IF	Initiation factor
K_M	Michaelis constant
LDL	Low-density lipoprotein
Mb	Myoglobin
NAD^+	Nicotinamide adenine dinucleotide (oxidized form)
NADH	Nicotinamide adenine dinucleotide (reduced form)
$NADP^+$	Nicotinamide adenine dinucleotide phosphate (oxidized form)
NADPH	Nicotinamide adenine dinucleotide phosphate (reduced form)
P_i	Phosphate ion
PAGE	Polyacrylamide gel electrophoresis
PCR	Polymerase chain reaction
PEP	Phosphoenolpyruvate
PIP_2	Phosphatidylinositol *bis*phosphate
PKU	Phenylketonuria
Pol	DNA polymerase
PP_i	Pyrophosphate ion
PRPP	Phosphoribosylpyrophosphate
PS	Photosystem
RF	Release factor
RFLPs	Restriction-fragment-length polymorphisms
RNA	Ribonucleic acid
RNase	Ribonuclease
mRNA	Messenger RNA
rRNA	Ribosomal RNA
tRNA	Transfer RNA
snRNP	Small nuclear ribounuclear protein
S	Svedberg unit
SCID	Severe combined immune deficiency
SSB	Single-strand binding protein
SV40	Simian virus 40
T	Thymine
TDP	Thymidine diphosphate
TMP	Thymidine monophosphate
TTP	Thymidine triphosphate
U	Uracil
UDP	Uridine diphosphate
UMP	Uridine monophosphate
UTP	Uridine triphosphate
V_{max}	Maximal velocity

Biochemistry ⬢ Now™

Take charge of your learning with BiochemistryNow™

http://now.brookscole.com/campbell5

Designed to maximize your time investment, **BiochemistryNow** helps you succeed by focusing your study time on the concepts that you're having the most difficulty mastering. **BiochemistryNow** gauges your unique study needs and provides you with a *Personalized Learning Plan* that enhances your problem-solving skills and conceptual understanding.

Integrated—chapter by chapter—with this text

Look for these references in the textbook. They direct you to the corresponding media-enhanced activities on **BiochemistryNow**. This precise page-by-page integration enables you to go beyond reading about biochemistry—you'll actually experience it in action!

> **Biochemistry ⬢ Now™**
> Test yourself on these Critical Questions at the BiochemistryNow website at **http://now.brookscole.com/campbell5**

How does it work?

The **BiochemistryNow** system includes three powerful assessment components:

▶ **What Do I Know?** This diagnostic *Pre-Test* based on the chapter-opening *Critical Questions* gives you an initial assessment of your knowledge.

▶ **What Do I Need to Learn?** Based on the automatically graded *Pre-Test*, you receive a *Personalized Learning Plan* that outlines key elements for review.

▶ **What Have I Learned?** After working through your *Personalized Learning Plan*, you can complete a *Post-Test* to assess your knowledge and e-mail results to your instructor.

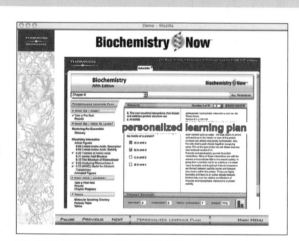

With a click of the mouse, the unique interactive activities contained at BiochemistryNow allow you to:

▶ Create a *Personalized Learning Plan* or review for an exam using the *Pre-Test* Web quizzes

▶ Explore biochemical concepts through simulations and tutorials with *Biochemistry Interactive*

▶ View *Active Figures* to interact with text illustrations and quiz yourself on the concepts

▶ Assess your mastery of core concepts and skills by completing the *Post-Test*

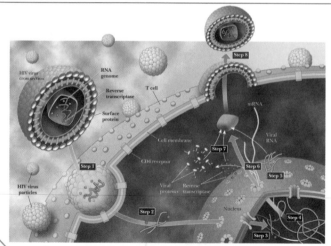

Biochemistry ⬢ Now™ ▲ ANIMATED FIGURE 14.29 HIV infection begins when the virus particle binds to CD4 receptors on the surface of the cell (Step 1). The viral core is inserted into the cell and partially disintegrates (Step 2). The reverse transcriptase catalyzes the production of DNA from the viral RNA. The viral DNA is integrated into the DNA of the host cell (Step 3). The DNA, including the integrated viral DNA, is transcribed to RNA (Step 4). Smaller RNAs are produced first, specifying the amino acid sequence of viral regulatory proteins (Step 5). Larger RNAs, ones that specify the amino acid sequences of viral enzymes and coat proteins, are made next (Step 6). The viral protease assumes particular importance in the budding of new virus particles (Step 7). Both the viral RNA and viral proteins are included in the budding virus, as is some of the membrane of the infected cell (Step 8). (*Adapted from AIDS and the Immune System, by Warner C. Green, illustration by Tomo Narachima, Sci. Amer. (1993).*) **See this figure animated at http://now.brookscole.com/campbell5**

Biochemistry ⬢ Now™ ▲ ANIMATED FIGURE 14.29 HIV infection begins when the virus particle binds to CD4 receptors on the surface of the cell (Step 1). The viral core is inserted into the cell and partially disintegrates (Step 2). The reverse transcriptase catalyzes the production of DNA from the viral RNA. The viral DNA is integrated into the DNA of the host cell (Step 3). The DNA, including the integrated viral DNA, is transcribed to RNA (Step 4). Smaller RNAs are produced first, specifying the amino acid sequence of viral regulatory proteins (Step 5). Larger RNAs, ones that specify the amino acid sequences of viral enzymes and coat proteins, are made next (Step 6). The viral protease assumes particular importance in the budding of new virus particles (Step 7). Both the viral RNA and viral proteins are included in the budding virus, as is some of the membrane of the infected cell (Step 8). (*Adapted from AIDS and the Immune System, by Warner C. Green, illustration by Tomo Narachima, Sci. Amer. (1993).*) **See this figure animated at http://now.brookscole.com/campbell5**

BIOCHEMISTRY

Mary K. Campbell
Mount Holyoke College

Shawn O. Farrell
Colorado State University

FIFTH EDITION

THOMSON

BROOKS/COLE

Australia · Canada · Mexico · Singapore
Spain · United Kingdom · United Stat

THOMSON
BROOKS/COLE

Biochemistry, Fifth Edition
Mary K. Campbell, Shawn O. Farrell

Publisher, Physical Sciences: David Harris

Development Editor: Jay Campbell

Assistant Editor: Ellen Bitter

Editorial Assistant: Candace Lum

Technology Project Manager: Donna Kelley

Marketing Manager: Amee Mosley

Marketing Assistant: Michele Colella

Advertising Project Manager: Nathaniel Bergson-Michelson

Project Manager, Editorial Production: Lisa Weber

Creative Director: Rob Hugel

Print/Media Buyer: Judy Inouye

Permissions Editor: Joohee Lee

Production Service: Lachina Publishing Services

Text Designer: Patrick Devine Design

Copy Editor: Gunder Hefta

Illustrators: J/B Woolsey and 2064design

Cover Designer: Lisa Devenish

Cover Image: © Digital Art/CORBIS

Cover Printer: Courier Corporation/Kendallville

Compositor: Lachina Publishing Services

Printer: Courier Corporation/Kendallville

For more information about our products, contact us at:

Thomson Learning Academic Resource Center
1-800-423-0563

For permission to use material from this text or product, submit a request online at **http://www.thomsonrights.com**.
Any additional questions about permissions can be submitted by email to **thomsonrights@thomson.com**.

COPYRIGHT 2006 Thomson Learning, Inc. All Rights Reserved. Thomson Learning WebTutor™ is a trademark of Thomson Learning, Inc.

Library of Congress Control Number: 2004111569

Student Edition: ISBN 0-534-40521-5

Instructor's Edition: ISBN 0-534-40523-1

International Student Edition: ISBN 0-534-39499-x
(Not for sale in the United States)

Thomson Brooks/Cole
10 Davis Drive
Belmont, CA 94002
USA

Asia (including India)
Thomson Learning
5 Shenton Way
#01-01 UIC Building
Singapore 068808

Australia/New Zealand
Thomson Learning Australia
102 Dodds Street
Southbank, Victoria 3006
Australia

Canada
Thomson Nelson
1120 Birchmount Road
Toronto, Ontario M1K 5G4
Canada

UK/Europe/Middle East/Africa
Thomson Learning
High Holborn House
50/51 Bedford Row
London WC1R 4LR
United Kingdom

Latin America
Thomson Learning
Seneca, 53
Colonia Polanco
11560 Mexico
D.F. Mexico

Spain (includes Portugal)
Thomson Paraninfo
Calle Magallanes, 25
28015 Madrid, Spain

DEDICATION

To all of those who made this text possible and especially to all of the students who will use it.

—Mary K. Campbell

To the returning adult students in my classes, especially those with children and a full-time job . . . my applause.

—Shawn O. Farrell

Mary K. Campbell

Mary K. Campbell is professor emeritus of chemistry at Mount Holyoke College, where she taught a one-semester biochemistry course, and advised undergraduates working on biochemical research projects. She frequently taught general chemistry and physical chemistry as well. At some point in her 36 years at Mount Holyoke, she taught every subfield of chemistry, except the lecture portion of organic chemistry. Her avid interest in writing led to the publication of the first four highly successful editions of this textbook. Originally from Philadelphia, Mary received her Ph.D. from Indiana University and did postdoctoral work in biophysical chemistry at Johns Hopkins University. Her area of interest includes researching the physical chemistry of biomolecules, specifically, spectroscopic studies of protein–nucleic acid interactions.

Mary enjoys traveling and has recently revisited favorite haunts in the United States from Atlantic (Newport, RI) to Pacific (San Francisco). She can frequently be seen hiking the Appalachian Trail.

Shawn O. Farrell

Shawn O. Farrell grew up in northern California and received a B.S. degree in biochemistry from the University of California, Davis, where he studied carbohydrate metabolism. He completed his Ph.D. in biochemistry at Michigan State University, where he studied fatty acid metabolism. For the last 18 years, Shawn has worked at Colorado State University teaching undergraduate biochemistry lecture and laboratory courses. Because of his interest in biochemical education, Shawn has written a number of scientific journal articles about teaching biochemistry. He is the coauthor (with Lynn E. Taylor) of *Experiments in Biochemistry: A Hands-On Approach*. Shawn became interested in biochemistry while in college because it coincided with his passion for bicycle racing. An active outdoorsman, Shawn raced competitively for 15 years and now officiates at bicycle races around the world. He is currently the Technical Director of USA Cycling, the national governing body of bicycle racing in the United States. He is also a distance runner and an avid fly fisherman, and recently achieved his third-degree black belt in Tae Kwon Do and first-degree black belt in combat hapkido. Shawn has also written articles on fly fishing for *Salmon Trout Steelheader* magazine. His other passions are soccer, chess, and foreign languages. He is fluent in Spanish and French, and is currently learning German and Italian.

Contents in Brief

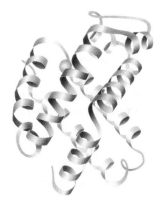

Table of Contents

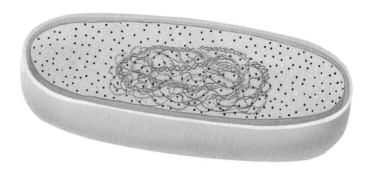

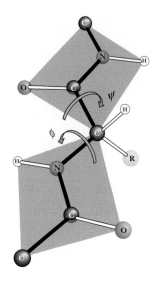

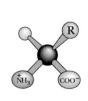
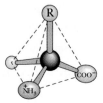

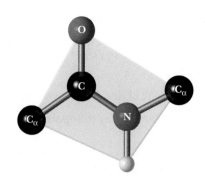

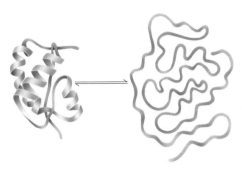

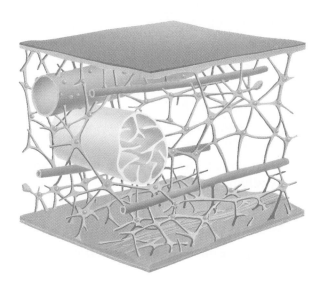

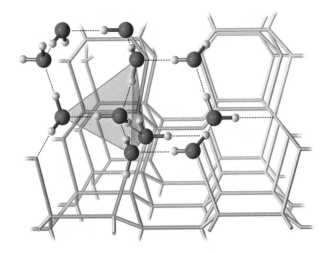

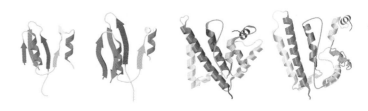

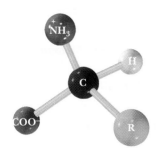
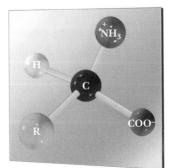

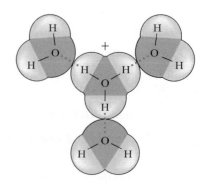

This text is intended for students in any field of science or engineering who want a one-semester introduction to biochemistry but who do not intend to be biochemistry majors. Our main goal in writing this book is to make biochemistry as clear and applied as possible and to familiarize science students with the major aspects of biochemistry. For students of biology, chemistry, physics, geology, nutrition, sports physiology, and agriculture, biochemistry impacts greatly on the content of their fields, especially in the areas of medicine and biotechnology. For engineers, studying biochemistry is especially important for those who hope to enter a career in biomedical engineering or some form of biotechnology.

Students who will use this text are at an intermediate level in their studies. A beginning biology course, general chemistry, and at least one semester of organic chemistry are assumed as preparation.

NEW TO THIS EDITION

All textbooks evolve to meet the interests and needs of students and instructors and to include the most current information. Several changes mark this edition.

Technology integration   First, and foremost, is the integration of **Biochemistry-Now,**™ the *first* assessment-centered student learning tool for biochemistry! This powerful and interactive online resource helps students gauge their unique study needs, then gives them a *Personalized Learning Plan* that focuses their study time on the concepts and problems that will most enhance their computational skills and understanding. **BiochemistryNow** gives students the resources and responsibility to manage their concept mastery. The system includes diagnostic tests to determine where students need help, online tutorials to help turn student weaknesses into strengths, Active and Animated Figures (which make extensive use of Java and MDL® Chime software) to make concepts come alive, and more. Register to access **BiochemistryNow** at **http://now.brookscole.com/campbell5.**

Biochemistry ⬥ Now™

Critical Question **framework**   We employ a new *Critical Question* framework for this edition to emphasize key biochemistry concepts. This focused approach guides students through each chapter by using section head questions, supporting concept statements, and summaries—and is enhanced by outstanding text and media integration through **BiochemistryNow.** The end-of-chapter summaries have been completely revised to reflect the *Critical Question* framework. At the end of each chapter the *Critical Questions* are restated and then the summary paragraphs are designed to highlight the concepts associated with the questions.

New chapter on advances in biochemistry   We have added a new chapter, Chapter 14, entitled *Hot Topics in Cell and Molecular Biology.* This chapter contains up-to-date material on new breakthroughs and topics in the area of biochemistry, like SARS, gene therapy, stem-cell research, AIDS, and cancer.

Early inclusion of thermodynamics   Select material on thermodynamics appears much earlier in the text. Chapter 1 includes sections on *Energy and Change, Spontaneity,* and the connection between *Thermodynamics and Life.* Also, Chapter 4 contains sections on the *Thermodynamics of Protein Folding* and *Predicting Protein Folding from Sequence.* We feel it is critical that students understand the driving force of biological processes and that so much of biology (protein folding, protein-protein interactions, small molecule binding, etc.) is driven by the favorable disordering of water molecules.

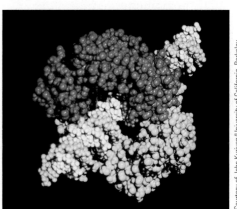

Courtesy of John Kuriyan/University of California, Berkeley

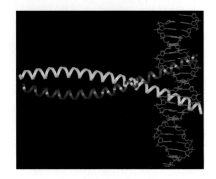

Expanded and updated coverage of select topics   We have increased the coverage of certain important topics in the text. Now included in Chapter 4 on the three-dimensional structure of proteins is an expanded description of prions and chaperonins. Chapter 13 now contains Section 13.12, covering bioinformatics, genomics, and proteomics. This material is included in the context that DNA sequences, protein sequences, etc. provide the database for these popular approaches. Also, the chapters on nucleic acids and biotechnology have been updated significantly due to the vast interest in the human genome project, cloning, and gene therapy, as well as proteomics.

Renumbering of chapters   In response to reviewer feedback, former Interchapters A (*Protein Purification and Characterization Techniques*) and B (*Nucleic Acid Biotechnology Techniques*) have been numbered Chapters 5 and 13, respectively. Reviewer feedback revealed that some felt labeling this material as "Interchapters" relegated them as optional or superfluous. Along with the addition of a new Chapter 14 and the deletion of former Interchapter C (*The Anabolism of Nitrogen-Containing Compounds*), you may notice the text now has 24 chapters compared to 21 chapters in the fourth edition. However, we should note that *the book has not grown*. In fact, *the book is **shorter** by **48 pages**!*

New format in problem sets   The end-of-chapter problem sets now are broken up by *Critical Question* and each problem is individually labeled according to its type (*Fact Check, Thought Question, Mathematical,* and *Biochemical Connections*). Also, where appropriate, we have added a few more problems that are more quantitative in nature. These carry the *Mathematical* label.

Strategy information added into Practice Session solutions   Where appropriate, we include suggestions on how to answer the questions asked in the Practice Sessions.

New design and art   To complement the integration of *BiochemistryNow* and the new *Critical Question* format, we have given the book an overhaul both in design and art. Approximately 25% of the art is new to this edition, and, as necessary, other figures have been "tuned up."

PROVEN FEATURES

The new elements in the text build upon many time-tested features found in previous editions.

Visual Impact   One of the most distinctive features of this text is its visual impact. Its extensive four-color art program includes artwork by the late Irving Geis, John and Bette Woolsey, and Greg Gambino of 2064 Design. The illustrations convey meaning so powerfully, it is certain that many of them will become standard presentations in the field.

Chapter Overviews   These chapter-opening paragraphs include overviews for each chapter. They transition together material from previous chapters with the topics to be discussed and serve as building blocks for new ideas.

Biochemical Connections   These boxes highlight special topics of particular interest to students. Topics frequently have clinical implications such as cancer, AIDS, and nutrition. These essays help students make the connection between biochemistry and the real world.

Practice Sessions   The *Practice Sessions* are interspersed within chapters and designed to give students problem-solving experience. The topics chosen are those areas of study where students usually have the most difficulty. *Solutions* and *problem-solving strategies* are now included, giving examples of the problem-solving approach for specific material.

Summaries and Questions   Each chapter closes with a concise summary, a broad selection of questions, and an annotated bibliography. As stated previously, the summaries have been completely revised to reflect the *Critical Question* framework. At the end of each chapter, the *Critical Questions* are restated and the summary paragraphs highlight the concepts associated with each question. The number of questions has been expanded in this edition to provide additional self-testing of content mastery and more homework material. These exercises fall into four categories: *Fact Check, Thought Question, Mathematical,* and *Biochemical Connections*. The *Fact Check* questions are designed for students to quickly assess their mastery of the material, while the *Thought Question* questions are for students to work through more thought-provoking questions. *Biochemical Connections* questions test students on the *Biochemical Connections* essays in that chapter. New to this edition are the *Mathematical* questions. These questions are quantitative in nature and focus on calculations.

Essential Information   These sidebars in each chapter highlight the key, important material. If a student flips through the chapter and reads the *Essential Information* boxes in the margins, even before reading the text, he or she will have a very good idea of the content of the chapter.

Glossary and Answers

The book ends with a glossary of important terms and concepts (including the section number where the term was first introduced), an answer section, and a detailed index.

Accuracy

The page proofs for this text were reviewed by the authors and Dr. Paul D. Adams of SUNY-Cortland.

ORGANIZATION

Because biochemistry is a multidisciplinary science, the first task in presenting it to students of widely varying backgrounds is to put it in context. Chapters 1 and 2 provide the necessary background and connect biochemistry to the other sciences. Chapters 3 through 8 focus on the structure and dynamics of important cellular components. Molecular biology is covered in Chapters 9 through 14. The final part of the book is devoted to intermediary metabolism. Some topics are discussed several times, such as the control of carbohydrate metabolism. Subsequent discussions make use of and build on information students have already learned. It is particularly useful to return to a topic after students have had time to assimilate and reflect on it.

The first two chapters of the book relate biochemistry to other fields of science. Chapter 1 deals with some of the less obvious relationships, such as the connections of biochemistry with physics, astronomy, and geology, mostly in the context of the origins of life. Functional groups on organic molecules are discussed from the point of view of their role in biochemistry. This chapter goes on to the more readily apparent linkage of biochemistry with biology, especially with respect to the distinction between prokaryotes and eukaryotes, as well as the role of organelles in eukaryotic cells. New to Chapter 1 for this edition are three sections of material on thermodynamics. Chapter 2 builds

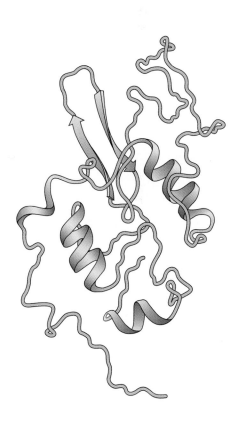

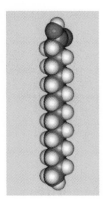

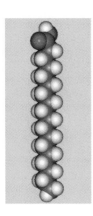

on material familiar from general chemistry, such as buffers and the solvent properties of water, but emphasizes the biochemical point of view toward such material.

The following six chapters (3 through 8), on the structure of cellular components, focus on the structure and dynamics of proteins and membranes in addition to giving an introduction to some aspects of molecular biology. Chapters 3, 4, 6, and 7 deal with amino acids, peptides, and the structure and action of proteins including enzyme catalysis. Chapter 4 includes more material on thermodynamics, like hydrophobic interactions. The discussion of enzymes is split into two chapters (Chapters 6 and 7) to give students more time to fully understand enzyme kinetics and enzyme mechanisms. Chapter 5 focuses on techniques for isolating and studying proteins. Chapter 8 treats the structure of membranes and their lipid components.

Chapters 9 through 14 explore the topics of molecular biology. Chapter 9 introduces the structure of nucleic acids. In Chapter 10, the replication of DNA is discussed. Chapter 11 focuses on transcription and gene regulation. This material on the biosynthesis of nucleic acids is split into two chapters to give students ample time to appreciate the workings of these processes. Chapter 12 finishes the topic with translation of the genetic message and protein synthesis. Chapters 13 and 14 cover topics often in the news today. Chapter 13 focuses on biotechnology techniques, and Chapter 14 deals with recent phenomena, like SARS, stem-cell research, and AIDS.

Chapters 15 through 24 explore intermediary metabolism. Chapter 15 opens the topic with chemical principles that provide some unifying themes. Thermodynamic concepts learned earlier in general chemistry and in Chapter 1 are applied specifically to biochemical topics such as coupled reactions. In addition, this chapter explicitly makes the connection between metabolism and electron transfer (oxidation–reduction) reactions.

Coenzymes are introduced in this chapter and are discussed in later chapters in the context of the reactions in which they play a role. Chapter 16 discusses carbohydrates. Chapter 17 begins the overview of the metabolic pathways by discussing glycolysis. Glycogen metabolism, gluconeogenesis, and the pentose phosphate pathway (Chapter 18) provide bases for treating control mechanisms in carbohydrate metabolism. Discussion of the citric acid cycle is followed by the electron transport chain and oxidative phosphorylation in Chapters 19 and 20. The catabolic and anabolic aspects of lipid metabolism are dealt with in Chapter 21. In Chapter 22, photosynthesis rounds out the discussion of carbohydrate metabolism. Chapter 23 completes the survey of the pathways by discussing the metabolism of nitrogen-containing compounds such as amino acids, porphyrins, and nucleobases. Chapter 24 is a summary chapter. It gives an integrated look at metabolism, including a treatment of hormones and second messengers. The overall look at metabolism includes a brief discussion of nutrition and a somewhat longer one of the immune system.

This text gives an overview of important topics of interest to biochemists and shows how the remarkable recent progress of biochemistry impinges on other sciences. The length is intended to provide instructors with a choice of favorite topics without being overwhelming for the limited amount of time available in one semester.

ALTERNATIVE TEACHING OPTIONS

The order in which individual chapters are covered can be changed to suit the needs of specific groups of students. Although we prefer an early discussion of thermodynamics, the portions of Chapters 1 and 4 that deal with thermodynamics can be covered at the beginning of Chapter 15, *The Importance of Energy Changes and Electron Transfer in Metabolism*. All of the molecular biology chapters (9–14) can precede metabolism or can follow it, depending on the instructor's choice. The order in which the material on molecular biology is treated can be varied according to the preference of the instructor.

SUPPLEMENTS

This fifth edition of Campbell and Farrell's *Biochemistry* is accompanied by the following rich array of web-based, electronic, and print supplements.

Web-Based Resources:

- **BiochemistryNow at http://now.brookscole.com/campbell5** This web-based, assessment-centered learning tool has been developed in concert with the text and is a natural extension of the *Critical Question* framework. Register to access **BiochemistryNow** at **http://now.brookscole.com/campbell5.**
- **WebTutor ToolBox for WebCT, WebTutor ToolBox for Blackboard** Preloaded with content and available via a free access code when packaged with this text, WebTutor ToolBox pairs all the content of this text's rich Book Companion Website at **http://now.brookscole.com/campbell5** with sophisticated course management functionality. Instructors can assign materials (including online quizzes) and have the results flow automatically to their gradebook. ToolBox is ready to use upon logging on—or instructors can customize its preloaded content by uploading images and other resources, adding weblinks, or creating their own practice materials. Students have access only to student resources on the website. Instructors can enter an access code for password-protected Instructor Resources. Contact your Thomson representative for information on packaging WebTutor ToolBox with this text.

Instructor Resources

Supporting materials are available to qualified adopters. Please consult your local Thomson Brooks/Cole sales representative for details. Visit the *BiochemistryNow* website at **http://now.brookscole.com/campbell5** to see samples of these materials, request a desk copy, locate your sales representative, or purchase a copy online.

- *Online Instructor's Manual and Test Bank* by Michael A. Sypes, Pennsylvania State University. Each chapter includes a chapter summary, lecture outline, answers to all the exercises in the text, and a bank of multiple-choice exam questions. Electronic files of the *Instructor's Manual and Test Bank* are available for download on the instructor's website.
- *iLrn Computerized Testing* With a balance of efficiency and high performance, simplicity and versatility, iLrn Testing lets instructors test the way they teach, giving them the power to transform the learning and teaching experience. iLrn Testing is a revolutionary, Internet-ready, cross-platform, text-specific testing suite that allows instructors to customize exams and track student progress in an accessible, browser-based format delivered via the web (at **http://www.iLrn.com**). Results flow automatically to instructors' gradebooks so that they are better able than ever to assess students' understanding of the material prior to class or an actual test.
- *Transparency Acetates* A set of 150 full-color overhead transparency acetates of text images are available for use in lectures.
- *Multimedia Manager Instructor CD-ROM* A dual-platform digital library and presentation tool that provides art, photos, and tables from the main text in a variety of electronic formats that are easily exported into other soft-

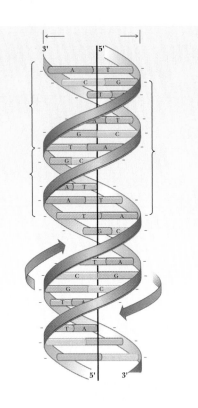

ware packages. Instructors can use Brooks/Cole's text-specific presentations or customize their own presentations by importing personal lecture slides or other selected materials.

Student Resources

- *Student Lecture Notebook* Contains all the instructor overhead transparency images printed in booklet format and includes pages for student notes. The *Student Lecture Notebook* can be packaged for free with each new copy of the text.
- *Experiments in Biochemistry: A Hands-On Approach* by Shawn O. Farrell and Lynn E. Taylor. This interactive manual for the introductory biochemistry laboratory course offers a great selection of classroom-tested experiments, each designed to be completed in a normal laboratory period.

ACKNOWLEDGMENTS

The help of many made this book possible. A grant from the Dreyfus Foundation made possible the experimental introductory course that was the genesis of many of the ideas for this text. Edwin Weaver and Francis DeToma from Mount Holyoke College gave much of their time and energy in initiating that course. Many others at Mount Holyoke were generous with their support, encouragement, and good ideas, especially Anna Harrison, Lilian Hsu, Dianne Baranowski, Sheila Browne, Janice Smith, Jeffrey Knight, Sue Ellen Frederick Gruber, Peter Gruber, Marilyn Pryor, Craig Woodard, Diana Stein, and Sue Rusiecki. Particular thanks go to Sandy Ward, science librarian, and to Rosalia Tungaraza, a biochemistry major in the class of 2004. Special thanks to Laurie Stargell, Marve Paule, and Steven McBryant at Colorado State University for their help and editorial assistance.

We thank the many biochemistry students who have used and commented on early versions of this text.

We would like to acknowledge colleagues who contributed their ideas and critiques of the manuscript. Some reviewers responded to specific queries regarding the text itself. We thank them for their efforts and their helpful suggestions.

- Denise Greathouse—University of Arkansas
- Charles C. Hardin—North Carolina State University
- Gavin MacBeath—Harvard University
- Dr. S. Madhavan—University of Nebraska at Lincoln
- Jamil Momand—California State University, Los Angeles
- Kazem Mostafapour—University of Michigan-Dearborn
- Thomas L. Selby—University of Central Florida
- David Smith—University of Wisconsin at Madison
- Dan M. Sullivan, Ph.D.—University of Nebraska at Omaha
- Martin Teintze—Montana State University
- Bryan A. White—University of Illinois at Urbana-Champaign
- John C. Wriston, Jr.—University of Delaware

We doubly thank Kazem Mostafapour for organizing his student evaluations of this book. His students' comments were insightful indeed.

There also were colleagues in the field who looked over our preliminary table of contents to aid us in judging whether our proposed shifting of material would be beneficial for students. We also thank them for their time.

- Dr. Paul D. Adams—State University of New York College at Cortland
- Arthur S. Brecher—Bowling Green State University

- Robert P. Cameron, Jr., Ph.D.—Samford University
- Jack Huang—Western Illinois University
- Dr. Theodore Jones—University of San Francisco
- William M. Scovell—Bowling Green State University
- Jeffrey Temple—Southeastern Louisiana University
- Paul Toom—Southwest Missouri State University
- Anthony P. Toste, Ph.D.—Southwest Missouri State University
- Lisa Wen—Western Illinois University

The efforts of Jay Campbell, Developmental Editor at Brooks/Cole Publishing, were essential to the development of this book. Lisa Weber, Senior Production Manager, directed production of this book with magnificent results. Ronn Jost of Lachina Publishing Services served diligently as our production editor. We feel privileged that the late Irving Geis contributed some of his classic illustrations; his passing in the summer of 1997 leaves a unique place in the sciences unfilled. Greg Gambino outdid himself at every turn with illustrations and turned crude sketches into works of art. Dena Digilio-Betz, photo researcher, found many splendid photographs, in some cases with considerable effort. We extend our most sincere gratitude to those listed here and to all others to whom we owe the opportunity to do this book. Instrumental in the direction given to this project was the late John Vondeling. John was a legend in the publishing field. His guidance and friendship shall be missed.

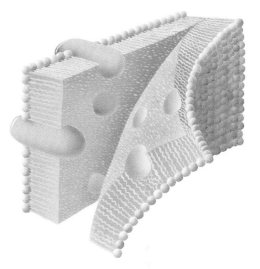

A Final Note from Mary Campbell

I thank my family and friends, whose moral support has meant so much to me in the course of my work. When I started this project years ago, I did not realize that it would become a large part of my life. It has been a thoroughly satisfying one.

and from Shawn Farrell

I cannot adequately convey how impossible this project would have been without my wonderful family who put up with a husband and father who became a hermit in the back office. My wife, Courtney, knows the challenge of living with me when I am working on 4 hours of sleep per night. It isn't pretty, and few would have been so understanding. I would also like to thank David Hall, book representative, for starting me down this path, and John Vondeling for giving me an opportunity to expand into other types of books and projects. Lastly, of course, I thank all of my students who have helped proofread the fifth edition, especially those who did it without getting extra credit for it.

Biochemistry and the Organization of Cells

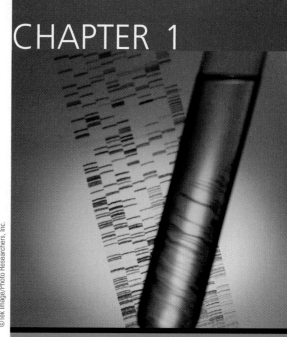

CHAPTER 1

©Tek Image/Photo Researchers, Inc.

Biochemistry unlocks the mysteries of the human body.

Complex living organisms originate from simple elements. Carbon, hydrogen, and oxygen combine to make up many different kinds of biomolecules, such as carbohydrates and fatty acids. The addition of nitrogen, as well as sulfur, makes possible the amino acids that combine to form proteins. In turn, added phosphorus provides the ingredients for making DNA, RNA, and complex lipids. Thus, there occurs a "building-up" from atoms to small molecular units to large biomolecules, such as proteins and the nucleic acids, DNA and RNA. A collection of interacting molecules, encased in a suitable membrane, becomes a cell—the basic unit of life. Cells have a central core of the hereditary material, DNA, which contains the information needed to make the complete organism. In one-celled prokaryotes, such as bacteria, the nuclear material is not enclosed in a membrane. The cells of plants and animals (called eukaryotes) are more highly organized, with the nucleus enclosed in a separate membrane. Fungi and protists are also classified as eukaryotes. Compartments specialized for particular functions are characteristic of eukaryotic cells. In plants, photosynthesis takes place in chloroplasts: Light energy is converted to chemical energy and stored as carbohydrates. In the mitochondria of eukaryotic cells, the stored energy of carbohydrates and lipids is recovered through respiration, a process in which carbon compounds are oxidized to carbon dioxide and water.

Critical Questions

1.1 What Are the Basic Themes for This Text?

1.2 What Is the Chemical Nature of Important Biomolecules?

1.3 What Can Biochemistry Say about Possible Origins of Life?

1.4 How Do Prokaryotes and Eukaryotes Differ in Levels of Organization?

1.5 What Are the Main Structural Features of Prokaryotic Cells?

1.6 What Are the Main Structural Features of Eukaryotic Cells?

1.7 How Do We Classify Organisms: Five Kingdoms or Three Domains?

1.8 Is There Common Ground for All Cells?

1.9 How Do Cells Use Energy?

1.10 What Is the Connection between Energy and Change?

1.11 What Is the Criterion for Spontaneity in Biochemical Reactions?

1.12 What Is the Connection between Thermodynamics and Life?

1.1 | What Are the Basic Themes for This Text?

Living organisms, and even the individual cells of which they are composed, are enormously complex and diverse. Nevertheless, certain unifying features are common to all living things. They all use the same types of *biomolecules*, and they all use energy. As a result, organisms can be studied via the methods of chemistry and physics. The belief in "vital forces" (forces thought to exist only in living organisms) held by 19th-century biologists has long since given way to awareness of an underlying unity throughout the natural world.

Disciplines that appear to be unrelated to biochemistry can provide answers to important biochemical questions. For example, physicists in the early 20th century discovered that X rays can be diffracted by crystals. As a result, the experimental method of X-ray diffraction was developed, and, with this methodology, three-dimensional structures of molecules as complex as proteins and nucleic acids could be determined. The field of biochemistry draws on many disciplines, and its multidisciplinary nature allows it to use results from many sciences to answer questions about the *molecular nature of life processes*. Important applications of this kind of knowledge are made in medically related fields; an understanding of health and disease at the molecular level leads to more effective treatment of illnesses of many kinds.

The activities within a cell are similar to the transportation system of a city. The cars, buses, and taxis correspond to the molecules involved in reactions (or series of reactions) within a cell. The routes traveled by vehicles likewise can be compared to the reactions that occur in the life of the cell. Note particularly that many vehicles travel more than one route—for instance, cars and taxis can go almost anywhere—whereas other, more specialized modes of

Biochemistry✏️Now™

Test yourself on these Critical Questions at the BiochemistryNow website at **http://now .brookscole.com/campbell5**

transportation, such as subways and streetcars, are confined to single paths. Similarly, some molecules play multiple roles, whereas others take part only in specific series of reactions. Also, *the routes operate simultaneously;* we shall see that this is true of the many reactions within a cell.

To continue the comparison, the transportation system of a large city has more kinds of transportation than does a smaller one. Whereas a small city may have only cars, buses, and taxis, a large city may have all of these plus others, such as streetcars or subways. Analogously, some reactions are found in all cells, and others are found only in specific kinds of cells. Also, more structural features are found in the larger, more complex cells of larger organisms than in the simpler cells of organisms such as bacteria.

An inevitable consequence of this complexity is the large quantity of terminology that is needed to describe it; learning considerable new vocabulary is an essential part of the study of biochemistry. You will also see many cross-references in this book, which are a reflection of the many connections among the processes that take place in the cell.

The fundamental similarity of cells of all types makes speculating on the origins of life interesting and illuminating. Even the structures of comparatively small biomolecules consist of several parts. Large biomolecules, such as proteins and nucleic acids, have complex structures, and living cells are enormously more complex. Even so, *both molecules and cells must have arisen ultimately from very simple molecules,* such as water, methane, carbon dioxide, ammonia, nitrogen, and hydrogen (Figure 1.1). In turn, these simple molecules must have arisen from atoms. The way in which the universe itself, and the atoms of which it is composed, came to be is a topic of great interest to astrophysicists as well as other scientists. Simple molecules were formed by combining atoms, and reactions of simple molecules led in turn to more complex molecules. The molecules that play a role in living cells today are the same molecules as those encountered in organic chemistry; they simply operate in a different context.

1.2 | What Is the Chemical Nature of Important Biomolecules?

Organic chemistry is the study of compounds of carbon and hydrogen and their derivatives. Because the cellular apparatus of living organisms is made up of carbon compounds, biomolecules are part of the subject matter of organic chemistry. Additionally, there are many carbon compounds that are not found in any organism, and many topics of importance to organic chemistry have little connection with living things.

Until the early part of the 19th century, there was a widely held belief in "vital forces," forces presumably unique to living things. This belief included the idea that the compounds found in living organisms could not be produced in the laboratory. German chemist Friedrich Wöhler performed the critical experiment that disproved this belief in 1828. Wöhler synthesized urea, a well-known waste product of animal metabolism, from ammonium cyanate, a compound obtained from mineral (i.e., nonliving) sources.

$$NH_4OCN \rightarrow H_2NCONH_2$$

Ammonium   Urea
cyanate

It has subsequently been shown that any compound that occurs in a living organism can be synthesized in the laboratory, although in many cases the synthesis represents a considerable challenge to even the most skilled organic chemist.

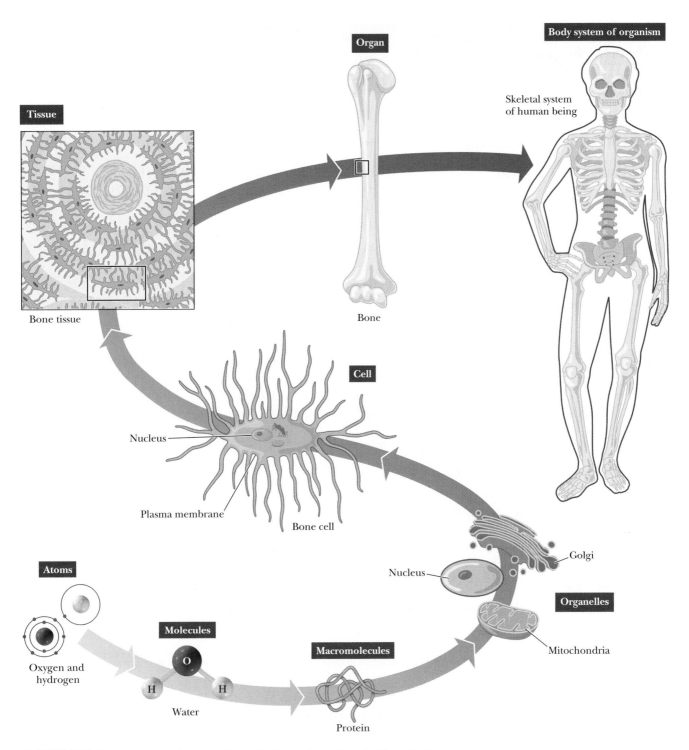

▲ **FIGURE 1.1** Levels of structural organization in the human body. Note the hierarchy from simple to complex.

The reactions of biomolecules can be described by the methods of organic chemistry, which requires the classification of compounds according to their **functional groups.** *The reactions of molecules are based on the reactions of their respective functional groups.* Table 1.1 lists some biologically important functional groups. Note that most of these functional groups contain oxygen and nitrogen, which are among the most electronegative elements. As a result,

Table 1.1

Functional Groups of Biochemical Importance

Class of Compound	General Structure	Characteristic Functional Group	Name of Functional Group	Example
Alkenes	$RCH{=}CH_2$ $RCH{=}CHR$ $R_2C{=}CHR$ $R_2C{=}CR_2$	$C{=}C$	Double bond	$CH_2{=}CH_2$
Alcohols	ROH	$-OH$	Hydroxyl group	CH_3CH_2OH
Ethers	ROR	$-O-$	Ether group	CH_3OCH_3
Amines	RNH_2 R_2NH R_3N	$-N\big\langle$	Amino group	CH_3NH_2
Thiols	RSH	$-SH$	Sulfhydryl group	CH_3SH
Aldehydes	$R-\overset{\overset{\displaystyle O}{\|\|}}{C}-H$	$-\overset{\overset{\displaystyle O}{\|\|}}{C}-$	Carbonyl group	$CH_3\overset{\overset{\displaystyle O}{\|\|}}{C}H$
Ketones	$R-\overset{\overset{\displaystyle O}{\|\|}}{C}-R$	$-\overset{\overset{\displaystyle O}{\|\|}}{C}-$	Carbonyl group	$CH_3\overset{\overset{\displaystyle O}{\|\|}}{C}\,CH_3$
Carboxylic acids	$R-\overset{\overset{\displaystyle O}{\|\|}}{C}-OH$	$-\overset{\overset{\displaystyle O}{\|\|}}{C}-OH$	Carboxyl group	$CH_3\overset{\overset{\displaystyle O}{\|\|}}{C}\,OH$
Esters	$R-\overset{\overset{\displaystyle O}{\|\|}}{C}-OR$	$-\overset{\overset{\displaystyle O}{\|\|}}{C}-OR$	Ester group	$CH_3\overset{\overset{\displaystyle O}{\|\|}}{C}\,OCH_3$
Amides	$R-\overset{\overset{\displaystyle O}{\|\|}}{C}-NR_2$ $R-\overset{\overset{\displaystyle O}{\|\|}}{C}-NHR$ $R-\overset{\overset{\displaystyle O}{\|\|}}{C}-NH_2$	$-\overset{\overset{\displaystyle O}{\|\|}}{C}-N\big\langle$	Amide group	$CH_3\overset{\overset{\displaystyle O}{\|\|}}{C}\,N(CH_3)_2$
Phosphoric acid esters	$R-O-\overset{\overset{\displaystyle O}{\|\|}}{\underset{\underset{\displaystyle OH}{\|}}{P}}-OH$	$-O-\overset{\overset{\displaystyle O}{\|\|}}{\underset{\underset{\displaystyle OH}{\|}}{P}}-OH$	Phosphoric ester group	$CH_3-O-\overset{\overset{\displaystyle O}{\|\|}}{\underset{\underset{\displaystyle OH}{\|}}{P}}-OH$
Phosphoric acid anhydrides	$R-O-\overset{\overset{\displaystyle O}{\|\|}}{\underset{\underset{\displaystyle OH}{\|}}{P}}-O-\overset{\overset{\displaystyle O}{\|\|}}{\underset{\underset{\displaystyle OH}{\|}}{P}}-OH$	$-\overset{\overset{\displaystyle O}{\|\|}}{\underset{\underset{\displaystyle OH}{\|}}{P}}-O-\overset{\overset{\displaystyle O}{\|\|}}{\underset{\underset{\displaystyle OH}{\|}}{P}}-$	Phosphoric anhydride group	$HO-\overset{\overset{\displaystyle O}{\|\|}}{\underset{\underset{\displaystyle OH}{\|}}{P}}-O-\overset{\overset{\displaystyle O}{\|\|}}{\underset{\underset{\displaystyle OH}{\|}}{P}}-OH$

The symbol R refers to any carbon-containing group. When there are several R groups in the same molecule, they may be different groups or they may be the same.

many of these functional groups are polar, and their polar nature plays a crucial role in their reactivity. Some groups that are of vital importance to organic chemists are missing from the table because molecules containing these groups, such as alkyl halides and acyl chlorides, do not have any particular applicability in biochemistry. Conversely, carbon-containing derivatives of phosphoric acid are mentioned infrequently in beginning courses on organic chemistry, but esters and anhydrides of phosphoric acid (Figure 1.2) are of vital importance in biochemistry. Adenosine triphosphate (ATP), a molecule that is the energy currency of the cell, contains both ester and anhydride linkages involving phosphoric acid.

Important classes of biomolecules have characteristic functional groups that determine their reactions. We shall discuss the reactions of the functional groups when we consider the compounds in which they occur.

Biochemistry ⓔ Now™
Go to BiochemistryNow and click on Biochemistry Interactive for a tutorial on functional groups.

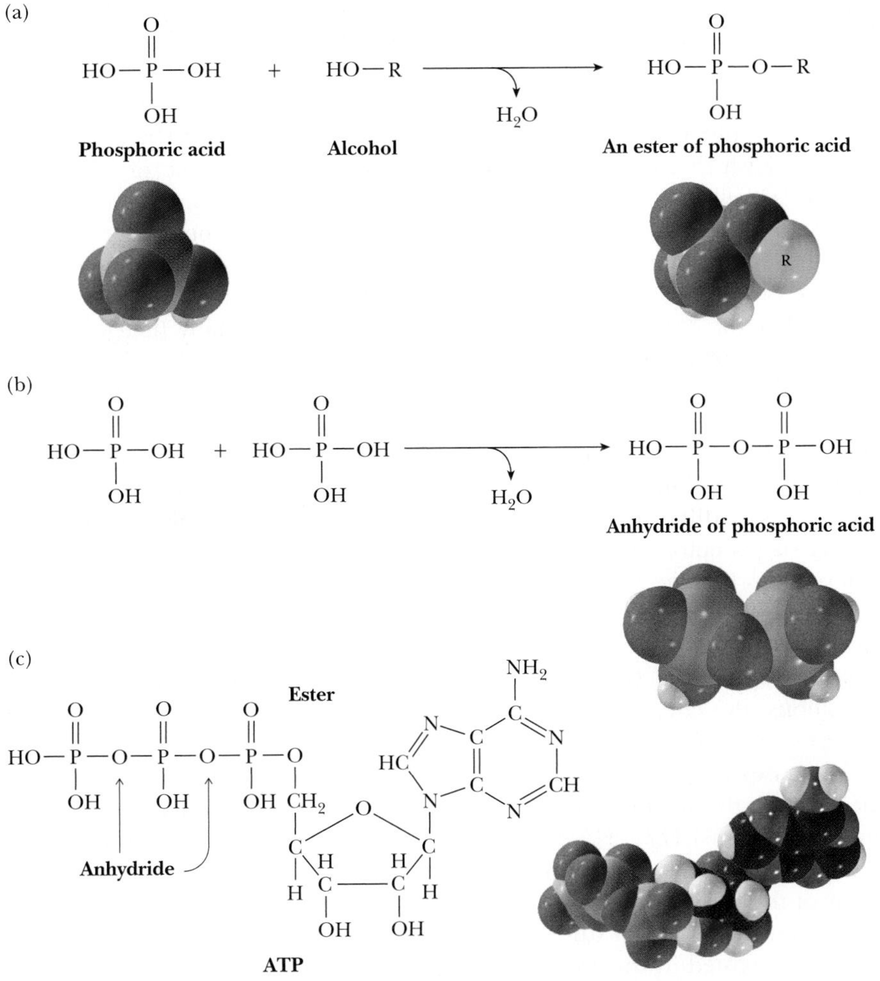

(a)

Phosphoric acid   Alcohol   An ester of phosphoric acid

(b)

Anhydride of phosphoric acid

(c)

ATP

◀ **FIGURE 1.2** ATP and the reactions for its formation. (a) Reaction of phosphoric acid with a hydroxyl group to form an ester, which contains a P–O–R linkage. Phosphoric acid is shown in its nonionized form in this figure. Space-filling models of phosphoric acid and its methyl ester are shown. The red spheres represent oxygen; the white, hydrogen; the green, carbon; and the orange, phosphorus. (b) Reaction of two molecules of phosphoric acid to form an anhydride, which contains a P–O–P linkage. A space-filling model of the anhydride of phosphoric acid is shown. (c) The structure of ATP (*a*denosine *trip*hosphate), showing two anhydride linkages and one ester.

1.3 | What Can Biochemistry Say about Possible Origins of Life?

The Earth and Its Age

To date, we are aware of only one planet that unequivocally supports life: our own. (The widely publicized reports of life on Mars are, at the moment, in the realm of conjecture rather than fact. See the article by Balter in the bibliography at the end of this chapter for more information about this point.) The Earth and its waters are universally understood to be the source and mainstay of life as we know it. A natural first question is how the Earth, along with the Universe of which it is a part, came to be.

Currently, the most widely accepted cosmological theory for the origin of the universe is the *big bang,* a cataclysmic explosion. According to big-bang cosmology, all the matter in the universe was originally confined to a comparatively small volume of space. As a result of a tremendous explosion, this "primordial fireball" started to expand with great force. Immediately after the big bang, the Universe was extremely hot, on the order of 15 billion (15×10^9) K. (Note that Kelvin temperatures are written without a degree symbol.) The average temperature of the Universe has been decreasing ever since as a result of expansion, and the lower temperatures have permitted the formation of stars and planets. In its earliest stages, the Universe had a fairly simple composition. Hydrogen, helium, and some lithium (the three smallest and simplest elements on the periodic table) were present, having been formed in the original big-bang explosion. The rest of the chemical elements are thought to have been formed in three ways: (1) by thermonuclear reactions that normally take place in stars, (2) in explosions of stars, and (3) by the action of cosmic rays outside the stars since the formation of the galaxy. The process by which the elements are formed in stars is a topic of interest to chemists as well as to astrophysicists. For our purposes, note that the most abundant isotopes of biologically important elements such as carbon, oxygen, nitrogen, phosphorus, and sulfur have *particularly stable nuclei.* These elements were produced by nuclear reactions in first-generation stars, the original stars produced after the beginning of the Universe (Table 1.2). Many first-generation stars were destroyed by explosions called *supernovas,* and their stellar material was recycled to produce second-generation stars, such as our own Sun, along with our solar system. Radioactive dating, which uses the decay of unstable nuclei, indicates that the age of the Earth (and the rest of the solar system) is 4 billion to 5 billion (4×10^9 to 5×10^9) years. The atmosphere of the early Earth was very different from the one we live in, and it probably went through several stages before reaching its current composition. The most important difference is that, according to most theories of the origins of the Earth, very little or no free oxygen (O_2) existed in the early stages (Figure 1.3). The early Earth was constantly irradiated with ultraviolet light from the Sun because there was no ozone (O_3) layer in the atmosphere to block it. Under these conditions, the chemical reactions that produced simple biomolecules took place.

The gases usually postulated to have been present in the atmosphere of the early Earth include NH_3, H_2S, CO, CO_2, CH_4, N_2, H_2, and (in both liquid and vapor forms) H_2O. However, there is no universal agreement on the relative amounts of these components, from which biomolecules ultimately arose. Many of the earlier theories of the origin of life postulated CH_4 as the carbon source, but more recent studies have shown that appreciable amounts of CO_2 must have existed in the atmosphere at least 3.8 billion (3.8×10^9) years ago. This conclusion is based on geological evidence: The earliest known rocks are 3.8 billion years old, and they are carbonates, which arise from CO_2. Any NH_3 originally present must have dissolved in the oceans, leaving N_2 in the atmo-

Table 1.2		
Abundance of Important Elements Relative to Carbon*		
Element	**Abundance in Organisms**	**Abundance in Universe**
Hydrogen	80–250	10,000,000
Carbon	1,000	1,000
Nitrogen	60–300	1,600
Oxygen	500–800	5,000
Sodium	10–20	12
Magnesium	2–8	200
Phosphorus	8–50	3
Sulfur	4–20	80
Potassium	6–40	0.6
Calcium	25–50	10
Manganese	0.25–0.8	1.6
Iron	0.25–0.8	100
Zinc	0.1–0.4	0.12

* Each abundance is given as the number of atoms relative to a thousand atoms of carbon.

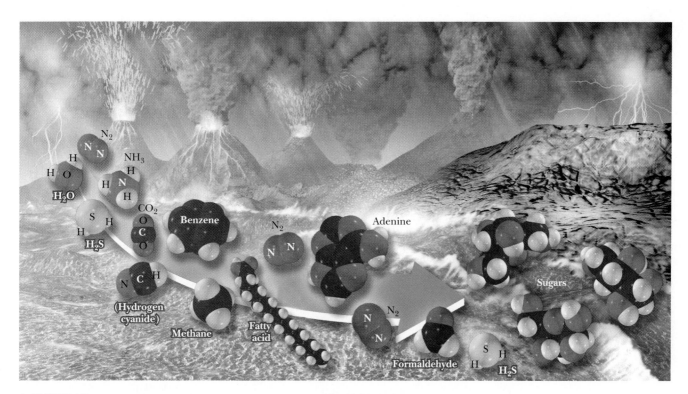

▲ **FIGURE 1.3** Conditions on early Earth would have been inhospitable for most of today's life. Very little or no oxygen (O_2) existed. Volcanoes erupted, spewing gases, and violent thunderstorms produced torrential rainfall that covered the Earth. The green arrow indicates the formation of biomolecules from simple precursors.

sphere as the nitrogen source required for the formation of proteins and nucleic acids.

Biomolecules

Experiments have been performed in which the simple compounds of the early atmosphere were allowed to react under the varied sets of conditions that might have been present on the early Earth. The results of such experiments indicate that these simple compounds react *abiotically* or, as the word indicates (*a*, "not" and *bios*, "life"), in the absence of life, to give rise to biologically important compounds such as the components of proteins and nucleic acids. Of historic interest is the well-known Miller–Urey experiment, shown schematically in Figure 1.4. In each trial, an electric discharge, simulating lightning, is passed through a closed system that contains H_2, CH_4, and NH_3, in addition to H_2O. Simple organic molecules, such as formaldehyde (HCHO) and hydrogen cyanide (HCN), are typical products of such reactions, as are amino acids, the building blocks of proteins. According to one theory, reactions such as these took place in the Earth's early oceans; other researchers postulate that such reactions occurred on the surfaces of clay particles that were present on the early Earth. It is certainly true that mineral substances similar to clay can serve as catalysts in many types of reactions. Both theories have their proponents, and more research will be needed to answer the many questions that remain.

Living cells as they exist today are assemblages that include very large molecules, such as proteins, nucleic acids, and polysaccharides. These molecules are larger by many powers of ten than the smaller molecules from which they are built. Hundreds or thousands of these smaller molecules, or **monomers,**

Biochemical Connections

Structure and Function of Biomolecules

A study of Table 1.2 shows clearly that the distribution of elements in living organisms is very different from that in the whole Universe (or in the Earth's crust, ocean, and atmosphere). Two of the most abundant elements in the Earth's crust are silicon and aluminum, 26% and 7.5% by weight, respectively. These two elements rarely occur in living organisms. Much of the hydrogen, oxygen, and nitrogen in the Universe is found in the gaseous, elemental form, not combined in complex compounds.

One important reason for this difference is that most living organisms depend on the nonmetals—that is, those elements that form complex molecules based on *covalent* bonding. Bio-molecules are frequently made up of only six elements—carbon, hydrogen, oxygen, nitrogen, sulfur, and phosphorus. Central to these biomolecules is *carbon*, which has the unique property of being able to bond to itself in long chains. This self-bonding is so important in living organisms because it allows many different compounds to be formed by mere rearrangement of the existing skeleton, not by having to reduce the compound to its different elements and then resynthesize them from scratch. For example, even a four-carbon chain has three different possible skeletons. Adding just one oxygen or double bond to this simple molecule can provide many different structures, each potentially with a different biological function.

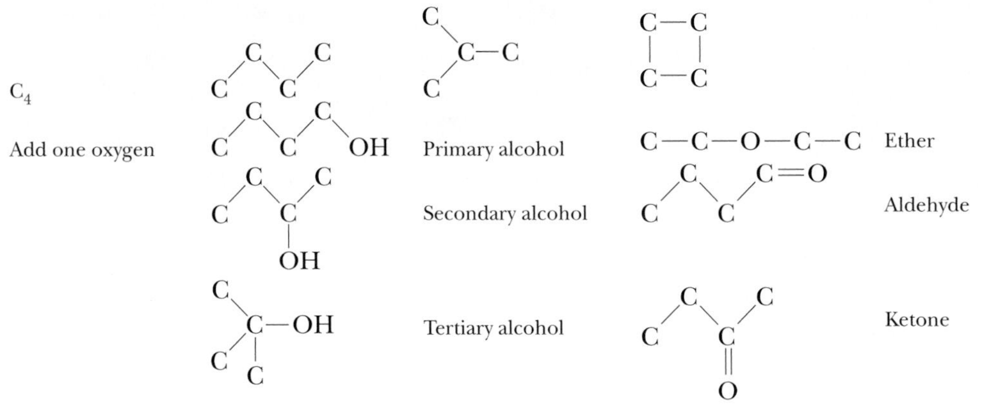

Two examples illustrate the difference that minor structural change can make. The simple sugars include glucose (a not-so-sweet aldehyde) and fructose (a very sweet ketone), both with the molecular formula $C_6H_{12}O_6$. The chemical differences between testosterone (a male sex hormone) and estrogen (a female sex hormone) are minor, although the biological difference is not.

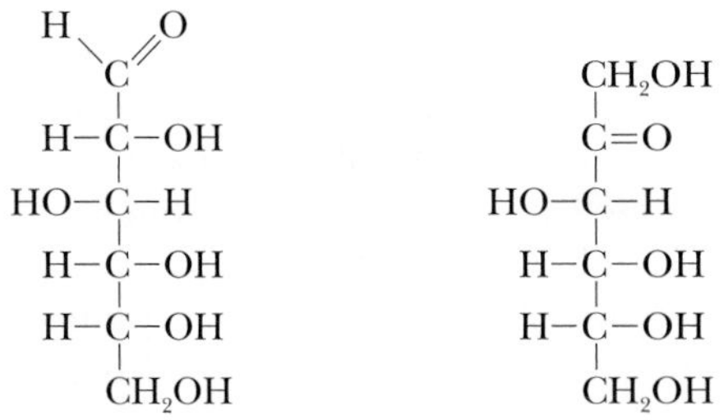

Glucose

Fructose

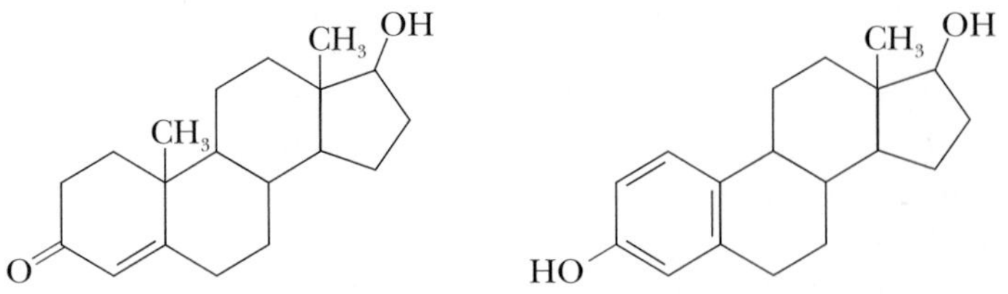

Testosterone

Estrogen (estradiol)

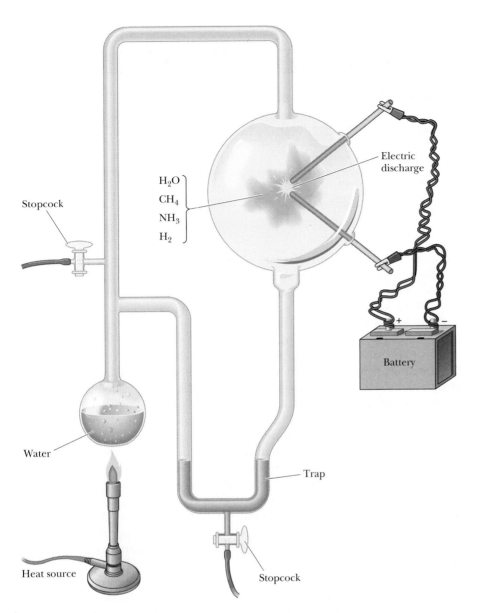

H_2O
CH_4
NH_3
H_2

Stopcock

Electric
discharge

Battery

Water

Heat source

Trap

Stopcock

◀ **FIGURE 1.4** An example of the Miller–Urey experiment. Water is heated in a closed system that also contains CH_4, NH_3, and H_2. An electric discharge is passed through the mixture of gases to simulate lightning. After the reaction has been allowed to take place for several days, organic molecules such as formaldehyde (HCHO) and hydrogen cyanide (HCN) accumulate. Amino acids are also frequently encountered as products of such reactions.

can be linked to produce macromolecules, which are also called **polymers.** The versatility of carbon is important here. Carbon is tetravalent and able to form bonds with itself and with many other elements, giving rise to different kinds of monomers, such as amino acids, nucleotides, and monosaccharides (sugar monomers). In present-day cells, amino acids (the monomers) combine by polymerization to form **proteins,** and nucleotides (also monomers) combine to form **nucleic acids;** the polymerization of sugar monomers produces polysaccharides. Polymerization experiments with amino acids carried out under early-Earth conditions have produced proteinlike polymers. Similar experiments have been done on the abiotic polymerization of nucleotides and sugars, which tends to happen less readily than the polymerization of amino acids.

The several types of amino acids and nucleotides can easily be distinguished from one another. When amino acids form polymers, with the loss of water accompanying this spontaneous process, the sequence of amino acids determines the properties of the polypeptide formed. Likewise, the genetic code lies in the sequence of monomeric nucleotides that polymerize to form nucleic acids (Figure 1.5). In polysaccharides, however, the order of

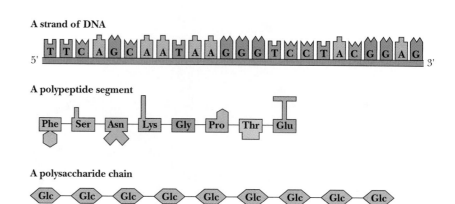

A strand of DNA

A polypeptide segment

A polysaccharide chain

Biochemistry☰Now™ ACTIVE FIGURE 1.5
Biological macromolecules are informational. The sequence of monomeric units in a biological polymer has the potential to contain information if the order of units is not overly repetitive. Nucleic acids and proteins are informational macromolecules; polysaccharides are not. **Watch this Active Figure at http://now.brookscole.com/campbell5**

monomers rarely has an important effect on the properties of the polymer, nor does the order of the monomers carry any genetic information. (Other aspects of the *linkage* between monomers are important in polysaccharides, as we shall see when we discuss carbohydrates in Chapter 16). Notice that all the building blocks have a "head" and a "tail," giving a sense of direction even at the monomer level (Figure 1.6).

The effect of monomer sequence on the properties of polymers can be illustrated by another example. Proteins of the class called *enzymes* display **catalytic activity,** which means that they increase the rates of chemical reactions compared with uncatalyzed reactions. In the context of the origin of life, catalytic molecules can facilitate the production of large numbers of complex molecules, allowing for the accumulation of such molecules. When a large group of related molecules accumulates, a complex system arises with some of the characteristics of living organisms. Such a system has a nonrandom organization, it tends to reproduce itself, and it competes with other systems for the simple organic molecules present in the environment. One of the most important functions of proteins is **catalysis,** and the catalytic effectiveness of a given enzyme depends on its amino acid sequence. The specific sequence of the amino acids present ultimately determines the properties of all types of proteins, including enzymes.

In present-day cells, the sequence of amino acids in proteins is determined by the sequence of nucleotides in nucleic acids. The process by which genetic information is translated into the amino acid sequence is very complex. *DNA (deoxyribonucleic acid),* one of the nucleic acids, serves as the coding material. The **genetic code** is the relationship between the nucleotide sequence in nucleic acids and the amino acid sequence in proteins. As a result of this relationship, the information for the structure and function of all living things is passed from one generation to the next. The workings of the genetic code are no longer completely mysterious, but they are far from completely understood. Theories on the origins of life consider how a coding system might have developed, and new insights in this area could shine some light on the present-day genetic code.

Molecules to Cells

A discovery with profound implications for discussions of the origin of life is that *RNA (ribonucleic acid),* another nucleic acid, is capable of catalyzing its own processing. Until this discovery, catalytic activity was associated exclusively with proteins. RNA, rather than DNA, is now considered by many scientists to have been the original coding material, and it still serves this function in some viruses. The idea that catalysis and coding both occur in one molecule has provided a point of departure for more research on the origins of life.

Essential Information

Several classes of molecules play a key role in life processes. Among the most important are proteins and nucleic acids. Both proteins and nucleic acids are polymers, very large molecules formed by linking together smaller units called monomers. In the case of proteins, the monomers are amino acids; in nucleic acids, the monomers are nucleotides.

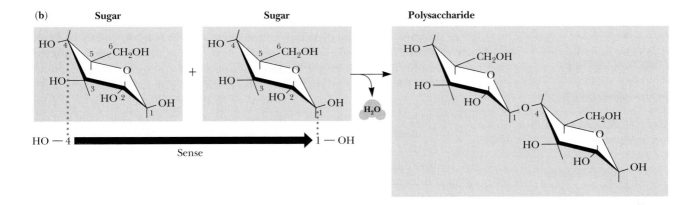

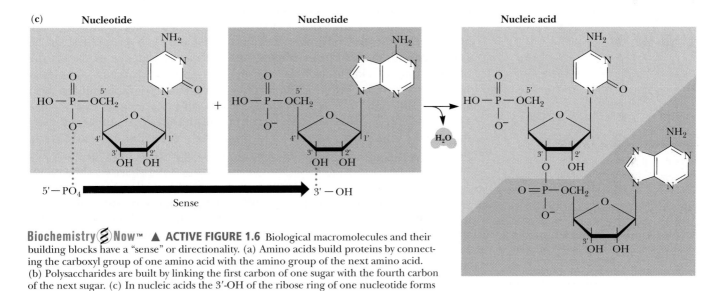

Biochemistry ⊛Now™ ▲ **ACTIVE FIGURE 1.6** Biological macromolecules and their building blocks have a "sense" or directionality. (a) Amino acids build proteins by connecting the carboxyl group of one amino acid with the amino group of the next amino acid. (b) Polysaccharides are built by linking the first carbon of one sugar with the fourth carbon of the next sugar. (c) In nucleic acids the 3′-OH of the ribose ring of one nucleotide forms a bond to the 5′-OH of the ribose ring of a neighboring nucleotide. All these polymerization reactions are accompanied by the elimination of water. **Watch this Active Figure at http://now.brookscole.com/campbell5**

(See the article by Cech in the bibliography at the end of this chapter.) The "RNA world" is the current conventional wisdom, but many unanswered questions exist regarding this point of view.

According to the RNA-world theory, the appearance of a form of RNA capable of coding for its own replication was the pivotal point in the origin of life. Polynucleotides can direct the formation of molecules whose sequence is an exact copy of the original. This process depends on a template mechanism (Figure 1.7), which is highly effective in producing exact copies but is a relatively slow process. A catalyst is required, which can be a polynucleotide, even the original molecule itself. Polypeptides, however, are more efficient catalysts than polynucleotides, but there is still the question whether they can direct

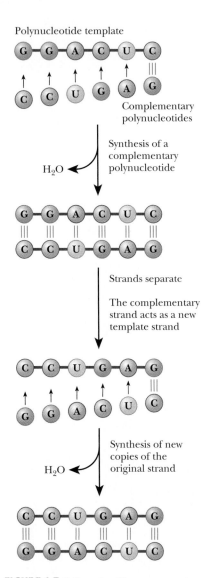

▲ **FIGURE 1.7** Polynucleotides use a template mechanism to produce exact copies of themselves: G pairs with C, and A pairs with U by a relatively weak interaction. The original strand acts as a template to direct the synthesis of a complementary strand. The complementary strand then acts as a template for the production of copies of the original strand. Note that the original strand can be a template for a number of complementary strands, each of which in turn can produce a number of copies of the original strand. This process gives rise to a many-fold amplification of the original sequence. (*Copyright © 1994 from* The Molecular Biology of the Cell, *3rd Edition by A. Alberts, D. Bray, J. Lewis, M. Raff, K. Roberts, and J. D. Watson. Reproduced by permission of Garland Science/Taylor & Francis Books, Inc.*)

the formation of exact copies of themselves. Recall that, in present-day cells, the genetic code is based on nucleic acids, and catalysis relies primarily on proteins. How did nucleic acid synthesis (which requires many protein enzymes) and protein synthesis (which requires the genetic code to specify the order of amino acids) come to be? According to this hypothesis, RNA (or a system of related kinds of RNA) originally played both roles, catalyzing and encoding its own replication. Eventually, the system evolved to the point of being able to encode the synthesis of more effective catalysts, namely proteins (Figure 1.8). Even later, DNA took over as the primary genetic material, relegating the more versatile RNA to an intermediary role in directing the synthesis of proteins under the direction of the genetic code residing in DNA. A certain amount of controversy surrounds this theory, but it has attracted considerable attention recently. Many unanswered questions remain about the role of RNA in the origin of life, but clearly that role must be important.

Another key point in the development of living cells is the formation of membranes that separate cells from their environment. The clustering of coding and catalytic molecules in a separate compartment brings molecules into closer contact with each other and excludes extraneous material. For reasons we shall explore in detail in Chapters 2 and 8, lipids are perfectly suited to form cell membranes (Figure 1.9).

Some theories on the origin of life focus on the importance of proteins in the development of the first cells. A strong piece of experimental evidence for the importance of proteins is that amino acids form readily under abiotic conditions, whereas nucleotides do so with great difficulty. Proteinoids are artificially synthesized polymers of amino acids, and their properties can be compared with those of true proteins. Although some evidence exists that the order of amino acids in artificially synthesized proteinoids is not completely random—a certain order is preferred—there is no definite amino acid sequence. In contrast, *a well-established, unique amino acid sequence exists for each protein produced by present-day cells.* According to the theory that gives primary importance to proteins, aggregates of proteinoids formed on the early Earth, probably in the oceans or at their edges. These aggregates took up other abiotically produced precursors of biomolecules to become *protocells,* the precursors of true cells. Several researchers have devised model systems for protocells. In one model, artificially synthesized proteinoids are induced to aggregate, forming structures called *microspheres.* Proteinoid microspheres are spherical in shape, as the name implies, and, in a given sample, they are approximately uniform in diameter. Such microspheres are certainly not cells, but they provide a model for protocells. Microspheres prepared from proteinoids with catalytic activity exhibit the same catalytic activity as the proteinoids. Furthermore, it is possible to construct such aggregates with more than one type of catalytic activity as a model for primitive cells. Note that these aggregates lack a coding system. Self-replication of peptides (coding and catalysis carried out by the same molecule) has been reported (see the article by Lee et al. in the bibliography at the end of this chapter), but that work was done on isolated peptides, not on aggregates.

Recently, attempts have been made to combine several lines of reasoning about the origin of life into a *double-origin theory.* According to this line of thought, the development of catalysis and the development of a coding system came about separately, and the combination of the two produced life as we know it. The rise of aggregates of molecules capable of catalyzing reactions was one origin of life, and the rise of a nucleic acid-based coding system was another origin.

A theory that life began on clay particles is a form of the double-origin theory. According to this point of view, coding arose first, but the coding material was the surface of naturally occurring clay. The pattern of ions on the clay

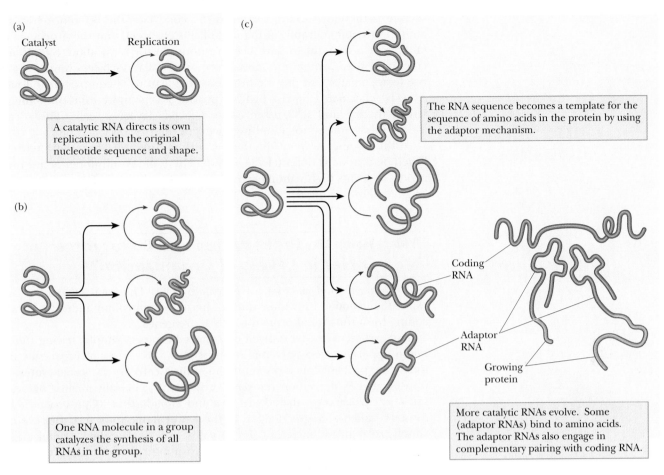

(a)

Catalyst   Replication

A catalytic RNA directs its own replication with the original nucleotide sequence and shape.

(b)

One RNA molecule in a group catalyzes the synthesis of all RNAs in the group.

(c)

The RNA sequence becomes a template for the sequence of amino acids in the protein by using the adaptor mechanism.

Coding RNA

Adaptor RNA

Growing protein

More catalytic RNAs evolve. Some (adaptor RNAs) bind to amino acids. The adaptor RNAs also engage in complementary pairing with coding RNA.

▲ **FIGURE 1.8** Stages in the evolution of a system of self-replicating RNA molecules. At each stage, more complexity appears in the group of RNAs, leading eventually to the synthesis of proteins as more effective catalysts. (*Copyright © 1994 from* The Molecular Biology of the Cell, *3rd Edition by A. Alberts, D. Bray, J. Lewis, M. Raff, K. Roberts, and J. D. Watson. Reproduced by permission of Garland Science/Taylor & Francis Books, Inc.*)

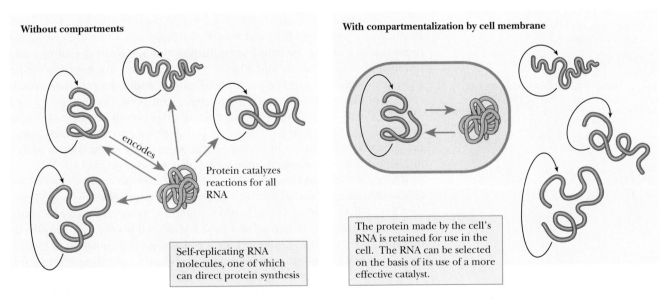

Without compartments

encodes

Protein catalyzes reactions for all RNA

Self-replicating RNA molecules, one of which can direct protein synthesis

With compartmentalization by cell membrane

The protein made by the cell's RNA is retained for use in the cell. The RNA can be selected on the basis of its use of a more effective catalyst.

▲ **FIGURE 1.9** The vital importance of a cell membrane in the origin of life. Without compartments, groups of RNA molecules must compete with others in their environment for the proteins they synthesize. With compartments, the RNAs have exclusive access to the more effective catalysts and are closer to each other, making it easier for reactions to take place. (*Copyright © 1994 from* The Molecular Biology of the Cell, *3rd Edition by A. Alberts, D. Bray, J. Lewis, M. Raff, K. Roberts, and J. D. Watson. Reproduced by permission of Garland Science/Taylor & Francis Books, Inc.*)

surface is thought to have served as the code (see the reference by Cairns-Smith in the bibliography at the end of this chapter), and the process of crystal growth is thought to have been responsible for replication. Simple molecules, and then protein enzymes, arose on the clay surface, eventually giving rise to aggregates that provided the essential feature of compartmentalization. At some later date, the rise of RNA provided a far more efficient coding system than clay, and RNA-based cells replaced clay-based cells. This scenario assumes that time is not a limiting factor in the process.

At this writing, none of the theories of the origin of life is definitely established, and none is definitely disproved. The topic is still under active investigation. It seems highly unlikely that we will ever know with certainty how life originated on this planet.

1.4 | How Do Prokaryotes and Eukaryotes Differ in Levels of Organization?

Both prokaryotic and eukaryotic cells contain DNA. The total DNA of a cell is called the **genome.** Individual units of heredity, controlling individual traits by coding for a functional protein or RNA, are **genes.**

The earliest cells that evolved must have been very simple, having the minimum apparatus necessary for life processes. The types of organisms living today that probably most resemble the earliest cells are the **prokaryotes.** This word, of Greek derivation (*karyon,* "kernel, nut"), literally means "before the nucleus." Prokaryotes include *bacteria* and *cyanobacteria.* (Cyanobacteria were formerly called blue-green algae; as the newer name indicates, they are more closely related to bacteria.) Prokaryotes are single-celled organisms, but groups of them can exist in association, forming colonies with some differentiation of cellular functions.

The word "eukaryote" means true nucleus. **Eukaryotes** are more complex organisms and can be multicellular or single-celled. A well-defined nucleus, set off from the rest of the cell by a membrane, is one of the chief features distinguishing a eukaryote from a prokaryote. A growing body of fossil evidence indicates that eukaryotes evolved from prokaryotes about 1.5 billion (1.5×10^9) years ago, about 2 billion years after life first appeared on Earth. Examples of single-celled eukaryotes include yeasts and *Paramecium* (an organism frequently discussed in beginning biology courses); all multicellular organisms (e.g., animals and plants) are eukaryotes. As might be expected, eukaryotic cells are more complex and usually much larger than prokaryotic cells. The diameter of a typical prokaryotic cell is on the order of 1 to 3 μm $(1 \times 10^{-6}$ to 3×10^{-6} m), whereas that of a typical eukaryotic cell is about 10 to 100 μm. The distinction between prokaryotes and eukaryotes is so basic that it is now a key point in the classification of living organisms; it is far more important than the distinction between plants and animals.

The main difference between prokaryotic and eukaryotic cells is the existence of organelles, especially the nucleus, in eukaryotes. An **organelle** is a part of the cell that has a distinct function; it is surrounded by its own membrane within the cell. In contrast, the structure of a prokaryotic cell is relatively simple, lacking membrane-enclosed organelles. Like a eukaryotic cell, however, a prokaryotic cell has a cell membrane, or plasma membrane, separating it from the outside world. The plasma membrane is the only membrane found in the prokaryotic cell. Both in prokaryotes and in eukaryotes, the cell membrane consists of a double layer (bilayer) of lipid molecules with a variety of proteins embedded in it.

Organelles have specific functions. A typical eukaryotic cell has a *nucleus* with a nuclear membrane. *Mitochondria* (respiratory organelles) and an inter-

Table 1.3

A Comparison of Prokaryotes and Eukaryotes

Organelle	Prokaryotes	Eukaryotes
Nucleus	No definite nucleus; DNA present but not separate from rest of cell	Present
Cell membrane (plasma membrane)	Present	Present
Mitochondria	None; enzymes for oxidation reactions located on plasma membrane	Present
Endoplasmic reticulum	None	Present
Ribosomes	Present	Present
Chloroplasts	None; photosynthesis (if present) is localized in chromatophores	Present in green plants

nal membrane system known as the *endoplasmic reticulum* are also common to all eukaryotic cells. Energy-yielding oxidation reactions take place in eukaryotic mitochondria. In prokaryotes, similar reactions occur on the plasma membrane. *Ribosomes* (particles consisting of RNA and protein), which are the sites of protein synthesis in all living organisms, are frequently bound to the endoplasmic reticulum in eukaryotes. In prokaryotes, ribosomes are found free in the cytosol. A distinction can be made between the cytoplasm and the cytosol. *Cytoplasm* refers to the portion of the cell outside the nucleus, and the *cytosol* is the aqueous portion of the cell that lies outside the membrane-bounded organelles. *Chloroplasts,* organelles in which photosynthesis takes place, are found in plant cells and green algae. In prokaryotes that are capable of photosynthesis, the reactions take place in layers called *chromatophores,* which are extensions of the plasma membrane, rather than in chloroplasts.

Table 1.3 summarizes the basic differences between prokaryotic and eukaryotic cells.

1.5 | What Are the Main Structural Features of Prokaryotic Cells?

Although no well-defined nucleus is present in prokaryotes, the DNA of the cell is concentrated in one region called the **nuclear region.** This part of the cell directs the workings of the cell very much as the eukaryotic nucleus does. The DNA of prokaryotes is not complexed with proteins in extensive arrays with specified architecture, as is the DNA of eukaryotes. In general, there is only a single, closed, circular molecule of DNA in prokaryotes. This circle of DNA, which is the genome, is attached to the cell membrane. Before a prokaryotic cell divides, the DNA replicates itself, and both DNA circles are bound to the plasma membrane. The cell then divides, and each of the two daughter cells receives one copy of the DNA (Figure 1.10).

In a prokaryotic cell, the cytosol (the fluid portion of the cell outside the nuclear region) frequently has a slightly granular appearance because of the presence of **ribosomes.** Because these consist of RNA and protein, they are also called *ribonucleoprotein particles;* they are the sites of protein synthesis in all organisms. The presence of ribosomes is the main visible feature of prokaryotic cytosol. (Membrane-bound organelles, characteristic of eukaryotes, are not found in prokaryotes.)

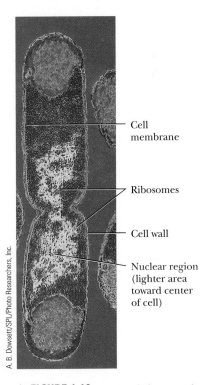

Cell membrane

Ribosomes

Cell wall

Nuclear region (lighter area toward center of cell)

A. B. Dowsett/SPL/Photo Researchers, Inc.

▲ **FIGURE 1.10** A colored electron microscope image of a typical prokaryote: the bacterium *Escherichia coli* (magnified 16,500×). The pair in the center shows that division into two cells is nearly complete.

Every cell is separated from the outside world by a **cell membrane,** or plasma membrane, an assemblage of lipid molecules and proteins. In addition to the cell membrane and external to it, a prokaryotic bacterial cell has a **cell wall,** which is made up mostly of polysaccharide material, a feature it shares with eukaryotic plant cells. The chemical natures of prokaryotic and eukaryotic cell walls differ somewhat, but a common feature is that the polymerization of sugars produces the polysaccharides found in both. Because the cell wall is made up of rigid material, it presumably serves as protection for the cell.

1.6	**What Are the Main Structural Features of Eukaryotic Cells?**

Multicellular plants and animals are eukaryotes, as are protista and fungi, but obvious differences exist among them. These differences are reflected on the cellular level. Plant cells, like bacteria, have cell walls. A plant cell wall is mostly made up of the polysaccharide cellulose, giving the cell its shape and mechanical stability. **Chloroplasts,** the photosynthetic organelles, are found in green plants and algae.

Animal cells have neither cell walls nor chloroplasts; the same is true of some protists. Figure 1.11 shows some of the important differences between typical plant cells, typical animal cells, and prokaryotes.

Important Organelles

The **nucleus** is perhaps the most important eukaryotic organelle. A typical nucleus exhibits several important structural features (Figure 1.12). It is surrounded by a *nuclear double membrane* (usually called the nuclear envelope). One of its prominent features is the **nucleolus,** which is rich in RNA. The RNA of a cell (with the exception of the small amount produced in such organelles as mitochondria and chloroplasts) is synthesized on a DNA template in the nucleolus for export to the cytoplasm through pores in the nuclear membrane. This RNA is ultimately destined for the ribosomes. Also visible in the nucleus, frequently near the nuclear membrane, is **chromatin,** an aggregate of DNA and protein. The main eukaryotic genome (its nuclear DNA) is duplicated before cell division takes place, as in prokaryotes. In eukaryotes, both copies of DNA, which are to be equally distributed between the daughter cells, are associated with protein. When a cell is about to divide, the loosely organized strands of chromatin become tightly coiled, and the resulting **chromosomes** can be seen under a microscope. The genes, responsible for the transmission of inherited traits, are part of the DNA found in each chromosome.

A second very important eukaryotic organelle is the **mitochondrion,** which, like the nucleus, has a double membrane (Figure 1.13). The outer membrane has a fairly smooth surface, but the inner membrane exhibits many folds called **cristae.** The space within the inner membrane is called the **matrix.** Oxidation processes that occur in mitochondria yield energy for the cell. Most of the enzymes responsible for these important reactions are associated with the inner mitochondrial membrane. Other enzymes needed for oxidation reactions, as well as DNA that differs from that found in the nucleus, are found in the internal mitochondrial matrix. Mitochondria also contain ribosomes similar to those found in bacteria. Mitochondria are approximately the size of many bacteria, typically about 1 μm in diameter and 2 to 8 μm in length. In theory, they may have arisen from the absorption of aerobic bacteria by larger host cells.

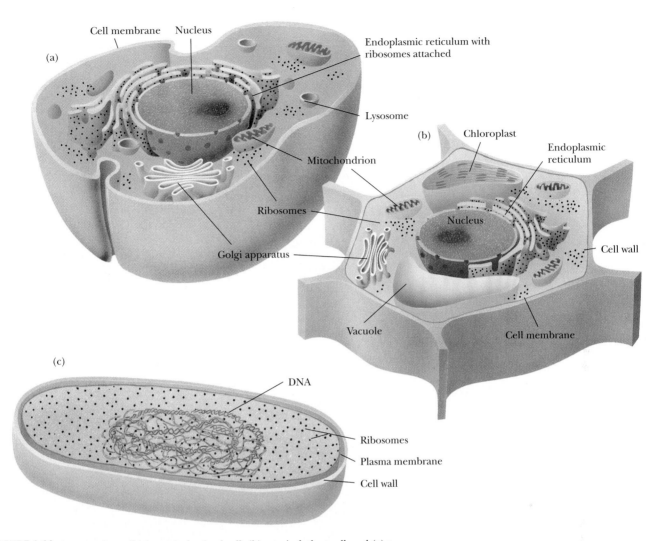

▲ FIGURE 1.11 A comparison of (a) a typical animal cell, (b) a typical plant cell, and (c) a prokaryotic cell.

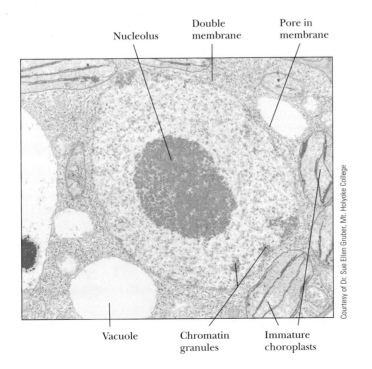

◄ FIGURE 1.12 The nucleus of a tobacco leaf cell (magnified 15,000×).

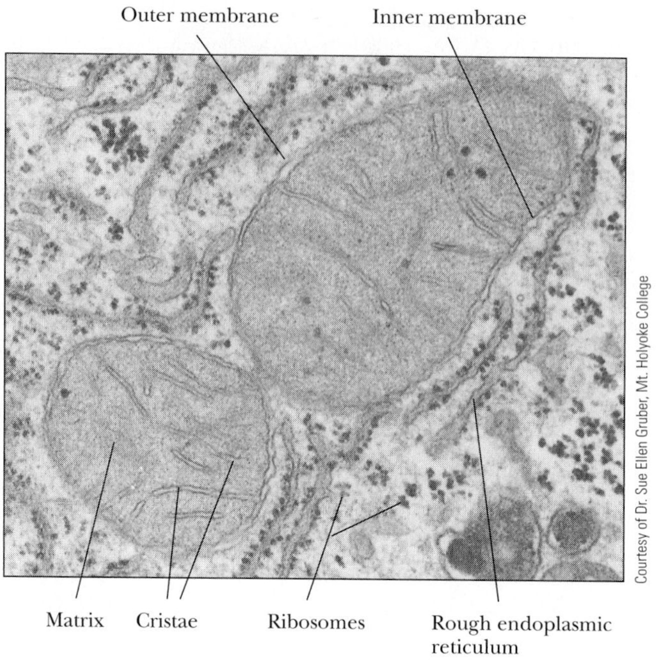

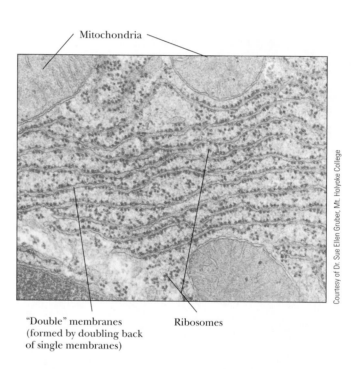

Outer membrane   Inner membrane

Matrix   Cristae   Ribosomes   Rough endoplasmic reticulum

Courtesy of Dr. Sue Ellen Gruber, Mt. Holyoke College

▶ **FIGURE 1.13** Mouse liver mitochondria (magnified 50,000×).

Mitochondria

"Double" membranes (formed by doubling back of single membranes)   Ribosomes

Courtesy of Dr. Sue Ellen Gruber, Mt. Holyoke College

▶ **FIGURE 1.14** Rough endoplasmic reticulum from mouse liver cells (magnified 50,000×).

The **endoplasmic reticulum (ER)** is part of a continuous single-membrane system throughout the cell; the membrane doubles back on itself to give the appearance of a double membrane in electron micrographs. The endoplasmic reticulum is attached to the cell membrane and to the nuclear membrane. It occurs in two forms, rough and smooth. The *rough endoplasmic reticulum* is studded with ribosomes bound to the membrane (Figure 1.14). Ribosomes, which can also be found free in the cytosol, are the sites of protein synthesis in all organisms. The *smooth endoplasmic reticulum* does not have ribosomes bound to it.

Chloroplasts are important organelles found only in green plants and green algae. Their structure includes membranes, and they are relatively large, typically up to 2 μm in diameter and 5 to 10 μm in length. The photosynthetic apparatus is found in specialized structures called *grana* (singular *granum*), membranous bodies stacked within the chloroplast. Grana are easily seen through an electron microscope (Figure 1.15). Chloroplasts, like mitochondria, contain a characteristic DNA that is different from that found in the nucleus. Chloroplasts and mitochondria also contain ribosomes similar to those found in bacteria.

Other Organelles and Cellular Constituents

Membranes are important in the structures of some less well-understood organelles. One, the **Golgi apparatus,** is separate from the endoplasmic reticulum but is frequently found close to the smooth endoplasmic reticulum. It is a series of membranous sacs (Figure 1.16). The Golgi apparatus is involved in secretion of proteins from the cell, but it also occurs in cells in which the primary function is not protein secretion. In particular, it is the site in the cell in which sugars are linked to other cellular components, such as proteins. The function of this organelle is still a subject of research.

Other organelles in eukaryotes are similar to the Golgi apparatus in that they involve single, smooth membranes and have specialized functions. **Lysosomes,** for example, are membrane-enclosed sacs containing hydrolytic enzymes that could cause considerable damage to the cell if they were not physically separated from the lipids, proteins, or nucleic acids that they are

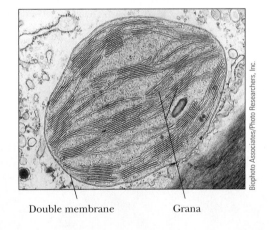

Double membrane   Grana

◀ **FIGURE 1.15** An electron microscope image of a chloroplast from the alga *Nitella* (magnified 60,000×).

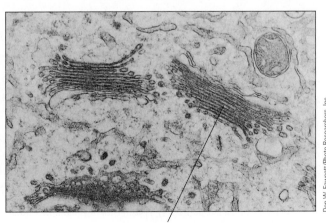

Stack of flattened membranous vesicles

◀ **FIGURE 1.16** Golgi apparatus from a mammalian cell (magnified 25,000×).

able to attack. Inside the lysosome, these enzymes break down target molecules, usually from outside sources, as a first step in processing nutrients for the cell. **Peroxisomes** are similar to lysosomes; their principal characteristic is that they contain enzymes involved in the metabolism of hydrogen peroxide (H_2O_2), which is toxic to the cell. The enzyme *catalase,* which occurs in peroxisomes, catalyzes the conversion of H_2O_2 to H_2O and O_2. **Glyoxysomes** are found in plant cells only. They contain the enzymes that catalyze the *glyoxylate cycle,* a pathway that converts some lipids to carbohydrate with glyoxylic acid as an intermediate.

The **cytosol** was long considered to be nothing more than a viscous liquid, but recent studies by electron microscopy have revealed that this part of the cell has some internal organization. The organelles are held in place by a lattice of fine strands that seem to consist mostly of protein. This **cytoskeleton,** or *microtrabecular lattice,* is connected to all organelles (Figure 1.17). Many questions remain about its function in cellular organization, but its importance in maintaining the infrastructure of the cell is not doubted.

The cell membrane of eukaryotes serves to separate the cell from the outside world. It consists of a double layer of lipids, with several types of proteins embedded in the lipid matrix. Some of the proteins transport specific substances across the membrane barrier. Transport can take place in both directions, with substances useful to the cell being taken in and others being exported.

Plant cells (and algae), but not animal cells, have cell walls external to the plasma membrane. The cellulose that makes up plant cell walls is a major component of plant material; wood, cotton, linen, and most types of paper are mainly cellulose. Also present in plant cells are large central **vacuoles,** sacs in the cytoplasm surrounded by a single membrane. Although vacuoles sometimes appear in animal cells, those in plants are more prominent. They tend to increase in number and size as the plant cell ages. An important function of vacuoles is to isolate waste substances that are toxic to the plant and are produced in greater amounts than the plant can secrete to the environment.

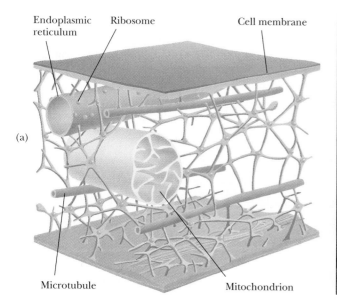

(a)

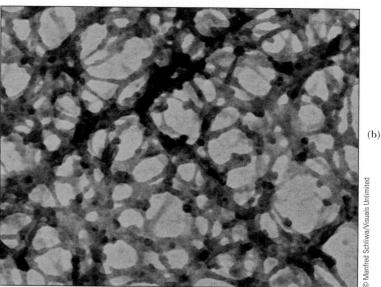

(b)

© Manfred Schliwa/Visuals Unlimited

▲ **FIGURE 1.17** The microtrabecular lattice. (a) This network of filaments, also called the cytoskeleton, pervades the cytosol. Some filaments, called microtubules, are known to consist of the protein tubulin. Organelles such as mitochondria are attached to the filaments. (b) An electron micrograph of the microtrabecular lattice (magnified 87,450×).

Table 1.4	
A Summary of Organelles and Their Functions	
Organelle	**Function**
Nucleus	Location of main genome; site of most DNA and RNA synthesis
Mitochondrion	Site of energy-yielding oxidation reactions; has its own DNA
Chloroplast	Site of photosynthesis in green plants and algae; has its own DNA
Endoplasmic reticulum	Continuous membrane throughout the cell; rough part studded with *ribosomes (the site of protein synthesis)**
Golgi apparatus	Series of flattened membranes; involved in secretion of proteins from cells and in reactions that link sugars to other cellular components
Lysosomes	Membrane-enclosed sacs containing hydrolytic enzymes
Peroxisomes	Sacs that contain enzymes involved in the metabolism of hydrogen peroxide
Cell membrane	Separates the cell contents from the outside world; contents include organelles (held in place by the *cytoskeleton**) and the *cytosol*
Cell wall	Rigid exterior layer of plant cells
Central vacuole	Membrane-bounded sac (plant cells)

* Because an organelle is defined as a portion of a cell enclosed by a membrane, ribosomes are not, strictly speaking, organelles. Smooth endoplasmic reticulum does not have ribosomes attached, and ribosomes also occur free in the cytosol. The definition of organelle also affects discussion of the cell membrane, cytosol, and cytoskeleton.

These waste products may be unpalatable or even poisonous enough to discourage herbivores (plant-eating organisms) from ingesting them and may thus provide some protection for the plant.

Table 1.4 summarizes organelles and their functions.

1.7 | How Do We Classify Organisms: Five Kingdoms or Three Domains?

The original biological classification scheme, established in the 18th century, divided all organisms into two kingdoms: the plants and the animals. In this scheme, plants are organisms that obtain food directly from the Sun, and animals are organisms that move about to search for food. It was discovered that some organisms, bacteria in particular, do not have an obvious relationship to either kingdom. It has also become clear that a more fundamental division of living organisms is actually not between plants and animals, but between prokaryotes and eukaryotes. In the 20th century, classification schemes that divide living organisms into more than the two traditional kingdoms have been introduced. The five-kingdom system takes into account the differences between prokaryotes and eukaryotes, and it also provides classifications for eukaryotes that appear to be neither plants nor animals.

The kingdom **Monera** consists only of prokaryotic organisms. Bacteria and cyanobacteria are members of this kingdom. The other four kingdoms are made up of eukaryotic organisms. The kingdom **Protista** includes unicellular organisms such as *Euglena, Volvox, Amoeba,* and *Paramecium.* Some protists, including algae, are multicellular. The three kingdoms that consist mainly of multicellular eukaryotes (with a few unicellular eukaryotes) are Fungi, Plantae,

▲ **FIGURE 1.18** The five-kingdom classification scheme.

and Animalia. The kingdom Fungi includes yeasts, molds, and mushrooms. Fungi, plants, and animals must have evolved from simpler eukaryotic ancestors, but the major evolutionary change was the development of eukaryotes from prokaryotes (Figure 1.18).

There is a group of organisms that can be classified as prokaryotes in the sense that the organisms lack a well-defined nucleus. These organisms are called **archaebacteria** (early bacteria) to distinguish them from **eubacteria** (true bacteria) because there are marked differences between the two kinds of organisms. Archaebacteria are found in extreme environments (see Biochemical Connections box) and, for this reason, are also called extremophiles. Most of the differences between archaebacteria and other organisms are biochemical features, such as the molecular structure of the cell walls, membranes, and some types of RNA. (The article by Woese listed in the bibliography at the end of this chapter makes biochemical comparisons between archaebacteria and other life forms.) Some biologists prefer a three-domain classification scheme—**Bacteria** (eubacteria), **Archaea** (archaebacteria), and **Eukarya** (eukaryotes)—to the five-kingdom classification (Figure 1.19). The basis for this preference is the emphasis on biochemistry as the basis for classification. The three-domain classification scheme will certainly become more important as time goes on. A complete genome of the archaebacterium *Methanococcus jannaschii* has been obtained (see the article by Morrell in the

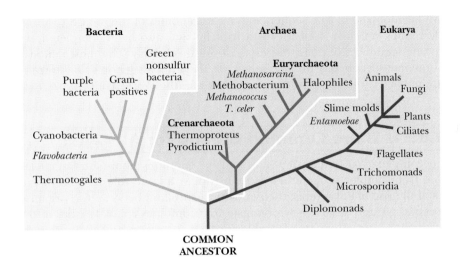

◄ FIGURE 1.19 The three-domain classification scheme. Two domains, Bacteria and Archaea, consist of prokaryotes. The third kingdom, Eukarya, consists of eukaryotes. All three domains have a common ancestor early in evolution. (*Reprinted with permission from Science 273, 1044. Copyright © 1996 AAAS.*)

bibliography at the end of this chapter). More than half the genes of this organism (56%) differ markedly from genes already known in both prokaryotes and eukaryotes, a piece of evidence that lends strong support to a three-domain classification scheme. Complete genomes are being obtained for organisms from all three domains. They include those of bacteria such as *Haemophilus influenzae* and *Escherichia coli,* the latter being a bacterium in which many biochemical pathways have been investigated. Complete sequences for eukaryotes such as *Saccharomyces cerevisiae* (brewer's yeast), *Arabidopsis thaliana* (mouse-ear cress), and *Caenorhabditis elegans* (a nematode)

Biochemical Connections

Extremophiles: The Toast of the Biotechnology Industry

Archaebacteria live in extreme environments and, therefore, are sometimes called extremophiles. The three groups of archaebacteria—methanogens, halophiles, and thermacidophiles—have specific preferences about the precise nature of their environment. *Methanogens* are strict anaerobes that produce methane (CH_4) from carbon dioxide (CO_2) and hydrogen (H_2). *Halophiles* require very high salt concentrations, such as those found in the Dead Sea, for growth. *Thermacidophiles* require high temperatures and acid conditions for growth—typically, 80°C–90°C and pH 2. These growth requirements may have resulted from adaptations to harsh conditions on the early Earth. Since these organisms can tolerate these conditions, the enzymes they produce must also be stable. Most enzymes isolated from eubacteria and eukaryotes are not stable under such conditions. Some of the reactions that are of greatest importance to the biotechnology industry are both enzyme-catalyzed and carried out under conditions that cause most enzymes to lose their catalytic ability in a short time. This difficulty can be avoided by using enzymes from extremophiles. An example is the DNA polymerase from *Thermus aquaticus* (Taq polymerase). Polymerase chain reaction (PCR) technology depends heavily on the properties of this enzyme (Section 13.6). Representatives of the biotechnology industry constantly search undersea thermal vents and hot springs for organisms that can provide such enzymes.

▲ A hot spring at Yellowstone National Park. Some bacteria can thrive even in this inhospitable environment.

have been obtained. The sequencing of the genomes of the mouse (*Mus musculus*) and *Drosophila melanogaster* (a fruit fly) has also been completed, with genome sequences of many more organisms on the way. The most famous of all genome-sequencing projects, that for the human genome, has received wide publicity, with the results now available on the World Wide Web.

1.8 | Is There Common Ground for All Cells?

The complexity of eukaryotes raises many questions about how such cells arose from simpler progenitors. Symbiosis plays a large role in current theories of the rise of eukaryotes; the symbiotic association between two organisms is seen as giving rise to a new organism that combines characteristics of both the original ones. The type of symbiosis called *mutualism* is a relationship that benefits both species involved, as opposed to *parasitic symbiosis,* in which one species gains at the other's expense. A classic example of mutualism (although it has been questioned from time to time) is the lichen, which consists of a fungus and an alga. The fungus provides water and protection for the alga; the alga is photosynthetic and provides food for both partners. Another example is the root-nodule system formed by a leguminous plant, such as alfalfa or beans, and anaerobic nitrogen-fixing bacteria (Figure 1.20). The plant gains useful compounds of nitrogen, and the bacteria are protected from oxygen, which is harmful to them. Still another example of mutualistic symbiosis, of great practical interest, is that between humans and bacteria, such as *Escherichia coli,* that live in the intestinal tract. The bacteria receive nutrients and protection from their immediate environment. In return, they aid our digestive process. Without beneficial intestinal bacteria, we would soon develop dysentery and other intestinal disorders. These bacteria are also a source of certain vitamins for us, since they can synthesize these vitamins and we cannot. The disease-causing strains of *E. coli* that have been in the news from time to time differ markedly from the ones that naturally inhabit the intestinal tract.

In hereditary symbiosis, a larger host cell contains a genetically determined number of smaller organisms. An example is the protist *Cyanophora paradoxa,* a eukaryotic host that contains a genetically determined number of cyanobacteria (blue-green algae). This relationship is an example of **endosymbiosis,** because the cyanobacteria are contained within the host organism. The cyanobacteria are aerobic prokaryotes and are capable of photosynthesis (Figure 1.21). The host cell gains the products of photosynthesis; in return, the cyanobacteria are protected from the environment and still have access to oxygen and sunlight because of the host's small size. In this model, with the passage of many generations, the cyanobacteria would have gradually lost the ability to exist independently and would have become organelles within a new and more complex type of cell. Such a situation in the past may well have given rise to chloroplasts, which are not capable of independent existence. Their autonomous DNA and their apparatus for synthesizing ribosomal proteins can no longer meet all their needs, but the very fact that these organelles have their own DNA and are capable of protein synthesis suggests that they may have existed as independent organisms in the distant past.

A similar model can be proposed for the origin of mitochondria. Consider this scenario: A large anaerobic host cell assimilates a number of smaller aerobic bacteria. The larger cell protects the smaller ones and provides them with nutrients. As in the example we used for the development of chloroplasts, the smaller cells still have access to oxygen. The larger cell is not itself capable of aerobic oxidation of nutrients, but some of the end products of its anaerobic oxidation can be further oxidized by the more efficient aerobic

▲ Like that of humans, the genome of *Caenorhabditis elegans* has been decoded. *C. elegans* is ideal for studying genetic blueprints because of its tendency to reproduce by self-fertilization. This results in offspring that are identical to the parent.

▲ **FIGURE 1.20** Leguminous plants live symbiotically with nitrogen-fixing bacteria in their root systems.

▲ **FIGURE 1.21** Stromatolite fossils. Stromatolites are large, stony, cushionlike masses, composed of numerous layers of cyanobacteria (blue-green algae) that have been preserved due to their ability to secrete calcium carbonate. They are among the oldest organic remains to have been found. This specimen dates from around 2.4 billion years ago. Stromatolite formation reached a peak during the late Precambrian period (4000–570 million years ago) but is still occurring today. This specimen was found in Argentina.

metabolism of the smaller cells. As a result, the larger cell can get more energy out of a given amount of food than it could without the bacteria. In time, the two associated organisms evolve to form a new aerobic organism, which contains mitochondria derived from the original aerobic bacteria.

The fact that both mitochondria and chloroplasts have their own DNA is an important piece of biochemical evidence in favor of this model. Additionally, both mitochondria and chloroplasts have their own apparatus for synthesis of RNA and proteins. The genetic code in mitochondria differs slightly from that found in the nucleus, which supports the idea of an independent origin. Thus, the remains of these systems for synthesis of RNA and protein could reflect the organelles' former existence as free-living cells. It is reasonable to conclude that large unicellular organisms that assimilated aerobic bacteria went on to evolve mitochondria from the bacteria and eventually gave rise to animal cells. Other types of unicellular organisms assimilated both aerobic bacteria and cyanobacteria and evolved both mitochondria and chloroplasts; these organisms eventually gave rise to green plants.

The proposed connections between prokaryotes and eukaryotes are not established with complete certainty, and they leave a number of questions unanswered. Still, they provide an interesting frame of reference from which to consider evolution and the origins of the reactions that take place in cells.

1.9 | How Do Cells Use Energy?

All cells require energy for a number of purposes. Many reactions that take place in the cell, particularly those involving synthesis of large molecules, cannot take place unless energy is supplied. The Sun is the ultimate source of energy for all life on Earth. Photosynthetic organisms trap light energy and use it to drive the energy-requiring reactions that convert carbon dioxide and water to carbohydrates and oxygen. (Note that these reactions involve the chemical process of **reduction.**) Nonphotosynthetic organisms, such as animals that consume these carbohydrates, use them as energy sources. (The reactions that release energy involve the chemical process of **oxidation.**) We shall discuss the roles that oxidation and reduction reactions play in cellular processes in Chapter 15, and you will see many examples of such reactions in subsequent chapters. For the moment, it is useful and sufficient to recall from general chemistry that oxidation is the loss of electrons and reduction is the gain of electrons.

One of the most important questions about any process is whether or not it is energetically favorable. **Thermodynamics** is the branch of science that deals with this question. The key point is that *processes that release energy are favored.* Conversely, processes that require energy are disfavored. The change in energy depends only on the state of the molecules present at the start of the process and the state of those present at the end of the process. This is true whether the process in question is the formation or breaking of a bond, the formation or disruption of an intermolecular interaction, or any possible process that requires or can release energy. We are going to discuss these points in some detail when we look at protein folding in Chapter 4 and at energy considerations in metabolism in Chapter 15. This material is of central importance, and it tends to be challenging for many. What we say about it now will make it easier to apply in later chapters.

A reaction that takes place as a part of many biochemical processes is the hydrolysis of the compound adenosine triphosphate, or ATP (Section 1.2).

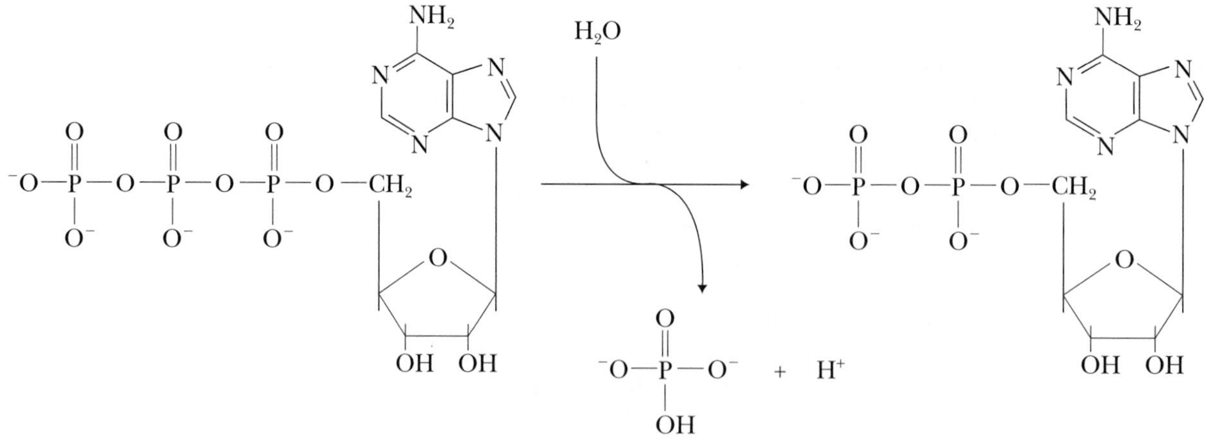

ATP
adenosine triphosphate

Phosphate ion
P_i

ADP
(adenosine diphosphate)

This is a reaction that releases energy (30.5 kJ mol^{-1} ATP = 7.3 kcal/mol ATP). More to the point, the energy released by this reaction allows energy-

(a)

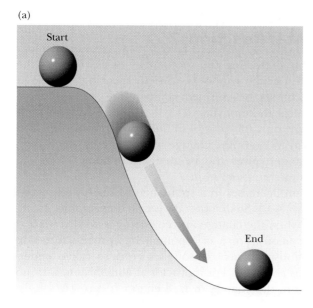

(b)

ATP

ADP + Phosphate ion

Biochemistry⊛Now™ ANIMATED FIGURE 1.22
Schematic representation of the lowering of energy.
(a) A ball rolls down a hill, releasing potential
energy. (b) ATP is hydrolyzed to produce ADP and
phosphate ion, releasing energy. The release of
energy when a ball rolls down a hill is analogous to
the release of energy in a chemical reaction. **See this
figure animated at http://now.brookscole.com/
campbell5**

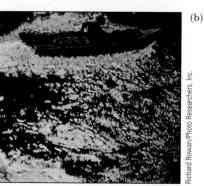

requiring reactions to proceed. Many ways are available to express energy
transfer. One of the most common is the free energy, *G*, which is discussed in
general chemistry. Also recall from general chemistry that a lowering
(release) of energy leads to a more stable state of the system under consideration. The lowering of energy is frequently shown in pictorial form as analogous to an object rolling down a hill (Figure 1.22) or over a waterfall. This
representation calls on common experience and aids understanding.

1.10 | What Is the Connection between Energy and Change?

Energy can take several forms, and it can be converted from one form to
another. All living organisms require and use energy in varied forms; for
example, motion involves mechanical energy, and maintenance of body temperature uses thermal energy. Photosynthesis requires light energy from the
Sun. Some organisms, such as several species of fish, are striking examples of
the use of chemical energy to produce electrical energy. The formation and
breakdown of biomolecules involve changes in chemical energy.

Any process that will actually take place with no outside intervention is
spontaneous in the specialized sense used in thermodynamics. *Spontaneous
does not mean "fast"; some spontaneous processes can take a long time to occur.* In the
last section, we used the term "energetically favorable" to indicate spontaneous processes. The laws of thermodynamics can be used to predict whether
any change involving transformations of energy will take place. An example of
such a change is a chemical reaction in which covalent bonds are broken and
new ones are formed. Another example is the formation of noncovalent interactions, such as hydrogen bonds, or hydrophobic interactions, when proteins
fold to produce their characteristic three-dimensional structures. The tendency of polar and nonpolar substances to exist in separate phases is a reflection of the energies of interaction between the individual molecules—in
other words, a reflection of the thermodynamics of the interaction.

▲ Two examples of transformations of energy in
biological systems. (a) This electric ray (a marine
fish in the family Torpedinidae) converts chemical
energy to electrical energy, and (b) phosphorescent
bacteria convert chemical energy into light energy.

▲ J. Willard Gibbs (1839–1903). The symbol G is given to free energy in his honor. His work is the basis of biochemical thermodynamics, and he is considered by some to have been the greatest scientist born in the United States.

1.11   What Is the Criterion for Spontaneity in Biochemical Reactions?

The most useful criterion for predicting the spontaneity of a process is the **free energy,** which is indicated by the symbol G. (Strictly speaking, the use of this criterion requires conditions of constant temperature and pressure, which are usual in biochemical thermodynamics.) It is not possible to measure absolute values of energy; only the *changes* in energy that occur during a process can be measured. The value of the change in free energy, ΔG (where the symbol Δ indicates change), gives the needed information about the spontaneity of the process under consideration.

The free energy of a system decreases in a spontaneous (energy-releasing) process, so ΔG is negative ($\Delta G < 0$). Such a process is called **exergonic,** meaning that energy is released. When the change in free energy is positive ($\Delta G > 0$), the process is nonspontaneous. For a nonspontaneous process to occur, energy must be supplied. Nonspontaneous processes are also called **endergonic,** meaning that energy is absorbed. For a process at **equilibrium,** with no net change in either direction, the change in free energy is zero ($\Delta G = 0$). *The sign of the change in free energy, ΔG, indicates the direction of the reaction:*

$$\Delta G < 0 \qquad \text{Spontaneous exergonic—energy released}$$

$$\Delta G = 0 \qquad \text{Equilibrium}$$

$$\Delta G > 0 \qquad \text{Nonspontaneous endergonic—energy required}$$

An example of a spontaneous process is the aerobic metabolism of glucose, in which glucose reacts with oxygen to produce carbon dioxide, water, and energy for the organism.

$$\text{Glucose} + 6\,O_2 \rightarrow 6CO_2 + 6H_2O \qquad \Delta G < 0$$

An example of a nonspontaneous process is the reverse of the reaction that we saw in section 1.9—namely, the phosphorylation of ADP (adenosine diphosphate) to give ATP (adenosine triphosphate). This reaction takes place in living organisms because metabolic processes supply energy.

$$\text{ADP} + {}^-\text{O}-\overset{\overset{\textstyle O}{\|}}{\underset{\underset{\textstyle OH}{|}}{P}}-\text{O}^- + H^+ \longrightarrow \text{ATP} + H_2O \qquad \Delta G > 0$$

Adenosine   Phosphate   Adenosine
diphosphate   triphosphate

1.12   What Is the Connection between Thermodynamics and Life?

From time to time, one encounters the statement that the existence of living things is a violation of the laws of thermodynamics, specifically of the second law. A look at the laws will clarify whether life is thermodynamically possible, and further discussion of thermodynamics will increase our understanding of this important topic.

The laws of thermodynamics can be stated in several ways. According to one formulation, the first law is "You can't win" and the second is "You can't break even." Put less flippantly, the first law states that it is impossible to convert energy from one form to another at greater than 100% efficiency. In other words, the first law of thermodynamics is the law of conservation of energy. The second law states that even 100% efficiency in energy transfer is impossible.

The two laws of thermodynamics can be related to the free energy by means of a well-known equation:

$$\Delta G = \Delta H - T\Delta S$$

In this equation, G is the free energy, as before; H stands for the **enthalpy,** and S for the entropy. Discussions of the first law focus on the change in enthalpy, ΔH, which is the **heat of a reaction at constant pressure.** This quantity is relatively easy to measure. Enthalpy changes for many important reactions have been determined and are available in tables in textbooks of general chemistry. Discussions of the second law focus on changes in entropy, ΔS, a concept that is less easily described and measured than changes in enthalpy. Entropy changes are particularly important in biochemistry.

One of the most useful definitions of entropy arises from statistical considerations. From a statistical point of view, an increase in the entropy of a system (the substance or substances under consideration) represents an increase in the number of possible arrangements of objects, such as individual molecules. Books have a higher entropy when they are scattered around the reading room of a library than when they are in their proper places on the shelves. Scattered books are clearly in a more dispersed state than books on shelves. The natural tendency of the universe is in the direction of increasing dispersion of energy, and living organisms put a lot of energy into maintaining order against this tendency. As all parents know, they can spend hours cleaning up a two-year-old's room, but the child can undo it all in seconds. Another statement of the second law is this: *in any spontaneous process, the entropy of the universe increases* ($\Delta S_{univ} > 0$). This statement is general, and it applies to any set of conditions. It is not confined to the special case of constant temperature and pressure, as is the statement that the free energy decreases in a spontaneous process. *Entropy changes are particularly important in determining the energetics of protein folding.*

▲ Ludwig Boltzmann (1844–1906). His equation for entropy in terms of the disorder of the universe was one of his supreme achievements; his equation is carved on his tombstone.

© Bettmann Archive/CORBIS

Biochemical Connections

Entropy and Probability

Let us consider a very simple system to illustrate the concept of entropy. We place four molecules in a container. There is an equal chance that each molecule will be on the left or on the right side of the container. Mathematically stated, the *probability* of finding a given molecule on one side is $1/2$. We can express any probability as a fraction ranging from 0 (impossible) to 1 (completely certain). We can see that 16 possible ways exist to arrange the four molecules in the container. In only one of these will all four molecules lie on the left side, but six possible arrangements exist with the four molecules evenly distributed between the two sides. *A less ordered (more dispersed) arrangement is more probable than a highly ordered arrangement.* Entropy is defined in terms of the number of possible arrangements of molecules. Boltzmann's equation for entropy, S, is $S = k \ln W$. In this equation, the term W represents the number of possible arrangements of molecules, ln is the logarithm to the base "e," and k is the constant universally referred to as Boltzmann's constant. It is equal to R/N where R is the gas constant and N is Avogadro's number (6.02×10^{23}), the number of molecules in a mole.

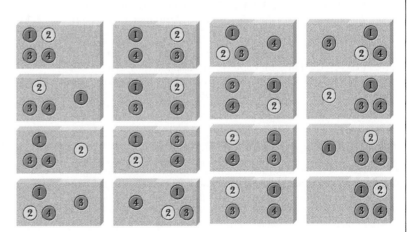

▲ The 16 possible states for a system of four molecules that may occupy either side of a container. In only one of these states are all four molecules on the left side.

Summary

1.1 What Are the Basic Themes for This Text? Biochemistry is a multidisciplinary field that addresses questions about the molecular nature of life processes. The fundamental biochemical similarities observed in all living organisms have engendered speculation about the origins of life.

1.2 What Is the Chemical Nature of Important Biomolecules? Both organic chemistry and biochemistry deal with the reactions of carbon-containing molecules. Both disciplines base their approaches on the behavior of functional groups, but their emphases differ because some functional groups important to organic chemistry do not play a role in biochemistry, and vice versa. Functional groups of importance in biochemistry include carbonyl groups, hydroxyl groups, carboxyl groups, amines, amides, and esters; derivatives of phosphoric acid such as esters and anhydrides are also important.

1.3 What Can Biochemistry Say about Possible Origins of Life? It has been shown that important biomolecules can be produced under abiotic (nonliving) conditions from simple compounds postulated to have been present in the atmosphere of the early Earth. These simple biomolecules can polymerize, also under abiotic conditions, to give rise to compounds resembling proteins and others having a less marked resemblance to nucleic acids.

All cellular activity depends on the presence of catalysts, which increase the rates of chemical reactions, and on the genetic code, which directs the synthesis of the catalysts. In present-day cells, catalytic activity is associated with proteins, and transmission of the genetic code is associated with nucleic acids, particularly with DNA. Both these functions may once have been carried out by a single biomolecule, RNA. It has been postulated that RNA was the original coding material, and it has recently been shown to have catalytic activity as well. The formation of peptide bonds in protein biosynthesis is catalyzed by the RNA portions of the ribosome.

1.4 How Do Prokaryotes and Eukaryotes Differ in Levels of Organization? Organisms are divided into two main groups based on their cell structures. *Prokaryotes* do not have internal membranes, whereas *eukaryotes* do.

1.5 What Are the Main Structural Features of Prokaryotic Cells? In *prokaryotes,* the cell lacks a well-defined nucleus and internal membrane; it has only a nuclear region, the portion of the cell that contains DNA, and a cell membrane that separates it from the outside world. The other principal feature of a prokaryotic cell's interior is the presence of ribosomes, the site of protein synthesis.

1.6 What Are the Main Structural Features of Eukaryotic Cells? In contrast, a *eukaryotic* cell has a well-defined nucleus, internal membranes as well as a cell membrane, and a considerably more complex internal structure. In eukaryotes, the nucleus is separated from the rest of the cell by a double membrane. Eukaryotic DNA in the nucleus is associated with proteins, particularly a class of proteins called histones. The combination of the two has specific structural motifs, which is not the case in prokaryotes. There is a continuous membrane system, called the endoplasmic reticulum, throughout the cell. Eukaryotic ribosomes are frequently bound to the endoplasmic reticulum, but some are also free in the cytosol. Membrane-enclosed organelles are characteristic of eukaryotic cells. Two of the most important are mitochondria, the sites of energy-yielding reactions, and chloroplasts, the sites of photosynthesis.

1.7 How Do We Classify Organisms: Five Kingdoms or Three Domains? Two ways of classifying organisms depend on the distinction between prokaryotes and eukaryotes. In the five-kingdom scheme, prokaryotes occupy the kingdom Monera. The other four kingdoms consist of eukaryotes: Protista, Fungi, Plantae, and Animalia. In the three-domain scheme, prokaryotes occupy two domains—Bacteria and Archaea—based on biochemical differences, and all eukaryotes occupy a single domain, Eukarya.

1.8 Is There Common Ground for All Cells? A good deal of research has gone into the question of how eukaryotes may have arisen from prokaryotes. Much of the thinking depends on the idea of *endosymbiosis,* in which larger cells may have absorbed aerobic bacteria, eventually giving rise to mitochondria, or photosynthetic bacteria, eventually giving rise to chloroplasts.

1.9 How Do Cells Use Energy? All cells require energy to carry out life processes. The Sun is the ultimate source of energy on Earth. Photosynthetic organisms trap light energy from the Sun as the chemical energy of the carbohydrates they produce. These carbohydrates serve as energy sources for other organisms in turn. Reactions that release energy are energetically favored, whereas those that require energy are disfavored.

1.10 What Is the Connection between Energy and Change? Thermodynamics deals with the changes in energy that determine whether a process will take place. A process that will take place without outside intervention is called *spontaneous.*

1.11 What Is the Criterion for Spontaneity in Biochemical Reactions? In a spontaneous process, the free energy decreases (ΔG is negative). In a nonspontaneous process, the free energy increases.

1.12 What Is the Connection between Thermodynamics and Life? In addition to the free energy, entropy is an important quantity in thermodynamics. The entropy of the Universe increases in any spontaneous process. Local decreases in entropy can take place within an overall increase in entropy. Living organisms represent local decreases in entropy.

Critical Questions to Review

The exercises at the end of each chapter are keyed to the critical questions that we ask in that chapter. To provide the benefit of more than one approach to review, the exercises are divided into two or more categories. *Fact Check* questions will allow you to test yourself about having important facts readily available to you. In some chapters, the material lends itself to quantitative calculations, and in those chapters you will see a *Mathematical* category. *Thought Questions* ask you to put those facts to use in questions that require use of the concepts in the chapter in moderately creative ways. A number of these exercises relate specifically to the questions we ask in this chapter, and those connections are explicitly indicated where they occur. Lastly, questions that relate specifically to Biochemical Connection boxes are labeled *Biochemical Connections.*

1.1 What Are the Basic Themes for This Text?

1. **Fact Check** State why the following terms are important in biochemistry: polymer, protein, nucleic acid, catalysis, genetic code.

1.2 What Is the Chemical Nature of Important Biomolecules?

2. Biochemical Connections Match each entry in Column a with one in Column b; Column a shows the names of some important functional groups, and Column b shows their structures.

Column a	Column b
Amino group	CH_3SH
Carbonyl group (ketone)	$CH_3CH \!=\! CHCH_3$
Hydroxyl group	$CH_3CH_2\overset{O}{\overset{\|}{C}}H$
Carboxyl group	$CH_3CH_2NH_2$
Carbonyl group (aldehyde)	$CH_3\overset{O}{\overset{\|}{C}}OCH_2CH_3$
Thiol group	$CH_3CH_2OCH_2CH_3$
Ester linkage	$CH_3\overset{O}{\overset{\|}{C}}CH_3$
Double bond	$CH_3\overset{O}{\overset{\|}{C}}OH$
Amide linkage	CH_3OH
Ether	$CH_3\overset{O}{\overset{\|}{C}}N(CH_3)_2$

3. Fact Check Identify the functional groups in the following compounds.

Glucose

A triglyceride

A peptide

Vitamin A

4. Thought Question In 1828, Wöhler was the first person to synthesize an organic compound (urea, from ammonium cyanate). How did this contribute, ultimately, to *bio*chemistry?

5. Thought Question A friend who is enthusiastic about health foods and organic gardening asks you whether urea is "organic" or "chemical." How do you reply to this question?

6. Thought Question Does biochemistry differ from organic chemistry? Explain your answer. (Consider such features as solvents, concentrations, temperatures, speed, yields, side reactions, and internal control.)

7. Biochemical Connections How many carbon skeletons can be created for a molecule with five carbon atoms? Assume that hydrogen atoms would fill out the rest of the bonds.

8. Biochemical Connections How many different structures are possible if you add just one oxygen atom to the structures in Question 7?

1.3 What Can Biochemistry Say about Possible Origins of Life?

9. Thought Question An earlier mission to Mars contained instruments that determined that amino acids were present on the surface of Mars. Why were scientists excited by this discovery?

10. Thought Question Common proteins are polymers of 20 different amino acids. How many subunits would be necessary to have an Avogadro number of possible sequences?

11. Thought Question Nucleic acids are polymers of just four different monomers in a linear arrangement. How many different sequences are available if one makes a polymer with only 40 monomers? How does this number compare with Avogadro's number?

12. Thought Question RNA is often characterized as being the first "biologically active" molecule. What two properties or activities does RNA display that are important to the evolution of life? *Hint:* Neither proteins nor DNA have *both* of these properties.

13. Thought Question Why is the development of catalysis important to the development of life?

14. Thought Question What are two major advantages of enzyme catalysts in living organisms when compared with other simple chemical catalysts such as acids or bases?

15. Thought Question Why was the development of a coding system important to the development of life?

16. Thought Question Comment on RNA's role in catalysis and coding in theories of the origin of life.

17. Thought Question Do you consider it a reasonable conjecture that cells could have arisen as bare cytoplasm without a cell membrane?

1.4 How Do Prokaryotes and Eukaryotes Differ in Levels of Organization?

18. Fact Check List five differences between prokaryotes and eukaryotes.

19. Fact Check Do the sites of protein synthesis differ in prokaryotes and eukaryotes?

1.5 What Are the Main Structural Features of Prokaryotic Cells?

20. Thought Question Assume that a scientist claims to have discovered mitochondria in bacteria. Is such a claim likely to prove valid?

1.6 What Are the Main Structural Features of Eukaryotic Cells?

21. Fact Check Draw an idealized animal cell, and identify the parts by name and function.

22. Fact Check Draw an idealized plant cell, and identify the parts by name and function.

23. **Fact Check** What are the differences between the photosynthetic apparatus of green plants and photosynthetic bacteria?

24. **Fact Check** Which organelles are surrounded by a double membrane?

25. **Fact Check** Which organelles contain DNA?

26. **Fact Check** Which organelles are the sites of energy-yielding reactions?

27. **Fact Check** State how the following organelles differ from each other in terms of structure and function: Golgi apparatus, lysosomes, peroxisomes, glyoxysomes. How do they resemble each other?

1.7 How Do We Classify Organisms: Five Kingdoms or Three Domains?

28. **Fact Check** List the five kingdoms into which living organisms are divided, and give at least one example of an organism belonging to each kingdom.

29. **Fact Check** Which of the five kingdoms consist of prokaryotes? Which consist of eukaryotes?

30. **Fact Check** List the three domains into which living organisms are divided, and indicate how this scheme differs from the five-kingdom classification scheme.

1.8 Is There Common Ground for All Cells?

31. **Thought Question** What are the advantages of being eukaryotic (as opposed to prokaryotic)?

32. **Thought Question** Mitochondria and chloroplasts contain some DNA, which more closely resembles prokaryotic DNA than (eukaryotic) nuclear DNA. Use this information to suggest how eukaryotes may have originated.

33. **Thought Question** Fossil evidence indicates that prokaryotes have been around for about 3.5 billion years, whereas the origin of eukaryotes has been dated at only about 1.5 billion years ago. Suggest why, in spite of the lesser time for evolution, eukaryotes are much more diverse (much larger number of species) than prokaryotes.

1.9 How Do Cells Use Energy?

34. **Fact Check** Which processes are favored: those that require energy or those that release energy?

1.10 What Is the Connection between Energy and Change?

35. **Fact Check** Does the thermodynamic term "spontaneous" refer to a process that takes place quickly?

1.11 What Is the Criterion for Spontaneity in Biochemical Reactions?

36. **Biochemical Connections** For the process

$$\text{Nonpolar solute} + H_2O \rightarrow \text{Solution},$$

what are the signs of ΔS_{univ}, ΔS_{sys}, and ΔS_{surr}? What is the reason for each answer? (ΔS_{surr} refers to the entropy change of the surroundings, all of the universe but the system.)

37. **Fact Check** Which of the following are spontaneous processes? Explain your answer for each process.
    (a) The hydrolysis of ATP to ADP and P_i
    (b) The oxidation of glucose to CO_2 and H_2O by an organism
    (c) The phosphorylation of ADP to ATP
    (d) The production of glucose and O_2 from CO_2 and H_2O in photosynthesis

38. **Thought Question** In which of the following processes does the entropy increase? In each case, explain why it does or does not increase.
    (a) A bottle of ammonia is opened. The odor of ammonia is soon apparent throughout the room.
    (b) Sodium chloride dissolves in water.
    (c) A protein is completely hydrolyzed to the component amino acids.

Hint: For Questions 39 through 41, consider the equation $\Delta G = \Delta H - T(\Delta S)$.

39. **Thought Question** Why is it necessary to specify the temperature when making a table listing ΔG values?

40. **Thought Question** Why is the entropy of a system dependent on temperature?

41. **Thought Question** A reaction at 23°C has $\Delta G = 1$ kJ mol^{-1}. Why might this reaction become spontaneous at 37°C?

42. **Thought Question** Urea dissolves very readily in water, but the solution becomes very cold as the urea dissolves. How is this possible? It appears that the solution is absorbing energy.

43. **Thought Question** Would you expect the reaction ATP $\rightarrow$ ADP + P_i to be accompanied by a decrease or increase in entropy? Why?

1.12 What Is the Connection between Thermodynamics and Life?

44. **Thought Question** The existence of organelles in eukaryotic cells represents a higher degree of organization than that found in prokaryotes. How does this affect the entropy of the universe?

45. **Thought Question** Why is it advantageous for a cell to have organelles? Discuss this concept from the standpoint of thermodynamics.

46. **Thought Question** Which would you expect to have a higher entropy: DNA in its well-known double-helical form, or DNA with the strands separated?

47. **Thought Question** How would you modify your answer to question 31 in light of the material on thermodynamics?

48. **Thought Question** Would it be more or less likely that cells of the kind we know would evolve on a gas giant such as the planet Jupiter?

49. **Thought Question** What thermodynamic considerations might enter into finding a reasonable answer to question 48?

50. **Thought Question** If cells of the kind we know were to have evolved on any other planet in our solar system, would it have been more likely to have happened on Mars or on Jupiter? Why?

51. **Thought Question** The process of protein folding is spontaneous in the thermodynamic sense. It gives rise to a highly ordered conformation that has a lower entropy than the unfolded protein. How can this be?

52. **Thought Question** In biochemistry, the exergonic process of converting glucose and oxygen to carbon dioxide and water in aerobic metabolism can be considered the reverse of photosynthesis in which carbon dioxide and water are converted to glucose and oxygen. Do you expect both processes to be exergonic, both endergonic, or one exergonic and one endergonic? Why? Would you expect both processes to take place in the same way? Why?

Biochemistry ⊜ Now™
Assess your understanding of this chapter's topics with additional quizzing and tutorials at **http://now.brookscole.com/campbell5**

Annotated Bibliography

Research progress is very rapid in biochemistry, and the literature in the field is vast and growing. Many books appear each year, and a large number of primary research journals and review journals report on original research. References to this body of literature are provided at the end of each chapter. A particularly useful reference is *Scientific American;* its articles include general overviews of the topics discussed. *Trends in Biochemical Sciences* and *Science* (a journal published weekly by the American Association for the Advancement of Science) are more advanced but can serve as primary sources of information about a given topic. In addition to material in print, a wealth of information has become available in electronic form. *Science* regularly covers websites of interest and has its own website at http://www.sciencemag.org. Journals now appear on the Internet. Some require subscriptions, and many college and university libraries have subscriptions, making the journals available to students and faculty in this form. Others are free of charge. One, PubMed, is a service of the U.S. government. It lists articles in the biomedical sciences and has links to them. Its URL is http://www.ncbi.nlm.nih.gov/PubMed. Databases provide instant access to structures of proteins and nucleic acids. References will be given to electronic resources as well.

Allen, R. D. The Microtubule as an Intracellular Engine. *Sci. Amer.* **256** (2), 42–49 (1987). [The role of the microtrabecular lattice and microtubules in the motion of organelles is discussed.]

Balter, M. Looking for Clues to the Mystery of Life on Earth. *Science* **273**, 870–872 (1996). [A report on proceedings of a conference about the origin of life. Read in conjunction with articles on pages 864 and 924 of the same issue about the discovery of putative microfossils on a meteorite that came from Mars.]

Barinaga, M. The Telomerase Picture Fills In. *Science* **276**, 528–529 (1997). [A Research News article about the identification of the catalytic component of telomerase, the enzyme that synthesizes telomeres (chromosome ends).]

Cairns-Smith, A. G. The First Organisms. *Sci. Amer.* **252** (6), 90–100 (1985). [A presentation of the point of view that the earliest life processes took place in clay rather than in the "primordial soup" of the early oceans.]

Cairns-Smith, A. G. *Genetic Takeover and the Mineral Origins of Life.* Cambridge, England: Cambridge Univ. Press, 1982. [A presentation of the idea that life began in clay.]

Cech, T. R. RNA as an Enzyme. *Sci. Amer.* **255** (5), 64–75 (1986). [A discussion of the ways in which RNA can cut and splice itself.]

de Duve, C. The Birth of Complex Cells. *Sci. Amer.* **274** (4), 50–57 (1996). [A Nobel laureate summarizes endosymbiosis and other aspects of cellular structure and function.]

Duke, R., D. Ojcius, and J. Young. Cell Suicide in Health and Disease. *Sci. Amer.* **275** (6), 80–87 (1996). [An article on cell death as a normal process in healthy organisms and the lack of it in cancer cells.]

Eigen, M., W. Gardiner, P. Schuster, and R. Winkler-Oswatitsch. The Origin of Genetic Information. *Sci. Amer.* **244** (4), 88–118 (1981). [A presentation of the case for RNA as the original coding material.]

Horgan, J. In the Beginning. . . . *Sci. Amer.* **264** (2), 116–125 (1991). [A report on new developments in the study of the origin of life.]

Knoll, A. The Early Evolution of Eukaryotes: A Geological Perspective. *Science* **256**, 622–627 (1992). [A comparison of biological and geological evidence on the subject.]

Lee, D., J. Granja, J. Martinez, K. Severin, and M. R. Ghadri. A Self-replicating Peptide. *Nature* **382**, 525–528 (1996). [An example of a research article, in this case one that offers evidence that coding and catalysis can be performed by peptides as well as by RNA.]

Madigan, M., and B. Marrs. Extremophiles. *Sci. Amer.* **276** (4), 82–87 (1997). [An account of various kinds of archaebacteria that live under extreme conditions and some of the useful enzymes that can be extracted from these organisms.]

Morell, V. Life's Last Domain. *Science* **273**, 1043–1045 (1996). [A Research News article about the genome of the archaebacterium *Methanococcus jannaschii.* This is the first genome sequence to be obtained for archaebacteria. Read in conjunction with the research article on pages 1058–1073 of the same issue.]

Pennisi, E. Laboratory Workhorse Decoded: Microbial Genomes Come Tumbling In. *Science* **277**, 1432–1434 (1997). [A Research News article about the genome of the bacterium *Escherichia coli.* This organism is widely used in the research laboratory, making its genome particularly important among the dozen bacterial genomes that have been obtained. Read in conjunction with the research article on pages 1453–1474 of the same issue.]

Robertson, H. How Did Replicating and Coding RNAs First Get Together? *Science* **274**, 66–67 (1996). [A short review on possible remains of an "RNA world."]

Rothman, J. E. The Compartmental Organization of the Golgi Apparatus. *Sci. Amer.* **253** (3), 74–89 (1985). [A description of the functions of the Golgi apparatus.]

Waldrop, M. Goodbye to the Warm Little Pond? *Science* **250**, 1078–1079 (1990). [Facts and theories on the role of meteorite impacts on the early Earth in the origin and development of life.]

Weber, K., and M. Osborn. The Molecules of the Cell Matrix. *Sci. Amer.* **253** (4), 100–120 (1985). [An extensive description of the cytoskeleton.]

Woese, C. R. Archaebacteria. *Sci. Amer.* **244** (6) 98–122 (1981). [A detailed description of the differences between archaebacteria and other types of organisms.]

CHAPTER 2

Water: The Solvent for Biochemical Reactions

Virtually all the chemical reactions of the cell involve water. They are the reactions of organic chemistry, using the same functional groups and operating in a cellular environment. Life has evolved around the special properties of water. Important structural considerations follow from the nature of the water molecule, which has a partial positive charge on each of its hydrogen atoms and a partial negative charge on its oxygen atom, which has two unshared pairs of electrons. This allows the water molecule to associate with four others of its kind. Four hydrogen bonds can be formed, pointing to the corners of a tetrahedron. Hydrogen bonds are important everywhere in biomolecular structures; they link parts of protein chains of enzymes and the two complementary chains of the DNA double helix. The tendency of nonpolar groups of biomolecules to sequester themselves from water gives rise to hydrophobic interactions, which are another key factor determining the structure of biomolecules. Another unique property of water is its role in the control of acidity within the cell by buffers. A cell's survival depends on strict control of its internal pH.

Life processes depend on the properties of water.

Critical Questions

Biochemistry⊘Now™

Test yourself on these Critical Questions at the BiochemistryNow website at **http://now** .brookscole.com/campbell5

2.1 | What Makes Water a Polar Molecule?

Water is the principal component of most cells. The geometry of the water molecule and its properties as a solvent play major roles in determining the properties of living systems.

When electrons are shared between atoms in a chemical bond, they need not be shared equally. Bonds that share electrons unequally are referred to as **polar.** The tendency of an atom to attract electrons to itself in a chemical bond (i.e., to become negative) is called **electronegativity.** Atoms of the same element, of course, share electrons equally in a bond—that is, they have equal electronegativity—but different elements do not necessarily have the same electronegativity. Oxygen and nitrogen are both highly electronegative, much more so than carbon and hydrogen (Table 2.1).

In the O—H bonds in water, oxygen is more electronegative than hydrogen, so there is a higher probability that the bonding electrons are closer to the oxygen. The difference in electronegativity between oxygen and hydrogen gives rise to a *partial* positive and negative charge, usually pictured as δ^+ and δ^-, respectively (Figure 2.1). The O—H bond is thus a polar bond. In situations in which the electronegativity difference is quite small, such as in the C—H bond in methane (CH_4), the sharing of electrons in the bond is very nearly equal, and the bond is essentially **nonpolar.**

A molecule may have polar bonds but still be nonpolar because of its geometry. Carbon dioxide is an example. The two C=O bonds are polar, but, because the CO_2 molecule is linear, the attraction of the oxygen for the electrons in one bond is cancelled out by the equal and opposite attraction for the electrons by the oxygen on the other side of the molecule.

$$\delta^- \quad 2\delta^+ \quad \delta^-$$

$$O=C=O$$

Water is a bent molecule with a bond angle of 104.3° (Figure 2.1), and the uneven sharing of electrons in the two bonds is not cancelled out as in CO_2. The result is that the bonding electrons are more likely to be found at the oxygen end of the molecule than at the hydrogen end. Bonds with positive and negative ends are called **dipoles.**

Solvent Properties of Water

The polar nature of water largely determines its solvent properties. *Ionic* compounds with full charges, such as potassium chloride (KCl, K^+ and Cl^- in solution), and *polar* compounds with partial charges (i.e., dipoles), such as ethyl alcohol (C_2H_5OH) or acetone [$(CH_3)_2C{=}O$], tend to dissolve in water (Figures 2.2 and 2.3). The underlying physical principle is electrostatic attraction between unlike charges. The negative end of a water dipole attracts a positive ion or the positive end of another dipole. The positive end of a water molecule attracts a negative ion or the negative end of another dipole. The aggregate of unlike charges, held in proximity to one another because of electrostatic attraction, has a lower energy than would be possible if this interaction could not take place. The lowering of energy makes the system more stable and more likely to exist. These *ion–dipole* and *dipole–dipole* interactions are similar to the interactions between water molecules themselves in terms of the quantities of energy involved. Examples of polar compounds that dissolve easily in water are small organic molecules containing one or more electronegative atoms (e.g., oxygen or nitrogen), including alcohols, amines, and carboxylic acids. The attraction between the dipoles of these molecules and the water dipoles makes them tend to dissolve. Ionic and polar substances are referred to as **hydrophilic** ("water-loving," from the Greek) because of this tendency.

Hydrocarbons (compounds that contain only carbon and hydrogen) are nonpolar. The favorable ion–dipole and dipole–dipole interactions responsible for the solubility of ionic and polar compounds do not occur for nonpolar compounds, so these compounds tend not to dissolve in water. The interactions between nonpolar molecules and water molecules are weaker than

Table 2.1

Electronegativities of Selected Elements

Element	Electronegativity*
Oxygen	3.5
Nitrogen	3.0
Sulfur	2.6
Carbon	2.5
Phosphorus	2.2
Hydrogen	2.1

* Electronegativity values are relative and are chosen to be positive numbers ranging from less than 1 for some metals to 4 for fluorine.

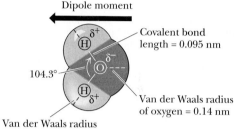

Biochemistry⬡Now™ ACTIVE FIGURE 2.1
The structure of water. Oxygen has a partial negative charge, and the hydrogens have a partial positive charge. The uneven distribution of charge gives rise to the large dipole moment of water. The dipole moment in this figure points in the direction from negative to positive, the convention used by physicists and physical chemists; organic chemists draw it pointing in the opposite direction. **Watch this Active Figure at http://now.brookscole.com/campbell5**

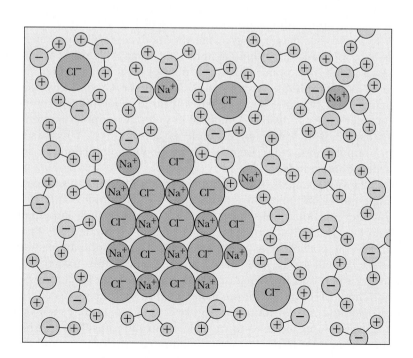

Biochemistry⬡Now™ ANIMATED FIGURE 2.2
Hydration shells surrounding ions in solution. Unlike charges attract. The partial negative charge of water is attracted to positively charged ions. Likewise, the partial positive charge on the other end of the water molecule is attracted to negatively charged ions. **See this figure animated at http://now.brookscole.com/campbell5**

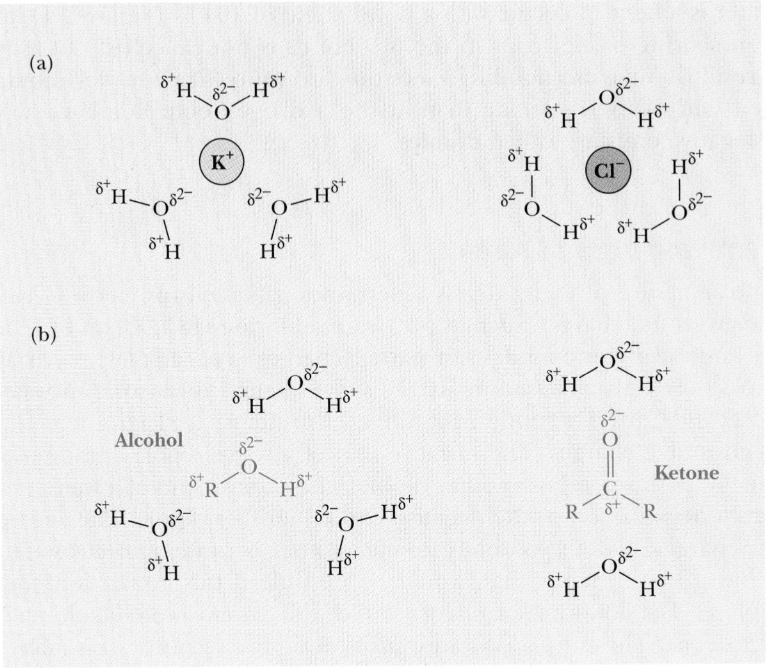

▶ **FIGURE 2.3** Ion–dipole and dipole–dipole interactions help ionic and polar compounds dissolve in water. (a) Ion–dipole interactions with water. (b) Dipole–dipole interactions of polar compounds with water. The examples shown here are an alcohol (ROH) and a ketone ($R_2C{=}O$).

dipolar interactions. The permanent dipole of the water molecule can induce a temporary dipole in the nonpolar molecule by distorting the spatial arrangements of the electrons in its bonds. Electrostatic attraction is possible between the induced dipole of the nonpolar molecule and the permanent dipole of the water molecule (a *dipole-induced dipole interaction*), but it is not as strong as that between permanent dipoles. Hence, its consequent lowering of energy is less than that produced by the attraction of the water molecules for one another. The association of nonpolar molecules with water is far less likely to occur than the association of water molecules with themselves.

A full discussion of why nonpolar substances are insoluble in water requires the thermodynamic arguments that we shall develop in Chapters 4 and 15. However, the points made here about intermolecular interactions will be useful background information for that discussion. For the moment, it is enough to know that it is less favorable thermodynamically for water molecules to be associated with nonpolar molecules than with other water molecules. As a result, nonpolar molecules do not dissolve in water and are referred to as **hydrophobic** ("water-hating," from the Greek). Hydrocarbons in particular tend to sequester themselves from an aqueous environment. A nonpolar solid leaves undissolved material in water. A nonpolar liquid forms a two-layer system with water; an example is an oil slick. The interactions between nonpolar molecules are called **hydrophobic interactions** or, in some cases, **hydrophobic bonds.**

Table 2.2 gives examples of hydrophobic and hydrophilic substances.

A single molecule may have both polar (hydrophilic) and nonpolar (hydrophobic) portions. Substances of this type are called **amphipathic.** A long-chain fatty acid having a polar carboxylic acid group and a long nonpolar hydrocarbon portion is a prime example of an amphipathic substance. The carboxylic acid group, the "head" group, contains two oxygen atoms in addition to carbon and hydrogen; it is very polar and can form a carboxylate anion at neutral pH. The rest of the molecule, the "tail," contains only carbon and hydrogen and is thus nonpolar (Figure 2.4). A compound such as this in the presence of water tends to form structures called **micelles,** in which the polar head groups are in contact with the aqueous environment and the nonpolar tails are sequestered from the water (Figure 2.5).

Table 2.2

Examples of Hydrophobic and Hydrophilic Substances

Hydrophilic	Hydrophobic
Polar covalent compounds [e.g., alcohols such as C_2H_5OH (ethanol) and ketones such as $(CH_3)_2C{=}O$ (acetone)]	Nonpolar covalent compounds [e.g., hydrocarbons such as C_6H_{14} (hexane)]
Sugars	Fatty acids, cholesterol
Ionic compounds (e.g., KCl)	
Amino acids, phosphate esters	

The sodium salt of palmitic acid: Sodium palmitate
$(Na^{+-}OOC(CH_2)_{14}CH_3)$

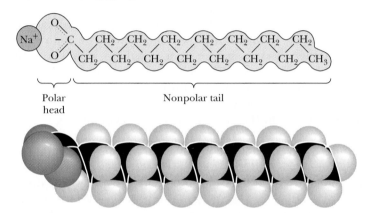

Polar head   Nonpolar tail

◀ **FIGURE 2.4** An amphiphilic molecule: sodium palmitate. Amphiphilic molecules are frequently symbolized by a ball and zigzag line structure, ●〰〰, where the ball represents the hydrophilic polar head and the zigzag line represents the nonpolar hydrophobic hydrocarbon tail.

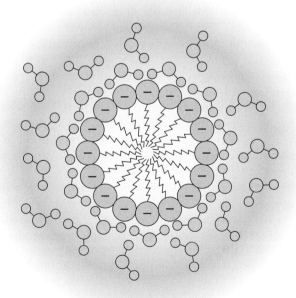

Biochemistry ⑤ Now™ ACTIVE FIGURE 2.5
Micelle formation by amphipathic molecules in aqueous solution. When micelles form, the ionized polar groups are in contact with the water, and the nonpolar parts of the molecule are protected from contact with the water. **Watch this Active Figure at http://now.brookscole.com/campbell5**

Interactions between nonpolar molecules themselves are very weak and depend on the attraction between short-lived temporary dipoles and the dipoles they induce. In a large sample of nonpolar molecules, there will always be some molecules with these temporary dipoles, which are caused by a momentary clumping of bonding electrons at one end of the molecule. A temporary dipole can induce another dipole in a neighboring molecule in the same way that a permanent dipole does. The interaction energy is low because the association is so short-lived. It is called a **van der Waals interaction** (also referred to as a van der Waals bond). The arrangement of molecules in cells strongly depends on the molecules' polarity, as we saw with micelles.

2.2 | What Is a Hydrogen Bond?

In addition to the interactions discussed in Section 2.1, there is another important type of noncovalent interaction: **hydrogen bonding.** Hydrogen bonding is of electrostatic origin and can be considered to be a special case of dipole–dipole interaction. When hydrogen is covalently bonded to an electronegative atom such as oxygen or nitrogen, it has a partial positive charge due to the polar bond, a situation that does not occur when hydrogen is covalently bonded to carbon. This partial positive charge on hydrogen can interact with an unshared (nonbonding) pair of electrons (a source of negative charge) on another electronegative atom. All three atoms lie in a straight line, forming a hydrogen bond. This arrangement allows for the greatest possible partial positive charge on the hydrogen and, consequently, for the strongest possible interaction with the unshared pair of electrons on the second electronegative atom (Figure 2.6). The group comprising the electronegative atom that is covalently bonded to hydrogen is called the *hydrogen-bond donor,* and the electronegative atom that contributes the unshared pair of electrons to the interaction is the *hydrogen-bond acceptor.* The hydrogen is not covalently bonded to the acceptor in the usual description of hydrogen bonding. Recent research has cast some doubt on this view, with experimental evidence to indicate some covalent character in the hydrogen bond. Some of this work is described in the article by Hellmans cited in the bibliography at the end of this chapter.

A consideration of the hydrogen-bonding sites in HF, H_2O, and NH_3 can yield some useful insights. Figure 2.7 shows that water constitutes an optimum situation in terms of the number of hydrogen bonds that each molecule can form. Water has two hydrogens to enter into hydrogen bonds and two unshared pairs of electrons on the oxygen to which other water molecules can be hydrogen-bonded. Each water molecule is involved in four hydrogen bonds—as a donor in two and as an acceptor in two. Hydrogen fluoride has only one hydrogen to enter into a hydrogen bond as a donor, but it has three unshared pairs of electrons on the fluorine that could bond to other hydrogens. Ammonia has three hydrogens to donate to a hydrogen bond but only one unshared pair of electrons, on the nitrogen.

The geometric arrangement of hydrogen-bonded water molecules has important implications for the properties of water as a solvent. The bond angle in water is 104.3°, as was shown in Figure 2.1, and the angle between the unshared pairs of electrons is similar. The result is a tetrahedral arrangement

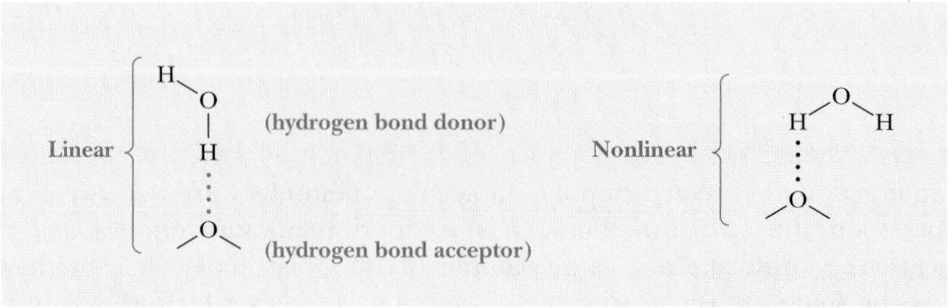

▶ **FIGURE 2.6** A comparison of linear and nonlinear hydrogen bonds. Nonlinear bonds are weaker than bonds in which all three atoms lie in a straight line.

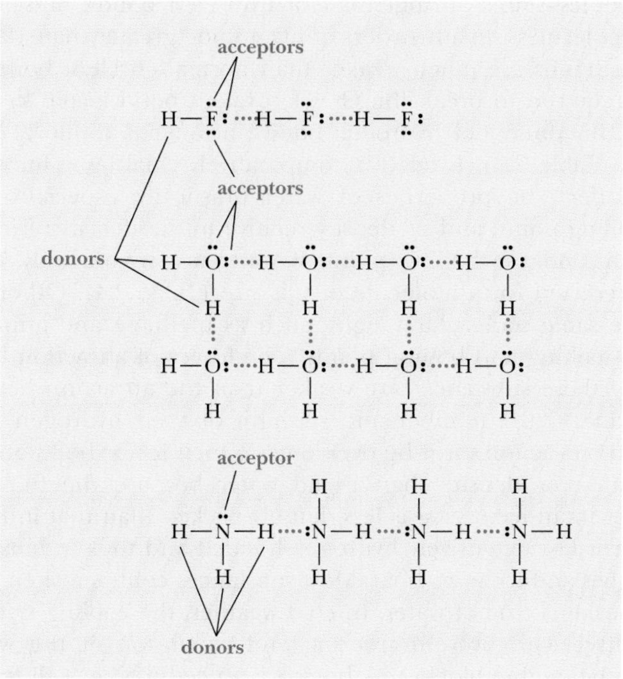

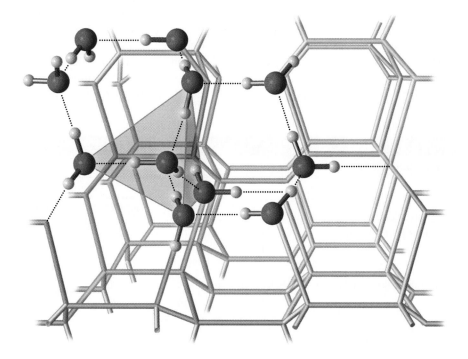

◀ **FIGURE 2.7** A comparison of the numbers of hydrogen bonding sites in HF, H_2O, and NH_3. (Actual geometries are not shown.) Each HF molecule has one hydrogen-bond donor and three hydrogen-bond acceptors. Each H_2O molecule has two donors and two acceptors. Each NH_3 molecule has three donors and one acceptor.

◀ **FIGURE 2.8** Tetrahedral hydrogen bonding in H_2O: an array of H_2O molecules in an ice crystal. Each H_2O molecule is hydrogen-bonded to four others.

of water molecules. Liquid water consists of hydrogen-bonded arrays that resemble ice crystals; each of these arrays can contain up to 100 water molecules. The hydrogen bonding between water molecules can be seen more clearly in the regular lattice structure of the ice crystal (Figure 2.8). There are several differences, however, between hydrogen-bonded arrays of this type in liquid water and the structure of ice crystals. In liquid water, hydrogen bonds are constantly breaking and new ones are constantly forming, with some molecules breaking off and others joining the cluster. A cluster can break up and re-form in 10^{-10} to 10^{-11} seconds in water at 25°C. An ice crystal, in contrast, has a more-or-less-stable arrangement of hydrogen bonds, and of course its number of molecules is many orders of magnitude greater than 100.

Hydrogen bonds are much weaker than normal covalent bonds. Whereas the energy required to break the O—H covalent bond is 460 kJ mol^{-1} (110 kcal mol^{-1}), the energy of hydrogen bonds in water is about 20 kJ mol^{-1} (5 kcal mol^{-1}) (Table 2.3). Even this comparatively small amount of energy is enough to affect the properties of water drastically, especially its melting point, its boiling point, and its density relative to the density of ice. Both the melting point and the boiling point of water are significantly higher than would be predicted for a molecule of this size (Table 2.4). Other substances of about the same molecular weight, such as methane and ammonia, have much lower melting and boiling points. The forces of attraction between the molecules of these substances are weaker than the attraction between water molecules, due to the number and strength of their hydrogen bonds. The energy of this attraction must be overcome to melt ice or boil water.

Ice has a lower density than liquid water because the fully hydrogen-bonded array in an ice crystal is less densely packed than that in liquid water. Liquid water is less extensively hydrogen-bonded and thus is denser than ice. Thus, ice cubes and icebergs float. Most substances contract when they freeze, but the opposite is true of water. In cold weather, the cooling systems of cars require antifreeze to prevent freezing and expansion of the water, which could crack the engine block. In laboratory procedures for cell fractionation, the same principle is used in a method of disrupting cells with several cycles of freezing and thawing. Finally, aquatic organisms can survive in cold climates because of the density difference between ice and liquid water; lakes and rivers freeze from top to bottom rather than vice versa.

Hydrogen bonding also plays a role in the behavior of water as a solvent. If a polar solute can serve as a donor or an acceptor of hydrogen bonds, not only can it form hydrogen bonds with water but it can also be involved in nonspe-

Table 2.3			
Some Bond Energies			
		Energy*	
	Type of Bond	**(kJ mol^{-1})**	**(kcal mol^{-1})**
Covalent Bonds (Strong)	O—H	460	110
	H—H	416	100
	C—H	413	105
Nonvalent Bonds (Weaker)	Hydrogen bond	20	5
	Ion–dipole interaction	20	5
	Hydrophobic interaction	4–12	1–3
	Van der Waals interactions	4	1

* Note that two units of energy are used throughout this text. The kilocalorie (kcal) is a commonly used unit in the biochemical literature. The kilojoule (kJ) is an SI unit and will come into wider use as time goes on.. The kcal is the same as the "Calorie" referred to on food labels.

Table 2.4			
Comparison of Properties of Water, Ammonia, and Methane			
Substance	**Molecular Weight**	**Melting Point (°C)**	**Boiling Point (°C)**
Water (H_2O)	18.02	0.0	100.0
Ammonia (NH_3)	17.03	−77.7	−33.4
Methane (CH_4)	16.04	−182.5	−161.5

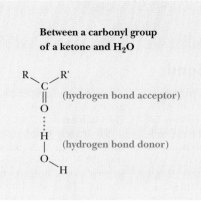

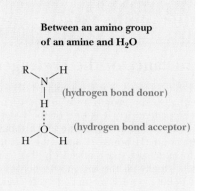

▲ **FIGURE 2.9** Examples of hydrogen bonding between polar groups and water.

cific dipole–dipole interactions. Figure 2.9 shows some examples. Alcohols, amines, carboxylic acids, and esters, as well as aldehydes and ketones, can all form hydrogen bonds with water, so they are soluble in water. It is difficult to overstate the importance of water to the existence of life on Earth, and it is difficult to imagine life based on another solvent. The following Biochemical Connections box explores some of the implications of this statement.

Biologically Important Hydrogen Bonds Other Than to Water Molecules

Hydrogen bonds have a vital involvement in stabilizing the three-dimensional structures of biologically important molecules, including DNA, RNA, and proteins.

The hydrogen bonds between complementary bases are one of the most striking characteristics of the double-helical structure of DNA (Section 9.3). Transfer RNA also has a complex three-dimensional structure characterized by hydrogen-bonded regions (Section 9.5). Hydrogen bonding in proteins gives rise to two important structures, the α-helix and β-pleated sheet conformations. Both types of conformation are widely encountered in proteins (Section 4.3). Table 2.5 summarizes some of the most important kinds of hydrogen bonds in biomolecules.

Table 2.5	
Examples of Major Types of Hydrogen Bonds Found in Biologically Important Molecules	
Bonding Arrangement	**Molecules Where the Bond Occurs**
—O—H••••••O— H	H bond formed in H_2O
—O—H••••••O=C<	Bonding of water to other molecules
N—H••••••O— H	
N—H••••••O=C<	Important in protein and nucleic acid structures
N—H••••••N<	
>N—H••••••N NH	

Biochemical Connections

The Importance of the Hydrogen Bond

Many noted biochemists have speculated that the hydrogen bond is essential to the evolution of life; just like carbon, polymers, and stereochemistry, it is one of the criteria that can be used to search for extraterrestrial life. Even though the individual hydrogen bond (H bond) is weak, the fact that so many H bonds can form means that collectively they can exert a *very* strong force. Virtually all the unique properties of water (high melting and boiling points, ice and density characteristics, and solvent potency) are a result of its ability to form many hydrogen bonds per molecule.

If we look at the solubility of a simple ion like Na^+ or Cl^-, we find that water is attracted to these ions by polarity. In addition, other water molecules form H bonds with those surrounding water molecules, typically 20 or more water molecules per dissolved ion. When we consider a simple biomolecule such as glyceraldehyde, the H bonds start at the molecule itself. At least eight water molecules bind directly to the glyceraldehyde molecule, and then more water molecules bind to those eight.

The orderly and repetitive arrangement of hydrogen bonds in polymers determines their shape. The extended structures of cellulose and of peptides in a β-sheet allow for the formation of strong fibers through intrachain H bonding. Single helices (as in starch) and the α-helices of proteins are stabilized by intrachain H bonds. Double and triple helices, as in DNA and collagen, involve H bonds between the two or three respective strands. Collagen contains several special amino acids that have an extra hydroxyl group; these allow for additional hydrogen bonds, which provide stability.

Hydrogen bonding is also fundamental to the specificity of transfer of genetic information. The complementary nature of the DNA double helix is assured by hydrogen bonds. The genetic code, both its specificity and its allowable variation, is a result of H bonds. Indeed, many compounds that cause genetic mutations work by altering the patterns of H bonding. For example, fluorouracil is often prescribed by dentists for cold sores (viral sores of the lip and mouth) because it causes mutations in the herpes simplex virus that causes the sores.

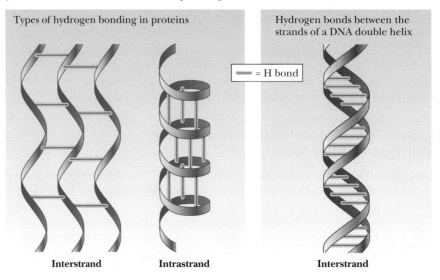

Types of hydrogen bonding in proteins

Hydrogen bonds between the strands of a DNA double helix

— = H bond

Interstrand   **Intrastrand**   **Interstrand**

2.3 | What Are Acids and Bases?

The biochemical behavior of many important compounds depends on their acid–base properties. A biologically useful definition of an acid is a molecule that acts as a proton (hydrogen ion) donor. A base is similarly defined as a proton acceptor. How readily acids or bases lose or gain protons depends on the chemical nature of the compounds under consideration. The degree of dissociation of acids in water, for example, ranges from essentially complete dissociation for a strong acid to practically no dissociation for a very weak acid, and any intermediate value is possible.

It is useful to derive a numerical measure of **acid strength,** which is the amount of hydrogen ion released when a given amount of acid is dissolved in water. Such an expression, called the **acid dissociation constant,** or K_a, can be written for any acid, HA, that reacts according to the equation

$$\underset{\text{Acid}}{HA} \rightleftharpoons \underset{\text{Conjugate base}}{H^+ + A^-}$$

$$K_a = \frac{[H^+][A^-]}{[HA]}$$

In this expression, the square brackets refer to molar concentration—that is, the concentration in moles per liter. For each acid, the quantity K_a has a fixed numerical value at a given temperature. This value is larger for more completely dissociated acids; the greater the K_a, the stronger the acid.

Strictly speaking, the preceding acid–base reaction is a proton-transfer reaction in which water acts as a base as well as the solvent.

$$\underset{\text{Acid}}{HA(aq)} + \underset{\text{Base}}{H_2O(\ell)} \rightleftharpoons \underset{\substack{\text{Conjugate} \\ \text{acid to } H_2O}}{H_3O^+(aq)} + \underset{\substack{\text{Conjugate} \\ \text{base to } HA}}{A^-(aq)}$$

The notation (aq) refers to solutes in aqueous solution, whereas (ℓ) refers to water in the liquid state. It is well established that there are no "naked protons" (free hydrogen ions) in solution; even the hydronium ion (H_3O^+) is an underestimate of the degree of hydration of hydrogen ion in aqueous solution. All solutes are extensively hydrated in aqueous solution. We will write the short form of equations for acid dissociation in the interest of simplicity, but the role of water should be kept in mind throughout our discussion.

2.4 | What Is pH, and What Does It Have to Do with the Properties of Water?

The acid–base properties of water play an important part in biological processes because of the central role of water as a solvent. The extent of self-dissociation of water to hydrogen ion and hydroxide ion

$$H_2O \rightleftharpoons H^+ + OH^-$$

is small, but the fact that it takes place determines important properties of many solutes (Figure 2.10). Both the hydrogen ion (H^+) and the hydroxide ion (OH^-) are associated with several water molecules, as are all ions in aqueous solution, and the water molecule in the equation is itself part of a cluster of such molecules (Figure 2.11). It is especially important to have a quantitative estimate of the degree of dissociation of water. We can start with the expression

$$K_a = \frac{[H^+][OH^-]}{[H_2O]}$$

The molar concentration of pure water, $[H_2O]$, is quite large compared with any possible concentrations of solutes and can be considered a constant. (The numerical value is 55.5 M, which can be obtained by dividing the number of grams of water in 1 liter, 1000 g, by the molecular weight of water, 18 g/mol; $1000/18 = 55.5$ M.) Thus,

$$K_a = \frac{[H^+][OH^-]}{55.5}$$

$$K_a \times 55.5 = [H^+][OH^-] = K_W$$

A new constant, K_w, the **ion product constant for water,** has just been defined, where the concentration of water has been included in its value.

The numerical value of K_w can be determined experimentally by measuring the hydrogen ion concentration of pure water. The hydrogen ion concentration is also equal, by definition, to the hydroxide ion concentration because water is a monoprotic acid (one that releases a single proton per molecule). At 25°C in *pure* water,

$$[H^+] = 10^{-7}\ M = [OH^-]$$

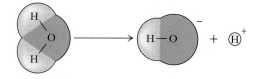

Biochemistry⬛Now™ ACTIVE FIGURE 2.10
The ionization of water. **Watch this Active Figure at http://now.brookscole.com/campbell5**

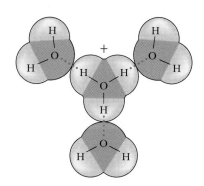

Biochemistry⬛Now™ ANIMATED FIGURE 2.11
The hydration of hydrogen ion in water. **See this figure animated at http://now.brookscole.com/campbell5**

Thus, at 25°C, the numerical value of K_w is given by the expression

$$K_w = [H^+][OH^-] = (10^{-7})(10^{-7}) = 10^{-14}$$

This relationship, which we have derived for pure water, is valid for *any* aqueous solution, whether neutral, acidic, or basic.

The wide range of possible hydrogen ion and hydroxide ion concentrations in aqueous solution makes it desirable to define a quantity for expressing these concentrations more conveniently than by exponential notation. This quantity is called pH and is defined as

$$pH = -\log_{10}[H^+]$$

with the logarithm taken to the base 10. Note that, because of the logarithms involved, a difference of one pH unit implies a tenfold difference in hydrogen ion concentration, $[H^+]$. The pH values of some typical aqueous samples can be determined by a simple calculation.

Practice Session

Since in pure water $[H^+] = 1 \times 10^{-7}$ M and pH = 7.0, you should be able to calculate the pH of the following aqueous solutions:
1. 1×10^{-3} M HCl   2. 1×10^{-4} M NaOH
Assume that the self-ionization of water makes a negligible contribution to the concentrations of hydronium ions and of hydroxide ions, which will typically be true unless the solutions are extremely dilute.

Solution

The key points in the approach to this problem are the definition of pH, which needs to be used in both parts, and the self-dissociation of water, needed in the second part.
1. For 1×10^{-3} M HCl, $[H_3O^+] = 1 \times 10^{-3}$ M; therefore, pH = 3.
2. For 1×10^{-4} M NaOH, $[OH^-] = 1 \times 10^{-4}$ M. Since $[OH^-][H_3O^+] = 1 \times 10^{-14}$, $[H_3O^+] = 1 \times 10^{-10}$ M; therefore, pH = 10.0.

Pure water with a pH of 7 is neutral, acidic solutions have pH values lower than 7, and basic solutions have pH values higher than 7.

A similar quantity, pK_a, can be defined by analogy with the definition of pH:

$$pK_a = -\log_{10}K_a$$

The pK_a is another numerical measure of acid strength; the smaller its value, the stronger the acid. This is the reverse of the situation with K_a, where larger values imply stronger acids (Table 2.6).

Monitoring Acidity

There is an equation that connects the K_a of any weak acid with the pH of a solution containing both that acid and its conjugate base. This relationship has wide use in biochemical practice, especially where it is necessary to control pH for optimum reaction conditions. Some reactions cannot take place if the pH varies from the optimum value. Important biological macromolecules lose activity at extremes of pH. Figure 2.12 shows how the activities of three enzymes are affected by pH. Note that each one has a peak activity that falls off rapidly as the pH is changed from the optimum. Also, some drastic physiological consequences can result from pH fluctuations in the body. Section 2.6 has more information about how pH can be controlled. To derive the

Table 2.6

Dissociation Constants of Some Acids

Acid	HA	A⁻	K_a	pK_a
Pyruvic acid	$CH_3COCOOH$	CH_3COCOO^-	3.16×10^{-3}	2.50
Formic acid	$HCOOH$	$HCOO^-$	1.78×10^{-4}	3.75
Lactic acid	$CH_3CHOHCOOH$	$CH_3CHOHCOO^-$	1.38×10^{-4}	3.86
Benzoic acid	C_6H_5COOH	$C_6H_5COO^-$	6.46×10^{-5}	4.19
Acetic acid	CH_3COOH	CH_3COO^-	1.76×10^{-5}	4.76
Ammonium ion	NH_4^+	NH_3	5.6×10^{-10}	9.25
Oxalic acid (1)	$HOOC—COOH$	$HOOC—COO^-$	5.9×10^{-2}	1.23
Oxalic acid (2)	$HOOC—COO^-$	$^-OOC—COO^-$	6.4×10^{-5}	4.19
Malonic acid (1)	$HOOC—CH_2—COOH$	$HOOC—CH_2—COO^-$	1.49×10^{-3}	2.83
Malonic acid (2)	$HOOC—CH_2—COO^-$	$^-OOC—CH_2—COO^-$	2.03×10^{-6}	5.69
Malic acid (1)	$HOOC—CH_2—CHOH—COOH$	$HOOC—CH_2—CHOH—COO^-$	3.98×10^{-4}	3.40
Malic acid (2)	$HOOC—CH_2—CHOH—COO^-$	$^-OOC—CH_2—CHOH—COO^-$	5.5×10^{-6}	5.26
Succinic acid (1)	$HOOC—CH_2—CH_2—COOH$	$HOOC—CH_2—CH_2—COO^-$	6.17×10^{-5}	4.21
Succinic acid (2)	$HOOC—CH_2—CH_2—COO^-$	$^-OOC—CH_2—CH_2—COO^-$	2.3×10^{-6}	5.63
Carbonic acid (1)	H_2CO_3	HCO_3^-	4.3×10^{-7}	6.37
Carbonic acid (2)	HCO_3^-	CO_3^{2-}	5.6×10^{-11}	10.20
Citric acid (1)	$HOOC—CH_2—C(OH)(COOH)—CH_2—COOH$	$HOOC—CH_2—C(OH)(COOH)—CH_2—COO^-$	8.14×10^{-4}	3.09
Citric acid (2)	$HOOC—CH_2—C(OH)(COOH)—CH_2—COO^-$	$^-OOC—CH_2—C(OH)(COOH)—CH_2—COO^-$	1.78×10^{-5}	4.75
Citric acid (3)	$^-OOC—CH_2—C(OH)(COOH)—CH_2—COO^-$	$^-OOC—CH_2—C(OH)(COO^-)—CH_2—COO^-$	3.9×10^{-6}	5.41
Phosphoric acid (1)	H_3PO_4	$H_2PO_4^-$	7.25×10^{-3}	2.14
Phosphoric acid (2)	$H_2PO_4^-$	HPO_4^{2-}	6.31×10^{-8}	7.20
Phosphoric acid (3)	HPO_4^{2-}	PO_4^{3-}	3.98×10^{-13}	12.40

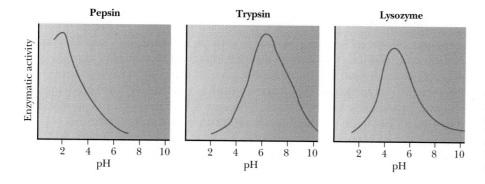

◀ **FIGURE 2.12** pH versus enzymatic activity. Pepsin, trypsin, and lysozyme all have steep pH optimum curves. Pepsin has maximum activity under very acidic conditions, as would be expected for a digestive enzyme that is found in the stomach. Lysozyme has its maximum activity near pH 5, while trypsin is most active near pH 6.

involved equation, it is first necessary to take the logarithm of both sides of the K_a equation.

$$K_a = \frac{[H^+][A^-]}{[HA]}$$

$$\log K_a = \log[H^+] + \log\frac{[A^-]}{[HA]}$$

$$-\log[H^+] = -\log K_a + \log\frac{[A^-]}{[HA]}$$

We then use the definitions of pH and pK_a:

$$pH = pK_a + \log \frac{[A^-]}{[HA]}$$

This relationship is known as the **Henderson–Hasselbalch equation** and is useful in predicting the properties of buffer solutions used to control the pH of reaction mixtures. When buffers are discussed in Section 2.6, we will be interested in the situation in which the concentration of acid, [HA], and the concentration of the conjugate base, [A⁻], are equal ([HA] = [A⁻]). The ratio [A⁻]/[HA] is then equal to 1, and the logarithm of 1 is equal to zero. Therefore, when a solution contains equal concentrations of a weak acid and its conjugate base, the pH of that solution equals the pK_a value of the weak acid.

2.5 | What Are Titration Curves?

When base is added to a sample of acid, the pH of the solution changes. A **titration** is an experiment in which measured amounts of base are added to a measured amount of acid. It is convenient and straightforward to follow the course of the reaction with a pH meter. The point in the titration at which the acid is exactly neutralized is called the **equivalence point.**

If the pH is monitored as base is added to a sample of acetic acid in the course of a titration, an inflection point in the titration curve is reached when the pH equals the pK_a of acetic acid (Figure 2.13). As we saw in our discussion of the Henderson–Hasselbalch equation, a pH value equal to the pK_a corresponds to a mixture with equal concentrations of the weak acid and its conjugate base—in this case, acetic acid and acetate ion, respectively. The pH at the inflection point is 4.76, which is the pK_a of acetic acid. The inflection point occurs when 0.5 mole of base has been added for each mole of acid present. Near the inflection point, the pH changes very little as more base is added.

When 1 mole of base has been added for each mole of acid, the equivalence point is reached, and essentially all the acetic acid has been converted to acetate ion. (See Question 42 at the end of this chapter.) Figure 2.13 also plots the relative abundance of acetic acid and acetate ion with increasing additions of NaOH. Notice that the percentage of acetic acid plus the percentage of acetate ion adds up to 100%. The acid (acetic acid) is progressively converted to its conjugate base (acetate ion) as more NaOH is added and the titration proceeds. It can be helpful to keep track of the percentages of a conjugate acid and base in this way to understand the full significance of the reaction taking place in a titration. The form of the curves in Figure 2.13 represents the behavior of any monoprotic weak acid, but the value of the pK_a for each individual acid determines the pH values at the inflection point and at the equivalence point.

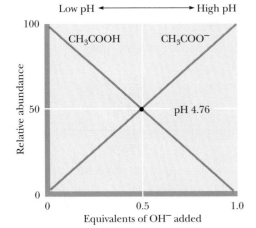

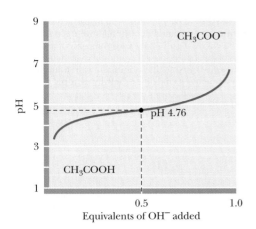

Biochemistry⊜Now™ ANIMATED FIGURE 2.13
Titration curve for acetic acid. Note that there is a region near the pK_a at which the titration curve is relatively flat. In other words, the pH changes very little as base is added in this region of the titration curve. **See this figure animated at http://now .brookscole.com/campbell5**

Practice Session

Calculate the relative amounts of acetic acid and acetate ion present at the following points when 1 mole of acetic acid is titrated with sodium hydroxide. Also use the Henderson–Hasselbalch equation to calculate the values of the pH at these points. Compare your results with Figure 2.13.
a. 0.1 mole of NaOH is added
b. 0.3 mole of NaOH is added
c. 0.5 mole of NaOH is added
d. 0.7 mole of NaOH is added
e. 0.9 mole of NaOH is added

Solution

We approach this problem as an exercise in stoichiometry. There is a 1:1 ratio of moles of acid reacted to moles of base added. The difference between the original number of moles of acid and the number reacted is the number of moles of acid remaining. These are the values to be used in the numerator and denominator, respectively, of the Henderson–Hasselbalch equation.

a. When 0.1 mol of NaOH is added, 0.1 mol of acetic acid reacts with it to form 0.1 mol of acetate ion, leaving 0.9 mol acetic acid. The composition is 90% acetic acid and 10% acetate ion.

$$pH = pK_a + \log \frac{0.1}{0.9}$$

$$pH = 4.76 + \log \frac{0.1}{0.9}$$

$$pH = 4.76 - 0.95$$

$$pH = 3.81$$

b. When 0.3 mol of NaOH is added, 0.3 mol of acetic acid reacts with it to form 0.3 mol of acetate ion, leaving 0.7 mol acetic acid. The composition is 70% acetic acid and 30% acetate ion.

$$pH = pK_a + \log \frac{0.3}{0.7}$$

$$pH = 4.39$$

c. When 0.5 mol of NaOH is added, 0.5 mol of acetic acid reacts with it to form 0.5 mol of acetate ion, leaving 0.5 mol acetic acid. The composition is 50% acetic acid and 50% acetate ion.

$$pH = pK_a + \log \frac{0.5}{0.5}$$

$$pH = 4.76$$

d. When 0.7 mol of NaOH is added, 0.7 mol of acetic acid reacts with it to form 0.7 mol of acetate ion, leaving 0.3 mol acetic acid. The composition is 30% acetic acid and 70% acetate ion.

$$pH = pK_a + \log \frac{0.7}{0.3}$$

$$pH = 5.13$$

e. When 0.9 mol of NaOH is added, 0.9 mol of acetic acid reacts with it to form 0.9 mol of acetate ion, leaving 0.1 mol acetic acid. The composition is 10% acetic acid and 90% acetate ion.

$$pH = pK_a + \log \frac{0.9}{0.1}$$

$$pH = 5.71$$

Table 2.6 lists values for the acid dissociation constant, K_a, and for the pK_a for a number of acids. Note that these acids are categorized in three groups. The first group consists of monoprotic acids, which release one hydrogen ion and have a single K_a and pK_a. The second group consists of diprotic acids, which can release two hydrogen ions and have two K_a values and two pK_a values. The third group consists of polyprotic acids, which can release more than two hydrogen ions. The two examples of polyprotic acids given here, citric acid and phosphoric acid, can release three hydrogen ions and have three K_a values and three pK_a values. Amino acids and peptides, the subject of Chapter 3, behave as diprotic and polyprotic acids; we shall see examples of their titration curves later. Here is a way to keep track of protonated and deprotonated forms of acids and their conjugate bases, and this can be particularly useful with diprotic and polyprotic acids. When the pH of a solution is less than the pK_a of an acid, the protonated form predominates. (Remember that the definition of pH includes a negative logarithm.) When the pH of a solution is greater than the pK_a of an acid, the deprotonated (conjugate base) form predominates.

$$pH < pK_a$$

H^+ on, substance protonated

$$pH > pK_a$$

H^+ off, substance deprotonated

2.6 | What Are Buffers, and Why Are They Important?

A **buffer solution** consists of a mixture of a weak acid and its conjugate base. Buffer solutions tend to resist a change in pH on the addition of moderate amounts of strong acid or base. Let us compare the changes in pH that occur on the addition of equal amounts of strong acid or strong base to pure water at pH 7 and to a buffer solution at pH 7. If 1.0 mL of 0.1 M HCl is added to 99.0 mL of pure water, the pH drops drastically. If the same experiment is conducted with 0.1 M NaOH instead of 0.1 M HCl, the pH rises drastically (Figure 2.14).

Practice Session

Calculate the pH value obtained when 1.0 mL of 0.1 M HCl is added to 99.0 mL of pure water. Also, calculate the pH observed when 1.0 mL of 0.1 M NaOH is added to 99.0 mL of pure water. *Hint:* Be sure to take the dilution of both acid and base to the final volume of 100 mL into account.

Solution

On dilution, we have 100 mL of 0.001 M HCl and 100 mL of 0.001 M NaOH.

Acid added, $[H_3O^+] = 10^{-3}$ M; therefore, pH = 3.

Base added, $[OH^-] = 10^{-3}$ M. Since $[OH^-][H_3O^+] = 1 \times 10^{-14}$, $[H_3O^+]$ $= 10^{-11}$ M; therefore, pH = 11.

The results are different when 99.0 mL of buffer solution is used instead of pure water. A solution that contains the monohydrogen phosphate and dihydrogen phosphate ions, HPO_4^{2-} and $H_2PO_4^-$, in suitable proportions can serve as such a buffer. The Henderson–Hasselbalch equation can be used to calculate the $HPO_4^{2-}/H_2PO_4^-$ ratio that corresponds to pH 7.0.

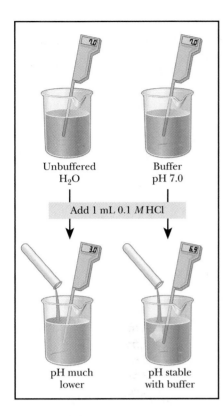

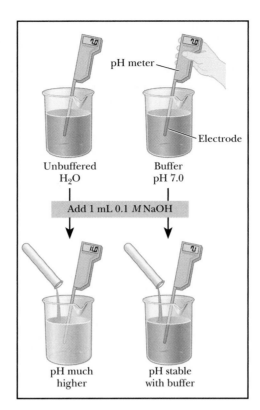

◀ **FIGURE 2.14** Buffering. Acid is added to the two beakers on the left. The pH of unbuffered H_2O drops dramatically while that of the buffer remains stable. Base is added to the two beakers on the right. The pH of the unbuffered water rises drastically while that of the buffer remains stable.

Practice Session

Convince yourself that the proper ratio for pH 7.00 is 0.63 parts HPO_4^{2-} to 1 part $H_2PO_4^-$ by doing the calculation now.

Solution

Use the Henderson–Hasselbalch equation with pH = 7.00 and pK_a = 7.20.

$$pH = pK_a + \log\frac{[A^-]}{[HA]}$$

$$7.00 = 7.20 + \log\frac{[HPO_4^{2-}]}{[H_2PO_4^-]}$$

$$-0.20 = \log\frac{[HPO_4^{2-}]}{[H_2PO_4^-]}$$

$$\frac{[HPO_4^{2-}]}{[H_2PO_4^-]} = 0.63$$

For purposes of illustration, let us consider a solution in which the concentrations are $[HPO_4^{2-}] = 0.063\ M$ and $[H_2PO_4^-] = 0.10\ M$; this gives the conjugate base/weak acid ratio of 0.63 seen above. If 1.0 mL of 0.10 M HCl is added to 99.0 mL of the buffer, the reaction

$$[HPO_4^{2-}] + H^+ \rightleftharpoons [H_2PO_4^-]$$

takes place, and almost all the added H^+ will be used up. The concentrations of $[HPO_4^{2-}]$ and $[H_2PO_4^-]$ will change, and the new concentrations can be calculated.

Concentrations (mol/L)

	$[HPO_4^{2-}]$	$[H^+]$	$[H_2PO_4^-]$
Before addition of HCl	0.063	1×10^{-7}	0.10
HCl added—no reaction yet	0.063	1×10^{-3}	0.10
After HCl reacts with HPO_4^{2-}	0.062	To be found	0.101

The new pH can then be calculated using the Henderson–Hasselbalch equation and the phosphate ion concentrations. The appropriate pK_a is 7.20 (Table 2.6).

$$pH = pK_a + \log \frac{[HPO_4^{2-}]}{[H_2PO_4^-]}$$

$$pH = 7.20 + \log \frac{0.062}{0.101}$$

$$pH = 6.99$$

The new pH is 6.99, a much smaller change than in the unbuffered pure water (Figure 2.14). Similarly, if 1.0 mL of 0.1 M NaOH is used, the same reaction takes place as in a titration:

$$H_2PO_4^- + OH^- \rightleftharpoons HPO_4^{2-}$$

Almost all the added OH^- is used up, but a small amount remains. Since this buffer is an aqueous solution, it is still true that $K_w = [H^+][OH^-]$. The increase in hydroxide ion concentration implies that the hydrogen ion concentration decreases and that the pH increases. Use the Henderson–Hasselbalch equation to calculate the new pH and to convince yourself that the result is pH = 7.01, again a much smaller change in pH than took place in pure water (Figure 2.14). Many biological reactions will not take place unless the pH remains within fairly narrow limits, and, as a result, buffers have great practical importance in the biochemistry laboratory.

A consideration of titration curves can give insight into how buffers work (Figure 2.15a). The pH of a sample being titrated changes very little in the

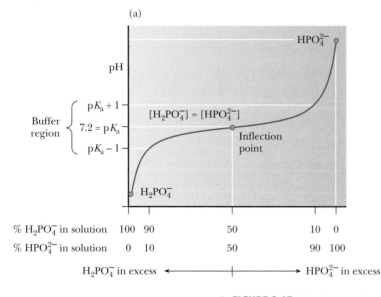

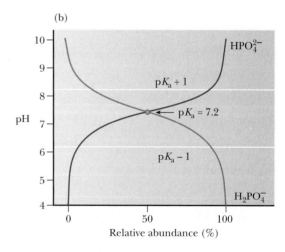

▲ **FIGURE 2.15** The relationship between the titration curve and buffering action in $H_2PO_4^-$. (a) The titration curve of $H_2PO_4^-$, showing the buffer region for the $H_2PO_4^-/HPO_4^{2-}$ pair. (b) Relative abundance of $H_2PO_4^-$ and HPO_4^{2-}.

vicinity of the inflection point of a titration curve. Also, at the inflection point, half the amount of acid originally present has been converted to the conjugate base. The second stage of ionization of phosphoric acid

$$H_2PO_4^- \rightleftharpoons H^+ + HPO_4^{2-}$$

was the basis of the buffer just used as an example. The pH at the inflection point of the titration is 7.20, a value numerically equal to the pK_a of the dihydrogen phosphate ion. At this pH, the solution contains equal concentrations of the dihydrogen phosphate ions and monohydrogen phosphate ions, the acid and base forms. Using the Henderson–Hasselbalch equation, we can calculate the ratio of the conjugate base form to the conjugate acid form for any pH when we know the pK_a. For example, if we choose a pH of 8.2 for a buffer composed of $H_2PO_4^-$ and HPO_4^{2-}, we can solve for the ratio

$$pH = pK_a + \log \frac{HPO_4^{2-}}{H_2PO_4^-}$$

$$8.2 = 7.2 + \log \frac{HPO_4^{2-}}{H_2PO_4^-}$$

$$1 = \log \frac{HPO_4^{2-}}{H_2PO_4^-}$$

$$\frac{HPO_4^{2-}}{H_2PO_4^-} = 10$$

Thus, when the pH is one unit higher than the pK_a, the ratio of the conjugate base form to the conjugate acid form is 10. When the pH is two units higher than the pK_a, the ratio is 100, and so on. Table 2.7 shows this relationship for several increments of pH value.

A buffer solution can maintain the pH at a relatively constant value because of the presence of appreciable amounts of both the acid and its conjugate base. This condition is met at pH values at or near the pK_a of the acid. If OH^- is added, an appreciable amount of the acid form of the buffer is present in solution to react with the added base. If H^+ is added, there is also an appreciable amount of the basic form of the buffer to react with the added acid.

The $H_2PO_4^-/HPO_4^{2-}$ pair is suitable as a buffer near pH 7.2, and the CH_3COOH/CH_3COO^- pair is suitable as a buffer near pH 4.76. At pH values below the pK_a, the acid form predominates, and at pH values above the pK_a, the basic form predominates. The plateau region in a titration curve, where the pH does not change rapidly, covers a pH range extending approximately one pH unit on each side of the pK_a. Thus, there is a range of about two pH

Table 2.7	
pH Values and Base/Acid Ratios for Buffers	
If the pH equals	**The ratio of base form/acid form equals**
$pK_a - 3$	1/1000
$pK_a - 2$	1/100
$pK_a - 1$	1/10
pK_a	1/1
$pK_a + 1$	10/1
$pK_a + 2$	100/1
$pK_a + 3$	1000/1

units in which the buffer is effective (Figure 2.15b). The condition that a buffer contains appreciable amounts of both a weak acid and its conjugate base applies both to the ratio of the two forms and to the absolute amount of each present in a given solution. If a buffer solution contained a suitable ratio of acid to base, but very low concentrations of both, it would take very little added acid to use up all the base form, and vice versa. A buffer solution with low concentrations of both the acid and base forms is said to have a low **buffering capacity.** A buffer that contains greater amounts of both acid and base has a higher buffering capacity. The Biochemical Connections box describes some of the considerations that go into the choice of a suitable buffer for a given application.

How We Make Buffers

When we study buffers in theory, we often use the Henderson–Hasselbalch equation and do many calculations concerning ratios of conjugate base form to conjugate acid form. In practice, however, making a buffer is much easier. To have a buffer, all that is necessary are the two forms of the buffer present in the solution at reasonable quantities. This situation can be obtained by adding predetermined amounts of the conjugate base form (A⁻) to the acid form (HA), or we could start with one and create the other. This is how it is done in practice. Remember that HA and A⁻ are interconverted by adding strong acid or strong base (Figure 2.16). To make a buffer, we could start with the HA form and add NaOH until the pH is correct, as determined by a pH meter. We could also start with A⁻ and add HCl until the pH is correct.

Biochemical Connections

Buffer Selection

Much of biochemistry is studied by carrying out enzymatic reactions in a test tube or *in vitro* (literally, in glass). Such reactions are usually buffered to maintain a constant pH. Similarly, virtually all methods for enzyme isolation, and even for growth of cells in tissue culture, use buffered solutions. The following criteria are typical for selecting a buffer for a biochemical reaction.

1. Suitable pK_a for the buffer.
2. No interference with the reaction or with the assay.
3. Suitable ionic strength of the buffer.
4. No precipitation of reactants or products due to presence of the buffer.
5. Nonbiological nature of the buffer.

The rule of thumb is that the pK_a should be ± 1 pH unit from the pH of the reaction; ± ½ pH unit is even better. Although the perfect generic buffer would have a pH equal to its pK_a, if the reaction is known to produce an acidic product, it is advantageous if the pK_a is below the reaction pH, because then the buffer capacity increases as the reaction proceeds.

Sometimes a buffer can interfere with a reaction or with the assay method. For example, a reaction that requires or produces phosphate or CO_2 may be inhibited if there is too much phosphate or carbonate in the reaction mixture. Even the counter-

ion may be important. Typically a phosphate or carbonate buffer is prepared from the Na⁺ or K⁺ salt. Since many enzymes that react with nucleic acids are activated by one of these two ions and inhibited by the other, the choice of Na⁺ or K⁺ for a counterion could be critical. A buffer can also affect the spectrophotometric determination of a colored assay product.

If a buffer has a poor buffering capacity at the desired pH, its efficiency can often be increased by increasing the concentration; however, many enzymes are sensitive to high salt concentration. Beginning students in biochemistry often have difficulty with enzyme isolations and assays because they fail to appreciate the sensitivity of many enzymes. Fortunately, to minimize this problem, most beginning biochemistry laboratory manuals call for the use of enzymes that are very stable.

A buffer may cause precipitation of an enzyme or even of a metallic ion that may be a cofactor for the reaction. For example, many phosphate salts of divalent cations are only marginally soluble.

Finally, it is often desirable to use a buffer that has no biological activity at all, so it can never interfere with the system being studied. TRIS is a very desirable buffer, since it rarely interferes with a reaction. Special buffers, such as HEPES and PIPES (Table 2.8), have been developed for growing cells in tissue culture.

Depending on the relationship of the pH we desire to the pK_a of the buffer, it may be more convenient to start with one than the other. For example, if we are making an acetic acid/acetate buffer at pH 5.7, it would make more sense to start with the A^- form and to add a small amount of HCl to bring the pH down to 5.7, rather than to start with HA and to add much more NaOH to bring the pH up past the pK_a.

Buffer Systems of Physiological Importance

Buffer systems in living organisms and in the laboratory are based on many types of compounds. Since physiological pH in most organisms stays around 7, it might be expected that the phosphate buffer system would be widely used in living organisms. This is the case where phosphate ion concentrations are high enough for the buffer to be effective, as in most intracellular fluids. The $H_2PO_4^-/HPO_4^{2-}$ pair is the principal buffer in cells. In blood, phosphate ion levels are inadequate for buffering, and a different system operates.

The buffering system in blood is based on the dissociation of carbonic acid (H_2CO_3):

$$H_2CO_3 \rightleftharpoons H^+ + HCO_3^-$$

where the pK_a of H_2CO_3 is 6.37. The pH of human blood, 7.4, is near the end of the buffering range of this system, but another factor enters into the situation.

Carbon dioxide can dissolve in water and in water-based fluids, such as blood. The dissolved carbon dioxide forms carbonic acid, which, in turn, reacts to produce bicarbonate ion:

$$CO_2(g) \rightleftharpoons CO_2(aq)$$

$$CO_2(aq) + H_2O(\ell) \rightleftharpoons H_2CO_3(aq)$$

$$H_2CO_3(aq) \rightleftharpoons H^+(aq) + HCO_3^-(aq)$$

$$\text{Net equation: } CO_2(g) + H_2O(\ell) \rightleftharpoons H^+(aq) + HCO_3^-(aq)$$

At the pH of blood, which is about one unit higher than the pK_a of carbonic acid, most of the dissolved CO_2 is present as HCO_3^-. The CO_2 being transported to the lungs to be expired takes the form of bicarbonate ion. There is a direct relationship between the pH of the blood and the pressure of carbon dioxide gas in the lungs. The properties of hemoglobin, the oxygen-carrying protein in the blood, also enter into the situation (see the Biochemical Connections box in Chapter 4).

The phosphate buffer system is common in the laboratory (*in vitro*, outside the living body) as well as in living organisms (*in vivo*). The buffer system based on TRIS [*tris*(hydroxymethyl)aminomethane] is also widely used in vitro. Other buffers that have come into wide use more recently are **zwitterions,** which are compounds that have both a positive charge and a negative charge. Zwitterions are usually considered less likely to interfere with biochemical reactions than some of the earlier buffers (Table 2.8).

Most living systems operate at pH levels close to 7. The pK_a values of many functional groups, such as the carboxyl and amino groups, are well above or well below this value. As a result, under physiological conditions, many important biomolecules exist as charged species to one extent or another. The practical consequences of this fact are explored in the following Biochemical Connections box.

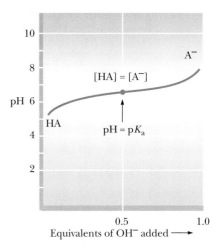

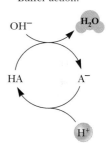

Buffer action:

Biochemistry ⊜ Now™ ACTIVE FIGURE 2.16
Two ways of looking at buffers. In the titration curve, we see that the pH varies only slightly near the region in which [HA] = [A⁻]. In the circle of buffers, we see that adding OH⁻ to the buffer converts HA to A⁻. Adding H⁺ converts A⁻ to HA. **Watch this Active Figure at http://now.brookscole.com/campbell5**

Table 2.8

Acid and Base Forms of Some Useful Biochemical Buffers

Acid Form		Base Form	pK_a
TRIS—H$^+$ (protonated form) $(HOCH_2)_3CNH_3^+$	N—*tris*[hydroxymethyl]aminomethane (TRIS) $\rightleftharpoons$	TRIS (free amine) $(HOCH_2)_3CNH_2$	8.3
$^-$TES—H$^+$ (zwitterionic form) $(HOCH_2)_3\overset{+}{C}NH_2CH_2CH_2SO_3^-$	N—*tris*[hydroxymethyl]methyl-2-aminoethane sulfonate (TES) $\rightleftharpoons$	$^-$TES (anionic form) $(HOCH_2)_3CNHCH_2CH_2SO_3^-$	7.55
$^-$HEPES—H$^+$ (zwitterionic form) $HOCH_2CH_2\overset{+}{N}\underset{H}{\bigcirc}NCH_2CH_2SO_3^-$	N—2—hydroxyethylpiperazine-N$'$-2-ethane sulfonate (HEPES) $\rightleftharpoons$	$^-$HEPES (anionic form) $HOCH_2CH_2N\bigcirc NCH_2CH_2SO_3^-$	7.55
$^-$MOPS—H$^+$ (zwitterionic form) $O\bigcirc\underset{H}{\overset{+}{N}}CH_2CH_2CH_2SO_3^-$	3—[N—morpholino]propane-sulfonic acid (MOPS) $\rightleftharpoons$	$^-$MOPS (anionic form) $O\bigcirc NCH_2CH_2CH_2SO_3^-$	7.2
$^{2-}$PIPES—H$^+$ (protonated dianion) $^-O_3SCH_2CH_2N\underset{H}{\overset{+}{\bigcirc}}NCH_2CH_2SO_3^-$	Piperazine—N,N$'$-*bis*[2-ethanesulfonic acid] (PIPES) $\rightleftharpoons$	$^{2-}$PIPES (dianion) $^-O_3SCH_2CH_2N\bigcirc NCH_2CH_2SO_3^-$	6.8

Biochemical Connections

Some Physiological Consequences of Blood Buffering

The process of respiration plays an important role in the buffering of blood. In particular, an increase in H$^+$ concentration can be dealt with by raising the rate of respiration. Initially, the added hydrogen ion binds to bicarbonate ion, forming carbonic acid.

$$H^+(aq) + HCO_3^-(aq) \rightleftharpoons H_2CO_3(aq)$$

An increased level of carbonic acid raises the levels of dissolved carbon dioxide and, ultimately, gaseous carbon dioxide in the lungs.

$$H_2CO_3(aq) \rightleftharpoons CO_2(aq) + H_2O(\ell)$$

$$CO_2(aq) \rightleftharpoons CO_2(g)$$

A high respiration rate removes this excess carbon dioxide from the lungs, starting a shift in the equilibrium positions of all the foregoing reactions. The removal of gaseous CO$_2$ decreases the amount of dissolved CO$_2$. Hydrogen ion reacts with HCO$_3^-$ and, in the process, lowers the H$^+$ concentration of blood back to its original level. In this way, the blood pH is kept constant.

In contrast, *hyperventilation* (excessively deep and rapid breathing) removes such large amounts of carbon dioxide from the lungs that it raises the pH of blood, sometimes to dangerously high levels that bring on weakness and fainting. Athletes, however, have learned how to use the increase in blood pH caused by hyperventilation. Short bursts of strenuous exercise produce high levels of lactic acid in the blood as a result of the breakdown of glucose. The presence of so much lactic acid tends to lower the pH of the blood, but a brief (30-second) period of hyperventilation before a short-distance event (say, a 400-m dash, 100-m swim, 1-km bicycle race, or any event that lasts between 30 seconds and about a minute) counteracts the effects of the added lactic acid and maintains the pH balance.

An increase in H$^+$ in blood can be caused by large amounts of any acid entering the bloodstream. Aspirin, like lactic acid, is an acid, and extreme acidity resulting from the ingestion of large doses of aspirin can cause *aspirin poisoning*. Exposure to *high altitudes* has an effect similar to hyperventilation at sea level. In response to the tenuous atmosphere, the rate of respiration increases. As with hyperventilation, more carbon dioxide is expired from the lungs, ultimately lowering the H$^+$ level in blood and raising the pH. When people who normally live at sea level are suddenly placed at a high elevation, their blood pH rises temporarily, until they become acclimated.

Summary

2.1 What Makes Water a Polar Molecule? The properties of the water molecule have a direct effect on the behavior of biomolecules. Water is a polar molecule, with a partial negative charge on the oxygen atom and partial positive charges on the hydrogen atoms. There are forces of attraction between the unlike partial charges. Polar substances tend to dissolve in water, but nonpolar substances do not.

2.2 What Is a Hydrogen Bond? A hydrogen bond is a special case of dipole–dipole interactions. In both the liquid state and the solid state, water molecules are extensively hydrogen-bonded to one another. Hydrogen bonding between water and polar solutes takes place in aqueous solutions. The three-dimensional structures of many important biomolecules, including proteins and nucleic acids, are stabilized by hydrogen bonds.

2.3 What Are Acids and Bases? Acids are proton donors, and bases are proton acceptors. Acid–base reactions involve proton transfer. Water can accept and donate protons. The degree of dissociation of acids in water can be characterized by an acid dissociation constant, K_a, which gives a numerical indication of the strength of the acid.

2.4 What Is pH, and What Does It Have to Do with the Properties of Water? The self-dissociation of water can be characterized by a similar constant, K_w. Since the hydrogen ion concentration of aqueous solutions can vary by many orders of magnitude, it is desirable to define a quantity, pH, that expresses the concentration of hydrogen ions conveniently. A similar quantity, pK_a, can be used as an alternative expression for the strength of any acid. The pH of a solution of a weak acid and its conjugate base can be related to the pK_a of that acid by the Henderson–Hasselbalch equation.

2.5 What Are Titration Curves? In an aqueous solution, the relative concentrations of a weak acid and its conjugate base can be related to the titration curve of that acid. In the region of the titration curve in which the pH changes very little upon addition of acid or base, the acid/base concentration ratio varies within a fairly narrow range (10:1 at one extreme and 1:10 at the other).

2.6 What Are Buffers, and Why Are They Important? The tendency to resist a change in pH on the addition of relatively small amounts of acid or base is characteristic of buffer solutions. The control of pH by buffers depends on the fact that their compositions reflect the acid/base concentration ratio in the region of the titration curve in which there is little change in pH.

Critical Questions to Review

2.1 What Makes Water a Polar Molecule?

1. **Thought Question** Why is water necessary for life?

2. **Thought Question** Contemplate biochemistry if atoms did not differ in electronegativity.

2.2 What Is a Hydrogen Bond?

3. **Fact Check** What are some macromolecules that have hydrogen bonds as a part of their structures?

4. **Biochemical Connections** How are hydrogen bonds involved in the transfer of genetic information?

5. **Thought Question** Rationalize the fact that hydrogen bonding has not been observed between CH_4 molecules.

6. **Thought Question** Draw three examples of types of molecules that can form hydrogen bonds.

7. **Thought Question** What are the requirements for molecules to form hydrogen bonds? (What atoms must be present and involved in such bonds?)

8. **Thought Question** Many properties of acetic acid can be rationalized in terms of a hydrogen-bonded dimer. Propose a structure for such a dimer.

9. **Thought Question** How many water molecules could hydrogen-bond *directly* to glucose? To sorbitol or ribitol?

Glucose

Sorbitol   Ribitol

10. **Thought Question** Both RNA and DNA have negatively charged phosphate groups as part of their structure. Would you expect ions that bind to nucleic acids to be positively or negatively charged? Why?

2.3 What Are Acids and Bases?

11. **Fact Check** Identify the conjugate acids and bases in the following pairs of substances:

$$(CH_3)_3NH^+/(CH_3)_3N$$
$$^+H_3N-CH_2COOH/^+H_3N-CH_2-COO^-$$
$$^+H_3N-CH_2-COO^-/H_2N-CH_2-COO^-$$
$$^-OOC-CH_2-COOH/^-OOC-CH_2-COO^-$$
$$^-OOC-CH_2-COOH/HOOC-CH_2-COOH$$

12. **Fact Check** Identify conjugate acids and bases in the following pairs of substances:

    (a) $(HOCH_2)_3 CNH_3^+$ $(HOCH_2)_3 CNH_2$

    (b) $HOCH_2 CH_2 N \overbrace{} N CH_2 CH_2 SO_3^-$

    $HOCH_2 CH_2 \overset{+}{\underset{H}{N}} \overbrace{} N CH_2 CH_2 SO_3^-$

    (c) $O_3^- SCH_2 CH_2 N \overbrace{} \overset{}{\underset{H}{N^+}} CH_2 CH_2 SO_3^-$

    $O_3^- SCH_2 CH_2 N \overbrace{} N CH_2 CH_2 SO_3^-$

13. **Thought Question** Aspirin is an acid with a pK_a of 3.5; its structure includes a carboxyl group. To be absorbed into the bloodstream, it must pass through the membrane lining the stomach and the small intestine. Electrically neutral molecules can pass through a membrane more easily than can charged molecules. Would you expect more aspirin to be absorbed in the stomach, where the pH of gastric juice is about 1, or in the small intestine, where the pH is about 6? Explain your answer.

2.4 What Is pH, and What Does It Have to Do with the Properties of Water?

14. **Fact Check** Why does the pH change by one unit if the hydrogen ion concentration changes by a factor of 10?

15. **Mathematical** Calculate the hydrogen ion concentration, $[H^+]$, for each of the following materials:
    (a) Blood plasma, pH 7.4
    (b) Orange juice, pH 3.5
    (c) Human urine, pH 6.2
    (d) Household ammonia, pH 11.5
    (e) Gastric juice, pH 1.8

16. **Mathematical** Calculate the hydrogen ion concentration, $[H^+]$, for each of the following materials:
    (a) Saliva, pH 6.5
    (b) Intracellular fluid of liver, pH 6.9
    (c) Tomato juice, pH 4.3
    (d) Grapefruit juice, pH 3.2

17. **Mathematical** Calculate the hydroxide ion concentration, $[OH^-]$, for each of the materials used in Question 16.

2.5 What Are Titration Curves?

18. **Fact Check** Define the following:
    (a) Acid dissociation constant
    (b) Acid strength
    (c) Amphipathic
    (d) Buffering capacity
    (e) Equivalence point
    (f) Hydrophilic
    (g) Hydrophobic
    (h) Nonpolar
    (i) Polar
    (j) Titration

2.6 What Are Buffers, and Why Are They Important?

19. **Biochemical Connections** List the criteria used to select a buffer for a biochemical reaction.

20. **Biochemical Connections** What is the relationship between pK_a and the useful range of a buffer?

21. **Mathematical** What is the $[CH_3COO^-]/[CH_3COOH]$ ratio in an acetate buffer at pH 5.00?

22. **Mathematical** What is the $[CH_3COO^-]/[CH_3COOH]$ ratio in an acetate buffer at pH 4.00?

23. **Mathematical** What is the ratio of TRIS/TRIS-H$^+$ in a TRIS buffer at pH 8.7?

24. **Mathematical** What is the ratio of HEPES/HEPES-H$^+$ in a HEPES buffer at pH 7.9?

25. **Mathematical** How would you prepare 1 liter of a 0.050 M phosphate buffer at pH 7.5 using crystalline K_2HPO_4 and a solution of 1.0 M HCl?

26. **Mathematical** The buffer needed for Exercise 25 can also be prepared using crystalline NaH_2PO_4 and a solution of 1.0 M NaOH. How would you do this?

27. **Mathematical** Calculate the pH of a buffer solution prepared by mixing 75 mL of 1.0 M lactic acid (see Table 2.6) and 25 mL of 1.0 M sodium lactate.

28. **Mathematical** Calculate the pH of a buffer solution prepared by mixing 25 mL of 1.0 M lactic acid and 75 mL of 1.0 M sodium lactate.

29. **Mathematical** Calculate the pH of a buffer solution that contains 0.10 M acetic acid (Table 2.6) and 0.25 M sodium acetate.

30. **Mathematical** A catalogue in the lab has a recipe for preparing 1 liter of a TRIS buffer at 0.0500 M and with pH = 8.0: dissolve 2.02 g of TRIS (free base, MW = 121.1 g/mol) and 5.25 g of TRIS hydrochloride (the acidic form, MW = 157.6 g/mol) in a total volume of 1 liter. Verify that this recipe is correct.

31. **Mathematical** If you mixed equal volumes of 0.1 M HCl and 0.20 M TRIS (free amine form; see Table 2.8), is the resulting solution a buffer? Why or why not?

32. **Mathematical** What would be the pH of the solution described in Question 31?

33. **Mathematical** If you have 100 mL of a 0.10 M TRIS buffer at pH 8.3 (Table 2.8) and you add 3.0 mL of 1 M HCl, what will be the new pH?

34. **Mathematical** What will be the pH of the solution in Question 33 if you were to add 3.0 mL more of 1 M HCl?

35. **Mathematical** Show that, for a pure weak acid in water, pH = $\frac{1}{2}$ (pK_a − log [HA]).

36. **Mathematical** What is the ratio of concentrations of acetate ion and undissociated acetic acid in a solution that has a pH of 5.12?

37. **Biochemical Connections** You need to carry out an enzymatic reaction at pH 7.5. A friend suggests a weak acid with a pK_a of 3.9 as the basis of a buffer. Will this substance and its conjugate base make a suitable buffer? Why or why not?

38. **Mathematical** If the buffer suggested in Question 37 were made, what would be the ratio of the conjugate base/conjugate acid?

39. **Biochemical Connections** Suggest a suitable buffer range for each of the following substances:
    (a) Lactic acid (pK_a = 3.86) and its sodium salt
    (b) Acetic acid (pK_a 4.76) and its sodium salt
    (c) TRIS (pK_a = 8.3; see Table 2.8) in its protonated form and its free amine form
    (d) HEPES (pK_a = 7.55; see Table 2.8) in its zwitterionic form and its anionic form

40. **Biochemical Connections** Which of the buffers shown in Table 2.8 would you choose to make a buffer with a pH of 7.3? Explain why.

41. **Thought Question** The solution in Question 25 is called 0.050 M, even though the concentration of neither the free base nor the

conjugate acid is 0.050 *M*. Why is 0.050 *M* the correct concentration to report?

42. **Thought Question** In Section 2.5 we said that, at the equivalence point of a titration of acetic acid, *essentially all* the acid has been converted to acetate ion. Why do we not say that *all* the acetic acid has been converted to acetate ion?

43. **Thought Question** Define buffering capacity. How do the following buffers differ in buffering capacity? How do they differ in pH?

    Buffer a: 0.01 *M* Na_2HPO_4 and 0.01 *M* NaH_2PO_4

    Buffer b: 0.10 *M* Na_2HPO_4 and 0.10 *M* NaH_2PO_4

    Buffer c: 1.0 *M* Na_2HPO_4 and 1.0 *M* NaH_2PO_4

44. **Biochemical Connections** If you wanted to make a HEPES buffer at pH 8.3, and you had both HEPES acid and HEPES base available, which would you start with, and why?

45. **Biochemical Connections** We usually say that a perfect buffer has its pH equal to its pK_a. Give an example of a situation in which it would be advantageous to have a buffer with a pH 0.5 units higher than its pK_a.

46. **Thought Question** What quality of zwitterions makes them desirable buffers?

47. **Thought Question** Many of the buffers used these days, such as HEPES and PIPES, were developed because they have desirable characteristics, such as resisting pH change with dilution. Why would resisting pH change with dilution be advantageous?

48. **Thought Question** Another characteristic of the modern buffers such as HEPES is that their pH changes little with changes in temperature. Why is this desirable?

49. **Thought Question** Identify the zwitterions in the list of substances in Question 11.

50. **Biochemical Connections** A frequently recommended treatment for hiccups is to hold one's breath. The resulting condition, hypoventilation, causes buildup of carbon dioxide in the lungs. Predict the effect on the pH of blood.

Biochemistry ⊘ Now™

Assess your understanding of this chapter's topics with additional quizzing and tutorials at **http://now.brookscole.com/campbell5**

Annotated Bibliography

Barrow, G. M. *Physical Chemistry for the Life Sciences,* 2nd ed. New York: McGraw-Hill, 1981. [Acid–base reactions are discussed in Chapter 4, with titration curves treated in great detail.]

Fasman, G. D., ed. *Handbook of Biochemistry and Molecular Biology: Physical and Chemical Data Section,* 2 vols., 3rd ed. Cleveland: The Chemical Rubber Company, 1976. [Includes a section on buffers and directions for preparation of buffer solutions (vol. 1, pp. 353–378). Other sections cover all important types of biomolecules.]

Ferguson, W. J., and N. E. Good. Hydrogen Ion Buffers. *Anal. Biochem.* **104,** 300–310 (1980). [A description of useful zwitterionic buffers.]

Gerstein, M., and M. Levitt. Simulating Water and the Molecules of Life. *Sci. Amer.* **279** (5), 101–105 (1998). [A description of computer modeling as a tool to investigate the interaction of water molecules with proteins and DNA.]

Hellmans, A. Getting to the Bottom of Water. *Science* **283,** 614–615 (1999). [Recent research indicates that the hydrogen bond may have some covalent character, affecting the properties of water.]

Jeffrey, G. A. *An Introduction to Hydrogen Bonding.* New York: Oxford Univ. Press, 1997. [An advanced, book-length treatment of hydrogen

bonding. Chapter 10 is devoted to hydrogen bonding in biological molecules.]

Olson, A., and D. Goodsell. Visualizing Biological Molecules. *Sci. Amer.* **268** (6), 62–68 (1993). [An account of how computer graphics can be used to represent molecular structure and properties.]

Pauling, L. *The Nature of the Chemical Bond,* 3rd ed. Ithaca, N.Y.: Cornell Univ. Press, 1960. [A classic. Chapter 12 is devoted to hydrogen bonding.]

Rand, R. Raising Water to New Heights. *Science* **256,** 618 (1992). [A brief perspective on the contribution of hydration to molecular assembly and protein catalysis.]

Westhof, E., ed. *Water and Biological Macromolecules.* Boca Raton, Fla.: CRC Press, 1993. [A series of articles about the role of water in hydration of biological macromolecules and the forces involved in macromolecular complexation and cell–cell interactions.]

CHAPTER 3

Amino Acids and Peptides

Proteins are long chains of amino acids linked together by peptide (amide) bonds with a positively charged nitrogen-containing amino group at one end and a negatively charged carboxyl group at the other end. Along the chain is a series of different side chains that differ for each of the 20 amino acids. A linkage of two amino acids is a dipeptide; three amino acids form a tripeptide. The sequence of the amino acids is of the utmost importance. Glycine–lysine–alanine is a different peptide from alanine–lysine–glycine, and it has a different chemical significance. (Similarly, the motto "Talk little, do much" has a different meaning from "Do little, talk much.") For a chain 20 amino acids long, there are more than a billion possible sequences. Literally, the sequence is the message. It determines exactly how the protein will fold up in a three-dimensional conformation to perform its precise biochemical function.

© Roger Ressmeyer/CORBIS

Stanley Miller's classic experiment used an electric discharge to produce amino acids.

Critical Questions

Biochemistry ⊜ Now™
Test yourself on these Critical Questions at the BiochemistryNow website at **http://now.brookscole.com/campbell5**

3.1 | What Are Amino Acids, and What Is Their Three-Dimensional Structure?

Among all the possible amino acids, only 20 are usually found in proteins. The general structure of amino acids includes an **amino group** and a **carboxyl group,** both of which are bonded to the α-carbon (the one next to the carboxyl group). The α-carbon is also bonded to a hydrogen and to the **side-chain group,** which is represented by the letter R. The R group determines the identity of the particular amino acid (Figure 3.1). The two-dimensional formula shown here can only partially convey the common structure of amino acids because one of the most important properties of these compounds is their three-dimensional shape, or **stereochemistry.**

Every object has a mirror image. Many pairs of objects that are mirror images can be superimposed on each other; two identical solid-colored coffee mugs are an example. In other cases, the mirror-image objects cannot be superimposed on one another but are related to each other as the right hand is to the left. Such nonsuperimposable mirror images are said to be **chiral** (from the Greek *cheir,* "hand"); many important biomolecules are chiral.

A frequently encountered chiral center in biomolecules is a carbon atom with four different groups bonded to it (Figure 3.1). Such a center occurs in all amino acids except glycine. Glycine has two hydrogen atoms bonded to the α-carbon; in other words, the side chain (R group) of glycine is hydrogen. Glycine is not chiral (or, alternatively, is **achiral**) because of this symmetry. In all the other commonly occurring amino acids, the α-carbon has four different groups bonded to it, giving rise to two nonsuperimposable mirror-image forms. Figure 3.2 shows perspective drawings of these two possibilities, or **stereoisomers,** for alanine, where the R group is —CH₃. The dashed wedges represent bonds directed away from the observer, and the solid triangles represent bonds directed out of the plane of the paper in the direction of the observer.

The two possible stereoisomers of another chiral compound, L- and D-glyceraldehyde, are shown for comparison with the corresponding forms of alanine. These two forms of glyceraldehyde are the basis of the classification of amino acids into L and D forms. The terminology comes from the Latin *laevus* and *dexter,* meaning "left" and "right," respectively, which comes from the ability of optically active compounds to rotate polarized light to the left or the right. The two stereoisomers of each amino acid are designated as **L- and D-amino acids** on the basis of their similarity to the glyceraldehyde standard. When drawn in a certain orientation, the L form of glyceraldehyde has the hydroxyl group on the left side of the molecule, and the D form has it on the right side, as shown in perspective in Figure 3.2 (a Fischer projection). To determine the L or D designation for an amino acid, it is drawn as shown. The position of the amino group on the left or right side of the α-carbon determines the L or D designation. The amino acids that occur in proteins are all of the L form. Although D-amino acids occur in nature, most often in bacterial cell walls and in some antibiotics, they are not found in proteins.

3.2 | What Are the Structures and Properties of the Individual Amino Acids?

The R groups, and thus the individual amino acids, are classified according to several criteria, two of which are particularly important. The first of these is the polar or nonpolar nature of the side chain. The second depends on the presence of an acidic or basic group in the side chain. Other useful criteria include the presence of functional groups other than acidic or basic ones in the side chains and the nature of those groups.

As mentioned, the side chain of the simplest amino acid, glycine, is a hydrogen atom, and in this case alone two hydrogen atoms are bonded to the α-carbon. In all other amino acids, the side chain is larger and more complex (Figure 3.3). Side-chain carbon atoms are designated with letters of the Greek alphabet, counting from the α-carbon. These carbon atoms are, in turn, the β-, γ-, δ-, and ε-carbons (see lysine in Figure 3.3); a terminal carbon atom is referred to as the ω-carbon, from the name of the last letter of the Greek alphabet. We frequently refer to amino acids by three-letter or one-letter abbreviations of their names, with the one-letter designations becoming much more prevalent these days; Table 3.1 lists these abbreviations.

Group 1—Amino Acids with Nonpolar Side Chains

One group of amino acids has nonpolar side chains. This group consists of alanine, valine, leucine, isoleucine, proline, phenylalanine, tryptophan, and methionine. (In some classification schemes, glycine is placed in this group because it does not have a polar side chain.) In several members of this

Text continues on page 62.

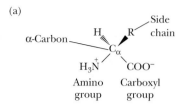

(a)

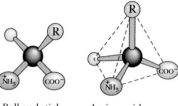

Ball-and-stick model   Amino acids are tetrahedral structures

(b)

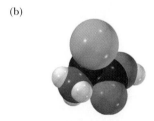

Biochemistry **Now**™ **ANIMATED FIGURE 3.1** The general formula of amino acids, showing the ionic forms that predominate at pH 7. **See this figure animated at http://now.brookscole.com/campbell5**

Essential Information

Proteins are polymers of α-amino acids. A carboxyl group and an amino group are bonded to the same carbon, the α-carbon. Two other groups are bonded to this carbon, so the common amino acids (with one exception) have an asymmetric center. They are chiral objects that cannot be superimposed on their mirror images.

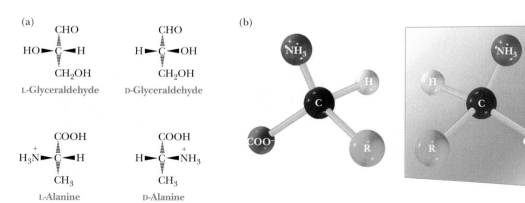

Biochemistry **Now**™ ▲ **ANIMATED FIGURE 3.2** Stereochemistry of alanine and glycine. The amino acids found in proteins have the same chirality as L-glyceraldehyde, which is opposite to that of D-glyceraldehyde. **See this figure animated at http://now.brookscole.com/campbell5**

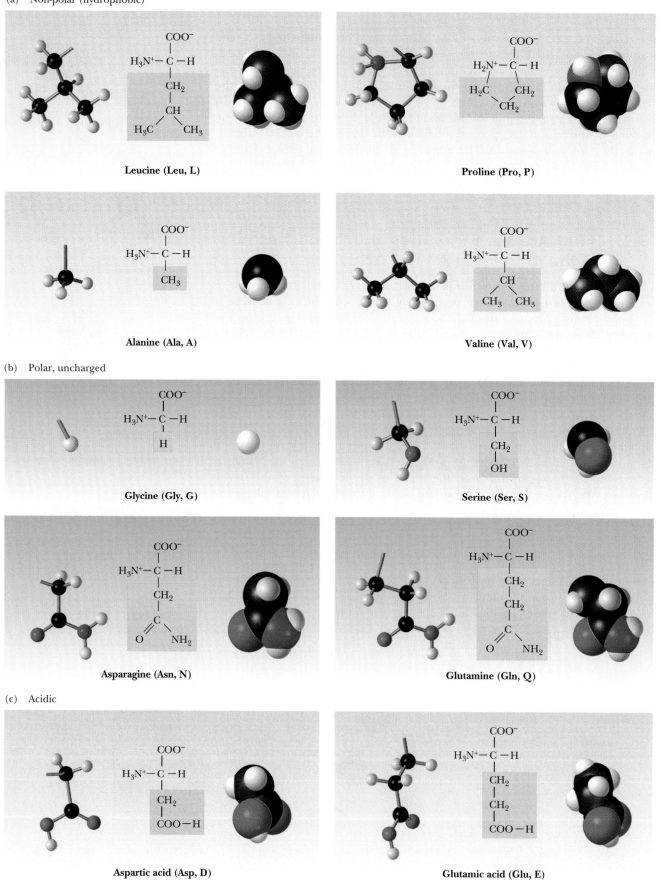

(a) Non-polar (hydrophobic)

Leucine (Leu, L)

Proline (Pro, P)

Alanine (Ala, A)

Valine (Val, V)

(b) Polar, uncharged

Glycine (Gly, G)

Serine (Ser, S)

Asparagine (Asn, N)

Glutamine (Gln, Q)

(c) Acidic

Aspartic acid (Asp, D)

Glutamic acid (Glu, E)

▲ **FIGURE 3.3** The 20 amino acids that are the building blocks of proteins can be classified as (a) nonpolar (hydrophobic), (b) polar, (c) acidic, or (d) basic. Also shown are the one-letter and three-letter codes used to denote amino acids. For each amino acid, the ball-and-stick model *(left)* and the space-filling model *(right)* show only the side chain. (*Illustration, Irving Geis. Rights owned by Howard Hughes Medical Institute. Not to be reproduced without permission.*)

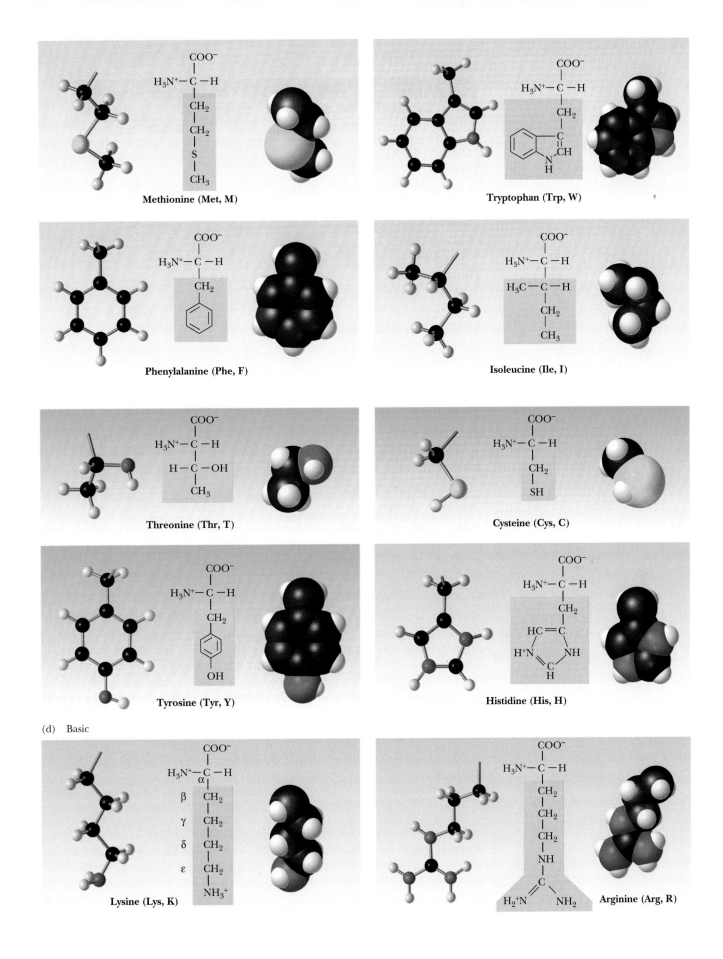

Methionine (Met, M)

Tryptophan (Trp, W)

Phenylalanine (Phe, F)

Isoleucine (Ile, I)

Threonine (Thr, T)

Cysteine (Cys, C)

Tyrosine (Tyr, Y)

Histidine (His, H)

(d)   Basic

Lysine (Lys, K)

Arginine (Arg, R)

Table 3.1

Names and Abbreviations of the Common Amino Acids

Amino Acid	Three-Letter Abbreviation	One-Letter Abbreviation
Alanine	Ala	A
Arginine	Arg	R
Asparagine	Asn	N
Aspartic acid	Asp	D
Cysteine	Cys	C
Glutamic acid	Glu	E
Glutamine	Gln	Q
Glycine	Gly	G
Histidine	His	H
Isoleucine	Ile	I
Leucine	Leu	L
Lysine	Lys	K
Methionine	Met	M
Phenylalanine	Phe	F
Proline	Pro	P
Serine	Ser	S
Threonine	Thr	T
Tryptophan	Trp	W
Tyrosine	Tyr	Y
Valine	Val	V

Note: One-letter abbreviations start with the same letter as the name of the amino acid where this is possible. When the names of several amino acids start with the same letter, phonetic names (occasionally facetious ones) are used, such as Rginine, asparDic, Fenylalanine, tWyptophan. Where two or more amino acids start with the same letter, it is the smallest one whose one-letter abbreviation matches its first letter.

group—namely alanine, valine, leucine, and isoleucine—each side chain is an aliphatic hydrocarbon group. (In organic chemistry, the term "aliphatic" refers to the absence of a benzene ring or related structure.) Proline has an aliphatic cyclic structure, and the nitrogen is bonded to two carbon atoms. In the terminology of organic chemistry, the amino group of proline is a secondary amine, and proline is often called an imino acid. In contrast, the amino groups of all the other common amino acids are primary amines. In phenylalanine, the hydrocarbon group is aromatic (it contains a cyclic group similar to a benzene ring) rather than aliphatic. In tryptophan, the side chain contains an indole ring, which is also aromatic. In methionine, the side chain contains a sulfur atom in addition to aliphatic hydrocarbon groupings. (See Figure 3.3.)

Group 2—Amino Acids with Electrically Neutral Polar Side Chains

Another group of amino acids has polar side chains that are electrically neutral (uncharged) at neutral pH. This group includes serine, threonine, tyrosine, cysteine, glutamine, and asparagine. Glycine is also included here for convenience because it lacks a nonpolar side chain, but, as mentioned before, some biochemists also put it in Group 1 because the C—H bond is nonpolar. In serine and threonine, the polar group is a hydroxyl (—OH) bonded to aliphatic hydrocarbon groups. The hydroxyl group in tyrosine is bonded to an aromatic hydrocarbon group, which eventually loses a proton at higher pH. (The hydroxyl group in tyrosine is a phenol, which is a stronger acid than an aliphatic alcohol. As a result, the side chain of tyrosine can lose a proton in a titration, whereas those of serine and threonine would require such a high pH that pK_a values are not normally listed for these side chains.) In cysteine, the polar side chain consists of a thiol group (—SH), which can react with other cysteine thiol groups to form disulfide (—S—S—) bridges in proteins in an oxidation reaction (Section 1.9). The thiol group can also lose a proton. The amino acids glutamine and asparagine have amide groups, which are derived from carboxyl groups, in their side chains. Amide bonds do not ionize in the range of pH usually encountered in biochemistry. Glutamine and asparagine can be considered to be derivatives of the Group 3 amino acids, glutamic acid and aspartic acid, respectively; those two amino acids have carboxyl groups in their side chains.

Group 3—Amino Acids with Carboxyl Groups in Their Side Chains

Two amino acids, glutamic acid and aspartic acid, have carboxyl groups in their side chains in addition to the one present in all amino acids. A carboxyl group can lose a proton, forming the corresponding carboxylate anion (Section 2.5)—glutamate and aspartate, respectively, in the case of these two amino acids. Because of the presence of the carboxylate, the side chain of each of these two amino acids is negatively charged at neutral pH.

Group 4—Amino Acids with Basic Side Chains

Three amino acids—histidine, lysine, and arginine—have basic side chains, and the side chain in all three is positively charged at or near neutral pH. In lysine, the side-chain amino group is attached to an aliphatic hydrocarbon tail. In arginine, the side-chain basic group, the guanidino group, is more complex in structure than the amino group, but it is also bonded to an aliphatic hydrocarbon tail. In free histidine, the pK_a of the side-chain imidazole group is 6.0, which is not far from physiological pH. The pK_a values for amino acids depend on the environment and can change significantly within the confines of a protein. Histidine can be found in the protonated or unprotonated forms in proteins, and the properties of many proteins depend on whether individual histidine residues are or are not charged.

> **Essential Information**
>
> Amino acids are classified according to two major criteria: the polarity of the side chains and the presence of an acidic or basic group in the side chain.

Biochemistry ⦵ Now™
Go to BiochemistryNow and click on Biochemistry Interactive to see how many amino acids you can recognize and name.

> **Practice Session**
>
> 1. In the following group, identify the amino acids with nonpolar side chains and those with basic side chains: alanine, serine, arginine, lysine, leucine, and phenylalanine.
> 2. The pK_a of the side-chain imidazole group of histidine is 6.0. What is the ratio of uncharged to charged side chains at pH 7.0?
>
> **Solution**
>
> Notice that, in the first part of this practice session, you are asked to do a fact check on material from this chapter, and, in the second part, you are asked to recall and apply concepts from an earlier chapter.
>
> 1. See Figure 3.3. Nonpolar: alanine, leucine, and phenylalanine; basic: arginine and lysine. Serine is not in either category because it has a polar side chain.
> 2. The ratio is 10:1 because the pH is one unit higher than the pK_a.

Amino Acids and Neurotransmitters

Two amino acids deserve some special notice because both are key precursors to many hormones and neurotransmitters (substances involved in the transmission of nerve impulses). The study of neurotransmitters is work in progress, but we do recognize that certain key molecules appear to be involved. Because many neurotransmitters have very short biological half-lives and function at very low concentrations, we also recognize that other derivatives of these molecules may be the actual biologically active forms.

Two of the neurotransmitter classes are simple derivatives of the two amino acids **tyrosine** and **tryptophan.** The active products are monoamine derivatives, which are themselves degraded or deactivated by monoamine oxidases (MAOs).

Tryptophan is converted to serotonin, more properly called 5-hydroxytryptamine.

Tyrosine, itself normally derived from phenylalanine, is converted to the class called catecholamines, which includes epinephrine, commonly known by its proprietary name, adrenalin.

Note that L-dihydroxyphenylalanine (L-dopa) is an intermediate in the conversion of tyrosine. Lower-than-normal levels of L-dopa are involved in Parkinson's disease. Tyrosine or phenylalanine supplements might increase the levels of dopamine, though L-dopa, the immediate precursor, is usually prescribed because L-dopa passes into the brain quickly through the blood–brain barrier.

Tyrosine and phenylalanine are precursors to norepinephrine and epinephrine, both of which are stimulatory. Epinephrine is commonly known as the "flight or fight" hormone. It causes the release of glucose and other nutrients into the blood and also stimulates brain function. People taking MAO inhibitors stay in a relatively high mental state, sometimes too high, because the epinephrine is not metabolized rapidly. Tryptophan is a precursor to serotonin, which has a sedative effect, giving a pleasant feeling. Very low levels of serotonin are associated with depression, while extremely high levels actually produce a manic state. Manic-depressive illness (also called bipolar disorder) can be managed by controlling the levels of serotonin and its further metabolites.

It has been suggested that tyrosine and phenylalanine may have unexpected effects in some people. For example, there is increasing evidence that some people get headaches from the phenylalanine in aspartame (a low-calorie sweetener), which is described in more detail in the Biochemical Connection box on page 75. It is also likely that many illegal psychedelic drugs, such as mescaline and psilocine, mimic and interfere with the effects of neurotransmitters. A recent Oscar-winning film, *A Beautiful Mind,* focused on the disturbing problems associated with schizophrenia. Until recently, the neurotransmitter dopamine was a major focus in the study of schizophrenia. More recently, it has been suggested that irregularities in the metabolism of glutamate, a neurotransmitter, can lead to the disease. (See the article by Javitt and Coyle cited in the bibliography at the end of this chapter.)

Some people insist that supplements of tyrosine give them a morning lift and that tryptophan helps them sleep at night. Milk proteins have high levels of tryptophan; a glass of warm milk before bed is widely believed to be an aid in inducing sleep. Cheese and red wines contain high amounts of tyramine, which mimics epinephrine; for many people a cheese omelet in the morning is a favorite way to start the day.

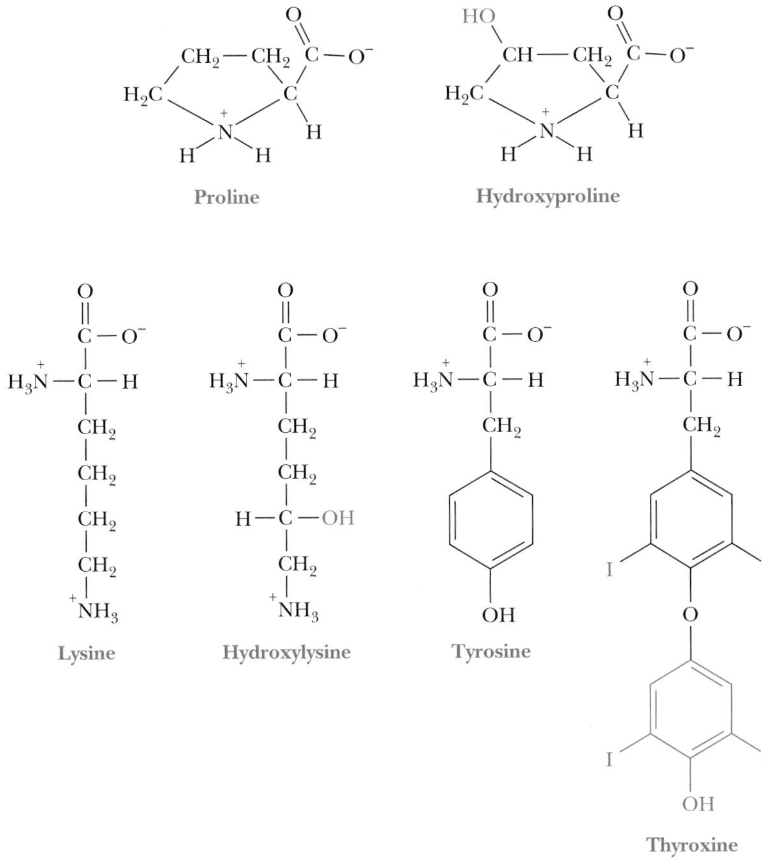

Proline   Hydroxyproline

Lysine   Hydroxylysine   Tyrosine

Thyroxine

◀ **FIGURE 3.4** Structures of hydroxyproline, hydroxylysine, and thyroxine. The structures of the parent amino acids—proline for hydroxyproline, lysine for hydroxylysine, and tyrosine for thyroxine—are shown for comparison. All amino acids are shown in their predominant ionic forms at pH 7.

Uncommon Amino Acids

Many other amino acids, in addition to the ones listed here, are known to exist. They occur in some, but by no means all, proteins. Figure 3.4 shows some examples of the many possibilities. They are derived from the common amino acids and are produced by modification of the parent amino acid after the protein is synthesized by the organism in a process called posttranslational modification. Hydroxyproline and hydroxylysine differ from the parent amino acids in that they have hydroxyl groups on their side chains; they are found only in a few connective-tissue proteins, such as collagen. Thyroxine differs from tyrosine in that it has an extra iodine-containing aromatic group on the side chain; it is produced only in the thyroid gland, formed by posttranslational modification of tyrosine residues in the protein thyroglobulin. Thyroxine is then released as a hormone by proteolysis of thyroglobulin.

3.3 | Do Amino Acids Have Specific Acid–Base Properties?

In a free amino acid, the carboxyl group and amino group of the general structure are charged at neutral pH—the carboxylate portion negatively and the amino group positively. Amino acids without charged groups on their side chains exist in neutral solution as zwitterions with no net charge. A zwitterion has equal positive and negative charges; in solution, it is electrically neutral. Neutral amino acids do not exist in the form NH_2—CHR—COOH (that is, without charged groups).

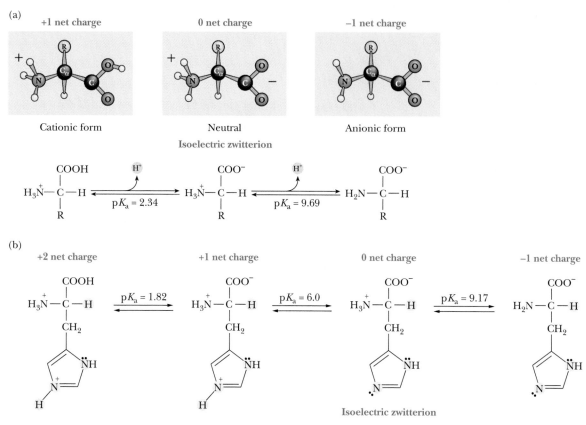

(a)

+1 net charge   0 net charge   −1 net charge

Cationic form   Neutral   Anionic form

Isoelectric zwitterion

$$\underset{R}{\overset{COOH}{H_3\overset{+}{N}-C-H}} \underset{pK_a = 2.34}{\overset{H^+}{\rightleftarrows}} \underset{R}{\overset{COO^-}{H_3\overset{+}{N}-C-H}} \underset{pK_a = 9.69}{\overset{H^+}{\rightleftarrows}} \underset{R}{\overset{COO^-}{H_2N-C-H}}$$

(b)

+2 net charge   +1 net charge   0 net charge   −1 net charge

$$\overset{COOH}{\underset{CH_2}{H_3\overset{+}{N}-C-H}} \underset{pK_a = 1.82}{\rightleftarrows} \overset{COO^-}{\underset{CH_2}{H_3\overset{+}{N}-C-H}} \underset{pK_a = 6.0}{\rightleftarrows} \overset{COO^-}{\underset{CH_2}{H_3\overset{+}{N}-C-H}} \underset{pK_a = 9.17}{\rightleftarrows} \overset{COO^-}{\underset{CH_2}{H_2N-C-H}}$$

Isoelectric zwitterion

Biochemistry❸Now™ ▲ **ANIMATED FIGURE 3.5** The ionization of amino acids. (a) The ionic forms of the amino acids, shown without consideration of any ionizations on the side chain. The cationic form is the low-pH form, and the titration of the cationic species with base yields the zwitterions and finally the anionic form. (b) The ionization of histidine (an amino acid with a titratable side chain). **See this figure animated at http://now.brookscole.com/campbell5**

Biochemistry❸Now™
Go to BiochemistryNow and click on Biochemistry Interactive to explore the titration behavior of amino acids.

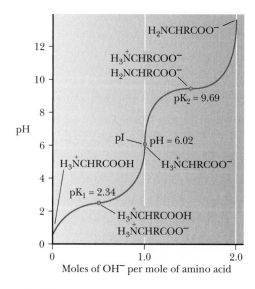

▲ **FIGURE 3.6** The titration curve of alanine.

When an amino acid is titrated, its titration curve indicates the reaction of each functional group with hydrogen ion. In alanine, the carboxyl and amino groups are the two titratable groups. At very low pH, alanine has a protonated (and thus uncharged) carboxyl group and a positively charged amino group that is also protonated. Under these conditions, the alanine has a net positive charge of 1. As base is added, the carboxyl group loses its proton to become a negatively charged carboxylate group (Figure 3.5a), and the pH of the solution increases. Alanine now has no net charge. As the pH increases still further with addition of more base, the protonated amino group (a weak acid) loses its proton, and the alanine molecule now has a negative charge of 1. The titration curve of alanine is that of a diprotic acid (Figure 3.6).

In histidine, the imidazole side chain also contributes a titratable group. At very low pH values, the histidine molecule has a net positive charge of 2 because both the imidazole and amino groups have positive charges. As base is added and the pH increases, the carboxyl group loses a proton to become a carboxylate as before, and the histidine now has a positive charge of 1 (Figure 3.5b). As still more base is added, the charged imidazole group loses its proton, and this is the point at which the histidine has no net charge. At still higher values of pH, the amino group loses its proton, as was the case with alanine, and the histidine molecule now has a negative charge of 1. The titration curve of histidine is that of a triprotic acid (Figure 3.7).

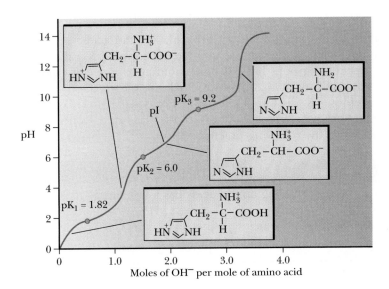

Biochemistry⚡Now™ **ACTIVE FIGURE 3.7**
The titration curve of histidine. The isoelectric pH (pI) is the value at which positive and negative charges are the same. The molecule has no net charge. **Watch this Active Figure at http://now .brookscole.com/campbell5**

Like the acids we discussed in Chapter 2, the titratable groups of each of the amino acids have characteristic pK_a values. The pK_a values of α-carboxyl groups are fairly low, around 2. The pK_a values of amino groups are much higher, with values ranging from 9 to 10.5. The pK_a values of side-chain groups, including side-chain carboxyl and amino groups, depend on the groups' chemical nature. Table 3.2 lists the pK_a values of the titratable groups of the amino acids. The classification of an amino acid as acidic or basic depends on the pK_a of the side chain as well as the chemical nature of the group. Histidine, lysine, and arginine are considered basic amino acids because each of their side chains has a nitrogen-containing group that can exist in either a protonated or deprotonated form. However, histidine has a pK_a in the acidic range. Aspartic acid and glutamic acid are considered to be acidic because each has a carboxylic acid side chain with a low pK_a value. These groups can still be titrated after the amino acid is incorporated into a peptide or protein, but the pK_a of the titratable group on the side chain is not necessarily the same in a protein as it is in a free amino acid. In fact, it can be very different. For example, a pK_a of 9 has been reported for an aspartate side chain in the protein thioredoxin. (For more information, see the article by Wilson et al. cited in the bibliography at the end of this chapter.)

The fact that amino acids, peptides, and proteins have different pK_a values gives rise to the possibility that they can have different charges at a given pH. Alanine and histidine, for example, both have net charges of −1 at high pH, above 10; the only charged group is the carboxylate anion. At lower pH, around 5, alanine is a zwitterion with no net charge, but histidine has a net charge of 1 at this pH because the imidazole group is protonated. This property is useful in **electrophoresis,** a common method for separating molecules in an electric field. This method is extremely useful in determining the important properties of proteins and nucleic acids. We shall see the applications to proteins in Chapter 5 and to nucleic acids in Chapter 14. The pH at which a molecule has no net charge is called the **isoelectric pH,** or isoelectric point (given the symbol **pI**). At its isoelectric pH, a molecule will not migrate in an electric field. This property can be put to use in separation methods. The pI of an amino acid can be calculated by the following equation:

$$pI = \frac{pK_{a1} + pK_{a2}}{2}$$

Table 3.2

pK_a Values of Common Amino Acids

Acid	α-COOH	α-NH$_3^+$	RH or RH$^+$
Gly	2.34	9.60	
Ala	2.34	9.69	
Val	2.32	9.62	
Leu	2.36	9.68	
Ile	2.36	9.68	
Ser	2.21	9.15	
Thr	2.63	10.43	
Met	2.28	9.21	
Phe	1.83	9.13	
Trp	2.38	9.39	
Asn	2.02	8.80	
Gln	2.17	9.13	
Pro	1.99	10.6	
Asp	2.09	9.82	3.86*
Glu	2.19	9.67	4.25*
His	1.82	9.17	6.0*
Cys	1.71	10.78	8.33*
Tyr	2.20	9.11	10.07
Lys	2.18	8.95	10.53
Arg	2.17	9.04	12.48

*For these amino acids, the R group ionization occurs before the α-NH$_3^+$ ionization.

For the majority of the amino acids, there are only two pK_a values, so this equation is easily used to calculate the pI. For the acidic and basic amino acids, however, we must be sure to average the correct pK_a values. The pK_{a1} is for the functional group that has dissociated at its isoelectric point. If there are two groups dissociated at isoelectric pH, the pK_{a1} is the higher pK_a of the two. Therefore, pK_{a2} is for the group that has not dissociated at isoelectric pH. If there are two groups that are not dissociated, the one with the lower pK_a is used. See the following practice session.

Practice Session

1. Which of the following amino acids has a net charge of $+2$ at low pH? Which has a net charge of -2 at high pH? Aspartic acid, alanine, arginine, glutamic acid, leucine, lysine.
2. What is the pI for histidine?

Solution

Notice that the first part of this practice session deals only with the qualitative description of the successive loss of protons by the titratable groups on the individual amino acids. In the second part, you need to refer to the titration curve as well to do a numerical calculation of pH values.

1. Arginine and lysine have net charges of $+2$ at low pH because of their basic side chains; aspartic acid and glutamic acid have net charges of -2 at high pH because of their carboxylic acid side chains. Alanine and leucine do not fall into either category because they do not have titratable side chains.
2. Draw or picture histidine at very low pH. It will have the formula shown in Figure 3.5b on the far left side. This form has a net charge of $+2$. To arrive at the isoelectric point, we must add some negative charge or remove some positive charge. This will happen in solution in order of increasing pK_a. Therefore, we begin by taking off the hydrogen from the carboxyl group because it has the lowest pK_a (1.82). This leaves us with the form shown second from the left in Figure 3.5. This form has a charge of $+1$, so we must remove yet another hydrogen to arrive at the isoelectric form. This hydrogen would come from the imidazole side chain because it has the next highest pK_a (6.0); this is the isoelectric form (second from right). Now we average the pK_a from the highest pK_a group that lost a hydrogen with that of the lowest pK_a group that still retains its hydrogen. In the case of histidine, the numbers to substitute in the equation for the pI are 6.0 [pK_{a1}] and 9.17 [pK_{a2}], which gives a pI of 7.58.

3.4 | What Is the Peptide Bond?

Individual amino acids can be linked together by forming covalent bonds. The bond is formed between the α-carboxyl group of one amino acid and the α-amino group of the next one. Water is eliminated in the process, and the linked amino acid **residues** remain after water is eliminated (Figure 3.8). A bond formed in this way is called a **peptide bond. Peptides** are compounds formed by linking small numbers of amino acids, ranging from two to several dozen. In a protein, many amino acids (usually more than a hundred) are linked by peptide bonds to form a **polypeptide chain** (Figure 3.9). Another

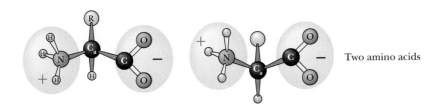

Two amino acids

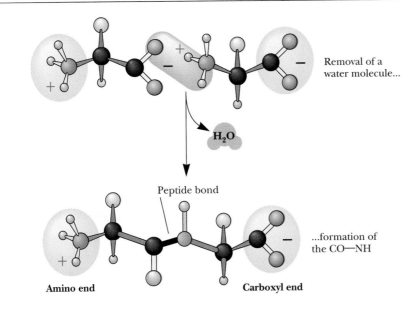

Removal of a water molecule...

H_2O

Peptide bond

...formation of the CO—NH

Amino end   **Carboxyl end**

BiochemistryNow™ **ANIMATED FIGURE 3.8**
Formation of the peptide bond. (*Illustration, Irving Geis. Rights owned by Howard Hughes Medical Institute. Not to be reproduced without permission.*) **See this figure animated at http://now.brookscole.com/campbell5**

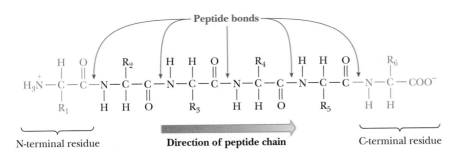

◀ **FIGURE 3.9** A small peptide showing the direction of the peptide chain (N-terminal to C-terminal).

name for a compound formed by the reaction between an amino group and a carboxyl group is an *amide.*

The carbon–nitrogen bond formed when two amino acids are linked in a peptide bond is usually written as a single bond, with one pair of electrons shared between the two atoms. With a simple shift in the position of a pair of electrons, it is quite possible to write this bond as a double bond. This shifting of electrons is well known in organic chemistry and results in **resonance structures,** structures that differ from one another only in the positioning of electrons. The positions of double and single bonds in one resonance structure are different from their positions in another resonance structure of the same compound. No single resonance structure actually represents the bonding in the compound; instead all resonance structures contribute to the bonding situation.

Text continues on page 72.

Essential Information

When the carboxyl group of one amino acid reacts with the amino group of another to give an amide linkage and eliminate water, a peptide bond is formed. In a protein, upward of a hundred amino acids are so joined to form a polypeptide chain.

Biochemical Connections

Amino Acid Functions Other Than in Peptides

Amino acids have biological functions other than as parts of proteins and oligopeptides. The following examples illustrate some of these functions for a few of the amino acids.

Glycine

As the simplest amino acid, glycine is among the most water soluble, and it is often added to other molecules to make them more water soluble, often so that they can be excreted in the urine. Many drugs and medications are oxidized in the liver to compounds that contain a hydroxyl group, which then are conjugated to glycine; the final product is then removed from the blood in the kidney. Benzoic acid, a byproduct of many aromatic substances, which is not water soluble, is conjugated via an amide bond to the amino group of glycine to form hippuric acid, a metabolic waste product.

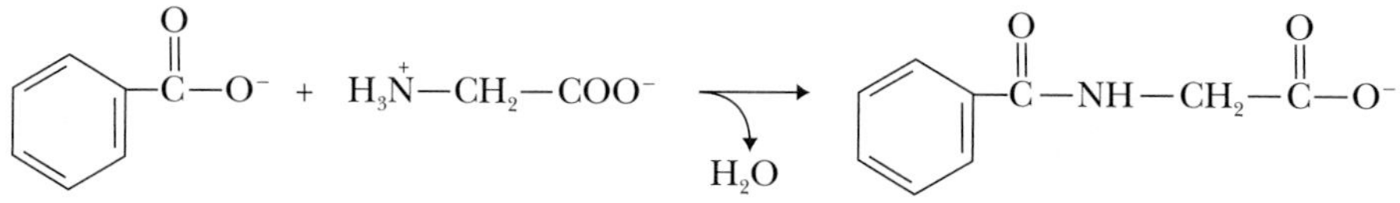

Benzoic acid
(as benzoate)

Glycine

Hippuric acid
(as hippurate)

Glycine is also added to cholic acid to form glycocholic acid, one of the two major bile salts, potent detergents used in the digestion of fats.

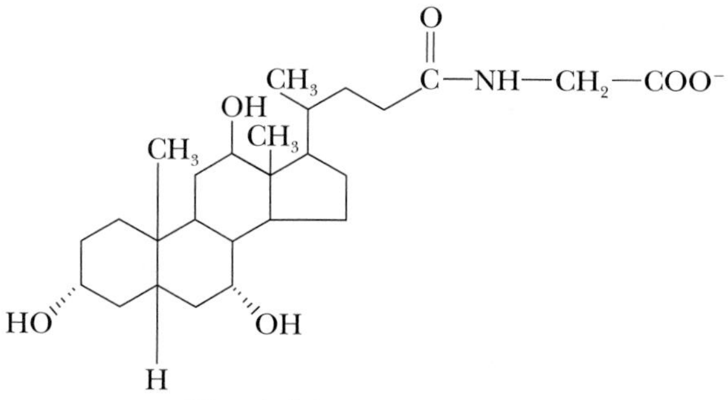

Glycocholate

Methionine

A derivative of this amino acid, S-adenosylmethionine, is the source of the methyl group in many methylation reactions. The corresponding compound that contains an ethyl group, ethionine, is a potent poison because it transfers an ethyl group, rather than the required methyl group.

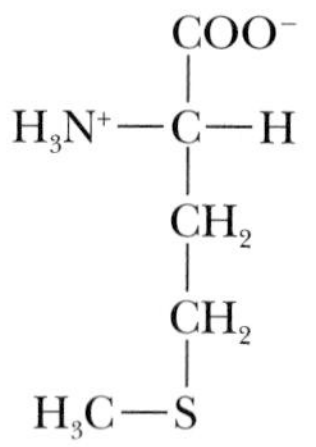

Methionine

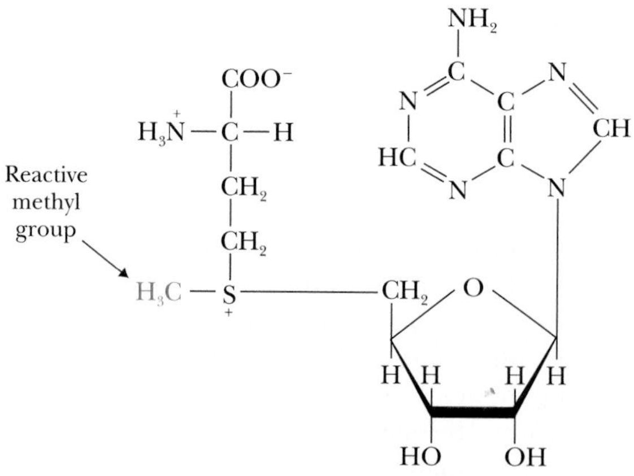

S-Adenosylmethionine

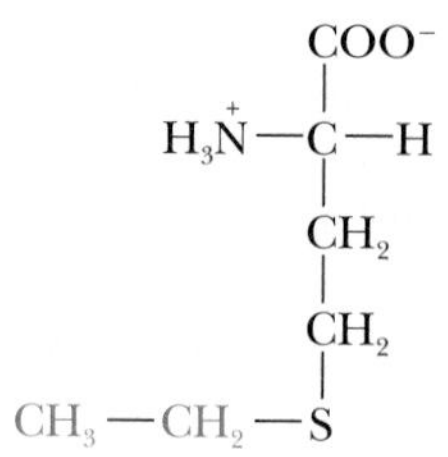

Ethionine

Glutamic Acid

Monosodium glutamate, or MSG, is a derivative of glutamic acid that finds wide use as a flavor enhancer. MSG causes a physiological reaction in some people, with chills, headaches, and dizziness resulting. Because many Asian foods contain significant amounts of MSG, this problem is often referred to as *Chinese restaurant syndrome.*

β-Alanine

The α-amino acids are not the only biologically important ones. This β-amino acid is found in the vitamin pantothenic acid and is an important part of the enzyme cofactor Coenzyme A.

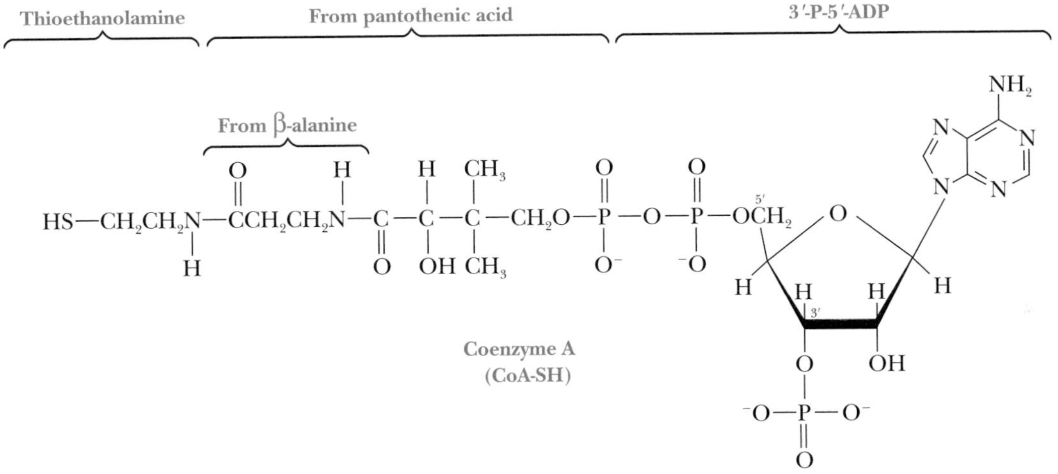

Coenzyme A
(CoA-SH)

Histidine

If the acid group of histidine is removed, it is converted to histamine, which is a potent vasodilator, increasing the diameter of blood vessels. Histamine, which is released as part of the immune response, increases the localized blood volume for white blood cells. This results in the swelling and stuffiness that are associated with a cold. Most cold medications contain antihistamines to overcome this stuffiness.

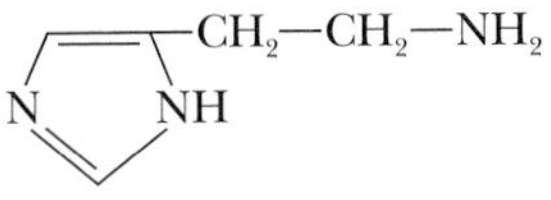

Histamine

Arginine

This basic amino acid is involved in the urea cycle, a series of reactions of fundamental importance in the use of nitrogen by living organisms.

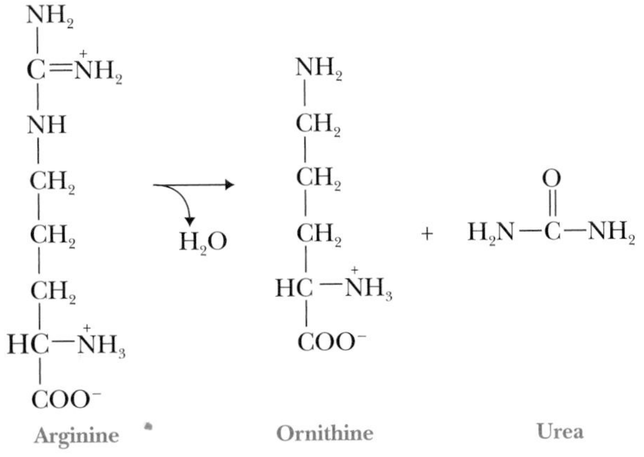

Arginine   Ornithine   Urea

Asparagine and Glutamine

These two amino acids can be considered derivatives of the acidic amino acids aspartate and glutamate. Like arginine, however, they play a role in the way living things use nitrogen. In animals, they are involved in detoxification of ammonia; in plants, they play a role in nitrogen storage.

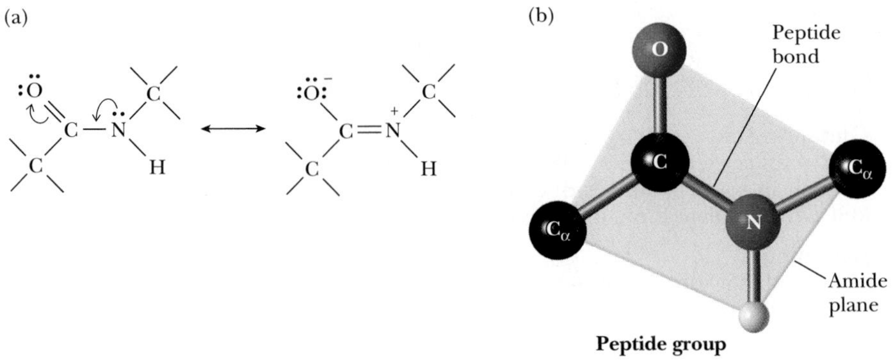

(a)   (b)

▶ **FIGURE 3.10** The resonance structures of the peptide bond lead to a planar group. (a) Resonance structures of the peptide group. (b) The planar peptide group. (*Illustration, Irving Geis. Rights owned by Howard Hughes Medical Institute. Not to be reproduced without permission.*)

The peptide bond can be written as a resonance hybrid of two structures (Figure 3.10), one with a single bond between the carbon and nitrogen and the other with a double bond between the carbon and nitrogen. The peptide bond has partial double bond character. As a result, the peptide group that forms the link between the two amino acids is planar. The peptide bond is also stronger than an ordinary single bond because of this resonance stabilization.

This structural feature has important implications for the three-dimensional conformations of peptides and proteins. There is free rotation around the bonds between the α-carbon of a given amino acid residue and the amino nitrogen and carbonyl carbon of that residue, but there is no significant rotation around the peptide bond. This stereochemical constraint plays an important role in determining how the protein backbone can fold.

3.5 | Are Small Peptides Physiologically Active?

The simplest possible covalently bonded combination of amino acids is a dipeptide, in which two amino acid residues are linked by a peptide bond. An example of a naturally occurring dipeptide is carnosine, which is found in muscle tissue. This compound, which has the alternative name β-alanyl-L-histidine, has an interesting structural feature. (In the systematic nomenclature of peptides, the **N-terminal** amino acid residue—the one with the free amino group—is given first; then other residues are given as they occur in sequence. The **C-terminal** amino acid residue—the one with the free carboxyl group—is given last.) The N-terminal amino acid residue, β-alanine, is structurally different from the α-amino acids we have seen up to now. As the name implies, the amino group is bonded to the third or β-carbon of the alanine (Figure 3.11).

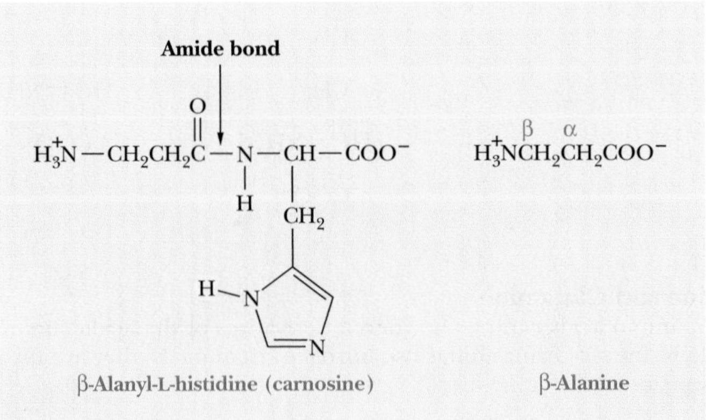

▶ **FIGURE 3.11** Structures of carnosine and its component amino acid β-alanine.

The peptide bond in this dipeptide is formed between the carboxyl group of the β-alanine and the amino group of the histidine, which is the C-terminal amino acid. The following Biochemical Connections box discusses another dipeptide of some interest, and the Biochemical Connections box on page 75 discusses health-related implications of the use of this same dipeptide.

Practice Session

Write an equation with structures for the formation of a dipeptide when alanine reacts with glycine to form a peptide bond. Is there more than one possible product for this reaction?

Solution

The main point here is to be aware of the possibility that amino acids can be linked together in more than one order when they form peptide bonds. Thus, there are two possible products when alanine and glycine react: alanylglycine, in which alanine is at the N-terminal end and glycine is at the C-terminal end, and glycylalanine, in which glycine is at the N-terminal end and alanine is at the C-terminal end.

Glutathione is a commonly occurring tripeptide; it has considerable physiological importance because it is a scavenger for oxidizing agents. Recall from Section 1.9 that oxidation is the loss of electrons; an oxidizing agent causes another substance to lose electrons. (It is thought that some oxidizing agents are harmful to organisms and play a role in the development of cancer.) In terms of its amino acid composition and bonding order, it is γ-glutamyl-L-cysteinylglycine (Figure 3.12a). The letter γ (gamma) is the third letter in the

---

Biochemical Connections

Aspartame, the Sweet Peptide

The dipeptide L-aspartyl-L-phenylalanine is of considerable commercial importance. The aspartyl residue has a free α-amino group, the N-terminal end of the molecule, and the phenylalanyl residue has a free carboxyl group, the C-terminal end. This dipeptide is about 200 times sweeter than sugar. A methyl ester derivative of this dipeptide is of even greater commercial importance than the dipeptide itself. The derivative has a methyl group at the C-terminal end in an ester linkage to the carboxyl group. The methyl ester derivative is called *aspartame* and is marketed as a sugar substitute under the trade name NutraSweet.

The consumption of common table sugar in the United States is about 100 pounds per person per year. Many people want to curtail their sugar intake in the interest of fighting obesity. Others must limit their sugar intake because of diabetes. One of the most common ways of doing so is by drinking diet soft drinks. The soft-drink industry is one of the largest markets for aspartame. The use of this sweetener was approved by the U.S. Food and Drug Administration in 1981 after extensive testing, although there is still considerable controversy about its safety. Diet soft drinks sweetened with aspartame

carry warning labels about the presence of phenylalanine. This information is of vital importance to people who have phenylketonuria, a genetic disease of phenylalanine metabolism. (See the Biochemical Connections box on page 75). Note that both amino acids have the L configuration. If a D-amino acid is substituted for either amino acid or for both of them, the resulting derivative is bitter rather than sweet.

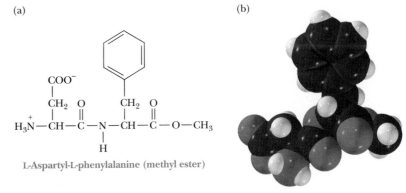

(a)

L-Aspartyl-L-phenylalanine (methyl ester)

(b)

▲ (a) Structure of aspartame. (b) Space-filling model of aspartame.

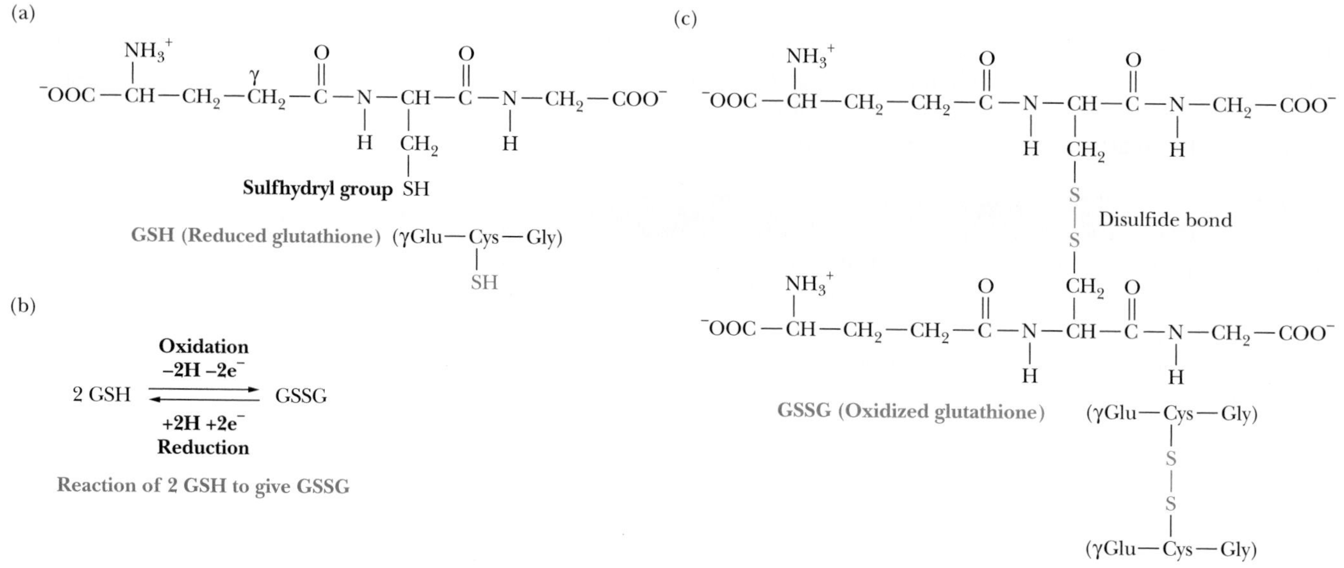

(a)

$$^-OOC-\underset{\overset{|}{NH_3^+}}{CH}-CH_2-\underset{\gamma}{CH_2}-\underset{\overset{\parallel}{O}}{C}-\underset{\overset{|}{H}}{N}-\underset{\overset{|}{CH_2}}{CH}-\underset{\overset{\parallel}{O}}{C}-\underset{\overset{|}{H}}{N}-CH_2-COO^-$$

Sulfhydryl group SH

GSH (Reduced glutathione) (γGlu—Cys—Gly)
|
SH

(b)

$$2\ GSH \underset{\underset{\textbf{Reduction}}{\textbf{+2H +2e}^-}}{\overset{\overset{\textbf{Oxidation}}{\textbf{−2H −2e}^-}}{\rightleftarrows}} GSSG$$

Reaction of 2 GSH to give GSSG

(c)

$$^-OOC-\underset{\overset{|}{NH_3^+}}{CH}-CH_2-CH_2-\underset{\overset{\parallel}{O}}{C}-\underset{\overset{|}{H}}{N}-\underset{\overset{|}{CH_2}}{CH}-\underset{\overset{\parallel}{O}}{C}-\underset{\overset{|}{H}}{N}-CH_2-COO^-$$

Disulfide bond

S
|
S

$$^-OOC-\underset{\overset{|}{NH_3^+}}{CH}-CH_2-CH_2-\underset{\overset{\parallel}{O}}{C}-\underset{\overset{|}{H}}{N}-\underset{\overset{|}{CH_2}}{CH}-\underset{\overset{\parallel}{O}}{C}-\underset{\overset{|}{H}}{N}-CH_2-COO^-$$

GSSG (Oxidized glutathione) (γGlu—Cys—Gly)
|
S
|
S
|
(γGlu—Cys—Gly)

▶ **FIGURE 3.12** The oxidation and reduction of glutathione. (a) The structure of reduced glutathione. (b) A schematic representation of the oxidation–reduction reaction. (c) The structure of oxidized glutathione.

Greek alphabet; in this notation, it refers to the third carbon atom in the molecule, counting the one bonded to the amino group as the first. Once again, the N-terminal amino acid is given first. In this case, the γ-carboxyl group (the side-chain carboxyl group) of the glutamic acid is involved in the peptide bond; the amino group of the cysteine is bonded to it. The carboxyl group of the cysteine is bonded, in turn, to the amino group of the glycine. The carboxyl group of the glycine forms the other end of the molecule, the C-terminal end. The glutathione molecule shown in Figure 3.12a is the reduced form. It scavenges oxidizing agents by reacting with them. The oxidized form of glutathione is generated from two molecules of the reduced peptide by forming a disulfide bond between the —SH groups of the two cysteine residues (Figure 3.12b). The full structure of oxidized glutathione is shown in Figure 3.12c.

Two pentapeptides found in the brain are known as enkephalins, naturally occurring analgesics (pain relievers). For molecules of this size, abbreviations for the amino acids are more convenient than structural formulas. The same notation is used for the amino acid sequence, with the N-terminal amino acid listed first and the C-terminal listed last. The two peptides in question, leucine enkephalin and methionine enkephalin, differ only in their C-terminal amino acids.

Tyr—Gly—Gly—Phe—Leu (three-letter abbreviations)

Y—G—G—F—L (one-letter abbreviations)
Leucine enkephalin

Tyr—Gly—Gly—Phe—Met

Y—G—G—F—M
Methionine enkephalin

It is thought that the aromatic side chains of tyrosine and phenylalanine in these peptides play a role in their activities. It is also thought that there are similarities between the three-dimensional structures of opiates, such as mor-

Biochemical Connections

Phenylketonuria and Inborn Errors of Metabolism

Mutations leading to deficiencies in enzymes are usually referred to as "inborn errors of metabolism," since they involve defects in the DNA of the affected individual. Errors in enzymes that catalyze reactions of amino acids frequently have disastrous consequences, many of them leading to severe forms of mental retardation. Phenylketonuria (PKU) is a well-known example. Phenylalanine, phenylpyruvate, phenyllactate, and phenylacetate all accumulate in the blood and urine. Available evidence suggests that phenylpyruvate, which is a phenylketone, causes mental retardation by interfering with the conversion of pyruvate to acetyl-CoA (an important intermediate in many biochemical reactions) in the brain. It is also likely that the accumulation of these products in the brain cells results in an osmotic imbalance in which water flows into the brain cells. These cells expand in size until they crush each other in the developing brain. In either case, the brain is not able to develop normally.

Fortunately, PKU can be easily detected in newborns, and all 50 states and the District of Columbia mandate that such a test be performed because it is cheaper to treat the disease with a modified diet than to cope with the costs of a mentally retarded individual who is usually institutionalized for life. The dietary changes are relatively simple. Phenylalanine must be limited to the amount needed for protein synthesis, and tyrosine must now be supplemented, since phenylalanine is no longer a source. You may have noticed that foods containing aspartame carry a warning about the phenylalanine portion of that artificial sweetener. A substitute for aspartame, which carries the trade name Alatame, contains alanine rather than phenylalanine. It has been introduced to retain the benefits of aspartame without the dangers associated with phenylalanine.

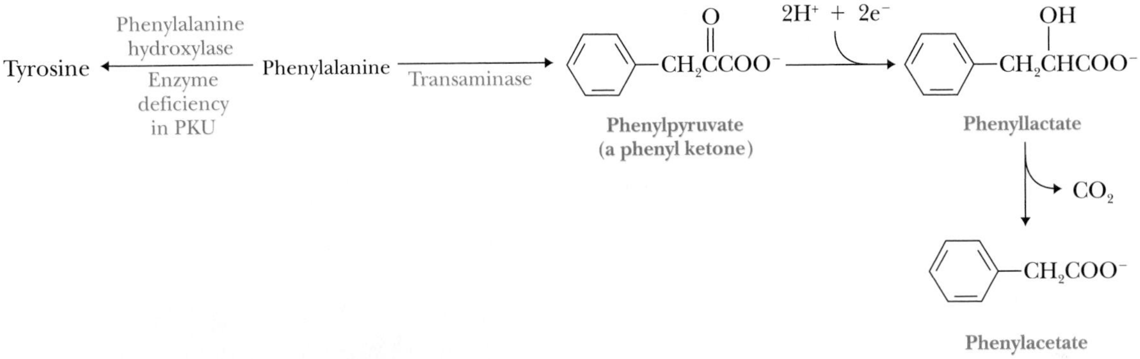

▲ Reactions involved in the development of phenylketonuria (PKU). A deficiency in the enzyme that catalyzes the conversion of phenylalanine to tyrosine leads to the accumulation of phenylpyruvate, a phenyl ketone.

phine, and those of the enkephalins. As a result of these structural similarities, opiates bind to the receptors in the brain intended for the enkephalins and thus produce their physiological activities.

Some important peptides have cyclic structures. Two well-known examples with many structural features in common are oxytocin and vasopressin (Figure 3.13). In each, there is an —S—S— bond similar to that in the oxidized form of glutathione. The disulfide bond is responsible for the cyclic structure. Each of these peptides contains nine amino acid residues, each has an amide group (rather than a free carboxyl group) at the C-terminal end, and each has a disulfide link between cysteine residues at positions 1 and 6. The difference between these two peptides is that oxytocin has an isoleucine residue at position 3 and a leucine residue at position 8, and vasopressin has a phenylalanine residue at position 3 and an arginine residue at position 8. Both of these peptides have considerable physiological importance as hormones (see the following Biochemical Connections box).

In some other peptides, the cyclic structure is formed by the peptide bonds themselves. Two cyclic decapeptides (peptides containing ten amino acid

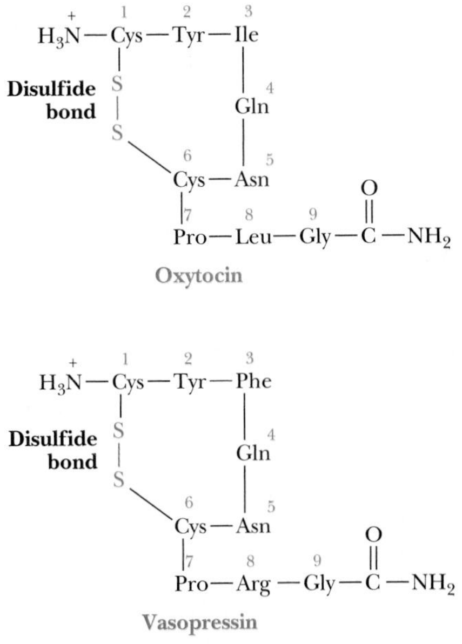

▲ **FIGURE 3.13** Structures of oxytocin and vasopressin.

$$\text{CH}_2 - \text{CH}_2 - \text{CH}_2 - \text{NH}_3{}^+$$
$$^+\text{NH}_3 - \text{CH} - \text{COO}^-$$

Ornithine (Orn)

L-Val — L-Orn — L-Leu — D-Phe — L-Pro
L-Pro — L-Phe — L-Leu — D-Orn — L-Val   **Direction of peptide bond**

Gramicidin S

L-Val — L-Orn — L-Leu — D-Phe — L-Pro
L-Tyr — L-Glu — L-Asp — D-Phe — L-Phe   **Direction of peptide bond**

Tyrocidine A

▶ **FIGURE 3.14** Structures of ornithine, gramicidin S, and tyrocidine A.

residues) produced by the bacterium *Bacillus brevis* are interesting examples. Both of these peptides, gramicidin S and tyrocidine A, are antibiotics, and both contain D-amino acids as well as the more usual L-amino acids (Figure 3.14). In addition, both contain the amino acid ornithine (Orn), which does not occur in proteins, but which does play a role as a metabolic intermediate in several common pathways (Section 23.6).

Biochemical Connections

Peptide Hormones

Both oxytocin and vasopressin are peptide hormones. Oxytocin induces labor in pregnant women and controls contraction of uterine muscle. During pregnancy, the number of receptors for oxytocin in the uterine wall increases. At term, the number of receptors for oxytocin is great enough to cause contraction of the smooth muscle of the uterus in the presence of small amounts of oxytocin produced by the body toward the end of pregnancy. The fetus moves toward the cervix of the uterus because of the strength and frequency of the uterine contractions. The cervix stretches, sending nerve impulses to the hypothalamus. When the impulses reach this part of the brain, positive feedback leads to the release of still more oxytocin by the posterior pituitary gland. The presence of more oxytocin leads to stronger contractions of the uterus so that the fetus is forced through the cervix and the baby is born. Oxytocin also plays a role in stimulating the flow of milk in a nursing mother. The process of suckling sends nerve signals to the hypothalamus of the mother's brain. Oxytocin is released and carried by the blood to the mammary glands. The presence of oxytocin causes the smooth muscle in the mammary glands to contract, forcing out the milk that is in them. As suckling continues, more hormone is released, producing still more milk.

Vasopressin plays a role in the control of blood pressure by regulating contraction of smooth muscle. Like oxytocin, vasopressin is released by the action of the hypothalamus on the posterior pituitary and is transported by the blood to specific receptors. Vasopressin stimulates reabsorption of water by the kidney, thus having an antidiuretic effect. More water is retained, and the blood pressure increases.

G&M David de Lossy/Image Bank/Getty Images

▲ Nursing stimulates the release of oxytocin, producing more milk.

Summary

3.1 What Are Amino Acids, and What Is Their Three-Dimensional Structure?
The amino acids that are the monomer units of proteins have a general structure in common, with an amino group and a carboxyl group bonded to the same carbon atom. The nature of the side chains, which are referred to as R groups, is the basis of the differences among amino acids. Except for glycine, amino acids can exist in two forms, designated L and D. These two stereoisomers are nonsuperimposable mirror images of each other. The amino acids found in proteins are of the L form, but some D-amino acids occur in nature.

3.2 What Are the Structures and Properties of the Individual Amino Acids?
A classification scheme for amino acids can be based on the properties of their side chains. Two particularly important criteria are the polar or nonpolar nature of the side chain and the presence of an acidic or basic group in the side chain.

3.3 Do Amino Acids Have Specific Acid–Base Properties?
In free amino acids at neutral pH, the carboxylate group is negatively charged and the amino group is positively charged. Amino acids without charged groups on their side chains exist in neutral solution as zwitterions, with no net charge. Titration curves of amino acids indicate the pH ranges in which titratable groups gain or lose a proton. Side chains of amino acids can also contribute titratable groups; the charge (if any) on the side chain must be taken into consideration in determining the net charge on the amino acid.

3.4 What Is the Peptide Bond?
Peptides are formed by linking the carboxyl group of one amino acid to the amino group of another amino acid in a covalent (amide) bond. Proteins consist of polypeptide chains; the number of amino acids in a protein is usually 100 or more. The peptide group is planar; this stereochemical constraint plays an important role in determining the three-dimensional structures of peptides and proteins.

3.5 Are Small Peptides Physiologically Active?
Small peptides, containing two to several dozen amino acid residues, can have marked physiological effects in organisms.

Critical Questions to Review

3.1 What Are Amino Acids, and What Is Their Three-Dimensional Structure?

1. **Fact Check** How do D-amino acids differ from L-amino acids? What biological roles are played by peptides that contain D-amino acids?

3.2 What Are the Structures and Properties of the Individual Amino Acids?

2. **Fact Check** Which amino acid is technically *not* an amino acid? Which amino acid contains no chiral carbon atoms?

3. **Fact Check** Name an amino acid in which the R group contains the following:

a hydroxyl group	a sulfur atom
a second chiral carbon atom	an amino group
an amide group	an acid group
an aromatic ring	a branched side chain

4. **Fact Check** Identify the polar amino acids, the aromatic amino acids, and the sulfur-containing amino acids, given a peptide with the following amino acid sequence:

   Val—Met—Ser—Ile—Phe—Arg—Cys—Tyr—Leu

5. **Fact Check** Identify the nonpolar amino acids and the acidic amino acids in the following peptide:

   Glu—Thr—Val—Asp—Ile—Ser—Ala

6. **Fact Check** Are amino acids other than the usual 20 amino acids found in proteins? If so, how are such amino acids incorporated into proteins? Give an example of such an amino acid and a protein in which it occurs.

3.3 Do Amino Acids Have Specific Acid–Base Properties?

7. **Mathematical** Predict the predominant ionized forms of the following amino acids at pH 7: glutamic acid, leucine, threonine, histidine, and arginine.

8. **Mathematical** Draw structures of the following amino acids, indicating the charged form that exists at pH 4: histidine, asparagine, tryptophan, proline, and tyrosine.

9. **Mathematical** Predict the predominant forms of the amino acids from question 8 at pH 10.

10. **Mathematical** Calculate the isoelectric point of each of the following amino acids: glutamic acid, serine, histidine, lysine, tyrosine, and arginine.

11. **Mathematical** Sketch a titration curve for the amino acid cysteine, and indicate the pK_a values for all titratable groups. Also indicate the pH at which this amino acid has no net charge.

12. **Mathematical** Sketch a titration curve for the amino acid lysine, and indicate the pK_a values for all titratable groups. Also indicate the pH at which the amino acid has no net charge.

13. **Mathematical** An organic chemist is generally happy with 95% yields. If you synthesized a polypeptide and realized a 95% yield with each amino acid residue added, what would be your overall yield after adding 10 residues (to the first amino acid)? After adding 50 residues? After 100 residues? Would these low yields be biochemically "satisfactory"? How are low yields avoided, biochemically?

14. **Mathematical** Sketch a titration curve for aspartic acid, and indicate the pK_a values of all titratable groups. Also indicate the pH range in which the conjugate acid–base pair +1 Asp and 0 Asp will act as a buffer.

15. **Thought Question** Suggest a reason why amino acids are usually more soluble at pH extremes than they are at neutral pH. (Note that this does not mean that they are insoluble at neutral pH.)

16. **Thought Question** Write equations to show the ionic dissociation reactions of the following amino acids: aspartic acid, valine, histidine, serine, and lysine.

17. **Thought Question** Based on the information in Table 3.2, is there any amino acid that could serve as a buffer at pH 8? If so, which one?

18. **Thought Question** If you were to have a mythical amino acid based on glutamic acid, but one in which the hydrogen that is attached to the γ-carbon were replaced by another amino group, what would be the predominant form of this amino acid at pH 4, 7, and 10, if the pK_a value were 10 for the unique amino group?

19. **Thought Question** What would be the pI for the mythical amino acid described in Question 18?

20. **Thought Question** Identify the charged groups in the peptide shown in Question 4 at pH l and at pH 7. What is the net charge of this peptide at these two pH values?

21. **Thought Question** Consider the following peptides: Phe—Glu—Ser—Met and Val—Trp—Cys—Leu. Do these peptides have different net charges at pH l? At pH 7? Indicate the charges at both pH values.

22. **Thought Question** In each of the following two groups of amino acids, which amino acid would be the easiest to distinguish from the other two amino acids in the group, based on a titration?

    (a) gly, leu, lys

    (b) glu, asp, ser

23. **Thought Question** Could the amino acid glycine serve as the basis of a buffer system? If so, in what pH range would it be useful?

3.4 What Is the Peptide Bond?

24. **Fact Check** Sketch resonance structures for the peptide group.

25. **Fact Check** How do the resonance structures of the peptide group contribute to the planar arrangement of this group of atoms?

26. **Biochemical Connections** Which amino acids or their derivatives are neurotransmitters?

27. **Biochemical Connections** What is a monoamine oxidase, and what function does it serve?

28. **Thought Question** Consider the peptides Ser—Glu—Gly—His—Ala and Gly—His—Ala—Glu—Ser. How do these two peptides differ?

29. **Thought Question** Would you expect the titration curves of the two peptides in Question 28 to differ? Why or why not?

30. **Thought Question** What are the sequences of all the possible tripeptides that contain the amino acids aspartic acid, leucine, and phenylalanine? Use the three-letter abbreviations to express your answer.

31. **Thought Question** Answer Question 30 using one-letter designations for the amino acids.

32. **Thought Question** Most proteins contain more than 100 amino acid residues. If you decided to synthesize a "100-mer," with 20 different amino acids available for each position, how many different molecules could you make?

33. **Biochemical Connections** What is the stereochemical basis of the observation that D-aspartyl-D-phenylalanine has a bitter taste, whereas L-aspartyl-L-phenylalanine is significantly sweeter than sugar?

34. **Biochemical Connections** Why might a glass of warm milk help you to sleep at night?

35. **Biochemical Connections** Which would be better to eat before an exam, a glass of milk or a piece of cheese? Why?

36. **Thought Question** What might you infer (or know) about the stability of amino acids, when compared with that of other building-block units of biopolymers (sugars, nucleotides, fatty acids, etc.)?

37. **Thought Question** If you knew everything about the properties of the 20 common (proteinous) amino acids, would you be able to predict the properties of a protein (or large peptide) made from them?

38. **Thought Question** Suggest a reason why the amino acids thyroxine and hydroxyproline are produced by posttranslational modification of the amino acids tyrosine and proline, respectively.

39. **Thought Question** Consider the peptides Gly—Pro—Ser—Glu—Thr (open chain) and Gly—Pro—Ser—Glu—Thr with a peptide bond linking the threonine and the glycine. Are these peptides chemically the same?

40. **Thought Question** Can you expect to separate the peptides in Question 39 by electrophoresis?

41. **Thought Question** Suggest a reason why biosynthesis of amino acids and of proteins would eventually cease in an organism with carbohydrates as its only food source.

42. **Thought Question** You are studying with a friend who draws the structure of alanine at pH 7. It has a carboxyl group (—COOH) and an amino group (—NH₂). What suggestions would you make?

43. **Thought Question** Suggest a reason (or reasons) why amino acids polymerize to form proteins that have comparatively few covalent crosslinks in the polypeptide chain.

44. **Thought Question** Suggest the effect on the structure of peptides if the peptide group were not planar.

45. **Thought Question** Speculate on the properties of proteins and peptides if none of the common amino acids were to contain sulfur.

46. **Thought Question** Speculate on the properties of proteins that would be formed if amino acids were not chiral.

3.5 Are Small Peptides Physiologically Active?

47. **Fact Check** What are the structural differences between the peptide hormones oxytocin and vasopressin? How do they differ in function?

48. **Fact Check** How do the oxidized and reduced forms of glutathione differ from each other?

49. **Fact Check** What is an enkephalin?

50. **Thought Question** The enzyme D-amino acid oxidase, which converts D-amino acids to their α-keto form, is one of the most potent enzymes in the human body. Suggest a reason why this enzyme should have such a high rate of activity.

Biochemistry ⑤ Now™
Assess your understanding of this chapter's topics with additional quizzing and tutorials at **http://now.brookscole.com/campbell5**

Annotated Bibliography

Barrett, G. C., ed. *Chemistry and Biochemistry of the Amino Acids.* New York: Chapman and Hall, 1985. [Wide coverage of many aspects of the reactions of amino acids.]

Javitt, D. C., and J. T. Coyle. Decoding Schizophrenia, *Scientific American,* **290 (1),** 48–55 (2004).

Larsson, A., ed. *Functions of Glutathione: Biochemical, Physiological, Toxicological and Chemical Aspects.* New York: Raven Press, 1983. [A collection of articles on the many roles of a ubiquitous peptide.]

McKenna, K. W., and V. Pantic, eds. *Hormonally Active Brain Peptides: Structure and Function.* New York: Plenum Press, 1986. [A discussion of the chemistry of enkephalins and related peptides.]

Siddle, K., and J. C. Hutton. *Peptide Hormone Action—A Practical Approach.* Oxford, England: Oxford University Press, 1990. [A book that concentrates on experimental methods for studying the actions of peptide hormones.]

Stegink, L. D., and L. J. Filer, Jr. *Aspartame—Physiology and Biochemistry*. New York: Marcel Dekker, 1984. [A comprehensive treatment of metabolism, sensory and dietary aspects, preclinical studies, and issues relating to human consumption (including ingestion by phenylketonurics and consumption during pregnancy).]

Wilson, N., E. Barbar, J. Fuchs, and C. Woodward. Aspartic Acid in Reduced *Escherichia coli* Thioredoxin Has a $pK_a > 9$. *Biochemistry* **34,** 8931–8939 (1995). [A research report on a remarkably high pK_a value for a specific amino acid in a protein.]

Wold, F. *In vivo* Chemical Modification of Proteins (Post-Translational Modification). *Ann. Rev. Biochem.* **50,** 788–814 (1981). [A review article on the modified amino acids found in proteins.]

CHAPTER 4

The Three-Dimensional Structure of Proteins

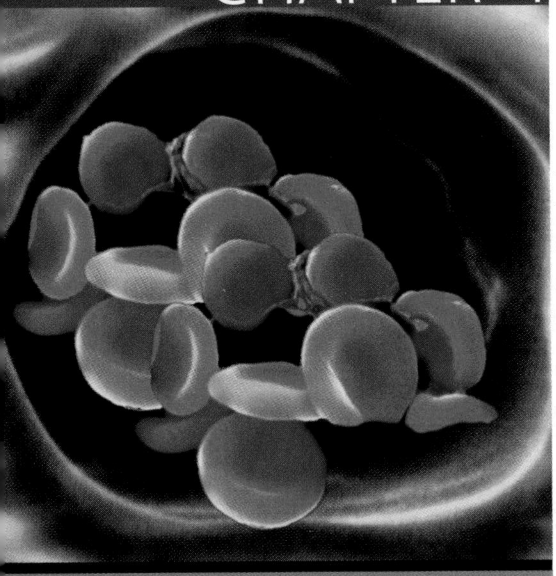

Amino acids joined together form a protein (polypeptide) chain. The repeating units are amide planes containing peptide bonds. These amide planes can twist about their connecting carbon atoms to create the three-dimensional conformations of proteins. More than 50 years ago, Linus Pauling predicted that linked amino acids could form an α-helix. Years later, his prediction was confirmed when myoglobin, an oxygen-binding protein, was found to be made from Pauling's α-helices. This type of local folding of the protein chain is called secondary structure, the linear sequence being the primary structure. The conformation of a complete protein chain is its tertiary structure. Myoglobin, a molecule that binds oxygen tightly, has a single protein chain. Hemoglobin, a protein with four myoglobin-like subunits fitted together, has a quaternary structure. This allows it to change from the oxy conformation, when it binds oxygen in the lungs, to the deoxy form, when it releases oxygen to working tissues. The discovery of structure–function relationships in hemoglobin led to an understanding of the way complex multisubunit enzymes regulate metabolic pathways.

Red blood cells contain hemoglobin, a classic example of protein structure.

© Dr. Philippa Uwins, Whistler Research Pty./Photo Researchers, Inc.

Critical Questions

Biochemistry ⓔ Now™

Test yourself on these Critical Questions at the BiochemistryNow website at **http://now .brookscole.com/campbell5**

| 4.1 | **How Does the Structure of Proteins Determine Their Function?** |

Levels of Structure in Proteins

Biologically active proteins are polymers consisting of amino acids linked by covalent peptide bonds. Many different conformations (three-dimensional structures) are possible for a molecule as large as a protein. Of these many structures, one or (at most) a few have biological activity; these are called the **native conformations.** Many proteins have no obvious regular repeating structure. As a consequence, these proteins are frequently described as having large segments of "random structure" (also referred to as random coil). The term "random" is really a misnomer, since the same nonrepeating structure is found in the native conformation of all molecules of a given protein, and this conformation is needed for its proper function. Because proteins are complex, they are defined in terms of four levels of structure.

Primary structure is the order in which the amino acids are covalently linked together. The peptide Leu—Gly—Thr—Val—Arg—Asp—His (recall that the N-terminal amino acid is listed first) has a different primary structure from the peptide Val—His—Asp—Leu—Gly—Arg—Thr, even though both have the same number and kinds of amino acids. Note that the order of amino acids can be written on one line. The primary structure is the one-dimensional first step in specifying the three-dimensional structure of a protein. Some biochemists define primary structure to include all covalent interactions, including the disulfide bonds that can be formed by cysteines; however, we shall consider the disulfide bonds to be part of the tertiary structure, which will be considered later.

Two three-dimensional aspects of a single polypeptide chain, called the secondary and tertiary structure, can be considered separately. **Secondary structure** is the arrangement in space of the atoms in the peptide backbone. The α-helix and β-pleated sheet arrangements are two different types of secondary structure. Secondary structures have repetitive interactions resulting from hydrogen bonding between the amide N—H and the carbonyl groups of the peptide backbone. The conformations of the side chains of the amino acids are not part of the secondary structure. In many proteins, the folding of parts of the chain can occur independently of the folding of other parts. Such independently folded portions of proteins are referred to as **domains** or **supersecondary structure.**

Tertiary structure includes the three-dimensional arrangement of all the atoms in the protein, including those in the side chains and in any **prosthetic groups** (groups of atoms other than amino acids).

A protein can consist of multiple polypeptide chains called **subunits.** The arrangement of subunits with respect to one another is the **quaternary structure.** Interaction between subunits is mediated by noncovalent interactions, such as hydrogen bonds, electrostatic attractions, and hydrophobic interactions.

We shall discuss secondary structure in more detail in Section 4.3, tertiary structure in Section 4.5, and quaternary structure in Section 4.7.

4.2 | What Is the Primary Structure of Proteins?

The amino acid sequence (the primary structure) of a protein determines its three-dimensional structure, which, in turn, determines its properties. In every protein, the correct three-dimensional structure is needed for correct functioning.

One of the most striking demonstrations of the importance of primary structure is found in the hemoglobin associated with *sickle-cell anemia.* In this genetic disease, red blood cells cannot bind oxygen efficiently. The red blood cells also assume a characteristic sickle shape, giving the disease its name. The sickled cells tend to become trapped in small blood vessels, cutting off circulation and thereby causing organ damage. These drastic consequences stem from a change in one amino acid residue in the sequence of the primary structure.

Considerable research is being done to determine the effects of changes in primary structure on the functions of proteins. Using molecular-biology techniques, such as site-directed mutagenesis (Section 14.7), it is possible to replace any chosen amino acid residue in a protein with another specific amino acid residue. The conformation of the altered protein, as well as its biological activity, can then be determined. The results of such amino acid substitutions range from negligible effects to complete loss of activity, depending on the protein and the nature of the altered residue.

Determining the sequence of amino acids in a protein is a routine, but not trivial, operation in classical biochemistry. It consists of several steps, which must be carried out carefully to obtain accurate results (Section 5.4).

The following Biochemical Connections box describes an important practical aspect of the amino acid composition of proteins. This property can differ markedly, depending on the source of the protein (plant or animal), with important consequences for human nutrition.

Essential Information

The primary structure of a protein is the sequence of amino acids. Determination of the sequence involves cleaving the protein to smaller peptides, determining the sequence of the individual peptides, and combining the peptide sequences to obtain that of the protein.

4.3 | What Is the Secondary Structure of Proteins?

The secondary structure of proteins is the hydrogen-bonded arrangement of the backbone of the protein, the polypeptide chain. The nature of the bonds in the peptide backbone plays an important role here. Within each amino acid residue are two bonds with reasonably free rotation. They are (1) the

Biochemical Connections

Complete Proteins and Nutrition

A **complete protein** is one that provides all essential amino acids (Section 23.5) in appropriate amounts for human survival. These amino acids cannot be synthesized by humans, but they are needed for the biosynthesis of proteins. Lysine and methionine are two essential amino acids that are frequently in short supply in plant proteins.

Because grains such as rice and corn are usually poor in lysine, and because beans are usually poor in methionine, vegetarians are at risk for malnutrition unless they eat grains and beans together. This leads to the concept of *complementary proteins,* mixtures that provide all the essential amino acids—for example, corn and beans in *succotash,* or a bean burrito made with a corn tortilla. The specific recommended dietary allowances for adult males follow. Adult females who are neither pregnant nor lactating need 20% less than the amounts indicated for adult males.

RDA		RDA	
Arg*	Unknown	Met	0.70 g
His*	Unknown	Phe	1.12 g (includes Tyr)
Ile	0.84 g	Thr	0.56 g
Leu	1.12 g	Trp	0.21 g
Lys	0.84 g	Val	0.96 g

*The inclusion of His and Arg is controversial. They appear to be required only by growing children and for the repair of injured tissue. Arg is required to maintain fertility in males.

The *p*rotein *e*fficiency *r*atio (PER) describes how well a protein supplies essential amino acids. This parameter is useful for deciding how much of a food you need to eat. Most college-age, nonpregnant females require 46 g (or about 1.6 oz) of complete protein, and males require 58 g (or about 2 oz) of complete protein per day. If one chooses to pick only a *single* source of protein for the diet, eggs are perhaps the best choice because they contain high-quality protein. For a female, the need for 1.6 oz of complete protein could be met with 10.7 oz of eggs, or about four whole extra-large eggs. For a male, 13.6 oz of eggs, or a little more than five eggs, would be needed. The same requirement could be met with a lean beef steak, but it would require 345 g, or about 0.75 lb, for a female (or 431 g, or nearly a full pound, for a male) because beef steak has a lower PER. If one ate only corn, it would require 1600 g/day for women and 2000 g/day for men (1600 g is about 3.6 pounds of fresh corn kernels—something in excess of 160 eight-inch ears per day). However, if you simply combine a small amount of beans or peas with the corn, it complements the low amount of lysine in the corn, and the protein is now complete. This can easily be done with normal food portions.

Protein	PER	% Protein
Whole egg	100	15
Beef muscle	84	16
Cow's milk	66	4 (largely H_2O)
Peanuts	45	28
Corn	32	9
Wheat	26	12

In an attempt to increase the nutritional value of certain crops that are grown as food for livestock, scientists have used genetic techniques to create strains of corn that are much higher in lysine than the wild-type corn. This has proven effective in increasing growth rates in pigs. Many vegetable crops are now being produced using biotechnology to increase shelf life, decrease spoilage, and give crops defenses against insects. These genetically modified foods are currently a hot spot of debate and controversy.

bond between the α-carbon and the amino nitrogen of that residue and (2) the bond between the α-carbon and the carboxyl carbon of that residue. The combination of the planar peptide group and the two freely rotating bonds has important implications for the three-dimensional conformations of peptides and proteins. A peptide-chain backbone can be visualized as a series of playing cards, each card representing a planar peptide group. The cards are linked at opposite corners by swivels, representing the bonds about which there is considerable freedom of rotation (Figure 4.1). The side chains also play a vital role in determining the three-dimensional shape of a protein, but only the backbone is considered in the secondary structure. The angles φ (phi) and ψ (psi), frequently called Ramachandran angles (after their originator, G. N. Ramachandran), are used to designate rotations around the C—N and C—C bonds, respectively. The conformation of a protein backbone can be described by specifying the values of φ and ψ for each residue (−180° to 180°). Two kinds of secondary structures that occur frequently in proteins are the repeating **α-helix** and **β-pleated sheet** (or β-sheet) hydrogen-

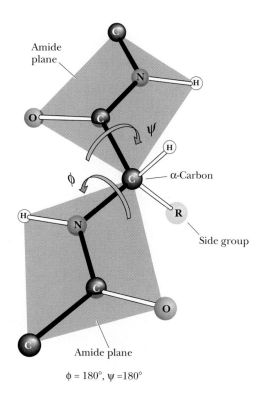

Amide plane

ψ

φ

α-Carbon

R

Side group

Amide plane

$\phi = 180°, \psi = 180°$

◀ **FIGURE 4.1** Definition of the angles that determine the conformation of a polypeptide chain. The rigid planar peptide groups (called "playing cards" in the text) are shaded. The angle of rotation around the C^α—N bond is designated φ (phi), and the angle of rotation around the C^α—C bond is designated ψ(psi). These two bonds are the ones around which there is freedom of rotation. (*Illustration, Irving Geis. Rights owned by Howard Hughes Medical Institute. Not to be reproduced without permission.*)

bonded structures. The φ and ψ angles repeat themselves in contiguous amino acids in regular secondary structures. The α-helix and β-pleated sheet are not the only possible secondary structures, but they are by far the most important and deserve a closer look.

Periodic Structures in Protein Backbones

The α-helix and β-pleated sheet are periodic structures; their features repeat at regular intervals. The α-helix is rodlike and involves only one polypeptide chain. The β-pleated sheet structure can give a two-dimensional array and can involve one or more polypeptide chains.

The α-Helix

The α-helix is stabilized by hydrogen bonds parallel to the helix axis within the backbone of a single polypeptide chain. Counting from the N-terminal end, the C—O group of each amino acid residue is hydrogen bonded to the N—H group of the amino acid four residues away from it in the covalently bonded sequence. The helical conformation allows a linear arrangement of the atoms involved in the hydrogen bonds, which gives the bonds maximum strength and thus makes the helical conformation very stable (Section 2.2). There are 3.6 residues for each turn of the helix, and the *pitch* of the helix (the linear distance between corresponding points on successive turns) is 5.4 Å (Figure 4.2).

The angstrom unit, $1 \text{ Å} = 10^{-8} \text{ cm} = 10^{-10} \text{ m}$, is convenient for interatomic distances in molecules, but it is not a Système International [SI] unit. Nanometers ($1 \text{ nm} = 10^{-9} \text{ m}$) and picometers ($1 \text{ pm} = 10^{-12} \text{ m}$) are the SI units used for interatomic distances. In SI units, the pitch of the α-helix is 0.54 nm or 540 pm.). Figure 4.3 shows the structures of two proteins with a high degree of α-helical content.

Essential Information

Two of the most important structural motifs in proteins are the α-helix and β-pleated sheet.

Biochemistry ⒺNow™
Go to BiochemistryNow and click on Biochemistry Interactive to explore the anatomy of the α-helix.

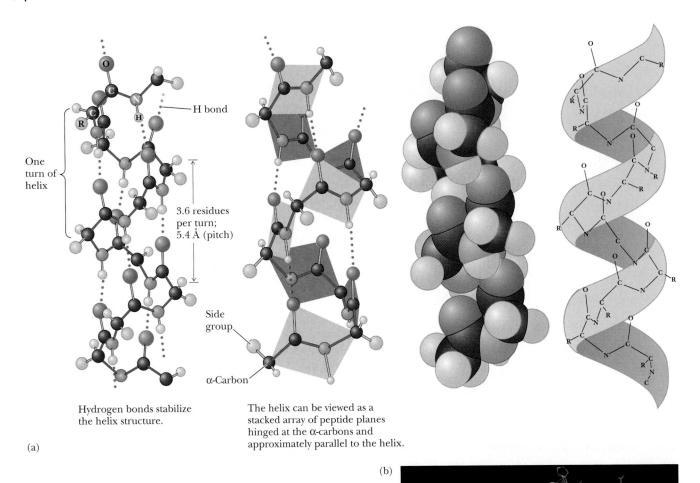

Hydrogen bonds stabilize the helix structure.

(a)

The helix can be viewed as a stacked array of peptide planes hinged at the α-carbons and approximately parallel to the helix.

(b)

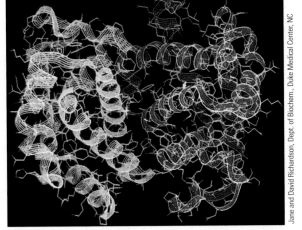

▲ **FIGURE 4.2** The α-helix. (a) From left to right, ball-and-stick model of the α-helix, showing terminology; ball-and-stick model with planar peptide groups shaded; computer-generated space-filling model of the α-helix; outline of the α-helix. (b) Model of the protein hemoglobin, showing the helical regions. (*Illustration, Irving Geis. Rights owned by Howard Hughes Medical Institute. Not to be reproduced without permission.*)

Proteins have varying amounts of α-helical structures, varying from a few percent to nearly 100%. Several factors can disrupt the α-helix. The amino acid proline creates a bend in the backbone because of its *cyclic* structure. It cannot fit into the α-helix because (1) rotation around the bond between the nitrogen and the α-carbon is severely restricted, and (2) proline's α-amino group cannot participate in intrachain hydrogen bonding. Other localized factors involving the side chains include strong electrostatic repulsion owing to the proximity of several charged groups of the same sign, such as groups of positively charged lysine and arginine residues or groups of negatively

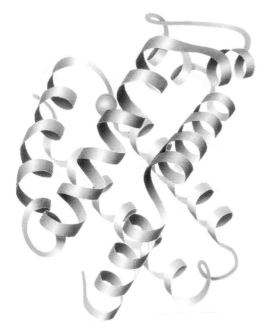

β-Hemoglobin subunit

Myohemerythrin

Biochemistry ⊜Now™ **ANIMATED FIGURE 4.3**
The three-dimensional structure of two proteins with
substantial amounts of α-helix in their structures.
The helices are represented by the regularly coiled
sections of the ribbon diagram. Myohemerythrin is
an oxygen-carrying protein in invertebrates. **See this
figure animated at http://now.brookscole.com/
campbell5** (*Jane Richardson.*)

charged glutamate and aspartate residues. Another possibility is crowding
(steric repulsion) caused by the proximity of several bulky side chains. In the
α-helical conformation, all the side chains lie outside the helix; there is not
enough room for them in the interior. The β-carbon is just outside the helix,
and crowding can occur if it is bonded to two atoms other than hydrogen, as
is the case with valine, isoleucine, and threonine.

The β-Sheet

The arrangement of atoms in the β-pleated sheet conformation differs
markedly from that in the α-helix. The peptide backbone in the β-sheet is
almost completely extended. Hydrogen bonds can be formed between different parts of a single chain that is doubled back on itself (*intrachain bonds*) or
between different chains (*interchain bonds*). If the peptide chains run in the
same direction (i.e., if they are all aligned in terms of their N-terminal and C-
terminal ends), a *parallel pleated sheet* is formed. When alternating chains run
in opposite directions, an *antiparallel* pleated sheet is formed (Figure 4.4).
The hydrogen bonding between peptide chains in the β-pleated sheet gives
rise to a repeated zigzag structure; hence, the name "pleated sheet" (Figure
4.5). Note that the hydrogen bonds are perpendicular to the direction of the
protein chain, not parallel to it as in the α-helix.

Irregularities in Regular Structures

Other helical structures are found in proteins. These are often found in
shorter stretches than with the α-helix, and they sometimes break up the regular nature of the α-helix. The most common is the 3_{10} helix, which has three
residues per turn and ten atoms in the ring formed by making the hydrogen
bond. Other common helices are designated 2_7 and 4.4_{16}, following the same
nomenclature as the 3_{10} helix.

Biochemistry ⊜Now™
Go to BiochemistryNow and click on Biochemistry
Interactive to explore β-sheets, one of the principal
types of secondary structure in proteins.

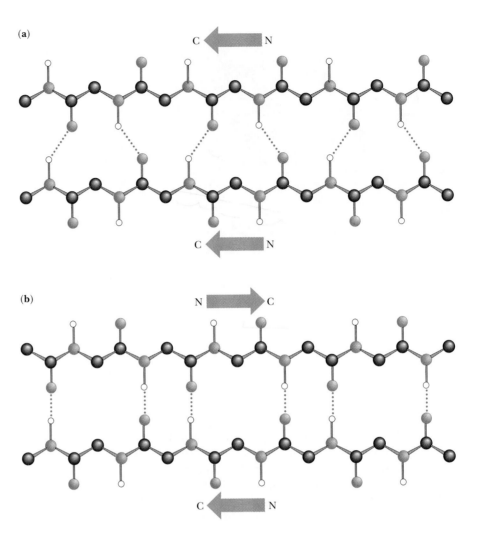

▶ **FIGURE 4.4** The arrangement of hydrogen bonds in (a) parallel and (b) antiparallel β-pleated sheets.

A **β-bulge** is a common nonrepetitive irregularity found in antiparallel β-sheets. It occurs between two normal β-structure hydrogen bonds and involves two residues on one strand and one on the other. Figure 4.6 shows typical β-bulges.

Protein folding requires that the peptide backbones and the secondary structures be able to change directions. Often a reverse turn marks a transition between one secondary structure and another. For steric (spatial) reasons, glycine is frequently encountered in **reverse turns,** at which the polypeptide chain changes direction; the single hydrogen of the side chain prevents crowding (Figures 4.7a and 4.7b). Because the cyclic structure of proline has the correct geometry for a reverse turn, this amino acid is also frequently encountered in such turns (Figure 4.7c).

Supersecondary Structures and Domains

The α-helix, β-pleated sheet, and other secondary structures are combined in many ways as the polypeptide chain folds back on itself in a protein. The combination of α- and β-strands produces various kinds of supersecondary structures in proteins. The most common feature of this sort is the βαβ *unit,* in which two parallel strands of β-sheet are connected by a stretch of α-helix (Figure 4.8a). An αα *unit* (helix-turn-helix) consists of two antiparallel α-helices

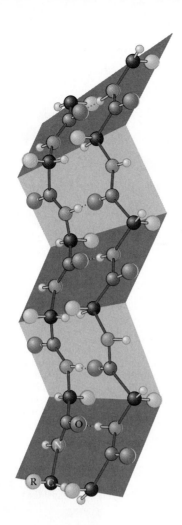

◀ **FIGURE 4.5** The three-dimensional form of the antiparallel β-pleated sheet arrangement. The chains do not fold back on each other but are in a fully extended conformation. (*Illustration, Irving Geis. Rights owned by Howard Hughes Medical Institute. Not to be reproduced without permission.*)

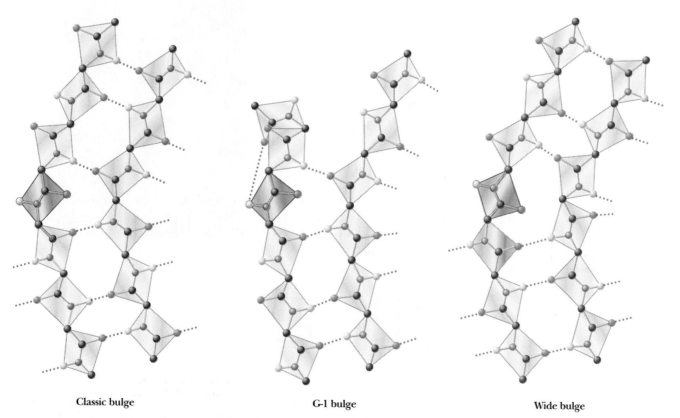

Classic bulge   **G-1 bulge**   **Wide bulge**

▲ **FIGURE 4.6** Three different β-bulge structures. Hydrogen bonds are shown as red dots.

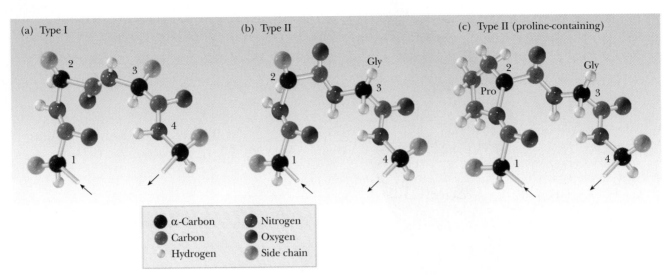

(a) Type I (b) Type II (c) Type II (proline-containing)

α-Carbon
Carbon
Hydrogen
Nitrogen
Oxygen
Side chain

▲ **FIGURE 4.7** Structures of reverse turns. Arrows indicate the directions of the polypeptide chains. (a) A type I reverse turn. In residue 3, the side chain (gold) lies outside the loop, and any amino acid can occupy this position. (b) A type II reverse turn. The side chain of residue 3 has been rotated 180° from the position in the type I turn and is now on the inside of the loop. Only the hydrogen side chain of glycine can fit into the space available, so glycine must be the third residue in a type II reverse turn. (c) The five-membered ring of proline has the correct geometry for a reverse turn; this residue normally occurs as the second residue of a reverse turn. The turn shown here is type II, with glycine as the third residue.

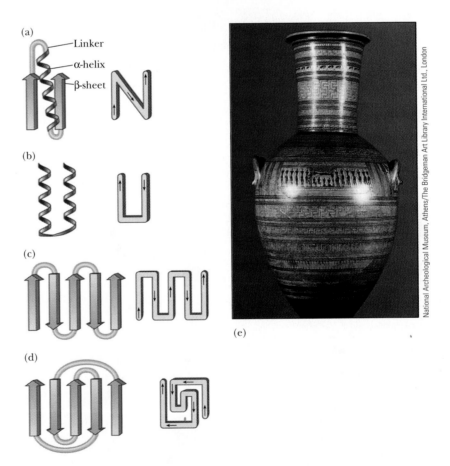

(a) Linker — α-helix — β-sheet
(b)
(c)
(d)
(e)

National Archeological Museum, Athens/The Bridgeman Art Library International Ltd., London

▲ **FIGURE 4.8** Schematic diagrams of supersecondary structures. Arrows indicate the directions of the polypeptide chains. (a) A βαβ unit, (b) an αα unit, (c) a β-meander, and (d) the Greek key. (e) The Greek key motif in protein structure resembles the geometric patterns on this ancient Greek vase, giving rise to the name.

(Figure 4.8b). In such an arrangement, energetically favorable contacts exist between the side chains in the two stretches of helix. In a β-*meander*, an antiparallel sheet is formed by a series of tight reverse turns connecting stretches of the polypeptide chain (Figure 4.8c). Another kind of antiparallel sheet is formed when the polypeptide chain doubles back on itself in a pattern known as the *Greek key*, named for a decorative design found on pottery from the classical period (Figure 4.8e). A **motif** is a repetitive supersecondary structure. Some of the common smaller motifs are shown in Figure 4.9. These smaller motifs can often be repeated and organized into larger motifs. Protein sequences that allow for a β-meander or Greek key can often be found arranged into a β-barrel in the tertiary structure of the protein (Figure 4.10). Motifs are important and tell us much about the folding of proteins. However, these motifs do not allow us to predict anything about the biological function of the protein because they are found in proteins and enzymes with very dissimilar functions.

Many proteins that have the same type of function have similar protein sequences; consequently, domains with similar conformations are associated with the particular function. Many types of domains have been identified, including three different types of domains by which proteins bind to DNA.

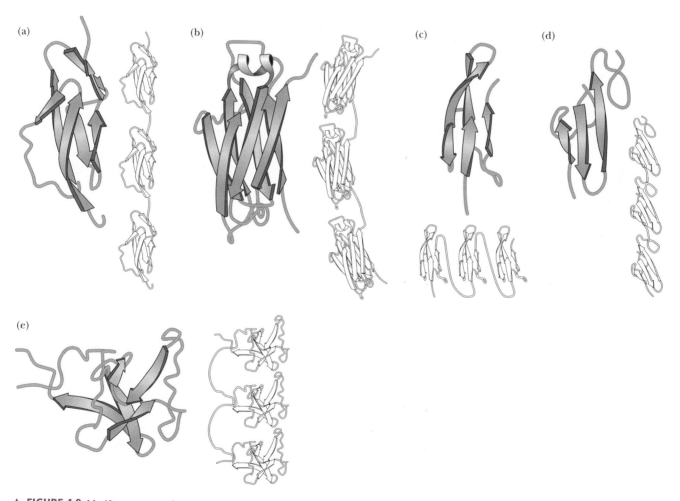

▲ **FIGURE 4.9** Motifs are repeated supersecondary structures, sometimes called modules. (a) The complement-control protein module. (b) The immunoglobulin module. (c) The fibronectin type I module. (d) The growth-factor module. (e) The kringle module. All of these have a particular secondary structure that is repeated in the protein. (*Reprinted from "Protein Modules," Trends in Biochemical Sciences, Vol. 16, p. 13–17, Copyright © 1991, with permission from Elsevier.*)

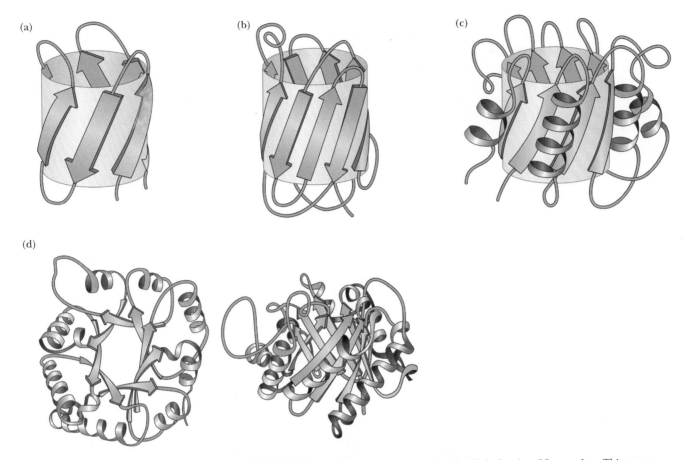

▲ **FIGURE 4.10** Some β-barrel arrangements. (a) A linked series of β-meanders. This arrangement occurs in the protein rubredoxin from *Clostridium pasteurianum*. (b) The Greek key pattern occurs in human prealbumin. (c) A β-barrel involving alternating βαβ units. This arrangement occurs in triose phosphate isomerase from chicken muscle. (d) Top and side views of the polypeptide backbone arrangement in triose phosphate isomerase. Note that the α-helical sections lie outside the actual β-barrel.

In addition, short polypeptide sequences within a protein direct the posttranslational modification and subcellular localization. For example, several sequences play a role in the formation of glycoproteins (ones that contain sugars in addition to the polypeptide chain). Other specific sequences indicate that a protein is to be bound to a membrane or secreted from the cell. Still other specific sequences mark a protein for phosphorylation by a specific enzyme.

The Collagen Triple Helix

Collagen, a component of bone and connective tissue, is the most abundant protein in vertebrates. It is organized in water-insoluble fibers of great strength. A collagen fiber consists of three polypeptide chains wrapped around each other in a ropelike twist, or triple helix. Each of the three chains has, within limits, a repeating sequence of three amino acid residues, X—Pro—Gly or X—Hyp—Gly, where Hyp stands for hydroxyproline, and any amino acid can occupy the first position, designated by X.

Proline and hydroxyproline can constitute up to 30% of the residues in collagen. Hydroxyproline is formed from proline by a specific hydroxylating enzyme after the amino acids are linked together. Hydroxylysine also occurs in collagen. In the amino acid sequence of collagen, every third position must be occupied by glycine. The triple helix is arranged so that every third residue on each chain is inside the helix. Only glycine is small enough to fit into the space available (Figure 4.11).

Hydroxylysine   Hydroxyproline

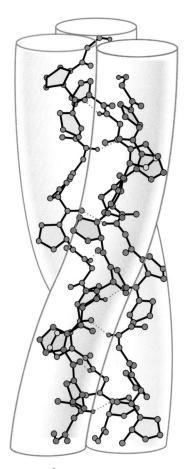

The three individual collagen chains are themselves helices that differ from the α-helix. They are twisted around each other in a superhelical arrangement to form a stiff rod. This triple helical molecule is called *tropocollagen;* it is 300 nm (3000 Å) long and 1.5 nm (15 Å) in diameter. The three strands are held together by hydrogen bonds involving the hydroxyproline and hydroxylysine residues. The molecular weight of the triple-stranded array is about 300,000; each strand contains about 800 amino acid residues. Collagen is both intramolecularly and intermolecularly linked by covalent bonds formed by reactions of lysine and histidine residues. The amount of cross-linking in a tissue increases with age. That is why meat from older animals is tougher than meat from younger animals.

Collagen in which the proline is not hydroxylated to hydroxyproline to the usual extent is less stable than normal collagen. Symptoms of scurvy, such as bleeding gums and skin discoloration, are the results of fragile collagen. The enzyme that hydroxylates proline and thus maintains the normal state of collagen requires ascorbic acid (vitamin C) to remain active. Scurvy is ultimately caused by a dietary deficiency of vitamin C. See the Biochemical Connections box in Chapter 16.

Biochemistry ⒺNow™ ACTIVE FIGURE 4.11
Poly (Gly—Pro—Pro), a collagen-like right-handed triple helix composed of three left-handed helical chains. (*Adapted from M. H. Miller and H. A. Scheraga, 1976, Calculation of the structures of collagen models. Role of interchain interactions in determining the triple-helical coiled-coil conformations. I. Poly(glycyl-prolyl-prolyl).* Journal of Polymer Science Symposium *54:171–200. © 1976 John Wiley & Sons, Inc. Reprinted by permission.*) **Watch this Active Figure at http://now.brookscole.com/campbell5**

Two Types of Protein Conformations: Fibrous and Globular

It is difficult to draw a clear separation between secondary and tertiary structures. The nature of the side chains in a protein (part of the tertiary structure) can influence the folding of the backbone (the secondary structure). Comparing collagen with silk and wool fibers can be illuminating. Silk fibers consist largely of the protein fibroin, which, like collagen, has a fibrous structure, but which, unlike collagen, consists largely of β-sheets. Fibers of wool consist largely of the protein keratin, which is largely α-helical. The amino acids of which collagen, fibroin, and keratin are composed determine which conformation they will adopt, but all are **fibrous proteins** (Figure 4.12a).

In other proteins, the backbone folds back on itself to produce a more or less spherical shape. These are called **globular proteins** (Figure 4.12b), and we shall see many examples of them. Their helical and pleated-sheet sections can be arranged so as to bring the ends of the sequence close to each other in three dimensions. Globular proteins, unlike fibrous proteins, are water-soluble and have compact structures; their tertiary and quaternary structures can be quite complex.

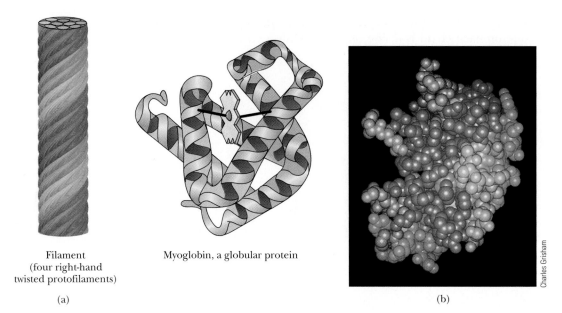

Filament
(four right-hand
twisted protofilaments)

(a)

Myoglobin, a globular protein

(b)

Charles Grisham

▲ **FIGURE 4.12** A comparison of the shapes of fibrous and globular proteins. (a) Schematic diagrams of a portion of a fibrous protein and of a globular protein. (b) Computer-generated model of a globular protein. The color-coding in this model differs from that of models of smaller molecules. The carbons are represented by light blue spheres, and the yellow spheres represent sulfur.

4.4 | What Can We Say about the Thermodynamics of Protein Folding?

The primary structure of a protein—the order of amino acids in the polypeptide chain—depends on the formation of peptide bonds, which are covalent. Higher-order levels of structure, such as the conformation of the backbone (secondary structure) and the positions of all the atoms in the protein (tertiary structure), depend on noncovalent interactions; if the protein consists of several subunits, the interaction of the subunits (quaternary structure) also depends on noncovalent interactions. Noncovalent stabilizing forces contribute to the most stable structure for a given protein, the one with the lowest energy.

Several types of hydrogen bonding occur in proteins. *Backbone* hydrogen bonding is a major determinant of secondary structure; hydrogen bonds *between the side chains of amino acids* are also possible in proteins. Nonpolar residues tend to cluster together in the interior of protein molecules as a result of *hydrophobic* interactions. *Electrostatic* attraction between oppositely charged groups, which frequently occurs on the surface of the molecule, results in such groups being close to one another. Several side chains can be *complexed* to a single metal ion. (Metal ions also occur in some prosthetic groups.)

In addition to these noncovalent interactions, *disulfide bonds* form covalent links between the side chains of cysteines. When such bonds form, they restrict the folding patterns available to polypeptide chains. There are specialized laboratory methods for determining the number and positions of disulfide links in a given protein. Information about the locations of disulfide links can then be combined with knowledge of the primary structure to give the *complete covalent structure* of the protein. Note the subtle difference here: The primary structure is the order of amino acids, whereas the complete covalent structure also specifies the positions of the disulfide bonds (Figure 4.13).

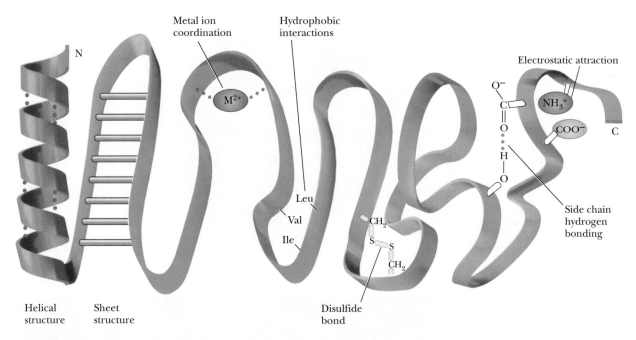

▲ FIGURE 4.13 Forces that stabilize the tertiary structure of proteins. Note that the helical structure and sheet structure are two kinds of backbone hydrogen bonding. Although backbone hydrogen bonding is part of secondary structure, the conformation of the backbone puts constraints on the possible arrangement of the side chains.

Recall that, as a result of this assortment of stabilizing forces, residues that are far apart in the primary sequence can be close to each other in the three-dimensional structure produced by the folding of the protein. When a polypeptide chain folds back on itself, it can assume a compact globular shape. A different polypeptide chain (or the same chain under different conditions) can assume a rodlike fibrous form.

The most stable form of the protein is the one with the lowest energy, representing a complex interplay of all the forces involved. Many of these forces involve bond formation, frequently the formation of a large number of weak, noncovalent bonds. Of these, hydrophobic interactions are a special case in the sense that the concept of entropy plays a large role in describing them. This is a good place to take a detailed look at hydrophobic interactions.

Hydrophobic Interactions: A Case Study in Thermodynamics

Hydrophobic interactions have important consequences in biochemistry. Large arrays of molecules can take on definite structures as a result of hydrophobic interactions. We have already seen the way in which phospholipid bilayers can form one such array. Recall (Chapter 2, Section 2.1) that phospholipids are molecules that have polar head groups and long nonpolar tails of hydrocarbon chains. These bilayers are less complex than a folded protein, but the interactions that lead to their formation also play a vital role in protein folding. Under suitable conditions, a double-layer arrangement is formed so that the polar head groups of many molecules face the aqueous environment, while the nonpolar tails are in contact with each other and are kept away from the aqueous environment. These bilayers form three-dimensional structures called **liposomes** (Figure 4.14). Such structures are useful model systems for biological membranes, which consist of similar bilayers with proteins embedded in them. The interactions between the bilayer and the

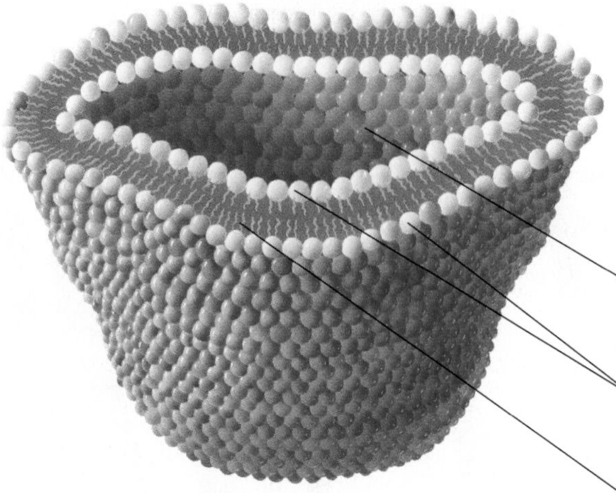

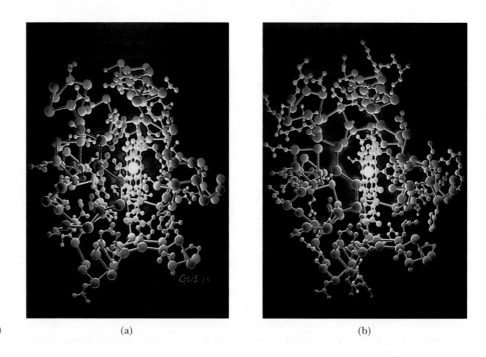

▶ **FIGURE 4.14** Schematic diagram of a liposome. This three-dimensional structure is arranged so that hydrophilic head groups of lipids are in contact with the aqueous environment. The hydrophobic tails are in contact with each other and are kept away from the aqueous environment.

Inner aqueous compartment

Hydrophilic surfaces

Hydrophobic tails

▶ **FIGURE 4.15** The three-dimensional structure of the protein cytochrome c. (a) The hydrophobic side chains (shown in red) are found in the interior of the molecule. (b) The hydrophilic side chains (shown in green) are found on the exterior of the molecule. (*Illustration, Irving Geis. Rights owned by Howard Hughes Medical Institute. Not to be reproduced without permission.*)

(a)   (b)

embedded proteins are also examples of hydrophobic interactions. The very existence of membranes depends on hydrophobic interactions. The same hydrophobic interactions play a crucial role in protein folding.

Hydrophobic interactions are a major factor in the folding of proteins into the specific three-dimensional structures required for their functioning as enzymes, oxygen carriers, or structural elements. The order of amino acids (i.e., the nature of the side chains) automatically determines the three-dimensional structure of the protein. It is known experimentally that proteins tend to be folded so that the nonpolar hydrophobic side chains are sequestered from water in the interior of the protein, while the polar hydrophilic side chains lie on the exterior of the molecule and are accessible to the aqueous environment (Figure 4.15). What makes hydrophobic interactions favorable?

Hydrophobic interactions are spontaneous processes. The entropy of the universe increases when hydrophobic interactions occur.

$$\Delta S_{universe} > 0$$

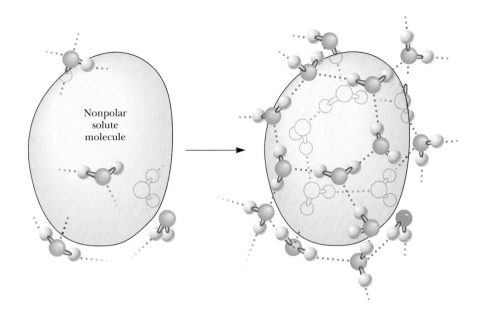

Biochemistry ❷ Now™ ANIMATED FIGURE 4.16
A "cage" of water molecules forms around a non-polar solute. **See this figure animated at http://now .brookscole.com/campbell5**

As an example, let us assume that we have tried to mix the liquid hydrocarbon hexane (C_6H_{14}) with water and have obtained not a solution but a two-layer system, one layer of hexane and one of water. Formation of a mixed solution is nonspontaneous, and the formation of two layers is spontaneous. Unfavorable entropy terms enter into the picture if solution formation requires the creation of ordered arrays of solvent, in this case water (Figure 4.16). The water molecules surrounding the nonpolar molecules can hydrogen bond with each other, but they have fewer possible orientations than if they were surrounded by other water molecules on all sides. This introduces a higher degree of order, preventing the dispersion of energy, more like the lattice of ice than liquid water, and thus a lower entropy. The required entropy decrease is too large for the process to take place. Therefore, nonpolar substances do not dissolve in water; rather, nonpolar molecules associate with one another by hydrophobic interactions and are excluded from water.

4.5 | What Is the Tertiary Structure of Proteins?

The tertiary structure of a protein is the three-dimensional arrangement of all the atoms in the molecule. The conformations of the side chains and the positions of any prosthetic groups are parts of the tertiary structure, as is the arrangement of helical and pleated-sheet sections with respect to one another. In a fibrous protein, the overall shape of which is a long rod, the secondary structure also provides much of the information about the tertiary structure. The helical backbone of the protein does not fold back on itself, and the only important aspect of the tertiary structure that is not specified by the secondary structure is the arrangement of the atoms of the side chains.

For a globular protein, considerably more information is needed. It is necessary to determine the way in which the helical and pleated-sheet sections fold back on each other, in addition to the positions of the side-chain atoms and any prosthetic groups. The interactions between the side chains play an important role in the folding of proteins. The folding pattern frequently brings residues that are separated in the amino acid sequence into proximity in the tertiary structure of the native protein.

Not every protein necessarily exhibits all possible structural features of the kinds we described in Section 4.4. For instance, there are no disulfide bridges in myoglobin and hemoglobin, which are oxygen-storage and transport proteins and classic examples of protein structure, but they both contain Fe(II) ions as part of a prosthetic group. In contrast, the enzymes trypsin and chymotrypsin do not contain complexed metal ions, but they do have disulfide bridges. Hydrogen bonds, electrostatic interactions, and hydrophobic interactions occur in most proteins.

The three-dimensional conformation of a protein is the result of the interplay of all the stabilizing forces. It is known, for example, that proline does not fit into an α-helix and that its presence can cause a polypeptide chain to turn a corner, ending an α-helical segment. The presence of proline is not, however, a *requirement* for a turn in a polypeptide chain. Other residues are routinely encountered at bends in polypeptide chains. The segments of proteins at bends in the polypeptide chain and in other portions of the protein that are not involved in helical or pleated-sheet structures are frequently referred to as "random" or "random coil." In reality, the forces that stabilize each protein are responsible for its conformation.

The experimental technique used to determine the tertiary structure of a protein is **X-ray crystallography.** Perfect crystals of some proteins can be grown under carefully controlled conditions. In such a crystal, all the individual protein molecules have the same three-dimensional conformation and the same orientation. Crystals of this quality can be formed only from proteins of very high purity, and it is not possible to obtain a structure if the protein cannot be crystallized.

When a suitably pure crystal is exposed to a beam of X rays, a *diffraction pattern* is produced on a photographic plate (Figure 4.17a) or a radiation counter. The pattern is produced when the electrons in each atom in the molecule scatter the X rays. The number of electrons in the atom determines the intensity of its scattering of X rays; heavier atoms scatter more effectively than lighter atoms. The scattered X rays from the individual atoms can reinforce each other or cancel each other (set up constructive or destructive interference), giving rise to the characteristic pattern for each type of molecule. A series of diffraction patterns taken from several angles contains the information needed to determine the tertiary structure. The information is extracted from the diffraction patterns through a mathematical analysis known as a *Fourier series*. Many thousands of such calculations are required to determine the structure of a protein, and even though they are performed by computer, the process is a fairly long one. Improving the calculation procedure is a subject of active research. The articles by Hauptmann and by Karle listed in the bibliography at the end of this chapter outline some of the accomplishments in the field.

Another technique that supplements the results of X-ray diffraction has come into wide use in recent years. It is a form of **nuclear magnetic resonance (NMR) spectroscopy.** In this particular application of NMR, called *2-D* (two-dimensional) *NMR*, large collections of data points are subjected to computer analysis (Figure 4.17b). Like X-ray diffraction, this method uses a Fourier series to analyze results. It is similar to X-ray diffraction in other ways: It is a long process, and it requires considerable amounts of computing power and milligram quantities of protein. One way in which 2-D NMR differs from X-ray diffraction is that it uses protein samples in aqueous solution rather than crystals. This environment is closer to that of proteins in cells, and thus it is one of the main advantages of the method. The NMR method most widely used in the determination of protein structure ultimately depends on the distances between hydrogen atoms, giving results independent of those obtained by X-ray crystallography. The NMR method is undergoing constant improvement and is being applied to larger proteins as these improvements progress.

Essential Information

The tertiary structure of a protein is the three-dimensional arrangement of all atoms in a protein chain. The secondary and tertiary structures of a protein can be determined simultaneously.

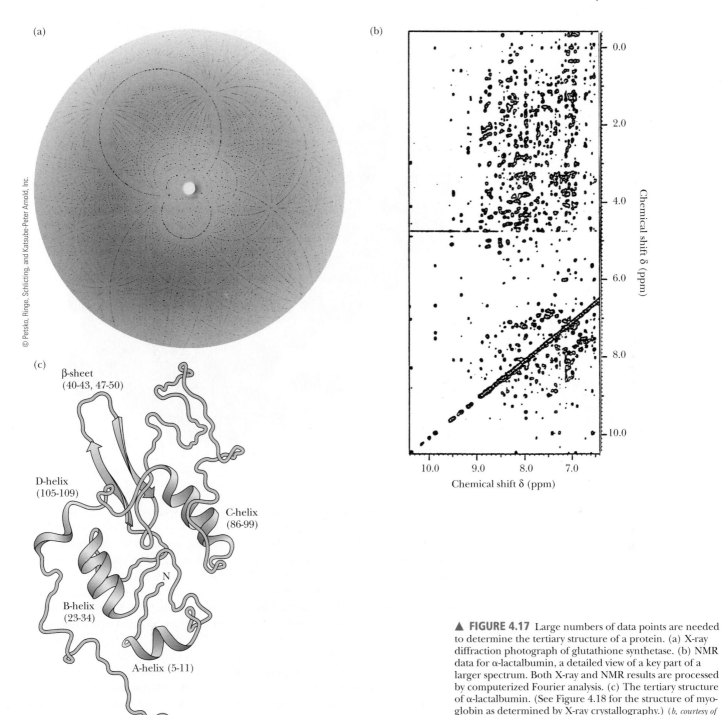

(a)

© Petsko, Ringe, Schlicting, and Katsube-Peter Arnold, Inc.

(b)

Chemical shift δ (ppm)

Chemical shift δ (ppm)

(c)

β-sheet
(40-43, 47-50)

D-helix
(105-109)

C-helix
(86-99)

N

B-helix
(23-34)

A-helix (5-11)

C

▲ **FIGURE 4.17** Large numbers of data points are needed to determine the tertiary structure of a protein. (a) X-ray diffraction photograph of glutathione synthetase. (b) NMR data for α-lactalbumin, a detailed view of a key part of a larger spectrum. Both X-ray and NMR results are processed by computerized Fourier analysis. (c) The tertiary structure of α-lactalbumin. (See Figure 4.18 for the structure of myoglobin as determined by X-ray crystallography.) (*b, courtesy of Professor C. M. Dobson, University of Oxford.*)

Myoglobin: An Example of Protein Structure

In many ways, myoglobin is the classic example of a globular protein. We shall use it here as a case study in tertiary structure. (We shall see the tertiary structures of many other proteins in context when we discuss their roles in biochemistry.) Myoglobin was the first protein for which the complete tertiary structure (Figure 4.18) was determined by X-ray crystallography. The complete myoglobin molecule consists of a single polypeptide chain of 153 amino acid residues and includes a prosthetic group, the **heme** group, which also occurs in hemoglobin. The myoglobin molecule (including the heme group) has a compact structure, with the interior atoms very close to each other. This

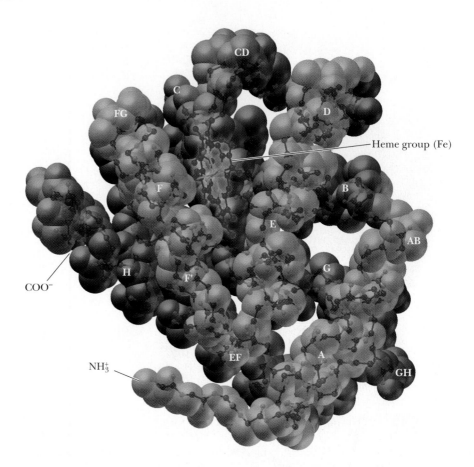

Heme group (Fe)

COO⁻

NH₃⁺

▶ **FIGURE 4.18** The structure of the myoglobin molecule, showing the peptide backbone and the heme group. The helical segments are designated by the letters A through H. The terms NH_3 and COO^- indicate the N-terminal and C-terminal ends, respectively.

structure provides examples of many of the forces responsible for the three-dimensional shapes of proteins.

In myoglobin, there are eight α-helical regions and no β-pleated sheet regions. Approximately 75% of the residues in myoglobin are found in these helical regions, which are designated by the letters A through H. Hydrogen bonding in the polypeptide backbone stabilizes the α-helical regions; amino acid side chains are also involved in hydrogen bonds. The polar residues are on the exterior of the molecule. The interior of the protein contains almost exclusively nonpolar amino acid residues. Two polar histidine residues are found in the interior; they are involved in interactions with the heme group and bound oxygen, and thus play an important role in the function of the molecule. The planar heme group fits into a hydrophobic pocket in the protein portion of the molecule and is held in position by hydrophobic attractions between heme's porphyrin ring and the nonpolar side chains of the protein. The presence of the heme group drastically affects the conformation of the polypeptide: The apoprotein (the polypeptide chain alone, without the prosthetic heme group) is not as tightly folded as the complete molecule.

The heme group consists of a metal ion, Fe(II), and an organic part, protoporphyrin IX (Figure 4.19). (The notation Fe(II) is preferred to Fe^{2+} when metal ions occur in complexes.) The porphyrin part consists of four five-membered rings based on the pyrrole structure; these four rings are linked by bridging methine (—CH═) groups to form a square planar structure. The Fe(II) ion has six coordination sites, and it forms six metal–ion complexation bonds. Four of the six sites are occupied by the nitrogen atoms of the four pyrrole-type rings of the porphyrin to give the complete heme group. The presence of the heme group is required for myoglobin to bind oxygen.

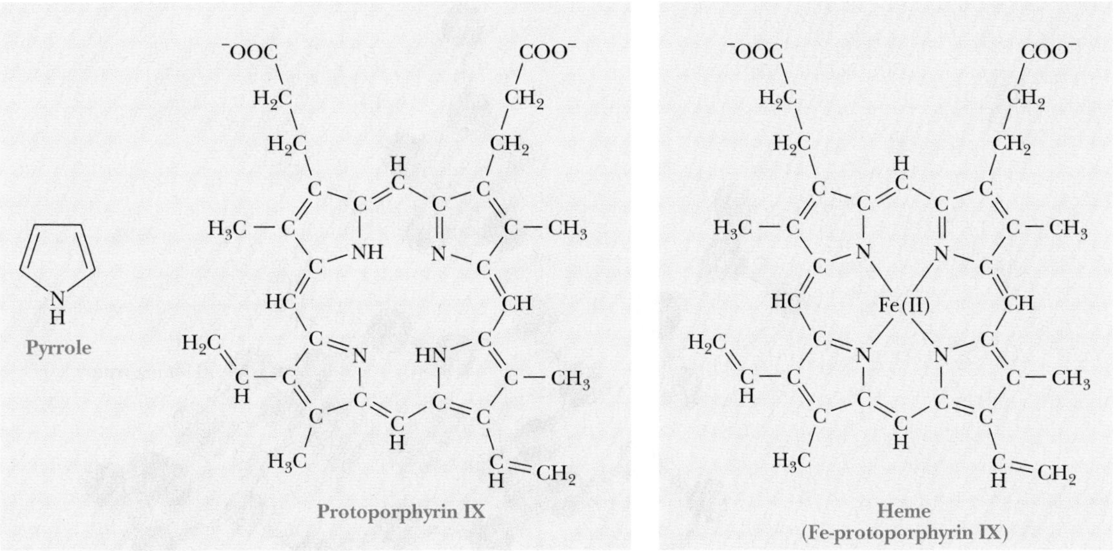

▲ **FIGURE 4.19** The structure of the heme group. Four pyrrole rings are linked by bridging groups to form a planar porphyrin ring. Several isomeric porphyrin rings are possible, depending on the nature and arrangement of the side chains. The porphyrin isomer found in heme is protoporphyrin IX. Addition of iron to protoporphyrin IX produces the heme group.

The fifth coordination site of the Fe(II) ion is occupied by one of the nitrogen atoms of the imidazole side chain of histidine residue F8 (the eighth residue in helical segment F). This histidine residue is one of the two in the interior of the molecule. The oxygen is bound at the sixth coordination site of the iron. The fifth and sixth coordination sites lie perpendicular to, and on opposite sides of, the plane of the porphyrin ring. The other histidine residue in the interior of the molecule, residue E7 (the seventh residue in helical segment E), lies on the same side of the heme group as the bound oxygen (Figure 4.20). This second histidine is not bound to the iron, or to any part of the heme group, but it acts as a gate that opens and closes as oxygen enters the hydrophobic pocket to bind to the heme. The E7 histidine sterically inhibits oxygen from binding perpendicularly to the heme plane, with biologically important ramifications. The affinity of free heme for carbon monoxide (CO) is 25,000 times greater than its affinity for oxygen. When carbon monoxide is forced to bind at an angle in myoglobin due to the steric block by His E7, its advantage over oxygen drops by two orders of magnitude (Figure 4.21). This guards against the possibility that traces of CO produced during metabolism would occupy all the oxygen-binding sites on the hemes. Nevertheless, CO is a potent poison in larger quantities because of its effect both on oxygen binding to hemoglobin and on the final step of the electron transport chain (Section 20.5).

In the absence of the protein, the iron of the heme group can be oxidized to Fe(III); the oxidized heme will not bind oxygen. Thus, the combination of both heme and protein is needed to bind O_2 for oxygen storage.

Denaturation and Refolding

The noncovalent interactions that maintain the three-dimensional structure of a protein are weak, and it is not surprising that they can be disrupted easily. The unfolding of a protein is called **denaturation.** Reduction of disulfide bonds leads (Section 3.5) to even more extensive unraveling of the tertiary

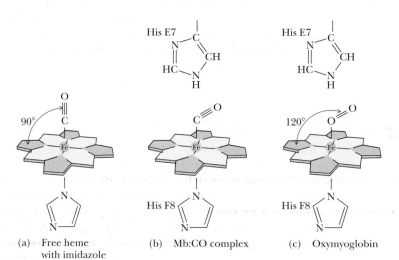

His E7

Binding site
for oxygen

Heme group

Fe

His F8

▶ **FIGURE 4.20** The oxygen-binding site of myoglobin. The porphyrin ring occupies four of the six coordination sites of the Fe(II). Histidine F8 (His F8) occupies the fifth coordination site of the iron (see text). Oxygen is bound at the sixth coordination site of the iron, and histidine E7 lies close to the oxygen. (*Leonard Lessin/Waldo Feng/Mt. Sinai CORE.*)

▶ **FIGURE 4.21** Oxygen and carbon monoxide binding to the heme group of myoglobin. The presence of the E7 histidine forces a 120° angle to the oxygen or CO.

(a) Free heme with imidazole

(b) Mb:CO complex

(c) Oxymyoglobin

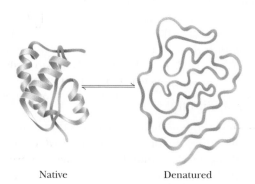

Biochemistry ❂ Now™ ANIMATED FIGURE 4.22
Denaturation of a protein. The native conformation can be recovered when denaturing conditions are removed. **See this figure animated at http://now.brookscole.com/campbell5**

Native

Denatured

structure. Denaturation and reduction of disulfide bonds are frequently combined when complete disruption of the tertiary structure of proteins is desired. Under proper experimental conditions, the disrupted structure can then be completely recovered. This process of denaturation and refolding is a dramatic demonstration of the relationship between the primary structure of the protein and the forces that determine the tertiary structure. For many proteins, various other factors are needed for complete refolding, but the important point is that the primary structure determines the tertiary structure.

Proteins can be denatured in several ways. One is *heat.* An increase in temperature favors vibrations within the molecule, and the energy of these vibrations can become great enough to disrupt the tertiary structure. At either high or low *extremes of pH,* at least some of the charges on the protein are missing, and so the electrostatic interactions that would normally stabilize the native, active form of the protein are drastically reduced. This leads to denaturation. The binding of *detergents,* such as sodium dodecyl sulfate (SDS), also denatures proteins. Detergents tend to disrupt hydrophobic interactions. If a detergent is charged, it can also disrupt electrostatic interactions within the protein. Other reagents, such as *urea* and *guanidine hydrochloride,* form hydrogen bonds with the protein that are stronger than those within the protein itself. These two reagents can also disrupt hydrophobic interactions in much the same way as detergents (Figure 4.22).

β-Mercaptoethanol (HS—CH$_2$—CH$_2$—OH) is frequently used to reduce disulfide bridges to two sulfhydryl groups. Urea is usually added to the reaction mixture to facilitate unfolding of the protein and to increase the accessibility of the disulfides to the reducing agent. If experimental conditions are properly chosen, the native conformation of the protein can be recovered when both mercaptoethanol and urea are removed (Figure 4.23). Experiments of this type provide some of the strongest evidence that the amino acid sequence of the protein contains all the information required to produce the complete three-dimensional structure. Protein researchers are pursuing with some interest the conditions under which a protein can be denatured—including reduction of disulfides—and its native conformation later recovered.

4.6 | Can We Predict Protein Folding from Sequence?

Since the sequence of amino acids determines the three-dimensional structure of a protein, a question that arises naturally is, "Can we predict the tertiary structure of a protein if we know its amino acid sequence?" The answer is that we can, within limits. Modern computing techniques greatly facilitate the operation, which requires processing large amounts of information. The encounter of biochemistry and computing has given rise to the burgeoning field of **bioinformatics.** Prediction of protein structure is one of the principal applications of bioinformatics. Another important application is the comparison of base sequences in nucleic acids, a topic we shall discuss in Chapter 14, along with other methods for working with nucleic acids.

The first step in predicting protein architecture is a search of databases of known structures for *sequence homology* between the protein whose structure is to be determined and proteins of known architecture, where the term **homology** refers to similarity of two or more sequences. If the sequence of the known protein is similar enough to that of the protein being studied, the known protein's structure becomes the point of departure for *comparative modeling.* Use of modeling algorithms that compare the protein being studied with known structures leads to a structure prediction. This method is most

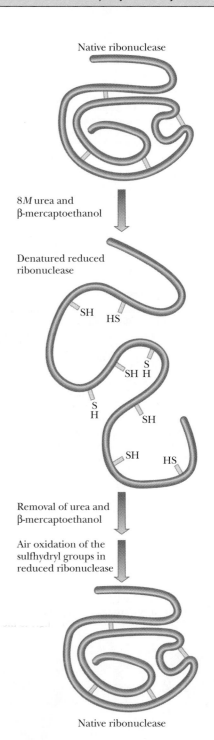

▲ **FIGURE 4.23** Denaturation and refolding in ribonuclease. The protein ribonuclease can be completely denatured by the actions of urea and mercaptoethanol. When denaturing conditions are removed, activity is recovered.

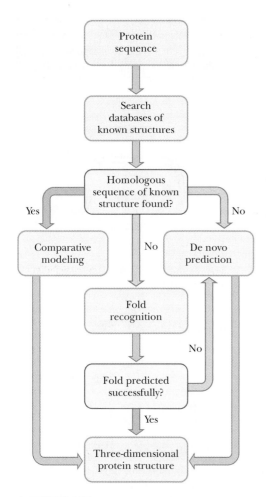

▲ **FIGURE 4.24** A flow chart showing the use of existing information from databases to predict protein conformation. (*Courtesy of Rob Russell, EMBL.*)

useful when the sequence homology is greater than 25–30%. If the sequence homology is less than 25–30%, other approaches are more useful. *Fold recognition* algorithms allow comparison with known folding motifs common to many secondary structures. We saw a number of these motifs in Section 4.3. Here is an application of that information. Yet another method is *de novo prediction*, based on first principles from chemistry, biology, and physics. This method too can give rise to structures subsequently confirmed by X-ray crystallography. The flow chart in Figure 4.24 shows how prediction techniques use existing information from databases. Figure 4.25 shows a comparison of the predicted structures of two proteins (right side) for the DNA repair protein MutS and the bacterial protein HI0817. The crystal structures of the two proteins are shown on the left.

A considerable amount of information about protein sequences and architecture is available on the World Wide Web. One of the most important resources is the Protein Data Bank operated under the auspices of the Research Collaboratory for Structural Bioinformatics (RCSB). Its URL is http://www.rcsb.org/pdb. This site, which has a number of mirror sites around the world, is the single repository of structural information about large molecules. It includes material about nucleic acids as well as proteins. Its home page has a button with links specifically geared to educational applications.

Results of structure prediction using the methods discussed in this section are available on the Web as well. One of the most useful URLs is http://predictioncenter.llnl.gov/casp5. Other excellent sources of information are available through the National Institutes of Health (http://pubmedcentral.nih.gov/tocrender.fcgi?iid=1005, and http://www.ncbi.nlm.nih.gov), and through the ExPASy (Expert Protein Analysis System) server (http://us.expasy.org).

Protein-Folding Chaperones

The primary structure conveys all the information necessary to produce the correct tertiary structure, but the folding process in vivo can be a bit trickier. In the protein-dense environment of the cell, proteins may begin to fold incorrectly as they are produced, or they may begin to associate with other proteins before completing their folding process. In eukaryotes, proteins may need to remain unfolded long enough to be transported across the membrane of a subcellular organelle. Special proteins called **chaperones** aid in the correct and timely folding of many other proteins (see the Biochemical Connections box in Chapter 12). The first such proteins discovered were a family called

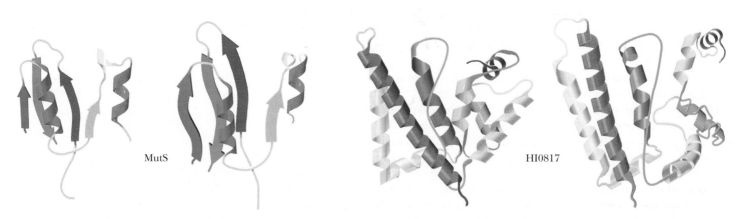

▲ **FIGURE 4.25** A comparison of the predicted structures of two proteins (right side) for the DNA repair protein MutS and the bacterial protein HI0817. The crystal structures of the two proteins are shown on the left. (*Courtesy of University of Washington, Seattle.*)

Biochemical Connections

Prions

It has been established that the causative agent of mad-cow disease, as well as the related diseases scrapie in sheep and spongiform encephalopathy (kuru and Creutzfeldt-Jakob disease) in humans, is a small (28-kD) protein called a **prion.** Prions are glycoproteins found in the cell membranes of nerve tissue. The diseases come about when the normal form of the prion protein, PrP (Figure a), folds into an incorrect form called PrP^{sc} (Figure b). The abnormal form of the prion protein is able to convert other, normal forms into abnormal forms. As recently discovered, this change can be propagated in nervous tissue. Scrapie had been known for years, but it had not been known to cross species barriers. Then an outbreak of mad-cow disease was shown to have followed the inclusion of sheep remains in cattle feed. It is now known that eating tainted beef from animals with mad-cow disease can cause spongiform encephalopathy, now known as new variant Creutzfeldt-Jakob disease, in humans. The normal prions have a large percentage of α-helix, but the abnormal forms have more β-pleated sheets. Notice that in this case the same protein (a single, well-defined sequence) can exist in alternative forms. These β-pleated sheets in the abnormal proteins interact between protein molecules and form insoluble plaques, a fate also seen in Alzheimer's disease. Ingested abnormal prions use macrophages from the immune system to travel in the body until they come in contact with nerve tissue. They can then propagate up the nerves until they reach the brain.

This mechanism was a subject of considerable controversy when it was first proposed. A number of scientists expected that a slow-acting virus would be found to be the ultimate cause of these neurological diseases. A susceptibility to these diseases can be inherited, so some involvement of DNA (or RNA) was also expected. Some went so far as to talk about "heresy" when Stanley Prusiner received the 1997 Nobel Prize in medicine for his discovery of prions. It now appears that genes for susceptibility to the incorrect form exist in all vertebrates, giving rise to the observed pattern of disease transmission, but many individuals with the genetic susceptibility never develop the disease if they do not come in contact with abnormal prions from another source. See the articles by Ferguson and Peretz in the bibliography of this chapter.

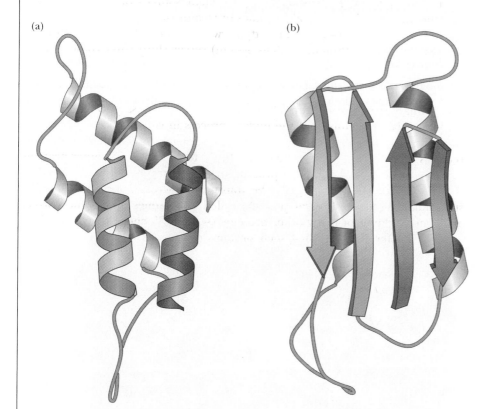

(a)

(b)

◀ (a) Normal prion structure (PrP).
(b) Abnormal prion (PrP^{sc}).

hsp70 (for 70,000 MW Heat-Shock Protein), which are proteins produced in *E. coli* grown above optimal temperatures. Chaperones exist in organisms from prokaryotes through humans, and their mechanisms of action are currently being studied. (See the article by Helfand in the bibliography of this chapter.) In recent years, it has become evident that protein-folding dynamics is crucial to protein function in vivo. The Biochemical Connections box on the previous page describes a particularly striking example of the importance of protein folding.

4.7 | What Is the Quaternary Structure of Proteins?

Quaternary structure is a property of proteins that consist of more than one polypeptide chain. Each chain is called a subunit. The number of chains can range from two to more than a dozen, and the chains may be identical or different. Commonly occurring examples are **dimers, trimers,** and **tetramers,** consisting of two, three, and four polypeptide chains, respectively. (The generic term for such a molecule, made up of a small number of subunits, is **oligomer.**) The chains interact with one another noncovalently via electrostatic attractions, hydrogen bonds, and hydrophobic interactions.

As a result of these noncovalent interactions, subtle changes in structure at one site on a protein molecule may cause drastic changes in properties at a distant site. Proteins that exhibit this property are called **allosteric.** Not all multisubunit proteins exhibit allosteric effects, but many do.

A classic illustration of the quaternary structure of proteins and its effect on properties is a comparison of hemoglobin, an allosteric protein, with myoglobin, which consists of a single polypeptide chain.

Hemoglobin

Hemoglobin is a tetramer, consisting of four polypeptide chains, two α-chains and two β-chains (Figure 4.26). (In oligomeric proteins, the types of polypep-

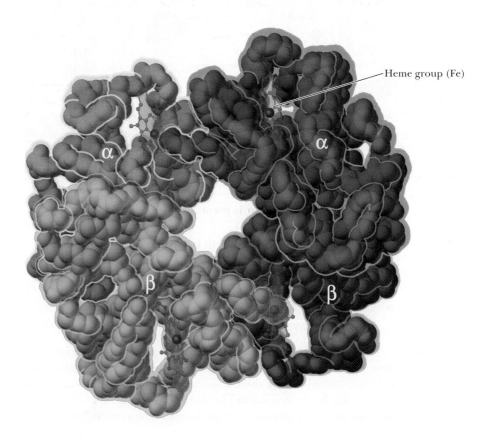

▶ **FIGURE 4.26** The structure of hemoglobin. Hemoglobin ($\alpha_2\beta_2$) is a tetramer consisting of four polypeptide chains (two α-chains and two β-chains).

tide chains are designated with Greek letters.) The two α-chains of hemoglobin are identical, as are the two β-chains. The overall structure of hemoglobin is $\alpha_2\beta_2$ in Greek-letter notation. Both the α- and β-chains of hemoglobin are very similar to the myoglobin chain. The α-chain is 141 residues long, and the β-chain is 146 residues long; for comparison, the myoglobin chain is 153 residues long. Many of the amino acids of the α-chain, the β-chain, and myoglobin are *homologous;* that is, the same amino acid residues are in the same positions. The heme group is the same in myoglobin and hemoglobin.

We have already seen that one molecule of myoglobin binds one oxygen molecule. Four molecules of oxygen can therefore bind to one hemoglobin molecule. Both hemoglobin and myoglobin bind oxygen reversibly, but the binding of oxygen to hemoglobin exhibits **positive cooperativity,** whereas oxygen binding to myoglobin does not. Positive cooperativity means that when one oxygen molecule is bound, it becomes easier for the next to bind. A graph of the oxygen-binding properties of hemoglobin and myoglobin is one of the best ways to illustrate this point (Figure 4.27).

When the degree of saturation of myoglobin with oxygen is plotted against oxygen pressure, a steady rise is observed until complete saturation is approached and the curve levels off. The oxygen-binding curve of myoglobin is thus said to be **hyperbolic.** In contrast, the shape of the oxygen-binding curve for hemoglobin is **sigmoidal.** This shape indicates that the binding of the first oxygen molecule facilitates the binding of the second oxygen, which facilitates the binding of the third, which in turn facilitates the binding of the fourth. This is precisely what is meant by the term "cooperative binding." However, note that even though cooperative binding means that binding of each subsequent oxygen is easier than the previous one, the binding curve is still lower than that of myoglobin at any oxygen pressure. In other words, at any oxygen pressure, myoglobin will have a higher percentage of saturation than hemoglobin.

The two types of behavior are also related to the functions of these proteins. Myoglobin has the function of oxygen *storage* in muscle. It must bind strongly to oxygen at very low pressures, and it is 50% saturated at 1 torr partial pressure of oxygen. (The **torr** is a widely used unit of pressure, but it is not an SI unit. One torr is the pressure exerted by a column of mercury 1 mm high at 0°C. One atmosphere is equal to 760 torr.) The function of hemoglobin is oxygen *transport,* and it must be able both to bind strongly to oxygen and to release oxygen easily, depending upon conditions. In the alveoli of lungs (where hemoglobin must bind oxygen for transport to the tissues), the oxygen pressure is 100 torr. At this pressure, hemoglobin is 100% saturated with oxygen. In the capillaries of active muscles, the pressure of oxygen is 20 torr, corresponding to less than 50% saturation of hemoglobin, which occurs at 26 torr. In other words, hemoglobin gives up oxygen easily in capillaries, where the need for oxygen is great.

Structural changes during binding of small molecules are characteristic of allosteric proteins such as hemoglobin. Hemoglobin has different quaternary structures in the bound (oxygenated) and unbound (deoxygenated) forms. The two β-chains are much closer to each other in oxygenated hemoglobin than in deoxygenated hemoglobin. The change is so marked that the two forms of hemoglobin have different crystal structures (Figure 4.28).

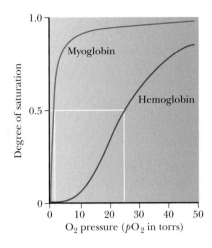

▲ **FIGURE 4.27** A comparison of the oxygen-binding behavior of myoglobin and hemoglobin. The oxygen-binding curve of myoglobin is hyperbolic, whereas that of hemoglobin is sigmoidal. Myoglobin is 50% saturated with oxygen at 1 torr partial pressure; hemoglobin does not reach 50% saturation until the partial pressure of oxygen reaches 26 torr.

Conformational Changes That Accompany Hemoglobin Function

Other ligands are involved in cooperative effects when oxygen binds to hemoglobin. Both H^+ and CO_2, which themselves bind to hemoglobin, affect the affinity of hemoglobin for oxygen by altering the protein's three-dimensional

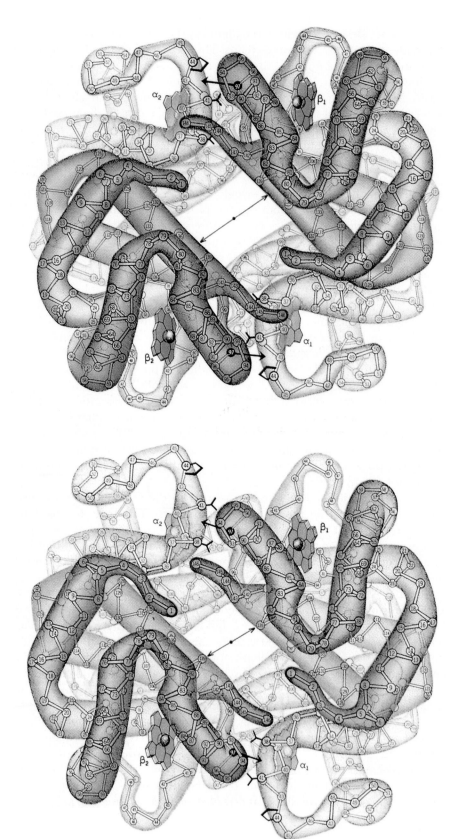

▶ **FIGURE 4.28** The structures of (a) deoxyhemoglobin and (b) oxyhemoglobin. Note the motions of subunits with respect to one another. There is much less room at the center of oxyhemoglobin. (*Illustration, Irving Geis. Rights owned by Howard Hughes Medical Institute. Not to be reproduced without permission.*)

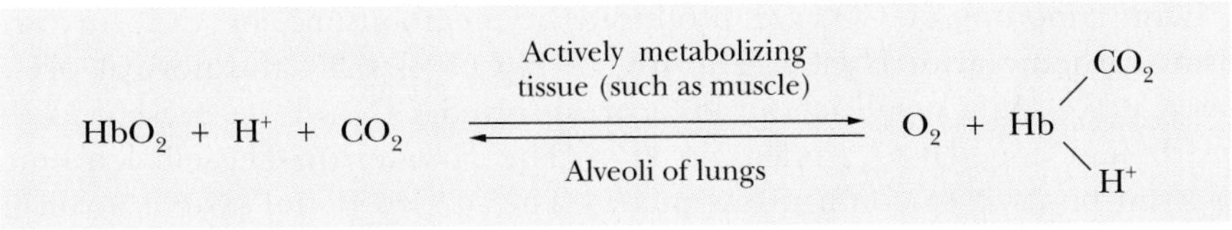

$$HbO_2 + H^+ + CO_2 \underset{\text{Alveoli of lungs}}{\overset{\substack{\text{Actively metabolizing}\\\text{tissue (such as muscle)}}}{\rightleftharpoons}} O_2 + Hb \diagdown^{CO_2}_{H^+}$$

◀ **FIGURE 4.29** The general features of the Bohr effect. In actively metabolizing tissue, hemoglobin releases oxygen and binds both CO_2 and H^+. In the lungs, hemoglobin releases both CO_2 and H^+ and binds oxygen.

structure in subtle but important ways. The effect of H^+ (Figure 4.29) is called the *Bohr effect,* after its discoverer, Christian Bohr (the father of physicist Niels Bohr). The oxygen-binding ability of myoglobin is not affected by the presence of H^+ or of CO_2.

An increase in the concentration of H^+ (i.e., a lowering of the pH) reduces the oxygen affinity of hemoglobin. Increasing H^+ causes the protonation of key amino acids, including the N-terminals of the α-chains and His[146] of the β-chains. The protonated histidine is attracted to, and stabilized by, a salt bridge to Asp[94]. This favors the deoxygenated form of hemoglobin. Actively metabolizing tissue, which requires oxygen, releases H^+, thus acidifying its local environment. Hemoglobin has a lower affinity for oxygen under these conditions, and it releases oxygen where it is needed (Figure 4.30). Hemoglobin's acid–base properties affect, and are affected by, its oxygen-binding properties. The oxygenated form of hemoglobin is a stronger acid (has a lower pK_a) than the deoxygenated form. In other words, deoxygenated hemoglobin has a higher affinity for H^+ than does the oxygenated form. Thus, changes in the quaternary structure of hemoglobin can modulate the buffering of blood through the hemoglobin molecule itself.

Table 4.1 summarizes the important features of the Bohr effect.

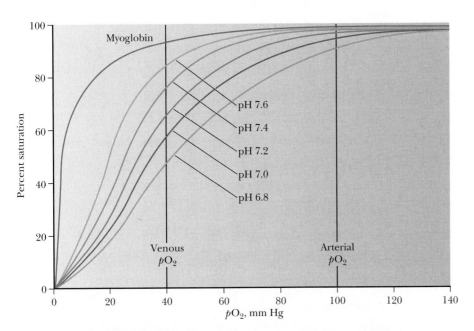

◀ **FIGURE 4.30** The oxygen saturation curves for myoglobin and for hemoglobin at five different pH values.

Table 4.1	
A Summary of the Bohr Effect	
Lungs	**Actively Metabolizing Muscle**
Higher pH than actively metabolizing tissue	Lower pH due to production of H^+
Hemoglobin binds O_2	Hemoglobin releases O_2
Hemoglobin releases H^+	Hemoglobin binds H^+

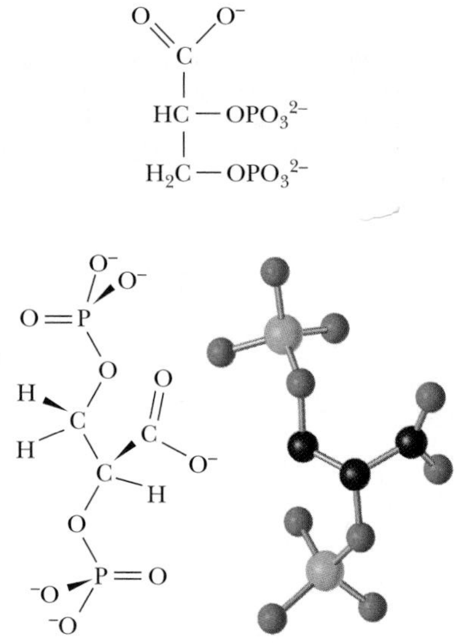

▲ FIGURE 4.31 The structure of BPG (2,3-*bis*phosphoglycerate), an important allosteric effector of hemoglobin.

Large amounts of CO_2 are produced by metabolism. The CO_2, in turn, forms carbonic acid, H_2CO_3. The pK_a of H_2CO_3 is 6.35; the normal pH of blood is 7.4. As a result, about 90% of dissolved CO_2 will be present as the bicarbonate ion, HCO_3^-, releasing H^+. (The Henderson–Hasselbalch equation can be used to confirm this point.) The in vivo buffer system involving H_2CO_3 and HCO_3^- in blood was discussed in Section 2.6. The presence of larger amounts of H^+ as a result of CO_2 production favors the quaternary structure that is characteristic of deoxygenated hemoglobin. Hence, the affinity of hemoglobin for oxygen is lowered. The HCO_3^- is transported to the lungs, where it combines with H^+ released when hemoglobin is oxygenated, producing H_2CO_3. In turn, H_2CO_3 liberates CO_2, which is then exhaled. Hemoglobin also transports some CO_2 directly. When the CO_2 concentration is high, it combines with the free α-amino groups to form carbamate:

$$R—NH_2 + CO_2 \rightleftharpoons R—NH—COO^- + H^+$$

This reaction turns the α-amino terminals into anions, which can then interact with the α-chain Arg[141], also stabilizing the deoxygenated form.

In the presence of large amounts of H^+ and CO_2, as in respiring tissue, hemoglobin releases oxygen. The presence of large amounts of oxygen in the lungs reverses the process, causing hemoglobin to bind O_2. The oxygenated

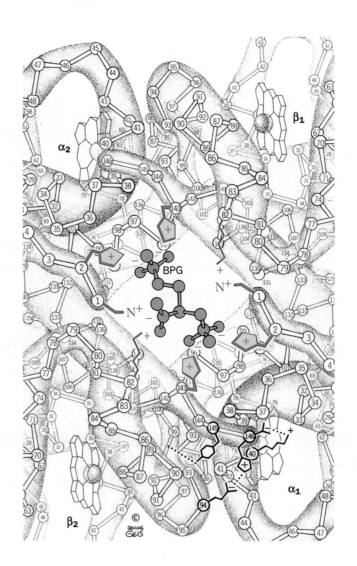

▶ FIGURE 4.32 The binding of BPG to deoxyhemoglobin. Note the electrostatic interactions between the BPG and the protein. (*Illustration, Irving Geis. Rights owned by Howard Hughes Medical Institute. Not to be reproduced without permission.*)

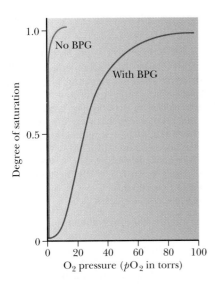

▲ **FIGURE 4.33** A comparison of the oxygen-binding properties of hemoglobin in the presence and absence of BPG. Note that the presence of the BPG markedly decreases the affinity of hemoglobin for oxygen.

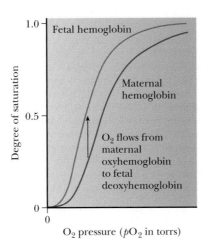

▲ **FIGURE 4.34** A comparison of the oxygen-binding capacity of fetal and maternal hemoglobins. Fetal hemoglobin binds less strongly to BPG and, consequently, has a greater affinity for oxygen than does maternal hemoglobin.

hemoglobin can then transport oxygen to the tissues. The process is complex, but it allows for fine tuning of pH as well as levels of CO_2 and O_2.

Hemoglobin in blood is also bound to another ligand, **2,3-*bis*phosphoglycerate (BPG)** (Figure 4.31), with drastic effects on its oxygen-binding capacity. The binding of BPG to hemoglobin is electrostatic; specific interactions take place between the negative charges on BPG and the positive charges on the protein (Figure 4.32). In the presence of BPG, the partial pressure at which 50% of hemoglobin is bound to oxygen is 26 torr. If BPG were not present in blood, the oxygen-binding capacity of hemoglobin would be much higher (50% of hemoglobin bound to oxygen at about 1 torr), and little oxygen would be released in the capillaries. "Stripped" hemoglobin, which is isolated from blood and from which the endogenous BPG has been removed, displays this behavior (Figure 4.33).

BPG also plays a role in supplying a growing fetus with oxygen. The fetus obtains oxygen from the mother's bloodstream via the placenta. Fetal hemoglobin (Hb F) has a higher affinity for oxygen than does maternal hemoglobin, allowing for efficient transfer of oxygen from the mother to the fetus (Figure 4.34). Two features of fetal hemoglobin contribute to this higher oxygen-binding capacity. One is the presence of two different polypeptide chains. The subunit structure of Hb F is $\alpha_2\gamma_2$, where the β-chains of adult hemoglobin (Hb A), the usual hemoglobin, have been replaced by the γ-chains, which are similar but not identical in structure. The second feature is that Hb F binds less strongly to BPG than does Hb A. In the β-chain of adult hemoglobin, His[143] makes a salt bridge to BPG. In the fetal hemoglobin, the γ-chain has an amino acid substitution of a serine for His[143]. This change of a positively charged amino acid for a neutral one diminishes the number of contacts between the hemoglobin and the BPG, effectively reducing the allosteric effect enough to give fetal hemoglobin a higher binding curve than adult hemoglobin.

Summary

4.1 How Does the Structure of Proteins Determine Their Function?
The structure of proteins is complex, with few obvious regular structures. Many three-dimensional conformations are possible for proteins, but only one, or at most a few, have biological activity; these are called the native conformations. To facilitate structure determination, it is customary to define four levels of organization.

4.2 What Is the Primary Structure of Proteins?
Primary structure is the order in which the amino acids are covalently linked. The primary structure of a protein can be determined by chemical methods. The amino acid sequence (the primary structure) of a protein determines its three-dimensional structure, which in turn determines its properties. A striking example of the importance of primary structure is sickle-cell anemia, a disease caused by a change in one amino acid in each of two of the four chains of hemoglobin.

4.3 What Is the Secondary Structure of Proteins?
Secondary structure is the hydrogen-bonded arrangement in space of the backbone, the polypeptide chain. Some of the most important backbone arrangements are the α-helix, the β-sheet, and the β-turn. They can be combined in a number of ways to produce structural motifs that occur in many proteins.

4.4 What Can We Say about the Thermodynamics of Protein Folding?
The higher-order (secondary and tertiary) levels of structure depend primarily on noncovalent interactions, including hydrogen bonds, hydrophobic interactions, electrostatic interactions, and complexation of metal ions. Hydrophobic interactions, which depend on the unfavorable entropy of the water of hydration surrounding nonpolar solutes, are particularly important determinants of protein folding.

4.5 What Is the Tertiary Structure of Proteins?
Tertiary structure includes the three-dimensional arrangement of *all* the atoms in the protein. The three-dimensional structures of proteins can be completely disrupted and, under proper experimental conditions, completely recovered. This process of denaturation and refolding is a dramatic example of the relationship between the primary structure of the protein and the forces that determine the tertiary structure. The secondary and tertiary structures of a protein can be determined simultaneously by X-ray crystallography. The oxygen-storage protein myoglobin was the first protein for which the complete tertiary structure was determined by crystallography.

4.6 Can We Predict Protein Folding from Sequence?
It is possible, to some extent, to predict the three-dimensional structure of a protein from its amino acid sequence. Computer algorithms are based on two approaches, one of which is based on comparison of sequences with those of proteins whose folding pattern is known. Another one is based on the folding motifs that occur in many proteins.

4.7 What Is the Quaternary Structure of Proteins?
Quaternary structure is the arrangement of subunits in multisubunit proteins. The individual polypeptide chains of multisubunit proteins interact with one another noncovalently. As a result, subtle changes in structure at one site on the molecule can cause drastic changes in properties at a distant site. Proteins that exhibit this property are referred to as allosteric. The properties of the allosteric protein hemoglobin can be contrasted with those of myoglobin, which is not allosteric. In hemoglobin, an oxygen-transport protein, the binding of oxygen is cooperative (as each oxygen is bound, it becomes easier for the next one to bind) and is modulated by such ligands as H^+, CO_2, and BPG. The binding of oxygen to myoglobin is not cooperative.

Critical Questions to Review

4.1 How Does the Structure of Proteins Determine Their Function?

1. **Fact Check** Match the following statements about protein structure with the proper levels of organization.

(a) Primary structure	(1) Three-dimensional arrangement of all atoms
(b) Secondary structure	(2) The order of amino acid residues in the polypeptide chain
(c) Tertiary structure	(3) The interaction between subunits in proteins that consist of more than one polypeptide chain
(d) Quaternary structure	(4) The hydrogen-bonded arrangement of the polypeptide backbone

2. **Fact Check** Define denaturation in terms of the effects of secondary, tertiary, and quaternary structure.

3. **Fact Check** What is the nature of "random" structure in proteins?

4.2 What Is the Primary Structure of Proteins?

4. **Thought Question** Suggest an explanation for the observation that, when proteins are chemically modified so that specific side chains have a different chemical nature, these proteins cannot be denatured reversibly.

5. **Thought Question** Rationalize the following observations.

   (a) Serine is the amino acid residue that can be replaced with the least effect on protein structure and function.

   (b) Replacement of tryptophan causes the greatest effect on protein structure and function.

   (c) Replacements such as Lys → Arg and Leu → Ile usually have very little effect on protein structure and function.

6. **Thought Question** Glycine is a highly conserved amino acid residue in proteins (i.e., it is found in the same position in the primary structure of related proteins). Suggest a reason why this might occur.

7. **Thought Question** A mutation that changes an alanine residue in a protein to an isoleucine leads to a loss of activity. Activity is regained when a further mutation at the same site changes the isoleucine to a glycine. Why?

8. **Thought Question** A biochemistry student characterizes the process of cooking meat as an exercise in denaturing proteins. Comment on the validity of this remark.

9. **Biochemical Connections** Severe combined immunodeficiency disease (SCID) is characterized by the complete lack of an immune system. Strains of mice have been developed that have SCID. When SCID mice that carry genetic predisposition to prion diseases are infected with PrP^{sc}, they do not develop prion diseases. How do these facts relate to the transmission of prion diseases?

10. **Biochemical Connections** An isolated strain of sheep was found in New Zealand. Most of these sheep carried the gene for predisposition to scrapie, yet none of them ever came down with the disease. How do these facts relate to the transmission of prion diseases?

4.3 What Is the Secondary Structure of Proteins?

11. **Fact Check** List three major differences between fibrous and globular proteins.

12. **Biochemical Connections** What is a protein efficiency ratio?

13. **Biochemical Connections** Which food has the highest PER?

14. **Biochemical Connections** What are the essential amino acids?

15. **Biochemical Connections** Why are scientists currently trying to create genetically modified foods?

16. **Fact Check** What are Ramachandran angles?

17. **Fact Check** What is a β-bulge?

18. **Fact Check** What is a reverse turn? Draw two types of reverse turns.

19. **Fact Check** List some of the differences between the α-helix and β-sheet forms of secondary structure.

20. **Fact Check** List some of the possible combinations of α-helices and β-sheets in supersecondary structures.

21. **Fact Check** Why is proline frequently encountered at the places in the myoglobin and hemoglobin molecules where the polypeptide chain turns a corner?

22. **Fact Check** Why must glycine be found at regular intervals in the collagen triple helix?

23. **Thought Question** You hear the comment that the difference between wool and silk is the difference between helical and pleated-sheet structures. Do you consider this a valid point of view? Why or why not?

24. **Thought Question** Woolen clothing shrinks when washed in hot water, but items made of silk do not. Suggest a reason, based on information from this chapter.

4.4 What Can We Say about the Thermodynamics of Protein Folding?

25. **Fact Check** List five forces that are responsible for maintaining the correct three-dimensional shapes of proteins. Specify which groups on the protein are involved in each type of interaction.

26. **Thought Question** Comment on the energetics of protein folding in light of the information in this chapter.

4.5 What Is the Tertiary Structure of Proteins?

27. **Fact Check** Draw two hydrogen bonds, one that is part of a secondary structure and another that is part of a tertiary structure.

28. **Fact Check** Draw a possible electrostatic interaction between two amino acids in a polypeptide chain.

29. **Fact Check** Draw a disulfide bridge between two cysteines in a polypeptide chain.

30. **Fact Check** Draw a region of a polypeptide chain showing a hydrophobic pocket containing nonpolar side chains.

31. **Fact Check** What is a chaperone?

32. **Thought Question** The terms *configuration* and *conformation* appear in descriptions of molecular structure. How do they differ?

33. **Thought Question** Theoretically, a protein could assume a virtually infinite number of configurations and conformations. Suggest several features of proteins that drastically limit the actual number.

34. **Thought Question** What is the highest level of protein structure found in collagen?

4.6 Can We Predict Protein Folding from Sequence?

35. **Thought Question** You have discovered a new protein, one whose sequence has about 25% homology with ribonuclease A. How would you go about predicting, rather than experimentally determining, its tertiary structure?

36. **Thought Question** Go to the RCSB site for the Protein Data Bank (http://www.rcsb.org/pdb). Give a brief description of the molecule prefoldin, which can be found under *chaperones*.

4.7 What Is the Quaternary Structure of Proteins?

37. **Biochemical Connections** What is a prion?

38. **Biochemical Connections** What are the known diseases caused by abnormal prions?

39. **Biochemical Connections** What are the protein secondary structures that differ between a normal prion and an infectious one?

40. **Fact Check** List two similarities and two differences between hemoglobin and myoglobin.

41. **Fact Check** What are the two critical amino acids near the heme group in both myoglobin and hemoglobin?

42. **Fact Check** What is the highest level of organization in myoglobin? In hemoglobin?

43. **Fact Check** Suggest a way in which the difference between the functions of hemoglobin and myoglobin is reflected in the shapes of their respective oxygen-binding curves.

44. **Fact Check** Describe the Bohr effect.

45. **Fact Check** Describe the effect of 2,3-*bis*phosphoglycerate on the binding of oxygen by hemoglobin.

46. **Fact Check** How does the oxygen-binding curve of fetal hemoglobin differ from that of adult hemoglobin?

47. **Fact Check** What is the critical amino acid difference between the β-chain and the γ-chain of hemoglobin?

48. **Thought Question** In oxygenated hemoglobin, $pK_a = 6.6$ for the histidines at position 146 on the β-chain. In deoxygenated hemoglobin, the pK_a of these residues is 8.2. How can this piece of information be correlated with the Bohr effect?

49. **Thought Question** You are studying with a friend who is in the process of describing the Bohr effect. She tells you that, in the lungs, hemoglobin binds oxygen and releases hydrogen ion; as a result, the pH increases. She goes on to say that, in actively metabolizing muscle tissue, hemoglobin releases oxygen and binds hydrogen ion and, as a result, the pH decreases. Do you agree with her reasoning? Why or why not?

50. **Thought Question** How does the difference between the β-chain and the γ-chain of hemoglobin explain the differences in oxygen binding between Hb A and Hb F?

51. **Thought Question** Suggest a reason for the observation that persons with sickle-cell trait sometimes have breathing problems during high-altitude flights.

52. **Thought Question** Does a fetus homozygous for Hb S have normal Hb F?

53. **Thought Question** Why is fetal Hb essential for the survival of placental animals?

54. **Thought Question** Why might you expect to find some Hb F in adults who are afflicted with sickle-cell anemia?

55. **Thought Question** When deoxyhemoglobin was first isolated in crystalline form, the researcher who did so noted that the crystals changed color from purple to red and also changed shape as he observed them under a microscope. What is happening on the molecular level? *Hint:* The crystals were mounted on a microscope slide with a *loosely* fitting cover slip.

Biochemistry ⊜ Now™
Assess your understanding of this chapter's topics with additional quizzing and tutorials at **http://now.brookscole.com/campbell5**

Annotated Bibliography

Ferguson, N. M., A. C. Ghan, C. A. Donnelly, T. J. Hagenaars, and R. M. Anderson. Estimating the Human Health Risk from Possible BSE Infection of the British Sheep Flock. *Nature* **415**, 420–424 (2002). [The title says it all.]

Gibbons, A., and M. Hoffman. New 3-D Protein Structures Revealed. *Science* **253**, 382–383 (1991). [Examples of the use of X-ray crystallography to determine protein structure.]

Gierasch, L. M., and J. King, eds. *Protein Folding: Deciphering the Second Half of the Genetic Code.* Waldorf, Md.: AAAS Books, 1990. [A collection of articles on recent discoveries about the processes involved in protein folding. Experimental methods for studying protein folding are emphasized.]

Hall, S. Protein Images Update Natural History. *Science* **267**, 620–624 (1995). [Combining X-ray crystallography and computer software to produce images of protein structure.]

Hauptmann, H. The Direct Methods of X-ray Crystallography. *Science* **233**, 178–183 (1986). [A discussion of improvements in methods of doing the calculations involved in determining protein structure; based on a Nobel Prize address. This article should be read in connection with the one by Karle, and it provides an interesting contrast with the articles by Perutz, both of which describe early milestones in protein crystallography.]

Helfand, S. L. Chaperones Take Flight. *Science* **295**, 809–810 (2002). [An article about using chaperones to combat Parkinson's disease.]

Holm, L., and C. Sander. Mapping the Protein Universe. *Science* **273**, 595–602 (1996). [An article on searching databases on protein structure to predict the three-dimensional structure of proteins. Part of a series of articles on computers in biology.]

Karle, J. Phase Information from Intensity Data. *Science* **232**, 837–843 (1986). [A Nobel Prize address on the subject of X-ray crystallography. See remarks on the article by Hauptmann.]

Kasha, K. J. Biotechnology and the World Food Supply. *Genome* **42** (4), 642–645 (1999). [Proteins are frequently in short supply in the diet of many people in the world, but biotechnology can help improve the situation.]

Mitten, D. D., R. MacDonald, and D. Klonus. Regulation of Foods Derived from Genetically Engineered Crops. *Curr. Opin. Biotechnol.* **10**, 298–302 (1999). [How genetic engineering can affect the food supply, especially that of proteins.]

O'Quinn, P. R., J. L. Nelssen, R. D. Goodband, D. A. Knabe, J. C. Woodworth, M. D. Tokach, and T. T. Lohrmann. Nutritional Value of a Genetically Improved High-Lysine, High-Oil Corn for Young Pigs. *J. Anim. Sci.* **78** (8), 2144–2149 (2000). [The availability of amino acids affects the proteins formed.]

Peretz, D., R. A. Williamson, K. Kaneko, J. Vergara, E. Leclerc, G. Schmitt-Ulms, I. R. Mehlhorn, G. Legname, M. R. Wormald, P. M. Rudd, R. A. Dwek, D. R. Burton, and S. B. Prusiner. Antibodies Inhibit Prion Propagation and Clear Cell Cultures of Prion Infectivity. *Nature* **412**, 739–742 (2001). [Description of a possible treatment for prion diseases.]

Perutz, M. The Hemoglobin Molecule. *Sci. Amer.* **211** (5), 64–76 (1964). [A description of work that led to a Nobel Prize.]

Perutz, M. The Hemoglobin Molecule and Respiratory Transport. *Sci. Amer.* **239** (6), 92–125 (1978). [The relationship between molecular structure and cooperative binding of oxygen.]

Ruibal-Mendieta, N. L., and F. A. Lints. Novel and Transgenic Food Crops: Overview of Scientific versus Public Perception. *Transgenic Res.* **7** (5), 379–386 (1998). [A practical application of protein structure research.]

Yam, P. Mad Cow Disease's Human Toll. *Sci. Amer.* **284** (5), 12–13 (2001). [An overview of mad-cow disease and how it has crossed over to infect people.]

Protein Purification and Characterization Techniques

Because a cell contains thousands of different protein molecules, the task of separating them and determining the structure of a single protein is exceedingly difficult. There are many techniques for purifying and characterizing a protein, ranging from strategies for determining such physical characteristics as molecular weight, isoelectric point, and number of subunits to discovering the number and type of its constituent amino acids and elucidating its complete amino acid sequence. When a protein has been degraded to its amino acids, they can be identified by chromatography according to their charge and polarity. The amino acids at the ends of a protein can be established by chemical labeling. The whole chain can be degraded by specific cleavage to give related peptide fragments. Each peptide can then be degraded one amino acid at a time to discover its sequence. In a final step of structure determination, a complete protein can be subjected to X-ray diffraction analysis to determine its three-dimensional conformation. However, the protein must first be purified, by such techniques as column chromatography and electrophoresis, and then crystallized.

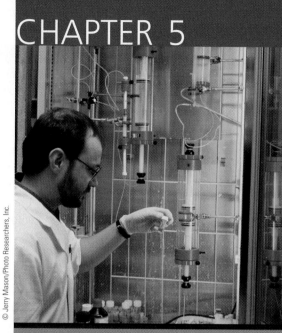

© Jerry Mason/Photo Researchers, Inc.

Column chromatography is widely used in working with proteins.

Critical Questions

5.1 How Do We Extract Pure Proteins from Cells?

5.2 What Is Column Chromatography?

5.3 What Is Electrophoresis?

5.4 How Do We Determine the Primary Structure of a Protein?

5.1 How Do We Extract Pure Proteins from Cells?

Many different proteins exist in a single cell. A detailed study of the properties of any one protein requires a homogeneous sample consisting of only one kind of molecule. The separation and isolation, or purification, of proteins constitutes an essential first step to further experimentation. In general, separation techniques focus on size, charge, and polarity—the sources of differences between molecules. Many techniques are performed to eliminate contaminants and to arrive at a pure sample of the protein of interest. As the purification steps are followed, we make a table of the recovery and purity of the protein to gauge our success. Table 5.1 shows a typical purification for an enzyme. The **percent recovery** column tracks how much of the protein of interest has been retained at each step. This number usually drops steadily during the purification; however, we hope that, by the time the protein is pure, sufficient product will be left for study and characterization. The **fold purification** column compares the purity of the protein at each step, and this value should go up if the purification is successful.

Isolation of Proteins from Cells

Before the real purification steps can begin, the protein must be released from the cells and subcellular organelles. The first step is called **homogenization** and involves the breaking open of the cells. This can be done with a wide variety of techniques. The simplest approach is grinding the tissue in a blender with a suitable buffer. The cells are broken open, releasing soluble proteins. This process also breaks many of the subcellular organelles, such as mitochondria, peroxisomes, and endoplasmic reticulum. A gentler technique

Table 5.1

Example of a Protein Purification Scheme:
Purification of the Enzyme Xanthine Dehydrogenase from a Fungus

Fraction	Volume (mL)	Total Protein (mg)	Total Activity	Specific Activity	Percent Recovery
1. Crude extract	3,800	22,800	2,460	0.108	100
2. Salt precipitate	165	2,800	1,190	0.425	48
3. Ion-exchange chromatography	65	100	720	7.2	29
4. Molecular-sieve chromatography	40	14.5	555	38.3	23
5. Immunoaffinity chromatography	6	1.8	275	152.108	11

is to use a Potter–Elvejhem homogenizer, a thick-walled test tube through which a tight-fitting plunger is passed. The squeezing of the homogenate around the plunger breaks open cells, but it leaves many of the organelles intact. Another technique, called sonication, involves using sound waves to break open the cells. Cells can also be ruptured by cycles of freezing and thawing. If the protein of interest is solidly attached to a membrane, detergents may have to be added to detach the proteins.

After the cells are homogenized, they are subjected to **differential centrifugation.** Spinning the sample at 600 times the force of gravity (600 × *g*) will result in a pellet of unbroken cells and nuclei. If the protein of interest is not found in the nuclei, this precipitate is discarded. The supernatant can then be centrifuged at higher speed, such as 15,000 × *g*, to bring down the mitochondria. Further centrifugation at 100,000 × *g* brings down the microsomal fraction, consisting of ribosomes and membrane fragments. If the protein of interest is soluble, the supernatant from this spin will be collected and will already be partially purified because the nuclei and mitochondria will have been removed. Figure 5.1 shows a typical separation via differential centrifugation.

After the proteins are solubilized, they are often subjected to a crude purification based on solubility. Ammonium sulfate is the most common reagent to use at this step, and this procedure is referred to as **salting out.** Proteins have varying solubilities in polar and ionic compounds. Proteins remain soluble due to their interactions with water. When ammonium sulfate is added to a protein solution, some of the water is taken away from the protein to make ion–dipole bonds with the salts. With less water available to hydrate the proteins, they begin to interact with each other through hydrophobic bonds. At a defined amount of ammonium sulfate, a precipitate that contains contaminating proteins forms. These proteins are centrifuged down and discarded. Then more salt is added, and a different set of proteins, which usually contains the protein of interest, will precipitate. This precipitate is collected by centrifugation and saved. The quantity of ammonium sulfate is usually measured in comparison with a 100% saturated solution. A common procedure involves bringing the solution to around 40% saturation and then spinning down the precipitate that forms. Next, more ammonium sulfate is added to the supernatant, often to a level of 60% to 70% saturation. The precipitate that forms often contains the protein of interest. These preliminary techniques will not generally give a sample that is very pure, but they serve the important task of preparing the crude homogenate for the more effective procedures that follow.

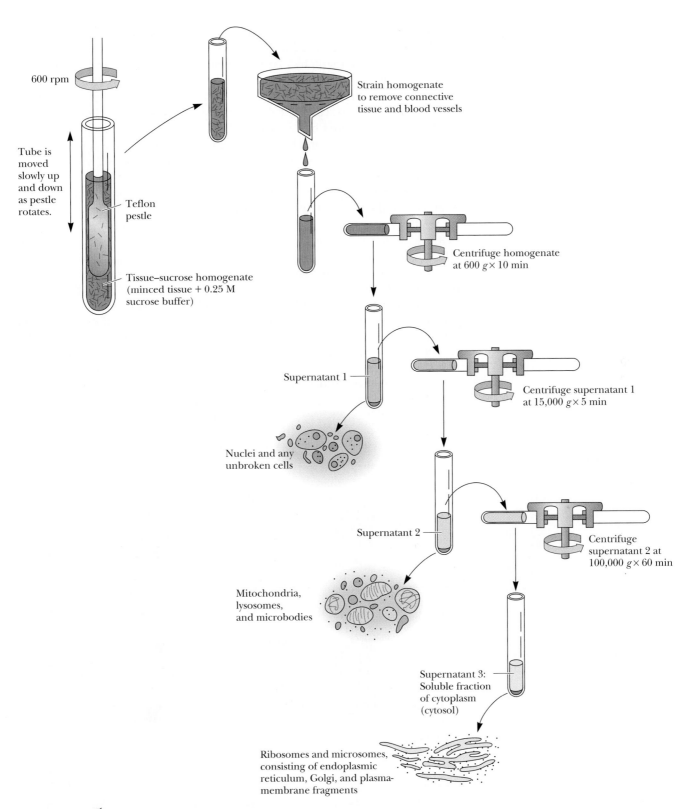

600 rpm

Tube is moved slowly up and down as pestle rotates.

Teflon pestle

Tissue–sucrose homogenate (minced tissue + 0.25 M sucrose buffer)

Strain homogenate to remove connective tissue and blood vessels

Centrifuge homogenate at 600 g × 10 min

Supernatant 1

Nuclei and any unbroken cells

Centrifuge supernatant 1 at 15,000 g × 5 min

Supernatant 2

Mitochondria, lysosomes, and microbodies

Centrifuge supernatant 2 at 100,000 g × 60 min

Supernatant 3: Soluble fraction of cytoplasm (cytosol)

Ribosomes and microsomes, consisting of endoplasmic reticulum, Golgi, and plasma-membrane fragments

Biochemistry&Now™ ▲ **ACTIVE FIGURE 5.1** Differential centrifugation is used to separate cell components. As a cell homogenate is subjected to increasing g forces, different cell components end up in the pellet. **Watch this Active Figure at http://now.brookscole.com/campbell5**

5.2 | What Is Column Chromatography?

The word "chromatography" comes from the Greek *chroma,* "color," and *graphein,* "to write"; the technique was first used around the beginning of the 20th century to separate plant pigments with easily visible colors. It has long since been possible to separate colorless compounds, as long as there are methods for detecting them. Chromatography is based on the fact that different compounds can distribute themselves to varying extents between different phases, or separable portions of matter. One phase is the **stationary phase,** and the other is the **mobile phase.** The mobile phase flows over the stationary material and carries the sample to be separated along with it. The components of the sample interact with the stationary phase to different extents. Some components interact relatively strongly with the stationary phase and are therefore carried along more slowly by the mobile phase than are those that interact less strongly. The differing mobilities of the components are the basis of the separation.

Many chromatographic techniques used for research on proteins are forms of **column chromatography,** in which the material that makes up the stationary phase is packed in a column. The sample is a small volume of concentrated solution that is applied to the top of the column; the mobile phase, called the *eluent,* is passed through the column. The sample is diluted by the eluent, and the separation process also increases the volume occupied by the sample. In a successful experiment, the entire sample eventually comes off the column. Figure 5.2 diagrams an example of column chromatography.

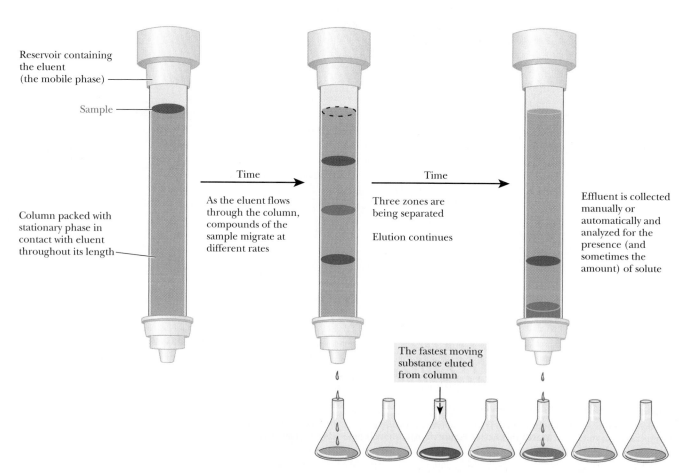

Reservoir containing the eluent (the mobile phase)

Sample

Column packed with stationary phase in contact with eluent throughout its length

Time

As the eluent flows through the column, compounds of the sample migrate at different rates

Time

Three zones are being separated

Elution continues

The fastest moving substance eluted from column

Effluent is collected manually or automatically and analyzed for the presence (and sometimes the amount) of solute

▲ **FIGURE 5.2** An example of column chromatography. A sample containing several components is applied to the column. The various components travel at different rates and can be collected individually.

Size-exclusion chromatography, also called **gel-filtration chromatography,** separates molecules on the basis of size, making it a useful way to sort proteins of varied molecular weights. It is a form of column chromatography in which the stationary phase consists of cross-linked gel particles. The gel particles are usually in bead form and consist of one of two kinds of polymers. The first is a carbohydrate polymer, such as **dextran** or **agarose;** these two polymers are often referred to by the trade names Sephadex® and Sepharose™, respectively (Figure 5.3). The second is based on **polyacrylamide** (Figure 5.4), which is sold under the trade name Bio-Gel®. The cross-linked structure of these polymers produces pores in the material. The extent of cross-linking can be controlled to select a desired pore size. When a sample is applied to the column, smaller molecules, which are able to enter the pores, tend to be delayed in their progress down the column, unlike the larger molecules. As a result, the larger molecules are eluted first, followed later by the smaller ones, after having escaped from the pores. Molecular-sieve chromatography is represented schematically in Figure 5.5. The advantages of this type of chromatography are (1) its convenience as a way to separate molecules on the basis of size and (2) the fact that it can be used to estimate molecular weight by comparing the sample with a set of standards. Each type of gel used has a specific range of sizes that will separate linearly with the log of the molecular weight. Each gel also has an exclusion limit, a size of protein that is too large to fit inside the pores. All proteins that size or larger will elute first and simultaneously.

Affinity chromatography uses the specific binding properties of many proteins. It is another form of column chromatography with a polymeric material used as the stationary phase. The distinguishing feature of affinity chromatography is that the polymer is covalently linked to some compound, called a *ligand,* that binds specifically to the desired protein (Figure 5.6). The other proteins in the sample do not bind to the column and can easily be eluted with buffer, while the bound protein remains on the column. The bound protein can then be eluted from the column by adding high concentrations of the ligand in soluble form, thus competing for the binding of the protein with

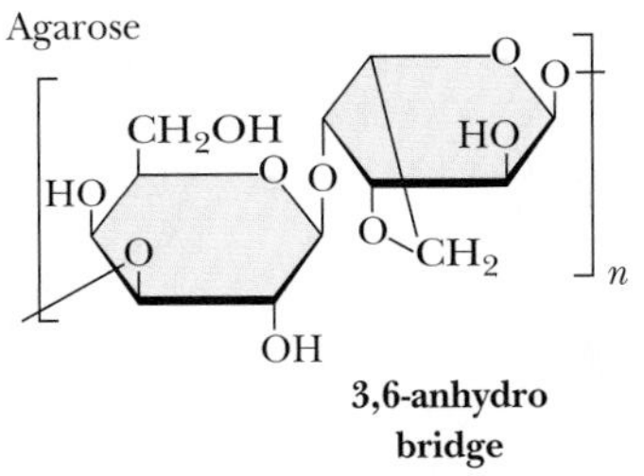

▲ FIGURE 5.3 The repeating disaccharide unit of agarose, which is used for column chromatography.

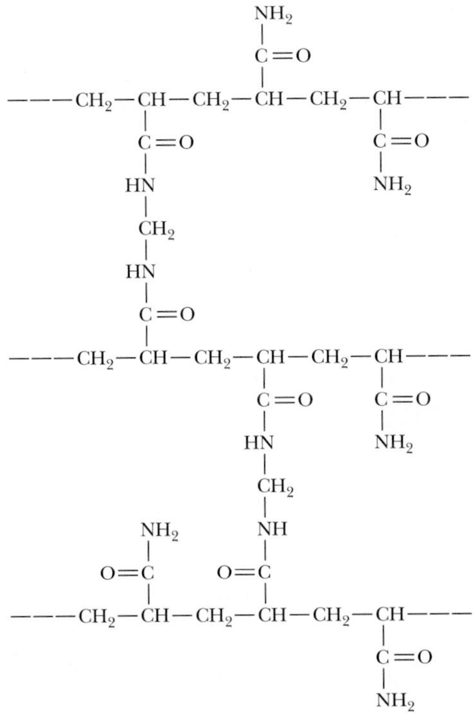

◀ FIGURE 5.4 The structure of cross-linked polyacrylamide, a polymer used in column chromatography.

(a)

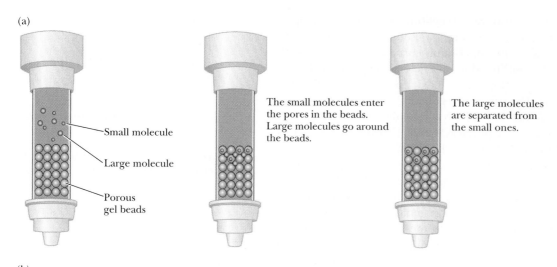

The small molecules enter the pores in the beads. Large molecules go around the beads.

The large molecules are separated from the small ones.

Small molecule

Large molecule

Porous gel beads

(b)

Protein concentration

Elution profile of a large macromolecule

A smaller macromolecule

V_o V_e V_t

Volume (mL) →

◀ **FIGURE 5.5** Gel-filtration chromatography. (a) Larger molecules are excluded from the gel and move more quickly through the column. Small molecules have access to the interior of the gel beads, so they take a longer time to elute. (b) V_0 is the void volume, the volume of elution for a molecule excluded from the gel bead. V_e is the elution volume for a particular molecule that can enter the bead. V_t is the total volume, the elution volume for a very small molecule that enters the bead unhindered.

Column with substance S covalently bonded to supporting material

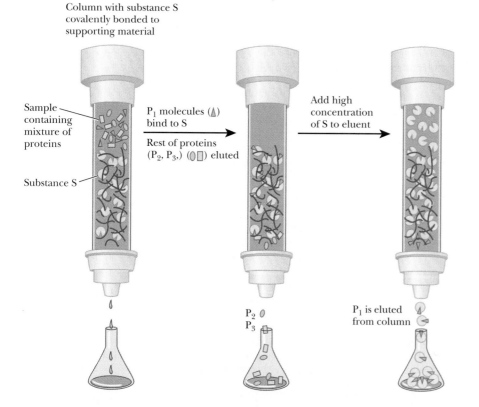

Sample containing mixture of proteins

Substance S

P_1 molecules (▲) bind to S

Rest of proteins (P_2, P_3,) (◯▢) eluted

Add high concentration of S to eluent

P_2
P_3

P_1 is eluted from column

◀ **FIGURE 5.6** The principle of affinity chromatography. In a mixture of proteins, only one (designated P_1) will bind to a substance (S) called the substrate. The substrate is attached to the column matrix. Once the other proteins (P_2 and P_3) have been washed out, P_1 can be eluted, either by adding a solution of high salt concentration or by adding free S.

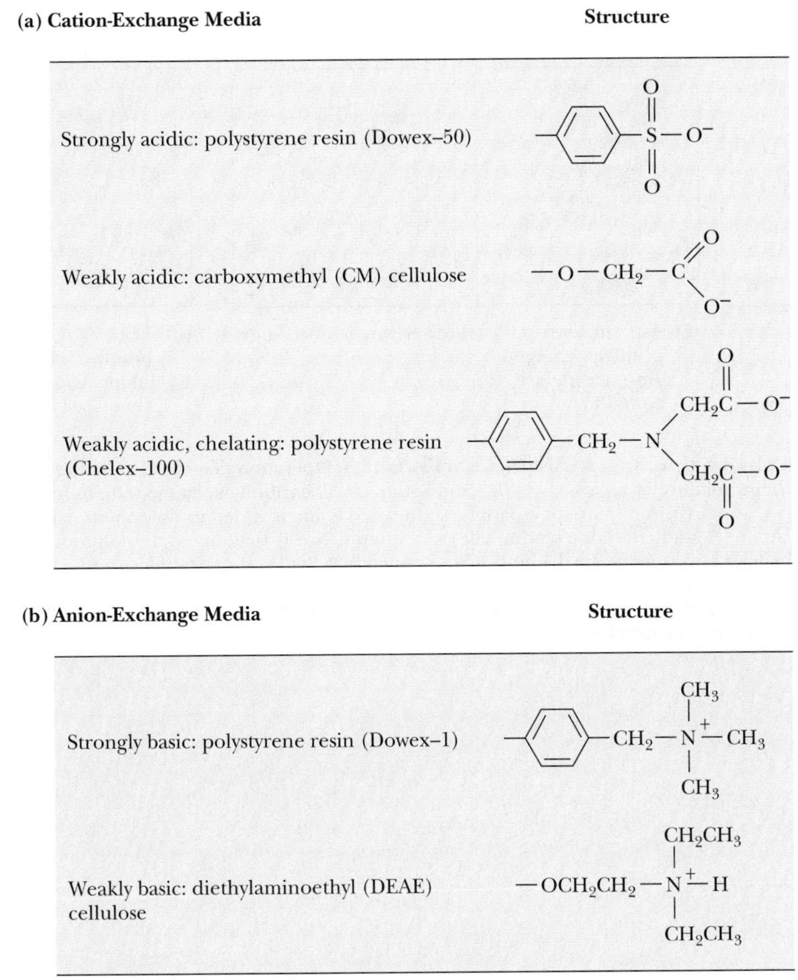

(a) Cation-Exchange Media **Structure**

Strongly acidic: polystyrene resin (Dowex–50)

Weakly acidic: carboxymethyl (CM) cellulose

Weakly acidic, chelating: polystyrene resin (Chelex–100)

(b) Anion-Exchange Media **Structure**

Strongly basic: polystyrene resin (Dowex–1)

Weakly basic: diethylaminoethyl (DEAE) cellulose

◀ **FIGURE 5.7** (a) Cation-exchange resins and (b) anion-exchange resins commonly used for biochemical separations.

the stationary phase. The protein binds to the ligand in the mobile phase and is recovered from the column. This protein–ligand interaction can also be disrupted with a change in pH or ionic strength. Affinity chromatography is a convenient separation method and has the advantage of producing very pure proteins. The Biochemical Connections box in Chapter 13 describes an interesting way in which affinity chromatography can be combined with molecular biological techniques to offer a one-step purification of a protein.

Ion-exchange chromatography is logistically similar to affinity chromatography. Both use a column resin that binds the protein of interest. With ion-exchange chromatography, however, the interaction is less specific and is based on net charge. An ion-exchange resin will have a ligand with a positive charge or a negative charge. A negatively charged resin is a **cation exchanger,** and a positively charged one is an **anion exchanger.** Figure 5.7 shows some typical ion-exchange ligands. Figure 5.8 illustrates their principle of operation with three amino acids of different charge. Figure 5.9 shows how cation-exchange chromatography would separate proteins. The column is initially equilibrated with a buffer of suitable pH and ionic strength. The exchange resin is bound to counterions. A cation-exchange resin is usually bound to Na^+ or K^+ ions, and an anion exchanger is usually bound to Cl^- ions. A mixture of proteins is loaded on the column and allowed to flow through it. Those proteins that have a net charge opposite to that of the exchanger will stick to the column, *exchanging* places with the bound counterions. Those proteins that have no net charge or have the same charge as the exchanger will elute. After all the nonbinding proteins are eluted, the eluent will be changed

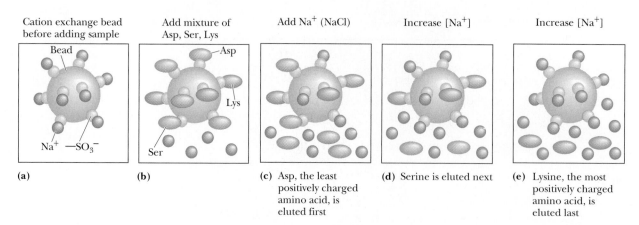

Cation exchange bead before adding sample (a)

Bead

Na^+ —SO_3^-

Add mixture of Asp, Ser, Lys (b)

Asp

Lys

Ser

Add Na^+ (NaCl)

(c) Asp, the least positively charged amino acid, is eluted first

Increase [Na^+]

(d) Serine is eluted next

Increase [Na^+]

(e) Lysine, the most positively charged amino acid, is eluted last

Biochemistry ⓢ Now™ ▲ **ANIMATED FIGURE 5.8** Operation of a cation-exchange column, separating a mixture of aspartate, serine, and lysine. (a) The cation-exchange resin in the beginning, Na^+ form. (b) A mixture of aspartate, serine, and lysine is added to the column containing the resin. (c) A gradient of the eluting salt (for example, NaCl) is added to the column. Aspartate, the least positively charged amino acid, is eluted first. (d) As the salt concentration increases, serine is eluted. (e) As the salt concentration is increased further, lysine, the most positively charged of the three amino acids, is eluted last. **See this figure animated at http://now .brookscole.com/campbell5**

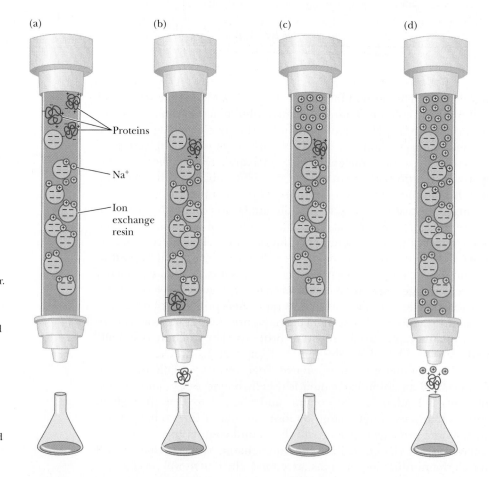

(a)   (b)   (c)   (d)

Proteins

Na^+

Ion exchange resin

▶ **FIGURE 5.9** Ion-exchange chromatography using a cation exchanger. (a) At the beginning of the separation, various proteins are applied to the column. The column resin is bound to Na^+ counterions (small red spheres). (b) Proteins that have no net charge or a net negative charge pass through the column. Proteins that have a net positive charge stick to the column, displacing the Na^+. (c) An excess of Na^+ ion is then added to the column. (d) The Na^+ ions outcompete the bound proteins for the binding sites on the resin, and the proteins elute.

either to a buffer that has a pH that will remove the charge on the bound proteins or to one with a higher salt concentration. The latter will outcompete the bound proteins for the limited binding space on the column. The once-bound molecules will then elute, having been separated from many of the contaminating ones.

5.3 | What Is Electrophoresis?

Electrophoresis is based on the motion of charged particles in an electric field toward an electrode of opposite charge. Macromolecules have differing mobilities based on their charge, shape, and size. Although many supporting media have been used for electrophoresis, including paper and liquid, the most common support is a polymer of agarose or acrylamide that is similar to those used for column chromatography. A sample is applied to wells that are formed in the supporting medium. An electric current is passed through the medium at a controlled voltage to achieve the desired separation (Figure 5.10). After the proteins are separated on the gel, the gel is stained to reveal the protein locations, as shown in Figure 5.11.

Agarose-based gels are most often used to separate nucleic acids and will be discussed in Chapter 13. For proteins, the most common electrophoretic support is polyacrylamide (Figure 5.4). The polyacrylamide gel is prepared and cast as a continuous cross-linked matrix, rather than being produced in the bead form employed in column chromatography. In one variation of polyacrylamide-gel electrophoresis, the protein sample is treated with the detergent sodium dodecyl sulfate (SDS) before it is applied to the gel. The structure of SDS is $CH_3(CH_2)_{10}CH_2OSO_3^-Na^+$. The anion binds strongly to proteins via nonspecific adsorption. The larger the protein, the more of the anion it will adsorb. SDS completely denatures proteins, breaking all the non-covalent interactions that determine tertiary and quaternary structure. This means that multisubunit proteins can be analyzed as the component polypeptide chains. All the proteins in a sample have a negative charge as a result of adsorption of the anionic SO_3^-. The proteins will also have roughly the same shape, which will be a random coil. In **SDS–polyacrylamide-gel electrophoresis (SDS–PAGE),** the acrylamide offers more resistance to large molecules than to small molecules. Because the shape and charge are approximately the same for all the proteins in the sample, the size of the protein becomes the determining factor in the separation: small proteins move faster than large ones. Like molecular-sieve chromatography, SDS–PAGE can be used to estimate the molecular weights of proteins by comparing the sample with standard samples. For most proteins, the log of the molecular weight is linearly related to its mobility on SDS–PAGE, as shown in Figure 5.12.

Isoelectric focusing is another variation of gel electrophoresis. Since different proteins have different titratable groups, they also have different isoelectric points. Recall (Section 3.3) that the isoelectric pH (pI) is the pH at which a protein (or amino acid or peptide) has no net charge. At the pI, the number of positive charges exactly balances the number of negative charges. In an isoelectric focusing experiment, the gel is prepared with a pH gradient that parallels the electric-field gradient. As proteins migrate through the gel under the influence of the electric field, they encounter regions of different pH, so the charge on the protein changes. Eventually each protein reaches the point at which it has no net charge—its isoelectric point—and no longer migrates. Each protein remains at the position on the gel corresponding to its pI, allowing for an effective method of separation.

An ingenious combination, known as two-dimension gel electrophoresis (2-D gels), allows for enhanced separation by using isoelectric focusing in one dimension and SDS–PAGE run at 90° to the first (Figure 5.13).

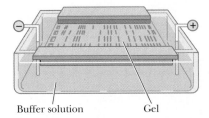

Buffer solution   Gel

▲ **FIGURE 5.10** The experimental setup for gel electrophoresis. The samples are placed on the left side of the gel. When the current is applied, the negatively charged molecules migrate toward the positive electrode.

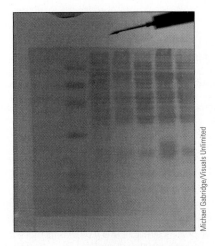

Michael Gabridge/Visuals Unlimited

▲ **FIGURE 5.11** Separation of proteins by gel electrophoresis. Each band seen in the gel represents a different protein. In the SDS–PAGE technique, the sample is treated with detergent before being applied to the gel. In isoelectric focusing, a pH gradient runs the length of the gel.

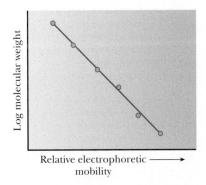

Relative electrophoretic mobility

▲ **FIGURE 5.12** A plot of the log of the molecular weight versus the relative electrophoretic mobility.

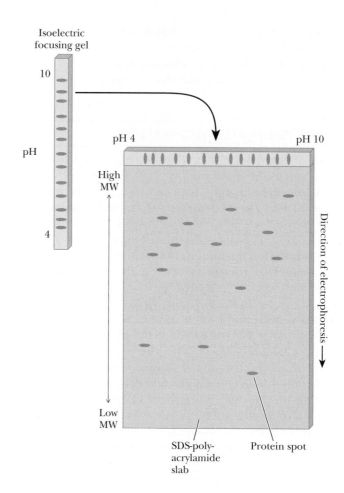

▶ **FIGURE 5.13** Two-dimensional electrophoresis. A mixture of proteins is separated by isoelectric focusing in one direction. The focused proteins are then run using SDS–PAGE perpendicular to the direction of the isoelectric focusing. Thus the bands that appear on the gel have been separated first by charge and then by size.

Essential Information

The primary structure of a protein is its sequence of amino acids. The sequence is determined by cleaving the protein into smaller peptides, verifying the sequence of the individual peptides, and combining overlapping peptide sequences to obtain that of the protein.

5.4 | How Do We Determine the Primary Structure of a Protein?

Determining the sequence of amino acids in a protein is a routine, but not trivial, operation in classical biochemistry. Its several parts must be carried out carefully to obtain accurate results (Figure 5.14).

Step 1 in determining the primary structure of a protein is to establish which amino acids are present and in what proportions. Breaking a protein down to its component amino acids is relatively easy: Heat a solution of the protein in acid, usually 6 M HCl, at 100°C to 110°C for 12 to 36 hours to hydrolyze the peptide bonds. Separation and identification of the products are somewhat more difficult and are best done by an amino acid analyzer. This automated instrument gives both qualitative information about the identities of the amino acids present and quantitative information about the relative amounts of those amino acids. Not only does it analyze amino acids, but it also allows informed decisions to be made about which procedures to choose later in the sequencing (see Steps 3 and 4 in Figure 5.14). An amino acid analyzer separates the mixture of amino acids either by ion-exchange chromatography or by **high-performance liquid chromatography (HPLC),** a chromatographic technique that allows high-resolution separations of many amino acids in a short time frame. Figure 5.15 shows a typical result of amino acid separation with this technique.

In Step 2, the identities of the N-terminal and C-terminal amino acids in a protein sequence are determined. This procedure is becoming less and less

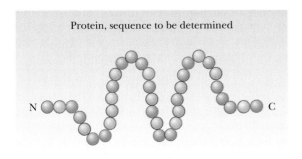

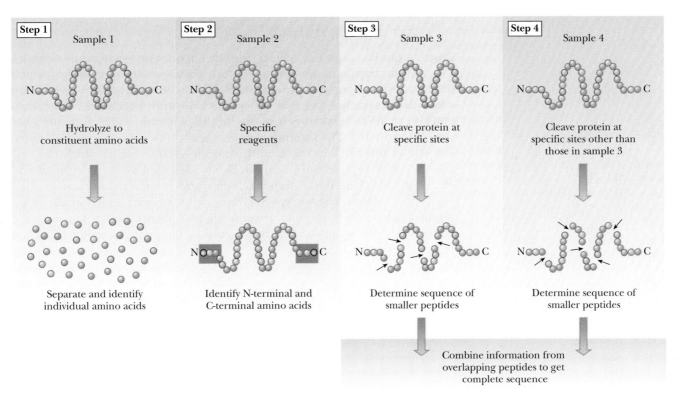

▲ **FIGURE 5.14** The strategy for determining the primary structure of a given protein. The amino acid sequence can be determined by four different analyses performed on four separate samples of the same protein.

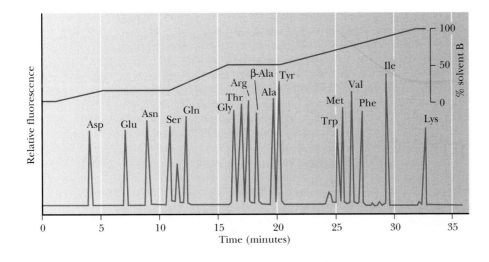

◀ **FIGURE 5.15** HPLC chromatogram of amino acid separation.

necessary as the sequencing of individual peptides improves, but it can be used to check whether a protein consists of one or two polypeptide chains.

In Steps 3 and 4, the amino acid sequence is determined. Automated instruments can perform a stepwise modification starting from the N-terminal end, followed by cleavage of each amino acid in the sequence and the subsequent identification of each modified amino acid as it is removed. The process (the Edman degradation method) becomes more difficult as the number of amino acids increases. In most proteins, the chain is more than 100 residues long. For sequencing, it is usually necessary to break a long polypeptide chain into fragments, ranging from 20 to 50 residues.

Cleavage of the Protein into Peptides

Proteins can be cleaved at specific sites by enzymes or by chemical reagents. The enzyme **trypsin** cleaves peptide bonds preferentially at amino acids that have positively charged R groups, such as lysine and arginine. The cleavage takes place in such a way that the amino acid with the charged side chain ends up at the C-terminal end of one of the peptides produced by the reaction (Figure 5.16). The C-terminal amino acid of the original protein can be any one of the 20 amino acids and is not necessarily one at which cleavage takes place. A peptide can be automatically identified as the C-terminal end of the original chain if its C-terminal amino acid is not a site of cleavage.

Another enzyme, **chymotrypsin,** cleaves peptide bonds preferentially at the aromatic amino acids: tyrosine, tryptophan, and phenylalanine. The aromatic amino acid ends up at the C-terminal ends of the peptides produced by the reaction (Figure 5.17).

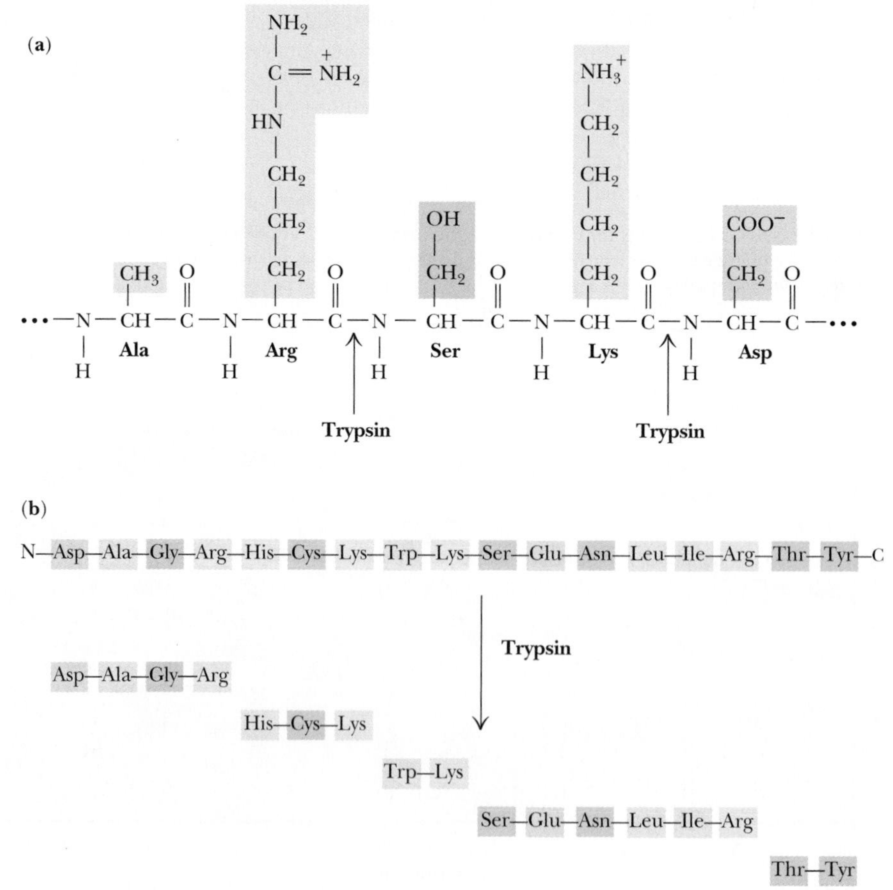

Biochemistry ⓔ Now™ ANIMATED FIGURE 5.16
(a) Trypsin is a proteolytic enzyme, or protease, that specifically cleaves only those peptide bonds in which arginine or lysine contributes the carbonyl function. (b) The products of the reaction are a mixture of peptide fragments with C-terminal Arg or Lys residues and a single peptide derived from the polypeptide's C-terminal end. **See this figure animated at http://now.brookscole.com/campbell5**

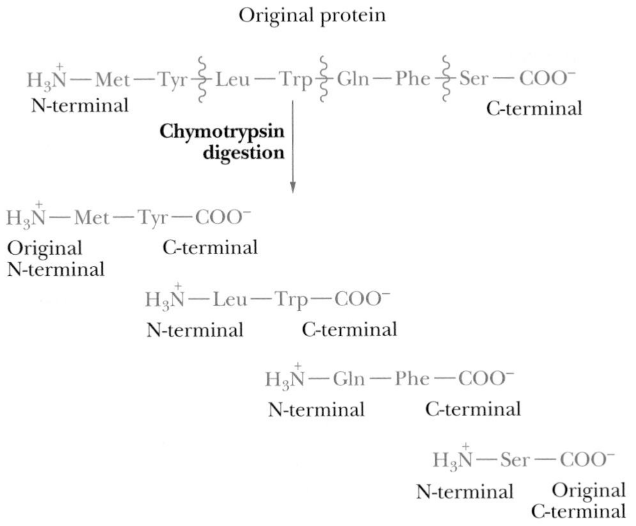

Original protein

$$H_3\overset{+}{N}—Met—Tyr\overset{\xi}{\xi}Leu—Trp\overset{\xi}{\xi}Gln—Phe\overset{\xi}{\xi}Ser—COO^-$$

N-terminal   C-terminal

Chymotrypsin digestion

$$H_3\overset{+}{N}—Met—Tyr—COO^-$$

Original   C-terminal
N-terminal

$$H_3\overset{+}{N}—Leu—Trp—COO^-$$

N-terminal   C-terminal

$$H_3\overset{+}{N}—Gln—Phe—COO^-$$

N-terminal   C-terminal

$$H_3\overset{+}{N}—Ser—COO^-$$

N-terminal   Original
C-terminal

◄ **FIGURE 5.17** Cleavage of proteins by enzymes. Chymotrypsin hydrolyzes proteins at aromatic amino acids.

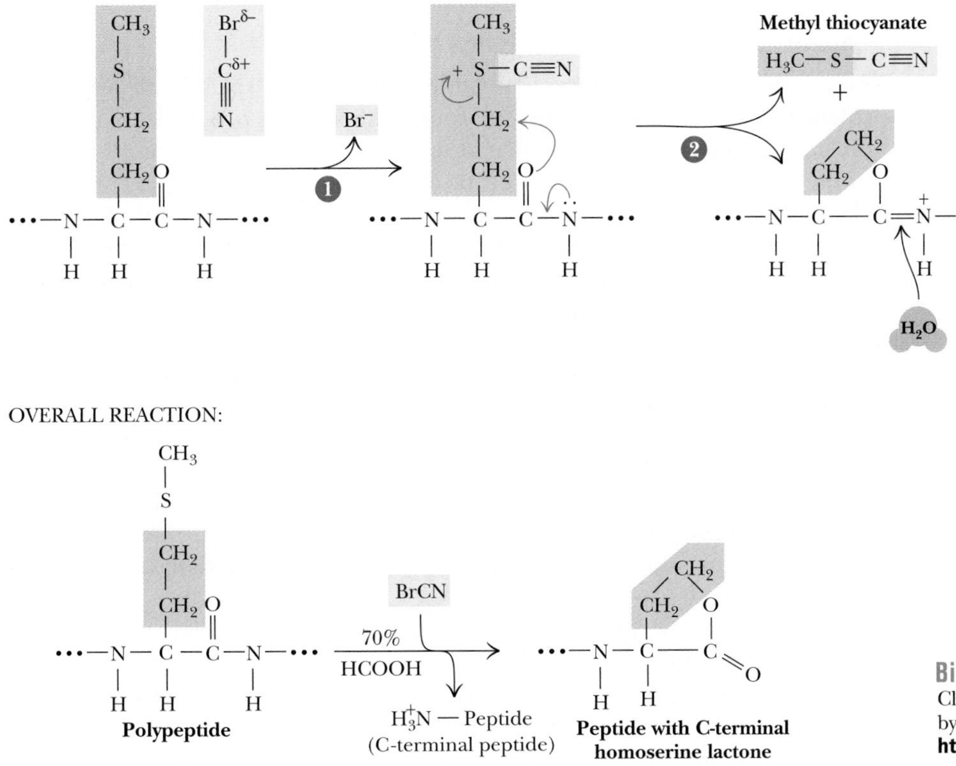

OVERALL REACTION:

Polypeptide   **Peptide with C-terminal homoserine lactone**

$H_3\overset{+}{N}—Peptide$
(C-terminal peptide)

Biochemistry🕭Now™ **ANIMATED FIGURE 5.18** Cleavage of proteins at internal methionine residues by cyanogen bromide. **See this figure animated at http://now.brookscole.com/campbell5**

In the case of the chemical reagent **cyanogen bromide** (CNBr), the sites of cleavage are at internal methionine residues. The sulfur of the methionine reacts with the carbon of the cyanogen bromide to produce a homoserine lactone at the C-terminal end of the fragment (Figure 5.18).

The cleavage of a protein by any of these reagents produces a mixture of peptides, which are then separated by high-performance liquid chromatography. The use of several such reagents on different samples of a protein to be sequenced produces different mixtures. The sequences of a set of peptides produced by one reagent will overlap the sequences produced by another reagent (Figure 5.19). As a result, the peptides can be arranged in the proper order after their own sequences have been determined.

Chymotrypsin	$H_3\overset{+}{N}$—Leu—Asn—Asp—Phe	
Cyanogen bromide	$H_3\overset{+}{N}$—Leu—Asn—Asp—Phe—His—Met	
Chymotrypsin		His—Met—Thr—Met—Ala—Trp
Cyanogen bromide		Thr—Met
Cyanogen bromide		Ala—Trp—Val—Lys—COO⁻
Chymotrypsin		Val—Lys—COO⁻
Overall sequence	$H_3\overset{+}{N}$—Leu—Asn—Asp—Phe—His—Met—Thr—Met—Ala—Trp—Val—Lys—COO⁻	

▲ **FIGURE 5.19** Use of overlapping sequences to determine protein sequence. Partial digestion was effected using chymotrypsin and cyanogen bromide. For clarity, only the original N-terminus and C-terminus of the complete peptide are shown.

Sequencing of Peptides: The Edman Method

The actual sequencing of each peptide produced by specific cleavage of a protein is accomplished by repeated application of a procedure called the **Edman degradation.** The sequence of a peptide containing 10 to 40 residues can be determined by this method in about 30 minutes using as little as 10 picomoles of material, with the range being based on the amount of purified fragment and the complexity of the sequence. For example, proline is more difficult to sequence than serine because of its chemical reactivity. (The amino acid sequences of the individual peptides in Figure 5.19 are determined by the Edman method after the peptides are separated from one another.) The overlapping sequences of peptides produced by different reagents provide the key to solving the puzzle. The alignment of like sequences on different peptides makes deducing the overall sequence possible. The Edman method has become so efficient that it is no longer considered necessary to identify the N-terminal and C-terminal ends of a protein by chemical or enzymatic methods. While interpreting results, however, it is necessary to keep in mind that a protein may consist of more than one polypeptide chain.

In the sequencing of a peptide, the Edman reagent, *phenyl isothiocyanate,* reacts with the peptide's N-terminal residue. The modified amino acid can be cleaved off, *leaving the rest of the peptide intact,* and can be detected as the phenylthiohydantoin derivative of the amino acid. The second amino acid of the original peptide can then be treated in the same way, as can the third. With an automated instrument called a **sequencer** (Figure 5.20), the process is repeated until the whole peptide is sequenced.

Another sequencing method uses the fact that the amino acid sequence of a protein reflects the base sequence of the DNA in the gene that coded for that protein. Using currently available methods, it is sometimes easier to obtain the sequence of the DNA than that of the protein. (See Section 13.11 for a discussion of sequencing methods for nucleic acids.) Using the genetic code (Section 12.2), one can immediately determine the amino acid sequence of the protein. Convenient though this method may be, it does not determine the positions of disulfide bonds or detect amino acids, such as hydroxyproline, that are modified after translation, nor does it take into account the extensive processing that occurs with eukaryotic genomes before the final protein is synthesized (Chapters 11 and 12).

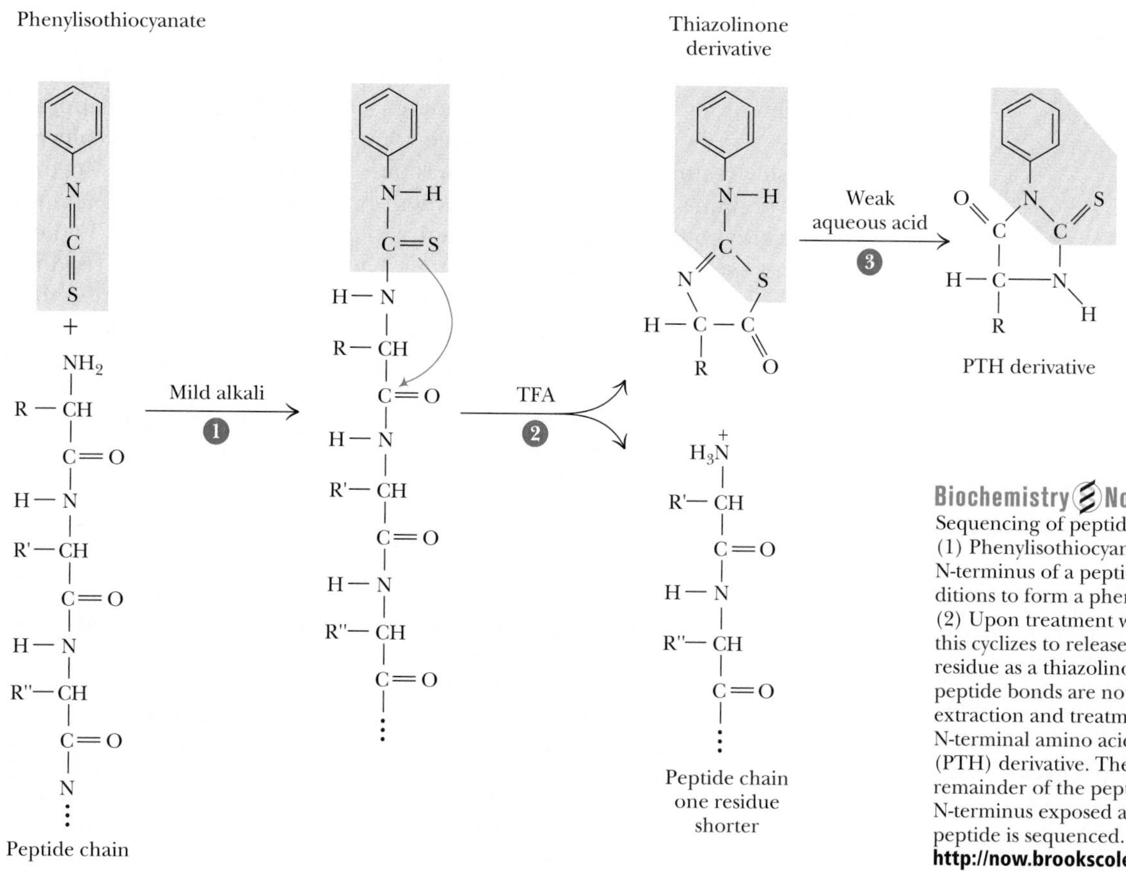

Phenylisothiocyanate

Thiazolinone
derivative

Biochemistry ⒺNow™ **ANIMATED FIGURE 5.20**
Sequencing of peptides by the Edman method.
(1) Phenylisothiocyanate combines with the
N-terminus of a peptide under mildly alkaline conditions to form a phenylthiocarbamoyl substitution.
(2) Upon treatment with TFA (trifluoroacetic acid),
this cyclizes to release the N-terminal amino acid
residue as a thiazolinone derivative, but the other
peptide bonds are not hydrolyzed. (3) Organic
extraction and treatment with aqueous acid yield the
N-terminal amino acid as a phenylthiohydantoin
(PTH) derivative. The process is repeated with the
remainder of the peptide chain to determine the
N-terminus exposed at each stage until the entire
peptide is sequenced. **Watch this Active Figure at
http://now.brookscole.com/campbell**

Practice Session

A solution of a peptide of unknown sequence was divided into two samples. One sample was treated with trypsin, and the other was treated with
chymotrypsin.

The smaller peptides obtained by trypsin treatment had the following
sequences:

Leu—Ser—Tyr—Ala—Ile—Arg
LSYAIR

and

Asp—Gly—Met—Phe—Val—Lys
DGMFVK

The smaller peptides obtained by chymotrypsin treatment had the following sequences:

Val—Lys—Leu—Ser—Tyr
VKLSY

Ala—Ile—Arg
AIR

and

Asp—Gly—Met—Phe
DGMF

Deduce the sequence of the original peptide.

Solution

The key point here is that the fragments produced by treatment with the two different enzymes have overlapping sequences. These overlapping sequences can be compared to give the complete sequence. The results of the trypsin treatment indicate that there are two basic amino acids in the peptide, arginine and lysine. One of them must be the C-terminal amino acid, because no fragment was generated with a C-terminal amino acid other than these two. If there had been an amino acid other than a basic residue at the C-terminal position, trypsin treatment alone would have provided the sequence. Treatment with chymotrypsin gives the information needed. The sequence of the peptide Val—Lys—Leu—Ser—Tyr (VKLSY) indicates that lysine is an internal residue. The complete sequence is Asp—Gly—Met—Phe—Val—Lys—Leu—Ser—Tyr—Ala—Ile—Arg (DGMFVKLSYAIR).

Summary

5.1 How Do We Extract Pure Proteins from Cells?
Disruption of cells is the first step in protein purification. The various parts of cells can be separated by centrifugation. This is a useful step because proteins tend to occur in given organelles. High salt concentrations will precipitate groups of proteins, which are then further separated by chromatography and electrophoresis.

5.2 What Is Column Chromatography?
Two of the most important methods for separating amino acids, peptides, and proteins are chromatography and electrophoresis. The various forms of chromatography rely on differences in charge, polarity, or size of the molecules to be separated, depending on the application.

5.3 What Is Electrophoresis?
In electrophoresis, differences in charge and in size are the criteria for separation. The sieving action of gel slabs is used in conjunction with the charge on proteins to achieve separation. The electrophoretic mobilities of proteins can be used to estimate their molecular weights.

5.4 How Do We Determine the Primary Structure of a Protein?
Determination of the N-terminal and C-terminal amino acids of proteins depends on the use of these separation methods after the ends of the molecule have been chemically labeled. Selective cleavage of the protein into peptides by enzymatic or chemical hydrolysis produces fragments of manageable size for sequencing. The amino acid sequence can then be determined by the Edman method.

Critical Questions to Review

5.1 How Do We Extract Pure Proteins from Cells?

1. **Fact Check** What are the types of homogenization techniques available for solubilizing a protein?

2. **Fact Check** When would you choose to use a Potter–Elvejhem homogenizer instead of a blender?

3. **Fact Check** What is meant by "salting out"? How does it work?

4. **Fact Check** What differences between proteins are responsible for their differential solubility in ammonium sulfate?

5. **Fact Check** How could you isolate mitochondria from liver cells using differential centrifugation?

6. **Fact Check** Can you separate mitochondria from peroxisomes using only differential centrifugation?

7. **Fact Check** Give an example of a scenario in which you could partially isolate a protein with differential centrifugation using only one spin.

8. **Fact Check** Describe a procedure for isolating a protein that is strongly embedded in the mitochondrial membrane.

9. **Thought Question** You are purifying a protein for the first time. You have solubilized it with homogenization in a blender followed by differential centrifugation. You wish to try ammonium sulfate precipitation as the next step. Knowing nothing beforehand about the amount of ammonium sulfate to add, design an experiment to find the proper concentration (% saturation) of ammonium sulfate to use.

10. **Thought Question** If you were to have a protein X, which is a soluble enzyme found inside the peroxisome, and you wished to separate it from a similar protein Y, which is an enzyme found embedded in the mitochondrial membrane, what would be your initial techniques for isolating those proteins?

5.2 What Is Column Chromatography?

11. **Fact Check** What is the basis for the separation of proteins by the following techniques?
    (a) gel-filtration chromatography
    (b) affinity chromatography
    (c) ion-exchange chromatography

12. **Fact Check** What is the order of elution of proteins on a gel-filtration column? Why is this so?

13. **Fact Check** What are two ways that a compound can be eluted from an affinity column? What could be the advantages or disadvantages of each?

14. **Fact Check** What are two ways that a compound can be eluted from an ion-exchange column? What could be the advantages or disadvantages of each?

15. **Fact Check** Why do most people elute bound proteins from an ion-exchange column by raising the salt concentration instead of changing the pH?

16. **Fact Check** What are two types of compounds that make up the resin for column chromatography?

17. **Fact Check** Draw an example of a compound that would serve as a cation exchanger. Draw one for an anion exchanger.

18. **Fact Check** How can gel-filtration chromatography be used to arrive at an estimate of the molecular weight of a protein?

19. **Thought Question** Sephadex® G-75 has an exclusion limit of 80,000 molecular weight for globular proteins. If you tried to use this column material to separate alcohol dehydrogenase (MW 150,000) from β-amylase (MW 200,000), what would happen?

20. **Thought Question** Referring to the question above, could you separate β-amylase from bovine serum albumin (MW 66,000) using this column?

21. **Thought Question** Design an experiment to purify protein X on an anion-exchange column. Protein X has an isoelectric point of 7.0.

22. **Thought Question** Referring to the problem above, how would you purify protein X using ion-exchange chromatography if it turns out the protein is only stable at a pH between 6 and 6.5?

23. **Thought Question** What could be an advantage of using an anion exchange column based on a quaternary amine [i.e., resin–$N^+(CH_2CH_3)_3$] as opposed to a tertiary amine [resin–$NH^+(CH_2CH_3)_2$]?

24. **Thought Question** You wish to separate and purify enzyme A from contaminating enzymes B and C. Enzyme A is found in the matrix of the mitochondria. Enzyme B is embedded in the mitochondrial membrane, and enzyme C is found in the peroxisome. Enzymes A and B have molecular weights of 60,000 daltons. Enzyme C has a molecular weight of 100,000. Enzyme A has a pI of 6.5. Enzymes B and C have pI values of 7.5. Design an experiment to separate enzyme A from the other two enzymes.

25. **Thought Question** An amino acid mixture consisting of lysine, leucine, and glutamic acid is to be separated by ion-exchange chromatography, using a cation-exchange resin at pH 3.5, with the eluting buffer at the same pH. Which of these amino acids will be eluted from the column first? Will any other treatment be needed to elute one of these amino acids from the column?

26. **Thought Question** An amino acid mixture consisting of phenyl-alanine, glycine, and glutamic acid is to be separated by HPLC. The stationary phase is aqueous and the mobile phase is a solvent less polar than water. Which of these amino acids will move the fastest? Which one will move the slowest?

27. **Thought Question** In reverse-phase HPLC, the stationary phase is nonpolar and the mobile phase is a polar solvent at neutral pH. Which of the three amino acids in Question 26 will move fastest on a reverse-phase HPLC column? Which one will move the slowest?

28. **Thought Question** Gel-filtration chromatography is a useful method for removing salts, such as ammonium sulfate, from protein solutions. Describe how such a separation is accomplished.

5.3 What Is Electrophoresis?

29. **Fact Check** What are the physical parameters of a protein that control its migration on electrophoresis?

30. **Fact Check** What are the types of compounds that make up the gels used in electrophoresis?

31. **Fact Check** Of the two principal polymers used in column chromatography and electrophoresis, which one would be most immune to contamination by bacteria and other organisms?

32. **Fact Check** What types of macromolecules are usually separated on agarose electrophoresis gels?

33. **Fact Check** If you had a mixture of proteins with different sizes, shapes, and charges and you separated them with electrophoresis, which proteins would move fastest toward the anode (positive electrode)?

34. **Fact Check** What does SDS–PAGE stand for? What is the benefit of doing SDS–PAGE?

35. **Fact Check** How does the addition of sodium dodecylsulfate to proteins affect the basis of separation on electrophoresis?

36. **Fact Check** Why is the order of separation based on size opposite for gel filtration and gel electrophoresis, even though they often use the same compound to form the matrix?

37. **Fact Check** The figure shown below is from an electrophoresis experiment using SDS–PAGE. The left lane has the following standards: Bovine Serum Albumin (MW 66,000), Ovalbumin (MW 45,000), Glyceraldehyde 3-Phosphate Dehydrogenase (MW 36,000), Carbonic Anhydrase (MW 24,000), and Trypsinogen (MW 20,000). The right lane is an unknown. Calculate the MW of the unknown.

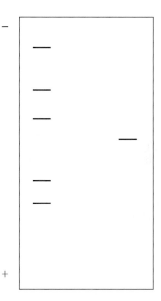

5.4 How Do We Determine the Primary Structure of a Protein?

38. **Fact Check** Why is it no longer considered necessary to determine the N-terminal amino acid of a protein as a separate step?

39. **Fact Check** What useful information might you get if you did determine the N-terminal amino acid as a separate step?

40. **Thought Question** Show by a series of equations (with structures) the first stage of the Edman method applied to a peptide that has leucine as its N-terminal residue.

41. **Thought Question** Why can the Edman degradation not be used effectively with very long peptides? (*Hint:* Think about the stoichiometry of the peptides and the Edman reagent and the percent yield of the organic reactions involving them.)

42. **Thought Question** What would happen during an amino acid sequencing experiment using the Edman degradation if you accidentally added twice as much Edman reagent (on a per-mole basis) as the peptide you were sequencing?

43. **Thought Question** A sample of an unknown peptide was divided into two aliquots. One aliquot was treated with trypsin, and the other with cyanogen bromide. Given the following sequences (N-terminal to C-terminal) of the resulting fragments, deduce the sequence of the original peptide.

Trypsin treatment

Asn—Thr—Trp—Met—Ile—Lys

Gly—Tyr—Met—Gln—Phe

Val—Leu—Gly—Met—Ser—Arg

Cyanogen bromide treatment

Gln—Phe

Val—Leu—Gly—Met

Ile—Lys—Gly—Tyr—Met

Ser—Arg—Asn—Thr—Trp—Met

44. **Thought Question** A sample of a peptide of unknown sequence was treated with trypsin; another sample of the same peptide was treated with chymotrypsin. The sequences (N-terminal to C-terminal) of the smaller peptides produced by trypsin digestion were

Met—Val—Ser—Thr—Lys

Val—Ile—Trp—Thr—Leu—Met—Ile

Leu—Phe—Asn—Glu—Ser—Arg

The sequences of the smaller peptides produced by chymotrypsin digestion were

Asn—Glu—Ser—Arg—Val—Ile—Trp

Thr—Leu—Met—Ile

Met—Val—Ser—Thr—Lys—Leu—Phe

Deduce the sequence of the original peptide.

45. **Thought Question** You are in the process of determining the amino acid sequence of a protein and must reconcile contradictory results. In one trial, you determine a sequence with glycine as the N-terminal amino acid and asparagine as the C-terminal amino acid. In another trial, your results indicate phenylalanine as the N-terminal aminio acid and alanine as the C-terminal amino acid. How do you reconcile this apparent contradiction?

46. **Thought Question** You are in the process of determining the amino acid sequence of a peptide. After trypsin digestion followed by the Edman degradation, you see the following peptide fragments:

Leu—Gly—Arg

Gly—Ser—Phe—Tyr—Asn—His

Ser—Glu—Asp—Met—Cys—Lys

Thr—Tyr—Glu—Val—Cys—Met—His

What is abnormal concerning these results? What might have been the problem that caused it?

47. **Thought Question** Amino acid compositions can be determined by heating a protein in 6 M HCl and running the hydrolysate

through an ion-exchange column. If you were going to do an amino acid sequencing experiment, why would you want to get an amino acid composition first?

48. **Thought Question** Assume that you are getting ready to do an amino acid sequencing experiment on a protein containing 100 amino acids, and amino acid analysis shows the following data:

Amino Acid	Number of Residues
Ala	7
Arg	23.7
Asn	5.6
Asp	4.1
Cys	4.7
Gln	4.5
Glu	2.2
Gly	3.7
His	3.7
Ile	1.1
Leu	1.7
Lys	11.4
Met	0
Phe	2.4
Pro	4.5
Ser	8.2
Thr	4.7
Trp	0
Tyr	2.0
Val	5.1

Which of the chemicals or enzymes normally used for cutting proteins into fragments would be the least useful to you?

49. **Thought Question** Which enzymes or chemicals would you choose to use to cut the protein from Question 48? Why?

50. **Thought Question** With which amino acid sequences would chymotrypsin be an effective reagent for sequencing the protein from Question 48? Why?

Biochemistry ⊜ Now™

Assess your understanding of this chapter's topics with additional quizzing and tutorials at **http://now.brookscole.com/campbell5**

Annotated Bibliography

Ahern, H. Chromatography, Rooted in Chemistry, Is a Boon for Life Scientists. *The Scientist* **10** (5), 17–19 (1996). [General treatise on chromatography.]

Boyer, R. F. *Modern Experimental Biochemistry.* Boston: Addison-Wesley, 1993. [Textbook specializing in biochemical techniques.]

Dayhoff, M. O., ed. *Atlas of Protein Sequence and Structure.* Washington, D.C.: National Biomedical Research Foundation, 1978. [A listing of all known amino acid sequences, updated periodically.]

Deutscher, M. P., ed. *Guide to Protein Purification.* Vol. 182, *Methods in Enzymology.* San Diego: Academic Press, 1990. [The standard reference for all aspects of research on proteins.]

Dickerson, R. E., and I. Geis. *The Structure and Action of Proteins,* 2nd ed. Menlo Park, Calif.: Benjamin Cummings, 1981. [A well-written

and particularly well-illustrated general introduction to protein chemistry.]

Farrell, S. O., and R. Ranallo. *Experiments in Biochemistry: A Hands-on Approach.* Philadelphia: Saunders College Publishing, 2000. [A laboratory manual for undergraduates that focuses on protein purification techniques.]

Robyt, J. F., and B. J. White. *Biochemical Techniques Theory and Practice.* Monterey, Calif.: Brooks/Cole Publishing Co., 1987. [An all-purpose review of techniques.]

Whitaker, J. R. Determination of Molecular Weights of Proteins by Gel Filtration on Sephadex®. *Analytical Chemistry* **35** (12), 1950–1953 (1963). [Classic paper describing gel filtration as an analytical tool.]

The Behavior of Proteins: Enzymes

Your automobile is powered by the oxidation of the hydrocarbon gasoline to carbon dioxide and water in a controlled explosion within an engine where hot gases can reach 4000°F. In contrast, the living cell gets energy by oxidizing the carbohydrate glucose to carbon dioxide and water at a temperature (in humans) of 98.6°F (37°C). The secret ingredient in living organisms is catalysis, a process performed by protein enzymes. Their three-dimensional architecture gives them exquisite specificity to select the substrate molecules to which they will bind and on which they will operate. Each enzyme has, in fact, a miniature "operating table" where the substrate is momentarily held in a predetermined position so that it can be cut or altered with surgical precision. The scene of the operation, called the active site, is usually a groove, cleft, or cavity on the surface of the protein. Enzyme surgery, such as cleaving molecules or "stitching" them together, frequently occurs many times (and in some cases many thousands of times) per second. The miracle of life is that a myriad chemical reactions in the cell occur simultaneously with great accuracy and at astonishing speed. Without the proper enzymes to process the food you eat, it might take you 50 years to digest your breakfast.

©BrandX Pictures/Getty Images

Traveling over a mountain pass is an analogy frequently used to describe the progress of a chemical reaction. Catalysts speed up the process.

Critical Questions

6.1 What Makes Enzymes Such Effective Biological Catalysts?

6.2 What Is the Difference between the Kinetic and the Thermodynamic Aspects of Reactions?

6.3 How Can We Describe Enzyme Kinetics in Mathematical Terms?

6.4 How Do Substrates Bind to Enzymes?

6.5 What Are Some Examples of Enzyme-Catalyzed Reactions?

6.6 What Is the Michaelis–Menten Approach to Enzyme Kinetics?

6.7 How Do Enzymatic Reactions Respond to Inhibitors?

6.1 | What Makes Enzymes Such Effective Biological Catalysts?

Of all the functions of proteins, **catalysis** is probably most important. In the absence of catalysis, most reactions in biological systems would take place far too slowly to provide products at an adequate pace for a metabolizing organism. The catalysts that serve this function in organisms are called **enzymes.** With the exception of some RNAs (ribozymes) that have catalytic activity (described in Section 11.7 and 12.4), all enzymes are proteins. Enzymes are the most efficient catalysts known; they can increase the rate of a reaction by a factor of up to 10^{20} over uncatalyzed reactions. Nonenzymatic catalysts, in contrast, typically enhance the rate of reaction by factors of 10^2 to 10^4. Enzymes are highly specific, even to the point of being able to distinguish stereoisomers of a given compound. In many cases, the actions of enzymes are fine-tuned by regulatory processes.

6.2 | What Is the Difference between the Kinetic and the Thermodynamic Aspects of Reactions?

The rate of a reaction and its thermodynamic favorability are two different topics, although they are closely related. This is true of all reactions, whether or not a catalyst is involved. The difference between the energies of the reactants (the initial state) and the energies of the products (the final state) of a reaction gives the energy change for that reaction, expressed as the **standard free energy change,** or $\Delta G°$. Energy changes can be described by several related

Biochemistry ⊘ Now™
Test yourself on these Critical Questions at the BiochemistryNow website at **http://now .brookscole.com/campbell5**

thermodynamic quantities. We shall use standard free energy changes for our discussion; the question whether a reaction is favored depends on $\Delta G°$ (see Sections 1.9 and 15.2). Enzymes, like all catalysts, speed up reactions, but they cannot alter the equilibrium constant or the free energy change. The reaction rate depends on the free energy of activation or **activation energy** ($\Delta G°^{\ddagger}$), the energy input required to initiate the reaction. The activation energy for an uncatalyzed reaction is higher than that for a catalyzed reaction; in other words, an uncatalyzed reaction requires more energy to get started. For this reason, its rate is slower than that of a catalyzed reaction.

The reaction of glucose and oxygen gas to produce carbon dioxide and water is an example of a reaction that requires a number of enzymatic catalysts:

$$\text{Glucose} + 6O_2 \rightarrow 6CO_2 + 6H_2O$$

This reaction is thermodynamically favorable (spontaneous in the thermodynamic sense) because its free energy change is negative ($\Delta G° = -2880$ kJ mol^{-1} = -689 kcal mol^{-1}). Note that the term "spontaneous" does not mean instantaneous. Glucose is stable in air with an unlimited supply of oxygen. The energy that must be supplied to start the reaction (which then proceeds with a release of energy)—the activation energy—is conceptually similar to the act of pushing an object to the top of a hill so that it can then slide down the other side.

Activation energy and its relationship to the free energy change of a reaction can best be shown graphically. In Figure 6.1a, the x coordinate shows the extent to which the reaction has taken place, and the y coordinate indicates free energy for an idealized reaction. The *activation energy profile* shows the intermediate stages of a reaction, those between the initial and final states. Activation energy profiles are essential in the discussion of catalysts. The activation energy directly affects the rate of reaction, and the presence of a catalyst speeds up a reaction by changing the mechanism and thus lowering the activation energy. Figure 6.1a plots the energies for an exergonic, spontaneous reaction, such as the complete oxidation of glucose. At the maximum of the curve connecting the reactants and the products lies the **transition state** with the necessary amount of energy and the correct arrangement of atoms to produce products. The activation energy can also be seen as the amount of free energy required to bring the reactants to the transition state.

The analogy of traveling over a mountain pass between two valleys is frequently used in discussions of activation energy profiles. The change in energy corresponds to the change in elevation, and the progress of the reaction corresponds to the distance traveled. The analogue of the transition state is the top of the pass. Considerable effort has gone into elucidating the intermediate stages in reactions of interest to chemists and biochemists and determining the pathway or reaction mechanism that lies between the initial and final states. Reaction dynamics, the study of the intermediate stages of reaction mechanisms, is currently a very active field of research.

The most important effect of a catalyst on a chemical reaction is apparent from a comparison of the activation energy profiles of the same reaction, catalyzed and uncatalyzed, as shown in Figure 6.1b. The standard free energy change for the reaction, $\Delta G°$, remains unchanged when a catalyst is added, but the activation energy, $\Delta G°^{\ddagger}$, is lowered. In the hill-and-valley analogy, the catalyst is a guide that finds an easier path between the two valleys. A similar comparison can be made between two routes from San Francisco to Los Angeles. The highest point on Interstate 5 is Tejon Pass (elevation 4400 feet) and is analogous to the uncatalyzed path. The highest point on U.S. Highway 101 is not much over 1000 feet. Thus, Highway 101 is an easier route and is analogous to the catalyzed pathway. The initial and final points of the trip are the same, but the paths between them are different, as are the mechanisms of catalyzed and uncatalyzed reactions. The presence of an enzyme lowers the

(a)

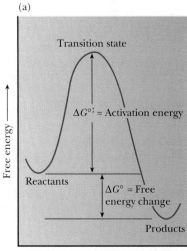

(b)

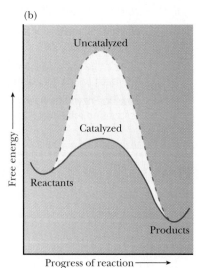

▲ **FIGURE 6.1** Activation energy profiles. (a) The activation energy profile for a typical reaction. The reaction shown here is exergonic (energy-releasing). Note the difference between the activation energy ($\Delta G°^{\ddagger}$) and the standard free energy of the reaction ($\Delta G°$). (b) A comparison of activation energy profiles for catalyzed and uncatalyzed reactions. The activation energy of the catalyzed reaction is much less than that of the uncatalyzed reaction.

Table 6.1

Lowering of the Activation Energy of Hydrogen Peroxide Decomposition by Catalysts

Reaction Conditions	Activation Free Energy		Relative Rate
	kJ mol^{-1}	kcal mol^{-1}	
No catalyst	75.2	18.0	1
Platinum surface	48.9	11.7	2.77×10^4
Catalase	23.0	5.5	6.51×10^8

Rates are given in arbitrary units relative to a value of 1 for the uncatalyzed reaction at 37°C.

activation energy needed for substrate molecules to reach the transition state. The concentration of the transition state increases markedly. As a result, the rate of the catalyzed reaction is much greater than the rate of the uncatalyzed reaction. Enzymatic catalysts enhance a reaction rate by many powers of 10.

The biochemical reaction in which hydrogen peroxide (H_2O_2) is converted to water and oxygen provides an example of the effect of catalysts on activation energy.

$$2\ H_2O_2 \rightarrow 2\ H_2O + O_2$$

The activation energy of this reaction is lowered if the reaction is allowed to proceed on platinum surfaces, but it is lowered even more by the enzyme catalase. Table 6.1 summarizes the energies involved.

> **Essential Information**
>
> Enzymes are biological catalysts. They increase the rates of reactions by lowering the free energy of activation, but they do not affect the thermodynamic aspects of reactions.

Biochemical Connections

Enzymes as Markers for Disease

Some enzymes are found only in specific tissues or in a limited number of such tissues. The enzyme lactate dehydrogenase (LDH) has two different types of subunits—one found primarily in heart muscle (H), and another found in skeletal muscle (M). The two different subunits differ slightly in amino acid composition; consequently, they can be separated electrophoretically or chromatographically on the basis of charge. Because LDH is a tetramer of four subunits, and because the H and M subunits can combine in all possible combinations, LDH can exist in five different forms, called **isozymes,** depending on the source. An increase of any form of LDH in the blood indicates some kind of tissue damage. A heart attack used to be diagnosed by an increase of LDH from heart muscle. Similarly, there are different forms of creatine kinase (CK), an enzyme that occurs in the brain, heart, and skeletal muscle. Appearance of the brain type can indicate a stroke or a brain tumor, whereas the heart type indicates a heart attack. After a heart attack, CK shows up more rapidly in the blood than LDH. Monitoring the presence of both enzymes extends the possibility of diagnosis, which is useful, since a very mild heart attack might be difficult to diagnose. An elevated level of the isozyme from heart muscle in blood is a definite indication of damage to the heart tissue.

A particularly useful enzyme to assay is acetylcholinesterase (ACE), which is important in controlling certain nerve impulses. Many pesticides interfere with this enzyme, so farm workers are often tested to be sure that they have not received inappropriate exposure to these important agricultural toxins. In fact, more

than 20 enzymes are typically used in the clinical lab to diagnose disease. There are highly specific markers for enzymes active in the pancreas, red blood cells, liver, heart, brain, prostate gland, and many of the endocrine glands. Because these enzymes are relatively easy to assay, even using automated techniques, they are part of the "standard" blood test your doctor is likely to request.

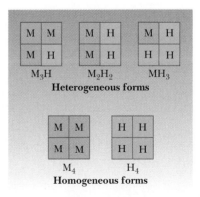

▲ The possible isozymes of lactate dehydrogenase. The symbol M refers to the dehydrogenase form that predominates in skeletal muscle, and the symbol H refers to the form that predominates in heart (cardiac) muscle.

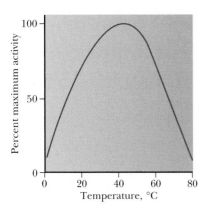

▲ **FIGURE 6.2** The effect of temperature on enzyme activity. The relative activity of an enzymatic reaction as a function of temperature. The decrease in activity above 50°C is due to thermal denaturation.

Raising the temperature of a reaction mixture increases the energy available to the reactants to reach the transition state. Consequently, the rate of a chemical reaction increases with temperature. One might be tempted to assume that this is universally true for biochemical reactions. In fact, increase of reaction rate with temperature occurs only to a limited extent with biochemical reactions. It is helpful to raise the temperature at first, but eventually there comes a point at which heat denaturation of the enzyme (Section 4.4) is reached. Above this temperature, adding more heat denatures more enzyme and slows down the reaction. Figure 6.2 shows a typical curve of temperature effect on an enzyme-catalyzed reaction. The Biochemical Connections box above describes another way in which the specificity of enzymes is of great use.

6.3 | How Can We Describe Enzyme Kinetics in Mathematical Terms?

The rate of a chemical reaction is usually expressed in terms of a change in the concentration of a reactant or of a product in a given time interval. Any convenient experimental method can be used to monitor changes in concentration. In a reaction of the form A + B → P, where A and B are reactants and P is the product, the rate of the reaction can be expressed either in terms of the rate of disappearance of one of the reactants or in terms of the rate of appearance of the product. The rate of disappearance of A is $-\Delta[A]/\Delta t$, where Δ symbolizes change, [A] is the concentration of A in moles per liter, and t is time. Likewise, the rate of disappearance of B is $-\Delta[B]/\Delta t$, and the rate of appearance of P is $\Delta[P]/\Delta t$. The rate of the reaction can be expressed in terms of any of these changes because the rates of appearance of product and disappearance of reactant are related by the stoichiometric equation for the reaction.

$$\text{Rate} = \frac{-\Delta[A]}{\Delta t} = \frac{-\Delta[B]}{\Delta t} = \frac{\Delta[P]}{\Delta t}$$

The negative signs for the changes in concentration of A and B indicate that A and B are being used up in the reaction, while P is being produced.

It has been established that the rate of a reaction at a given time is proportional to the product of the concentrations of the reactants raised to the appropriate powers,

$$\text{Rate} \propto [A]^f[B]^g$$

or, as an equation,

$$\text{Rate} = k[A]^f[B]^g$$

where k is a proportionality constant called the **rate constant.** The exponents f and g *must be determined experimentally.* They are *not necessarily* equal to the coefficients of the balanced equation, but frequently they are. The square brackets, as usual, denote molar concentration. When the exponents in the rate equation have been determined experimentally, a mechanism for the reaction—a description of the detailed steps along the path between reactants and products—can be proposed.

The exponents in the rate equation are usually small whole numbers, such as 1 or 2. (There are also some cases in which the exponent 0 occurs.) The values of the exponents are related to the number of molecules involved in the detailed steps that constitute the mechanism. The *overall order* of a reaction is the sum of all the exponents. If, for example, the rate of a reaction A → P is given by the rate equation

$$\text{Rate} = k[\text{A}]^1 \tag{6.1}$$

where k is the rate constant and the exponent for the concentration of A is 1, then the reaction is **first order** with respect to A and first order overall. The rate of radioactive decay of the widely used tracer isotope phosphorus 32 (^{32}P; atomic weight = 32) depends only on the concentration of ^{32}P present. Here we have an example of a first-order reaction. Only the ^{32}P atoms are involved in the mechanism of the radioactive decay, which, as an equation, takes the form

$$^{32}\text{P} \rightarrow \text{decay products}$$

$$\text{Rate} = k[^{32}\text{P}]^1 = k[^{32}\text{P}]$$

If the rate of a reaction A + B → C + D is given by

$$\text{Rate} = k[\text{A}]^1[\text{B}]^1 \tag{6.2}$$

where k is the rate constant, the exponent for the concentration of A is 1, and the exponent for the concentration of B is 1, then the reaction is said to be first order with respect to A, first order with respect to B, and **second order** overall. In the reaction of glycogen$_n$ (a polymer of glucose with n glucose residues) with inorganic phosphate, P_i, to form glucose 1-phosphate + glycogen$_{n-1}$, the rate of reaction depends on the concentrations of both reactants.

$$\text{Glycogen}_n + P_i \rightarrow \text{Glucose 1-phosphate} + \text{Glycogen}_{n-1}$$

$$\text{Rate} = k[\text{Glycogen}]^1[P_i]^1 = k[\text{Glycogen}][P_i]$$

where k is the rate constant. Both the glycogen and the phosphate take part in the reaction mechanism. The reaction of glycogen with phosphate is first order with respect to glycogen, first order with respect to phosphate, and second order overall.

Many common reactions are first or second order. After the order of the reaction is determined experimentally, proposals can be made about the mechanism of a reaction.

The possibility exists that exponents in a rate equation may be equal to zero, with the rate for a reaction A → B given by the equation

$$\text{Rate} = k[\text{A}]^0 = k \tag{6.3}$$

Such a reaction is called **zero order,** and its rate, which is constant, depends not on concentrations of reactants but on other factors, such as the presence of catalysts. Enzyme-catalyzed reactions can exhibit zero-order kinetics when the concentrations of reactants are so high that the enzyme is completely saturated with reactant molecules. This point will be discussed in more detail later in this chapter, but, for the moment, we can consider the situation analogous to a traffic bottleneck in which six lanes of cars are trying to cross a two-lane bridge. The rate at which the cars cross is not affected by the number of waiting cars, only by the number of lanes available on the bridge.

6.4 | How Do Substrates Bind to Enzymes?

In an enzyme-catalyzed reaction, the enzyme binds to the **substrate** (one of the reactants) to form a complex. The formation of the complex leads to the formation of the transition-state species, which then forms the product. The nature of transition states in enzymatic reactions is a large field of research in itself, but some general statements can be made on the subject. A substrate

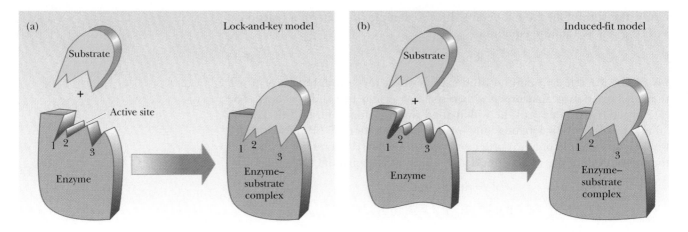

▲ **FIGURE 6.3** Two models for the binding of a substrate to an enzyme. (a) In the lock-and-key model, the shape of the substrate and the conformation of the active site are complementary to one another. (b) In the induced-fit model, the enzyme undergoes a conformational change upon binding to substrate. The shape of the active site becomes complementary to the shape of the substrate only after the substrate binds to the enzyme.

binds, usually by noncovalent interactions, to a small portion of the enzyme called the **active site,** frequently situated in a cleft or crevice in the protein and consisting of certain amino acids that are essential for enzymatic activity (Figure 6.3). The catalyzed reaction takes place at the active site, usually in several steps. The first step is the binding of substrate to the enzyme, which occurs because of highly specific interactions between the substrate and the side chains and backbone groups of the amino acids making up the active site. Two important models have been developed to describe the binding process. The first, the **lock-and-key model,** assumes a high degree of similarity between the shape of the substrate and the geometry of the binding site on the enzyme (Figure 6.3a). The substrate binds to a site whose shape complements its own, like a key in a lock or the correct piece in a three-dimensional jigsaw puzzle. This model is now largely of historical interest because it does not take into account an important property of proteins—namely, their conformational flexibility. The second model takes into account the fact that proteins have some three-dimensional flexibility. According to this **induced-fit model,** the binding of the substrate induces a conformational change in the enzyme that results in a complementary fit after the substrate is bound (Figure 6.3b). The binding site has a different three-dimensional shape before the substrate is bound. The induced-fit model is also more attractive when we consider the nature of the transition state and the lowered activation energy that occurs with an enzyme-catalyzed reaction. The enzyme and substrate must bind to form the ES complex before anything else can happen. What would happen if this binding were too perfect? Figure 6.4 shows what happens when E and S bind. There must be an attraction between E and S for them to bind. This attraction will cause the ES complex to be lower on an energy diagram than the E + S at the start. Then the bound ES must attain the conformation of the transition state $EX^{\ddagger}$. If the binding of E and S to form ES were a perfect fit, the ES would be at such a low energy that the difference between ES and $EX^{\ddagger}$ would be very large. This would slow down the rate of reaction. Many studies have shown that enzymes increase the rate of reaction by lowering the energy of the transition state, $EX^{\ddagger}$, while raising the energy of the ES complex. The induced-fit model certainly supports this last consideration better than the lock-and-key model; in fact, the induced-fit model mimics the transition state.

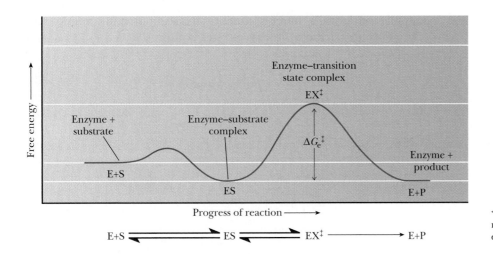

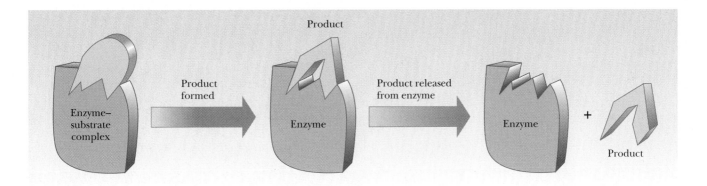

◄ **FIGURE 6.4** The activation energy profile of a reaction with strong binding of the substrate to the enzyme to form an enzyme–substrate complex.

▲ **FIGURE 6.5** Formation of product from substrate (bound to the enzyme), followed by release of the product.

After the substrate is bound and the transition state is subsequently formed, catalysis can occur. This means that bonds must be rearranged. In the transition state, the substrate is bound close to atoms with which it is to react. Furthermore, the substrate is placed in the correct orientation with respect to those atoms. Both effects, proximity and orientation, speed up the reaction. As bonds are broken and new bonds are formed, the substrate is transformed into product. The product is released from the enzyme, which can then catalyze the reaction of more substrate to form more product (Figure 6.5). Each enzyme has its own unique mode of catalysis, which is not surprising in view of enzymes' great specificity. Even so, there are some general modes of catalysis in enzymatic reactions. Two enzymes, chymotrypsin and aspartate transcarbamoylase, are good examples of these general principles.

6.5 What Are Some Examples of Enzyme-Catalyzed Reactions?

Chymotrypsin is an enzyme that catalyzes the hydrolysis of peptide bonds, with some specificity for residues containing aromatic side chains. Chymotrypsin also cleaves peptide bonds at other sites, such as leucine, histidine, and glutamine, but with a lower frequency than at aromatic amino acid residues. It also catalyzes the hydrolysis of ester bonds.

Reactions catalyzed by chymotrypsin

Peptide   + H_2O ⇌   Acid   +   Amine

Ester   + H_2O ⇌   Acid   +   Alcohol

p-Nitrophenylacetate $\xrightarrow[\textbf{Basic conditions}]{\textbf{H}_2\textbf{O}}$ p-Nitrophenolate (yellow)   + $2H^+$ +

Although ester hydrolysis is not important to the physiological role of chymotrypsin in the digestion of proteins, it is a convenient model system for investigating the enzyme's catalysis of hydrolysis reactions. The usual laboratory procedure is to use p-nitrophenyl esters as the substrate and to monitor the progress of the reaction by the appearance of a yellow color in the reaction mixture caused by the production of p-nitrophenolate ion.

In a typical reaction in which a p-nitrophenyl ester is hydrolyzed by chymotrypsin, the experimental rate of the reaction depends on the concentration of the substrate—in this case, the p-nitrophenyl ester. At low substrate concentrations, the rate of reaction increases as more substrate is added. At higher substrate concentrations, the rate of the reaction changes very little with the addition of more substrate, and a maximum rate is reached. When these results are presented in a graph, the curve is hyperbolic (Figure 6.6).

Another enzyme-catalyzed reaction is the one catalyzed by the enzyme **aspartate transcarbamoylase** (ATCase). This reaction is the first step in a pathway leading to the formation of cytidine triphosphate (CTP) and uridine triphosphate (UTP), which are ultimately needed for the biosynthesis of RNA and DNA. In this reaction, carbamoyl phosphate reacts with aspartate to produce carbamoyl aspartate and phosphate ion.

$$\text{Carbamoyl phosphate} + \text{Aspartate} \rightarrow \text{Carbamoyl aspartate} + HPO_4^{2-}$$
Reaction catalyzed by aspartate transcarbamoylase

The rate of this reaction also depends on substrate concentration—in this case, the concentration of aspartate (the carbamoyl phosphate concentration

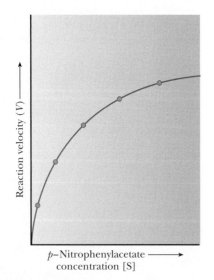

▲ **FIGURE 6.6** Dependence of reaction velocity, V, on p-nitrophenylacetate concentration, [S], in a reaction catalyzed by chymotrypsin. The shape of the curve is hyperbolic.

is kept constant). Experimental results show that, once again, the rate of the reaction depends on substrate concentration at low and moderate concentrations, and, once again, a maximum rate is reached at high substrate concentrations. There is, however, one very important difference. For this reaction, a graph showing the dependence of reaction rate on substrate concentration has a sigmoidal rather than hyperbolic shape (Figure 6.7).

The results of experiments on the reaction kinetics of chymotrypsin and aspartate transcarbamoylase are representative of experimental results obtained with many enzymes. The overall kinetic behavior of many enzymes resembles that of chymotrypsin, while other enzymes behave similarly to aspartate transcarbamoylase. We can use this information to draw some general conclusions about the behavior of enzymes. The comparison between the kinetic behaviors of chymotrypsin and ATCase is reminiscent of the relationship between the oxygen-binding behaviors of myoglobin and hemoglobin, discussed in Chapter 4. ATCase and hemoglobin are allosteric proteins; chymotrypsin and myoglobin are not. (Recall, from Section 4.7, that allosteric proteins are the ones in which subtle changes at one site affect structure and function at another site. Cooperative effects, such as the fact that the binding of the first oxygen molecule to hemoglobin makes it easier for other oxygen molecules to bind, are a hallmark of allosteric proteins.) The differences in behavior between allosteric and nonallosteric proteins can be understood in terms of models based on structural differences between the two kinds of proteins. We shall need a model that explains the hyperbolic plot of kinetic data for nonallosteric enzymes and another model that explains the sigmoidal plot for allosteric enzymes, when we encounter the mechanisms of the many enzyme-catalyzed reactions in subsequent chapters. The Michaelis–Menten model is widely used for nonallosteric enzymes, and several models are used for allosteric enzymes.

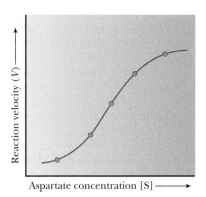

▲ **FIGURE 6.7** Dependence of reaction velocity, V, on aspartate concentration, $[S]$, in a reaction catalyzed by aspartate transcarbamoylase. The shape of the curve is sigmoidal.

6.6 | What Is the Michaelis–Menten Approach to Enzyme Kinetics?

A particularly useful model for the kinetics of enzyme-catalyzed reactions was devised in 1913 by Leonor Michaelis and Maud Menten. It is still the basic model for nonallosteric enzymes and is widely used, even though it has undergone many modifications.

A typical reaction might be the conversion of some substrate, S, to a product, P. The stoichiometric equation for the reaction is

$$S \rightarrow P$$

The mechanism for an enzyme-catalyzed reaction can be summarized in the form

$$E + S \underset{k_{-1}}{\overset{k_1}{\rightleftharpoons}} ES \overset{k_2}{\rightarrow} E + P \tag{6.4}$$

Note the assumption that the product is not converted to substrate to any appreciable extent. In this equation, k_1 is the rate constant for the formation of the enzyme–substrate complex, ES, from the enzyme, E, and the substrate, S; k_{-1} is the rate constant for the reverse reaction, dissociation of the ES complex to free enzyme and substrate; and k_2 is the rate constant for the conversion of the ES complex to product P and the subsequent release of product from the enzyme. The enzyme appears explicitly in the mechanism, and the concentrations of both free enzyme, E, and enzyme–substrate complex, ES, therefore, appear in the rate equations. Catalysts characteristically are regenerated at the end of the reaction, and this is true of enzymes.

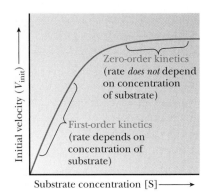

▲ **FIGURE 6.8** The rate and the observed kinetics of an enzymatic reaction depend on substrate concentration. The concentration of enzyme, [E], is constant.

Essential Information

The main feature of the Michaelis–Menten model for enzymatic reactions is the formation of an enzyme–substrate complex. The concentration of enzyme–substrate complex is low, but it remains unchanged to any appreciable extent over the course of the reaction. The substrate is converted to product, which is released from the enzyme. Like all catalysts, the enzyme is regenerated at the end of the reaction.

When we measure the rate (also called the velocity) of an enzymatic reaction at varying substrate concentrations, we see that the rate depends on the substrate concentration, [S]. We measure the initial rate of the reaction (the rate measured immediately after the enzyme and substrate are mixed) so that we can be certain that the product is not converted to substrate to any appreciable extent. This velocity is sometimes written V_{init} or V_0 to indicate this initial velocity, but it is important to remember that all the calculations involved in enzyme kinetics assume that the velocity measured is the initial velocity. We can graph our results as in Figure 6.8. In the lower region of the curve (at low levels of substrate), the reaction is first order (Section 6.3), implying that the velocity, V, depends on substrate concentration [S]. In the upper portion of the curve (at higher levels of substrate), the reaction is zero order; the rate is independent of concentration. The active sites of all of the enzyme molecules are saturated. At infinite substrate concentration, the reaction would proceed at its maximum velocity, written V_{max}.

The substrate concentration at which the reaction proceeds at one-half its maximum velocity has a special significance. It is given the symbol K_M, which can be considered an inverse measure of the affinity of the enzyme for the substrate. The lower the K_M, the higher the affinity.

Let us examine the mathematical relationships among the quantities [E], [S], V_{max}, and K_M. The general mechanism of the enzyme-catalyzed reaction involves binding of the enzyme, E, to the substrate to form a complex, ES, which then forms the product. The rate of formation of the enzyme–substrate complex, ES, is

$$\text{Rate of formation} = \frac{\Delta[ES]}{\Delta t} = k_1[E][S] \tag{6.5}$$

where $\Delta[ES]/\Delta t$ means the change in the concentration of the complex, $\Delta[ES]$, during a given time Δt, and k_1 is the rate constant for the formation of the complex.

The ES complex breaks down in two reactions, by returning to enzyme and substrate or by giving rise to product and releasing enzyme. The rate of disappearance of complex is the sum of the rates of the two reactions.

$$\text{Rate of breakdown} = \frac{-\Delta[ES]}{\Delta t} = k_{-1}[ES] + k_2[ES] \tag{6.6}$$

The negative sign in the term $-\Delta[ES]/\Delta t$ means that the concentration of the complex decreases as the complex breaks down. The term k_{-1} is the rate constant for the dissociation of complex to regenerate enzyme and substrate, and k_2 is the rate constant for the reaction of the complex to give product and enzyme.

Enzymes are capable of processing the substrate very efficiently, and a **steady state** is soon reached in which the rate of formation of the enzyme–substrate complex equals the rate of its breakdown. Very little complex is present, and it turns over rapidly, but its concentration stays the same with time. According to the *steady-state theory*, then, the rate of formation of the enzyme–substrate complex equals the rate of its breakdown,

$$\frac{\Delta[ES]}{\Delta t} = \frac{-\Delta[ES]}{\Delta t} \tag{6.7}$$

and

$$k_1[E][S] = k_{-1}[ES] + k_2[ES] \tag{6.8}$$

To solve for the concentration of the complex, ES, it is necessary to know the concentration of the other species involved in the reaction. The initial concentration of substrate is a known experimental condition and does not change significantly during the initial stages of the reaction. The substrate concentration is much greater than the enzyme concentration. The total concentration of the enzyme, $[E]_T$, is also known, but a large proportion of it may be involved in the complex. The concentration of free enzyme, $[E]$, is the difference between $[E]_T$, the total concentration, and $[ES]$, which can be written as an equation:

$$[E] = [E]_T - [ES] \qquad (6.9)$$

Substituting for the concentration of free enzyme, $[E]$, in Equation 6.8,

$$k_1([E]_T - [ES])\,[S] = k_{-1}\,[ES] + k_2\,[ES] \qquad (6.10)$$

Collecting all the rate constants for the individual reactions,

$$\frac{([E]_T - [ES])[S]}{[ES]} = \frac{k_{-1} + k_2}{k_1} = K_M \qquad (6.11)$$

where K_M is called the **Michaelis constant.** It is now possible to solve Equation 6.11 for the concentration of enzyme–substrate complex, $[ES]$,

$$\frac{([E]_T - [ES])\,[S]}{[ES]} = K_M$$

$$[E]_T[S] - [ES][S] = K_M\,[ES]$$

$$[E]_T[S] = [ES](K_M + [S])$$

or

$$[ES] = \frac{[E]_T[S]}{K_M + [S]} \qquad (6.12)$$

In the initial stages of the reaction, so little product is present that no reverse reaction of product to complex need be considered. Thus the initial rate determined in enzymatic reactions depends on the rate of breakdown of the enzyme–substrate complex into product and enzyme. In the Michaelis–Menten model, the initial rate, V, of the formation of product depends only on the rate of the breakdown of the ES complex,

$$V = k_2[ES] \qquad (6.13)$$

and on the substitution of the expression for $[ES]$ from Equation 6.12,

$$V = \frac{k_2[E]_T[S]}{K_M + [S]} \qquad (6.14)$$

If the substrate concentration is so high that the enzyme is completely saturated with substrate ($[ES] = [E]_T$), the reaction proceeds at its maximum possible rate (V_{max}). Substituting $[E]_T$ for $[ES]$ in Equation 6.13,

$$V = V_{max} = k_2[E]_T \qquad (6.15)$$

The total concentration of enzyme is a constant, which means that

$$V_{max} = \text{Constant}$$

This expression for V_{max} resembles that for a zero-order reaction given in Equation 6.3:

$$\text{Rate} = k[A]_0 = k$$

Note that the concentration of substrate, $[A]$, appears in Equation 6.3 rather than the concentration of enzyme, $[E]$, as in Equation 6.15. When the enzyme is saturated with substrate, zero-order kinetics with respect to substrate are observed.

Substituting the expression for V_{max} into Equation 6.14 enables us to relate the observed velocity at any substrate concentration to the maximum rate of an enzymatic reaction:

$$V = \frac{V_{max}[S]}{K_M + [S]} \qquad (6.16)$$

Figure 6.8 shows the effect of increasing substrate concentration on the observed rate. In such an experiment, the reaction is run at several substrate concentrations, and the rate is determined by following the disappearance of reactant, or the appearance of product, by way of any convenient method. At low-substrate concentrations, first-order kinetics are observed. At higher substrate concentrations (well beyond $10 \times K_M$), when the enzyme is saturated, the constant reaction rate characteristic of zero-order kinetics is observed.

This constant rate, when the enzyme is saturated with substrate, is the V_{max} for the enzyme, a value that can be roughly estimated from the graph. The value of K_M can also be estimated from the graph. From Equation 6.16,

$$V = \frac{V_{max}[S]}{K_M + [S]}$$

When experimental conditions are adjusted so that $[S] = K_M$,

$$V = \frac{V_{max}[S]}{[S] + [S]}$$

and

$$V = \frac{V_{max}}{2}$$

In other words, when the rate of the reaction is half its maximum value, the substrate concentration is equal to the Michaelis constant (Figure 6.9). This fact is the basis of the graphical determination of K_M.

Note that the reaction used to generate the Michaelis–Menten equation was the simplest enzyme equation possible, that with a single substrate going to a single product. Most enzymes catalyze reactions containing two or more substrates. This does not invalidate our equations, however. For enzymes with multiple substrates, the same equations can be used, but only one substrate can be studied at a time. If, for example, we had the enzyme-catalyzed reaction

$$A + B \rightarrow P + Q$$

we could still use the Michaelis–Menten approach. If we hold A at saturating levels and then vary the amount of B over a broad range, the curve of velocity versus [B] will still be a hyperbola, and we can still calculate the K_M for B. Conversely, we could hold the level of B at saturating levels and vary the amount of A to determine the K_M for A. There are even enzymes that have two substrates where, if we plot V versus [substrate A], we will see the Michaelis–Menten hyperbola, but, if we plot V versus [substrate B], we will see the sigmoidal curve shown for aspartate transcarbamoylase in Figure 6.7. Technically the term K_M is only appropriate for those enzymes that exhibit a hyperbolic curve of velocity versus [substrate].

Michaelis–Menten equation

Biochemistry ⓔ Now™
Go to BiochemistryNow and click on Biochemistry Interactive to review Michaelis–Menten kinetics.

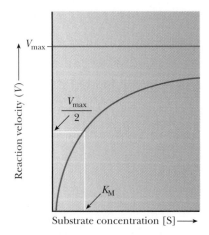

▲ **FIGURE 6.9** Graphical determination of V_{max} and K_M from a plot of reaction velocity, V, against substrate concentration, [S]. V_{max} is the constant rate reached when the enzyme is completely saturated with substrate, a value that frequently must be estimated from such a graph.

Linearizing the Michaelis–Menten Equation

The curve that describes the rate of a nonallosteric enzymatic reaction is hyperbolic. It is quite difficult to estimate V_{max} because it is an asymptote, and the value is never reached with any finite substrate concentration. This, in

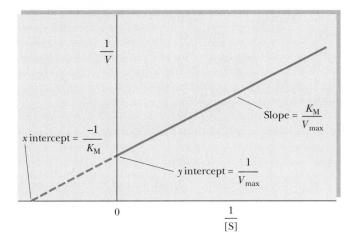

$$\frac{1}{V} = \frac{K_M}{V_{max}} \left(\frac{1}{[S]}\right) + \frac{1}{V_{max}}$$

Biochemistry Now™ ACTIVE FIGURE 6.10
A Lineweaver–Burk double reciprocal plot of enzyme kinetics. The reciprocal of reaction velocity, $1/V$, is plotted against the reciprocal of the substrate concentration, $1/[S]$. The slope of the line is K_M/V_{max}, and the y intercept is $1/V_{max}$. The x intercept is $-1/K_M$. **Watch this Active Figure at http://now .brookscole.com/campbell5**

turn, makes it difficult to determine the K_M of the enzyme. It is considerably easier to work with a straight line than a curve. One can transform the equation for a hyperbola (Equation 6.16) into an equation for a straight line by taking the reciprocals of both sides:

$$\frac{1}{V} = \frac{K_M + [S]}{V_{max}[S]}$$

$$\frac{1}{V} = \frac{K_M}{V_{max}[S]} + \frac{[S]}{V_{max}[S]}$$

$$\frac{1}{V} = \frac{K_M}{V_{max}} \times \frac{1}{[S]} + \frac{1}{V_{max}} \tag{6.17}$$

The equation now has the form of a straight line, $y = mx + b$, where $1/V$ takes the place of the y coordinate and $1/[S]$ takes the place of the x coordinate. The slope of the line, m, is $K_M/V_{max,}$ and the intercept, b, is $1/V_{max}$. Figure 6.10 presents this information graphically as a **Lineweaver–Burk double-reciprocal plot.** It is usually easier to draw the best straight line through a set of points than to estimate the best fit of points to a curve. There are convenient computer methods for drawing the best straight line through a series of experimental points. Such a line can be extrapolated to high values of [S], ones that might be unattainable due to solubility limits or the cost of the substrate. The extrapolated line can be used to obtain V_{max}.

Practice Session

The following data were obtained for the hydrolysis of carbobenzoxyglycyl-L-tryptophan catalyzed by the enzyme carboxypeptidase (R. Lumry, E. L. Smith, and R. R. Glantz, *J. Amer. Chem. Soc.* **73,** 4330, 1951). The reaction in question is

carbobenzoxyglycyl-L-tryptophan + H_2O →
carbobenzoxyglycine + L-tryptophan

Plot these results using the Lineweaver–Burk method, and determine values for K_M and V_{max}. The symbol mM represents millimoles per liter;

$1 \text{ m}M = 1 \times 10^{-3} \text{ mol L}^{-1}$. (The concentration of the enzyme is the same in all experiments.)

Substrate Concentration (mM)	Velocity (mM sec⁻¹)
2.5	0.024
5.0	0.036
10.0	0.053
15.0	0.060
20.0	0.064

Solution

The reciprocal of substrate concentration and of velocity gives the following results:

$1/[S]$ (mM⁻¹)	$1/V$ (mM sec⁻¹)⁻¹
0.400	41.667
0.200	27.778
0.100	18.868
0.067	16.667
0.050	15.625

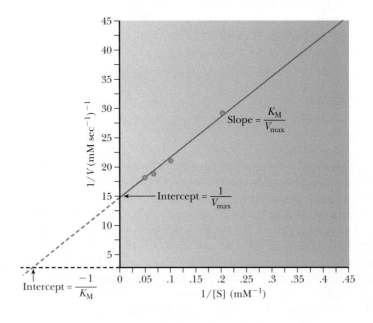

Plotting the results gives a straight line; the best fit to the experimental points is $1/V = 75.431 \, (1/[S]) + 11.8$. The reciprocal of the y intercept is V_{max}, and the slope is K_M/V_{max}. Hence, $V_{max} = 0.0847 \text{ m}M \text{ sec}^{-1}$; $K_M = 6.39 \text{ m}M$.

Significance of K_M and V_{max}

We have already seen that, when the rate of a reaction, V, is equal to half the maximum rate possible, $V = V_{max}/2$, then $K_M = [S]$. One interpretation of the Michaelis constant, K_M, is that it equals the concentration of substrate at which 50% of the enzyme active sites are occupied by substrate. The Michaelis constant has the units of concentration.

Another interpretation of K_M relies on the assumptions of the original Michaelis–Menten model of enzyme kinetics. Recall Equation 6.4:

$$E + S \underset{k_{-1}}{\overset{k_1}{\rightleftharpoons}} ES \overset{k_2}{\rightarrow} E + P \tag{6.4}$$

As before, k_1 is the rate constant for the formation of the enzyme–substrate complex, ES, from the enzyme and substrate; k_{-1} is the rate constant for the reverse reaction, dissociation of the ES complex to free enzyme and substrate; and k_2 is the rate constant for the formation of product P and the subsequent release of product from the enzyme. Also recall from Equation 6.11 that

$$K_M = \frac{k_{-1} + k_2}{k_1}$$

Consider the case in which the reaction $E + S \rightarrow ES$ takes place more frequently than $E + S \rightarrow E + P$. In kinetic terms, this means that the dissociation rate constant k_{-1} is greater than the rate constant for the formation of product, k_2. If k_{-1} is *much* larger than k_2 ($k_{-1} \gg k_2$), as was originally assumed by Michaelis and Menten, then approximately

$$K_M = \frac{k_{-1}}{k_1}$$

It is informative to compare the expression for the Michaelis constant with the equilibrium constant expression for the dissociation of the ES complex,

$$ES \underset{k_1}{\overset{k_{-1}}{\rightleftharpoons}} E + S$$

The k values are the rate constants, as before. The equilibrium constant expression is

$$K_{eq} = \frac{[E][S]}{[ES]} = \frac{k_{-1}}{k_1}$$

This expression is the same as that for K_M and makes the point that, when the assumption that $k_{-1} \gg k_2$ is valid, K_M is simply the dissociation constant for the ES complex. The K_M is a measure of how tightly the substrate is bound to the enzyme. The greater the value of the K_M, the less tightly the substrate is bound to the enzyme. Note that, in the steady-state approach, k_2 is not assumed to be small compared with k_{-1}; therefore, K_M is not technically a dissociation constant, even though it is often used to estimate the affinity of the enzyme for the substrate.

V_{max} is related to the **turnover number** of an enzyme, a quantity equal to the catalytic constant, k_2. This constant is also referred to as k_{cat} or k_p:

$$\frac{V_{max}}{[E_T]} = \text{turnover number} = k_{cat}$$

The turnover number is the number of moles of substrate that react to form product per mole of enzyme per unit time. This statement assumes that the enzyme is fully saturated with substrate and thus that the reaction is proceeding at the maximum rate. Table 6.2 lists turnover numbers for typical enzymes, where the units are *per second.*

Turnover numbers are a particularly dramatic illustration of the efficiency of enzymatic catalysis. Catalase is an example of a particularly efficient enzyme. In Section 6.1, we encountered catalase in its role in converting hydrogen peroxide to water and oxygen. As Table 6.2 indicates, it can transform 40 million moles of substrate to product every second. The following Biochemical Connections describes some practical information available from the kinetic parameters we have discussed in this section.

Table 6.2

Turnover Numbers and K_M for Some Typical Enzymes

Enzyme	Function	k_{cat} = Turnover Number*	K_M**
Catalase	Conversion of H_2O_2 to H_2O and O_2	4×10^7	25
Carbonic Anhydrase	Hydration of CO_2	1×10^6	12
Acetylcholinesterase	Regenerates acetylcholine, an important substance in transmission of nerve impulses, from acetate and choline	1.4×10^4	9.5×10^{-2}
Chymotrypsin	Proteolytic enzyme	1.9×10^2	6.6×10^{-1}
Lysozyme	Degrades bacterial cell-wall polysaccharides	0.5	6×10^{-3}

*The definition of turnover number is the moles of substrate converted to product per mole of enzyme per second. The units are $\sec^{-1}$.

**The units of K_M are millimolar.

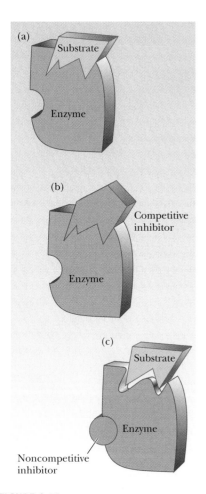

▲ **FIGURE 6.11** Modes of action of inhibitors. The distinction between competitive and noncompetitive inhibitors is that a competitive inhibitor prevents binding of the substrate to the enzyme, whereas a noncompetitive inhibitor does not. (a) An enzyme–substrate complex in the absence of inhibitor. (b) A competitive inhibitor binds to the active site; the substrate cannot bind. (c) A noncompetitive inhibitor binds at a site other than the active site. The substrate still binds, but the enzyme cannot catalyze the reaction because of the presence of the bound inhibitor.

6.7 | How Do Enzymatic Reactions Respond to Inhibitors?

An **inhibitor,** as the name implies, is a substance that interferes with the action of an enzyme and slows the rate of a reaction. A good deal of information about enzymatic reactions can be obtained by observing the changes in the reaction caused by the presence of inhibitors. There are two ways in which inhibitors can affect an enzymatic reaction. A reversible inhibitor can bind to the enzyme and subsequently be released, leaving the enzyme in its original condition. An irreversible inhibitor reacts with the enzyme to produce a protein that is not enzymatically active and from which the original enzyme cannot be regenerated.

Two major classes of reversible inhibitors can be distinguished on the basis of the sites on the enzyme to which they bind. One class consists of compounds very similar in structure to the substrate. In this case, the inhibitor can bind to the active site and block the substrate's access to it. This mode of action is called **competitive inhibition** because the inhibitor competes with the substrate for the active site on the enzyme. The other major class of reversible inhibitors includes any inhibitor that binds to the enzyme at a site other than the active site and, as a result of binding, causes a change in the structure of the enzyme, especially around the active site. The substrate is still able to bind to the active site, but the enzyme cannot catalyze the reaction when the inhibitor is bound to it. This mode of action is called **noncompetitive inhibition** (Figure 6.11).

The two kinds of inhibition can be distinguished from one another in the laboratory. The reaction is carried out in the presence of inhibitor at several substrate concentrations, and the rates obtained are compared with those of the uninhibited reaction. The differences in the Lineweaver–Burk plots for the inhibited and uninhibited reactions provide the basis for the comparison.

Kinetics of Competitive Inhibition

In the presence of a competitive inhibitor, the slope of the Lineweaver–Burk plot changes, but the *y* intercept does not. (The *x* intercept also changes.) The V_{max} is unchanged, but the K_M increases. More substrate is needed to get to a given rate in the presence of inhibitor than in its absence. This point specifically applies to the specific value $V_{max}/2$ (recall that at $V_{max}/2$, the substrate concentration, [S], equals K_M) (Figure 6.12). Competitive inhibition can be overcome by a sufficiently high substrate concentration.

Biochemical Connections

Practical Information from Kinetic Data

The mathematics of enzyme kinetics can certainly look challenging. In fact, an understanding of kinetic parameters can often provide key information about the role of an enzyme within a living organism. Four aspects are useful: comparison of K_M, comparison of k_{cat} or turnover number, comparison of k_{cat}/K_M ratios, and specific locations of enzymes within an organism.

Comparison of K_M

Let us start by comparing the values of the K_M for two enzymes that catalyze an early step in the breakdown of sugars: hexokinase and glucokinase. Both enzymes catalyze the formation of a phosphate ester linkage to a hydroxyl group of a sugar. Hexokinase can use any one of several six-carbon sugars, including glucose and fructose, the two components of sucrose (common table sugar), as substrates. Glucokinase is an isozyme of hexokinase that is primarily involved in glucose metabolism. The K_M for hexokinase is 0.15 mM for glucose and 1.5 mM for fructose. The K_M for glucokinase, a liver-specific enzyme, is 20 mM. (We shall use the expression K_M here, even though some hexokinases studied do not follow Michaelis–Menten kinetics, and the term $[S]_{0.5}$ might be more appropriate. Not all enzymes have a K_M, but they do all have a substrate concentration that gives rise to $\frac{1}{2}V_{max}$.)

Comparison of these numbers tells us a lot about sugar metabolism. Because the resting level for blood glucose is about 5 mM, hexokinase would be expected to be fully active for all body cells. The liver would not be competing with the other cells for glucose. However, after a carbohydrate-rich meal, the blood glucose levels often exceed 10 mM, and, at that concentration, the liver glucokinase would have reasonable activity. Furthermore, since the enzyme is found only in the liver, the excess glucose will be preferentially taken into the liver, where it can be stored as glycogen until it is needed. Also, the comparison of the two sugars for hexokinase indicates clearly that glucose is preferred over fructose as a nutrient.

Similarly, if one compares the form of the enzyme lactate dehydrogenase found in heart muscle to the type found in skeletal muscle, one can see that there are small differences in amino acid composition. These differences in turn affect the reaction catalyzed by this enzyme, the conversion of pyruvate to lactate. The heart type has a high K_M, or a *low* affinity for pyruvate, and the muscle type has a low K_M, or a *high* affinity for pyruvate. This means that the pyruvate will be preferentially converted to lactate in the muscle but will be preferentially used for aerobic metabolism in the heart, rather than being converted to lactate. These conclusions are consistent with the known biology and metabolism of these two tissues.

Comparison of Turnover Number

As can be seen from Table 6.2, the first two enzymes are very reactive; catalase has one of the highest turnover numbers of all known enzymes. These high numbers allude to their importance in detoxifying hydrogen peroxide and preventing formation of CO_2 bubbles in the blood; these are their respective reactions. The values for chymotrypsin and acetylcholinesterase are within the range for "normal" metabolic enzymes. Lysozyme is an enzyme that degrades certain polysaccharide components of bacterial cell walls. It is present in many body tissues. Its low catalytic efficiency indicates that it operates well enough to catalyze polysaccharide degradation under normal conditions.

Comparison of k_{cat}/K_M

Even though the k_{cat} alone is indicative of the catalytic efficiency under saturating substrate conditions, [S] is rarely saturating under physiological conditions for many enzymes. The in vivo ratio of $[S]/K_M$ is often in the range of 0.01 to 1, meaning that active sites are not filled with substrate. Under these conditions, the level of substrate is small, and the amount of free enzyme approximates the level of total enzyme, because most of it is not bound to substrate. The Michaelis–Menten equation can be rewritten in the following form:

$$V = \frac{V_{max}[S]}{K_M + [S]} = \frac{k_{cat}[E_T][S]}{K_M + [S]}$$

If we then replace E_T with E and assume that the [S] is negligible compared with K_M, we can rewrite the equation as follows:

$$V = (k_{cat}/K_M)[E][S]$$

Thus, under these conditions, the ratio of k_{cat} to K_M is a second-order rate constant and provides a measure of the catalytic efficiency of the enzyme under nonsaturating conditions. The ratio of k_{cat} to K_M is much more constant between different enzymes than either the K_M or k_{cat} alone. Looking at the first three enzymes in Table 6.2, we can see that the k_{cat} values vary over a range of nearly 3000. The K_M values vary over a range of nearly 300. When the ratio of k_{cat} to K_M is compared, however, the range is only 4. The upper limit of a second-order rate constant is dependent on the diffusion-controlled limit of how fast the E and S can come together. The diffusion limit in an aqueous environment is in the range of 10^8 to 10^9. Many enzymes have evolved to have k_{cat} to K_M ratios that do indeed allow reactions to proceed at these limiting rates. This is referred to as being catalytically perfect.

Specific Enzyme Locations

We have already seen an important example here. Because the liver is the only organ in the human body with glucokinase, it must be the major organ for storage of excess dietary sugar as glycogen. Similarly, to replenish blood glucose levels, the glucose produced in the tissue must have its phosphate group removed by an enzyme called glucose phosphatase. Because this enzyme is found only in the liver and, to a lesser extent, in the kidney, we now know that the liver has the primary role of maintaining blood glucose levels.

In the presence of a competitive inhibitor, the equation for an enzymatic reaction becomes

$$EI \rightleftharpoons E \underset{+I}{\overset{+S}{\rightleftharpoons}} ES \rightarrow E + P$$

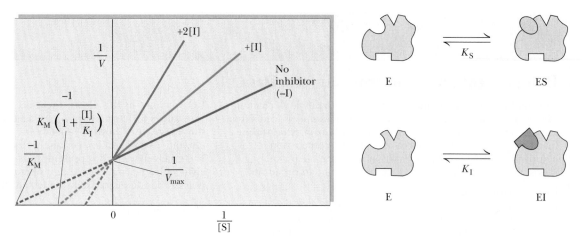

Biochemistry ⓈNow™ ▲ ACTIVE FIGURE 6.12 A Lineweaver–Burk double-reciprocal plot of enzyme kinetics for competitive inhibition. **Watch this Active Figure at http://now.brookscole .com/campbell5**

Essential Information

The action of enzymes can be inhibited reversibly or irreversibly. Irreversible inhibition normally involves formation or breaking of covalent bonds in the enzyme. In reversible inhibition, some substance can bind to the enzyme and subsequently be released. These reversible inhibitors can be divided into two major groups: competitive inhibitors that bind at the active site and prevent binding of substrate and noncompetitive inhibitors that bind at a site other than the active site, changing the structure of the enzyme and preventing catalysis.

Competitive inhibition

where EI is the enzyme–inhibitor complex. The dissociation constant for the enzyme–inhibitor complex can be written

$$EI \rightleftharpoons E + I$$

$$K_I = \frac{[E][I]}{[EI]}$$

It can be shown algebraically (although we shall not do it here) that, in the presence of inhibitor the value of K_M increases by the factor

$$1 + \frac{[I]}{K_I}$$

If we substitute $K_M (1 + [I]/K_I)$ for K_M in Equation 6.17, we obtain

$$\frac{1}{V} = \frac{K_M}{V_{max}} \times \frac{1}{[S]} + \frac{1}{V_{max}}$$

$$\frac{1}{V} = \frac{K_M}{V_{max}}\left(1 + \frac{[I]}{K_I}\right) \times \frac{1}{[S]} + \frac{1}{V_{max}}$$

$$y = \qquad m \qquad \times \quad x \;+\; b \qquad\qquad (6.18)$$

Here the term $1/V$ takes the place of the y coordinate, and the term $1/[S]$ takes the place of the x coordinate, as was the case in Equation 6.17. The intercept $1/V_{max}$, the b term in the equation for a straight line, has not changed from the earlier equation, but the slope K_M/V_{max} in Equation 6.17 has increased by the factor $(1 + [I]/K_I)$. The slope, the m term in the equation for a straight line, is now

$$\frac{K_M}{V_{max}}\left(1 + \frac{[I]}{K_I}\right)$$

accounting for the changes in the slope of the Lineweaver–Burk plot. Note that the y intercept does not change. This algebraic treatment of competitive inhibition agrees with experimental results, validating the model, just as experimental results validate the underlying Michaelis–Menten model for enzyme action. It is important to remember that the most distinguishing characteristic of a competitive inhibitor is that substrate or inhibitor can bind the enzyme, but not both. Because both are vying for the same location, suffi-

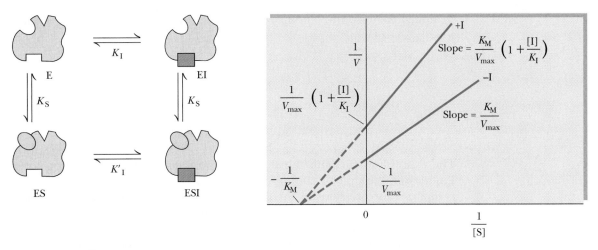

Biochemistry ⓔ Now™ ▲ **ACTIVE FIGURE 6.13** A Lineweaver–Burk plot of enzyme kinetics for noncompetitive inhibition. **Watch this Active Figure at http://now.brookscole.com/campbell5**

ciently high substrate will "outcompete" the inhibitor. This is why the V_{max} does not change; it is a measure of the velocity at infinite [substrate].

Kinetics of Noncompetitive Inhibition

The kinetic results of noncompetitive inhibition differ from those of competitive inhibition. The Lineweaver–Burk plots for a reaction in the presence and absence of a noncompetitive inhibitor show that both the slope and the y intercept change for the inhibited reaction (Figure 6.13), without changing the x intercept. The value of V_{max} decreases, but that of K_M remains the same; the inhibitor does not interfere with the binding of substrate to the active site. Increasing the substrate concentration cannot overcome noncompetitive inhibition because the inhibitor and substrate are not competing for the same site.

The reaction pathway has become considerably more complicated, and several equilibria must be considered.

$$
\begin{array}{c}
+S \\
E \rightleftharpoons ES \rightarrow E + P \\
+I \Updownarrow \qquad \Updownarrow +I \\
EI \leftrightharpoons ESI \\
+S
\end{array}
$$

In the presence of a noncompetitive inhibitor, I, the maximum velocity of the reaction, V^I_{max}, has the form (we shall not do the derivation here)

$$
V^I_{max} = \frac{V_{max}}{1 + [I]/K_I}
$$

where K_I is again the dissociation constant for the enzyme–inhibitor complex, EI. Recall that the maximum rate, V_{max}, appears in the expressions for both the slope and the intercept in the equation for the Lineweaver–Burk plot (Equation 6.17):

$$
\frac{1}{V} = \frac{K_M}{V_{max}} \times \frac{1}{[S]} + \frac{1}{V_{max}}
$$

$$
y = m \times x + b
$$

In noncompetitive inhibition, we replace the term V_{max} with the expression for V^I_{max}, to obtain

$$\underbrace{\frac{1}{V}}_{y\,=} = \underbrace{\frac{K_M}{V_{max}}\left(1 + \frac{[I]}{K_I}\right)}_{m} \times \underbrace{\frac{1}{[S]}}_{x} + \underbrace{\frac{1}{V_{max}}\left(1 + \frac{[I]}{K_I}\right)}_{b}$$

(6.19)

Noncompetitive inhibition

The expressions for both the slope and the intercept in the equation for a Lineweaver–Burk plot of an uninhibited reaction have been replaced by more complicated expressions in the equation that describes noncompetitive inhibition. This interpretation is borne out by the observed results. With a pure, noncompetitive inhibitor, the binding of substrate does not affect the binding of inhibitor, and vice versa. Since the K_M is a measure of the affinity of the enzyme and substrate, and since the inhibitor does not affect the binding, the K_M does not change with noncompetitive inhibition.

The two types of inhibition presented here are the two extreme cases. There are many other types of inhibition. **Uncompetitive inhibition** is seen when an inhibitor can bind to the ES complex but not to free E. A Lineweaver–Burk plot of an uncompetitive inhibitor shows parallel lines. The V_{max} decreases and the apparent K_M decreases as well. Noncompetitive inhibition is actually a limiting case of a more general inhibition type called **mixed inhibition.** With a mixed inhibitor, the same binding diagram is seen as in the equilibrium equations above, but, in this case, the binding of inhibitor does affect the binding of substrate and vice versa. A Lineweaver–Burk plot of an enzyme plus mixed inhibitor gives lines that intersect in the left-hand quadrant of the graph. The K_M increases, and the V_{max} decreases.

Practice Session

Sucrose (common table sugar) is hydrolyzed to glucose and fructose (Section 16.3) in a classic experiment in kinetics. The reaction is catalyzed by the enzyme invertase. Using the following data, determine, by the Lineweaver–Burk method, whether the inhibition of this reaction by 2 *M* urea is competitive or noncompetitive.

Sucrose Concentration (mol L^{-1})	V, no inhibitor (arbitrary units)	V, Inhibitor Present (same arbitrary units)
0.0292	0.182	0.083
0.0584	0.265	0.119
0.0876	0.311	0.154
0.117	0.330	0.167
0.175	0.372	0.192

Solution:

Plot the data with the reciprocal of the sucrose concentration on the *x* axis and the reciprocals of the two reaction velocities on the *y* axis. Note that the two plots have different slopes and different *y* axis intercepts, typical of noncompetitive inhibition. Note the same intercept on the negative *x* axis, which gives $-1/K_M$.

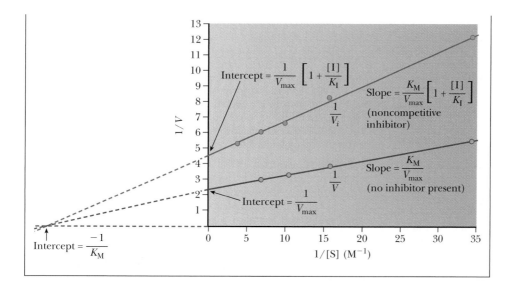

$$\text{Intercept} = \frac{1}{V_{max}} \left[1 + \frac{[I]}{K_I}\right]$$

$$\text{Slope} = \frac{K_M}{V_{max}} \left[1 + \frac{[I]}{K_I}\right]$$
(noncompetitive inhibitor)

$$\frac{1}{V_i}$$

$$\text{Slope} = \frac{K_M}{V_{max}}$$
(no inhibitor present)

$$\frac{1}{V}$$

$$\text{Intercept} = \frac{1}{V_{max}}$$

$$\text{Intercept} = \frac{-1}{K_M}$$

$1/V$

$1/[S]$ (M^{-1})

Biochemical Connections

Enzyme Inhibition in the Treatment of AIDS

A key strategy in the treatment of acquired immunodeficiency syndrome, or AIDS, has been to develop specific inhibitors that selectively block the actions of enzymes unique to the human immunodeficiency virus (HIV), which causes AIDS. Many laboratories are working on this approach to the development of therapeutic agents.

One of the most important target enzymes is HIV protease, an enzyme essential to the production of new virus particles in infected cells. HIV protease is unique to this virus. It catalyzes the processing of viral proteins in an infected cell. Without these proteins, viable virus particles cannot be released to cause further infection. The structure of HIV protease, including its active site, was known from the results of X-ray crystallography. With this structure in mind, scientists have designed and synthesized compounds to bind to the active site. Improvements were made in the drug design by obtaining structures of a series of

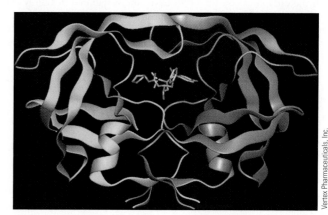

▲ Active site of VX-478 complexed with HIV-1 protease.

Vertex Pharmaceuticals, Inc.

inhibitors bound to the active site of HIV protease. These structures were also elucidated by X-ray crystallography. This process eventually led to several compounds marketed by several different pharmaceutical companies. These HIV protease inhibitors include saquinavir from Hoffman-LaRoche, ritonavir from Abbott Laboratories, indinavir from Merck, Viracept from Pfizer, and amprenavir from Vertex Pharmaceuticals. (These companies maintain highly informative home pages on the World Wide Web.)

Treatment of AIDS is most effective when a combination of drug therapies is used, and HIV protease inhibitors play an important role. Especially promising results (e.g., lowering of levels of the virus in the bloodstream) are obtained when HIV protease inhibitors are part of drug therapies for AIDS.

▲ Structure of amprenavir (VX-478), an HIV protease inhibitor developed by Vertex Pharmaceuticals. (*Vertex Pharmaceuticals, Inc.*)

Summary

6.1 What Makes Enzymes Such Effective Biological Catalysts?
Probably the most important function of proteins is catalysis. Biological catalysts are called enzymes. With the exception of some recently discovered RNAs that have catalytic activity, all enzymes are globular proteins. Enzymes are the most efficient catalysts known.

6.2 What Is the Difference between the Kinetic and the Thermodynamic Aspects of Reactions?
Catalysts speed up a reaction by lowering the activation energy, a kinetic parameter. They do not affect the thermodynamics of the reaction.

6.3 How Can We Describe Enzyme Kinetics in Mathematical Terms?
We can describe the kinetic aspects of reactions by rate equations. In such equations, we define reaction rate in terms of the disappearance of reactants or the appearance of products. We can describe rates in terms of changes in the concentrations of reactants or products as a function of time.

6.4 How Do Substrates Bind to Enzymes?
The first step in an enzyme-catalyzed reaction is the binding of the enzyme to the substrate to form a complex. The formation of the complex leads to formation of the transition-state species, which, in turn, forms the product. A substrate binds to a small portion of the enzyme called the active site. Two models have been proposed to describe enzyme–substrate binding: the lock-and-key model, in which there is an exact fit between the enzyme and substrate, and the induced-fit model, in which the enzyme is considered to have conformational flexibility and there is an exact fit only when the substrate is bound. The active site of an enzyme forces the substrate to mimic the transition state of the reaction, which is the primary way the activation energy of the reaction is lowered.

6.5 What Are Some Examples of Enzyme-Catalyzed Reactions?
In some enzyme-catalyzed reactions, the rate of reaction rises as substrate concentration increases and then levels off. When this behavior is shown on a graph, the curve is hyperbolic. Chymotrypsin is a digestive enzyme that shows this behavior. In other reactions, the shape of the curve is sigmoidal. Aspartate transcarbamoylase is an enzyme involved in the production of pyrimidines, and it shows sigmoidal kinetics.

6.6 What Is the Michaelis–Menten Approach to Enzyme Kinetics?
The kinetics of many enzyme-catalyzed reactions can be described by the Michaelis–Menten model. In this model, the concept of the steady state, with a constant concentration of the enzyme–substrate complex, plays a vital role. Much can be learned about the nature of an enzyme-catalyzed reaction by determining the kinetic constants, K_M and k_{cat}, for the enzyme.

6.7 How Do Enzymatic Reactions Respond to Inhibitors?
Inhibitors can give a considerable amount of information about enzymatic reactions. A reversible inhibitor can bind to the enzyme and subsequently be released. An irreversible inhibitor reacts with the enzyme to produce a protein that is not enzymatically active. Two major kinds of reversible inhibitors are competitive and noncompetitive inhibitors. Competitive inhibitors bind to the active site and block access of the substrate to the active site. Noncompetitive inhibitors bind to the enzyme at a site other than the active site and cause a change in the structure of the enzyme, especially around the active site, as a result of binding. In the Michaelis–Menten model, competitive inhibitors increase the K_M but leave the V_{max} unchanged; noncompetitive inhibitors change the V_{max} but leave the K_M unchanged.

Critical Questions to Review

6.1 What Makes Enzymes Such Effective Biological Catalysts?

1. **Fact Check** How does the catalytic effectiveness of enzymes compare with that of nonenzymatic catalysts?

2. **Fact Check** Are all enzymes proteins?

3. **Mathematical** Catalase breaks down hydrogen peroxide about 10^7 times faster than the uncatalyzed reaction. If the latter required one year, how much time would be needed by the catalase catalyzed reaction?

4. **Thought Question** Give two reasons why enzyme catalysts are 10^3 to 10^5 more effective than reactions that are catalyzed by, for example, simple H^+ or OH^-.

6.2 What Is the Difference between the Kinetic and Thermodynamic Aspects of Reactions?

5. **Fact Check** For the reaction of glucose with oxygen to produce carbon dioxide and water,

$$\text{Glucose} + 6\,O_2 \rightarrow 6\,CO_2 + 6\,H_2O$$

the $\Delta G°$ is –2880 kJ mol^{-1}, a strongly exergonic reaction. However, a sample of glucose can be maintained indefinitely in an oxygen-containing atmosphere. Reconcile these two statements.

6. **Thought Question** Would nature rely on the same enzyme to catalyze a reaction either way (forward or backward) if the $\Delta G°$ were -0.8 kcal mol^{-1}? If it were -5.3 kcal mol^{-1}?

7. **Thought Question** Suggest a reason why heating a solution containing an enzyme markedly decreases its activity. Why is the decrease of activity frequently much less when the solution contains high concentrations of the substrate?

8. **Thought Question** A model is proposed to explain the reaction catalyzed by an enzyme. Experimentally obtained rate data fit the model to within experimental error. Do these findings prove the model or not?

9. **Thought Question** Does the presence of a catalyst alter the standard free energy change of a chemical reaction?

10. **Thought Question** What effect does a catalyst have on the activation energy of a reaction?

11. **Thought Question** An enzyme catalyzes the formation of ATP from ADP and phosphate ion. What is its effect on the rate of hydrolysis of ATP to ADP and phosphate ion?

12. **Thought Question** Can the presence of a catalyst increase the amount of product obtained in a reaction?

6.3 How Can We Describe Enzyme Kinetics in Mathematical Terms?

13. **Fact Check** For the hypothetical reaction

$$3\,A + 2\,B \rightarrow 2\,C + 3\,D$$

the rate was experimentally determined to be

$$\text{Rate} = k[A]^1\,[B]^1$$

What is the order of the reaction with respect to A? With respect to B? What is the overall order of the reaction? Suggest how many molecules each of A and B are likely to be involved in the detailed mechanism of the reaction.

14. **Thought Question** The enzyme lactate dehydrogenase catalyzes the reaction

$$\text{Pyruvate} + NADH + H^+ \rightarrow \text{lactate} + NAD^+$$

NADH absorbs light at 340 nm in the near ultraviolet region of the electromagnetic spectrum, but NAD^+ does not. Suggest an experimental method for following the rate of this reaction, assuming that you have available a spectrophotometer capable of measuring light at this wavelength.

15. **Thought Question** Would you use a pH meter to monitor the progress of the reaction described in Question 14? Why or why not?

16. **Thought Question** Suggest a reason for carrying out enzymatic reactions in buffer solutions.

6.4 How Do Substrates Bind to Enzymes?

17. **Fact Check** Distinguish between the lock-and-key and induced-fit models for binding of a substrate to an enzyme.

18. **Fact Check** Using an energy diagram, show why the lock-and-key model could lead to an inefficient enzyme mechanism (*Hint:* Remember that the distance to the transition state must be minimized for an enzyme to be an effective catalyst).

19. **Thought Question** Other things being equal, what is a potential disadvantage of an enzyme having a very high affinity for its substrate?

20. **Thought Question** Amino acids that are far apart in the amino acid sequence of an enzyme can be essential for its catalytic activity. What does this suggest about its active site?

21. **Thought Question** If only a few of the amino acid residues of an enzyme are involved in its catalytic activity, why does the enzyme need such a large number of amino acids?

22. **Thought Question** A chemist synthesizes a new compound that may be structurally analogous to the transition-state species in an enzyme-catalyzed reaction. The compound is experimentally shown to inhibit the enzymatic reaction strongly. Is it likely that this compound is indeed a transition-state analog?

6.5 What Are Some Examples of Enzyme-Catalyzed Reactions?

23. **Fact Check** Show graphically the dependence of reaction velocity on substrate concentration for an enzyme that follows Michaelis–Menten kinetics and for an allosteric enzyme.

24. **Fact Check** Do all enzymes display kinetics that obey the Michaelis–Menten equation? Which ones do not?

25. **Fact Check** How can you recognize an enzyme that does not display Michaelis–Menten kinetics?

6.6 What Is the Michaelis–Menten Approach to Enzyme Kinetics?

26. **Fact Check** Show graphically how the reaction velocity depends on the enzyme concentration. Can a reaction be saturated with enzyme?

27. **Fact Check** Define *steady state,* and comment on the relevance of this concept to theories of enzyme reactivity.

28. **Fact Check** How is the turnover number of an enzyme related to V_{max}?

29. **Mathematical** For an enzyme that displays Michaelis–Menten kinetics, what is the reaction velocity, V (as a percentage of V_{max}), observed at (a) $[S] = K_M$; (b) $[S] = 0.5K_M$; (c) $[S] = 0.1K_M$; (d) $[S] = 2K_M$; (e) $[S] = 10K_M$?

30. **Mathematical** Determine the values of K_M and V_{max} for the decarboxylation of a β-keto acid given the following data.

Substrate Concentration (mol L^{-1})	Velocity (mM min^{-1})
2.500	0.588
1.000	0.500
0.714	0.417
0.526	0.370
0.250	0.256

31. **Mathematical** The kinetic data in the following table were obtained for the reaction of carbon dioxide and water to produce bicarbonate and hydrogen ion catalyzed by carbonic anhydrase

$$CO_2 + H_2O \rightarrow HCO_3^- + H^+$$

(H. De Voe and G. B. Kistiakowsky, *J. Am. Chem. Soc.* **83,** 274, 1961). From these data, determine K_M and V_{max} for the reaction.

Carbon Dioxide Concentration (mmol L^{-1})	1/Velocity (M^{-1} sec)
1.25	36×10^3
2.5	20×10^3
5.0	12×10^3
20.0	6×10^3

32. **Mathematical** The enzyme β-methylaspartase catalyzes the deamination of β-methylaspartate

mesaconate
absorbs at
240 nm

(V. Williams and J. Selbin, *J. Biol. Chem.* **239,** 1636, 1964). The rate of the reaction was determined by monitoring the absorbance of the product at 240 nm (A_{240}). From the data in the following table, determine K_M for the reaction. How does the method of calculation differ from that in Exercises 30 and 31?

Substrate Concentration (mol L^{-1})	Velocity (ΔA_{240} min^{-1})
0.002	0.045
0.005	0.115
0.020	0.285
0.040	0.380
0.060	0.460
0.080	0.475
0.100	0.505

33. **Mathematical** The hydrolysis of a phenylalanine-containing peptide is catalyzed by α-chymotrypsin with the following results. Calculate K_M and V_{max} for the reaction.

Peptide Concentration (M)	Velocity (M min^{-1})
2.5×10^{-4}	2.2×10^{-6}
5.0×10^{-4}	3.8×10^{-6}
10.0×10^{-4}	5.9×10^{-6}
15.0×10^{-4}	7.1×10^{-6}

34. **Mathematical** For the V_{max} obtained in Question 30, calculate the turnover number (catalytic rate constant) assuming that 1×10^{-4} mol of enzyme were used.

35. **Mathematical** You do an enzyme kinetic experiment and calculate a V_{max} of 100 μmol product per minute. If each assay used 0.1 mL of an enzyme solution that had a concentration of 0.2 mg/mL, what would be the turnover number if the enzyme had a molecular weight of 128,000 g/mol?

36. **Thought Question** The enzyme D-amino acid oxidase has a very high turnover number because the D-amino acids are potentially toxic. The K_M for the enzyme is in the range of 1 to 2 mM for the aromatic amino acids and in the range of 15 to 20 mM for such amino acids as serine, alanine, and the acidic amino acids. Which of these amino acids are the preferred substrates for the enzyme?

37. **Thought Question** Why is it useful to plot rate data for enzymatic reactions as a straight line rather than as a curve?

38. **Thought Question** Under what conditions can we make the assumption that K_M is an indication of the binding affinity between substrate and enzyme?

6.7 How Do Enzymatic Reactions Respond to Inhibitors?

39. **Fact Check** How can competitive and noncompetitive inhibition be distinguished in terms of K_M?

40. **Fact Check** Why does a competitive inhibitor not change the V_{max}?

41. **Fact Check** Why does a noncompetitive inhibitor not change the observed K_M?

42. **Fact Check** Distinguish between the molecular mechanisms of competitive and noncompetitive inhibition.

43. **Fact Check** Can enzyme inhibition be reversed in all cases?

44. **Fact Check** Why is a Lineweaver–Burk plot useful in analyzing kinetic data from enzymatic reactions?

45. **Fact Check** Where do lines intersect on a Lineweaver–Burk plot showing competitive inhibition? On a Lineweaver–Burk plot showing noncompetitive inhibition?

46. **Mathematical** Draw Lineweaver–Burk plots for the behavior of an enzyme for which the following experimental data are available.

[S] (mM)	V, No Inhibitor (mmol min^{-1})	V, Inhibitor Present (mmol min^{-1})
3.0	4.58	3.66
5.0	6.40	5.12
7.0	7.72	6.18
9.0	8.72	6.98
11.0	9.50	7.60

What are the K_M and V_{max} values for the inhibited and uninhibited reactions? Is the inhibitor competitive or noncompetitive?

47. **Mathematical** For the following aspartase reaction (see Question 32) in the presence of the inhibitor hydroxymethylaspartate, determine the K_M and whether the inhibition is competitive or noncompetitive.

[S] (molarity)	V, No Inhibitor (arbitrary units)	V, Inhibitor Present (same arbitrary units)
1×10^{-4}	0.026	0.010
5×10^{-4}	0.092	0.040
1.5×10^{-3}	0.136	0.086
2.5×10^{-3}	0.150	0.120
5×10^{-3}	0.165	0.142

48. **Thought Question** Is it good (or bad) that enzymes can be reversibly inhibited? Why?

49. **Thought Question** Noncompetitive inhibition is a limiting case in which the effect of binding inhibitor has no effect on the affinity for the substrate and vice versa. Suggest what a Lineweaver–Burk plot would look like for an inhibitor that had a reaction scheme similar to that on page 149 [noncompetitive inhibition reaction], but where binding inhibitor lowered the affinity of EI for the substrate.

50. **Biochemical Connections** You have been hired by a pharmaceutical company to work on development of drugs to treat AIDS. What information from this chapter will be useful to you?

51. **Thought Question** Would you expect an irreversible inhibitor of an enzyme to be bound by covalent or by noncovalent interactions? Why?

52. **Thought Question** Would you expect the structure of a noncompetitive inhibitor of a given enzyme to be similar to that of its substrate?

Biochemistry ⊘Now™

Assess your understanding of this chapter's topics with additional quizzing and tutorials at **http://now.brookscole.com/campbell5**

Annotated Bibliography

Althaus, I., J. Chou, A. Gonzales, M. Deibel, K. Chou, F. Kezdy, D. Romero, J. Palmer, R. Thomas, P. Aristoff, W. Tarpley, and F. Reusser. Kinetic Studies with the Non-nucleoside HIV-1 Reverse Transcriptase Inhibitor U-88204E. *Biochemistry* **32**, 6548–6554 (1993). [How enzyme kinetics can play a role in AIDS research.]

Bachmair, A., D. Finley, and A. Varshavsky. *In Vivo* Half-Life of a Protein Is a Function of Its Amino Terminal Residue. *Science* **234**, 179–186 (1986). [A particularly striking example of the relationship between structure and stability in proteins.]

Bender, M. L., R. L. Bergeron, and M. Komiyama. *The Bioorganic Chemistry of Enzymatic Catalysis.* New York: Wiley, 1984. [A discussion of mechanisms in enzymatic reactions.]

Danishefsky, S. Catalytic Antibodies and Disfavored Reactions. *Science* **259**, 469–470 (1993). [A short review of chemists' use of antibodies as the basis of "tailor-made" catalysts for specific reactions.]

Dressler, D., and H. Potter. *Discovering Enzymes.* New York: Scientific American Library, 1991. [A well-illustrated book that introduces important concepts of enzyme structure and function.]

Dugas, H., and C. Penney. *Bioorganic Chemistry: A Chemical Approach to Enzyme Action.* New York: Springer-Verlag, 1981. [Discusses model systems as well as enzymes.]

Fersht, A. *Enzyme Structure and Mechanism,* 2nd ed. New York: W. H. Freeman, 1985. [A thorough coverage of enzyme action.]

Hammes, G. *Enzyme Catalysis and Regulation.* New York: Academic Press, 1982. [A good basic text on enzyme mechanisms.]

Kraut, J. How Do Enzymes Work? *Science* **242**, 533–540 (1988). [An advanced discussion of the role of transition states in enzymatic catalysis.]

Lerner, R., S. Benkovic, and P. Schultz. At the Crossroads of Chemistry and Immunology: Catalytic Antibodies. *Science* **252**, 659–667 (1991). [A review of how antibodies can bind to almost any molecule of interest and then catalyze some reaction of that molecule.]

Marcus, R. Skiing the Reaction Rate Slopes. *Science* **256**, 1523–1524 (1992). [A brief, advanced-level look at reaction transition states.]

Moore, J. W., and R. G. Pearson. *Kinetics and Mechanism,* 3rd ed. New York: John Wiley Interscience, 1980. [A classic, quite advanced treatment of the use of kinetic data to determine mechanisms.]

Rini, J., U. Schulze-Gahmen, and I. Wilson. Structural Evidence for Induced Fit as a Mechanism for Antibody–Antigen Recognition. *Science* **255,** 959–965 (1992). [The results of structure determination by X-ray crystallography.]

Sigman, D., ed. *The Enzymes. Vol. 20. Mechanisms of Catalysis.* San Diego: Academic Press, 1992. [Part of a definitive series on enzymes and their structures and functions.]

Sigman, D., and P. Boyer, eds. *The Enzymes. Vol. 19. Mechanisms of Catalysis.* San Diego: Academic Press, 1990. [Part of a definitive series on enzymes and their structures and functions.]

CHAPTER 7

The Behavior of Proteins: Enzymes, Mechanisms, and Control

©Macduff Everton/CORBIS

Signals regulate the flow of traffic in much the same fashion as control mechanisms in chemical reactions.

Biochemistry ⋛ Now™
Test yourself on these Critical Questions at the BiochemistryNow website at **http://now** **.brookscole.com/campbell5**

A number of control mechanisms combine to regulate enzymatic pathways. The conformational changes that take place in allosteric enzymes can shut down long synthetic pathways at their first steps, frequently saving considerable amounts of energy for the cell. Conformational changes combined with covalent modification of enzymes give rise to a higher level of control. Enzymes can bind covalently to phosphate groups, affecting both their activity and their allosteric interactions. This process is reversible because phosphate groups can be removed by hydrolysis; however, the covalent modification in zymogen activation is irreversible. It involves the cleavage of bonds, followed by conformational change. No matter how enzyme activity is controlled, the net result is to ensure that the three-dimensional arrangement of the active site puts essential amino acid residues into position for optimum catalytic activity. Concepts from organic chemistry play an important role in catalysis, even though they operate in an unfamiliar environment. Nucleophilic substitution reactions with specific stereochemistry occur frequently in the active sites of enzymes. A number of other well-known kinds of reaction mechanisms occur in enzymatic reactions. In addition, reactions of enzymes may involve cofactors that are not amino acids but are compounds called vitamins, or metabolites of vitamins.

7.1 | Does the Michaelis–Menten Model Describe the Behavior of Allosteric Enzymes?

The behavior of many well-known enzymes can be described quite adequately by the Michaelis–Menten model, but allosteric enzymes behave very differently. In the last chapter, we saw that there are similarities between the reaction kinetics of an enzyme such as chymotrypsin, which does not display allosteric behavior, and the binding of oxygen by myoglobin, which is also an example of nonallosteric behavior. The analogy extends to show the similarity in the kinetic behavior of an allosteric enzyme such as aspartate transcarbamoylase (ATCase) and the binding of oxygen by hemoglobin. Both ATCase and hemoglobin are allosteric proteins; the behaviors of both exhibit cooperative effects caused by subtle changes in quaternary structure. (Recall that *quaternary structure* is the arrangement in space that results from the interaction of subunits through noncovalent forces, and that *positive cooperativity* refers to the fact that the binding of low levels of substrate facilitates the action of the protein at higher levels of substrate, whether the action is catalytic or some other kind of binding.) In addition to displaying cooperative kinetics, allosteric enzymes have a different response to the presence of inhibitors from that of nonallosteric enzymes, as characterized by the Michaelis–Menten model.

Control Mechanisms That Affect Allosteric Enzymes

ATCase catalyzes the first step in a series of reactions in which the end product is cytidine triphosphate (CTP), a nucleoside triphosphate needed to make RNA and DNA (Chapter 9). The pathways that produce nucleotides are energetically costly and involve many steps. The reaction catalyzed by aspartate transcarbamoylase is a good example of how such a pathway is controlled to avoid overproduction of such compounds. For DNA and RNA synthesis, the levels of several nucleotide triphosphates are controlled. CTP is an inhibitor of ATCase, the enzyme that catalyzes the first reaction in the pathway. This behavior is an example of **feedback inhibition** (also called end-product inhibition), in which the end product of the sequence of reactions inhibits the first reaction in the series (Figure 7.1). Feedback inhibition is an efficient control mechanism because the entire series of reactions can be shut down when an excess of the final product exists, thus preventing the accumulation of intermediates in the pathway. Feedback inhibition is a general feature of metabolism and is not confined to allosteric enzymes. However, the observed kinetics of the ATCase reaction, including the mode of inhibition, are typical of allosteric enzymes.

When ATCase catalyzes the condensation of aspartate and carbamoyl phosphate to form carbamoyl aspartate, the graphical representation of the rate as a function of increasing substrate concentration (aspartate) is a sigmoidal curve rather than the hyperbola obtained with nonallosteric enzymes (Figure 7.2a). The sigmoidal curve is indicative of the cooperative behavior of allosteric enzymes. In this two-substrate reaction, aspartate is the substrate for

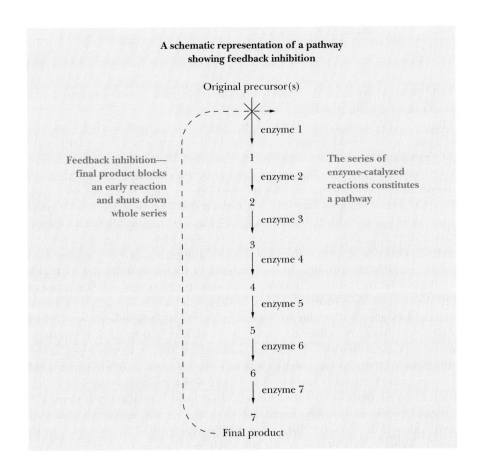

◀ FIGURE 7.1 Schematic representation of a pathway, showing feedback inhibition.

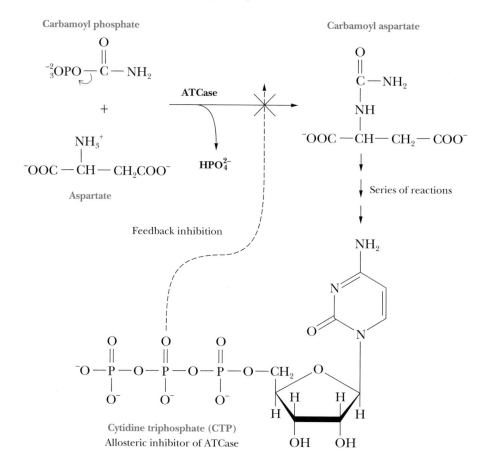

The reaction catalyzed by ATCase
leads eventually to the production of CTP

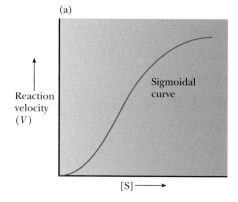

(a)

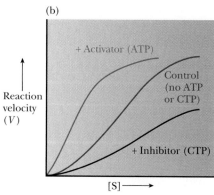

(b)

▲ **FIGURE 7.2** (a) Plot of velocity vs. substrate concentration (aspartate) for aspartate trans-carbamoylase. (b) The effect of inhibitors and activators on an allosteric enzyme.

which the concentration is varied, while the concentration of carbamoyl phosphate is kept constant at high levels.

Figure 7.2b compares the rate of the uninhibited reaction of ATCase with the reaction rate in the presence of CTP. In the latter case, a sigmoidal curve still describes the rate behavior of the enzyme, but the curve is shifted to higher substrate levels; a higher concentration of aspartate is needed for the enzyme to achieve the same rate of reaction. At high substrate concentrations, the same maximal rate, V_{max}, is observed in the presence and absence of **inhibitor.** (Recall this from Section 6.7.) Because in the Michaelis–Menten scheme the V_{max} changes when a reaction takes place in the presence of a noncompetitive inhibitor, noncompetitive inhibition cannot be the case here. The same Michaelis–Menten model associates this sort of behavior with competitive inhibition, but that part of the model still does not provide a reasonable picture. Competitive inhibitors bind to the same site as the substrate because they are very similar in structure. The CTP molecule is very *different* in structure from the substrate, aspartate, and it is bound to a different site on the ATCase molecule. ATCase is made up of two different types of subunits. One of them is the catalytic subunit, which consists of six protein subunits organized into two trimers. The other is the regulatory subunit, which also consists of six protein subunits organized into three dimers (Figure 7.3). The catalytic subunits can be separated from the regulatory subunits by treatment with *p*-hydroxymercuribenzoate, which reacts with the cysteines in the protein. When so treated, ATCase still catalyzes the reaction, but it loses its allosteric control by CTP, and the curve becomes hyperbolic.

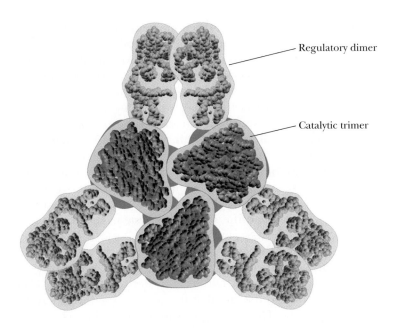

Regulatory dimer

Catalytic trimer

◀ **FIGURE 7.3** Organization of aspartate transcarbamoylase, showing the two catalytic trimers and the three regulatory dimers.

The situation becomes "curiouser and curiouser" when the ATCase reaction takes place not in the presence of CTP, a pyrimidine nucleoside triphosphate, but in the presence of adenosine triphosphate (ATP), a purine nucleoside triphosphate. The structural similarities between CTP and ATP are apparent, but ATP is not a product of the pathway that includes the reaction of ATCase and that produces CTP. Both ATP and CTP are needed for the synthesis of RNA and DNA. The relative proportions of ATP and CTP are specified by the needs of the organism. If there is not enough CTP relative to the amount of ATP, the enzyme requires a signal to produce more. In the presence of ATP, the rate of the enzymatic reaction is increased at lower levels of aspartate, and the shape of the rate curve becomes less sigmoidal and more hyperbolic (Figure 7.2b). In other words, there is less cooperativity in the reaction. The binding site for ATP on the enzyme molecule is the same as that for CTP (which is not surprising in view of their structural similarity), but ATP is an activator rather than an inhibitor like CTP. When CTP is in short

Adenosine triphosphate (ATP)
a purine nucleotide;
activator of
ATCase

supply in an organism, the ATCase reaction is not inhibited, and the binding of ATP increases the activity of the enzyme still more.

Even though it is tempting to consider inhibition of allosteric enzymes in the same fashion as nonallosteric enzymes, much of the terminology is not appropriate. "Competitive inhibition" and "noncompetitive inhibition" are terms reserved for the enzymes that behave in line with Michaelis–Menten kinetics. With allosteric enzymes, the situation is more complex. In general, two types of enzyme systems exist, called **K systems** and **V systems.** A K system is an enzyme where the substrate concentration that yields one-half V_{max} is altered by the presence of inhibitors or activators. ATCase is an example of a K system. Because we are not dealing with a Michaelis–Menten type of enzyme, the term K_M is not applicable. For an allosteric enzyme, the substrate level at one-half V_{max} is called the $K_{0.5}$. In a V system, the effect of inhibitors and activators changes the V_{max}, but not the $K_{0.5}$.

The key to allosteric behavior, including cooperativity and modifications of cooperativity, is the existence of multiple forms for the quaternary structures of allosteric proteins. The word "allosteric" is derived from *allo,* "other," and *steric,* "shape," referring to the fact that the possible conformations affect the behavior of the protein. The binding of substrates, inhibitors, and activators changes the quaternary structure of allosteric proteins, and the changes in structure are reflected in the behavior of those proteins. A substance that modifies the quaternary structure, and thus the behavior, of an allosteric protein by binding to it is called an **allosteric effector.** The term "effector" can apply to substrates, inhibitors, or activators. Several models for the behavior of allosteric enzymes have been proposed, and it is worthwhile to compare them.

Let us first define two terms. **Homotropic** effects are allosteric interactions that occur when several identical molecules are bound to a protein. The binding of substrate molecules to different sites on an enzyme, such as the binding of aspartate to ATCase, is an example of a homotropic effect. **Heterotropic** effects are allosteric interactions that occur when different substances (such as inhibitor and substrate) are bound to the protein. In the ATCase reaction, inhibition by CTP and activation by ATP are both heterotropic effects.

7.2 | What Are the Models for the Behavior of Allosteric Enzymes?

The two principal models for the behavior of allosteric enzymes are the concerted model and the sequential model. They were proposed in 1965 and 1966, respectively, and both are currently used as a basis for interpreting experimental results. The concerted model has the advantage of comparative simplicity, and it describes the behavior of some enzyme systems very well. The sequential model sacrifices a certain amount of simplicity for a more realistic picture of the structure and behavior of proteins; it also deals very well with the behavior of some enzyme systems.

The Concerted Model for Allosteric Behavior

In 1965, Jacques Monod, Jeffries Wyman, and Jean-Pierre Changeux proposed the **concerted model** for the behavior of allosteric proteins in a paper that has become a classic in the biochemical literature. (It is listed in the bibliography at the end of this chapter.) In this picture, the protein has two conformations, the active R (relaxed) conformation, which binds substrate tightly, and the inactive T (tight, also called taut) conformation, which binds

(a)   A dimeric protein can exist in either of two
        conformational states at equilibrium.

(b)   Substrate binding shifts equilibrium in favor of R.

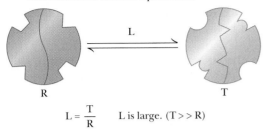

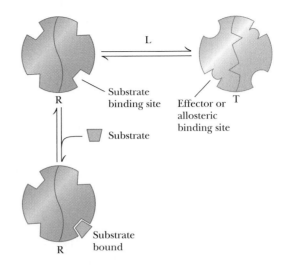

$$L = \frac{T}{R} \qquad \text{L is large. } (T >> R)$$

▲ **FIGURE 7.4** Monod–Wyman–Changeux (MWC) model for allosteric transitions, also called
the concerted model. (a) A dimeric protein can exist in either of two conformational states at
equilibrium, the T (taut) form or the R (relaxed) form. L is the ratio of the T form to the R form.
With most allosteric systems, L is large, so there is more enzyme present in the T form than in the
R form. (b) By Le Chatelier's principle, substrate binding shifts the equilibrium in favor of the
relaxed state (R) by removing unbound R. The dissociation constant for the enzyme–substrate
complex is K_R for the relaxed form and K_T for the taut form. $K_R < K_T$, so the substrate binds better to the relaxed form. The ratio of K_R/K_T is called c. This figure shows a limiting case in which
the taut form does not bind substrate at all, in which case K_T is infinite and $c = 0$.

substrate less tightly. The distinguishing feature of this model is that the conformations of *all* subunits change simultaneously. Figure 7.4a shows a hypothetical protein with two subunits. Both subunits change conformation from
the inactive T conformation to the active R conformation at the same time;
that is, a concerted change of conformation occurs. The equilibrium ratio of
the T/R forms is called L and is assumed to be high—that is, there is more of
the unbound T form present than the unbound R form. The binding of substrate to either form can be described by the dissociation constant of the
enzyme and substrate, *K*, with the affinity for substrate higher in the R form
than in the T form. Thus, $K_R \ll K_T$. The ratio of K_R/K_T is called *c*. Figure
7.4b shows a limiting case in which K_T is infinitely greater than K_R ($c = 0$). In
other words, substrate will not bind to the T form at all. The allosteric effect
is explained by this model based on perturbing the equilibrium between the
T and R forms. Although initially the amount of enzyme in the R form is
small, when substrate binds to the R form, it removes free R form. This causes
the production of more R form to reestablish the equilibrium, which makes
binding more substrate possible. This shifting of the equilibrium is responsible for the observed allosteric effects. The Monod–Wyman–Changeux model
has been shown mathematically to explain the sigmoidal effects seen with
allosteric enzymes. The shape of the curve will be based on the L and *c* values. As L increases (free T form more highly favored), the shape becomes
more sigmoidal (Figure 7.5). As the value for *c* decreases (higher affinity
between substrate and R form), the shape also becomes more sigmoidal.

In the concerted model, the effects of inhibitors and activators can also be
considered in terms of shifting the equilibrium between the T and R forms of
the enzyme. The binding of inhibitors to allosteric enzymes is cooperative;
allosteric inhibitors bind to and stabilize the T form of the enzyme. The binding of activators to allosteric enzymes is also cooperative; allosteric activators

bind to and stabilize the R form of the enzyme. When an activator, A, is present, the cooperative binding of A shifts the equilibrium between the T and R forms, with the R form favored (Figure 7.6). As a result, there is less need for substrate, S, to shift the equilibrium in favor of the R form, and less cooperativity in the binding of S is seen.

When an inhibitor, I, is present, the cooperative binding of I also shifts the equilibrium between the T and R forms, but this time the T form is favored

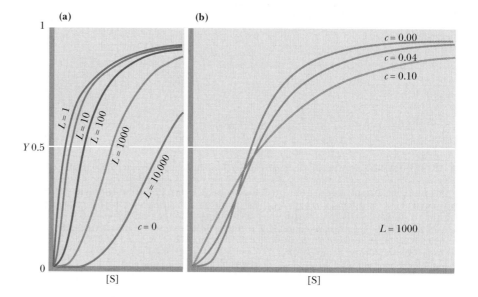

Biochemistry ⊛ Now™   **ANIMATED FIGURE 7.5**
The Monod–Wyman–Changeux (or concerted) model. (a) As L (the ratio of the T/R form) increases, the shape becomes more sigmoidal. (b) The level of cooperativity is also based on the affinity of the substrates for the T or R form. When K_T is infinite (zero affinity), cooperativity is high, as shown in the blue line, where $c = 0$ ($c = K_R/K_T$). As c increases, the difference in binding between the T and R forms decreases, and the lines become less sigmoidal. (*Adapted from Monod, J., Wyman, J., and Changeux, J.-P., 1965. On the nature of allosteric transitions: A plausible model.* Journal of Molecular Biology *12:92.*) **See this figure animated at http://now.brookscole.com/campbell5**

A dimeric protein that can exist in either of two states: R_0 or T_0. This protein can bind three ligands:

1) Substrate (S) ▆ : A positive homotropic effector that binds only to R at site S

2) Activator (A) ▲ : A positive heterotropic effector that binds only to R at site F

3) Inhibitor (I) ◥ : A negative heterotropic effector that binds only to T at site F

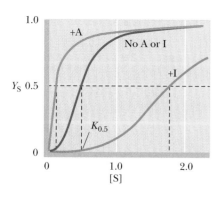

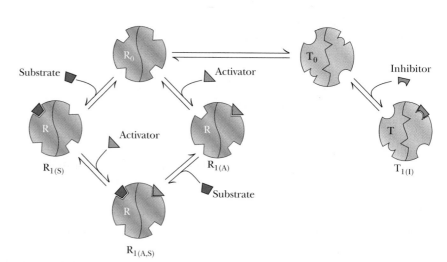

Effects of A:
$A + R_0 \longrightarrow R_{1(A)}$
Increase in number of R-conformers shifts $R_0 \rightleftharpoons T_0$ so that $T_0 \longrightarrow R_0$

(1) More binding sites for S made available.

(2) Decrease in cooperativity of substrate saturation curve. Effector A lowers the apparent value of L.

Effects of I:
$I + T_0 \longrightarrow T_{1(I)}$
Increase in number of T-conformers (decrease in R_0 as $R_0 \longrightarrow T_0$ to restore equilibrium)

Thus, I inhibits association of S and A with R by lowering R_0 level. I increases cooperativity of substrate saturation curve. I raises the apparent value of L.

Biochemistry ⊛ Now™   ▲ **ACTIVE FIGURE 7.6** Effects of binding activators and inhibitors with the concerted model. An activator is a molecule that stabilizes the R form. An inhibitor stabilizes the T form. **Watch this Active Figure at http://now.brookscole.com/campbell5**

(Figure 7.6). More substrate is needed to shift the T-to-R equilibrium in favor of the R form. A greater degree of cooperativity is seen in the binding of S.

The Sequential Model for Allosteric Behavior

The name Daniel Koshland is associated with the direct **sequential model** of allosteric behavior. The distinguishing feature of this model is that the binding of substrate induces the conformational change from the T form to the R form—the type of behavior postulated by the induced-fit theory of substrate binding. (The reference to the original article describing this model is given in the bibliography at the end of this chapter.) A conformational change from T to R in one subunit makes the same conformational change easier in another subunit, and this is the form in which cooperative binding is expressed in this model (Figure 7.7a).

In the sequential model, the binding of activators and inhibitors also takes place by the induced-fit mechanism. The conformational change that begins with binding of inhibitor or activator to one subunit affects the conformations of other subunits. The net result is to favor the R state when activator is present and to favor the T form when inhibitor, I, is present (Figure 7.7b). Binding I to one subunit causes a conformational change such that the T form is even less likely to bind substrate than before. This conformational change is passed along to other subunits, making them also more likely to bind inhibitor and less likely to bind substrate. This is an example of cooperative behavior that leads to more inhibition of the enzyme. Likewise, binding an activator causes a conformational change that favors substrate binding, and this effect is passed from one subunit to another.

The sequential model for binding effectors of all types, including substrates, to allosteric enzymes has a unique feature, not seen in the concerted model. The conformational changes thus induced can make the enzyme less likely to bind more molecules of the same type. This phenomenon, called **negative cooperativity,** has been observed in a few enzymes. One is tyrosyl tRNA synthetase, which plays a role in protein synthesis. In the reaction catalyzed by this enzyme, the amino acid tyrosine forms a covalent bond to a

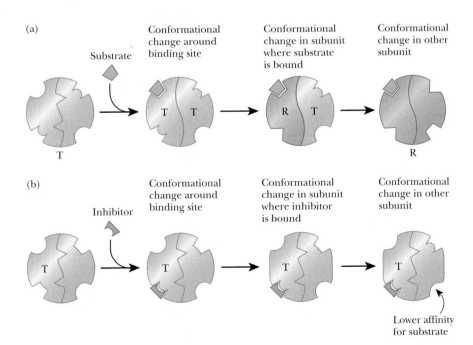

(a)

Substrate

Conformational change around binding site

Conformational change in subunit where substrate is bound

Conformational change in other subunit

T

T   T

R   T

R

(b)

Inhibitor

Conformational change around binding site

Conformational change in subunit where inhibitor is bound

Conformational change in other subunit

T

T

T

T

Lower affinity for substrate

◀ **FIGURE 7.7** (a) Sequential model of cooperative binding of substrate S to an allosteric enzyme. Binding substrate to one subunit induces the other subunit to adopt the R state, which has a higher affinity for substrate. (b) Sequential model of cooperative binding of inhibitor I to an allosteric enzyme. Binding inhibitor to one subunit induces a change in the other subunit to a form that has a lower affinity for substrate.

molecule of transfer RNA (tRNA). In subsequent steps, the tyrosine is passed along to its place in the sequence of the growing protein. The tyrosyl tRNA synthetase consists of two subunits. Binding of the first molecule of substrate to one of the subunits inhibits binding of a second molecule to the other subunit. The sequential model has successfully accounted for the negative cooperativity observed in the behavior of tyrosyl tRNA synthetase. The concerted model makes no provision for negative cooperativity.

| 7.3 | How Does Phosphorylation of Specific Residues Regulate Enzyme Activity? |

The side-chain hydroxyl groups of serine, threonine, and tyrosine can all form phosphate esters. The presence of the phosphate can convert an inactive precursor into an active enzyme, or vice versa. Transport across membranes provides an important example, such as the sodium–potassium ion pump, which moves potassium into the cell and sodium out (Section 8.6). The source of the phosphate group for the protein component of the sodium–potassium ion pump and for many enzyme phosphorylations is the ubiquitous ATP. When ATP is hydrolyzed to adenosine diphosphate (ADP), enough energy is released to allow a number of otherwise energetically unfavorable reactions to take place. In the case of the Na^+/K^+ pump, ATP donates a phosphate to aspartate 369 as part of the mechanism, causing a conformation change in the enzyme (Figure 7.8). Proteins that catalyze these phosphorylation reactions are called **protein kinases.** Kinase refers to an

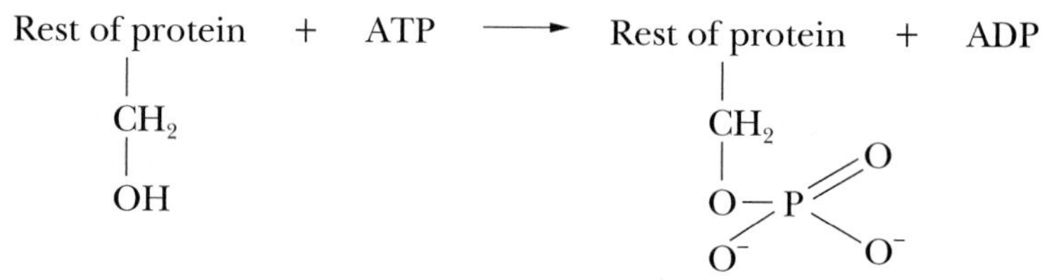

Serine residue · Phosphorylated serine residue

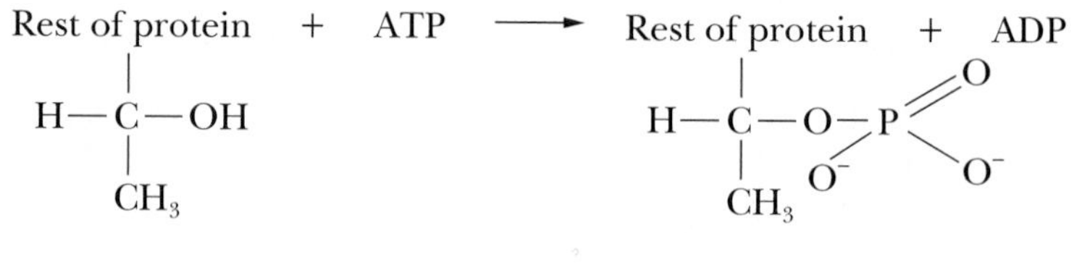

Threonine residue · Phosphorylated threonine residue

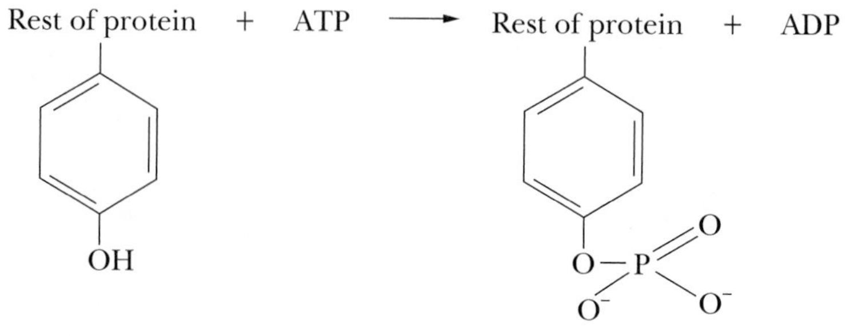

Tyrosine residue · Phosphorylated tyrosine residue

enzyme that catalyzes transfer of a phosphate group, almost always from ATP, to some substrate. These enzymes play an important role in metabolism.

Many examples appear in processes involved in generating energy, as is the case in carbohydrate metabolism. Glycogen phosphorylase, which catalyzes the initial step in the breakdown of stored glycogen (Section 18.1), exists in two forms—the phosphorylated glycogen phosphorylase *a* and the dephosphorylated glycogen phosphorylase *b* (Figure 7.9). The *a* form is more active than the *b* form, and the two forms of the enzyme respond to different allosteric effectors, depending on tissue type. Glycogen phosphorylase is thus subject to two kinds of control—allosteric regulation and covalent modification. The net result is that the *a* form is more abundant and active when phosphorylase is needed to break down glycogen to provide energy.

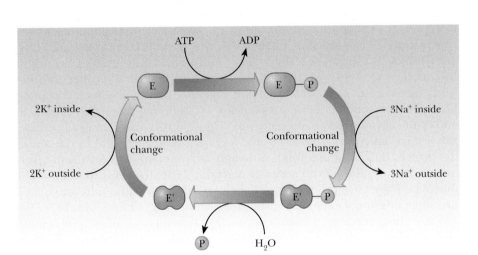

◀ **FIGURE 7.8** Phosphorylation of the sodium–potassium pump is involved in cycling the membrane protein between the form that binds to sodium and the form that binds to potassium.

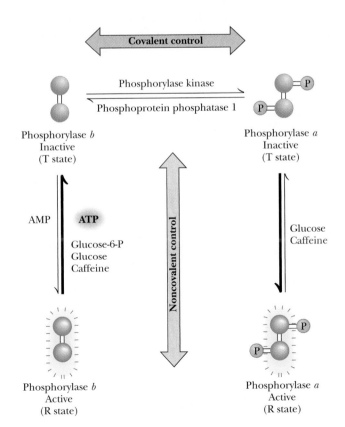

7.4 | What Are Zymogens, and How Do They Control Enzyme Activity?

Allosteric interactions control the behavior of proteins through reversible changes in quaternary structure, but this mechanism, effective though it may be, is not the only one available. A **zymogen,** an inactive precursor of an enzyme, can be irreversibly transformed into an active enzyme by cleavage of covalent bonds.

The proteolytic enzymes trypsin and chymotrypsin provide a classic example of zymogens and their activation. Their inactive precursor molecules, trypsinogen and chymotrypsinogen, respectively, are formed in the pancreas, where they would do damage if they were in an active form. In the small intestine, where their digestive properties are needed, they are activated by cleavage of specific peptide bonds. The conversion of chymotrypsinogen to chymotrypsin is catalyzed by trypsin, which in turn arises from trypsinogen as a result of a cleavage reaction catalyzed by the enzyme enteropeptidase. Chymotrypsinogen consists of a single polypeptide chain 245 residues long, with five disulfide (—S—S—) bonds. When chymotrypsinogen is secreted into the small intestine, trypsin present in the digestive system cleaves the peptide bond between arginine 15 and isoleucine 16, counting from the N-terminal end of the chymotrypsinogen sequence (Figure 7.10). The cleavage produces active π-chymotrypsin. The 15-residue fragment remains bound to the rest of the protein by a disulfide bond. Although π-chymotrypsin is fully active, it is not the end product of this series of reactions. It acts on itself to remove two dipeptide fragments, producing α-chymotrypsin, which is also fully active. The two dipeptide fragments cleaved off are Ser 14—Arg 15 and Thr 147—Asn 148; the final form of the enzyme, α-chymotrypsin, has three polypeptide chains held together by two of the five original, and still intact, disulfide bonds. (The other three disulfide bonds remain intact as well; they link portions of single polypeptide chains.) When the term "chymotrypsin" is used without specifying the π or the α form, the final α form is meant.

The changes in primary structure that accompany the conversion of chymotrypsinogen to α-chymotrypsin bring about changes in the tertiary struc-

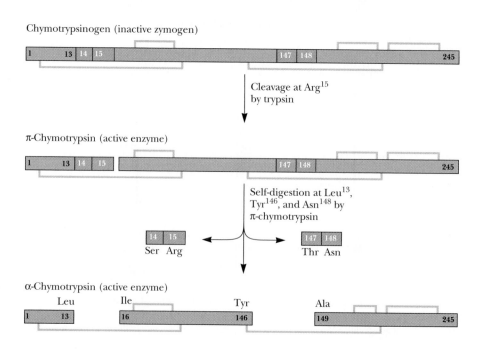

Biochemistry ⊜ Now™ ANIMATED FIGURE 7.10
The proteolytic activation of chymotrypsinogen. **See this figure animated at http://now.brookscole.com/campbell5**

ture. The enzyme is active because of its tertiary structure, just as the zymogen is inactive because of its tertiary structure. The three-dimensional structure of chymotrypsin has been determined by X-ray crystallography. The protonated amino group of the isoleucine residue exposed by the first cleavage reaction is involved in an ionic bond with the carboxylate side chain of aspartate residue 194. This ionic bond is necessary for the active conformation of the enzyme because it is near the active site. Chymotrypsinogen lacks this bond; therefore, it does not have the active conformation and cannot bind substrate.

Blood clotting also requires a series of proteolytic activations involving several proteins, particularly the conversions of prothrombin to thrombin and of fibrinogen to fibrin. Blood clotting is a complex process; for this discussion, it is sufficient to know that activation of zymogens plays a crucial role. In the final, best-characterized step of clot formation, the soluble protein fibrinogen is converted to the insoluble protein fibrin as a result of the cleavage of four peptide bonds. The cleavage occurs as the result of action of the proteolytic enzyme thrombin, which, in turn, is produced from a zymogen called prothrombin. The conversion of prothrombin to thrombin requires Ca^{2+} as well as a number of proteins called *clotting factors*.

Some of the Processes Involved in Blood Clotting

The early stages of blood clotting consist of an elaborate multistep mechanism in which the action of one clotting factor affects the behavior of many molecules of the next factor. This cascade effect allows for fine-tuning of the process but can also cause great problems if something goes wrong with one of the steps. The molecular disease *hemophilia*, for example, is typically caused by a lack of one of the clotting factors. A hemophiliac can bleed to death from a very small cut that would not trouble another person.

7.5 | How Do Active-Site Events of an Enzyme Affect the Reaction Mechanism?

We can ask several questions about the mode of action of an enzyme. Here are some of the most important:

1. Which amino acid residues on the enzyme are in the active site (recall this term from Chapter 6) and catalyze the reaction? In other words, which are the critical amino acid residues?
2. What is the spatial relationship of the critical amino acid residues in the active site?
3. What is the mechanism by which the critical amino acid residues catalyze the reaction?

Answers to these questions are available for chymotrypsin, and we shall use its mechanism as an example of enzyme action. Information on well-known systems such as chymotrypsin can lead to general principles that are applicable to all enzymes. Enzymes catalyze chemical reactions in many ways, but all reactions have in common the requirement that some reactive group on the enzyme interact with the substrate. In proteins, the α-carboxyl and α-amino groups of the amino acids are no longer free because they have formed peptide bonds. Thus, the side-chain reactive groups are the ones involved in the action of the enzyme. Hydrocarbon side chains do not contain reactive groups and are not involved in the process. Functional groups that can play a catalytic role include the imidazole group of histidine, the hydroxyl group of serine, the carboxyl side chains of aspartate and glutamate, the sulfhydryl group of cysteine, the amino side chain of lysine, and the phenol group of tyrosine.

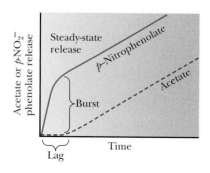

▲ **FIGURE 7.11** The kinetics observed in the chymotrypsin reaction. An initial burst of *p*-nitrophenolate is seen, followed by a slower, steady-state release that matches the appearance of the other product, acetate.

Biochemistry ⬙ Now™
Go to BiochemistryNow and click on Biochemistry Interactive for more information about chymotrypsin.

Chymotrypsin catalyzes the hydrolysis of peptide bonds adjacent to aromatic amino acid residues in the protein being hydrolyzed; other residues are attacked at a lower frequency. In addition, chymotrypsin catalyzes the hydrolysis of esters in model studies in the laboratory. The use of model systems is common in biochemistry because a model provides the essential features of a reaction in a simple form that is easier to work with than the one found in nature. The amide (peptide) bond and the ester bond are similar enough that the enzyme can accept both types of compounds as substrates. Model systems based on the hydrolysis of esters are frequently used to study the peptide hydrolysis reaction.

A typical model compound is *p*-nitrophenyl acetate, which is hydrolyzed in two stages. The acetyl group is covalently attached to the enzyme at the end of the first stage (Step 1) of the reaction, but the *p*-nitrophenolate ion is released. In the second stage (Step 2), the acyl-enzyme intermediate is hydrolyzed, releasing acetate and regenerating the free enzyme. The kinetics observed when *p*-nitrophenyl acetate is first mixed with chymotrypsin shows an initial burst and then a slower phase (Figure 7.11). This reaction is consistent with an enzyme that has two phases, one often forming an acylated-enzyme intermediate.

Step 1

$$E \;+\; O_2N-\langle\text{ring}\rangle-O-\overset{\overset{\displaystyle O}{\|}}{C}-CH_3 \longrightarrow E-\overset{\overset{\displaystyle O}{\|}}{C}-CH_3 \;+\; O_2N-\langle\text{ring}\rangle-O^-$$

Enzyme   *p*-Nitrophenyl acetate   Acyl-enzyme intermediate   *p*-Nitrophenolate

Step 2

$$E-\overset{\overset{\displaystyle O}{\|}}{C}-CH_3 \;\;\xrightarrow{\;H_2O\;}\;\; E \;+\; {}^-O-\overset{\overset{\displaystyle O}{\|}}{C}-CH_3$$

Acyl-enzyme intermediate   Acetate

Determining the Essential Amino Acid Residues

The serine residue at position 195 is required for the activity of chymotrypsin; in this respect, chymotrypsin is typical of a class of enzymes known as **serine proteases.** Trypsin and thrombin, mentioned previously, are also serine proteases (see the Biochemical Connections on p. 176). The enzyme is completely inactivated when this serine reacts with diisopropylphosphofluoridate (DIPF), forming a covalent bond that links the serine side chain with DIPF. The formation of covalently modified versions of specific side chains on proteins is called **labeling;** it is widely used in laboratory studies. The other serine residues of chymotrypsin are far less reactive and are not labeled by DIPF (Figure 7.12).

Biochemistry ⬙ Now™ ACTIVE FIGURE 7.12
Diisopropylphosphofluoridate (DIPF) labels the active-site serine of chymotrypsin. **Watch this Active Figure** at http://now.brookscole.com/campbell5

$$\underset{\text{OH}}{\overset{\textbf{E}}{|}} + \;\; H-\overset{CH_3}{\underset{CH_3}{\overset{|}{\underset{|}{C}}}}-O-\overset{F}{\underset{O}{\overset{|}{\underset{\|}{P}}}}-O-\overset{CH_3}{\underset{CH_3}{\overset{|}{\underset{|}{C}}}}-H \;\;\xrightarrow{\;F^-\;}\;\; H-\overset{CH_3}{\underset{CH_3}{\overset{|}{\underset{|}{C}}}}-O-\overset{\overset{\displaystyle\textbf{E}}{|}\;O}{\underset{O}{\overset{|}{\underset{\|}{P}}}}-O-\overset{CH_3}{\underset{CH_3}{\overset{|}{\underset{|}{C}}}}-H$$

Diisopropylphosphofluoridate   **Diisopropylphosphoryl derivative of chymotrypsin**

Histidine 57 is another critical amino acid residue in chymotrypsin. Chemical labeling again provides the evidence for involvement of this residue in the activity of chymotrypsin. In this case, the reagent used to label the critical amino acid residue is *N*-tosylamido-L-phenylethyl chloromethyl ketone (TPCK), also called tosyl-L-phenylalanine chloromethyl ketone. The phenylalanine moiety is bound to the enzyme because of the specificity for aromatic amino acid residues at the active site, and the active site histidine residue reacts because the labeling reagent is similar to the usual substrate.

The labeling of the active-site histidine of chymotrypsin by TPCK

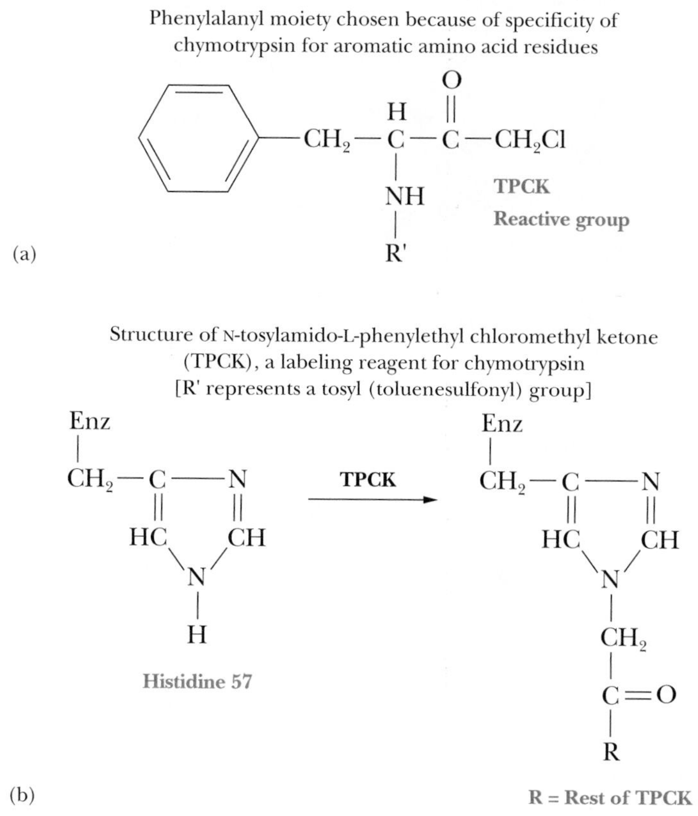

Phenylalanyl moiety chosen because of specificity of chymotrypsin for aromatic amino acid residues

(a)

Structure of N-tosylamido-L-phenylethyl chloromethyl ketone (TPCK), a labeling reagent for chymotrypsin [R' represents a tosyl (toluenesulfonyl) group]

(b)

R = Rest of TPCK

The Architecture of the Active Site

Both serine 195 and histidine 57 are required for the activity of chymotrypsin; therefore, they must be close to each other in the active site. The determination of the three-dimensional structure of the enzyme by X-ray crystallography provides evidence that the active-site residues do indeed have a close spatial relationship. The folding of the chymotrypsin backbone, mostly in an antiparallel pleated-sheet array, positions the essential residues around an active-site pocket (Figure 7.13). Only a few residues are directly involved in the active site, but the whole molecule is necessary to provide the correct three-dimensional arrangement for those critical residues.

Other important pieces of information about the three-dimensional structure of the active site emerge when a complex is formed between chymotrypsin and a substrate analogue. When one such substrate analog, formyl-L-tryptophan, is bound to the enzyme, the tryptophan side chain fits into a hydrophobic pocket near serine 195. This type of binding is not surprising, in

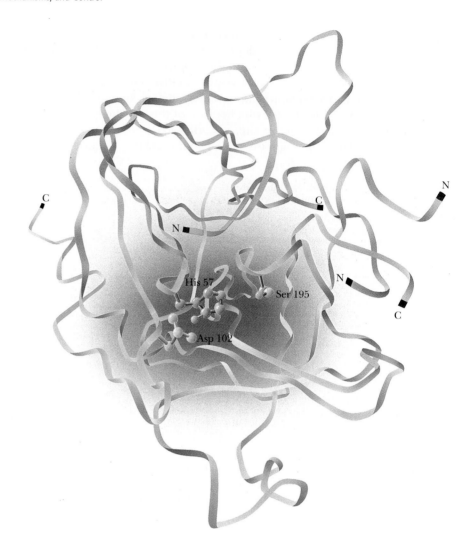

► **FIGURE 7.13** The tertiary structure of chymotrypsin places the essential amino acid residues close to one another. They are shown in blue and red. (*Abeles, R., Frey, P., Jencks, W.* Biochemistry © *Boston: Jones and Bartlett, Publishers, 1992, reprinted by permission.*)

Formyl-L-tryptophan

view of the specificity of the enzyme for aromatic amino acid residues at the cleavage site.

The results of X-ray crystallography show, in addition to the binding site for aromatic amino acid side chains of substrate molecules, a definite arrangement of the amino acid side chains that are responsible for the catalytic activity of the enzyme. The residues involved in this arrangement are serine 195 and histidine 57.

The Mechanism of Chymotrypsin Action

Any postulated reaction mechanism must be modified or discarded if it is not consistent with experimental results. There is consensus, but not total agreement, on the main features of the mechanism discussed in this section.

The critical amino acid residues, serine 195 and histidine 57, are involved in the mechanism of catalytic action. In the terminology of organic chemistry, the oxygen of the serine side chain is a **nucleophile,** or nucleus-seeking substance. A nucleophile tends to bond to sites of positive charge or polarization (electron-poor sites) in contrast to an **electrophile,** or electron-seeking substance, which tends to bond to sites of negative charge or polarization (electron-rich sites). The nucleophilic oxygen of the serine attacks the carbonyl carbon of the peptide group. The carbon now has four single bonds, and a tetrahedral intermediate is formed; the original —C=O bond becomes a single bond, and the carbonyl oxygen becomes an oxyanion. The acyl-enzyme intermediate is formed from the tetrahedral species (Figure 7.14). The histidine and the amino portion of the original peptide group are involved in this part of the reaction as the amino group hydrogen bonds to the imidazole portion of the histidine. Note that the imidazole is already protonated and that the proton came from the hydroxyl group of the serine. The histidine behaves as a base in abstracting the proton from the serine; in the terminology of the physical organic chemist, the histidine acts as a general base catalyst. The carbon—nitrogen bond of the original peptide group breaks, leaving the acyl-enzyme intermediate. The proton abstracted by the histidine has been donated to the leaving amino group. In donating the proton, the histidine has acted as an acid in the breakdown of the tetrahedral intermediate, although it acted as a base in its formation.

In the deacylation phase of the reaction, the last two steps are reversed, with water acting as the attacking nucleophile. In this second phase, the water is hydrogen-bonded to the histidine. The oxygen of water now performs the nucleophilic attack on the acyl carbon that came from the original peptide group. Once again, a tetrahedral intermediate is formed. In the final step of

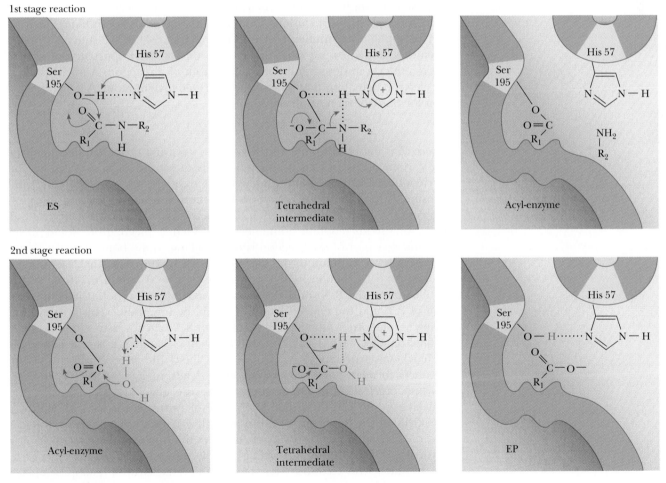

Biochemistry ⑤Now™ ▲ ANIMATED FIGURE 7.14 The mechanism of chymotrypsin action. In the first stage of the reaction, the nucleophile serine 195 attacks the carbonyl carbon of the substrate. In the second stage, water is the nucleophile that attacks the acyl-enzyme intermediate. Note the involvement of histidine 57 in both stages of the reaction. (*From Hammes, G.:* Enzyme Catalysis and Regulation, *New York: Academic Press, 1982.*) **See this figure animated at http://now .brookscole.com/campbell5**

the reaction, the bond between the serine oxygen and the carbonyl carbon breaks, releasing the product with a carboxyl group where the original peptide group used to be and regenerating the original enzyme. Note that the serine is hydrogen-bonded to the histidine. This hydrogen bond increases the nucleophilicity of the serine, whereas, in the second part of the reaction, the hydrogen bond between the water and the histidine increased the nucleophilicity of the water.

The mechanism of chymotrypsin action is particularly well studied and, in many respects, typical. Numerous types of reaction mechanisms for enzyme action are known, and we shall discuss them in the contexts of the reactions catalyzed by the enzymes in question. To lay the groundwork, it is useful to discuss some general types of catalytic mechanisms and how they affect the specificity of enzymatic reactions.

7.6 | What Types of Chemical Reactions Are Involved in Enzyme Mechanisms?

The overall mechanism for a reaction may be fairly complex, as we have seen in the case of chymotrypsin, but the individual parts of a complex mechanism can themselves be fairly simple. Concepts such as nucleophilic attack and acid catalysis commonly enter into discussions of enzymatic reactions. We can draw quite a few general conclusions from these two general descriptions.

Nucleophilic substitution reactions play a large role in the study of organic chemistry, and they are excellent illustrations of the importance of kinetic measurements in determining the mechanism of a reaction. A nucleophile is an electron-rich atom that attacks an electron-deficient atom. A general equation for this type of reaction is

$$R:X + :Z \rightarrow R:Z + X$$

where :Z is the nucleophile and X is called a leaving group. In biochemistry, the carbon of a carbonyl group (C=O) is often the atom attacked by the nucleophile. Common nucleophiles are the oxygens of serine, threonine, and tyrosine. If the rate of the reaction shown here is found to depend solely on the concentration of the R:X, then the nucleophilic reaction is called an S_N1 (substitution nucleophilic unimolecular). Such a mechanism would mean that the slow part of the reaction is the breaking of the bond between R and X, and that the addition of the nucleophile Z happens very quickly compared to that. An S_N1 reaction follows first-order kinetics (Chapter 6). If the nucleophile attacks the R:X while the X is still attached, then both the concentration of R:X and the concentration of :Z will be important. This reaction will follow second-order kinetics and is called an S_N2 reaction (substitution nucleophilic bimolecular). The difference between S_N1 and S_N2 is very important to biochemists because it explains much about the stereospecificity of the products formed. An S_N1 reaction often leads to loss of stereospecificity. Because the leaving group is gone before the attacking group enters, the attacking group can often end up in one of two orientations, although the specificity of the active site can also limit this. With an S_N2 reaction, the fact that the leaving group is still attached forces the nucleophile to attack from a particular side of the bond, leading to only one possible stereospecificity in the product. The chymotrypsin nucleophilic attacks were examples of S_N2 reactions, although no stereochemistry is noted because the carbonyl that was attacked became a carbonyl group again at the end of the reaction and was, therefore, not chiral.

To discuss acid–base catalysis, it is helpful to recall the definitions of acids and bases. In the Brønsted–Lowry definition, an acid is a proton donor and a

Biochemical Connections

Enzymes Catalyze Familiar Reactions of Organic Chemistry

Biochemical reactions are those reactions described in organic chemistry textbooks. Important compounds such as alcohols, aldehydes, and ketones appear many times. Carboxylic acids are involved in many other reactions, frequently as their derivatives, esters, and amides. Still other reactions, called condensations, form new carbon—carbon bonds. Reverse condensations break carbon—carbon bonds, as their name implies. The breakdown of sugars provides examples of this. Glucose, a six-carbon compound, is converted to pyruvate, a three-carbon compound, in glycolysis (Chapter 17). A reverse condensation reaction cleaves the six-carbon glucose derivative fructose-1,6-*bis*phosphate to two three-carbon fragments, glyceraldehyde-3-phosphate and dihydroxyacetone phosphate.

Fructose-1,6-*bis*phosphate $\rightleftharpoons$ dihydroxyacetone phosphate + D-glyceraldehyde-3-phosphate

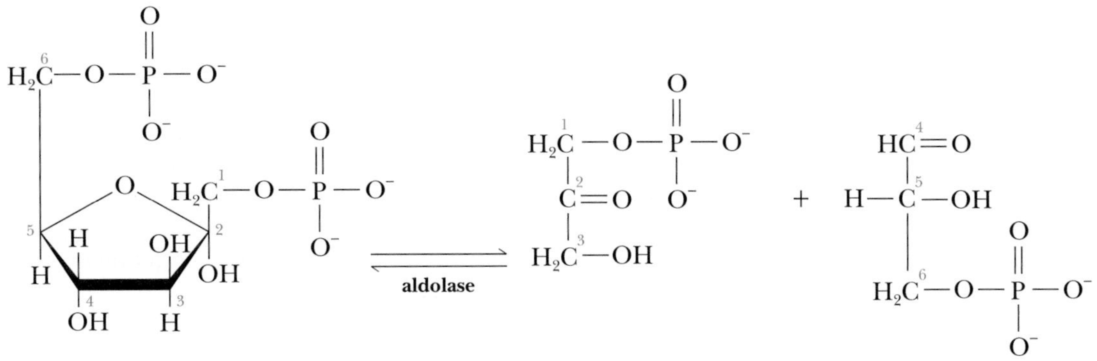

| Fructose-1,6-*bis*phosphate | Dihydroxyacetone phosphate | D-Glyceraldehyde-3-phosphate |

Glyceraldehyde-3-phosphate, in turn, is converted to 1,3-*bis*phosphoglycerate in a reaction that converts the aldehyde to carboxylic acid involved in a mixed anhydride linkage to phosphoric acid.

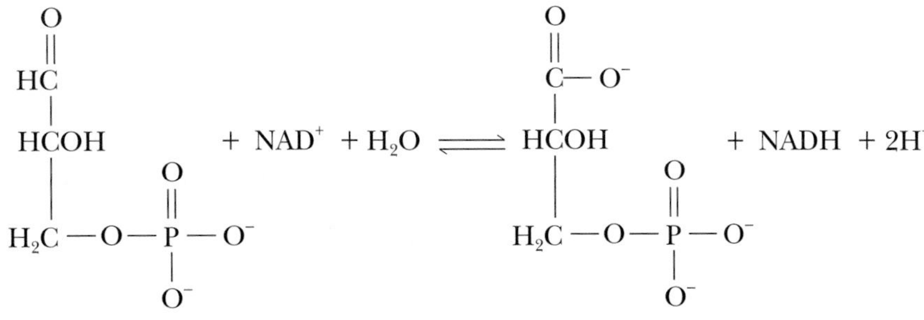

glyceraldehyde-3-phosphate + NAD$^+$ + H$_2$O $\rightleftharpoons$ 3-phosphoglycerate + NADH + 2H$^+$

The conversion of an aldehyde to a carboxylic acid is an oxidation with the compound NAD$^+$ as the oxidizing agent. These reactions, like all biochemical reactions, are catalyzed by specific enzymes. In many cases, the catalytic mechanism is known, as is the case with both reactions. Several common organic mechanisms appear repeatedly in biochemical mechanisms.

base is a proton acceptor. The concept of **general acid–base catalysis** depends on donation and acceptance of protons by groups such as the imidazole, hydroxyl, carboxyl, sulfhydryl, amino, and phenolic side chains of amino acids; all these functional groups can act as acids or bases. The donation and acceptance of protons gives rise to the bond breaking and re-formation that constitute the enzymatic reaction. If the enzyme mechanism involves an amino acid donating a hydrogen ion, as in the reaction

$$R—H^+ + R—O^- \rightarrow R + R—O—H$$

then that part of the mechanism would be called general acid catalysis. If an amino acid takes a hydrogen ion from one of the substrates, such as in the reaction

$$R + R—OH \rightarrow R—H^+ + R—O^-$$

then that part is called general base catalysis. Histidine is an amino acid that often takes part in both reactions, since it has a reactive hydrogen on the imidazole side chain that dissociates near physiological pH. In the chymotrypsin mechanism, we saw both acid and base catalysis by histidine.

A second form of acid–base catalysis reflects another, more general definition of acids and bases. In the Lewis formulation, an acid is an electron-pair acceptor, and a base is an electron-pair donor. Metal ions, including such biologically important ones as Mn^{2+}, Mg^{2+}, and Zn^{2+}, are Lewis acids. Thus, they can play a role in **metal–ion catalysis** (also called Lewis acid–base catalysis). The involvement of Zn^{2+} in the enzymatic activity of carboxypeptidase A is an example of this type of behavior. This enzyme catalyzes the hydrolysis of C-terminal peptide bonds of proteins. The Zn(II), which is required for the activity of the enzyme, is complexed to the imidazole side chains of histidines 69 and 196 and to the carboxylate side chain of glutamate 72. The zinc ion is also complexed to the substrate.

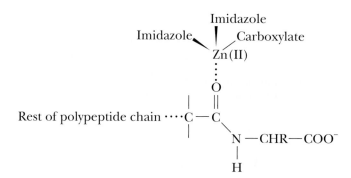

A zinc ion is complexed to three side chains of carboxypeptidase and to a carbonyl group on the substrate.

The type of binding involved in the complex is similar to the binding that links iron to the large ring involved in the heme group. Binding the substrate to the zinc ion polarizes the carbonyl group, making it susceptible to attack by water and allowing the hydrolysis to proceed more rapidly than it does in the uncatalyzed reaction.

A definite connection exists between the concepts of acids and bases and the idea of nucleophiles and their complementary substances, electrophiles. A Lewis acid is an electrophile, and a Lewis base is a nucleophile. Catalysis by enzymes, including their remarkable specificity, is based on these well-known chemical principles operating in a complex environment.

The nature of the active site plays a particularly important role in the specificity of enzymes. An enzyme that displays *absolute specificity*, catalyzing the reaction of one, and only one, substrate to a particular product, is likely to have a fairly rigid active site that is best described by the lock-and-key model of substrate binding. The many enzymes that display *relative specificity*, catalyzing the reactions of structurally related substrates to related products,

The reaction catalyzed by carboxypeptidase A.

$$Rest\ of\ polypeptide\ chain -C-C-N-CHR-COO^-$$

$$Rest\ of\ polypeptide\ chain -C-C-O^-$$

$$+$$

$$H_3\overset{+}{N}-CHR-COO^-$$

apparently have more flexibility in their active sites and are better characterized by the induced-fit model of enzyme–substrate binding; chymotrypsin is a good example. Finally, there are *stereospecific* enzymes with specificity in which optical activity plays a role. The binding site itself must be asymmetric in this situation (Figure 7.15). If the enzyme is to bind specifically to an optically active substrate, the binding site must have the shape of the substrate and not its mirror image. There are even enzymes that introduce a center of optical activity into the product. The substrate itself is not optically active in this case. There is only one product, which is one of two possible isomers, not a mixture of optical isomers.

Essential Information

The catalytic behavior of enzymes frequently involves a series of relatively simple reactions. Substitution reactions and acid–base reactions are frequently encountered in the detailed processes of enzymatic reactions.

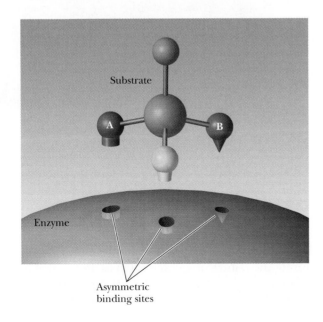

Substrate

A   B

Enzyme

Asymmetric binding sites

◀ **FIGURE 7.15** An asymmetric binding site on an enzyme can distinguish between identical groups, such as A and B. Note that the binding site consists of three parts, giving rise to asymmetric binding because one part is different from the other two.

Biochemical Connections

Families of Enzymes: Proteases

Large numbers of enzymes catalyze similar functions. Many oxidation–reduction reactions take place, each catalyzed by a specific enzyme. We have already seen that kinases transfer phosphate groups. Still other enzymes catalyze hydrolytic reactions. Enzymes that have similar functions may have widely varying structures. The important feature that they have in common is that they have an active site that can catalyze the reaction in question. A number of different enzymes catalyze the hydrolysis of proteins. Chymotrypsin is one example of the class of serine proteases, but many others are known, including elastase, which catalyzes the degradation of the connective tissue protein elastin and the digestive enzyme trypsin. (Recall that we first saw trypsin in its role in protein sequencing.) All these enzymes are similar in structure. Other proteases employ other essential amino acid residues as the nucleophile in the active site. Papain, the basis of commercial meat tenderizers, is a proteolytic enzyme derived from papayas. It, however, has a cysteine rather than a serine as the nucleophile in its active site. Aspartyl proteases differ still more widely in structure from the common serine proteases. A pair of aspartate side chains, sometimes on different subunits, participates in the reaction mechanism. A number of aspartyl proteases, such as the digestive enzyme pepsin, are known. However, the most notorious aspartyl protease is the one necessary for the maturation of the human immunodeficiency virus, HIV-1 protease.

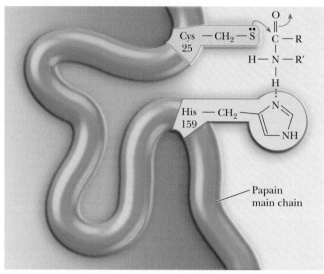

▲ Papain is a cysteine protease. A critical cysteine residue is involved in the nucleophilic attack on the peptide bonds it hydrolyzes.

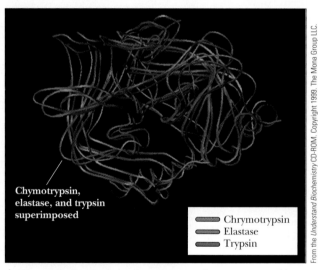

▲ Chymotrypsin, elastin, and trypsin are serine proteases and have similar structures.

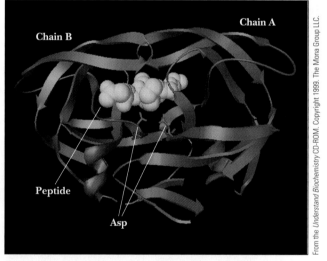

▲ HIV-1 protease is a member of the class of enzymes called the aspartic proteases. Two aspartates are involved in the reaction.

Biochemistry⊘Now™
Go to BiochemistryNow and click on Biochemistry Interactive for more information about HIV-1 protease.

7.7 | What Is the Connection between the Active Site and Transition States?

Now that we have spent some time looking at mechanisms and the active site, it is worth revisiting the nature of enzyme catalysis. Recall that an enzyme lowers the activation energy by lowering the energy necessary to reach the transition state (Figure 6.1). The true nature of the transition state is a chemical species that is intermediate in structure between the substrate and the product. This transition state often has a very different shape from either the substrate or the product. In the case of chymotrypsin, the substrate has the carbonyl group that is attacked by the reactive serine. The carbon of the carbonyl group has three bonds, and the orientation is planar. After the serine performs the nucleophilic attack, the carbon has four bonds and a tetrahedral arrangement. This tetrahedral shape is the transition state of the reaction, and the active site must make this change more likely.

The fact that the enzyme stabilizes the transition state has been shown many times by the use of **transition-state analogs,** which are molecules with a shape that mimics the transition state of the substrate. Proline racemase catalyzes a reaction that converts L-proline to D-proline. In the progress of the reaction, the α-carbon must change from a tetrahedral arrangement to a planar form, and then back to tetrahedral, but with the orientation of two bonds reversed (Figure 7.16). An inhibitor of the reaction is pyrrole-2-carboxylate, a chemical that is structurally similar to what proline would look like at its transition state because it is always planar at the equivalent carbon. This inhibitor binds to proline racemase 160 times more strongly than proline does. Transition-state analogs have been used with many enzymes to help verify a suspected mechanism and structure of the transition state as well as to inhibit an enzyme selectively. Back in 1969, William Jencks proposed that an immunogen (a molecule that elicits an antibody response) would elicit antibodies with catalytic activity if the immunogen mimicked the transition state of the reaction. Richard Lerner and Peter Schultz, who created the first catalytic antibodies, verified this hypothesis in 1986. Because an antibody is a protein designed to bind to specific molecules on the immunogen, the antibody will, in essence, be a fake active site. For example, the reaction of pyridoxal phosphate and an amino acid to form the corresponding α-keto acid and pyridoxamine phosphate is a very important reaction in amino acid metabolism. The molecule, N^{α}-(5'-phosphopyridoxyl)-L-lysine serves as a transition-state analog for this reaction. When this antigen molecule was used to elicit antibodies, these antibodies, or **abzymes,** had catalytic activity (Figure 7.17). Thus, in addition to helping us verify the nature of the transition state or making an inhibitor, transition-state analogs now offer the possibility of making designer enzymes to catalyze a wide variety of reactions.

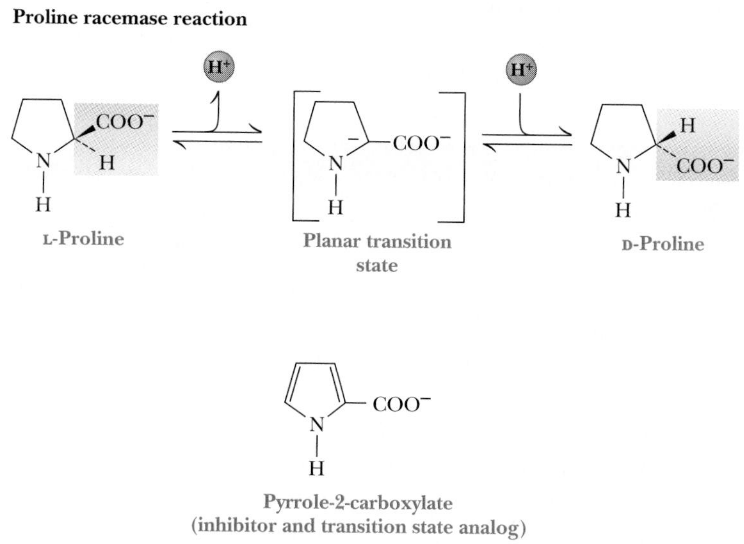

Proline racemase reaction

Pyrrole-2-carboxylate
(inhibitor and transition state analog)

◀ **FIGURE 7.16** The proline racemase reaction. Pyrrole-2-carboxylate and Δ-1-pyrroline-2-carboxylate mimic the planar transition state of the reaction.

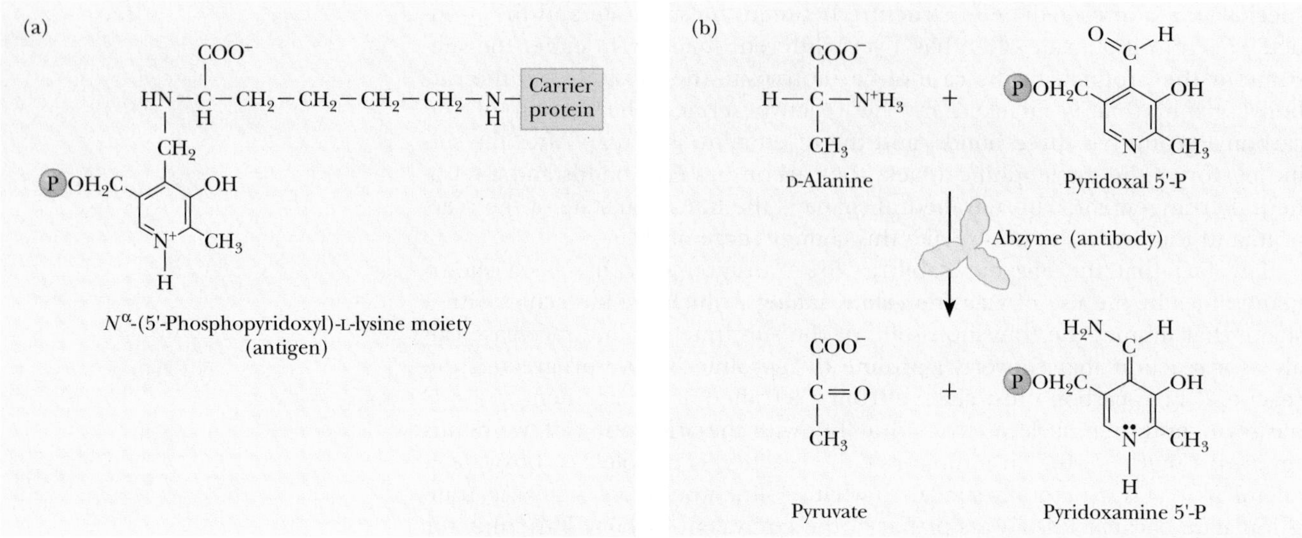

(a)

$$N^\alpha\text{-}(5'\text{-Phosphopyridoxyl)-L-lysine moiety}$$
(antigen)

(b)

D-Alanine
Pyridoxal 5'-P

Abzyme (antibody)

Pyruvate
Pyridoxamine 5'-P

▲ **FIGURE 7.17** (a) N^α-(5′-phosphopyridoxyl)-L-lysine moiety is a transition-state analog for the reaction of an amino acid with pyridoxal 5′-phosphate. When this moiety is attached to a protein and injected into a host, it acts like an antigen, and the host then produces antibodies that have catalytic activity (abzymes). (b) The abzyme is then used to catalyze the reaction.

7.8 What Are Coenzymes?

Cofactors are nonprotein substances that take part in enzymatic reactions and are regenerated for further reaction. Metal ions frequently play such a role, and they make up one of two important classes of cofactors. The other important class (**coenzymes**) is a mixed bag of organic compounds; many of them are vitamins or are metabolically related to vitamins.

Because metal ions are Lewis acids (electron-pair acceptors), they can act as Lewis acid–base catalysts. They can also form coordination compounds by behaving as Lewis acids, while the groups to which they bind act as Lewis bases. Coordination compounds are an important part of the chemistry of metal ions in biological systems, as shown by Zn(II) in carboxypeptidase and by Fe(II) in hemoglobin. The coordination compounds formed by metal ions tend to have quite specific geometries, which aid in positioning the groups involved in a reaction for optimum catalysis.

Some of the most important organic coenzymes are vitamins and their derivatives, especially B vitamins. Many of these coenzymes are involved in oxidation–reduction reactions, which provide energy for the organism. Others serve as group-transfer agents in metabolic processes (Table 7.1). We shall

Table 7.1

Coenzymes, Their Reactions, and Their Vitamin Precursors

Coenzyme	Reaction Type	Vitamin Precursor	See Section
Biotin	Carboxylation	Biotin	18.2, 21.6
Coenzyme A	Acyl transfer	Pantothenic acid	15.7, 19.3, 21.6
Flavin coenzymes	Oxidation–reduction	Riboflavin (B_2)	15.7, 19.3
Lipoic acid	Acyl transfer	—	19.3
Nicotinamide adenine coenzymes	Oxidation–reduction	Niacin	15.7, 17.3, 19.3
Pyridoxal phosphate	Transamination	Pyridoxine (B_6)	23.4
Tetrahydrofolic acid	Transfer of one-carbon units	Folic acid	23.4
Thiamine pyrophosphate	Aldehyde transfer	Thiamine (B_1)	17.4, 18.4

Catalytic Antibodies against Cocaine

Many addictive drugs, such as heroin, operate by binding to a particular receptor in the neurons, mimicking the action of a neurotransmitter. When a person is addicted to such a drug, a common way to attempt to treat the addiction is to use a compound to block the receptor, thereby denying the drug's access to it. Cocaine addiction has always been difficult to treat, due primarily to its unique modus operandi. As shown, cocaine blocks the reuptake of the neurotransmitter dopamine. Thus, dopamine stays in the system longer, overstimulating the neuron and leading to the reward signals in the brain that lead to addiction. Using a drug to block a receptor would be of no use with cocaine addiction and would probably just make removal of dopamine even more unlikely. Cocaine can be degraded by a specific esterase, an enzyme that hydrolyzes an ester bond that is part of cocaine's structure. In the process of this hydrolysis, the cocaine must pass through a transition state that changes its shape. Catalytic antibodies to the transition state of the hydrolysis of cocaine were created (see articles by Landry in the bibliography at the end of this chapter). When administered to patients suffering from cocaine addiction, the antibodies successfully hydrolyzed cocaine to two harmless degradation products—benzoic acid and ecgonine methyl ester. When degraded, the cocaine cannot block dopamine reuptake. No prolongation of the neuronal stimulus occurs, and the addictive effects of the drug vanish over time.

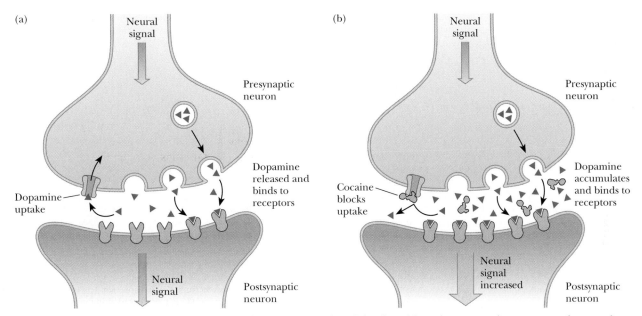

▲ The mechanism of action of cocaine. (a) Dopamine acts as a neurotransmitter. It is released from the presynaptic neuron, travels across the synapse, and bonds to dopamine receptors on the postsynaptic neuron. It is later released and taken up into vesicles in the presynaptic neuron. (b) Cocaine increases the amount of time that dopamine is available to the dopamine receptors by blocking its uptake. (*From* Scientific American, *Vol. 276(2), pp. 42–45. Reprinted by permission of Tomoyuki Narashima.*)

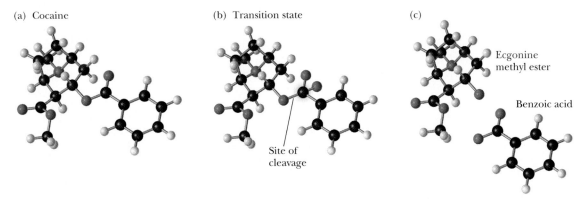

▲ Degradation of cocaine by esterases or catalytic antibodies. Cocaine (a) passes through a transition state (b) on its way to being hydrolyzed to benzoic acid and ecgonine methyl ester (c). Transition-state analogs are used to generate catalytic antibodies for this reaction. (*From* Scientific American, *Vol. 276(2), pp. 42–45. Reprinted by permission of Tomoyuki Narashima.*)

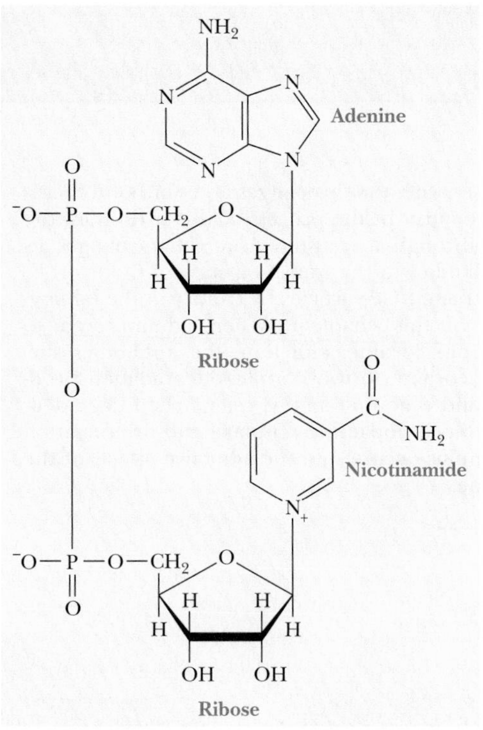

▲ **FIGURE 7.18** The structure of nicotinamide adenine dinucleotide (NAD$^+$).

see these coenzymes again when we discuss the reactions in which they are involved. For the present, we shall investigate one particularly important oxidation–reduction coenzyme and one group-transfer coenzyme.

Nicotinamide adenine dinucleotide (NAD$^+$) is a coenzyme in many oxidation–reduction reactions. Its structure (Figure 7.18) has three parts—a nicotinamide ring, an adenine ring, and two sugar—phosphate groups linked together. The nicotinamide ring contains the site at which oxidation and reduction reactions occur (Figure 7.19). Nicotinic acid is another name for the vitamin niacin. The adenine—sugar—phosphate portion of the molecule is structurally related to nucleotides.

The B$_6$ vitamins (pyridoxal, pyridoxamine, and pyridoxine and their phosphorylated forms, which are the coenzymes) are involved in the transfer of amino groups from one molecule to another, an important step in the biosynthesis of amino acids (Figure 7.20). In the reaction, the amino group is transferred from the donor to the coenzyme and then from the coenzyme to the ultimate acceptor (Figure 7.21).

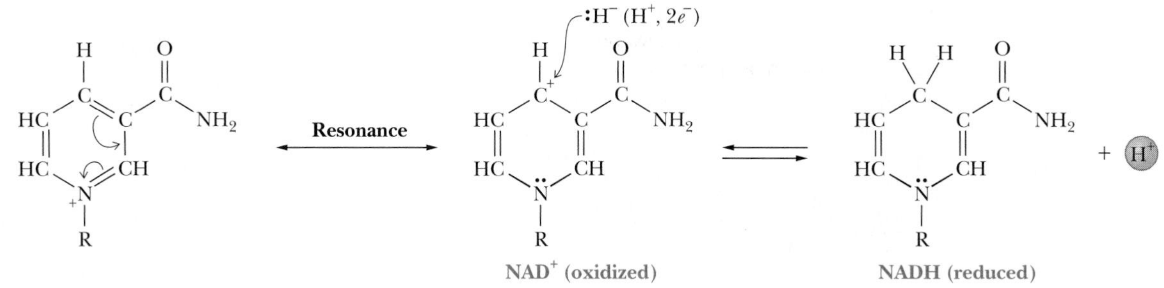

▲ **FIGURE 7.19** The role of the nicotinamide ring in oxidation–reduction reactions. R is the rest of the molecule. In reactions of this sort, an H$^+$ is transferred along with the two electrons.

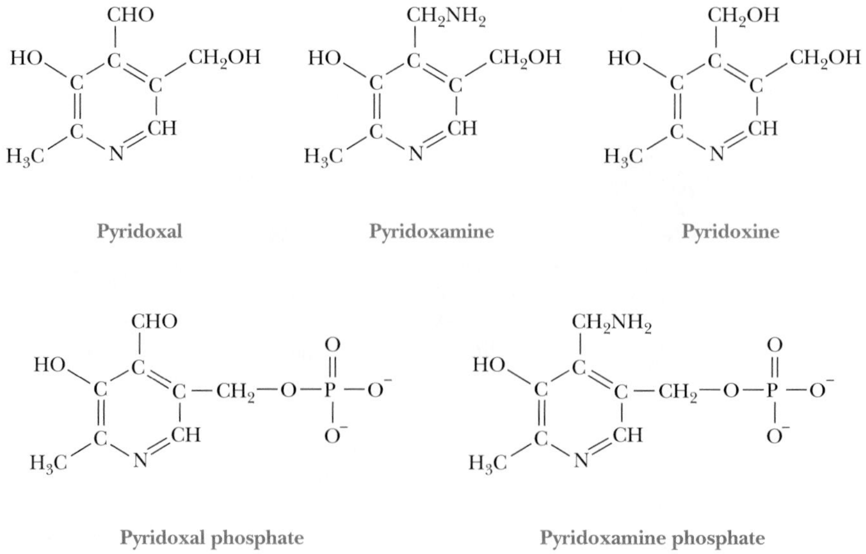

▶ **FIGURE 7.20** Forms of vitamin B$_6$. The first three structures are vitamin B$_6$ itself, and the last two structures show the modifications that give rise to the metabolically active coenzyme.

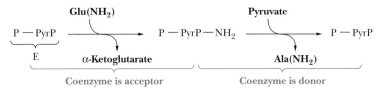

This amino (NH_2) group transfer reaction occurs in two stages:

FIGURE 7.21 The role of pyridoxal phosphate as a coenzyme in a transamination reaction. PyrP is pyridoxal phosphate, P is the apoenzyme (the polypeptide chain alone), and E is the active holoenzyme (polypeptide plus coenzyme).

Summary

7.1 Does the Michaelis–Menten Model Describe the Behavior of Allosteric Enzymes? The Michaelis–Menten model does not describe the behavior of allosteric enzymes. Changes in quaternary structure on binding of substrates, inhibitors, and activators all affect the observed kinetics of such enzymes.

7.2 What Are the Models for the Behavior of Allosteric Enzymes? In the concerted model for allosteric behavior, the binding of substrate, inhibitor, or activator to one subunit shifts the equilibrium between an active form of the enzyme, which binds substrate strongly, and an inactive form, which does not bind substrate strongly. The conformational change takes place in all subunits at the same time. In the sequential model, the binding of substrate induces the conformational change in one subunit, and the change is subsequently passed along to other subunits. Both models are useful; they may eventually be incorporated in a single, more inclusive model.

7.3 How Does Phosphorylation of Specific Residues Regulate Enzyme Activity? Still other enzymes are activated or inactivated, depending on the presence or absence of phosphate groups. This kind of covalent modification can be combined with allosteric interactions to allow for a high degree of control over enzymatic pathways.

7.4 What Are Zymogens, and How Do They Control Enzyme Activity? Another type of control mechanism in enzyme action is zymogen activation, in which an inactive precursor of an enzyme is transformed into an active enzyme by cleavage of covalent bonds. For example, the proteolytic enzymes trypsin and chymotrypsin arise from the zymogens trypsinogen and chymotrypsinogen, respectively. Similar protein activations take place in blood clotting.

7.5 How Do Active-Site Events of an Enzyme Affect the Reaction Mechanism? Several questions arise about the events that occur at the active site of an enzyme in the course of a reaction. Some of the most important of these questions address the nature of the critical amino acid residues, their spatial arrangement, and the mechanism of the reaction. Chymotrypsin is a good example of an enzyme for which most of the questions about its mechanism of action have been answered. Its critical amino acid residues have been determined to be serine 195 and histidine 57. The complete three-dimensional structure of chymotrypsin, including the architecture of the active site, has been determined by X-ray crystallography. Nucleophilic attack by serine is the main feature of the mechanism, with histidine hydrogen-bonded to serine in the course of the reaction.

7.6 What Types of Chemical Reactions Are Involved in Enzyme Mechanisms? Common organic reaction mechanisms, such as nucleophilic substitution and general acid–base catalysis, are known to play roles in enzymatic catalysis.

7.7 What Is the Connection between the Active Site and Transition States? The nature of catalysis has been aided by the use of transition-state analogs, molecules that mimic the transition state. The compounds usually bind to the enzyme better than the natural substrate and help to verify the mechanism. They can also be used to develop potent inhibitors or to create antibodies with catalytic activity, called abzymes.

7.8 What Are Coenzymes? Coenzymes are nonprotein substances that take part in enzymatic reactions and are regenerated for further reaction. Metal ions can serve as coenzymes, frequently by acting as Lewis acids. There are also many organic coenzymes, most of which are vitamins or are structurally related to vitamins.

Critical Questions to Review

7.1 Does the Michaelis–Menten Model Describe the Behavior of Allosteric Enzymes?

1. **Fact Check** What features distinguish enzymes that undergo allosteric control from those that obey the Michaelis–Menten equation?

2. **Fact Check** What is the metabolic role of aspartate transcarbamoylase?

3. **Fact Check** What molecule acts as a positive effector (activator) of ATCase? What molecule acts as an inhibitor?

4. **Fact Check** Is the term K_M used with allosteric enzymes? What about competitive and noncompetitive inhibition? Explain.

5. **Fact Check** What is a K system?

6. **Fact Check** What is a V system?

7. **Fact Check** What is a homotropic effect? What is a heterotropic effect?

8. **Fact Check** What is the structure of ATCase?

9. **Fact Check** How is the cooperative behavior of allosteric enzymes reflected in a plot of reaction rate against substrate concentration?

10. **Fact Check** Does the behavior of allosteric enzymes become more or less cooperative in the presence of inhibitors?

11. **Fact Check** Does the behavior of allosteric enzymes become more or less cooperative in the presence of activators?

12. **Fact Check** Explain what is meant by $K_{0.5}$.

13. **Thought Question** Explain the experiment used to determine the structure of ATCase. What happens to the activity and regulatory activities when the subunits are separated?

7.2 What Are the Models for the Behavior of Allosteric Enzymes?

14. **Fact Check** Distinguish between the concerted and sequential models for the behavior of allosteric enzymes.

15. **Fact Check** Which allosteric model can explain negative cooperativity?

16. **Fact Check** With the concerted model, what conditions favor greater cooperativity?

17. **Fact Check** With respect to the concerted model, what is the L value? What is the c value?

18. **Thought Question** Is it possible to envision models for the behavior of allosteric enzymes other than the ones that we have seen in this chapter?

7.3 How Does Phosphorylation of Specific Residues Regulate Enzyme Activity?

19. **Fact Check** What is the function of a protein kinase?

20. **Fact Check** What amino acids are often phosphorylated by kinases?

21. **Thought Question** What are some possible advantages to the cell in combining phosphorylation with allosteric control?

22. **Thought Question** Explain how phosphorylation is involved in the function of the sodium—potassium ATPase.

23. **Thought Question** Explain how glycogen phosphorylase is controlled allosterically and by covalent modification.

7.4 What Are Zymogens, and How Do They Control Enzyme Activity?

24. **Fact Check** Name three proteins that are subject to the control mechanism of zymogen activation.

25. **Biochemical Connection** List three proteases and their substrates.

26. **Fact Check** How is blood clotting related to zymogens?

27. **Thought Question** Explain why cleavage of the bond between arginine 15 and isoleucine 16 of chymotrypsinogen activates the zymogen.

28. **Thought Question** Why is it necessary or advantageous for the body to make zymogens?

29. **Thought Question** Why is it necessary or advantageous for the body to make inactive hormone precursors?

7.5 How Do Active-Site Events of an Enzyme Affect the Reaction Mechanism?

30. **Fact Check** What are the two essential amino acids in the active site of chymotrypsin?

31. **Fact Check** Why does the enzyme reaction for chymotrypsin proceed in two phases?

32. **Thought Question** Briefly describe the role of nucleophilic catalysis in the mechanism of the chymotrypsin reaction.

33. **Thought Question** Explain the function of histidine 57 in the mechanism of chymotrypsin.

34. **Thought Question** Explain why the second phase of the chymotrypsin mechanism is slower than the first phase.

35. **Thought Question** Explain how the pKa for histidine 57 is important to its role in the mechanism of chymotrypsin action.

36. **Thought Question** An inhibitor that specifically labels chymotrypsin at histidine 57 is N-tosylamido-L-phenylethyl chloromethyl ketone. How would you modify the structure of this inhibitor to label the active site of trypsin?

7.6 What Types of Chemical Reactions Are Involved in Enzyme Mechanisms?

37. **Thought Question** What properties of metal ions make them useful cofactors?

38. **Biochemical Connection** Is the following statement true or false? Why? "The mechanisms of enzymatic catalysis have nothing in common with those encountered in organic chemistry."

39. **Thought Question** What is meant by general acid catalysis with respect to enzyme mechanisms?

40. **Thought Question** Explain the difference between an S_N1 reaction mechanism and an S_N2 reaction mechanism.

41. **Thought Question** Which of the two reaction mechanisms in Question 40 is likely to cause the loss of stereospecificity? Why?

42. **Thought Question** An experiment is performed to test a suggested mechanism for an enzyme-catalyzed reaction. The results fit the model exactly (to within experimental error). Do the results prove that the mechanism is correct? Why or why not?

7.7 What Is the Connection between the Active Site and Transition States?

43. **Thought Question** What would be the characteristics of a transition-state analog for the chymotrypsin reaction?

44. **Thought Question** What is the relationship between a transition-state analog and the induced-fit model of enzyme kinetics?

45. **Thought Question** Explain how a researcher makes an abzyme. What is the purpose of an abzyme?

46. **Biochemical Connection** Why can cocaine addiction not be treated with a drug that blocks the cocaine receptor?

47. **Biochemical Connection** Explain how abzymes can be used to treat cocaine addiction.

7.8 What Are Coenzymes?

48. **Fact Check** List three coenzymes and their functions.

49. **Fact Check** How are coenzymes related to vitamins?

50. **Fact Check** What type of reaction uses vitamin B_6?

51. **Thought Question** Suggest a role for coenzymes based on reaction mechanisms.

52. **Thought Question** An enzyme uses NAD^+ as a coenzyme. Using Figure 7.19, predict whether a radiolabeled $H:^-$ ion would tend to appear preferentially on one side of the nicotinamide ring as opposed to the other side.

Biochemistry ⓔNow™

Assess your understanding of this chapter's topics with additional quizzing and tutorials at **http://now.brookscole.com/campbell5**

Annotated Bibliography

Danishefsky, S. Catalytic Antibodies and Disfavored Reactions. *Science* **259,** 469–470 (1993). [A short review of chemists' use of antibodies as the basis of "tailor-made" catalysts for specific reactions.]

Dressler, D., and H. Potter. *Discovering Enzymes.* New York: Scientific American Library, 1991. [A well-illustrated book that introduces important concepts of enzyme structure and function.]

Koshland, D., G. Nemethy, and D. Filmer. Comparison of Experimental Binding Data and Theoretical Models in Proteins Containing Subunits. *Biochemistry* **5,** 365–385 (1966).

Kraut, J. How Do Enzymes Work? *Science* **242,** 533–540 (1988). [An advanced discussion of the role of transition states in enzymatic catalysis.]

Landry, D. W. Immunotherapy for Cocaine Addiction. *Sci. Amer.,* **276**(2), 42–45 (1997). [How catalytic antibodies have been used to treat cocaine addiction.]

Landry, D. W., K. Zhao, G. X. Q. Yang, M. Glickman, and T. M. Georgiadis. Antibody Catalyzed Degradation of Cocaine. *Science* **259,** 1899–1901 (1993). [How antibodies can degrade an addictive drug.]

Lerner, R., S. Benkovic, and P. Schultz. At the Crossroads of Chemistry and Immunology: Catalytic Antibodies. *Science* **252,** 659–667 (1991). [A review of how antibodies can bind to almost any molecule of interest and then catalyze some reaction of that molecule.]

Marcus, R. Skiing the Reaction Rate Slopes. *Science* **256,** 1523–1524 (1992). [A brief, advanced-level look at reaction transition states.]

Monod, J., J. Wyman, and J.-P. Changeux. On the Nature of Allosteric Transitions: A Plausible Model. *J. Mol. Biol.* **12,** 88–118 (1965).

Sigman, D., ed. *The Enzymes. Vol. 20. Mechanisms of Catalysis.* San Diego: Academic Press, 1992. [Part of a definitive series on enzymes and their structures and functions.]

Sigman, D., and P. Boyer, eds. *The Enzymes. Vol. 19. Mechanisms of Catalysis.* San Diego: Academic Press, 1990. [Part of a definitive series on enzymes and their structures and functions.]

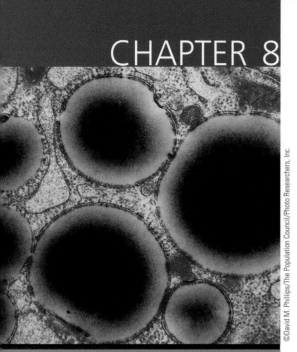

Electron micrograph of a fat cell. Much of the cell volume is taken up by lipid droplets.

CHAPTER 8

Lipids and Proteins Are Associated in Biological Membranes

The most striking feature of lipids is their nonpolar nature, which leads to their insolubility in water. A fatty acid is a lipid that contains a carboxyl head group attached to a hydrocarbon "tail." With three long-chain fatty acids, the triacylglycerols (also referred to as fats) are ideal reservoirs for energy storage in the cell. Some lipids have large, charged polar heads in addition to their uncharged hydrocarbon tails.

The chief ingredients of biological membranes are the phospholipids. In water, they form lipid bilayers, with their flexible tails in the hydrophobic interior of the membrane and their polar heads on exterior surfaces in contact with water. About half of the membrane consists of protein molecules associated with the lipid bilayer. Some small molecules can migrate through the membrane, from a high concentration on one side to a low concentration on the other side, by simple diffusion. Some proteins form pores that allow specified ions and small molecules to pass through the membrane. Heart (cardiac) muscle cells, which act in close synchrony, are connected by gap junctions—gated tubes that join the cells through their outer membranes. On the surfaces of cells are glycoproteins and lipoproteins that recognize other molecules, as well as receptors that act as gates for the passage of ions and molecules into the cell.

Critical Questions

Biochemistry ⓔNow™
Test yourself on these Critical Questions at the BiochemistryNow website at **http://now .brookscole.com/campbell5**

8.1 | What Is the Definition of a Lipid?

Lipids are compounds that occur frequently in nature. They are found in places as diverse as egg yolks and the human nervous system and are an important component of plant, animal, and microbial membranes. The definition of a lipid is based on solubility. Lipids are marginally soluble (at best) in water but readily soluble in organic solvents, such as chloroform or acetone. Fats and oils are typical lipids in terms of their solubility, but that fact does not really define their chemical nature. In terms of chemistry, lipids are a mixed bag of compounds that share some properties based on structural similarities, mainly a preponderance of nonpolar groups.

Classified according to their chemical nature, lipids fall into two main groups. One group, which consists of open-chain compounds with polar head groups and long nonpolar tails, includes *fatty acids, triacylglycerols, sphingolipids, phosphoacylglycerols,* and *glycolipids.* The second major group consists of fused-ring compounds, the *steroids;* an important representative of this group is cholesterol.

8.2 | What Are the Chemical Natures of the Lipid Types?

Fatty Acids

A fatty acid has a carboxyl group at the polar end and a hydrocarbon chain at the nonpolar tail. Fatty acids are **amphipathic** compounds because the car-

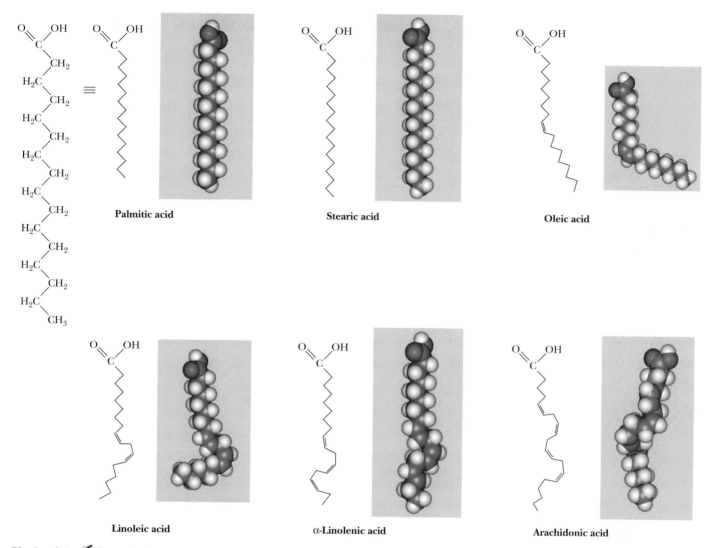

Biochemistry ⊛**Now**™ ▲ **ANIMATED FIGURE 8.1** The structures of some typical fatty acids. Note that most naturally occurring fatty acids contain even numbers of carbon atoms and that the double bonds are nearly always *cis* and rarely conjugated. **See this figure animated at http://now.brookscole.com/campbell5**

boxyl group is hydrophilic and the hydrocarbon tail is hydrophobic. The carboxyl group can ionize under the proper conditions.

A fatty acid that occurs in a living system normally contains an even number of carbon atoms, and the hydrocarbon chain is usually unbranched (Figure 8.1). If there are carbon–carbon double bonds in the chain, the fatty acid is *unsaturated;* if there are only single bonds, the fatty acid is *saturated.* Tables 8.1 and 8.2 list a few examples of the two classes. In unsaturated fatty acids, the stereochemistry at the double bond is usually *cis* rather than *trans.* The difference between *cis* and *trans* fatty acids is very important to their overall shape. A *cis* double bond puts a kink in the long-chain hydrocarbon tail, whereas the shape of a *trans* fatty acid is like that of a saturated fatty acid in its fully extended conformation. Note that the double bonds are isolated from one another by several singly bonded carbons; fatty acids do not normally have conjugated double-bond systems. The notation used for fatty acids indicates the number of carbon atoms and the number of double bonds. In this system, 18:0 denotes an 18-carbon saturated fatty acid with no double bonds, and 18:1 denotes an 18-carbon fatty acid with one double bond. Note that, in the unsaturated fatty acids in Table 8.2 (except arachidonic acid), there is a

Table 8.1

Typical Naturally Occurring Saturated Fatty Acids

Acid	Number of Carbon Atoms	Formula	Melting Point (°C)
Lauric	12	$CH_3(CH_2)_{10}CO_2H$	44
Myristic	14	$CH_3(CH_2)_{12}CO_2H$	58
Palmitic	16	$CH_3(CH_2)_{14}CO_2H$	63
Stearic	18	$CH_3(CH_2)_{16}CO_2H$	71
Arachidic	20	$CH_3(CH_2)_{18}CO_2H$	77

Table 8.2

Typical Naturally Occurring Unsaturated Fatty Acids

Acid	Number of Carbon Atoms	Degree of Unsaturation	Formula	Melting Point (°C)
Palmitoleic	16	$16{:}1—\Delta^9$	$CH_3(CH_2)_5CH{=}CH(CH_2)_7CO_2H$	−0.5
Oleic	18	$18{:}1—\Delta^9$	$CH_3(CH_2)_7CH{=}CH(CH_2)_7CO_2H$	16
Linoleic	18	$18{:}2—\Delta^{9,12}$	$CH_3(CH_2)_4CH{=}CH(CH_2)CH{=}CH(CH_2)_7CO_2H$	−5
Linolenic	18	$18{:}3—\Delta^{9,\,12,\,15}$	$CH_3(CH_2CH{=}CH)_3(CH_2)_7CO_2H$	−11
Arachidonic	20	$20{:}4—\Delta^{5,\,8,\,11,\,14}$	$CH_3(CH_2)_4CH{=}CHCH_2)_4(CH_2)_2CO_2H$	−50

* Degree of unsaturation refers to the number of double bonds. The superscript indicates the position of double bonds. For example, Δ^9 refers to a double bond at the ninth carbon atom from the carboxyl end of the molecule.

double bond at the ninth carbon atom from the carboxyl end. The position of the double bond results from the way unsaturated fatty acids are synthesized in organisms (Section 21.6). Unsaturated fatty acids have lower melting points than saturated ones. Plant oils are liquid at room temperature because they have higher proportions of unsaturated fatty acids than do animal fats, which tend to be solids. Conversion of oils to fats is a commercially important process. It involves hydrogenation, the process of adding hydrogen across the double bond of unsaturated fatty acids to produce the saturated counterpart. Oleomargarine, in particular, uses partially hydrogenated vegetable oils, which tend to include *trans* fatty acids (see the Biochemical Connections box on page 195).

Fatty acids are rarely found free in nature, but they form parts of many commonly occurring lipids.

Triacylglycerols

Glycerol is a simple compound that contains three hydroxyl groups (Figure 8.2). When all three of the alcohol groups form ester linkages with fatty acids, the resulting compound is a **triacylglycerol;** an older name for this type of compound is *triglyceride.* Note that the three ester groups are the polar part of the molecule, whereas the tails of the fatty acids are nonpolar. It is usual for three different fatty acids to be esterified to the alcohol groups of the same glycerol molecule. Triacylglycerols do not occur as components of membranes (as do other types of lipids), but they accumulate in adipose tissue (primarily fat cells) and provide a means of storing fatty acids, particularly in animals. They serve as concentrated stores of metabolic energy. Complete oxidation of fats yields about 9 kcal g^{-1}, in contrast with 4 kcal g^{-1} for carbohydrates and proteins (see Section 21.3 and 24.2).

When an organism uses fatty acids, the ester linkages of triacylglycerols are hydrolyzed by enzymes called **lipases.** The same hydrolysis reaction can take

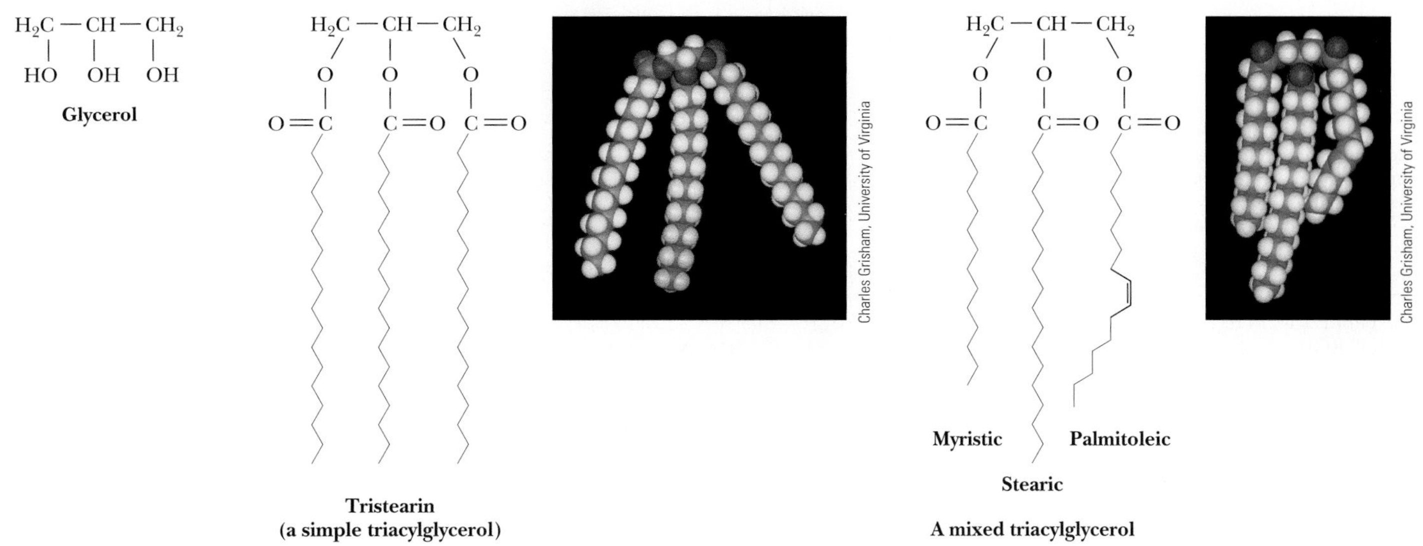

H₂C — CH — CH₂
| | |
HO   OH   OH
Glycerol

H₂C — CH — CH₂
| | |
O   O   O
| | |
O=C   C=O C=O

Tristearin
(a simple triacylglycerol)

H₂C — CH — CH₂
| | |
O   O   O
| | |
O=C   C=O C=O

Myristic   **Palmitoleic**

Stearic

A mixed triacylglycerol

Charles Grisham, University of Virginia

▲ **FIGURE 8.2** Triacylglycerols are formed from glycerol and fatty acids.

place outside organisms, with acids or bases as catalysts. When a base such as sodium hydroxide or potassium hydroxide is used, the products of the reaction, which is called *saponification* (Figure 8.3), are glycerol and the sodium or potassium salts of the fatty acids. These salts are soaps. When soaps are used with hard water, the calcium and magnesium ions in the water react with the fatty acids to form a precipitate—the characteristic scum left on the insides of sinks and bathtubs. The other product of saponification, glycerol, is used in creams and lotions as well as in the manufacture of nitroglycerin.

Phosphoacylglycerols (Phospholipids)

It is possible for one of the alcohol groups of glycerol to be esterified by a phosphoric acid molecule rather than by a carboxylic acid. In such lipid molecules, two fatty acids are also esterified to the glycerol molecule. The resulting compound is called a **phosphatidic acid** (Figure 8.4a). Fatty acids are usually monoprotic acids with only one carboxyl group able to form an ester bond, but phosphoric acid is triprotic and thus can form more than one ester linkage. One molecule of phosphoric acid can form ester bonds both to glycerol and to some other alcohol, creating a *phosphatidyl ester* (Figure 8.4b). Phosphatidyl esters are classed as **phosphoacylglycerols.** The natures of the fatty acids vary widely, as they do in triacylglycerols. As a result, the names of the types of lipids (such as triacylglycerols and phosphoacylglycerols) that contain fatty acids must be considered generic names.

The classification of a phosphatidyl ester depends on the nature of the second alcohol esterified to the phosphoric acid. Some of the most important lipids in this class are *phosphatidyl ethanolamine* (cephalin), *phosphatidyl serine*, *phosphatidyl choline* (lecithin), *phosphatidyl inositol, phosphatidyl glycerol,* and *diphosphatidyl glycerol* (cardiolipin) (Figure 8.5). In each of these types of compounds, the nature of the fatty acids in the molecule can vary widely. All these compounds have long, nonpolar, hydrophobic tails and polar, highly hydrophilic head groups and thus are markedly amphipathic. (We have already seen this characteristic in fatty acids.) In a phosphoacylglycerol, the polar head group is charged, since the phosphate group is ionized at neutral pH. There is frequently also a positively charged amino group contributed by an amino alcohol esterified to the phosphoric acid. Phosphoacylglycerols are important components of biological membranes.

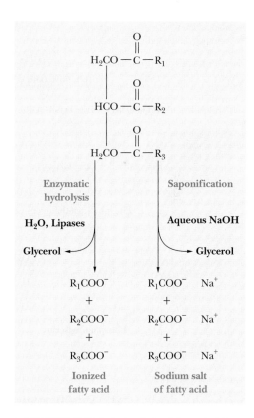

▲ **FIGURE 8.3** Hydrolysis of triacylglycerols. The term "saponification" refers to the reactions of glyceryl ester with sodium or potassium hydroxide to produce a soap, which is the corresponding salt of the long-chain fatty acid.

(a)

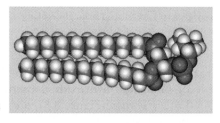

Phosphatidic acid

(b)

Stearyl group

Linoleyl group

Phosphatidyl ester

▶ **FIGURE 8.4** The molecular architecture of phosphoacylglycerols. (a) A phosphatidic acid, in which glycerol is esterified to phosphoric acid and to two different carboxylic acids. R_1 and R_2 represent the hydrocarbon chains of the two carboxylic acids. (b) A phosphatidyl ester (phosphoacylglycerol). Glycerol is esterified to two carboxylic acids, stearic acid and linoleic acid, as well as to phosphoric acid. Phosphoric acid, in turn, is esterified to a second alcohol, ROH.

Phosphatidylcholine

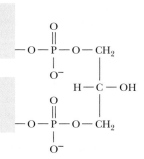

GLYCEROLIPIDS WITH OTHER HEAD GROUPS:

Phosphatidylethanolamine

Phosphatidylserine

Phosphatidylglycerol

Diphosphatidylglycerol (Cardiolipin)

Phosphatidylinositol

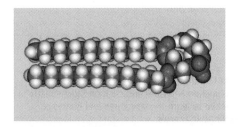

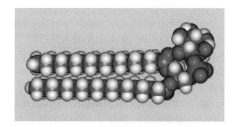

Biochemistry⊜Now™ ANIMATED FIGURE 8.5
Structures of some phosphoacylglycerols and space-filling models of phosphatidylcholine, phosphatidylglycerol, and phosphatidylinositol. **See this figure animated at http://now.brookscole.com/campbell5**

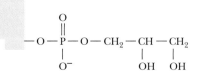

Waxes

Waxes are complex mixtures of esters of long-chain carboxylic acids and long-chain alcohols. They frequently serve as protective coatings for both plants and animals. In plants, they coat stems, leaves, and fruit; in animals, they are found on fur, feathers, and skin. Myricyl cerotate (Figure 8.6a), the principal component of carnauba wax, is produced by the Brazilian wax palm. Carnauba wax is extensively used in floor wax and automobile wax. The principal component of spermaceti, a wax produced by whales, is cetyl palmitate (Figure 8.6a). The use of spermaceti as a component of cosmetics made it one of the most highly prized products of 19th-century whaling efforts.

Sphingolipids

Sphingolipids do not contain glycerol, but they do contain the long-chain amino alcohol sphingosine, from which this class of compounds takes its name (Figure 8.6b). Sphingolipids are found in both plants and animals; they are particularly abundant in the nervous system. The simplest compounds of this class are the ceramides, which consist of one fatty acid linked to the amino group of sphingosine by an amide bond (Figure 8.6b). In **sphingomyelins,** the primary alcohol group of sphingosine is esterified to phosphoric acid, which, in turn, is esterified to another amino alcohol, choline (Figure 8.6b). Note the structural similarities between sphingomyelin and other phospholipids. Two long hydrocarbon chains are attached to a backbone that contains alcohol groups. One of the alcohol groups of the backbone is esterified to phosphoric acid. A second alcohol—choline, in this case—is also esterified to the phosphoric acid. We have already seen that choline occurs in phosphoacylglycerols. Sphingomyelins are amphipathic; they occur in cell membranes in the nervous system (see the following Biochemical Connections box).

Glycolipids

If a carbohydrate is bound to an alcohol group of a lipid by a glycosidic linkage (see Section 16.3 for a discussion of glycosidic linkages), the resulting compound is a **glycolipid.** Quite frequently, **ceramides** (see Figure 8.6) are the parent compounds for glycolipids, and the glycosidic bond is formed between the primary alcohol group of the ceramide and a sugar residue. The resulting compound is called a **cerebroside.** In most cases, the sugar is glucose or galactose; for example, a glucocerebroside is a cerebroside that contains glucose (Figure 8.7). As the name indicates, cerebrosides are found in nerve and brain cells, primarily in cell membranes. The carbohydrate portion of these compounds can be very complex. Gangliosides are examples of glycolipids with a complex carbohydrate moiety that contains more than three sugars. One of them is always a sialic acid (Figure 8.8). These compounds are also referred to as acidic glycosphingolipids due to their net negative charge at neutral pH. Glycolipids are often found as markers on cell membranes and play a large role in tissue and organ specificity. Gangliosides are also present in large quantities in nerve tissues. Their biosynthesis and breakdown are discussed in Section 21.7 and in the Biochemical Connections box on page XXX in Chapter 21.

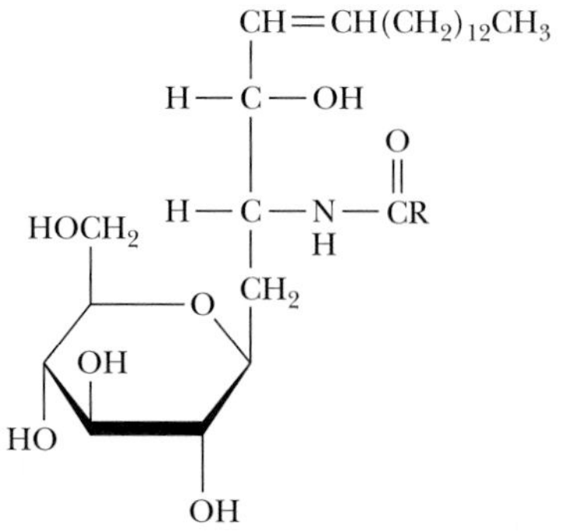

Steroids

Many compounds of widely differing functions are classified as **steroids** because they have the same general structure: a fused-ring system consisting of three six-membered rings (the A, B, and C rings) and one five-membered ring (the D ring). There are many important steroids, including sex hormones.

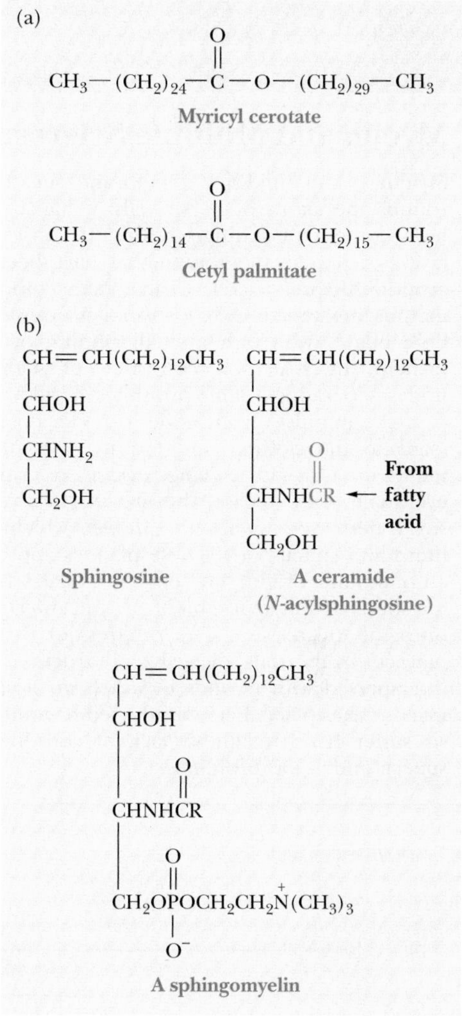

▲ **FIGURE 8.6** Structures of some waxes and sphingolipids.

A Glucocerebroside

▲ **FIGURE 8.7** Structure of a glucocerebroside.

Biochemical Connections

Myelin and Multiple Sclerosis

Myelin is the lipid-rich membrane sheath that surrounds the axons of nerve cells; it has a particularly high content of sphingomyelins. It consists of many layers of plasma membrane that have been wrapped around the nerve cell. Unlike many other types of membranes (Section 8.5), myelin is essentially an all-lipid bilayer with only a small amount of embedded protein. Its structure, consisting of segments with nodes separating them, promotes rapid transmission of nerve impulses from node to node. Loss of myelin leads to the slowing and eventual cessation of the nerve impulse. In *multiple sclerosis,* a crippling and eventually fatal disease, the myelin sheath is progressively destroyed by *sclerotic plaques,* which affect the brain and spinal cord. These plaques appear to be of autoimmune origin, but epidemiologists have raised questions about involvement of viral infections in the onset of the disease. The progress of the disease is marked by periods of active destruction of myelin interspersed with periods in which no destruction of myelin takes place. Persons affected by multiple sclerosis suffer from weakness, lack of coordination, and speech and vision problems.

Reuters/Corbis-Bettmann

◀ Annette Funicello enjoyed a successful career in television and films before she was stricken with multiple sclerosis. She started to display the lack of coordination characteristic of the early stages of this disease, causing concern among those who knew her. To end speculation, she announced that she had developed multiple sclerosis.

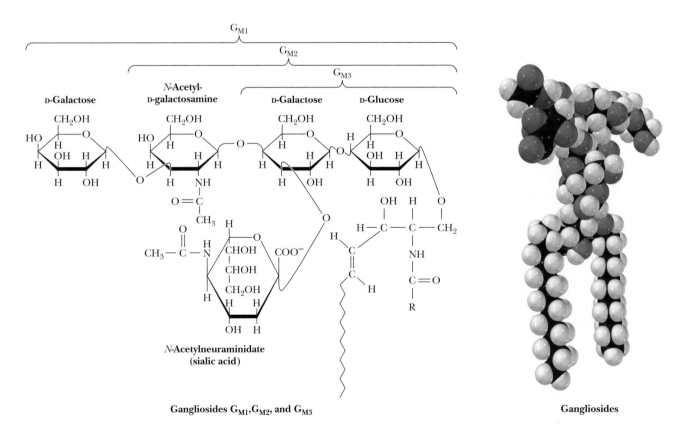

Gangliosides G_{M1}, G_{M2}, and G_{M3}

Gangliosides

▲ **FIGURE 8.8** The structures of several important gangliosides. Also shown is a space-filling model of ganglioside G_{M1}.

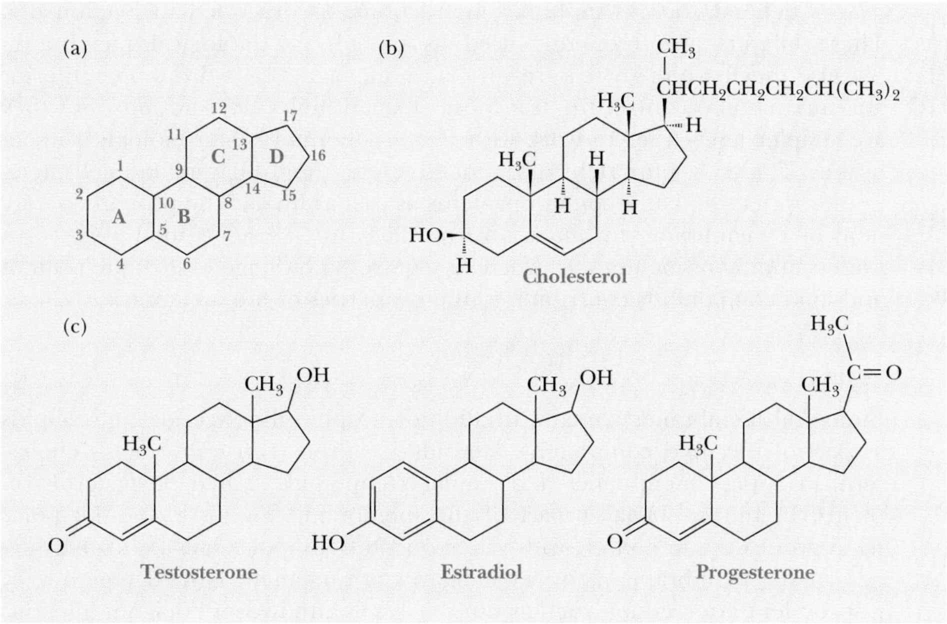

▲ **FIGURE 8.9** Structures of some steroids. (a) The fused-ring structure of steroids. (b) Cholesterol. (c) Some steroid sex hormones.

(See Section 24.3 for more steroids of biological importance.) The steroid that is of most interest in our discussion of membranes is **cholesterol** (Figure 8.9). The only hydrophilic group in the cholesterol structure is the single hydroxyl group. As a result, the molecule is highly hydrophobic. Cholesterol is widespread in biological membranes, especially in animals, but it does not occur in prokaryotic cell membranes. The presence of cholesterol in membranes can modify the role of membrane-bound proteins. Cholesterol has a number of important biological functions, including its role as a precursor of other steroids and of vitamin D_3. We will see a five-carbon structural motif (the isoprene unit) that is common to steroids and to fat-soluble vitamins, which is an indication of their biosynthetic relationship (Sections 8.7 and 21.8). However, cholesterol is best known for its harmful effects on health when it is present in excess in the blood. It plays a role in the development of *atherosclerosis*, a condition in which lipid deposits block the blood vessels and lead to heart disease (see Section 21.8).

> **Essential Information**
>
> Lipids are compounds with a preponderance of nonpolar groups. They can be open-chain molecules with a polar head group and a long nonpolar tail. Glycerol, fatty acids, and phosphoric acid can frequently be obtained as degradation products of these compounds. Another class of lipids consists of fused-ring compounds, the steroids.

8.3 | What Is the Nature of Biological Membranes?

Every cell has a cell membrane (also called a plasma membrane); eukaryotic cells also have membrane-enclosed organelles, such as nuclei and mitochondria. The molecular basis of the membrane's structure lies in its lipid and protein components. Now it is time to see how the interaction between the lipid bilayer and membrane proteins determines membrane function. Membranes not only separate cells from the external environment but also play important roles in transport of specific substances into and out of cells. In addition, a number of important enzymes are found in membranes and depend on this environment for their function.

Phosphoglycerides are prime examples of amphipathic molecules, and they are the principal lipid components of membranes. The existence of *lipid bilayers* depends on hydrophobic interactions, as described in Section 4.4. These bilayers are frequently used as models for biological membranes because they have many features in common, such as a hydrophobic interior and an ability to control the transport of small molecules and ions, but they are simpler and easier to work with in the laboratory than biological membranes. The most important difference between lipid bilayers and cell membranes is that the latter contain proteins as well as lipids. The protein component of a membrane can make up from 20% to 80% of its total weight. An understanding of membrane structure requires knowledge of how the protein and lipid components contribute to the properties of the membrane.

Lipid Bilayers

Biological membranes contain, in addition to phosphoglycerides, glycolipids as part of the lipid component. Steroids are present in eukaryotes—cholesterol in animal membranes and similar compounds, called phytosterols, in plants. In the **lipid-bilayer** part of the membrane (Figure 8.10), the polar head groups are in contact with water, and the nonpolar tails lie in the interior of the membrane. The whole bilayer arrangement is held together by noncovalent interactions, such as van der Waals and hydrophobic interactions (Section 2.1). The surface of the bilayer is polar and contains charged groups. The nonpolar hydrocarbon interior of the bilayer consists of the saturated and unsaturated chains of fatty acids and the fused-ring system of cholesterol. Both the inner and outer layers of the bilayer contain mixtures of lipids, but their compositions differ and can be used to distinguish the inner and outer layers from each other (Figure 8.11). Bulkier molecules tend to occur in the outer layer, and smaller molecules tend to occur in the inner layer.

The arrangement of the hydrocarbon interior of the bilayer can be ordered and rigid or disordered and fluid. The bilayer's fluidity depends on its composition. In saturated fatty acids, a linear arrangement of the hydro-

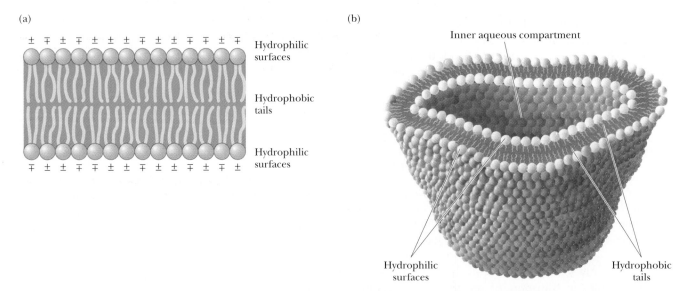

(a)

Hydrophilic surfaces

Hydrophobic tails

Hydrophilic surfaces

(b)

Inner aqueous compartment

Hydrophilic surfaces

Hydrophobic tails

▲ **FIGURE 8.10** Lipid bilayers. (a) Schematic drawing of a portion of a bilayer consisting of phospholipids. The polar surface of the bilayer contains charged groups. The hydrocarbon "tails" lie in the interior of the bilayer. (b) Cutaway view of a lipid bilayer vesicle. Note the aqueous inner compartment and the fact that the inner layer is more tightly packed than the outer layer. (*From Bretscher, M. S. The Molecules of the Cell Membrane.* Scientific American, *October 1985, p. 103. Art by Dana Burns-Pizer.*)

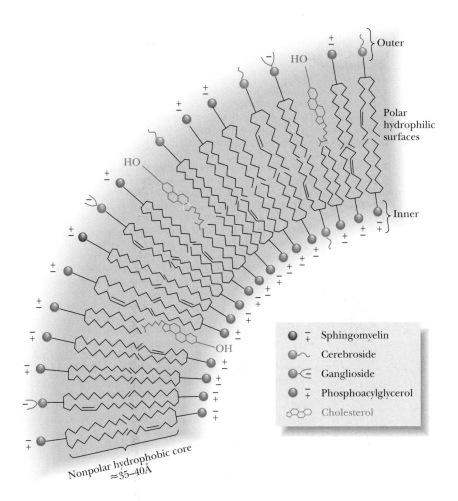

Sphingomyelin
Cerebroside
Ganglioside
Phosphoacylglycerol
Cholesterol

◀ **FIGURE 8.11** Lipid bilayer asymmetry. The compositions of the outer and inner layers differ; the concentration of bulky molecules is higher in the outer layer, which has more room.

carbon chains leads to close packing of the molecules in the bilayer, and thus to rigidity. In unsaturated fatty acids, there is a kink in the hydrocarbon chain that does not exist in saturated fatty acids (Figure 8.12). The kinks cause disorder in the packing of the chains, which makes for a more open structure than would be possible for straight saturated chains (Figure 8.13). In turn, the disordered structure caused by the presence of unsaturated fatty acids with *cis* double bonds (and therefore kinks) in their hydrocarbon chains causes greater fluidity in the bilayer. The lipid components of a bilayer are always in motion, to a greater extent in more fluid bilayers and to a lesser extent in more rigid ones.

The presence of cholesterol may also enhance order and rigidity. The fused-ring structure of cholesterol is itself quite rigid, and the presence of cholesterol stabilizes the extended straight-chain arrangement of saturated fatty acids by van der Waals interactions (Figure 8.14). The lipid portion of a plant membrane has a higher percentage of unsaturated fatty acids, especially polyunsaturated (containing two or more double bonds) fatty acids, than does the lipid portion of an animal membrane. Furthermore, the presence of cholesterol is characteristic of animal, rather than plant, membranes. As a result, animal membranes are less fluid (more rigid) than plant membranes, and the membranes of prokaryotes, which contain no appreciable amounts of steroids, are the most fluid of all. Research suggests that plant sterols can act as natural cholesterol blockers, interfering with the uptake of dietary cholesterol.

With heat, ordered bilayers become less ordered; bilayers that are comparatively disordered become even more disordered. This cooperative transition

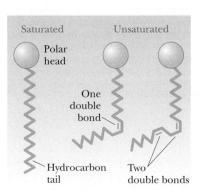

▲ **FIGURE 8.12** The effect of double bonds on the conformations of the hydrocarbon tails of fatty acids. Unsaturated fatty acids have kinks in their tails.

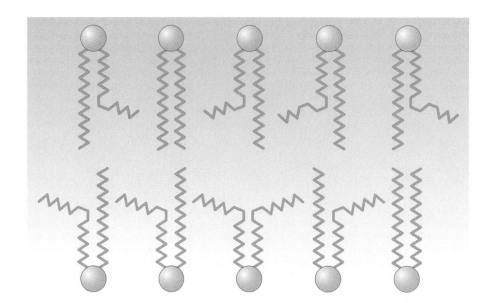

▶ **FIGURE 8.13** Schematic drawing of a portion of a highly fluid phospholipid bilayer. The kinks in the unsaturated side chains prevent close packing of the hydrocarbon portions of the phospholipids.

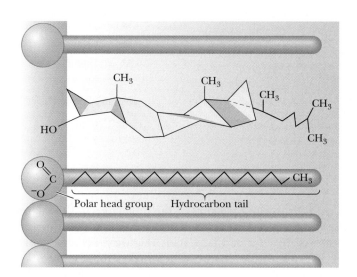

▶ **FIGURE 8.14** Stiffening of the lipid bilayer by cholesterol. The presence of cholesterol in a membrane reduces fluidity by stabilizing extended chain conformations of the hydrocarbon tails of fatty acids, as a result of van der Waals interactions.

Polar head group   Hydrocarbon tail

Biochemistry ⊘ Now™ **ANIMATED FIGURE 8.15** An illustration of the gel-to-liquid crystalline phase transition, which occurs when a membrane is warmed through the transition temperature, T_m. Notice that the surface area must increase and the thickness must decrease as the membrane goes through a phase transition. The mobility of the lipid chains increases dramatically. **See this figure animated at http://now.brookscole.com/campbell5**

Gel   Heat   Liquid crystal

takes place at a characteristic temperature, like the melting of a crystal, which is also a cooperative transition (Figure 8.15). The transition temperature is higher for more rigid and ordered membranes than it is for relatively fluid and disordered membranes. The following Biochemical Connections box looks at some connections between the fatty acid composition of bilayers and membranes and how they behave at different temperatures.

Biochemical Connections

Butter Versus Margarine—Which Is Healthier?

We use the terms animal "fats" and plant "oils" because of the solid and fluid nature of these two groups of lipids. The major difference between fats and oils is the percentage of unsaturated fatty acids in the triglycerides and the phosphoglycerides of membranes. This difference is far more important than the fact that the length of the fatty acid chain can affect the melting points. Butter is an exception; it has a high proportion of short-chain fatty acids and thus can "melt in your mouth."

Membranes must maintain a certain degree of fluidity to be functional. Consequently, unsaturated fats are distributed in varying proportions in different parts of the body. The membranes of internal organs of warm-blooded mammals have a higher percentage of saturated fats than do the membranes of skin tissues, which helps to keep the membrane more solid at the higher temperature of the internal organ. An extreme example of this is found in the legs and the body of reindeer, where there are marked differences in the percentages of saturated fatty acids. When bacteria are grown at different temperatures, the fatty acid composition of the membranes changes to reflect more unsaturated fatty acids at lower temperatures and more saturated fatty acids at higher temperatures. The same type of difference can be seen in eukaryotic cells grown in tissue culture.

Even if we look at plant oils alone, we find different proportions of saturated fats in different oils. The following table gives the distribution for a tablespoon (14 g) of different oils.

Because cardiovascular disease is correlated with diets high in saturated fats, a diet of more unsaturated fats may reduce the risk of heart attacks and strokes. Canola oil is an attractive dietary choice because it has a high ratio of unsaturated fatty acids to saturated fatty acids. Since the 1960s, we have known that foods higher in polyunsaturated fats were healthier. Unfortunately, even though olive oil is popular in cooking Italian food and canola oil is trendy for other cooking, pouring oil on bread or toast is not appealing. Thus companies began to market butter substitutes that were based on unsaturated fatty acids but that would also have the physical characteristics of butter, such as being solid at room temperature. They accomplished this task by partially hydrogenating the double bonds in the unsaturated fatty acids making up the oils. The irony here is that, to avoid eating the saturated fatty acids in butter, butter substitutes were created from polyunsaturated oils by removing some of the double bonds, thus making them more saturated. In addition, many of the soft spreads that are marketed as being healthy (safflower-oil spread and canola-oil spread) may indeed pose new health risks. In the hydrogenation process, some double bonds are converted to the *trans* form. Studies now show that *trans* fatty acids raise the ratio of LDL (low-density lipoprotein) cholesterol compared to HDL (high-density lipoprotein) cholesterol, a positive correlator of heart disease. Thus the effects of *trans* fatty acids are similar to those of saturated fatty acids. In the last few years, however, new butter substitutes have been marketed that advertise "no *trans* fatty acids."

Type of Oil or Fat	Example	Saturated (g)	Monounsaturated (g)	Polyunsaturated (g)
Tropical oils	Coconut oil	13	0.7	0.3
Semitropical oils	Peanut oil	2.4	6.5	4.5
	Olive oil		10.3	1.3
Temperate oils	Canola oil	1	8.2	4.1
	Safflower oil	1.3	1.7	10.4
Animal fat	Lard	5.1	5.9	1.5
	Butter	9.2	4.2	0.6

Recall that the distribution of lipids is not the same in the inner and outer portions of the bilayer. Because the bilayer is curved, the molecules of the inner layer are more tightly packed (refer to Figure 8.11). Bulkier molecules, such as cerebrosides (see Section 8.2), tend to be located in the outer layer. There is very little tendency for "flip-flop" migration of lipid molecules from one layer of the bilayer to another, but it does occur occasionally. Lateral motion of lipid molecules within one of the two layers frequently takes place, however, especially in more fluid bilayers. Several methods exist for monitoring the motions of molecules within a lipid bilayer. These methods depend on labeling some part of the lipid component with an easily detected "tag." The tags are usually fluorescent compounds, which can be detected with highly sensitive equipment. Another kind of labeling method depends on the fact that some nitrogen compounds have unpaired electrons. These compounds are used as labels and can be detected by magnetic measurements.

8.4 | What Are Some Common Types of Membrane Proteins?

Proteins in a biological membrane can be associated with the lipid bilayer in either of two ways—as **peripheral proteins** on the surface of the membrane or as **integral proteins** within the lipid bilayer (Figure 8.16). Peripheral proteins are usually bound to the charged head groups of the lipid bilayer by polar interactions, electrostatic interactions, or both. They can be removed by such mild treatment as raising the ionic strength of the medium. The relatively numerous charged particles present in a medium of higher ionic strength undergo more electrostatic interactions with the lipid and with the protein, "swamping out" the comparatively fewer electrostatic interactions between the protein and the lipid.

Removing integral proteins from membranes is much more difficult. Harsh conditions, such as treatment with detergents or extensive sonication (exposure to ultrasonic vibrations), are usually required. Such measures frequently denature the protein, which often remains bound to lipids in spite of all efforts to obtain it in pure form. The denatured protein is of course inactive, whether or not it remains bound to lipids. Fortunately, nuclear magnetic resonance techniques are now enabling researchers to study proteins of this sort in living tissue. The structural integrity of the whole membrane system appears to be necessary for the activities of most membrane proteins.

Proteins can be attached to the membrane in a variety of ways. When a protein completely spans the membrane, it is often in the form of an α-helix or β-sheet. These structures minimize contact of the polar parts of the peptide backbone with the nonpolar lipids in the interior of the bilayer (Figure 8.17). Proteins can also be anchored to the lipids via covalent bonds from cysteines or free amino groups on the protein to one of several lipid anchors. Myristoyl and palmitoyl groups are common anchors (Figure 8.17).

Membrane proteins have a variety of functions. Most, but not all, of the important functions of the membrane as a whole are those of the protein component. **Transport proteins** help move substances in and out of the cell, and **receptor proteins** are important in the transfer of extracellular signals, such as those carried by hormones or neurotransmitters, into the cell. In addition, some enzymes are tightly bound to membranes; examples include many of the enzymes responsible for aerobic oxidation reactions, which are found in specific parts of mitochondrial membranes. Some of these enzymes are on the inner surface of the membrane, and some are on the outer surface. There is an uneven distribution of proteins of all types on the inner and outer layers of all cell membranes, just as there is an asymmetric distribution of lipids.

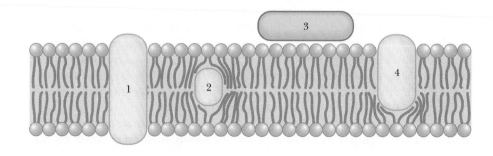

▶ **FIGURE 8.16** Some types of associations of proteins with membranes. The proteins marked 1, 2, and 4 are integral proteins, and protein 3 is a peripheral protein. Note that the integral proteins can be associated with the lipid bilayer in several ways. Protein 1 tranverses the membrane, protein 2 lies entirely within the membrane, and protein 4 projects into the membrane.

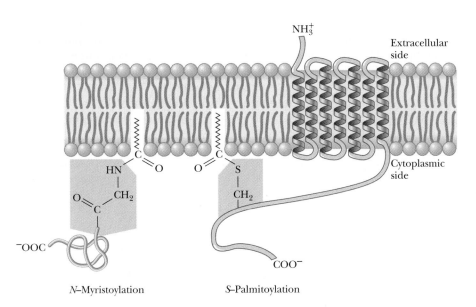

◀ **FIGURE 8.17** Certain proteins are anchored to biological membranes by lipid anchors. Particularly common are the *N*-myristoyl- and *S*-palmitoyl-anchoring motifs shown here. *N*-myristoylation always occurs at an *N*-terminal glycine residue, whereas thioester linkages occur at cysteine residues within the polypeptide chain. G-protein–coupled receptors, with seven transmembrane segments, may contain one (and sometimes two) palmitoyl anchors in thioester linkage to cysteine residues in the *C*-terminal segment of the protein.

8.5 | What Is the Fluid-Mosaic Model of Membrane Structure?

We have seen that biological membranes have both lipid and protein components. How do these two parts combine to produce a biological membrane? Currently, the **fluid-mosaic model** is the most widely accepted description of biological membranes. The term "mosaic" implies that the two components exist side by side without forming some other substance of intermediate nature. The basic structure of biological membranes is that of the lipid bilayer, with the proteins embedded in the bilayer structure (Figure 8.18). These proteins tend to have a specific orientation in the membrane. The

Essential Information

Biological membranes consist of lipid bilayers combined with proteins. Peripheral proteins are loosely attached to one surface of the membrane via hydrogen bonds or electrostatic attractions. Integral proteins are embedded more solidly in the membrane and may be covalently attached to lipid anchors.

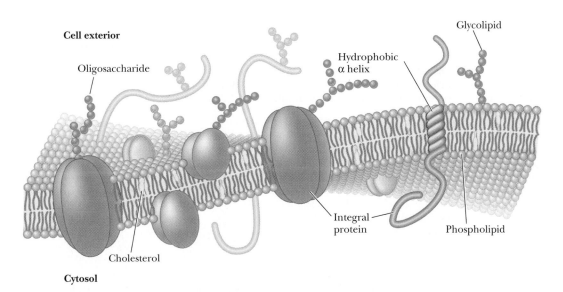

▲ **FIGURE 8.18** Fluid-mosaic model of membrane structure. Membrane proteins can be seen embedded in the lipid bilayer. (*From Singer, S. J., in G. Weissman and R. Claiborne, Eds.,* Cell Membranes: Biochemistry, Cell Biology, and Pathology, *New York: HP Pub., 1975, p. 37.*)

term "fluid mosaic" implies that the same sort of lateral motion that we have already seen in lipid bilayers also occurs in membranes. The proteins "float" in the lipid bilayer and can move along the plane of the membrane.

Electron micrographs can be made of membranes that have been frozen and then fractured along the interface between the two layers. The outer layer is removed, exposing the interior of the membrane. The interior has a granular appearance because of the presence of the integral membrane proteins (Figures 8.19 and 8.20).

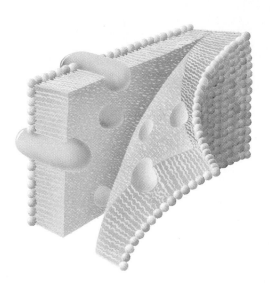

▶ **FIGURE 8.19** Replica of a freeze-fractured membrane. In the freeze-fracture technique, the lipid bilayer is split parallel to the surface of the membrane. The hydrocarbon tails of the two layers are separated from each other, and the proteins can be seen as "hills" in the replica shown. In the other layer, seen edge on, there are "valleys" where the proteins were. (*From Singer, S. J., in G. Weissman and R. Claiborne, Eds.,* Cell Membranes: Biochemistry, Cell Biology, and Pathology, *New York: HP Pub., 1975, p. 37.*)

Biochemical Connections

Membranes in Medicine

Because the driving force behind the formation of lipid bilayers is the exclusion of water from the hydrophobic region of lipids, and not some enzymatic process, artificial membranes can be created in the lab. **Liposomes** are stable structures based on a lipid bilayer that form a spherical vesicle. These vesicles can be prepared with therapeutic agents on the inside and then used to deliver the agent to a target tissue.

Every year, more than a million Americans are diagnosed with skin cancer, most often caused by long-term exposure to ultraviolet light. The ultraviolet (UV) light damages DNA in several ways, with one of the most common being the production of dimers between two pyrimidine bases (Section 9.5). For a species with little body hair and a fondness for sunshine, humans are poorly equipped to fight damaged DNA in their skin. Of the 130 known human DNA repair enzymes, only one system is designed to repair the main DNA lesions caused by exposure to UV. Several lower species have repair enzymes that we lack.

Researchers have developed a skin lotion to counteract the effects of UV light. The lotion contains liposomes filled with a DNA-repair enzyme from a virus, called T4 endonuclease V. The liposomes penetrate the skin cells. Once inside, the enzymes

make their way to the nucleus, where they attack pyrimidine dimers and start a DNA-repair mechanism that the normal cellular processes can complete. The skin lotion, marketed by AGI Dermatics, is currently undergoing clinical trials. Check out the AGI Dermatics website (http://www.agiderm.com) for information on the results of the clinical trials.

(a) Bilayer   (b) Unilamellar vesicle

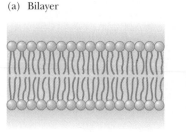

▲ Schematic drawing of a bilayer and a unilamellar vesicle. Because exposure of the edges of a bilayer to solvent is highly unfavorable, extensive bilayers usually wrap around themselves to form closed vesicles.

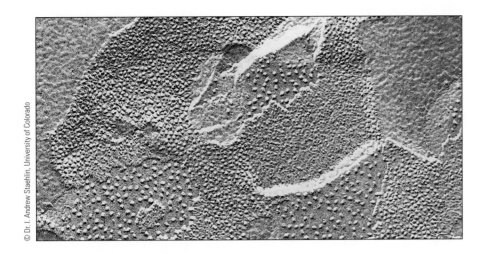

© Dr. I. Andrew Staehlin, University of Colorado

◀ **FIGURE 8.20** Electron micrograph of a freeze-fractured thylakoid membrane of a pea (magnified 110,000×). The grains protruding from the surface are integral membrane proteins.

8.6	**What Are Some of the Functions of Membranes?**

As already mentioned, three important functions take place in or on membranes (in addition to the structural role of membranes as the boundaries and containers of all cells and of the organelles within eukaryotic cells). The first of these functions is *transport*. Membranes are semipermeable barriers to the flow of substances into and out of cells and organelles. Transport through the membrane can involve the lipid bilayer as well as the membrane proteins. The other two important functions primarily involve the membrane proteins. One of these functions is *catalysis*. As we have seen, enzymes can be bound—in some cases very tightly—to membranes, and the enzymatic reaction takes place on the membrane. The third significant function is the *receptor property,* in which proteins bind specific biologically important substances that trigger biochemical responses in the cell. We shall discuss enzymes bound to membranes in subsequent chapters (especially in our treatment of aerobic oxidation reactions in Chapters 19 and 20). The other two functions we now consider in turn.

Membrane Transport

The most important question about transport of substances across biological membranes is whether the process requires the cell to expend energy. In **passive transport,** a substance moves from a region of higher concentration to one of lower concentration. In other words, the movement of the substance is in the same direction as a *concentration gradient,* and the cell does not expend energy. In **active transport,** a substance moves from a region of lower concentration to one of higher concentration (against a concentration gradient), and this process requires the cell to expend energy.

The process of passive transport can be subdivided into two categories—simple diffusion and facilitated diffusion. In **simple diffusion,** a molecule moves directly through the membrane without interacting with another molecule. Small, uncharged molecules, such as O_2, N_2, H_2O, and CO_2, can pass through membranes via simple diffusion. The rate of movement through the membrane is controlled solely by the concentration difference across the membrane (Figure 8.21). Larger molecules (especially polar ones) and ions cannot pass through a membrane by simple diffusion. The process of moving a molecule passively through a membrane using a carrier protein, to which molecules bind, is called **facilitated diffusion.** A good example is the movement of glucose into erythrocytes. The concentration of glucose in the blood is about 5 mM. The

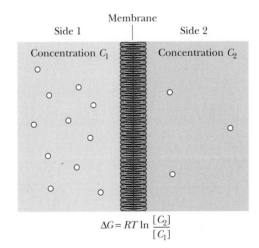

$$\Delta G = RT \ln \frac{[C_2]}{[C_1]}$$

Biochemistry ⊗ Now™ ACTIVE FIGURE 8.21
Passive diffusion of an uncharged species across a membrane depends only on the concentrations (C_1 and C_2) on the two sides of the membrane. **Watch this Active Figure at http://now.brookscole .com/campbell5**

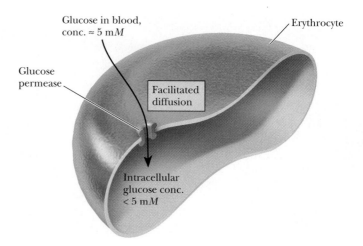

Glucose in blood,
conc. ≈ 5 mM

Erythrocyte

Glucose
permease

Facilitated
diffusion

Intracellular
glucose conc.
< 5 mM

▶ **FIGURE 8.22** Glucose passes into an erythrocyte via glucose permease by facilitated diffusion. Glucose flows using its concentration gradient via passive transport. (*Adapted from Lehninger,* Principles of Biochemistry, *Third Edition, by David L. Nelson and Michael M. Cox. © 1982, 1992, 2000 by Worth Publishers. Used with permission of W. H. Freeman and Company.*)

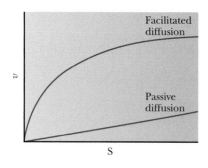

▲ **FIGURE 8.23** Passive diffusion and facilitated diffusion may be distinguished graphically. The plots for facilitated diffusion are similar to plots of enzyme-catalyzed reactions (Chapter 6), and they display saturation behavior. The value v stands for velocity of transport. S is the concentration of the substrate being transported.

glucose concentration in the erythrocyte is less than 5 mM. Glucose passes through a carrier protein called glucose permease (Figure 8.22). This process is labeled as facilitated diffusion because no energy is expended and a protein carrier is used. In addition, facilitated diffusion is identified by the fact that the rate of transport, when plotted against the concentration of the molecule being transported, gives a hyperbolic curve similar to that seen in Michaelis–Menten enzyme kinetics (Figure 8.23). In a carrier protein, a pore is created by folding the backbone and side chains. Many of these proteins have several α-helical portions that span the membrane; in others, a β-barrel forms the pore. In one example, the helical portion of the protein spans the membrane. The exterior, which is in contact with the lipid bilayer, is hydrophobic, whereas the interior, through which ions pass, is hydrophilic. Note that this orientation is the inverse of that observed in water-soluble globular proteins.

Active transport requires moving substances against a concentration gradient. It is identified by the presence of a carrier protein and the need for an energy source to move solutes against a gradient. In **primary active transport,** the movement of molecules against a gradient is directly linked to the hydrolysis of a high-energy molecule, such as ATP. The situation is so markedly similar to pumping water uphill that one of the most extensively studied examples of active transport, moving potassium ions into a cell and simultaneously moving sodium ions out of the cell, is referred to as the **sodium–potassium ion pump** (or Na^+/K^+ pump).

Under normal circumstances, the concentration of K^+ is higher inside a cell than in extracellular fluids ($[K^+]_{inside} > [K^+]_{outside}$), but the concentration of Na^+ is lower inside the cell than out ($[Na^+]_{inside} < [Na^+]_{outside}$). The energy required to move these ions against their gradients comes from an exergonic (energy-releasing) reaction, the hydrolysis of ATP to ADP and P_i (phosphate ion). There can be no transport of ions without hydrolysis of ATP. The same protein appears to serve both as the enzyme that hydrolyzes the ATP (the ATPase) and as the transport protein; it consists of several subunits. The reactants and products of this hydrolysis reaction—ATP, ADP, and P_i— remain within the cell, and the phosphate becomes covalently bonded to the transport protein for part of the process.

The Na^+/K^+ pump operates in several steps (Figure 8.24). One subunit of the protein hydrolyzes the ATP and transfers the phosphate group to an aspartate side chain on another subunit (Step 1). (The bond formed here is a mixed anhydride; see Section 1.2.) Simultaneously, binding of three Na^+ ions from the interior of the cell takes place. The phosphorylation of one subunit causes a conformational change in the protein, which opens a channel or pore through which the three Na^+ ions can be released to the extracellular

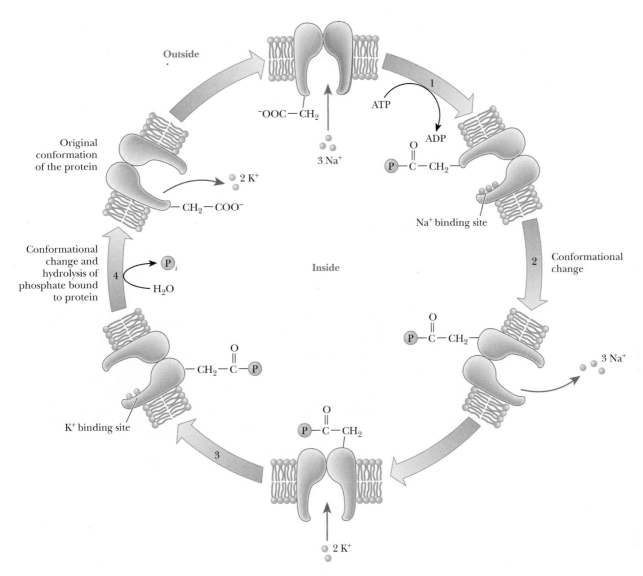

▲ **FIGURE 8.24** The sodium–potassium ion pump (see text for details).

fluid (Step 2). Outside the cell, two K^+ ions bind to the pump enzyme, which is still phosphorylated (Step 3). Another conformational change occurs when the bond between the enzyme and the phosphate group is hydrolyzed. This second conformational change regenerates the original form of the enzyme and allows the two K^+ ions to enter the cell (Step 4). The pumping process transports three Na^+ ions out of the cell for every two K^+ ions transported into the cell (Figure 8.25).

The operation of the pump can be reversed when there is no K^+ and a high concentration of Na^+ in the extracellular medium; in this case, ATP is produced by the phosphorylation of ADP. The actual operation of the Na^+/K^+ pump is not completely understood and probably is even more complicated than we now know. There is also a calcium ion (Ca^{2+}) pump, which is a subject of equally active investigation. Unanswered questions about the detailed mechanism of active transport provide opportunities for future research.

Another type of transport is called **secondary active transport.** An example is the galactoside permease in bacteria (Figure 8.26). The lactose concentration inside the bacterial cell is higher than the concentration outside, so moving lactose into the cell requires energy. The galactoside permease does not directly hydrolyze ATP, however. Instead, it harnesses the energy by letting hydrogen ions flow through the permease into the cell with their concentration gradient.

Biochemistry ⊜ Now™ **ANIMATED FIGURE 8.25**
A mechanism for Na⁺/K⁺ ATPase (the sodium–potassium ion pump). The model assumes two principal conformations, E_1 and E_2. Binding of Na⁺ ions to E_1 is followed by phosphorylation and release of ADP. Na⁺ ions are transported and released, and K⁺ ions are bound before dephosphorylation of the enzyme. Transport and release of K⁺ ions complete the cycle. **See this figure animated at http://now .brookscole.com/campbell5**

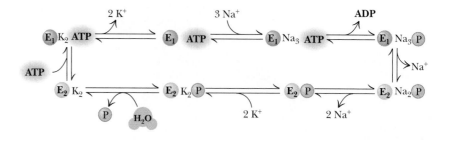

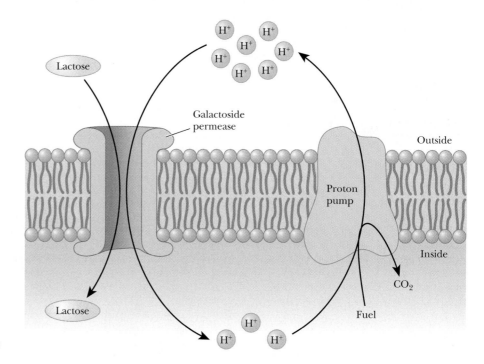

▶ **FIGURE 8.26** An example of secondary active transport. Galactoside permease uses the higher concentration of H⁺ outside the cell to drive the concentration of lactose inside the cell. (*Adapted from Lehninger,* Principles of Biochemistry, *Third Edition, by David L. Nelson and Michael M. Cox.* © *1982, 1992, 2000 by Worth Publishers. Used with permission of W. H. Freeman and Company.*)

Essential Information

Membrane proteins play key roles in transport of a number of substances across membranes. Proteins also serve as receptors for substances that bind to cell surfaces.

As long as more energy is available allowing the hydrogen ions to flow $(-\Delta G)$ than is required to concentrate the lactose $(+\Delta G)$, the process is possible. However, to arrive at a situation in which there is a higher concentration of hydrogen ions on the outside than on the inside, some other primary active transporter must establish the hydrogen ion gradient. Active transporters that create hydrogen ion gradients are called **proton pumps.**

Membrane Receptors

The first step in producing the effects of some biologically active substances is binding the substance to a protein receptor site on the exterior of the cell. The interaction between receptor proteins and the active substances which bind to them has features in common with enzyme–substrate recognition. There is a requirement for essential functional groups that have the correct three-dimensional conformation with respect to each other. The binding site, whether on a receptor or an enzyme, must provide a good fit for the substrate. In receptor binding, as in enzyme behavior, inhibition of the action of the protein by some sort of "poison" or inhibitor is possible. The study of receptor proteins is less advanced than the study of enzymes because many receptors are tightly bound integral proteins, and their activity depends on the membrane environment. Receptors are often large oligomeric proteins (ones with several subunits), with molecular weights on the order of hundreds of thousands. Also, quite frequently, the receptor has very few molecules in each cell, adding to the difficulties of isolating and studying this type of protein.

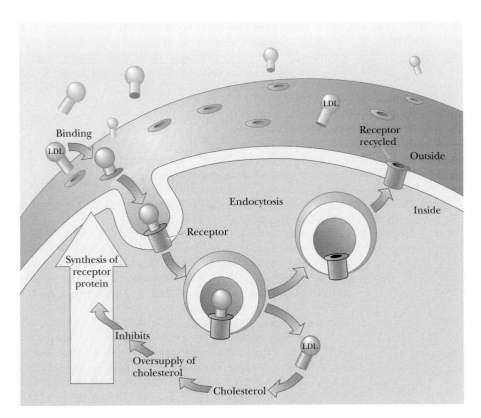

◀ **FIGURE 8.27** The mode of action of the LDL receptor. A portion of the membrane with LDL receptor and bound LDL is taken into the cell as a vesicle. The receptor protein releases LDL and is returned to the cell surface when the vesicle fuses to the membrane. LDL releases cholesterol in the cell. An oversupply of cholesterol inhibits synthesis of the LDL receptor protein. An insufficient number of receptors leads to elevated levels of LDL and cholesterol in the bloodstream. This situation increases the risk of heart attack.

An important type of receptor is that for low-density lipoprotein (LDL), the principal carrier of cholesterol in the bloodstream. LDL is a particle that consists of various lipids—in particular, cholesterol and phosphoglycerides—as well as a protein. The protein portion of the LDL particle binds to the LDL receptor of a cell. The complex formed between the LDL and the receptor is pinched off into the cell in a process called *endocytosis*. (This important aspect of receptor action is described in detail in the articles by Brown and Goldstein and by Dautry-Varsat and Lodish listed in the bibliography at the end of this chapter.) The receptor protein is then recycled back to the surface of the cell (Figure 8.27). The cholesterol portion of the LDL is used in the cell, but an oversupply of cholesterol causes problems. Excess of cholesterol inhibits the synthesis of LDL receptor. If there are too few receptors for LDL, the level of cholesterol in the bloodstream increases. Eventually, the excess cholesterol is deposited in the arteries, blocking them severely. This blocking of arteries, called atherosclerosis, can eventually lead to heart attacks and strokes. In many industrialized countries, typical blood cholesterol levels are high, and the incidence of heart attacks and strokes is correspondingly high. (We will say more about this subject after we have seen the pathway by which cholesterol is synthesized in the body in Section 21.8.)

8.7 | Which Are the Lipid-Soluble Vitamins, and What Are Their Functions?

Some vitamins, having a variety of functions, are of interest in this chapter because they are soluble in lipids. These lipid-soluble vitamins are hydrophobic, which accounts for their solubility (Table 8.3).

Vitamin A

The extensively unsaturated hydrocarbon **β-carotene** is the precursor of **vitamin A,** which is also known as **retinol.** As the name suggests, β-carotene is

Table 8.3	
Lipid-Soluble Vitamins and Their Functions	
Vitamin	**Function**
Vitamin A	Serves as the site of the primary photochemical reaction in vision
Vitamin D	Regulates calcium (and phosphorus) metabolism
Vitamin E	Serves as an antioxidant; necessary for reproduction in rats and may be necessary for reproduction in humans
Vitamin K	Has a regulatory function in blood clotting

abundant in carrots, but it also occurs in other vegetables, particularly the yellow ones. When an organism requires vitamin A, β-carotene is converted to the vitamin (Figure 8.28).

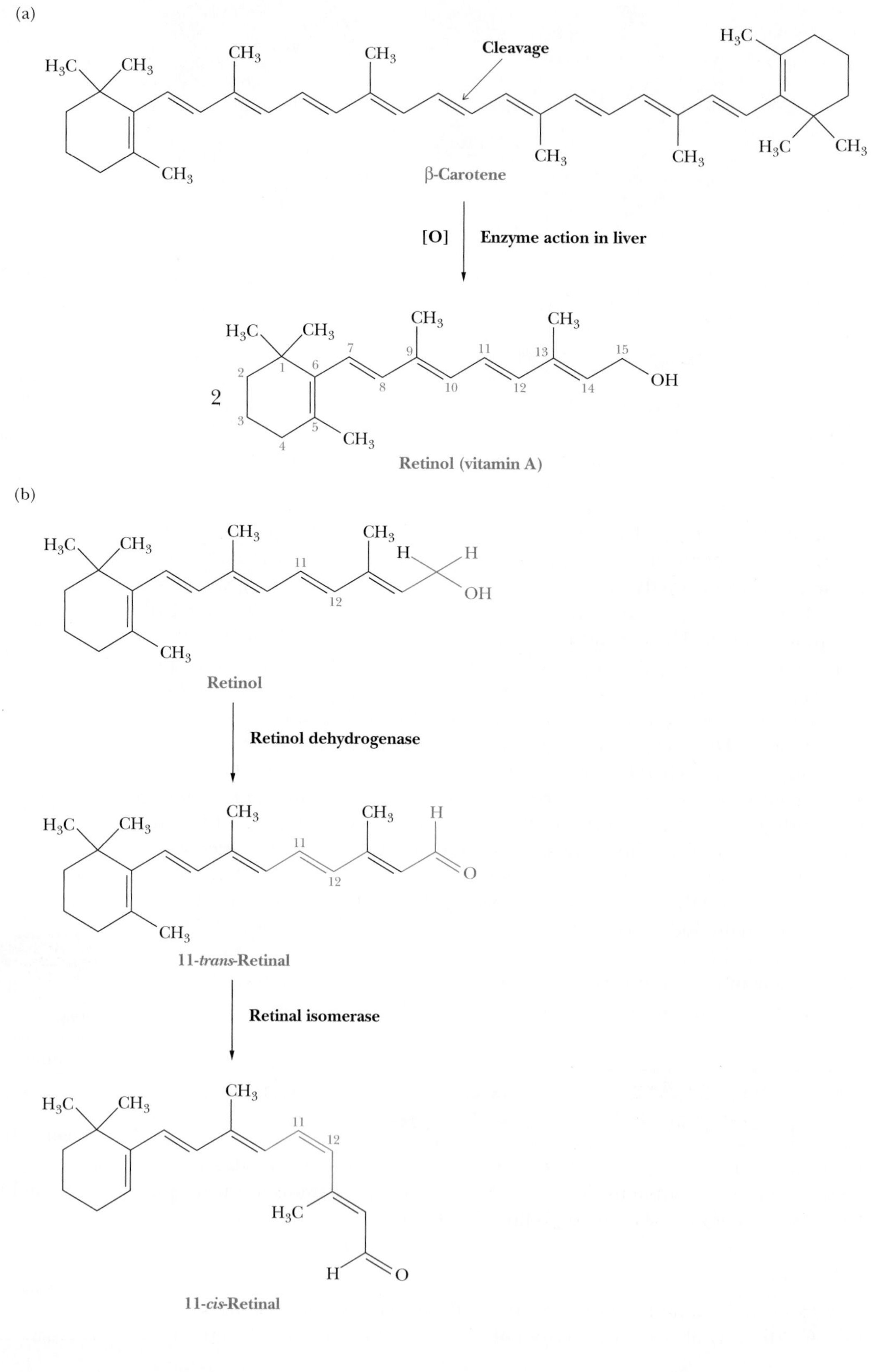

(a)

β-Carotene

[O] | **Enzyme action in liver**

Retinol (vitamin A)

(b)

Retinol

Retinol dehydrogenase

11-*trans*-Retinal

Retinal isomerase

11-*cis*-Retinal

▶ **FIGURE 8.28** Reactions of vitamin A. (a) The conversion of β-carotene to vitamin A. (b) The conversion of vitamin A to 11-*cis*-retinal.

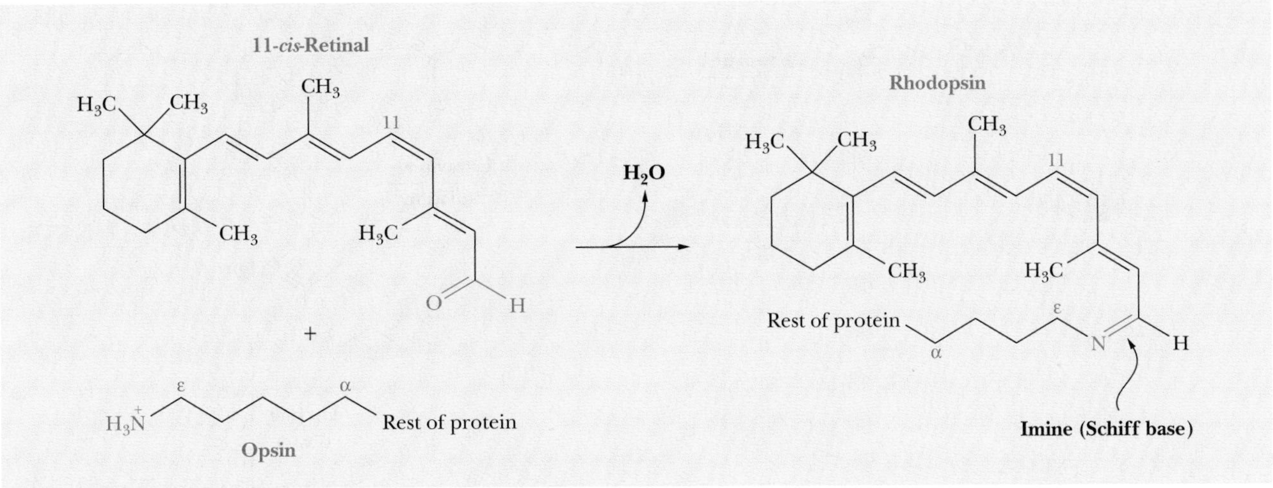

▲ **FIGURE 8.29** The formation of rhodopsin from 11-*cis*-retinal and opsin.

A derivative of vitamin A plays a crucial role in vision when it is bound to a protein called *opsin*. The cone cells in the retina of the eye contain several types of opsin and are responsible for vision in bright light and for color vision. The rod cells in the retina contain only one type of opsin; they are responsible for vision in dim light. The chemistry of vision has been more extensively studied in rod cells than in cone cells, and we shall discuss events that take place in rod cells.

Vitamin A has an alcohol group that is enzymatically oxidized to an aldehyde group, forming **retinal** (Figure 8.28b). Two isomeric forms of retinal, involving *cis–trans* isomerization around one of the double bonds, are important in the behavior of this compound in vivo. The aldehyde group of retinal forms an imine (also called a Schiff base) with the side-chain amino group of a lysine residue in rod-cell opsin (Figure 8.29).

The product of the reaction between retinal and opsin is **rhodopsin.** The outer segment of rod cells contains flat membrane-bounded discs, the membrane consisting of about 60% rhodopsin and 40% lipid. (For more details about rhodopsin, see the following Biochemical Connections box.)

Vitamin D

The several forms of **vitamin D** play a major role in the regulation of calcium and phosphorus metabolism. One of the most important of these compounds, vitamin D_3 (cholecalciferol), is formed from cholesterol by the action of ultraviolet radiation from the sun. Vitamin D_3 is further processed in the body to form hydroxylated derivatives, which are the metabolically active form of this vitamin (Figure 8.30). The presence of vitamin D_3 leads to increased synthesis of a Ca^{2+}-binding protein, which increases the absorption of dietary calcium in the intestines. This process results in calcium uptake by the bones.

A deficiency of vitamin D can lead to *rickets,* a condition in which the bones of growing children become soft, resulting in skeletal deformities. Children, especially infants, have higher requirements for vitamin D than do adults. Milk with vitamin D supplements is available to most children. Adults who are exposed to normal amounts of sunlight do not usually require vitamin D supplements.

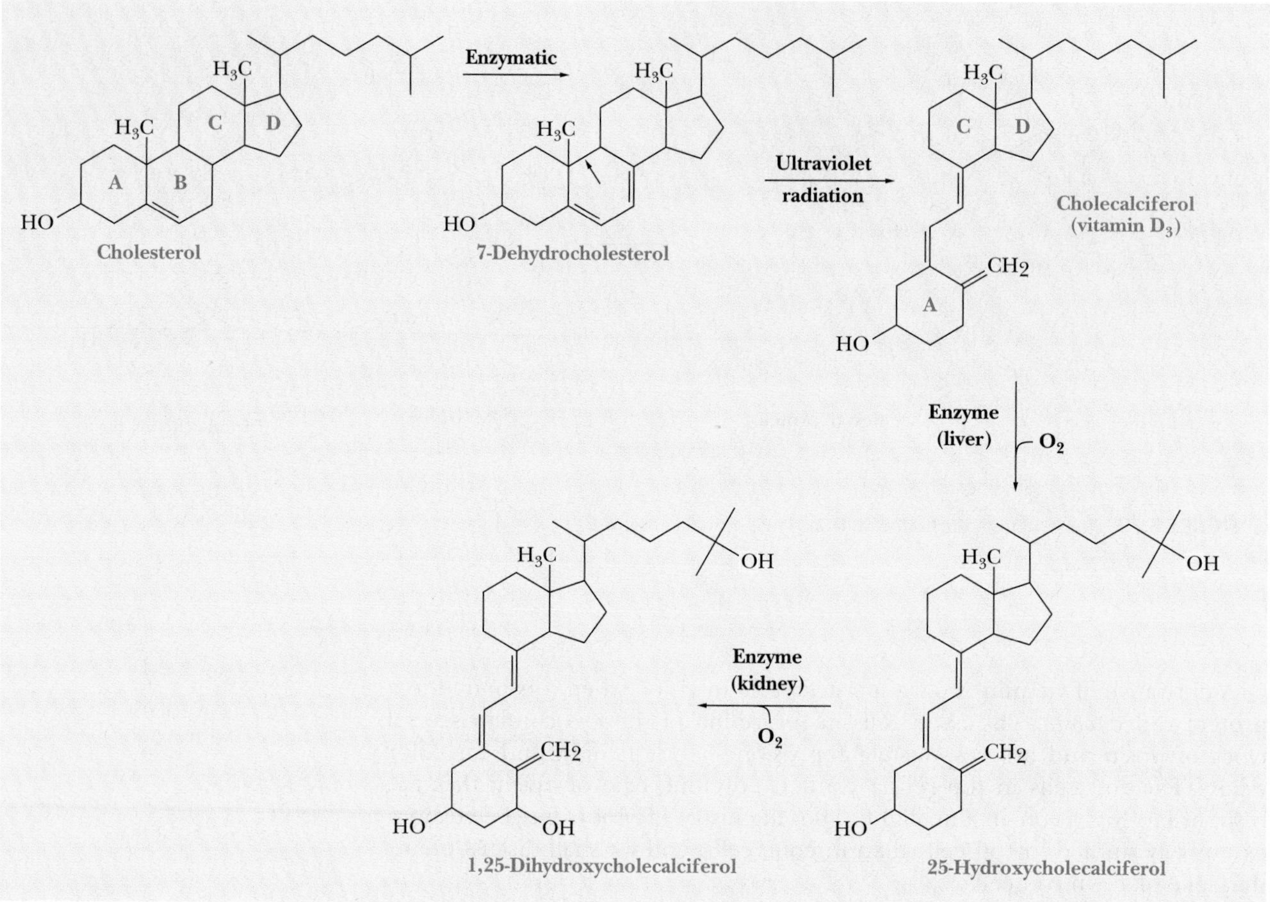

▲ **FIGURE 8.30** Reactions of vitamin D. The photochemical cleavage occurs at the bond shown by the arrow; electron rearrangements after the cleavage produce vitamin D_3. The final product, 1,25-dihydrocholecalciferol, is the form of the vitamin that is most active in stimulating the intestinal absorption of calcium and phosphate and in mobilizing calcium for bone development.

Vitamin E

The most active form of **vitamin E** is **α-tocopherol.** In rats, vitamin E is required for reproduction and for the prevention of the disease *muscular dystrophy*. It is not known whether this requirement exists in humans. A well-established chemical property of vitamin E is that it is an **antioxidant**—that is,

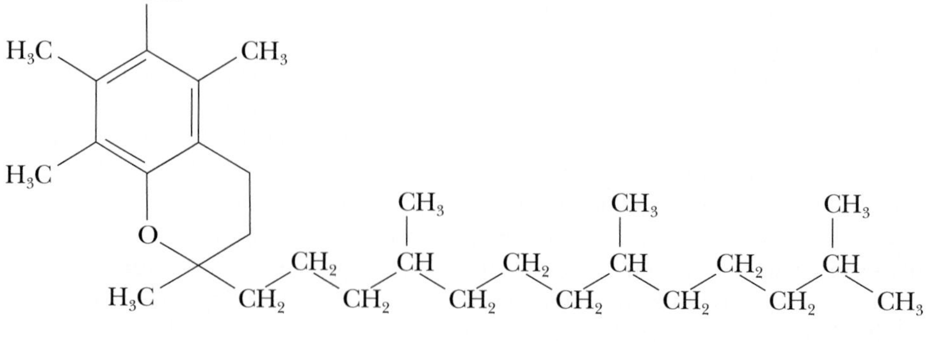

Vitamin E (α-tocopherol)

▲ The most active form of vitamin E is α-tocopherol.

Biochemical Connections

The Chemistry of Vision

The primary chemical reaction in vision, the one responsible for generating an impulse in the optic nerve, involves *cis*–*trans* isomerization around one of the double bonds in the retinal portion of rhodopsin. When rhodopsin is active (that is, when it can respond to visible light), the double bond between carbon atoms 11 and 12 of the retinal (11-*cis*-retinal) has the *cis* orientation. Under the influence of light, an isomerization reaction occurs at this double bond, producing all-*trans*-retinal. Because the all-*trans* form of retinal cannot bind to opsin, all-*trans*-retinal and free opsin are released. As a result of this reaction, an electrical impulse is generated in the optic nerve and transmitted to

the brain to be processed as a visual event. The active form of rhodopsin is regenerated by enzymatic isomerization of the all-*trans*-retinal back to the 11-*cis* form and subsequent re-formation of the rhodopsin.

Vitamin A deficiency can have drastic consequences, as would be predicted from its importance in vision. Night blindness—and even total blindness—can result, especially in children. On the other hand, an excess of vitamin A can have harmful effects, such as bone fragility. Lipid-soluble compounds are not excreted as readily as water-soluble substances, and it is possible for excessive amounts of lipid-soluble vitamins to accumulate in adipose tissue.

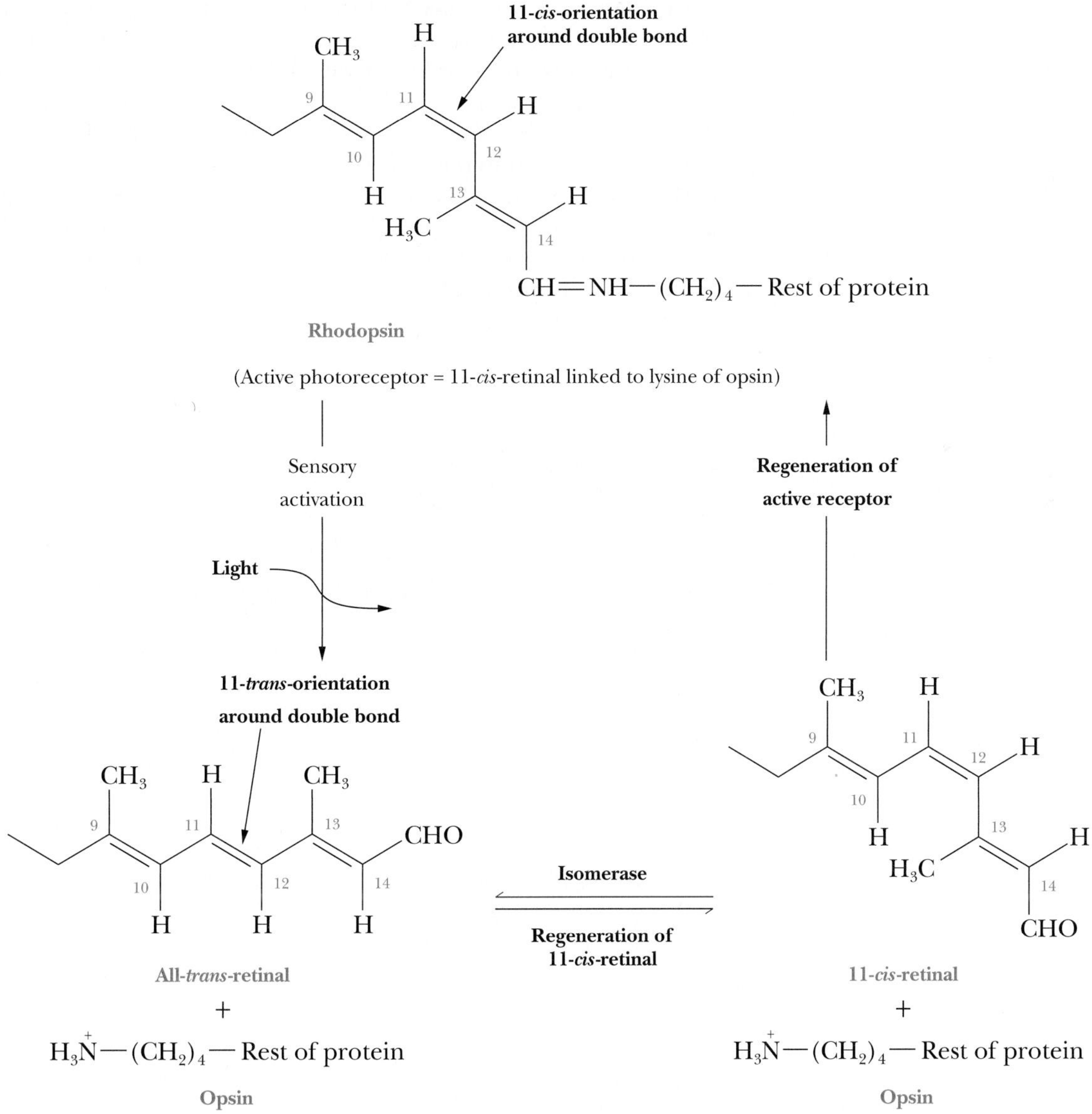

▲ The primary chemical reaction of vision.

a good reducing agent—so it reacts with oxidizing agents before they can attack other biomolecules. The antioxidant action of vitamin E has been shown to protect important compounds, including vitamin A, from degradation in the laboratory; it probably also serves this function in organisms. Recent research has shown that the interaction of vitamin E with membranes enhances its effectiveness as an antioxidant. Another function of antioxidants such as vitamin E is to react with, and thus to remove, the very reactive and highly dangerous substances known as **free radicals.** A free radical has at least one unpaired electron, which accounts for its high degree of reactivity. Free radicals may play a part in the development of cancer and in the aging process.

Vitamin K

The name of **vitamin K** comes from the Danish *Koagulation* because this vitamin is an important factor in the blood-clotting process. The bicyclic ring system contains two carbonyl groups, the only polar groups on the molecule (Figure 8.31). A long unsaturated hydrocarbon side chain consists of repeating *isoprene* units, the number of which determines the exact form of vitamin K. Several forms of this vitamin can be found in a single organism, but the reason for this variation is not well understood. Vitamin K is not the first vitamin we have encountered that contains isoprene units, but it is the first one in which the number of isoprene units and their degree of saturation make a difference. (Can you pick out the isoprene-derived portions of the structures of vitamins A and E?) It is also known that the steroids are biosynthetically derived from isoprene units, but the structural relationship is not immediately obvious (Section 21.8).

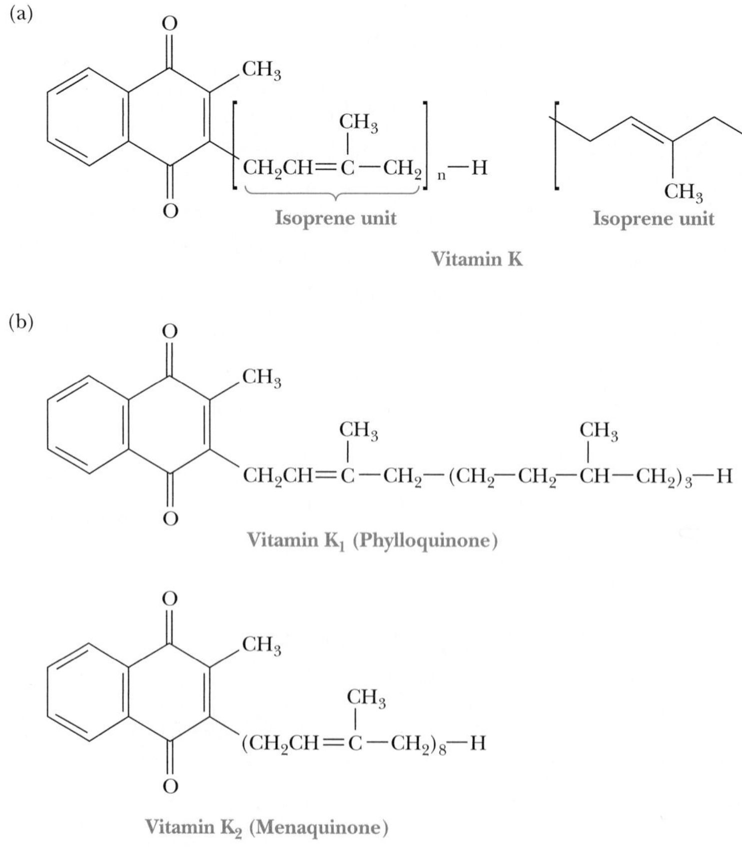

▶ **FIGURE 8.31** (a) The general structure of vitamin K, which is required for blood clotting. The value of *n* is variable, but it is usually <10. (b) Vitamin K_1 has one unsaturated isoprene unit; the rest are saturated. Vitamin K_2 has eight unsaturated isoprene units.

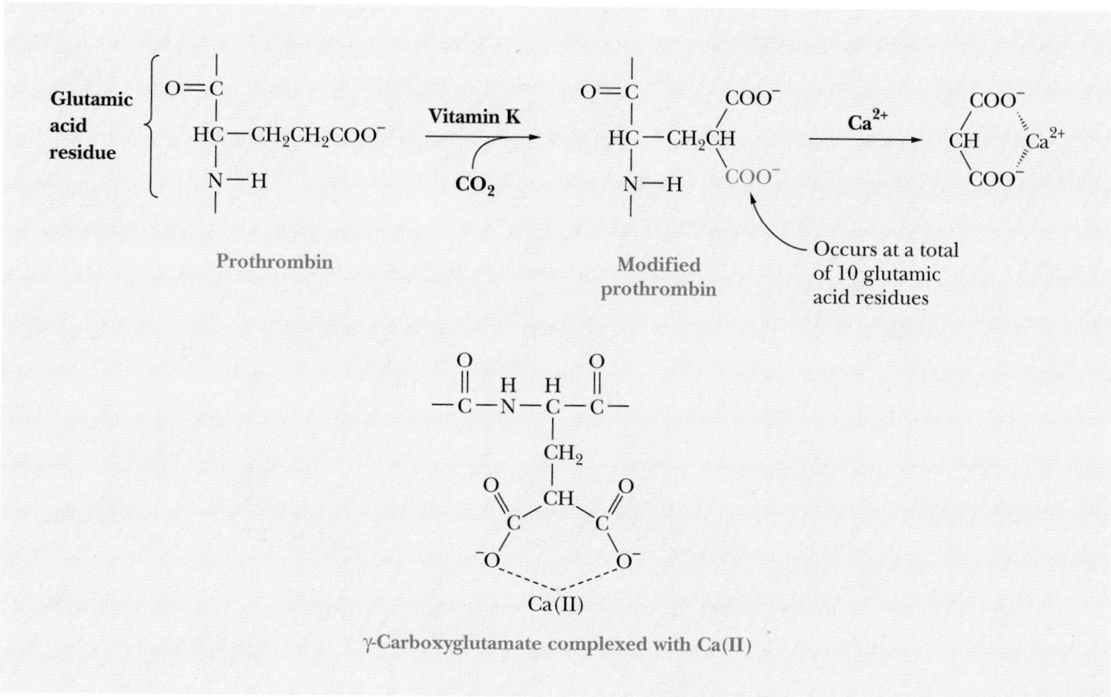

▲ **FIGURE 8.32** The role of vitamin K in the modification of prothrombin. The detailed structure of the γ-carboxyglutamate at the calcium complexation site is shown at the bottom.

The presence of vitamin K is required in the complex process of blood clotting, which involves many steps and many proteins and has stimulated numerous unanswered questions. It is known definitely that vitamin K is required to modify prothrombin and other proteins involved in the clotting process. Specifically, with prothrombin, the addition of another carboxyl group alters the side chains of several glutamate residues of prothrombin. This modification of glutamate produces γ-carboxyglutamate residues (Figure 8.32). The two carboxyl groups in proximity form a *bidentate* ("two teeth") *ligand,* which can bind calcium ion (Ca^{2+}). If prothrombin is not modified in this way, it does not bind Ca^{2+}. Even though there is a lot more to be learned about blood clotting and the role of vitamin K in the process, this point, at least, is well established, because Ca^{2+} is required for blood clotting. (Two well-known anticoagulants, dicumarol and warfarin (a rat poison), are vitamin-K antagonists.)

8.8 What Are Prostaglandins and Leukotrienes, and What Do They Have to Do with Lipids?

A group of compounds derived from fatty acids has a wide range of physiological activities; they are called **prostaglandins** because they were first detected in seminal fluid, which is produced by the prostate gland. It has since been shown that they are widely distributed in a variety of tissues. The metabolic precursor of all prostaglandins is **arachidonic acid,** a fatty acid that contains 20 carbon atoms and four double bonds. The double bonds are not conjugated. The production of the prostaglandins from arachidonic acid takes place in several steps, which are catalyzed by enzymes. The prostaglandins themselves each have a five-membered ring; they differ from one

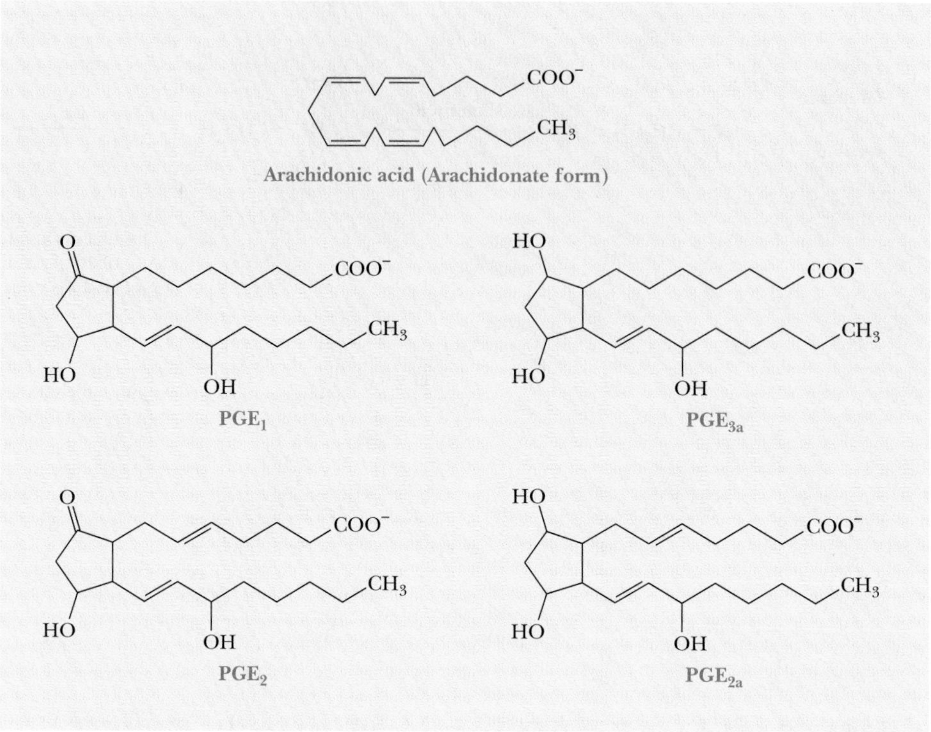

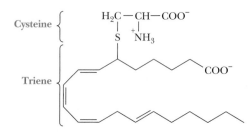

▲ **FIGURE 8.33** Arachidonic acid and some prostaglandins.

another in the numbers and positions of double bonds and oxygen-containing functional groups (Figure 8.33).

The structures of prostaglandins and their laboratory syntheses have been topics of great interest to organic chemists, largely because of the many physiological effects of these compounds and their possible usefulness in the pharmaceutical industry. Some of the functions of prostaglandins are control of blood pressure, stimulation of smooth-muscle contraction, and induction of inflammation. Aspirin inhibits the synthesis of prostaglandins, particularly in blood platelets, a property that accounts for its anti-inflammatory and fever-reducing properties. Cortisone and other steroids also have anti-inflammatory effects because of their inhibition of prostaglandin synthesis.

Prostaglandins are known to inhibit the aggregation of platelets. They may thus be of therapeutic value by preventing the formation of blood clots, which can cut off the blood supply to the brain or the heart and cause certain types of strokes and heart attacks. Even if this behavior were the only useful property of prostaglandins, it would justify considerable research effort. Heart attacks and strokes are two of the leading causes of death in industrialized countries. More recently, the study of prostaglandins has been a topic of great interest because of their possible antitumor and antiviral activity.

Leukotrienes are compounds that, like prostaglandins, are derived from arachidonic acid. They are found in leukocytes (white blood cells) and have three conjugated double bonds; these two facts account for the name. (Fatty acids and their derivatives do not normally contain conjugated double bonds.) Leukotriene C (Figure 8.34) is a typical member of this group; note the 20 carbon atoms in the carboxylic acid backbone, a feature that relates this compound structurally to arachidonic acid. (The 20-carbon prostaglan-

▲ **FIGURE 8.34** Leukotriene C.

J. S. Reid/Custom Medical Stock

◀ Research on leukotrienes may provide new treatments for asthma, perhaps eliminating the need for inhalers, such as the one shown here.

dins and leukotrienes are also called eicosinoids.) An important property of leukotrienes is their constriction of smooth muscle, especially in the lungs. Asthma attacks may result from this constricting action because the synthesis of leukotriene C appears to be facilitated by allergic reactions, such as a reaction to pollen. Drugs that inhibit the synthesis of leukotriene C are now being used in the treatment of asthma, as are other drugs designed to block leukotriene receptors. In the United States, the incidence of asthma increased 46% between 1982 and 1993, providing considerable incentive to find new treatments. (The National Asthma Education and Prevention Program has released "Guidelines for the Diagnoses and Management of Asthma." This document can be accessed on the Internet at http://www .nhlgbi.nih.gov.) Leukotrienes may also have inflammatory properties and may be involved in rheumatoid arthritis.

Thromboxanes are a third class of derivatives of arachidonic acid. They contain cyclic ethers as part of their structures. The most widely studied member of the group, thromboxane A_2 (TxA_2), is known to induce platelet aggregation and smooth-muscle contraction.

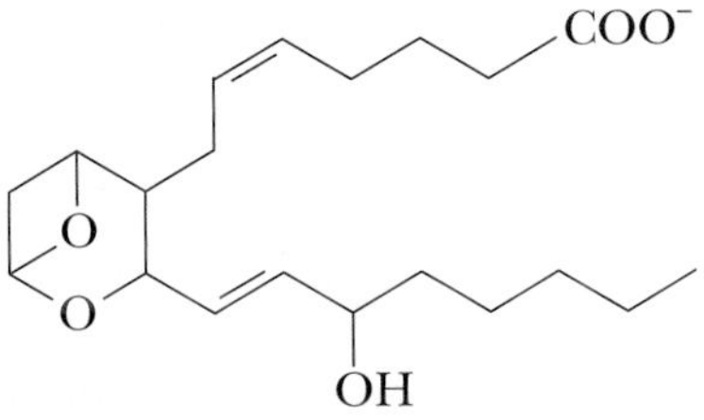

COO⁻

O

O

OH

Thromboxane A_2
(TxA_2)

▲ Thromboxane A_2

The following Biochemical Connections box explores some connections among topics we have discussed in this chapter.

Biochemical Connections

Omega-3 Fatty Acids and Platelets in Heart Disease

Platelets are elements in the blood that initiate blood clotting and tissue repair by releasing clotting factors and platelet-derived growth factor (PDGF). Turbulence in the bloodstream may cause platelets to rupture. Fat deposits and bifurcations of arteries lead to such turbulence, so platelets and PDGF are implicated in blood clotting and growth of atherosclerotic plaque. Furthermore, the anaerobic conditions that exist under a large plaque deposit may lead to weakness and dead cells in the arterial wall, aggravating the problem.

In cultures that depend on fish as a major food source, including some Eskimo tribes, very little heart disease is diagnosed, even though people in these groups eat high-fat diets and have high levels of blood cholesterol. Analysis of the their diet led to the discovery that certain highly unsaturated fatty acids are found in the oils of fish and diving mammals. One class of these fatty acids is called omega-3 (ω_3), an example of which is eicosapentenoic acid (EPA).

$$CH_3CH_2(CH{=}CHCH_2)_5(CH_2)_2COOH$$

Eicosapentenoic acid (EPA)

Note the presence of a double bond at the third carbon atom from the end of the hydrocarbon tail. The omega system of nomenclature is based on numbering the double bonds from the last carbon in the fatty acid instead of the carbonyl group [the delta (Δ) system]. Omega is the last letter in the Greek alphabet.

The omega-3 fatty acids inhibit the formation of certain prostaglandins and thromboxane A, which is similar in structure to prostaglandins. Thromboxane released by ruptured arteries causes other platelets to clump in the immediate area and to increase the size of the blood clot. Any disruption in thromboxane synthesis will result in a lower tendency to form blood clots and, thus, in a lower potential for artery damage.

It is interesting to note that aspirin is also an inhibitor of prostaglandin synthesis, although it is less potent than EPA. Aspirin inhibits the synthesis of the prostaglandins responsible for inflammation and the perception of pain. Aspirin has been implicated in reducing the incidence of heart disease, probably by a mechanism similar to that of EPA. However, people who are being treated with blood thinners or who are prone to easy bleeding should not take aspirin.

Summary

8.1 What Is the Definition of a Lipid? Lipids are compounds that are insoluble in water but soluble in nonpolar organic solvents. Their chemical structures consist primarily of nonpolar moieties.

8.2 What Are the Chemical Natures of the Lipid Types? One group of lipids consists of open-chain compounds, each with a polar head group and a long nonpolar tail; this group includes fatty acids, triacylglycerols, phosphoacylglycerols, sphingolipids, and glycolipids. A second major group consists of fused-ring compounds, the steroids. Triacylglycerols are the storage forms of fatty acids, and phosphoacylglycerols are important components of biological membranes, as are sphingolipids and glycolipids.

8.3 What Is the Nature of Biological Membranes? A biological membrane consists of a lipid part and a protein part. The lipid part is a bilayer, with the polar head groups in contact with the aqueous interior and exterior of the cell, and the nonpolar portions of the lipid in the interior of the membrane. Lateral motion of lipid molecules within one layer of a membrane occurs frequently.

8.4 What Are Some Common Types of Membrane Proteins? The proteins that occur in membranes can be peripheral proteins, which are found on the surface of the membrane, or integral proteins, which lie within the lipid bilayer. Various structural motifs, such as bundles of seven α-helices, occur in proteins that span membranes.

8.5 What Is the Fluid-Mosaic Model of Membrane Structure? The fluid-mosaic model describes the interaction of lipids and proteins in biological membranes. The proteins "float" in the lipid bilayer.

8.6 What Are Some of the Functions of Membranes? Three important functions take place in or on membranes. The first, transport across the membrane, can involve the lipid bilayer as well as the membrane proteins. The second, catalysis, is carried out by enzymes bound to the membrane. Finally, receptor proteins in the membrane bind biologically important substances that trigger a biochemical response in the cell. The most important question about transport of substances across biological membranes is whether the process requires expenditure of energy by the cell. In passive transport, a substance moves from a region of higher concentration to one of lower concentration, requiring no expenditure of energy by the cell. Active transport requires moving substances against a concentration gradient, a situation similar to pumping water up a hill. Energy, as well as a carrier protein, is required for active transport. The sodium–potassium ion pump is an example of active transport. The first step in the effects of some biologically active substances is binding to a protein receptor site on the exterior of the cell. The interaction between receptor proteins and the active substances to which they bind is very similar to enzyme–substrate recognition. The action of a receptor frequently depends on a conformational change in the receptor protein. Receptors can be ligand-gated channel proteins, in which the binding of ligand transiently opens a channel protein through which substances such as ions can flow in the direction of a concentration gradient.

8.7 Which Are the Lipid-Soluble Vitamins, and What Are Their Functions? Lipid-soluble vitamins are hydrophobic, accounting for their solubility properties. A derivative of vitamin A plays a crucial role in vision. Vitamin D controls calcium and phosphorus metabolism, affecting the structural integrity of bones. Vitamin E is known to be an antioxidant; its other metabolic functions are not definitely established. The presence of vitamin K is required in the blood-clotting process.

8.8 What Are Prostaglandins and Leukotrienes, and What Do They Have to Do with Lipids? The unsaturated fatty acid arachidonic acid is the precursor of prostaglandins and leukotrienes, compounds that have a wide range of physiological activities. Stimulation of smooth-muscle contraction and induction of inflammation are common to both classes of compounds. Prostaglandins are also involved in control of blood pressure and inhibition of blood-platelet aggregation.

Critical Questions to Review

8.1 What Is the Definition of a Lipid?

1. **Fact Check** Proteins, nucleic acids, and carbohydrates are grouped by common structural features found within their group. What is the basis for grouping substances as lipids?

8.2 What Are the Chemical Natures of the Lipid Types?

2. **Fact Check** What structural features do a triacylglycerol and a phosphatidyl ethanolamine have in common? How do the structures of these two types of lipids differ?

3. **Fact Check** Draw the structure of a phosphoacylglycerol that contains glycerol, oleic acid, stearic acid, and choline.

4. **Fact Check** What structural features do a sphingomyelin and a phosphatidyl choline have in common? How do the structures of these two types of lipids differ?

5. **Fact Check** You have just isolated a pure lipid that contains only sphingosine and a fatty acid. To what class of lipids does it belong?

6. **Fact Check** What structural features does a sphingolipid have in common with proteins? Are there functional similarities?

7. **Fact Check** Write the structural formula for a triacylglycerol, and name the component parts.

8. **Fact Check** How does the structure of steroids differ from that of the other lipids discussed in this chapter?

9. **Fact Check** What are the structural features of waxes? What are some common uses of compounds of this type?

10. **Thought Question** Which is more hydrophilic, cholesterol or phospholipids? Defend your answer.

11. **Thought Question** Write an equation, with structural formulas, for the saponification of the triacylglycerol in Question 7.

12. **Thought Question** Succulent plants from arid regions generally have waxy surface coatings. Suggest why such a coating is valuable for the survival of the plant.

13. **Thought Question** In the produce department of supermarkets, vegetables and fruits (cucumbers are an example) have been coated with wax for shipping and storage. Suggest a reason why this is done.

14. **Thought Question** Egg yolks contain a high amount of cholesterol, but they also contain a high amount of lecithin. From a diet and health standpoint, how do these two molecules complement each other?

15. **Thought Question** In the preparation of sauces that involve mixing water and melted butter, egg yolks are added to prevent separation. How do the egg yolks prevent separation? *Hint:* Egg yolks are rich in phosphatidylcholine (lecithin).

16. **Thought Question** When water birds have had their feathers fouled with crude oil after an oil spill, they are cleaned by rescuers to remove the spilled oil. Why are they not released immediately after they are cleaned?

8.3 What Is the Nature of Biological Membranes?

17. **Fact Check** Which of the following lipids are *not* found in animal membranes?

    (a) Phosphoglycerides

    (b) Cholesterol

    (c) Triacylglycerols

    (d) Glycolipids

    (e) Sphingolipids

18. **Fact Check** Which of the following statements is (are) consistent with what is known about membranes?

    (a) A membrane consists of a layer of proteins sandwiched between two layers of lipids.

    (b) The compositions of the inner and outer lipid layers are the same in any individual membrane.

    (c) Membranes contain glycolipids and glycoproteins.

    (d) Lipid bilayers are an important component of membranes.

    (e) Covalent bonding takes place between lipids and proteins in most membranes.

19. **Thought Question** Why might some food companies find it economically advantageous to advertise their product (for example, triacylglycerols) as being composed of polyunsaturated fatty acids with *trans*-double bonds?

20. **Thought Question** Suggest a reason why partially hydrogenated vegetable oils are used so extensively in packaged foods.

21. **Biochemical Connections** Crisco is made from vegetable oils, which are usually liquid. Why is Crisco a solid? *Hint:* Read the label.

22. **Biochemical Connections** Why does the American Heart Association recommend the use of canola oil or olive oil rather than coconut oil in cooking?

23. **Thought Question** In lipid bilayers, there is an order–disorder transition similar to the melting of a crystal. In a lipid bilayer in which most of the fatty acids are unsaturated, would you expect this transition to occur at a higher temperature, a lower temperature, or the same temperature as it would in a lipid bilayer in which most of the fatty acids are saturated? Why?

24. **Biochemical Connections** Briefly discuss the structure of myelin and its role in the nervous system.

25. **Thought Question** Suggest a reason why the cell membranes of bacteria grown at 20°C tend to have a higher proportion of unsaturated fatty acids than the membranes of bacteria of the same species grown at 37°C. In other words, the bacteria grown at 37°C have a higher proportion of saturated fatty acids in their cell membranes.

26. **Thought Question** Suggest a reason why animals that live in cold climates tend to have higher proportions of polyunsaturated fatty acid residues in their lipids than do animals that live in warm climates.

27. **Thought Question** What is the energetic driving force for the formation of phospholipid bilayers?

8.4 What Are Some Common Types of Membrane Proteins?

28. **Fact Check** Define glycoprotein and glycolipid.

29. **Fact Check** Do all proteins associated with membranes span the membrane from one side to another?

30. **Thought Question** A membrane consists of 50% protein by weight and 50% phosphoglycerides by weight. The average molecular weight of the lipids is 800 daltons, and the average molecular weight of the proteins is 50,000 daltons. Calculate the molar ratio of lipid to protein.

31. **Thought Question** Suggest a reason why the same protein system moves both sodium and potassium ions into and out of the cell.

32. **Thought Question** Suppose that you are studying a protein involved in transporting ions in and out of cells. Would you expect to find the nonpolar residues in the interior or the exterior? Why? Would you expect to find the polar residues in the interior or the exterior? Why?

8.5 What Is the Fluid-Mosaic Model of Membrane Structure?

33. **Thought Question** Which statements are consistent with the fluid-mosaic model of membranes?

    (a) All membrane proteins are bound to the interior of the membrane.

(b) Both proteins and lipids undergo transverse (flip-flop) diffusion from the inside to the outside of the membrane.

(c) Some proteins and lipids undergo lateral diffusion along the inner or outer surface of the membrane.

(d) Carbohydrates are covalently bonded to the outside of the membrane.

(e) The term "mosaic" refers to the arrangement of the lipids alone.

8.6 What Are Some of the Functions of Membranes?

34. Thought Question Suggest a reason why inorganic ions, such as K^+, Na^+, Ca^{2+}, and Mg^{2+}, do not cross biological membranes by simple diffusion.

35. Thought Question Which statements are consistent with the known facts about membrane transport?

(a) Active transport moves a substance from a region in which its concentration is lower to one in which its concentration is higher.

(b) Transport does not involve any pores or channels in membranes.

(c) Transport proteins may be involved in bringing substances into cells.

8.7 Which Are the Lipid-Soluble Vitamins, and What Are Their Functions?

36. Fact Check What is the structural relationship between vitamin D_3 and cholesterol?

37. Fact Check List an important chemical property of vitamin E.

38. Fact Check What are isoprene units? What do they have to do with the material of this chapter?

39. Fact Check List the fat-soluble vitamins, and give a physiological role for each.

40. Biochemical Connections What is the role in vision of the *cis–trans* isomerization of retinal?

41. Thought Question Why is it possible to argue that vitamin D is not a vitamin?

42. Thought Question Give a reason for the toxicity that can be caused by overdoses of lipid-soluble vitamins.

43. Thought Question Why can some vitamin-K antagonists act as anticoagulants?

44. Thought Question Why are many vitamin supplements sold as antioxidants? How does this relate to material in this chapter?

45. Thought Question A health-conscious friend asks whether eating carrots is better for the eyesight or for preventing cancer. What do you tell your friend? Explain.

8.8 What Are Prostaglandins and Leukotrienes, and What Do They Have to Do with Lipids?

46. Biochemical Connections Define omega-3 fatty acid.

47. Fact Check What are the main structural features of leukotrienes?

48. Fact Check What are the main structural features of prostaglandins?

49. Thought Question List two classes of compounds derived from arachidonic acid. Suggest some reasons for the amount of biomedical research devoted to these compounds.

50. Biochemical Connections Outline a possible connection between the material in this chapter and the integrity of blood platelets.

Biochemistry ⊛ Now™

Assess your understanding of this chapter's topics with additional quizzing and tutorials at **http://now.brookscole.com/campbell5**

Annotated Bibliography

Barinaga, M. Forging a Path to Cell Death. *Science* **273**, 735–737 (1996). [A Research News article describing a process apparently missing in cancer cells, and that depends on interactions among receptor proteins on cell surfaces.]

Bayley, H. Building Doors into Cells. *Sci. Amer.* **277** (3), 62–67 (1997). [Protein engineering can create artificial pores in membranes for drug delivery.]

Bretscher, M. S. The Molecules of the Cell Membrane. *Sci. Amer.* **253** (4), 100–108 (1985). [A particularly well-illustrated description of the roles of lipids and proteins in cell membranes.]

Brown, M. S., and J. L. Goldstein. A Receptor-Mediated Pathway for Cholesterol Homeostasis. *Science* **232**, 34–47 (1986). [A description of the role of cholesterol in heart disease.]

Dautry-Varsat, A., and H. F. Lodish. How Receptors Bring Proteins and Particles into Cells. *Sci. Amer.* **250** (5), 52–58 (1984). [A detailed description of endocytosis.]

Engelman, D. Crossing the Hydrophobic Barrier: Insertion of Membrane Proteins. *Science* **274**, 1850–1851 (1996). [A short review of the processes by which transmembrane proteins become associated with lipid bilayers.]

Hajjar, D., and A. Nicholson. Atherosclerosis. *Amer. Scientist* **83**, 460–467 (1995). [The cellular and molecular basis of lipid deposition in arteries.]

Karow, J. Skin So Fixed. *Sci. Amer.* **284** (3), 21 (2001). [A discussion of liposomes used to deliver DNA repair enzymes to skin cells.]

Keuhl, F. A., and R. W. Egan. Prostaglandins, Arachidonic Acid and Inflammation. *Science* **210**, 978–984 (1980). [A discussion of the chemistry of these compounds and their physiological effects.]

Wood, R. D., M. Mitchell, J. Sgouros, and T. Lindahl. Human DNA repair genes. *Science* **291** (5507), 1284–1289 (2001).

Nucleic Acids: How Structure Conveys Information

Genes, the hereditary material within the chromosomes, are essentially long stretches of double-helical DNA. In a process mediated by RNA (the other kind of nucleic acid) the sequence of DNA bases specifies the sequence of amino acids in a single polypeptide (protein) chain. The protein's amino acid sequence, in turn, determines its structure and function. Thus, the base sequence of the DNA ultimately determines the activities of proteins, the essential machinery of life. Each cell carries in its DNA the instructions for making the complete organism. When the cell divides, each new cell bears a copy of the original DNA. Replication of the hereditary material is made possible by the complementary nature of the DNA bases. Adenine on one strand pairs with thymine on the opposite strand of the double helix. The same is true for the other two bases: guanine on one strand pairs with cytosine on the opposite strand. Thus, one strand of DNA is a template for the other strand. It is now possible to control some aspects of genetic coding. Starting in the 1970s, techniques were introduced for manipulating DNA by cutting and splicing it in a manner that both mimics and transcends natural processes. These techniques will provide valuable insight into the manner in which proteins interact with DNA molecules to control gene activation and repression.

© Steve Oh, M.S./Phototake

Determination of the double-helical structure of DNA has illuminated molecular biology for more than half a century.

Critical Questions

9.1 What Are the Levels of Structure in Nucleic Acids?

9.2 What Is the Covalent Structure of Polynucleotides?

9.3 What Is the Structure of DNA?

9.4 How Does the Denaturation of DNA Take Place?

9.5 What Are the Principal Kinds of RNA and Their Structures?

9.1	What Are the Levels of Structure in Nucleic Acids?

In Chapter 4, we identified four levels of structure—primary, secondary, tertiary, and quaternary—in proteins. Nucleic acids can be viewed in the same way. The *primary structure* of nucleic acids is the order of bases in the polynucleotide sequence, and the *secondary structure* is the three-dimensional conformation of the backbone. The *tertiary structure* is specifically the supercoiling of the molecule.

DNA (deoxyribonucleic acid) and RNA (ribonucleic acid) are the two kinds of nucleic acids. Important differences between them appear in their secondary and tertiary structures, and so we shall describe these structural features separately for DNA and for RNA. Even though nothing in nucleic acid structure is directly analogous to the quaternary structure of proteins, the interaction of nucleic acids with other classes of macromolecules (for example, proteins) to form complexes is similar to the interactions of the subunits in an oligomeric protein. One well-known example is the association of RNA and proteins in **ribosomes** (the polypeptide-generating machinery of the cell); another is the self-assembly of tobacco mosaic virus, in which the nucleic acid strand winds through a cylinder of coat-protein subunits.

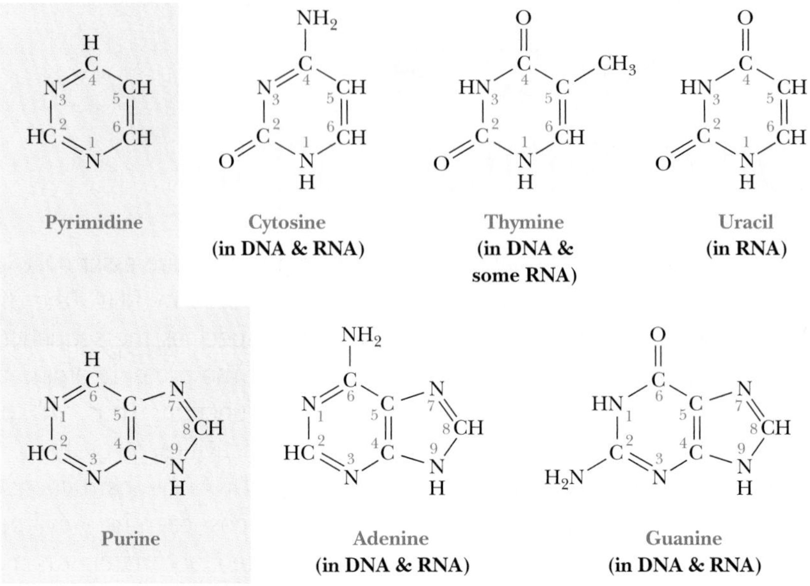

► **FIGURE 9.1** Structures of the common nucleobases. The structures of pyrimidine and purine are shown for comparison.

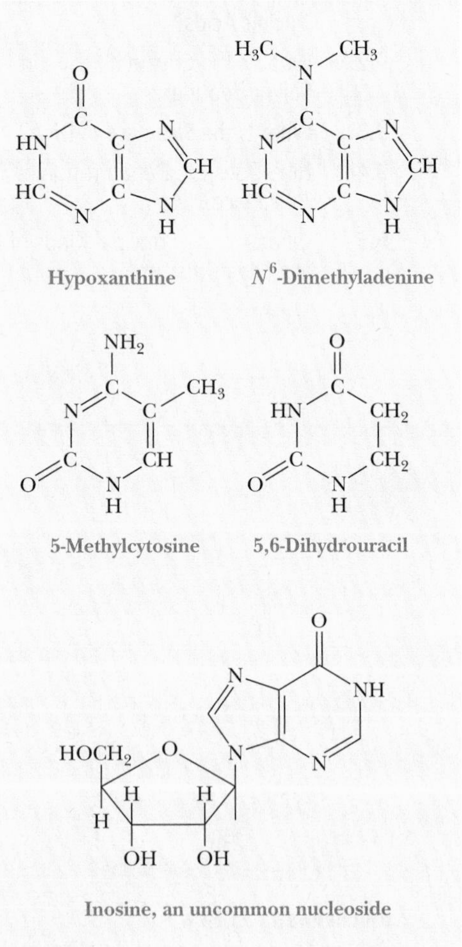

▲ **FIGURE 9.2** Structures of some of the less common nucleobases. When hypoxanthine is bonded to a sugar, the corresponding compound is called inosine.

9.2 | What Is the Covalent Structure of Polynucleotides?

The monomers of nucleic acids are **nucleotides.** An individual nucleotide consists of three parts—a nitrogenous base, a sugar, and a phosphoric acid residue—all of which are covalently bonded together.

The order of bases in the nucleic acids of DNA contains the information necessary to produce the correct amino acid sequence in the cell's proteins. The **nucleic acid bases** (also called **nucleobases**) are of two types—*pyrimidines* and *purines* (Figure 9.1). In this case, the word "base" does not refer to an alkaline compound, such as NaOH; rather, it refers to a one- or two-ring nitrogenous aromatic compound. Three **pyrimidine bases** (single-ring aromatic compounds)—*cytosine, thymine,* and *uracil*—commonly occur. Cytosine is found both in RNA and in DNA. Uracil occurs only in RNA. In DNA, thymine is substituted for uracil; thymine is also found to a small extent in some forms of RNA. The common **purine bases** (double-ring aromatic compounds) are *adenine* and *guanine,* both of which are found in RNA and in DNA (Figure 9.1). In addition to these five commonly occurring bases, there are "unusual" bases, with slightly different structures, that are found principally, but not exclusively, in transfer RNA (Figure 9.2). In many cases, the base is modified by methylation.

A **nucleoside** is a compound that consists of a base and a sugar covalently linked together. It differs from a nucleotide by lacking a phosphate group in its structure. In a nucleoside, a base forms a glycosidic linkage with the sugar. Glycosidic linkages and the stereochemistry of sugars are discussed in detail in Section 16.2. If you wish to look now at the material on the structure of sugars, you will find that it does not depend on material in the intervening chapters. For now, it is sufficient to say that a *glycosidic bond* is one that links a sugar and some other moiety. When the sugar is β-D-ribose, the resulting compound is a **ribonucleoside;** when the sugar is β-D-deoxyribose, the resulting compound is a **deoxyribonucleoside** (Figure 9.3). The glycosidic linkage is from the C-1′ carbon of the sugar to the N-1 nitrogen of pyrimidines or to the N-9 nitrogen of purines. The ring atoms of the base and the carbon atoms of the sugar are both numbered, with the numbers of the sugar atoms primed to prevent confusion. Note that the sugar is linked to a nitrogen in both cases (an *N*-glycosidic bond).

When phosphoric acid is esterified to one of the hydroxyl groups of the sugar portion of a nucleoside, a nucleotide is formed (Figure 9.4). A

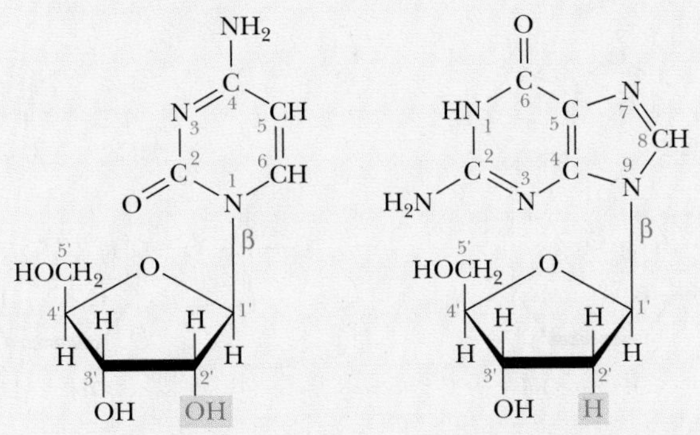

◀ **FIGURE 9.3** A comparison of the structures of a ribonucleoside and a deoxyribonucleoside. (A nucleoside does not have a phosphate group in its structure.)

nucleotide is named for the parent nucleoside, with the suffix "monophosphate" added; the position of the phosphate ester is specified by the number of the carbon atom at the hydroxyl group to which it is esterified—for instance, adenosine 3'-monophosphate or deoxycytidine 5'-monophosphate. The 5' nucleotides are most commonly encountered in nature. If additional phosphate groups form anhydride linkages to the first phosphate, the corresponding nucleoside diphosphates and triphosphates are formed. Recall this point from Section 2.2. These compounds are also nucleotides.

The polymerization of nucleotides gives rise to nucleic acids. The linkage between monomers in nucleic acids involves formation of two ester bonds by phosphoric acid. The hydroxyl groups to which the phosphoric acid is esterified are those bonded to the 3' and 5' carbons on adjacent residues. The resulting repeated linkage is a **3',5'-phosphodiester bond.** The nucleotide residues of nucleic acids are numbered from the 5' end, which normally carries a phosphate group, to the 3' end, which normally has a free hydroxyl group.

Figure 9.5 shows the structure of a fragment of an RNA chain. The *sugar–phosphate backbone* repeats itself down the length of the chain. The most important features of the structure of nucleic acids are the identities of the bases. Abbreviated forms of the structure can be written to convey this essential information. In one system of notation, single letters, such as A, G, C, U, and T, represent the individual bases. Vertical lines show the positions of the sugar moieties to which the individual bases are attached, and a diagonal line through the letter "P" represents a phosphodiester bond (Figure 9.5). However, an even more common system of notation uses only the single letters to show the order of the bases. When it is necessary to indicate the position on the sugar to which the phosphate group is bonded, the letter "p" is written to the left of the single-letter code for the base to represent a 5' nucleotide and to the right to represent a 3' nucleotide. For example, pA signifies 5'-adenosine monophosphate (5'-AMP), and Ap signifies 3'-AMP. The sequence of an oligonucleotide can be represented as pGpApCpApU or, even more simply, as GACAU, with the phosphates understood.

A portion of a DNA chain differs from the RNA chain just described only in the fact that the sugar is 2'-deoxyribose rather than ribose (Figure 9.6). In abbreviated notation, the deoxyribonucleotide is specified in the usual manner. Sometimes a "d" is added to indicate a deoxyribonucleotide residue; for example, dG is substituted for G, and the deoxy analogue of the ribooligonucleotide in the preceding paragraph would be d(GACAT). However, given that the sequence must refer to DNA because of the presence of thymine, the sequence GACAT is not ambiguous and would also be a suitable abbreviation.

Biochemistry ⑤ Now™
Go to BiochemistryNow and click on Biochemistry Interactive to explore the structures of purines and pyrimidines.

Essential Information

Both DNA and RNA consist of nucleotides joined by phosphodiester bonds to form a sugar–phosphate backbone. The sugar moiety is deoxyribose in DNA and ribose in RNA. Two kinds of nitrogen-containing nucleobases, pyrimidines and purines, are bonded to the sugar portion of the backbone. The sequence of bases is a very important feature of the primary structure of nucleic acids, because the sequence is the genetic information that ultimately leads to the sequence of RNA and protein.

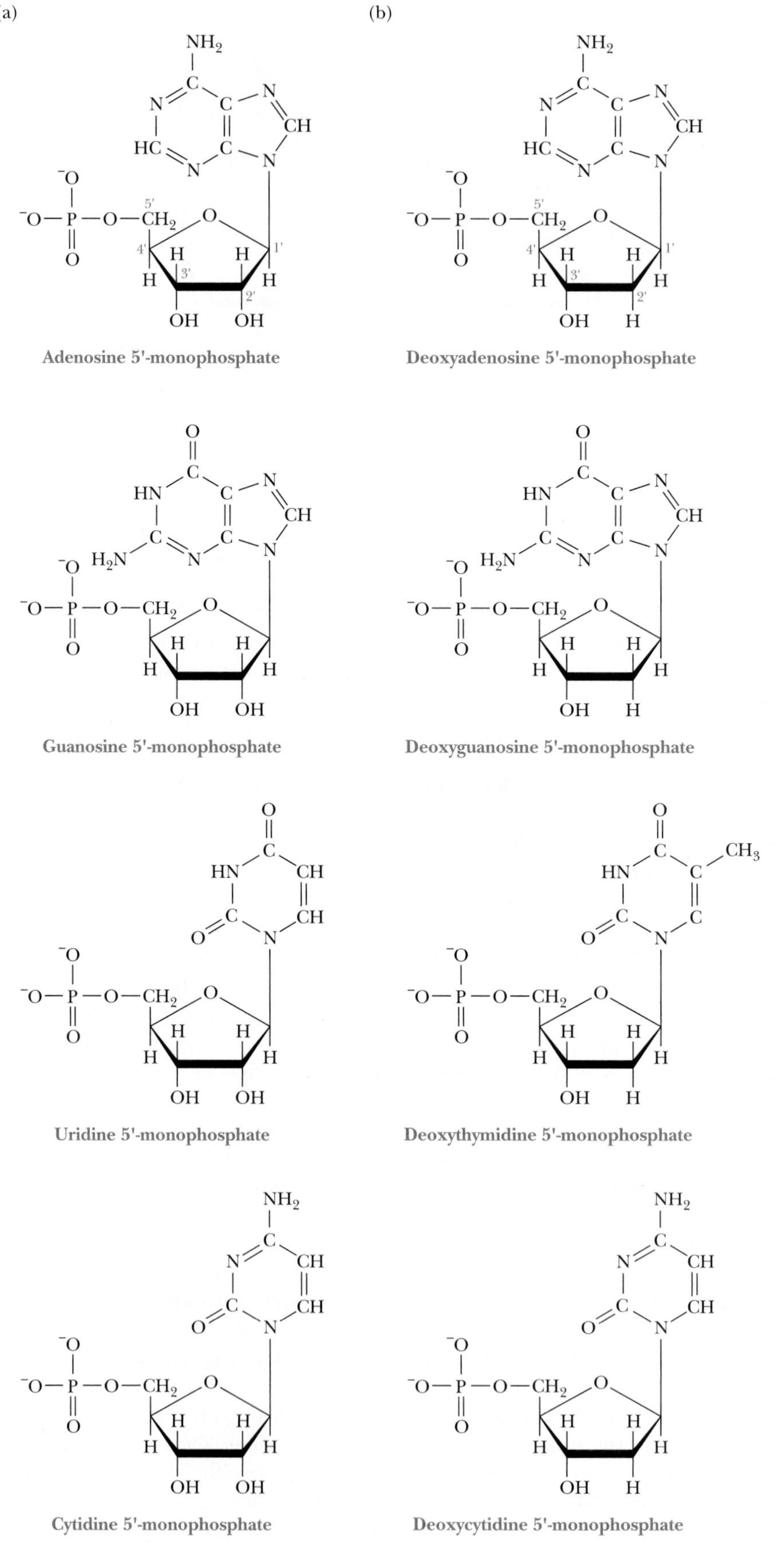

(a)

(b)

Adenosine 5'-monophosphate

Deoxyadenosine 5'-monophosphate

Guanosine 5'-monophosphate

Deoxyguanosine 5'-monophosphate

Uridine 5'-monophosphate

Deoxythymidine 5'-monophosphate

Cytidine 5'-monophosphate

Deoxycytidine 5'-monophosphate

▶ **FIGURE 9.4** The structures and names of the commonly occurring nucleotides. Each nucleotide has a phosphate group in its structure. All structures are shown in the forms that exist at pH 7. (a) Ribonucleotides. (b) Deoxyribonucleotides.

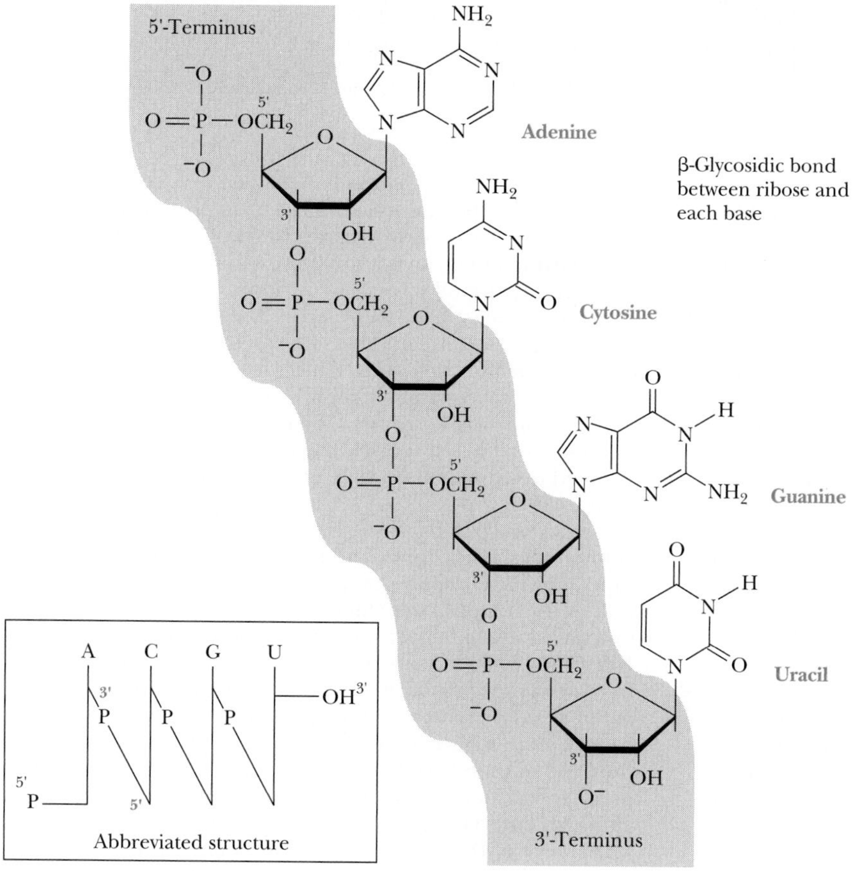

◀ **FIGURE 9.5** A fragment of an RNA chain.

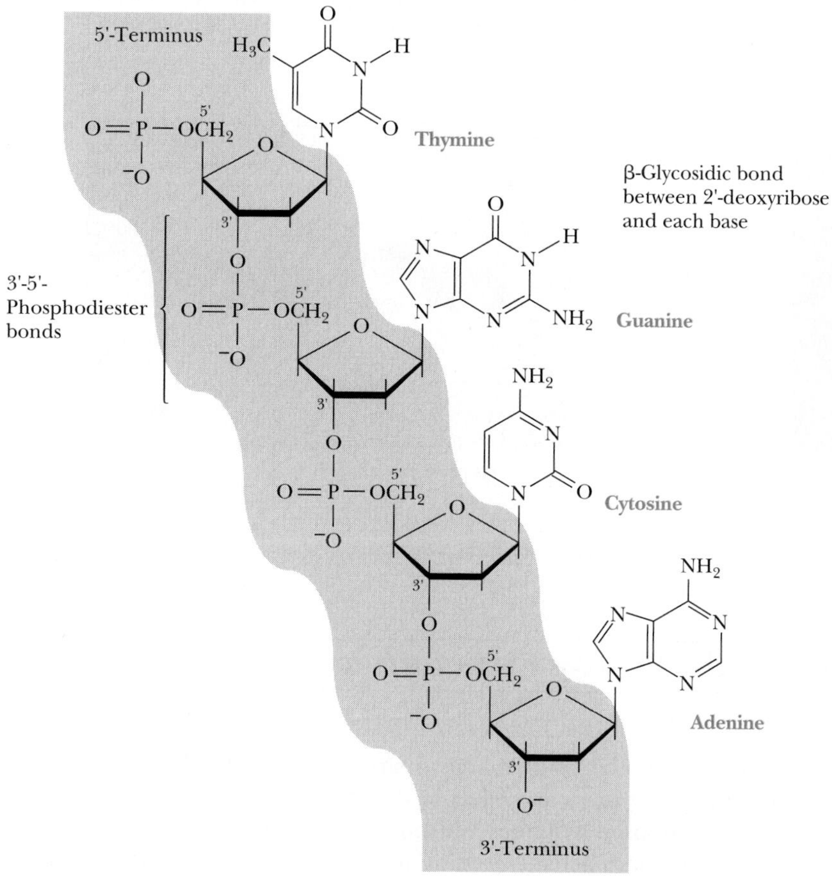

◀ **FIGURE 9.6** A portion of a DNA chain.

Biochemical Connections

The DNA Family Tree

Because it is easy to determine the sequence of DNA, even using automated and robotic systems that require little supervision, the amount of DNA sequence data available has virtually exploded. Many scientific journals no longer report full sequences; the information is just incorporated into the so-called gene banks, large computer systems that store the data. The sequence information, for proteins as well as for DNA, is readily available to anyone with a web search program. See http://www.tigr.org (the Institute for Genomic Research) for genomic databases and http://expasy.hcuge.ch/ (the ExPASy molecular biology server maintained by Geneva University Hospital and the University of Geneva). The ExPASy site is a repository for information about protein sequences as well as DNA sequences. A particularly useful site is http://www.ncbi.nlm.nih.gov/ (the National Center for Biotechnology Information). This site has a gene-sequence database (GenBank), molecular databases for protein sequence and structures, and a literature databank for searching for publications online. So much information is entered into the databanks that it has become necessary to develop new and more efficient computer technology to search and to compare such sequences. We are just beginning to appreciate the usefulness of so much information (Section 13.12). Many new applications, not even thought of at this time, will undoubtedly be developed. Here are two applications that give molecular information about evolution, the "family tree" of all living things.

1. *Molecular taxonomy*. In ways never before possible, we can compare the sequences not just from existing organisms but also, when DNA is available from fossil specimens, from extinct ancestors of living organisms. Within given genetic families of limited size, this information has enabled very detailed evolutionary trees to be developed. It has been possible to show that, in some areas, all plants are clones of one another. The largest living organism is a soil fungus that spreads over several acres. Redwood trees grow as clones from a central root system after forest fires. Sadly, many endangered species have such small remaining numbers that all living specimens are closely related to each other. This is true for all nene geese, which are native to Hawaii; for all California condors; and even for some whale species. The lack of genetic diversity in these endangered species may mean that the species are doomed to extinction, in spite of human attempts to ensure their survival.

2. *Ancient DNA*. DNA has been isolated from human fossils, such as mummies, bog people, and the frozen man found in the Alps, allowing comparisons of modern humans to recent relatives. Mitochondrial sequencing has shown that all humans now alive radiated out from one region in Africa some 100,000 to 200,000 years ago. More ancient DNA sequences from insect specimens preserved in amber have been compared to their modern counterparts. The film *Jurassic Park* is based on the suggestion that dinosaurs might be cloned from the DNA in their blood, which survived in the gut of an insect preserved in amber—certainly a far-fetched possibility, although entertaining (and profitable to filmmakers). The acceptance of DNA sequence data from ancient DNA is still controversial because of the likelihood of DNA degradation over time, contamination with modern DNA, and damage due to the initial chemical treatment of the samples.

▲ The nene geese, native to Hawaii, are an endangered species. Even those in European zoos are related to the ones left in Hawaii.

▲ Insects preserved in amber.

9.3 | What Is the Structure of DNA?

Secondary Structure of DNA: The Double Helix

Representations of the double-helical structure of DNA have become common in the popular press as well as in the scientific literature. When the **double helix** was proposed by James Watson and Francis Crick in 1953, it touched off a flood of research activity, leading to great advances in molecular biology. The determination of the double-helical structure was based primarily on model building and X-ray diffraction patterns. Information from X-ray patterns was added to information from chemical analyses that showed that the amount of A was always the same as the amount of T, and that the amount of G always equaled the amount of C. Both of these lines of evidence were used to conclude that DNA consists of two polynucleotide chains wrapped around each other to form a helix. Hydrogen bonds between bases on opposite chains determine the alignment of the helix, with the paired bases lying in planes perpendicular to the helix axis. The sugar–phosphate backbone is the outer part of the helix (Figure 9.7). The chains run in antiparallel directions, one 3′ to 5′ and the other 5′ to 3′.

The X-ray diffraction pattern of DNA demonstrated the helical structure and the diameter. The combination of evidence from X-ray diffraction and chemical analysis led to the conclusion that the base pairing is *complementary,* meaning that adenine pairs with thymine and that guanine pairs with cytosine. Because complementary base pairing occurs along the entire double helix, the two chains are also referred to as *complementary strands.* By 1953, studies of the base composition of DNA from many species had already shown that, to within experimental error, the mole percentages of adenine and thymine (moles of these substances as percentages of the total) were equal; the same was found to be the case with guanine and cytosine. An adenine–thymine (A–T) base pair has two hydrogen bonds between the bases; a guanine–cytosine (G–C) base pair has three (Figure 9.8).

The inside diameter of the sugar–phosphate backbone of the double helix is about 11 Å (1.1 nm). The distance between the points of attachment of the

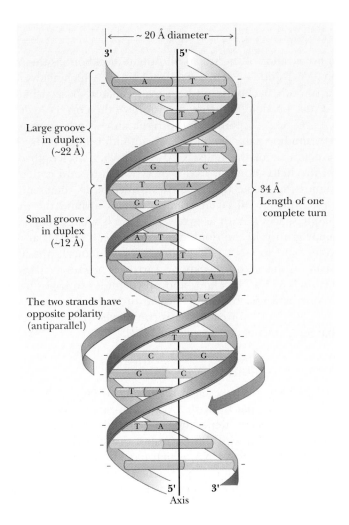

◀ **FIGURE 9.7** The double helix. A complete turn of the helix spans ten base pairs, covering a distance of 34 Å (3.4 nm). The individual base pairs are spaced 34 Å (3.4 nm) apart. The places where the strands cross hide base pairs that extend perpendicular to the viewer. The inside diameter is 11 Å (1.1 nm), and the outside diameter is 20 Å (2.0 nm). Within the cylindrical outline of the double helix are two grooves, a small one and a large one. Both are large enough to accommodate polypeptide chains. The minus signs alongside the strands represent the many negatively charged phosphate groups along the entire length of each strand.

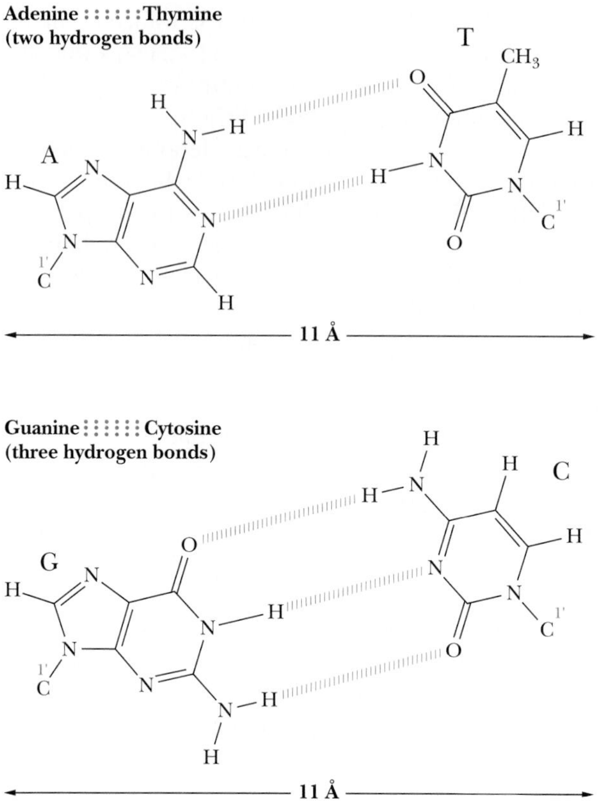

Adenine : : : : : :Thymine
(two hydrogen bonds)

► **FIGURE 9.8** Base pairing. The adenine–thymine (A–T) base pair has two hydrogen bonds, whereas the guanine–cytoside (G–C) base pair has three hydrogen bonds.

bases to the two strands of the sugar–phosphate backbone is the same for the two base pairs (A–T and G–C), about 11 Å (1.1 nm), which allows for a double helix with a smooth backbone and no overt bulges. Base pairs other than A–T and G–C are possible, but they do not have the correct hydrogen bonding pattern (A–C or G–T pairs) or the right dimensions (purine–purine or pyrimidine–pyrimidine pairs) to allow for a smooth double helix (Figure 9.8). The outside diameter of the helix is 20 Å (2 nm). The length of one complete turn of the helix along its axis is 34 Å (3.4 nm) and contains ten base pairs. The atoms that make up the two polynucleotide chains of the double helix do not completely fill an imaginary cylinder around the double helix; they leave empty spaces known as grooves. There is a large **major groove** and a smaller **minor groove** in the double helix; both can be sites at which drugs or polypeptides bind to DNA (see Figure 9.7). At neutral, physiological pH, each phosphate group of the backbone carries a negative charge. Positively charged ions, such as Na^+ or Mg^{2+}, and polypeptides with positively charged side chains must be associated with DNA in order to neutralize the negative charges. Eukaryotic DNA, for example, is complexed with histones, which are positively charged proteins, in the cell nucleus.

Conformational Variations in DNA

The form of DNA that we have been discussing so far is called **B-DNA.** It is thought to be the principal form that occurs in nature. However, other secondary structures can occur, depending on conditions such as the nature of the positive ion associated with the DNA and the specific sequence of bases. One of those other forms is **A-DNA,** which has 11 base pairs for each turn of the helix. Its base pairs are not perpendicular to the helix axis but lie at an angle of about 20° to the perpendicular (Figure 9.9). An important shared

Essential Information

The double helix is the predominant secondary structure of DNA. The sugar–phosphate backbones, which run in antiparallel directions on the two strands, lie on the outside of the helix. Pairs of bases, one on each strand, are held in alignment by hydrogen bonds. The base pairs lie in a plane perpendicular to the helix axis.

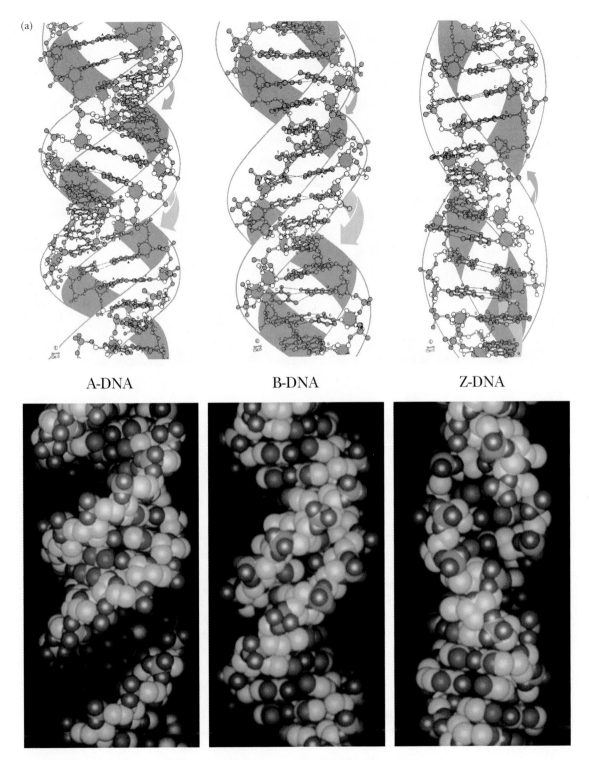

(a)

A-DNA   B-DNA   Z-DNA

▲ **FIGURE 9.9** Comparison of the A, B, and Z forms of DNA. (a) Side views.

(Figure 9.9 is continued on next page)

feature of A-DNA and B-DNA is that both are right-handed helices; that is, the helix winds upward in the direction in which the fingers of the right hand curl when the thumb is pointing upward (Figure 9.10). The A form of DNA was originally found in dehydrated DNA samples, and many researchers believed that the A form was an artifact of DNA preparation. DNA:RNA

(b)

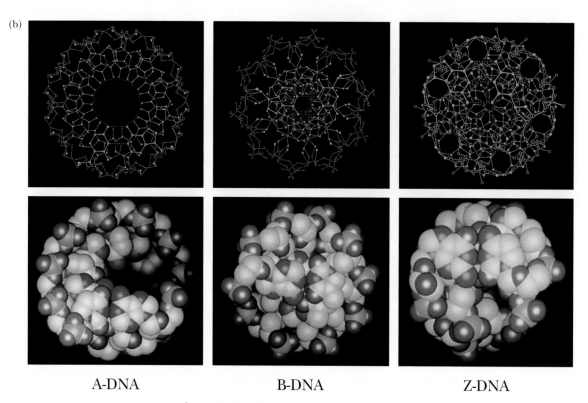

A-DNA   B-DNA   Z-DNA

▲ **FIGURE 9.9–cont'd** (b) Top views. Both parts include computer-generated space-filling models (bottom). The top half of each part shows corresponding ball-and-stick drawings. In the A form, the base pairs have a marked propeller-twist with respect to the helix axis. In the B form, the base pairs lie in a plane that is close to perpendicular to the helix axis. Z-DNA is a left-handed helix and in this respect differs from A-DNA and B-DNA, both of which are right-handed helices. (*Robert Stodala, Fox Chase Cancer Research Center. Illustration, Irving Geis. Rights owned by Howard Hughes Medical Institute. Not to be reproduced without permission.*)

▲ **FIGURE 9.10** Right- and left-handed helices are related to each other in the same way as right and left hands.

hybrids can adopt an A formation because the 2′-hydroxyl on the ribose prevents an RNA helix from adopting the B form; RNA:RNA hybrids may also be found in the A form.

Another variant form of the double helix, **Z-DNA,** is left-handed; it winds in the direction of the fingers of the left hand (Figure 9.10). Z-DNA is known to occur in nature, most often when there is a sequence of alternating purine–pyrimidine, such as dCpGpCpGpCpG. Sequences with cytosine methylated at the number 5 position of the pyrimidine ring can also be found in the Z form. It may play a role in the regulation of gene expression. The Z form of DNA is also a subject of active research among biochemists. The Z form of DNA can be considered a derivative of the B form of DNA, produced by flipping one side of the backbone 180° without having to break either the backbone or the hydrogen bonding of the complementary bases. Figure 9.11 shows how this might occur. The Z form of DNA gets its name from the zigzag look of the phosphodiester backbone when viewed from the side.

The B form of DNA has long been considered the normal, physiological DNA form. It was predicted from the nature of the hydrogen bonds between purines and pyrimidines and later found experimentally. Although it is easy to focus completely on the base pairing and the order of bases in DNA, other features of DNA structure are just as important. The ring portions of the DNA bases are very hydrophobic and interact with each other via hydrophobic bonding of their pi-cloud electrons. This process is usually referred to as **base stacking,** and even single-stranded DNA has a tendency to form struc-

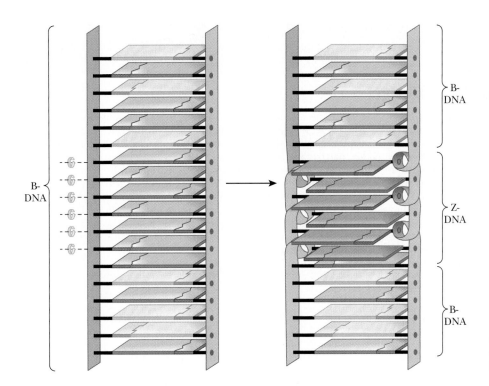

◀ **FIGURE 9.11** A Z-DNA section can form in the middle of a section of B-DNA by rotation of the base pairs, as indicated by the curved arrows.

tures in which the bases can stack. In standard B-DNA, each base pair is rotated 32° with respect to the preceding one (Figure 9.12). This form is perfect for maximal base pairing, but it is not optimal for maximal overlap of the bases. In addition, the edges of the bases that are exposed to the minor groove must come in contact with water in this form. Many of the bases twist in a characteristic way, called *propeller-twist* (Figure 9.13). In this form, the base-pairing distances are less optimal, but the base stacking is more optimal, and water is eliminated from the minor groove contacts with the bases. Besides twisting, bases also slide sideways, allowing them to interact better with the bases above and below them. The twist and slide depends on which bases are present, and researchers have identified that a basic unit for studying DNA structure is actually a dinucleotide with its complementary pairs. This is called a *step* in the nomenclature of DNA structure. For example, in Figure 9.13, we see an AG/CT step, which tends to adopt a different structure than a GC/GC step. As more and more is learned about DNA structure, it is evident that the standard B-DNA structure, while a good model, does not really describe local regions of DNA very well. Many DNA-binding proteins recognize the overall structure of a sequence of DNA, which depends upon the sequence but is not the DNA sequence itself.

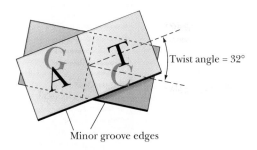

▲ **FIGURE 9.12** Two base pairs with 32° of right-handed helical twist; the minor-groove edges are drawn with heavy shading.

Tertiary Structure of DNA: Supercoiling

The DNA molecule has a length considerably greater than its diameter; it is not completely stiff and can fold back on itself in a manner similar to that of proteins as they fold into their tertiary structures. The double helix we have discussed so far is relaxed, which means that it has no twists in it, other than the helical twists themselves. Further twisting and coiling, or **supercoiling,** of the double helix is possible. The first example of supercoiling we shall consider is the case of prokaryotic DNA.

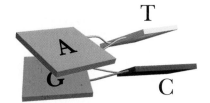

▲ **FIGURE 9.13** Propeller-twisted base pairs. Note how the hydrogen bonds between bases are distorted by this motion, yet remain intact. The minor-groove edges of the bases are shaded.

Biochemical Connections

Triple-Helical DNA: A Tool for Drug Design

Triple-helical DNA was first observed in 1957 in the course of an investigation of synthetic polynucleotides, but for decades it remained a laboratory curiosity. Recent studies have shown that synthetic oligonucleotides (usually about 15 nucleotide residues long) will bind to specific sequences of naturally occurring double-helical DNA. The oligonucleotides are chemically synthesized to have the correct base sequence for specific binding. The oligonucleotide that forms the third strand fits into the major groove of the double helix and forms specific hydrogen bonds. When the third strand is in place, the major groove is inaccessible to proteins that might otherwise bind to that site—specifically, proteins that activate or repress expression of that portion of DNA as a gene. This behavior suggests a possible in vivo role for triple helices, especially in view of the fact that hybrid triplexes with a short RNA strand bound to a DNA double helix are particularly stable.

In another aspect of this work, researchers who have studied triple helices have synthesized oligonucleotides with reactive sites that can be positioned in definite places in DNA sequences. Such a reactive site can be used to modify or cleave DNA at a chosen point in a given sequence. This kind of specific cutting of DNA is crucial to recombinant DNA technology and to genetic engineering.

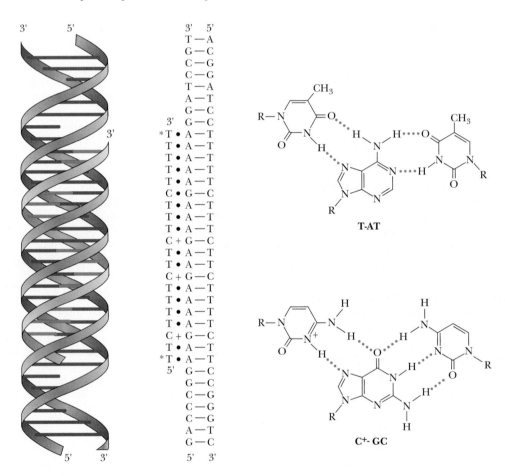

▲ (*left*) Model of a triple helix. (*middle*) Schematic diagram of a triple helix complex. C^+ is protonated cytosine. T^* indicates the site of attachment of the third helix. (*right*) The hydrogen-bonding scheme for a triple-helix formation.

Supercoiling in Prokaryotic DNA

If the sugar–phosphate backbone of a prokaryotic DNA forms a covalently bonded circle, the structure is still relaxed. Some extra twists are added if the DNA is unwound slightly before the ends are joined to form the circle. A

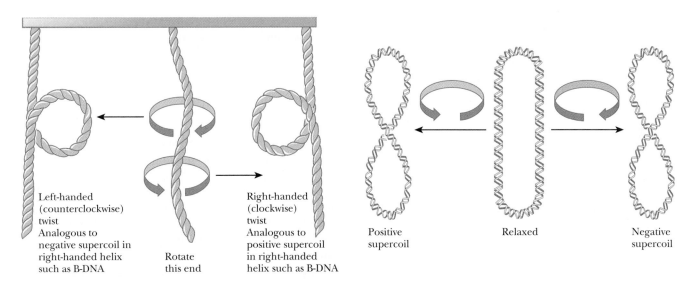

▲ **FIGURE 9.14** Supercoiled DNA topology. The DNA double helix can be approximated as a two-stranded, right-handed coiled rope. If one end of the rope is rotated counterclockwise, the strands begin to separate (negative supercoiling). If the rope is twisted clockwise (in a right-handed fashion), the rope becomes overwound (positive supercoiling). Get a piece of right-handed multistrand rope, and carry out these operations to convince yourself.

strain is introduced in the molecular structure, and the DNA assumes a new conformation to compensate for the unwinding. If, because of unwinding, a right-handed double helix acquires an extra left-handed helical twist (a super-coil), the circular DNA is said to be *negatively* supercoiled (Figure 9.14). Under different conditions, it is possible to form a right-handed, or *positively* supercoiled, structure in which there is overwinding of the closed-circle dou-ble helix. The difference between the positively and negatively supercoiled forms lies in their right- and left-handed natures, which, in turn, depend on the overwinding or underwinding of the double helix.

Enzymes that affect the supercoiling of DNA have been isolated from a variety of organisms. Naturally occurring circular DNA is negatively super-coiled except during replication, when it becomes positively supercoiled. It is critical for the cell to regulate this process. Enzymes that are involved in changing the supercoiled state of DNA are called **topoisomerases,** and they fall into two classes. Class I topoisomerases cut the phosphodiester backbone of one strand of DNA, pass the other end through, and then reseal the back-bone. Class II topoisomerases cut both strands of DNA, pass some of the remaining DNA helix between the cut ends, and then reseal. In either case, supercoils can be added or removed. As we shall see in upcoming chapters, these enzymes play an important role in replication and transcription, where separation of the helix strands causes supercoiling. **DNA gyrase** is a bacterial topoisomerase that introduces negative supercoils into DNA. The mechanism is shown in Figure 9.15. The enzyme is a tetramer. It cuts both strands of DNA, so it is a class II topoisomerase.

Supercoiling has been observed experimentally in naturally occurring DNA. Particularly strong evidence has come from electron micrographs that clearly show coiled structures in circular DNA from a number of different sources, including bacteria, viruses, mitochondria, and chloroplasts. Ultracen-trifugation can be used to detect supercoiled DNA because it sediments more rapidly than the relaxed form. (See Section 9.5 for a discussion of ultracen-trifugation.) Scientists have known for some time that prokaryotic DNA is

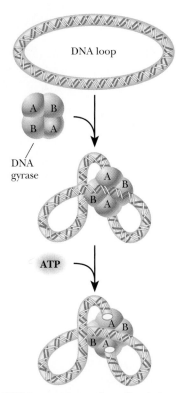

DNA loop

DNA gyrase

ATP

DNA is cut and a conformational change allows the DNA to pass through. Gyrase rejoins the DNA ends and then releases it.

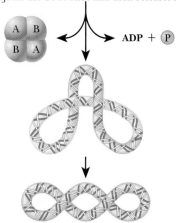

ADP + P

▲ **FIGURE 9.15** A model for the action of bacterial DNA gyrase (topoisomerase II).

normally circular, but supercoiling is a relatively recent subject of research. Computer modeling has helped scientists to visualize many aspects of the twisting and knotting of supercoiled DNA by obtaining "stop-action" images of very fast changes.

Supercoiling in Eukaryotic DNA

The supercoiling of the nuclear DNA of eukaryotes (such as plants and animals) is more complicated than the supercoiling of the circular DNA from prokaryotes. Eukaryotic DNA is complexed with a number of proteins, especially with basic proteins that have abundant positively charged side chains at physiological (neutral) pH. Electrostatic attraction between the negatively charged phosphate groups on the DNA and the positively charged groups on the proteins favors the formation of complexes of this sort. The resulting material is called **chromatin.** Thus, topological changes induced by supercoiling must be accommodated by the histone-protein component of chromatin.

The principal proteins in chromatin are the **histones,** of which there are five main types, called H1, H2A, H2B, H3, and H4. All these proteins contain large numbers of basic amino acid residues, such as lysine and arginine. In the chromatin structure, the DNA is tightly bound to all the types of histone except H1. The H1 protein is comparatively easy to remove from chromatin, but dissociating the other histones from the complex is more difficult. Proteins other than histones are also complexed with the DNA of eukaryotes, but they are neither as abundant nor as well studied as histones.

In electron micrographs, chromatin resembles beads on a string (Figure 9.16). This appearance reflects the molecular composition of the protein–DNA complex. Each "bead" is a **nucleosome,** consisting of DNA wrapped around a histone core. This protein core is an octamer, which includes two molecules of each type of histone but H1; the composition of the octamer is $(H2A)_2(H2B)_2(H3)_2(H4)_2$. The "string" portions are called *spacer regions;* they consist of DNA complexed to some H1 histone and nonhistone proteins. As the DNA coils around the histones in the nucleosome, about 150 base pairs are in contact with the proteins; the spacer region is about 30 to 50 base pairs long. Histones can be modified by acetylation, methylation, phosphorylation, and ubiquitinylation. Ubiquitin is a protein involved in the degradation of other proteins. It will be studied further in Chapter 12. Modifying histones changes their DNA and protein-binding characteristics, and how these changes affect transcription and replication is a subject of active research (Chapter 11). See the article by Jenuwein listed in the bibliography of this chapter.

9.4 | How Does the Denaturation of DNA Take Place?

We have already seen that the hydrogen bonds between base pairs are an important factor in holding the double helix together. The amount of stabilizing energy associated with the hydrogen bonds is not great, but the hydrogen bonds hold the two polynucleotide chains in the proper alignment. However, the stacking of the bases in the native conformation of DNA contributes the largest part of the stabilization energy. Energy must be added to a sample of DNA to break the hydrogen bonds and to disrupt the stacking interactions. This is usually carried out by heating the DNA in solution.

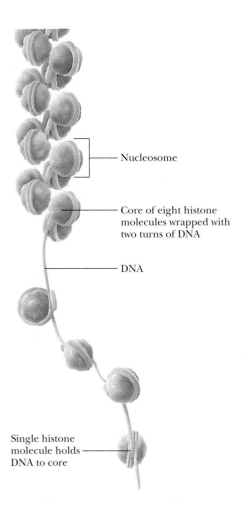

Nucleosome

Core of eight histone molecules wrapped with two turns of DNA

DNA

Single histone molecule holds DNA to core

◄ **FIGURE 9.16** The structure of chromatin. DNA is associated with histones in an arrangement that gives the appearance of beads on a string. The "string" is DNA, and each of the "beads" (nucleosomes) consists of DNA wrapped around a protein core of eight histone molecules. Further coiling of the DNA spacer regions produces the compact form of chromatin found in the cell.

The heat denaturation of DNA, also called *melting*, can be monitored experimentally by observing the absorption of ultraviolet light. The bases absorb light in the 260-nm wavelength region. As the DNA is heated and the strands separate, the wavelength of absorption does not change, but the amount of light absorbed increases (Figure 9.17). This effect is called *hyperchromicity*. It is based on the fact that the bases, which are stacked on top of one another in native DNA, become unstacked as the DNA is denatured. Because the bases interact differently in the stacked and unstacked orientations, their absorbance changes. Heat denaturation is a way to obtain single-stranded DNA (Figure 9.18), which has many uses. Some of these uses are discussed in Chapter 14. When DNA is replicated, it first becomes single-stranded so that the complementary bases can be aligned. This same principle is seen during a chemical reaction used to determine the DNA sequence (Chapter 14). A most ambitious example of this reaction is described in the following Biochemical Connections box.

Under a given set of conditions, there is a characteristic midpoint of the melting curve (the transition temperature, or melting temperature, written T_m) for DNA from each distinct source. The underlying reason for this property is that each type of DNA has a given, well-defined base composition. A G–C base pair has three hydrogen bonds, and an A–T base pair has only two. The higher the percentage of G–C base pairs, the higher the melting temperature of a DNA molecule. In addition to the effect of the base pairs, G–C

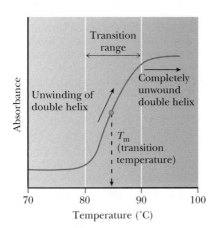

▲ **FIGURE 9.17** The experimental determination of DNA denaturation. This is a typical melting-curve profile of DNA, depicting the hyperchromic effect observed on heating. The transition (melting) temperature, T_m, increases as the guanine and cytosine (the G–C content) increase. The entire curve would be shifted to the right for a DNA with higher G–C content and to the left for a DNA with lower G–C content.

Biochemical Connections

The Human Genome Project: Prospects and Possibilities

The Human Genome Project (HGP) is a massive attempt to sequence the entire human genome, some 3.3 billion base pairs spread over 23 pairs of chromosomes. This project, started formally in 1990, is a worldwide effort driven forward by two groups. One is a private company called Celera Genomics, and its preliminary results were published in *Science* in February, 2001. The other is a publicly funded group of researchers called the International Human Genome Sequencing Consortium. Their preliminary results were published in *Nature* in February, 2001. Researchers were surprised to find that there are only about 30,000 genes in the human genome (although this is still debated), which is similar to many other eukaryotes, including some as simple as the roundworm *Caenorhabditis elegans*.

What does one do with the information? From this information, we will eventually be able to identify all human genes and to determine which sets of genes are likely to be involved in all human genetic traits, including diseases that have a genetic basis. There is an elaborate interplay of genes, so it may never be possible to say that a defect in a given gene will ensure that the individual will develop a particular disease. Nevertheless, some forms of genetic screening will certainly become a routine part of medical testing in the future. It would be beneficial, for example, if someone more susceptible to heart disease than the average person were to have this information at an early age. This person could then decide on some minor adjustments in lifestyle and diet that might make heart disease much less likely to develop.

Many people are concerned that the availability of genetic information could lead to genetic discrimination. For that reason, HGP is a rare example of scientific project in which definite percentages of financial support and research effort have been devoted to the ethical, legal, and social implications (ELSI) of the research. The question is often posed in this form: Who has a right to know your genetic information? You? Your doctor? Your potential spouse or employer? An insurance company? These questions are not trivial, but they have not yet been answered definitively. The 1997 movie *GATTACA* depicted a society in which one's social and economic classes are established at birth based on one's genome. Many citizens have expressed concern that genetic screening would lead to a new type of prejudice and bigotry aimed against "genetically challenged" people. Many people have suggested that there is no point in screening for potentially disastrous genes if there is no meaningful therapy for the disease they may "cause." However, couples often want to know in advance if they are likely to pass on a potentially lethal disease to their children.

Three specific examples are pertinent here:

1. There is no advantage in testing for the breast-cancer gene if a woman is *not* in a family at high risk for the disease. The presence of a "normal" gene in such a low-risk individual tells nothing about whether a mutation might occur in the future. The risk of breast cancer is not changed if a low-risk person has the normal gene, so mammograms and monthly self-examination are in order. (See the articles by Levy-Lahad and Couzin in the bibliography at the end of this chapter.)

2. Couples whose offspring are at risk for Tay–Sachs disease (see the Biochemical Connections box on page 591 in Chapter 21) often choose never to have children, rather than to have children who die at a tragically early age. The availability of a good genetic test for the disease has actually increased the birth rate and decreased the abortion rate for such people. Parents can now be assured that their children will not suffer from Tay–Sachs disease.

3. The presence of a gene has not always predicted the development of the disease. Some individuals who have been shown to be carriers of the gene for Huntington's disease have lived to old age without developing the disease. Some males who are functionally sterile have been found to have cystic fibrosis, which carries a side effect of sterility due to the improper chloride-channel function that is a feature of that disease (see Section 13.8). They learn this when they go to a clinic to assess the nature of their fertility problem, even though they may never have shown true symptoms of the disease as a child, other than perhaps a high occurrence of respiratory ailments.

Another major area for concern about the HGP is the possibility of gene therapy, which many people fear is akin to "playing God." Some people envision an era of so-called designer babies, with attempts made to create the "perfect" human. A more moderate view has been that gene therapy may be useful in correcting diseases that impair life or are lethal. Tests with human subjects are already underway for cystic fibrosis, the "bubble boy" type of immune deficiency, and some other diseases. Current guidelines in the United States allow for gene therapy of somatic cells, but they do not allow for genetic modifications that would be passed on to the next generation (see Section 14.4).

A new science called "behavioral genomics" has been born in the wake of the Human Genome Project. Besides looking at the genetic causes of physical diseases, the genetics of behavior have also been studied. Because behavior has both inheritable as well as environmental aspects, this is a very difficult task, but many genes for behavioral abnormalities have been isolated, including genes correlated with Alzheimer's disease, dyslexia, attention-deficit disorder, schizophrenia, and certain types of aggression (see the article by McGuffin, Riley, and Plomin in the bibliography at the end of this chapter).

pairs are more hydrophobic than A–T pairs, so they stack better, which also affects the melting curve.

Renaturation of denatured DNA is possible on slow cooling (Figure 9.18). The separated strands can recombine and form the same base pairs responsible for maintaining the double helix.

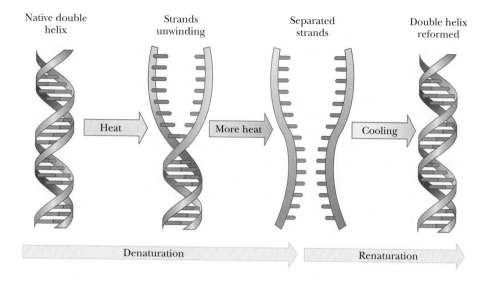

◄ **FIGURE 9.18** The double helix unwinds when DNA is denatured, with eventual separation of the strands. The double helix is re-formed on renaturation with slow cooling and annealing.

9.5 | What Are the Principal Kinds of RNA and Their Structures?

Six kinds of RNA—**transfer RNA (tRNA), ribosomal RNA (rRNA), messenger RNA (mRNA), small nuclear RNA (snRNA), micro RNA (miRNA),** and **small interfering RNA (siRNA)**—play an important role in the life processes of cells. Figure 9.19 shows the process of information transfer. The various kinds

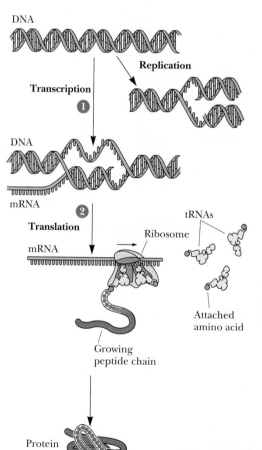

Replication
DNA replication yields two DNA molecules identical to the original one, ensuring transmission of genetic information to daughter cells with exceptional fidelity.

Transcription
The sequence of bases in DNA is recorded as a sequence of complementary bases in a single-stranded mRNA molecule.

Translation
Three-base codons on the mRNA corresponding to specific amino acids direct the sequence of building a protein. These codons are recognized by tRNAs (transfer RNAs) carrying the appropriate amino acids. Ribosomes are the "machinery" for protein synthesis.

◄ **FIGURE 9.19** The fundamental process of information transfer in cells. (1) Information encoded in the nucleotide sequence of DNA is transcribed through synthesis of an RNA molecule whose sequence is dictated by the DNA sequence. (2) As the sequence of this RNA is read (as groups of three consecutive nucleotides) by the protein-synthesis machinery, it is translated into the sequence of amino acids in a protein. This information transfer system is encapsulated in the dogma: DNA → RNA → protein.

of RNA participate in the synthesis of proteins in a series of reactions ultimately directed by the base sequence of the cell's DNA. *The base sequences of all types of RNA are determined by that of DNA.* The process by which the order of bases is passed from DNA to RNA is called **transcription** (Chapter 11).

Ribosomes, in which rRNA is associated with proteins, are the sites for assembly of the growing polypeptide chain in protein synthesis. Amino acids are brought to the assembly site covalently bonded to tRNA, as aminoacyl-tRNAs. The order of bases in mRNA specifies the order of amino acids in the growing protein; this process is called **translation** of the genetic message. A sequence of three bases in mRNA directs the incorporation of a particular amino acid into the growing protein chain. (We shall discuss the details of protein synthesis in Chapter 12.) We are going to see that the details of the process will differ in prokaryotes and in eukaryotes (Figure 9.20). In prokaryotes, there is no nuclear membrane, so mRNA can direct the synthesis of pro-

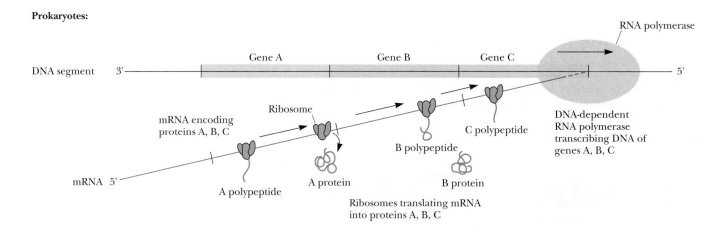

Prokaryotes:

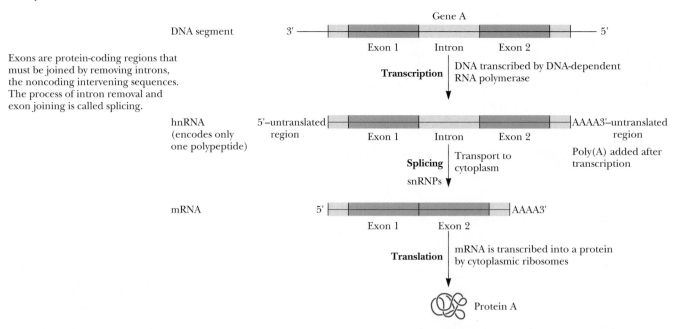

Eukaryotes:

Exons are protein-coding regions that must be joined by removing introns, the noncoding intervening sequences. The process of intron removal and exon joining is called splicing.

Biochemistry❸Now™ ▲ **ACTIVE FIGURE 9.20** The properties of mRNA molecules in prokaryotic versus eukaryotic cells during transcription and translation. **Watch this Active Figure at http://now.brookscole.com/campbell5**

Table 9.1		
The Roles of Different Kinds of RNA		
RNA Type	**Size**	**Function**
Transfer RNA	Small	Transports amino acids to site of protein synthesis
Ribosomal RNA	Several kinds— variable in size	Combines with proteins to form ribosomes, the site of protein synthesis
Messenger RNA	Variable	Directs amino acid sequence of proteins
Small nuclear RNA	Small	Processes initial mRNA to its mature form in eukaryotes
Small interfering RNA	Small	Affects gene expression; used by scientists to knock out a gene being studied
Micro RNA	Small	Affects gene expression; important in growth and development

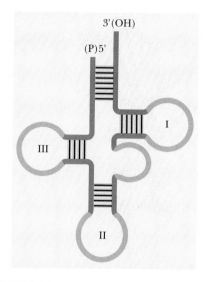

▲ **FIGURE 9.21** The cloverleaf depiction of transfer RNA. Double-stranded regions (shown in red) are formed by folding the molecule and stabilized by hydrogen bonds (∭) between complementary base pairs. Peripheral loops are shown in yellow. There are three major loops (numbered) and one minor loop of variable size (not numbered).

teins while it is still in the process of being transcribed. Eukaryotic mRNA, on the other hand, undergoes considerable processing. One of the most important parts of the process is splicing out intervening sequences (introns), so that the parts of the mRNA that will be expressed (exons) are contiguous to each other.

Small nuclear RNAs are found only in the nucleus of eukaryotic cells, and they are distinct from the other three RNA types. They are involved in processing of initial mRNA transcription products to a mature form suitable for export from the nucleus to the cytoplasm for translation. Micro RNAs and small interfering RNAs are the most recent discoveries. SiRNAs are the main players in **RNA interference (RNAi),** a process that was first discovered in plants and later in mammals, including humans. RNAi causes the suppression of certain genes (see Chapter 11). It is also being used extensively by scientists who wish to eliminate the effect of a gene to help discover its function (see Chapter 13). Table 9.1 summarizes the types of RNA.

Transfer RNA

The smallest of the three important kinds of RNA is tRNA. Different types of tRNA molecules can be found in every living cell because at least one tRNA bonds specifically to each of the amino acids that commonly occur in proteins. Frequently there are several tRNA molecules for each amino acid. A tRNA is a single-stranded polynucleotide chain, between 73 and 94 nucleotide residues long, that generally has a molecular mass of about 25,000 Da. (Note that biochemists tend to call the unit of atomic mass the *dalton,* for which the abbreviation is Da.)

Intrachain hydrogen bonding occurs in tRNA, forming A–U and G–C base pairs similar to those that occur in DNA except for the substitution of uracil for thymine. The duplexes thus formed have the A-helical form, rather than the B-helical form, which is the predominant form in DNA (Section 9.3). The molecule can be drawn as a *cloverleaf structure,* which can be considered the secondary structure of tRNA because it shows the hydrogen bonding between certain bases (Figure 9.21). The hydrogen-bonded portions of the molecule are called *stems,* and the non-hydrogen-bonded portions are *loops.* Some of these loops contain modified bases (Figure 9.22). During protein synthesis, both tRNA and mRNA are bound to the ribosome in a definite spatial arrangement that ultimately ensures the correct order of the amino acids in the growing polypeptide chain.

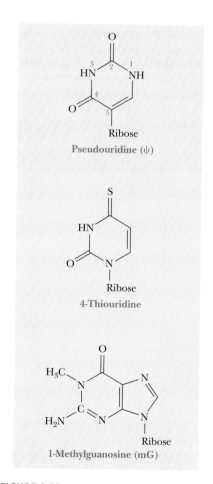

▲ **FIGURE 9.22** Structures of some modified bases found in transfer RNA. Note that the pyrimidine in pseudouridine is linked to ribose at C-5 rather than at the usual N-1.

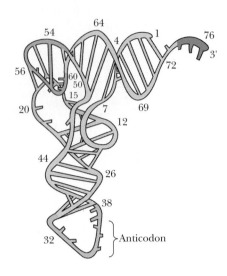

▲ **FIGURE 9.23** The three-dimensional structure of yeast phenylalanine tRNA as deduced from X-ray diffraction studies of its crystals. The tertiary folding is illustrated, and the ribose–phosphate backbone is presented as a continuous ribbon; H bonds are indicated by crossbars. Unpaired bases are shown as short, unconnected rods. The anticodon loop is at the bottom and the —CCA 3′—OH acceptor end is at the top right.

Essential Information

Four kinds of RNA—transfer RNA, ribosomal RNA, messenger RNA, and small nuclear RNA—are involved in protein synthesis. Transfer RNA transports amino acids to the sites of protein synthesis on ribosomes, which consist of ribosomal RNAs and proteins. Messenger RNA directs the amino acid sequence of proteins. Small nuclear RNA is used to help process eukaryotic mRNA to its final form.

A particular tertiary structure is necessary for tRNA to interact with the enzyme that covalently attaches the amino acid to the 2′ or 3′ end. To produce this tertiary structure, the tRNA folds into an L-shaped conformation that has been determined by X-ray diffraction (Figure 9.23).

Ribosomal RNA

In contrast with tRNA, rRNA molecules tend to be quite large, and only a few types of rRNA exist in a cell. Because of the intimate association between rRNA and proteins, a useful approach to understanding the structure of rRNA is to investigate ribosomes themselves.

The RNA portion of a ribosome accounts for 60%–65% of the total weight, and the protein portion constitutes the remaining 35%–40% of the weight. Dissociation of ribosomes into their components has proved to be a useful way of studying their structure and properties. A particularly important endeavor has been to determine both the number and the kind of RNA and protein molecules that make up ribosomes. This approach has helped to elucidate the role of ribosomes in protein synthesis. In both prokaryotes and eukaryotes, a ribosome consists of two subunits, one larger than the other. In turn, the smaller subunit consists of one large RNA molecule and about 20 different proteins; the larger subunit consists of two RNA molecules in prokaryotes (three in eukaryotes) and about 35 different proteins in prokaryotes (about 50 in eukaryotes). The subunits are easily dissociated from one another in the laboratory by lowering the Mg^{2+} concentration of the medium. Raising the Mg^{2+} concentration to its original level reverses the process, and active ribosomes can be reconstituted by this method.

A technique called *analytical ultracentrifugation* has proved very useful for monitoring the dissociation and reassociation of ribosomes. Figure 9.24 shows an analytical ultracentrifuge. We need not consider all the details of this technique, as long as it is clear that its basic aim is the observation of the motion of ribosomes, RNA, or protein in a centrifuge. The motion of the particle is characterized by a *sedimentation coefficient*, expressed in *Svedberg units* (S), which are named after Theodor Svedberg, the Swedish scientist who invented the ultracentrifuge. The S value increases with the molecular weight of the sedimenting particle, but it is not directly proportional to it because the particle's shape also affects its sedimentation rate.

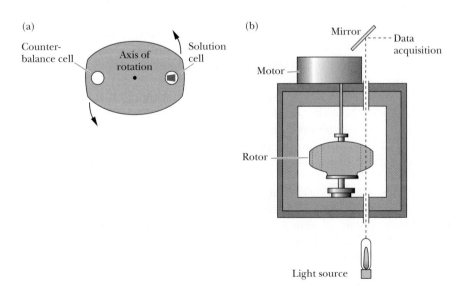

▶ **FIGURE 9.24** The analytical ultracentrifuge. (a) Top view of an ultracentrifuge rotor. The solution cell has optical windows; the cell passes through a light path once each revolution. (b) Side view of an ultracentrifuge rotor. The optical measurement taken as the solution cell passes through the light path makes it possible to monitor the motion of sedimenting particles.

Ribosomes and ribosomal RNA have been studied extensively via sedimentation coefficients. Most research on prokaryotic systems has been done with the bacterium *Escherichia coli*, which we shall use as an example here. An *E. coli* ribosome typically has a sedimentation coefficient of 70S. When an intact 70S bacterial ribosome dissociates, it produces a light 30S subunit and a heavy 50S subunit. Note that the values of sedimentation coefficients are not additive, showing the dependence of the S value on the shape of the particle. The 30S subunit contains a 16S rRNA and 21 different proteins. The 50S subunit contains a 5S rRNA, a 23S rRNA, and 34 different proteins (Figure 9.25). For comparison, eukaryotic ribosomes have a sedimentation coefficient of 80S, and the small and large subunits are 40S and 60S, respectively. The small subunit of eukaryotes contains an 18S rRNA, and the large subunit contains three types of rRNA molecules: 5S, 5.8S, and 28S.

The 5S rRNA has been isolated from many different types of bacteria, and the nucleotide sequences have been determined. A typical 5S rRNA is about 120 nucleotide residues long and has a molecular mass of about 40,000 Da. Some sequences have also been determined for the 16S and 23S rRNA molecules. These larger molecules are about 1500 and 2500 nucleotide residues long, respectively. The molecular mass of 16S rRNA is about 500,000 Da, and that of 23S rRNA is about one million Da. The degrees of secondary and tertiary structure in the larger RNA molecules appear to be substantial.

A secondary structure has been proposed for 16S rRNA (Figure 9.26), and suggestions have been made about the way in which the proteins associate with the RNA to form the 30S subunit.

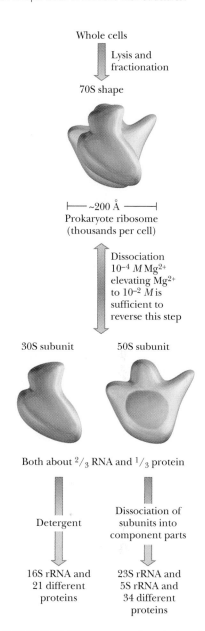

▲ FIGURE 9.25 The structure of a typical prokaryotic ribosome. The individual components can be mixed, producing functional subunits. Reassociation of subunits gives rise to an intact ribosome.

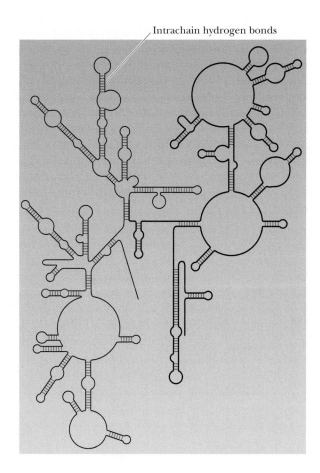

◀ FIGURE 9.26 A schematic drawing of a proposed secondary structure for 16S rRNA. The intrachain folding pattern includes loops and double-stranded regions. Note the extensive intrachain hydrogen bonding.

The *self-assembly of ribosomes* takes place in the living cell, but the process can be duplicated in the laboratory. Elucidation of ribosomal structure is an active field of research. The binding of antibiotics to bacterial ribosomal subunits so as to prevent self-assembly of the ribosome is one focus of the investigation. The structure of ribosomes is also one of the points used to compare and contrast eukaryotes, eubacteria, and archaebacteria (Chapter 1). For more information on this subject, see the articles by Lake, especially the review article, listed in the bibliography at the end of this chapter. The study of RNA became much more exciting in 1986, when Thomas Cech showed that certain RNA molecules exhibited catalytic activity (Section 11.7). Equally exciting was the recent discovery that the ribosomal RNA, and not protein, is the part of a ribosome that catalyzes the formation of peptide bonds in bacteria (Chapter 12). See the article by Cech in the bibliography at the end of this chapter for more on this development.

Messenger RNA

The least abundant of the main types of RNA is mRNA. In most cells, it constitutes no more than 5%–10% of the total cellular RNA. The sequences of bases in mRNA specify the order of the amino acids in proteins. In rapidly growing cells, many different proteins are needed within a short time interval. Fast turnover in protein synthesis becomes essential. Consequently, it is logical that mRNA is formed when it is needed, directs the synthesis of proteins, and then is degraded so that the nucleotides can be recycled. Of the main types of RNA, mRNA is the one that usually turns over most rapidly in the cell. Both tRNA and rRNA (as well as ribosomes themselves) can be recycled intact for many rounds of protein synthesis.

The sequence of mRNA bases that directs the synthesis of a protein reflects the sequence of DNA bases in the gene that codes for that protein, although this mRNA sequence is often altered after it is produced from the DNA. Messenger RNA molecules are heterogeneous in size, as are the proteins whose sequences they specify. Less is known about possible intrachain folding in mRNA, with the exception of folding that occurs during termination of transcription (Chapter 11). It is also likely that several ribosomes are associated with a single mRNA molecule at some time during the course of protein synthesis. In eukaryotes, mRNA is initially formed as a larger precursor molecule called **heterogeneous nuclear RNA (hnRNA).** These contain lengthy portions of intervening sequences called **introns** that do not encode a protein. These introns are removed by posttranscriptional splicing. In addition, protective units called *5'-caps* and *3' poly(A) tails* are added before the mRNA is complete (Section 11.5).

Small Nuclear RNA

The most recently discovered RNA molecule is the small nuclear RNA (snRNA), which is found, as the name implies, in the nucleus of eukaryotic cells. This type of RNA is small, about 100 to 200 nucleotides long, but it is not a tRNA molecule nor a small subunit of rRNA. In the cell, it is complexed with proteins forming **small nuclear ribonucleoprotein particles,** which are usually abbreviated **snRNPs** (pronounced "snurps"). These particles have a sedimentation coefficient of 10S. Their function is to help with the processing of the initial mRNA transcribed from DNA into a mature form that is ready for export out of the nucleus. In eukaryotes, transcription occurs in the nucleus, but because most protein synthesis occurs in the cytosol, the mRNA must first be exported. Many researchers are working on the processes of RNA splicing, which will be described further in Section 11.5.

RNA Interference

The process called RNA interference was heralded as the breakthrough of the year in 2002 in *Science* magazine. Short stretches of RNA (20–30 nucleotides long) have been found to have an enormous control over gene expression. This process has been found to be a protection mechanism in many species, with the siRNAs being used to eliminate expression of an undesirable gene, such as one that is causing uncontrolled cell growth or a gene that came from a virus. These small RNAs are also being used by scientists who wish to study gene expression. In what has become an explosion of new biotechnology, many companies have been created to produce and to market designer siRNA to knock out hundreds of known genes. This technology also has medical applications: siRNA has been used to protect mouse liver from hepatitis and to help clear infected liver cells of the disease (see articles by Couzin, Gitlin, and Lau). The biotech applications of RNA interference will be discussed more in Chapter 13.

Summary

9.1 What Are the Levels of Structure in Nucleic Acids? The primary structure of nucleic acids is the order of bases in the polynucleotide sequence, and the secondary structure is the three-dimensional conformation of the backbone. The tertiary structure is specifically the supercoiling of the molecule.

9.2 What Is the Covalent Structure of Polynucleotides? The monomers of nucleic acids are nucleotides. An individual nucleotide consists of three parts—a nitrogenous base, a sugar, and a phosphoric acid residue—all of which are covalently bonded together. The bases are bonded to the sugars, forming nucleosides. Nucleosides are linked by ester bonds to phosphoric acid to form the phosphodiester backbone.

9.3 What Is the Structure of DNA? The double helix originally proposed by Watson and Crick is the most striking feature of DNA structure. The two coiled strands run in antiparallel directions with hydrogen bonds between complementary bases. Adenine pairs with thymine, and guanine pairs with cytosine. Supercoiling is a feature of DNA structure both in prokaryotes and in eukaryotes. Eukaryotic DNA is complexed with histones and other basic proteins, but less is known about proteins bound to prokaryotic DNA.

9.4 How Does the Denaturation of DNA Take Place? When DNA is denatured, the double-helical structure breaks down; the progress of this phenomenon can be followed by monitoring the absorption of ultraviolet light. The temperature at which DNA becomes denatured by heat depends on its base composition; higher temperatures are needed to denature DNA rich in G–C base pairs.

9.5 What Are the Principal Kinds of RNA and Their Structures? The six kinds of RNA—transfer RNA (tRNA), ribosomal RNA (rRNA), messenger RNA (mRNA), small nuclear RNA (snRNA), micro RNA (miRNA), and small interfering RNA (siRNA)—differ in structure and function. Transfer RNA is relatively small, about 80 nucleotides long. It exhibits extensive intrachain hydrogen bonding, represented in two dimensions by a cloverleaf structure. Ribosomal RNA molecules tend to be quite large and are complexed with proteins to form ribosomal subunits. Ribosomal RNA also exhibits extensive internal hydrogen bonding. The sequence of bases in a given mRNA determines the sequence of amino acids in a specified protein. The size of mRNA molecules varies with the size of the protein. Eukaryotic mRNA is processed in the nucleus by a fourth type of RNA, small nuclear RNA, which is complexed with proteins to give small nuclear ribonuclear protein particles (snRNPs). Eukaryotic mRNA is initially produced in an immature form that must be processed by removing introns and adding protective units at the 5′ and 3′ ends. Micro RNA and small interfering RNA are both very small, about 20–30 bases long. They function in the control of gene expression and were the most recent discoveries in RNA research.

Critical Questions to Review

9.1 What Are the Levels of Structure in Nucleic Acids?

1. **Thought Question** Consider the following in light of the concept of levels of structure (primary, secondary, tertiary, quaternary) as defined for proteins.
   (a) What level is shown by double-stranded DNA?
   (b) What level is shown by tRNA?
   (c) What level is shown by mRNA?

9.2 What Is the Covalent Structure of Polynucleotides?

2. **Fact Check** What is the structural difference between thymine and uracil?

3. **Fact Check** What is the structural difference between adenine and hypoxanthine?

4. **Fact Check** Give the name of the base, the ribonucleoside or deoxyribonucleoside, and the ribonucleoside triphosphate for A, G, C, T, and U.

5. **Fact Check** What is the difference between ATP and dATP?

6. **Fact Check** Give the sequence on the opposite strand for ACGTAT, AGATCT, and ATGGTA (all read 5′ → 3′).

7. **Fact Check** Are the sequences shown in Question 6 those of RNA or DNA? How can you tell?

8. **Thought Question** (a) Is it biologically advantageous that DNA is stable? Why or why not? (b) Is it biologically advantageous that RNA is unstable? Why or why not?

9. **Thought Question** A friend tells you that only four different kinds of bases are found in RNA. What would you say in reply?

10. **Thought Question** In the early days of molecular biology, some researchers speculated that RNA, but not DNA, might have a branched rather than linear covalent structure. Why might this speculation have come about?

11. **Thought Question** Why is RNA more vulnerable to alkaline hydrolysis than DNA?

9.3 What Is the Structure of DNA?

12. **Fact Check** In what naturally occurring nucleic acids would you expect to find A form helices, B form helices, Z form helices, nucleosomes, and circular DNA?

13. **Fact Check** Draw a G–C base pair. Draw an A–T base pair.

14. **Fact Check** Which of the following statements is (are) true?

    (a) Bacterial ribosomes consist of 40S and 60S subunits.

    (b) Prokaryotic DNA is normally complexed with histones.

    (c) Prokaryotic DNA normally exists as a closed circle.

    (d) Circular DNA is supercoiled.

15. **Biochemical Connections** Binding sites for the interaction of polypeptides and drugs with DNA are found in the major and minor grooves. True or false?

16. **Fact Check** How do the major and minor grooves in B-DNA compare to those in A-DNA?

17. **Fact Check** Which of the following statements is (are) true?

    (a) The two strands of DNA run parallel from their 5′ to their 3′ ends.

    (b) An adenine–thymine base pair contains three hydrogen bonds.

    (c) Positively charged counterions are associated with DNA.

    (d) DNA base pairs are always perpendicular to the helix axis.

18. **Fact Check** Define supercoiling, positive supercoil, topoisomerase, and negative supercoil.

19. **Fact Check** What is propeller-twist?

20. **Fact Check** What is an AG/CT step?

21. **Fact Check** Why does propeller-twist occur?

22. **Fact Check** What is the difference between B-DNA and Z-DNA?

23. **Fact Check** If circular B-DNA is positively supercoiled, will these supercoils be left- or right-handed?

24. **Fact Check** Briefly describe the structure of chromatin.

25. **Biochemical Connections** Draw the interactions between bases that make triple-helical DNA possible.

26. **Thought Question** List three mechanisms that relax the twisting stress in helical DNA molecules.

27. **Thought Question** Explain how DNA gyrase works.

28. **Thought Question** Explain, and draw a diagram to show, how acetylation or phosphorylation could change the binding affinity between DNA and histones.

29. **Thought Question** Would you expect to find adenine–guanine or cytosine–thymine base pairs in DNA? Why?

30. **Thought Question** One of the original structures proposed for DNA had all the phosphate groups positioned at the center of a long fiber. Give a reason why this proposal was rejected.

31. **Thought Question** What is the complete base composition of a double-stranded eukaryotic DNA that contains 22 percent guanine?

32. **Thought Question** Why was it necessary to specify that the DNA in Question 31 is double-stranded?

33. **Thought Question** What would be the most obvious characteristic of the base distribution of a single-stranded DNA molecule?

34. **Biochemical Connections** What is the purpose of the Human Genome Project? Why do researchers want to know the details of the human genome?

35. **Biochemical Connections** Explain the legal and ethical considerations involved in human gene therapy.

36. **Biochemical Connections** A recent commercial for a biomedical company talked about a future in which every individual would have a card that told his or her complete genotype. What would be some advantages and disadvantages of this?

37. **Thought Question** A technology called PCR is used for replicating large quantities of DNA in forensic science (Chapter 13). With this technique, DNA is separated by heating with an automated system. Why is information about the DNA sequence needed to use this technique?

9.4 How Does the Denaturation of DNA Take Place?

38. **Thought Question** Why does DNA with a high A–T content have a lower transition temperature, T_m, than DNA with a high G–C content?

9.5 What Are the Principal Kinds of RNA and Their Structures?

39. **Fact Check** Sketch a typical cloverleaf structure for transfer RNA. Point out any similarities between the cloverleaf pattern and the proposed structures of ribosomal RNA.

40. **Fact Check** What is the purpose of small nuclear RNA? What is a snRNP?

41. **Fact Check** Which type of RNA is the biggest? Which is the smallest?

42. **Fact Check** Which type of RNA has the least amount of secondary structure?

43. **Fact Check** Why does the absorbance increase when a DNA sample unwinds?

44. **Fact Check** What is RNA interference?

45. **Thought Question** Would you expect tRNA or mRNA to be more extensively hydrogen bonded? Why?

46. **Thought Question** The structures of tRNAs contain several unusual bases in addition to the typical four. Suggest a purpose for the unusual bases.

47. **Thought Question** Would you expect mRNA or rRNA to be degraded more quickly in the cell? Why?

48. **Thought Question** Which would be more harmful to a cell, a mutation in DNA or a transcription mistake that leads to an incorrect mRNA? Why?

49. **Thought Question** Explain briefly what happens to eukaryotic mRNA before it can be translated to protein.

50. **Thought Question** Explain why a 50S ribosomal subunit and a 30S ribosomal subunit combine to form a 70S subunit, instead of an 80S subunit.

Biochemistry ⊘ Now™

Assess your understanding of this chapter's topics with additional quizzing and tutorials at **http://now.brookscole.com/campbell5**

Annotated Bibliography

Most textbooks of organic chemistry have a chapter on nucleic acids.

Baltimore, D. Our Genome Unveiled. *Nature* **409,** 814–816 (2001). [A Nobel Prize winner's guide to the special issue describing human genome sequencing.]

Berg, P., and M. Singer. *Dealing with Genes: The Language of Heredity.* Mill Valley, CA: University Science Books, 1992. [Two leading biochemists have produced an eminently readable book on molecular genetics; highly recommended.]

Cech, T. R. The Ribosome Is a Ribozyme. *Science* **289** (5481), 878–879 (2000). [The title says it all.]

Claverie, J. M. What If There Are Only 30,000 Human Genes? *Science,* **252** (5507), 1255–1257 (2001). [Implications of the low gene number for human molecular biology.]

International Human Genome Sequencing Consortium (F. Collins et al.). Initial Sequencing and Analysis of the Human Genome. *Nature* **409,** 860–921 (2001). [One of two simultaneous publications of the sequence of the human genome.]

Couzin, J. Mini RNA Molecules Shield Mouse Liver from Hepatitis. *Science* **299,** 995 (2003). [An example of RNA interference.]

Couzin, J. Small RNAs Make Big Splash. *Science* **298,** 2296–2297 (2002). [A description of the small, recently discovered forms of RNA.]

Couzin, J. The Twists and Turns in BRCA's Path. *Science,* **302,** 591–593 (2003). [Genes involved in breast cancer have given researchers some big surprises and continue to do so.]

Gitlin, L., S. Karelsky, and R. Andino. Short interfering RNA confers intracellular antiviral immunity in human cells. *Nature,* **418,** 430–434 (2002). [An example of RNA interference.]

Jeffords, J. M., and T. Daschle. Political Issues in the Genome Era. *Science* **252** (5507), 1249–1251 (2001). [Comments on the Human Genome Project by two members of the U.S. Senate.]

Jenuwein, T., and C. D. Allis. Translating the Histone Code. *Science* **293,** 1074–1079 (2001). [An in-depth article about chromatin, histones, and methylation.]

Lake, J. A. Evolving Ribosome Structure: Domains in Archaebacteria, Eubacteria, Eocytes and Eukaryotes. *Ann. Rev. Biochem.* **54,** 507–530 (1985). [A review of the evolutionary implications of ribosome structure.]

Lake, J. A. The Ribosome. *Scientific American* **245** (2), 84–97 (1981). [A look at some of the complexities of ribosome structure.]

Lau, N. C., and D. P. Bartel. Censors of the Genome. *Scientific American* **289** (2), 34–41 (2003). [An article primarily about RNA interference.]

Levy-Lahad, E., and S. E. Plon. A Risky Business—Assessing Breast Cancer Risk. *Science* **302,** 574–575 (2003). [A discussion of risk factors and probabilities for *BRCA* gene carriers.]

McGuffin, P., B. Riley, and R. Plomin. Toward Behavioral Genomics. *Science* **291** (5507), 1232–1249 (2001). [Discussion of genetic basis for behavioral disorders.]

Moffat, A. Triplex DNA Finally Comes of Age. *Science* **252,** 1374–1375 (1991). [Triple helices as "molecular scissors."]

Paabo, S. The Human Genome and Our View of Ourselves. *Science* **252** (5507), 1219–1220 (2001). [A look at human DNA and its comparison with the DNA of other species.]

Peltonen, L., and V. A. McKusick. Dissecting Human Disease in the Postgenomic Era. *Science* **252** (5507), 1224–1229 (2001). [How diseases may be studied in the genomic era.]

Scovell, W. M. Supercoiled DNA. *J. Chem. Ed.* **63,** 562–565 (1986). [A discussion focused mainly on the topology of circular DNA.]

Venter, J. C., et al. The Sequence of the Human Genome. *Science* **291** (2001), 1304–1351. [One of two simultaneous publications of the sequence of the human genome.]

Watson, J. D., and F. H. C. Crick. Molecular Structure of Nucleic Acid. A Structure for Deoxyribose Nucleic Acid. *Nature* **171,** 737–738 (1953). [The original article describing the double helix. Of historical interest.]

Wolfsberg, T., J. McEntyre, and G. Schuler. Guide to the Draft Human Genome. *Nature* **409,** 824–826 (2001). [How to analyze the results of the Human Genome Project.]

CHAPTER 10 Biosynthesis of Nucleic Acids: Replication

CNRI/Photo Researchers, Inc.

Prokaryotic cells divide by pinching in two.

Critical Questions

10.1 What Is the Flow of Genetic Information in the Cell?

10.2 What Are the General Considerations in the Replication of DNA?

10.3 How Does the DNA Polymerase Reaction Take Place?

10.4 Which Proteins Are Required for DNA Replication?

10.5 How Do Proofreading and Repair Take Place?

10.6 How Is DNA Replicated in Eukaryotes?

Biochemistry Now™
Test yourself on these Critical Questions at the BiochemistryNow website at **http://now.brookscole.com/campbell5**

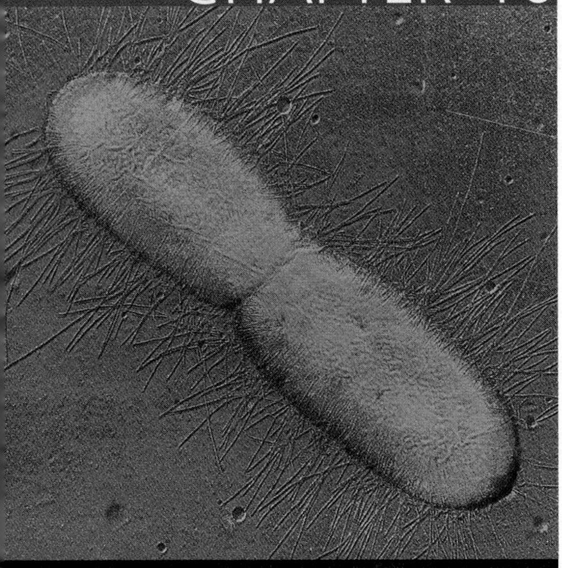

Before double-helical DNA can be replicated, helical sections of DNA must be unwound so that the two parental strands can serve as templates for the synthesis of new daughter strands, thus making a precise copy of the original double helix. DNA polymerases promote the synthesis of DNA by aligning nucleotides complementary to those on the exposed single-stranded DNA template and catalyzing their addition to a growing second strand. The fidelity of DNA synthesis is of utmost importance because errors of replication will be passed to future generations. The polymerases have "proofreading" powers capable of self-correction. In the use of genetic information, the sequence of DNA bases is transcribed into a complementary sequence of RNA bases called messenger RNA. The RNA message differs from DNA in one respect: the DNA base thymine (T) is replaced by the RNA base uracil (U). In eukaryotes, messenger RNA carries the genetic code from the nucleus to the ribosomes in the cytosol where the sequence of RNA bases is translated into the amino acid sequence of proteins. Numerous mRNA transcripts can be made from a single gene. This is a powerful way to amplify the production of protein molecules. Proteins, in turn, are the workhorses of the cell. They play a structural role as well as serving as antibodies and receptors on membranes. Above all, they are catalysts, a function they share with only a few kinds of RNA, and, in a rather circular mechanism, the proteins control the manipulation of the DNA that ultimately leads to their production.

10.1 What Is the Flow of Genetic Information in the Cell?

The sequence of bases in DNA encodes genetic information. The duplication of DNA, giving rise to a new DNA molecule with the same base sequence as the original, is necessary whenever a cell divides to produce daughter cells. This duplication process is called **replication.** The actual formation of gene products requires RNA; the production of RNA on a DNA template is called **transcription,** which will be studied in Chapter 11. The base sequence of DNA is reflected in the base sequence of RNA.

Three kinds of RNA are involved in the biosynthesis of proteins; of the three, messenger RNA (mRNA) is of particular importance. A sequence of three bases in mRNA specifies the identity of one amino acid in a manner directed by the genetic code. The process by which the base sequence directs the amino acid sequence is called **translation,** which will be studied in Chapter 12. In nearly all organisms, the flow of genetic information is DNA → RNA → protein. The only major exceptions are some viruses (called retroviruses) in which RNA, rather than DNA, is the genetic material. In those viruses, RNA can direct its own synthesis as well as that of DNA; the enzyme *reverse transcriptase* catalyzes this process. (Not all viruses in which RNA is the genetic material are retroviruses, but all retroviruses have a reverse transcrip-

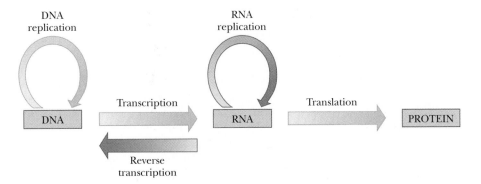

DNA replication

RNA replication

Transcription

DNA → RNA

Translation

RNA → PROTEIN

Reverse transcription

◀ **FIGURE 10.1** Mechanisms for transfer of information in the cell. The yellow arrows represent general cases, and the blue arrows represent special cases (mostly in RNA viruses).

tase. In fact, that is the origin of the term "retrovirus," referring to the reverse of the usual situation with transcription. See the article by Varmus listed in the bibliography at the end of this chapter.) In cases of infection by retroviruses, such as HIV, reverse transcriptase is a target for drug design. Figure 10.1 shows ways in which information is transferred in the cell. This scheme has been called the "Central Dogma" of molecular biology.

> **Essential Information**
>
> The base sequence of DNA contains the genetic code. It undergoes the process of *replication* when a cell divides and a new copy of DNA is produced. The base sequence of DNA determines the base sequence of RNA in the process of *transcription*. The base sequence of messenger RNA, in turn, determines the amino acid sequence of proteins in the *translation* of the genetic message.

10.2 | What Are the General Considerations in the Replication of DNA?

Naturally occurring DNA exists in many forms. Single- and double-stranded DNAs are known, and both can exist in linear and circular forms. As a result, it is difficult to generalize about all possible cases of DNA replication. Since many DNAs are double-stranded, we can present some general features of the replication of double-stranded DNA, features that apply both to linear and to circular DNA. Most of the details of the process that we shall discuss here were first investigated in prokaryotes, particularly in the bacterium *Escherichia coli*. We shall use information obtained by experiments on this organism for most of our discussion of the topic. Section 10.6 will discuss differences between prokaryotic and eukaryotic replication.

The process by which one double-helical DNA molecule is duplicated to produce two such double-stranded molecules is complex. The very complexity allows for a high degree of fine-tuning, which, in turn, ensures considerable fidelity in replication. The cell faces three important challenges in carrying out the necessary steps. The first challenge is *how to separate the two DNA strands*. The two strands of DNA are wound around each other in such a way that they must be unwound if they are to be separated. In addition to achieving continuous unwinding of the double helix, the cell also must protect the unwound portions of DNA from the action of **nucleases** that preferentially attack single-stranded DNA. The second task involves *the synthesis of DNA from the 5′ to the 3′ end*. Two antiparallel strands must be synthesized in the same direction on antiparallel templates. In other words, the template has one 5′ → 3′ strand and one 3′ → 5′ strand, as does the newly synthesized DNA. The third task is *how to guard against errors in replication*, ensuring that the correct base is added to the growing polynucleotide chain. Finding the answers to these challenges requires an understanding of the material in this section and the three following sections.

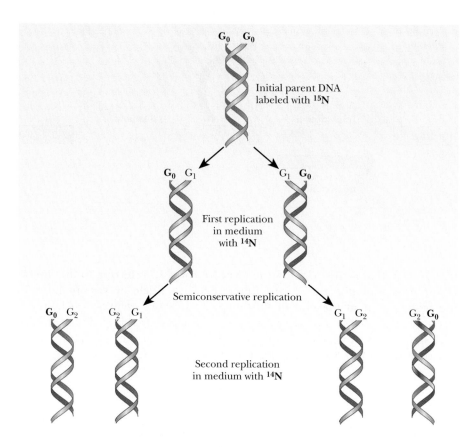

▶ **FIGURE 10.2** The labeling pattern of ^{15}N strands in semiconservative replication. (G_0 indicates original strands; G_1 indicates new strands after first generation; G_2 indicates new strands after second generation.)

Semiconservative Replication

DNA replication involves separation of the two original strands and production of two new strands with the original strands as templates. Each new DNA molecule contains one strand from the original DNA and one newly synthesized strand. This situation is what is called **semiconservative replication** (Figure 10.2). The details of the process differ in prokaryotes and eukaryotes, but the semiconservative nature of replication is observed in all organisms.

Semiconservative replication of DNA was established unequivocally in the late 1950s by experiments performed by Matthew Meselson and Franklin Stahl. *E. coli* bacteria were grown with $^{15}NH_4Cl$ as the sole nitrogen source, ^{15}N being a heavy isotope of nitrogen. (The usual isotope of nitrogen is ^{14}N.) In such a medium, all newly formed nitrogen compounds, including purine and pyrimidine nucleobases, become labeled with ^{15}N. The ^{15}N-labeled DNA has a higher density than unlabeled DNA, which contains the usual isotope, ^{14}N. In this experiment, the ^{15}N-labeled cells were then transferred to a medium that contained only ^{14}N. The cells continued to grow in the new medium. With every new generation of growth, a sample of DNA was extracted and analyzed by the technique of **density-gradient centrifugation** (Figure 10.3). This technique depends on the fact that heavy ^{15}N DNA (DNA that contains ^{15}N alone) will form a band at the bottom of the tube; light ^{14}N DNA (containing ^{14}N alone) will appear at the top of the tube. DNA containing a 50–50 mixture of ^{14}N and ^{15}N will appear at a position halfway between the two bands. In the actual experiment, this 50–50 hybrid DNA was observed after one generation, a result to be expected with semiconservative replication. After two generations in the lighter medium, half of the DNA in the cells should be the 50–50 hybrid and half should be the lighter ^{14}N DNA. This prediction of the kind and amount of DNA that should be observed was confirmed by the experiment.

Essential Information

In the process of DNA replication, a new strand is formed on a template strand (semiconservative replication). Synthesis of new DNA takes place in both directions from an origin of replication.

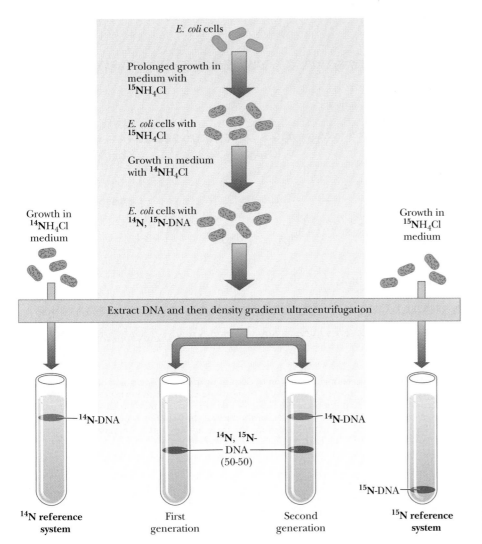

◀ **FIGURE 10.3** The experimental evidence for semiconservative replication. Heavy DNA labeled with ^{15}N forms a band at the bottom of the tube, and light DNA with ^{14}N forms a band at the top. DNA that forms a band at an intermediate position has one heavy strand and one light strand.

Bidirectional Replication

During replication, the DNA double helix unwinds at a specific point called the **origin of replication** (OriC in *E. coli*). New polynucleotide chains are synthesized using each of the exposed strands as a template. Two possibilities exist for the growth of the new strands: synthesis can take place in both directions from the origin of replication, or in one direction only. It has been established that DNA synthesis is bidirectional in most organisms, with the exception of a few viruses and plasmids. (Plasmids are rings of DNA that are found in bacteria and that replicate independently from the regular bacterial genome. They are discussed in Section 13.3). For each origin of replication, there are two points **(replication forks)** at which new polynucleotide chains are formed. A "bubble" (also called an "eye") of newly synthesized DNA between regions of the original DNA is a manifestation of the advance of the two replication forks in opposite directions. This feature is also called a θ structure because of its resemblance to the lowercase Greek letter theta. There is one such bubble (and one origin of replication) in the circular DNA of prokaryotes (Figure 10.4a). In eukaryotes, several origins of replication, and thus several bubbles, exist (Figure 10.4b). The bubbles grow larger and eventually merge, giving rise to two complete daughter DNAs. This bidirectional growth of both new polynucleotide chains represents *net chain growth*. Both new polynucleotide chains are synthesized in the 5'-to-3' direction.

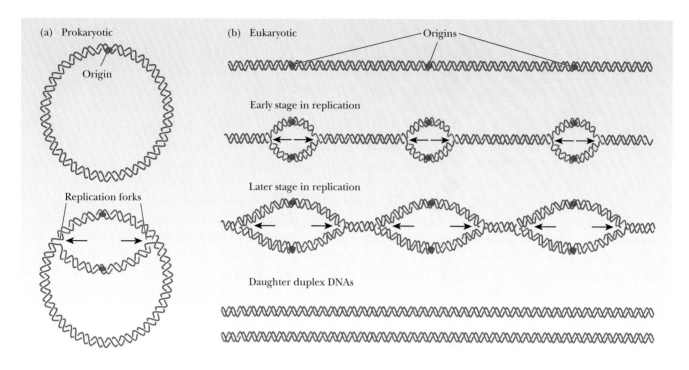

▲ **FIGURE 10.4** Bidirectional replication of DNA in prokaryotes (one origin of replication) and in eukaryotes (several origins). Bidirectional replication refers to overall synthesis (compare this with Figure 10.5). (a) Replication of the chromosome of *E. coli*, a typical prokaryote. There is one origin of replication, and there are two replication forks. (b) Replication of a eukaryotic chromosome. There are several origins of replication, and there are two replication forks for each origin. The "bubbles" that arise from each origin eventually coalesce.

| 10.3 | **How Does the DNA Polymerase Reaction Take Place?** |

One Strand of DNA Is Synthesized Semidiscontinuously

A major challenge for the cell in DNA replication is how to achieve $5' \rightarrow 3'$ polymerization in the opposite direction from the template strand, which is itself exposed in the 5'-to-3' direction. (There is no problem with the other strand, which is exposed by unwinding from the 3' end to the 5' end.)

The problem is solved by different modes of polymerization for the two growing strands. One newly formed strand (the **leading strand**) is formed continuously from its 5' end to its 3' end at the replication fork on the exposed 3'-to-5' template strand. The other strand (the **lagging strand**) is formed semidiscontinuously in small fragments (typically 1000 to 2000 nucleotides long), sometimes called **Okazaki fragments,** after the scientist who first studied them (Figure 10.5). The 5' end of each of these fragments is closer to the replication fork than the 3' end. The fragments of the lagging strand are then linked together by an enzyme called **DNA ligase.**

DNA Polymerase from *E. coli*

The first DNA polymerase discovered was found in *E. coli*. A universal feature of DNA replication is that the **nascent chain** (the new one being synthesized) grows from the 5' to the 3' end; there is a 5'-phosphate on the sugar at one end and a free 3'-hydroxyl on the sugar at the other end. **DNA polymerase** catalyzes the successive addition of each new nucleotide to the growing chain. The 3'-hydroxyl group at the end of the growing chain is a nucleophile. It

Essential Information

The two growing strands of DNA follow two different modes of polymerization in the replication process. The leading strand is formed continuously from the 5' to the 3' end. The lagging strand is formed from the 5' to the 3' end in small fragments that are then linked together.

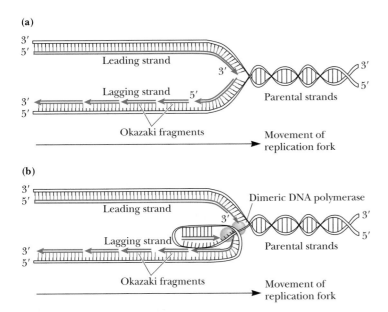

Biochemistry☾Now™ ◄ **ANIMATED FIGURE 10.5** The semidiscontinuous model for DNA replication. Newly synthesized DNA is shown in red. Because DNA polymerases only polymerize nucleotides $5' \rightarrow 3'$, both strands must be synthesized in the $5' \rightarrow 3'$ direction. Thus, the copy of the parental $3' \rightarrow 5'$ strand is synthesized continuously; this newly made strand is designated the leading strand. (a) As the helix unwinds, the other parental strand (the $5' \rightarrow 3'$ strand) is copied in a discontinuous fashion through synthesis of a series of fragments 1000 to 2000 nucleotides in length, called the Okazaki fragments; the strand constructed from the Okazaki fragments is called the lagging strand. (b) Because both strands are synthesized in concert by a dimeric DNA polymerase situated at the replication fork, the $5' \rightarrow 3'$ parental strand must wrap around in trombone fashion so that the unit of the dimeric DNA polymerase replicating it can move along it in the $3' \rightarrow 5'$ direction. This parental strand is copied in a discontinuous fashion because the DNA polymerase must occasionally dissociate from this strand and rejoin it further along. The Okazaki fragments are then covalently joined by DNA ligase to form an uninterrupted DNA strand. **See this figure animated at http://now.brookscole.com/campbell5**

attacks the phosphorus adjacent to the sugar in the nucleotide to be added to the growing chain, leading to the elimination of the pyrophosphate and the formation of a new phosphodiester bond (Figure 10.6). We discussed nucleophilic attack by a hydroxyl group at length in the case of serine proteases (Section 7.5); here we see another instance of this kind of mechanism. It is helpful to always keep this mechanism in mind. The further in depth we study DNA, the more the directionality of $5' \rightarrow 3'$ can lead to confusion over which strand of DNA we are discussing. If you always remember that all synthesis of nucleotides occurs in the $5' \rightarrow 3'$ direction from the perspective of the growing chain, it will be much easier to understand the processes to come.

Practice Session

A nucleoside derivative that has been very much in the news is 3′-azido-3′-deoxythymidine (AZT). This compound has been widely used in the treatment of AIDS (acquired immune deficiency syndrome), as has 2′-3′-dideoxyinosine (DDI).

Propose a reason for the effectiveness of these two compounds. *Hint:* How might these two compounds fit into a DNA chain?

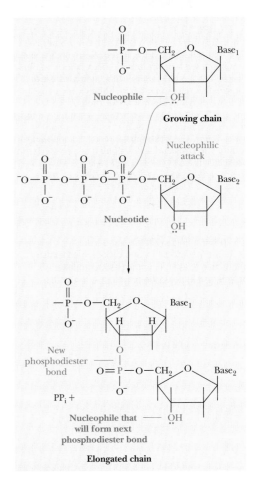

▲ **FIGURE 10.6** The addition of a nucleotide to a growing DNA chain. The 3′-hydroxyl group at the end of the growing DNA chain is a nucleophile. It attacks at the phosphorus adjacent to the sugar in the nucleotide, which will be added to the growing chain. Pyrophosphate is eliminated, and a new phosphodiester bond is formed.

Solution

Both compounds lack a hydroxyl group at the 3′-position of the sugar moiety. They cannot form the phosphodiester linkages found in nucleic acids. Thus, they interfere with the replication of the AIDS virus by preventing nucleic acid synthesis.

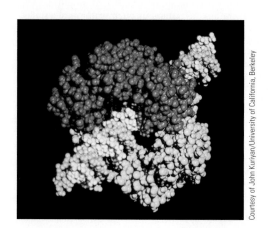

Courtesy of John Kuriyan/University of California, Berkeley

▲ **FIGURE 10.7** The dimer of β-subunits of DNA polymerase III bound to DNA. One monomer is shown in yellow, the other in red. Note that the dimer forms a closed loop around the DNA (shown in blue). The rest of the polymerase III holoenzyme is not shown. The remainder of the holoenzyme consists of the core enzyme responsible for the polymerization and the 3′ exonuclease activity (α-, ε-, and θ-subunits) and the γ-complex (γ-, δ-, δ′-, χ-, and ψ-subunits), which allows the β-subunits to form a clamp that surrounds the DNA and slides along it as polymerization proceeds. (*Adapted from Kong, X. P., et al. Three-Dimensional Structure of the β Subunit of E. Coli DNA Polymerase Holoenzyme: A Sliding DNA Clamp Cell* **69**, *425–437 (1992)*).

Biochemistry ⊘ Now™
Go to BiochemistryNow and click on Biochemistry Interactive to learn how the β-subunit dimer of polymerase III holds the polymerase to the DNA.

There are at least five DNA polymerases in *E. coli*. Three of them have been studied more extensively, and some of their properties are listed in Table 10.1. DNA polymerase I (Pol I) was discovered first, with the subsequent discovery of polymerases II (Pol II) and polymerase III (Pol III). Polymerase I consists of a single polypeptide chain, but polymerases II and III are multisubunit proteins that share some common subunits. Polymerase II is not required for replication; rather, it is strictly a repair enzyme. Recently, two more polymerases, Pol IV and Pol V, were discovered. They, too, are repair enzymes, and both are involved in a unique repair mechanism called the SOS response (see the Biochemical Connections box on page 255.) Two important considerations regarding the effect of any of the polymerases are the speed of the synthetic reaction (turnover number) and the **processivity,** which is the number of nucleotides joined before the enzyme dissociates from the template (Table 10.1).

Polymerase III consists of a core enzyme responsible for the polymerization and the 3′ exonuclease activity—consisting of α-, ε-, and θ-subunits—and a number of other subunits, including a dimer of α-subunits responsible for DNA binding, and the γ-complex—consisting of γ-, δ-, δ′, χ-, and ψ-subunits—which allows the β-subunits to form a clamp that surrounds the DNA and slides along it as polymerization proceeds (Figure 10.7). Table 10.2 gives the subunit composition of the DNA polymerase III complex. All these polymerases add nucleotides to a growing polynucleotide chain but have different roles in the overall replication process. As can be seen in Table 10.1, DNA polymerase III has the highest turnover number and a huge processivity compared to polymerases I and II.

If DNA polymerases are added to a single-stranded DNA template with all the deoxynucleotide triphosphates necessary to make a strand of DNA, no reaction will occur. It was discovered that DNA polymerases cannot catalyze de novo synthesis. All three enzymes require the presence of a **primer,** a short oligonucleotide strand to which the growing polynucleotide chain is cova-

Table 10.1			
Properties of DNA Polymerases of *E. coli*			
Property	**Pol I**	**Pol II**	**Pol III**
Mass (kDa)	103	90	830
Turnover number (min⁻¹)	600	30	1200
Processivity	200	1500	≥500,000
Number of subunits	1	≥4	≥10
Structural gene	*polA*	*polB**	*polC**
Polymerization 5′ → 3′	Yes	Yes	Yes
Exonuclease 5′ → 3′	Yes	No	No
Exonuclease 3′ → 5′	Yes	Yes	Yes

* Polymerization subunit only. These enzymes have multiple subunits, and some of them are shared between both enzymes.

Table 10.2

The Subunits of *E. coli* DNA Polymerase III Holoenzyme

Subunit	Mass (kDa)	Structural Gene	Function
α	130.5	*polC (dnaE)*	Polymerase
ε	27.5	*dnaQ*	3′-exonuclease
θ	8.6	*holE*	α, ε assembly?
τ	71	*dnaX*	Assembly of holoenzyme on DNA
β	41	*dnaN*	Sliding clamp, processivity
γ	47.5	*dnaX(Z)*	Part of the γ complex*
δ	39	*holA*	Part of the γ complex*
δ′	37	*holB*	Part of the γ complex*
χ	17	*holC*	Part of the γ complex*
ψ	15	*holD*	Part of the γ complex*

* Subunits γ-, δ-, δ′-, χ-, and ψ form the so-called γ complex, which is responsible for the placement of the β-subunits (the sliding clamp) on the DNA. The γ complex is referred to as the clamp loader. The δ and τ subunits are encoded by the same gene.

lently attached in the early stages of replication. In essence, DNA polymerases must have a nucleotide with a free 3′-hydroxyl already in place so that they can add the first nucleotide as part of the growing chain. In natural replication, this primer is RNA.

The DNA polymerase reaction requires all four deoxyribonucleoside triphosphates—dTTP, dATP, dGTP, and dCTP (Figure 10.8). Mg^{2+} and a DNA template are also necessary. Because of the requirement for an RNA primer, all four ribonucleoside triphosphates—ATP, UTP, GTP, and CTP—are needed as well; they are incorporated into the primer. The primer (RNA) is hydrogen-bonded to the template (DNA); the primer provides a stable framework on which the nascent chain can start to grow. The newly synthesized DNA strand begins to grow by forming a covalent linkage to the free 3′-hydroxyl group of the primer.

It is now known that DNA polymerase I has a specialized function in replication—repairing and "patching" DNA—and that DNA polymerase III is the enzyme primarily responsible for the polymerization of the newly formed DNA strand. The major function of DNA polymerases II, IV, and V is as repair enzymes. The exonuclease activities listed in Table 10.1 are part of the proofreading-and-repair functions of DNA polymerases, a process by which incorrect nucleotides are removed from the polynucleotide so that the correct nucleotides can be incorporated. The $3′ \rightarrow 5′$ exonuclease activity, which all three polymerases possess, is part of the **proofreading** function; incorrect nucleotides are removed in the course of replication and are replaced by the correct ones. Proofreading is done one nucleotide at a time. The $5′ \rightarrow 3′$ exonuclease activity clears away short stretches of nucleotides during **repair,** usually involving several nucleotides at a time. This is also how the RNA primers are removed. The proofreading-and-repair function is less effective in some DNA polymerases.

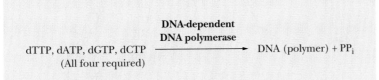

◀ **FIGURE 10.8** The requirements for the DNA polymerase reaction. Template DNA, Mg^{2+}, and an RNA primer are also required. Because of the need for an RNA primer, there is also an implicit requirement for all four ribonucleotide triphosphates (ATP, UTP, GTP, and CTP) for formation of the primer.

| 10.4 | **Which Proteins Are Required for DNA Replication?** |

Unwinding the Double Helix

Two questions arise in separating the two strands of the original DNA so that it can be replicated. The first is how to achieve continuous unwinding of the double helix. This question is complicated by the fact that prokaryotic DNA exists in a supercoiled, closed-circular form (see "Tertiary Structure of DNA: Supercoiling" in Section 9.3). The second related question is how to protect single-stranded stretches of DNA that are exposed to intracellular nucleases as a result of the unwinding.

An enzyme called **DNA gyrase** (class II topoisomerase) catalyzes the conversion of relaxed, circular DNA with a nick in one strand to the supercoiled form with the nick sealed (Figure 10.9). A slight unwinding of the helix before the nick is sealed introduces the supercoiling. The energy required for the process is supplied by the hydrolysis of ATP. Some evidence exists that DNA gyrase causes a double-strand break in DNA in the process of converting the relaxed, circular form to the supercoiled form. In replication, the role of the gyrase is somewhat different. The prokaryotic DNA is negatively supercoiled in its natural state; however, opening the helix during replication would introduce positive supercoils ahead of the replication fork. To see this phenomenon for yourself, try straightening out a section of a phone cord and watch what happens to the coils ahead. If the replication fork continued to move, the torsional strain of the positive supercoils would eventually make further replication impossible. DNA gyrase acts to fight these positive supercoils by putting negative supercoils ahead of the replication fork (Figure 10.10). A helix-destabilizing protein, called a **helicase,** promotes unwinding by binding at the replication fork. A number of helicases are known, including the *DnaB protein* and the *rep protein*. Another protein, called the **single-strand binding protein (SSB),** stabilizes the single-stranded regions by binding tightly to these portions of the molecule. The presence of this DNA-binding protein protects the single-stranded regions from hydrolysis by nucleases.

The Primase Reaction

One of the great surprises in studies of DNA replication was the discovery that *RNA serves as a primer in DNA replication*. In retrospect, it is not surprising at all, because RNA can be formed de novo without a primer, even though DNA synthesis requires a primer. This finding lends support to theories of the origin of life in which RNA, rather than DNA, was the original genetic mate-

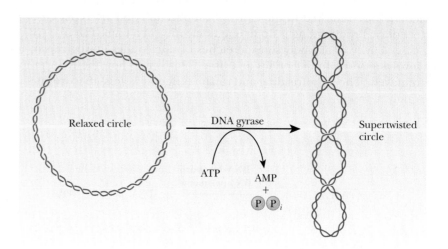

▶ **FIGURE 10.9** DNA gyrase introduces supertwisting in circular DNA.

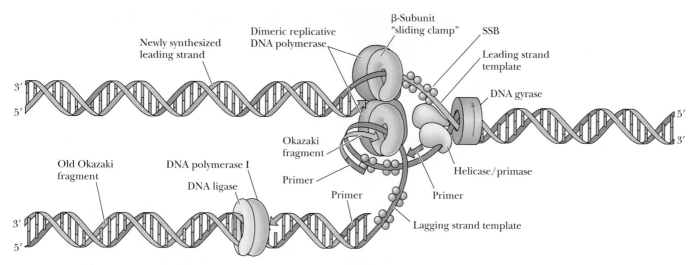

Biochemistry❀Now™ ▲ **ACTIVE FIGURE 10.10** General features of a replication fork. The DNA duplex is unwound by the action of DNA gyrase and helicase, and the single strands are coated with SSB (ssDNA-binding protein). Primase periodically primes synthesis on the lagging strand. Each half of the dimeric replicative polymerase is a holoenzyme bound to its template strand by a β-subunit sliding clamp. DNA polymerase I and DNA ligase act downstream on the lagging strand to remove RNA primers, replace them with DNA, and ligate the Okazaki fragments. **Watch this Active Figure at http://now.brookscole.com/campbell5**

rial. The fact that RNA has been shown to have catalytic ability in several cases has added support to that theory (Chapter 11). A primer in DNA replication must have a free 3′-hydroxyl to which the growing chain can attach, and both RNA and DNA can provide this group. The primer activity of RNA was first observed in vivo. In some of the original in vitro experiments, DNA was used as a primer because a primer consisting of DNA was expected. Living organisms are, of course, far more complex than isolated molecular systems and, as a result, can be full of surprises for researchers. It has subsequently been found that a separate enzyme, called **primase,** is responsible for copying a short stretch of the DNA template strand to produce the RNA primer sequence. The first primase was discovered in *E. coli.* The enzyme consists of a single polypeptide chain, with a molecular weight of about 60,000. There are 50 to 100 molecules of primase in a typical *E. coli* cell. The primer and the protein molecules at the replication fork constitute the **primosome.** The general features of DNA replication, including the use of an RNA primer, appear to be common to all prokaryotes (Figure 10.10).

Synthesis and Linking of New DNA Strands

The synthesis of two new strands of DNA is begun by DNA polymerase III. The newly formed DNA is linked to the 3′-hydroxyl of the RNA primer, and synthesis proceeds from the 5′ end to the 3′ end on both the leading and the lagging strands. Two molecules of Pol III, one for the leading strand and one for the lagging strand, are physically linked to the *primosome.* The resulting multiprotein complex is called the **replisome.** As the replication fork moves, the RNA primer is removed by polymerase I, using its exonuclease activity. The primer is replaced by deoxynucleotides, also by DNA polymerase I, using its polymerase activity. (The removal of the RNA primer and its replacement with the missing portions of the newly formed DNA strand by polymerase I are the repair function we mentioned earlier.) None of the DNA polymerases can seal the nicks that remain; DNA ligase is the enzyme responsible for the final linking of the new strand. Table 10.3 summarizes the main points of DNA replication in prokaryotes.

Table 10.3

A Summary of DNA Replication in Prokaryotes

1. DNA synthesis is bidirectional. Two replication forks advance in opposite directions from an origin of replication.

2. The direction of DNA synthesis is from the 5′ end to the 3′ end of the newly formed strand. One strand (the leading strand) is formed continuously, while the other strand (the lagging strand) is formed discontinuously. On the lagging strand, small fragments of DNA (Okazaki fragments) are subsequently linked.

3. Five DNA polymerases have been found in *E. coli*. Polymerase III is primarily responsible for the synthesis of new strands. The first polymerase enzyme discovered, polymerase I, is involved in synthesis, proofreading, and repair. Polymerases II, IV, and V function as repair enzymes under unique conditions.

4. DNA gyrase introduces a swivel point in advance of the movement of the replication fork. A helix-destabilizing protein, a helicase, binds at the replication fork and promotes unwinding. The exposed single-stranded regions of the template DNA are stabilized by a DNA-binding protein.

5. Primase catalyzes the synthesis of an RNA primer.

6. The synthesis of new strands is catalyzed by Pol III. The primer is removed by Pol I, which also replaces the primer with deoxynucleotides. DNA ligase seals the remaining nicks.

10.5 | How Do Proofreading and Repair Take Place?

DNA replication takes place only once each generation in each cell, unlike other processes, such as RNA and protein synthesis, which occur many times. It is essential that the fidelity of the replication process be as high as possible to prevent **mutations,** which are errors in replication. Mutations are frequently harmful, even lethal, to organisms. Nature has devised several ways to ensure that the base sequence of DNA is copied faithfully.

Errors in replication occur spontaneously only once in every 10^9 to 10^{10} base pairs. Proofreading refers to the removal of incorrect nucleotides immediately after they are added to the growing DNA during the replication process. DNA polymerase I has three active sites, as demonstrated by Hans Klenow. Pol I can be cleaved into two major fragments. One of them (the Klenow fragment) contains the polymerase activity and the proofreading activity. The other contains the 5′ → 3′ repair activity. Figure 10.11 shows the proofreading activity of Pol I. Errors in hydrogen bonding lead to the incorporation of an incorrect nucleotide into a growing DNA chain once in every 10^4 to 10^5 base pairs. DNA polymerase I uses its 3′ exonuclease activity to remove the incorrect nucleotide. Replication resumes when the correct

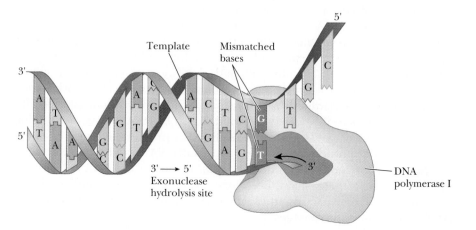

▶ **FIGURE 10.11** The 3′ → 5′ exonuclease activity of DNA polymerase I removes nucleotides from the 3′ end of the growing DNA chain.

Biochemical Connections

Why Does DNA Contain Thymine and Not Uracil?

Given that both uracil and thymine base-pair with adenine, why does RNA contain uracil and DNA contain thymine? Scientists now believe that RNA was the original hereditary molecule, and that DNA developed later. If we compare the structure of uracil and thymine, the only difference is the presence of a methyl group at C-5 of thymine. This group is not on the side of the molecule involved in base pairing. Because carbon sources and energy are required to methylate a molecule, there must be a reason for DNA developing with a base that does the same thing as uracil but that requires more energy to produce. The answer is that thymine helps to guarantee replication fidelity. One of the most common spontaneous mutations of bases is the natural deamination of cytosine.

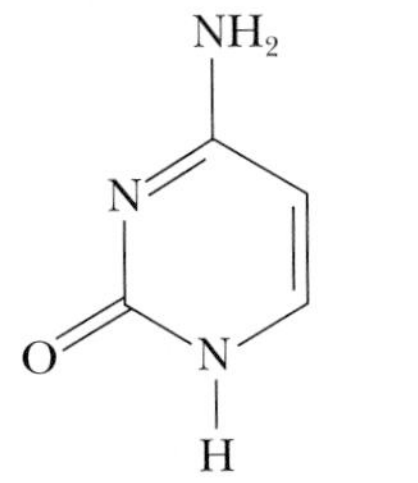

Cytosine
(2-oxy-4-amino
pyrimidine)

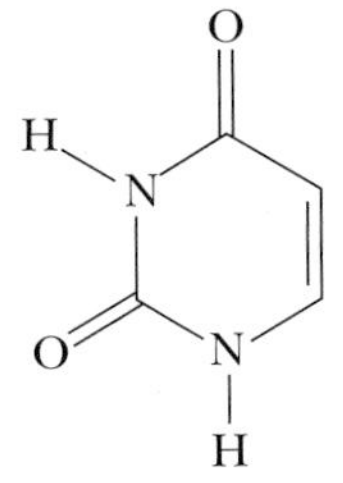

Uracil
(2-oxy-4-oxy
pyrimidine)

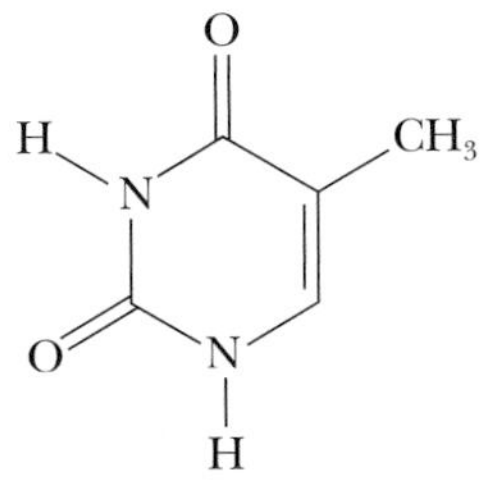

Cytosine

Uracil

Thymine
(2-oxy-4-oxy
5-methyl pyrimidine)

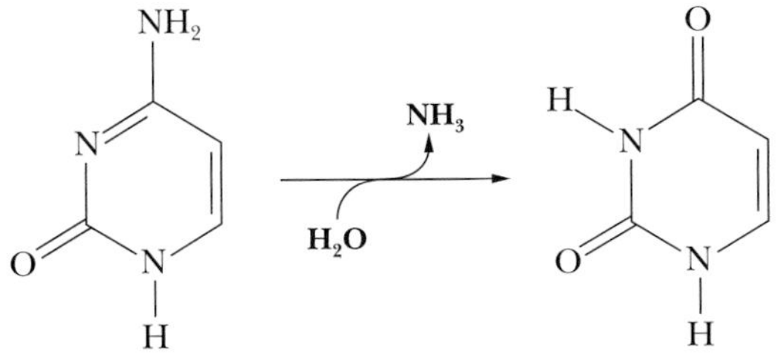

At any moment, a small but finite number of cytosines lose their amino groups to become uracil. Imagine that, during replication, a C–G base pair separates. If, at that moment, the C deaminates to U, it would have a tendency to base-pair to A instead of to G. If U were a natural base in DNA, the DNA polymerases would just line up an adenine across from the uracil, and there would be no way to know that the uracil was a mistake. This would lead to a much higher level of mutation during replication. Because uracil is an unnatural base in DNA, DNA polymerases can recognize it as a mistake and can replace it. Thus, the incorporation of thymine into DNA, while energetically more costly, helps to ensure that the DNA is replicated faithfully.

nucleotide is added, also by DNA polymerase I. Although the specificity of hydrogen-bonded base pairing accounts for one error in every 10^4 to 10^5 base pairs, the proofreading function of DNA polymerase improves the fidelity of replication to one error in every 10^9 to 10^{10} base pairs.

During replication, a *cut-and-patch process* catalyzed by polymerase I takes place. The cutting is the removal of the RNA primer by the 5′ exonuclease function of the polymerase, and the patching is the incorporation of the required deoxynucleotides by the polymerase function of the same enzyme. Note that this part of the process takes place after polymerase III has produced the new polynucleotide chain. Existing DNA can also be repaired by polymerase I, using the cut-and-patch method, if one or more bases have been damaged by an external agent, or if a mismatch was missed by the proofreading activity. DNA polymerase I is able to use its 5′ → 3′ exonuclease activity to remove RNA primers or DNA mistakes as it moves along the DNA. It then fills in behind it with its polymerase activity. This process is called **nick translation** (Figure 10.12). In addition to experiencing those spontaneous mutations caused by misreading the genetic code, organisms are frequently exposed to **mutagens,** agents that produce mutations. Common mutagens include ultraviolet light, ionizing radiation (radioactivity), and various chemical agents, all of which lead to changes in DNA over and above those produced by spontaneous

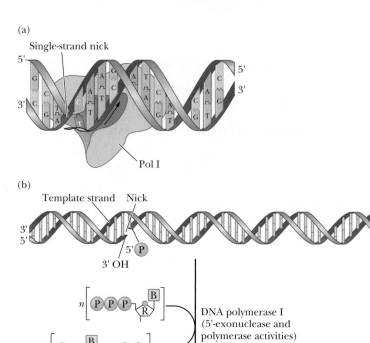

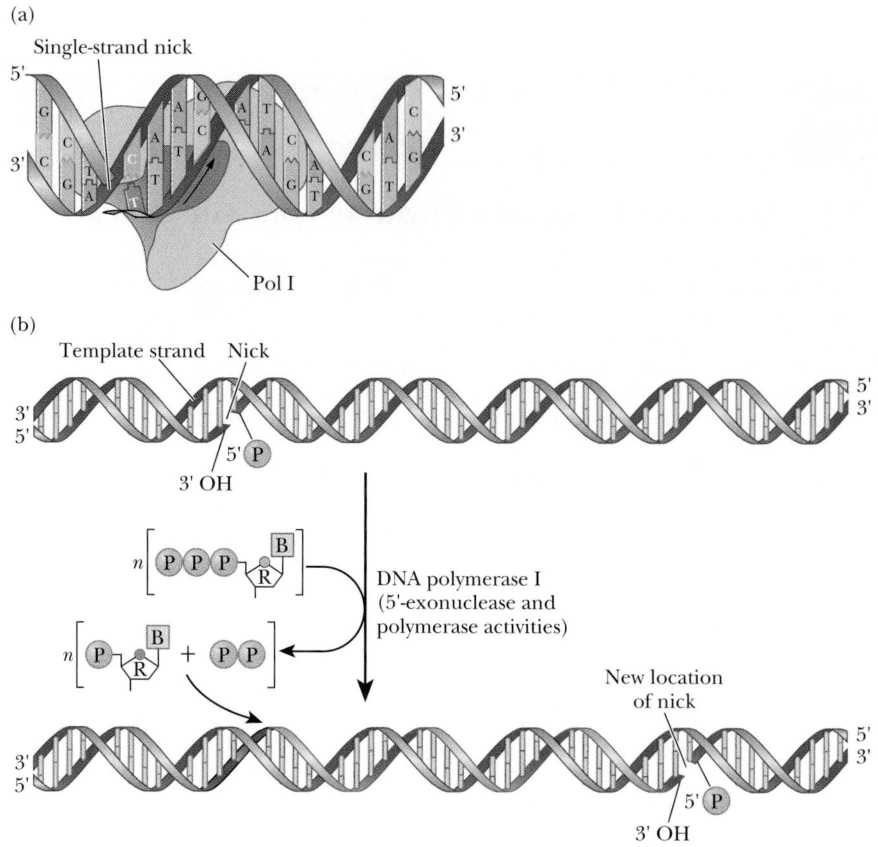

▶ **FIGURE 10.12** (a) The 5′ → 3′ exonuclease activity of DNA polymerase I can remove up to 10 nucleotides in the 5′ direction downstream from a 3′-OH single-strand nick. (b) If the 5′ → 3′ polymerase activity fills in the gap, the net effect is nick translation by DNA polymerase.

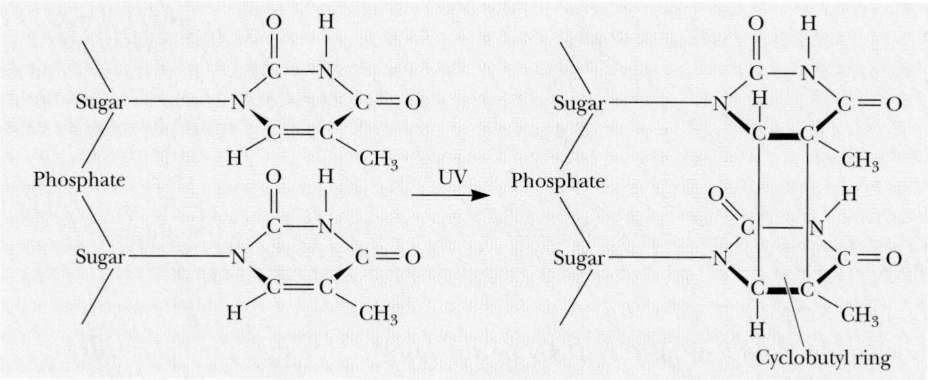

▶ **FIGURE 10.13** UV irradiation causes dimerization of adjacent thymine bases. A cyclobutyl ring is formed between carbons 5 and 6 of the pyrimidine rings. Normal base pairing is disrupted by the presence of such dimers.

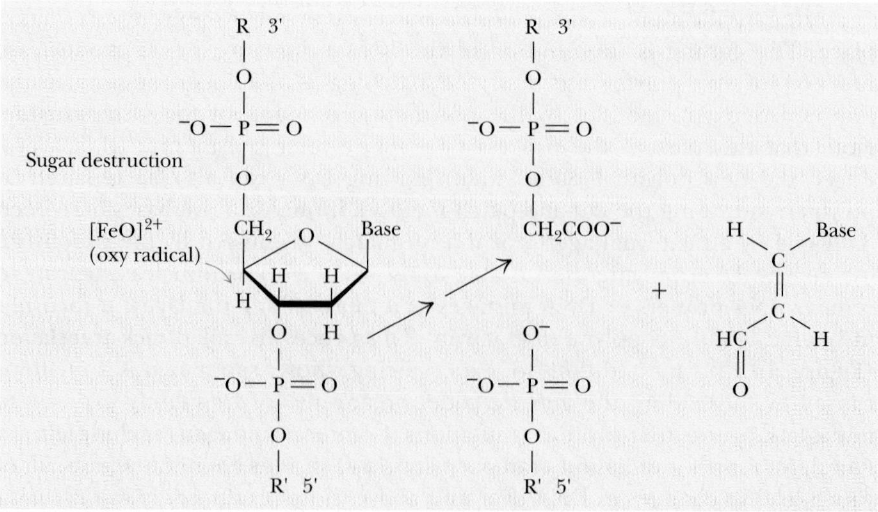

▶ **FIGURE 10.14** Oxygen radicals, in the presence of metal ions such as Fe²⁺, can destroy sugar rings in DNA, breaking the strand.

mutation. The most common effect of ultraviolet light is the creation of pyrimidine dimers (Figure 10.13). The π electrons from two carbons on each of two pyrimidines form a cyclobutyl ring, which distorts the normal shape of the DNA and interferes with replication and transcription. Chemical damage, which is often caused by free radicals (Figure 10.14), can lead to a break in the phosphodiester backbone of the DNA strand. This is one of the primary reasons that antioxidants are so popular as dietary supplements these days.

When damage has managed to escape the normal exonuclease activities of DNA polymerases I and III, prokaryotes have a variety of other repair mechanisms at their disposal. In **mismatch repair,** enzymes recognize that two bases are incorrectly paired. The area with the mismatch is removed, and DNA polymerases replicate the area again. If there is a mismatch, the challenge for the repair system is to know which of the two strands is the correct one. This is possible only because prokaryotes alter their DNA at certain locations (Chapter 13) by modifying bases with added methyl groups. This **methylation** occurs shortly after replication. Thus, immediately after replication, there is a window of opportunity for the mismatch-repair system. Figure 10.15 shows

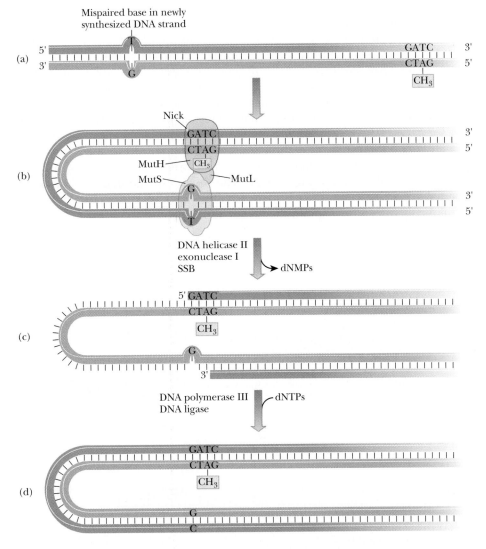

▲ **FIGURE 10.15** Mismatch repair in *E.coli.* (a) The newly synthesized DNA (shown in red) has a mismatch (G–T). (b) MutH, MutS, and MutL link the mismatch with the nearest methylation site, which identifies the blue strand as the parental (correct) strand. (c) An exonuclease removes DNA from the red strand between the proteins. (d) DNA polymerases replace the removed DNA with the correct sequence. (*Adapted from Lehninger,* Principles of Biochemistry, *Third Edition, by David L. Nelson and Michael M. Cox. © 1982, 1992, 2000 by Worth Publishers. Used with permission of W. H. Freeman and Company.*)

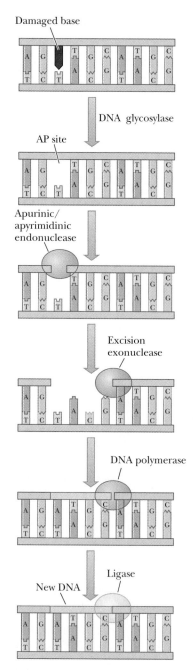

▲ **FIGURE 10.16** Base-excision repair. A damaged base (❚) is excised from the sugar–phosphate backbone by DNA glycosylase, creating an AP site. Then, an apurinic/apyrimidinic endonuclease severs the DNA strand, and an excision nuclease removes the AP site and several nucleotides. DNA polymerase I and DNA ligase then repair the gap.

▶ **FIGURE 10.17** Nucleotide-excision repair. When a serious lesion, such as a pyrimidine dimer, is detected, ABC exinuclease binds to the region and cuts out a large piece of DNA, including the lesion. DNA polymerase I and DNA ligase then resynthesize and seal the DNA. (*Adapted from Lehninger,* Principles of Biochemistry, *Third Edition, by David L. Nelson and Michael M. Cox. © 1982, 1992, 2000 by Worth Publishers. Used with permission of W. H. Freeman and Company.*)

how this works. Assume that a bacterial species methylates adenines that are part of a unique sequence. Originally, both parental strands are methylated. When the DNA is replicated, a mistake is made, and a T is placed opposite a G (Figure 10.15a). Because the parental strand contained methylated adenines, the enzymes can distinguish the parental strand from the newly synthesized daughter strand without the modified bases. Thus, the T is the mistake and not the G. Several proteins and enzymes are then involved in the repair process. *MutH, MutS,* and *MutL* form a loop between the mistake and a methylation site. DNA helicase II helps unwind the DNA. *Exonuclease I* removes the section of DNA containing the mistake (Figure 10.15b). Single-stranded binding proteins protect the template (blue) strand from degradation. DNA polymerase III then fills in the missing piece (Figure 10.15c).

Another repair system is called **base-excision repair** (Figure 10.16). A base that has been damaged by oxidation or chemical modification is removed by *DNA glycosylase,* leaving an *AP site,* so called because it is apurinic or apyrimidinic (without purine or pyrimidine). An *AP endonuclease* then removes the sugar and phosphate from the nucleotide. An *excision exonuclease* then removes several more bases. Finally, DNA polymerase I fills in the gap, and DNA ligase seals the phosphodiester backbone.

Nucleotide-excision repair is common for DNA lesions caused by ultraviolet or chemical means, which often lead to deformed DNA structures. Figure 10.17 demonstrates how a large section of DNA containing the lesion is removed by *ABC excinuclease.* DNA polymerase I and DNA ligase then work to fill in the gap. This type of repair is also the most common repair for ultraviolet damage in mammals. Defects in DNA repair mechanisms can have drastic consequences. One of the most remarkable examples is the disease *xeroderma pigmentosum.* Affected individuals develop numerous skin cancers at an early age because they do not have the repair system to correct damage caused by ultraviolet light. The endonuclease that nicks the damaged portion of the DNA is probably the missing enzyme. The repair enzyme that recognizes the lesion has been named XPA protein after the disease. The cancerous lesions eventually spread throughout the body, causing death.

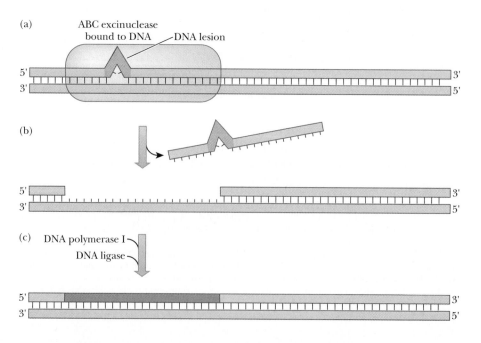

Biochemical Connections

The SOS Response in *E. coli*

When bacteria are subjected to extreme conditions and a great deal of DNA damage occurs, the normal repair mechanisms are not up to the task of repairing the damage. Prolonged exposure to ultraviolet light can do much damage to bacterial DNA. However, bacteria have one last card to play, which is called, appropriately, the SOS response. At least 15 proteins are activated as part of this response, including the mysterious DNA polymerase II. Another important protein is called *recA*. It gets its name from the fact that it is involved in a recombination event. Homologous DNA can recombine by a variety of mechanisms that we will not go into in this book. Suffice it to say that there are DNA sequences that can be used to cross one strand over another and replace it. Part (a) of the figure shows how this might work. If there were a lesion too complex for the normal repair enzymes to function, a gap would be left behind during replication because DNA polymerases could not synthesize new DNA over the lesion. However, the other replicating strand (shown in blue) should have the correct complement. RecA and many other proteins act to recombine this section of DNA to the lower strand. This would leave the upper strand without a piece of DNA, but it, too, has its correct complement (shown in red), so DNA polymerases can replicate it.

If the damaged strand has too many lesions, DNA polymerase II becomes involved in **error-prone repair.** In this case, the DNA polymerase continues to replicate over the damaged area, although it can't really match bases directly over the lesions. Thus, it inserts bases without a template, in essence "guessing." This goes against the idea of fidelity of replication, but it is better than nothing for the damaged cells. Many of the replication attempts produce mutations that are lethal, and many cells die. However, some may survive, which is better than the alternative.

▶ Recombination can be used to repair infrequent lesions. (a) The parental DNA (blue) has a lesion on the bottom strand. The newly synthesized top strand (light blue) has the correct sequence. Through recombination, the top blue strand can cross over and pair with the bottom blue strand that has the lesion. The top light blue strand can then be replicated to give the product shown in (b). In (c), the lesions are too numerous for this system to work. Instead, error-prone replication using DNA polymerase II patches over the lesion as best it can. Many mistakes are made in the process. (*Adapted from Lehninger,* Principles of Biochemistry, *Third Edition, by David L. Nelson and Michael M. Cox. © 1982, 1992, 2000 by Worth Publishers. Used with permission of W. H. Freeman and Company.*)

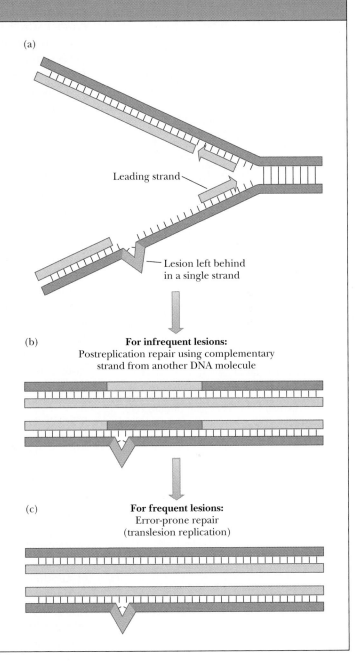

(a)

Leading strand

Lesion left behind in a single strand

(b) **For infrequent lesions:**
Postreplication repair using complementary strand from another DNA molecule

(c) **For frequent lesions:**
Error-prone repair (translesion replication)

10.6 | How Is DNA Replicated in Eukaryotes?

Our understanding of replication in eukaryotes is not as extensive as that in prokaryotes, owing to the higher level of complexity in eukaryotes and the consequent difficulty in studying the processes. Even though many of the principles are the same, eukaryotic replication is more complicated in three basic ways: there are multiple origins of replication, the timing must be controlled to that of cell divisions, and more proteins and enzymes are involved. (See the article by Gilbert in the bibliography at the end of this chapter.)

In a human cell, a few billion base pairs of DNA must be replicated once, and only once, per cell cycle. Cell growth and division are divided into

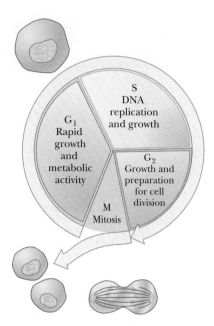

▲ **FIGURE 10.18** The eukaryotic cell cycle. The stages of mitosis and cell division define the M phase ("M" for *mitosis*). G_1 ("G" for *gap,* not *growth*) is typically the longest part of the cell cycle; G_1 is characterized by rapid growth and metabolic activity. Cells that are quiescent—that is, not growing and dividing (such as neurons)—are said to be in G_0. The S phase is the time of DNA synthesis. S is followed by G_2, a relatively short period of growth in which the cell prepares for division. Cell cycle times vary from less than 24 hours (rapidly dividing cells, such as the epithelial cells lining the mouth and gut) to hundreds of days.

phases—M, G_1, S, and G_2 (Figure 10.18). DNA replication takes place during a few hours in the S phase, and pathways exist to make sure that the DNA is replicated only once per cycle. Eukaryotic chromosomes accomplish this DNA synthesis by having replication begin at multiple origins of replication, also called **replicators.** These are specific DNA sequences that are usually between gene sequences. An average human chromosome may have several hundred replicators. The zones where replication is proceeding are called **replicons,** and the size of these varies with the species. In higher mammals, replicons may span 500 to 50,000 base pairs.

Cell-Cycle Control of Replication

The best understood model for control of eukaryotic replication is from yeast cells (Figure 10.19). Only chromosomes from cells that have reached the G_1 phase are competent to initiate DNA replication. Many proteins are involved in the control of replication and its link to the cell cycle. As usual, these proteins are usually given an abbreviation that makes them easier to say, but more difficult for the uninitiated to comprehend at first glance. The first proteins involved are seen during a window of opportunity that occurs between the early and late G_1 phase (see Figure 10.19 top). Replication is initiated by a multisubunit protein called the **origin recognition complex (ORC),** which binds to the origin of replication. This protein complex appears to be bound to the DNA throughout the cell cycle, but it serves as an attachment site for several proteins that help control replication. The next protein to bind is an activation factor called the **replication activator protein (RAP).** After the activator protein is bound, **replication licensing factors (RLFs)** can bind. In yeast, there are at least six different RLFs. They get their name from the fact that replication cannot proceed until they are bound. One of the keys to linking replication to cell division is that some of the RLF proteins have been found to be cytosolic. Thus, they have access to the chromosome only when the nuclear membrane dissolves during mitosis. Until they are bound, replication cannot occur. After RLFs bind, the DNA is then competent for replication. The combination of the DNA, the ORC, RAP, and RLFs constitutes what researchers call the **pre-replication complex (pre-RC).**

The next step involves other proteins and protein kinases. In Chapter 7, we learned that many processes are controlled by kinases phosphorylating target proteins. One of the great discoveries in this field was the existence of **cyclins,** which are proteins that are produced in one part of a cell cycle and degraded in another. Cyclins are able to combine with specific protein kinases, called **cyclin-dependent protein kinases** (CDKs). When these cyclins combine with CDKs, they are able to activate DNA replication and also to block reassembly of a pre-RC after initiation. The state of activity of the CDKs and the cyclins determines the window of opportunity for DNA synthesis. Cyclin–CDK complexes phosphorylate sites on RAP, the RLFs, and the ORC itself. Once phosphorylated, RAP dissociates from the pre-RC, as do the RLFs. Once phosphorylated and released, RAP and the RLFs are degraded (Figure 10.19, middle). Thus, the activation of cyclin–CDKs serves both to initiate DNA replication and to prevent formation of another pre-RC. In the G_2 phase, the DNA has been replicated. During mitosis, the DNA is separated into the daughter cells. At the same time, the dissolved nuclear membrane allows entrance of the licensing factors that are produced in the cytosol so that each daughter cell can initiate a new round of replication.

Eukaryotic DNA Polymerases

Five different DNA polymerases have been isolated from animal systems (Table 10.4). The use of animals rather than plants for study avoids the complication

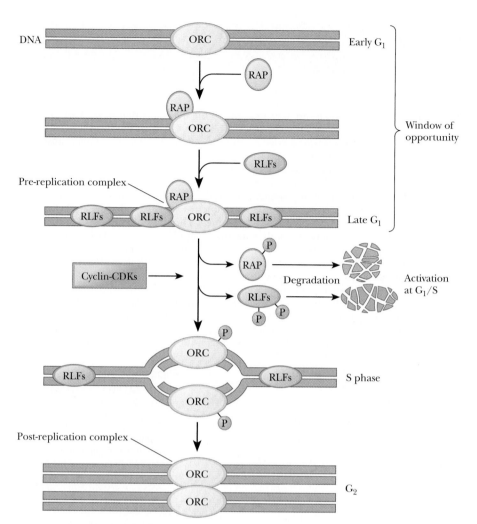

◀ **FIGURE 10.19** Model for initiation of the DNA replication cycle in eukaryotes. ORC is present at the replicators throughout the cell cycle. The pre-replication complex (pre-RC) is assembled through the sequential addition of RAP (replication activator protein) and RLFs (replication licensing factors) during a window of opportunity defined by the state of cyclin–CDKs. Phosphorylation of the RAP, ORC, and RLFs triggers replication. After initiation, a post-RC state is established, and the RAP and RLFs are degraded. (*Adapted from Figure 2 in Stillman, B., 1996. Cell Cycle Control of DNA Replication.* Science ***274:** 1659–1663. © 1996 AAAS. Used by permission.*)

Table 10.4

The Biochemical Properties of Eukaryotic DNA Polymerases

	α	δ	ε	β	γ
Mass (kDa)					
Native	>250	170	256	36–38	160–300
Catalytic core	165–180	125	215	36–38	125
Other subunits	70, 50, 60	48	55	None	35, 47
Location	Nucleus	Nucleus	Nucleus	Nucleus	Mitochondria
Associated functions					
$3' \rightarrow 5'$ exonuclease	No	Yes	Yes	No	Yes
Primase	Yes	No	No	No	No
Properties					
Processivity	Low	High	High	Low	High
Fidelity	High	High	High	Low	High
Replication	Yes	Yes	Yes	No	Yes
Repair	No	?	Yes	Yes	No

Source: Adapted from Kornberg. A., and Baker, T. A., 1992. *DNA Replication,* 2nd ed. New York: W. H. Freeman and Co.

Biochemical Connections

Telomerase and Cancer

Replication of linear DNA molecules poses particular problems at the ends of the molecules. Remember that, at the 5′ end of a strand of DNA being synthesized, there will initially be a short RNA primer, which must later be removed and replaced by DNA. This is never a problem with a circular template because the DNA polymerase I that is coming from the 5′ side of the primer (the previous Okazaki fragment) can then patch over the RNA with DNA. However, with a linear chromosome, this is not possible. At each end, there will be a 3′ and a 5′ DNA chain. The 5′-end template strand is not a problem because a DNA polymerase copying it will be moving from 5′ to 3′ and will be able to proceed to the end of the chromosome from the last RNA primer. The 3′ end template strand does pose a problem, however—see part (a) of the figure. The RNA primer at the 5′ end of the new strand (shown in green on the opposite page) will not have any way of being replaced. Remember that all DNA polymerases require a primer and, because there is nothing upstream (to the 5′ side),

there is no way to replace the RNA primer with DNA. RNA is unstable, and the RNA primer will be degraded in time. In effect, unless some special mechanism is created, the linear molecule gets shorter each time it is replicated.

The ends of eukaryotic chromosomes have a special structure called a telomere, which is a series of repeated DNA sequences. In human sperm-cell and egg-cell DNA, the sequence is 5′TTAGGG3′, and this sequence is repeated over 1000 times at the end of the chromosomes. This repetitive DNA is noncoding and acts as a buffer against degradation of the DNA sequence at the ends, which would occur with each replication as the RNA primers are degraded. There has been some evidence that a relationship exists between longevity and telomere length, and some researchers have suggested that the loss of the telomere DNA with age is part of the natural aging process. Eventually, the DNA would become nonviable and the cell would die.

However, even with long telomeres, cells will eventually die when their DNA gets shorter with each replication unless there is

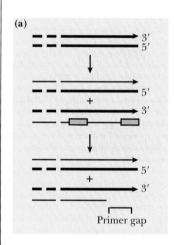

(a)

Primer gap

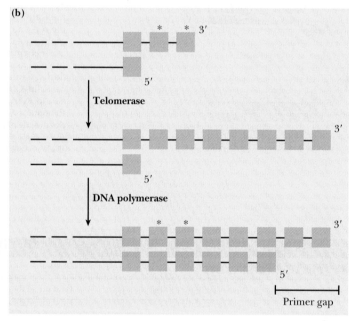

(b)

Telomerase

DNA polymerase

Primer gap

Biochemistry ⑤ Now™

◀ **ANIMATED FIGURE** Telomere replication. (a) In replication of the lagging strand, short RNA primers are added (pink) and extended by DNA polymerase. When the RNA primer at the 5′ end of each strand is removed, there is no nucleotide sequence to read in the next round of DNA replication. The result is a gap (primer gap) at the 5′ end of each strand (only one end of a chromosome is shown in this figure). (b) Asterisks indicate sequences at the 3′ end that cannot be copied by conventional DNA replication. Synthesis of telomeric DNA by telomerase extends the 5′ ends of DNA strands, allowing the strands to be copied by normal DNA replication. **See this figure animated at http://now.brookscole .com/campbell5**

(Continued)

of any DNA synthesis in chloroplasts. The various polymerases are called α, β, γ, δ, and ε. The α, β, δ, and ε enzymes are found in the nucleus, and the γ form occurs in mitochondria. Polymerase α was the first discovered, and it has the most subunits. It also has the ability to make primers, but it lacks a 3′ → 5′ proofreading activity and has low processivity. Thus, Pol α is not the main DNA synthesizer. It has been found to be active primarily in lagging-strand replication, for which it makes short RNA and/or DNA primers. Polysome compensatory mechanism. The creative solution is an enzyme called telomerase, which provides a mechanism for synthesis of the telomeres—see part (b) of the figure. The enzyme telomerase is a ribonuclear protein, containing a section of RNA that is the complement of the telomere. In humans, this sequence is 5′CCCUAA3′. Telomerase binds to the 5′ strand at the chromosome end and uses a **reverse transcriptase** activity to synthesize DNA (shown in red) on the 3′ strand, using its own RNA as the template. This allows the template strand (shown in purple) to be elongated, effectively lengthening the telomere. When the nature of telomerase was discovered, it was originally believed that it was a "fountain of youth" and that, if we could figure out how to keep it going, cells (and perhaps individuals) would never die. Very recent work has shown that, even though the enzyme telomerase *does* remain active in rapidly growing tissues such as blood cells, the intestinal lumen, skin, and others, it is *not* active in most adult tissues. When the cells of most adult tissues divide, for replacement or for repair, they do not preserve the chromosome ends. Eventually, enough DNA is gone, a vital gene is lost, and the cell dies. This may be a part of the normal aging and death process.

The big surprise was the discovery that telomerase is reactivated in cancer cells, explaining, in part, their immortality and their ability to keep dividing rapidly. This observation has opened a new possibility for cancer therapy: if we can prevent the reactivation of telomerase in cancerous tissues, the cancer might die of natural causes. The study of telomerase is just the tip of the iceberg. Other mechanisms must exist to protect the integrity of chromosomes besides telomerase. Using techniques described in Chapter 13, mice have been genetically engineered to lack telomerase. These mice did show continued shortening of their telomeres with successive replication and generations, but, eventually, the chromo-

some shortening did stop, indicating that some other process was also able to conserve the length of the chromosomes. Currently, the relationship between telomeres, recombination, and DNA repair is being studied (see the articles by Wu and Kucherlapati listed in the bibliography at the end of this chapter).

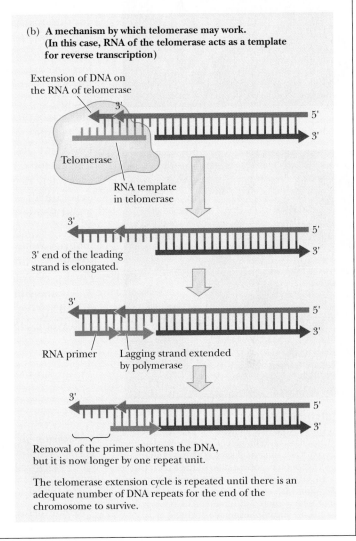

(b) **A mechanism by which telomerase may work. (In this case, RNA of the telomerase acts as a template for reverse transcription)**

Extension of DNA on the RNA of telomerase

Telomerase

RNA template in telomerase

3′ end of the leading strand is elongated.

RNA primer   Lagging strand extended by polymerase

Removal of the primer shortens the DNA, but it is now longer by one repeat unit.

The telomerase extension cycle is repeated until there is an adequate number of DNA repeats for the end of the chromosome to survive.

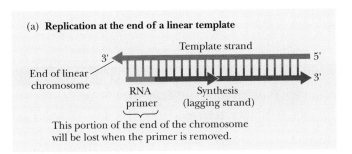

(a) **Replication at the end of a linear template**

Template strand

End of linear chromosome

RNA primer   Synthesis (lagging strand)

This portion of the end of the chromosome will be lost when the primer is removed.

merase δ is the principal DNA polymerase in eukaryotes. It interacts with a special protein called *PCNA* (for *proliferating cell nuclear antigen*). PCNA is the eukaryotic equivalent of the part of Pol III that functions as a sliding clamp (β). It is a trimer of three identical proteins that surround the DNA (Figure 10.20). DNA polymerase ε plays a role in replication, but its function is less clear. It may replace polymerase δ in certain situations, such as DNA repair, and it may function at the replication fork to remove primers on the lagging

(a)

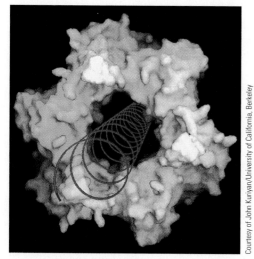

Courtesy of John Kuriyan/University of California, Berkeley

(b)

Courtesy of John Kuriyan/University of California, Berkeley

▲ **FIGURE 10.20** Structure of the PCNA homotrimer. Note that the trimeric PCNA ring of eukaryotes is remarkably similar to its prokaryotic counterpart, the dimeric β sliding clamp (Figure 10.7). (a) Ribbon representation of the PCNA trimer with an axial view of a B-form DNA duplex in its center. (b) Molecular surface of the PCNA trimer with each monomer colored differently. The red spiral represents the sugar–phosphate backbone of a strand of B-form DNA. (*Adapted from Figure 3 in Krishna, T. S., et al., 1994. Crystal Structure of the Eukaryotic DNA Polymerase Processivity Factor PCNA. Cell* **79**: *1233–1243.*)

Biochemistry ⓔNow™

Go to BiochemistryNow and click on Biochemistry Interactive to discover how PCNA is a eukaryotic analog of the prokaryotic β-subunit dimer sliding clamp.

strand. DNA polymerase β appears to be a repair enzyme. DNA polymerase γ carries out DNA replication in mitochondria. Several of the DNA polymerases isolated from animals lack exonuclease activity (the α and β enzymes). In this regard, the animal enzymes differ from prokaryotic DNA polymerases. Separate exonucleolytic enzymes exist in animal cells.

The Eukaryotic Replication Fork

The general features of DNA replication in eukaryotes are similar to those in prokaryotes. Table 10.5 summarizes the differences. As with prokaryotes, DNA replication in eukaryotes is semiconservative. There is a leading strand with continuous synthesis in the $5' \rightarrow 3'$ direction and a lagging strand with discontinuous synthesis in the $5' \rightarrow 3'$ direction. An RNA primer is formed by a specific enzyme in eukaryotic DNA replication, as is the case with prokaryotes, but, in this case, the primase activity is associated with Pol α. The structures involved at the eukaryotic replication fork are shown in Figure 10.21. The formation of Okazaki fragments (typically 150 to 200 nucleotides long in eukaryotes) is initiated by Pol α. After the RNA primer is made and a few nucleotides are added by Pol α, the polymerase dissociates and is replaced by Pol δ and its attached PCNA protein. Another protein, called *RFC* (replication factor C), is involved in attaching PCNA to Pol δ. The RNA primer is eventually degraded, but, in the case of eukaryotes, the polymerases do not have the $5' \rightarrow 3'$ exonuclease activity to do it. Instead, separate enzymes, FEN-1 and RNase H1, degrade the RNA. Continued movement of Pol δ fills in the gaps made by primer removal. As with prokaryotic replication, topoisomerases relieve the torsional strain from unwinding the helix, and a single-strand binding protein, called RPA, protects the DNA from degradation. Finally, DNA ligase seals the nicks that separate the fragments.

Another important difference between DNA replication in prokaryotes and in eukaryotes is that prokaryotic DNA is not complexed to histones, as is eukaryotic DNA. Histone biosynthesis occurs at the same time and at the

Table 10.5	
Differences in DNA Replication in Prokaryotes and Eukaryotes	
Prokaryotes	**Eukaryotes**
Five polymerases (I, II, III, IV, V)	Five polymerases (α, β, γ, δ, ε)
Functions of polymerase:	Functions of polymerases:
I is involved in synthesis, proofreading, repair, and removal of RNA primers	α: a polymerizing enzyme
II is also a repair enzyme	β: is a repair enzyme
III is main polymerizing enzyme	γ: mitochondrial DNA synthesis
IV, V are repair enzymes under unusual conditions	δ: main polymerizing enzyme ε: function unknown
Polymerases are also exonucleases	Not all polymerases are exonucleases
One origin of replication	Several origins of replication
Okazaki fragments 1000–2000 residues long	Okazaki fragments 150–200 residues long
No proteins complexed to DNA	Histones complexed to DNA

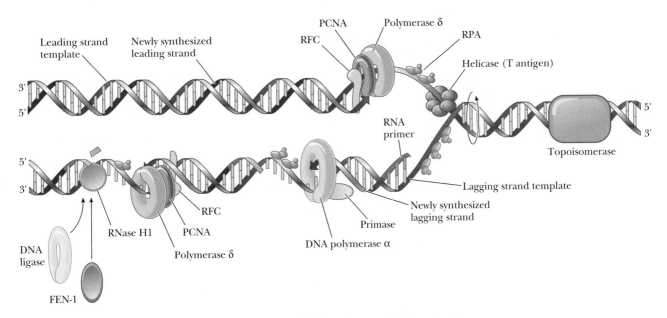

▲ **FIGURE 10.21** The basics of the eukaryotic replication fork. The primase activity is associated with DNA polymerase α. After a few nucleotides are incorporated, DNA polymerase δ, with its associated proteins called PCNA and RFC, bind and do the majority of the synthesis. The enzymes FEN-1 and RNase H1 degrade the RNA primers in eukaryotic replication. (*From* Cellular and Molecular Biology *by Karp, Figure 13-22. Used by permission of John Wiley & Sons, Inc.*)

same rate as DNA biosynthesis. In eukaryotic replication, histones are associated with DNA as it is formed.

An important aspect of DNA replication in eukaryotes, specifically affecting humans, is described in the Biochemical Connections box on pages 258 and 259.

Summary

10.1 What Is the Flow of Genetic Information in the Cell? In all organisms except RNA viruses, the flow of genetic information is DNA → RNA → protein. The duplication of DNA is called replication, and the production of RNA on a DNA template is called transcription. Translation is the process of protein synthesis, in which the sequence of amino acids is directed by the sequence of bases in the RNA transcript.

10.2 What Are the General Considerations in the Replication of DNA? Replication of DNA is semiconservative and bidirectional. Two replication forks advance in opposite directions from an origin of replication. Both new polynucleotide chains are synthesized in the 5′ to 3′ direction. One strand (the leading strand) is synthesized continuously, while the other (the lagging strand) is synthesized discontinuously in fragments that are subsequently linked together.

10.3 How Does the DNA Polymerase Reaction Take Place? Two DNA polymerases play important roles in replication in *E. coli*, a typical prokaryote. Polymerase III is primarily responsible for the synthesis of new strands. The first polymerase enzyme discovered, polymerase I, is mainly a repair enzyme.

10.4 Which Proteins Are Required for DNA Replication? DNA gyrase introduces a swivel point in advance of the movement of the replication fork. A helix-destabilizing protein, a heli-case, binds at the replication fork and promotes unwinding. The exposed single-stranded regions of the template DNA are protected from nuclease digestion by a DNA-binding protein. Primase catalyzes the synthesis of an RNA primer. The synthesis of new strands linked to the primer is catalyzed by Pol III. The primer is removed by Pol I, which also replaces the primer with deoxynucleotides. DNA ligase seals any remaining nicks.

10.5 How Do Proofreading and Repair Take Place? DNA replication takes place only once each generation in each cell. It is essential that the fidelity of the replication process be as high as possible to prevent mutations, which are errors in replication. Pol III does proofreading in the course of replication. In addition, Pol I carries out a cut-and-patch process, removing the RNA primer and replacing it with deoxyribonucleotides during replication. Pol I uses the same cut-and-patch process to repair existing DNA. Several other mechanisms exist to repair damaged DNA after replication is over, including mismatch repair, base-excision repair, and nucleotide-excision repair.

10.6 How Is DNA Replicated in Eukaryotes? Replication in eukaryotes follows the same general outline as replication in prokaryotes, with the most important difference being the presence of histone proteins complexed to eukaryotic DNA. Different proteins are used, and the system is more complex than it is in prokaryotes. Replication is controlled so that it occurs only once during a cell-division cycle, during the S phase.

Critical Questions to Review

10.1 What Is the Flow of Genetic Information in the Cell?

1. **Fact Check** Define replication, transcription, and translation.

2. **Thought Question** Is the following statement true or false? Why? "The flow of genetic information in the cell is always DNA → RNA → protein."

3. **Thought Question** Why is it more important for DNA to be replicated accurately than transcribed accurately?

10.2 What Are the General Considerations in the Replication of DNA?

4. **Fact Check** Why is the replication of DNA referred to as a semiconservative process? What is the experimental evidence for the semiconservative nature of the process? What experimental results would you expect if replication of DNA were a conservative process?

5. **Fact Check** What is a replication fork? Why is it important in replication?

6. **Fact Check** Describe the structural features of an origin of replication.

7. **Fact Check** Why is it necessary to unwind the DNA helix in the replication process?

8. **Thought Question** In the Meselson–Stahl experiment that established the semiconservative nature of DNA replication, the extraction method produced short fragments of DNA. What sort of results might have been obtained with longer pieces of DNA?

9. **Thought Question** Suggest a reason why it would be unlikely for replication to take place without unwinding the DNA helix.

10.3 How Does the DNA Polymerase Reaction Take Place?

10. **Fact Check** Do DNA-polymerase enzymes also function as exonucleases?

11. **Fact Check** Compare and contrast the properties of the enzymes DNA polymerase I and polymerase III from *E. coli*.

12. **Fact Check** Define processivity, and indicate the importance of this concept in DNA replication.

13. **Thought Question** Comment on the dual role of the monomeric reactants in replication.

14. **Thought Question** What is the importance of pyrophosphatase in the synthesis of nucleic acids?

15. **Thought Question** DNA synthesis always takes place from the 5′ to the 3′ end. The template strands have opposite directions. How does nature deal with this situation?

16. **Thought Question** What would happen to the replication process if the growing DNA chain did not have a free 3′ end?

17. **Thought Question** Suggest a reason for the rather large energy "overkill" in inserting a deoxyribonucleotide into a growing DNA molecule. (About 15 kcal mol^{-1} is used in forming a phosphate ester bond that actually requires only about a third as much energy.)

18. **Thought Question** Why is it not surprising that the addition of nucleotides to a growing DNA chain takes place by nucleophilic substitution?

19. **Thought Question** Is it unusual that the β-subunits of DNA polymerase III that form a sliding clamp along the DNA do not contain the active site for the polymerization reaction? Explain your answer.

10.4 Which Proteins Are Required for DNA Replication?

20. **Fact Check** List the substances required for replication of DNA catalyzed by DNA polymerase.

21. **Fact Check** Describe the discontinuous synthesis of the lagging strand in DNA replication.

22. **Fact Check** What are the functions of the gyrase, primase, and ligase enzymes in DNA replication?

23. **Fact Check** Single-stranded regions of DNA are attacked by nucleases in the cell, yet portions of DNA are in a single-stranded form during the replication process. Explain.

24. **Fact Check** Describe the role of DNA ligase in the replication process.

25. **Fact Check** What is the primer in DNA replication?

26. **Thought Question** How does the replication process take place on a supercoiled DNA molecule?

27. **Thought Question** Why is a short RNA primer needed for replication?

10.5 How Do Proofreading and Repair Take Place?

28. **Fact Check** How does proofreading take place in the process of DNA replication?

29. **Fact Check** Does proofreading always take place by the same process in replication?

30. **Fact Check** Describe the excision repair process in DNA, using the excision of thymine dimers as an example.

31. **Thought Question** Of what benefit is it for DNA to have thymine rather than uracil?

32. **Thought Question** Your book contains about 2 million characters (letters, spaces, and punctuation marks). If you could type with the accuracy with which the prokaryote *E. coli* incorporates, proofreads, and repairs bases in replication (about one uncorrected error in 10^9 to 10^{10} bases), how many such books would you have to type before an uncorrected error is "permitted"? (Assume that the error rate is one in 10^{10} bases.)

33. **Thought Question** *E. coli* incorporates deoxyribonucleotides into DNA at a rate of 250 to 1000 bases per second. Using the higher value, translate this into typing speed in words per minute. (Assume five characters per word, using the typing analogy from Question 32.)

34. **Thought Question** Given the typing speed from Question 33, how long must you type, nonstop, at the fidelity shown by *E. coli* (see Question 32) before an uncorrected error would be permitted?

35. **Thought Question** Can methylation of nucleotides play a role in DNA replication? If so, what sort of role?

36. **Thought Question** How can breakdown in DNA repair play a role in the development of human cancers?

37. **Biochemical Connections** Can prokaryotes deal with drastic DNA damage in ways that are not available to eukaryotes?

10.6 How Is DNA Replicated in Eukaryotes?

38. **Fact Check** Do eukaryotes have fewer origins of replication than prokaryotes, or more origins, or the same number?

39. **Fact Check** How does DNA replication in eukaryotes differ from the process in prokaryotes?

40. **Fact Check** What role do histones play in DNA replication?

41. Thought Question (a) Eukaryotic DNA replication is more complex than prokaryotic. Give one reason why this should be so. (b) Why might eukaryotic cells need more kinds of DNA polymerases than bacteria?

42. Thought Question How do the DNA polymerases of eukaryotes differ from those of prokaryotes?

43. Thought Question What is the relationship between control of DNA synthesis in eukaryotes and the stages of the cell cycle?

44. Biochemical Connections What would be the effect on DNA synthesis if the telomerase enzyme were inactivated?

45. Thought Question Would it be advantageous to a eukaryotic cell to have histone synthesis take place at a faster rate than DNA synthesis?

46. Thought Question What are replication licensing factors? How did they get their name?

47. Thought Question Is DNA synthesis likely to be faster in prokaryotes or in eukaryotes?

48. Thought Question Outline a series of steps by which reverse transcriptase produces DNA on an RNA template.

49. Biochemical Connections Name an important difference in the replication of circular DNA versus linear double-stranded DNA.

50. Thought Question Why is it reasonable that eukaryotes have a DNA polymerase (Pol γ) that operates only in mitochondria?

Biochemistry ⊗ Now™

Assess your understanding of this chapter's topics with additional quizzing and tutorials at **http://now.brookscole.com/campbell5**

Annotated Bibliography

Botchan, M. Coordinating DNA Replication with Cell Division: Current Status of the Licensing Concept. *Proc. Nat. Acad. Sci.* **93,** 9997–10,000 (1996). [An article about control of replication in eukaryotes.]

Buratowski, S. DNA Repair and Transcription: The Helicase Connection. *Science* **260,** 37–38 (1993). [How repair and transcription are coupled.]

Gilbert, D. M. Making Sense of Eukaryotic DNA Replication Origins. *Science* **294,** 96–100 (2001). [The latest information on replication origins in eukaryotes].

Kornberg, A., and T. Baker. *DNA Replication,* 2nd ed. New York: W. H. Freeman and Co., 1991. [Most aspects of DNA biosynthesis are covered. The first author received a Nobel Prize for his work in this field.]

Kucherlapati, R., and R. A. DePinho. Telomerase meets its mismatch. *Nature* **411,** 647–648 (2001). [An article about a possible relationship between telomerase and mismatch repair.]

Radman, M., and R. Wagner. The High Fidelity of DNA Duplication. *Sci. Amer.* **259** (1), 40–46 (1988). [A description of replication, concentrating on the mechanisms for minimizing errors.]

Stillman, B. Cell Cycle Control of DNA Replication. *Science* **274,** 1659–1663 (1996). [A description of how eukaryotic replication is controlled and linked to cell division.]

Varmus, H. Reverse Transcription. *Sci. Amer.* **257** (3), 56–64 (1987). [A description of RNA-directed DNA synthesis. The author was one of the recipients of the 1989 Nobel Prize in medicine for his work on the role of reverse transcription in cancer.]

Wu, L., and D. Hickson. DNA Ends RecQ-uire Attention. *Science* **292,** 229–230 (2001). [An article describing various ways that the ends of chromosomes are protected.]

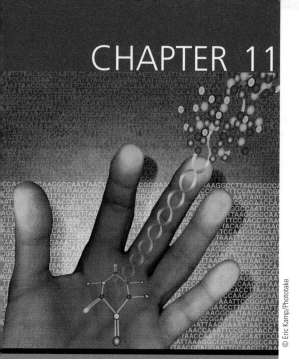

CHAPTER 11

In transcription, the template strand of DNA is used to produce a complementary strand of RNA.

Critical Questions

11.1 How Does Transcription Take Place in Prokaryotes?

11.2 How Is Transcription Regulated in Prokaryotes?

11.3 How Does Transcription Take Place in Eukaryotes?

11.4 How Is Transcription Regulated in Eukaryotes?

11.5 What Are Some Structural Motifs in DNA-Binding Proteins?

11.6 How Is RNA Modified after Transcription?

11.7 How Does RNA Act as an Enzyme?

Biochemistry ⊘ Now™

Test yourself on these Critical Questions at the BiochemistryNow Web site at **http://now.brookscole.com/campbell5**

Transcription of the Genetic Code: The Biosynthesis of RNA

In the use of genetic information, one of the strands of the double-stranded DNA molecule is transcribed into a complementary sequence of RNA. The RNA sequence differs from DNA in one respect: The DNA base thymine (T) is replaced by the RNA base uracil (U). Of all the DNA in a cell, only some is transcribed. Transcription produces all the types of RNA—mRNA, tRNA, rRNA, snRNA, miRNA, and siRNA. In prokaryotes, where there is no cell compartmentalization, messenger RNA can be, and frequently is, translated at one end while it is still being transcribed at the other end. In eukaryotes, messenger RNA carries the genetic code from the nucleus to the ribosomes in the cytosol where the sequence of RNA bases is translated into the amino acid sequence of proteins. The process is much more complicated in eukaryotes than in prokaryotes, involving a number of transcription factors. Copying the genetic message is a powerful way to amplify the production of protein molecules. Proteins, in turn, are the workhorses of the cell. They not only play a structural role but also serve as antibodies and receptors on membranes. Above all, they are catalysts, a function that they share with only a few kinds of RNA.

As we saw in Chapter 10, the central dogma of molecular biology is that DNA makes RNA, and RNA makes proteins. The process of making RNA from DNA is called **transcription,** and it is the major control point in the expression of genes and the production of proteins.

The details of RNA transcription differ somewhat in prokaryotes and eukaryotes. Most of the research on the subject has been done in prokaryotes, especially *E. coli,* but some general features are found in all organisms except in the case of cells infected by RNA viruses. Table 11.1 summarizes the main features of the process.

11.1	**How Does Transcription Take Place in Prokaryotes?**

RNA Polymerase in *Escherichia coli*

The most extensively studied RNA polymerase is that isolated from *E. coli*. The molecular weight of this enzyme is about 470,000, and it has a multisubunit structure. Five different types of subunits, designated α, ω, β, β', and σ, have been identified. The actual composition of the enzyme is $\alpha_2\omega\beta\beta'\sigma$. The σ-subunit is rather loosely bound to the rest of the enzyme (the $\alpha_2\omega\beta\beta'$ portion), which is called the **core enzyme.** The **holoenzyme** consists of all the subunits, including the σ-subunit. The σ-subunit is involved in the recognition of specific promoters, whereas the β-, β'-, α-, and ω-subunits combine to make the active site for polymerization.

Figure 11.1 shows the basics of information transfer from DNA to protein. Of the two strands of DNA, one of them is the template for RNA synthesis.

Table 11.1
General Features of RNA Synthesis

1. RNA is initially synthesized using a DNA template in the process called transcription; the enzyme that catalyzes the process is **DNA-dependent RNA polymerase.**
2. All four ribonucleoside triphosphates (ATP, GTP, CTP, and UTP) are required, as is Mg^{2+}.
3. A primer is not needed in RNA synthesis, but a DNA template is required.
4. As is the case with DNA biosynthesis, the RNA chain grows from the 5′ to the 3′ end. The nucleotide at the 5′ end of the chain retains its triphosphate group (abbreviated ppp).
5. The enzyme uses one strand of the DNA as the template for RNA synthesis. The base sequence of the DNA contains signals for initiation and termination of RNA synthesis. The enzyme binds to the template strand and moves along it in the 3′-to-5′ direction.
6. The template is unchanged.

RNA polymerase reads it from 3′ to 5′. This strand has several names. The most common is the **template strand,** because it is the strand that will direct the synthesis of the RNA. It is also called the **antisense strand,** because its code is the complement of the RNA that will be produced. It is sometimes called the **(−) strand** by convention. The other strand is called the **coding strand** because its sequence of DNA will be the same as the RNA sequence that is produced (with the exception of U replacing T). It is also called the **sense strand,** since the RNA sequence is the sequence that we use to determine what amino acids will be produced in the case of mRNA. It is also called the **(+) strand** by convention, or even the **nontemplate strand.** For our purposes, we will use the terms *template strand* and *coding strand* throughout. Because the DNA in the coding strand has the same sequence as the RNA that is produced, it is used when discussing the sequence of genes for proteins or for promoters and controlling elements on the DNA.

The core enzyme of RNA polymerase is catalytically active but lacks specificity. The core enzyme alone would transcribe both strands of DNA, when only one strand contains the information in the gene. The holoenzyme of RNA polymerase binds to specific DNA sequences and transcribes only the correct strand. The essential role of the σ-subunit is recognition of the **promoter locus** (a DNA sequence that signals the start of RNA transcription; see Section 11.2). The loosely bound σ-subunit is released after transcription

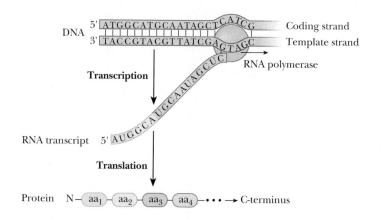

◄ FIGURE 11.1 The basics of transcription. RNA polymerase uses the template strand of DNA to make an RNA transcript that has the same sequence as the nontemplate DNA strand, with the exception that T is replaced by U. If this RNA is mRNA, it can later be translated to protein.

begins and about 10 nucleotides have been added to the RNA chain. Prokaryotes can have more than one type of σ-subunit. The nature of the σ-subunit can direct RNA polymerases to different promoters and cause the transcription of various genes to reflect different metabolic conditions.

Promoter Structure

Even the simplest organisms contain a great deal of DNA that is not transcribed. RNA polymerase must have a way of knowing which of the two strands is the template strand, which part of the strand is to be transcribed, and where the first nucleotide of the gene to be transcribed is located. Promoters are DNA sequences that provide this direction for RNA polymerase. The promoter region to which RNA polymerase binds is closer to the 3′ end of the template strand than is the actual gene for the RNA to be synthesized. The RNA is formed from the 5′ end to the 3′ end, so the polymerase moves along the template strand from the 3′ end to the 5′ end. However, by convention, all control sequences are given for the coding strand, which is 5′ to 3′. The binding site for the polymerase is said to lie *upstream* of the start of transcription, which is farther to the 5′ side of the coding strand.

Most bacterial promoters have at least three components. Figure 11.2 shows some typical promoter sequences for *E. coli* genes. The component closest to the first nucleotide to be incorporated is about 10 bases upstream. Also by convention, the first base to be incorporated into the RNA chain is said to be at position +1 and is called the **transcription start site (TSS).** All the nucleotides upstream from this start site are given negative numbers. Because the first promoter element is about 10 bases upstream, it is called the −10 region, but is also called the **Pribnow box** after its discoverer. After the Pribnow box, there are 16 to 18 bases that are completely variable. The next promoter element is about 35 bases upstream of the TSS and is simply called the **−35 region** or **−35 element.** An element is a general term for a DNA

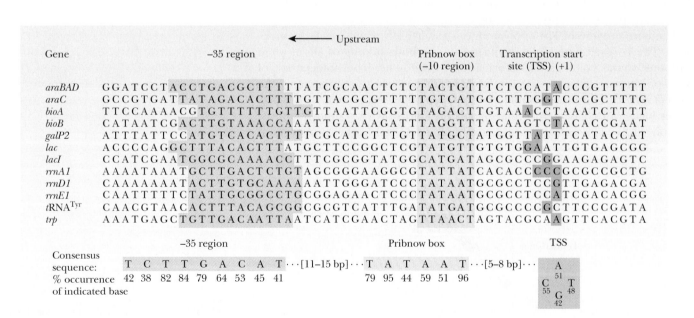

▲ **FIGURE 11.2** Sequences of representative promoters from *E. coli*. By convention, these are given as the sequence that would be found on the coding strand going from left to right as the 5′ to 3′ direction. The numbers below the consensus sequences indicate the percentage of the time that a certain position is occupied by the indicated nucleotide.

sequence that is somehow important in controlling transcription. The area from the −35 element to the TSS is called the **core promoter.** Upstream of the core promoter can be an **UP element,** which enhances the binding of RNA polymerase. UP elements usually extend from −40 to −60. The region from the end of the UP element to the transcription start site is known as the **extended promoter.**

The base sequence of promoter regions has been determined for a number of prokaryotic genes, and a striking feature is that they contain many bases in common. These are called **consensus sequences.** Promoter regions are A–T rich, with two hydrogen bonds per base pair; consequently, they are more easily unwound than G–C-rich regions, which have three hydrogen bonds per base pair. Figure 11.2 shows the consensus sequences for the −10 and −35 regions.

Even though the −10 and −35 regions of many genes are similar, there are also some significant variations that are important to the metabolism of the organism. Besides directing the RNA polymerase to the correct gene, the promoter base sequence controls the frequency with which the gene is transcribed. Some promoters are strong, and others are weak. A strong promoter will bind RNA polymerase tightly, and the gene will therefore be transcribed more often. In general, as a promoter sequence varies from the consensus sequence, the binding of RNA polymerase becomes weaker.

Chain Initiation

The process of transcription (and translation as well, as we will see in Chapter 12) is usually broken down into phases for easier studying. The first phase of transcription is called **chain initiation,** and it is the part of transcription that has been studied the most. It is also the part that is the most controlled.

Chain initiation begins when RNA polymerase (RNA pol) binds to the promoter and forms what is called the **closed complex** (Figure 11.3). The σ-subunit directs the polymerase to the promoter. It bridges the −10 and −35 regions of the promoter to the RNA polymerase core via a flexible "flap" in the β-subunit. Core enzymes lacking the σ-subunit will bind to areas of DNA that lack promoters. The holoenzyme may bind to "promoterless" DNA, but it will dissociate without transcribing.

Chain initiation requires formation of the **open complex,** and prematurely terminated initiation of RNA chains is common. The polymerase is not released but reinitiates transcription until the open complex is formed and incorporation of nucleoside triphosphates proceeds. Recent studies show that it is a portion of the β′ and the σ-subunits that initiate strand separation (melting) of the DNA starting at about −10 from the start site. Once the DNA is separated, RNA polymerase binds to the nontemplate strand. A purine ribonucleoside triphosphate is the first base in RNA, and it binds to its complementary DNA base at position +1. Of the purines, A tends to occur more often than G. This first residue retains its 5′-triphosphate group (indicated by ppp in Figure 11.3). (See the articles by deHaseth and Nisen and by Young et al. for the most current information on how RNA polymerase initiates transcription.)

Chain Elongation

After the strands have separated, a transcription bubble of about 17 base pairs moves down the DNA sequence to be transcribed (Figure 11.3), and RNA polymerase catalyzes the formation of the phosphodiester bonds between the incorporated ribonucleotides. When about 10 nucleotides have been incorporated, the σ-subunit dissociates and is later recycled to bind to another RNA polymerase core enzyme.

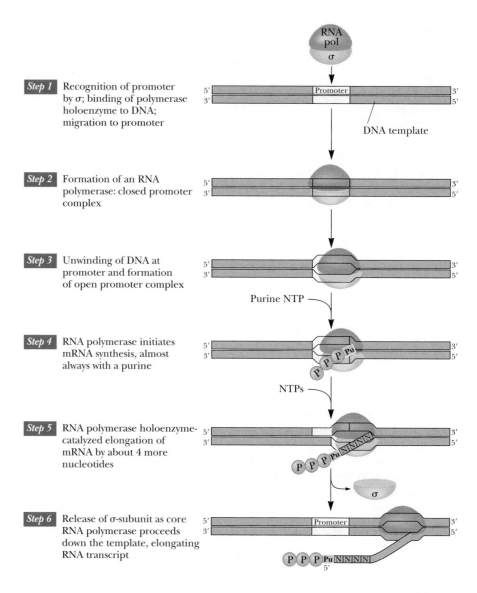

Step 1 Recognition of promoter by σ; binding of polymerase holoenzyme to DNA; migration to promoter

DNA template

Step 2 Formation of an RNA polymerase: closed promoter complex

Step 3 Unwinding of DNA at promoter and formation of open promoter complex

Purine NTP

Step 4 RNA polymerase initiates mRNA synthesis, almost always with a purine

NTPs

Step 5 RNA polymerase holoenzyme-catalyzed elongation of mRNA by about 4 more nucleotides

Step 6 Release of σ-subunit as core RNA polymerase proceeds down the template, elongating RNA transcript

Biochemistry ⊜ **Now**™ **ACTIVE FIGURE 11.3**
Sequence of events in the initiation and elongation phases of transcription as it occurs in prokaryotes. Nucleotides in this region are numbered with reference to the base at the transcription start site, which is designated +1. **Watch this Active Figure at http://now.brookscole.com/campbell5**

The transcription process supercoils DNA, with negative supercoiling upstream of the transcription bubble and positive supercoiling downstream, as shown in Figure 11.4. Topoisomerases relax the supercoils in front of and behind the advancing transcription bubble.

The rate of chain elongation is not constant. The RNA polymerase moves quickly through some DNA regions and slowly through others. It may pause for as long as one minute before continuing.

Chain Termination

Termination of RNA transcription also involves specific sequences *downstream* of the actual gene for the RNA to be transcribed. There are two types of termination mechanisms. The first is called **intrinsic termination,** and it is controlled by specific sequences called **termination sites.** The termination sites are characterized by two inverted repeats spaced by a few other bases (Figure 11.5). Inverted repeats are sequences of bases that are complementary, such that they can loop back on themselves. The DNA will then encode a series of uracils. When the RNA is created, the inverted repeats will form a hairpin loop. This will tend to stall the advancement of RNA polymerase. At the same

(a)

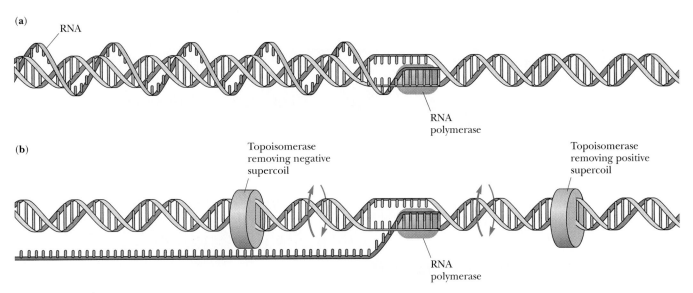

Biochemistry Now™ ▲ ACTIVE FIGURE 11.4 Two models for transcription elongation.
(a) If the RNA polymerase followed the template strand around the axis of the DNA duplex, there
would be no strain, and no supercoiling of the DNA would occur, but the RNA chain would be
wrapped around the double helix once every 10 base pairs. This possibility seems unlikely because
it would be difficult to disentangle the transcript from the DNA duplex. (b) Alternatively, topoisomerases could remove the supercoils. A topoisomerase capable of relaxing positive supercoils
situated ahead of the advancing transcription bubble would "relax" the DNA. A second topoisomerase behind the bubble would remove the negative supercoils. (*Adapted from Futcher, B., 1988.
Supercoiling and transcription, or vice versa?* Trends in Genetics **4**, *271–272. Used by permission of Elsevier Science.*)
Watch this Active Figure at http://now.brookscole.com/campbell5

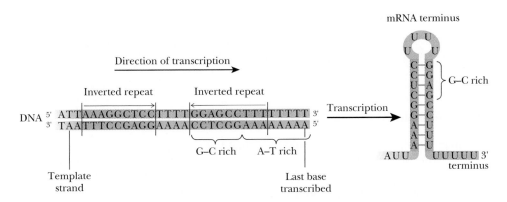

▲ FIGURE 11.5 Inverted repeats in the DNA sequence being transcribed can lead to an mRNA
molecule that forms a hairpin loop. This is often used to terminate transcription.

time, the presence of the uracils causes a series of A–U base pairs between the
template strand and the RNA. A–U pairs are weakly hydrogen-bonded compared with G–C pairs, and the RNA dissociates from the transcription bubble,
ending transcription.

The other type of termination involves a special protein called *rho* (ρ). Rho-
dependent termination sequences also cause a hairpin loop to form. In this
case, the ρ protein binds to the RNA and chases the polymerase, as shown in
Figure 11.6. When the polymerase transcribes the RNA that forms a hairpin
loop (not shown in figure), it stalls, giving the ρ protein a chance to catch up.
When the ρ protein reaches the termination site, it facilitates the dissociation
of the transcription machinery. The movement of the ρ protein and the dissociation require ATP.

(a)

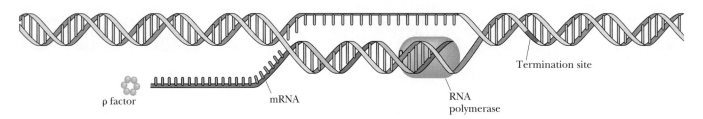

ρ factor   mRNA   RNA
polymerase   Termination site

(b)

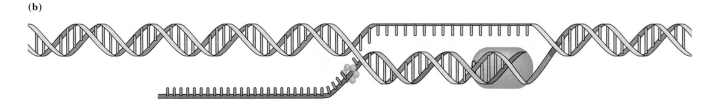

(c)

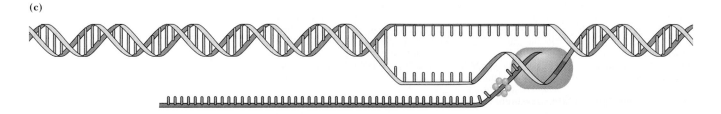

(d)

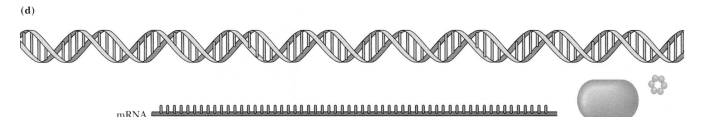

mRNA

Biochemistry ⚛ Now™ ▲ **ANIMATED FIGURE 11.6** The rho-factor mechanism of transcription termination. Rho factor (a) attaches to a recognition site on mRNA and (b) moves along it behind RNA polymerase. (c) When RNA polymerase pauses at the termination site, rho factor unwinds the DNA:RNA hybrid in the transcription bubble, releasing the nascent mRNA (d). **See this figure animated at http://now.brookscole.com/campbell5**

11.2	How Is Transcription Regulated in Prokaryotes?

Transcription is controlled in prokaryotes in several ways. The control of transcription is largely responsible for controlling the level of protein production. In fact, many equate transcription control with gene expression.

Alternative σ Factors

Viruses and bacteria can exert some control over which genes are expressed by producing different σ-subunits that will direct the RNA polymerase to different genes. A classic example of how this works is the action of phage SPO1, a virus that infects the bacteria *Bacillus subtilis*. The virus has a set of genes called the *early genes,* which are transcribed by the host's RNA polymerase, using its regular σ-subunit (Figure 11.7). One of the viral early genes codes for a protein called *gp28*. This protein is actually another σ-subunit, which directs the RNA polymerase to transcribe preferentially more of the viral genes during the *middle phase*. Products of the middle phase transcription are *gp33* and *gp34*, which together make up another σ factor that directs the transcription of the *late genes*. Remember that σ factors are recycled. Initially, the *B. subtilis* uses the standard σ factor. As more and more of the gp28 is produced, it competes for binding with standard σ for the RNA polymerase, eventually subverting the transcription machinery for the virus instead of the bacterium.

Another example of alternative σ factors is seen in the response of *E. coli* to heat shock. The normal σ-subunit in this species is called σ^{70} because it has a molecular weight of 70,000. When *E. coli* are grown at higher temperatures than their optimum, they produce another set of proteins in response. Another σ factor, called σ^{32}, is produced. It directs the RNA polymerase to bind to different promoters that are not normally recognized by σ^{70}.

Enhancers

In certain *E. coli* genes, there are sequences upstream of the extended promoter region. The genes for ribosomal RNA production have three upstream sites, called *Fis sites* because they are binding sites for the protein called Fis (Figure 11.8). These sites extend from the end of the UP element at −60 to −150. RNA polymerase does not bind to the Fis sites, so they cannot be considered part of the promoter. Instead, they are examples of a class of DNA sequences called **enhancers.** Enhancers are sequences that can be bound by proteins called **transcription factors,** a class of molecule we will see a lot of in Sections 11.3 and 11.4. When enhancers allow a response to changing metabolic conditions, such as temperature shock, they are usually referred to as **response elements.** When binding the transcription factor increases the level of transcription, the element is said to be an enhancer. When binding the transcription factor decreases transcription, the element is said to be a **silencer.** The position and orientation of enhancers is less important than for sequences that are part of the promoter. Molecular biologists can study the nature of control elements by making changes to them. When enhancer sequences are moved from one place on the DNA to another or have their sequences reversed, they still function as enhancers. The study of the number and nature of transcription factors is the most common research in molecular biology these days.

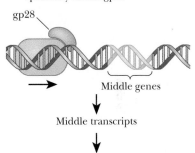

(a) Early transcription; specificity factor: host σ

RNA polymerase

Early genes

Early transcripts

Early proteins, including gp28

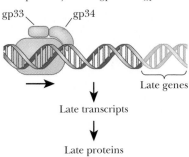

(b) Middle transcription; specificity factor: gp28

gp28

Middle genes

Middle transcripts

Middle proteins, including gp33 and gp34

(c) Late transcription; specificity factor: gp33 and gp34

gp33   gp34

Late genes

Late transcripts

Late proteins

▲ **FIGURE 11.7** Control of transcription via different σ subunits. (a) When the phage SPO1 infects *B. subtilis,* the host RNA polymerase (tan) and σ-subunit (blue) transcribe the early genes of the infecting viral DNA. One of the early gene products is gp28 (green) an alternative σ-subunit. (b) The gp28 directs the RNA polymerase to transcribe the middle genes, which produces gp33 (purple) and gp34 (red). (c) The gp33 and gp34 direct the host's RNA polymerase to transcribe the late genes. (*Adapted by permission from* Molecular Biology, *by R. F. Weaver, McGraw-Hill, 1999.*)

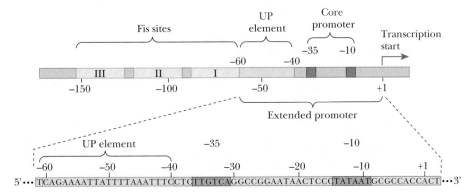

◀ **FIGURE 11.8** Schematic representation of elements of a bacterial promoter. The core promoter includes the −10 and −35 regions. The extended promoter includes the UP element. Upstream of the UP element, there may be enhancers, such as the Fis sites seen in the promoters for genes that code for ribosomal RNA in *E. coli.* The protein Fis is a transcription factor. (*Adapted by permission from* Molecular Biology, *by R. F. Weaver, McGraw-Hill, 1999.*)

Operons

In prokaryotes, genes that encode enzymes of certain metabolic pathways are often controlled as a group, with the genes encoding the proteins of the pathway being close together and under the control of a common promoter. Such a group of genes is called an **operon.** Usually the genes are not transcribed all the time. Rather, the production of these proteins can be triggered by the presence of a suitable substance called an **inducer.** This phenomenon is called **induction.** A particularly well-studied example of an inducible protein is the enzyme β-*galactosidase* in *E. coli.*

The disaccharide *lactose* (a β-galactoside; Section 16.3) is the substrate of β-galactosidase. The enzyme hydrolyzes the glycosidic linkage between galactose and glucose, the monosaccharides that are the component parts of lactose. *E. coli* can survive with lactose as its sole carbon source. To do so, the bacterium needs β-galactosidase to catalyze the first step in lactose degradation. The production of β-galactosidase takes place only in the presence of lactose, not in the presence of other carbon sources, such as glucose. A metabolite of lactose, allolactose, is the actual inducer, and β-galactosidase is an *inducible enzyme.*

β-Galactosidase is coded for by a **structural gene** (*lacZ*) (Figure 11.9). Structural genes encode the gene products that are involved in the biochemical pathway of the operon. Two other structural genes are part of the operon. One is *lacY,* which encodes the enzyme lactose permease, which allows lactose to enter the cell. The other is *lacA,* which encodes an enzyme called transacetylase. The purpose of this last enzyme is not known, but some hypothesize that its role is to inactivate certain antibiotics that may enter the cell through the lactose permease. The expression of these structural genes is in turn under control of a **regulatory gene** (*lacI*), and the mode of operation of the regulatory gene is the most important part of the *lac* operon mechanism. The regulatory gene is responsible for the production of a protein, the **repressor.** As the name indicates, the repressor inhibits the expression of the structural genes. In the presence of the inducer, this inhibition is removed. This is an example of **negative regulation** because the *lac* operon is turned on unless something is present to turn it off, which is the repressor in this case.

The repressor protein that is made by the *lacI* gene forms a tetramer when it is translated. It then binds to a portion of the operon called the **operator** (O) (Figure 11.9). When the repressor is bound to the operator, RNA polymerase cannot bind to the adjacent promoter region (p_{lac}), which facilitates the expression of the structural genes. The operator and promoter together constitute the **control sites.**

In induction, the inducer binds to the repressor, producing an inactive repressor that cannot bind to the operator (Figure 11.9). Because the repressor is no longer bound to the operator, RNA polymerase can now bind to the promoter, and transcription and translation of the structural genes can take place. The *lacI* gene is adjacent to the structural genes in the *lac* operon, but this need not be the case. Many operons are known in which the regulatory gene is far removed from the structural genes.

The *lac* operon is induced when *E. coli* has lactose, and no glucose, available to it as a carbon source. When both glucose and lactose are present, the cell does not make the *lac* proteins. The repression of the synthesis of the *lac* proteins by glucose is called **catabolite repression.** The mechanism by which *E. coli* recognizes the presence of glucose involves the promoter. The promoter has two regions. One is the binding site for RNA polymerase, and the other is the binding site for another regulatory protein, the **catabolite activator protein (CAP)** (Figure 11.10). The binding site for RNA polymerase also overlaps the binding site for the repressor in the operator region.

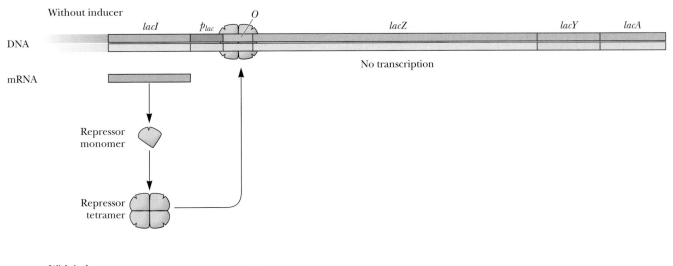

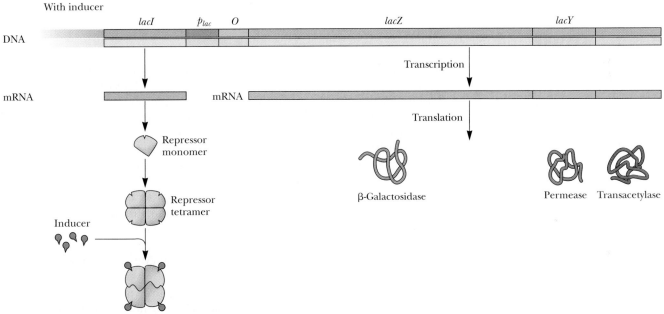

BiochemistryⒺNow™ ▲ **ACTIVE FIGURE 11.9** The mode of action of the *lac* repressor. The *lacI* gene produces a protein that represses the *lac* operon by binding to the operator. In the presence of an inducer, the repressor cannot bind, and the operon genes are transcribed. **Watch this Active Figure at http://now.brookscole.com/campbell5**

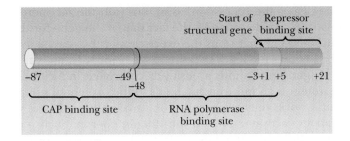

◄ **FIGURE 11.10** Binding sites in the *lac* operon. Numbering refers to base pairs. Negative numbers are assigned to base pairs in the regulatory sites. Positive numbers indicate the structural gene, starting with base pair +1. The CAP binding site is seen next to the RNA polymerase binding site.

The binding of CAP to the promoter depends on the presence or absence of 3′,5′-cyclic AMP (cAMP). When glucose is not present, cAMP is formed, serving as a "hunger signal" for the cell. CAP forms a complex with cAMP. The complex binds to the CAP site in the promoter region. When the complex is bound to the CAP site on the promoter, the RNA polymerase can bind

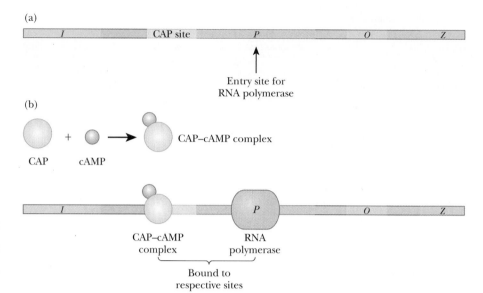

▶ **FIGURE 11.11** Catabolite repression. (a) The control sites of the *lac* operon. The CAP–cAMP complex, not CAP alone, binds to the CAP site of the *lac* promoter. When the CAP site on the promoter is not occupied, RNA polymerase does not bind. (b) In the absence of glucose, cAMP forms a complex with CAP. The complex binds to the CAP site, allowing RNA polymerase to bind to the entry site on the promoter and to transcribe the structural genes.

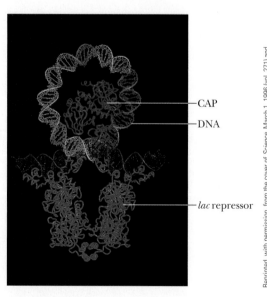

▲ The *lac* repressor and CAP bound to DNA.

Reprinted, with permission, from the cover of *Science*. March 1, 1996 (vol. 271) and from Dr. Mitchell Lewis, University of Pennsylvania School of Medicine.

at the binding site available to it and proceed with transcription (Figure 11.11). The *lac* promoter is particularly weak, and RNA polymerase binding is minimal in the absence of the CAP–cAMP complex bound to the CAP site. The CAP site is an example of an enhancer element, and the CAP–cAMP complex is a transcription factor. The modulation of transcription by CAP is a type of **positive regulation.**

When the cell has an adequate supply of glucose, the level of cAMP is low. CAP binds to the promoter only when it is complexed to cAMP. The combination of positive and negative regulation with the *lac* operon means that the presence of lactose is necessary, but not sufficient, for transcription of the operon structural genes. It takes the presence of lactose *and* the absence of glucose for the operon to be active. As we shall see later, many transcription factors and response elements involve the use of cAMP, a common messenger in the cell.

Operons can be controlled by positive or negative regulation mechanisms. They are also classified as **inducible, repressible,** or both, depending on how they respond to the molecules that control their expression. There are four general possibilities, as shown in Figure 11.12. The top left figure shows a negative control system with induction. It is negative control because a repressor protein stops transcription when it binds to the promoter. It is an inducible system because the presence of the inducer or **co-inducer,** as it is often called, releases the repression, as we saw with the *lac* operon. Negative control systems can be identified by the fact that, if the gene for the repressor is mutated in some way that stops the expression of the repressor, the operon will always be expressed. Genes that are always expressed are called **constitutive.** The top right figure shows a positively controlled inducible system. The controlling protein is an inducer that binds to the promoter, stimulating transcription, but it works only when bound to its co-inducer. This is what is seen with the catabolite activator protein with the *lac* operon. Such positively controlled systems can be identified by the fact that, if the gene for the inducer is mutated, it cannot be expressed—that is, it is **uninducible.** The bottom left figure shows a negatively controlled repressible system. A repressor stops transcription, but this repressor functions only in the presence of a **co-repressor.** The bottom right figure shows a positively controlled repressible system. An inducer protein binds to the promoter, stimulating transcription; but, in the presence of the co-repressor, the inducer is inactivated.

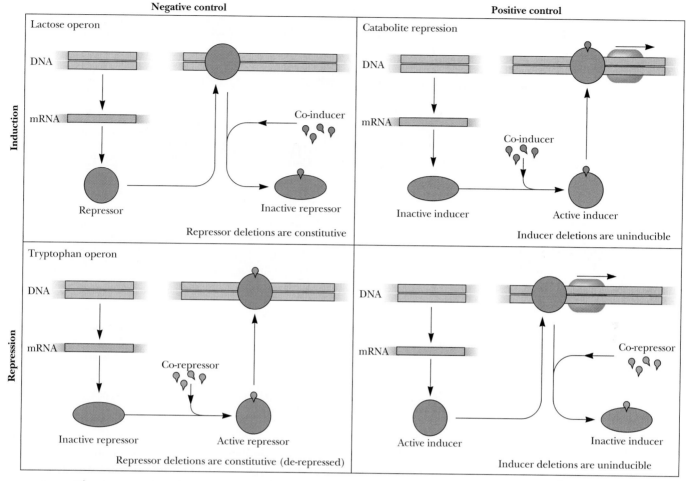

Negative control

Positive control

Induction

Lactose operon

Catabolite repression

Repressor deletions are constitutive

Inducer deletions are uninducible

Repression

Tryptophan operon

Repressor deletions are constitutive (de-repressed)

Inducer deletions are uninducible

Biochemistry ⑤Now™ ▲ **ANIMATED FIGURE 11.12** Basic control mechanisms seen in the control of genes. They may be inducible or repressible, and they may be positively or negatively controlled. **See this figure animated at http://now.brookscole.com/campbell5**

The *trp* operon of *E. coli* codes for a leader sequence (*trpL*) and five polypeptides, *trpE* through *trpA,* as shown in Figure 11.13. The five proteins make up four different enzymes (shown in the three boxes near the bottom of the figure). These enzymes catalyze the multistep process that converts chorismate to tryptophan. Control of the operon is via a repressor protein that binds to two molecules of tryptophan. When tryptophan is plentiful, this repressor–tryptophan complex binds to the *trp* operator that is next to the *trp* promoter. This binding prevents the binding of RNA polymerase, so the operon is not transcribed. When tryptophan levels are reduced, the repression is lifted because the repressor will not bind to the operator in the absence of the co-repressor, tryptophan. This is an example of a system that is repressible and under negative regulation, as shown in Figure 11.12. The *trp* repressor protein is itself produced by the *trpR* operon and also represses that operon. It is an example of **autoregulation,** because the product of the *trpR* operon regulates its own production.

Transcription Attenuation

In addition to repression, the *trp* operon is regulated by transcription **attenuation.** This control mechanism works by altering transcription *after* it has begun via transcription termination or pausing. Prokaryotes have no separation of

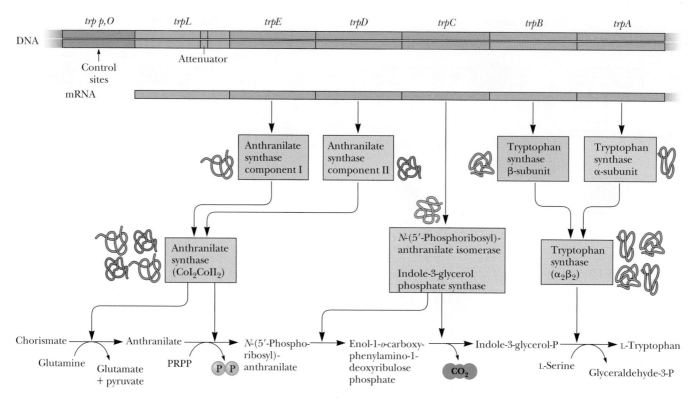

Biochemistry Now™ ▲ **ANIMATED FIGURE 11.13** The *trp* operon of *E. coli*. **See this figure animated at http://now.brookscole.com/campbell5**

transcription and translation as eukaryotes do, so the ribosomes are attached to the mRNA while it is being transcribed. The *trp* operon's first gene is the *trpL* sequence that codes for a leader peptide. This leader peptide has two key tryptophan residues in it. Translation of the mRNA leader sequence depends on having an adequate supply of tryptophan-charged tRNA (Chapters 9 and 12). When tryptophan is scarce, the operon is translated normally. When it is plentiful, transcription is terminated prematurely after only 140 nucleotides of the leader sequence have been transcribed. Secondary structures formed in the mRNA of the leader sequence are responsible for this effect (Figure 11.14). Three possible hairpin loops can form in this RNA—the **1·2 pause structure,** the **3·4 terminator,** or the **2·3 antiterminator.** Transcription begins normally and proceeds until position 92, at which point the 1·2 pause structure can form. This causes RNA polymerase to pause in its RNA synthesis. A ribosome begins to translate the leader sequence, which releases the RNA polymerase from its pause and allows transcription to resume. The ribosome follows closely behind the RNA polymerase shown in Figure 11.15. The ribosome stops over the UGA stop codon of the mRNA, which prevents the 2·3 antiterminator hairpin from forming and allows instead the 3·4 terminator hairpin to form. This hairpin has the series of uracils characteristic of rho-independent termination. The RNA polymerase ceases transcription when this terminator structure forms.

If tryptophan is limiting, the ribosome stalls out over the tryptophan codons on the mRNA of the leader sequence. This leaves the mRNA free to form the 2·3 antiterminator hairpin, which stops the 3·4 terminator sequence from forming, so that the RNA polymerase continues to transcribe the rest of the operon. Transcription is attenuated in several other operons dealing with amino acid synthesis. In these cases, there are always codons for the amino acid, which is the product of the pathway that acts in the same way as the tryptophan codons in this example.

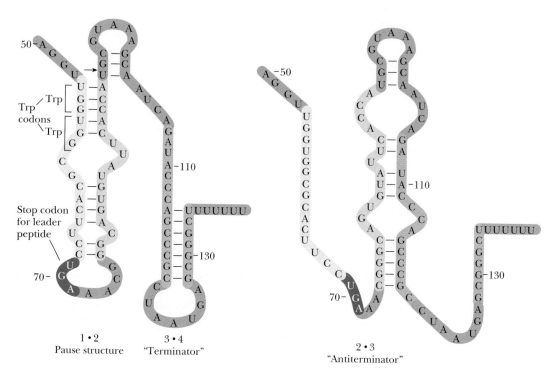

▲ FIGURE 11.14 Alternative secondary structures can form in the leader sequence of mRNA for the *trp* operon. Binding between regions 1 and 2 (yellow and tan) is called a pause structure. Regions 3 and 4 (purple) then form a terminator hairpin loop. Alternative binding between regions 2 and 3 forms an antiterminator structure.

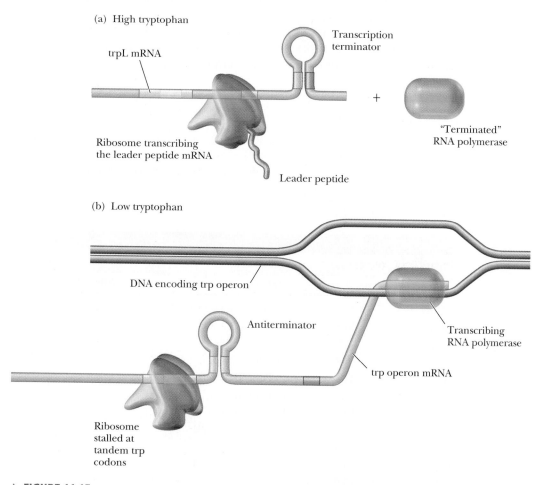

▲ FIGURE 11.15 The attenuation mechanism in the *trp* operon. The pause structure forms when the ribosome passes over the Trp codons quickly when tryptophan levels are high. This causes premature abortion of the transcript as the terminator loop is allowed to form. When tryptophan is low, the ribosome stalls at the Trp codons, allowing the antiterminator loop to form, and transcription continues.

11.3 | How Does Transcription Take Place in Eukaryotes?

We have seen that prokaryotes have a single RNA polymerase that is responsible for the synthesis of all three kinds of prokaryotic RNA—mRNA, tRNA, and rRNA. The polymerase can switch σ factors to interact with different promoters, but the core polymerase stays the same. The transcription process is predictably more complex in eukaryotes than in prokaryotes. Three RNA polymerases with different activities are known to exist. Each one transcribes a different set of genes and recognizes a different set of promoters:

1. RNA polymerase I is found in the nucleolus and synthesizes precursors of most, but not all, ribosomal RNAs.

2. RNA polymerase II is found in the nucleoplasm and synthesizes mRNA precursors.
3. RNA polymerase III is found in the nucleoplasm and synthesizes the tRNAs, precursors of 5S ribosomal RNA, and a variety of other small RNA molecules involved in mRNA processing and protein transport.

All three of the eukaryotic RNA polymerases are large (500–700 kDa), complex proteins consisting of ten or more subunits. Their overall structures differ, but they all have a few subunits in common. They all have two larger subunits that share sequence homology with the β- and β′-subunits of prokaryotic RNA polymerase that make up the catalytic unit. There are no σ-subunits to direct polymerases to promoters. The detection of a gene to be transcribed is accomplished in a different way in eukaryotes, and the presence of transcription factors, of which there are hundreds, plays a larger role. We shall restrict our discussion to transcription by Pol II.

Structure of RNA Polymerase II

Of the three RNA polymerases, RNA polymerase II is the most extensively studied, and the yeast *Saccharomyces cerevisiaie* is the most common model system. Yeast RNA polymerase II consists of 12 subunits, as shown in Table 11.2.

Table 11.2			
Yeast RNA Polymerase II Subunits			
Subunit	Size (kDa)	Features	*E. coli* Homologue
RPB1	191.6	Phosphorylation site	β′
RPB2	138.8	NTP binding site	β
RPB3	35.3	Core assembly	α
RPB4	25.4	Promoter recognition	σ
RPB5	25.1	In Pol I, II, and III	
RPB6	17.9	In Pol I, II, and III	
RPB7	19.1	Unique to Pol II	
RPB8	16.5	In Pol I, II, and III	
RPB9	14.3		
RPB10	8.3	In Pol I, II, and III	
RPB11	13.6		
RPB12	7.7	In Pol I, II, and III	

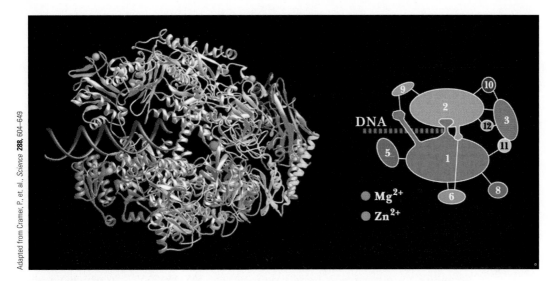

▲ **FIGURE 11.16** Architecture of yeast RNA polymerase II. The backbone models for the 10 subunits are shown as ribbon diagrams. B-DNA is shown in blue. Zinc atoms are shown as turquoise spheres, and magnesium is shown as pink spheres. The box on the right is a key to the subunit color codes.

Adapted from Cramer, P., et. al., *Science* **288**, 604–649

The subunits are called RPB1 through RPB12. **RPB** stands for **RNA polymerase B** because another nomenclature system refers to the polymerases as A, B, and C, instead of I, II, and III.

The function of many of the subunits is not known. The core subunits, RBP1 through RBP3, seem to play a role similar to their homologues in prokaryotic RNA polymerase. Five of them are present in all three RNA polymerases. RPB1 has a repeated sequence of PTSPSYS in the **C-terminal domain (CTD),** which, as the name applies, is found at the C-terminal region of the protein. Threonine, serine, and tyrosine are all substrates for phosphorylation, which is important in the control of transcription initiation.

X-ray crystallography has been used to determine the structure of RNA polymerase II (see the article by Cramer et al. listed in the Annotated Bibliography at the end of this chapter). Notable features include a pair of jaws formed by subunits RPB1, RPB5, and RPB9, which appear to grip the DNA downstream of the active site. A clamp near the active site is formed by RPB1, RPB2, and RPB6, which may be involved in locking the DNA:RNA hybrid to the polymerase, increasing the stability of the transcription unit. Figure 11.16 shows a ribbon diagram of the structure of RNA polymerase II.

The recent structural work on RNA polymerases from prokaryotes and eukaryotes has led to some exciting conclusions regarding their evolution. There is extensive homology between the core regions of RNA polymerases from bacteria, yeast, and humans, leading researchers to speculate that RNA polymerase evolved eons ago, at a time when only prokaryotes existed. As more complex organisms developed, layers of other subunits were added to the core polymerase to reflect the more complicated metabolism and compartmentalization of eukaryotes.

Biochemistry Ⓔ**Now**™

Go to BiochemistryNow and click on Biochemistry Interactive to explore the RNA polymerase II as the machine of transcription.

Pol II Promoters

There are four elements to Pol II promoters (Figure 11.17). The first includes a variety of **upstream elements,** which act as **enhancers** and **silencers.** Specific binding proteins either activate transcription above basal levels, in the case of

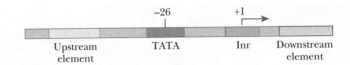

▶ **FIGURE 11.17** Four elements of Pol II promoters.

enhancers, or suppress it, in the case of silencers. Two common elements that are close to the core promoter are the GC box (−40), which has a **consensus sequence** of GGGCGG, and the CAAT box (extending to −110), which has a consensus sequence of GGCCAATCT.

The second element, found at position −25, is the **TATA box,** which has a consensus sequence of TATAA(T/A).

The third element includes the transcription start site at position +1, but, in the case of eukaryotes, it is surrounded by a sequence called the **initiator element** (*Inr*). This sequence is not well conserved. For instance, the sequence for a particular gene type may be $_{-3}$YYCAYYYYY$_{+6}$, in which Y indicates either pyrimidine, and A is the purine at the **transcription start site (TSS).**

The fourth element is a possible downstream regulator, although these are more rare than upstream regulators. Many natural promoters lack at least one of the four elements. The initiator plus the TATA box make up the core promoter and are the two most consistent parts across different species and genes. Some genes do not have TATA boxes; they are called "TATA-less" promoters. In some genes, the TATA box is necessary for transcription, and deletion of the TATA box causes a loss of transcription. In others, the TATA box serves to orient the RNA polymerase correctly. Elimination of the TATA box in these genes causes transcription at random starting points. Whether a particular regulatory element is considered to be part of the promoter or not is often a judgment call. Those that are considered part of the promoter are close to the TSS (50–200 bp) and show specificity with regard to distance and orientation of the sequence. Those regulatory sequences that are not considered to be part of the promoter can be far removed from the TSS, and their orientation is irrelevant. Experiments have shown that, when such sequences are reversed, they still work, and when they are moved several thousand base pairs upstream, they still work.

Initiation of Transcription

The biggest difference between transcription in prokaryotes and eukaryotes is the sheer number of proteins associated with the eukaryotic version of the process. Any protein that regulates transcription but that is not itself a subunit of RNA polymerase is a **transcription factor.** There are many transcription factors for eukaryotic transcription, as we shall see. The molecular mass of the entire complex of Pol II and all of the associated factors exceeds 2.5 million Da.

Transcription initiation begins by the formation of a **preinitiation complex,** and the vast majority of the control of transcription occurs at this step. This complex normally contains RNA polymerase II and six **general transcription factors (GTFs)—TFIIA, TFIIB, TFIID, TFIIE, TFIIF,** and **TFIIH.** These GTFs are required for all promoters. Much work is still going on to determine the structure and function of each of the parts of the preinitiation complex. Each of the GTFs has a specific function, and each is added to the complex in a defined order. Table 11.3 is a summary of the components of the preinitiation complex.

Table 11.3

General Transcription Initiation Factors

Factor	Subunits	Size (kDa)	Function
TFIID-TBP	1	27	TATA box recognition, positioning of TATA box DNA around TFIIB and Pol II
TFIID-TAF$_{II}$s	14	15–250	Core promoter recognition (non-TATA elements), positive and negative regulation
TFIIA	3	12, 19, 35	Stabilization of TBP binding; stabilization of TAF–DNA binding
TFIIB	1	38	Recruitment of Pol II and TFIIF; start-site recognition for Pol II
TFIIF	3	156 total	Promoter targeting of Pol II
TFIIE	2	92 total	TFIIH recruitment; modulation of TFIIH helicase ATPase, and kinase activities; promoter melting
TFIIH	9	525 total	Promoter melting; promoter clearance via phosphorylation of CTD

Figure 11.18 shows the sequence of events in Pol II transcription. The first step in the formation of the preinitiation complex is the recognition of the TATA box by TFIID. This transcription factor is actually a combination of several proteins. The primary protein is called **TATA-binding protein (TBP).** Associated with TBP are many **TBP-associated factors (TAF$_{II}$s).** Because TBP is also present and required for Pol I and Pol III, it is a universal transcription factor. TBP is highly conserved. From species as different as yeast, plants, fruit flies, and humans, the TBPs have more than 80 percent identical amino acids. The TBP protein binds to the minor groove of the DNA at the TATA box via the last 180 amino acids of its C-terminal domain. As shown in Figure 11.19, the TBP sits on the TATA box like a saddle. The minor groove of the DNA is opened, and the DNA is bent to an 80° angle.

As shown in Figure 11.18, once TFIID is bound, TFIIA binds, and TFIIA also has interactions with both the DNA and TFIID. TFIIB also binds to TFIID, bridging the TBP and Pol II. TFIIA and TFIIB can actually bind in either order, and they do not interact with each other. TFIIB is critical for the assembly of the initiation complex and for the location of the correct transcription start site. TFIIF then binds tightly to Pol II and suppresses nonspecific binding. Pol II and TFIIF then bind stably to the promoter. TFIIF interacts with Pol II, TBP, TFIIB, and the TAF$_{II}$s. It also regulates the activity of the CTD phosphatase.

The last two factors to be added are TFIIE and TFIIH. TFIIE interacts with unphosphorylated Pol II. These two factors have been implicated in the phosphorylation of polymerase II. TFIIH also has helicase activity. After all these GTFs have bound to unphosphorylated Pol II, the preinitiation complex is complete. TFIIH has been found to have other functions as well, such as DNA repair (see the Biochemical Connections box on page 284).

Before transcription can begin, the preinitiation complex must form the *open complex*. In the open complex, the Pol II CTD is phosphorylated, and the DNA strands are separated (Figure 11.18).

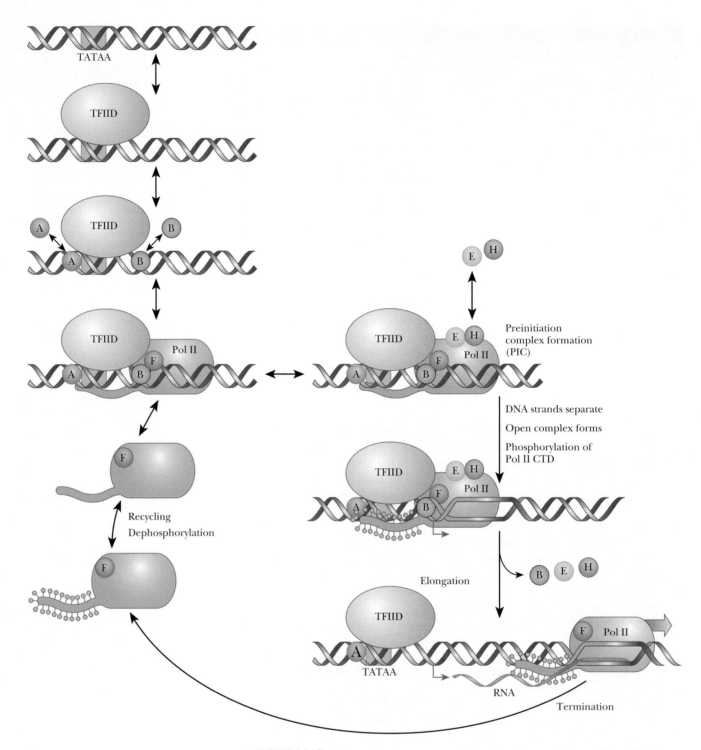

▲ **FIGURE 11.18** A schematic representation of the order of events of transcription. TFIID (which contains the TATA-box binding protein, TBP) binds to the TATA box. TFIIA and TFIIB then bind, followed by recruitment of RNA polymerase II and TFIIF. TFIIH and TFIIE then bind to form the preinitiation complex (PIC). Kinases phosphorylate the C-terminal domain of Pol II, leading to the open complex in which the DNA strands are separated. RNA is produced during elongation as Pol II and TFIIF leave the promoter and the other general transcription factors behind. Pol II dissociates during the termination phase, and the CTD is dephosphorylated. Pol II/TFIIF is then recycled to bind to another promoter.

Elongation and Termination

Less is known about elongation and termination in eukaryotes than in prokaryotes. Most of the research efforts have focused on the preinitiation complex and on the regulation by enhancers and silencers. As shown in Figure 11.18, the phosphorylated Pol II synthesizes RNA and leaves the promoter region behind. At the same time, the GTFs either are left at the promoter or dissociate from Pol II.

Pol II does not elongate efficiently when alone in vitro. Under those circumstances, it can synthesize only 100–300 nucleotides per minute, whereas the in vivo rates are between 1500 and 2000 nucleotides per minute. The difference is due to elongation factors. One is TFIIF, which, in addition to its role in the formation of the preinitiation complex, also has a separate stimulatory effect on elongation. A second elongation factor, which was named *TFIIS*, was more recently discovered.

Elongation is controlled in several ways. There are sequences called *pause sites*, where the RNA polymerase will hesitate. This is very similar to the transcription attenuation we saw with prokaryotes. Elongation can also be aborted, leading to premature termination. Finally, elongation can proceed past the normal termination point. This is called *antitermination*. The TFIIF class of elongation factors promotes a rapid read-through of pause sites, perhaps locking the Pol II into an elongation-competent form that will not pause and dissociate.

The TFIIS class of elongation factors are called *arrest release factors*. They act to help the RNA polymerase to move again after it has paused. A third class of elongation factors consists of the *P-TEF* and *N-TEF* proteins (Positive-Transcription Elongation Factor and Negative-Transcription Elongation Factor). They increase the productive form of transcription and decrease the abortive form, or vice versa. At some point during either elongation or termination, TFIIF dissociates from Pol II.

Termination begins by stopping the RNA polymerase. There is a eukaryotic consensus sequence for termination, which is AAUAAA. This sequence may be 100–1000 bases away from the actual end of the mRNA. After termination occurs, the transcript is released, and the Pol II open form (phosphorylated) is released from the DNA. The phosphates are removed by phosphatases, and the Pol II/TFIIF complex is recycled for another round of transcription (Figure 11.18).

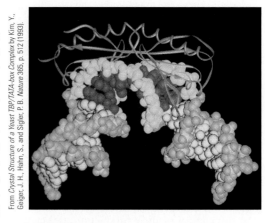

From *Crystal Structure of a Yeast TBP/TATA-box Complex* by Kim, Y., Geiger, J. H., Hahn, S., and Sigler, P. B. *Nature* 365, p. 512 (1993).

▲ **FIGURE 11.19** Model of yeast TATA-binding protein (TBP) binding to DNA. The DNA backbone of the TATA box is shown in yellow, and the TATA bases are shown in red. Adjacent DNA sequences are shown in turquoise. The TBP, which is shown in green, sits on the minor groove of the DNA like a saddle.

11.4 | How Is Transcription Regulated in Eukaryotes?

In the last section, we saw how the general transcription machinery, consisting of the RNA polymerase and general transcription factors, functions to initiate transcription. This is the general case that is consistent for all transcription of mRNA. However, this machinery alone produces only a low level of transcription called the **basal level.** The actual transcription level of some genes may be many times the basal level. The difference is gene-specific transcription factors, otherwise known as **activators.** Recall that eukaryotic DNA is complexed to histone proteins in chromatin. The DNA is wound tightly around the histone proteins, and many of the promoters and other regulatory DNA sequences may be inaccessible much of the time.

Biochemical Connections

TFIIH—Making the Most Out of the Genome

The dogma for decades had always been that humans were more complex than other species, and this complexity was supposedly due to our having a larger amount of DNA and a greater number of genes. With the preliminary data from the Human Genome Project just in, it is now clear that we are not that much more complicated in terms of gene number. How, then, can very different structures and metabolisms between humans and nematodes, for example, be explained? Scientists must now look both at the effects of the proteins produced and at the control of their production, rather than simply counting the number of genes that encode proteins. A complex organism must get a lot of bang for the buck out of its gene products. This is seen clearly in the field of transcription. Eukaryotes have three RNA polymerases, but they all share some common sub-

units. Each polymerase has a unique organization of subunits and transcription factors, but many of these are shared among the multiple polymerases. Transcription factor TFIIH is particularly versatile. Besides its role in initiation of transcription of Pol II, it also has a cyclin-dependent kinase activity. Cyclins are proteins that are involved in the control of the cell cycle. Thus TFIIH is involved not only in tying transcription and cell division together but also in repairing DNA, as seen in Chapter 10.

Two human genetic diseases, xeroderma pigmentosum (XP) and Cockayne syndrome, are characterized by extreme skin sensitivity to sunlight. Several genes are involved in the former disease, and most of the mutations lead to missing or defective DNA polymerases that act as repair enzymes. However, in a couple of the XP mutations and in Cockayne syndrome, there is a defect in the TFIIH protein. Besides its role in general transcription, it has been implicated in a DNA repair mechanism called *transcription-coupled repair (TCR)*. The figure here shows the model of the function of TFIIH. When RNA polymerase is attempting to transcribe DNA and it encounters a lesion, it cannot continue. The polymerase is released. TFIIH and one of the protein products of the XP family, XPG, bind to the DNA. It is believed that these factors recruit the particular repair enzymes that are needed to correct the damage.

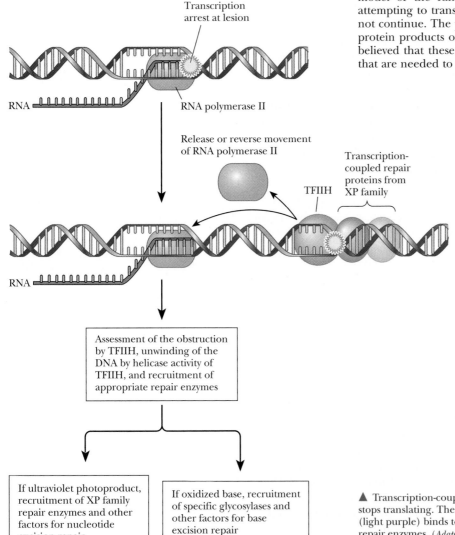

▲ Transcription-coupled repair. When RNA Pol II encounters a lesion, it stops translating. The polymerase must release or back up, while TFIIH (light purple) binds to the lesion and helps to recruit the correct DNA-repair enzymes. (*Adapted by permission from Hanawalt, P. C. DNA Repair: The Bases for Cockayne Syndrome.* Nature **405,** *415 [2000].*)

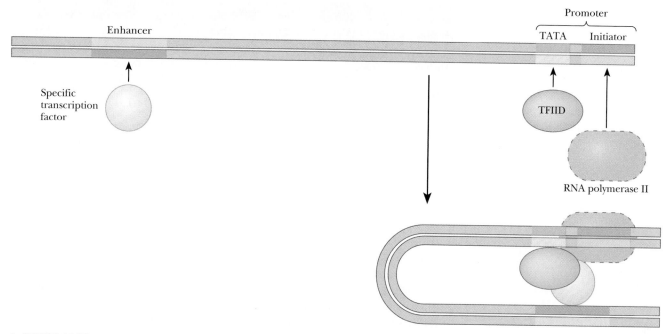

▲ **FIGURE 11.20** DNA looping brings enhancers in contact with transcription factors and RNA polymerase.

Enhancers and Silencers

Enhancers and silencers are regulatory sequences that augment or diminish transcription, respectively. They can be upstream or downstream from the transcription initiator, and their orientation doesn't matter. They act through the intermediary of a gene-specific transcription-factor protein. As shown in Figure 11.20, the DNA must loop back so that the enhancer element and its associated transcription factor can contact the preinitiation complex.

Response Elements

Some transcription control mechanisms can be categorized based on a common response to certain metabolic factors. Enhancers that are responsive to these factors are called **response elements.** Examples include the **heat-shock element (HSE),** the **glucocorticoid-response element (GRE),** the **metal-response element (MRE),** and the **cyclic-AMP-response element (CRE).** These response elements all bind proteins (transcription factors) that are produced under certain cell conditions, and several related genes are activated. This is not the same as an operon because the genes are not linked in sequence and are not controlled by a single promoter. Several different genes, all with unique promoters, may all be affected by the same transcription factor binding the response element.

In the case of HSE, elevated temperatures lead to the production of specific heat-shock transcription factors that activate the associated genes. Glucocorticoid hormones bind to a steroid receptor. Once bound, this becomes the transcription factor that binds to the GRE. Table 11.4 summarizes some of the best-understood response elements.

We will look more closely at the cyclic-AMP-response element as an example of eukaryotic control of transcription. Hundreds of research papers deal with this topic as more and more genes are found to have this response element as part of their control. Remember that cAMP was also involved in the control of prokaryotic operons via the CAP protein.

Table 11.4				
Response Elements and Their Characteristics				
Response Element	Physiological Signal	Consensus Sequence	Transcription Factor	Size (kDa)
CRE	cAMP-dependent activation of protein kinase A	TGACGTCA	CREB, CREM, ATF1	43
GRE	Presence of glucocorticoids	TGGTACAAA TGTTCT	Glucocorticoid receptor	94
HSE	Heat shock	CNNGAANNT CCNNG*	HSTF	93
MRE	Presence of cadmium	CGNCCCGGN CNC*	?	?

*N stands for any nucleotide.

Cyclic AMP is produced as a second messenger from several hormones, such as epinephrine and glucagon (see Chapter 21). When the levels of cAMP rise, the activity of **cAMP-dependent protein kinase** (protein kinase A) is stimulated. This enzyme phosphorylates many other proteins and enzymes inside the cell and is usually associated with switching the cell to a catabolic mode, in which macromolecules will be broken down for energy. Protein kinase A phosphorylates a protein called **cyclic-AMP-response-element binding protein (CREB),** which binds to the cyclic-AMP-response element and activates the associated genes (see the Biochemical Connections box on page 288). The CREB does not directly contact the basal transcription machinery (RNA polymerase and GTFs), however, and the activation requires another protein. **CREB-binding protein (CBP)** binds to CREB after it has been phosphorylated and bridges the response element and the promoter region, as shown in Figure 11.21. After this bridge is made, transcription is activated above basal levels. CBP is called a *mediator* or *coactivator.* Many abbreviations are used in the language of transcription, and Table 11.5 summarizes the more important ones.

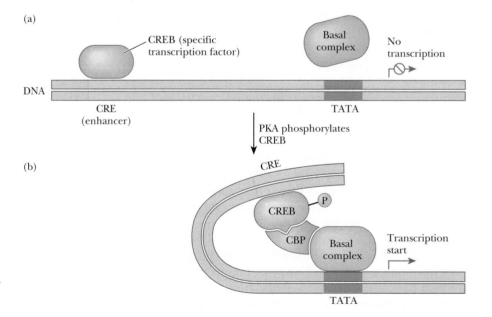

▶ **FIGURE 11.21** Activation of transcription via CREB and CBP. (a) Unphosphorylated CREB does not bind to CREB binding protein, and no transcription occurs. (b) Phosphorylation of CREB causes binding of CREB to CBP, which forms a complex with the basal complex (RNA polymerase and GTFs), thereby activating transcription. (*Adapted by permission from* Molecular Biology, *by R. F. Weaver, McGraw-Hill, 1999.*)

Table 11.5

Abbreviations Used in Transcription

bZIP	Basic-region leucine zipper	NTD	N-terminal domain
CAP	Catabolite activator protein	N-TEF	Negative transcription elongation factor
CBP	CREB-binding protein		
CRE	Cyclic-AMP-response element	Pol II	RNA polymerase II
		P-TEF	Positive transcription elongation factor
CREB	Cyclic-AMP-response-element binding protein	RPB	RNA polymerase B (Pol II)
CREM	Cyclic-AMP-response-element modulating protein	RNP	Ribonucleoprotein particle
		snRNP	Small nuclear ribonucleoprotein particle ("snurps")
CTD	C-terminal domain		
GRE	Glucocorticoid-response element	TAF	TBP-associated factor
		TATA	Consensus promoter element in eukaryotes
GTF	General transcription factor		
HSE	Heat-shock-response element	TBP	TATA-box binding protein
HTH	Helix–turn–helix	TCR	Transcription-coupled repair
Inr	Initiator element	TF	Transcription factor
MRE	Metal-response element	TSS	Transcription start site
MAPK	Mitogen-activated protein kinase	XP	Xeroderma pigmentosum

The CBP protein and a similar one called p300 are a major bridge to several different hormone signals, as can be seen in Figure 11.22. Several hormones that act through cAMP cause the phosphorylation and binding of CREB to CPB. Steroid and thyroid hormones and some others act upon receptors in the nucleus to bind to CBP/p300. Growth factors and stress signals cause *mitogen-activated protein kinase (MAPK)* to phosphorylate transcription factors *AP-1 (activating protein 1)* and *Sap-1a*, both of which bind to CBP. See the article by Brivanlou in the Annotated Bibliography of this chapter for a review of transcription factors.

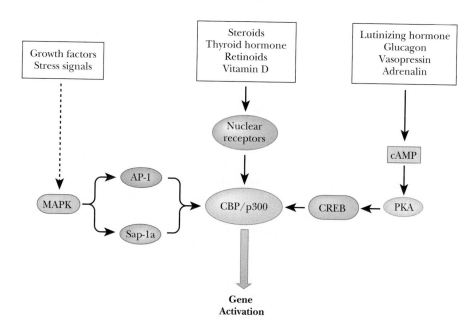

◀ **FIGURE 11.22** Multiple ways in which CREB-binding protein (CBP) and p300 are involved in gene expression. MAPK is mitogen-activated protein kinase. It acts on two other transcription factors, AP-1 and Sap-1a, which bind to CBP. Steroid hormones affect nuclear receptors, which then bind to CBP. Other hormones activate a cAMP cascade, leading to phosphorylation of CREB, which then binds to CBP. (*Adapted by permission from* Molecular Biology, *by R. F. Weaver, McGraw-Hill, 1999.*)

Biochemical Connections

CREB—The Most Important Protein You Have Never Heard Of?

Hundreds of genes are controlled by the cyclic-AMP-response element. CREs are bound by a family of transcription factors that include CREB, cyclic-AMP-response-element modulating protein (**CREM**), and activating transcription factor 1 (**ATF-1**). All these proteins share a high degree of homology, and all belong to the basic-region leucine zipper class of transcription factors (see Section 11.5). CREB itself is a 43-kDa protein with a critical serine at position 133 that can be phosphorylated. Transcription is activated only when CREB is phosphorylated at this site. CREB can be phosphorylated by a variety of mechanisms. The classical mechanism is via protein kinase A, which is stimulated by cAMP release. Protein kinase C, which is stimulated by Ca^{2+} release, and MAPK also phosphorylate CREB. The ultimate signals for these processes can be peptide hormones, growth or stress factors, or neuronal activity. Phosphorylated CREB does not act alone to stimulate transcription of its target genes. It works in concert with a 265-kDa protein, the CREB-binding protein (CBP), which connects CREB and the basal transcription machinery. More than 100 known transcription factors also bind to CBP. To add to the diversity of transcriptional control, CREB and CREM are both synthesized in alternate forms due to differ-ent posttranscriptional splicing mechanisms (see Section 11.6). In the case of CREM, some of the isoforms are stimulatory while others are inhibitory.

Although the research is still ongoing, CREB-mediated transcription has been implicated in a tremendous variety of physiological processes, such as *cell proliferation, cell differentiation,* and *spermatogenesis.* It controls *release of somatostatin,* a hormone that inhibits growth hormone secretion. It has been shown to be critical for *development of mature T-lymphocytes* (immune-system cells), and has been shown to confer *protection to nerve cells* in the brain under hypoxic conditions. It is involved in *metabolism of the pineal gland* and *control of circadian rhythms.* CREB levels have been shown to be elevated during the body's adaptation to strenuous physical exercise. It is involved in the *regulation of gluconeogenesis* by the peptide hormones, glucagon and insulin, and it directly affects *transcription of metabolic enzymes,* such as phosphoenolpyruvate carboxykinase (PEPCK) and lactate dehydrogenase. Most interestingly, CREB has been shown to be *critical in learning and storage in long-term memory,* and low levels of CREB have been found in brain tissue of those suffering from Alzheimer's disease.

11.5 | What Are Some Structural Motifs in DNA-Binding Proteins?

Proteins that bind to DNA during the course of transcription do so by the same types of interactions that we have seen in protein structures and enzymes—hydrogen bonding, electrostatic attractions, and hydrophobic interactions. Most proteins that activate or inhibit transcription by RNA polymerase II have two functional domains. One of them is the **DNA-binding domain,** and the other is the **transcription-activation domain.**

DNA-Binding Domains

Most DNA-binding proteins have domains that fall into one of three categories, **helix–turn–helix (HTH), zinc fingers,** and **basic-region leucine zipper (bZIP).** These domains interact with DNA in either the major or minor groove, with the major groove being more common.

Helix–Turn–Helix Motifs

A common feature seen in proteins that bind to DNA is the presence of a segment of α-helix that fits into the major groove. The width of the major groove and the α-helix are similar, so the protein helix can fit snugly. This is the most common motif because the standard form of DNA, B-DNA, has the major groove of the correct size, and no alterations in its topology are necessary. Such binding proteins are often dimers with two regions of HTH, as shown in Figure 11.23.

The HTH motif is a sequence of 20 amino acids that is relatively conserved in many different DNA-binding proteins. Table 11.6 shows the sequence for the HTH region of several transcription factors. The first helical region is

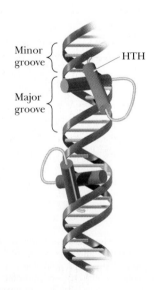

Minor groove
Major groove
HTH

▲ **FIGURE 11.23** The helix–turn–helix motif. Proteins containing the HTH motif bind to DNA via the major groove. (*With permission, from the* Annual Review of Biochemistry, *Volume 58* © *1989 by Annual Reviews.* www.annualreviews.org.)

Table 11.6

Amino Acid Sequences in the HTH Regions of Selected Transcription Regulatory Proteins

434 *Rep* and *Cro* are bacteriophage 434 proteins; *Lam Rep* and *Cro* are bacteriophage λ proteins; CAP, *trp Rep*, and *Lac Rep* are catabolite activator protein, Trp repressor, and *lac* repressor of *E. coli*, respectively. *Antp* is the homeodomain protein of the *Antennapedia* gene of the fruit fly *Drosophila melanogaster*. The numbers in each sequence indicate the location of the HTH within the amino acid sequences of the various polypeptides.

	Helix							Turn				Helix								
	1	2	3	4	5	6	7	8	9	10	11	12	13	14	15	16	17	18	19	20
434 Rep	17-Gln	Ala	Glu	**Leu**	Ala	Gln	Lys	**Val**	**Gly**	Thr	Thr	Gln	Gln	Ser	**Ile**	Glu	Gln	**Leu**	Glu	Asn-36
434 Cro	17-Gln	Thr	Glu	**Leu**	Ala	Thr	Lys	**Ala**	**Gly**	Val	Lys	Gln	Gln	Ser	**Ile**	Gln	Leu	**Ile**	Glu	Ala-36
Lam Rep	33-Gln	Glu	Ser	**Val**	Ala	Asp	Lys	**Met**	**Gly**	Met	Gly	Gln	Ser	Gly	**Val**	Gly	Ala	**Leu**	Phe	Asn-52
Lam Cro	16-Gln	Thr	Lys	**Thr**	Ala	Lys	Asp	**Leu**	**Gly**	Val	Tyr	Gln	Ser	Ala	**Ile**	Asn	Lys	**Ala**	Ile	His-35
CAP	169-Arg	Gln	Glu	**Ile**	**Gly**	Glu	Ile	**Val**	**Gly**	Cys	Ser	Arg	Glu	Thr	**Val**	Gly	Arg	**Ile**	Leu	Lys-18
Trp Rep	68-Gln	Arg	Glu	**Leu**	**Lys**	Asn	Glu	**Leu**	**Gly**	Ala	Gly	Ile	Ala	Thr	**Ile**	Thr	Arg	**Gly**	Ser	Asn-87
Lac Rep	6-Leu	Tyr	Asp	**Val**	Ala	Arg	Leu	**Ala**	**Gly**	Val	Ser	Tyr	Gln	Thr	**Val**	Ser	Arg	**Val**	Val	Asn-25
Antp	31-Arg	Ile	Glu	**Ile**	Ala	His	Ala	**Leu**	**Cys**	Leu	Thr	Glu	Arg	Gln	**Ile**	Lys	Ile	**Trp**	Phe	Gln-50

Source: Adapted from Harrison, S. C., and Aggarwal, A. K., 1990, DNA recognition by proteins with the helix–turn–helix motif. *Annual Review of Biochemistry* **59**, 933–969.

composed of the first eight residues of the region. A sequence of three or four amino acids separates it from the second helical region. Position 9 is a glycine involved in a β-turn (Chapter 4).

Proteins that recognize DNA with specific base sequences are more likely to bind to the major groove. The orientation of the bases in the standard base pairings puts more of the unique structure into the major groove. Figure 11.24 shows how glutamine and arginine can interact favorably with adenine and guanine, respectively. Some interactions, however, including many in the minor groove, only read the DNA indirectly. As discussed in Chapter 9, the B form of DNA is not as constant as was once thought. Local variations in helix structure occur based on the actual sequence, especially when there are A–T-rich areas. The bases undergo extensive propeller-twist. Many proteins bind to the edges of the bases that protrude into the minor groove. Studies have shown that artificial molecules that mimic the base protrusions into the minor groove are equally capable of binding to many transcription factors.

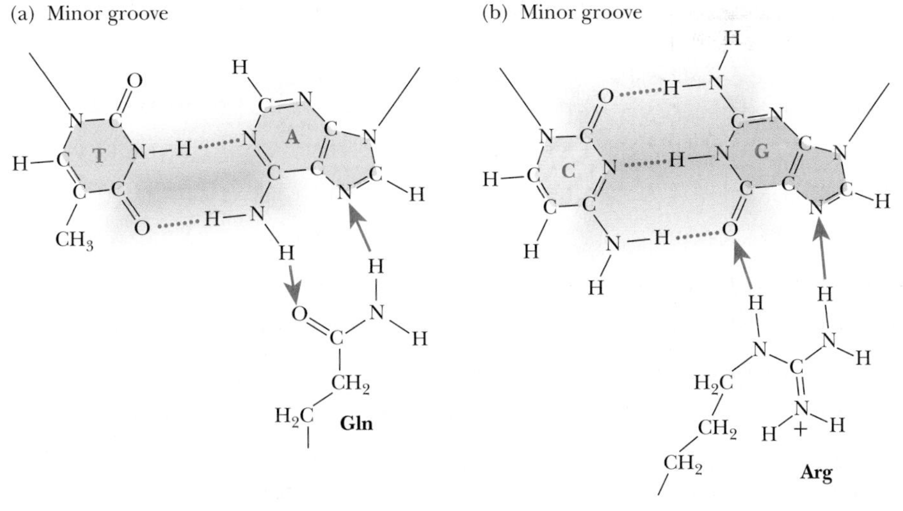

(a) Minor groove   (b) Minor groove

◀ **FIGURE 11.24** (a) Hydrogen-bonding interactions between glutamine and adenine. (b) Hydrogen-bonding interactions between arginine and guanine.

(a) (b)

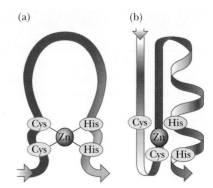

▲ **FIGURE 11.25** Cys$_2$His$_2$ zinc-finger motifs.
(a) The coordination between zinc and cysteine and histidine residues. (b) The secondary structure. (*Adapted from Evans, R. M., and Hollenberg, S. M., 1988.* Cell **52**, *1, Figure 1.*)

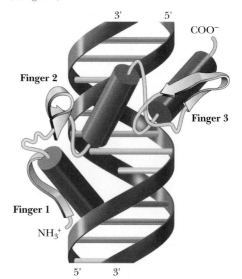

▲ **FIGURE 11.26** Zinc-finger proteins follow the major groove of DNA. (*Adapted with permission from Pavletich, N., and Pabo, C. O., 1991,* Science **252**, *809, Figure 2. Copyright © 1991 AAAS.*)

Thus, while the base pairing is clearly important to DNA, sometimes the overall shape of the bases is important for other reasons. A particular binding protein might not be recognizing the part of the base that is involved in hydrogen bonds but, rather, a part that is protruding into the grooves.

Zinc Fingers

In 1985, it was discovered that a transcription factor of RNA polymerase III, TFIIIA, had nine repeating structures of 30 amino acids each. Each repeat contained two closely spaced cysteines and two closely spaced histidines 12 amino acids later. It was also found that this factor had enough associated zinc ions to bind to each of the repeats. This led to the discovery of the zinc-finger domain in DNA-binding proteins, which is represented in Figure 11.25. The motif gets its name from the shape adopted by the 12 amino acids that are looped out from the intersection of the two cysteines and two histidines with the zinc ion. When TFIIIA binds to DNA, the repeated zinc fingers follow the major groove around the DNA, as shown in Figure 11.26.

Basic-Region Leucine Zipper Motif

The third major class of sequence dependent DNA-binding proteins is called the basic-region leucine zipper motif. Many transcription factors are known to contain this motif, including CREB (see the Biochemical Connections box on page 288). Figure 11.27 shows the sequence homology of several such transcription factors. Half of the protein is composed of the basic region with many conserved residues of lysine, arginine, and histidine. The second half contains a series of leucines every seven residues. The significance of the spacing of the leucines is clear. It takes 3.6 amino acids to make a turn of an α-helix. With a seven-residue spacing, the leucines will all line up on one side of an α-helix, as shown in Figure 11.28. The motif gets the name, "zipper," from the fact that the line of hydrophobic residues will interact with a second analogous protein fragment via hydrophobic bonds, interweaving themselves like a zipper. DNA-binding proteins with leucine zippers bind the DNA in the major groove via the strong electrostatic interactions between the basic region and the sugar phosphates. Protein dimers form, and the leucine half interacts

Protein	Basic region A	Basic region B	Leucine zipper
C/EBP	278–D K N S N E Y R V R R E R N N I A V R K S H D K A K Q R N V E T Q Q K V L E L T S D N D R L R K R V E Q L S R E L D T L R G–341		
Jun	257–S Q E R I K A E R K R M R N R I A A S K C H K R K L E R I A R L E E K V K T L K A Q N S E L A S T A N M L T E Q V A Q L K O–320		
Fos	233–E E R R R I R R I R R E R N K M A A A K C R N R R R E L T D T L Q A E T D Q L E D K K S A L Q T E I A N L L K E K E K L E F–296		
GCN4	221–P E S S D P A A L K R A R N T E A A R R S R A R K L Q R M K O L E D K V E E L L S K N Y H L E N E V A R L K K L V G E R–COOH		
YAP1	60–D L D P E T K Q K R T A Q N R A A Q R A F H E R K E R K M K E L E K K V Q S L E S I Q Q Q N E V E A T F L R D Q L I T L V N–123		
CREB	279–E E A A R K R E V R L M K N R E A A R E C R R K K K E Y V K C L E N R V A V L E N Q N K T L I E E L K A L K D L Y C H K S D–342		
Cys-3	95–A S R L A A E E D K R K R N T A A S A R F R I K K K Q R E Q A L E K S A K E M S E K V T Q L E G R I Q A L E T E N K Y L K G–148		
CPC1	211–E D P S D V V A M K R A R N T L A A R K S B E R K A Q R L E E L E A K I E E L I A E R D R Y K N L A L A H G A S T E–COOH		
HBP1	176–W D E R E L K K Q K R L S N R E S A R R S R L R K Q A E C E E L G Q R A E A L K S E N S S L R I E L D R I K K E Y E E L L S–239		
TGA1	68–S K P V E K V L R R L A Q R N E A A R K S R L R K K A Y V Q Q L E N S K L K L I Q L E Q E L E R A R K Q G M C V G G G V D A–131		
Opaque2	223–M P T E E R V R K R K E S N R E S A R R S R Y R K A A H L K E L E D Q V A Q L K A E N S C L L R R I A A L N Q K Y N D A N V–286		

▲ **FIGURE 11.27** Comparisons of the amino acid sequences of several DNA-binding proteins with basic-region leucine zippers. (*Adapted with permission from Vinson, C. R., Sigler, P. B., and McKnight, S. L. 1989.* Science **246**, *912, Figure 1. Copyright © 1989 AAAS.*)

with the other subunit, while the basic part interacts with the DNA, as shown in Figure 11.29.

Transcription-Activation Domains

The three motifs mentioned above are involved in the binding of transcription factors to DNA. Not all transcription factors bind directly to DNA, however. Some bind to other transcription factors and never contact the DNA. An example is CBP, which bridges CREB and the RNA polymerase II transcription-initiation complex. The motifs whereby transcription factors recognize other proteins can be broken down into three categories:

1. *Acidic domains* are regions rich in acidic amino acids. *Gal4* is a transcription factor in yeast that activates the genes for metabolizing galactose. It has a domain of 49 amino acids, 11 of which are acidic.

2. *Glutamine-rich domains* are seen in several transcription factors. *Sp1* is an upstream transcription factor that activates transcription in the presence of an additional promoter element called a *GC box*. It has two glutamine-rich domains, one of which contains 39 glutamines in 143 amino acids. CREB and CREM (see the Biochemical Connections box on page 288) also have this domain.

3. A *proline-rich domain* is seen in the activator *CTF-1*. It has a domain of 84 amino acids, 19 of which are prolines. CTF-1 is a member of a class of transcription factors that bind to an extended promoter element called a CCAAT box. The N-terminal domain has been shown to regulate transcription of certain genes. The C-terminal end is a transcription regulator and is known to bind to histone proteins via the proline repeats. An active area of study is how transcription is linked to the acetylation of histones. The coactivator CBP, which was discussed in the previous section, is also a histone acetyl transferase. See the article by Struhl in the Annotated Bibliography of this chapter.

Despite the seemingly overwhelming complexity of transcription factors, their elucidation has been made more manageable by the similarities in the motifs described in this section. For example, if a new protein is discovered or a new DNA sequence is elucidated, evidence of its role as a transcription factor can be determined by locating the DNA-binding protein motifs discussed in this section.

11.6 | How Is RNA Modified after Transcription?

The three principal kinds of RNA—tRNA, rRNA, and mRNA—are all modified enzymatically after transcription to give rise to the functional form of the RNA in question. The type of processing in prokaryotes can differ greatly from that in eukaryotes, especially in the case of mRNA. The initial size of the RNA transcripts is greater than the final size because of leader sequences at the 5′ end and trailer sequences at the 3′ end. The leader and trailer sequences must be removed, and other forms of *trimming* are also possible. *Terminal sequences* can be added after transcription, and *base modification* is frequently observed, especially in tRNA.

Transfer RNA and Ribosomal RNA

The precursor of several tRNA molecules is frequently transcribed in one long polynucleotide sequence. All three types of modification—trimming, addition of terminal sequences, and base modification—take place in the

Biochemistry ⊛ Now™
Go to BiochemistryNow and click on Biochemistry Interactive to learn more about leucine zippers.

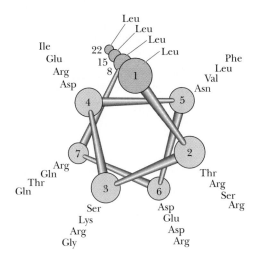

▲ **FIGURE 11.28** The helical wheel structure of a basic-region leucine zipper of a typical DNA-binding protein. The amino acids listed show the progression down the helix. Note that the leucines line up along one side, forming a hydrophobic spine. (*Adapted with permission from Landschulz, W. H., Johnson, P. F., and McKnight, S. L., 1988,* Science **240**, *1759–1764, Figure 1. Copyright © 1988 AAAS.*)

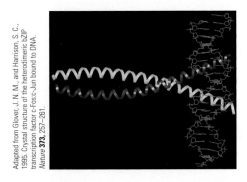

Adapted from Glover, J. N. M. and Harrison, S. C., 1995. Crystal structure of the heterodimeric bZIP transcription factor c-Fos:c-Jun bound to DNA. *Nature* **373**, 257–261.

▲ **FIGURE 11.29** Crystal structure of the bZIP transcription factor c-Fos:c-Jun bound to a DNA oligomer containing the AP-1 consensus target sequence TGACTCA. The basic region binds to the DNA while the leucine regions of the two helices bind via hydrophobic interactions.

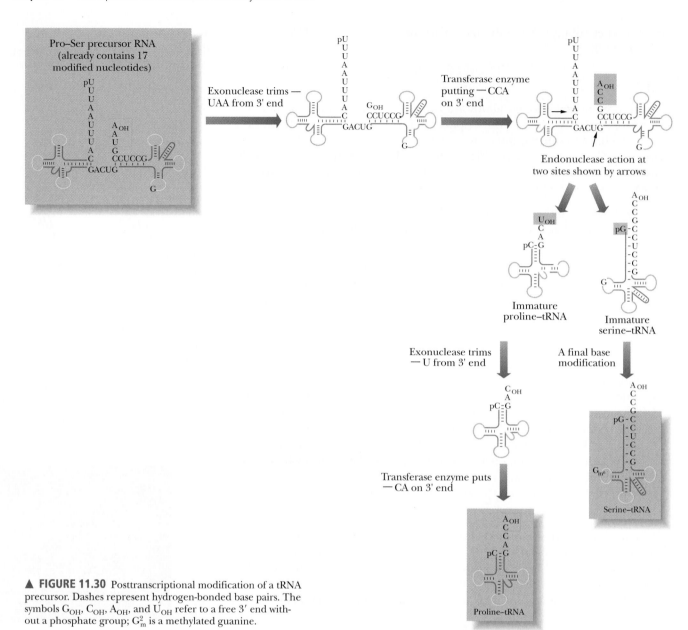

▲ **FIGURE 11.30** Posttranscriptional modification of a tRNA precursor. Dashes represent hydrogen-bonded base pairs. The symbols G_{OH}, C_{OH}, A_{OH}, and U_{OH} refer to a free 3′ end without a phosphate group; G_m^2 is a methylated guanine.

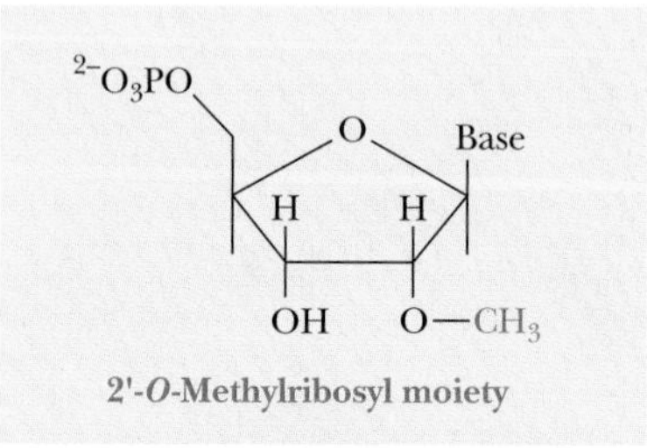

▲ **FIGURE 11.31** The structure of a nucleotide containing a 2′-*O*-methylribosyl group.

2′-*O*-Methylribosyl moiety

transformation of the initial transcript to the mature tRNAs (Figure 11.30). (The enzyme responsible for generating the 5′ ends of all *E. coli* tRNAs, *RNase P*, consists of both RNA and protein.) The RNA moiety is responsible for the catalytic activity. This was one of the first examples of catalytic RNA (Section 11.7). Some base modifications take place before trimming, and some occur after. Methylation and substitution of sulfur for oxygen are two of the more usual types of base modification. (See Section 9.2 and "Transfer RNA" in Section 9.5 for the structures of some of the modified bases.) One type of methylated nucleotide found only in eukaryotes contains a 2′-*O*-methylribosyl group (Figure 11.31).

The trimming and addition of terminal nucleotides produce tRNAs with the proper size and base sequence. Every tRNA contains a CCA sequence at the 3′ end. The presence of this portion of the molecule is of great importance in protein synthesis because the 3′ end is the acceptor for amino acids to be added to a growing protein chain (Chapter 12). Trimming of large precursors of eukaryotic tRNAs takes place in the nucleus, but most methylating enzymes occur in the cytosol.

The processing of rRNAs is primarily a matter of methylation and of trimming to the proper size. In prokaryotes, there are three rRNAs in an intact ribosome, which has a sedimentation coefficient of 70S. (Sedimentation coefficients and some aspects of ribosomal structure are discussed in "Ribosomal RNA" in Section 9.5.) In the smaller subunit, which has a sedimentation coefficient of 30S, one RNA molecule has a sedimentation coefficient of 16S. The 50S subunit contains two kinds of RNA, with sedimentation coefficients of 5S and 23S. The ribosomes of eukaryotes have a sedimentation coefficient of 80S, with 40S and 60S subunits. The 40S subunit contains an 18S RNA, and the 60S subunit contains a 5S RNA, a 5.8S RNA, and a 28S RNA. Base modifications in both prokaryotic and eukaryotic rRNA are accomplished primarily by methylation.

Messenger RNA

Extensive processing takes place in eukaryotic mRNA. Modifications include **capping** of the 5′ end, **polyadenylating** (adding a poly-A sequence to) the 3′ end, and **splicing** of coding sequences. Such processing is not a feature of the synthesis of prokaryotic mRNA.

The cap at the 5′ end of eukaryotic mRNA is a guanylate residue that is methylated at the N-7 position. This modified guanylate residue is attached to the neighboring residue by a 5′-5′ triphosphate linkage (Figure 11.32). The 2′-hydroxyl group of the ribosyl portion of the neighboring residue is frequently methylated, and sometimes that of the next nearest neighbor is as well. The **polyadenylate tail** (abbreviated poly-A, or $poly[r(A)_n]$) at the 3′ end of a message (typically 100 to 200 nucleotides long) is added before the mRNA leaves the nucleus. It is thought that the presence of the tail protects the mRNA from nucleases and phosphatases, which would degrade it. According to this point of view, the adenylate residues would be cleaved off before the portion of the molecule that contains the actual message is attacked. The presence of the 5′ cap also protects the mRNA from exonuclease degradation.

The presence of the poly-A tail has been very fortuitous for researchers. By designing an affinity chromatography column (Chapters 5 and 13) with a **poly-T tail** (or poly[d(T)] tail), the isolation of mRNA from a cell lysate can be quickly accomplished. This enables the study of transcription by looking at which genes are being transcribed at a particular time under various cell conditions.

The genes of prokaryotes are continuous; every base pair in a continuous prokaryotic gene is reflected in the base sequence of mRNA. The genes of eukaryotes are not necessarily continuous; eukaryotic genes frequently contain intervening sequences that do not appear in the final base sequence of the mRNA for that gene product. The DNA sequences that are expressed (the ones actually retained in the final mRNA product) are called **exons.** The intervening sequences, which are not expressed, are called **introns.** Such genes are often referred to as **split genes.** The expression of a eukaryotic gene involves not just its transcription but also the processing of the primary transcript into its final form. Figure 11.33 shows how a split gene might be processed. When the gene is transcribed, the mRNA transcript contains regions at the 5′ and 3′ ends that are not translated and several introns shown in green. The introns are removed, linking the exons together. The 3′ end is modified by adding a poly-A tail and a 7-mG cap to yield the mature mRNA.

Some genes have very few introns, while others have many. There is one intron in the gene for the muscle protein actin; there are two for both the

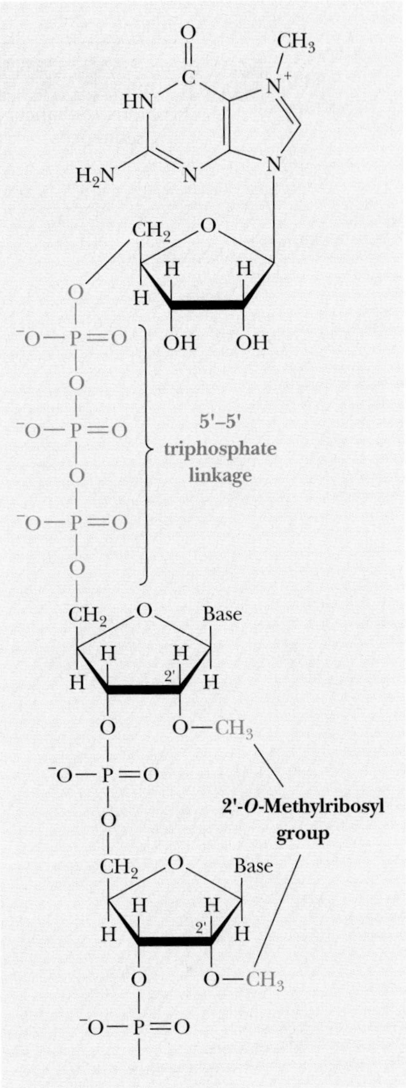

▲ **FIGURE 11.32** The structures of some typical mRNA caps.

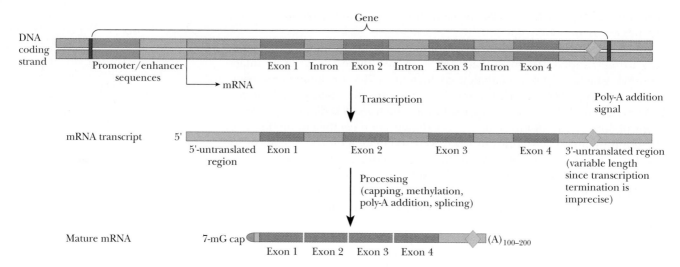

▲ **FIGURE 11.33** The organization of split genes in eukaryotes.

α- and β-chains of hemoglobin, three for lysozyme, and so on, up to as many as 50 introns in a single gene. The pro α-2 collagen gene in chickens is about 40,000 base pairs long, but the actual coding regions amount to only 5000 base pairs spread out over 51 exons. With so much splicing needed, the splicing mechanisms must be very accurate. Splicing is a little easier because the genes have the exons in the correct order, even if they are separated by introns. Also, the primary transcript is usually spliced in the same positions in all tissues of the organism.

A major exception to this is the splicing that occurs with immunoglobulins, in which antibody diversity is maintained by having multiple ways of splicing mRNA. In the last few years, more eukaryotic proteins that are the products of alternative splicing have been discovered. The need for this was also demonstrated by the preliminary data from the Human Genome Project. Differential splicing would be necessary to explain the fact that the known number of proteins exceeds the number of human genes found.

The Splicing Reaction: Lariats and Snurps

The removal of intervening sequences takes place in the nucleus, where RNA forms **ribonucleoprotein particles (RNPs)** through association with a set of nuclear proteins. These proteins interact with RNA as it is formed, keeping it in a form that can be accessed by other proteins and enzymes. The substrate for splicing is the capped, polyadenylated pre-mRNA. Splicing requires cleavage at the 5′ and 3′ end of introns and the joining of the two ends. This process must be done with great precision to avoid shifting the sequence of the mRNA product. Specific sequences make up the *splice sites* for the process, with GU at the 5′ end and AG at the 3′ end of the introns in higher eukaryotes. A *branch site* within the intron also has a conserved sequence. This site is found 18 to 40 bases upstream from the 3′ splice site. The branch site sequence in higher eukaryotes is PyNPyPuAPy, where Py represents any pyrimidine and Pu any purine. N can be any nucleotide. The A is invariant.

Figure 11.34 shows how splicing occurs. The G that is always present on the 5′ end of the intron loops back in close contact with the invariant A from the branch point. The 2′ hydroxyl of the A performs a nucleophilic attack on the phosphodiester backbone at the 5′ splice site, forming a *lariat* structure and releasing exon 1. The AG at the 3′ end of the exon then does the same

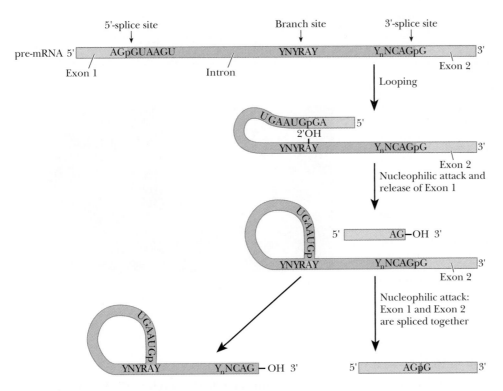

▲ **FIGURE 11.34** Splicing of mRNA precursors. Exon 1 and exon 2 are separated by the intervening sequence (intron) shown in green. In the splicing together of the two exons, a lariat forms in the intron. (*Adapted from Sharp, P. A., 1987,* Science **235,** *766, Figure 1.*)

to the G at the 3′ splice site, fusing the two exons. These lariat structures can be seen with an electron microscope, although the structure is inherently unstable and soon is linearized.

The splicing also depends on **small nuclear ribonucleoproteins, or snRNPs** (pronounced "snurps"), to mediate the process. This snRNP is a fourth basic type of RNA, separate from mRNA, tRNA, and rRNA. The snRNPs, as their name implies, contain both RNA and proteins. The RNA portion is between 100 and 200 nucleotides in higher eukaryotes, and there are ten or more proteins. With more than 100,000 copies of some snRNPs in eukaryotic cells, snRNPs are one of the most abundant gene products. They are enriched in uridine residues and are therefore often given names like U1 and U2. snRNPs also have an internal consensus sequence of AUUUUG. The snRNPs bind to the RNAs being spliced via complementary regions between the snRNP and the branch and splice sites. The actual splicing involves a 50S to 60S particle called the **spliceosome,** which is a large multisubunit particle similar in size to a ribosome. Several different snRNPs are involved, and there is an ordered addition of them to the complex. In addition to their role in splicing, certain snRNPs have been found to stimulate transcription elongation. It is now widely recognized that some RNAs can catalyze their own self-splicing, as will be discussed in Section 11.7. The present process involving ribonucleoproteins may well have evolved from the self-splicing of RNAs. An important similarity between the two processes is that both proceed via a lariat mechanism by which the splice sites are brought together. (For additional information, see the article by Steitz listed in the Annotated Bibliography at the end of this chapter.) The following Biochemical Connections box describes an autoimmune disease that develops when the body makes antibodies to one of these snRNPs.

Biochemical Connections

Lupus: An Autoimmune Disease Involving RNA Processing

Systemic lupus erythematosus (SLE) is an autoimmune disease that can have fatal consequences. It starts, usually in late adolescence or early adulthood, with a rash on the forehead and cheekbones, giving the wolflike appearance from which the disease takes its name. (*Lupus* means "wolf" in Latin.) Severe kidney damage may follow, along with arthritis, accumulation of fluid around the heart, and inflammation of the lungs. About 90% of lupus patients are women. It has been established that this disease is of autoimmune origin, specifically from the production of antibodies to one of the snRNPs, *U1-snRNP.* This snRNP is so designated because it contains a uracil-rich RNA, U1-snRNA, which recognizes the 5′ splice junction of mRNA. Because the processing of mRNA affects every tissue and organ in the body, this disease affects widely dispersed target areas and can spread easily.

▲ A characteristic rash is frequently seen on the cheekbones and foreheads of victims of systemic lupus erythematosus.

Alternative RNA Splicing

Gene expression can also be controlled at the level of RNA splicing. Many proteins are always spliced in the same way, but many others can be spliced in different ways to give different **isoforms** of the protein to be produced. In humans, 5% of the proteins produced have isoforms based on alternative splicing. These differences may be seen by having two forms of the mRNA in the same cell, or there might be only one form in one tissue, but a different form in another tissue. Regulatory proteins can affect the recognition of splice sites and direct the alternative splicing.

It has been found that a protein called Tau accumulates in the brain of people afflicted with Alzheimer's disease. This protein has six isoforms generated by differential splicing, with the forms appearing during specific developmental stages. The human troponin T gene produces a muscle protein that has many isoforms due to differential splicing. Figure 11.35 shows the complexity of this gene. There are 18 exons that can be linked together to make

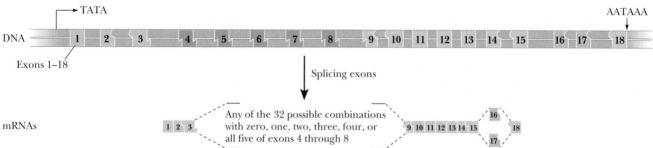

▲ **FIGURE 11.35** Organization of the fast skeletal muscle troponin T gene and the 64 possible mRNAs that can be generated from it. Exons shown in orange are constitutive, appearing in all mRNAs produced. Exons in green are combinatorial, giving rise to all possible combinations from zero to all five. The exons in blue and red are mutually exclusive: Only one or the other may be used. (*With permission, from the* Annual Review of Biochemistry, *Volume 58,* © *1989 by Annual Reviews. www.annualreviews.org.*)

the mature mRNA. Some of them are always present, such as exons 1–3 and 9–15, which are always linked together in their respective orders. However, exons 4 through 8 can be added in any combination group of 32 possible combinations. On the right side, either exon 16 or 17 is used, but not both. This leads to a total of 64 possible troponin molecules, which highlights the tremendous diversity in protein structure and function that can come from splicing mRNA.

11.7 How Does RNA Act as an Enzyme?

There was a time when proteins were considered to be the only biological macromolecules capable of catalysis. The discovery of the catalytic activity of RNA has thus had a profound impact on the way biochemists think. A few enzymes with RNA components had been discovered, such as telomerase (Chapter 10) and RNase P, an enzyme that cleaves extra nucleotides off the 5′ ends of tRNA precursors. It was later shown that the RNA portion of RNase P has the catalytic activity. The field of catalytic RNA (ribozymes) was launched in earnest by the discovery of RNA that catalyzes its own self-splicing. It is easy to see a connection between this process and the splicing of mRNA by snRNPs. More recently, it has been shown that RNAs can catalyze reactions involved in protein synthesis, as will be explained further in Chapter 12. The catalytic efficiency of catalytic RNAs is less than that of protein enzymes, and the catalytic efficiency of currently existing RNA systems is greatly enhanced by the presence of protein subunits in addition to the RNA. Recall that many important coenzymes include an adenosine phosphate moiety in their structure (Section 7.8). Compounds of such central importance in metabolism must be of ancient origin, another piece of evidence in support of the idea of an RNA-based world, where RNA was the original genetic molecule and the original catalytic one as well.

Several groups of ribozymes are known to exist. In **Group I ribozymes,** there is a requirement for an external guanosine, which becomes covalently bonded to the splice site in the course of excision. An example is the self-splicing that takes place in pre-rRNA of the ciliate protist *Tetrahymena* (Figure 11.36). The transesterification (of phosphoric acid esters) that takes place here releases one end of the intron. The free 3′-OH end of the exon attacks the 5′ end of the other exon, splicing the two exons and releasing the intron. The free 3′-OH end of the intron then attacks a nucleotide 15 residues from the 5′ end, cyclizing the intron and releasing a 5′ terminal sequence. The precision of this sequence of reactions depends on the folded conformation of the RNA, which remains internally hydrogen bonded throughout the process. In vitro, this catalytic RNA can act many times, being regenerated in the usual way for a true catalyst. In vivo, however, it appears to act only once by splicing itself out. **Group II ribozymes** display a lariat mechanism of operation similar to the mechanism seen in Section 11.6 that was facilitated by snRNPs. There is no requirement for an external nucleotide; the 2′-OH of an internal adenosine attacks the phosphate at the 5′ splice site. Clearly, DNA cannot self-splice in this fashion because it does not have a 2′-OH.

The folding of the RNA is crucial to its catalytic activity, as is the case with protein catalysts. A divalent cation (Mg^{2+} or Mn^{2+}) is required; it is quite

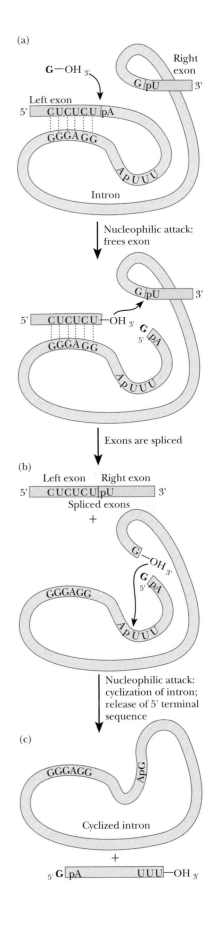

▶ **FIGURE 11.36** The self-splicing of pre-rRNA of the ciliate protist *Tetrahymena,* a Class I ribozyme. (a) A guanine nucleotide attacks at the splice site of the exon on the left, giving a free 3′-OH end. (b) The free 3′-OH end of the exon attacks the 5′ end of the exon on the right, splicing the two exons and releasing the intron. (c) The free 3′-OH end of the intron then attacks a nucleoside 15 residues from the 5′ end, cyclizing the intron and releasing a 5′ terminal sequence.

likely that metal ions stabilize the folded structure by neutralizing some of the negative charges on the phosphate groups of the RNA. A divalent cation is essential for the functioning of the smallest ribozymes known, the hammerhead ribozymes, which can be catalytically active with as few as 43 nucleotides. (The name comes from the fact that their structures resemble the head of a hammer when shown in conventional representations of hydrogen-bonded secondary structure.) (See the article by Doudna listed in the Annotated Bibliography at the end of this chapter.) The folding of RNA is such that large-scale conformational changes can take place with great precision. Similar large-scale changes take place in the ribosome in protein synthesis and in the spliceosome in the processing of mRNA. Note that they remain RNA machines when proteins have taken over much of the catalytic functioning of the cell. The ability of RNA to undergo the requisite large-scale conformational changes may well play a role in the process. A recently proposed clinical application of ribozymes has been suggested. If a ribozyme can be devised that can cleave the RNA genome of HIV, the virus that causes AIDS (Section 14.6), it will be a great step forward in the treatment of this disease. Research on this topic is in progress in several laboratories. (See the article by Barinaga listed in the Annotated Bibliography at the end of this chapter.)

Summary

11.1 How Does Transcription Take Place in Prokaryotes? RNA synthesis is the transcription of the base sequence of DNA to that of RNA. All RNAs are synthesized on a DNA template; the enzyme that catalyzes the process is DNA-dependent RNA polymerase. All four ribonucleoside triphosphates—ATP, GTP, CTP, and UTP—are required, as is Mg^{2+}. There is no need for a primer in RNA synthesis. As is the case with DNA biosynthesis, the RNA chain grows from the 5′ to the 3′ end. The enzyme uses one strand of the DNA (the antisense strand, or template strand) as the template for RNA synthesis. In prokaryotic transcription, RNA polymerase is directed to the gene to be transcribed by the interactions between the polymerase's σ-subunit and sequences of DNA near the start site called promoters. Consensus sequences have been established for prokaryotic promoters, and the key elements are sequences at −35 and −10, the latter called the Pribnow box.

11.2 How Is Transcription Regulated in Prokaryotes? Frequency of transcription is controlled by the promoter sequence. Additional sequences upstream can also be involved in regulating prokaryotic transcription. These sequences are called enhancers or silencers, and they stimulate or inhibit transcription, respectively. Proteins called transcription factors can bind to these enhancer or silencer elements. Many prokaryotic genes that produce proteins that are part of a pathway are controlled in groups called operons.

11.3 How Does Transcription Take Place in Eukaryotes? Eukaryotic transcription is much more complicated. Three RNA polymerases exist to handle the different RNA types. RNA polymerase II transcribes messenger RNA. The promoters are very important in eukaryotic transcription, and a key promoter element is the TATA box at −25 from the transcription start site. RNA polymerase II requires many general transcription factors for proper binding and initiation of transcription. In addition, there are also many enhancer elements.

11.4 How Is Transcription Regulated in Eukaryotes? Many eukaryotic genes are controlled in clusters by enhancers that are responsive to cellular conditions. These are called response elements. The cyclic-AMP-response element (CRE) is one of the most important.

11.5 What Are Some Structural Motifs in DNA-Binding Proteins? Many DNA-binding proteins are involved in eukaryotic transcription, and most of them can be categorized by a few common structural motifs, such as zinc fingers, leucine zippers, and helix–turn–helix.

11.6 How Is RNA Modified after Transcription? Post-transcriptional processing takes place in RNA. Base modification frequently occurs, as does trimming of long polynucleotide chains. In eukaryotes, the removal of intervening sequences, which reflect the base sequence of a portion of the DNA not expressed in the mature RNA, is an important step in the processing of RNA.

11.7 How Does RNA Act as an Enzyme? Some forms of RNA, called ribozymes, have catalytic activity. The mechanism by which such RNAs catalyze their own self-splicing may have an evolutionary relationship to the processing of eukaryotic mRNA.

Critical Questions to Review

11.1 How Does Transcription Take Place in Prokaryotes?

1. **Fact Check** What is the difference in the requirement for a primer in RNA transcription compared to DNA replication?

2. **Fact Check** List three important properties of RNA polymerase from *E. coli.*

3. **Fact Check** What is the subunit composition of *E. coli* RNA polymerase?

4. **Fact Check** What is the difference between the core enzyme and the holoenzyme?

5. **Fact Check** What are the different terms used to describe the two strands of DNA involved in transcription?

6. **Fact Check** Define promoter region, and list three of its properties.

7. **Fact Check** Put the following in linear order: UP element, Pribnow box, TSS, −35 region, Fis site.

8. **Fact Check** Distinguish between rho-dependent termination and intrinsic termination.

9. **Thought Question** Diagram a section of DNA being transcribed. Give the various names for the two strands of DNA.

11.2 How Is Transcription Regulated in Prokaryotes?

10. **Fact Check** Define inducer and repressor.

11. **Fact Check** What is a σ factor? Why is it important in transcription?

12. **Fact Check** What is the difference between σ^{70} and σ^{32}?

13. **Fact Check** What is the function of the catabolite activator protein?

14. **Fact Check** What is transcription attenuation?

15. **Thought Question** What role does an operon play in the synthesis of enzymes in prokaryotes?

16. **Thought Question** Diagram a termination of transcription showing how inverted repeats can be involved in releasing the RNA transcript.

17. **Thought Question** Give an example of a system in which alternative σ factors can control which genes are transcribed. Explain how this works.

18. **Thought Question** Explain, with diagrams, how transcription attenuation works in the *trp* operon.

11.3 How Does Transcription Take Place in Eukaryotes?

19. **Fact Check** Define exon and intron.

20. **Fact Check** What are some of the main differences between transcription in prokaryotes and in eukaryotes?

21. **Fact Check** What are the products of the reactions of the three eukaryotic RNA polymerases?

22. **Fact Check** List the components of eukaryotic Pol II promoters.

23. **Fact Check** List the Pol II general transcription factors.

24. **Thought Question** What are the functions of TFIIH?

11.4 How Is Transcription Regulated in Eukaryotes?

25. **Fact Check** Describe the function of three eukaryotic response elements.

26. **Fact Check** What is the purpose of CREB?

27. **Thought Question** How does regulation of transcription in eukaryotes differ from regulation of transcription in prokaryotes?

28. **Thought Question** What is the mechanism of transcription attenuation?

29. **Thought Question** How do the roles of enhancers and silencers differ from each other?

30. **Thought Question** How do response elements modulate RNA transcription?

31. **Thought Question** Diagram a gene that is affected by CRE and CREB, showing which proteins and nucleic acids contact each other.

32. **Thought Question** Explain the relationship between TFIID, TBP, and TAFs.

33. **Thought Question** Defend or attack this statement: "All eukaryotic promoters have TATA boxes."

34. **Thought Question** Explain the different ways in which eukaryotic transcription elongation is controlled.

35. **Thought Question** Explain the importance of CREB, giving examples of genes activated by it.

36. **Thought Question** Give examples of structural motifs found in transcription factors that interact with other proteins instead of DNA.

11.5 What Are Some Structural Motifs in DNA-Binding Proteins?

37. **Fact Check** List three important structural motifs in DNA-binding proteins.

38. **Thought Question** Give examples of the major structural motifs in DNA-binding proteins, and explain how they bind.

11.6 How Is RNA Modified after Transcription?

39. **Fact Check** List several ways in which RNA is processed after transcription.

40. **Fact Check** How is lupus related to snRNPs?

41. **Fact Check** What do the proteins Tau and Troponin have in common?

42. **Thought Question** Why is a trimming process important in converting precursors of tRNA and rRNA to the active forms?

43. **Thought Question** List three molecular changes that take place in the processing of eukaryotic mRNA.

44. **Thought Question** What are snRNPs? What is their role in the processing of eukaryotic mRNAs?

45. **Thought Question** What roles can RNA play, other than that of transmission of the genetic message?

46. **Thought Question** Diagram the formation of a lariat in RNA processing.

47. **Thought Question** Explain how differential splicing of RNA is thought to be relevant to the information gathered from the Human Genome Project.

11.7 How Does RNA Act as an Enzyme?

48. **Fact Check** What is a ribozyme? List some examples of ribozymes.

49. **Thought Question** Outline a mechanism by which RNA can catalyze its own self-splicing.

50. **Thought Question** Why are proteins more effective catalysts than RNA molecules?

Biochemistry ⓔNow™

Assess your understanding of this chapter's topics with additional quizzing and tutorials at **http://now.brookscole.com/campbell5**

Annotated Bibliography

Barinaga, M. Ribozymes: Killing the Messenger. *Science* **262**, 1512–1514 (1993). [A report on research designed to use ribozymes to attack the RNA genome of HIV.]

Bentley, D. RNA Processing: A Tale of Two Tails. *Nature* **395**, 21–22 (1998). [The relationship between RNA processing and the structure of the RNA polymerase that made it.]

Brivanlou, A. H., and J. E. Darnell. Signal Transduction and the Control of Gene Expression. *Science* **295**, 813–818 (2002). [A comprehensive review of eukaryotic transcription factors.]

Bushnell, D. A., K. D. Westover, R. E. Davis, and R. D. Kornberg. Structural Basis of Transcription: An RNA Polymerase II-TFIIB Cocrystal at 4.5 Angstroms. *Science* **303**, 983–988 (2004).

Cammarota, M., et al. Cyclic AMP-Responsive Element Binding Protein in Brain Mitochondria. *J. Neurochem.* **72** (6), 2272–2277 (1999). [An article about the suspected relationship of CREB to memory.]

Cech, T. R. RNA as an Enzyme. *Sci. Amer.* **255** (5), 64–75 (1986). [A description of the discovery that some RNAs can catalyze their own self-splicing. The author was a recipient of the 1989 Nobel Prize in chemistry for this work.]

Cramer, P., et al. Architecture of RNA Polymerase II and Implications for the Transcription Mechanism. *Science* **288**, 640–649 (2000). [Recent structural analysis of RNA polymerase to very high resolution has shed more light on transcription mechanisms.]

De Cesare, D., and P. Sassone-Corsi. Transcriptional Regulation by Cyclic AMP-Responsive Factors. *Prog. Nucleic Acid Res. Mol. Biol.* **64**, 343–369 (2000). [A review of cAMP response elements.]

DeHaseth, P. L., and T. W. Nisen. When a Part Is as Good as the Whole. *Science* **303**, 1307–1308 (2004). [An article describing the structure and function of RNA polymerase.]

Doudna, J. A. RNA Structure: A Molecular Contortionist. *Nature* **388**, 830–831 (1997). [An article about RNA structure and its relationship to transcription.]

Fong, Y. W., and Q. Zhou. Stimulatory Effect of Splicing Factors on Transcriptional Elongation. *Nature* **414**, 929–933 (2001). [An article about the link between transcription and splicing.]

Grant, P. A., and J. L. Workman. Transcription: A Lesson in Sharing? *Nature* **396**, 410–411 (1998). [A review of TBP and TAFs that discusses the requirements for transcription.]

Hanawalt, P. C. DNA Repair: The Bases for Cockayne Syndrome. *Nature* **405**, 415 (2000). [The relationship between DNA repair and transcription factors.]

Kuras, L., and K. Struhl. Binding of TBP to Promoters *in vivo* Is Stimulated by Activators and Requires Pol II Holoenzyme. *Nature* **399**, 609–613 (1999). [The nature of the TATA box and binding proteins.]

Kuznedelov, K., L. Minakhin, A. Niedzuda-Majka, S. L. Dove, D. Rogulja, B. E. Nickels, A. Hochschild, T. Heyduk, and K. Severinov. A Role for Interaction of the RNA Polymerase Flap Domain with the σ-Subunit in Promoter Recognition. *Science* **295**, 855–857 (2002). [The most recent structural information on bacterial RNA polymerase binding.]

Mandelkow, E. Alzheimer's Disease: The Tangled Tale of Tau. *Nature* **402**, 588–589 (1999). [Phosphorylation of a neuron protein leads to neurofibrillary tangles that are associated with Alzheimer's dementia.]

Montminy, M. Transcriptional Activation: Something New to Hang Your HAT On. *Nature* **387**, 654–655 (1997). [Transcription in eukaryotes requires opening the DNA/histone complex, which can be controlled by acetylation.]

Rhodes, D., and A. Klug. Zinc Fingers. *Sci. Amer.* **268** (2), 56–65 (1993). [How the structure of these zinc-containing proteins enables them to play a role in regulating the activity of genes.]

Riccio, A., et al. Mediation by a CREB Family Transcription Factor of NGF-Dependent Survival of Sympathetic Neurons. *Science* **286**, 2358–2361 (1999). [How CREB may help protect neurons in times of stress.]

Steitz, J. A. Snurps. *Sci. Amer.* **258** (6), 56–63 (1988). [A discussion of the role of small nuclear ribonucleoproteins, or snRNPs, in the removal of introns from mRNA.]

Struhl, K. A Paradigm for Precision. *Science* **293**, 1054–1055 (2001). [A recent article discussing how transcription factors, coactivators, and histone acetyltransferases work together to enhance transcription.]

Tupler, R., G. Perini, and M. R. Green. Expressing the Human Genome. *Nature* **409**, 832–833 (2001). [An excellent review of transcription and the use of data from the Human Genome Project to search for transcription factors.]

Westover, K. D., D. A. Bushnell, and R. D. Kornberg. Structural Basis of Transcription: Separation of RNA from DNA by RNA Polymerase II. *Science* **303**, 1014–1016 (2004). [An article describing the mechanism of separation of the DNA and RNA during transcription.]

Young, B. A., T. M. Gruber, and C. A. Gross. Minimal Machinery of RNA Polymerase Holoenzyme Sufficient for Promoter Melting. *Science* **303**, 1382–1384 (2004).

Protein Synthesis: Translation of the Genetic Message

After the DNA base sequence has been transcribed to RNA, the genetic code is needed to translate the RNA sequence into the amino acid sequence of a protein. In eukaryotic cells, DNA is transcribed in the nucleus but typically translated in the cytosol. The transcript is exported from the nucleus in the form of messenger RNA, which is read and translated at the ribosome. Molecules of transfer RNA, one for each amino acid, are required to collect activated amino acids and deliver them, one at a time, to the ribosome. There they are sequentially joined to synthesize the polypeptide chain of a protein. The sequence of amino acids, derived from the sequence of DNA bases, is specified by the genetic code, using the four RNA bases A, U, G, and C, taken three at a time. In the triplet code, there are 64 possible "code words," called codons, of which 3 are stop signals and 61 specify the 20 amino acids with considerable redundancy. In the actual mechanism of translation, messenger RNA is temporarily bonded to transfer RNA. As tRNA molecules deliver their amino acids to the ribosome, in succession, peptide bonds covalently join the amino acids to form the growing polypeptide chain. A stop signal on the messenger RNA terminates the protein chain, which is then released from the ribosome. This whole mechanism is subject to outside takeover when a virus infects a cell, sometimes with drastic consequences.

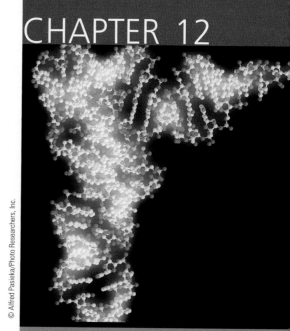

© Alfred Pasieka/Photo Researchers, Inc.

Transfer RNA brings amino acids to the site at which they are incorporated into a growing polypeptide chain.

Critical Questions

12.1 What Is the Overall Process of Translating the Genetic Message?

12.2 What Is the Genetic Code?

12.3 What Is the Role of Aminoacyl-tRNA Synthetases in Amino Acid Activation?

12.4 How Does Translation Take Place in Prokaryotes?

12.5 How Does Translation Take Place in Eukaryotes?

12.6 How Does Posttranslational Modification of Proteins Take Place?

12.7 How Are Proteins Degraded?

12.1 | What Is the Overall Process of Translating the Genetic Message?

Protein biosynthesis is a complex process requiring ribosomes, messenger RNA (mRNA), transfer RNA (tRNA), and a number of protein factors. The ribosome is the site of protein synthesis. The mRNA and tRNA, which are bound to the ribosome in the course of protein synthesis, are responsible for the correct order of amino acids in the growing protein chain.

Before an amino acid can be incorporated into a growing protein chain, it must first be **activated,** a process involving both tRNA and a specific enzyme of the class known as **aminoacyl-tRNA synthetases.** The amino acid is covalently bonded to the tRNA in the process, forming an aminoacyl-tRNA. The actual formation of the polypeptide chain occurs in three steps. In the first step, **chain initiation,** the first aminoacyl-tRNA is bound to the mRNA at the site that encodes the start of polypeptide synthesis. In this complex, the mRNA and the ribosome are bound to each other. The next aminoacyl-tRNA forms a complex with the ribosome and with mRNA. The binding site for the second aminoacyl-tRNA is close to that for the first aminoacyl-tRNA. A peptide bond is formed between the amino acids in the second step, called **chain elongation.** The chain-elongation process repeats itself until the polypeptide chain is complete. Finally, in the third step, **chain termination** takes place. Each of these steps has many distinguishing features (Figure 12.1), and we shall look at each of them in detail.

Biochemistry⊘Now™
Test yourself on these Critical Questions at the BiochemistryNow website at **http://now.brookscole.com/campbell5**

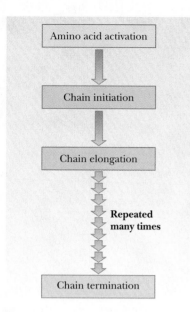

▲ **FIGURE 12.1** A flow chart showing the steps in protein biosynthesis.

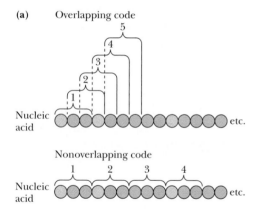

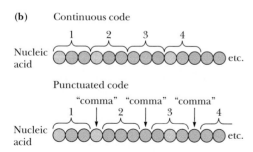

▲ **FIGURE 12.2** (a) An overlapping versus a nonoverlapping code. (b) A continuous versus a punctuated code.

12.2 | What Is the Genetic Code?

Some of the most important features of the code can be specified by saying that the genetic message is contained in a *triplet, nonoverlapping, commaless, degenerate, universal code.* Each of these terms has a definite meaning that describes the way in which the code is translated.

A **triplet** code means that a sequence of three bases (called a **codon**) is needed to specify one amino acid. The genetic code must translate the language of DNA, which contains four bases, into the language of the 20 common amino acids that are found in proteins. If there were a one-to-one relationship between bases and amino acids, then the four bases could only encode four amino acids, and all proteins would have to be combinations of these four. If it took two bases to make a codon, then there would be 4^2 possibilities, or 16 possible amino acids, which is still not enough. Thus, one could have guessed that a codon would have to be at least three bases long. With three bases, there are 4^3 possibilities, or 64 possible codons, which is more than enough to encode the 20 amino acids. The term *nonoverlapping* indicates that no bases are shared between consecutive codons; the ribosome moves along the mRNA three bases at a time rather than one or two at a time (Figure 12.2). If the ribosome moved along the mRNA more than three bases at a time, this situation would be referred to as "a punctuated code." Since no intervening bases exist between codons, the code is *commaless.* In a *degenerate* code, more than one triplet can encode the same amino acid. There are 64 ($4 \times 4 \times 4$) possible triplets of the four bases that occur in RNA, and all are used to encode the 20 amino acids or one of the three stop signals. Note that there is a big difference between a degenerate code and an ambiguous one. Each amino acid may have more than one codon, so the genetic code is a little redundant, but no codon can encode more than one amino acid. If it did, the code would be ambiguous, and the protein-synthesizing machinery would not know which amino acid should be inserted in the sequence. All 64 codons have been assigned meanings, with 61 of them coding for amino acids and the remaining 3 serving as the termination signals (Table 12.1).

Two amino acids, tryptophan and methionine, have only one codon each, but the rest have more than one. A single amino acid can have as many as six codons, as is the case with leucine and with arginine. Originally, the genetic code was thought to be a random selection of bases encoding amino acids. More recently, it is becoming clear why the code has withstood billions of years of natural selection. Multiple codons for a single amino acid are not randomly distributed in Table 12.1 but have one or two bases in common. The bases that are common to several codons are usually the first and second bases, with more room for variation in the third base, which is called the "wobble" base. The degeneracy of the code acts as a buffer against deleterious mutations. For example, for eight of the amino acids (L, V, S, P, T, A, G, and R), the third base is completely irrelevant. Thus, any mutation in the third base of these codons would not change the amino acid at that location. In addition, the second base of the codon also appears to be very important for determining the type of amino acid. For example, when the second base is U, all the amino acids generated from the codon possibilities are hydrophobic. Thus, if the first or third base were mutated, the damage would not be as great because one hydrophobic amino acid would be replaced with another. Codons sharing the same first letter often code for amino acids that are products of one another or precursors of one another. A recent paper in *Scientific American* looked at the error rate for other hypothetical genetic codes and calculated that, of one million possible genetic codes that could be conceived of, only 100 of them would have the effect of reducing errors in protein func-

Table 12.1

The Genetic Code

First Position (5'-end)	Second Position				Third Position (3'-end)
	U	C	A	G	
U	UUU Phe	UCU Ser	UAU Tyr	UGU Cys	U
	UUC Phe	UCC Ser	UAC Tyr	UGC Cys	C
	UUA Leu	UCA Ser	UAA Stop	UGA Stop	A
	UUG Leu	UCG Ser	UAG Stop	UGG Trp	G
C	CUU Leu	CCU Pro	CAU His	CGU Arg	U
	CUC Leu	CCC Pro	CAC His	CGC Arg	C
	CUA Leu	CCA Pro	CAA Gln	CGA Arg	A
	CUG Leu	CCG Pro	CAG Gln	CGG Arg	G
A	AUU Ile	ACU Thr	AAU Asn	AGU Ser	U
	AUC Ile	ACC Thr	AAC Asn	AGC Ser	C
	AUA Ile	ACA Thr	AAA Lys	AGA Arg	A
	AUG Met*	ACG Thr	AAG Lys	AGG Arg	G
G	GUU Val	GCU Ala	GAU Asp	GGU Gly	U
	GUC Val	GCC Ala	GAC Asp	GGC Gly	C
	GUA Val	GCA Ala	GAA Glu	GGA Gly	A
	GUG Val	GCG Ala	GAG Glu	GGG Gly	G

Third-Base Degeneracy Is Color-Coded

Third-Base Relationship	Third Bases with Same Meaning	Number of Codons
Third-base irrelevant	U, C, A, G	32 (8 families)
Purines	A or G	12 (6 pairs)
Pyrimidines	U or C	14 (7 pairs)
Three out of four	U, C, A	3 (AUX = Ile)
Unique definitions	G only	2 (AUG = Met)
		(UGG = Trp)
Unique definition	A only	1 (UGA = Stop)

*AUG signals translation initiation as well as coding for Met residues.

tion when compared with the real code. Indeed, it seems that the genetic code has withstood the test of time because it is one of the best ways to protect an organism from DNA mutations. (See the article by Freeland and Hurst in the Annotated Bibliography for this chapter.)

The assignment of triplets in the genetic code was based on several types of experiments. One of the most significant experiments involved the use of synthetic polyribonucleotides as messengers.

When homopolynucleotides (polyribonucleotides that contain only one type of base) are used as a *synthetic mRNA* for polypeptide synthesis in laboratory systems, homopolypeptides (polypeptides that contain only one kind of amino acid) are produced. When poly U is the messenger, the product is polyphenylalanine. With poly A as the messenger, polylysine is formed. The product for poly C is polyproline, and the product for poly G polyglycine is specified. This procedure was used to establish the code for the four possible homopolymers quickly. When an alternating copolymer (a polymer with an alternating sequence of two bases) is the messenger, the product is an alternating polypeptide (a polypeptide with an alternating sequence of two amino acids). For example, when the sequence of the polynucleotide is –ACACACA-CACACACACACACAC–, the polypeptide produced has alternating threonines and histidines. There are two types of coding triplets in this polynucleotide, ACA and CAC, but this experiment cannot establish which one codes for threonine and which one codes for histidine. More information is needed for an unambiguous assignment, but it is interesting that this result proves that the code is a triplet code. If it were a doublet code, the product would be a mixture of two homopolymers, one specified by the codon AC and the other by the codon CA. (The terminology for the different ways of reading this message as a doublet is to say that they have different **reading frames,** /AC/AC/ and /CA/CA/. In a triplet code, only one reading frame is possible, namely, /ACA/CAC/ACA/CAC/, which gives rise to an alternating polypeptide.) Use of other synthetic polynucleotides can yield other coding assignments, but, as in our example here, many questions remain.

Other methods are needed to answer the remaining questions about codon assignment. One of the most useful methods is the **filter-binding assay** (Figure 12.3). In this technique, various tRNA molecules, one of which is radioactively labeled with carbon-14 (^{14}C), are mixed with ribosomes and synthetic trinucleotides that are bound to a filter. The mixture of tRNAs is passed through the filter, and some will bind and others will pass through. If the radioactive label is detected on the filter, then it is known that the particular tRNA did bind. If the radioactive label is found in a solution that flowed through the filter, then the tRNA did not bind. This technique depends on the fact that aminoacyl-tRNAs bind strongly to ribosomes in the presence of the correct trinucleotide. In this situation, the trinucleotide plays the role of an mRNA codon. The possible trinucleotides are synthesized by chemical methods, and binding assays are repeated with each type of trinucleotide. For example, if the aminoacyl-tRNA for histidine binds to the ribosome in the presence of the trinucleotide CAU, the sequence CAU is established as a codon for histidine. About 50 of the 64 codons were identified by this method.

Essential Information

In the genetic code, a sequence of three bases in mRNA (a codon) specifies each amino acid. There are 64 possible codons, each of which specifies one of the 20 amino acids or a stop signal.

Codon–Anticodon Pairing and Wobble

A codon forms base pairs with a complementary **anticodon** of a tRNA when an amino acid is incorporated during protein synthesis. Because there are 64 possible codons, one might expect to find 64 types of tRNA but, in fact, the number is less than 64 in all cells. Some tRNAs bond to one codon exclusively, but many of them can recognize more than one codon because of variations in the allowed pattern of hydrogen bonding. This variation is called **"wobble"** (Figure 12.4), and it applies to the first base of an anticodon, the one at the 5′ end, but not to the second or the third base. Recall that mRNA is read from the 5′ to the 3′ end. The first (wobble) base of the anticodon hydrogen-bonds to the third base of the codon, the one at the 3′ end. The base in the wobble position of the anticodon can base-pair with several different bases in the codon, not just the base specified by Watson–Crick base pairing (Table 12.2).

When the wobble base of the anticodon is uracil, it can base-pair not only with adenine, as expected, but also with guanine, the other purine base. When the wobble base is guanine, it can base-pair with cytosine, as expected, and also with uracil, the other pyrimidine base. The purine base hypoxanthine frequently occurs in the wobble position in many tRNAs, and it can

---

Table 12.2

Base-Pairing Combinations in the Wobble Scheme

Base at 5′ End of Anticodon	Base at 3′End of Codon
I*	A, C, or U
G	C or U
U	A or G
A	U
C	G

* I = hypoxanthine.

Note that there are no variations in base pairing when the wobble position is occupied by A or C.

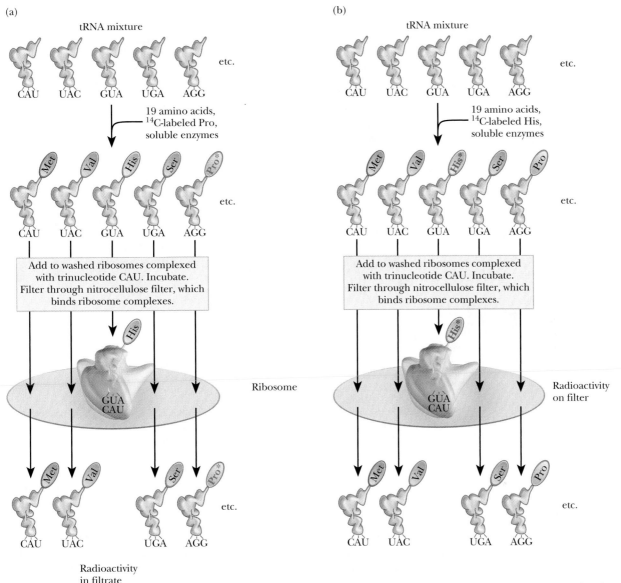

▲ **FIGURE 12.3** The filter-binding assay for elucidation of the genetic code. A reaction mixture combines washed ribosomes, Mg^{2+}, a particular trinucleotide, and all 20 aminoacyl-tRNAs, one of which is radioactively (^{14}C) labeled. (a) ^{14}C-labeled prolyl-tRNA. (b) ^{14}C-labeled histidine-tRNA. Only the aminoacyl-tRNA whose binding is directed by the trinucleotide codon will become bound to the ribosomes and retained on the nitrocellulose filter. The amount of radioactivity retained by the filter is a measure of trinucleotide-directed binding of a particular labeled aminoacyl-tRNA by ribosomes. Use of this binding assay to test the 64 possible codon trinucleotides against the 20 different amino acids quickly enabled researchers to assign triplet code words to the individual amino acids. The genetic code was broken. (*Adapted from Nirenberg, M. W., and Leder, P., 1964. RNA Codewords and Protein Synthesis.* Science **145,** *1399–1407.*)

base-pair with adenine, cytosine, and uracil in the codon (Figure 12.5). Adenine and cytosine do not form any base pairs other than the expected ones with uracil and guanine, respectively (Table 12.2). To summarize, when the wobble position is occupied by I (from inosine, the nucleoside made up of ribose and hypoxanthine), G, or U, variations in hydrogen bonding are allowed; when the wobble position is occupied by A or C, these variations do not occur.

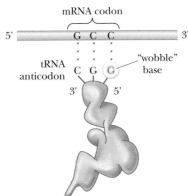

▲ **FIGURE 12.4** "Wobble" base pairing. The wobble base of the anticodon is the one at the 5′ end; it forms hydrogen bonds with the last base of the mRNA codon, the one at the 3′ end of the codon. (*Adapted by permission from Crick, F. H. C., 1966. Codon–anticodon pairing: The wobble hypothesis.* Journal of Molecular Biology **19,** *548–555.*)

The guanine–adenine base pair

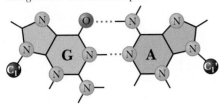

The two possible uracil–uracil base pairs

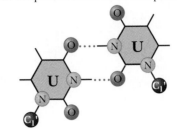
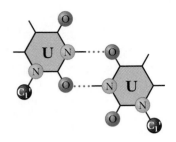

The uracil–cytosine base pair

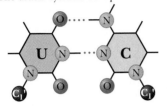

The guanine–uracil and inosine–uracil base pairs are similar

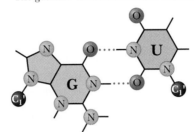
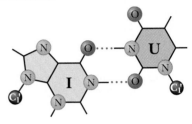

The inosine–adenine base pair

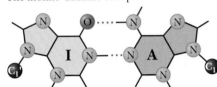

▲ **FIGURE 12.5** Various base-pairing alternatives. G:A is unlikely, because the 2-NH$_2$ of G cannot form one of its H bonds; even water is sterically excluded. U:C may be possible, even though the two C═O are juxtaposed. Two U:U arrangements are feasible. G:U and I:U are both possible and somewhat similar. The purine pair I:A is also possible. (*Adapted from Crick, F. H. C., 1966. Codon–anticodon pairing: The wobble hypothesis.* Journal of Molecular Biology **19**, *548–555.*)

The wobble hypothesis provides insight into some aspects of the degeneracy of the code. In many cases, the degenerate codons for a given amino acid differ in the third base, the one that pairs with the wobble base of the anticodon. Fewer different tRNAs are needed because a given tRNA can base-pair with several codons. As a result, a cell would have to invest less energy in the synthesis of needed tRNAs. The existence of wobble also minimizes the damage that can be caused by misreading of the code. If, for example, a leucine codon, CUU, were to be misread as CUC, CUA, or CUG during transcription of mRNA, this codon would still be translated as leucine during protein synthesis; no damage to the organism would occur. We saw in earlier chapters that drastic consequences can result from misreading the genetic code in other codon positions, but here we see that such effects are not inevitable.

A *universal* code is one that is the same in all organisms. The universality of the code has been observed in viruses, prokaryotes, and eukaryotes. However, there are some exceptions. Some codons seen in mitochondria are different from those seen in the nucleus. There are also at least 16 organisms that have code variations. For example, the marine alga *Acetabularia* translates the standard stop codons, UAG and UAA, as a glycine rather than as a stop. Fungi of the genus *Candida* translate the codon CUG as a serine, where that codon would specify leucine in most organisms. The evolutionary origin of these differences is not known at this writing, but many researchers believe that understanding these code variations is important to understanding evolution.

12.3 What Is the Role of Aminoacyl-tRNA Synthetases in Amino Acid Activation?

The activation of the amino acid and the formation of the aminoacyl-tRNA take place in two separate steps, both of which are catalyzed by the aminoacyl-tRNA synthetase (Figure 12.6). First, the amino acid forms a covalent bond to an adenine nucleotide, producing an aminoacyl-AMP. The free energy of hydrolysis of ATP provides energy for bond formation. The aminoacyl moiety is then transferred to tRNA, forming an aminoacyl-tRNA.

$$\text{Amino acid} + \text{ATP} \rightarrow \text{aminoacyl-AMP} + \text{PP}_i$$
$$\underline{\text{Aminoacyl-AMP} + \text{tRNA} \rightarrow \text{aminoacyl-tRNA} + \text{AMP}}$$
$$\text{Amino acid} + \text{ATP} + \text{tRNA} \rightarrow \text{aminoacyl-tRNA} + \text{AMP} + \text{PP}_i$$

Aminoacyl-AMP is a mixed anhydride of a carboxylic acid and a phosphoric acid. Because anhydrides are reactive compounds, the free-energy change for the hydrolysis of aminoacyl-AMP favors the second step of the overall reaction. Another point that favors the process is the energy released when pyrophosphate (PP$_i$) is hydrolyzed to orthophosphate (P$_i$) to replenish the phosphate pool in the cell.

In the second part of the reaction, an ester linkage is formed between the amino acid and either the 3'-hydroxyl or the 2'-hydroxyl of the ribose at the 3' end of the tRNA. There are two classes of aminoacyl-tRNA synthetases. Class I loads the amino acid onto the 2' hydroxyl. Class II uses the 3' hydroxyl. These two classes of enzyme appear to be unrelated and indicate a convergent evolution. Several tRNAs can exist for each amino acid, but a given tRNA will not bond to more than one amino acid. The synthetase enzyme requires Mg^{2+} and is highly specific both for the amino acid and for the tRNA. A separate synthetase exists for each amino acid, and this synthetase functions for all the different tRNA molecules for that amino acid. The specificity of the enzyme contributes to the accuracy of the translation process. A student who used an earlier edition of this book compared the mode of action of the aminoacyl-tRNA synthetases to a "dating service" for amino acids and tRNAs. The synthetase assures that the right amino acid pairs up with the right tRNA, and this is its primary function. The synthetase has another level of activity as well. An extra level of proofreading by the synthetase is part of what is sometimes called the "second genetic code."

The two-stage reaction allows for selectivity to operate at two levels: that of the amino acid and that of the tRNA. The specificity of the first stage uses the fact that the aminoacyl-AMP remains bound to the enzyme. For example, isoleucyl-tRNA synthetase can form an aminoacyl-AMP of isoleucine or the structurally similar valine. If the valyl moiety is then transferred to the tRNA for isoleucine, it is detected by an editing site in the tRNA synthetase, which then hydrolyzes the incorrectly acylated aminoacyl-tRNA. The selectivity resides in the tRNA, not in the amino acid.

The second aspect of selectivity depends on the selective recognition of tRNAs by aminoacyl-tRNA synthetases. Specific binding sites on tRNA are recognized by aminoacyl-tRNA synthetases. The exact position of the recognition site varies with different synthetases, and this feature, in and of itself, is a source of greater specificity. Contrary to what one might expect, the anticodon is not always the part of the tRNA that is recognized by the aminoacyl-tRNA synthetase, although it frequently is involved. Figure 12.7 shows the locations of the recognition sites for the tRNAs for various amino acids.

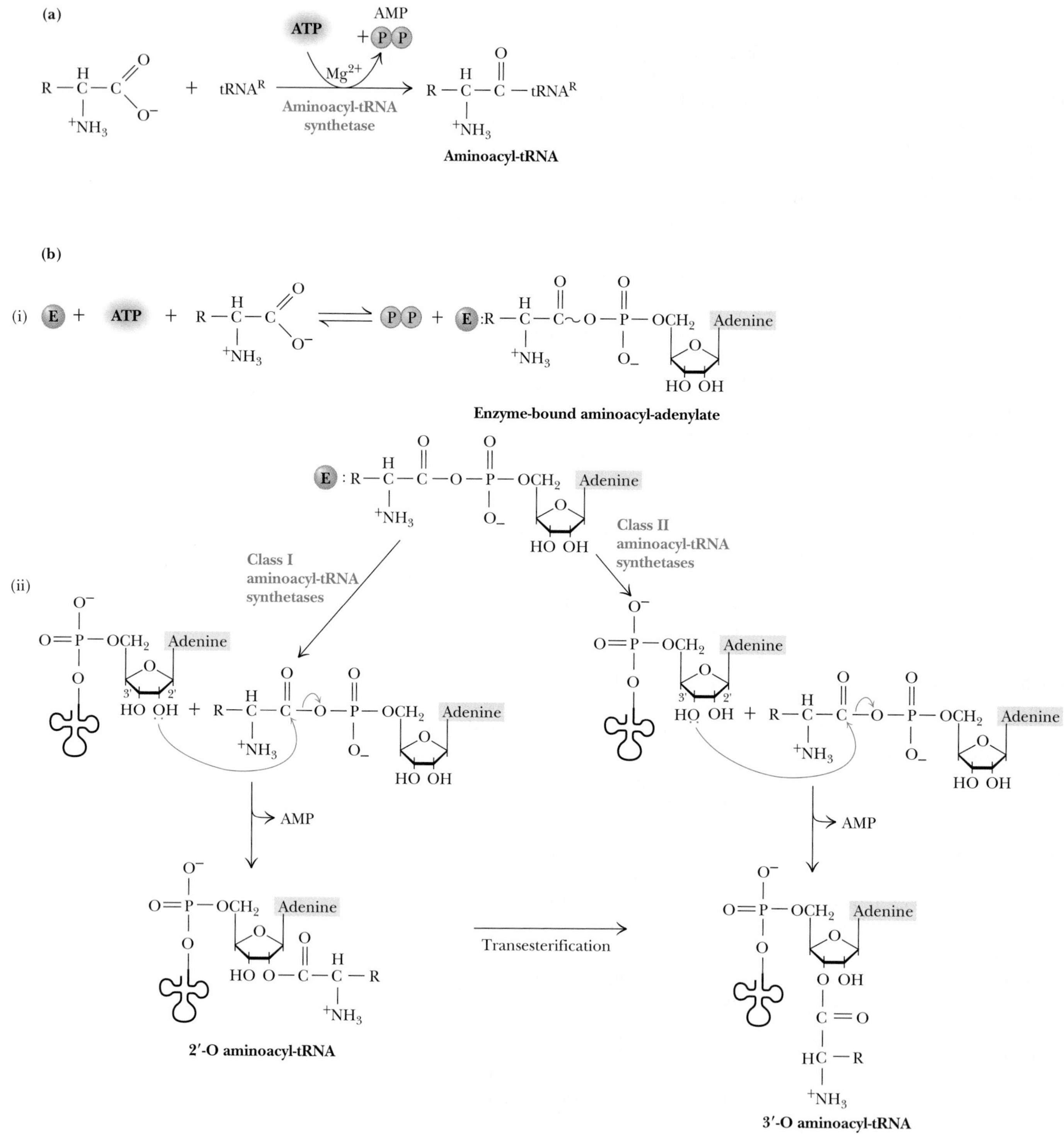

Biochemistry Now™ ▲ ACTIVE FIGURE 12.6 The aminoacyl-tRNA synthetase reaction. (a) The overall reaction. Ever-present pyrophosphatases in cells quickly hydrolyze the PP$_i$ produced in the aminoacyl-tRNA synthetase reaction, rendering aminoacyl-tRNA synthesis thermodynamically favorable and essentially irreversible. (b) The overall reaction commonly proceeds in two steps: (i) formation of an aminoacyl-adenylate and (ii) transfer of the activated amino acid moiety of the mixed anhydride to either the 2'-OH (class I aminoacyl-tRNA synthetases) or 3'-OH (class II aminoacyl-tRNA synthetases) of the ribose on the terminal adenylic acid at the 3'-OH terminus common to all tRNAs. Those aminoacyl-tRNAs formed as 2'-OH esters undergo a transesterification that moves the aminoacyl group to the 3'-OH of tRNA. Only the 3'-esters are substrates for protein synthesis. **Watch this Active Figure at http://now.brookscole.com/campbell5**

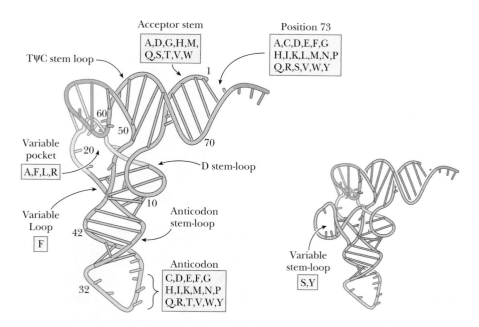

◀ **FIGURE 12.7** Ribbon diagram of the tRNA tertiary structure. Numbers represent the consensus nucleotide sequence. The locations of nucleotides recognized by the various aminoacyl-tRNA synthetases are indicated; shown within the boxes are one-letter designations of the amino acids whose respective aminoacyl-tRNA synthetases interact at the discriminator base (position 73), acceptor stem, variable pocket and/or loop, or anticodon. The inset shows additional recognition sites in those tRNAs having a variable loop that forms a stem-loop structure. (*Adapted with permission from Saks, M. E., Sampson, J. R., and Abelson, J. N., 1994. The transfer RNA problem: A search for rules. Science* **263**, *191–197, Figure 2. Copyright © 1994 AAAS.*)

The recognition of the correct tRNA by the synthetase is vital to the fidelity of translation because most of the final proofreading occurs at this step. See the articles by LaRiviere et al. and Ibba in the Annotated Bibliography at the end of this chapter for the latest information on this topic.

12.4 | How Does Translation Take Place in Prokaryotes?

The details of the chain of events in translation differ somewhat in prokaryotes and eukaryotes. Like DNA and RNA synthesis, this process has been more thoroughly studied in prokaryotes. We shall use *Escherichia coli* as our principal example, because all aspects of protein synthesis have been most extensively studied in this bacterium. As was the case with replication and transcription, translation can be divided into stages—chain initiation, chain elongation, and chain termination.

Ribosomal Architecture

Protein synthesis requires the specific binding of mRNA and aminoacyl-tRNAs to the ribosome. Ribosomes have a specific architecture that facilitates the binding. As Figure 12.8 shows, the smaller subunit has a head, a base, and a platform. A cleft lies between the platform and the other two portions. The larger subunit is a concave structure with a central protuberance, a wing, and a stalk. When the two subunits fit together, a cavity remains. This cavity is large enough to hold at least two tRNAs. This central region is also considered to contain the mRNA and the growing peptide chain. Elucidation of the details of ribosomal structure is a recent triumph of X-ray crystallography.

Chain Initiation

In all organisms, the synthesis of polypeptide chains starts at the N-terminal end; the chain grows from the N-terminal end to the C-terminal end. This is one of the reasons that scientists chose to record DNA sequences from 5′ to 3′ and to focus on the coding strand of DNA and the mRNA. The coding

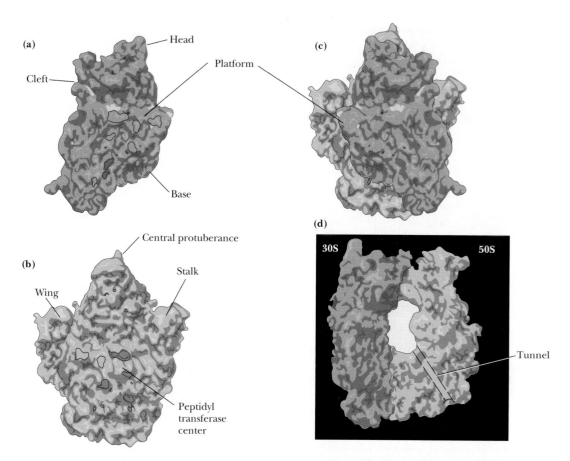

▲ **FIGURE 12.8** Structure of the *E. coli* ribosomal subunits and 70S ribosome, as deduced by X-ray crystallography. Prominent structural features are labeled. Parts (a) and (b) present views of the 30S and 50S subunits, respectively. These views show the sides of these two subunits that form the interface between them when they come together to form a 70S subunit (c). Part (d) is a side view of the 70S ribosome; the white area represents the region in which mRNA and tRNAs are bound and peptide bond formation occurs. The tunnel through the 50S subunit that the growing peptide chain transits is shown as a dashed line. (*Adapted from Figures 2 and 3 in Cate, J. H., et al., 1999. X-ray crystal structures of 70S ribosomal functional complexes.* Science **285**, *2095–2104.*)

Biochemistry⊜Now™ ANIMATED FIGURE 12.9
Formation of the *N*-formylmethionine-tRNA^fmet (first reaction). Methionine must be bound to tRNA^fmet to be formylated. **See this figure animated at** http://now.brookscole.com/campbell5

strand sequences are read from 5′ to 3′, the mRNA is read from 5′ to 3′, and the proteins are built from the N-terminus to the C-terminus. In prokaryotes, the initial N-terminal amino acid of all proteins is *N*-formylmethionine (fmet) (Figure 12.9). However, this residue often is removed by posttranslational processing after the polypeptide chain is synthesized. There are two different tRNAs for methionine in *E. coli*, one for unmodified methionine and one for

N-formylmethionine. These two tRNAs are called tRNA^{met} and tRNA^{fmet}, respectively (the superscript identifies the tRNA). The aminoacyl-tRNAs that they form with methionine are called met-tRNA^{met} and met-tRNA^{fmet}, respectively (the prefix identifies the bound amino acid). In the case of met-tRNA^{fmet}, a formylation reaction takes place after methionine is bonded to the tRNA, producing **N-formylmethionine-tRNA^{fmet}** (fmet-tRNA^{fmet}). The source of the formyl group is N^{10}-formyltetrahydrofolate (see "One-Carbon Transfers and the Serine Family" in Section 23.4). Methionine bound to tRNA^{met} is not formylated.

Both tRNAs (tRNA^{met} and tRNA^{fmet}) contain a specific sequence of three bases (a triplet), 3′-UAC-5′, which base-pairs with the sequence **5′-AUG-3′** in the mRNA sequence. The tRNA^{fmet} triplet in question, 3′-UAC-5′, recognizes the AUG triplet, which is the **start signal** when it occurs at the beginning of the mRNA sequence that directs the synthesis of the polypeptide. The same 3′-UAC-5′ triplet in tRNA^{met} recognizes the AUG triplet when it is found in an internal position in the mRNA sequence.

The start of polypeptide synthesis requires the formation of an **initiation complex** (Figure 12.10). At least eight components enter into the formation of the initiation complex, including mRNA, the 30S ribosomal subunit, fmet-tRNA^{fmet}, GTP, and three protein initiation factors, called IF-1, IF-2, and IF-3. The IF-3 protein facilitates the binding of mRNA to the 30S ribosomal subunit. It also appears to prevent premature binding of the 50S subunit, which takes place in a subsequent step of the initiation process. IF-2 binds GTP and aids in the selection of the initiator tRNA (fmet-tRNA^{fmet}) from all the other aminoacylated tRNAs available. The function of IF-1 is less clear; it appears to bind to IF-3 and to IF-2, and it facilitates the action of both. It also catalyzes the separation of the 30S and the 50S ribosomal subunits being recycled for another round of translation. The resulting combination of mRNA, the 30S ribosomal subunit, and fmet-tRNA^{fmet} is the **30S initiation complex** (Figure 12.10). A 50S ribosomal subunit binds to the 30S initiation complex to produce the **70S initiation complex.** The hydrolysis of GTP to GDP and P$_i$ favors the process by providing energy; the initiation factors are released at the same time. The correct positioning of the initiator tRNA is maintained as a result of a small difference between it and tRNA for an internal methionine. A single C–A mismatched base pair near the acceptor stem allows the 30S subunit to recognize the initiator tRNA.

For the mRNA to be translated correctly, the ribosome must be placed at the correct start location. The start signal is preceded by a purine-rich leader segment of mRNA, called the **Shine–Dalgarno sequence** (5′-GGAGGU-3′) (Figure 12.10), which usually lies about ten nucleotides upstream of the AUG start signal (also known as the initiation codon) and acts as a ribosomal binding site. Figure 12.11 gives some characteristic Shine–Dalgarno sequences. This purine-rich area binds to a pyrimidine-rich sequence on the 16S ribosomal RNA part of the 30S subunit and aligns it for proper translation beginning with the AUG start codon.

Chain Elongation

The elongation phase of prokaryotic protein synthesis (Figure 12.12) uses the fact that three binding sites for tRNA are present on the 50S subunit of the 70S ribosome. The three tRNA binding sites are called the **P (peptidyl) site,** the **A (aminoacyl) site,** and the **E (exit) site.** The P site binds a tRNA that carries a peptide chain, and the A site binds an incoming aminoacyl-tRNA. The E site carries an uncharged tRNA that is about to be released from the ribosome. Chain elongation begins with the addition of the second amino acid

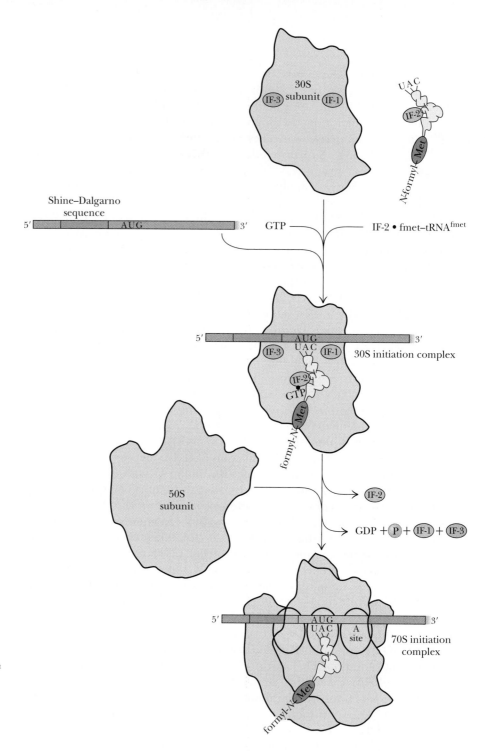

The formation of an initiation complex. The 30S ribosomal subunit binds to mRNA and fmet-tRNA^{fmet} in the presence of GTP and the three initiation factors, IF-1, IF-2, and IF-3, forming the 30S initiation complex. The 50S ribosomal subunit is added, forming the 70S initiation complex. **Watch this Active Figure at http://now.brookscole.com/campbell5**

▶ **FIGURE 12.11** Various Shine–Dalgarno sequences recognized by *E. coli* ribosomes. These sequences lie about ten nucleotides upstream from their respective AUG initiation codon and are complementary to the UCCU core sequence element of *E. coli* 16S rRNA. G:U as well as canonical G:C and A:U base pairs are involved here.

Initiation codon

araB	– U U U G G A U G G A G U G A A A C G A U G G C G A U U –
galE	– A G C C U A A U G G A G C G A A U U A U G A G A G U U –
lacI	– C A A U U C A G G G U G G U G A U U G U G A A A C C A –
lacZ	– U U C A C A C A G G A A A C A G C U A U G A C C A U G –
Q β phage replicase	– U A A C U A A G G A U G A A A U G C A U G U C U A A G –
φX174 phage A protein	– A A U C U U G G A G G C U U U U U U A U G G U U C G U –
R17 phage coat protein	– U C A A C C G G G G U U U G A A G C A U G G C U U C U –
ribosomal protein S12	– A A A A C C A G G A G C U A U U U A A U G G C A A C A –
ribosomal protein L10	– C U A C C A G G A G C A A A G C U A A U G G C U U U A –
trpE	– C A A A A U U A G A G A A U A A C A A U G C A A A C A –
trpL leader	– G U A A A A A G G G U A U C G A C A A U G A A A G C A –

3'-end of 16S rRNA   3' ₕₒ A U U C C U C C A C U A G – 5'

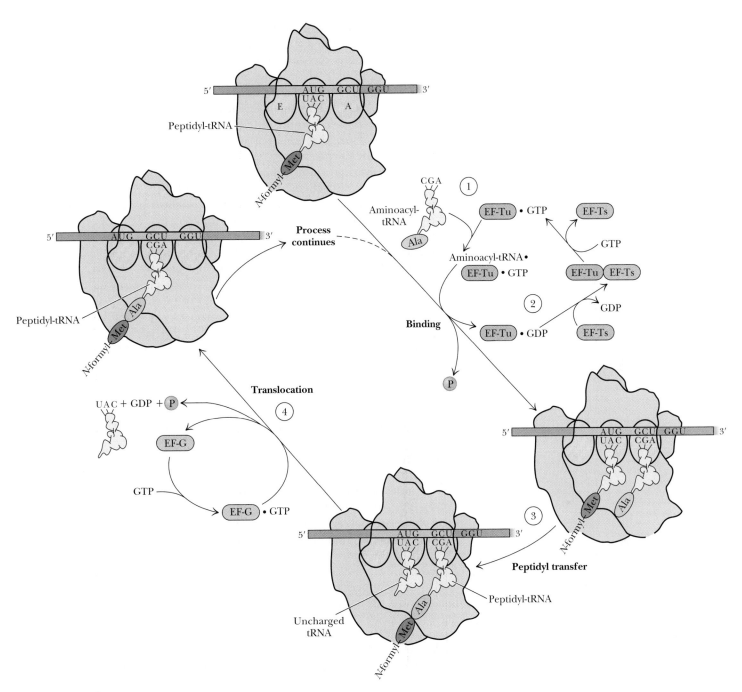

Biochemistry⊗Now™ ▲ ACTIVE FIGURE 12.12 A summary of the steps in chain elongation. Step 1: An aminoacyl-tRNA is bound to the A site on the ribosome. Elongation factor EF-Tu
(Tu) and GTP are required. The P site on the ribosome is already occupied. Step 2: Elongation
factor EF-Tu is released from the ribosome and regenerated in a process requiring elongation
factor EF-Ts (Ts) and GTP. Step 3: The peptide bond is formed, leaving an uncharged tRNA at
the P site. Step 4: In the translocation step, the uncharged tRNA is released. The peptidyl-tRNA
is translocated to the P site, leaving an empty A site. The uncharged tRNA is translocated to the
E site and subsequently released. Elongation factor EF-G and GTP are required. **Watch this Active
Figure at http://now.brookscole.com/campbell5**

specified by the mRNA to the 70S initiation complex (Step 1). The P site on
the ribosome is the one initially occupied by the fmet-tRNA^{fmet} in the 70S
initiation complex. The second aminoacyl-tRNA binds at the A site. A triplet
of tRNA bases (the anticodon AGC in our example) forms hydrogen bonds
with a triplet of mRNA bases (GCU, the codon for alanine, in this example). In addition, GTP and two protein elongation factors, EF-Tu and EF-Ts

(temperature-unstable and temperature-stable elongation factors, respectively), are required (Step 2). EF-Tu guides the aminoacyl-tRNA into part of the A site and aligns the anticodon with the mRNA codon. Only when the match is found to be correct is the aminoacyl-tRNA inserted completely into the A site. GTP is hydrolyzed and EF-Tu dissociates. EF-Ts is involved in regeneration of EF-Tu-GTP. This small EF-Tu protein (43 kDa) is the most abundant protein in *E. coli*, comprising 5% of the dry weight of the cell. It has recently been shown that EF-Tu is involved in another level of translation fidelity. When the correct amino acid is bound to the correct tRNA, EF-Tu is efficient at delivering the activated tRNA to the ribosome. If the tRNA and amino acid are mismatched, then either the EF-Tu does not bind the activated tRNA very well, in which case it does not deliver it well to the ribosome, or it binds the activated tRNA too well, in which case it will not release it from the ribosome. See the articles by LaRiviere et al. and Ibba in the Annotated Bibliography at the end of this chapter for more on this topic.

A **peptide bond** is then formed in a reaction catalyzed by *peptidyl transferase*, which is a part of the 50S subunit (Step 3). The mechanism for this reaction is shown in Figure 12.13. The α-amino group of the amino acid in the A site performs a nucleophilic attack on the carbonyl group of the amino acid linked to the tRNA in the P site. There is now a dipeptidyl-tRNA at the A site and a tRNA with no amino acid attached (an "*uncharged tRNA*") at the P site.

A **translocation** step then takes place before another amino acid can be added to the growing chain (Figure 12.12, Step 4). In the process, the uncharged tRNA moves from the P site to the E site, from which it is subsequently released; the peptidyl-tRNA moves from the A site to the vacated P site. In addition, the mRNA moves with respect to the ribosome. Another elongation factor, EF-G, also a protein, is required at this point, and once again GTP is hydrolyzed to GDP and P_i.

The three steps of the chain elongation process are aminoacyl-tRNA binding, peptide bond formation, and translocation (Steps 1, 3, and 4 in Figure 12.12). They are repeated for each amino acid specified by the genetic message of the mRNA until the stop signal is reached. Step 2 in Figure 12.12 shows the regeneration of aminoacyl-tRNA.

Much of the information about this phase of protein synthesis has been gained from the use of inhibitors. Puromycin is a structural analog for the 3′ end of an aminoacyl-tRNA, making it a useful probe to study chain elongation (Figure 12.14). In an experiment of this sort, puromycin binds to the A site, and a peptide bond is formed between the C-terminus of the growing polypeptide and the puromycin. The peptidyl puromycin is weakly bound to the ribosome and dissociates from it easily, resulting in premature termination and a defective protein. Puromycin also binds to the P site and blocks the translocation process, although it does not react with peptidyl-tRNA in

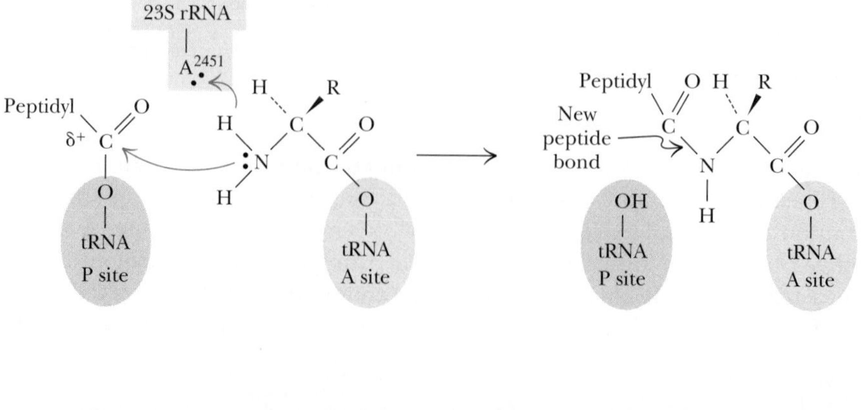

Biochemistry ⊛ Now™ ANIMATED FIGURE 12.13
Peptide bond formation in protein synthesis. Nucleophilic attack by the α-amino group of the A-site aminoacyl-tRNA on the carbonyl-C of the P-site peptidyl-tRNA is facilitated when a purine moiety of rRNA abstracts a proton. **See this figure animated at http://now.brookscole.com/campbell5**

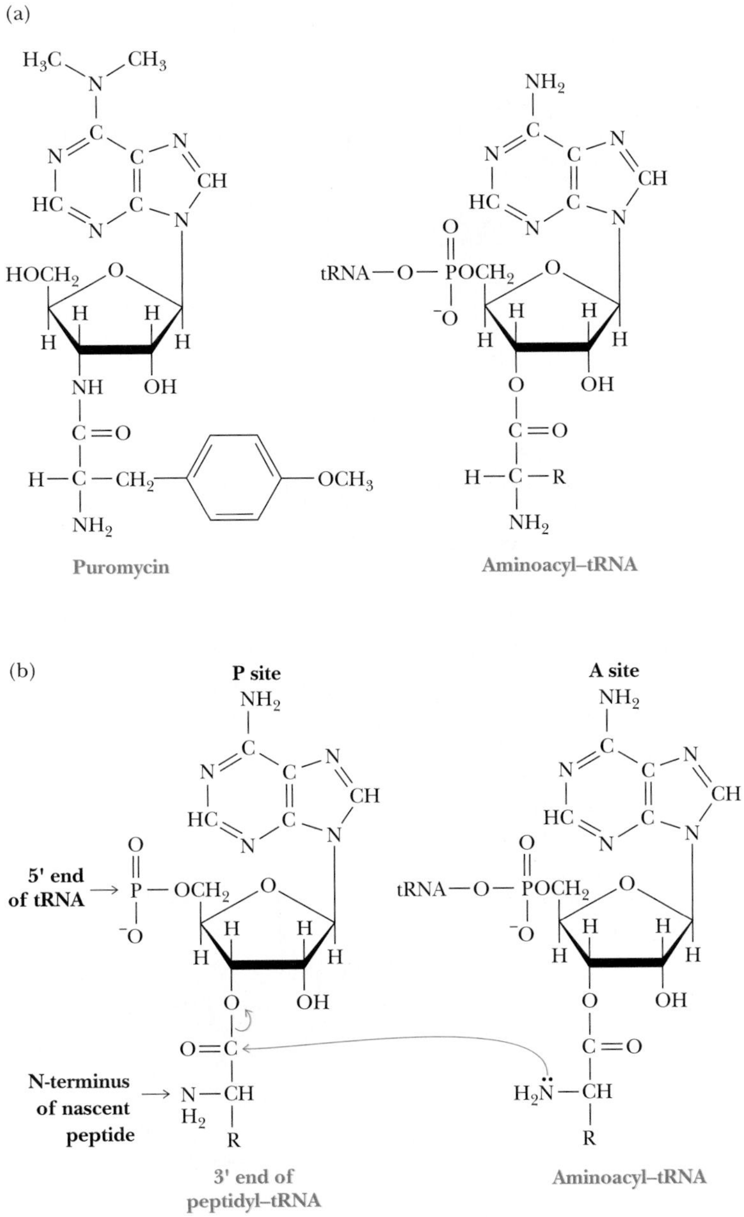

(a)

Puromycin

Aminoacyl–tRNA

(b)

◀ **FIGURE 12.14** The mode of action of puromycin. (a) A comparison of the structures of puromycin and the 3′ end of an aminoacyl-tRNA. (b) Formation of a peptide bond between a peptidyl-tRNA bound at the P site of a ribosome and an aminoacyl-tRNA bound at the A site.

this case. The existence of A and P sites was determined by these experiments with puromycin.

Chain Termination

A stop signal is required for the termination of protein synthesis. The codons UAA, UAG, and UGA are the stop signals. These codons are not recognized by any tRNAs, but they are recognized by proteins called release factors (Figure 12.15). One of two protein release factors (RF-1 or RF-2) is required, as is GTP, which is bound to a third release factor, RF-3. RF-1 binds to UAA and UAG, and RF-2 binds to UAA and UGA. RF-3 does not bind to any codon, but it does facilitate the activity of the other two release factors. Either RF-1 or RF-2 is bound near the A site of the ribosome when one of the termination codons is reached. The release factor not only blocks the binding of a new aminoacyl-tRNA but also affects the activity of the peptidyl transferase so that

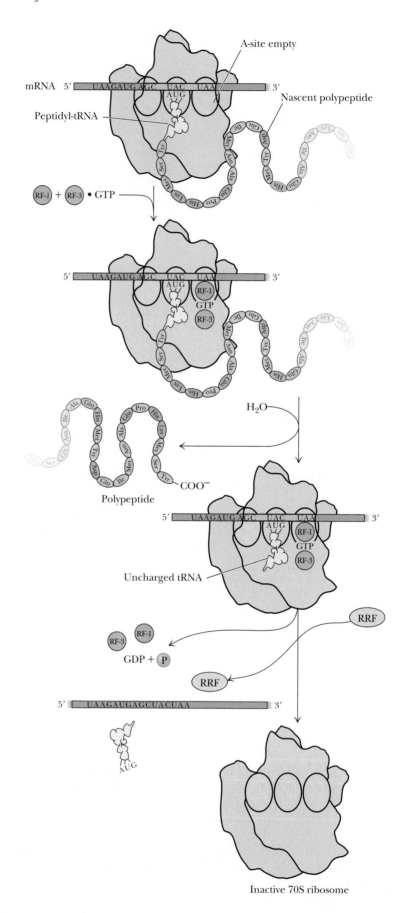

Table 12.3	
Components Required for Each Step of Protein Synthesis in *Escherichia coli*	
Step	**Components**
Amino acid activation	Amino acids
	tRNAs
	Aminoacyl-tRNA synthetases
	ATP, Mg^{2+}
Chain initiation	fmet-tRNA^{fmet}
	Initiation codon (AUG) of mRNA
	30S ribosomal subunit
	50S ribosomal subunit
	Initiation factors (IF-1, IF-2, and IF-3)
	GTP, Mg^{2+}
Chain elongation	70S ribosome
	Codons of mRNA
	Aminoacyl-tRNAs
	Elongation factors (EF-Tu, EF-Ts, and EF-G)
	GTP, Mg^{2+}
Chain termination	70S ribosome
	Termination codons (UAA, UAG, and UGA) of mRNA
	Release factors (RF-1, RF-2, and RF-3)
	GTP, Mg^{2+}

the bond between the carboxyl end of the peptide and the tRNA is hydrolyzed. GTP is hydrolyzed in the process. The whole complex dissociates, setting free the release factors, tRNA, mRNA, and the 30S and 50S ribosomal subunits. All these components can be reused in further protein synthesis. Table 12.3 summarizes the steps in protein synthesis and the components required for each step. The following Biochemical Connections box describes an interesting variation on stop codons.

Biochemical Connections

The 21st Amino Acid?

Many amino acids, such as citrulline and ornithine (which are found in the urea cycle), are not building blocks of proteins. Other nonstandard amino acids, such as hydroxyproline, are formed after translation by posttranslational modification. When discussing amino acids and translation, the magic number was always 20. Only 20 standard amino acids were put onto tRNA molecules for protein synthesis. In the late 1980s, another amino acid was found in proteins from eukaryotes and prokaryotes alike, including humans. It is selenocysteine, a cysteine residue in which the sulfur atom has been replaced by a selenium atom.

It was later determined that the formation of a selenocysteine beings with a serine molecule being bound to a special tRNA molecule called tRNA^{sec}. Once the serine is bound, the oxygen in the serine side chain is replaced by selenium. This tRNA molecule has an anticodon that matches the UGA stop codon. In special cases, the UGA is not read as a stop; rather, the selenocysteine-tRNA^{sec} is loaded into the A site and translation continues. Some are therefore calling selenocysteine the 21st amino acid. The methods by which the cell knows when to put selenocysteine into the protein instead of reading UGA as a stop codon are still being investigated.

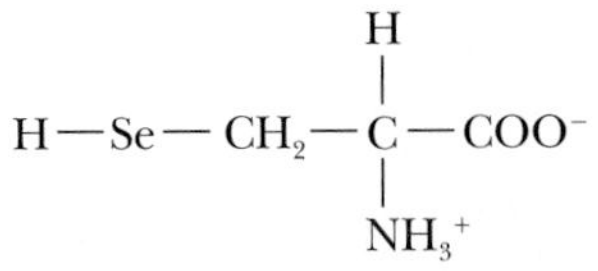

$$H-Se-CH_2-\overset{\overset{\displaystyle H}{|}}{\underset{\underset{\displaystyle NH_3^+}{|}}{C}}-COO^-$$

Selenocysteine

▲ Selenocysteine

The Ribosome Is a Ribozyme

Until recently, proteins were thought to be the only molecules with catalytic ability. Then the self-splicing ability of the *Tetrahymena* snRNP showed that RNA can also catalyze reactions. In 2000, the complete structure of the large ribosomal subunit was determined by X-ray crystallography to 2.4-Å (0.24-nm) resolution (Figure 12.16). Ribosomes had been studied for 40 years, but the complete structure had been elusive. When the active sites for peptidyl transferase were looked at, it turned out that there is no protein in the vicinity of the new peptide bond, proving once again that RNA has catalytic ability. This is an exciting finding because it answers questions that have been plaguing scientists for decades. It was assumed that RNA was the first genetic material, and RNA can encode proteins that act as catalysts; but, because it takes proteins to do the translation, how could the first proteins have been created? With the discovery of an RNA-based peptidyl transferase, it was suddenly possible to imagine an "RNA world" in which the RNA both carried the message and processed it. This discovery is very intriguing, but it has not yet been accepted by many researchers, and some evidence questions the nature of catalytic RNA. One study showed that mutations of the putative RNA bases involved in the catalytic mechanism do not significantly reduce the efficiency of peptidyl transferase, throwing into question whether the RNA is chemically involved in the catalysis (see the article by Polacek et al. in the Annotated Bibliography at the end of this chapter).

Polysomes

In our description of protein synthesis, we have considered, up to now, the reactions that take place at one ribosome. It is, however, not only possible but quite usual for several ribosomes to be attached to the same mRNA. Each of these ribosomes will bear a polypeptide in one of various stages of completion, depending on the position of the ribosome as it moves along the mRNA (Figure 12.17). This complex of mRNA with several ribosomes is called a **polysome;** an alternative name is polyribosome. In prokaryotes, translation begins very soon after mRNA transcription. It is possible for a molecule of mRNA that is still being transcribed to have a number of ribosomes attached to it that are in various stages of translating that mRNA. It is also possible for

Essential Information

The biosynthesis of proteins depends on four important steps. (1) In the activation step, amino acids are bonded to tRNAs in a reaction catalyzed by aminoacyl-tRNA synthetases. (2) The initiation step requires assembly of ribosomes, mRNA, and aminoacyl-tRNAs into a functional unit. (3) In chain elongation, the ribosome moves along the mRNA, and the protein is assembled as new amino acids are added. (4) Chain termination requires protein release factors as well as stop signals on the mRNA.

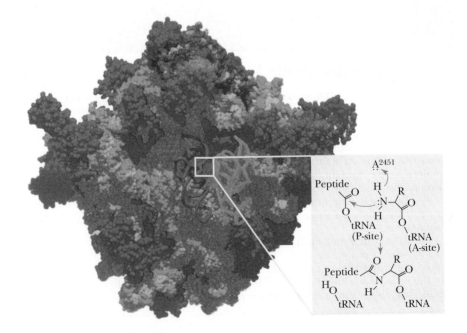

▶ **FIGURE 12.16** The large subunit of the ribosome seen from the viewpoint of the small subunit, with proteins in purple, 23S rRNA in orange and white, 5S rRNA (at the top) in burgundy and white, and A-site tRNA (green) and P-site tRNA (red) docked. In the box, the peptidyl transfer mechanism is catalyzed by RNA. The general base (adenine 2451 in *E. coli* 23S rRNA) is rendered unusually basic by its environment within the folded structure; it could abstract a proton at any of several steps, one of which is shown here. (*Reprinted by permission of Thomas Cech. The ribosome is a ribozyme.* Science **289**, *p. 878. Copyright © 2000 AAAS.*)

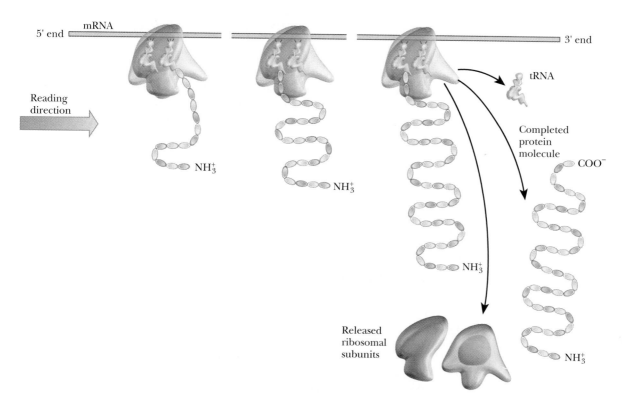

▲ **FIGURE 12.17** Simultaneous protein synthesis on polysomes. A single mRNA molecule is translated by several ribosomes simultaneously. Each ribosome produces one copy of the polypeptide chain specified by the mRNA. When the protein has been completed, the ribosome dissociates into subunits that are used in further rounds of protein synthesis.

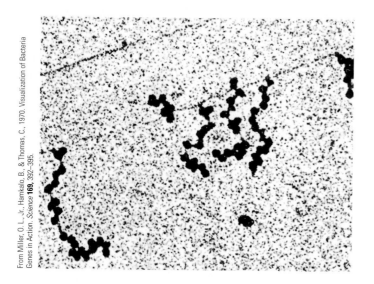

◀ **FIGURE 12.18** Electron micrograph showing coupled translation. The dark spots are ribosomes, arranged in clusters on a strand of mRNA. Several mRNAs have been transcribed from one strand of DNA (diagonal line from center left to upper right).

DNA to be in various stages of being transcribed. In this situation, several molecules of RNA polymerase are attached to a single gene, giving rise to several mRNA molecules, each of which has a number of ribosomes attached to it. The prokaryotic gene is being simultaneously transcribed and translated. This process, which is called *coupled translation* (Figure 12.18), is possible in prokaryotes because of the lack of cell compartmentalization. In eukaryotes, mRNA is produced in the nucleus, and the majority of protein synthesis takes place in the cytosol.

12.5 | How Does Translation Take Place in Eukaryotes?

The main features of translation are the same in prokaryotes and eukaryotes, but the details differ. The messenger RNAs of eukaryotes are characterized by two major posttranscriptional modifications. The first is the 5′ cap, and the second is the 3′ poly-A tail (Figure 12.19). Both modifications are essential to eukaryotic translation.

Chain Initiation

This is the part of eukaryotic translation that is the most different from that in prokaryotes. Thirteen more initiation factors are given the designation **eIF,** for **eukaryotic initiation factor.** Many of them are multisubunit proteins. Table 12.4 summarizes pertinent information about these initiation factors.

Step 1 in chain initiation involves the assembly of a 43S preinitiation complex (Figure 12.20). The initial amino acid is methionine, which is attached to a special tRNA$_i$ that serves only as the initiator tRNA. There is no fmet in eukaryotes. The met-tRNA$_i$ is delivered to the 40S ribosomal subunit as a complex with GTP and eIF2. The 40S ribosome is also bound to eIF1A and eIF3. This order of events is different from that in prokaryotes in that the first tRNA binds to the ribosome without the presence of the mRNA. In Step 2, the mRNA is recruited. There is no Shine–Dalgarno sequence for location of the start codon. The **5′ cap** orients the ribosome to the correct AUG via what is called a *scanning mechanism,* which is driven by ATP hydrolysis. The eIF4E is also a cap-binding protein, which forms a complex with several other eIFs. A **poly A binding protein (Pab1p)** links the **poly A tail** to eIF4G. The eIF-40S complex is initially positioned upstream of the start codon (Figure 12.21). It moves downstream until it encounters the first AUG in the correct context. The context is determined by a few bases surrounding the start codon, called the **Kozak sequence.** It is characterized by the consensus sequence −3ACCAUGG+4. The ribosome may skip the first AUG it finds if the next one has the Kozak sequence. Another factor is the presence of mRNA secondary structure. If hairpin loops form downstream of an AUG, an earlier AUG may be chosen. The mRNA and the seven eIFs constitute the 48S

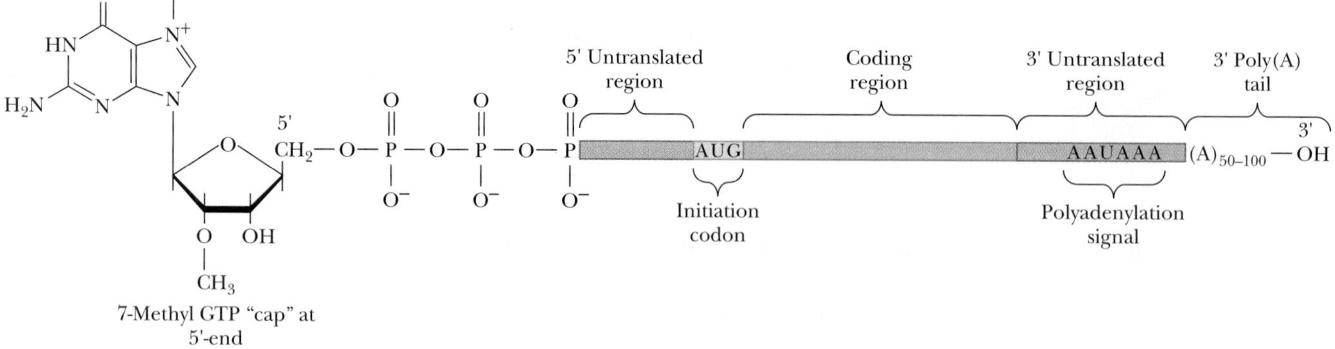

7-Methyl GTP "cap" at 5′-end

▲ **FIGURE 12.19** The characteristic structure of eukaryotic mRNAs. Untranslated regions ranging between 40 and 150 bases in length occur at both the 5′ and 3′ ends of the mature mRNA. An initiation codon at the 5′ end, invariably AUG, signals the translation start site.

Table 12.4			
Properties of Eukaryotic Translation Initiation Factors			
Factor	**Subunit**	**Size (kDa)**	**Function**
eIF1		15	Enhances initiation complex formation
eIF1A		17	Stabilizes Met-tRNA$_i$ binding of 40S ribosomes
eIF2		125	GTP-dependent Met-tRNA$_i$ binding to 40S ribosomes
	α	36	Regulated by phosphorylation
	β	50	Binds Met-tRNA$_i$
	γ	55	Binds GTP, Met-tRNA$_i$
eIF2B		270	Promotes guanine nucleotide exchange on eIF2
	α	26	Binds GTP
	β	39	Binds ATP
	γ	58	Binds ATP
	δ	67	Regulated by phosphorylation
	ε	82	
eIF2C		94	Stabilizes ternary complex in presence of RNA
eIF3		550	Promotes Met-tRNA$_i$ and mRNA binding
	p35	35	
	p36	36	
	p40	40	
	p44	44	
	p47	47	
	p66	66	Binds RNA
	p115	115	Major phosphorylated subunit
	p170	170	
eIF4A		46	Binds RNA; ATPase; RNA helicase; promotes mRNA binding to 40S ribosomes
eIF4B		80	Binds mRNA; promotes RNA helicase activity and mRNA binding to 40S ribosomes
eIF4E		25	Binds to mRNA caps
eIF4G		153.4	Binds eIF4A, eIF4E, and eIF3
eIF4F			Complex binds to mRNA caps; RNA helicase activity; promotes mRNA binding to 40S
eIF5		48.9	Promotes GTPase of eIF2, ejection of eIF
eIF6			Dissociates 80S; binds to 60S

Adapted from Clark, B. F. C., et al., eds. 1996. Prokaryotic and eukaryotic translation factors. *Biochimie* **78**, 1119–1122.

preinitiation complex. In Step 3, the 60S ribosome is recruited, forming the 80S initiation complex. GTP is hydrolyzed, and the initiation factors are released.

Chain Elongation

Peptide chain elongation in eukaryotes is very similar to that of prokaryotes. The same mechanism of peptidyl transferase and ribosome translocation is seen. The structure of the eukaryotic ribosome is different in that there is no E site, only the A and P sites. There are two eukaryotic elongation factors, eEF1 and eEF2. The eEF1 consists of two subunits, eEF1A and eEF1B. The 1A subunit is the counterpart of EF-Tu in prokaryotes. The 1B subunit is the equivalent of the EF-Ts in prokaryotes. The eEF2 protein is the counterpart of the prokaryotic EF-G, which causes translocation.

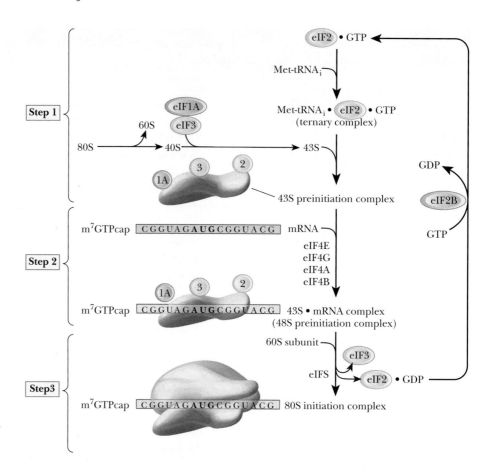

▶ **FIGURE 12.20** The three stages in the initiation of translation in eukaryotic cells. See Table 12.4 for a description of the functions of the eukaryotic initiation factors (eIFs).

▶ **FIGURE 12.21** Initiation factor eIF4G serves as a multipurpose adapter to engage the ^{7}methyl-G cap:eIF4E complex, the Pab1p:poly(A) tract, and the 40S ribosomal subunit in eukaryotic translation initiation. (*Adapted with permission from Heutze, H. W., 1997. eIF4G: A multipurpose ribosome adapter?* Science **275,** *500–501. Copyright © 1997 AAAS.*)

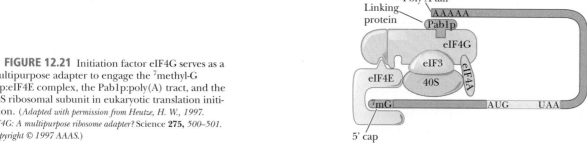

Many of the differences between translation in prokaryotes and eukaryotes can be seen in the response to inhibitors of protein synthesis and to toxins. The antibiotic chloramphenicol (a trade name is Chloromycetin) binds to the A site and inhibits peptidyl transferase activity in prokaryotes, but not in eukaryotes. This property has made chloramphenicol useful in treating bacterial infections. In eukaryotes, diphtheria toxin is a protein that interferes with protein synthesis by decreasing the activity of the eukaryotic elongation factor eEF2.

Chain Termination

As in prokaryotic termination, the ribosome will encounter a stop codon, either UAG, UAA, or UGA, and these will not be recognized by a tRNA molecule. In prokaryotes, three different release factors—RF1, RF2, and RF3—

were used, with two of them alternating, depending on which stop codon was found. In eukaryotes, only one release factor binds to all three stop codons and catalyzes the hydrolysis of the bond between the C-terminal amino acid and the tRNA.

There is a special tRNA called a **suppressor tRNA,** which allows translation to continue through a stop codon (see the Biochemical Connections box on p. 317 for a description of a unique tRNA that inserts a selenocysteine residue because the anticodon of the corresponding tRNA binds to the stop codon). Suppressor tRNAs tend to be found in cells in which a mutation has introduced a stop codon.

Coupled Transcription and Translation in Eukaryotes?

Until recently, the dogma of eukaryotic translation was that it was physically separated from transcription. Transcription occurred in the nucleus, and mRNA was then exported to the cytosol for translation. While this system is accepted as the normal process, recent evidence has shown that the nucleus has all of the components (mRNA, ribosomes, protein factors) necessary for translation. In addition, evidence shows that, in isolated test systems, proteins are translated in the nucleus. The authors of the most recent work suggest that 10–15% of the cell's protein synthesis occurs in the nucleus. See the articles by Hentze and by Iborra et al. in the Annotated Bibliography of this chapter.

12.6 | How Does Posttranslational Modification of Proteins Take Place?

Newly synthesized polypeptides are frequently processed before they reach the form in which they have biological activity. We have already mentioned that, in prokaryotes, *N*-formylmethionine is cleaved off. Specific bonds in precursors can be hydrolyzed, as in the cleavage of preproinsulin to proinsulin and of proinsulin to insulin (Figure 12.22). Proteins destined for export to specific parts of the cell or from the cell have leader sequences at their N-terminal ends. These leader sequences, which direct the proteins to their proper destination, are recognized and removed by specific proteases associated with the *endoplasmic reticulum*. The finished protein then enters the *Golgi apparatus*, which directs it to its final destination.

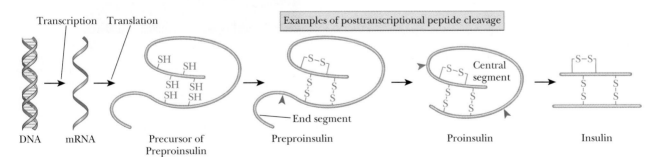

▲ **FIGURE 12.22** Some examples of posttranslational modification of proteins. After a precursor of preproinsulin is formed by the transcription–translation process, it is transformed into preproinsulin by formation of three disulfide bonds. Specific cleavage that removes an end segment converts preproinsulin to proinsulin. Finally, two further specific cleavages remove a central segment, with insulin as the end result.

Biochemical Connections

Molecular Chaperones: Preventing Unsuitable Associations

It is sometimes said that a chaperone's task is to prevent unsuitable associations. The class of proteins known as molecular chaperones operate in this way by preventing aggregation of newly formed proteins until they fold into their active forms. The information necessary for protein folding is present in the amino acid sequence, and many proteins will fold correctly without any outside help, as shown in part (a) of the figure. However, some proteins may form aggregates with other proteins, or may fold with incorrect secondary and tertiary structures, unless they interact first with a chaperone. Well-known examples include heat-shock proteins, which are produced by cells as a result of heat stress. The prime examples of this are the *Hsp70* class of proteins, named after the 70-kDa heat-shock protein that occurs in mammalian cytosol, shown in part (b) of the figure. The Hsp70 protein binds to the nascent polypeptide and prevents it from interacting with other proteins or from folding into an unproductive form. Completion of correct folding requires release from the chaperone and is driven by ATP hydrolysis. All proteins in this class, which were first studied as a response to heat stress in cells of all types, have highly conserved primary structures, in both prokaryotes and eukaryotes.

About 85% of proteins fold as shown in parts (a) and (b) of the figure. Another group, the **chaperonins** (also called the *Hsp60* proteins from their 60-kDa molecular weight) are known to be involved in the folding of the other 15% of proteins. A large multisubunit protein forms a cage of 60-kDa subunits around the nascent protein to protect it during the folding process, shown in part (c) of the figure. **GroEL** and **GroES** are the best characterized chaperonins from *E. coli*. GroEL is formed by two stacked seven-membered rings of 60-kDa subunits with a central cavity, as shown in the figure below. Protein folding occurs in the central cavity and is dependent upon ATP hydrolysis. GroES is a single seven-membered ring of 10-kDa subunits that sits on top of GroEL. During protein folding, the polypeptide chain goes through cycles of binding and unbinding to the surface of the central cavity. In some cases, more than 100 ATP molecules must be hydrolyzed before protein folding is complete.

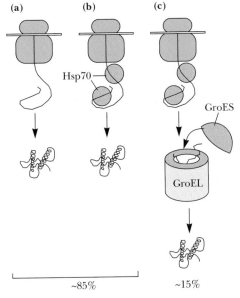

▲ Protein folding pathways. (a) Chaperone-independent folding. (b) Folding facilitated by a chaperone (gray)—in this case, the Hsp70 protein. About 85% of proteins fold by one of the two mechanisms shown in (a) and (b). (c) Folding facilitated by chaperonins—in this case, GroEL and GroES. (*Adapted from Netzer, W. J., and Hartl, F. U., 1998. Protein folding in the cytosol: Chaperonin-dependent and -independent mechanisms.* Trends in Biochemical Sciences, **23,** *68–73, Figure 2.*)

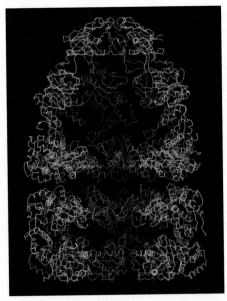

▲ Structure of the GroEL–GroES complex, emphasizing the central cavity. The protein is shown in orange. Bound ADP molecules are shown in green, and their associated magnesium ions are shown in red. (*Adapted by permission from Figure 1 in Xu, Z., Horwich, A. L., and Sigler, P. B., 1997,* Nature **388,** *741–750. Molecular graphics courtesy of Paul B. Sigler, Yale University.*)

In addition to the processing of proteins by breaking bonds, other substances can be linked to the newly formed polypeptide. Various cofactors, such as heme groups, are added, and disulfide bonds are formed (Figure 12.22). Some amino acid residues are also covalently modified, as in the conversion of proline to hydroxyproline. Other covalent modifications can take place, an example being the addition of carbohydrates or lipids to yield an active final form of the protein in question. Proteins can also be methylated, phosphorylated, and ubiquitinylated (Section 12.7).

A highly important question concerns the proper folding of the newly synthesized protein. In principle, the primary structure of the protein conveys enough information to specify its three-dimensional structure. In the cell, the complexity of the process and the number of possible conformations make it less likely that a protein would spontaneously fold into the correct conformation. The Biochemical Connections box on the previous page describes the processes involved in protein folding in vivo.

12.7 | How Are Proteins Degraded?

One of the most often overlooked controls of gene expression occurs at the level of the degradation of proteins. Proteins are in a dynamic state in which they are turned over often. Athletes are painfully aware of this because it means that they must work very hard to get in shape, but then they get out of shape very quickly. For some classes of proteins, there is a 50% turnover every three days. In addition, abnormal proteins that were formed from errors in either transcription or translation are degraded quickly. It is believed that a single break in the peptide backbone of a protein is enough to trigger the rapid degradation of the pieces, because breakdown products from natural proteins are rarely seen in vivo.

If protein degradation is so quick, clearly it is a process that must be heavily controlled to avoid destruction of the wrong polypeptides. The degradation pathways are restricted to degradative subcellular organelles, such as lysosomes, or to macromolecular structures called **proteasomes** (see the Biochemical Connections box on page 326). Proteins are directed to lysosomes by specific signal sequences, often added in a posttranslational modification step. Once in the lysosome, the destruction is nonspecific. Proteasomes are found in both prokaryotes and eukaryotes, and specific pathways exist to target a protein so that it will complex with a proteasome and be degraded.

In eukaryotes, the most common mechanism for targeting protein for destruction in a proteasome is by **ubiquitinylation.** Ubiquitin is a small polypeptide (76 amino acids) that is highly conserved in eukaryotes. There is a high degree of homology between the sequences in species as widespread as yeast and humans. When ubiquitin is linked to a protein, it condemns that protein to destruction in a proteasome. Figure 12.23 shows the mechanism of ubiquitinylation. Three enzymes are involved—*ubiquitin-activating enzyme (E1), ubiquitin-carrier protein (E2),* and *ubiquitin-protein ligase (E3).* The ligase transfers the ubiquitin to free amino groups on the targeted protein, either the N-terminus or lysine side chains. Proteins must have a free α-amino group to be susceptible, so those proteins that are modified at the N-terminus—with an acetyl group, for instance—are protected from ubiquitin-mediated degradation. The nature of the N-terminal amino acid also influences its susceptibility to ubiquitinylation. Proteins with Met, Ser, Ala, Thr, Val, Gly, or Cys at the N-terminus are resistant. Those with Arg, Lys, His, Phe, Tyr, Trp, Leu, Asn, Gln, Asp, or Glu at the N-terminus have very short half-lives, between 2 and 30 minutes. Proteins with an acidic residue at the N-terminus have a requirement for tRNA as part of their destruction pathway. The tRNA for arginine, Arg-tRNA^arg, is used to transfer arginine to the N-terminus, making the protein much more susceptible to the ubiquitin ligase (Figure 12.24). The following Biochemical Connections box gives an interesting example of how transcription regulation and protein degradation work together to control the process of acclimation to high altitude.

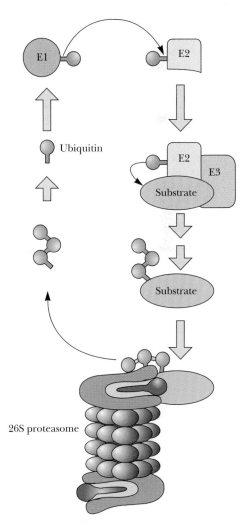

ACTIVE FIGURE 12.23
Diagram of the ubiquitin-proteasome degradation pathway. Pink "lollipop" structures symbolize ubiquitin molecules. (*Adapted with permission from Hilt, W., and Wolf, D. H., 1996. Proteasomes: Destruction as a program.* Trends in Biochemical Sciences **21,** *96–102, Figure 1.*) **Watch this Active Figure at http://now.brookscole.com/ campbell5**

Biochemical Connections

How Do We Adapt to High Altitude?

Those of us who live at low altitudes are quite aware of the sensations of oxygen deprivation and the associated physiological changes that occur with prolonged exposure to high altitudes. Physiologists have studied this phenomenon for many years, and biochemists have tried to find the mechanism that explains how cells sense the oxygen partial pressure and make the adaptive changes. Two of the major changes are the increase in red blood cells, which is stimulated by the hormone **erythropoietin (EPO)**, and *angiogenesis,* the stimulation of formation of new capillaries, which is stimulated by **vascular endothelial growth factor (VEGF).** Researchers have learned a great deal about how cells respond to low oxygen pressure, or **hypoxia,** and the results have many applications, including the production of drugs to treat inflammation, heart disease, and cancer.

A family of transcription factors called *hypoxia inducible factors (HIFs)* is the key to these processes (see figure). Heterodimers composed of HIFα and HIFβ subunits bind to DNA and up-regulate a variety of genes when the partial pressure of oxygen in the blood is low. Oxygen can be low for many reasons, such as a person being at high altitude, or a tissue that has to work unusually hard. During a heart attack or stroke, the oxygen partial pressure can drop, and these transcription factors can help reduce the damage.

The genes that are controlled by HIF are responsible for production of EPO and angiogenesis, as well as production of glycolytic enzymes that can provide energy for the cells when aerobic metabolism is compromised. In addition, many types of cancers are found associated with elevated levels of HIF. This may be involved in the tumor's ability to grow, which requires a large oxygen supply. As it turns out, if cell growth is uncontrolled, the dimerization of the two HIF subunits is unchecked, which would lead to constant expression of the adaptive genes. When this hap-

pens, overgrowth of endothelial cells occurs, leading to tumors. Any given cell maintains relatively constant levels of the HIFβ subunit, but the level of the HIFα subunit is regulated. The system is mainly controlled by the degradation of the HIFα subunit (see the right side of the figure). Proline 564 in the HIFα subunit can be hydroxylated by an enzyme called *proline hydroxylase (PH).* After it is hydroxylated, it binds to a protein called von Hippel–Lindau protein (pVHL), which was first discovered to be a tumor suppressor (see Section 14.8 for more on tumor suppressors). After the pVHL is bound, it stimulates the organization of a complex with *ubiquitin ligase (UL),* which ubiquitinylates the HIFα subunit. Ubiquitin is a 76-residue polypeptide that is very abundant and conserved in eukaryotes. When ubiquitin is put onto a protein, it targets the protein for transport to a proteasome, where the protein is degraded. Searching for how this pathway is related to the body's ability to sense the oxygen partial pressure is currently an area of active research. Evidence exists that the proline hydroxylase has a requirement for iron and oxygen. It is possible that, in the absence of sufficient oxygen, the HIF proline hydroxylase cannot function. Thus, the HIFα subunit is not targeted for destruction and will be available to bind to the HIFβ subunit, thus forming the active dimer and stimulating the adaptive effects of reduced oxygen pressure.

A second control point was recently discovered to be mediated by another hydroxylase. This time the target is an asparagine residue on HIFα. When the HIFα subunit is hydroxylated, it cannot bind to the transcription mediator, p300 (see Chapter 11), and can therefore not induce transcription. Thus, when oxygen is low, two different hydroxylations proceed at reduced levels. Reduction of the asparagine's hydroxylase reaction allows the HIFα to do its job, and reduction of the prolyl hydroxylase reac-

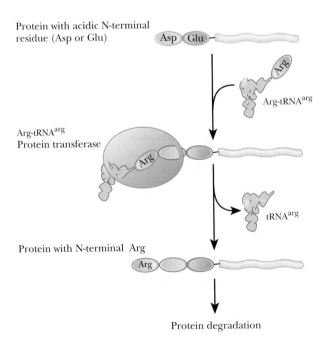

▶ **FIGURE 12.24** Proteins with acidic N-termini show a tRNA requirement for degradation. Arginyl-tRNA^{arg}:protein transferase catalyzes the transfer of Arg to the free α-NH$_2$ of proteins with Asp or Glu N-terminal residues. Arg-tRNA^{arg}:protein transferase serves as part of the protein degradation recognition system.

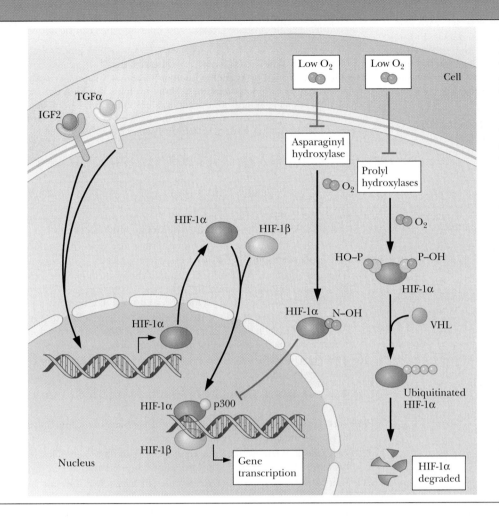

tion stops degradation of the HIFα by the ubiquitin-linked pathway. Many researchers believe that these two reactions are the mechanism for oxygen sensing, but others believe they are just incidental to the true mechanism, which has yet to be discovered.

HIF is also controlled in a positive way by **insulin growth factor 2 (IGF2)** and by **transforming growth factor α (TGFα)**. These are common growth factors that are active in several growth and differentiation pathways, and they are not related to oxygen availability.

Researchers are looking at control of HIF as a potential therapy against cancer because so many cancers are found to be associated with high HIF levels. Since tumors need oxygen to grow, stopping the tumor's activation of HIF-mediated transcription could effectively choke out the tumor before it progresses.

◀ The control of the level and function of hypoxia inducible fctor α by the level of oxygen. (*Adapted with permission from Marx, J., 2004. How cells endure low oxygen. Science* **303**, *1454–1456. Illustration by Carin Cain. Copyright © 2004 AAAS.*)

Summary

12.1 What Is the Overall Process of Translating the Genetic Message?
Protein biosynthesis requires ribosomes, messenger RNA, transfer RNA, and a number of protein factors. The ribosome is the site of protein synthesis. The mRNA and tRNA, which are bound to the ribosome in the course of protein synthesis, are responsible for maintaining the correct order of amino acids in the growing protein chain.

12.2 What Is the Genetic Code?
The genetic message is contained in a triplet, nonoverlapping, commaless, degenerate, universal code. A codon—in other words, a series of three bases adjacent to one another in sequence (nonoverlapping and commaless)—specifies a given amino acid. Several codons can, and usually do, specify the same amino acid (degeneracy of the code). The same code has been observed in viruses, prokaryotes, and eukaryotes (universality of the code). All 64 possible codons have been assigned meanings, with 61 of them coding for amino acids and the remaining three serving as the termination signals. However, there are a few organisms that have code variations, including certain green algae and fungi. In addition, mitochondria exhibit other variations. For example, the green alga *Acetabularia* translates the standard "stop" codons, UAG and UAA, as glycine.

12.3 What Is the Role of Aminoacyl-tRNA Synthetases in Amino Acid Activation?
Before an amino acid can be incorporated into a growing protein chain, it must first be activated. A covalent bond is formed between the amino acid and a tRNA, yielding an aminoacyl-tRNA.

12.4 How Does Translation Take Place in Prokaryotes?
The actual formation of the polypeptide chain takes place in three steps. In the initiation step, the first aminoacyl-tRNA is bound to the ribosome and to mRNA. A second aminoacyl-tRNA forms a complex with the ribosome and with mRNA. The binding site for the second aminoacyl-tRNA is close to that for the first aminoacyl-tRNA. A peptide bond is formed between the amino acids (chain elongation). The chain elongation process—which involves translocation of the ribosome along the mRNA, in addition to peptide-bond formation—repeats itself until the polypeptide chain is complete. Finally, chain termination takes place. Newly synthesized polypeptides frequently undergo posttranslational modification to produce the final active form of the protein. In the actual translation process, it is usual for several ribosomes to be bound to the same mRNA. Such a complex is called a polysome. Each of the ribosomes in the polysome has a

polypeptide in one of various stages of completion, depending on the position of the ribosome as it moves along the mRNA and transcribes the genetic message.

12.5 How Does Translation Take Place in Eukaryotes?
In eukaryotes, the process differs in several details. Eukaryotic mRNA undergoes a lot of processing that is not observed in prokaryotes. The number of initiation factors and elongation factors is higher in eukaryotes than in prokaryotes. Stop codons in eukaryotes are recognized by suppressor tRNAs, which allow insertion of nonstandard amino acids, such as selenocysteine.

12.6 How Does Posttranslational Modification of Proteins Take Place?
Newly synthesized polypeptides are processed before they reach the form in which they have biological activity. In prokaryotes, *N*-formylmethionine is cleaved off. Specific bonds in precursors can be hydrolyzed, as in the cleavage of preproinsulin to proinsulin and of proinsulin to insulin. Leader sequences, which direct the proteins to specific destinations, are recognized and removed by specific proteases. Sugar and lipid moieties can be added as well.

12.7 How Are Proteins Degraded?
Proteins are degraded in subcellular organelles, such as lysosomes, or in macromolecular structures called proteasomes. Many proteins are targeted for destruction by being bound to a protein called ubiquitin. The nature of the amino acid sequence at the N-terminus is often very important to control of the timing of destruction of a protein.

Critical Questions to Review

12.1 What Is the Overall Process of Translating the Genetic Message?

1. **Fact Check** Prepare a flow chart showing the stages of protein synthesis.

12.2 What Is the Genetic Code?

2. **Fact Check** A genetic code in which two bases encode a single amino acid is not adequate for protein synthesis. Give a reason why.

3. **Fact Check** Define degenerate code.

4. **Fact Check** How can the binding assay technique be used to assign coding triplets to the corresponding amino acids?

5. **Fact Check** Which nucleotides will break the rules of Watson–Crick base pairing when they are found at the wobble position of the anticodon? Which ones will not?

6. **Fact Check** Describe the role of stop codons in the termination of protein synthesis.

7. **Thought Question** Consider a three-base sequence in the template of DNA: $5' \ldots 123 \ldots 3'$, in which 1, 2, and 3 refer to the relative positions of deoxyribonucleotides. Comment on the probable effect on the resulting protein if the following point mutations (one-base substitutions) occurred.

   (a) Changing one purine for another in position 1.

   (b) Changing one pyrimidine for another in position 2.

   (c) Changing a purine to a pyrimidine in position 2.

   (d) Changing one purine for another in position 3.

8. **Thought Question** It is possible for the codons for a single amino acid to have the first two bases in common and to differ in the third base. Why is this experimental observation consistent with the concept of wobble?

9. **Thought Question** The nucleoside inosine frequently occurs as the third base in codons. What role does inosine play in wobble base pairing?

10. **Thought Question** Is it reasonable that codons for the same amino acid have one or two nucleotides in common? Why or why not?

11. **Thought Question** How would protein synthesis be affected if a single codon could specify the incorporation of more than one amino acid (an ambiguous code)?

12. **Thought Question** Comment on the evolutionary implications of the differences in the genetic code observed in mitochondria.

12.3 What Is the Role of Aminoacyl-tRNA Synthetases in Amino Acid Activation?

13. **Fact Check** What is the role of ATP in amino acid activation?

14. **Fact Check** Outline the proofreading processes in amino acid activation.

15. **Fact Check** What ensures fidelity in protein synthesis? How does this compare with the fidelity of replication and transcription?

16. **Fact Check** Can the same enzyme esterify more than one amino acid to its corresponding tRNA?

17. **Thought Question** A friend tells you that she is starting a research project on aminoacyl esters. She asks you to describe the biological role of this class of compounds. What do you tell her?

18. **Thought Question** Suggest a reason why the proofreading step in protein synthesis takes place at the level of amino acid activation rather than that of codon–anticodon recognition.

19. **Thought Question** Is amino acid activation energetically favored? Why or why not?

12.4 How Does Translation Take Place in Prokaryotes?

20. **Fact Check** Identify the following by describing their functions: EF-G, EF-Tu, EF-Ts, and peptidyl transferase.

21. **Fact Check** What are the components of the initiation complex in protein synthesis? How do they interact with one another?

22. **Fact Check** What is the role of the 50S ribosomal subunit in prokaryotic protein synthesis?

23. **Fact Check** What are the A site and the P site? How are their roles in protein synthesis similar? How do they differ? What is the E site?

24. **Fact Check** How does puromycin function as an inhibitor of protein synthesis?

25. **Fact Check** Describe the role of the stop signals in protein synthesis.

26. **Fact Check** Does mRNA bind to one or to both ribosomal subunits in the course of protein synthesis?

27. **Fact Check** What is the Shine–Dalgarno sequence? What role does it play in protein synthesis?

28. **Thought Question** You are studying with a friend who says that the hydrogen-bonded portions of tRNA play no important role in its function. What is your reply?

29. **Thought Question** *E. coli* has two tRNAs for methionine. What is the basis for the distinction between the two?

30. **Thought Question** In prokaryotic protein synthesis, formylmethionine (fmet) is the first amino acid incorporated, whereas (normal) methionine is incorporated in eukaryotes. The same codon (AUG) serves both. What prevents methionine from being inserted into the beginning and formethionine in the interior?

31. **Thought Question** Describe the recognition process by which the tRNA for *N*-formylmethionine interacts with the portion of mRNA that specifies the start of transcription.

32. **Thought Question** The fidelity of protein synthesis is assured twice during protein synthesis. How and when?

33. **Thought Question**
    (a) How many activation cycles are needed for a protein with 150 amino acids?
    (b) How many initiation cycles are needed for a protein with 150 amino acids?
    (c) How many elongation cycles are needed for a protein with 150 amino acids?
    (d) How many termination cycles are needed for a protein with 150 amino acids?

34. **Thought Question** What is the energy cost per amino acid in prokaryotic protein synthesis? Relate this to low entropy.

35. **Thought Question** Would it be possible to calculate the cost of protein synthesis, including the cost of making mRNA and DNA?

36. **Thought Question** Suggest a possible conclusion from the fact that peptidyl transferase is one of the most conserved sequences in all of biology.

37. **Thought Question** In the early days of research on protein synthesis, some scientists observed that their most highly purified ribosome preparations, containing almost exclusively single ribosomes, were less active than preparations that were less highly purified. Suggest an explanation for this observation.

38. **Thought Question** Suggest a scenario for the origin and development of peptidyl transferase as an integral part of the ribosome.

39. **Thought Question** Would you expect electron microscopy to give detailed information about ribosomal structure? *Hint:* Look at Figure 12.18.

40. **Thought Question** How does it improve the efficiency of protein synthesis to have several binding sites for tRNA close to each other on the ribosome?

41. **Thought Question** A virus does not contain ribosomes. How does it manage to ensure the synthesis of its proteins?

12.5 How Does Translation Take Place in Eukaryotes?

42. **Fact Check** What are two major similarities between protein synthesis in bacteria compared to eukaryotes? What are two major differences?

43. **Thought Question** Why do amino acids other than methionine occur in the N-terminal position of proteins from eukaryotes?

44. **Thought Question** Would puromycin be useful for the treatment of a virus infection? Why or why not? Would chloramphenicol be useful?

45. **Thought Question** Protein synthesis takes place much more slowly in eukaryotes than in prokaryotes. Suggest a reason why this is so.

46. **Thought Question** Why is it advantageous to have a mechanism to override the effect of stop codons in protein synthesis?

12.6 How Does Posttranslational Modification of Proteins Take Place?

47. **Thought Question** The amino acid hydroxyproline is found in collagen. There is no codon for hydroxyproline. Explain the occurrence of this amino acid in a common protein.

12.7 How Are Proteins Degraded?

48. **Fact Check** What role does ubiquitin play in the degradation of proteins?

49. **Thought Question** Consider protein degradation in the absence of ubiquitinylation. Is the process likely to be more or less efficient?

50. **Thought Question** Is it reasonable to expect that protein degradation can take place at any location in a cell?

Biochemistry ⊛ Now™

Assess your understanding of this chapter's topics with additional quizzing and tutorials at **http://now.brookscole.com/campbell5**

Annotated Bibliography

Ban, N. The Complete Atomic Structure of the Large Ribosomal Subunit at 2.4 Ångstrom Resolution. *Science* **289**, 905–920. (2000). [Current information on the structure of the ribosome.]

Cech, T. R. The Ribosome Is a Ribozyme. *Science* **289**, 878–879 (2000). [A classic paper describing the RNA-based catalysis of the ribosome.]

Fabrega, C., M. A. Farrow, B. Mukhopadhyay, V. de Crecy-Lagard, A. R. Ortiz, and P. Schimmel. An Aminoacyl tRNA Synthetase Whose Sequence Fits into Neither of the Two Known Classes. *Nature* **411**, 110–114 (2001). [An article describing a tRNA synthetase discovered to not belong to one of the two known classes.]

Freeland, S. J., and L. D. Hurst. Evolution Encoded. *Sci. Amer.* **290** (4), 84–91 (2004). [An article on the evolutionary fitness of the genetic code.]

Goldberg, A. Functions of the Proteasome: The Lysis at the End of the Tunnel. *Science* **268**, 522–523 (1995). [A perspective view of the multisubunit proteins involved in protein degradation.]

Hartl, F. Molecular Chaperones in Cellular Protein Folding. *Nature* **381**, 571–579 (1996). [A review article on the various classes of molecular chaperones.]

Hentze, M. eIF4G: A Multipurpose Ribosome Adapter? *Science* **275**, 500–501 (1997). [A short review of translation initiation in eukaryotes.]

Hentze, M. W. Believe It or Not—Translation in the Nucleus. *Science* **293**, 1058–1059 (2001). [A recent article introducing the research that showed nuclear translation.]

Ibba, M. Discriminating Right from Wrong. *Science* **294**, 70–71 (2001). [A summary of the latest research showing a type of proofreading ability by EF-Tu.]

Iborra, F. J., D. A. Jackson, and P. R. Cook. Coupled Transcription and Translation within Nuclei of Mammalian Cells. *Science* **293**, 1139–1142 (2001). [The primary research showing translation in the nucleus.]

LaRiviere, F. J., A. D. Wolfson, and O. C. Uhlenbeck. Uniform Binding of Aminoacyl-tRNAs to Elongation Factor Tu by Thermodynamic Compensation. *Science* **294**, 154–168 (2001). [An in-depth article about the newest evidence that EF-Tu provides another level of fidelity in translation.]

Polacek, N., M. Gaynor, A. Yassin, and A. S. Mankin. Ribosomal Peptidyl Transferase Can Withstand Mutations at the Putative Catalytic Nucleotide. *Nature* **411**, 498–501 (2001). [An article questioning the ribozyme theory of peptidyl transferase.]

Zhu, H., and H. F. Bunn. How do cells sense oxygen? *Science* **292** (5516), 449–451 (2001). [An article about protein degradation and gene expression in the system that allows adaptation to high altitude.]

CHAPTER 13 Nucleic Acid Biotechnology Techniques

Gopal Murti/CNRI/Phototake, NYC

Human chromosomes viewed through a scanning electron microscope.

Critical Questions

- **13.1** How Do We Purify and Detect Nucleic Acids?
- **13.2** What Makes Restriction Endonucleases an Important Tool for DNA Research?
- **13.3** What Is Cloning?
- **13.4** What Is Genetic Engineering, and Why Do We Do It?
- **13.5** What Are DNA Libraries?
- **13.6** What Is the Polymerase Chain Reaction?
- **13.7** What Is Site-Directed Mutagenesis?
- **13.8** What Is DNA Fingerprinting?
- **13.9** How Can We Study DNA–Protein Interactions?
- **13.10** What Are Some Methods for Studying Transcription?
- **13.11** How Do We Determine the Base Sequences of Nucleic Acids?
- **13.12** How Can We Use Bioinformatics to Study Genomics and Proteomics?

Biochemistry Now™
Test yourself on these Critical Questions at the BiochemistryNow website at **http://now.brookscole.com/campbell5**

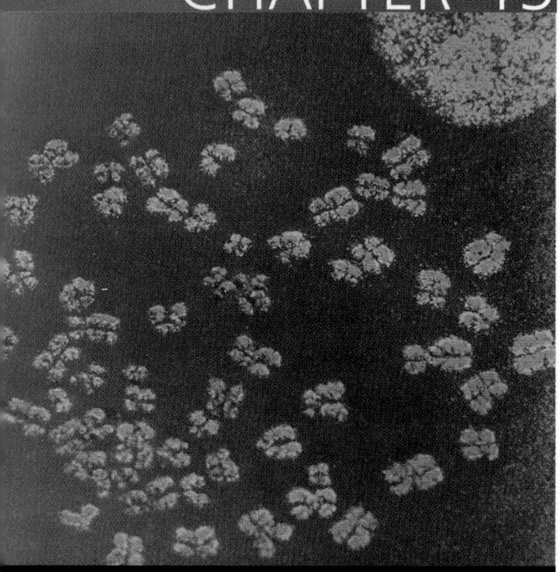

Methods of manipulating nucleic acids take advantage of their unique properties, particularly those of DNA. Methods of determining the base sequences of DNA make use of the way in which it is replicated. Interactions with specific proteins play a pivotal role in nucleic acid research. The discovery of restriction endonucleases made it possible to produce recombinant DNA in the laboratory, mimicking a process that takes place extensively in nature. The ability to pick out specific genes and to increase the available supply of their DNA, whether by cloning or by chain-reaction amplification, allows a degree of manipulation of living organisms far beyond that achieved by selective breeding. These methods have made it possible to contemplate and carry out the Human Genome Project, in which human DNA is completely mapped and the base sequences determined. Medical applications of this technology to genetic diseases are widely discussed in the media. Another application that potentially touches everyone's life is agriculture. "Not everybody gets sick, but everybody has to eat." The results of genetic engineering are on the way to your dinner table. Analytical techniques involving nucleic acids have wide applications in pure science, such as determining the types of DNA: protein interactions that control replication and transcription, as well as in applied science, such as the use of DNA samples for forensic identification of individuals.

13.1 How Do We Purify and Detect Nucleic Acids?

In early 1997, headlines around the world reported the successful cloning of a sheep by Scottish scientists, followed at intervals by reports of cloning of other animals. These striking examples of the power of the techniques for manipulating DNA sparked enormous amounts of discussion. In this chapter, we will focus our attention on some of the most important methods used in biotechnology.

Experiments on nucleic acids frequently involve extremely small quantities of materials of widely varying molecular size. Two of the primary necessities are to separate the components of a mixture and to detect the presence of nucleic acids; fortunately, powerful methods exist for accomplishing both goals.

Separation Techniques

Any separation method depends on the differences between the items to be separated. Charge and size are two properties of molecules that are frequently used for separation. One of the most widely used techniques in molecular biology, **gel electrophoresis,** uses both these properties. Electrophoresis is based on the motion of charged particles in an electric field. For our purposes, it is enough to know that the motion of a charged molecule in an electric field depends on the ratio of its charge to its mass. A sample is applied to a supporting medium. With the use of electrodes, an electric current is passed through the medium to achieve the desired separation. Polymeric gels, such as agarose and polyacrylamide, are frequently used as supporting media for electrophoresis (Figure 13.1). They are prepared and cast as a continuous cross-linked matrix. The cross-linking gives rise to pores, and the choice of agarose versus polyacrylamide gels depends on the size of the molecules to be separated—agarose for larger fragments (thousands of oligonucleotides) and polyacrylamide for smaller (hundreds of oligonucleotides).

The charge on the molecules to be separated leads them to move through the gel toward an electrode of opposite charge. Nucleic acids and oligonucleotide fragments are negatively charged at neutral pH because of the presence of the phosphate groups. When these negatively charged molecules are placed in an electric field between two electrodes, they all migrate toward the positive electrode. In nucleic acids, each nucleotide residue contributes a negative charge from the phosphate to the overall charge of the fragment, but the mass of the nucleic acid or oligonucleotide increases correspondingly. Thus, the ratio of charge to mass remains approximately the same regardless of the size of the molecule in question. As a result, the separation takes place simply on the basis of size and is due to the sieving action of the gel. In a given amount of time, with a sample consisting of a mixture of oligonucleotides, a smaller oligonucleotide moves farther than a larger one in an electrophoretic separation. The oligonucleotides move in the electric field because of their charges; the distances they move in a given time depend on their sizes.

Most separations are done with an agarose gel in a horizontal position, called a submarine gel because it is actually underneath the buffer in the chamber. However, when DNA sequencing is done (see Section 13.12), a polyacrylamide gel is run in a vertical position. Many different samples can be separated on a single gel. Each sample is loaded at a given place (a distinct well) at the negative-electrode end of the gel, and an electric current flows until the separation is complete (Figure 13.2).

Detection Methods

After the DNA pieces have been separated, they must be treated in some way that will allow them to be seen. Some of these techniques will allow all of the DNA to be seen, but others are more specific for certain DNA pieces.

The original method for detecting the separated products is based on radioactive labeling of the sample. A label, or tag, is an atom or molecule that allows visualization of another molecule. The isotope of phosphorus of mass number 32 (^{32}P, spoken as "P-thirty-two") was widely used in the past for this purpose. More recently, ^{35}S, or the isotope of sulfur of mass number 35 (spoken as "S-thirty-five"), has been used extensively. The DNA molecules undergo a reaction that incorporates the radioactive isotope into the DNA. When the labeled oligonucleotides have been separated, the gel is placed in contact with a piece of X-ray film. The radioactively labeled oligonucleotides expose the portions of the film with which they are in contact. When the film is developed, the positions of the labeled substances show up as dark bands. This technique is called **autoradiography,** and the resulting film image is an autoradiograph (Figure 13.3).

Many examples of autoradiographs can be seen in the scientific literature, but, as time goes on, autoradiography is being replaced by detection methods that do not use radioactive materials and their associated hazards. Many of these methods depend on emission of light (**luminescence**) by a chemical label attached to the fragments, and they can detect amounts of substances measured in picomoles. The way in which the label emits light depends on

▲ The late Dolly, the most famous sheep in the world, produced by cloning techniques.

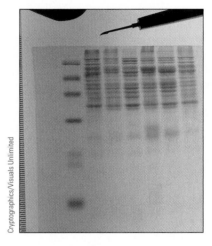

▲ **FIGURE 13.1** Separation of oligonucleotides by gel electrophoresis. Each band seen in the gel represents a different oligonucleotide.

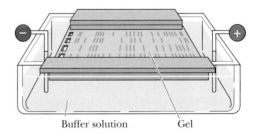

▲ **FIGURE 13.2** The experimental setup for gel electrophoresis. The samples are placed on the left side of the gel. When the current is applied, the negatively charged oligonucleotides migrate toward the positive electrode.

Hank Morgan/Photo Researchers, Inc.

▶ **FIGURE 13.3** An example of an autoradiogram.

the application. When the base sequence of DNA is to be determined, the label is a series of four fluorescent compounds, one for each base. The gel with the separated products is irradiated with a laser; the wavelength of the laser light is one that is absorbed by each of the four labels. Each of the four labeled compounds re-emits light at a different, characteristic, longer wavelength. This is called **fluorescence.** Another detection method that uses fluorescence involves the compound ethidium bromide. Its molecular structure includes a planar portion that can slip between the bases of DNA, giving ethidium bromide fluorescence properties when it binds to DNA that differ from those observed when it is free in solution. An ethidium bromide solution is used as a stain for DNA in a gel. The solution soaks into the gel, and the DNA fragments in the gel can be seen as orange bands by shining ultraviolet light on the gel.

Essential Information

Manipulation of nucleic acids requires methods for separating polynucleotides and oligonucleotides of different sizes. This is done by gel electrophoresis. Molecules of different sizes appear as separate bands on the gels. The bands can be seen by labeling the molecules with radioactive or luminescent "tags."

13.2 | What Makes Restriction Endonucleases an Important Tool for DNA Research?

The methods designed specifically for manipulating DNA depend on the availability of specific enzymes to produce DNA pieces of manageable size. These enzymes are called **restriction endonucleases.**

Many enzymes act on nucleic acids. A group of specific enzymes acts in concert to ensure the faithful replication of DNA, and another group directs the transcription of the base sequence of DNA into that of RNA. (We needed all of Chapters 10 and 11 to describe the manner in which these enzymes operate.) Other enzymes, called **nucleases,** catalyze the hydrolysis of the phosphodiester backbones of nucleic acids. Some nucleases are specific for DNA; others are specific for RNA. Cleavage from the ends of the molecule (by *exo*nucleases) is known, as is cleavage in the middle of the chain (by *endo*nucleases). Some enzymes are specific for single-stranded nucleic acids, and others cleave double-stranded ones. One group of nucleases, restriction endonucleases, has played a crucial role in the development of recombinant DNA technology (see the Biochemical Connections box on page 335).

Essential Information

Restriction endonucleases hydrolyze phosphodiester linkages in both strands of DNA at highly specific sites. These restriction enzymes are useful in sequencing nucleic acids because they produce DNA pieces of manageable size.

Many Restriction Endonucleases Produce "Sticky Ends"

Each restriction endonuclease hydrolyzes only a specific bond of a specific sequence in DNA. The sequences recognized by restriction endonucleases—their sites of action—read the same from left to right as they do from right to left (on the complementary strand). The term for such a sequence is a **palindrome.** ("Able was I ere I saw Elba" and "Madam, I'm Adam" are well-known linguistic palindromes.) A typical restriction endonuclease called *Eco*RI is isolated from *E. coli* (each restriction endonuclease is designated by an abbreviation of the name of the organism in which it occurs). The *Eco*RI site in DNA is 5′-GAATTC-3′, where the base sequence on the other strand is 3′-CTTAAG-5′. The sequence from left to right on one strand is the same as the sequence from right to left on the other strand. The phosphodiester bond between G and A is the one hydrolyzed. This same break is made on both strands of the DNA. There are four nucleotide residues—two adenines and two thymines in each strand—between the two breaks on opposite strands, leaving **sticky ends,** which can still be joined by hydrogen bonding between the complementary bases. With the ends held in place by the hydrogen bonds, the two breaks can then be resealed covalently by the action of DNA ligases (Figure 13.4). If no ligase is present, the ends can remain separated, and the hydrogen bonding at the sticky ends holds the molecule together until gentle

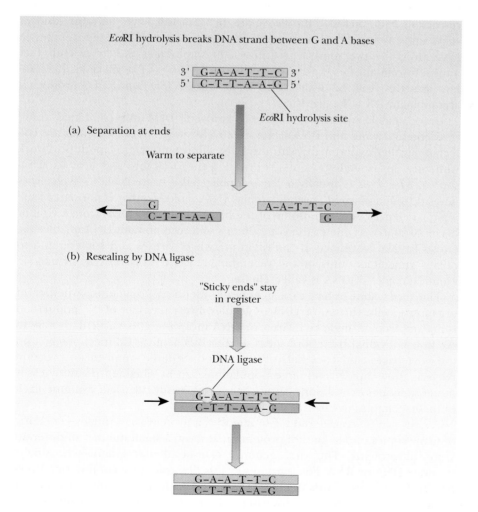

◀ **FIGURE 13.4** Hydrolysis of DNA by restriction endonucleases. (a) Separation of ends. (b) Resealing of ends by DNA ligase.

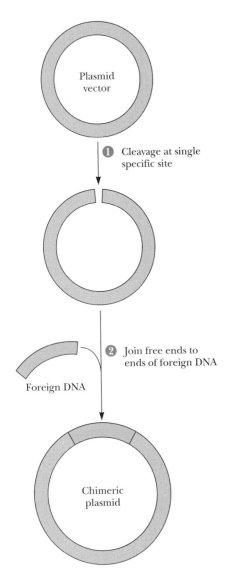

Biochemistry ⓔ**Now**™ **ACTIVE FIGURE 13.5**
(1) Foreign DNA sequences can be inserted into plasmid vectors by opening the circular plasmid with a restriction endonuclease. (2) The ends of the linearized plasmid DNA are then joined with the ends of a foreign sequence, reclosing the circle to create a chimeric plasmid. **Watch this Active Figure at http://now.brookscole.com/campbell5**

Essential Information

Recombinant DNA can be produced in the laboratory when DNA samples from different sources are treated with the same restriction endonuclease. The two different kinds of DNA have ends with complementary sequences, which can base-pair to one another. DNA ligases then form covalent bonds in the polynucleotide chain to give the recombinant DNA.

---

warming or vigorous stirring effects a separation. Some enzymes, such as *Hae*III, cut in a way that leave a blunt end.

$$5'\text{-GGCC-}3' \qquad\qquad 5'\text{-GG} \qquad \text{CC-}3'$$
$$\xrightarrow[]{\textit{Hae}\text{III}}$$
$$3'\text{-CCGG-}5' \qquad\qquad 3'\text{-CC} \qquad \text{GG-}5'$$
$$\textbf{Blunt-end cut}$$

To make life more challenging for molecular biologists, some enzymes can also cut with less than absolute specificity. This is called *star (*) activity* and can often be seen if the enzyme concentration is too high or the enzyme is incubated with the DNA too long.

13.3 | What Is Cloning?

DNA molecules containing covalently linked segments derived from two or more DNA sources are called **recombinant DNA.** (Another name for recombinant DNA is **chimeric DNA,** named after the chimera, a monster in Greek mythology that had the head of a lion, the body of a goat, and the tail of a serpent.) The production of recombinant DNA was made possible by the isolation of restriction endonucleases.

Using "Sticky Ends" to Construct Recombinant DNA

If DNA from two different organisms has recognition sites for the same restriction endonuclease, the two kinds of DNA will have the same kind of sticky ends as a result of treatment with that enzyme. If digested samples of DNA from the two sources are mixed, in some cases the sticky ends that anneal to one another will be from different sources. The nicks in the covalent structure can be sealed with **DNA ligases** (Section 10.4), producing recombinant DNA (Figure 13.5).

Unfortunately, when two different kinds of DNA are combined using restriction enzymes and DNA ligase, relatively few product molecules are collected. Further experiments with the DNA will require large amounts to work with, and this is made possible by inserting the DNA into a viral or bacterial source. The virus is usually a bacteriophage; the bacterial DNA typically is derived from a **plasmid,** a small circular DNA molecule that is not part of the main circular DNA chromosome of the bacterium. Using DNA from a viral or bacterial source as one of the components of a recombinant DNA enables scientists to take advantage of the rapid growth of viruses and bacteria and to obtain greater amounts of the recombinant DNA. This process of making identical copies of DNA is called **cloning.**

The term **clone** refers to a genetically identical population, whether of organisms, cells, viruses, or DNA molecules. Every member of the population is derived from a single cell, virus, or DNA molecule. It is particularly easy to see how individual bacteriophages (see the Biochemical Connections box on the next page) and bacterial cells can produce large numbers of progeny. Bacteria grow rapidly, and large populations can be obtained relatively easily under laboratory conditions. Viruses also grow easily. We shall examine each of these examples in turn.

A virus can be considered a genome with a protein coat, usually consisting of many copies of one kind of protein or, at most, a small number of different kinds of proteins. The viral genome can be double-stranded or single-stranded DNA or RNA. For purposes of this discussion, we confine our attention to DNA viruses with double-stranded DNA. In the cloning of bacteriophages, a "lawn" of bacteria covering a petri dish is infected with the phage.

Biochemical Connections

Restriction Endonucleases: "Molecular Scissors"

This class of enzymes was discovered in the course of genetic investigations of bacteria and **bacteriophages** (**phages** for short; from the Greek *phagein*, "to eat"), the viruses that infect bacteria. The researchers noted that bacteriophages that grew well in one strain of the bacterial species they infected frequently grew poorly (had *restricted* growth) in another strain of the same species. Further work showed that this phenomenon arises from a subtle difference between the phage DNA and the DNA of the strain of bacteria in which phage growth is restricted. This difference is the presence of methylated bases at certain sequence-specific sites in the host DNA and not in the viral DNA.

The growth-restricting host cells contain cleavage enzymes, the restriction endonucleases, that produce double chain breaks at the unmethylated specific sequences in phage DNA; the cells' own corresponding DNA sequences, in which methylated bases occur, are not attacked (see figure). These cleavage enzymes consequently degrade DNA from any source *but* the host cell. The most immediate consequence is a slowing of the growth of the phage in that bacterial strain, but the important thing for our discussion is that DNA from any source can be cleaved by such an enzyme if it contains the target sequence. More than 800 restriction endonucleases have been discovered in a variety of bacterial species. More than 100 specific sequences are recognized by one or more of these enzymes. Table 13.1 shows several target sequences.

Table 13.1

Restriction Endonucleases and Their Cleavage Sites

Enzyme*	Recognition and Cleavage Site
*Bam*HI	↓ 5′-GGATCC-3′ 3′-CCTAGG-5′ ↑
*Eco*RI	↓ 5′-GAATTC-3′ 3′-CTTAAG-5′ ↑
*Hae*III	↓ 5′-GGCC-3′ 3′-CCGG-5′ ↑
*Hind*III	↓ 5′-AAGCTT-3′ 3′-TTCGAA-5′ ↑
*Hpa*II	↓ 5′-CCGG-3′ 3′-GGCC-5′ ↑
*Not*I	↓ 5′-GCGGCCGC-3′ 3′-CGCCGGCG-5′ ↑
Pst	↓ 5′-CTGCAG-3′ 3′-GACGTC-5′ ↑

Arrows indicate the phosphodiester bonds cleaved by the restriction endonucleases.

* The name of the restriction endonuclease consists of a three-letter abbreviation of the bacterial species from which it is derived—for example *Eco* for *Escherichia coli*.

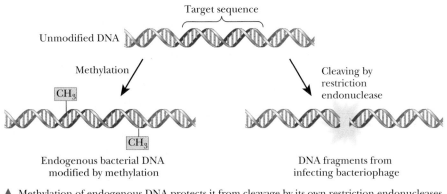

▲ Methylation of endogenous DNA protects it from cleavage by its own restriction endonucleases.

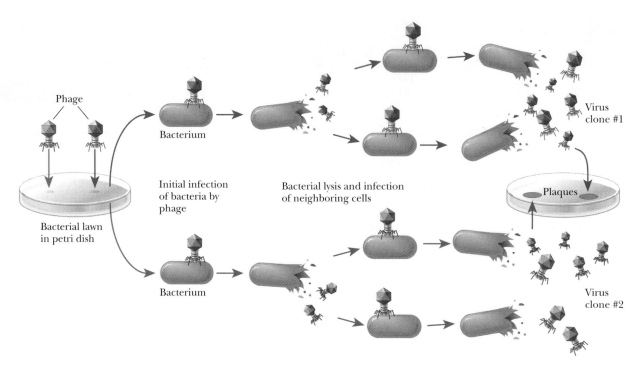

▲ **FIGURE 13.6** The cloning of a virus. The progeny of each individual phage (bacterial virus) infects and destroys bacteria on the petri dish, leaving clear spots known as plaques. Each plaque indicates the presence of a clone. (*Adapted with permission from* Dealing with Genes: The Language of Heredity, *by Paul Berg and Maxine Singer,* © *1992 by University Science Books.*)

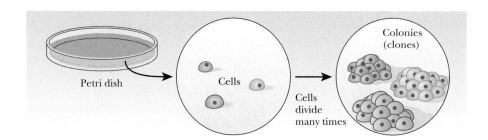

▶ **FIGURE 13.7** The cloning of cells. Each individual cell divides many times, producing a colony of progeny. Each colony is a clone. (*Adapted with permission from* Dealing with Genes: The Language of Heredity, *by Paul Berg and Maxine Singer,* © *1992 by University Science Books.*)

Each individual virus infects a bacterial cell and reproduces, as do its progeny when they infect and destroy other bacterial cells. As the virus multiplies, a clear spot, called a "plaque," appears on the petri dish, marking the area in which the bacterial cells have been killed (Figure 13.6). The plaque consists of the progeny viruses that are clones of the original.

To clone individual cells, whether from a bacterial or a eukaryotic source, a small number of cells is spread thinly over a suitable growth medium in a dish. Spreading the cells thinly ensures that each cell will multiply in isolation from the others. Each colony of cells that appears on the dish will then be a clone derived from a single cell (Figure 13.7). Since large quantities of bacteria and bacteriophages can be grown in short time intervals under laboratory conditions, it is useful to introduce DNA from a larger, slower-growing organism into bacteria or phages and to produce more of the desired DNA by cloning. If, for example, we want to take a portion of human DNA, which would be hard to acquire, and clone it in a virus, we can treat the human DNA and the virus DNA with the same restriction endonuclease, mix the two, and allow the sticky ends to anneal. (Recall our discussion of restriction endonucleases from the previous Biochemical Connections box.) If we then treat the mixture with DNA ligase, we have produced recombinant DNA. To clone it, we incorporate the chimeric DNA into virus particles by adding viral coat protein and allowing the virus to assemble itself. The virus particles are spread on a lawn of bacteria, and the cloned segments in each plaque can then be identified (Figure 13.8). The bacteriophage is called a **vector,** the carrier for the gene of interest that was cloned. The gene of interest is called many things, such as the *"foreign DNA,"* the *"insert," "geneX,"* or even *"YFG,"* for "your favorite gene."

The other principal vector is a plasmid—bacterial DNA that is not part of the main circular DNA chromosome of the bacterium. This DNA, which usually exists as a closed circle, replicates independently of the main bacterial genome and can be transferred from one strain of a bacterial species to another by cell-to-cell contact. The foreign gene can be inserted into the plasmid by the successive actions of restriction endonucleases and DNA ligase, as was seen in Figure 13.5. When the plasmid is taken up by a bacterium, the

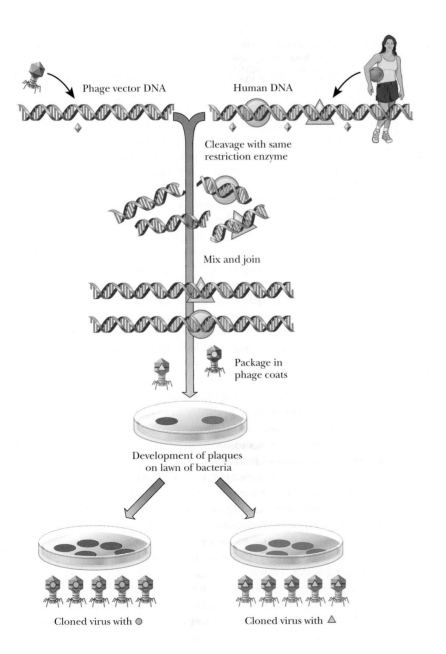

Phage vector DNA

Human DNA

Cleavage with same
restriction enzyme

Mix and join

Package in
phage coats

Development of plaques
on lawn of bacteria

Cloned virus with ⬤

Cloned virus with △

◀ **FIGURE 13.8** The cloning of human DNA fragments with a viral vector. Human DNA is inserted into viral DNA and then cloned. (*Adapted with permission from* Dealing with Genes: The Language of Heredity, *by Paul Berg and Maxine Singer, © 1992 by University Science Books.*)

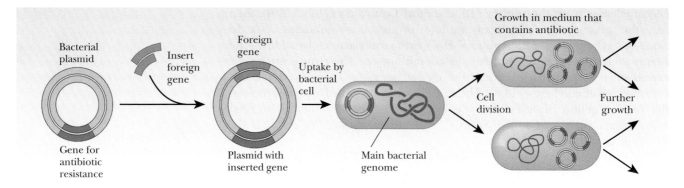

▲ FIGURE 13.9 Selecting for recombinant DNA in a bacterial plasmid. The plasmid also contains a gene for antibiotic resistance. When bacteria are grown in a medium that contains the antibiotic, those that have acquired a plasmid will grow. Bacteria without a plasmid cannot grow in this medium. (*Adapted with permission from* Dealing with Genes: The Language of Heredity, *by Paul Berg and Maxine Singer,* © 1992 by University Science Books.)

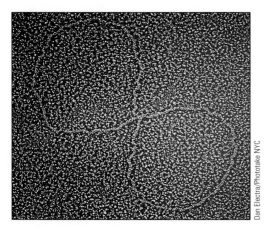

▲ DNA plasmids—extrachromosomal self-replicating genetic elements of a bacterial cell.

DNA insert goes along for the ride (Figure 13.9). The bacteria that contain the DNA insert can then be grown in fermentation tanks under conditions that allow them to divide rapidly, amplifying the inserted gene many thousandfold.

While the theory of cloning DNA into a plasmid is straightforward, there are several considerations for a successful experiment. When bacteria take up a plasmid, we say they have been *transformed*. **Transformation** is the process whereby new DNA is incorporated into a host. Bacteria are encouraged to take up foreign DNA by a couple of methods. One is to heat-shock the bacteria at 42°C, followed by placing them on ice. Another is to place them in an electric field, a technique called *electroporation*.

How are we to know which of the bacteria have taken up the plasmid? Since bacteria divide quickly, we would not want all the bacteria to grow— rather, only the ones that have the plasmid. This process is called **selection.** Each plasmid chosen for cloning must have some type of **selectable marker** that lets us know that the growing bacterial colonies contain the plasmid. These markers are usually genes that confer resistance to antibiotics. After transformation, the bacteria are plated on a medium containing the antibiotic to which the plasmid carries resistance. In that way, only the bacteria that took up the plasmids will grow (Figure 13.9). One of the first plasmids used for cloning is pBR322 (Figure 13.10). This simple plasmid was created from a naturally occurring one found in *E. coli*. Like all plasmids, it has an origin of replication, so it can replicate independently of the rest of the genome. It has genes that confer resistance to two antibiotics, tetracycline and ampicillin. The genes are indicated *tet*^{r} and *amp*^{r}. The pBR322 plasmid has several restriction enzyme sites. The number and location of restriction sites is very important to a cloning experiment. The foreign DNA must be inserted at unique restriction sites so that the use of restriction enzymes opens up the plasmid at only one point. Also, if the restriction site chosen is inside one of the selection markers, the resistance to the antibiotic will be lost upon inserting the foreign DNA. This was, in fact, the original way selection was done with plasmids. Foreign DNA was inserted using restriction sites in the *tet*^{r} gene. Selection was achieved by noting the loss of the ability of bacteria to grow on a medium containing tetracycline.

One of the early stumbling blocks to cloning was finding the right plasmid, one that had restriction sites that matched those enzymes needed to cut out the foreign DNA. As the technology to design plasmids improved, they were

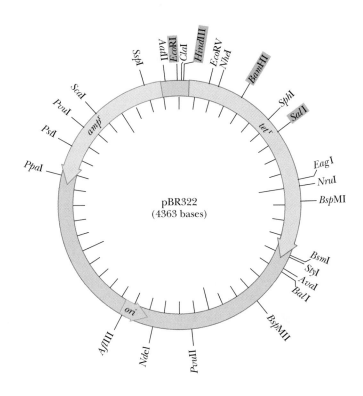

◀ **FIGURE 13.10** One of the first widely used cloning vectors, the plasmid pBR322. This 4363-base-pair plasmid contains an origin of replication (*ori*) and genes encoding resistance to the drugs ampicillin (*amp*[r]) and tetracycline (*tet*[r]). The locations of restriction endonuclease cleavage sites are indicated.

◀ **FIGURE 13.11** A vector cloning site containing multiple restriction sites. This is called a polylinker or multiple cloning site (MCS). The colored amino acids are from the *lacZ* gene that is part of the plasmid. The MCS does not disrupt the normal reading frame of this sequence, so this plasmid can be used for blue/white screening (see text). (*Adapted from Figure 1.14.2, in Ausubel, F. M., et al., 1987,* Current Protocols in Molecular Biology. *New York: John Wiley and Sons. Used by permission.*)

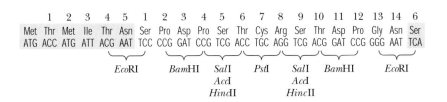

created with regions that had many different restriction sites in a small space. This region was called a **multiple cloning site (MCS)** or a **polylinker** (Figure 13.11). A popular cloning vector series is based on the pUC plasmids (Figure 13.12). The acronym pUC stands for *universal cloning plasmid*. Each of these cloning vectors has an extensive MCS, which helps solve another problem with cloning—the directionality of the inserted DNA. Depending on what is to be done with the cloned DNA, it may be important to control its orientation in the vector. If only one restriction enzyme is used, such as *Bam*H1, then the foreign DNA will be able to enter the plasmid in either of two directions. However, if the foreign DNA is cut out of its source at one end with *Bam*H1 and at the other end with *Hind*III, and if these same two restriction enzymes are used to open up the plasmid, then the ends will match in only one direction (Figure 13.12).

The use of the pUC plasmids also aids in the selection procedure. The older plasmids that were based on pBR322 had the shortcoming that the foreign DNA was inserted into the tetracycline resistance gene. This meant that the only way to spot bacteria that had taken up a plasmid that had also taken up the insert was that the bacteria would *not* grow on a medium containing tetracycline. This lack of growth by the successful clone made it challenging to go back and find the proper bacterial colonies. The pUC plasmids, however, have a characteristic that alleviates this procedure—they contain the *lacZ*

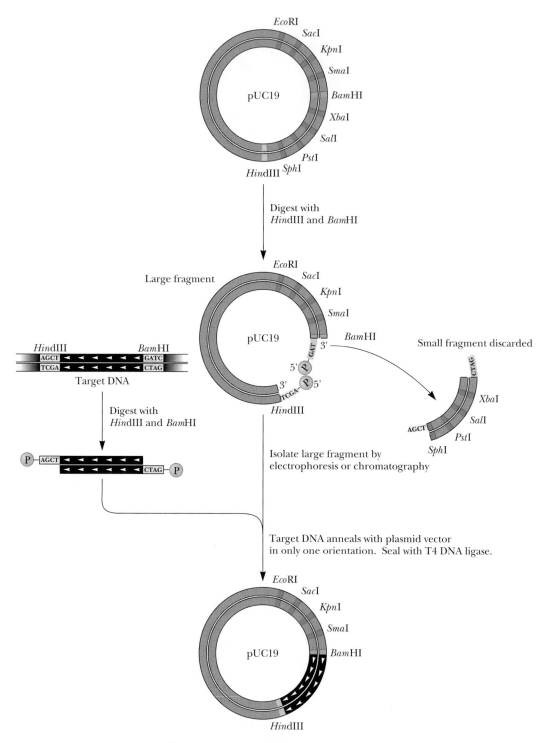

Biochemistry ⊗ Now™ ▲ **ANIMATED FIGURE 13.12** The pUC series of plasmids is very popular. They have extensive multiple cloning sites. Here we see an example of directional cloning. Two different restriction enzymes are used to cut open the MCS and to cut out a piece of DNA to be cloned. As a result, the DNA that is to be inserted can be incorporated in only one orientation. **See this figure animated at http://now.brookscole.com/campbell5**

gene, which is the basis for a selection technique called **blue/white screening.** The *lac*Z gene codes for the α-subunit of the enzyme β-galactosidase, which is used to cleave disaccharides, such as lactose (Chapter 16). The MCS is located inside the *lac*Z gene, so that, when foreign DNA is inserted, it will inactivate the gene. Figure 13.13 shows how this characteristic is useful. The

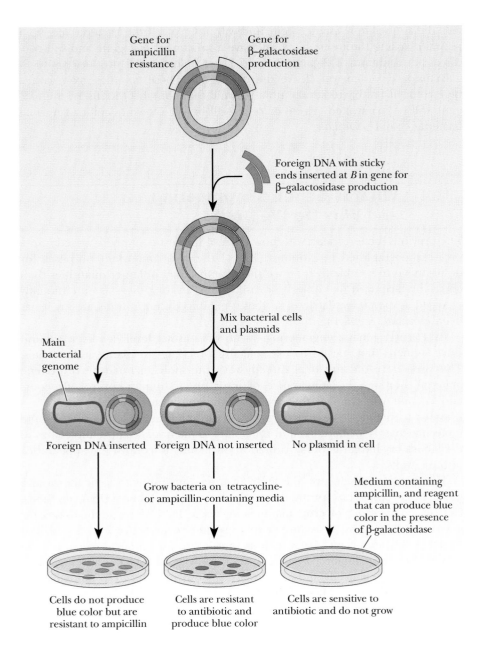

Gene for ampicillin resistance

Gene for β–galactosidase production

Foreign DNA with sticky ends inserted at *B* in gene for β–galactosidase production

Mix bacterial cells and plasmids

Main bacterial genome

Foreign DNA inserted

Foreign DNA not inserted

No plasmid in cell

Grow bacteria on tetracycline- or ampicillin-containing media

Medium containing ampicillin, and reagent that can produce blue color in the presence of β-galactosidase

Cells do not produce blue color but are resistant to ampicillin

Cells are resistant to antibiotic and produce blue color

Cells are sensitive to antibiotic and do not grow

◀ **FIGURE 13.13** Clone selection via blue/white screening. The pUC plasmid contains a gene for ampicillin resistance and the *lacZ* gene. The latter produces the α-subunit of β-galactosidase. Transformed cells are plated on an agar medium containing ampicillin and a dye called X-gal. The *lacZ* gene is found inside the multiple cloning site of the plasmid. When the plasmid and the DNA to be cloned are cut with a restriction enzyme and then mixed together, two possibilities result. The DNA insert can be incorporated as shown (red insert seen inside plasmid), or the plasmid can recircularize without the insert. When this mixture is used to transform bacteria, there can be three products. *Left side:* The bacteria take up a plasmid that has the insert. This plasmid confers ampicillin resistance to the cells, but the *lacZ* gene is inactivated by the presence of the insert. These cells grow and are the normal off-white color of bacterial colonies. *Middle:* The bacteria take up the recircularized plasmid. This plasmid confers ampicillin resistance, so the cells grow. The plasmid makes the α-subunit of β-galactosidase. The β-galactosidase cleaves the X-gal, causing the dye to turn blue, so these cells grow with a blue color. *Right side:* Bacteria take up no plasmid at all. These cells do not grow, due to their sensitivity to ampicillin. (*Adapted with permission from* Dealing with Genes: The Language of Heredity, *by Paul Berg and Maxine Singer,* © *1992 by University Science Books.*)

pUC plasmid is cut open in its MCS by restriction enzymes, and the foreign DNA (red) is cut out of its source with the same enzymes. These are then combined and joined together with DNA ligase to yield two products in the ligation reaction. The desired product is the plasmid that now contains the foreign DNA. The other is a plasmid that has reclosed upon itself without the inserted DNA. This type is much less common when two different restriction enzymes are used to open the plasmid, but it still occurs infrequently. When this mixture is used to transform the bacteria, there will be three possible products: (1) bacteria that took up the plasmid with the insert, (2) bacteria that took up the plasmid without the insert, and (3) bacteria that took up no plasmid at all. The mixture of bacteria from the transformation is plated on a medium containing ampicillin and a dye, such as X-gal. β-Galactosidase hydrolyzes a bond in the X-gal molecule, turning it blue. The bacteria to be transformed are mutants that make a defective version of the β-galactosidase that lacks the α-subunit. If the bacteria take up no plasmid at all, they will lack the ampicillin-resistance gene and will not grow. If the bacteria take up a plasmid lacking the insert, they will have a functional *lacZ* gene and will produce the α-subunit of β-galactosidase. These colonies will produce active β-galactosidase, which will cleave the dye, X-gal. These colonies grow with a blue color. If the bacteria take up a plasmid that contains the insert, the *lacZ* gene will be inactivated. These colonies will be the off-white color of normal bacterial colonies on agar.

13.4 What Is Genetic Engineering, and Why Do We Do It?

The previous sections dealt with how DNA of interest could be inserted into a vector and amplified by cloning. One of the most important purposes for doing this is to be able to produce the gene product in larger quantities than could be acquired by other means. When an organism is intentionally changed at the molecular level so that it exhibits different traits, we say it has been *genetically engineered.*

In a sense, **genetic engineering** on an organismal level has been around since humans first started to use selective breeding on plants and animals. This procedure did not deal directly with the molecular nature of genetic material, nor was the appearance of traits under human control. Breeders had to cope with changes that arose spontaneously, and the only choice was whether to breed for a trait or to let it die out. An understanding of the molecular nature of heredity and the ability to manipulate those molecules in the laboratory have, of course, added to our ability to control the appearance of these traits.

The practice of selective alteration of organisms for both agricultural and medical purposes has profited greatly from recombinant DNA methods. Genetic engineering of crop plants is an active field of research. Genes for increased yields, frost resistance, and resistance to pests are introduced into commercially important plants such as strawberries, tomatoes, and corn. Similarly, animals of commercial importance—mostly mammals, but also including fish—are also genetically altered. Some variations introduced in animals have medical implications. Mice with altered genetic makeup are used in the research laboratory. In another medically related field, researchers working with insect-borne diseases, such as malaria, are trying to engineer strains of insects, such as the mosquito, *Anopheles gambiae,* that can no longer transmit the infection to humans (Figure 13.14). In all cases, the focus of the research is to introduce *traits that can be inherited* by the descendants of the treated organisms. In the treatment of human genetic disease, however, the aim is not to produce heritable changes. Serious ethical questions arise with the manipulation of human genetics; consequently, the focus of research has been on forms of **gene therapy** in which cells of specific tissues in a living person are altered in a way that alleviates the effects of the disease. Examples of diseases that may someday be treated in this way include cystic fibrosis, hemophilia, Duchenne muscular dystrophy, and severe combined immune deficiency (SCID). The last of these is also known as the bubble-boy syndrome, because its victims must live in isolation (in a large "bubble") to avoid infection.

James Gathany and the Centers for Disease Control

▲ **FIGURE 13.14** Two adult female *Anopheles gambiae* mosquitoes (ventral view). The one on the left is a mutant. Scientists are attempting to produce strains of these mutant mosquitoes, which are unable to transmit malaria to humans, in hopes that they will replace the malaria carriers.

DNA Recombination Occurs in Nature

When recombinant-DNA technology was in its early stages in the 1970s, considerable concern arose both about safety and about ethical questions. Some of the ethical questions are still matters of concern. One that has definitely

Biochemical Connections

Genetic Engineering in Agriculture

The idea of genetic engineering in humans often upsets people because they are fearful of "playing God." Genetic manipulations in plants appear to cause less controversy, although many people are still concerned about the practice. Nevertheless, many types of modifications have been made, and some have been introduced with little fanfare and have shown some signs of success. Several examples are listed here. It is important to realize that many of the modifications that have been made using genetic engineering are just controlled versions of the selective breeding used for centuries to improve crop and animal production.

1. *Disease resistance.* Because most high-production crops involve specialized strains, many are more susceptible to fungal disease and insect damage. As a consequence, many herbicides and insecticides are applied liberally during the growing season. In many cases, other plants have natural resistance to these pests. When the gene that produces the resistance can be isolated, it can be transferred into other plants. There has been limited success in transferring such resistance to crop species. In 2000, the public became more aware of genetically modified (GM) food crops because of news about Bt corn, a crop carrying a bacterial gene that produces a toxin poisonous to certain caterpillars. The Bt gene, from *Bacillus thuringensis,* has been put into corn and cotton to increase crop yields by killing the caterpillars that would otherwise eat those crops. It also reduces the amount of pesticides needed to grow the crops. On the negative side is the potential harm to other species. For example, environmental groups oppose the planting of Bt corn because of its effect on the monarch butterfly in lab tests.

2. *Nitrogen fixation.* Nitrogen fixation is most easily accomplished by bacteria that grow in nodules on the roots of certain legumes, such as beans, peas, and alfalfa. We now realize that the genes of the nitrogen-fixing bacteria actually become shared with the genes of the host plant. There is much research going on to determine if these genes could be incorporated into other plants, which would reduce the amount of nitrogen fertilizer needed for maximal plant growth and crop production.

3. *Frost-free plants.* Many marine organisms, such as fish found in the Antarctic, produce a so-called antifreeze protein, a protein characterized by its hydrophobic surface, which prevents the formation of ice crystals at low temperature. In plants exposed to freezing temperatures, ice crystals in their tissues actually cause the frost damage. Insertion of the gene for the antifreeze protein into strawberries and potatoes, for example, has resulted in crops that are stable during late spring frost or in areas with very short growing seasons. There has been much controversy over these crops, because, in each case, a foreign gene from a nonplant species was introduced into a plant. This type of cross-species gene modification causes widespread concern, even though the taste and texture of the modified product is indistinguishable from those of foods produced by unmodified plants.

4. *Tomatoes with a long shelf life.* The Flavr-Savr tomato has a genetic modification, but one in which no new gene has been introduced. Rather, one of the plant's own genes has been deactivated. The gene in questions is the one that allows the tomato plant to produce ethylene, a key compound in the ripening process. Because this gene has been deactivated, the tomatoes mature on the plant until they just begin to show some pink color—the exact stage at which tomatoes are picked and sent to market. Typically, fields of unmodified tomatoes need to be picked as many as nine times in a season, since the tomatoes all ripen at different rates. When an unmodified tomato ripens, it continues to make ethylene, which leads to overripening, softening, and deterioration. Fields of the modified tomatoes can be picked once or twice, and the harvested fruits can be made to ripen as needed by exposing them to exogenous ethylene gas. Since the Flavr-Savr tomato does not make its own ethylene, the shelf life of the ripe, ready-to-eat tomato is increased. The financial saving made possible by less time spent picking tomatoes and the extended shelf life results in a cheaper and better product. Best of all, the taste is indistinguishable from that of unmodified tomatoes. (One of the authors of this book has eaten Flavr-Savr tomatoes many times.)

5. *Increased milk production.* There is much controversy over providing cows with supplemental bovine somatotropin (BST), also called growth hormone, a hormone that increases metabolism and milk production in dairy cows. The controversy centers on the human consumption of the hormone. However, BST is a peptide hormone, so it is hydrolyzed in the digestive tract and is not absorbed directly into the human bloodstream (Section 24.2). Furthermore, all milk must contain some of this hormone, because the cow cannot produce milk without it. A more soundly based concern about BST supplementation in dairy cows is that the cows given extra BST often develop mastitis, an inflammation of the udder caused by bacterial infection. Mastitis is frequently treated with high doses of antibiotics, and there is a possibility that some of these antibiotics could end up in the milk, causing problems for people with certain food sensitivity.

Courtesy Monsanto Company

▲ Roundup Ready® soybeans. A gene has been introduced into these plants that allows them to tolerate the herbicide Roundup.® The weeds are killed, but the soybeans are not. Crop yields are improved, and the amount of herbicide can be reduced.

been laid to rest is the question of whether the process of cutting and splicing DNA is an unnatural process. Indeed, DNA recombination is a common part of the crossing over of chromosomes. There are many, varied reasons for in vivo recombination of DNA, two of which are the maintenance of genetic diversity and the repair of damaged DNA (Section 10.5).

Until recently, heritable changes in organisms were solely those that arose from mutations. Researchers in the field took advantage of both spontaneous mutations and those produced by exposure of organisms to radioactive materials and other substances known to induce mutations. Selective breeding was then used to increase the population of desired mutants. It was not possible to produce "custom-tailored" changes in genes.

Since the advent of recombinant-DNA technology, it is possible (within limits) to change specific genes, and even to change specific DNA sequences within those genes, to alter the inherited characteristics of organisms. Bacteria can be altered to produce large amounts of medically and economically important proteins. Animals can be manipulated to cure, or to alleviate the symptoms of, their genetic diseases, and agriculturally important plants can be made to produce greater crop yields or be given increased resistance to pests. The Biochemical Connections box on page 343 gives some examples of agricultural applications of genetic engineering.

Bacteria as "Protein Factories"

We can use the reproductive power of bacteria to express large quantities of a mammalian protein of interest; however, the process is often more complicated than it might seem because most mammalian proteins are heavily processed after their initial transcription and translation (Section 12.7). Because bacteria have little posttranslational modification of their proteins, they lack the enzymes necessary for this processing.

An application of genetic engineering that is of considerable practical importance is the production of human insulin by *E. coli*. This was one of the very first human proteins produced through genetic engineering, and its production eliminated the problems related to harvesting insulin from large numbers of laboratory animals and giving humans a peptide from another species. The process is far from straightforward, however. A significant problem is that the insulin gene is split. It contains an **intron,** a DNA sequence that codes for RNA that will eventually be deleted in the processing of the mRNA that directs the synthesis of the protein (see Section 11.6). Only the RNA transcribed from DNA sequences called **exons** will appear in mature mRNA (Figure 13.15). Bacteria do not have the cellular apparatus for splicing introns out of RNA transcripts to give functional mRNA. One might think that the problem could be solved by using cDNA (Section 13.6) obtained from the mRNA for insulin in a reaction catalyzed by reverse transcriptase. The problem here is that the polypeptide encoded by this mRNA contains an end peptide and a central peptide, that is to be removed from it by further processing in insulin-producing cells to yield two polypeptide chains, designated A and B (Figure 13.15).

The approach to this problem is to use two synthetic DNAs, one encoding the A chain of insulin and the other encoding the B chain. These synthetic DNAs are produced in the laboratory using methods that were developed by synthetic organic chemists. Each DNA is inserted into a separate plasmid vector (Figure 13.16). The vectors are taken up by two different populations of *E. coli*. The two groups are then cloned separately; each group of bacteria produces one of the two polypeptide chains of insulin. The A and B chains

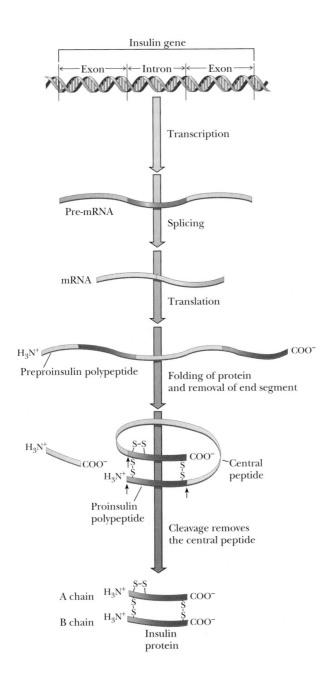

Insulin gene

Exon—Intron—Exon

Transcription

Pre-mRNA

Splicing

mRNA

Translation

H_3N^+ COO⁻

Preproinsulin polypeptide

Folding of protein and removal of end segment

S–S

H_3N^+
COO⁻
COO⁻ Central peptide

H_3N^+

Proinsulin polypeptide

Cleavage removes the central peptide

A chain H_3N^+ S–S COO⁻

B chain H_3N^+ COO⁻

Insulin protein

◀ **FIGURE 13.15** Synthesis of insulin in humans. The insulin gene is a split gene. The intervening sequence (intron) encodes an RNA transcript that is spliced out of the mRNA. Only the portions of the gene called exons are reflected in the base sequence of mRNA. Once protein synthesis takes place, the polypeptide is folded, cut, and spliced. The end product, active insulin, has two polypeptide chains as a result. (*Adapted with permission from* Dealing with Genes: The Language of Heredity, *by Paul Berg and Maxine Singer,* © *1992 by University Science Books.*)

are extracted and mixed, finally producing functional human insulin. The Biochemical Connections box on page 347 describes some other successes in the production of human proteins for which the demand greatly exceeds the supply.

Protein Expression Vectors

The plasmid vectors pBR322 and pUC are referred to as cloning vectors. They are used to insert the foreign DNA and to amplify it. However, if the goal is to produce the protein product from the foreign DNA, they are not suitable. An **expression vector** is needed. An expression vector will have many of the same attributes as a cloning vector, such as the origin of replication, a

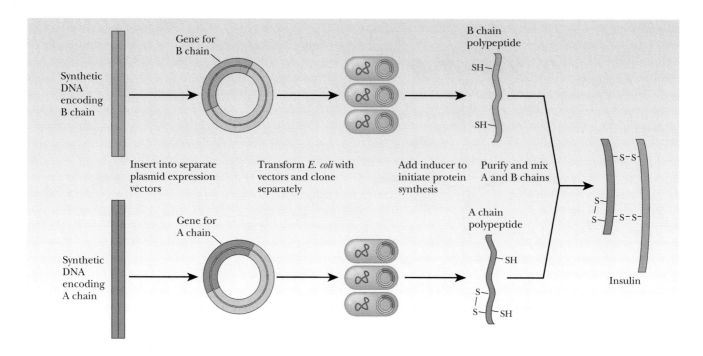

▲ **FIGURE 13.16** Active human insulin can be produced in bacteria by the use of two separate batches of *E. coli*. Each batch produces one of the two chains, the A chain or the B chain. The two chains are mixed to produce active insulin. (*Adapted with permission from* Dealing with Genes: The Language of Heredity, *by Paul Berg and Maxine Singer,* © *1992 by University Science Books.*)

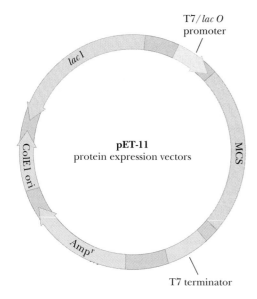

▲ **FIGURE 13.17** The pET expression vectors. These plasmids have the usual plasmid components, such as an origin of replication, MCS, and an antibiotic resistance gene (which provides resistance to ampicillin). In addition, their MCS is between the promoter for binding T7 RNA polymerase and a termination site for T7 RNA polymerase. When these vectors carry inserts, the insert DNA can be transcribed in the cell by T7 RNA polymerase. The cell then translates the mRNA into protein. (*Courtesy of Stratagene.*)

multiple cloning site, and at least one selectable marker. In addition, it must be able to be transcribed by the genetic machinery of the bacteria into which it is transformed. This means that it must have a promoter for RNA polymerase, and the RNA transcribed must have a ribosomal binding site so that it can be translated. It must also have a transcription termination sequence; otherwise, the entire plasmid will be transcribed instead of just the inserted gene. Figure 13.17 shows a schematic of an expression vector. Upstream of the site where the foreign DNA is inserted is the transcription promoter. Often this is the promoter for a viral RNA polymerase called **T7 polymerase.** There will also be a T7 terminator at the other end of the MCS. After the insert is successfully ligated, the plasmid is transformed into an expression strain of bacteria, such as *E. coli* JM109 DE3. What makes this strain unique is that it has a gene that produces T7 RNA polymerase, but the gene is under the control of the *lac* operon (Chapter 11). Once the bacteria are growing well with the plasmid, the cells are given a lactose analogue, IPTG (isopropylthiogalactoside). This stimulates the *lac* operon in the bacteria, which then produce T7 RNA polymerase, which then binds to the plasmid T7 promoter and transcribes the gene. The bacterial cells then translate the mRNA into protein. This selective control of the expression is important because many foreign proteins are toxic to the cells; expression must be timed carefully. The plasmid shown in Figure 13.17 also has the *lac*I gene, although it is transcribed in the opposite direction. This produces the repressor for the *lac* operon to help make sure that none of the foreign proteins are transcribed unless the system is induced by IPTG. The Biochemical Connections box on page 348 gives an example of how protein expression can be linked to a novel purification scheme.

Biochemical Connections

Human Proteins through Genetic Recombination Techniques

Genetic engineering techniques have made it possible to prepare many proteins with the amino acid sequences found in humans. This is done by isolating the gene that encodes the protein and then incorporating the gene into a bacterium or into eukaryotic cells grown in tissue culture. Recent work in the cloning of cows and sheep has been justified in part as a potential source of human proteins that might be produced in the animals' milk. In the process of establishing systems for the commercial production of any protein product, it is always easier to isolate and purify the proteins if they are present in the extracellular fluid surrounding the cells that produced them.

Frequently, bacterial systems are used to produce human proteins. Bacteria are very cheap to grow and easy to work with. Experimental systems using eukaryotic cells are much more difficult and expensive, in spite of the fact that eukaryotic cells offer the advantage of being able to add sugar residues to glycoproteins and perhaps help the proteins fold into their biologically active conformations. Still, using bacteria and recombinant DNA for protein production has its complications. Growing proteins in bacteria requires the addition of the Shine–Dalgarno sequence (Section 12.4) to the mRNA to ensure its binding to the ribosome. The genetic manipulations necessary to make the product in a foreign host often result in a protein that is slightly modified, usually by having a few extra amino acids on the N-terminus. The presence of the extra amino acids complicates the process of approval by the FDA, since there is more concern about side effects.

Some notable successes in human protein production include the following:

1. Insulin (see text) is traditionally isolated from horses and swine, and it has been used to treat diabetes mellitus type I. However, after many years of use, about 5% of diabetics develop a severe allergy to the foreign protein. These people can now be treated with bacteria-derived human insulin, which costs only about 10% more than the animal-derived hormones.
2. In the past, human growth hormone (HGH) has been obtained only by extraction from the pituitary glands of cadavers, a practice that carries the risk of the cells being contaminated with HIV or other diseases. HGH is used for therapy in genetic dwarfs and for muscle-wasting diseases, including AIDS. HGH is a relatively large protein hormone, with more than 300 amino acids, but it is a simple protein with no sugar residues, so it was relatively easy to clone into bacteria. It is interesting to note that the availability of increased amounts of safe HGH has resulted in the emergence of a black market in the sale of the hormone for muscle building in athletes.
3. Two proteins, tissue plasminogen activator (TPA) and enterokinase (EK), are known to dissolve blood clots. If they are injected into the body within a critical period after a heart attack or stroke, either of these two proteins can prevent or minimize the disastrous effects of blood clots in the heart or brain. Without genetic recombination, there would never be enough of these proteins available for this treatment for blood clots to be of practical use. Currently, two different companies are producing the two proteins.
4. Erythropoietin (EPO) is a hormone that stimulates the bone marrow to produce erythrocytes, commonly referred to as red blood cells (RBCs). This relatively small protein is lost during kidney dialysis; a healthy kidney also filters this hormone from the blood but then reabsorbs it back into the body. People with chronic kidney failure, who are on dialysis while awaiting a kidney transplant, thus suffer an additional problem of being chronically anemic from having too few RBCs. Such individuals must receive regular blood transfusions, which carries the risk of disease and possible allergic reaction. A genetically engineered erythropoietin, Epogen, is now available from Amgen; it is the most commercially successful example of a human hormone coming from genetic research. EPO is also one of the most successful examples of a recombinant protein that has had its original purpose subverted. Because some endurance athletes have used EPO to boost their levels of RBCs, giving them a big advantage over their competitors, there now exists a black market for EPO in the sports industry. The situation came to a head during the 1998 Tour de France when a team doctor was arrested crossing the border into France while carrying countless vials of EPO. The EPO was eventually traced to a theft from a hospital. The team was ejected from the race, and the international governing body of cycling has invested hundreds of thousands of dollars to combat EPO usage, including developing an assay for recombinant EPO.

Genetic Engineering in Eukaryotes

When the target organism for genetic engineering is an animal or a plant, one must consider that these are multicellular organisms with multiple kinds of tissues. In bacteria, altering the genetic makeup of a cell implies a change in the whole single-celled organism. In multicellular organisms, one possibility is to change a gene in a specific tissue, one that contains only one kind of differentiated cell. In other words, the change is *somatic,* affecting only the body tissues of the altered organism. In contrast, changes in germ cells (egg and sperm cells), called *germ-line* changes, are passed on to succeeding generations. If germ cells are to be modified, the change must be made at an early stage in development, before the germ cells are sequestered from the rest of

Biochemical Connections

Fusion Proteins and Fast Purifications

Affinity chromatography was introduced in Chapter 5 as an example of a powerful technique for protein purification. Molecular biologists have taken the idea one step further and have incorporated affinity chromatography ligand-binding sites directly into a protein to be expressed. A protein is created that contains not only the amino acid sequence of the desired polypeptide but also some extra amino acids at the N-terminus or C-terminus. These new proteins are called **fusion proteins.** The figure indicates how this might work. An expression vector that has a promoter for T7 polymerase, followed by a start sequence ATG, is used. A *his-tag* sequence that follows the ATG will code for six histidine residues. Following the his-tag is a sequence that is specific for a proteolytic enzyme called *enterokinase*. Finally comes the MCS, where the gene of interest can be cloned. After the desired gene is cloned into this vector, it is transformed into bacteria and expressed. The fusion proteins that will be translated will have the initial methionine, six histidines, the enterokinase-specific amino acid sequence, and then the desired protein. Remember from Chapter 5 that an affinity-chromatography resin uses a ligand that binds specifically to a protein of interest. This technique is used with a his-tagged fusion protein. A nickel affinity column is set up, which is very specific for histidine residues. The cells are lysed and passed over the column. All the proteins will pass through except the fusion protein, which binds tightly to the nickel column. The fusion protein can be eluted with imidazole, a histidine analog. Enterokinase is then added to cleave off the his-tag, leaving the desired protein. Under ideal circumstances, this can be an almost perfect, one-step purification of the protein.

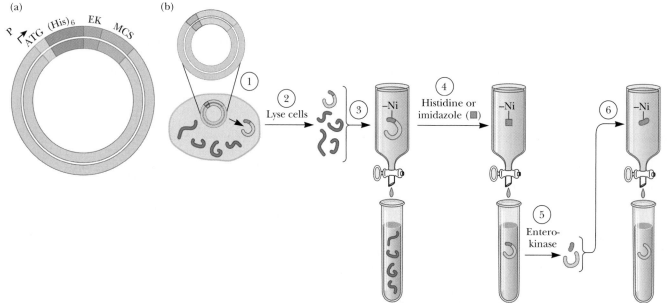

▲ Affinity chromatography and fusion proteins combine to purify a protein efficiently. (a) An expression vector is used that inserts six histidines at the N-terminus just after the ATG start site. Following the histidines is a sequence that will code for an enterokinase site, after which will be the protein you are trying to purify. (b) In Step 1, the plasmid is cloned and expressed to form cell proteins, including the fusion protein (red and yellow). In Step 2, the cells are lysed and, in Step 3, the lysate is run over a nickel affinity column. Nickel attracts the histidine tag on the fusion protein, and the other cell components are washed out. In Step 4, high concentrations of histidine or imidazole (red rectangles) are used to elute the fusion protein from the nickel affinity column. In Step 5, enterokinase is then used to cut out the histidines, leaving (in Step 6) the desired protein. (*Adapted from* Molecular Biology, *by R. F. Weaver, McGraw-Hill, 1999.*)

the organism. Attempts to produce such changes have succeeded in comparatively few organisms, such as plants, fruit flies, and some other animals, such as mice. Genetic engineering in plants frequently uses a vector based on a bacterial plasmid from the crown gall bacterium, *Agrobacterium tumefaciens.* Cells of this bacterium bind to wounded plant tissue, allowing plasmids to move from the bacterial cells into the plant cells. Some of the plasmid DNA inserts itself into the DNA of the plant cells in the only known natural transfer of genes from a bacterial plasmid to a eukaryotic genome. Expression of plasmid genes in the plant gives rise to a tumor called a crown gall. Whole,

Courtesy Calgene LLC

◀ **FIGURE 13.18** A transgenic tomato plant. Recombinant DNA methods have produced plants that resist defoliation by caterpillars. Tomatoes with a longer shelf life are another result of this research.

healthy plants can grow from gall cells, even though they are not germ cells. (This process, of course, does not take place in animals.) The plants that grow from the gall cells can produce fertile seeds, allowing the gene that has been transferred to be continued in a new strain of the plant. Genes from any desired source can be incorporated into the *A. tumefaciens* plasmid and then transferred to a plant.

This method was used to genetically engineer tomato plants that resist defoliation by caterpillars (Figure 13.18). A gene that encodes a protein toxic to caterpillars was taken from the bacterium *Bacillus thuringensis* to bring about this modification. Work is continuing on other useful modifications of food crops. Many observers of this whole line of research have raised questions about both the safety and the ethics of the process. The public became more aware of the extent to which genetically modified (GM) foods were in circulation in the year 2000, when corn that had been modified with the gene from *Bacillus thuringensis* (Bt corn) showed up in taco shells. This had been an accident as the Bt corn had been approved only for animal feed and not for human consumption, pending studies of potential allergenic effects. (See the article by Hopkin in the Annotated Bibliography at the end of this chapter.) Environmentalists are also concerned about the effect of GM crops for two reasons. The first concern is the effect on nontargeted insects, such as the monarch butterfly, which may be particularly sensitive to the toxin produced by the Bt gene. Second is the potential to create a super breed of insect accidentally that is immune to the effect of the toxin. (See the article by Brown in the Annotated Bibliography at the end of this chapter.) On the positive side, fields planted with Bt cotton plants can sometimes use up to 80% less pesticide than fields planted with ordinary cotton.

13.5 | What Are DNA Libraries?

Since methods exist for selecting regions of DNA from the genome of an organism of interest and cloning those regions in suitable vectors, a question that immediately comes to mind is whether we can take *all* the DNA of an organism (the total genome) and clone it in chunks of reasonable size. The answer is that we can do this, and the result is a **DNA library.**

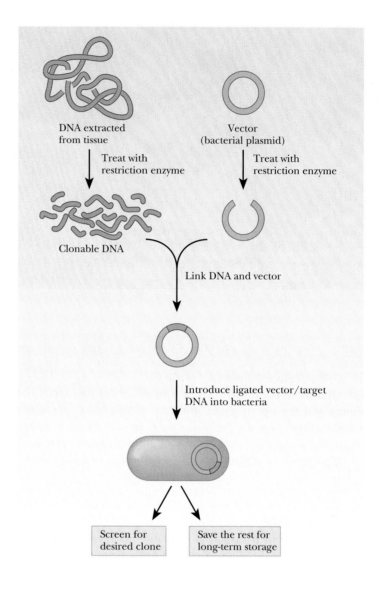

▶ **FIGURE 13.19** Steps involved in the construction of a DNA library. All the DNA of a given organism is extracted and treated with a restriction enzyme. The DNA fragments are incorporated into bacterial plasmids. Specific clones can be selected. The remaining clones are saved for future use. (*Adapted with permission from* Dealing with Genes: The Language of Heredity, *by Paul Berg and Maxine Singer, © 1992 by University Science Books.*)

Let us say that we want to construct a library of the human genome. There are six billion base pairs in a diploid human cell (a cell that has a set of chromosomes from both parents). If we consider that 20,000 base pairs is a reasonable size for a cloned insert, we will need a minimum of 300,000 different recombinant DNAs. It is quite a feat to achieve this number of different recombinants, and, in practice, we would need several times this minimum to ensure full representation and to account for the vector molecules that do not acquire an insert. For the purposes of this discussion, we shall use a bacterial plasmid as the vector to construct a suitable number of recombinant DNA molecules by the methods described in Section 13.3 (Figure 13.19).

The next step is to separate the individual members of the population of plasmid DNA molecules by cloning them. The group of clones that has acquired a plasmid with an insert constitutes the recombinant library. The whole library can be stored for future use, or a single clone can be selected for further study. The process of constructing a DNA library can be quite laborious, leading many researchers to obtain previously constructed libraries from other laboratories or from commercial sources. Some journals require that libraries and individual clones that have been discussed in articles they publish be freely available to other laboratories.

Finding an Individual Clone in a DNA Library

Imagine that, after a DNA library has been constructed, someone wants to find a single clone (for example, one that contains a gene responsible for an inherited disease) out of the hundreds of thousands, or possibly millions, in the library. This degree of selectivity requires specialized techniques. One of the most useful of these techniques depends on separating and annealing complementary strands. An imprint is taken of the petri dish on which the bacterial colonies (or phage plaques) have grown. A nitrocellulose disc is placed on the dish and then removed. Some of each colony or plaque is transferred to the disc, and *the position of each is the same as it was on the dish.* The rest of the original colonies or plaques remain on the dish and can be stored for future use (Figure 13.20).

The nitrocellulose disc is treated with a denaturing agent to unwind all the DNA on it. [The DNA has become accessible by disruption (lysis) of the bacterial cells or phage.] After it is denatured, the DNA is permanently fixed to the disc by treatment with heat or ultraviolet light. The next step is to expose the disc to a solution that contains a single-stranded DNA (or RNA) probe that has a sequence complementary to one of the strands in the clone of interest (Figure 13.20). The probe anneals to the DNA of interest and only to that DNA. Any excess solution is washed off the nitrocellulose disc. In the case of radioactive probes, the disc is placed in contact with X-ray film (Section 13.1). Only those spots on the disc in which some of the probe has annealed to the DNA already there are radioactive, and only those spots expose the X-ray film. Because the original petri dish has been saved, the desired clone can be picked off the plate and allowed to reproduce.

If the nucleotide sequence of the desired DNA segment is not known and no probe is available, a complication arises. If the gene of interest directs the synthesis of a given protein, one chooses a vector that will allow cloned genes to be transcribed and translated. If the presence of the desired protein can be detected by its function, that serves as the basis for detecting it. Alternatively, labeled antibodies can be used as a basis for protein detection.

RNA libraries are not constructed and cloned as such. Rather, the RNA of interest (usually mRNA) is used as the template for synthesis of complementary DNA (cDNA) in a reaction catalyzed by reverse transcriptase. The cDNA is incorporated into a vector (Figure 13.21). Ligating the cDNA to a vector requires the use of a synthetic linker. From this point on, the process of producing a **cDNA library** is virtually identical to that for constructing a genomic DNA library. For a given organism, a genomic DNA library is the same no matter what the tissue source, and the DNA represents both expressed and

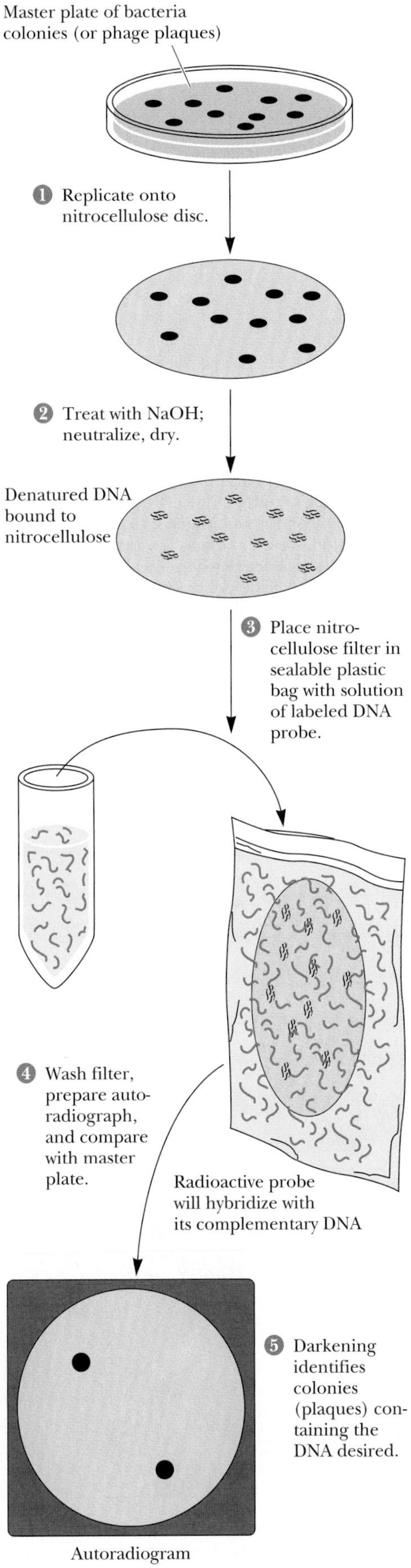

Master plate of bacteria colonies (or phage plaques)

1 Replicate onto nitrocellulose disc.

2 Treat with NaOH; neutralize, dry.

Denatured DNA bound to nitrocellulose

3 Place nitro-cellulose filter in sealable plastic bag with solution of labeled DNA probe.

4 Wash filter, prepare auto-radiograph, and compare with master plate.

Radioactive probe will hybridize with its complementary DNA

5 Darkening identifies colonies (plaques) con-taining the DNA desired.

Autoradiogram

Biochemistry ⊗Now™ ▶ **ACTIVE FIGURE 13.20** Screening a genomic library by colony hybridization (or plaque hybridization). Host bacteria transformed with a plasmid-based genomic library or infected with a bacteriophage-based genomic library are plated on a petri dish and incubated overnight to allow bacterial colonies (or phage plaques) to form. A replica of the bacterial colonies (or phage plaques) is then obtained by overlaying the plate with a nitrocellulose disc (1). Nitrocellulose strongly binds nucleic acids; single-stranded nucleic acids are bound more tightly than double-stranded nucleic acids. Once the nitrocellulose disc has taken up an impression of the bacterial colonies (or phage plaques), it is removed and the petri dish is set aside and saved. The disc is treated with 2 M NaOH, neutralized, and dried (2). NaOH both lyses any bacteria (or phage particles) and dissociates the DNA strands. When the disc is dried, the DNA strands become immobilized on the filter. The dried disc is placed in a sealable plastic bag, and a solution containing heat-denatured (single-stranded), labeled probe is added (3). The bag is incubated to allow annealing of the probe DNA to any target DNA sequences that might be present on the nitrocellulose. The filter is then washed, dried, and placed on a piece of X-ray film to obtain an autoradiogram (4). The position of any spots on the X-ray film reveals where the labeled probe has hybridized with target DNA (5). The location of these spots can be used to recover the genomic clone from the bacteria (or phage plaques) on the original petri dish. **Watch this Active Figure at http://now.brookscole.com/campbell5**

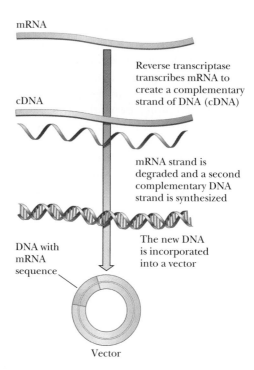

mRNA

Reverse transcriptase transcribes mRNA to create a complementary strand of DNA (cDNA)

cDNA

mRNA strand is degraded and a second complementary DNA strand is synthesized

DNA with mRNA sequence

The new DNA is incorporated into a vector

Vector

▶ **FIGURE 13.21** Reverse transcriptase catalyzes the synthesis of a strand of complementary DNA (cDNA) on a template of mRNA. The cDNA directs the synthesis of a second strand, which is then incorporated into a vector. (*Adapted with permission from* Dealing with Genes: The Language of Heredity, *by Paul Berg and Maxine Singer, © 1992 by University Science Books.*)

unexpressed DNA. In contrast, a cDNA library will be different depending on the tissue used and the expression profile of the cells.

13.6 | What Is the Polymerase Chain Reaction?

It is possible to increase the amount of a given DNA many times over without cloning that DNA. The method that makes this amplification possible is the **polymerase chain reaction (PCR).** Any chosen DNA can be amplified, and it need not be separated from the rest of the DNA in a sample before the procedure is applied. PCR copies both complementary strands of the desired DNA sequence. Scientists had long wished for a cell-free, automated way of synthesizing DNA, but any system that could be automated would need to function at high temperatures so that the DNA strands could be separated physically without the need for the many enzymes found in DNA replication, such as topoisomerases and helicases. Unfortunately, such temperatures (around 90°C) would denature and inactivate the DNA polymerases that would be making the DNA strands. What made the process possible was the discovery of bacteria that live around deep-sea hydrothermal vents, under extreme pressures, and at temperatures higher than 100°C. If bacteria can live under those conditions, then their enzymes must be able to function at those temperatures. The bacterium, *Thermus aquaticus,* from which a heat-resistant polymerase is extracted, is one of those that live in hot springs. The enzyme is called *Taq* polymerase. We saw, in Chapter 1, that the biotechnology industry eagerly searches for organisms that live under extreme conditions, and here we have an example of why they do so.

At the start of the process, the two DNA strands are separated by heating, after which short oligonucleotide primers are added in large excess and, via cooling, are allowed to anneal to the DNA strands. These primers are complementary to the ends of the DNA chosen for amplification and serve the same purpose as the RNA primers in normal replication. Once the primers

Wide World Photos

▲ Kary B. Mullis, inventor of the polymerase chain reaction and 1993 Nobel laureate in chemistry.

have annealed to the DNA, the temperature is raised again to optimize the activity of *Taq* polymerase, which begins synthesizing the new DNA from the 3′ end of the primer. The two complementary strands grow in the 5′ to 3′ direction (Figure 13.22), and the *Taq* polymerase is allowed to work until the desired length of DNA has been synthesized. This first round doubles the amount of the desired DNA. The process of unwinding the two strands,

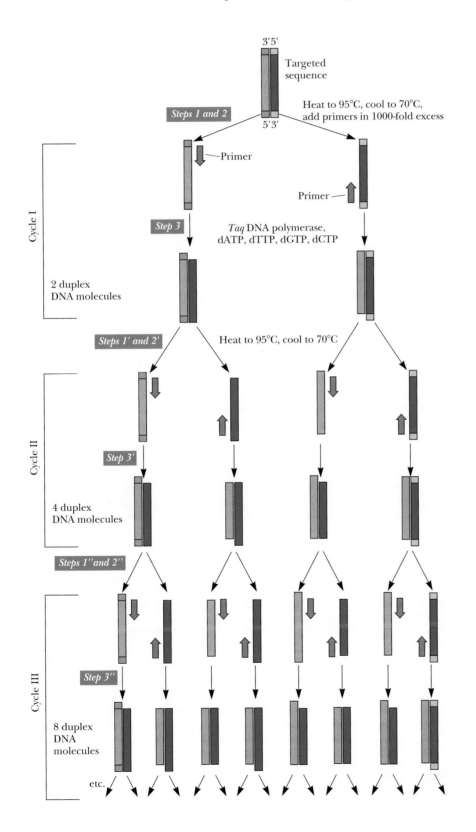

Biochemistry⊛Now™ ANIMATED FIGURE 13.22 Polymerase chain reaction (PCR). Oligonucleotides complementary to a given DNA sequence prime the synthesis of only that sequence. Heat-stable *Taq* DNA polymerase survives many cycles of heating. Theoretically, the amount of the specific primed sequence is doubled in each cycle. **See this figure animated at http://now.brookscole.com/campbell5**

annealing the primers, and copying the complementary strand is repeated, bringing about a second doubling of the selected double-stranded DNA. It is not necessary to add more primer because it is present in large excess. The whole process is automated. Control of the temperature to which the strands are heated to separate them is crucial, as is the temperature chosen for annealing the primers.

The amount of DNA continues to double in subsequent rounds of amplification. After about an hour, and 25 to 40 cycles of replication, one obtains millions to hundreds of millions of copies of the desired DNA segment, usually a few hundred to a few thousand base pairs in length (Figure 13.22). Other DNA sequences are not amplified and do not interfere with the reaction or subsequent use of the amplified DNA.

The most important part of the science behind PCR is the design of the primers. They have to be sufficiently long to be specific for the target sequence but not so long that they are too expensive. Usually, the primers are 18 to 30 bases long. They must also have optimal binding properties, such as an amount of G and C that is sufficient to allow them to anneal before the entire DNA renatures. In addition, the two primers should contain similar amounts of G and C so that they have the same melting temperature. The sequence of the primer must not lead to secondary structures within a primer or between the two different primers; otherwise, the primers will bind to themselves instead of to the DNA being amplified. For example, if a primer had the sequence, AAAAATTTTT, it would form a hairpin loop with itself and would not be available to bind the DNA. Section 13.7 explains how the primer sequence can be controlled to make changes in the DNA that is being amplified.

Amplification of the amounts of DNA in extremely small samples has made it possible to obtain accurate analyses that were not possible earlier. Forensic applications of the technique have resulted in positive identifications of crime victims and suspects. Even minuscule amounts of ancient DNA, such as those available from Egyptian mummies, can now be researched after amplification. The Biochemical Connections box describes some forensic uses of DNA technology.

Biochemical Connections

Forensic Uses of DNA Testing

It has been suggested that tissue samples be taken from all convicted felons to allow for identification of their presence at future crime sites by DNA fingerprinting (Section 13.8). This may seem very desirable, but some ethical and constitutional questions must first be answered. Nevertheless, DNA testing has been used to determine if people currently in jail might have been involved in unsolved crimes. One crime frequently examined in this way is rape, since body fluids are usually left behind. Many unsolved crimes are being solved, and, in some cases, some convicted prisoners are being found innocent. In at least one bizarre case, a felon was found innocent of the crime for which he was in jail, only to be rearrested in a few weeks because his DNA matched samples found on rape victims in three other cases.

The power of DNA testing cannot be overemphasized. In the trial of O. J. Simpson, the identity of the DNA was never questioned, although the defense successfully raised questions as to whether the evidence had been planted. In the vast majority of cases, DNA evidence results in the release of innocent suspects in a crime. When the DNA matches, however, the rate of plea-bargaining goes up, and the crimes are solved without long and expensive court trials.

Establishing paternity is a natural application for DNA testing. The DNA markers found in a child must arise from either the mother or the father. A man who is suspected of being the father of a particular child is immediately excluded from consideration if that child's DNA contains markers that are not found in the DNA of the mother or the suspect father. It is more difficult to prove conclusively that a person is the father of a particular child because it is usually too expensive to test enough markers to provide conclusive proof. However, in cases in which there are only two or three candidate fathers, it is usually possible to determine the correct one.

13.7 | **What Is Site-Directed Mutagenesis?**

Much of what we know about genetics has come from studying bacteria, viruses, and other simple organisms that divide quickly. Mutations from wild-type genes have enabled researchers to study the nature of DNA, RNA, and proteins. Until recently, however, scientists were limited by having to wait for naturally occurring mutations. The invention of PCR has changed all that. PCR has led naturally into the technique known as **site-directed mutagenesis.** With this technique, a PCR reaction is used to amplify DNA, but the primer is changed in such a way that the resulting DNA has base changes. The effect of these changes can then be studied. Remember that a single base change can, and often will, change the amino acid sequence. We shall look at how site-directed mutagenesis can be used for two different purposes—changing or adding restriction sites for cloning, and studying the effect of a mutation on an enzyme's activity.

Let's assume that you are doing a cloning-and-expression experiment. You have a 1-kb piece of DNA that encodes your protein of interest (geneX). At the 5′ end of the coding strand, there is an *Nde*I restriction site. At the 3′ end, there is an *Eco*RI site. Your current protocol calls for expressing the gene in a pET 5 vector, which has a multiple cloning site containing the same restriction sites and a promoter for T7 polymerase. This is an acceptable vector, but it lacks the blue/white screening capability that some other vectors have, and you would like to find a more convenient vector. Another company markets an expression plasmid called pBlue, and this plasmid does have the *lacZ* gene as part of its multiple cloning region. Unfortunately, when you look at the restriction sites found in the MCS, you see that there is an *Eco*RI site, but no *Nde*I site. *Nde*I reorganizes the sequence CATATG, and one of its advantages is that it has the ATG start signal, which, when turned into RNA, would code for a methionine. Thus, many PCR-amplified genes have the *Nde*I site at the 5′ end of the coding strand, so that the start signal is retained and so that, when the gene is cut out with *Nde*I, little extra DNA goes along for the ride. However, you notice that there is an *Nco*I site, which uses the sequence CCATGG. How will you use this restriction site to insert your geneX? The answer is, with site-directed mutagenesis. A possible sequence of the geneX DNA might look like the following, with the restriction sites shown in blue and the rest of the sequence shown in green:

5′-CATATGGTTGCATTG——————//—TGTCAAGCGGAATTC3′

3′-GTATACCAACGTAAC——————//—ACAGTTCGCCTTAAG5′

Remember that one of the strands—in this case, the top one—is the sense strand or coding strand (Chapter 11). To use site-directed mutagenesis to replace the *Nde*I site for an *Nco*I site, you must construct a primer with the sequence 5′CCATGGTTGCATTG3′. In the first round of PCR, this primer (shown in red) will anneal to the bottom strand from above, as shown here:

3′-GTATACCAACGTAAC——————//—ACAGTTCGCCTTAAG5′

5′-CCCATGGTTGCATTG

Note that at the 5′ end of the primer, two bases (the second and third cytosines) do not match correctly. However, this won't hinder the process because the primer has enough complementary bases that it will recognize and anneal to the opposite strand. During the synthesis step of PCR, this primer will be extended to copy the rest of the DNA (with new DNA shown in purple):

3′-GTATACCAACGTAAC——————//—ACAGTTCGCCTTAAG5′

5′-CCCATGGTTGCATTG——————//—TGTCAAGCGGAATTC-→3′

The bottom strand now has the desired sequence since the primer was designed that way. At the 5′ end there is now an *Nco*I site instead of the *Nde*I site, but the job isn't quite done. In the next cycle, the top strand (3′ to 5′) will be replicated in the same way. The bottom strand (5′ to 3′) will be replicated starting with the primer that will anneal to the right-hand side of the DNA. The primer for that will be the one that has the *Eco*RI sequence:

5′—CCCATGGTTGCATTG————————//—TGTCAAGCGGAATTC3′

ACAGTTCGCCTTAAG5′

During the synthesis phase of the reaction, this primer will be extended from the 5′ to 3′ direction as follows:

5′—CCCATGGTTGCATTG——————//—TGTCAAGCGGAATTC3′

3′←—GGGTACCAACGTAAC—————//—ACAGTTCGCCTTAAG5′

As you can see, the synthesis of the bottom strand used the 5′ end primer for a template, making DNA that has the correct base pairing for the *Nco*I site. From this point forward in the PCR cycles, all the DNA will have the new sequence and the geneX DNA now can be inserted into the new vector using *Nco*I.

This same technique can be applied to study the effect of amino acid substitutions on protein function directly. After a gene for a protein is isolated and cloned, PCR can be used to change a specific base and to look at the resulting effects on the protein. If you had an enzyme that was suspected to be a tyrosine protease (a proteolytic enzyme with a tyrosine at its active site, Chapter 7) and you had isolated and cloned the gene, you would be able to find the DNA sequence that coded for the amino acids at the active site. Such a sequence might be the following:

Ser Asp Gly Tyr Val His Ile

5′—//—TCA GAC GGT TAC GTA CAT ATC————//—3′

3′—//—AGT CTG CCA ATG CAT GTA TAG————//—5′

To help prove your hypothesis that the enzyme is a tyrosine protease and that the tyrosine shown here is the active-site tyrosine, you can use site-directed mutagenesis to change the amino acids at the active site. A primer can be made to this region of the DNA and can be annealed to the preceding bottom strand:

3′—//—AGT CTG CCA ATG CAT GTA TAG————//—5′

5′—//—TCA GAC GGT TTC GTA CAT ATC————//—3′

The thymine (T) shown in purple is the one that was changed (the mutated base). This bottom DNA piece with the primer would be extended 5′ to 3′ to complete the gene. In the next PCR cycle, when this fragment is used as the template for DNA synthesis, the complementary fragment will then match the primer sequence perfectly,

3′—//—AGT CTG CCA AAG CAT GTA TAG————//—5′

5′—//—TCA GAC GGT TTC GTA CAT ATC————//—3′

and, from that point forward, the DNA sequence will have been permanently changed. The codon TTC specifies a phenylalanine instead of a tyrosine. This gene can then be cloned into an expression vector and expressed, and the function of the protein can be compared with the wild type. If the hypothesis (that the enzyme is a tyrosine protease) is true, and this is the active-site tyrosine, the enzyme will have no activity with a phenylalanine replacing the native tyrosine.

13.8 | What Is DNA Fingerprinting?

DNA samples can be studied and compared using a technique called DNA fingerprinting. The DNA is digested with restriction enzymes and then run on an agarose gel (Figure 13.23). The DNA fragments can be seen directly on

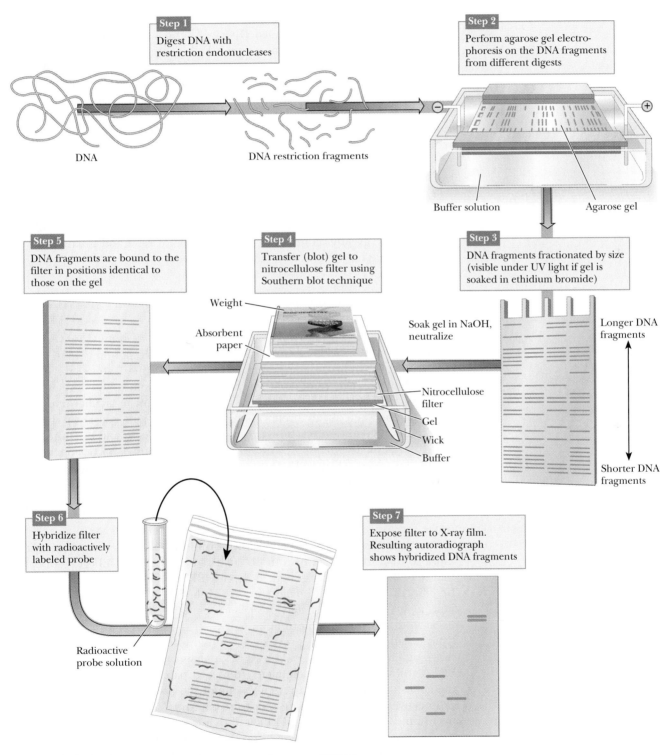

▲ **FIGURE 13.23** The Southern blot. Electrophoretically separated DNA fragments are transferred to a nitrocellulose sheet. A radioactively labeled probe for a DNA sequence of interest is bound to the nitrocellulose, and bands are visualized with an autoradiogram.

the gel if it is soaked in ethidium bromide and viewed under ultraviolet light (Section 13.1). As shown in Figure 13.23, Step 3, this will give bands of varying sizes, depending on the nature of the DNA and the restriction enzymes used. If greater sensitivity is needed, or if the number of fragments would be too great to distinguish the bands, this technique can be modified to visualize only selected DNA sequences. The first step would be to transfer the DNA to a nitrocellulose membrane in a procedure called a **Southern blot,** after its inventor, E. M. Southern. The agarose gel is soaked in NaOH to denature the DNA because only single-stranded DNA will bind to the nitrocellulose. The membrane is placed on the agarose gel, which is on top of a filter-paper wick placed in buffer. Dry absorbent paper is placed on top of the nitrocellulose. Wicking action carries the buffer from the buffer chamber up through the gel and nitrocellulose into the dry paper. The DNA bands move out of the gel and stick to the nitrocellulose. This is similar to the procedure for proteins called western blotting that we saw in Chapter 5. (In fact, the name western blotting is a take-off on Southern blotting, as is northern blotting for RNA, which is described in Section 13.10.) The next step is to visualize the bands on the nitrocellulose. A specific DNA probe that is labeled with ^{32}P is incubated with the nitrocellulose membrane. The DNA probe will bind to DNA fragments that are complementary. The membrane is then placed on photographic paper to produce an autoradiogram (Section 13.1). The use of the specific probe greatly reduces the number of bands seen and isolates desired DNA sequences.

Restriction-Fragment Length Polymorphisms: A Powerful Method for Forensic Analysis

In organisms (such as humans) with two sets of chromosomes, a given gene on one chromosome may differ slightly from the corresponding gene on the paired chromosome. In the language of genetics, these genes are **alleles.** When they are the same on the paired chromosomes, the organism is **homozygous** for that gene; when they differ, the organism is **heterozygous.** A difference between alleles, even a change in one base pair, can mean that one allele has a recognition site for a restriction endonuclease and the other does not. Restriction fragments of different sizes are obtained on treatment with the endonuclease (Figure 13.24); they are called **restriction-fragment length polymorphisms,** or **RFLPs** (pronounced "riflips") for short. These polymorphisms (a word meaning "many shapes") are analyzed via gel electrophoresis to separate the fragments by size, followed by blotting and the annealing of a probe for a specific sequence.

Research has shown that such polymorphisms are quite common, much more so than the mutations in traits such as eye color and inherited diseases that were used for earlier genetic mapping. RFLPs can be used as markers for heredity in the same way as mutations in visible traits because they are inherited in the manner predicted by classical genetics (Figure 13.25). However, because they are much more abundant than mutations that lead to phenotypic variations, they have provided many more markers for detailed genetic mapping. Figure 13.25 shows how RFLP analysis can be used in paternity testing. A child gets one of each allele from each parent, so every fragment that the child has must also be present in one of the parents. Thus, if there was a prospective father undergoing a test for paternity, it would be easy to eliminate him. If the child had a RFLP band that the mother did not have and the prospective father did not have, then the prospective father is excluded. The same kinds of analyses are done with evidence found at crime scenes. Suspects in criminal cases may be exonerated if their DNA samples do not match

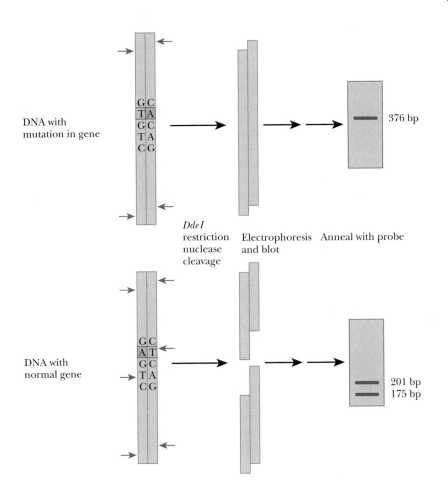

DNA with
mutation in gene

G C
T A
G C
T A
C G

DdeI
restriction
nuclease
cleavage

Electrophoresis
and blot

Anneal with probe

376 bp

DNA with
normal gene

G C
A T
G C
T A
C G

201 bp
175 bp

◀ **FIGURE 13.24** The basis for restriction-fragment length polymorphism. A change of one base pair eliminates a restriction nuclease cleavage site. A portion of DNA that codes for a protein has a cleavage site for the restriction nuclease *DdeI*. The corresponding DNA with a mutation does not have this cleavage site. The difference can be detected by electrophoresis, followed by blotting and the annealing of a probe specific for this fragment. (The abbreviation bp stands for base pairs.) (*Adapted with permission from* Dealing with Genes: The Language of Heredity, *by Paul Berg and Maxine Singer, © 1992 by University Science Books.*)

those found at the scene. RFLP analysis was used extensively in the process of locating the altered gene that causes cystic fibrosis, a prevalent genetic disease. Once the gene was located on chromosome 7, a series of RFLP markers was used to help map its exact position (Figure 13.26). Then the gene was isolated from restriction endonuclease digests and cloned, and its protein product was then characterized. The protein in question is involved in the transport of chloride ion (Cl^-) through membranes. If this protein is defective, chloride ions remain in the cells and take up water by osmosis from the surrounding mucus. The mucus thickens as a result. In the lungs, the thickened mucus favors infections, particularly pneumonia. The results of this disease can be tragic, leading to a short life span in those who are affected by it. This information deepens our insight into the nature of cystic fibrosis and provides approaches to new treatment.

13.9 | How Can We Study DNA-Protein Interactions?

In Chapters 10 and 11, we saw that proteins control many of the processes involving DNA replication and transcription. New transcription factors are discovered almost daily. One of the ways in which the interaction of DNA and proteins has been studied is with a technique called a **DNA footprint.** The DNA to be studied is labeled at one end with ^{32}P. It is then digested with an

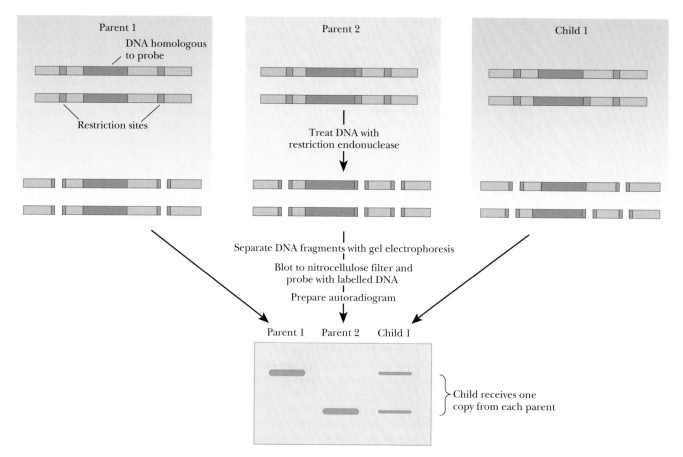

▲ **FIGURE 13.25** Restriction-fragment length polymorphisms (RFLPs) can be detected by probes for homologous DNA. Parent 1 has DNA with two restriction sites near the DNA homologous to the probe. Because Parent 1 has the same pattern of restriction sites on both DNA strands, one large restriction fragment will be detected. Parent 2 has three restriction sites on each DNA strand. A single smaller restriction fragment will be detected by the homologous probe for Parent 2. Their child has inherited one copy of each DNA strand from each parent. The child's DNA will produce one fragment of each size. (*Adapted with permission from* Dealing with Genes: The Language of Heredity, *by Paul Berg and Maxine Singer,* © *1992 by University Science Books.*)

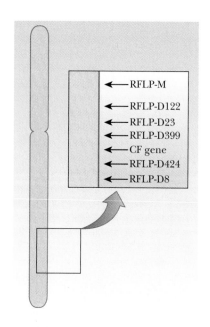

▲ **FIGURE 13.26** Localization of the gene associated with cystic fibrosis (CF) on human chromosome 7. The RFLP markers are given arbitrary names. The location of the CF gene is given relative to the RFLP markers. (*Adapted with permission from* Dealing with Genes: The Language of Heredity, *by Paul Berg and Maxine Singer,* © *1992 by University Science Books.*)

enzyme such as DNase I, which cuts DNA at random locations (Figure 13.27). This will generate a series of fragments based on where the enzyme cuts and how often. If the conditions are controlled correctly, multiple fragments that represent all the possible fragment lengths will be generated. Then the original DNA being studied is allowed to bind to a protein, and the DNA/protein complex is digested with the enzyme. This might be RNA polymerase or one of the many transcription factors (Chapter 11). If the protein binds to a particular DNA sequence, then that region will no longer be susceptible to cutting with DNase I. The result will be a gel with bands that shows a gap where no fragments appear. This gap will indicate the region of DNA that was protected by the protein.

Another method for studying DNA-protein interactions takes advantage of the fact that DNA that is bound to protein will move more slowly through a gel than naked DNA. Figure 13.28 shows how this procedure works. A sample of DNA containing the suspected target sequence for protein binding is electrophoresed in lane 1 of a gel. In lane 2, the DNA is mixed with a protein suspected of binding to the target sequence. Lane 3 contains the DNA, the original protein, and perhaps a second protein, if it is suspected that two proteins bind in that region. If the DNA binds to the target sequence, a band of lower

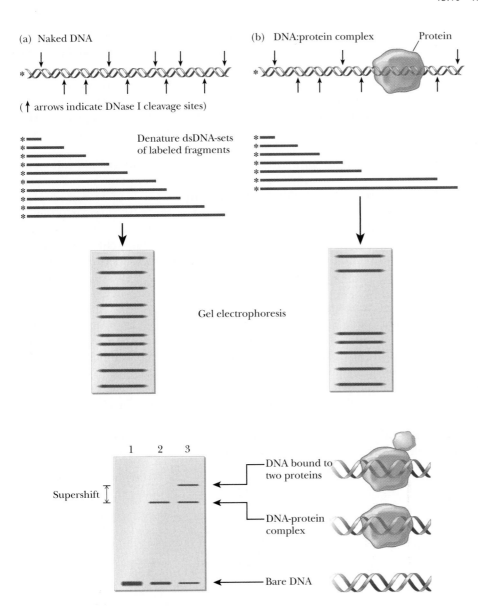

(a) Naked DNA

(↑ arrows indicate DNase I cleavage sites)

(b) DNA:protein complex   Protein

Denature dsDNA-sets of labeled fragments

Gel electrophoresis

◀ **FIGURE 13.27** DNA footprinting. DNA is labeled on one end with ^{32}P. DNA is then treated with DNase I, an enzyme that cuts randomly on the DNA. A mixture of DNA pieces is generated that can be separated by electrophoresis. If the same DNA is bound to protein, such as RNA polymerase or a transcription factor, the bound protein will protect the DNA from the nuclease. A gap will appear in the gel electrophoresis, indicating the location on the DNA at which the protein was bound. (*From D. Rhodes and L. Fairall, Analysis of sequence-specific DNA-binding proteins.* © *Oxford University Press, 1997. Reprinted from* Protein Function: A Practical Approach, *2/e by T. E. Creighton by permission of Oxford University Press.*)

Supershift

DNA bound to two proteins

DNA-protein complex

Bare DNA

◀ **FIGURE 13.28** An electrophoretic mobility-shift assay, or EMSA. DNA is separated by electrophoresis. Lane 1 contains naked DNA. Lane 2 contains the same DNA, but some of it is complexed to protein, which makes the sample larger, so it does not travel as far in the gel. This is called a gel shift. Lane 3 contains naked DNA, DNA bound to the protein, and DNA bound to two proteins, shifting the band even higher. (*Adapted from* Molecular Biology, *Figure 5.31, by R. F. Weaver, 1999.*)

mobility will appear on the gel. This shift in the mobility of the band is called a *gel shift* or *band shift*. The technique is called an **electrophoretic mobility-shift assay (EMSA).**

13.10	**What Are Some Methods for Studying Transcription?**

While much has been made recently about sequencing the human genome, the fact is that just knowing the DNA sequence is rarely enough to understand the cell's metabolism. Genes are not active all the time, and researchers would like to understand which genes are active in which tissues and under what conditions. The amount and type of mRNA in a cell is a better indication of its metabolic state. Fortunately, most eukaryotic mRNA has a poly-A tail at the 3′ end that makes it fairly easy to isolate. A cell lysate can be passed over an affinity-chromatography column (Chapter 5) containing poly-T. The mRNA will stick to the column and then can be separated from the rest of the cell components. The mRNA can be run on an agarose gel and then blotted

Days of Growth

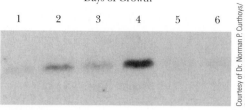

1   2   3   4   5   6

▲ **FIGURE 13.29** A northern blot. RNA was isolated from rat tissues at various times. Equal amounts
of RNA from each time point were electrophoresed
and transferred to nitrocellulose. The RNA on the
blot was then hybridized to a labeled probe for the
rat glyceraldehyde-3-phosphate dehydrogenase
(G3PDH) gene, and the blot was then exposed to
X-ray film. The bands indicate the presence of the
G3PDH gene, and the intensity indicates the relative
amount of the RNA at each time point.

to nitrocellulose in a procedure called a **northern blot.** A specific radioactive
probe can then be used to spot a particular mRNA on the blot. Figure 13.29
shows an experiment where the mRNA from cultured pig cells was isolated,
separated on a gel, and subjected to northern blot. The probe used was
cDNA for glutaminase. The intensity of each band indicates the relative level
of the transcription of the enzyme after the specified days of cell growth. The
following Biochemical Connections box discusses how some recent technology can greatly expand on this theme.

To study transcription promoters and how subtle changes might influence
transcription rates, it is convenient to link the transcription to a **reporter
gene.** The reporter gene produces something that is very easy to assay, so that
many reactions can be compared quickly. One of the most popular reporter
genes is the bacterial gene *cat,* which encodes the enzyme *chloramphenicol
acetyltransferase (CAT).* This is an enzyme that acetylates chloramphenicol
(CAM), an antibiotic that inhibits bacterial translation, rendering it harmless
to the bacteria. Figure 13.30 shows how the CAT assay works. The gene you
are studying, *gene*X, is under the control of its promoter. The coding portion
of the gene is removed with restriction enzymes (1), replaced with the *cat*
gene (2), and transformed into eukaryotic cells (3). Cells are extracted (4)
and incubated with radioactive chloramphenicol and acetyl-CoA, the source
of the acetyl groups (5). When CAT is active, the chloramphenicol will be
acetylated and will run higher on a thin-layer chromatogram (6). The intensity of the acetylated form indicates the transcription activity of those cells.
Duplicate trials with mutated promoters would give valuable information
about the nature of transcription promoters.

Recently, more direct methods have been developed to quantify transcription both in vitro and in vivo. The promoter of interest is cloned upstream of

Text continues on page 365.

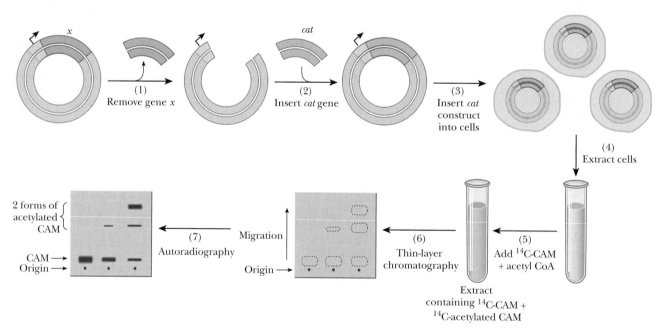

▲ **FIGURE 13.30** Using a reporter gene. A plasmid carrying a gene *x* (blue) under the control
of its own promoter (yellow) is removed (1) with restriction enzymes and replaced with the chloramphenicol acetyltransferase (CAT) gene, *cat* (2). The recombinant plasmid is transformed into
cells (3). The cells are extracted (4), and radioactively labeled chloramphenicol (CAM) is added
along with acetyl-CoA (5). If CAT is active, the CAM will become acetylated. The extracts are run
on thin-layer chromatography (6) and visualized with autoradiography (7). The level of acetylated
CAM reveals the activity of the transcription of the CAT gene and the effectiveness of the gene *x*
promoter.

DNA Chips—Robotic Technology Meets Biochemistry

Thousands of genes and their products (i.e., RNA and proteins) in a given living organism function in a complicated and harmonious way. Unfortunately, traditional methods in molecular biology have always focused on analyzing one gene per experiment. In the past several years, a new technology, called **DNA microarray** (**DNA chip** or **gene chip**), has attracted tremendous interest among molecular biologists. Microarrays allow for the analysis of an entire genome in one experiment and are used to study gene expression, the transcription rates of the genome in vivo. The genes that are being transcribed at any particular time are known as the **transcriptome.** The principle behind the microarray is the placement of specific nucleotide sequences in an ordered array, which then will base-pair with complementary sequences of DNA or RNA that have been labeled with fluorescent markers of different colors. The locations where binding occurred and the colors observed are then used to quantify the amount of DNA or RNA bound. This would be similar to getting data from thousands of northern-blot assays simultaneously.

Microarray chips are manufactured by high-speed robotics, which can put thousands of samples on a glass slide with an area of about 1 cm². The diameter of an individual sample might be 200 μm or less. There are several different methods for implanting the DNA to be studied on the chip, and many companies make microarray chips.

One comprehensive study used yeast *(Saccharomyces cerevisiae)*, the genome of which is completely known, and microarray chips have been produced commercially with the entire genome on them. (See the article by DeRisi in the Annotated Bibliography

at the end of this chapter.) Yeast cells were grown anaerobically, their mRNA was collected, and fluorescently labeled cDNA was produced using reverse transcriptase. The cells were then switched to an oxygen environment, and, at various times, the mRNA was collected and converted to cDNA with a fluorescent label of a different color. By annealing the cDNA to the microarray chips, the changes in the binding of the cDNA can be seen, which correlates with changes in the genes being transcribed (see figure). The green dots (first fluorescent marker) are the background genes that were active under anaerobic conditions. The red dots (second fluorescent marker) are the genes that were turned on by the switch to an oxygen environment. Specific genes were identified in the figure. Data such as these allowed the researchers to better understand the changes in metabolism associated with the shift from anaerobic to aerobic growth. This same procedure can easily be adapted to check the metabolism of yeast with mutant genes or to study transcription when a suspected transcription factor is either missing or overproduced.

The same robotic technology has also allowed the production of protein chips, which follow the same idea as DNA chips but have thousands of bound proteins, which can then be probed for protein:protein interactions.

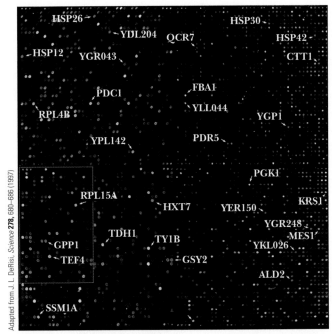

▲ A yeast genome microarray. The actual size of the microarray is 18 mm by 18 mm. The microarray contains the entire yeast genome. A fluorescently labeled cDNA probe was prepared from mRNA isolated from cells at the beginning of the experiment. The probe used a green fluorescent marker. After switching the yeast to an aerobic environment, another cDNA probe was made to mRNA isolated from cells in the aerobic environment, this time using a red fluorescent marker. The dots showing up as red indicate the genes for which RNA was actively produced under aerobic conditions. Those showing up as green were those active under anaerobic conditions.

▲ A microarray chip. Thousands of DNA samples are placed (by robotics) on a slide with an area of about 1 cm². These are then bound to complementary DNA or RNA and visualized using fluorescent markers. The location and intensity of the fluorescent signal is analyzed by computer to establish which genes are currently being transcribed and to what extent.

RNA Interference—The Newest Way to Study Genes

RNA interference (RNAi) was first discovered in a nematode worm (*Caenorhabditis elegans*). Double-stranded RNA (dsRNA) was found to cause gene silencing in a sequence-specific way. Researchers had long thought that RNA would be the perfect way to control gene expression since the right sequence of RNA should bind to DNA and interfere with its transcription. While checking on the efficiency of antisense RNA (Section 11.1) as a suppressor of gene expression, it was discovered that dsRNA was more than ten times as effective at shutting off transcription of a gene. In addition, suppression by RNAi was shown to be transmissible to other cells of the organism and to the progeny in *C. elegans*. RNAi has since been found as a natural phenomenon in plants as a way of targeting viral RNAs for destruction. It has also been seen in *Drosophila* and, most recently, in mammals. It is now believed that RNAi is a natural regulatory mechanism for controlling gene expression. It may also be a protection mechanism against oncogenes that produce too much of a harmful product.

The process starts with an enzyme that is a member of the RNase III class, called **Dicer.** This enzyme binds to dsRNA and cleaves it into small interfering RNA (siRNA) of between 22 and 25 nucleotides, as shown in the figure. The siRNAs then bind to a protein complex called **RNA-induced silencing complex (RISC).** The siRNA–RISC complex then binds to target mRNA that has the same sequence as part of the siRNA and degrades it. In this way, the sequence of the dsRNA controls the degradation of an mRNA target. We are still learning how this may have arisen naturally. Perhaps an endogenous RNA-dependent RNA polymerase senses that there is too much of a particular mRNA being made. It could then create the opposite strand, thereby forming the dsRNA. This would then trigger the eventual destruction of the mRNA. The same logic could apply to a defense mechanism against viruses.

Regardless of RNAi's natural purpose, it has become the fastest growing new field in molecular biology. Many companies have sprung up overnight that produce RNAi kits and dsRNAs to use to initiate reactions. The technique is quickly becoming the newest way to knock out specific genes to see what then happens to the organism. All it takes is a knowledge of the gene sequence, and the correct dsRNA could then be used to produce siRNAs to shut off the gene. Researchers using RNAi have been better able to map the thousands of genes in certain organisms, such as *C. elegans*. As an example of the power of the technique, researchers just mapped the gene for an enzyme involved in the metabolism of vitamin K. This enzyme, **vitamin K epoxide reductase (VKOR),** is the target of an widely prescribed anticoagulant. The enzyme has been known for 40 years now, but attempts to purify it and to locate it on the human chromosome had met with no success. Using information from the Human Genome Project, researchers were able to map human chromosomes and to come up with thirteen genes that had characteristics that could mean that they were the VKOR gene. Using RNAi, the correct one was isolated in a few weeks, a major breakthrough for researchers that had been studying this enzyme for decades.

Medical researchers are very hopeful about the possibilities for this technique. Given the specificity of RNAi, mutant alleles that are the basis of diseases could be knocked out with the right sequence of dsRNA given as a trigger for the Dicer enzyme to process. For example, if one allele is mutated but the other is normal, RNAi could knock out the mutated one while not affecting the normal one. See the article by Hannon in the Annotated Bibliography for this chapter for a complete review of RNAi.

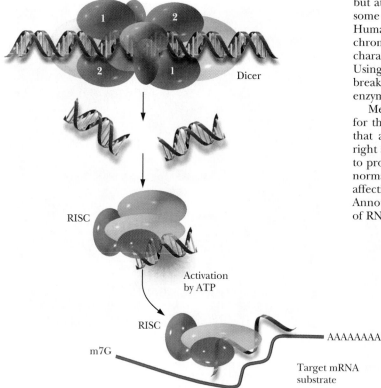

▲ RNA interference. An enzyme complex called Dicer cleaves double-stranded RNA into small interfering RNAs (siRNA). These then bind to a protein complex called RNA-induced silencing complex (RISC). RISC unwinds the siRNA and one strand binds to the complementary mRNA, which is subsequently degraded.

a synthetic gene of fixed length that is devoid of guanine nucleotides (called G-less). Transcription occurs in the presence of all four deoxyribonucleotides, including trace amounts of ^{32}P-UTP. Following digestion with a nuclease specific for guanine nucleotides, which effectively trims the RNA to a fixed, expected length, a radiolabeled transcript is electrophoresed and visualized with an autoradiogram. The directness of the measurement renders this technique very quantitative, and the ^{32}P-UTP allows for great sensitivity.

RNA transcripts can also be measured in a living cell. The promoter of interest is cloned upstream of a firefly (*Photinus pyralis*) luciferase gene. In cultured cells, a lipid-based solution allows the incorporation of the reporter vector. Naturally occurring transcription factors, or similarly incorporated vectors that direct the expression of one or more transcription factors, act on the promoter in vivo. These transcription factors direct the expression of the luciferase enzyme RNA, and the cell translates the RNA into protein. The cells are then harvested and lysed, and the substrates D-luciferin and ATP are added. Luciferase acts on the substrates to produce light, which is measured with an instrument called a luminometer. The amount of light detected is proportional to the amount of luciferase enzyme made, which is, in turn, proportional to the amount of RNA made. This technique is 30–1000 times more sensitive than the CAT reporter system, and shows a linear response over eight orders of magnitude.

13.11 | How Do We Determine the Base Sequences of Nucleic Acids?

We have already seen that the primary structure of a protein determines its secondary and tertiary structures. The same is true of nucleic acids; the nature and order of monomer units determine the properties of the whole molecule. Base pairing in both RNA and DNA depends on a series of complementary bases, whether these bases are on different polynucleotide strands, as in DNA, or on the same strand, as is frequently the case in RNA. Sequencing of nucleic acids is now fairly routine, and this relative ease would have amazed the scientists of the 1950s and 1960s.

The method devised by Sanger and Coulson for determining the base sequences of nucleic acids depends on selective interruption of oligonucleotide synthesis. A single-stranded DNA fragment whose sequence is to be determined is used as a template for the synthesis of a complementary strand. The new strand grows from the 5′ end to the 3′ end. This unique direction of growth is true for all nucleic acid synthesis (Chapter 10). The synthesis is interrupted at every possible site in a population of molecules. The interruption of

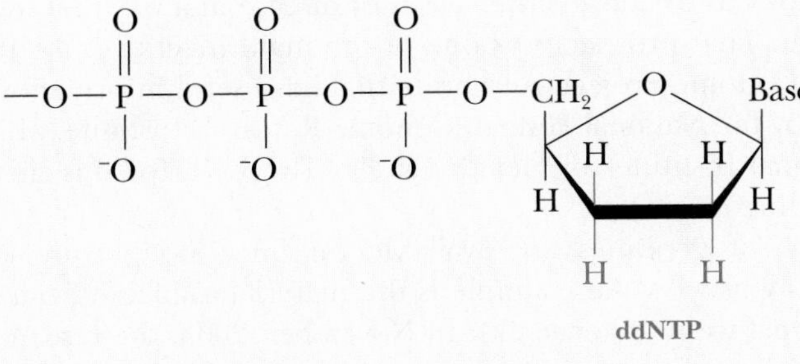

ddNTP

synthesis depends on the presence of 2′,3′-dideoxyribonucleoside triphosphates (ddNTPs).

The 3′-hydroxyl group of deoxyribonucleoside triphosphates (the usual monomer unit for DNA synthesis) has been replaced by a hydrogen. These ddNTPs can be incorporated in a growing DNA chain, but they lack a 3′-hydroxyl group to form a bond to another nucleoside triphosphate. The incorporation of a ddNTP into the growing chain causes termination at that point. The presence of small amounts of ddNTPs in a replicating mixture causes random termination of chain growth.

The DNA to be sequenced is mixed with a short oligonucleotide that serves as a primer for synthesis of the complementary strand. The primer is hydrogen-bonded toward the 3′ end of the DNA to be sequenced. The DNA with primer is divided into four separate reaction mixtures. Each reaction mixture contains all four deoxyribonucleoside triphosphates (dNTPs), one of which is labeled to allow the newly synthesized fragments to be visualized by autoradiography or by fluorescence, as described in Section 13.1. In addition, each of the reaction mixtures contains one of the four ddNTPs. Synthesis of the chain is allowed to proceed in each of the four reaction mixtures. In each mixture, chain termination occurs at all possible sites for that nucleotide.

When gel electrophoresis is performed on each reaction mixture, a band corresponding to each position of chain termination appears. The sequence of the newly formed strand, which is complementary to that of the template DNA, can be "read" directly from the sequencing gel (Figure 13.31). A variation on this method is to use a single reaction mixture with a different fluorescent label on each of the four ddNTPs. Each fluorescent label can be detected by its characteristic spectrum, requiring only a single gel electrophoresis experiment. The use of fluorescent labels makes it possible to automate DNA sequencing, with the whole process under computer control. Commercial kits are available for these sequencing methods (Figure 13.32).

When RNA is to be sequenced, the method of choice is not to analyze the RNA itself but to use the methods of DNA sequencing on a DNA complementary (cDNA) to the RNA in question. The cDNA, in turn, is generated by using the enzyme reverse transcriptase, which catalyzes the synthesis of DNA from an RNA template.

13.12 How Can We Use Bioinformatics to Study Genomics and Proteomics?

With more and more full DNA sequences becoming available, it is tempting to compare those sequences to see whether patterns emerge from genes that encode proteins with similar functions. The amount of data makes use of a computer essential for the process. Databases on genome and protein sequences are so extensive as to require information technology at its best to solve problems. Knowing the full DNA sequence of the human genome, for example, allows us to address the causes of disease in a way that was not possible until now. That prospect was one of the main incentives for undertaking the Human Genome Project. A web site that has useful information is the one maintained by the National Human Genome Research Institute, which is a part of the National Institutes of Health (NIH). The URL for this site is (http://www.nhgri.nih.gov/).

A number of genomes are available on line, along with software for sequence comparisons. An example is the material available from the Sanger Institute (http://www.sanger.ac.uk). In November, 2003, the researchers at this institute announced that they had sequenced two billion bases from the DNA of several organisms (human, mouse, zebrafish, yeasts, and the roundworm

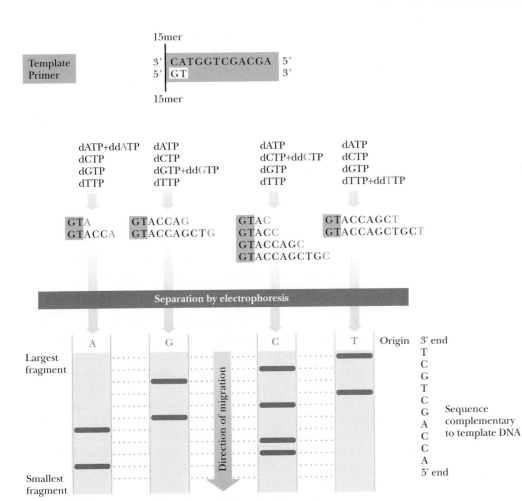

▲ **FIGURE 13.31** The Sanger–Coulson method for sequencing DNA. A primer at least 15 residues long is hydrogen-bonded to the 3′ end of the DNA to be sequenced. Four reaction mixtures are prepared; each contains the four dNTPs and one of the four possible ddNTPs. In each reaction mixture, synthesis takes place; but, in a given population of molecules, synthesis is interrupted at every possible site. A mixture of oligonucleotides of varying lengths is produced. The components of the mixture are separated by gel electrophoresis.

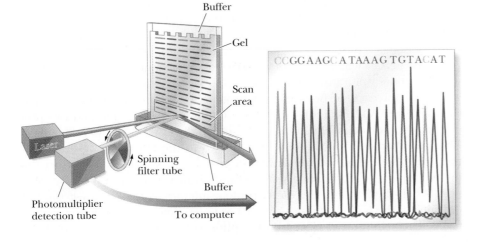

◀ **FIGURE 13.32** A schematic diagram of the methodology used in fluorescent labeling and automated sequencing of DNA. Four reactions are set up, one for each base, and the primer in each is end-labeled with one of four different fluorescent dyes; the dyes serve to color-code the base-specific sequencing protocol (a unique dye is used in each dideoxynucleotide reaction). The four reaction mixtures are then combined and run in one lane. Thus, each lane in the gel represents a different sequencing experiment. As the differently sized fragments pass down the gel, a laser beam excites the dye in the scan area. The emitted energy passes through a rotating color filter and is detected by a fluorometer. The color of the emitted light identifies the final base in the fragment. (*Applied Biosystems, Inc., Foster City, CA.*)

Caenorhabiditis elegans, among others). If this amount of DNA were the size of a spiral staircase, it would reach from the Earth to the Moon.

A question implicit in the determination of the genome of any organism is that of assignment of sequences to the chromosome in which they belong. This is a challenging task, and only suitable computer algorithms make it possible. Once this has been achieved, one can compare genomes to see what changes have occurred in the DNA of complex organisms compared with those of simpler ones.

Beyond this application, challenging though it may be, lies the application to medicine, which is leading to a number of surprises. Two closely related genes (*BRCA1* and *BRCA2*) involved in the development of breast cancer interact with other genes and proteins, and this is a topic of feverish research. The connection between these genes and a number of seemingly unrelated cancers is only starting to be unraveled. (See the article by Couzin referred to in the second Editors of *Science* entry in the Annotated Bibliography at the end of this chapter.) Clearly, there is need to determine not only the genetic blueprint but the manner in which an organism puts it into action.

The **proteome** is the protein version of the genome. In all organisms for which sequence information is available, **proteomics** (the study of interactions among all the proteins in a cell) is assuming an important place in the life sciences. If the genome is the script, the proteome puts the play on stage. The genetically determined amino acid sequence of proteins determines their structure and how they will interact with each other. Those interactions determine how they will behave in a living organism. The potential medical applications of the human proteome are apparent, but these have not yet been realized. Proteomic information does exist, for eukaryotes such as yeast and the fruit fly *Drosophila melanogaster,* and the methods that have been developed for those experiments will be of use in unraveling the human proteome. A two-hybrid method of determining protein:protein interactions was developed for yeast and further developed for *Drosophila.* It will be useful to see how this method works, because it involves application of a number of techniques that we have already seen. The method involves dividing a gene that encodes a transcription activator into two parts and cloning those parts into separate plasmids for expression. Each part of the divided gene has the gene for another protein fused to it. Transcription activation (Sections 11.3 and 11.4) can be used to determine whether two proteins interact. In some cases, the added proteins, such as those marked X and Y in Figure 13.33, interact with each other. Because of this interaction, the two parts of the protein whose gene was divided, shown as DB and TA in Figure 13.33, also interact with each other. As a result of the interaction, the activity of the protein with the divided gene is reconstituted. The gene to be activated is the yeast GAL-4 gene, whose transcriptional activator has two domains. One of the domains binds to the DNA (DB), and the other is the activator (TA) that cannot bind to the DNA itself. The two proteins have to interact with each other to ensure transcriptional activation. The gene for the activator is divided between two plasmids, one of which encodes the DNA-binding domain and the other of which encodes the transcription activator.

A gene for each of two different proteins to be tested for interaction is spliced to the two different parts of the GAL-4 gene. Each plasmid can produce a fusion protein. The one that is fused to the GAL-4 activation domain is called the "target," and the one that is fused to the DNA-binding domain is the "bait." Both plasmids are allowed to grow in the same cell. This is accomplished in practice by mating yeast cells, each of which has a different plasmid. When the bait and the target protein interact with each other, the two domains of the GAL-4 activator interact as well, allowing transcription to take place. The activation of GAL-4 transcription is detected by use of a reporter

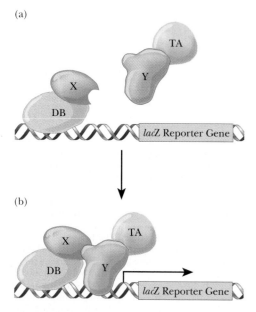

(a)

(b)

▲ **FIGURE 13.33** The two-hybrid system for detecting gene interaction.

gene, the *lac-Z* gene, for which we can use the blue/white screening that we met in Section 13.3. A similar two-hybrid approach was used for *Drosophila*, with high-throughput screening and computer analysis of the results. While it is not quite true that "humans are big flies without wings," this is a milestone in systems biology. The human proteome is on its way.

Summary

13.1 How Do We Purify and Detect Nucleic Acids?
Two of the primary necessities for successful experiments with nucleic acids are to separate the components of a mixture and to detect the presence of nucleic acids. Gel electrophoresis is widely used as a separation method; labeling with radioactive or luminescent materials allows for detection of small amounts of nucleic acids.

13.2 What Makes Restriction Endonucleases an Important Tool for DNA Research?
Restriction endonucleases play a large role in the manipulation of DNA. These enzymes produce short, single-stranded stretches, called sticky ends, at the ends of cleaved DNA. The sticky ends provide a way to link DNAs from different sources, even to the point of inserting eukaryotic DNA into bacterial genomes.

13.3 What Is Cloning?
DNA samples from different sources can be selectively cut by using restriction endonucleases and then spliced together with DNA ligase to produce recombinant DNA. DNA from another source can be introduced into the genome of a virus or a bacterium. In bacteria, the foreign DNA is usually introduced into a plasmid, a smaller circular DNA separate from the main bacterial chromosome. The growth of the virus or bacterium also produces large amounts of the other DNA by the process of cloning.

13.4 What Is Genetic Engineering, and Why Do We Do It?
Once the DNA is successfully cloned, it can be expressed using an expression vector and cell line. This allows for the production of eukaryotic proteins quickly and cheaply in bacterial hosts. Genetically altered organisms, such as mice and corn, have been engineered for both pure and applied scientific purposes, and many more changes are to come.

13.5 What Are DNA Libraries?
This technology has advanced to the point and which it is possible to cut the entire genome of a eukaryotic organism into fragments and to clone the fragments. The assemblage of cloned fragments is called a DNA library.

13.6 What Is the Polymerasae Chain Reaction?
An alternative method of producing large amounts of a given DNA, called the polymerase chain reaction, depends only on enzymatic reactions and does not require viral or bacterial hosts. The procedure is automated and relies on a heat-stable form of DNA polymerase.

13.7 What Is Site-Directed Mutagenesis?
This technique has also led to the possibility of changing specific bases in genes being studied by the process of site-directed mutagenesis. It is often used to change the restriction sites on a plasmid or DNA insert for the purposes of cloning. Another application involves selectively changing the base sequence of a gene so that the protein product has an altered amino acid at a known location.

13.8 What Is DNA Fingerprinting?
The ability to analyze DNA fragments is important for basic research and for forensic science. DNA fingerprinting allows for the identification of individuals from their DNA samples. DNA is digested with restriction enzymes, and a banding pattern is seen upon electrophoresis of the digest. No two individuals will have the same pattern, just as no two people have the same fingerprints. This technique is often used for paternity tests and for identification of criminals.

13.9 How Can We Study DNA–Protein Interactions?
The two most common techniques are called DNA footprinting and electrophoretic mobility-shift assay (EMSA). With a DNA footprint, DNA (with or without a suspected binding protein) is digested with DNase I. The banding pattern then seen with electrophoresis can show which part of the DNA is bound to protein. EMSA compares migration distances of naked DNA with DNA bound to protein.

13.10 What Are Some Methods for Studying Transcription?
A number of methods allow us to study which genes are transcribed at any given time. Detection methods similar to those used for DNA can be applied to the study of mRNA. Microarray chips provide the opportunity for simultaneous determination of the way in which many possible DNA or RNA sequences respond to a given set of conditions. The newest technique involves RNA interference to check the function of a particular gene.

13.11 How Do We Determine the Base Sequences of Nucleic Acids?
The ultimate way to study DNA is to sequence it. The Sanger–Coulson method allows for the determination of the entire DNA sequence. These procedures have contributed to the data on the sequence of the human genome, preliminary results of which were published in early 2001.

13.12 How Can We Use Bioinformatics to Study Genomics and Proteomics?
As more DNA sequences become available, it becomes possible to compare those sequences. Of particular interest is any pattern that may emerge from genes that encode proteins with similar functions. Important medical applications are emerging, and new methods are making it possible to analyze large quantities of data. Complete protein:protein interaction maps are now available for eukaryotes.

Critical Questions to Review

13.1 How Do We Purify and Detect Nucleic Acids?

1. **Fact Check** What advantages does fluorescent labeling offer over radioactive methods of labeling DNA?

2. **Fact Check** What methods are used to visualize radioactively labeled nucleic acids?

3. **Thought Question** When proteins are separated using native gel electrophoresis, size, shape, and charge control their rate of migration on the gel. Why is it that DNA separates based on size, and why do we not worry much about shape or charge?

13.2 What Makes Restriction Endonucleases an Important Tool for DNA Research?

4. **Fact Check** How does the use of restriction endonucleases of different specificities aid in the sequencing of DNA?

5. **Fact Check** What is the importance of methylation in the activity of restriction endonucleases?

6. **Fact Check** Why do restriction endonucleases not hydrolyze DNA from the organism that produces it?

7. **Fact Check** What role did restriction endonucleases play in localizing the gene associated with cystic fibrosis?

8. **Fact Check** Where did restriction endonucleases get their name?

9. **Fact Check** What do the following have in common? MOM; POP; NOON; MADAM, I'M ADAM; A MAN, A PLAN, A CANAL: PANAMA.

10. **Fact Check** Give three examples of DNA palindromes.

11. **Fact Check** What are three differences between the sites recognized by *Hae*III and those recognized by *Bam*HI?

12. **Fact Check** What are sticky ends? What is their importance in recombinant DNA technology?

13. **Fact Check** What would be an advantage of using *Hae*III for a cloning experiment? What would be a disadvantage?

13.3 What Is Cloning?

14. **Fact Check** Describe the cloning of DNA.

15. **Fact Check** What vectors can be used for cloning?

16. **Fact Check** Describe the method you would use to test for the uptake of a plasmid with a DNA insert.

17. **Fact Check** What is blue/white screening? What is the key feature of a plasmid that is used for it?

18. **Thought Question** What are some general "requirements" for recombinant DNA technology?

19. **Thought Question** What are some of the dangers of (and precautions against) recombinant DNA technology?

13.4 What Is Genetic Engineering, and Why Do We Do It?

20. **Fact Check** What are the purposes of genetic engineering in agriculture?

21. **Fact Check** What human proteins have been produced by genetic engineering?

22. **Fact Check** You go for a drive in the country with some friends and pass a cornfield with a sign. They do not understand the cryptic message on the sign, with the letters "Bt" followed by some numbers. You are able to enlighten them on the basis of information from this chapter. What is that information?

23. **Thought Question** Using information we have seen about lactate dehydrogenase, how could you clone and express human lactate dehydrogenase 3 (LDH 3) in bacteria?

24. **Thought Question** What are the requirements for an expression vector?

25. **Thought Question** What is a fusion protein? How are fusion proteins involved in cloning and expression?

26. **Thought Question** A friend tells you that she doesn't want to feed her baby high-production milk because she is afraid that the BST will interfere with the baby's growth by overstimulation. What do you tell her?

27. **Thought Question** The genes for both the α- and β-globin chains of hemoglobin contain introns (i.e., they are split genes). How would this fact affect your plans if you wanted to introduce the gene for α-globin into a bacterial plasmid and have the bacteria produce α-globin?

28. **Thought Question** Outline the methods you would use to produce human growth hormone (a substance used in the treatment of dwarfism) in bacteria.

29. **Thought Question** Bacteria and yeast are known not to have prions (Chapter 4). What does this fact have to do with the popularity of expressing mammalian proteins using bacterial vectors?

13.5 What Are DNA Libraries?

30. **Fact Check** What are the differences between a DNA library and a cDNA library?

31. **Thought Question** Why is it a large undertaking to construct a DNA library?

32. **Thought Question** Why do some journals require that the authors of articles describing DNA libraries make those libraries available to other researchers?

13.6 What Is the Polymerase Chain Reaction?

33. **Fact Check** Why is temperature control so important in the polymerase chain reaction?

34. **Fact Check** Why is the use of temperature-stable DNA polymerase an important factor in the polymerase chain reaction?

35. **Fact Check** What are the criteria for "good" primers in a PCR reaction?

36. **Thought Question** What difficulties arise in the polymerase chain reaction if there is contamination of the DNA that is to be copied?

37. **Thought Question** Each of the following pairs of primers has a problem with it. Tell why the primers would not work well.

(a) Forward primer

5′ GCCTCCGGAGACCCATTGG 3′

Reverse primer

5′ TTCTAAGAAACTGTTAAGG 3′

(b) Forward primer

5′ GGGGCCCCTCACTCGGGGCCCC 3′

Reverse primer

5′ TCGGCGGCCGTGGCCGAGGCAG 3′

(c) Forward primer

5′ TCGAATTGCCAATGAAGGTCCG 3′

Reverse primer

5′ CGGACCTTCATTGGCAATTCGA 3′

13.7 What Is Site-Directed Mutagenesis?

38. **Thought Question** Describe how you would use site-directed mutagenesis to change a *Bam*HI restriction site into an *Eco*RI site.

39. **Thought Question** Describe how you would use site-directed mutagenesis to help confirm the suspected mechanism of chymotrypsin (Chapter 7).

13.8 What Is DNA Fingerprinting?

40. **Thought Question** Suppose that you are a prosecuting attorney. How has the introduction of the polymerase chain reaction changed your job?

41. **Thought Question** Why is DNA evidence more useful as exclusionary evidence than for positive identification of a suspect?

13.9 How Can We Study DNA–Protein Interactions?

42. **Fact Check** What is the difference between a DNA footprint and an electrophoretic mobility-shift assay (EMSA)?

43. **Thought Question** How could EMSA be used to study the interactions of DNA, CREB, and CBP (see Chapter 11)?

13.10 What Are Some Methods for Studying Transcription?

44. **Thought Question** Explain how a microarray chip could be used to determine which cellular proteins were being expressed under two different metabolic conditions.

13.11 How Do We Determine the Base Sequences of Nucleic Acids?

45. **Thought Question** Give the DNA sequence for the template strand that gives rise to the following sequence gel, prepared using the Sanger method with a radioactive label at the 5′ end of the primer.

A	C	G	T

46. **Thought Question** Although techniques are available for determining the sequences of amino acids in proteins, it is becoming more and more common to sequence proteins indirectly by determining the base sequence of the gene for the protein and then inferring the amino acid sequence from the genetic-code relationships. Suggest why the latter technique is being used for proteins.

47. **Thought Question** Sometimes knowing the DNA sequence of a gene that codes for a protein does not tell you the amino acid sequence. Suggest several reasons why this is so.

48. **Thought Question** This is a conjectural question—there is no single "right" answer—that is good for discussion over tea and crumpets. In what ways might it be possible to prevent genetic discrimination due to information made available by the Human Genome Project?

49. **Thought Question** A recent television commercial featuring six-time Tour de France winner Lance Armstrong talked about the possibility of people carrying a DNA genotype card with them that would contain all of the information necessary to predict future diseases. This could, therefore, be used to help prescribe drugs to stop a medical condition before it became apparent. Give a couple of specific examples of how this ability could be used for the benefit or the detriment of humankind.

13.12 How Can We Use Bioinformatics to Study Genomics and Proteomics?

50. **Fact Check** What is the difference between the genome and the proteome?

51. **Fact Check** Has proteomic analysis been done on multicellular eukaryotes?

52. **Thought Question** Describe how the yeast two-hybrid system is used to determine interactions between two proteins in proteomic analysis.

Biochemistry 🧬 Now™
Assess your understanding of this chapter's topics with additional quizzing and tutorials at **http://now.brookscole.com/campbell5**

Annotated Bibliography

Berg, P., and M. Singer. *Dealing with Genes: The Language of Heredity.* Mill Valley, Calif.: University Science Books, 1992. [Two leading biochemists have produced an eminently readable book on molecular genetics. Highly recommended.]

Brown, K. Seeds of Concern. *Sci. Amer.* **284** (4), 52–57 (2000). [Discussion of potential environmental problems associated with genetically modified foods.]

Butler, D., and T. Reichhardt. Long-Term Effects of GM Crops Serves Up Food for Thought. *Nature* **398**, 651–653 (1999). [Reflections on possible problems with GM foods.]

DeRisi, J. L. Exploring the Metabolic and Genetic Control of Gene Expression on a Genomic Scale. *Science* **278**, 680–686 (1997). [Comprehensive discussion of microarray use for mapping transcription changes in yeast under anaerobic or aerobic conditions.]

Editors of *Science* et al. Genome Issue. *Science* **274**, 533–567 (1996). [A series of articles on genomes of organisms from yeast to humans, including a map of the human genome. The articles include an editorial about policy issues. There is also a web feature associated with this issue.]

Editors of *Science* et al. Genome Issue. *Science* **302**, 587–608 (2003). [A series of articles on progress in applying genomic information to medicine. The article by Couzin on the *BRCA* genes appears on pp. 591–593.]

Hannon, G. J. RNA Interference. *Nature* **418**, 244–251 (2002). [A description of one of the most important topics in nucleic acid manipulation.]

Hopkin, K. The Risks on the Table. *Sci. Amer.* **284** (4), 60–61 (2000). [Review of genetically modified crops, pros and cons.]

Li, T., C. Y. Chang, D. Y. Jin, P. J. Lin, A. Khvorova, and D. W. Stafford. Identification of the Gene for Vitamin K Epoxide Reductase. *Nature* **427**, 541–544 (2004). [A report on an important use of RNA interference.]

MacBeath, G. Printing Proteins as Microarrays for High-Throughput Function Determination. *Science* **289**, 1760–1763 (2000). [A research article on protein microchips.]

Marx, J. DNA Arrays Reveal Cancer in Its Many Forms. *Science* **289**, 1670–1672 (2000). [The use of DNA chips in the study of common cancers.]

O'Brien, S., and M. Dean. In Search of AIDS-Resistance Genes. *Sci. Amer.* **277** (3), 44–51 (1997). [Genetic resistance to HIV infection may provide the basis of new approaches to prevention and therapy of AIDS.]

Pennisi, E. Laboratory Workhorse Decoded: Microbial Genomes Come Tumbling In. *Science* **277**, 1432–1434 (1997). [A Research News article on the determination of the complete genome of the bacterium *Escherichia coli*, with a discussion of information available on genomes of other organisms. To be read in conjunction with the research article on pages 1453–1474 of the same issue.]

Reichhardt, T. Will Souped Up Salmon Sink or Swim. *Nature* **406**, 10–12 (2000). [A description of genetically modified oversized salmon.]

Ronald, P. Making Rice Disease-Resistant. *Sci. Amer.* **277** (5), 100–105 (1997). [Genetic engineering to improve yields of one of the world's most important food crops.]

Service, R. F. Protein Arrays Step Out of DNA's Shadow. *Science* **289**, 1673 (2000). [A single-page review of protein microchips.]

CHAPTER 14

Hot Topics in Cell and Molecular Biology

Students in Kowloon wear face masks as protection against the spread of SARS.

A. Givon/AP/Wide World Photos

Biochemistry ⊘ Now™
Test yourself on these Critical Questions at the BiochemistryNow website at **http://now .brookscole.com/campbell5**

Viruses are bits of genetic material wrapped in a protein coat. They are commonly thought of only for the considerable number of diseases they cause. Viruses are often specific for a host species and perhaps even for a particular tissue type or cell type. Most viruses have DNA as their genetic material, but some have RNA. Sometimes the DNA of a virus is incorporated into the host's DNA and other times it simply replicates independently and makes new virus particles. In 2003, a new, fast-spreading virus was discovered. This new virus, called SARS, was spread among people in China, Canada, and many places in between. Through a concerted effort on the part of the world's top scientists, it was quickly isolated, identified, and nearly eradicated. The most infamous virus, HIV, has not had the same fate, unfortunately. This virus persists and is still spreading today, causing the debilitating and fatal disease known as AIDS. Viruses have also been used for scientific and medical achievements. The growing but controversial field of gene therapy involves using viruses to transport genes into cells in an attempt to cure diseases. Diseases such as severe combined immunodeficiency syndrome (SCID), muscular dystrophy, cystic fibrosis, and several others have been treated with gene therapy.

To understand the effects of viruses, one must also understand the nature of the immune system, the body's defense against invasion. There are two kinds of immunity, innate and acquired. Among the latter are a host of cellular responses, such as T cells and B cells, and a humoral response that releases antibodies. One of the reasons that the HIV virus is so effective is that the immune cells called T cells are its target. Immune cells, like all cells, originate with a precursor cell called a stem cell. Stem cells have become the hottest topic in cellular biology today. Stem cells can differentiate into any type of cell, and scientists are trying to find ways to exploit this ability to cure problems once thought incurable, such as paralysis due to a broken spinal cord and replacement of dead brain cells after a stroke.

Having seen how DNA replicates and how this is tied to cell division, we can look at what happens when something goes awry. Uncontrolled cell growth and cell division leads to tumors and cancers, which are one of the leading causes of death. Many of the topics in this chapter are related. For example, viruses have also been implicated in cancer.

14.1 | What Are Viruses?

Viruses have always been difficult to classify according to normal taxonomy. Many have argued whether or not they should be considered to be living things. They cannot reproduce independently, and they cannot make proteins or generate energy independently, so they do not meet all of the requirements for life as we have traditionally defined it. But, if they are not life forms, what are they? The simplest definition would be a relatively small amount of genetic material surrounded by a protein envelope. The vast

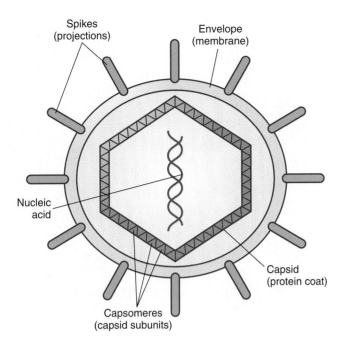

Spikes (projections)

Envelope (membrane)

Nucleic acid

Capsid (protein coat)

Capsomeres (capsid subunits)

◀ **FIGURE 14.1** The architecture of a typical virus particle. The nucleic acid is in the middle, surrounded by a protein coat called the capsid. Many viruses also have an envelope membrane that is usually covered with protein spikes.

majority of the viruses have only one type of nucleic acid, either DNA or RNA. Depending on the virus, this nucleic acid could be either single-stranded or double-stranded.

Viruses are known for the diseases they cause. They are pathogens of bacteria, plants, and animals. Some viruses are deadly, such as the fast-acting **Ebola virus,** which can have a mortality rate of more than 85%, and the slow-acting but equally deadly **human immunodeficiency virus (HIV),** which causes **acquired immunodeficiency syndrome (AIDS).** Other viruses might be simply annoying, such as **rhinovirus,** which causes common colds.

Virus Structure

Viruses are very small particles composed of nucleic acid and protein. The entire virus particle is called the **virion.** At the center of the virion is the nucleic acid. Surrounding this is the **capsid,** which is a protein coat. The combination of the nucleic acid and the capsid is called the **nucleocapsid,** and, for some viruses, such as the rhinovirus, that is the extent of the particle. Many other viruses, including HIV, have a **membrane envelope** surrounding the nucleocapsid. Many viruses also have **protein spikes** that help it attach to its host cell. Figure 14.1 shows the main features of a virus.

The overall shape of a virus varies. The classic viral shape most often seen in the literature has a hexagonal capsid with a rod sticking out of it that attaches to the host cell and acts like a syringe to inject the nucleic acid. Figure 14.2 shows the T2 bacteriophage of *E. coli,* a classic example of a virus of this shape. Tobacco mosaic virus (TMV), on the other hand, has a rod shape, as shown in Figure 14.3.

Families of Viruses

While there are many characteristics that distinguish viruses, most are organized by whether they have a genome of DNA or of RNA and whether or not they have an envelope. In addition, the nature of the nucleic acid (linear vs. circular, small vs. large, single-stranded vs. double-stranded) and the mode of incorporation (nucleic acid remains separate vs. nucleic acid joins with host

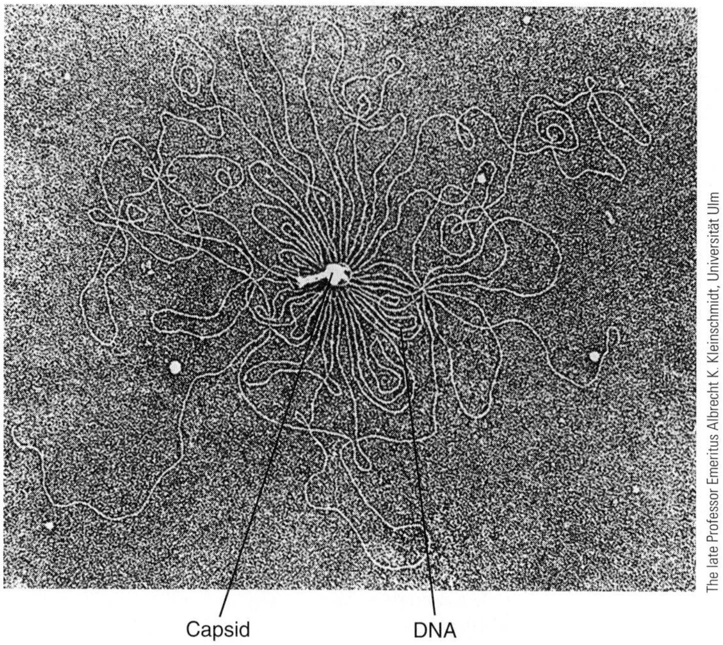

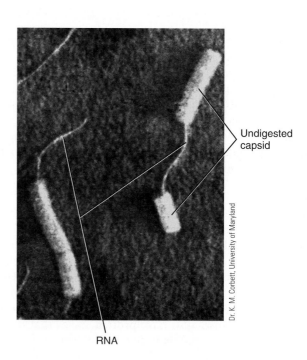

The late Professor Emeritus Albrecht K. Kleinschmidt, Universität Ulm

Capsid   DNA

▶ **FIGURE 14.2** An electron micrograph of a hexagonal virus. The bacteriophage T2 virus was gently disrupted, releasing the DNA, which can be seen as many loops outside the virus.

Undigested capsid

Dr. K. M. Corbett, University of Maryland

RNA

▶ **FIGURE 14.3** An electron micrograph of the rod-shaped tobacco mosaic virus.

chromosome) distinguishes the different virus types. Table 14.1 shows some of the known viral diseases and the families of viruses that cause them.

Virus Life Cycles

Most viruses cannot survive for long periods outside of cells, so they must quickly gain access to a cell. There are several mechanisms for gaining access, and preventing access to the cell has been a major focus of pharmaceutical companies trying to develop antiviral drugs. Figure 14.4 shows a generic example of a virus infecting a cell. The virus binds to the cell membrane and releases its DNA into the cell. The DNA is then replicated by host DNA polymerases and transcribed by host RNA polymerases. The transcription and

Table 14.1

Vertebrate Viruses and Diseases They Cause

DNA-Containing Viruses, Nonenveloped	
Adenoviruses	Respiratory and gastrointestinal diseases
Circoviruses	Anemia in chickens
Iridoviruses	Various diseases of insects, fish, and frogs
Papovaviruses	
Papillomaviruses	Warts
Human papillomavirus	Cervical cancer
Parvoviruses	
Human parvovirus B19	Fifth disease (a childhood rash)
Canine parvovirus	Viral gastroenteritis in dogs

DNA-Containing Viruses, Enveloped	
African swine fever virus	African swine fever (rarely in humans)
Hepadnaviruses	Hepatitis B
Herpesviruses	
Cytomegalovirus	Birth defects
Epstein–Barr virus (EBV)	Infectious mononucleosis and Burkitt's lymphoma
Herpes simplex virus (HSV) type 1	Cold sores
Herpes simplex virus (HSV) type 2	Genital herpes
Varicella-zoster virus (herpes zoster virus)	Chickenpox and shingles
Poxviruses	
Monkeypox virus	Monkeypox
Variola major virus	Smallpox

RNA-Containing Viruses, Nonenveloped	
Astroviruses	Gastroenteritis
Birnaviruses	Various diseases of birds, fish, and insects
Calciviruses	Norwalk gastroenteritis
Picornaviruses	
Hepatitis A virus	Hepatitis A
Polioviruses	Poliomyelitis
Rhinoviruses	Common cold
Reoviruses	
Rotaviruses	Infantile gastroenteritis

RNA-Containing, Enveloped	
Arenaviruses	
Lassa virus	Lassa fever
Arteriviruses	
Equine arteritis virus	Equine viral arteritis
Bunyaviruses	
California encephalitis virus	California encephalitis
Hantaviruses	Epidemic hemorrhagic fever or pneumonia
Coronaviruses	Respiratory disease, possibly gastroenteritis
Filoviruses	
Ebola virus	Ebola disease
Marburg virus	Marburg disease
Flaviviruses	
Dengue virus	Dengue fever
Hepatitis C virus	Hepatitis C
St. Louis encephalitis virus	St. Louis encephalitis
Yellow fever virus	Yellow fever
Orthomyxoviruses	Influenza
Paramyxoviruses	
Measles virus	Measles
Mumps virus	Mumps
Respiratory syncytial virus (RSV)	Pneumonia, bronchitis
Retroviruses	
Human T-cell lymphotropic viruses (HTLVs)	Leukemia, lymphoma
Human immuno-deficiency virus (HIV)	Acquired immune deficiency syndrome (AIDS)
Rhabdoviruses	
Rabies virus	Rabies
Togaviruses	
Rubella virus	Rubella (German measles)
Eastern equine encephalomyelitis (EEE) virus	Encephalomyelitis

translation of the mRNA leads to the proteins that are necessary to make the coat proteins of the capsid. New virions are produced and then released from the cell. This is called the **lytic** pathway, as the host cells are lysed by this process.

Viruses do not always lyse their host cells, however. A separate process called **lysogeny** involves the incorporation of the viral DNA into the host chromosome. **Simian virus 40 (SV40)** is an example of a DNA virus. It appears to be spherical, but it is actually an icosahedron, a geometric shape with 20 faces that are equilateral triangles, as shown in Figure 14.5. The genome of

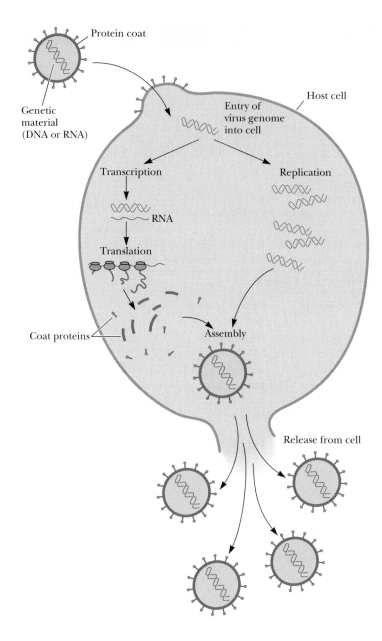

Biochemistry ⑤ Now™ ACTIVE FIGURE 14.4
The virus life cycle. Viruses are mobile bits of genetic
information encapsulated in a protein coat. The
genetic material may be either DNA or RNA. Once
this genetic material gains entry into the host cell, it
takes over the host machinery for macromolecular
synthesis and subverts it to the synthesis of viral-
specific nucleic acids and proteins. These virus
components are then assembled into mature virus
particles, which are then released from the cell.
Often, this parasitic cycle of virus infection leads to
cell death and disease. **Watch this Active Figure at
http://now.brookscole.com/campbell5**

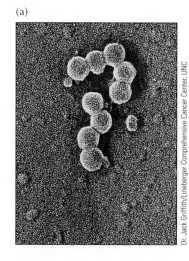

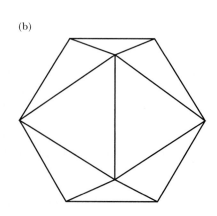

▶ **FIGURE 14.5** The architecture of simian virus 40
(SV40). (a) Virus particles appear almost spherical in
electron micrographs, but, on closer examination,
they can be seen to have an icosahedral shape.
(b) The geometry of an icosahedron. This regular
polyhedron has 20 faces, all of which are equilateral
triangles of identical size.

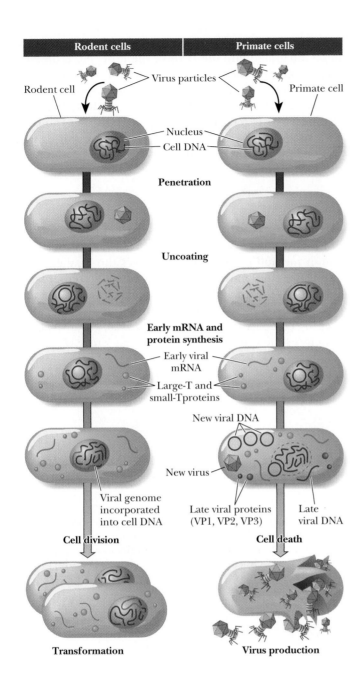

Rodent cells	Primate cells

Virus particles

Rodent cell

Primate cell

Nucleus
Cell DNA

Penetration

Uncoating

**Early mRNA and
protein synthesis**

Early viral
mRNA

Large-T and
small-Tproteins

New viral DNA

New virus

Viral genome
incorporated
into cell DNA

Late viral proteins
(VP1, VP2, VP3)

Late
viral DNA

Cell division

Cell death

Transformation

Virus production

◀ **FIGURE 14.6** The outcome of infection of cells by simian virus 40 depends on the nature of the cells. When primate cells are infected, the large-T protein is produced, and its presence ultimately leads to the production of new viral DNA and coat proteins. New virus particles are assembled and released: the death of the host cell takes place when the new virions are released. When rodent cells are infected, the viral genome is incorporated into the cell DNA. (*Adapted from* Dealing with Genes: The Language of Heredity, *by Paul Berg and Maxine Singer, © 1992 by University Science Books.*)

this virus is a closed circle of double-stranded DNA, with genes that encode the amino acid sequences of five proteins. Three of the five proteins are coat proteins. Of the remaining two proteins, one, the large-T protein, is involved in the development of the virus when it infects a cell. The function of the fifth protein, the small-T protein, is not known.

The outcome of infection by SV40 depends on the organism infected. When simian cells are infected, the virus enters the cell and loses its protein coat. The viral DNA is expressed first as mRNAs and then as proteins. The large-T protein is the first one made (Figure 14.6), triggering the replication of viral DNA, followed by viral coat proteins. The virus takes over the cellular machinery for both replication of DNA and protein synthesis. New virus particles are assembled, and eventually the infected cell bursts, releasing the new virus particles to infect other cells.

The results are different when SV40 infects rodent cells. The process is the same as far as the production of the large-T protein, but replication of the

---

Essential Information

Viruses are made from a small amount of nucleic acid enclosed in a protein shell. In some viruses, the nucleic acid is DNA; in others, it is RNA. The protein coat may or may not be surrounded by a membrane. There are many kinds of viruses that are known for a wide variety of diseases that they cause. Viruses enter host cells and conscript their enzymes to allow them to reproduce their own genetic material. Sometimes the virus lyses the host cell. At other times, viral DNA becomes incorporated into the host DNA.

viral DNA does not take place. The SV40 DNA already present in the cell can be lost or can be integrated into the DNA of the host cell. If the SV40 DNA is lost, there is no apparent result of the infection. If it becomes integrated into the DNA of the host cell, the infected cell loses control of its own growth. As a result of the accumulation of large-T protein, the infected cell behaves like a cancer cell. The large-T gene is an **oncogene,** one that causes cancer. Its mechanism is a subject of active research. The relationship between viruses and cancer will be looked at in more detail in Section 14.4 and Section 14.8.

Viral Attachment

A virus must attach to a host cell before it can penetrate, which is why so much research is involved in studying the exact mechanisms of viral attachment. A common method of attachment involves the binding of one of the spike proteins on the envelope of the virus to a specific receptor on the host cell. Figure 14.7 shows an example of HIV attachment. A specific spike protein called gp120 binds to a CD4 receptor on helper T cells. After this happens, a coreceptor complexes with CD4 and gp120. Another spike protein, gp41, then punctures the cell so that the capsid can enter.

Biochemical Connections

Influenza—The Virus That Won't Go Away

Certainly anyone reading this book has had the flu, a disease that most people take for granted as an annoying fact of life. The flu is caused by the influenza virus, which has been with us for a long time and has never been controlled by modern medicine. There are annual epidemics of influenza, and the first written record of such an epidemic was made by Hippocrates, the father of medicine, in 412 BC. In 1918, there was an influenza pandemic that killed more than 20 million people and was one of the worst plagues in history. (By comparison, as of the beginning of 2004, more than 40 million people have developed AIDS.) Between flu epidemics, new strains of the virus develop to which we have little resistance.

Influenza is caused by an enveloped virus with a single-stranded RNA template-strand genome. There are three major types, designated A, B, and C, depending on differences in the nucleocapsid proteins. Influenza viruses cause infections of the upper respiratory tract that lead to fever, muscle pain, headaches, nasal congestion, sore throat, and coughing. One of the biggest problems is that people stricken with the flu often get secondary infections, including pneumonia, which is what makes the flu potentially lethal.

The most prominent features of the virus envelope are two spike proteins. One is called **hemagglutinin (HA),** which gets its name because it causes erythrocytes to clump together. The second is **neuraminidase (NA),** an enzyme that catalyzes the hydrolysis of a linkage of sialic acid to galactose or galactosamine (see Chapter 16). The figure to the right shows the structural features of the influenza virus. HA is believed to help the virus in recognizing target cells. NA is believed to help the virus get through mucous membranes. As HA is the major protein on the surface of the virus, it is also the major site of attack of antibodies against influenza. It is also the protein that shows a lot of genetic drift, which is responsible for the fact that new strains are always developing, and it is difficult to develop a vaccine that remains effective for very long. That is also why there is a new flu vaccine every year that is supposed to be specific for the current strains that are being found.

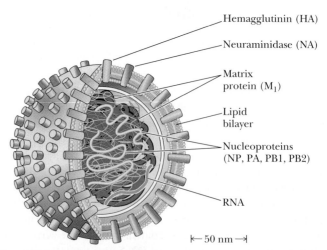

▲ A cutaway diagram of the influenza virion. The HA and NA spikes are embedded in a lipid bilayer that forms the virion's outer envelope. A matrix protein, M1, coats the inside of this membrane. The virion core contains the eight single-stranded segments that constitute its genome in a complex with the proteins NP, PA, PB1, and PB2 to form helical structures called neocapsids. (*Reprinted with permission from the Estate of Bunji Tagawa.*)

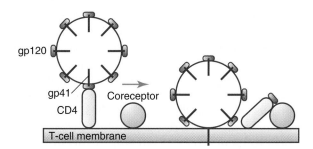

◀ **FIGURE 14.7** HIV attachment to a helper T cell. A specific spike protein called gp120 binds to a CD4 receptor on helper T cells. After this happens, a coreceptor complexes with CD4 and gp120. Another spike protein, gp41, then punctures the cell so that the capsid can enter.

14.2	What Virus Causes Severe Acute Respiratory Syndrome (SARS)?

In the fall of 2002, a new disease was seen in Guangdong Province in China. It was first referred to as "atypical pneumonia," and it was later renamed **severe acute respiratory syndrome (SARS).** In the first week, the disease produced fever, but few other symptoms. In the second week, pneumonia spread through the lungs. In the third week, destruction of blood vessels occurred, which was fatal in many cases. The overall mortality was 3%–6% of people infected, but mortality went up to 40%–50% in certain areas, when people older than 60 were infected. The study of SARS was a fascinating exercise in epidemiology on a worldwide scale, from which researchers have learned a lot about tracking down the roots of a disease. By early April of 2003, 2600 cases of SARS had been confirmed in 17 countries. By May of 2003, 8200 cases had been confirmed in 26 countries, with 700 deaths attributed to the disease. China was the hardest hit. The **World Health Organization (WHO)** issued an alert in March, and the possibility of a pandemic was feared. Fortunately, SARS wasn't nearly as contagious as previously thought. By comparison, in 1968, there was an outbreak of influenza that killed 700,000 people throughout the world in just 8 weeks.

SARS was found to have certain "hot spots," including Hong Kong and Toronto. The spread of the disease was eventually traced back to a few key incidents. A patient in Prince of Wales Hospital in Hong Kong was treated for supposed pneumonia by using a nebulizer to help him breathe. Unfortunately, nebulizers atomize droplets of respiration, enabling them to float further by a cough or sneeze. A week later, more than one hundred patients in the hospital came down with the disease. Then there was an outbreak in an apartment complex where 321 residents were stricken. Toronto became another site of the disease, which was eventually traced to a doctor who had been in the hospital in Hong Kong during the SARS outbreak there.

The first hypothesis for the origins of the disease came out of China, where a prominent scientist had noticed that many of the early samples from people with the disease contained the bacterium *Chlamydia*. In Beijing, a group of scientists had already identified another anomaly with the samples, a virus with a distinctive halo of spikes, which would have put it in a category of viruses never before known to kill humans, the **coronaviruses.** Figure 14.8 shows the structure of a coronavirus. Coronaviruses are members of a family of enveloped RNA viruses. This novel coronavirus was later shown to be the causative agent of SARS. In less than two months from its identification, the complete genome sequence had been simultaneously determined in several laboratories. The SARS coronavirus, identified as **SARS-HCoV,** is approximately 30,000 nucleotides of coding-strand RNA with a 5′ CAP and a 3′ poly-A tail. Once inside the cell, the RNA is translated into a large polypeptide that is cleaved by a virally encoded protease into several smaller proteins, including RNA-dependent RNA polymerase and an ATPase helicase. These proteins are responsible for replicating the viral genome and making viral proteins.

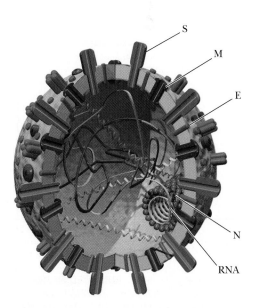

▲ **FIGURE 14.8** The structure of a coronavirus. The RNA genome is protected by a protein coat called N. The genome is further protected by an envelope containing glycoproteins called spike (S), envelope (E), and membrane (M). (*Adapted with permission from "The SARS Coronavirus: A Postgenomic Era," by K. V. Holmes and L. Enjuanes, Science **300**, 1377. Illustration by Katharine Sutliff. Copyright © 2003 AAAS.*)

Although the SARS virus is morphologically a coronavirus, its sequence is not related to any of the three known classes of human coronaviruses that cause nonlethal respiratory infections. Therefore, it has been used to define a fourth class of these viruses. Understanding the structure of the virus was important to allow scientists to think up countermeasures, such as antibody treatments or drugs. However, it was the old-fashioned technique of patient isolation that got the upper hand. In July of 2003, WHO declared that the SARS threat was over. In the end, a disease originally thought to be the potential cause of a new pandemic turned out not to spread fast enough to foil modern medicine.

Even after the SARS threat was declared over, scientists were left with questions about the origin of the disease. Some suspected that SARS was a biological agent that had been released by terrorists, but the evidence does not support that suspicion. Others believed that SARS was a mutation of one of the other human coronaviruses, but the data do not support that idea either, because there are too many base substitutions between SARS and any of the known human coronaviruses. A large number of the early patients were found to be working in the food industry in Southern China, which led scientists to theorize that SARS was a mutated form of a previously unknown animal coronavirus. In May, Chinese researchers showed that the virus could be found in Asian civets, weasel-like animals that are considered a delicacy in China. The genome of the virus in civets was similar to the genome of the human virus, except that the civet version of the virus had 29 extra nucleotides. It is currently not known if the civet was the natural host of a virus that eventually mutated to a form that could infect humans. Interestingly, of the 20 human coronaviruses sequenced, one patient who came from the province of Guangdong did have viruses with the sequence from civets with the extra 29 nucleotides, but the rest did not. At present, researchers do not believe that we have found the natural reservoir of the SARS virus, and many other questions about its origin and potential for mutation are still unanswered.

While the major threat from SARS seems over, China had another breakout in December of 2003, again in the province of Guangdong, when four more people were found with the disease. All of those patients were found to have the same coronavirus as the civets. The government is currently organizing slaughters of the civets in an attempt to eliminate the virus reservoir.

We may not have seen the end of the SARS virus, but it appears that it is largely contained. The lessons learned from the process have helped to bring the world together in a time of crisis. Laboratories that were adversaries worked together to prevent what could have become a devastating epidemic. One researcher for WHO said that it was fortunate that we had had the chance to practice on a virus that was not as contagious and as deadly as it might have been.

14.3 | What Is Unique about Retroviruses?

A retrovirus gets its name from the fact that its replication is backwards, compared to the central dogma of molecular biology: it makes DNA from RNA. The genome of a retrovirus is single-stranded RNA. Once it infects the cell, this RNA is used as a template to make a double-stranded DNA. The enzyme that does this is a virally encoded **reverse transcriptase.** One of the unique features of the retrovirus lifecycle is that the DNA produced by reverse transcription must be incorporated into the host DNA. This occurs because the ends of the DNA produced contain **long terminal repeats (LTRs).** LTRs are well known in DNA recombination events, and they allow the viral DNA to combine with the host's DNA. Figure 14.9 shows the replication cycle of a retrovirus.

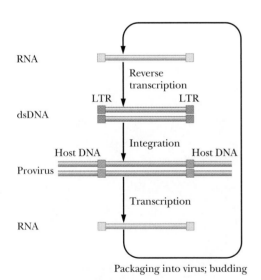

Biochemistry⊛Now™ ANIMATED FIGURE 14.9
The life cycle of a retrovirus. Viral RNA is released into the host cell, where the viral reverse transcriptase synthesizes double-stranded DNA from it. The DNA then incorporates into the host's DNA via recombination using the long terminal repeat (LTR) sequences. Eventually, the DNA is transcribed to RNA, which is then packaged into new virus particles. **See this figure animated at http://now.brookscole .com/campbell5**

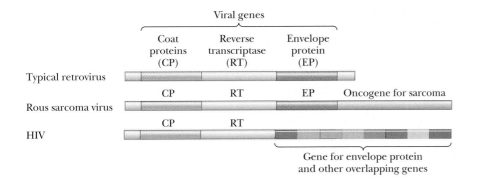

◀ **FIGURE 14.10** The RNA genomes of all retro-viruses have genes for coat proteins (CP), for reverse transcriptase (RT), and for envelope protein (EP). In addition to these essential genes, the Rous sarcoma virus carries the sarcoma oncogene. The HIV genome is more complex, with a number of overlapping genes for envelope proteins and other proteins. (*Adapted from* Dealing with Genes: The Language of Heredity, *by Paul Berg and Maxine Singer, © 1992 by University Science Books.*)

Retroviruses are the subject of extensive research in virology these days for three reasons. The first is that retroviruses have been linked to cancer. The second is that **human immunodeficiency virus (HIV)** is a retrovirus. The third is that retroviruses can be used in gene therapy.

All retroviruses have certain genes in common. There is a gene for proteins of the nucleocapsid, often called **coat proteins (CP).** They all have a gene for **reverse transcriptase (RT),** and they all have genes for **envelope proteins (EP).** Figure 14.10 shows a schematic of the RNA genomes of common retroviruses. In the case of the Rous sarcoma virus, the genome also contains an oncogene that causes tumors (see Section 14.8). The HIV genome is even more complex with overlapping genes for envelope proteins. All of these topics will be studied further in this chapter.

14.4 | How Are Viruses Used in Gene Therapy?

While viruses have usually been seen as problems for humans, there is one field now in which they are being used for good. Viruses can be used to make alterations in somatic cells, where a genetic disease is treated by the introduction of a gene for a missing protein. This is called **gene therapy.** The most successful form of gene therapy to date involves the gene for **adenosine deaminase (ADA),** an enzyme involved in purine catabolism (Section 23.8). If this enzyme is missing, dATP builds up in tissues, inhibiting the action of the enzyme ribonucleotide reductase. This results in a deficiency of the other three deoxyribonucleoside triphosphates (dNTPs). The dATP (in excess) and the other three dNTPs (deficient) are precursors for DNA synthesis. This imbalance particularly affects DNA synthesis in lymphocytes, on which much of the immune response depends. Individuals who are homozygous for adenosine deaminase deficiency develop **severe combined immune deficiency (SCID),** the "bubble-boy" syndrome. These individuals are highly prone to infection because of their highly compromised immune systems. The ultimate goal of the planned gene therapy is to take bone marrow cells from affected individuals; introduce the gene for adenosine deaminase into the cells using a virus as a vector; and then reintroduce the bone marrow cells in the body, where they will produce the desired enzyme. This procedure has been worked out in mice (Figure 14.11). The first clinical trials for ADA⁻ SCID were simple enzyme-replacement therapies begun in 1982. The patients were given injections of ADA. Later clinical trials focused on correction of the gene in mature T cells. In 1990, transformed T cells were given to recipients via transfusions. In trials at the National Institutes of Health (NIH), two girls, aged 4 and 9 at the start of treatment, showed improvement to the extent that they could attend regular public schools and have no more than the average number of infections. Administration of bone-marrow stem cells in addition to T

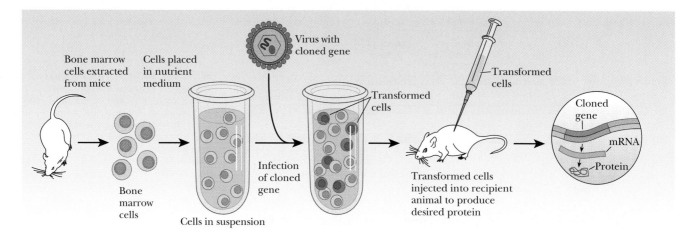

▲ **FIGURE 14.11** Gene therapy in bone-marrow cells. A cloned gene that directs the synthesis of a missing protein is introduced into bone-marrow cells from the mouse. The transformed cells are replaced in the mouse's body, where they produce the desired protein. (*Adapted from* Dealing with Genes: The Language of Heredity, *by Paul Berg and Maxine Singer,* © 1992 by University Science Books.)

cells was the next step; clinical trials of this procedure were undertaken with two infants, aged 4 months and 8 months in the year 2000. After 10 months, the children were healthy and had restored immune systems.

There are two types of delivery methods in human gene therapy. The first is called ex vivo, and is the type used to combat SCID. Ex vivo delivery means that somatic cells are removed from the patient, altered with the gene therapy, and then given back to the patient. The most common vector for this is **Maloney murine leukemia virus (MMLV).** Figure 14.12 shows how the virus is used for gene therapy. Some of the MMLV is altered to remove the *gag, pol,* and *env* genes, rendering the virus unable to replicate. These genes are replaced with an **expression cassette,** which contains the gene being administered, such as the ADA gene, along with a suitable promoter. This mutated virus is used to infect a packaging cell line. Normal MMLV is also used to infect the packaging cell line, which is not susceptible to the MMLV. The normal MMLV will not replicate in the packaging cell line, but its *gag, pol,* and *env* genes will restore the mutated virus's ability to replicate, but only in this cell line. These controls are necessary to keep mutant viruses from escaping to other tissues. The mutated virus particles are collected from the packaging cell line and used to infect the target cells, such as bone marrow cells in the case of SCID. MMLV is a retrovirus, so it infects the target cell and produces DNA from its RNA genome, and this DNA can then incorporate into the host genome, along with the promoter and ADA gene. In this way, the target cells that were collected have been transformed, and they will produce ADA. These cells are then put back into the patient.

The second delivery method is called in vivo, and means that the virus is used to directly infect the patient's tissues. The most common vector for this delivery is the **adenovirus** (which is a DNA virus). A particular vector can be chosen, based on specific receptors on the target tissue. Adenovirus has receptors in lung and liver cells, and it has been used in clinical trials for gene therapy of cystic fibrosis and ornithine transcarbamoylase deficiency. Figure 14.13 shows how in vivo delivery works. Adenovirus requires several genes for replication, including those called E1 through E4 (the E signifies early genes) and L1 through L5 (the L signifies late genes). A replication-deficient adenovirus is created by removing the E1 and E3 genes. The

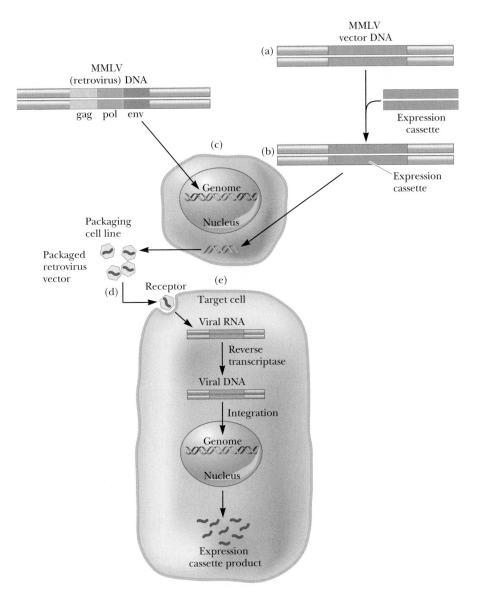

◄ **FIGURE 14.12** Gene therapy via retroviruses. The Maloney murine leukemia virus (MMLV) is used for ex vivo gene therapy. (a) Essential genes *(gag, pol, env)* are removed from the virus and (b) replaced with an expression cassette containing the gene being replaced with gene therapy. Removal of the essential viral genes renders the viruses unable to replicate. (c) The altered virus is then grown in a packaging cell line that will allow replication. (d) Viruses are collected and then used to infect cultured target cells from the patient needing the gene therapy. (e) The altered virus produces RNA, which then produces DNA via reverse transcriptase. The DNA then integrates in the patient's cells' genome, and then his or her cells produce the desired protein. The cultured cells are then given back to the patient. (*Adapted from Figure 1 in Crystal, R. G., 1995. Transfer of genes to humans: Early lessons and obstacles to success.* Science **270,** *404.*)

removed section of DNA is replaced with an expression cassette, as it was in the previous example. The adenovirus vector/expression cassette hybrids are grown in a complementing cell line that has been infected by wild-type adenovirus. Here the vectors are replicated. They are then collected and used to infect the patient's tissues.

Clinical trials using gene therapy to combat cystic fibrosis and certain tumors in humans are under way. In mice, gene therapy has been successful in fighting diabetes. The field of gene therapy is exciting and full of promise, but there are many obstacles to success in humans. There are also many risks, such as a dangerous immunological response to the vector carrying the gene, or the danger of a gene becoming incorporated into the host chromosome at a location that activates a cancer-causing gene. This possibility will be discussed further in Section 14.8. Gene therapy using in vivo delivery by adenovirus was suspended in early 2000, after the death of an 18-year-old following treatment for ornithine transcarbamoylase deficiency, another cause of SCID.

Essential Information

Many diseases are caused by the lack of a particular protein. With gene therapy, doctors attempt to give the patient back the missing protein by altering his or her cells. Viruses are used to carry in the gene for the missing protein. With ex vivo gene therapy, viruses are used to infect cells that have been isolated from the patient. Once these cells have been infected with the therapeutic gene, they are given back to the patient. With in vivo gene therapy, the virus is used to infect the patient directly.

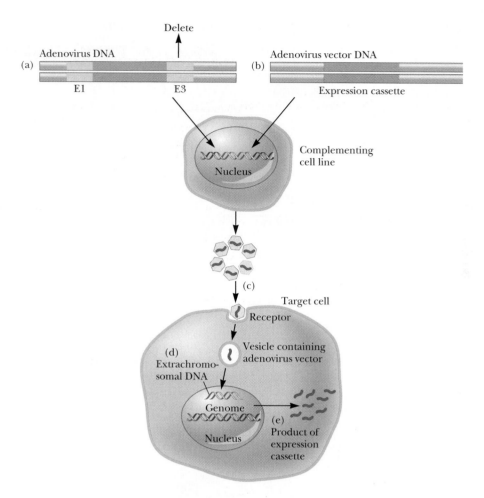

▶ **FIGURE 14.13** Gene therapy in vivo. Adenovirus is used for in vivo delivery of recombinant DNA. (a) Adenovirus vectors are generated by deleting gene E1 and E3, rendering the virus incapable of replicating unless introduced into a packaging cell line containing the E1 gene. (b) An expression cassette containing the therapeutic gene replaces the E1 and E3 genes. (c) Viruses grown in the packaging cell line are used to infect the patient directly. (d) The adenovirus enters the cells, and its DNA is replicated along with the patient's genome, although the DNA does not incorporate into the genome. (e) The therapeutic gene is transcribed and translated into product.

14.5 | How Does the Immune System Defend the Body?

One distinctive characteristic of the immune system is its ability to *distinguish self from nonself*. It is this ability that enables the cells and molecules responsible for immunity to recognize and destroy pathogens (disease-causing agents, such as viruses and bacteria) when they invade the body—or even one's own cells when they become cancerous. Since infectious diseases can be fatal, the operation of the immune system can be a matter of life and death. Striking confirmation of this last point is apparent in the lives of those who have AIDS (acquired immune deficiency syndrome). This disease so weakens the immune system that those who suffer from it become prey to infections that proceed unchecked, with ultimately fatal consequences. Suppression of the immune system can save lives as well as take them. The development of drugs that suppress the immune system has made *organ transplants* possible. Recipients of hearts, lungs, kidneys, or livers tolerate the transplanted organs without rejecting them because these drugs have thwarted the way in which the immune system tries to attack the grafts. However, the immune suppression also makes transplant recipients more susceptible to infections.

It is also possible for the immune system to go awry in distinguishing self from nonself. The result is **autoimmune disease,** in which the immune system attacks the body's own tissues. Examples include rheumatoid arthritis, insulin-dependent diabetes, and multiple sclerosis (Chapter 8, Biochemical Connections p. 190). A significant portion of research on the immune system is directed toward developing approaches for treating these diseases. **Allergies** are another example of improper functioning of the immune system. Millions suffer from asthma as a result of allergies to plant pollens and to other allergens (substances that trigger allergic attacks). Food allergies can evoke violent reactions that may be life-threatening.

Over the years, researchers have unraveled some of the mysteries of the immune system and have used its properties as a therapeutic aid. The first **vaccine,** that against smallpox, was developed about 200 years ago. Since that time it has been used so effectively as a preventive measure that smallpox has been eradicated. The action of vaccines of this sort depends on exposure to the infectious agent in a weakened form. The immune system mounts an attack, and *the immune system retains "memory" of the exposure.* In subsequent encounters with the same pathogen, the immune system can mount a quick and effective defense. This ability to retain "memory" is another major characteristic of the immune system. It is hoped that current research can be carried to the point of developing vaccines that can treat AIDS in persons already infected. Other strategies are directed at finding treatments for autoimmune diseases. Still others are attempting to use the immune system to attack and destroy cancer cells.

We need an understanding of how the immune system operates to go into more detail about how some of these goals might be achieved. There are two important aspects to the process: those that operate on the cellular level and those that operate on the molecular level. In addition, we have to look at whether the immune system is acquired or whether it is always present. We shall discuss these two aspects in turn.

A major component of the immune system is the class of cells called **leukocytes,** otherwise known as white blood cells. Like all blood cells, they arise from common precursor cells (stem cells) in the bone marrow. Unlike other blood cells, however, they can leave the blood vessels and circulate in the lymphatic system. Lymphoid tissues (such as lymph nodes, the spleen, and, above all, the thymus gland) play important roles in the workings of the immune system.

Innate Immunity—The Front Lines of Defense

When one considers the tremendous numbers of bacteria, viruses, parasites, and toxins that our bodies have to deal with, it is a wonder that we are not continually sick. Most students learn about antibodies in high school, and these days everyone learns about T cells, due to their relationship to AIDS. However, when discussing immunity, there are many more weapons of defense than T cells and antibodies. In reality, you only discover that you are sick once a pathogen has managed to beat your front-line defense, which is called **innate immunity.**

There are several parts to innate immunity. One part includes physical barriers, such as skin, mucus, and tears. All of these act to hinder penetration by pathogens and do not require specialized cells to fight the pathogen. However, if a pathogen, be it a bacterium, virus, or parasite, is able to breach this outer layer of defense, the cellular warriors of the innate system come into play. The cells of the innate immune system that we will discuss are **dendritic cells, macrophages,** and **natural killer (NK) cells.** One of the first and most important cells to join the fight are the dendritic cells, so called because of their dendrites, which are long, tentacle-like projections (see Figure 14.14). Dendritic cells are found in the skin, mucous membranes, the lungs, and the spleen, and are the first cells of the innate system that will have a crack at any virus or bacterium that wanders across their path. Using suction-cup-like receptors, they grab onto invaders and then engulf them by endocytosis. These cells then chop up the devoured pathogens and bring parts of their proteins to the surface. Here the protein fragments are displayed on a protein called a **major histocompatibility complex (MHC).** The dendritic cells travel through the lymph to the spleen, where they present these antigens to

▲ Allergic reactions arise when the immune system attacks innocuous substances. Allergies to plant pollens are common, producing well-known symptoms such as sneezing.

▲ Edward Jenner developed the world's first vaccine in 1796. It was a safe and effective way to prevent smallpox and has led to eradication of this disease.

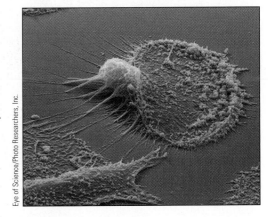

▲ **FIGURE 14.14** Dendritic cells get their name from their tentacle-like arms. The one shown is from a human.

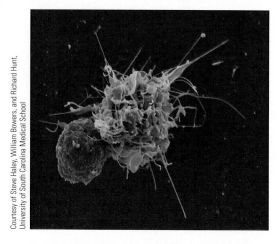

▲ FIGURE 14.15 Dendritic cells and the other cells of the immune system. This figure shows a rat dendritic cell interacting with a T cell. Through these interactions, the dendritic cells teach the acquired-immunity system what to attack.

other cells of the immune system, the **helper T cells (T$_H$ cells)**. Dendritic cells are members of a class of cells referred to as **antigen-presenting cells (APCs)**, and they are the starting point in most of the responses that are traditionally associated with the immune system. Once the dendritic cells present their antigens to the helper T cells, the latter release chemicals called **cytokines** that stimulate other members of the immune system, such as **killer T cells** (also called cytotoxic T cells or T$_C$ cells) and **B cells.** Figure 14.15 shows the basics of the relationship between dendritic cells and other immune cells. There are two classes of MHC proteins (I and II), based on their structure and on what they bind to. MHC I binds to killer T cells, while MHC II binds to helper T cells. Besides the link between the two cells based on the MHC, there is always another link (or perhaps two) that is necessary before cell proliferation occurs. The double signal is a trademark of most immune-cell responses, and it is thought to be one mechanism for making sure that the immune system is not activated in error.

Besides their basic role in presenting antigens to T cells and B cells, dendritic cells have recently been very popular with companies that are trying to generate antibodies to help fight cancer. Dendritic cells do have a downside, however. It was recently found that HIV uses a receptor on dendritic cells to hitch a ride in the lymph system until it can find a T$_H$ cell. Some labs are working on chemicals to block this interaction in the hopes of slowing down HIV's travel through the body. See the articles by Serbina and Pamer and by Banchereau in the Annotated Bibliography at the end of the chapter for more details about these fascinating cells.

Another important cell type in the innate immunity system is the **natural killer cell (NK)** (Figure 14.16), which is a member of a class of leukocytes

(a)

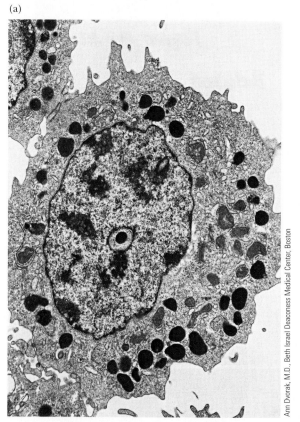

(b)

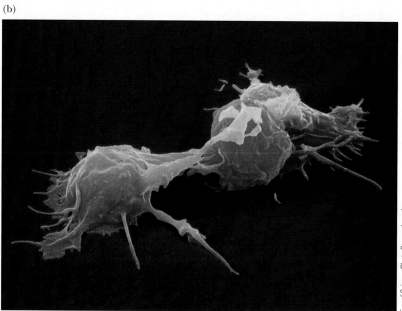

▲ FIGURE 14.16 Natural killer cells (NK) are among the first cells to be involved in an immune response. They are nonphagocytic cells that can interact with and destroy other cells, such as those infected with a virus, or cancer cells. (a) An electron micrograph. (b) A high-resolution photomicrograph.

called **lymphocytes,** because it is derived from a type of stem cell called a lymphoid stem cell. NK cells kill off cells that have been infected by viruses or that are cancerous, and they secrete cytokines that call up other cells, such as macrophages, another innate-immunity cell type that destroys microbes. They also work with dendritic cells, in a sense. If an infection is small, the NK cells may end up killing off the infected dendritic cells before the rest of the immune system is activated. Thus, NK cells help to decide whether the acquired-immunity system needs to be activated or not. NK cells are also important in fighting cancer. They are stimulated by interferon, an antiviral glycoprotein, which was employed as one of the first treatments for cancer and the first protein to be cloned and expressed for human use.

Acquired Immunity: Cellular Aspects

Acquired immunity is dependent on two other types of lymphocytes: T cells and B cells. **T cells** develop primarily in the thymus gland and **B cells** develop primarily in the bone marrow, accounting for their names (Figure 14.17). Much of the cellular aspect of acquired immunity is the province of the T cells, whereas much of the molecular aspect depends on the activities of the B cells.

T-Cell Functions

T cells can have a number of functions. As T cells differentiate, each becomes specialized for one of the possible functions. The first of these possibilities, that of the **killer T cells,** involves **T-cell receptors (TCRs)** on their surfaces that recognize and bind to **antigens,** the foreign substances that trigger the immune response. The antigens are presented to the T cell by antigen-presenting cells (APCs), such as macrophages and dendritic cells. The APCs ingest and process antigens, and then present them to the T cells. The processed antigen takes the form of a short peptide bound to an MHC I protein on the surface of the APC. Figure 14.18 shows how this works for macrophages. The macrophage also presents another molecule, a protein of a family known as B7, which binds to another T cell surface protein called CD28; the exact nature of this B7 protein is a subject of active research. (See the article by Cohen listed in the Annotated Bibliography at the end of this chapter.) The combination of the two signals leads to T-cell growth and differentiation, producing killer T cells. Proliferation of killer T cells is also triggered when macrophages bound to T cells produce small proteins called

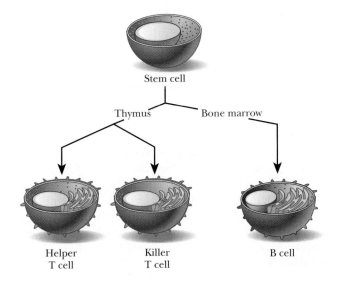

Stem cell

Thymus   Bone marrow

Helper   Killer   B cell
T cell   T cell

▲ **FIGURE 14.17** The development of lymphocytes. All lymphocytes are ultimately derived from the stem cells of the bone marrow. In the thymus, two kinds of T cells develop: helper T cells and killer T cells. B cells develop in the bone marrow.

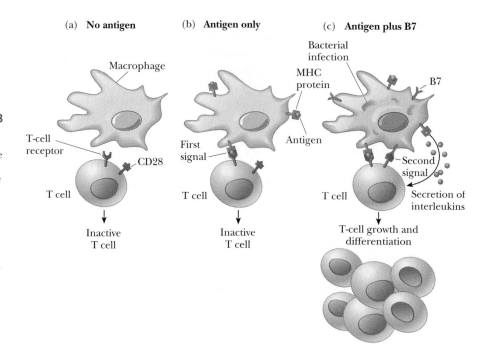

(a) **No antigen** (b) **Antigen only** (c) **Antigen plus B7**

Bacterial infection

Macrophage

MHC protein

B7

T-cell receptor

First signal

Antigen

Second signal

CD28

T cell

T cell

Secretion of interleukins

T cell

Inactive T cell

Inactive T cell

T-cell growth and differentiation

Biochemistry Now™ ANIMATED FIGURE 14.18
A two-stage process leads to the growth and differentiation of T cells. (a) In the absence of antigen, proliferation of T cells does not take place. (b) In the presence of antigen alone, the T cell receptor binds to antigen presented on the surface of a macrophage cell by the MHC protein. There is still no proliferation of T cells because the second signal is missing. In this way the body can avoid an inappropriate response to its own antigens. (c) When an infection takes place, a B7 protein is produced in response to the infection. The B7 protein on the surface of the infected cell binds to a CD28 protein on the surface of the immature T cell, giving the second signal that allows it to grow and proliferate. (*Adapted from "How the Immune System Recognizes Invaders," by Charles A. Janeway, Jr.; illustration by Ian Warpole.* Sci. Amer. *269 (3) (1993).*) **See this figure animated at http://now.brookscole.com/campbell5**

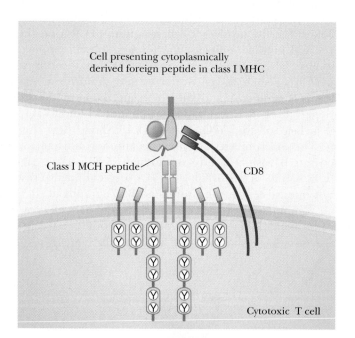

Cell presenting cytoplasmically derived foreign peptide in class I MHC

Class I MCH peptide

CD8

Cytotoxic T cell

▶ **FIGURE 14.19** Interaction between cytotoxic T cells (killer T cells) and antigen-presenting cells. Foreign peptides that are derived from the cytoplasm of infected cells are displayed on the surface by MHC I proteins. These bind to the T-cell receptor of a killer T cell. A docking protein called CD8 helps link the two cells together.

interleukins. The T cells make an interleukin-receptor protein as long as they are bound to the macrophage but do not do so when they are no longer bound. Interleukins are part of a class of substances called cytokines. When we discussed innate immunity, we saw that this term refers to soluble protein factors produced by one cell that specifically affect another cell. In this way, T cells do not proliferate in uncontrolled fashion. A killer T cell also has another membrane protein called CD8, which helps it dock to the MHC of the antigen-presenting cell, as shown in Figure 14.19. In fact, the CD8 protein is such a distinguishing characteristic that many researchers use the term CD8 cells instead of killer T cells.

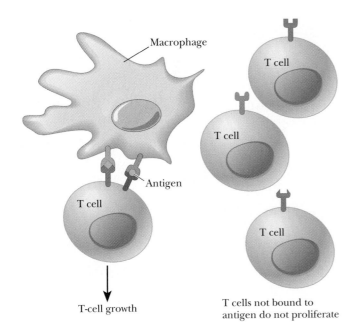

Macrophage

T cell

T cell

Antigen

T cell

T cell

T cell

T-cell growth

T cells not bound to
antigen do not proliferate

◀ **FIGURE 14.20** Clonal selection allows the immune system to be both versatile and efficient in responding to a wide range of possible antigens. Many different types of cells can be produced by the immune system, allowing it to deal with almost any possible challenge. Only those cells that respond to an antigen that is actually present are produced in quantity; this is an efficient use of resources.

T cells that bind to a given antigen and *only to that antigen* grow when these conditions are fulfilled. Note the specificity of which the immune system is capable. Many substances, including ones that do not exist in nature, can be antigens. The remarkable adaptability of the immune system in dealing with so many possible challenges is another of its main features. The process by which only those cells that respond to a given antigen grow in preference to other T cells is called **clonal selection** (Figure 14.20). The immune system can thus be versatile in its responses to the challenges it meets. This clonal selection is the basis of the definition of acquired immunity. The majority of the response of T cells stems from the rapid proliferation of cells once they are selected—the organism, in essence, acquiring these cells only when necessary. However, it must be pointed out that there must be at least one cell with the proper TCR to recognize the antigen and to bind to it. These receptors are not generated because there is a need; rather, they are generated randomly when stem cells differentiate into T cells. Fortunately, the diversity of T-cell receptors is so great that there are millions of TCR specificities.

The division of T cells during the peak of the immune response is very rapid, often reaching three or four divisions per day. This would lead to more than a thousandfold increase in the number of selected T cells in a few days.

As their name implies, killer T cells destroy antigen-infected cells. They do so by binding to them and by releasing a protein that perforates the plasma membranes of the infected cells. This aspect of the immune system is particularly effective in preventing the spread of viral infection by killing virus-infected host cells. In a situation such as this, the antigen can be considered to be all or part of the coat protein of the virus. When the infection subsides, some memory cells remain, conferring immunity against later attacks from the same virus.

T cells play another role in the immune system. Another class of T cells develops receptors for a different group of antigen-presenting MHC proteins, in this case MHC II. These become **helper T cells,** which develop in much the same way as killer T cells. Helper T cells are also referred to CD4 cells, due to the presence of that unique membrane protein. CD4 helps the cell dock to the MHC of the antigen-presenting cell, as shown in Figure 14.21. The function of helper T cells is primarily to aid in the stimulation of B cells. Maturing

Essential Information

The immune system allows an organism to distinguish self from nonself and can do so effectively by operating on the cellular and on the molecular level. Lymphocytes (a kind of white blood cell) are important in the immune response. Cells of the innate-immunity system, such as macrophages, natural killer cells, and dendritic cells, are the first to encounter pathogens. They chop up pieces of the pathogen, present them on the surfaces of the major histocompatibility proteins, and deliver them to the cells of the acquired immunity system, T cells and B cells.

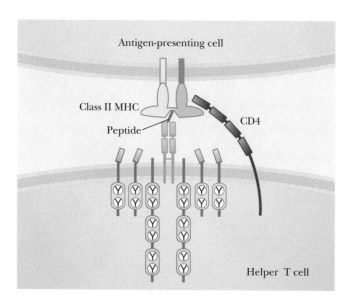

▶ **FIGURE 14.21** Interaction between helper T cells and antigen-presenting cells. Foreign peptides are displayed on the surface by MHC II proteins. These bind to the T-cell receptor of a helper T cell. A docking protein called CD4 helps link the two cells together.

B cells display the MHC II protein, with processed antigen, on their surfaces. Note particularly that the MHC proteins play a key role in the immune system. This property has led to a considerable amount of research to determine their structure, including determination by X-ray crystallography. The MHC II of the B cells is the binding site for helper T cells. The binding of helper T cells to B cells releases interleukins (IL-2 and IL-4) and triggers the development of B cells to plasma cells (Figure 14.22). Both B cells and plasma cells produce **antibodies** (also known as **immunoglobulins**), the proteins that will occupy most of our time as we discuss the molecular aspects of the immune response. B cells display antibodies on their surfaces in addition to the MHC II proteins. The antibodies recognize and bind to antigens. This property allows B cells to absorb antigens for processing. Plasma cells release circulating antibodies into the bloodstream, where they bind to antigen, marking it for destruction by the immune system. Helper T cells also help stimulate killer T cells and antigen-presenting cells via release of interleukins.

T-Cell Memory

One of the major characteristics of the acquired-immunity system is that it exhibits memory. While the system is slow to respond the first time it encounters an antigen, it is much quicker the next time. The process of generating T-cell memory involves the death of most of the T cells that were generated by the first infection with a particular antigen. Only a small percentage (5%–10%) of the original cells survive as memory cells. Still, this represents a much larger number than was present before the initial encounter with the antigen. These memory cells have a higher reproductive rate even in the absence of antigen than does a naïve T cell (one that has never encountered the antigen).

It is known that several of the interleukins play key roles in these processes. Interleukin 7 is involved in maintenance of naïve killer T cells at low levels. When stimulated by antigens, the proliferation of T_C cells is stimulated by interleukin 2. Memory T_C cells, on the other hand, are maintained by interleukin 15.

T-cell memory is one place where killer T cells and helper T cells come together. It has recently been shown by several researchers that CD8 cells will expand when confronted with the correct antigen in the absence of CD4

(a)

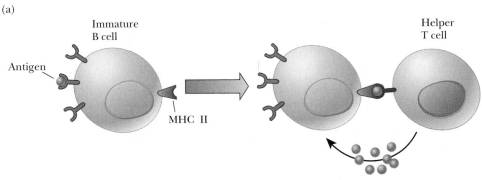

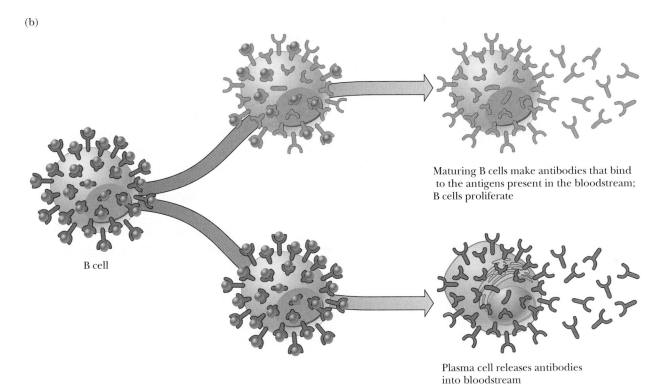

▲ **FIGURE 14.22** Helper T cells aid in the development of B cells. (a) A helper T cell has a receptor for the MHC II protein on the surfaces of immature B cells. When the helper T cells bind to the processed antigen presented by the MHC II protein, they release interleukins and trigger the maturation and proliferation of B cells. (b) B cells have antibodies on their surfaces, which allow them to bind to antigens. The B cells with antibodies for the antigens present grow and develop. When B cells develop into plasma cells, they release circulating antibodies into the bloodstream. (*Adapted from "How the Immune System Develops," by Irving L. Weissman and Max D. Cooper; illustrated by Jared Schneidman*. Sci. Amer. *September (1993).*)

cells. However, CD8 cells that were clonally expanded without CD4 cells were unable later to make memory cells that were as active.

The Immune System: Molecular Aspects

Antibodies are Y-shaped molecules, consisting of two identical heavy chains and two identical light chains held together by disulfide bonds (Figure 14.23). They are glycoproteins, with oligosaccharides linked to their heavy chains. There are different classes of antibodies, based on differences in the

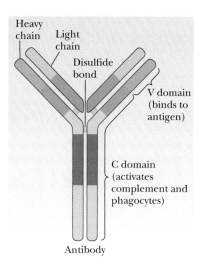

Heavy
chain
Light
chain
Disulfide
bond
V domain
(binds to
antigen)
C domain
(activates
complement and
phagocytes)
Antibody

▲ **FIGURE 14.23** A typical antibody molecule is a Y-shaped molecule consisting of two identical light chains and two identical heavy chains linked by disulfide bonds. Each light chain and each heavy chain has a variable region and a constant region. The variable region, which is at the prongs of the Y, binds to antigen. The constant region, toward the stem of the Y, activates phagocytes and complement, the parts of the immune system that destroy antibody-bound antigen. (*Adapted from "How the Immune System Recognizes Invaders," by Charles A. Janeway, Jr.; illustration by Ian Warpole. Sci. Amer. September (1993).*)

heavy chains. In some of these classes, heavy chains are linked to form dimers, trimers, or pentamers. Each light chain and each heavy chain has a constant region and a variable region. The variable region (also called the V domain) is found at the prongs of the Y and is the part of the antibody that binds to the antigen (Figure 14.24). The binding sites for the antibody on the antigen are called **epitopes.** The Biochemical Connections box describes an anticancer vaccine based on antigen–antibody recognition.

Most antigens have several such binding sites, so that the immune system will have several possible avenues of attack for naturally occurring antigens. Each antibody can bind to two antigens, and each antigen usually has several binding sites for antibody, giving rise to a precipitate that is the basis of experimental methods for immunological research. The constant region (the C domain) is located at the hinge and the stem of the Y; it is this part of the antibody that is recognized by phagocytes and by the complement system (the portion of the immune system that destroys antibody-bound antigen).

How does the body produce so many highly diverse antibodies to respond to essentially any possible antigen? The number of possible antibodies is virtually unlimited, as is the number of words in the English language. In a language, the letters of the alphabet can be arranged in countless ways to give a variety of words, and the same possibility for enormous numbers of rearrangements exists with the gene segments that code for portions of antibody chains. Antibody genes are inherited as small fragments that join together to form a complete gene in individual B cells as they develop (Figure 14.25). When gene segments are joined, the enzymes that catalyze the process add random DNA bases to the ends of segments being spliced, allowing for the wide variety observed experimentally. This rearrangement process takes place in the genes for both the light and the heavy chains. (Bear in mind that the exon splicing and mRNA processing that we discussed in Chapter 11 still takes place as well.) In addition to these factors, it is well known that B lymphocytes have a particularly high rate of somatic mutation, in which changes in the base sequence of DNA occur as the cell develops. Changes outside the germ cells apply only to the organism in which they take place and are not passed on to succeeding generations.

Each B cell (and each progeny plasma cell) produces only one kind of antibody. In principle, each such cell should be a source of a supply of homogeneous antibody by cloning. This is not possible in practice because lymphocytes do not grow continuously in culture. In the late 1970s, Georges Köhler

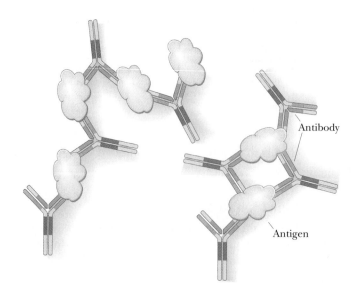

▶ **FIGURE 14.24** An antigen–antibody reaction forms a precipitate. An antigen, such as a bacterium or virus, typically has several binding sites for antibodies. Each variable region of an antibody (each prong of the Y) can bind to a different antigen. The aggregate thus formed precipitates and is attacked by phagocytes and the complement system.

Antibody

Antigen

Biochemical Connections

A Carbohydrate-Based Anticancer Vaccine

An anticancer vaccine based on a synthetically prepared oligosaccharide is being tested in clinical trials. The synthetic hexasaccharide is identical in structure to globo H, an epitope that has been isolated from breast, prostate, colon, and pancreatic cancer cells. The vaccine itself consists of the oligosaccharide linked to a carrier protein. When mice were injected with this vaccine, they produced antibodies specific for the hexasaccharide. These antibodies also recognize the antigen on tumor cells in mice. The clinical trials will show whether the same effect will be seen in humans. (For more information on this topic, see the article by Ragupathi et al. in the Annotated Bibliography at the end of this chapter.)

Epitope

▲ Structure of the epitope of an anticancer vaccine.

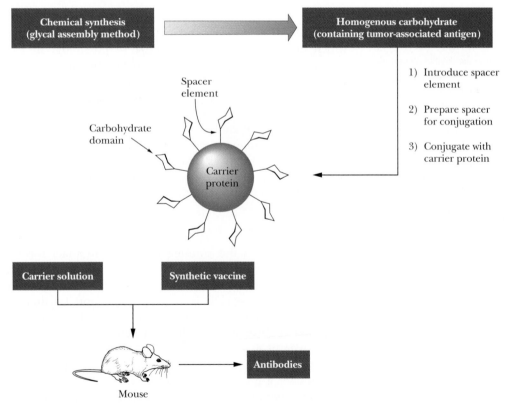

▲ The general strategy for the preparation of carbohydrate-based antitumor vaccines. (*Adapted from G. T. Ragupathi, et. al.*, Angew. Chem. Int. Ed. Engl. *36 (1/2), 125–128 (1997).*)

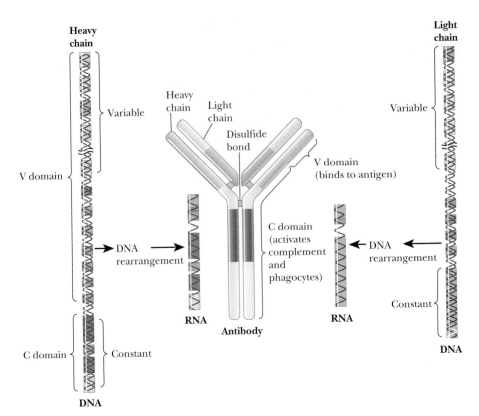

▶ **FIGURE 14.25** The heavy and light chains of antibodies are encoded by genes that consist of a number of DNA segments. These segments rearrange and, in the process, give rise to genes for different chains in each B cell. Since the joining is highly variable, comparatively few gene segments give rise to millions of distinct antibodies. (*Adapted from "How the Immune System Recognizes Invaders," by Charles A. Janeway, Jr.; illustration by Ian Warpole. Sci. Amer. September (1993).*)

Essential Information

Helper T cells are involved in the development of B cells and the plasma cells that arise from them. Antibodies in turn are produced by B cells and plasma cells. These Y-shaped glycoproteins circulate in the bloodstream. Antibodies have two binding sites for antigens, and antigens have several binding sites for antibodies, giving rise to cross-linked precipitates that remove antigens from the bloodstream for subsequent degradation. In antibodies, the prongs of the Y are highly variable in composition, allowing for specificity for a virtually limitless number of antigens.

and César Milstein developed a method to circumvent this problem, a feat for which they received the Nobel Prize in physiology in 1984. The technique requires fusing lymphocytes that make the desired antibody with mouse myeloma cells. The resulting **hybridoma** (hybrid myeloma), like all cancer cells, can be cloned in culture (Figure 14.26) and produces the desired antibody. Since the clones are the progeny of a single cell, they produce homogeneous **monoclonal antibodies.** In this way, it is possible to produce antibodies to almost any antigen in quantity. Monoclonal antibodies can be used to assay for biological substances that can act as antigens. A striking example of their usefulness is in testing blood for the presence of HIV; this procedure has become routine to protect the public blood supply.

Distinguishing Self from Nonself

With all the power the immune system has to attack foreign invaders, it must also do so with discretion, because we have our own cells that display proteins and other macromolecules on their surfaces. How the immune system knows not to attack these cells is a complicated and fascinating topic. When the body makes a mistake and does attack one of its own cells, the result is an **autoimmune disease,** examples of which are rheumatoid arthritis, lupus, and some forms of diabetes.

T cells and B cells have a wide variety of receptors on their surfaces. The affinities for a given antigen will vary greatly. Below a certain threshold, an encounter between a lymphocyte receptor and an antigen will not be sufficient to trigger that cell to become active and begin to multiply. These same cells also have stages of development. They mature in the bone marrow or the thymus and go through an early stage in which receptors first begin to appear on their surfaces.

In the case of T cells, there is a precursor form called a **DP cell** that has both the CD4 and the CD8 protein. This cell is the turning point for the fate of its progeny. If the receptors of the DP cell do not recognize anything,

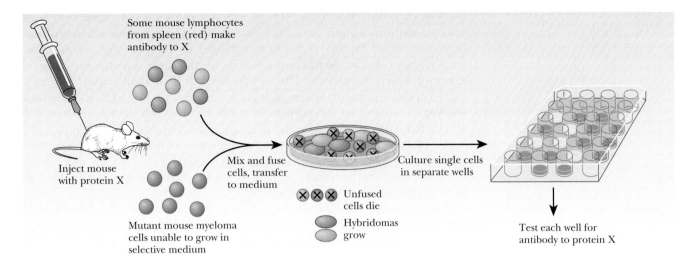

▲ FIGURE 14.26 A procedure for producing monoclonal antibodies against a protein antigen X. A mouse is immunized against the antigen X, and some of its spleen lymphocytes produce antibody. The lymphocytes are fused with mutant myeloma cells that cannot grow in a given medium because they lack an enzyme found in the lymphocytes. Unfused cells die because lymphocytes cannot grow in culture, and the mutant myeloma cells cannot survive in this medium. The individual cells are grown in culture in separate wells and are tested for antibody to protein X.

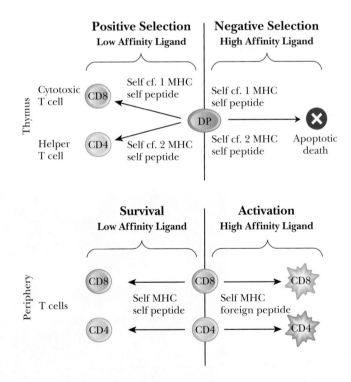

◀ FIGURE 14.27 Differentiation of T cells. A precursor to T cells called a DP cell is the turning point in the fate of T-cell progeny. If the DP cell reacts to nothing, including self-antigens or MHCs, then it dies by neglect (not shown). If it recognizes self-antigens or MHCs with high affinity, then it is programmed for apoptosis to avoid an autoimmune response. If it recognizes self-antigens or MHCs with low affinity, it differentiates into killer T cells and helper T cells. (*Reprinted with permission from "Signaling Life and Death in the Thymus: Timing Is Everything," by G. Werlen, B. Hausmann, D. Naeher, and E. Palmer. Science* **299,** *1859–1863. Copyright © 2003 AAAS.*)

including self-antigens or self-MHC proteins, then it dies by neglect. If the receptors recognize self-antigens or MHC but with low affinity, then the cell undergoes positive selection and differentiates into a killer T cell or a helper T cell, as shown in Figure 14.27. On the other hand, if the cell's receptors encounter self-antigens that are recognized with high affinity, it undergoes a process called **negative selection** and is programmed for apoptosis, or cell death.

By the time the lymphocytes leave their tissue of origin, they have therefore already been stripped of the most dangerous individual cells that would tend to react to self-antigens. There will still be some individual cells that have a receptor with very low affinity for a self-antigen. If these slip out of the bone marrow or thymus, they will not initiate an immune response because their affinity is below the minimum threshold, and there is always the requirement for a secondary signal. They would need to have another cell, such as a macrophage, also present them with an antigen. In the case of B cells, besides binding an antigen to its receptor, it would need to receive an interleukin 2 from a helper T cell that had also been stimulated by the same antigen.

All of these safeguards lead to the delicate balance that must be maintained by the immune system, a system that simultaneously has the diversity to bind to almost any molecule in the universe but does not react to the myriad proteins that are recognized as self.

14.6 | How Does Human Immunodeficiency Virus Cause AIDS?

Human immunodeficiency virus (HIV) is the most infamous of the retroviruses because it is the causative agent of acquired immunodeficiency syndrome, or AIDS. This disease affects more than 40 million people worldwide and has thwarted attempts to eradicate it. The best medicines today can slow it down, but nothing has been able to stop it.

Its genome is a single-stranded RNA molecule that has a number of proteins packed around it, including the virus-specific reverse transcriptase and protease. There is a protein coat around the RNA–protein assemblage, giving the overall shape of a truncated cone. Finally, there is a membrane envelope around the protein coat. The envelope consists of a phospholipid bilayer formed from the plasma membrane of cells infected earlier in the life cycle of the virus, as well as some specific glycoproteins, such as gp41 and gp120, as shown in Figure 14.28.

(a)

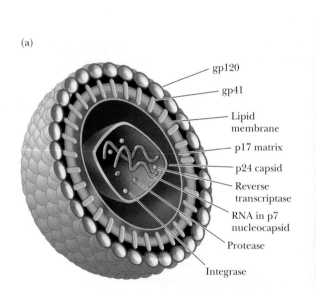

gp120
gp41
Lipid membrane
p17 matrix
p24 capsid
Reverse transcriptase
RNA in p7 nucleocapsid
Protease
Integrase

(b)

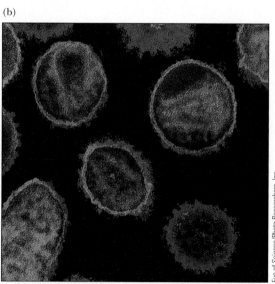

Eye of Science/Photo Researchers, Inc.

▲ **FIGURE 14.28** The architecture of HIV. (a) The RNA genome is surrounded by P7 nucleocapsid proteins and by several viral enzymes—namely, reverse transcriptase, integrase, and protease. The truncated cone consists of P24 capsid protein subunits. The P17 matrix (another layer of protein) lies inside the envelope, which consists of a lipid bilayer and glycoproteins, such as gp41 and gp120. (b) An electron micrograph shows both mature virus particles, in which the core (the truncated cone) is visible, and immature virus particles, in which it is not.

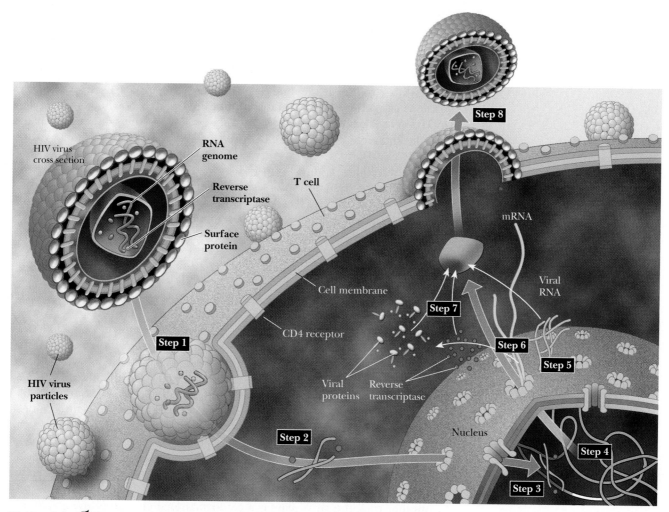

Biochemistry⑤Now™ ▲ ANIMATED FIGURE 14.29 HIV infection begins when the virus particle binds to CD4 receptors on the surface of the cell (Step 1). The viral core is inserted into the cell and partially disintegrates (Step 2). The reverse transcriptase catalyzes the production of DNA from the viral RNA. The viral DNA is integrated into the DNA of the host cell (Step 3). The DNA, including the integrated viral DNA, is transcribed to RNA (Step 4). Smaller RNAs are produced first, specifying the amino acid sequence of viral regulatory proteins (Step 5). Larger RNAs, ones that specify the amino acid sequences of viral enzymes and coat proteins, are made next (Step 6). The viral protease assumes particular importance in the budding of new virus particles (Step 7). Both the viral RNA and viral proteins are included in the budding virus, as is some of the membrane of the infected cell (Step 8). (*Adapted from* AIDS and the Immune System, *by Warner C. Green, illustration by Tomo Narachima*, Sci. Amer. *(1993).*) **See this figure animated at http://now.brookscole .com/campbell5**

The mode of action of HIV is a classic example of the mode of operation of retroviruses. It is known that the HIV infection begins when the virus particle binds to receptors on the surface of the cell (Figure 14.29). The viral core is inserted into the cell and partially disintegrates. The reverse transcriptase catalyzes the production of DNA from the viral RNA. The viral DNA is integrated into the DNA of the host cell. The DNA, including the integrated viral DNA, is transcribed to RNA. Smaller RNAs are produced first, specifying the amino acid sequences of viral regulatory proteins. Larger RNAs, ones that specify the amino acid sequences of viral enzymes and coat proteins, are made next. The viral protease (Chapter 6, Biochemical Connections, p. 151) assumes particular importance in the budding of new virus particles. Both the viral RNA and the viral proteins are included in the budding virus, as is some of the membrane of the infected cell.

HIV Confounds Our Immune Systems

Why is this virus so deadly and so hard to stop? We have seen examples of viruses, such as adenovirus, that cause nothing more than the common cold, while others, such as the SARS virus, can be deadly. At the same time, we have seen the complete eradication of the deadly SARS virus, whereas adenovirus is still with us. HIV has several characteristics that lead to its persistence and eventual deadliness. Ultimately, it is deadly because of its target, the helper T cell. The immune system is under constant attack by the virus, and millions of helper T cells and killer T cells are called up to fight billions of virus particles. Through degradation of the T-cell membrane via budding and the activation of caspases that lead to cell death, the number of T cells diminishes to the point at which the infected person is no longer able to mount a suitable immune response, eventually succumbing to pneumonia or another opportunistic disease.

There are many reasons that the disease is so persistent. One of them is that it is slow acting. The main reason SARS was eradicated so quickly was that the virus was quick to act, making it easy to find infected people before they had a chance to spread the disease. This is far from the truth with HIV, where people can go for years before they are aware that they have the infection. However, this is a small part of what makes HIV so difficult to kill.

HIV is difficult to kill because it is difficult to find. For an immune system to fight a virus, it needs to be able to locate specific macromolecules that can be bound to antibodies or to T-cell receptors. The reverse transcriptase of HIV is very inaccurate in its replication. The result is the rapid mutation of HIV, a situation that presents a considerable challenge to those who want to devise treatments for AIDS. The virus mutates so rapidly that there may be many strains of HIV in a single individual.

Another trick the virus plays is a conformational change of the gp120 protein when it binds to the CD4 receptor on the T cell. The normal shape of the gp120 monomer may elicit an antibody response, but these antibodies are largely ineffective. The gp120 forms a complex with gp41 and changes shape when it binds to CD4. It also binds to a secondary site on the T cell that normally binds to a cytokine. This change exposes a part of the gp120 that was previously hidden and therefore unable to elicit antibodies.

HIV is also good at evading the innate immunity system. Natural killer cells attempt to attack the virus, but HIV binds a particular cell protein, called cyclophilin, to its capsid, which blocks the antiviral agent known as restriction factor-1. Another of HIV's proteins blocks the viral inhibitor called CEM-15, which normally disrupts the viral life cycle.

Lastly, HIV hides from the immune system by cloaking its outer membrane in sugars that are very close to the natural sugars found on most of its host's cells, rendering the immune system blind to it.

The Search for a Vaccine

The attempt to find a vaccine for HIV is akin to the search for the Holy Grail, and it has met with about as much success. One strategy for using a vaccine to stimulate the body's immunity to HIV is shown in Figure 14.30. DNA for a unique HIV gene, such as the *gag* gene, is injected into muscle. The *gag* gene leads to the gag protein, which is taken up by antigen-presenting cells and displayed on their cell surfaces. This then elicits the cellular immune response, stimulating killer and helper T cells. It also stimulates the humoral

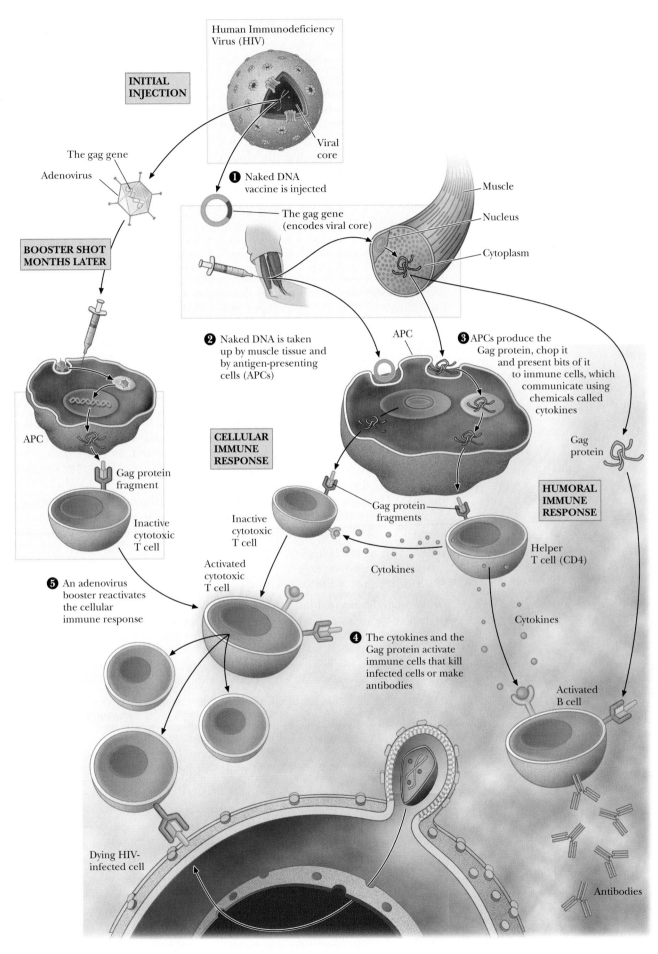

INITIAL INJECTION

Human Immunodeficiency Virus (HIV)

Viral core

The gag gene

Adenovirus

BOOSTER SHOT MONTHS LATER

❶ Naked DNA vaccine is injected

The gag gene (encodes viral core)

Muscle

Nucleus

Cytoplasm

❷ Naked DNA is taken up by muscle tissue and by antigen-presenting cells (APCs)

APC

❸ APCs produce the Gag protein, chop it and present bits of it to immune cells, which communicate using chemicals called cytokines

Gag protein

APC

CELLULAR IMMUNE RESPONSE

Gag protein fragment

Inactive cytotoxic T cell

Inactive cytotoxic T cell

Gag protein fragments

HUMORAL IMMUNE RESPONSE

Helper T cell (CD4)

Cytokines

❺ An adenovirus booster reactivates the cellular immune response

Activated cytotoxic T cell

❹ The cytokines and the Gag protein activate immune cells that kill infected cells or make antibodies

Cytokines

Activated B cell

Dying HIV-infected cell

Antibodies

▲ **FIGURE 14.30** One strategy for an AIDS vaccine. (© *2003 Terese Winslow.*)

immune response, stimulating production of antibodies. Figure 14.30 also shows a second part of the treatment, which is a booster shot of an altered adenovirus that carries the *gag* gene. Unfortunately, most attempts at making antibodies have been unsuccessful. The most thorough attempt was made by the VaxGen company, which continued the research through the third stage of clinical trials, testing the vaccine on more than a thousand high-risk people and comparing them with a thousand that did not receive the vaccine. 5.7% of the people that received the vaccine eventually became infected, compared with 5.8% of the placebo group. Many people analyzed the data and, despite attempts to show a better response in certain ethnic groups, the trials had to be declared a failure. The vaccine, called AIDSVAX, was a gp120 vaccine.

Antiviral Therapy

While the search for an effective AIDS vaccine continued with little or no success, pharmaceutical companies flourished by designing drugs that would inhibit retroviruses. By 1996 there were 16 drugs used to inhibit either the HIV reverse transcriptase or the protease, some of which we have seen before, such as AZT and saquinavir. Several others are in clinical trials, including drugs that target the gp41 and the gp120 in an attempt to prevent entry of the virus. Table 14.2 shows some of the newer drugs that are undergoing trials. A combination of drugs to inhibit retroviruses has been dubbed **highly active antiretroviral therapy (HAART).** Initial attempts at HAART were very successful, driving the viral load down almost to the point of being undetectable, with the concomitant rebounding of the CD4 cell population. However, as always seems to be the case with HIV, it later turned out that, although the virus was knocked down, it was not knocked out. HIV remained in hiding in the body and would bounce back as soon as the therapy was stopped. Thus, the best-case scenario for an AIDS patient was a lifetime of expensive drug therapies. In addition, long-term exposure to HAART was found to cause constant nausea and anemia, as well as diabetes symptoms, brittle bones, and heart disease.

Antibodies Get a Second Chance

In the wake of the realization that patients could not stay on the HAART program indefinitely, several researchers attempted to use a combination of

Table 14.2				
Anti-HIV Drugs in Clinical Trials				
Drug	**Manufacturer**	**Target**	**Stage**	**Attributes**
T-20	Trimeris/Hoffmann-La Roche	Entry (gp41)	Phase III	Novel target
Atazanavir	Bristol-Myers Squibb	Protease	Phase III	Low lipid tox.,1 pill/day
FTC (emtricitabine)	Triangle Pharmaceuticals	RT	Phase III	1 pill/day
Tipranavir	Boehringer Ingelheim	Protease	Phase II/III	Resistance
DPC-083	Bristol-Myers Squibb	RT	Phase II	Resistance
DAPD	Triangle Pharmaceuticals	RT	Phase I/II	Resistance
T-1249	Trimeris	Entry (gp41)	Phase I/II	Novel target
TMC 125	Tibotec-Virco	RT	Phase I	Ultrapotent
L-870, 810	Merck & Co. Inc.	Integrase	Phase I	Novel target
S-1360	Shionogi/GlaxoSmithKline	Integrase	Phase I	Novel target
SCH-C	Schering-Plough	Entry (CCR5)	Phase I	Novel target
BMS-806	Bristol-Myers Squibb	Entry (gp120/CD4)	Phase I	Earliest entry stage

HAART and vaccination. Even though most of the vaccines were not found to be effective when given alone, they were more effective in combination with HAART. In addition, once on the vaccine, patients were able to take a rest from the other drugs, giving their bodies and minds time to recover from the side effects of the antiviral therapy.

The Future of Antibody Research

Early attempts at creating a vaccine against HIV appear to have failed because the vaccine elicited too many useless antibodies. What patients need is a **neutralizing antibody,** one capable of completely eliminating its target. Researchers discovered a patient who had been infected with HIV for six years but had never developed AIDS. They then studied his blood and found a rare antibody, which they labeled **b12.** In lab trials, b12 was found to stop most strains of HIV. What made b12 different from the other antibodies? Structural analysis showed that antibody b12 has a different shape from a normal immunoglobulin. It has sections of long tendrils that fit into a fold in gp120. This fold in gp120 cannot mutate very much, otherwise the protein would not be able to dock properly with the CD4 receptor. Another antibody was found in another patient that seemed resistant to HIV. This antibody was actually a dimer and had a shape more like an "I" than the traditional "Y". This antibody, called **2G12,** recognizes some of the sugars on the HIV outer membrane that are unique to HIV.

By finding a few such antibodies, researchers have been able to search for a vaccine in the opposite direction from the normal way. This was called **retrovaccination,** because researchers already had the antibody and needed to find a vaccine to elicit it, instead of injecting vaccines and looking to see what antibodies they produced.

14.7	**Why Are Stem Cells Special?**

Stem cells are the precursors of all the other cell types. They are undifferentiated cells that have the ability to form any cell type as well as to replicate into more stem cells. Stem cells are often called **progenitor** cells, due to their ability to differentiate into many cell types. A **pluripotent** stem cell is one that is able to give rise to all cell types in an embryo or in an adult. Some cells are called **multipotent** because they can differentiate into more than one cell type, but not into all cell types. The farther from a zygote a cell is in the course of development, the less the potency of the cell type. The use of stem cells, especially **embryonic stem cells,** has been an exciting field of research for several years.

History of Stem-Cell Research

Stem-cell research began in the 1970s with studies on teratocarcinoma cells, which are found in testicular cancers. These cells are bizarre blends of differentiated and undifferentiated cells. They were referred to as **embryonal carcinoma (EC) cells.** They were found to be pluripotent, which led to the idea of using them for therapy. However, such research was suspended because the cells had come from tumors, which made their use dangerous, and because they were **aneuploid,** which means they had the wrong number of chromosomes.

Early work with embryonic stem (ES) cells came from cells that were grown in culture after being taken from embryos. It was found that these stem cells could be grown in culture and maintained for long periods. Most differentiated cells, on the other hand, will not grow for extended periods in culture. Stem cells are maintained in culture by the addition of certain factors, such as

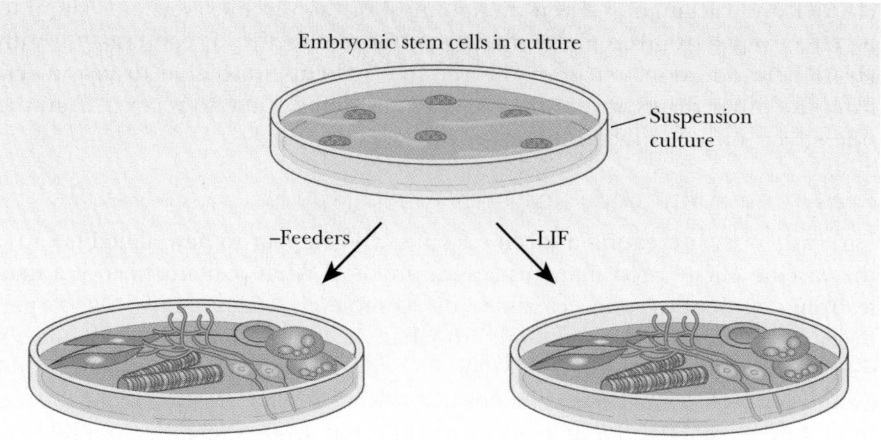

▶ **FIGURE 14.31** Pluripotent embryonic stem cells can be grown in cell culture. They can be maintained in an undifferentiated state by growing them on certain feeder cells, such as fibroblasts, or by using leukemia inhibitory factor (LIF). When removed from the feeder cells or when the LIF is removed, they begin to differentiate in a wide variety of tissue types, which could then be harvested and grown for tissue therapy. (*Taken from Donovan, P. J., and Gearhart, J. Nature,* **414,** *92–97 (2001).*)

leukemia inhibiting factor or feeder cells (non-mitotic cells such as fibroblasts). Once released from these controls, ES cells will differentiate into all kinds of cells, as shown in Figure 14.31.

Stem Cells Offer Hope

Stem cells placed into a particular tissue, such as blood, differentiate and grow into blood cells. Others placed into brain tissue grow into brain cells. This is a very exciting discovery because it had been believed that there was little hope for patients with spinal-cord and other nerve damage, because nerve cells do not normally regenerate. In theory, neurons could be produced to treat neurodegenerative diseases, such as Alzheimer's disease or Parkinson's disease. Muscle cells could be produced to treat muscular dystrophies and heart disease. In one study, mouse stem cells were injected into a mouse heart that had undergone a myocardial infarction. The cells spread from an unaffected region into the infarcted zone and began to grow new heart tissue. Human pluripotent stem cells have been used to regenerate nerve tissue in rats with nerve injuries and have been shown to improve motor and cognitive ability in rats that suffered strokes. (See the articles by Sussman, by Aldhous, and by Donovan in the Annotated Bibliography for this chapter.) Results such as these have led some scientists to claim that stem-cell technology will be the most important advancement since cloning.

Truly pluripotent stem cells have been harvested primarily from embryonic tissue, and these cells show the greatest ability to differentiate into various tissues and to reproduce in cell culture. Stem cells have also been taken from adult tissues, since there are always some stem cells in an organism even at the adult stage. These cells are usually multipotent, as they can form several different cell types, but they are not as versatile as embryonic stem (**ES**) cells. Many scientists believe that the ES cells represent a better source for tissue therapy than adult stem cells for this reason.

The acquisition and use of stem cells can also be related to a technique called **cell reprogramming,** which is a necessary component of whole-mammal cloning, such as the cloning that produced the world's most famous sheep, Dolly. Most somatic cells in an organism contain the same genes, but the cells develop as different tissues with extremely different patterns of gene expression. A mechanism that alters expression of genes without changing the actual DNA sequence is called an **epigenetic** mechanism. An epigenetic state of the DNA in a cell is a heritable trait that allows a "molecular memory" to exist in the cells. In essence, a liver cell remembers where it came from and will continue to divide and to remain a liver cell. These epigenetic states involve methylation of cytosine–guanine dinucleotides and interactions with proteins of chromatin. Mammalian genes have an additional level of epigenetic information called **imprinting,** which allows the DNA to retain a molecular memory of its germ-line origin. The paternal DNA is imprinted differently from the maternal DNA. In normal development, only DNA that came from both parents would be able to combine and to lead to a viable offspring. Normally, the epigenetic states of somatic cells are locked in a way that the differentiated tissues remain stable. The key to whole-organism cloning was the ability to erase the epigenetic state and to return to the state of a fertilized egg, which has the potential to produce all cell types. It has been shown that, if the nucleus of a somatic cell is injected into a recipient oocyte (see Figure 14.32), the epigenetic state of the DNA can be reprogrammed, or at least partially reprogrammed. The molecular memory is erased, and the cell begins to behave like a true zygote. This can be used to derive pluripotent stem cells or to transfer a blastocyst into a mother-carrier for growth and development. In November of 2001, the first cloned human blastocyst was created in this way, with the aim of growing enough cells to harvest pluripotent stem cells for research.

There is currently a controversy raging worldwide over the use of embryonic stem cells. The issue is one of ethics and the definition of life. Embryonic stem cells come from many sources, including aborted fetuses, umbilical cords, and embryos from in vitro fertilization clinics. The report about the cloned human embryonic cells added to the controversy. The U.S. government has banned government funding for stem-cell research, but it allows research to continue on all existing embryonic cell lines. Some big questions that people will have to answer are the following: Do a few cells created by therapeutic cloning of your own somatic cells constitute life? If these cells do constitute life, do they have the same rights as a human being conceived naturally? If it were possible, should someone be allowed to grow his or her own therapeutic clone into an adult?

14.8 | What Is the Biochemistry of Cancer?

Cancer is one of the leading causes of death in humans. It is characterized by cells that grow and divide out of control, often spreading to other tissues and causing them to become cancerous. Some estimates suggest that a third of all humans will get cancer during their lifetimes, so it is clearly a disease that is important for everyone to understand. However, the older a person gets, the more likely she or he is to get cancer. A 70-year-old is about 100 times more likely to get cancer than a 20-year-old.

The Mark of a Cancer Cell

All life-threatening cancers have at least six characteristics in common, and multiple problems must occur in a cell before it becomes cancerous. That may be why, even though cancer is common, most people will still grow to old age and not get cancer. First, cancer cells will continue to grow and divide in situations in which normal cells would not. Most cells must receive a growth-chemical signal, but cancer cells manage to keep growing without such signals. Second, cancer cells will continue to grow even when the neighboring cells send out "stop-growth" signals. For example, normal cells stop growing when compressed by other cells. Somehow, tumors manage to avoid this. (Figure 14.33 shows a tumor cell expanding and squeezing against neighboring tissue.) Third, cancer cells manage to keep going and avoid a "self-destruct"

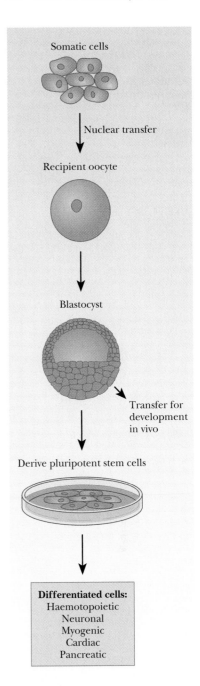

▲ FIGURE 14.32 Reprogramming a somatic nucleus. When transplanted into an oocyte, a somatic nucleus may respond to the cytoplasmic factors and be reprogrammed back to totipotency. These cytoplasmic factors erase the molecular memory of the somatic cells. Such cells can then be used to harvest pluripotent stem cells or to transfer a blastocyst into a carrier and develop an organism in vivo. (*Taken from Surani, M. A. Nature,* **414,** *122–127 (2001).*)

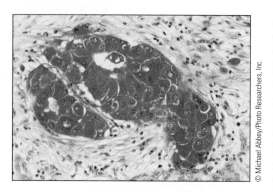

▲ FIGURE 14.33 Tumor cells. Normal cells stop growing when they are squeezed by other cells. This tumor cell continues dividing and growing even though it is being squeezed by the adjacent tissue.

signal that usually occurs when DNA damage has occurred. Fourth, they are able to co-opt the body's vascular system, causing the growth of new blood vessels to supply the cancerous cells with nutrients. Fifth, they are essentially immortal. Normal cells can divide only for a finite number of times, usually in the 50–70 range. However, cancer cells and tumors are able to divide far more often than that. The sixth characteristic is the most lethal: While cells that exhibit the first five characteristics can be a problem, it is the fact that cancer cells have the ability to break loose, to travel to other parts of the body, and to create new tumors that makes them lethal. This process is called **metastasis.** Stationary tumors can often be removed by surgery. However, once a cancer starts spreading, it is almost impossible to stop. Of every ten deaths due to cancer—including a high percentage of lung, colon, and breast cancers—nine of them are due to cancers that metastasized.

What Causes Cancer?

One often hears of many things that cause cancer. Smoking causes cancer. Radiation causes cancer. Asbestos or grilling meat causes cancer. However, these things cannot truly be the ultimate cause, although they may play a role. The real cause may be a combination of insults to the cell that leads to it turning malignant. Cancer is ultimately a DNA disease. It has its roots in changes in the DNA inside a cell. Somehow these changes cause the loss of control of division and the other characteristics described above.

The changes in the DNA cause changes to specific proteins that are responsible for controlling the cell cycle. Most mutations of DNA affect two types of genes. The first is called a **tumor suppressor,** a gene that makes a protein that restricts the cell's ability to divide. If a mutation damages the gene for a tumor suppressor, then the cell will have lost its brakes and will divide out of control. The second type of gene, called an **oncogene,** is one whose protein product stimulates growth and cell division. Mutations of the oncogene cause it to be permanently active. Scientists are still looking for changes to genes that are direct causes of cancer. So far, more than 100 oncogenes and 15 tumor-suppressor genes have been found that have been linked to cancer.

Oncogenes

An oncogene is a gene that has been implicated in cancer. The root word, *"onco,"* means cancer. In 1911, a scientist named Peyton Rous demonstrated that solutions taken from chicken carcinomas could infect other cells. This was the first discovery of tumor viruses, and Rous was given the Nobel prize in 1966 for his discovery. The virus was called the **Rous sarcoma virus,** and it was the first retrovirus shown to cause cancer. The gene that was specific to the cancer is called *v-src,* for viral sarcoma. It was found that this gene encodes a protein that causes transformation of the host cell into a cancer cell. Thus, the gene was given the name *oncogene.* The protein was called **pp60**^{src}, which stands for a phosphoprotein of 60,000 molecular weight from the sarcoma virus (*src*).

However, it was later found that the sequence of the gene was very similar to that of a normal gene in eukaryotes. These genes are called **proto-oncogenes.** Many proto-oncogenes are normal and necessary for proper growth and development in eukaryotic cells. However, some transforming event causes the proto-oncogene to lose control. Sometimes this is due to a viral infection. In other cases, the event that causes a proto-oncogene to become an oncogene is not known. Table 14.3 shows some proto-oncogenes implicated in human tumors. Many of these genes are involved in signal-transduction pathways that affect the transcription of genes that speed up cell division. In Chapter 11, we looked at the control of transcription in eukaryotes and noted that there were many signaling pathways that were routed through the CBP/P300

Table 14.3

A Representative List of Proto-Oncogenes Implicated in Human Tumors

Proto-Oncogene	Neoplasm(s)
abl	Chronic myelogenous leukemia
*erb*B-1	Squamous cell carcinoma; astrocytoma
*erb*B-2 *(neu)*	Adenocarcinoma of breast, ovary, and stomach
myc	Burkitt's lymphoma carcinoma of lung, breast, and cervix
H-*ras*	Carcinoma of colon, lung, and pancreas; melanoma
N-*ras*	Carcinoma of genitourinary tract and thyroid; melanoma
ros	Astrocytoma
src	Carcinoma of colon
jun	⎫
fos	⎭ Several

Adapted from Bishop, J. M. 1991, Molecular themes in oncogenesis, *Cell* **64**: 235–248.

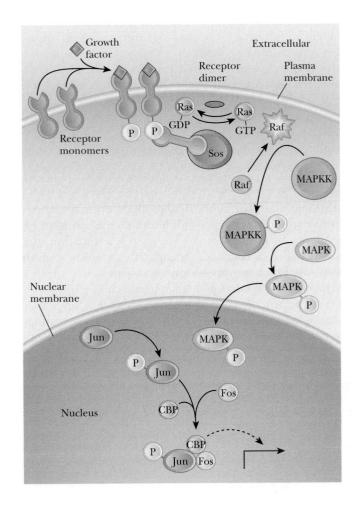

◀ **FIGURE 14.34** MAP kinase signal transduction. Signal transduction starts when a growth factor (blue) binds to a receptor monomer (red) on the cell membrane. The receptor is a tyrosine kinase, which then phosphorylates its partner receptor. The phosphorylated receptor is then recognized by GRB2 (light purple), which binds to the Ras exchanger Sos (blue). Sos is activated to exchange GDP for GTP on Ras (pink), activating it. Ras moves Raf (tan) to the cell membrane, where it becomes active. Raf phosphorylates MAP kinase kinase, which then phosphorylates MAP kinase (yellow). MAP kinase (MAPK) enters the nucleus and phosphorylates Jun (light green). Phosphorylated Jun binds to Fos and CBP and transcription is activated. (*Reprinted by permission from* Molecular Biology, *by R. F. Weaver, 2nd ed., p. 375, McGraw-Hill.*)

coactivator (see Figure 11.22). One of these pathways involved **mitogen-activated protein kinase (MAPK)** and a transcription factor called **AP-1.** To understand the nature of many of the oncogenes shown in Table 14.3, we must take another look at this pathway.

The process starts when an extracellular signal binds to a receptor on the cell membrane (see Figure 14.34). This receptor is a tyrosine kinase that dimerizes, and then each part phosphorylates the other. Once phosphorylated, the receptors are bound by an adaptor molecule, a protein called **GRB2** (pronounced *"grab two"*), which has a phosphotyrosine binding domain that is very similar to a domain found in the pp60^{src} protein. The other end of GRB2 binds to a protein called **Sos.**

At this point, there is an interaction with a very important 21-kilodalton protein. This protein, called **p21^{ras}** or just **Ras,** is involved in about 30% of human tumors. The designation Ras comes from *Rat sarcoma*, the original tissue in which it was discovered. The Ras family of proteins are GTP-binding proteins. In their resting state, they are bound to GDP. After the cell signal, the Sos replaces the GDP for GTP. Intrinsic hydrolysis of the GTP returns the protein to its inactive state, but this process is slow. Proteins known as GTPase-activating proteins (GAPs) speed up this hydrolysis and are involved in the control of the Ras proteins. GAPs inactivate Ras by accelerating the hydrolysis of GTP. Oncogenic forms of Ras have impaired GTPase activity and are insensitive to GAPs, thus leaving them bound to GTP, which causes them to stimulate cell division continually.

Although Ras mutations have been some of the most studied mutations leading to cancer, we can see that Ras is found rather early in the process that ultimately leads to cell division. Activated Ras attracts another protein called **Raf,** which then phosphorylates serines and threonines on **mitogen-activated protein kinase kinase (MAPKK).** As one can guess from its name, this enzyme then phosphorylates mitogen-activated protein kinase (MAPK). This enzyme enters the nucleus and phosphorylates a transcription factor called **Jun.** Jun binds to another transcription factor called **Fos.** Together, Jun and Fos make up the transcription factor that we saw before called AP-1, which binds to CBP and stimulates the transcription of genes that lead to rapid cell division. As we can see in the table, *jun* and *fos* oncogenes code for these proteins.

Tumor Suppressors

There are also many human genes that produce proteins called **tumor suppressors.** Tumor suppressors act as inhibitors to transcription of genes that would cause increased replication. When there is a mutation in any of these suppressors, replication and division become uncontrolled and tumors result. Table 14.4 lists some human tumor-suppressor genes.

A 53-kDa protein designated **p53** has become the focus of feverish activity in cancer research. Mutations in the gene that codes for p53 are found in more than half of all human cancers. When the gene is operating normally, it acts as a tumor suppressor; when it is mutated, it is involved in a wide variety

Table 14.4	
Representative Tumor-Suppressor Genes Implicated in Human Tumors	
Tumor-Suppressor Gene	**Neoplasm(s)**
RBI	Retinoblastoma; osteosarcoma; carcinoma of breast, bladder, and lung
p53	Astrocytoma; carcinoma of breast, colon, and lung; osteosarcoma
WT1	Wilms' tumor
DCC	Carcinoma of colon
NF1	Neurofibromatosis type 1
FAP	Carcinoma of colon
MEN-1	Tumors of parathyroid, pancreas, pituitary, and adrenal cortex

Adapted from Bishop, J. M. 1991. Molecular themes in oncogenesis. *Cell* **64**, 235–248.

of cancers. By the end of 1993, mutations in the *p53* gene had been found in 51 types of human tumors. The role of p53 is to slow down cell division and to promote cell death (apoptosis) under certain circumstances, including when DNA is damaged or when cells are infected by viruses.

It is known that p53 binds to the basal transcription machinery (one of the TAFs bound to TFIID; see Chapter 11). When cancer-causing mutations occur in p53, it can no longer bind to DNA in a normal fashion. The mode of action of p53 as a tumor suppressor is twofold. It is an activator of RNA transcription; it "turns on" the transcription and translation of several genes. One of them, *Pic1*, encodes a 21-kDa protein, **P21,** that is a key regulator of DNA synthesis and thus of cell division. The P21 protein, which is present in normal cells but is missing from (or mutated in) cancer cells, binds to the enzymes known as cyclin-dependent protein kinases (CDKs), which, as their name implies, become active only when they associate with proteins called cyclins. Recall, from Section 10.6, that cell division depends on the activity of cyclin-dependent kinases. Some of the oncogenes seen above work in such a way that the result is an overproduction of the CDK proteins, which keeps the cells dividing continuously. Normal levels of p53 protein cannot turn these genes off in cancer cells, but they could do this in normal cells. In normal cells, the result is that the cell cycle remains in the state between mitosis (in which cells divide) and the replication of DNA for the next cell division. DNA repair can take place at this stage. If the attempts at DNA repair fail, the p53 protein may trigger apoptosis, the programmed cell death characteristic of normal cells, but not of cancer cells.

The important point is that two different mechanisms are operating here. One is analogous to the brakes failing in your car (inadequate or defective p53 protein) and the other (overproduction of CDKs) is equivalent to the accelerator sticking in the open position—two opposite mechanisms with the same result: the car crashes.

A number of factors come together in explaining the variety of diseases we call cancer. Mutations of DNA lead to changes in the proteins that control cell growth, either by directly causing cell division or allowing it to occur by default. Still other mutations interfere with DNA repair. The possibility of finding new cancer therapies—and perhaps even cancer cures—is enhanced by understanding these contributing factors and how they affect each other.

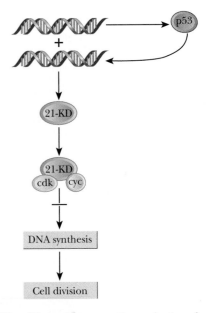

▲ The p53 protein turns on the production of a 21-kDa protein. This protein binds to complexes of cyclin-dependent kinases (CDKs) and cyclins. The result of binding is inhibition of DNA synthesis and cell growth. (*Adapted from* Science, *Figure 1, Vol. 262, 1993, p. 1644, by K. Sutliff,* © 1993 *by the AAAS.*)

Viruses and Cancer

The original work by Rous showed how viruses could cause cancer in certain situations. The close homology between the oncogene sequence found in some

Text continues on page 410.

Biochemical Connections

If It Isn't One Thing, It's Another

The number of cancers and tumors related to a mutation of the *p53* gene has risen every year to the point that most researchers will say that "most" human cancers are associated with problems in the p53 pathway. Therefore, it was only a matter of time before scientists tried to use p53 protein for a preemptive strike against cancer. The logic is clear: if a lack of p53 protein causes cancer, having more p53 might even prevent it.

To study this possibility, researchers created a strain of mice that had an extra copy of the *p53* gene. These "super *p53*" mice

did, in fact, show a greater resistance to DNA damage, and they were less susceptible to carcinogen-induced tumors than a wild-type mouse. Ideas about a new fountain of health were short-lived, however, when it turned out that these mice also aged much more quickly than their wild-type litter mates. This raises the possibility that aging is a necessary evil that is a safeguard against cancer. (See the articles by Ferbeyere and by Straus in the Annotated Bibliography at the end of this chapter.)

Biochemical Connections

Viruses Helping Cure Cancer

As we have seen in this chapter, viruses come in many types and cause many diseases. Viruses can be very specific to a single cell type because they rely upon a protein receptor on the cell to gain entry. Liver cells display receptors that nerve cells do not, and vice versa. Oncologists (doctors who treat cancer) have treated cancer for years with techniques such as radiation therapy and chemotherapy. These techniques attempt to target cancer cells, but, in the end, are very destructive to other cells as well. In some sense, the goal of chemotherapy is to kill the cancer before the treatment kills the patient. If doctors could come up with a treatment that would be completely specific to cancer cells, it would go a long way, both toward stopping the cancer and toward making the patient's life more comfortable during the treatment. This was another chance for researchers to find something helpful about viruses.

In the 1990s a new type of treatment for cancer, called **virotherapy,** was begun. This technique was shown to target human

tumor cells grafted onto mice. The treatment eliminated the human tumors. The virus of choice was an adenovirus, which we saw in the section on gene therapy. There are two strategies for virotherapy. One is to use the virus to attack and kill the cancer cell directly. The second is to have the virus ferry in a gene to the cancer cell that will make the cell more susceptible to a chemotherapy agent.

One of the biggest challenges in virotherapy is to make sure the virus specifically targets the cancer cell. The common adenovirus is not specific for cancer cells, so, in order to use it for virotherapy, other techniques must also be employed. One of these is called **transductional targeting.** In this technique, antibodies are attached to the virus. These antibodies are created so that they will target the cancer cell (see figure below). In this way, the normally indiscriminant adenovirus will attack only cancer cells. Once inside, the virus reproduces and eventually lyses the cell.

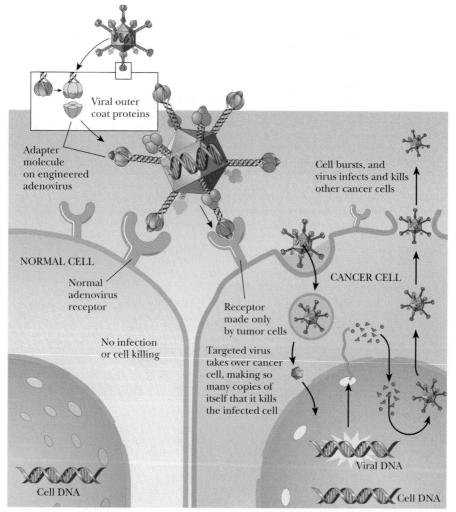

▲ Transductional targeting in virotherapy. Viruses, such as adenovirus, are used to infect and destroy cancer cells selectively. Spikes on the adenovirus are mutated so that they recognize unique receptors on cancer cells. The virus selectively infects and lyses the cancer cells. (© 2003 *Terese Winslow.*)

Another approach is called **transcriptional targeting.** With this technique, the replication genes for adenovirus are placed after a promoter that is specific for a cancer cell. For example, skin cells make much more of the pigment melanin than other cells. Therefore, the genes for enzymes that make melanin are turned on more often in skin cells than in other cells (see figure below). Adenovirus can be engineered to have the promoter for the melanin-producing enzyme near the genes for virus replication. In skin cells that are cancerous, these promoters are triggered more often, so adenovirus replicates much quicker in skin-cancer cells, killing them specifically. Similar techniques have been used to target liver-cancer cells and prostate-cancer cells.

The other basic strategy is to have the virus ferry in a gene that will make the cancer cell more susceptible to chemotherapy. One such system uses a virus that targets rapidly dividing cells. Inside these cells, and only in these cells, the gene carried by the virus converts an innocuous pro-drug into an anticancer drug. These viruses are sometimes called "smart viruses" for their ability to select only the cancer cells. They then allow drugs to be used that are not harmful to normal cells.

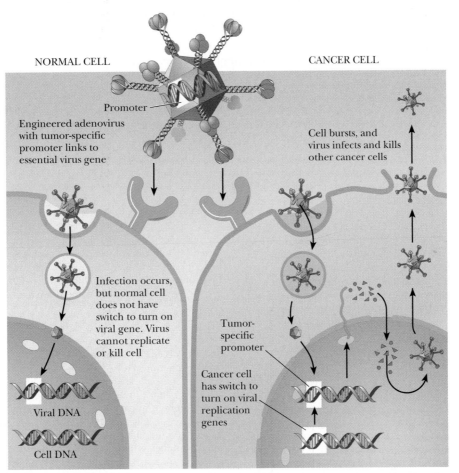

▲ Transcriptional targeting in virotherapy. A tumor-specific promoter is engineered into the adenovirus. The adenovirus will infect many cells, but it is only activated to replicate in the cancer cell. (© 2003 Terese Winslow.)

Essential Information

One of the most common causes of death in humans is cancer. Cancer cells grow and divide in an uncontrolled manner. There are many control points in cell division, and, when enough of them go wrong, the cell becomes cancerous. Many cancers have been linked to oncogenes, genes that cause cancer when they are overexpressed. Many cancers have also been linked to defects in genes that make tumor suppressors. A single protein, p53, a tumor suppressor, has been linked to almost half of the known human cancers.

viruses with the proto-oncogene sequences in the mammalian genome has led many researchers to theorize that the oncogenes may have been of mammalian origin. It is possible that, in the course of repeated infections and travels, the virus picks up pieces of DNA from a host and delivers another piece of DNA to a host. In the course of the rapid mutation that occurs in retroviruses, these proto-oncogenes could be mutated to a form that is oncogenic.

Retroviruses that cause cancer in humans are known; some forms of leukemia (caused by HTLV-I and HTLV-II, which infect T cells of the immune system) are well-known examples, as well as cervical cancer caused by cervical papillomavirus. Theoretically, any retrovirus that inserts its DNA into the host chromosome could accidentally disable a tumor-suppressor gene or enable an oncogene by insertion of a strong promoter sequence near a proto-oncogene. One of the biggest fears of using in vivo delivery techniques for human gene therapy (see Section 14.4) is that the viral DNA inserted into a human chromosome might become incorporated into an otherwise healthy tumor-suppressor gene. This would potentially solve one of the individual's problems by giving him or her a functional gene he or she was missing while causing an even greater problem. This happened, unfortunately, in 2003, where researchers in France were using viral-gene therapy to treat patients with X-linked SCID (see Section 14.4). In nine of eleven cases, the viral-gene therapy was able to restore the immune systems of the patients. However, in two cases, patients developed leukemia. It was later found that the virus had inserted itself, in each case, near a gene that has been found to be a leukemia oncogene. This was a tragic setback in viral-gene therapy, and now many government agencies are discussing the future of such therapy.

Summary

14.1 What Are Viruses? Viruses are simple genes, made up of RNA or DNA, that infect cells and take over their replication, transcription, and translation machinery. Viruses are known to cause many diseases, and they may be very specific to a particular species and cell type. Viruses enter the cell by binding to specific receptors on the cell. Once inside the cell, the virus may replicate, form new viruses, and burst the cell. The virus may also hide its DNA by incorporating it into the host's DNA. Viruses are characterized by their structure, their type of nucleic acid, whether it is single- or double-stranded, and their mode of infection.

14.2 What Virus Causes Severe Acute Respiratory Syndrome (SARS)? In 2003, a new virus was discovered, called the SARS virus, which is a coronavirus. Coronaviruses have a unique membrane with spike-like protuberances, which gives them their name. They had been known for a long time, but they had never been known to infect humans. SARS was spreading rapidly and was very dangerous, leading to hundreds of deaths in a short period. Fortunately, the world's scientists worked together to nearly eradicate the disease in just a few months. The natural host appears to be the Asian civet, but this conclusion is still tentative. A few individuals came down with the virus after it was thought to have been eliminated.

14.3 What Is Unique about Retroviruses? Retroviruses have a genome based on RNA. When they infect a cell, their RNA is turned into DNA. The DNA is then incorporated into the host's DNA genome as part of the replication cycle for the virus.

14.4 How Are Viruses Used in Gene Therapy? Gene therapy is a technique that makes a potentially good use of viruses. They are used as vectors to carry DNA into cells that are missing a functional copy of a necessary gene. The virus is constructed to remove its ability to replicate independently, and therapeutic genes

are inserted. The viruses are replicated in special cells and then collected for infection of the patient's cells. Several diseases are currently being treated with gene therapy. While many scientists are optimistic, they must also be very careful. A few patients have died as a result of the therapy, and two have come down with leukemia as a result of the treatment.

14.5 How Does the Immune System Defend the Body? Vertebrates have a complicated and elegant system of defense called the immune system. One type of immunity, called innate immunity, consists of physical barriers, such as skin, and cellular warriors, such as dendritic cells. This system is always present and waiting to attack invading organisms or even cancerous cells. Another type of immunity, called acquired immunity, is based on two types of T cells (killer T cells and helper T cells) and on B cells. These cells are generated randomly with receptors that can be specific for an unimaginable number of antigens. When these cells encounter their specific antigens, they are stimulated to multiply, exponentially increasing the number of cells that can fight the invading organism. Acquired immune cells also leave behind memory cells so that, if the same pathogen is seen again, the body is faster to eliminate it.

Immune cells must also be able to recognize self from nonself. T cells and B cells are conditioned, in their early stages of development, not to recognize proteins from that individual. In some cases, this system breaks down, and a person may be attacked by his or her own immune system, which may lead to an autoimmune disease.

14.6 How Does Human Immunodeficiency Virus Cause AIDS? Much work has gone into the study of human immunodeficiency virus (HIV), the cause of AIDS. HIV is a retrovirus, so it must incorporate its genetic information as DNA inside the host's DNA. Its target is the helper T cell, which it weakens and eventually kills. With fewer T cells, the infected person slowly loses the ability to fight off infections, eventually succumbing to an opportunistic infection, such as pneumonia. HIV is particularly difficult to fight because it changes frequently and has many tricks to thwart the immune system.

14.7 Why Are Stem Cells Special?

Stem cells are undeveloped cells that can turn into any type of cell in the body. While they can be found in adults, they are present in much higher concentrations in embryos and fetuses. They have shown the ability to become a specific type of cell if placed in a tissue of that type. This has led to hope for the cure of many diseases, such as heart disease, stroke, muscular dystrophy, and even spinal-cord damage. They are controversial because most of the world's stem-cell lines came from umbilical cords or aborted fetuses.

14.8 What Is the Biochemistry of Cancer?

One in three people will develop cancer in the course of a lifetime. All potentially fatal cancers have several things in common, such as having cells that are immortal, that divide despite "stop growth" signals from nearby cells, that stimulate blood-vessel formation near to themselves, and that spread to other parts of the body. The development of cancer requires multiple breakdowns in normal metabolism. Most cancers have been linked to specific genes called oncogenes or to tumor-suppressor genes. When these genes are mutated, the cell loses the ability to control its replication. There are many classical ways to fight cancer, such as radiation therapy and chemotherapy. Both of these are very hard on healthy cells and, therefore, on the patient. Novel techniques using viruses are now being tried to target cancer cells more directly, and some of these are showing tremendous promise.

Critical Questions to Review

14.1 What Are Viruses?

1. **Fact Check** What is the genetic material of a virus?
2. **Fact Check** Define the following:
   a) virion
   b) capsid
   c) nucleocapsid
   d) protein spike
3. **Fact Check** What determines the family in which a virus is categorized?
4. **Fact Check** How does a virus infect a cell?
5. **Fact Check** What is the difference between the lytic pathway and the lysogenic pathway?
6. **Biochemical Connections** For how long have humans had to endure the flu virus?
7. **Biochemical Connections** What are some of the structural features of a flu virus?
8. **Thought Question** Is there a correlation between the speed of a viral infection and its potential mortality rate? Explain.
9. **Thought Question** If you were going to design a drug to fight a virus, what would be likely targets for the drug design?
10. **Thought Question** Some viruses can undergo lysis or lysogeny even in the same host. What might be a reason for this? Under what conditions might the virus favor the one strategy over the other?
11. **Thought Question** What might be the characteristics of cells of a human who is immune to HIV infection?

14.2 What Virus Causes Severe Acute Respiratory Syndrome (SARS)?

12. **Fact Check** What structural feature distinguishes the SARS virus?
13. **Fact Check** Where did SARS come from?
14. **Thought Question** The SARS genome is about 30,000 bases of coding-strand RNA with a $5'$ cap and a poly-A tail. What might be the advantage of these components in the genome?
15. **Thought Question** Compare SARS with HIV in terms of our ability to fight it. Why has SARS been contained but not HIV?

14.3 What Is Unique about Retroviruses?

16. **Fact Check** What is unique about the lifecycle of a retrovirus?
17. **Fact Check** What enzyme is responsible for the production of viral DNA from a retrovirus?
18. **Fact Check** What are three reasons that retroviruses are studied so much these days?

14.4 How Are Viruses Used in Gene Therapy?

19. **Fact Check** What is meant by gene therapy?
20. **Fact Check** What are the two types of gene therapy?
21. **Fact Check** What types of viruses are used for gene therapy, and how are they manipulated to make them useful?
22. **Fact Check** What are the potential hazards of gene therapy?
23. **Thought Question** What are the considerations for choice of a vector in gene therapy?
24. **Thought Question** Both ADA⁻ SCID and type 1 diabetes are diseases based upon lack of a particular protein. Why is it that the pioneering work on gene therapy has focused on SCID instead of on diabetes?

14.5 How Does the Immune System Defend the Body?

25. **Fact Check** What health conditions are linked to malfunctioning immune systems?
26. **Fact Check** What is innate immunity? What is acquired immunity?
27. **Fact Check** What are the components of innate immunity?
28. **Fact Check** What are the components of acquired immunity?
29. **Fact Check** What is the purpose of a major histocompatibility complex?
30. **Fact Check** What is clonal selection?
31. **Thought Question** Describe the relationship between the innate-immunity system and the acquired-immunity system.
32. **Thought Question** One of the first human proteins cloned was interferon. Why would it be important to be able to produce interferon in a lab?
33. **Thought Question** Describe how the cells of the acquired-immunity system develop so that they do not recognize self-antigens but do recognize foreign antigens.

14.6 How Does Human Immunodeficiency Virus Cause AIDS?

34. **Fact Check** How does HIV cause AIDS?
35. **Fact Check** How does HIV confound the human immune system?
36. **Fact Check** What types of therapy are used to fight AIDS?
37. **Thought Question** Why have vaccines been relatively unsuccessful in stopping AIDS?
38. **Thought Question** What are the structural features of the two types of neutralizing antibodies that have been the most successful at combating AIDS? What makes these antibodies more effective?

14.7 Why Are Stem Cells Special?

39. Fact Check What characteristics are demonstrated by stem cells?

40. Fact Check What are the sources of stem cells?

41. Thought Question What are the arguments for and against the production of embryonic stem cells?

42. Thought Question Why is the epigenetic state of a cell important to whole-animal cloning?

14.8 What Is the Biochemistry of Cancer?

43. Fact Check What characteristics are shown by cancer cells?

44. Fact Check What is a tumor suppressor? What is an oncogene?

45. Fact Check Why are the proteins called p53 and Ras studied so much these days?

46. Fact Check How are viruses related to cancer?

47. Biochemical Connections What is virotherapy?

48. Thought Question Why is it inaccurate to say, "Smoking causes cancer"?

49. Thought Question Describe the difference between a tumor suppressor and an oncogene with respect to the actual causes of cancer.

50. Thought Question Describe the relationships between Ras, Jun, and Fos.

Biochemistry☰Now™

Assess your understanding of this chapter's topics with additional quizzing and tutorials at **http://now.brookscole.com/campbell5**

Annotated Bibliography

Aldhous, P. Can they rebuild us? *Nature* **410**, 622–625 (2001). [An article about how stem cells might lead to therapies to rebuild tissues.]

Bakker, T. C. M, and M. Zbinden. Counting on Immunity. *Nature* **414**, 262–263 (2001). [An article about how, in some species, individuals select mates that have major histocompatibility proteins as dissimilar to their own as possible.]

Banchereau, J. The Long Arm of the Immune System. *Sci. Amer.* **187(5)**, 52–59 (2002). [An in-depth article about dendritic cells.]

Batzing, Barry. *Microbiology: An Introduction.* Brooks/Cole Publishing, Pacific Grove, Calif. (2002). [Basic textbook in microbiology, including chapters on viruses.]

Check, E. Trial Suggests Vaccines Could Aid HIV Therapy. *Nature* **422**, 650 (2003). [An article about the efficacy of using antibodies to fight HIV.]

Check, E. Back to Plan A. *Nature* **423**, 912–914 (2003). [A discussion of the different strategies for using antibodies in the fight against AIDS.]

Cohen, J. Confronting the Limits of Success. *Science* **296**, 2320–2324 (2002). [An article about the problems associated with finding vaccines and other treatments for AIDS.]

Cohen, J. Escape Artist par Excellence. *Science* **299**, 1505–1508 (2003). [An article about how HIV confounds the immune system.]

Donovan, P. J., and G. Gearhart. The End of the Beginning for Pluripotent Stem Cells. *Nature* **414**, 92–97 (2001). [A review of the status of stem-cell research.]

Enserink, M. China's Missed Chance. *Science* **301**, 294–296 (2003). [An article about the search for the cause of SARS and how the process got delayed in China.]

Ezzel, C. *Sci. Amer.* **286 (6)**, 40–45 (2002). [An article about potential AIDS vaccines.]

Ferbeyere, G., and S. W. Lowe. The Price of Tumour Suppression. *Nature* **415**, 26–27 (2002). [An article about the tradeoffs between aging and tumor suppression.]

Gibbs, W. W. Roots of Cancer. *Sci. Amer.* **289 (1)**, 57–65 (2003). [An in-depth article about the many causes of cancer.]

Greene, W. AIDS and the Immune System. *Sci. Amer.* **269 (3)**, 98–105 (1993). [A description of the HIV virus and its life cycle in T cells.]

Holmes, K. V., and L. Enjuanes. The SARS Coronavirus: A Postgenomic Era. *Science* **300**, 1377–1378 (2003). [A review of the status of the research on the SARS virus.]

Janssen, E. M., E. E. Lemmens, T. Wolfe, U. Christen, M. G. von Herrath, and S. P. Schoenberger. CD4+ T Cells Are Required for Secondary Expansion and Memory in CD8+ T Lymphocytes. *Nature* **421**, 852–855 (2003). [An in-depth article about the research that led to the understanding of the CD4/CD8 cell relationship in memory.]

Jardetzky, T. Conformational Camouflage. *Nature* **420**, 623–624 (2002). [An article about how HIV can hide from antibodies.]

Kaech, S. M., and R. Ahmed. CD8 T Cells Remember with a Little Help. *Science* **300**, 263–265 (2003). [An article about how memory develops in immune cells.]

Kaiser, J. Seeking the Cause of Induced Leukemias in X-SCID Trial. *Science* **299**, 495 (2003). [An article about the search for information about why two patients receiving gene therapy developed leukemia.]

Knight, J. Researchers Get to Grips with Cause of Pneumonia Epidemic. *Nature* **422**, 547–548 (2003). [A progress report on the characterization of the SARS virus.]

Marra, M. A., et. al. The Genome Sequence of the SARS-Associated Coronavirus. *Science* **300**, 1399–1404 (2003). [The complete sequence information on the SARS virus.]

McCune, J. M. The Dynamics of CD4+ T-Cell Depletion in HIV Disease. *Nature* **410**, 974–979 (2001). [An in-depth article about T cells and how they are affected by HIV.]

McMichael, A. J., and S. L. Rowland-Jones. Cellular Immune Responses to HIV. *Nature* **410**, 980–987 (2001). [An in-depth review of the immune response to HIV infection.]

Nettelbeck, D. M., and D. T. Curiel. Tumor-Busting Viruses. *Sci. Amer.* **289 (4)**, 68–75 (2003). [An article describing a new use for viruses as specific weapons against cancer.]

Normile, D. Battling SARS on the Frontlines. *Science* **300**, 714–715 (2003). [A news brief on the fight against SARS.]

Normile, D., and M. Eserink. Tracking the Roots of a Killer. *Science* **301**, 297–299 (2003). [An article about the search for the origins of SARS in China.]

Nossal, G. J. V. A Purgative Mastery. *Nature* **412**, 685–686 (2001). [An article about the diversity of the immune system.]

Parham, P. The Unsung Heroes. *Nature* **423**, 20 (2003). [An article about the innate immune system.]

Piot, P. et al. The Global Impact of HIV/AIDS. *Nature* **410**, 968–973 (2001). [A summary of the social and economic impacts of AIDS worldwide.]

Ragupathi, G., T. Park, S. Zhang, I. Kim, L. Graber, S. Adluri, K. Lloyd, S. Danishefsky, and P. Livingston. Immunization of Mice with a Fully Synthetic Globo H Antigen Results in Antibodies against Human Cancer Cells: A Combined Chemical–Immunological Approach to the Fashioning of an Anticancer Vaccine. *Angew. Chem. Int. Ed. Engl.* **36** (1/2), 125–128 (1997). [Synthetic organic chemists and immunologists have teamed up to produce an anticancer vaccine.]

Reusch, T. B. H, M. A. Haberli, P. B. Aeschliman, and M. Milinski. *Nature* **414,** 300–302 (2001). [An article about how female sticklebacks select mates based on diversity of MHC proteins.]

Serbina, N. V., and E. G. Pamer. Giving Credit Where Credit Is Due. *Science* **301,** 1856–1857 (2003). [An article about the importance of dendritic cells.]

Soares, C. Caught Off Guard. *Sci. Amer.* **288 (6),** 18–19 (2003). [In-depth article about how China was not prepared for an epidemic.]

Sprent, J., and D. F. Tough. T Cell Death and Memory. *Science* **293,** 245–247 (2001). [An article about how T cells are selected.]

Straus, E. Cancer-Stalling System Accelerates Aging. *Science* **295,** 28–29 (2002). [An article about the apparent link between aging and cancer protection.]

Sussman, M. Cardiovascular Biology: Hearts and Bones. *Nature* **410,** 640–641 (2001). [An article about how stem cells can be used to regenerate tissue.]

Weiss, R. A. Guilliver's Travels in HIVland. *Nature* **410,** 963–967 (2001). [An excellent review of the current information on HIV and AIDS.]

Werlen, G., B. Hausmann, D. Naeher, and E. Palmer. Signaling Life and Death in the Thymus: Timing Is Everything. *Science* **299,** 1859–1863 (2003). [An article about how immune cells must be selected that will recognize foreign molecules but not self.]

CHAPTER 15

© David Muench

The potential energy of the water at the top of a waterfall is transformed into kinetic energy in spectacular fashion.

Critical Questions

Biochemistry⊘Now™

Test yourself on these Critical Questions at the BiochemistryNow website at **http://now .brookscole.com/campbell5**

The Importance of Energy Changes and Electron Transfer in Metabolism

Life processes require that molecules taken in as nutrients be broken down to extract energy and also to provide the building blocks to create new molecules. To maintain a steady state, a living organism needs a constant supply of energy from without to bring order to the constant turmoil within. The energy-extraction process takes place in a series of many small steps in which electron donors transfer energy to electron acceptors. These oxidation–reduction reactions are fundamental to the extraction of energy from molecules such as glucose. The principal electron carriers are NADH (the reduced form of nicotinamide adenine dinucleotide) and NAD^+, its oxidized form. NADH is oxidized to NAD^+ when it loses two electrons, and NAD^+ is reduced to NADH when it accepts two electrons. Two electrons from each NADH and two protons join an oxygen atom to form H_2O in the complete oxidation of glucose. Energy generated in this reaction is conserved by transforming "lower-energy" ADP to "higher-energy" ATP. The ADP–ATP system is like a very active checking account in which deposits and withdrawals are in a steady state. The energy from ATP is never used up, only transferred in the cell's myriad chemical reactions that require energy.

15.1 | What Are Standard States for Free-Energy Changes?

We have already seen how the lowering of energy, which really means dispersal on the molecular level, is spontaneous in the thermodynamic sense. In this chapter, we are going to see how energetic considerations apply to metabolism. We are going to be comparing so many different processes that it will be useful to have a benchmark against which to make those comparisons.

We can define *standard conditions* for any process and then use those standard conditions as the basis for comparing reactions. The choice of standard conditions is arbitrary. For a process under standard conditions, all substances involved in the reaction are in their **standard states,** in which case they are also said to be at *unit activity*. For pure solids and pure liquids, the standard state is the pure substance itself. For gases, the standard state is usually taken as a pressure of 1.00 atmosphere of that gas. For solutes, the standard state is usually taken as 1.00 molar concentration. Strictly speaking, these definitions for gases and for solutes are approximations, but they are valid for all but the most exacting work.

For any general reaction

$$a\text{A} + b\text{B} \rightarrow c\text{C} + d\text{D}$$

we can write an equation that relates the free-energy change (ΔG) for the reaction under *any* conditions to the free-energy change under *standard* conditions ($\Delta G°$); the superscript ° refers to standard conditions. This equation is

$$\Delta G = \Delta G° + RT \ln\frac{[\text{C}]^c[\text{D}]^d}{[\text{A}]^a[\text{B}]^b}$$

In this equation, the square brackets indicate molar concentrations, R is the gas constant (8.31 J mol^{-1} K^{-1}), and T is the absolute temperature. The notation ln refers to natural logarithms (to the base e) rather than logarithms to the base 10, for which the notation is log. This equation holds under all circumstances; the reaction does not have to be at equilibrium. The value of ΔG under a given set of conditions depends on the value of $\Delta G°$ and on the concentration of reactants and products (given by the second term in the equation). Most biochemical reactions are described in terms of $\Delta G°$, which is the ΔG under standard conditions (1.00 M concentration for solutes). There is only one $\Delta G°$ for a reaction at a given temperature.

When the reaction is at equilibrium, $\Delta G = 0$, and thus

$$0 = \Delta G° + RT \ln \frac{[C]^c[D]^d}{[A]^a[B]^b}$$

$$\Delta G° = -RT \ln \frac{[C]^c[D]^d}{[A]^a[B]^b}$$

The concentrations are now equilibrium concentrations, and this equation can be rewritten

$$\Delta G° = -RT \ln K_{eq}$$

where K_{eq} is the equilibrium constant for the reaction. We now have a relationship between the equilibrium concentrations of reactants and products and the standard free-energy change. After we have determined the equilibrium concentrations of reactants by any convenient method, we can calculate the equilibrium constant, K_{eq}. We can then calculate the standard free-energy change, $\Delta G°$, from the equilibrium constant.

15.2 What Is a Modified Standard State for Biochemical Applications?

We have just seen that the calculation of standard free-energy changes includes the stipulation that all substances be in standard states, which for solutes can be approximated as a concentration of 1 M. If the hydrogen-ion concentration of a solution is 1 M, the pH is zero. (Recall that the logarithm of 1 to any base is zero.) The interior of a living cell is, in many respects, an aqueous solution of the cellular components, and the pH of such a system is normally in the neutral range. Biochemical reactions in the laboratory are usually carried out in buffers that are also at or near neutral pH. For this reason, it is convenient to define, for biochemical practice, a modified standard state, one that differs from the original standard state only by the change in hydrogen-ion concentration from 1 M to 1×10^{-7} M, implying a pH of 7. When free-energy changes are calculated on the basis of this modified standard state, they are designated by the symbol $\Delta G°'$ (spoken "delta G zero prime"). The Biochemical Connections box describes other specific applications of thermodynamics to living organisms.

Practice Session

Use of Equilibrium Constants to Determine $\Delta G°'$

Let us assume that the relative concentrations of reactants have been determined for a reaction carried out at pH 7 and 25°C (298 K). Such concentrations can be used to calculate an equilibrium constant, K_{eq}, which, in turn, can be used to determine the standard free-energy change, $\Delta G°'$, for

Biochemical Connections

Biochemical Thermodynamics

Gibbs free energy, ΔG, is perhaps the most suitable way to measure energy changes in living systems because it measures *the energy available to do work at constant temperature and pressure,* which describes the living state. Even cold-blooded organisms are at constant temperature and pressure at any given point in time; any temperature and pressure changes are slow enough not to affect measurements of ΔG.

Spontaneity and Reversibility

The concept of spontaneity can be confusing, but it merely means that a reaction can occur without added energy. This is similar to water held behind a dam at the top of a hill, which has the potential energy to flow downhill, but it will not do so unless someone opens the dam. Because water flows only downhill, that is the direction with a negative value of the free-energy change $(-\Delta G)$; pumping water uphill is nonspontaneous (requires energy) and has a positive value of the free-energy change $(+\Delta G)$. If the free-energy change is only 1 kcal mol^{-1} (about 4 kJ mol^{-1}) in either direction, then the reaction is considered to be freely reversible. The reaction can readily go in either direction. If one adds reactants or removes products, the reaction shifts to the right; if one removes reactants or adds products, the reaction shifts to the left. This is a key aspect of a number of metabolic pathways; many reactions in the middle of the pathway are likely to be freely reversible. This means that the same enzymes can be used whether the pathway is in the process of breaking down a substance or of forming the substance. In reversible metabolic pathways, it is often just the reactions at the ends that are irreversible, and these reactions can be turned on or off to turn the whole pathway on or off, or even to reverse it.

Driving Endergonic Reactions

Reactions can sometimes be coupled together. This occurs when the phosphorylation of glucose is coupled to the hydrolysis of one phosphate group of ATP. Of course, there are not really two reactions going on; the enzyme merely transfers the phosphate from the ATP directly to the glucose (see Section 15.6). We can think of the phosphorylation of glucose and the hydrolysis of ATP as two parts of the same reaction. We can then add them together to determine the overall energy change and make sure that, overall, it is exergonic.

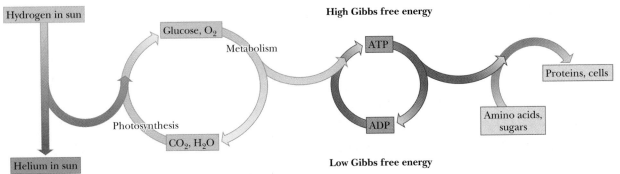

▲ The synthesis of glucose and other sugars in plants, the production of ATP from ADP, and the elaboration of proteins and other biological molecules are all processes in which the Gibbs free energy of the system must increase. They occur only through coupling to other processes in which the Gibbs free energy decreases by an even larger amount. There is a local decrease in entropy at the expense of higher entropy of the universe.

the reaction. A typical reaction to which this kind of calculation can be applied is the hydrolysis of ATP at pH 7, yielding ADP, monohydrogen phosphate ion (written as P_i), and H^+ (the reverse of a reaction we have already seen):

$$ATP + H_2O \rightleftharpoons ADP + P_i + H^+$$

$$K'_{eq} = \frac{[ADP][P_i][H^+]}{[ATP]} \quad \text{pH 7, 25 °C}$$

The concentrations of the solutes are used to approximate their activities, and the activity of the water is one. The value for K_{eq} for this reaction is determined in the laboratory; it is 2.23×10^5. Once we have this information, we can determine the standard free-energy change by substituting in the equation $\Delta G^\circ = -RT \ln K_{eq}$. The key point is to choose the correct quantities to substitute and to keep track of units. Substituting $R = 8.31$ J mol^{-1} K^{-1}, $T = 298$ K, and $\ln K_{eq} = 12.32$,

$$\Delta G^\circ = -RT \ln K'_{eq}$$

$$\Delta G^{\circ\prime} = (8.31 \text{ J mol}^{-1} \text{ K}^{-1})(298 \text{ K})(12.32)$$

$$\Delta G^{\circ\prime} = -3.0500 \times 10^4 \text{ J mol}^{-1} = -30.5 \text{ kJ mol}^{-1} = -7.29 \text{ kcal mol}^{-1}$$

1 kJ = 0.239 kcal

In addition to illustrating the usefulness of a modified standard state for biochemical work, the negative value of $\Delta G^{\circ\prime}$ indicates that the reaction of hydrolysis of ATP to ADP is a spontaneous process in which energy is released.

15.3 | What Is Metabolism?

Until now, we have discussed some basic chemical principles and investigated the natures of the molecules of which living cells are composed. We have yet to discuss the bulk of chemical reactions of biomolecules themselves, which constitute **metabolism,** the biochemical basis of all life processes. The molecules of carbohydrates, fats, and proteins taken into an organism are processed in a variety of ways (Figure 15.1). The breakdown of larger molecules to smaller ones is called **catabolism.** Small molecules are used as the starting points of a variety of reactions to produce larger and more complex molecules, including proteins and nucleic acids; this process is called **anabolism.** Catabolism and anabolism are separate pathways; they are not simply the reverse of each other.

Catabolism is an oxidative process that releases energy; anabolism is a reductive process that requires energy. We shall need several chapters to explore some of the implications of this statement. In this chapter, we discuss oxidation and reduction (electron-transfer reactions) and their relation to the use of energy by living cells. The Biochemical Connections box will deal with another aspect of the unique energetics of living things.

Essential Information

Metabolism is the sum total of the chemical reactions of biomolecules in an organism. In catabolism, large molecules are broken down to smaller products, releasing energy and transferring electrons to acceptor molecules of various sorts. In anabolism, small molecules react to give rise to larger ones; this process requires energy and involves acceptance of electrons from a variety of donors.

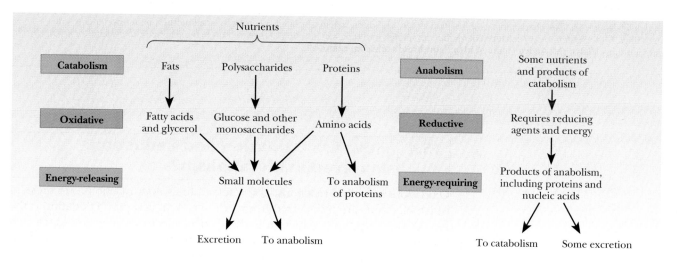

▲ **FIGURE 15.1** A comparison of catabolism and anabolism.

Biochemical Connections

Living Things Are Unique Thermodynamic Systems

Questions arise frequently about whether living organisms obey the laws of thermodynamics. The short answer is that they most definitely do. Most classical treatments of thermodynamics deal with closed systems at equilibrium. A closed system can exchange energy, but not matter, with its surroundings. A living organism is obviously not a closed system, but an open system that can exchange both matter and energy with its surroundings. Because living organisms are open systems, they cannot be at equilibrium as long as they are alive, as shown in the figure below. They can, however, achieve a *steady state*, which is a stable condition. It is the state in which living things can operate at maximum thermodynamic efficiency. This point was established by Ilya Prigogine, winner of the 1977 Nobel Prize for chemistry for his work on nonequilibrium thermodynamics. He showed that, for systems not at equilibrium, ordered structures can arise from disordered ones. This treatment of thermodynamics is quite advanced and highly mathematical, but the results are more directly applicable to biological systems than those of classical thermodynamics. This approach applies not only to living organisms but to the growth of cities and to predictions of auto traffic.

▲ Ilya Prigogine (1917–2003). Ilya Prigogine was born in Moscow in 1917. His family moved to Germany to escape the Russian revolution and subsequently moved to Belgium. He studied at the Université Libre in Brussels and remained there as a faculty member to conduct research on nonequilibrium thermodynamics. He was also associated with the University of Texas, which found a unique way to mark his receiving the Nobel Prize: A tower on the Texas campus is illuminated when one of the university's sports teams wins a championship. It was also illuminated at the time of the announcement of his Nobel Prize.

Isolated system:
No exchange of matter or energy

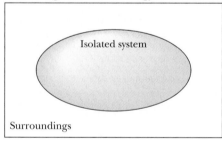

Closed system:
Energy exchange may occur

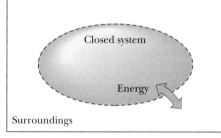

Open system:
Energy exchange and/or matter exchange may occur

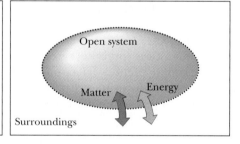

Biochemistry ⑤ Now™ ▲ The characteristics of isolated, closed, and open systems. Isolated systems exchange neither matter nor energy with their surroundings. Closed systems may exchange energy, but not matter, with their surroundings. Open systems may exchange either matter or energy with the surroundings. **Watch this Active Figure at http://now.brookscole.com/campbell5**

15.4	**How Are Oxidation and Reduction Involved in Metabolism?**

Oxidation–reduction reactions, also referred to as *redox* reactions, are those in which electrons are transferred from a donor to an acceptor. **Oxidation** is the loss of electrons, and **reduction** is the gain of electrons. The substance that loses electrons (the electron donor)—that is, the one that is oxidized—is called the **reducing agent** or reductant. The substance that gains electrons (the electron acceptor)—the one that is reduced—is called the **oxidizing agent** or oxidant. Both an oxidizing agent and a reducing agent are necessary for the transfer of electrons (an oxidation–reduction reaction) to take place.

An example of an oxidation–reduction reaction is the one that occurs when a strip of metallic zinc is placed in an aqueous solution containing copper ions. Although both zinc and copper ions play roles in life processes, this particular reaction does not occur in living organisms. However, it is a good place to start our discussion of electron transfer because, in this comparatively simple reaction, it is fairly easy to follow where the electrons are going. (It is not always quite as easy to keep track of the details in biological redox reactions.) The experimental observation is that the zinc metal disappears and zinc ions go into solution, while copper ions are removed from the solution and copper metal is deposited. The equation for this reaction is

$$Zn(s) + Cu^{2+}(aq) \rightarrow Zn^{2+}(aq) + Cu(s)$$

The notation (s) signifies a solid and (aq) signifies a solute in aqueous solution.

In the reaction between zinc metal and copper ion, the Zn lost two electrons to become the Zn^{2+} ion and was oxidized. A separate equation can be written for this part of the overall reaction, and it is called the **half reaction** of oxidation:

$$Zn \rightarrow Zn^{2+} + 2e^-$$

Zn is the reducing agent (it loses electrons; it is an electron donor; it is oxidized).

Likewise, the Cu^{2+} ion gained two electrons to form Cu and was reduced. An equation can also be written for this part of the overall reaction and is called the half reaction of reduction.

$$Cu^{2+} + 2e^- \rightarrow Cu$$

Cu^{2+} is the oxidizing agent (it gains electrons; it is an electron acceptor; it is reduced).

If the two equations for the half reactions are combined, the result is an equation for the overall reaction:

$$
\begin{array}{ll}
Zn \rightarrow Zn^{2+} + 2e^- & \text{Oxidation} \\
\underline{Cu^{2+} + 2e^- \rightarrow Cu} & \text{Reduction} \\
Zn + Cu^{2+} \rightarrow Zn^{2+} + Cu & \text{Overall reaction}
\end{array}
$$

This reaction is a particularly clear example of electron transfer. It will be useful to keep these basic principles in mind when we examine the flow of electrons in the more complex redox reactions of aerobic metabolism. In many of the biological redox reactions we will encounter, the oxidation state of a carbon atom changes. Figure 15.2 shows the changes that occur as carbon in its most reduced form (an alkane) becomes oxidized to an alcohol, an aldehyde, a carboxylic acid, and ultimately carbon dioxide. Each of these oxidations requires the loss of two electrons.

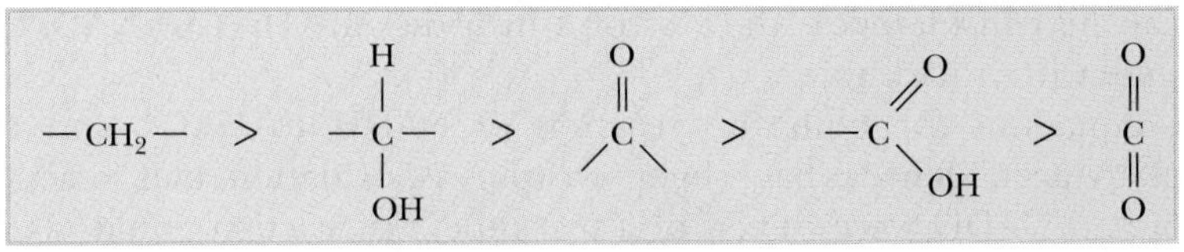

◄ **FIGURE 15.2** Comparison of the state of reduction of carbon atoms in biomolecules: —CH_2— (fats) > —CHOH— (carbohydrates) > —C=O (carbonyls) > —COOH (carboxyls) > CO_2 (carbon dioxide, the final product of catabolism).

15.5	How Are Coenzymes Used in Biologically Important Oxidation–Reduction Reactions?

Oxidation–reduction reactions are discussed at length in textbooks of general and inorganic chemistry, but the oxidation of nutrients by living organisms to provide energy requires its own special treatment. The description of redox reactions in terms of oxidation numbers, which is widely used with inorganic compounds, can be used to deal with the oxidation of carbon-containing molecules. However, our discussion will be more pictorial and easier to follow if we write equations for the half reactions and then concentrate on the functional groups of the reactants and products and on the number of electrons transferred. An example is the oxidation half reaction for the conversion of ethanol to acetaldehyde.

**The half reaction of oxidation
of ethanol to acetaldehyde**

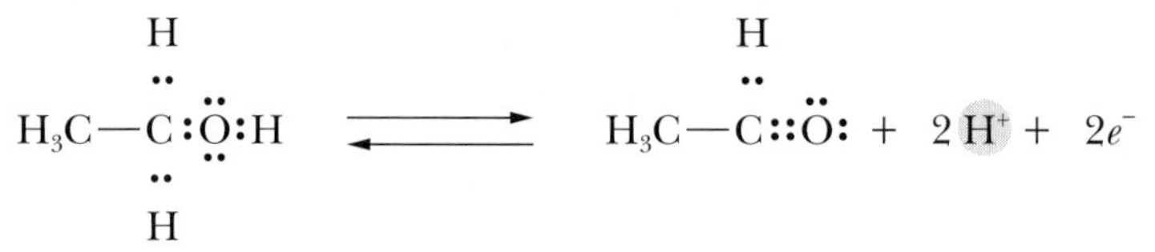

Ethanol (12 electrons in
groups involved in reaction)   Acetaldehyde (10 electrons
in groups involved in reaction)

Writing the Lewis electron-dot structures for the functional groups involved in the reaction helps us to keep track of the electrons being transferred. In the oxidation of ethanol, there are 12 electrons in the part of the ethanol molecule involved in the reaction and 10 electrons in the corresponding part of the acetaldehyde molecule; two electrons are transferred to an electron acceptor (an oxidizing agent). This type of "bookkeeping" is useful for dealing with biochemical reactions. Many biological oxidation reactions, like this example, are accompanied by the transfer of a proton (H^+). The oxidation half reaction has been written as a reversible reaction because the occurrence of oxidation or reduction depends on the other reagents present.

Another example of an oxidation half reaction is that for the conversion of NADH, the reduced form of nicotinamide adenine dinucleotide, to the oxidized form, NAD^+. This substance is an important **coenzyme** in many reactions.

Figure 15.3 shows the structure of NAD^+ and NADH; the nicotinamide portion, the functional group involved in the reaction, is indicated in red and blue. Nicotinamide is a derivative of nicotinic acid (also called niacin), one of the B-complex vitamins (see Section 7.8). A similar compound is NADPH (for which the oxidized form is $NADP^+$). It differs from NADH by having an additional phosphate group; the site of attachment of this phosphate group to ribose is also indicated in Figure 15.3. To simplify writing the equation for the oxidation of NADH, only the nicotinamide ring is shown explicitly, with the rest of the molecule designated as R. The two electrons that are lost when NADH is converted to NAD^+ can be considered to come from the bond between carbon and the lost hydrogen, with the nitrogen lone-pair electrons becoming involved in a bond. Note that the loss of a hydrogen and two electrons can be considered as the loss of a hydride ion ($H:^-$) by NADH and is sometimes written that way.

The equations for both the reaction of NADH to NAD^+ and that of ethanol to acetaldehyde have been written as oxidation half reactions. If ethanol and NADH were mixed in a test tube, no reaction could take place because there would be no electron acceptor. If, however, NADH were mixed

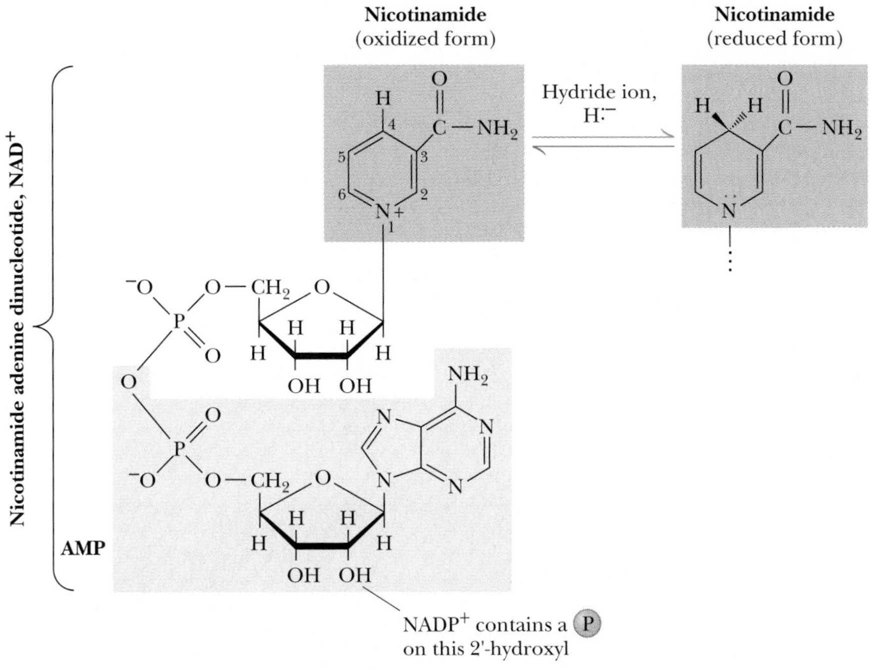

Nicotinamide
(oxidized form)

Nicotinamide
(reduced form)

Hydride ion,
H:⁻

Nicotinamide adenine dinucleotide, NAD⁺

AMP

NADP⁺ contains a ⓟ
on this 2'-hydroxyl

with acetaldehyde, which is an oxidized species, a transfer of electrons could take place, producing ethanol and NAD⁺. (This reaction would take place very slowly in the absence of an enzyme to catalyze it. Here we have an excellent example of the difference between the thermodynamic and kinetic aspects of reactions. The reaction is spontaneous in the thermodynamic sense but very slow in the kinetic sense.)

$$\text{NADH} \rightarrow \text{NAD}^+ + \text{H}^+ + 2e^- \qquad \text{Half reaction of oxidation}$$
$$\underline{\text{CH}_3\text{CHO} + 2\text{H}^+ + 2e^- \rightarrow \text{CH}_3\text{CH}_2\text{OH}} \qquad \text{Half reaction of reduction}$$
$$\text{NADH} + \text{H}^+ + \text{CH}_3\text{CHO} \rightarrow \text{NAD}^+ + \text{CH}_3\text{CH}_2\text{OH} \quad \text{Overall reaction}$$

Acetaldehyde   Ethanol

Such a reaction does take place in some organisms as the last step of alcoholic fermentation. The NADH is oxidized while the acetaldehyde is reduced.

Another important electron acceptor is FAD (flavin adenine dinucleotide) (Figure 15.4), which is the oxidized form of FADH$_2$. The symbol FADH$_2$ explicitly recognizes that protons (hydrogen ions) as well as electrons are accepted by FAD. The structures shown in this equation again point out the electrons that are transferred in the reaction. Several other coenzymes contain the flavin group; they are derived from the vitamin riboflavin (vitamin B$_2$).

The half reaction of reduction of FAD to FADH$_2$

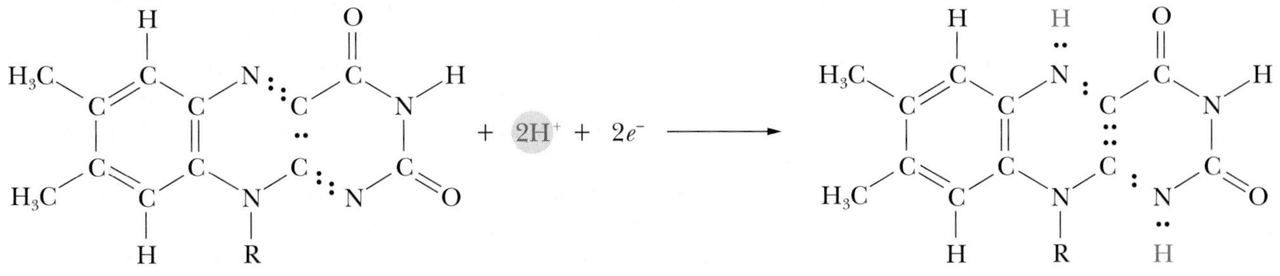

+ 2H⁺ + 2e⁻ ⟶

FAD oxidized form

FADH$_2$ reduced form

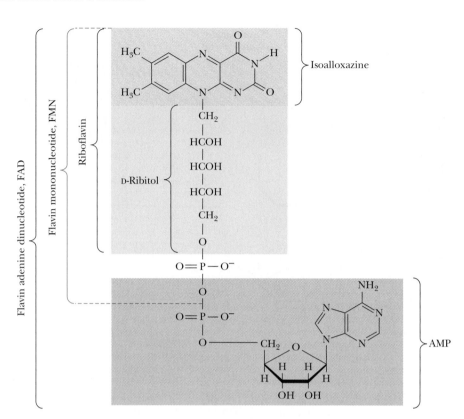

Biochemistry⚡Now™ **ANIMATED FIGURE 15.4**
The structures of riboflavin, flavin mononucleotide (FMN), and flavin adenine dinucleotide (FAD). Even in organisms that rely on the nicotinamide coenzymes (NADH and NADPH) for many of their oxidation–reduction cycles, the flavin coenzymes fill essential roles. Flavins are stronger oxidizing agents than NAD$^+$ and NADP. They can be reduced by both one-electron and two-electron pathways and can be reoxidized easily by molecular oxygen. Enzymes that use flavins to carry out their reactions—flavoenzymes—are involved in many kinds of oxidation–reduction reactions. **See this figure animated at http://now.brookscole.com/campbell5**

Oxidation of nutrients to provide energy for an organism cannot take place without reduction of some electron acceptor. The ultimate electron acceptor in aerobic oxidation is oxygen; we shall encounter intermediate electron acceptors as we discuss metabolic processes. Reduction of metabolites plays a significant role in living organisms in anabolic processes. Important biomolecules are synthesized in organisms by many reactions in which a metabolite is reduced while the reduced form of a coenzyme is oxidized.

15.6 How Are Production and Use of Energy Coupled?

Another important question about metabolism is: "How is the energy released by the oxidation of nutrients trapped and used?" This energy cannot be used directly; it must be shunted into an easily accessible form of chemical energy. In Section 1.11, we saw that several phosphorus-containing compounds, such as ATP, can be hydrolyzed easily, and that the reaction releases energy. Formation of ATP is intimately linked with the release of energy from oxidation of nutrients. The coupling of energy-producing reactions and energy-requiring reactions is a central feature in the metabolism of all organisms.

The phosphorylation of ADP (adenosine diphosphate) to produce ATP (adenosine triphosphate) requires energy, which can be supplied by the oxidation of nutrients Conversely, the hydrolysis of ATP to ADP releases energy (Figure 15.5).

The forms of ADP and ATP shown in this section are in their ionization states for pH 7. The symbol P$_i$ for phosphate ion comes from its name in biochemical jargon, "inorganic phosphate." Note that there are four negative charges on ATP and three on ADP; electrostatic repulsion makes ATP less stable than ADP. Energy must be expended to put an additional negatively charged phosphate group on ADP by forming a covalent bond to the phos-

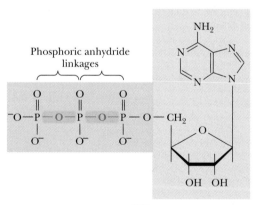

ATP
(adenosine-5'-triphosphate)

Biochemistry⚡Now™ **ACTIVE FIGURE 15.5**
The phosphoric anhydride bonds in ATP are "high-energy" bonds, referring to the fact that they require or release convenient amounts of energy, depending on the direction of the reaction. **Watch this Active Figure at http://now.brookscole.com/campbell5**

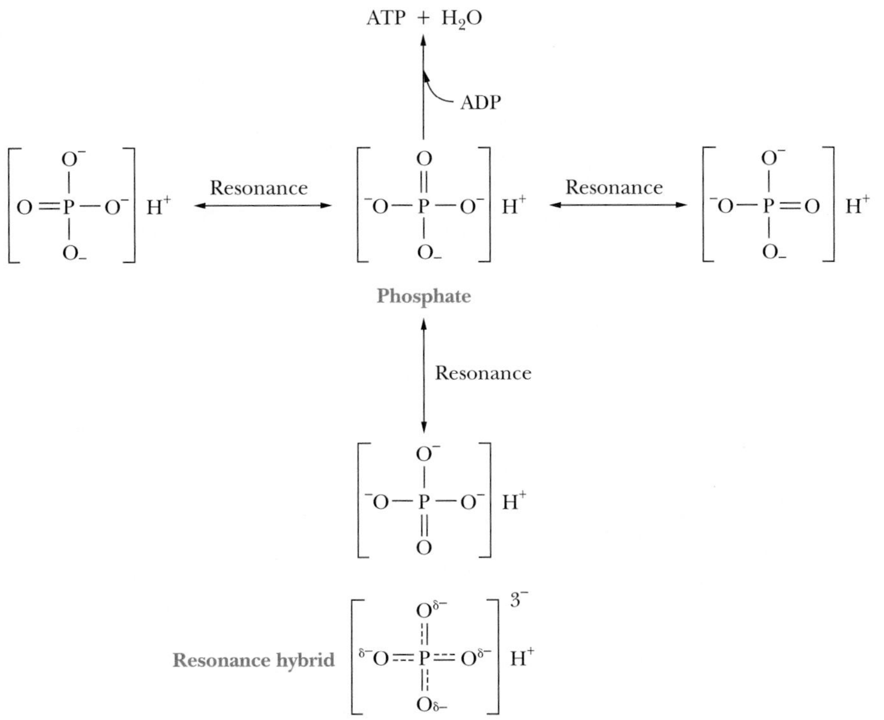

$$\text{ATP} + \text{H}_2\text{O}$$

$$\left[\begin{array}{c} \text{O}^- \\ | \\ \text{O}{=}\text{P}{-}\text{O}^- \\ | \\ \text{O}_- \end{array} \right] \text{H}^+ \quad \xleftrightarrow{\text{Resonance}} \quad \left[\begin{array}{c} \text{O} \\ || \\ {}^-\text{O}{-}\text{P}{-}\text{O}^- \\ | \\ \text{O}_- \end{array} \right] \text{H}^+ \quad \xleftrightarrow{\text{Resonance}} \quad \left[\begin{array}{c} \text{O}^- \\ | \\ {}^-\text{O}{-}\text{P}{=}\text{O} \\ | \\ \text{O}_- \end{array} \right] \text{H}^+$$

Phosphate

Resonance

$$\left[\begin{array}{c} \text{O}^- \\ | \\ {}^-\text{O}{-}\text{P}{-}\text{O}^- \\ || \\ \text{O} \end{array} \right] \text{H}^+$$

Resonance hybrid $\left[\begin{array}{c} \text{O}^{\delta-} \\ ||| \\ {}^{\delta-}\text{O}{=}\!\!=\!\!\text{P}{=}\!\!=\!\!\text{O}^{\delta-} \\ ||| \\ \text{O}_{\delta-} \end{array} \right]^{3-} \text{H}^+$

◀ **FIGURE 15.6** When ADP is phosphorylated to ATP, there is a loss of the resonance-stabilized phosphate ion, resulting in a decrease in entropy. (δ^- denotes a partial negative charge.)

phate group being added. In addition, there is an entropy loss when ADP is phosphorylated to ATP. Inorganic phosphate can adopt multiple resonance structures, and the loss of these potential structures results in a decrease in entropy when the phosphate is attached to ADP (Figure 15.6). The $\Delta G^{\circ\prime}$ for the reaction refers to the usual biochemical convention of pH 7 as the standard state for hydrogen ion (Section 15.2). Note, however, that there is a marked decrease in electrostatic repulsion on phosphorylation of ADP to ATP (Figure 15.7).

The reverse reaction, the hydrolysis of ATP to ADP and phosphate ion, releases 30.5 kJ mol^{-1} (7.3 kcal mol^{-1}) when energy is needed:

$$\text{ATP} + \text{H}_2\text{O} \rightarrow \text{ADP} + \text{P}_i + \text{H}^+$$

$$\Delta G^{\circ\prime} = -30.5 \text{ kJ mol}^{-1} = -7.3 \text{ kcal mol}^{-1}$$

Practice Session

In the following reactions, identify the substance oxidized, the substance reduced, the oxidizing agent, and the reducing agent.

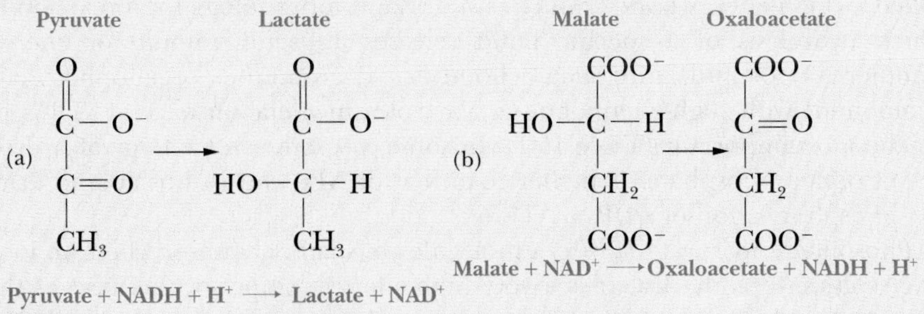

(a) Pyruvate $\longrightarrow$ Lactate

Pyruvate + NADH + H$^+$ $\longrightarrow$ Lactate + NAD$^+$

(b) Malate $\longrightarrow$ Oxaloacetate

Malate + NAD$^+$ $\longrightarrow$ Oxaloacetate + NADH + H$^+$

Solution

The way to approach this question is to recall that NADH is a reduced form of the coenzyme. It will be oxidized and will serve as a reducing

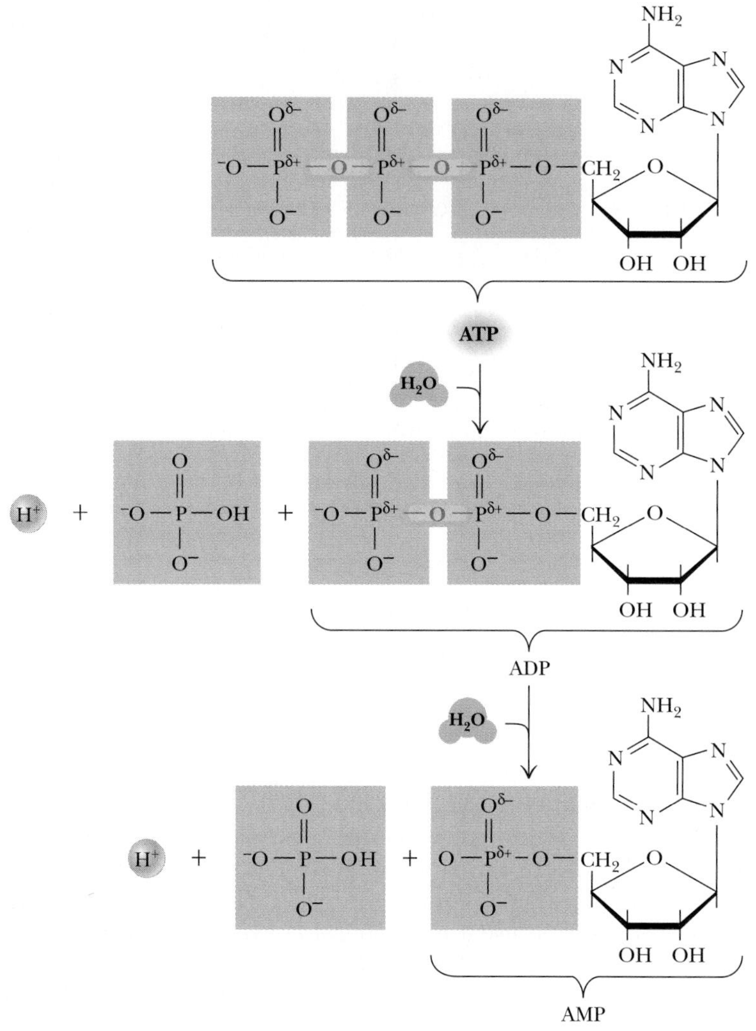

agent. NAD^+ is the oxidized form. It will be reduced and thus will serve as
the oxidizing agent. In the first reaction, pyruvate is reduced and NADH is
oxidized; pyruvate is the oxidizing agent, and NADH is the reducing agent.
In the second reaction, malate is oxidized, and NAD^+ is reduced; NAD^+ is
the oxidizing agent, and malate is the reducing agent.

The bond that is hydrolyzed when this reaction takes place is sometimes
called a "high-energy bond," which is shorthand terminology for a reaction in
which hydrolysis of a specific bond releases a useful amount of energy.
Another way of indicating such a bond is ~P. Numerous organophosphate
compounds with high-energy bonds play roles in metabolism, but ATP is by
far the most important (Table 15.1). In some cases, the free energy of hydrolysis of organophosphates is higher than that of ATP and is thus able to drive
the phosphorylation of ADP to ATP.

Phosphoenolpyruvate (PEP), a molecule we shall encounter when we look
at glycolysis, tops the list. It is a very high-energy compound because of the
resonance stabilization of the liberated phosphate when it is hydrolyzed (the
same effect as that seen with ATP) and because keto–enol tautomerization of
pyruvate is a possibility. Both effects increase the entropy upon hydrolysis
(Figure 15.8).

Table 15.1

Free Energies of Hydrolysis of Selected Organophosphates

Compound	$\Delta G^{\circ\prime}$	
	kJ mol^{-1}	kcal mol^{-1}
Phosphoenolpyruvate	−61.9	−14.8
Carbamoyl phosphate	−51.4	−12.3
Creatine phosphate	−43.1	−10.3
Acetyl phosphate	−42.2	−10.1
ATP (to ADP)	−30.5	−7.3
Glucose-1-phosphate	−20.9	−5.0
Glucose-6-phosphate	−12.5	−3.0
Glycerol-3-phosphate	−9.7	−2.3

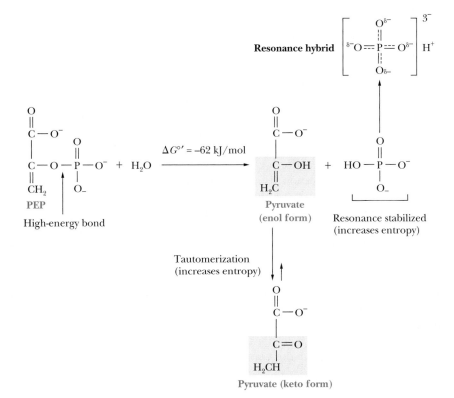

Biochemistry Now™ ANIMATED FIGURE 15.8
When phosphoenolpyruvate is hydrolyzed to pyruvate and phosphate, it results in an increase in entropy. Both the formation of the keto form of pyruvate and the resonance structures of phosphate lead to the increase in entropy. **See this figure animated at http://now.brookscole.com/campbell5**

The energy of hydrolysis of ATP is not stored energy, just as an electric current does not represent stored energy. Both ATP and electric current must be produced when they are needed—by organisms or by a power plant, as the case may be. The cycling of ATP and ADP in metabolic processes is a way of shunting energy from its production (by oxidation of nutrients) to its uses (in processes such as biosynthesis of essential compounds or muscle contraction) when it is needed. The oxidation processes take place when the organism needs the energy that can be generated by the hydrolysis of ATP. When chemical energy is stored, it is usually in the form of fats and carbohydrates, which are metabolized as needed. Certain small biomolecules, such as creatine phosphate, can also serve as vehicles for storing chemical energy. The energy that must be supplied for the many endergonic reactions in life processes comes directly from the hydrolysis of ATP and indirectly from the oxidation

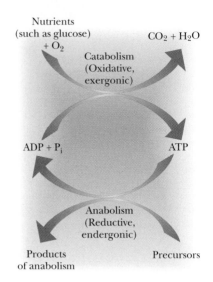

▲ **FIGURE 15.9** The role of ATP as energy currency in processes that release energy and processes that use energy.

of nutrients. The latter produces the energy needed to phosphorylate ADP to ATP (Figure 15.9).

Let us examine some biological reactions that release energy and see how some of that energy is used to phosphorylate ADP to ATP. The multistep conversion of glucose to lactate ions is an exergonic and anaerobic process. Two molecules of ADP are phosphorylated to ATP for each molecule of glucose metabolized. The basic reactions are the production of lactate, which is exergonic,

$$\text{Glucose} \rightarrow 2 \text{ Lactate ions} \qquad \Delta G^{\circ\prime} = -184.5 \text{ kJ mol}^{-1} = -44.1 \text{ kcal mol}^{-1}$$

and the phosphorylation of two moles of ADP for each mole of glucose, which is endergonic,

$$2 \text{ ADP} + 2 \text{ P}_i \rightarrow 2 \text{ ATP}$$

$$\Delta G^{\circ\prime} = 61.0 \text{ kJ m mol}^{-1} = 14.6 \text{ kcal mol}^{-1}$$

(In the interest of simplicity, we shall write the equation for phosphorylation of ADP in terms of ADP, P_i, and ATP only.) The overall reaction is

$$\text{Glucose} + 2 \text{ ADP} + 2 \text{ P}_i \rightarrow 2 \text{ Lactate ions} + 2 \text{ ATP}$$

$$\Delta G^{\circ\prime} \text{ overall} = -184.5 + 61.0 = -123.5 \text{ kJ mol}^{-1} = -29.5 \text{ kcal mol}^{-1}$$

Not only can we add the two chemical reactions to obtain an equation for the overall reaction, we can also add the free-energy changes for the two reactions to find the overall free-energy change. We can do this because the free-energy change is a **state function;** it depends only on the initial and final states of the system under consideration, not on the path between those states. The exergonic reaction provides energy, which drives the endergonic reaction. This phenomenon is called **coupling.** The percentage of the released energy that is used to phosphorylate ADP is the efficiency of energy use in anaerobic metabolism; it is (61.0/184.5) × 100, or about 33%. The number 61.0 comes from the number of kilojoules required to phosphorylate 2 moles of ADP to ATP, and the number 184.5 is the number of kilojoules released when 1 mole of glucose is converted to 2 moles of lactate.

The breakdown of glucose goes farther under aerobic conditions than under anaerobic conditions. The end products of aerobic oxidation are six molecules of carbon dioxide and six molecules of water for each molecule of glucose. Up to 32 molecules of ADP can be phosphorylated to ATP when one molecule of glucose is broken down completely to carbon dioxide and water. The exergonic reaction for the complete oxidation of glucose is

$$\text{Glucose} + 6 \text{ O}_2 \rightarrow 6 \text{ CO}_2 + 6 \text{ H}_2\text{O}$$

$$\Delta G^{\circ\prime} = -2867 \text{ kJ mol}^{-1} = -685.9 \text{ kcal mol}^{-1}$$

The endergonic reaction for phosphorylation is

$$32 \text{ ADP} + 32 \text{ P}_i \rightarrow 32 \text{ ATP}$$

$$\Delta G^{\circ\prime} = 976 \text{ kJ} = 233.5 \text{ kcal}$$

The net reaction is

$$\text{Glucose} + 6 \text{ O}_2 + 32 \text{ ADP} + 32 \text{P}_i \rightarrow 6 \text{ CO}_2 + 6 \text{ H}_2\text{O} + 32 \text{ ATP}$$

$$\Delta G^{\circ\prime} = -2867 + 976 = -1891 \text{ kJ mol}^{-1} = -452.4 \text{ kcal mol}^{-1}$$

Note that, once again, we add the two reactions and their respective free-energy changes to obtain the overall reaction and its free-energy change. The efficiency of aerobic oxidation of glucose is (976/2867) × 100, about 34%. (We performed this calculation in the same way that we did with the example of anaerobic oxidation of glucose.) More ATP is produced by the coupling process in aerobic oxidation of glucose than by the coupling process in anaerobic oxidation. The hydrolysis of ATP produced by breakdown (aerobic or anaerobic) of glucose can be coupled to endergonic processes, such as muscle contraction in exercise. As any jogger or long-distance swimmer knows, aerobic metabolism involves large quantities of energy, processed in a highly efficient fashion. We have now seen two examples of coupling of exergonic and endergonic processes, aerobic oxidation of glucose and anaerobic fermentation of glucose, involving different amounts of energy.

Practice Session

Calculations of Free Energies

We shall use values from Table 15.1 to calculate $\Delta G^{\circ\prime}$ for the following reaction and decide whether or not it is spontaneous. The most important point here is that we add algebraically. In particular, we have to remember to change the sign of the $\Delta G^{\circ\prime}$ when we reverse the direction of the reaction.

$$\text{ADP} + \text{Phosphoenolpyruvate} \rightarrow \text{ATP} + \text{Pyruvate}$$

From Table 15.1,

$$\text{Phosphoenolpyruvate} + \text{H}_2\text{O} \rightarrow \text{Pyruvate} + \text{P}_i$$

$$\Delta G^{\circ\prime} = -61.9 \text{ kJ mol}^{-1} = -14.8 \text{ kcal mol}^{-1}$$

Also,

$$\text{ATP} + \text{H}_2\text{O} \rightarrow \text{ADP} + \text{P}_i \qquad \Delta G^{\circ\prime} = -30.5 \text{ kJ mol}^{-1} = -7.3 \text{ kcal mol}^{-1}$$

We want the reverse of the second reaction:

$$\text{ADP} + \text{P}_i \rightarrow \text{ATP} + \text{H}_2\text{O} \qquad \Delta G^{\circ\prime} = +30.5 \text{ kJ mol}^{-1} = +7.3 \text{ kcal mol}^{-1}$$

We now add the two reactions and their free-energy changes:

$$\text{Phosphoenolpyruvate} + \text{H}_2\text{O} \rightarrow \text{Pyruvate} + \text{P}_i$$
$$\underline{\text{ADP} + \text{P}_i \rightarrow \text{ATP} + \text{H}_2\text{O}}$$
$$\text{Phosphoenolpyruvate} + \text{ADP} \rightarrow \text{Pyruvate} + \text{ATP} \qquad \textbf{Net reaction}$$

$$\Delta G^{\circ\prime} = -61.9 \text{ kJ mol}^{-1} + 30.5 \text{ kJ mol}^{-1} = -31.4 \text{ kJ mol}^{-1}$$

$$\Delta G^{\circ\prime} = -14.8 \text{ kcal mol}^{-1} + 7.3 \text{ kcal mol}^{-1} = -7.5 \text{ kcal mol}^{-1}$$

The reaction is spontaneous, as indicated by the $\Delta G^{\circ\prime}$. However, remember that, even though this reaction is spontaneous, nothing would happen if you simply put these chemicals in a test tube. Biochemical reactions require enzymes to catalyze them.

Essential Information

In the coupling of biochemical reactions, the energy released by one reaction provides energy for another.

15.7 How Is Coenzyme A Involved in Activation of Metabolic Pathways?

The metabolic oxidation of glucose that we saw in Section 15.6 does not take place in one step. The anaerobic breakdown of glucose requires many steps, and the complete aerobic oxidation of glucose to carbon dioxide and water has still more steps. One of the most important points about the multistep nature of all metabolic processes, including the oxidation of glucose, is that the many stages allow for efficient production and use of energy. The electrons produced by the oxidation of glucose are passed along to oxygen, the ultimate electron acceptor, by intermediate electron acceptors. Many of the intermediate stages of the oxidation of glucose are coupled to ATP production by phosphorylation of ADP.

A step frequently encountered in metabolism is the process of **activation.** In a reaction of this sort, a metabolite (a component of a metabolic pathway) is bonded to some other molecule, such as a coenzyme, and the free-energy change for breaking this new bond is negative. In other words, the next reaction in the metabolic pathway is exergonic. For example, if substance A is the metabolite and reacts with substance B to give AB, the following series of reactions might take place:

A + Coenzyme → A—Coenzyme   **Activation step**

A—Coenzyme + B → AB + Coenzyme $\Delta G^{\circ\prime} < 0$ **Exergonic reaction**

The formation of a more reactive substance in this fashion is called activation. There are many examples of activation in metabolic processes. We can discuss one of the most useful examples now. It involves forming a covalent bond to a compound known as coenzyme A (CoA).

The structure of CoA is complex. It consists of several smaller components linked together covalently (Figure 15.10). One part is 3′-P-5′-ADP, a derivative of adenosine with phosphate groups esterified to the sugar, as shown in the structure. Another part is derived from the vitamin pantothenic acid, and the part of the molecule involved in activation reactions contains a thiol group. In fact, coenzyme A is frequently written as CoA-SH to emphasize that the thiol group is the reactive portion of the molecule. For example, carboxylic acids form thioester linkages with CoA-SH. The metabolically active form of a carboxylic acid is the corresponding acyl-CoA thioester, in which the thioester linkage is a high-energy bond (Figure 15.11). Thioesters are high-energy compounds due to the possible dissociation of the products after hydrolysis and

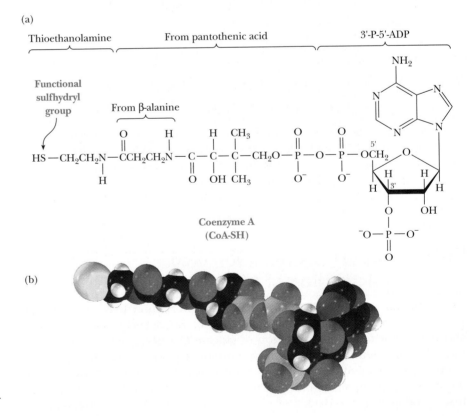

▶ **FIGURE 15.10** (a) The structure of coenzyme A. (b) Space-filling model of coenzyme A.

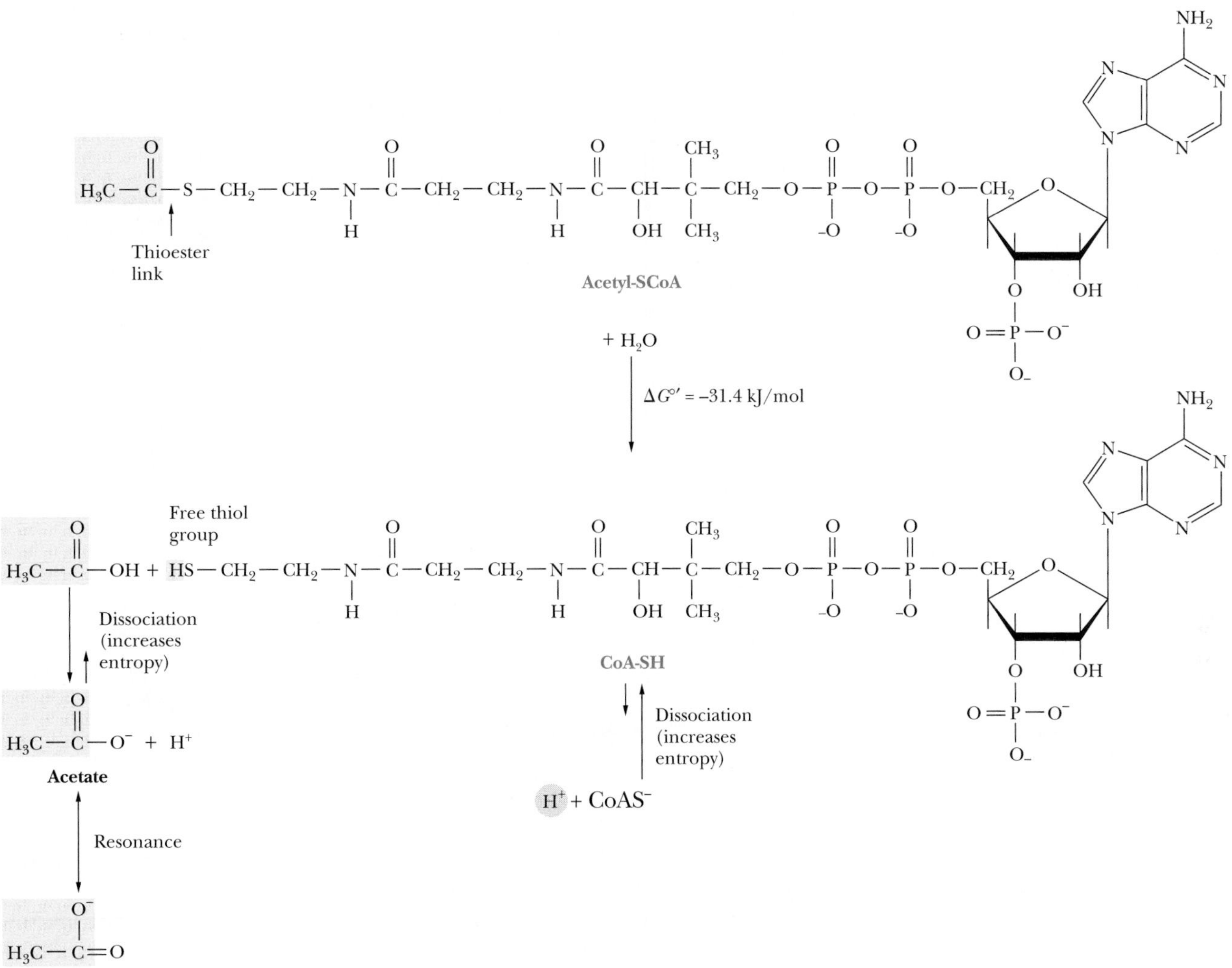

▲ **FIGURE 15.11** The hydrolysis of acetyl-CoA. The products are stabilized by resonance and by dissociation.

resonance structures of the products. For example, when acetyl-CoA is hydrolyzed, the —SH at the end of the molecule can dissociate slightly to form H^+ and CoA-S$^-$. The acetate released by the hydrolysis is stabilized by resonance. Acetyl-CoA is a particularly important metabolic intermediate; other acyl-CoA species figure prominently in lipid metabolism.

The important coenzymes we have met in this chapter—NAD$^+$, NADP$^+$, FAD, and coenzyme A—share an important structural feature: all contain ADP. In NADP$^+$, there is an additional phosphate group at the 2′ position of the ribose group of ADP. In CoA, the additional phosphate group is at the 3′ position.

Like catabolism, anabolism proceeds in stages. Unlike catabolism, which releases energy, anabolism requires energy. The ATP produced by catabolism is hydrolyzed to release the needed energy. Reactions in which metabolites are reduced are part of anabolism; they require reducing agents, such as NADH, NADPH, and FADH$_2$, all of which are the reduced forms of coenzymes mentioned in this chapter. In their oxidized forms, these coenzymes serve as the intermediate oxidizing agents needed in catabolism. In their reduced form, the same coenzymes provide the "reducing power" needed for the anabolic processes of biosynthesis; in this case, the coenzymes act as reducing agents.

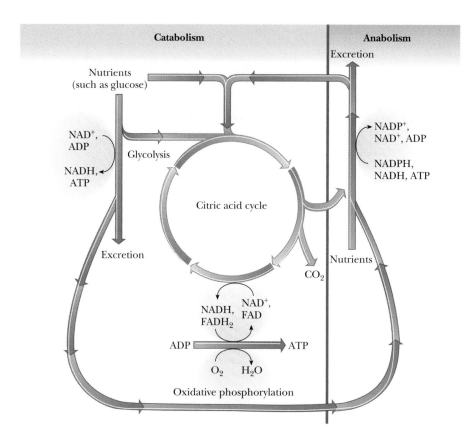

▶ **FIGURE 15.12** The role of electron transfer and ATP production in metabolism. NAD⁺, FAD, and ATP are constantly recycled.

We are now in a position to expand on our earlier statements about the natures of anabolism and catabolism. Figure 15.12 is an outline of metabolic pathways that explicitly points out two important features of metabolism: the role of electron transfer and the role of ATP in the release and utilization of energy. Even though this outline is more extended than the one in Figure 15.1, it is still very general. The more important specific pathways have been studied in detail, and some are still the subjects of active research. We shall discuss some of the most important metabolic pathways in the remainder of this textbook.

Summary

15.1 What Are Standard States for Free-Energy Changes? Thermodynamics deals with the changes in energy that determine whether a process will take place. In a spontaneous process, the free-energy decreases (ΔG is negative). In a nonspontaneous process, the free energy increases. The symbol for change in free energy is ΔG. A change in free energy under any set of conditions can be compared to the free-energy change under standard conditions ($\Delta G°$), in which the concentration of all reactants in solution is set at 1 M. Free-energy changes under standard conditions can be related to the equilibrium constant of a reaction by the equation

$$\Delta G° = -RT \ln K_{eq}.$$

15.2 What Is a Modified Standard State for Biochemical Applications? Since biochemical reactions do not occur naturally at a hydrogen-ion concentration of 1 M, a biochemical standard state ($\Delta G°'$) is often used, where the [H⁺] is set at 1×10^{-7} M (pH = 7.0).

15.3 What Is Metabolism? The reactions of the biomolecules in the cell constitute metabolism. The breakdown of larger molecules to smaller ones is called catabolism. The reaction of small molecules to produce larger and more complex molecules is called anabolism. Catabolism and anabolism are separate pathways, not the reverse of each other. Metabolism is the biochemical basis of all life processes.

15.4 How Are Oxidation and Reduction Involved in Metabolism? Catabolism is an oxidative process that releases energy; anabolism is a reductive process that requires energy. Oxidation–reduction (redox) reactions are those in which electrons are transferred from a donor to an acceptor. Oxidation is the loss of electrons, and reduction is the gain of electrons.

15.5 How Are Coenzymes Used in Biologically Important Oxidation–Reduction Recations? Many biologically important redox reactions involve coenzymes, such as

NADH and $FADH_2$. These coenzymes appear in many reactions as one of the half-reactions that can be written for a redox reaction.

15.6 How Are the Production and Use of Energy Coupled?
The coupling of energy-producing reactions and energy-requiring reactions is a central feature in the metabolism of all organisms. In catabolism, oxidative reactions are coupled to the endergonic production of ATP by phosphorylation of ADP. Aerobic metabolism is a more efficient means of using the chemical energy of nutrients than is anaerobic metabolism. In anabolism, the exergonic hydrolysis of the high-energy bond of ATP releases the energy needed to drive endergonic reductive reactions.

15.7 How Is Coenzyme A Involved in Activation of Metabolic Pathways?
Metabolism proceeds in stages, and the many stages allow for the efficient production and use of energy. The process of activation, producing high-energy intermediates, occurs in many metabolic pathways. The formation of thioester linkages by the reaction of carboxylic acids with coenzyme A is an example of the activation process that occurs in several pathways.

Critical Questions to Review

15.1 What Are Standard States for Free-Energy Changes?

1. **Fact Check** Is there a connection between the free-energy change for a reaction and its equilibrium constant? If there is a connection, what is it?

2. **Thought Question** What do the following indicators tell you about whether a reaction can proceed as written?
   (a) The standard free-energy change is positive.
   (b) The free-energy change is positive.
   (c) The reaction is exergonic.

3. **Thought Question** Consider the reaction

$$\text{Glucose-6-phosphate} + H_2O \rightarrow \text{glucose} + P_i$$

$$K_{eq} = \frac{[\text{glucose}][P_i]}{[\text{glucose-6-P}]}$$

   The K_{eq} at pH 8.5 and 38°C is 122. Can you determine the rate of the reaction from this information?

15.2 What Is a Modified Standard State for Biochemical Applications?

4. **Fact Check** Why is it necessary to define a modified standard state for biochemical applications of thermodynamics?

5. **Fact Check** Which of the following statements is (are) true about the modified standard state for biochemistry? For each, explain why or why not.
   (a) $[H^+] = 1 \times 10^{-7}$ M, not 1 M.
   (b) The concentration of any solute is 1×10^{-7} M.

6. **Fact Check** How can you tell if the standard Gibbs free energy given for a reaction is for chemical standard states or biological standard states?

7. **Fact Check** Can the thermodynamic property, $\Delta G°$, be used to predict the speed of a reaction in a living organism? Why or why not?

8. **Mathematical** Calculate $\Delta G°'$ for the following values of K_{eq}: 1×10^4, 1, 1×10^{-6}.

9. **Mathematical** For the hydrolysis of ATP at 25°C (298 K) and pH 7, ATP + $H_2O \rightarrow$ ADP + P_i + H^+, the standard free energy of hydrolysis ($\Delta G°'$) is -30.5 kJ mol^{-1} (-7.3 kcal mol^{-1}), and the standard enthalpy change ($\Delta H°'$) is -20.1 kJ mol^{-1} (-4.8 kcal mol^{-1}). Calculate the standard entropy change ($\Delta S°'$) for the reaction, in both joules and calories. Why is the positive sign of the answer to be expected in view of the nature of the reaction? Hint: You may want to review some material from Chapter 1.

10. **Mathematical** Consider the reaction $A \rightleftharpoons B + C$, in which $\Delta G° = 0.00$.
    (a) What is the value of ΔG (not $\Delta G°$) when the initial concentrations of A, B, and C are 1 M, 10^{-3} M, and 10^{-6} M?
    (b) Try the same calculations for the reaction $D + E \rightleftharpoons F$, for the same relative order of concentrations.

(c) Try the same calculations for the reaction $G \rightleftharpoons H$, if the concentrations are 1 M and 10^{-3} M for G and H, respectively.

11. **Mathematical** Compare your answers for parts (a) and (b) with that for part (c) in Question 10. What do your answers to parts (a), (b), and (c) say about the influence of concentrations of reactants and products on reactions?

12. **Mathematical** The $\Delta G°'$ for the reaction citrate $\rightarrow$ isocitrate is $+6.64$ kJ $mol^{-1} = +1.59$ kcal mol^{-1}. The $\Delta G°'$ for the reaction isocitrate $\rightarrow$ α-ketoglutarate is -267 kJ $mol^{-1} = -63.9$ kcal mol^{-1}. What is the $\Delta G°'$ for the conversion of citrate to α-ketoglutarate? Is that reaction exergonic or endergonic, and why?

13. **Mathematical** If a reaction can be written $A \rightarrow B$, and the $\Delta G°'$ is 20 kJ mol^{-1}, what would the substrate/product ratio have to be for the reaction to be thermodynamically favorable?

14. **Mathematical** All the organophosphate compounds listed in Table 15.1 undergo hydrolysis reactions in the same way as ATP. The following equation illustrates the situation for glucose-l-phosphate.

$$\text{Glucose-l-phosphate} + H_2O \rightarrow \text{Glucose} + P_i$$

$$\Delta G°' = -20.9 \text{ kJ } mol^{-1}$$

Using the free-energy values in Table 15.1, predict whether the following reactions will proceed in the direction written, and calculate the $\Delta G°'$ for the reaction, assuming that the reactants are initially present in a 1:1 molar ratio.
    (a) ATP + Creatine $\rightarrow$ Creatine phosphate + ADP
    (b) ATP + Glycerol $\rightarrow$ Glycerol-3-phosphate + ADP
    (c) ATP + Pyruvate $\rightarrow$ Phosphoenolpyruvate + ADP
    (d) ATP + Glucose $\rightarrow$ Glucose-6-phosphate + ADP

15. **Thought Question** Can you use the equation $\Delta G°' = -RT \ln K_{eq}$ to get the $\Delta G°'$ from the information in Question 3?

16. **Thought Question** Why are $\Delta G°'$ values not rigorously applicable to biochemical systems?

15.3 What Is Metabolism?

17. **Fact Check** Organize the following words into two related groups: catabolism, energy-requiring, reductive, anabolism, oxidative, energy-yielding.

18. **Biochemical Connections** Comment on the statement that the existence of life is a violation of the second law of thermodynamics, adding concepts from this chapter to those we saw in Chapter 1.

19. **Thought Question** Would you expect the production of sugars by plants in photosynthesis to be an exergonic or endergonic process? Give the reason for your answer.

20. **Thought Question** Would you expect the biosynthesis of a protein from the constituent amino acids in an organism to be an exergonic or endergonic process? Give the reason for your answer.

21. **Thought Question** Adult humans synthesize large amounts of ATP in the course of a day, but their body weights do not change significantly. In the same time period, the structures and compositions

of their bodies also do not change appreciably. Explain this apparent contradiction.

15.4 How Are Oxidation and Reduction Involved in Metabolism?

22. Fact Check Identify the molecules oxidized and reduced in the following reactions and write the half reactions.

(a) $CH_3CH_2CHO + NADH \rightarrow CH_3CH_2CH_2OH + NAD^+$

(b) $Cu^{2+}(aq) + Fe^{2+}(aq) \rightarrow Cu^+(aq) + Fe^{3+}(aq)$

23. Fact Check For each of the reactions in Exercise 22, give the oxidizing agent and reducing agents.

15.5 How Are Coenzymes Used in Biologically Important Oxidation–Reduction Reactions?

24. Fact Check What structural feature do NAD^+, $NADP^+$, and FAD have in common?

25. Fact Check What is the structural difference between NADH and NADPH?

26. Fact Check How does the difference between NADH and NADPH affect the reactions in which they are involved?

27. Thought Question Which of the following statements are true? For each, explain why or why not.

(a) All coenzymes are electron-transfer agents.

(b) Coenzymes do not contain phosphorus or sulfur.

(c) Generating ATP is a way of storing energy.

28. Thought Question A biochemical reaction transfers 60 kJ mol^{-1} (15 kcal mol^{-1}) of energy. What general process most likely would be involved in this transfer? What cofactor (or cosubstrate) likely would be used? Which cofactor probably would not be used?

29. Thought Question The following half reactions play important roles in metabolism.

$$\tfrac{1}{2} O_2 + 2 H^+ + 2e^- \rightarrow H_2O$$

$$NADH + H^+ \rightarrow NAD^+ + 2H^+ + 2e$$

Which of these two is a half reaction of oxidation? Which one is a half reaction of reduction? Write the equation for the overall reaction. Which reagent is the oxidizing agent (electron acceptor)? Which reagent is the reducing agent (electron donor)?

30. Thought Question Draw NAD^+ and FAD showing where the electrons and hydrogens go when the molecules are reduced.

31. Thought Question There is a reaction in carbohydrate metabolism in which glucose-6-phosphate reacts with $NADP^+$ to give 6-phosphoglucono-δ-lactone and NADPH.

Glucose-6-phosphate → 6-Phosphoglucono-δ-lactone (NADPH + H⁺ ; NADP⁺)

In this reaction, which substance is oxidized, and which is reduced? Which substance is the oxidizing agent, and which is the reducing agent?

32. Thought Question There is a reaction in which succinate reacts with FAD to give fumarate and $FADH_2$.

Succinate → Fumarate (FADH₂ ; FAD)

In this reaction, which substance is oxidized, and which is reduced? Which substance is the oxidizing agent, and which is the reducing agent?

33. Thought Question Suggest a reason that catabolic pathways generally produce NADH and $FADH_2$, whereas anabolic pathways generally use NADPH.

15.6 How Are the Production and Use of Energy Coupled?

34. Mathematical What substrate concentrations would be necessary to make the reaction in Question 14(c) a favorable reaction?

35. Mathematical Using the data in Table 15.1, calculate the value of $\Delta G^{\circ\prime}$ for the following reaction.

Creatine phosphate + Glycerol → Creatine + Glycerol-3-phosphate

Hint: This reaction proceeds in stages. ATP is formed in the first step, and the phosphate group is transferred from ATP to glycerol in the second step.

36. Mathematical Using information from Table 15.1, calculate the value of $\Delta G^{\circ\prime}$ for the following reaction.

Glucose-l-phosphate → Glucose-6-phosphate

37. Mathematical Show that the hydrolysis of ATP to AMP and 2 P_i releases the same amount of energy by either of the two following pathways.

Pathway 1

$$ATP + H_2O \rightarrow ADP + P_i$$

$$ADP + H_2O \rightarrow AMP + P_i$$

Pathway 2

$$ATP + H_2O \rightarrow AMP + PP_i \quad \text{(Pyrophosphate)}$$

$$PP_i + H_2O \rightarrow 2 P_i$$

38. Mathematical The standard free-energy change for the reaction

Arginine + ATP → Phosphoarginine + ADP

is +1.7 kJ mol^{-1}. From this information and that in Table 15.1, calculate the $\Delta G^{\circ\prime}$ for the reaction

Phosphoarginine + H_2O → Arginine + P_i

39. Thought Question What are the usual ionic forms of ATP and ADP in typical cells? Does this information have any bearing on the free-energy change for the conversion of ATP to ADP?

40. Thought Question Comment on the free energy of hydrolysis of the phosphate bond of ATP (-30.5 kJ mol^{-1}; -7.3 kcal mol^{-1}) relative to those of other organophosphates (e.g., sugar phosphates, creatine phosphate).

41. Thought Question A friend has seen creatine supplements for sale in a health-food store and asks why. What do you tell your friend?

42. Thought Question Would you expect an increase or a decrease of entropy to accompany the hydrolysis of phosphatidylcholine to the constituent parts (glycerol, two fatty acids, phosphoric acid, and choline)? Why?

43. Thought Question Explain and show why phosphoenolpyruvate is a high-energy compound.

44. **Thought Question** A very favorable reaction is the production of ATP and pyruvate from ADP and phosphoenolpyruvate. Given the standard free-energy change for this coupled reaction, why does the following reaction not occur?

$$PEP + 2\,ADP \rightarrow Pyruvate + 2\,ATP$$

45. **Thought Question** Short periods of exercise, such as sprints, are characterized by lactic acid production and the condition known as oxygen debt. Comment on this fact in light of the material discussed in this chapter.

15.7 How Is Coenzyme A Involved in Activation of Metabolic Pathways?

46. **Thought Question** Why is the process of activation a useful strategy in metabolism?

47. **Thought Question** What is the molecular logic that makes a pathway with a number of comparatively small energy changes more likely than a single reaction with a large energy change?

48. **Thought Question** Why are thioesters considered high-energy compounds?

49. **Thought Question** Explain why several biochemical pathways start by putting a coenzyme A onto the molecule that initiates the pathway.

50. **Thought Question** This is a conjectural question: If the reactive part of coenzyme A is the thioester, why is the molecule so complicated?

Biochemistry ⊗ Now™

Assess your understanding of this chapter's topics with additional quizzing and tutorials at **http://now.brookscole.com/campbell5**

Annotated Bibliography

Two standard multivolume references cover specific aspects of metabolism in detail. One of these, the third edition of *The Enzymes* (P. D. Boyer, ed. New York: Academic Press), is a series that has been in production since 1970. The other, *Comprehensive Biochemistry* (M. Florkin and E. H. Stotz, eds. New York: Elsevier), has been in production since 1962.

Atkins, P. W. *The Second Law.* San Francisco: W. H. Freeman, 1984. [A highly readable nonmathematical discussion of thermodynamics.]

Chang, R. *Physical Chemistry with Applications to Biological Systems,* 2nd ed. New York: Macmillan, 1981. [Chapter 12 contains a detailed treatment of thermodynamics.]

Fasman, G. D., ed. *Handbook of Biochemistry and Molecular Biology,* 3rd ed. Sec. D, *Physical and Chemical Data.* Cleveland: CRC Press, 1976.

[Volume 1 contains data on the free energies of hydrolysis of many important compounds, especially organophosphates.]

Harold, F. M. *The Vital Force: A Study of Bioenergetics.* New York: W. H. Freeman, 1986. [Energetic aspects of many important life processes.]

Hinkle, P. C., and R. E. McCarty. How Cells Make ATP. *Sci. Amer.* **238** (3), 104–125 (1978). [Getting old, but a particularly good treatment of energy coupling.]

Prigogine, I., and I. Stengers. *Order Out of Chaos.* Toronto: Bantam Press, 1984. [A comparatively accessible treatment of the thermodynamics of biological systems. Prigogine won the 1977 Nobel Prize in chemistry for his pioneering work on the thermodynamics of complex systems.]

CHAPTER 16

Carbohydrates

More than half of all the organic carbon on planet Earth is stored in just two carbohydrate molecules—starch and cellulose. Both are polymers of the sugar monomer, glucose. The only difference between them is the manner in which the glucose units are joined together. Glucose is made by green plants and stored in the form of starch as the plants' energy reserves. Animals (including humans) have an enzyme that recognizes the helical conformation of starch and can degrade it into its glucose units. Glucose, oxidized to carbon dioxide and water, is our primary energy source. Cellulose, a major component of plant cell walls and, therefore, of cotton and wood, is a polymer of glucose monomers, all of which lie in the same plane. Humans don't possess the enzyme cellulase, which is needed to break down cellulose, but termites do. A protozoan in their intestines contains cellulase, enabling them to digest wood. We learn this the hard way when wooden houses are damaged by termites. Modified carbohydrates are constituents of bacterial cell walls, where they are cross-linked by short polypeptide chains. By combining specific sugar monomers with amino acids, glycoproteins can be formed that function as cell-surface markers to be recognized by other biomolecules. This recognition feature is of life-or-death importance in blood transfusions, where compatibility of blood types depends on glycoproteins.

Like bread and pasta, fruits and vegetables are sources of carbohydrates, which provide energy.

Critical Questions

16.1 What Are the Structures and the Stereochemistry of Monosaccharides?

16.2 How Do Monosaccharides React?

16.3 What Are Some Important Oligosaccharides?

16.4 What Are the Structures and Functions of Polysaccharides?

16.5 What Are Glycoproteins?

Biochemistry ⊘ Now™

Test yourself on these Critical Questions at the BiochemistryNow website at **http://now .brookscole.com/campbell5**

When the word "carbohydrate" was coined, it originally referred to compounds of the general formula $C_n(H_2O)_n$. However, only the simple sugars, or **monosaccharides,** fit this formula exactly. The other types of carbohydrates, oligosaccharides and polysaccharides, are based on the monosaccharide units and have slightly different general formulas. **Oligosaccharides** are formed when a few (Greek *oligos*) monosaccharides are linked; polysaccharides are formed when many (Greek *polys*) monosaccharides are bonded together. The reaction that adds monosaccharide units to a growing carbohydrate molecule involves the loss of one H_2O for each new link formed, accounting for the difference in the general formula.

Many commonly encountered carbohydrates are polysaccharides, including glycogen, which is found in animals, and starch and cellulose, which occur in plants. Carbohydrates play a number of important roles in biochemistry. First, they are major energy sources (Chapters 17 through 20 are devoted to carbohydrate metabolism). Second, oligosaccharides play a key role in processes that take place on the surfaces of cells, particularly in cell–cell interactions and immune recognition. In addition, polysaccharides are essential structural components of several classes of organisms. Cellulose is a major component of grass and trees, and other polysaccharides are major components of bacterial cell walls.

16.1 | What Are the Structures and the Stereochemistry of Monosaccharides?

The building blocks of all carbohydrates are the simple sugars called **monosaccharides.** A monosaccharide can be a polyhydroxy aldehyde (**aldose**) or a polyhydroxy ketone **(ketose).** The simplest monosaccharides contain three carbon atoms and are called trioses (*tri* meaning three). *Glyceraldehyde* is the aldose with three carbons (an aldotriose), and *dihydroxyacetone* is the ketose with three carbon atoms (a ketotriose). Figure 16.1 shows these molecules. Aldoses with four, five, six, and seven carbon atoms are called aldotetroses, aldopentoses, aldohexoses, and aldoheptoses, respectively. The corresponding ketoses are ketotetroses, ketopentoses, ketohexoses, and ketoheptoses. Six-carbon sugars are the most abundant in nature, but two five-carbon sugars, ribose and deoxyribose, occur in the structures of RNA and DNA, respectively. Four-carbon and seven-carbon sugars play roles in photosynthesis and other metabolic pathways.

We have already seen (in Section 3.1) that some molecules are not superimposable on their mirror images and that these mirror images are **optical isomers (stereoisomers)** of each other. A chiral (asymmetric) carbon atom is the usual source of optical isomerism, as was the case with amino acids. The simplest carbohydrate that contains a chiral carbon is glyceraldehyde, which can exist in two isomeric forms that are mirror images of each other [Figure 16.1(b) and (c)]. Note that the two forms differ in the position of the hydroxyl group bonded to the central carbon. (Dihydroxyacetone does not contain a chiral carbon atom and does not exist in nonsuperimposable mirror-image forms.) The two forms of glyceraldehyde are designated D-glyceraldehyde and L-glyceraldehyde. Mirror-image stereoisomers are also called **enantiomers,** and D-glyceraldehyde and L-glyceraldehyde are enantiomers of each other. Certain conventions are used for two-dimensional drawings of the three-dimensional structures of stereoisomers. The dashed wedges represent bonds directed away from the viewer, below the plane of the paper, and the solid wedges represent bonds directed oppositely, toward the viewer and out of the plane of the paper. The **configuration** is the three-dimensional arrangement of groups around a chiral carbon atom, and stereoisomers differ from each other in configuration. The *D,L* system to denote stereochemistry is widely used by biochemists. Organic chemists tend to use a more recent one, designated the *R,S* system. There is not a one-to-one correspondence between the two systems. For example, some D-isomers are *R*, and some are *S*.

The two enantiomers of glyceraldehyde are the only possible stereoisomers of three-carbon sugars, but the possibilities for stereoisomerism increase as the number of carbon atoms increases. To show the structures of the resulting molecules, we need to say more about the convention for a two-dimensional perspective of the molecular structure, which is called the **Fischer-projection** method, after the German chemist Emil Fischer, who established the structures of many sugars. We shall use some common six-carbon sugars for purposes of illustration. In a Fischer projection, bonds written vertically on the two-dimensional paper represent bonds directed *behind* the paper in three dimensions, whereas bonds written horizontally represent bonds directed *in front of* the paper in three dimensions. Figure 16.2 shows that the most highly oxidized carbon—in this case, the one involved in the aldehyde group—is written at the "top" and is designated carbon 1, or C-1. In the ketose shown, the ketone group becomes C-2, the carbon atom next to the "top." Most common sugars are aldoses rather than ketoses, so our discussion will focus mainly on aldoses. The other carbon atoms are numbered in sequence from the "top." The designation of the configuration as L or D depends on the arrangement at the chiral carbon with the highest number. In the cases of both glucose and fructose, this is C-5. In the Fischer projection of the D configuration, the hydroxyl group is on the right of the highest-numbered chiral carbon, whereas the hydroxyl group is on the left of the highest-numbered chiral carbon in the L configuration. Let us see what happens as another

(a)

CH$_2$OH—CHOH—CH (with O double-bonded to CH)

Glyceraldehyde

CH$_2$OH—C—CH$_2$OH (with O double-bonded to C)

Dihydroxyacetone

(b)

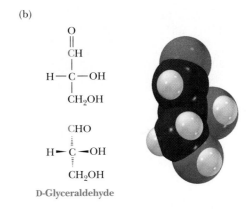

D-Glyceraldehyde

(c)

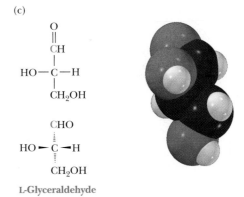

L-Glyceraldehyde

▲ **FIGURE 16.1** The structures of the simplest carbohydrates, the trioses. (a) A comparison of glyceraldehyde (an aldotriose) and dihydroxyacetone (a ketotriose). (b) The structure of D-glyceraldehyde and a space-filling model of D-glyceraldehyde. (c) The structure of L-glyceraldehyde and a space-filling model of L-glyceraldehyde. (*Leonard Lessin/Waldo Feng/Mt. Sinai CORE*)

▲ Emil Fischer (1852–1919) was a German-born scientist who won the Nobel prize in chemistry in 1902 for his studies on sugars, purine derivatives, and peptides.

Biochemistry ⑤ Now™

Go to BiochemistryNow and click on Biochemistry Interactive to learn how to identify the structures of simple sugars.

▶ **FIGURE 16.2** (a) Examples of an aldose (D-glucose) and a ketose (D-fructose), showing the numbering of carbon atoms. (b) A comparison of the structures of D-glucose and L-glucose.

carbon is added to glyceraldehyde to give a four-carbon sugar. In other words, what are the possible stereoisomers for an aldotetrose? The aldotetroses (Figure 16.3) have two chiral carbons, C-2 and C-3, and there are 2^2, or four, possible stereoisomers. Two of the isomers have the D configuration, and two have the L configuration. The two D isomers have the same configuration at C-3, but they differ in configuration (arrangement of the —OH group) at the other chiral carbon, C-2. These two isomers are called D-erythrose and D-threose. They are not superimposable on each other, but neither are they mirror images of each other. Such nonsuperimposable, nonmirror-image stereoisomers are called **diastereomers.** The two L isomers are L-erythrose and L-threose. L-Erythrose is the enantiomer (mirror image) of D-erythrose, and L-threose is the enantiomer of D-threose. L-Threose is a diastereomer of both D- and L-erythrose, and L-erythrose is a diastereomer of both D- and L-threose. Diastereomers that differ from each other in the configuration at only one chiral carbon are called **epimers;** D-erythrose and D-threose are epimers.

Aldopentoses have three chiral carbons, and there are 2^3, or 8, possible stereoisomers—four D forms and four L forms. Aldohexoses have four chiral carbons and 2^4, or 16, stereoisomers—eight D forms and eight L forms (Figure 16.4). Some of the possible stereoisomers are much more common in nature than others, and most biochemical discussion centers on the common, naturally occurring sugars. For example, D sugars, rather than L sugars, predominate in nature. Most of the sugars we encounter in nature, especially in foods, contain either five or six carbon atoms. We shall discuss D-glucose (an

(a)

		Carbon number
H⟍C⟋O	CH₂OH	1
H—C—OH	C=O	2
HO—C—H	HO—C—H	3
H—C—OH	H—C—OH	4
H—C—OH	H—C—OH	5
CH₂OH	CH₂OH	6
D-Glucose (an aldose)	D-Fructose (a ketose)	

(b)

		Carbon number
H⟍C⟋O	H⟍C⟋O	1
H—C—OH	HO—C—H	2
HO—C—H	H—C—OH	3
H—C—OH	HO—C—H	4
H—C—OH	HO—C—H	5
CH₂OH	CH₂OH	6
D-Glucose	L-Glucose	

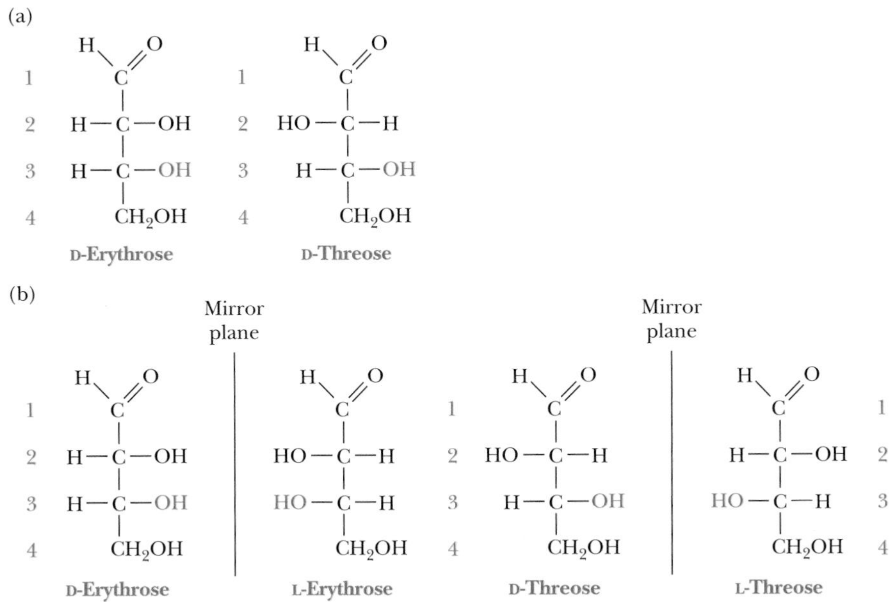

(a)

D-Erythrose   D-Threose

(b)

D-Erythrose   L-Erythrose   D-Threose   L-Threose

◄ **FIGURE 16.3** Stereoisomers of an aldotetrose. (a) Diastereomers D-erythrose and D-threose. (b) Enantiomers D- and L-erythrose and D- and L-threose. The carbons are numbered. The designation of D or L depends on the configuration at the highest-numbered chiral carbon atom.

aldohexose) and D-ribose (an aldopentose) far more than many other sugars. Glucose is a ubiquitous energy source, and ribose plays an important role in the structure of nucleic acids.

Cyclic Structures: Anomers

Sugars, especially those with five or six carbon atoms, normally exist as cyclic molecules rather than as the open-chain forms we have shown so far. The cyclization takes place as a result of interaction between the functional groups on distant carbons, such as C-l and C-5, to form a cyclic **hemiacetal** (in aldohexoses). Another possibility (Figure 16.5) is interaction between C-2 and C-5 to form a cyclic **hemiketal** (in ketohexoses). In either case, the carbonyl carbon becomes a new chiral center called the **anomeric carbon.** The cyclic sugar can take either of two different forms, designated α and β, and are called **anomers** of each other.

The Fischer projection of the α-anomer of a D sugar has the anomeric hydroxyl group to the right of the anomeric carbon (C—OH), and the β-anomer of a D sugar has the anomeric hydroxyl group to the left of the anomeric carbon (Figure 16.6). The free carbonyl species can readily form either the α- or β-anomer, and the anomers can be converted from one form to another through the free carbonyl species. In some biochemical molecules, any anomer of a given sugar can be used, but, in other cases, only one anomer occurs. For example, in living organisms, only β-D-ribose and β-D-deoxyribose are found in RNA and DNA, respectively.

Fischer projection formulas are useful for describing the stereochemistry of sugars, but their long bonds and right-angle bends do not give a realistic picture of the bonding situation in the cyclic forms, nor do they accurately represent the overall shape of the molecules. **Haworth projection formulas** are more useful for those purposes. In Haworth projections, the cyclic structures of sugars are shown in perspective drawings as planar five- or six-membered rings viewed nearly edge on. A five-membered ring is called a **furanose** because of its resemblance to furan; a six-membered ring is called a **pyranose**

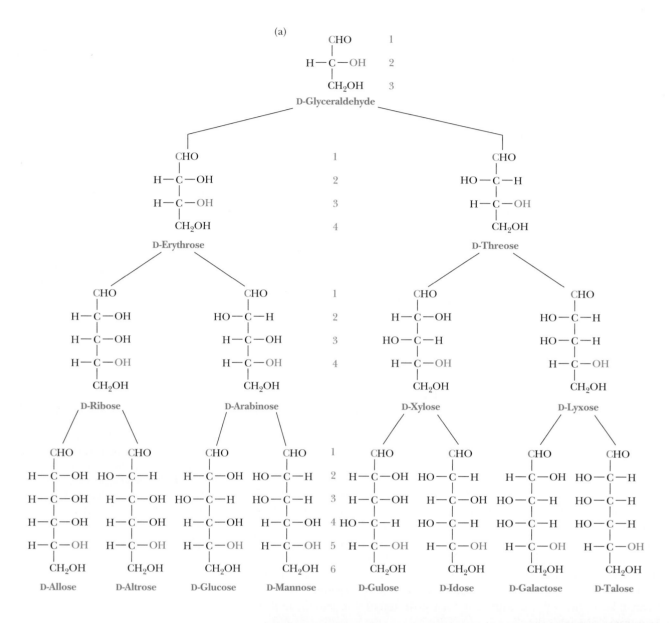

▲ **FIGURE 16.4** Stereochemical relationships among monosaccharides. (a) Aldoses containing from three to six carbon atoms, with the numbering of the carbon atom shown. Note that the figure shows only half the possible isomers. For each isomer shown, there is an enantiomer that is not shown, the L series. (b) The relationship between mirror images is of interest to mathematicians as well as to chemists. Lewis Carroll (C. L. Dodgson), the author of *Alice's Adventures in Wonderland*, was a contemporary of Emil Fischer.

Courtesy of Beverlee Tito Simboli

Alcohol   **Aldehyde**   **Hemiacetal**

D-Glucose

Pyran

Cyclization

α-D-Glucopyranose

β-D-Glucopyranose

HAWORTH PROJECTION FORMULAS

α-D-Glucopyranose

β-D-Glucopyranose

FISCHER PROJECTION FORMULAS

Biochemistry☯Now™ ▲ ANIMATED FIGURE 16.5 The linear form of D-glucose undergoes an intramolecular reaction to form a cyclic hemiacetal. **See this figure animated at http://now .brookscole.com/campbell5**

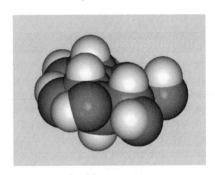

β-D-Glucopyranose

α-Configuration
OH to right of
anomeric carbon

β-Configuration
OH to left of
anomeric carbon

α-D-Glucose

*Reacts with CH=O
to form hemiacetal
Open chain form

β-D-Glucose

◀ **FIGURE 16.6** Fischer projection formulas of three forms of glucose. Note that the α and β forms can be converted to each other through the open-chain form. The configuration at carbon 5 determines the D designation.

because of its resemblance to pyran [Figure 16.7(a) and (b)]. These cyclic formulas approximate the shapes of the actual molecules better for furanoses than for pyranoses. The five-membered rings of furanoses are in reality very nearly planar, but the six-membered rings of pyranoses actually exist in solu- tion in the chair conformation [Figure 16.7(c)]. The chair conformation is

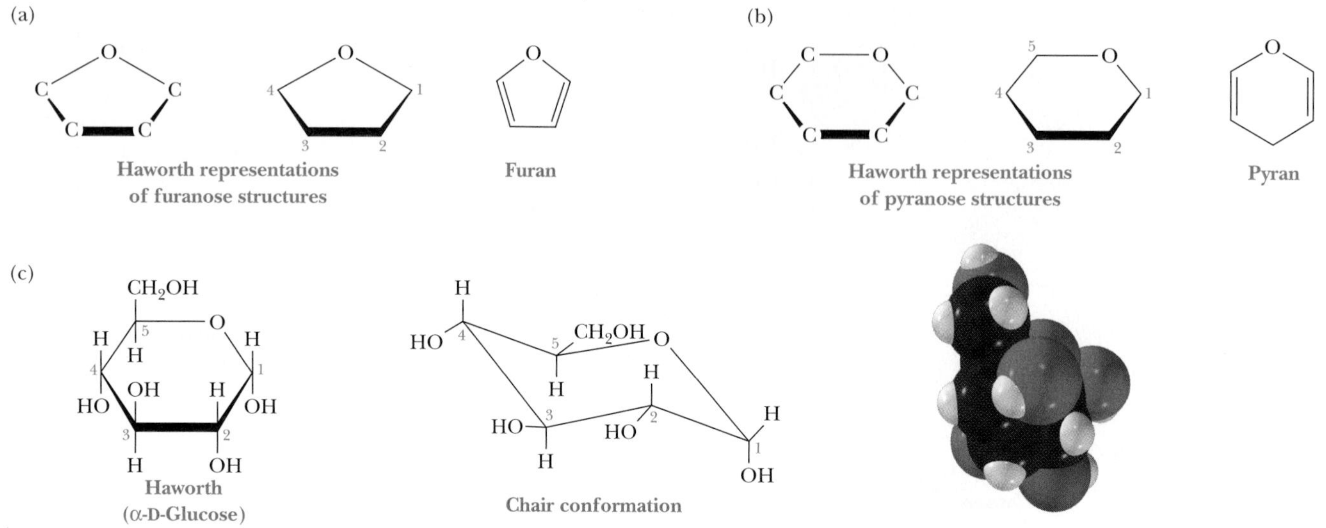

(a)

Haworth representations
of furanose structures

Furan

(b)

Haworth representations
of pyranose structures

Pyran

(c)

Haworth
(α-D-Glucose)

Chair conformation

▲ **FIGURE 16.7** Haworth representations of sugar structures. (a) A comparison of the structure of furan with Haworth representations of furanoses. (b) A comparison of the structure of pyran with Haworth representations of pyranoses. (c) α-D-Glucopyranose in the Haworth representation (left), in the chair conformation (middle), and as a space-filling model (right). (*Leonard Lessin/ Waldo Feng/Mt. Sinai CORE*)

Essential Information

Most sugars exist in cyclic forms with five- or six-membered rings. The cyclization process involves the carbonyl group and gives rise to another chiral center in addition to the ones already present in the sugar molecule. The two possible cyclic isomers, called *anomers*, are designated α and β.

widely shown in textbooks of organic chemistry. This kind of structure is particularly useful in discussions of molecular recognition. The chair conformation and the Haworth projections are alternative ways of expressing the same information. Even though the Haworth formulas are approximations, they are useful shorthand for the structures of reactants and products in many reactions that we are going to see. The Haworth projections represent the stereochemistry of sugars more realistically than do the Fischer projections, and the Haworth scheme is adequate for our purposes. That is why biochemists use them, even though organic chemists prefer the chair form. We shall continue to use Haworth projections in our discussion of sugars.

For a D sugar, any group that is written to the *right* of the carbon in a Fischer projection has a *downward* direction in a Haworth projection; any group that is written to the *left* in a Fischer projection has an *upward* direction in a Haworth projection. The terminal —CH$_2$OH group, which contains the carbon atom with the highest number in the numbering scheme, is shown in an upward direction. The structures of α- and β-D-glucose, which are both pyranoses, and of β-D-ribose, which is a furanose, illustrate this point (Figure 16.8). Note that, in the α-anomer, the hydroxyl on the anomeric carbon is on the opposite side of the ring from the terminal —CH$_2$OH group (i.e., pointing down). In the β-anomer, it is on the same side of the ring (pointing up). The same convention holds for α- and β-anomers of furanoses.

16.2 | How Do Monosaccharides React?

Oxidation–Reduction Reactions

Oxidation and reduction reactions of sugars play key roles in biochemistry. Oxidation of sugars provides energy for organisms to carry out their life processes; the highest yield of energy from carbohydrates occurs when sugars are completely oxidized to CO$_2$ and H$_2$O in aerobic processes. The reverse of

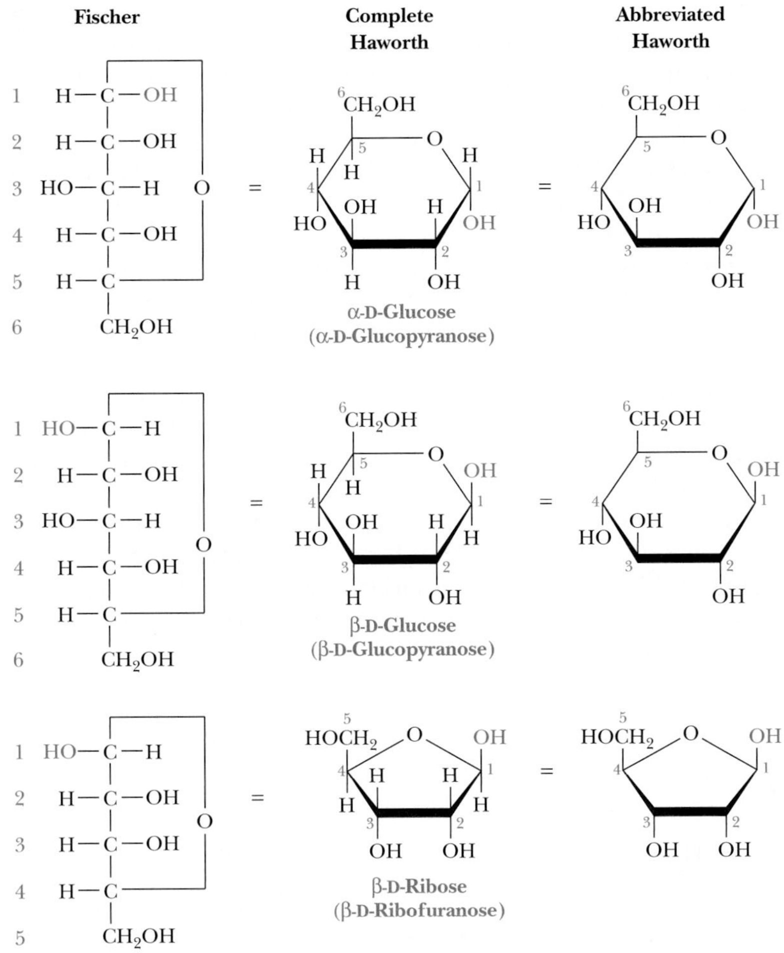

Fischer	Complete Haworth	Abbreviated Haworth

α-D-Glucose
(α-D-Glucopyranose)

β-D-Glucose
(β-D-Glucopyranose)

β-D-Ribose
(β-D-Ribofuranose)

◀ **FIGURE 16.8** A comparison of the Fischer, complete Haworth, and abbreviated Haworth representations of α- and β-D-glucose (glucopyranose) and β-D-ribose (ribofuranose). In the Haworth representation, the α-anomer is represented with the OH group (red) downward, and the β-anomer is represented with the OH group (red) upward.

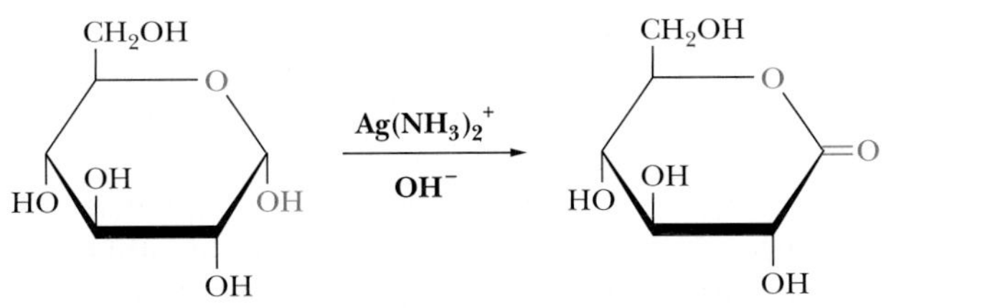

◀ **FIGURE 16.9** An example of an oxidation reaction of sugars: oxidation of α-D-glucose hemiacetal to give a lactone. Deposition of free silver as a silver mirror indicates that the reaction has taken place.

complete oxidation of sugars is the reduction of CO_2 and H_2O to form sugars, a process that takes place in photosynthesis.

Several oxidation reactions of sugars are of some importance in laboratory practice because they can be used to identify sugars. Aldehyde groups can be oxidized to give the carboxyl group that is characteristic of acids, and this reaction is the basis of a test for the presence of aldoses. When the aldehyde is oxidized, some oxidizing agent must be reduced. Aldoses are called **reducing sugars** because of this type of reaction; ketoses can also be reducing sugars because they isomerize to aldoses. In the cyclic form, the compound produced by oxidation of an aldose is a *lactone* (a cyclic ester linking the carboxyl group and one of the sugar alcohols, as shown in Figure 16.9). A lactone of considerable importance to humans is discussed in the Biochemical Connections box on page 443.

▲ By adding Tollens reagent to an aldehyde, a silver mirror has been deposited in the inside of this flask.

Two types of reagent are used in the laboratory to detect the presence of reducing sugars. The first of these is Tollens reagent, which uses the silver ammonia complex ion, $Ag(NH_3)_2^+$ as the oxidizing agent. A silver mirror is deposited on the wall of the test tube if a reducing sugar is present, as a result of the Ag^+ in the complex ion being reduced to free silver metal. A more recent method for detection of glucose, but not other reducing sugars, is based on the use of the enzyme glucose oxidase, which is specific for glucose.

In addition to oxidized sugars, there are some important reduced sugars. In *deoxy sugars*, a hydrogen atom is substituted for one of the hydroxyl groups of the sugar. One of these deoxy sugars is L-fucose (L-6-deoxygalactose), which is found in the carbohydrate portions of some glycoproteins (Figure 16.10), including the ABO blood-group antigens. The name "glycoprotein" indicates that these substances are conjugated proteins that contain some carbohydrate group (*glykos* is Greek for "sweet") in addition to the polypeptide chain. An even more important example of a deoxy sugar is D-2-deoxyribose, the sugar found in DNA (Figure 16.10).

When the carbonyl group of a sugar is reduced to a hydroxyl group, the resulting compound is one of the polyhydroxy alcohols known as *alditols*. Two compounds of this kind, xylitol and sorbitol, derivatives of the sugars xylulose and sorbose, respectively, have commercial importance as sweeteners in sugar-less chewing gum and candy.

D-Sorbose	D-Sorbitol	D-Xylitol	D-Xylulose
CH_2OH	CH_2OH	CH_2OH	CH_2OH
C=O	H—C—OH	H—C—OH	C=O
H—C—OH	H—C—OH	HO—C—H	HO—C—H
HO—C—H	HO—C—H	H—C—OH	H—C—OH
H—C—OH	H—C—OH	CH_2OH	CH_2OH
CH_2OH	CH_2OH		

Esterification Reactions

The hydroxyl groups of sugars behave exactly like all other alcohols in the sense that they can react with acids and derivatives of acids to form esters. The phosphate esters are particularly important because they are the usual

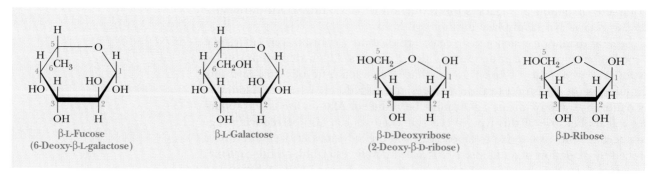

β-L-Fucose
(6-Deoxy-β-L-galactose)

β-L-Galactose

β-D-Deoxyribose
(2-Deoxy-β-D-ribose)

β-D-Ribose

▲ **FIGURE 16.10** Structures of two deoxy sugars. The structures of the parent sugars are shown for comparison.

Biochemical Connections

Vitamin C Is Related to Sugars

Vitamin C (ascorbic acid) is an unsaturated lactone with a five-membered ring structure. Each carbon is bonded to a hydroxyl group, except for the carboxyl carbon that is involved in the cyclic ester bond. Most animals can synthesize vitamin C; the exceptions are guinea pigs and primates, including humans. As a result, guinea pigs and primates must acquire vitamin C in their diet. Air oxidation of ascorbic acid, followed by hydrolysis of the ester bond, leads to loss of activity as a vitamin. Consequently, a lack of fresh food can cause vitamin C deficiencies, which, in turn, can lead to the disease scurvy (Section 4.3). In this disease, defects in collagen structure cause skin lesions and fragile blood vessels. The presence of hydroxyproline is necessary for collagen stability because of hydrogen-bonded cross-links between collagen strands. Ascorbic acid, in turn, is essential for the activity of prolyl hydroxylase, which converts proline residues in collagen to hydroxyproline. Lack of ascorbic acid eventually leads to the fragile collagen responsible for the symptoms of scurvy.

The British navy introduced citrus fruit into the diet of sailors in the 18th century to prevent scurvy during long sea voyages, and citrus fruit is still consumed by many for its vitamin C. Potatoes are another important source of vitamin C, not because potatoes contain a *high* concentration of ascorbic acid but because we eat so many potatoes.

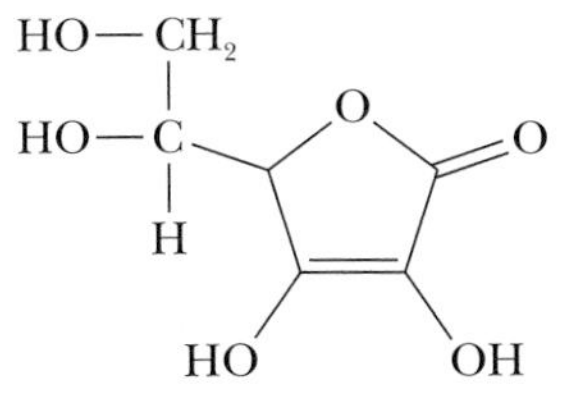

Ascorbic acid
(Vitamin C)

intermediates in the breakdown of carbohydrates to provide energy. Phosphate esters are frequently formed by transfer of a phosphate group from ATP (adenosine triphosphate) to give the phosphorylated sugar and ADP (adenosine diphosphate), as shown in Figure 16.11. Such reactions play an important role in the metabolism of sugars (Section 17.2).

The Formation of Glycosides

It is possible for a sugar hydroxyl group (ROH) bonded to the anomeric carbon to react with another hydroxyl (R′—OH) to form a glycosidic linkage (R′—O—R). A glycosidic linkage is *not* an ether (the R′—O—R notation is misleading) because glycosides can be hydrolyzed to the original alcohols. This type of reaction involves the anomeric carbon of the sugar in its cyclic form. (Recall that the anomeric carbon is the carbonyl carbon of the open-chain form of the sugar and is the one that becomes a chiral center in the cyclic form.) Stated in a slightly different way, a hemiacetal carbon can react

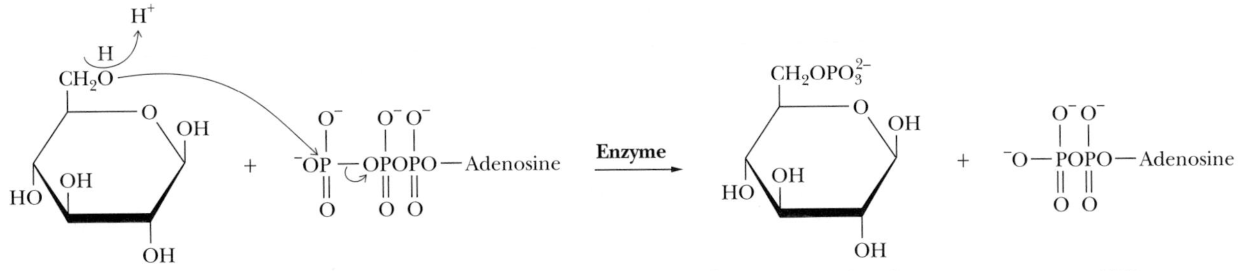

▲ **FIGURE 16.11** The formation of a phosphate ester of glucose. ATP is the phosphate group donor. The enzyme specifies the interaction with —CH₂OH on carbon 6.

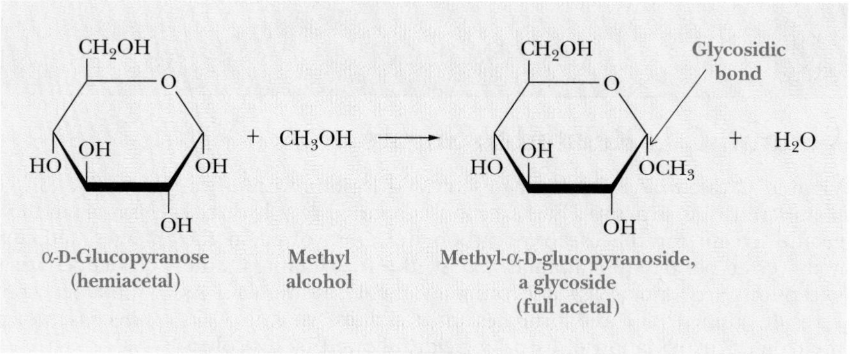

► **FIGURE 16.12** An example of the formation of a glycoside. Methyl alcohol (CH_3OH) and an α-D-glucopyranose react to form the corresponding glycoside.

with an alcohol such as methyl alcohol to give a *full acetal,* or **glycoside** (Figure 16.12). The newly formed bond is called a **glycosidic bond.** The glycosidic bonds discussed in this chapter are *O*-glycosides, with each sugar bonded to an oxygen atom of another molecule. (We encountered *N*-glycosides in Chapter 9 when we discussed nucleosides and nucleotides, in which the sugar is bonded to a nitrogen atom of a base.) Glycosides derived from furanoses are called **furanosides,** and those derived from pyranoses are called **pyranosides.**

Glycosidic bonds between monosaccharide units are the basis for the formation of oligosaccharides and polysaccharides. Glycosidic linkages can take various forms; the anomeric carbon of one sugar can be bonded to any one of the —OH groups on a second sugar to form an α- or β-glycosidic linkage. Many different combinations are found in nature. The —OH groups are numbered so that they can be distinguished, and the numbering scheme follows that of the carbon atoms. The notation for the glycosidic linkage between the two sugars specifies which anomeric form of the sugar is involved in the bond and also specifies which carbon atoms of the two sugars are linked together. Two ways in which two α-D-glucose molecules can be linked together are α(1→4) and α(1→6). In the first example, the α anomeric carbon (C-l) of the first glucose molecule is joined in a glycosidic bond to the fourth carbon atom (C-4) of the second glucose molecule; the C-1 of the first glucose molecule is linked to the C-6 of the second glucose molecule in the second example (Figure 16.13). Another possibility of a glycosidic bond, this time between two β-D-glucose molecules, is a β,β(1→1) linkage. The anomeric forms at both C-l carbons must be specified because the linkage is between the two anomeric carbons, each of which is C-l (Figure 16.14).

When oligosaccharides and polysaccharides form as a result of glycosidic bonding, their chemical natures depend on which monosaccharides are linked together and also on the particular glycosidic bond formed (i.e., which anomers and which carbon atoms are linked together). The difference between cellulose and starch depends on the glycosidic bond formed between glucose monomers. Because of the variation in glycosidic linkages, both linear and branched-chain polymers can be formed. If the internal monosaccharide

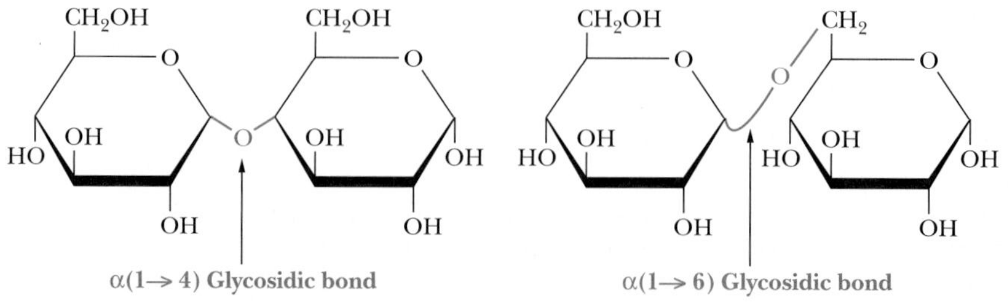

▲ **FIGURE 16.13** Two different disaccharides of α-D-glucose. These two chemical compounds have different properties because one has an α(1→4) linkage and the other has an α(1→6) linkage.

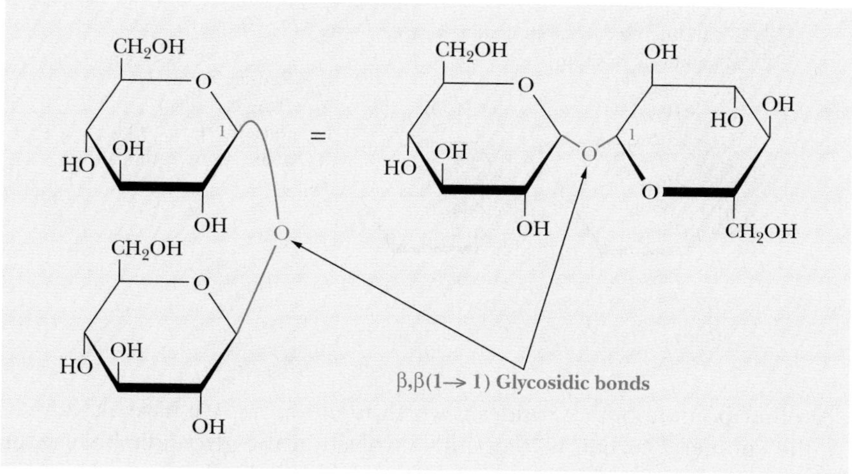

◀ **FIGURE 16.14** A disaccharide of β-D-glucose. Both anomeric carbons (C-1) are involved in the glycosidic linkage.

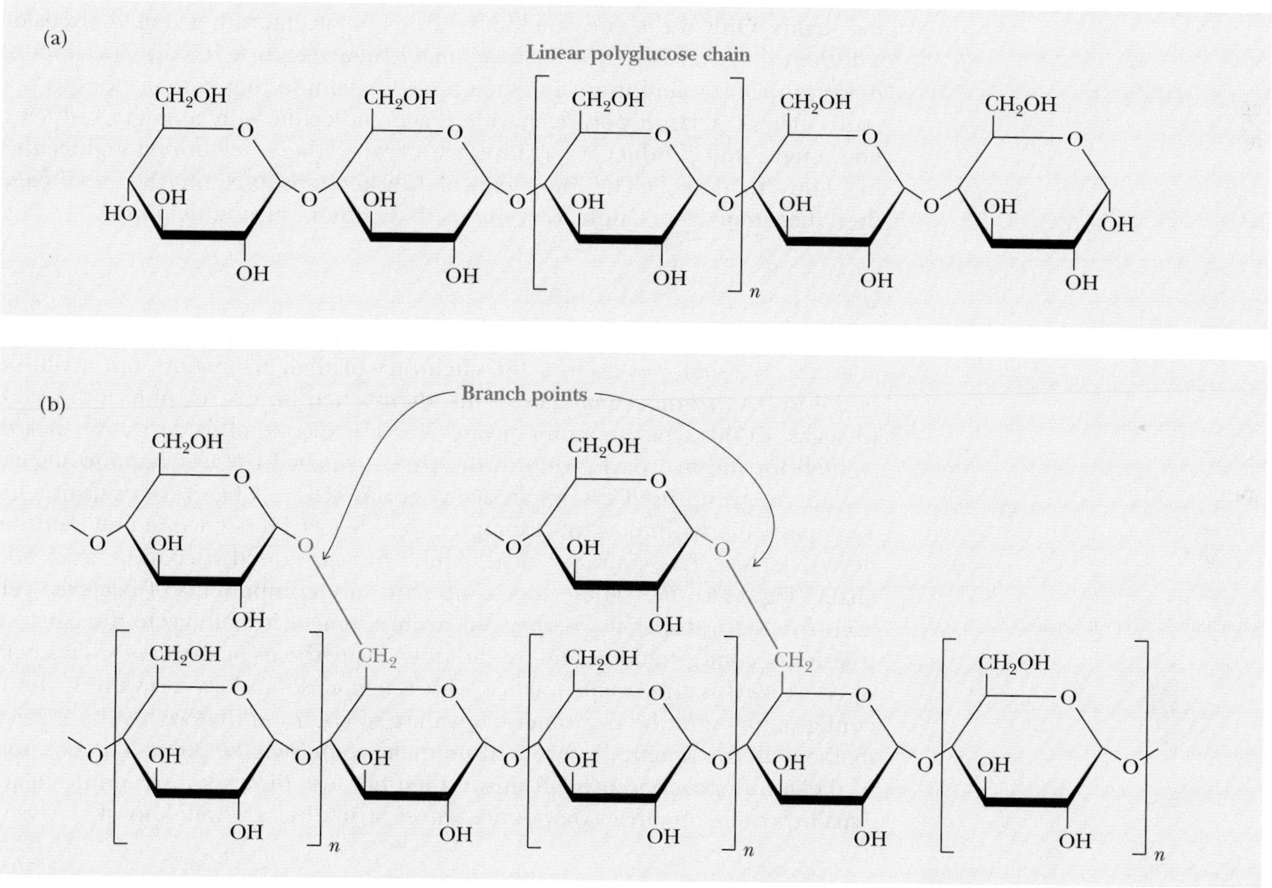

▲ **FIGURE 16.15** Linear and branched-chain polymers of α-D-glucose. The linear polyglucose chain (a) occurs in amylose, and the branched-chain polymer (b) occurs in amylopectin and glycogen (Section 13.4). All glycosidic bonds in (a) are α(1→4). Branched-polyglucose-chain glycosidic bonds in (b) are α(1→6) at branched points, but all glycosidic bonds along the chain are α(1→4).

residues that are incorporated in a polysaccharide form only two glycosidic bonds, the polymer will be linear. (Of course, the end residues will be involved in only one glycosidic linkage.) Some internal residues can form three glycosidic bonds, leading to the formation of branched-chain structures (Figure 16.15).

Essential Information

Glycosidic linkages are responsible for the bonding of monosaccharides to form oligosaccharides and polysaccharides. The reaction in question takes place when one sugar hydroxyl group forms a bond with another sugar hydroxyl, usually one on an anomeric carbon. Different stereochemical forms are possible in glycosidic linkages, having important consequences for the function of the substances thus formed.

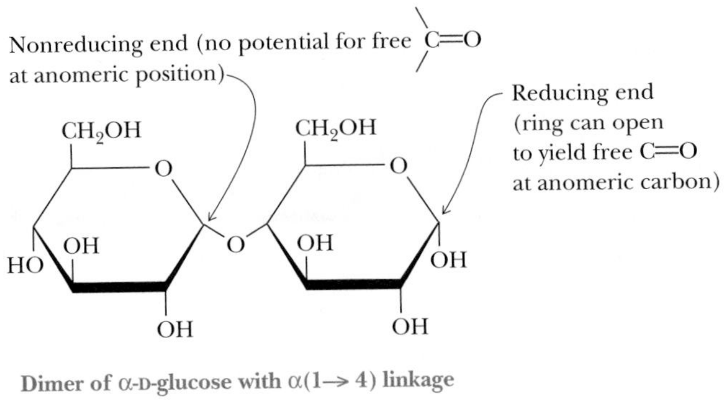

Dimer of α-D-glucose with α(1→ 4) linkage

► **FIGURE 16.16** A disaccharide with a free hemiacetal end is a reducing sugar because of the presence of a free anomeric aldehyde carbonyl or potential aldehyde group.

Another point about glycosides is worth mentioning. We have already seen that the anomeric carbon is frequently involved in the glycosidic linkage, and also that the test for the presence of sugars—specifically for reducing sugars—requires a reaction of the group at the anomeric carbon. The internal anomeric carbons in oligosaccharides are not free to give the test for reducing sugars. Only if the end residue is a free hemiacetal rather than a glycoside will there be a positive test for a reducing sugar (Figure 16.16). The level of detection can be important for such a test. A sample that contains only a few molecules of a large polysaccharide, each molecule with a single reducing end, might well produce a negative test because there are not enough reducing ends to detect. The Biochemical Connections box on the next page describes some interesting compounds that contain glycosidic bonds.

Other Derivatives of Sugars

Amino sugars are an interesting class of compounds related to the monosaccharides. We shall not go into the chemistry of their formation, but it will be useful to have some acquaintance with them when we discuss polysaccharides. In sugars of this type, an amino group (—NH$_2$) or one of its derivatives is substituted for the hydroxyl group of the parent sugar. In N-acetyl amino sugars, the amino group itself carries an acetyl group (CH$_3$—CO—) as a substituent. Two particularly important examples are N-acetyl-β-D-glucosamine and its derivative N-acetyl-β-muramic acid, which has an added carboxylic acid side chain (Figure 16.17). These two compounds are components of bacterial cell walls. We did not specify whether N-acetylmuramic acid belongs to the L or the D series of configurations, and we did not specify the α- or β-anomer. This type of shorthand is the usual practice with β-D-glucose and its derivatives; the D configuration and the β-anomeric form are so common that we need not specify them all the time unless we want to make some specific point. The position of the amino group is also left unspecified because discussion of amino sugars usually centers on a few compounds whose structures are well known.

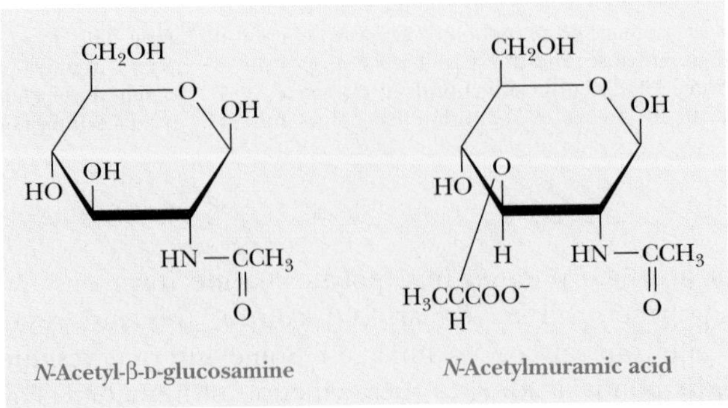

N-Acetyl-β-D-glucosamine   N-Acetylmuramic acid

► **FIGURE 16.17** The structures of N-acetyl-β-D-glucosamine and N-acetylmuramic acid.

Biochemical Connections

Glycosides, Fruits, and Flowers

In sucrose, starches, and other sugar polymers, the *O*-glycoside bonds attach sugars to sugars. Other major categories of glycosides are known in which the sugar binds to some other type of molecule. Probably the most common example is the structure of nucleotides (Section 9.2), *N*-glycosides, in which the sugar binds to the nitrogenous, aromatic base, as found in ATP, many vitamins, DNA, and RNA. In glycolipids (Section 8.2) and glycoproteins (Section 16.5), carbohydrates are attached to both lipids and proteins, respectively, by glycoside linkages.

The red and blue colors of some flowers are sugar derivatives, often called anthocyanins. These pigments involve various sugars bonded to the compound cyanidin and its derivatives. These compounds are water soluble because of the polar groups they possess. You may have done an acid–base titration of the pigment from red cabbage or from blueberry juice in a chemistry lab. In contrast, orange, yellow, and green plant pigments tend to be lipid in composition and insoluble in water.

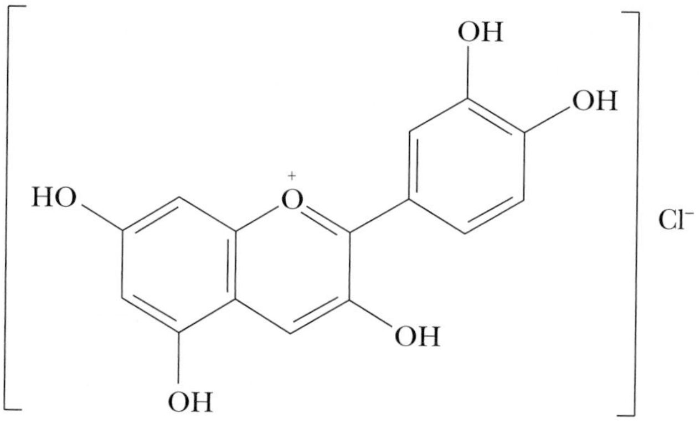

Cyanidin chloride

Many flavors involve sugar glycosides. Two familiar ones are cinnamon and vanilla, in which the sugars bond to cinnamaldehyde (3-phenyl-2-propenal) and vanillin, respectively. Both of these compounds are aromatic aldehydes. The distinctive taste of the kernel in a peach or apricot pit (a bitter-almond flavor) is due to laetrile, a controversial substance suggested as a cancer treatment by some.

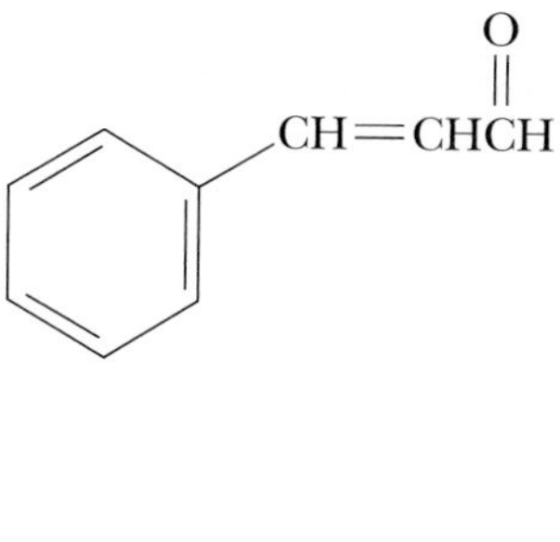

Cinnamaldehyde

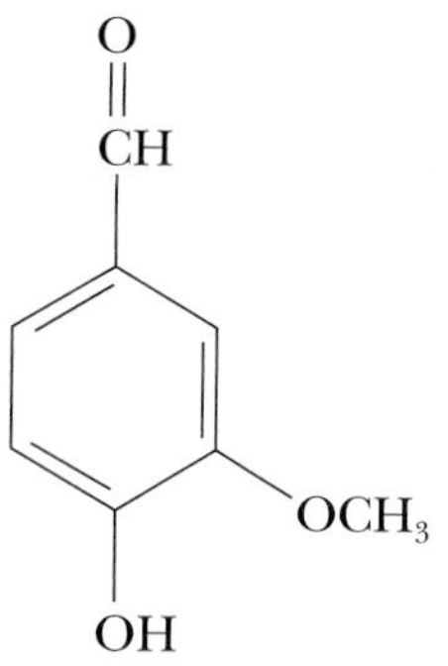

Vanillin

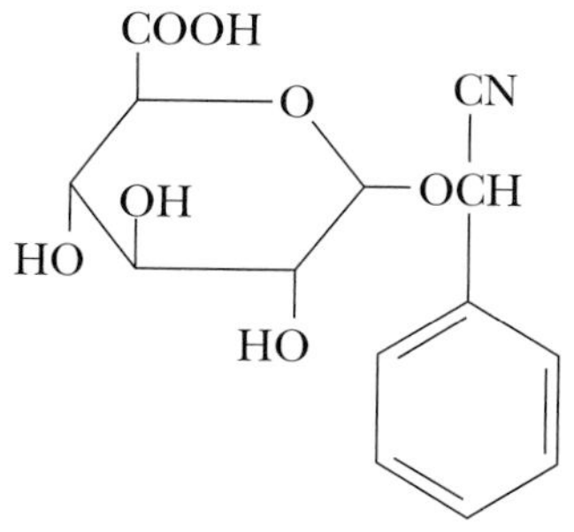

Laetrile

Many medically important substances have a glycosidic linkage as a part of their structure. Digitalis, prescribed for irregular heartbeat, is a mixture of several steroid complexes with sugars attached. Laetrile, a benzaldehyde derivative with a glycosidic linkage to glucuronic acid, was once thought to fight cancer, possibly because the cyanide moiety would poison the fast-growing cancer cells. This treatment is not approved in the United States, and it is likely that the cyanide causes more problems than it solves. The National Cancer Institute maintains a website, for which the URL is http://www.cancer.gov. Once there, use the search function to find information about laetrile.

Photodisc Green/Getty Images

▲ The foxglove plant produces the important cardiac medication digitalis.

16.3 | What Are Some Important Oligosaccharides?

Oligomers of sugars frequently occur as **disaccharides,** formed by linking two monosaccharide units by glycosidic bonds. Three of the most important examples of oligosaccharides are disaccharides. They are sucrose, lactose, and maltose (Figure 16.18). Two other disaccharides, isomaltose and cellobiose, are shown for comparison.

Sucrose is common table sugar, which is extracted from sugarcane and sugar beets. The monosaccharide units that make up sucrose are α-D-glucose and β-D-fructose. Glucose (an aldohexose) is a pyranose, and fructose (a keto-hexose) is a furanose. The α C-l carbon of the glucose is linked to the β C-2 carbon of the fructose (Figure 16.18) in a glycosidic linkage that has the notation α,β(1→2). Sucrose is not a reducing sugar because both anomeric groups are involved in the glycosidic linkage. Free glucose is a reducing sugar, and free fructose can also give a positive test, even though it is a ketone rather than an aldehyde in the open-chain form. Fructose and ketoses in general can act as reducing sugars because they can isomerize to aldoses in a rather complex rearrangement reaction. (We need not concern ourselves with the details of this isomerization.)

When animals consume sucrose, it is hydrolyzed to glucose and fructose, which are then degraded by metabolic processes to provide energy. Humans consume large quantities of sucrose, and excess consumption can contribute to health problems, which fact has led to a search for other sweetening

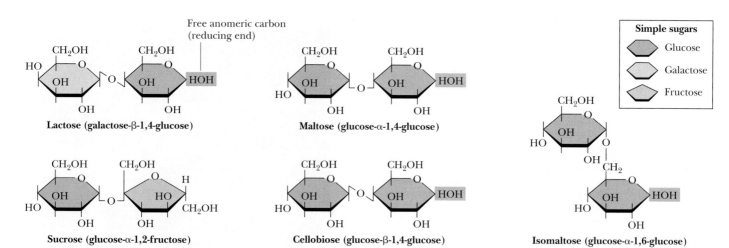

Lactose (galactose-β-1,4-glucose)

Maltose (glucose-α-1,4-glucose)

Sucrose (glucose-α-1,2-fructose)

Cellobiose (glucose-β-1,4-glucose)

Isomaltose (glucose-α-1,6-glucose)

Free anomeric carbon (reducing end)

Simple sugars
Glucose
Galactose
Fructose

Biochemistry Now™ ▲ **ACTIVE FIGURE 16.18** The structures of several important disaccharides. Note that the notation —HOH means that the configuration can be either α or β. When a D sugar is drawn in this orientation, if the —OH group is above the ring, the configuration is termed β. The configuration is termed α if the —OH group is below the ring. Also note that sucrose has no free anomeric carbon atoms. **Watch this Active Figure at http://now.brookscole .com/campbell5**

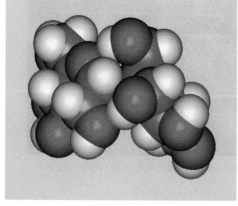

Sucrose

agents. One that has been proposed is fructose itself. It is sweeter than sucrose; therefore, a smaller amount (by weight) of fructose than sucrose can produce the same sweetening effect with fewer calories. Consequently, high-fructose corn syrup is frequently used in food processing. The presence of fructose changes the texture of food, and the reaction to the change tends to depend on the preference of the consumer. Artificial sweeteners have been produced in the laboratory and have frequently been suspected of having harmful side effects; the ensuing controversies bear eloquent testimony to the human craving for sweets. Saccharin, for example, has been found to cause cancer in laboratory animals, as have cyclamates, but the applicability of these results to human carcinogenesis has been questioned by some. Aspartame (NutraSweet; Section 3.5) has been suspected of causing neurological problems, especially in individuals whose metabolisms cannot tolerate phenylalanine (see the Biochemical Connections box on page 73).

Another artificial sweetener is a derivative of sucrose. This substance, sucralose, which is marketed under the trade name Splenda, differs from sucrose in two ways. The first difference is that three of the hydroxyl groups have been replaced with three chlorine atoms. The second is that the configuration at carbon atom number four of the six-membered pyranose ring of glucose has been inverted, producing a galactose derivative. The three hydroxyl groups that have been replaced by chlorine atoms are those bonded to carbon atoms one and six of the fructose moiety and to carbon atom four of the galactose moiety. Sucralose is not metabolized by the body, and, consequently, it does not provide calories. Tests conducted so far, as well as anecdotal evidence, indicate that it is a safe sugar substitute. It is likely to find wide use in the near future. It is safe to predict that the search for nonfattening sweeteners will continue and that it will be accompanied by controversy.

Lactose (see the Biochemical Connections box on page 450) is a disaccharide made up of β-D-galactose and D-glucose. Galactose is the C-4 epimer of glucose. In other words, the only difference between glucose and galactose is inversion of configuration at C-4. The glycosidic linkage is β(1→4), between the anomeric carbon C-1 of the β form of galactose and the C-4 carbon of glucose (Figure 16.18). Since the anomeric carbon of glucose is not involved in the glycosidic linkage, it can be in either the α or the β form. The two anomeric forms of lactose can be specified, and the designation refers to the glucose residue; galactose must be present as the β-anomer, since the β form of galactose is required by the structure of lactose. Lactose is a reducing sugar because the group at the anomeric carbon of the glucose portion is not involved in a glycosidic linkage, so it is free to react with oxidizing agents.

Maltose is a disaccharide obtained from the hydrolysis of starch. It consists of two residues of D-glucose in an α(1→4) linkage. Maltose differs from *cellobiose*, a disaccharide that is obtained from the hydrolysis of cellulose, only in the glycosidic linkage. In cellobiose, the two residues of D-glucose are bonded together in a β(1→4) linkage (Figure 16.18). Mammals can digest maltose, but not cellobiose. Yeast, specifically brewer's yeast, contains enzymes that hydrolyze the starch in sprouted barley (barley malt) first to maltose and then to glucose, which is fermented in the brewing of beer. Maltose is also used in other beverages, such as malted milk.

16.4 What Are the Structures and Functions of Polysaccharides?

When many monosaccharides are linked together, the result is a **polysaccharide.** Polysaccharides that occur in organisms are usually composed of a very few types of monosaccharide components. A polymer that consists of only one type of monosaccharide is a *homopolysaccharide;* a polymer that consists of

Biochemical Connections

Lactose Intolerance

Humans can be intolerant of milk and milk products for several reasons. Sugar intolerance results from the inability either to digest or to metabolize certain sugars. This problem differs from a food allergy, which involves an immune response (Section 14.5). A negative reaction to sugars in the diet usually involves intolerance, whereas proteins, including those found in milk, tend to cause allergies. Most sugar intolerance is due to missing or defective enzymes, so this is another example of inborn errors of metabolism.

Lactose is sometimes referred to as milk sugar because it occurs in milk. In some adults, a deficiency of the enzyme lactase in the intestinal villi causes a buildup of the disaccharide when milk products are ingested. This is because lactase is necessary to degrade lactose to galactose and glucose so that it can be absorbed into the bloodstream from the villi. Without the

Charles D. Winters

▲ These products help those with lactose intolerance to meet their calcium needs.

enzyme, an accumulation of lactose in the intestine can be acted on by the lactase of intestinal bacteria (as opposed to the desirable lactase of the villi), producing hydrogen gas, carbon dioxide, and organic acids. The products of the bacterial lactase reaction lead to digestive problems, such as bloating and diarrhea, as does the presence of undegraded lactose. In addition, the by-products of the extra bacterial growth draw water into the intestine, thus aggravating the diarrhea. This disorder affects only about one-tenth of the Caucasian population of the United States, but it is more common among African-Americans, Asians, Native Americans, and Hispanics.

Even if the enzyme lactase is present so that lactose can be broken down by the body, other problems can occur. A different but related problem can occur in the further metabolism of galactose. If the enzyme that catalyzes a subsequent reaction in the pathway is missing and galactose builds up, a condition known as galactosemia can result. This is a severe problem in infants because the nonmetabolized galactose accumulates within cells and is converted to the hydroxy sugar galactitol, which cannot escape. Water is drawn into these cells and the swelling and edema causes damage. The critical tissue is the brain, which is not fully developed at birth. The swelling cells can crush the brain tissue, resulting in severe and irreversible retardation. The clinical test for this disorder is inexpensive and is required by law in all states.

The dietary therapy for these two problems is quite different. Lactose-intolerant individuals must avoid lactose throughout their lives. Fortunately, tablets like Lactaid are available to add to regular milk, as are lactose- and galactose-free formulas for feeding infants. True fermented food products such as yogurt and many cheeses (especially aged ones) have had their lactose degraded during fermentation. However, many foods are not processed in this way, so lactose-intolerant individuals need to exercise caution in their food choices.

There is no way to treat milk to make it safe for people who have galactosemia, so affected individuals must avoid milk during childhood. Fortunately, a galactose-free diet is easy to achieve simply by avoiding milk. After puberty, the development of other metabolic pathways for galactose alleviates the problem in most afflicted individuals. For people who want to avoid milk, there are plenty of milk substitutes, such as soy milk or rice milk. You can even get your latte or mocha made with soy milk at Starbucks nowadays.

Essential Information

Polysaccharides are formed by linking monomeric sugars through glycosidic linkages. These polymers—such as cellulose, starch, and glycogen—play important roles in the structure of organisms such as plants and bacteria and act as vehicles for energy storage.

more than one type of monosaccharide is a *heteropolysaccharide*. Glucose is the most common monomer. When there is more than one type of monomer, frequently only two types of molecules occur in a repeating sequence. A complete characterization of a polysaccharide includes specification of which monomers are present and, if necessary, the sequence of monomers. It also requires that the type of glycosidic linkage be specified. We shall see the importance of the type of glycosidic linkage as we discuss different polysaccharides, since the nature of the linkage determines function. Cellulose and chitin are polysaccharides with β-glycosidic linkages, and both are structural materials. Starch and glycogen, also polysaccharides, have α-glycosidic linkages, and they serve as carbohydrate-storage polymers in plants and animals, respectively.

Cellulose and Starch

Cellulose is the major structural component of plants, especially of wood and plant fibers. It is a linear homopolysaccharide of β-D-glucose, and all residues are linked in β(1→4) glycosidic bonds (Figure 16.19). Individual polysaccharide chains are hydrogen-bonded together, giving plant fibers their mechanical strength. Animals lack the enzymes, called *cellulases*, that hydrolyze cellulose to glucose. Such enzymes attack the β-linkages between glucoses, which is common to structural polymers; the α-linkage between glucoses, which animals can digest, is characteristic of energy-storage polymers such as starch (Figure 16.20). Cellulases are found in certain bacteria, including the bacteria that inhabit the digestive tracts of insects, such as termites, and grazing animals, such as cattle and horses. The presence of these bacteria explains why cows and horses can live on grass and hay but humans cannot. The damage done by termites to the wooden parts of buildings arises from their ability to use cellulose in wood as a nutrient—owing to the presence of suitable bacteria in their digestive tracts.

The Forms of Starch

The importance of carbohydrates as energy sources suggests that there is some use for polysaccharides in metabolism. We shall now discuss in more detail some polysaccharides, such as starches, that serve as vehicles for storage of glucose.

Starches are polymers of α-D-glucose that occur in plant cells, usually as starch granules in the cytosol. Note that there is an α-linkage in starch, in contrast with the β-linkage in cellulose. The types of starches can be distinguished from one another by their degrees of chain branching. Amylose is a

▲ Termites can digest the cellulose in wood, and cattle can digest the cellulose in grass, because bacteria in their digestive tracts produce the enzyme cellulase, which hydrolyzes the β-glycosidic linkage in cellulose.

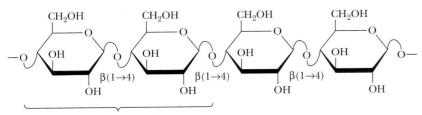

Repeating disaccharide
in cellulose
(β-cellobiose)

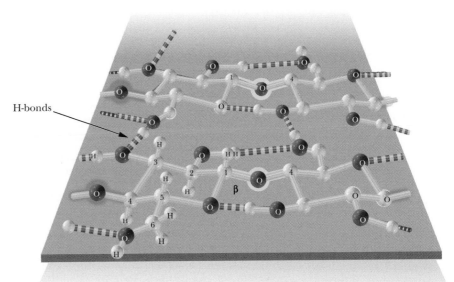

▲ **FIGURE 16.19** The polymeric structure of cellulose. β-Cellobiose is the repeating disaccharide. The monomer of cellulose is the β-anomer of glucose, which gives rise to long chains that can hydrogen-bond to one another.

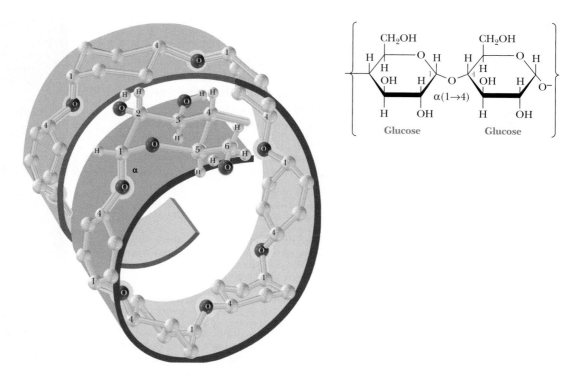

▲ **FIGURE 16.20** The monomer of starch is the α-anomer of glucose, which gives rise to a chain that folds into a helical form. The repeating dimer has α(1→4) linkages throughout.

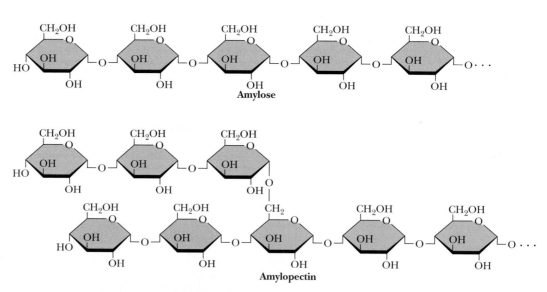

Biochemistry⒮**Now**™ ▲ **ANIMATED FIGURE 16.21** Amylose and amylopectin are the two forms of starch. Note that the linear linkages are α(1→4), but the branches in amylopectin are α(1→6). Branches in polysaccharides can involve any of the hydroxyl groups on the monosaccharide components. Amylopectin is a highly branched structure, with branches occurring every 12 to 30 residues. **See this figure animated at http://now.brookscole.com/campbell5**

linear polymer of glucose, with all the residues linked together by α(1→4) bonds. Amylopectin is a branched chain polymer, with the branches starting at α(1→6) linkages along the chain of α(1→4) linkages (Figure 16.21). The most usual conformation of amylose is a helix with six residues per turn. Iodine molecules can fit inside the helix to form a starch–iodine complex, which has a characteristic dark-blue color (Figure 16.22). The formation of this complex is a well-known test for the presence of starch. If there is a preferred conformation for amylopectin, it is not yet known. (It *is* known that the color of the product obtained when amylopectin and glycogen react with iodine is red-brown, not blue.)

Because starches are storage molecules, there must be a mechanism for releasing glucose from starch when the organism needs energy. Both plants and animals contain enzymes that hydrolyze starches. Two of these enzymes, known as α- and β-*amylase* (the α and β do not signify anomeric forms in this case), attack α(1→4) linkages. β-amylase is an *exoglycosidase* that cleaves from the nonreducing end of the polymer. Maltose, a dimer of glucose, is the product of reaction. The other enzyme, α-amylase, is an *endoglycosidase,* which can hydrolyze a glycosidic linkage anywhere along the chain to produce glucose and maltose. Amylose can be completely degraded to glucose and maltose by the two amylases, but amylopectin is not completely degraded because the branching linkages are not attacked. However, *debranching enzymes* occur in both plants and animals; they degrade the α(1→6) linkages. When these enzymes are combined with the amylases, they contribute to the complete degradation of both forms of starch.

Glycogen

Although starches occur only in plants, there is a similar carbohydrate storage polymer in animals. **Glycogen** is a branched-chain polymer of α-D-glucose, and in this respect it is similar to the amylopectin fraction of starch. Like amylopectin, glycogen consists of a chain of α(1→4) linkages with α(1→6) linkages at the branch points. The main difference between glycogen and amylopectin is that glycogen is more highly branched (Figure 16.23). Branch points occur about every 10 residues in glycogen and about every 25 residues in amylopectin. In glycogen, the average chain length is 13 glucose residues, and there are 12 layers of branching. At the heart of every glycogen molecule is a protein called glycogenin, which is discussed in Section 18.1. Glycogen is found in animal cells in granules similar to the starch granules in plant cells. Glycogen granules are observed in well-fed liver and muscle cells, but they are not seen in some other cell types, such as brain and heart cells under normal conditions. Some athletes, particularly long-distance runners, try to build up

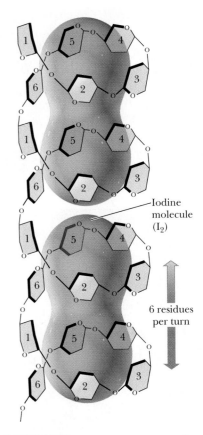

▲ **FIGURE 16.22** The starch–iodine complex. Amylose occurs as a helix with six residues per turn. In the starch–iodine complex, the iodine molecules are parallel to the long axis of the helix. Four turns of the helix are shown here. Six turns of the helix, containing 36 glycosyl residues, are required to produce the characteristic blue color of the complex.

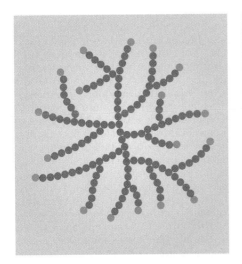

Amylopectin

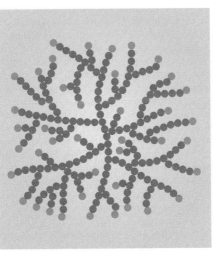

Glycogen

◀ **FIGURE 16.23** A comparison of the degrees of branching in amylopectin and glycogen.

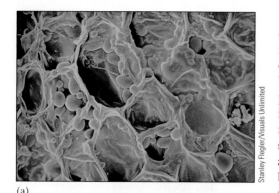

(a)

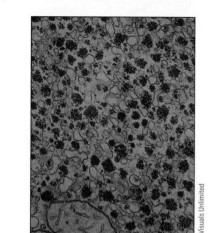

(b)

▲ Electron micrographs of starch granules in a plant and glycogen granules in an animal.

their glycogen reserves before a race by eating large amounts of carbohydrates. When the organism needs energy, various degradative enzymes remove glucose units (Section 18.1). Glycogen phosphorylase is one such enzyme; it cleaves one glucose at a time from the nonreducing end of a branch to produce glucose-1-phosphate, which then enters the metabolic pathways of carbohydrate breakdown. Debranching enzymes also play a role in the complete breakdown of glycogen. The number of branch points is significant for two reasons. First, a more branched polysaccharide is more water soluble. This may not be as important for a plant, but, for a mammal, the amount of glycogen in solution is. There are glycogen-storage diseases caused by lower-than-normal levels of branching enzymes. The glycogen products resemble starch and can fall out of solution, forming glycogen crystals in the muscles and liver. Second, when an organism needs energy quickly, the glycogen phosphorylase will have more potential targets if there are more branches, allowing a quicker mobilization of glucose. Again, this is not as important to a plant, so there was no evolutionary pressure to make starch highly branched.

Chitin

A polysaccharide that is similar to cellulose in both structure and function is chitin, which is also a linear homopolysaccharide with all the residues linked in β(1→4) glycosidic bonds. Chitin differs from cellulose in the nature of the monosaccharide unit; in cellulose, the monomer is β-D-glucose; in chitin, the monomer is N-acetyl-β-D-glucosamine. The latter compound differs from glucose only in the substitution of the N-acetylamino group (—NH—CO—CH₃) for the hydroxyl group (—OH) on carbon C-2 (Figure 16.24). Like cellulose, chitin plays a structural role and has a fair amount of mechanical strength because the individual strands are held together by hydrogen bonds. It is a major structural component of the exoskeletons of invertebrates such as insects and crustaceans (a group that includes lobsters and shrimp), and it also occurs in cell walls of algae, fungi, and yeasts.

Biochemical Connections

Dietary Fiber

Fiber in the diet is colloquially called roughage. It is principally made of complex carbohydrates, may have some protein components, and is moderately to fully insoluble. The health benefits of fiber are just beginning to be fully realized. We have known for a long time that roughage stimulates peristaltic action and thus helps move the digested food through the intestines, decreasing the transit time through the gut.

Potentially toxic substances in food and in bile fluid bind to fiber and are exported from the body, thus preventing them from doing damage to the lower intestine or being reabsorbed there. Statistical evidence indicates that high fiber also reduces colon and other cancers, precisely because fiber binds suspected carcinogens. It is also plausible that the benefit is due to a lack of other items in the high-fiber diet. People on high-fiber diets also tend to take in less fat and fewer calories. Any difference in heart disease or cancer may be due to these other differences. There has been much publicity about fiber in the diet reducing cholesterol. Fiber does bind cholesterol, and it certainly causes

some decrease in the amount in the blood. The reduction, expressed as a percentage, is higher in those cases in which the original level of cholesterol is higher. There is, however, no definitive evidence that lowering cholesterol via the ingestion of fiber will result in less heart disease.

Fiber comes in two forms: soluble and insoluble. The most common insoluble fiber is cellulose, which is found in lettuce, carrots, bean sprouts, celery, brown rice, most other vegetables, many fruit skins, and pumpernickel bread. Insoluble fiber binds various molecules but otherwise merely forms bulk in the lower intestine. Soluble fibers include amylopectin and other pectins, as well as complex starches. There will be a higher proportion of this type of fiber in uncooked and mildly processed foods. Because of increased surface area, these fibers seem to be more beneficial. Good sources include bran (especially oat bran), barley, and fresh fruits (with skin), Brussels sprouts, potatoes with skin, beans, and zucchini. Soluble fiber binds water very well, increasing satiety by helping to fill the stomach.

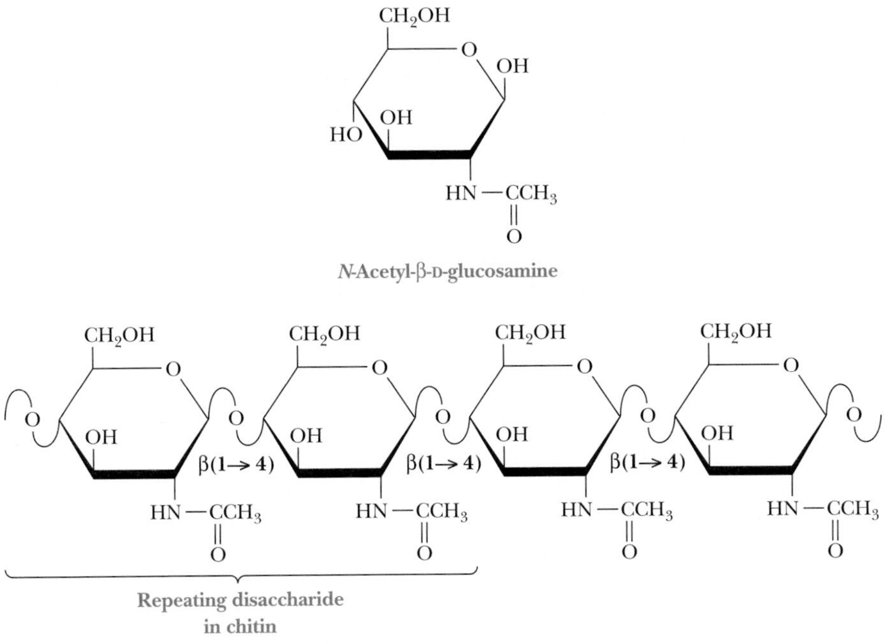

N-Acetyl-β-D-glucosamine

Repeating disaccharide
in chitin

◀ **FIGURE 16.24** The polymeric structure of chitin. *N*-Acetylglucosamine is the monomer, and a dimer of *N*-acetylglucosamine is the repeating disaccharide.

The Role of Polysaccharides in the Structure of Cell Walls

In organisms that have cell walls, such as bacteria and plants, the walls consist largely of polysaccharides. The cell walls of bacteria and plants have biochemical differences, however.

Bacterial Cell Walls. Heteropolysaccharides are major components of bacterial cell walls. A distinguishing feature of prokaryotic cell walls is that the polysaccharides are cross-linked by peptides. The repeating unit of the polysaccharide consists of two residues held together by β(1→4) glycosidic links, as was the case in cellulose and chitin. One of the two monomers is *N*-acetyl-D-glucosamine, which occurs in chitin, and the other monomer is *N*-acetylmuramic acid (Figure 16.25a). The structure of *N*-acetylmuramic acid differs from that of *N*-acetylglucosamine by the substitution of a lactic acid side chain [—O—CH(CH$_3$)—COOH] for the hydroxyl group (—OH) on carbon 3. *N*-Acetylmuramic acid is found only in prokaryotic cell walls; it does not occur in eukaryotic cell walls.

The cross-links in bacterial cell walls consist of small peptides. We shall use one of the best-known examples as an illustration. In the cell wall of the bacterium *Staphylococcus aureus*, an oligomer of four amino acids (a tetramer) is bonded to *N*-acetylmuramic acid, forming a side chain (Figure 16.25b). The tetrapeptides are themselves cross-linked by another small peptide, in this case consisting of five amino acids.

The carboxyl group of the lactic acid side chain of *N*-acetylmuramic acid forms an amide bond with the N-terminal end of a tetrapeptide that has the sequence L-Ala—D-Gln—L-Lys—D-Ala. Recall that bacterial cell walls are one of the few places where D-amino acids occur in nature. The occurrence of D-amino acids and *N*-acetylmuramic acid in bacterial cell walls but not in plant cell walls shows a biochemical as well as structural difference between prokaryotes and eukaryotes.

The tetrapeptide forms two cross-links, both of them to a pentapeptide that consists of five glycine residues, (Gly)$_5$. The glycine pentamers form peptide bonds to the C-terminal end and to the side-chain ε-amino group of the

(a)

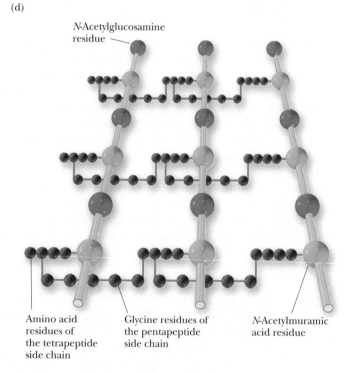

NAM
N-Acetylmuramic
acid

NAG
N-Acetylglucosamine

(b)

(c)

L-Ala

D-Gln

L-Lys — ε-NH — C — (Gly)₅ — NH —   To tetrapeptide
side chain

D-Ala

H — N

(Gly)₅

C=O

**To tetrapeptide
side chain**

H — N

L-Ala

D-Gln

L-Lys — ε-NH₃⁺

D-Ala

C=O

O⁻

(d)

N-Acetylglucosamine
residue

Amino acid
residues of
the tetrapeptide
side chain

Glycine residues of
the pentapeptide
side chain

N-Acetylmuramic
acid residue

▲ **FIGURE 16.25** The structure of the peptidoglycan of the bacterial cell wall of *Staphylococcus aureus*. (a) The repeating disaccharide. (b) The repeating disaccharide with the tetrapeptide side chain (shown in red). (c) Adding the pentaglycine cross-links (shown in red). (d) Schematic diagram of the peptidoglycan. The sugars are the larger spheres. The red spheres are the amino acid residues of the tetrapeptide, and the blue spheres are the glycine residues of the pentapeptide.

lysine in the tetrapeptide [Figure 16.25(c)]. This extensive cross-linking produces a three-dimensional network of considerable mechanical strength, which is why bacterial cell walls are extremely difficult to disrupt. The material that results from the cross-linking of polysaccharides by peptides is a **peptidoglycan** [Figure 16.25(d)], so named because it has both peptide and carbohydrate components.

Plant Cell Walls. Plant cell walls consist largely of **cellulose.** The other important polysaccharide component found in plant cell walls is **pectin,** a polymer made up mostly of D-galacturonic acid, a derivative of galactose in which the hydroxyl group on carbon C-6 has been oxidized to a carboxyl group.

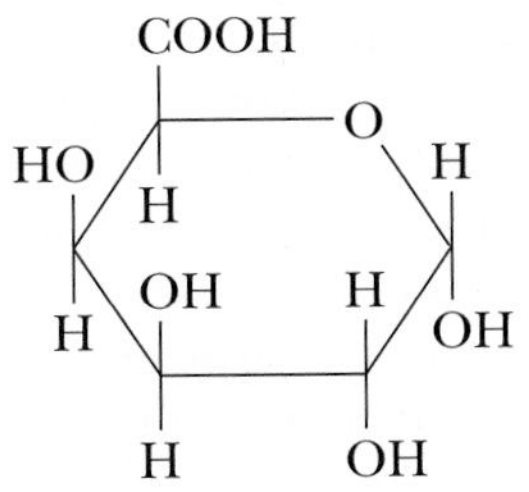

D-Galacturonic acid

Pectin is extracted from plants because it has commercial importance in the food-processing industry as a gelling agent in yogurt, fruit preserves, jams, and jellies. The major nonpolysaccharide component in plant cell walls, especially in woody plants, is **lignin** (Latin *lignum*, "wood"). Lignin is a polymer of coniferyl alcohol, and it is a very tough and durable material (Figure 16.26). Unlike bacterial cell walls, plant cell walls contain comparatively little peptide or protein.

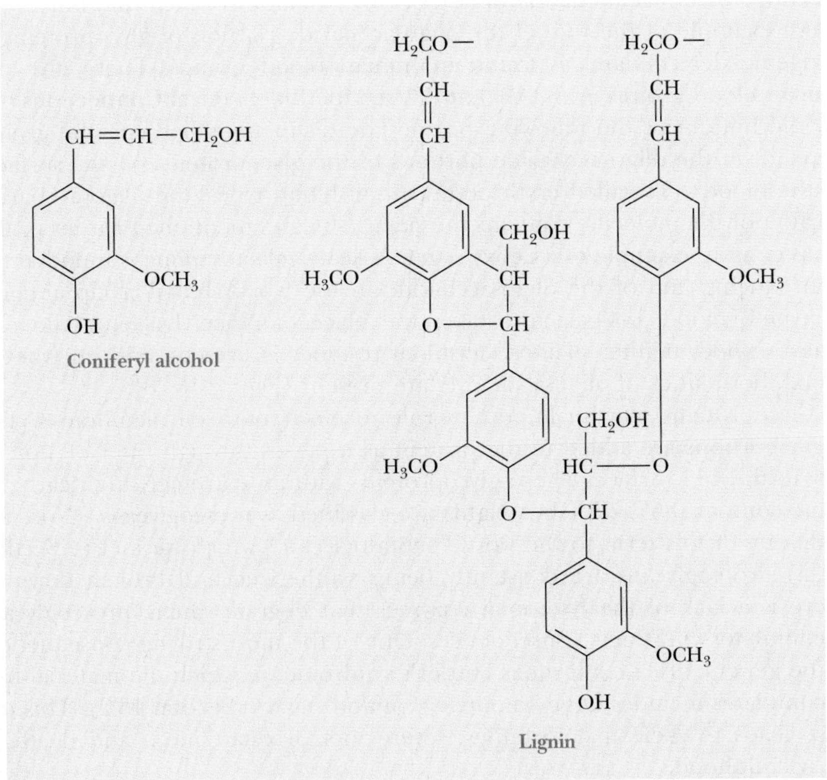

Lignin

◀ **FIGURE 16.26** The structure of lignin, a polymer of coniferyl alcohol.

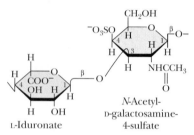

Chondroitin-4-sulfate

Chondroitin-6-sulfate

Dermatan sulfate

Heparin

Hyaluronate

Keratan sulfate

◀ **FIGURE 16.27** Glycosaminoglycans, which are formed from repeating disaccharide units, often occur as components of the proteoglycans.

Glycosaminoglycans

Glycosaminoglycans are a type of polysaccharide based on a repeating disaccharide in which one of the sugars is an amino sugar and at least one of them has a negative charge owing to the presence of a sulfate group or a carboxyl group. These polysaccharides are involved in a wide variety of cellular functions and tissues. Figure 16.27 shows the disaccharide structure of the most common ones. Heparin is a natural anticoagulant that helps prevent blood clots. Hyaluronic acid is a component of the vitreous humor of the eye and of the lubricating fluid of joints. The chondroitin sulfates and keratan sulfate are components of connective tissue. Glucosamine sulfate and chondroitin sulfate are sold in large quantities as over-the-counter drugs used to help repair frayed or otherwise damaged cartilage, especially in knees. Many people who are advised that they need knee surgery for damaged ligaments look for improvement first with a two- or three-month regimen of these glycosaminoglycans. Questions exist about the efficacy of this treatment, so it will be interesting to see what future it may have.

16.5 | What Are Glycoproteins?

Glycoproteins contain carbohydrate residues in addition to the polypeptide chain. Some of the most important examples of glycoproteins are involved in the immune response; for example, **antibodies,** which bind to and immobilize antigens (the substances attacking the organism), are glycoproteins. Carbohydrates also play an important role as **antigenic determinants,** the portions of an antigenic molecule that antibodies recognize and to which they bind.

An example of the role of the oligosaccharide portion of glycoproteins as antigenic determinants is found in human blood groups. There are four human blood groups, A, B, AB, and O (see the Biochemical Connections box on the bottom of the following page). The distinctions between the groups depend on the oligosaccharide portions of the glycoproteins on the surfaces of the blood cells called erythrocytes. In all blood types, the oligosaccharide portion of the molecule contains the sugar L-fucose, mentioned earlier in this chapter as an example of a deoxy sugar. N-Acetylgalactosamine is found at the nonreducing end of the oligosaccharide in the type-A blood-group antigen. In type-B blood, α-D-galactose takes the place of N-acetylgalactosamine. In type-O blood, neither of these terminal residues is present, and, in type-AB blood, both kinds of oligosaccharide are present (Figure 16.28).

Glycoproteins play an important role in eukaryotic cell membranes. The sugar portions are added to the protein as it passes through the Golgi on its way to the cell surface. Those glycoproteins with an extremely high carbohydrate content (85%–95% by weight) are classified as **proteoglycans.** (Note the similarity of this term to the word "peptidoglycan," which we met in Section 16.4.) Proteoglycans are constantly being synthesized and broken down. If there is a lack of the lysosomal enzymes that degrade them, proteoglycans accumulate, with tragic consequences. One of the most striking consequences is the genetic disease known as Hurler's syndrome, in which the material that accumulates includes large amounts of amino sugars (Section 16.2). This disease leads to skeletal deformities, severe mental retardation, and death in early childhood.

Biochemical Connections

Low-Carbohydrate Diets

In the 1970s, the diets that were supposed to be the healthiest were low in fat and high in carbohydrates. "Carboloading" was the craze for athletes of all types, genders, and ages, as well as for the average, sedentary person. Thirty years later, things have changed considerably. Now you can go to Burger King® and buy a burger wrapped in a piece of lettuce instead of a bun. Why did a macromolecule once thought to be healthy become something people want to avoid? The answer has to do with how glucose, the primary monosaccharide of life, is metabolized. When glucose levels rise in the blood, it causes a subsequent rise in levels of the hormone insulin. Insulin stimulates cells to take up glucose from the blood so that the cells get the energy and blood-glucose levels remain stable. We now know that insulin also has the unfortunate effect of stimulating fat synthesis and storage and inhibiting fat burning.

Some of the diets that have become most popular lately, such as the Zone Diet and the Atkins Diet, are based on keeping the carbohydrate levels low so that insulin levels do not rise and stimulate this fat storage. As with any popular diet, the evidence in support of it is not 100% conclusive, but many doctors are suggesting these diets for their patients wishing to lose weight. In the case of athletes, however, there is still little evidence to suggest that a low-carbohydrate diet will be effective for athletic performance, due to the extended time that it takes to replenish muscle and liver glycogen when the athlete is not on a high-carbohydrate diet.

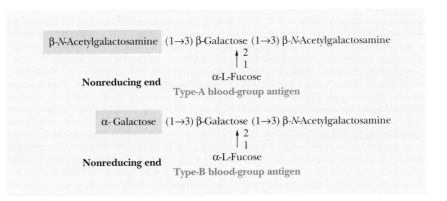

◀ **FIGURE 16.28** The structures of the blood-group antigenic determinants.

Biochemical Connections

Glycoproteins and Blood Transfusions

If a blood transfusion is attempted with incompatible blood types, as when blood from a type-A donor is given to a type-B recipient, an antigen–antibody reaction takes place because the type-B recipient has antibodies to the type-A blood. The characteristic oligosaccharide residues of type-A blood cells serve as the antigen. A cross-linking reaction occurs between antigens and antibodies, and the blood cells clump together. In the case of a transfusion of type-B blood to a type-A recipient, antibodies to type-B blood produce the same result. Type-O blood has neither antigenic determinant, and so people with type-O blood are considered universal donors. However, these people have antibodies to both type-A and type-B blood, and so they are not universal acceptors. Type-AB persons have both antigenic determi-

nants. As a result, they do not produce either type of antibody; they are universal acceptors.

Transfusion Relationships			
Blood Type	Makes Antibodies Against	Can Receive From	Can Donate To
O	A, B	O	O, A, B, AB
A	B	O, A	A, AB
B	A	O, B	B, AB
AB	None	O, A, B, AB	AB

Summary

16.1 What Are the Structures and the Stereochemistry of Monosaccharides?
The simplest examples of carbohydrates are monosaccharides, compounds that each contain a single carbonyl group and two or more hydroxyl groups. Monosaccharides frequently encountered in biochemistry are sugars that contain from three to seven carbon atoms. Sugars contain one or more chiral centers; the configurations of the possible stereoisomers can be represented by Fischer projection formulas. Sugars exist predominantly as cyclic molecules rather than in an open-chain form. Haworth projection formulas are more realistic representations of the cyclic forms of sugars than are Fischer projection formulas. Many stereoisomers are possible for five- and six-carbon sugars, but only a few of the possibilities are encountered frequently in nature.

16.2 How Do Monosaccharides React?
Monosaccharides can undergo various reactions, including oxidation and esterification, but the most important reaction by far is the formation of glycosidic linkages, which give rise to oligosaccharides and polysaccharides.

16.3 What Are Some Important Oligosaccharides?
Three important examples of oligosaccharides are the disaccharides sucrose, lactose, and maltose. Sucrose is common table sugar, lactose occurs in milk, and maltose is obtained via the hydrolysis of starch.

16.4 What Are the Structures and Functions of Polysaccharides?
In polysaccharides, the repeating unit of the polymer is frequently limited to one or two kinds of monomer. Cellulose and chitin are polymers based on single kinds of monomer units—glucose and N-acetylglucosamine, respectively.

Both polymers play structural roles in organisms. Starch, found in plants, and glycogen, which occurs in animals, are energy-storage polymers of glucose. They differ from each other in the degree of branching in the polymer structure, and they differ from cellulose in the stereochemistry of the glycosidic linkage between monomers.

16.5 What Are Glycoproteins?
In glycoproteins, carbohydrate residues are covalently linked to the polypeptide chain. Glycoproteins play a role in the recognition sites of antigens.

Critical Questions to Review

16.1 What Are the Structures and the Stereochemistry of Monosaccharides?

1. **Fact Check** Define the following terms: polysaccharide, furanose, pyranose, aldose, ketose, glycosidic bond, oligosaccharide, glycoprotein.

2. **Fact Check** Name which, if any, of the following are epimers of D-glucose: D-mannose, D-galactose, D-ribose.

3. **Fact Check** Name which, if any, of the following groups are *not* aldose–ketose pairs: D-ribose and D-ribulose, D-glucose and D-fructose, D-glyceraldehyde and dihydroxyacetone.

4. **Fact Check** What is the difference between an enantiomer and a diastereomer?

5. **Fact Check** How many possible epimers of D-glucose exist?

6. **Fact Check** Why are furanoses and pyranoses the most common cyclic forms of sugars?

7. **Fact Check** How many chiral centers are there in the open-chain form of glucose? In the cyclic form?

8. **Thought Question** Following are Fischer projections for a group of five-carbon sugars, all of which are aldopentoses. Identify the pairs that are enantiomers and the pairs that are epimers. (The sugars shown here are not all of the possible five-carbon sugars.)

9. **Thought Question** The sugar alcohol often used in "sugarless" gums and candies is L-sorbitol. Much of this alcohol is prepared by reduction of D-glucose. Compare these two structures and explain how this can be.

10. **Thought Question** Consider the structures of arabinose and ribose. Explain why nucleotide derivatives of arabinose, such as ara-C and ara-A, are effective metabolic poisons.

D-Ribose   D-Arabinose

11. **Thought Question** Two sugars are epimers of each other. Is it possible to convert one to the other without breaking covalent bonds?

12. **Thought Question** How does the cyclization of sugars introduce a new chiral center?

16.2 How Do Monosaccharides React?

13. **Fact Check** What is unusual about the structure of *N*-acetylmuramic acid (Figure 16.17), compared with the structures of other carbohydrates?

14. **Fact Check** What is the chemical difference between a sugar phosphate and a sugar involved in a glycosidic bond?

15. **Fact Check** Define the term *reducing sugar*.

16. **Biochemical Connections** What are the structural differences between vitamin C and sugars? Do these structural differences play a role in the susceptibility of this vitamin to air oxidation?

16.3 What Are Some Important Oligosaccharides?

17. **Fact Check** Name two differences between sucrose and lactose. Name two similarities.

18. **Thought Question** Draw a Haworth projection for the disaccharide gentibiose, given the following information:
    (a) It is a dimer of glucose.
    (b) The glycosidic linkage is $\beta(1\rightarrow6)$.
    (c) The anomeric carbon not involved in the glycosidic linkage is in the α configuration.

19. **Biochemical Connections** What is the metabolic basis for the observation that many adults cannot ingest large quantities of milk without developing gastric difficulties?

20. **Thought Question** Draw Haworth projection formulas for dimers of glucose with the following types of glycosidic linkages:
    (a) A $\beta(1\rightarrow4)$ linkage (both molecules of glucose in the β form)
    (b) An $\alpha,\alpha(1\rightarrow1)$ linkage
    (c) A $\beta(1\rightarrow6)$ linkage (both molecules of glucose in the β form)

21. **Biochemical Connections** A friend asks you why some parents at her child's school want a choice of beverages served at lunch, rather than milk alone. What do you tell your friend?

16.4 What Are the Structures and Functions of Polysaccharides?

22. **Fact Check** What are some of the main differences between the cell walls of plants and those of bacteria?

23. **Fact Check** How does chitin differ from cellulose in structure and function?

24. **Fact Check** How does glycogen differ from starch in structure and function?

25. **Fact Check** What is the main structural difference between cellulose and starch?

26. **Fact Check** What is the main structural difference between glycogen and starch?

27. **Fact Check** How do the cell walls of bacteria differ from those of plants?

28. **Thought Question** Pectin, which occurs in plant cell walls, exists in nature as a polymer of D-galacturonic acid methylated at carbon 6 of the monomer. Draw a Haworth projection for a repeating disaccharide unit of pectin with one methylated and one unmethylated monomer unit in $\alpha(1\rightarrow4)$ linkage.

29. **Thought Question** Advertisements for a food supplement to be taken by athletes claimed that the energy bars contained the two best precursors of glycogen. What were they?

30. **Thought Question** Explain how the minor structural difference between α- and β-glucose is related to the differences in structure and function in the polymers formed from these two monomers.

31. **Thought Question** All naturally occurring polysaccharides have one terminal residue, which contains a free anomeric carbon. Why do these polysaccharides *not* give a positive chemical test for a reducing sugar?

32. **Thought Question** An amylose chain is 5000 glucose units long. At how many places must it be cleaved to reduce the average length to 2500 units? To 1000 units? To 200 units? What percentage of the glycosidic links are hydrolyzed in each case? (Even partial hydrolysis can drastically alter the physical properties of polysaccharides and thus affect their structural role in organisms.)

33. **Thought Question** Suppose that a polymer of glucose with alternating $\alpha(1\rightarrow4)$ and $\beta(1\rightarrow4)$ glycosidic linkages has just been discovered. Draw a Haworth projection for a repeating tetramer (two repeating dimers) of such a polysaccharide. Would you expect this polymer to have primarily a structural role or an energy-storage role in organisms? What sort of organisms, if any, could use this polysaccharide as a food source?

34. **Thought Question** Glycogen is highly branched. What advantage, if any, does this provide an animal?

35. **Thought Question** No animal is able to digest cellulose. Reconcile this statement with the fact that many animals are herbivores that depend heavily on cellulose as a food source.

36. **Thought Question** How does the presence of α-bonds versus β-bonds influence the digestibility of glucose polymers by humans? *Hint:* there are *two* effects.

37. **Thought Question** How do the sites of cleavage of starch differ from one another when the cleavage reaction is catalyzed by α-amylase and β-amylase?

38. **Biochemical Connections** What is the benefit of fiber in the diet?

39. **Thought Question** How would you expect the active site of a cellulase to differ from the active site of an enzyme that degrades starch?

40. **Thought Question** Would you expect cross-linking to play a role in the structure of polysaccharides? If so, how would the cross-links be formed?

41. **Thought Question** Compare the information in the sequence of monomers in a polysaccharide with that in the sequence of amino acid residues in a protein.

42. **Thought Question** Why is it advantageous that polysaccharides can have branched chains? How do they achieve this structural feature?

43. **Thought Question** Why is the polysaccharide chitin a suitable material for the exoskeleton of invertebrates such as lobsters? What other sort of material can play a similar role?

44. **Thought Question** Could bacterial cell walls consist largely of protein? Why or why not?

45. **Thought Question** Some athletes eat diets high in carbohydrates before an event. Suggest a biochemical basis for this practice.

46. **Thought Question** You are a teaching assistant in a general chemistry lab. The next experiment is to be an oxidation–reduction titration involving iodine. You get a starch indicator from the stockroom. Why do you need it?

47. **Thought Question** Blood samples for research or medical tests sometimes have heparin added. Why is this done?

48. **Thought Question** Based on what you know about glycosidic bonds, propose a scheme for formation of covalent bonds between the carbohydrate and protein portions of glycoproteins.

16.5 What Are Glycoproteins?

49. **Fact Check** What are glycoproteins? What are some of their biochemical roles?

50. **Biochemical Connections** Briefly indicate the role of glycoproteins as antigenic determinants for blood groups.

Biochemistry ⊛ Now™

Assess your understanding of this chapter's topics with additional quizzing and tutorials at **http://now.brookscole.com/campbell5**

Annotated Bibliography

Most organic chemistry textbooks have one or more chapters on the structures and reactions of carbohydrates.

Kritchevsky, K., C. Bonfield, and J. Anderson, eds. *Dietary Fiber: Chemistry, Physiology, and Health Effects.* New York: Plenum Press, 1990. [A topic of considerable current interest, with explicit connections to the biochemistry of plant cell walls.]

Sharon, N. Carbohydrates. *Sci. Amer.* **243** (5), 90–102 (1980). [A good overview of structures.]

Sharon, N., and H. Lis. Carbohydrates in Cell Recognition. *Sci. Amer.* **268** (1), 82–89 (1993). [The development of drugs to stop infection and inflammation by targeting cell-surface carbohydrates.]

Takahashi, N., and T. Muramatsu. *Handbook of Endoglycosidases and Glyco-amidases.* Boca Raton, Fla.: CRC Press, 1992. [A source of practical information on how to manipulate biologically important carbohydrates.]

Glycolysis

The first stage of glucose metabolism is called glycolysis, and it was the first biochemical pathway elucidated. Glycolysis is an anaerobic process that, by itself, yields only two molecules of ATP. The complete aerobic oxidation of glucose to carbon dioxide and water (involving glycolysis, the citric acid cycle, and oxidative phosphorylation) yields the energy equivalent of 32 molecules of ATP. During strenuous physical activity, the body aerobically metabolizes carbohydrates, fats, and proteins for fuel; however, more carbohydrates are used as the intensity of physical activity increases. In sudden bursts of energy, such as in a 400-meter dash, the body uses carbohydrates faster than it can process them aerobically. Glucose will be metabolized via glycolysis, with pyruvate being the end product. This pyruvate will be converted to lactic acid, which will eventually be exported from the muscle to the liver. Thus, the two ATPs from anaerobic glycolysis will be an additional energy source under these conditions. Under aerobic conditions, the main purpose of glycolysis is to feed pyruvate into the citric acid cycle, where further metabolic steps will give rise to considerably more ATP.

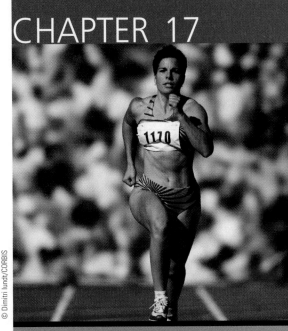

© Dimitri Iundt/CORBIS

For athletes, efficient use of carbohydrates can provide the margin of victory.

Critical Questions

17.1 What Is the Overall Pathway in Glycolysis?

17.2 How Is the 6-Carbon Glucose Converted to the 3-Carbon Glyceraldehyde-3-Phosphate?

17.3 How Is Glyceraldehyde-3-Phosphate Converted to Pyruvate?

17.4 How Is Pyruvate Metabolized Anaerobically?

17.5 How Much Energy Can Be Produced by Glycolysis?

17.1 | What Is the Overall Pathway in Glycolysis?

In **glycolysis,** one molecule of glucose (a six-carbon compound) is converted to fructose-1,6-*bis*phosphate (also a six-carbon compound), which eventually gives rise to two molecules of pyruvate (a three-carbon compound) (Figure 17.1). The glycolytic pathway (also called the Embden–Meyerhoff pathway) involves many steps, including the reactions in which metabolites of glucose are oxidized. Each reaction in the pathway is catalyzed by an enzyme specific for that reaction. In each of two reactions in the pathway, one molecule of ATP is hydrolyzed for each molecule of glucose metabolized; the energy released in the hydrolysis of these two ATP molecules makes coupled endergonic reactions possible. In each of two other reactions, two molecules of ATP are produced by phosphorylation of ADP for each molecule of glucose, giving a total of four ATP molecules produced. A comparison of the number of ATP molecules used by hydrolysis (two) and the number produced (four) shows that there is a net gain of two ATP molecules for each molecule of glucose processed in glycolysis (Section 15.10). Glycolysis plays a key role in the way organisms extract energy from nutrients.

When pyruvate is formed, it can have one of several fates (Figure 17.1). In aerobic metabolism (in the presence of oxygen), pyruvate loses carbon dioxide. The remaining two carbon atoms become linked to coenzyme A (Section 15.11) as an acetyl group to form acetyl-CoA, which then enters the citric acid cycle (Chapter 19). There are two fates for pyruvate in anaerobic metabolism (in the absence of oxygen). In organisms capable of alcoholic fermentation, pyruvate loses carbon dioxide, this time producing acetaldehyde, which, in turn, is reduced to produce ethanol (Section 17.4). The more common fate of pyruvate in anaerobic metabolism is reduction to lactate, called **anaerobic glycolysis** to distinguish it from conversion of glucose to pyruvate, which is simply called glycolysis. Anaerobic metabolism is the only energy source in mammalian red blood cells, as well as in several species of bacteria, such as *Lactobacillus* in sour milk and *Clostridium botulinum* in tainted canned foods. The Biochemical Connections box on page 466 gives some early history of research on fermentation.

Biochemistry ⊘ Now™
Test yourself on these Critical Questions at the BiochemistryNow website at **http://now .brookscole.com/campbell5**

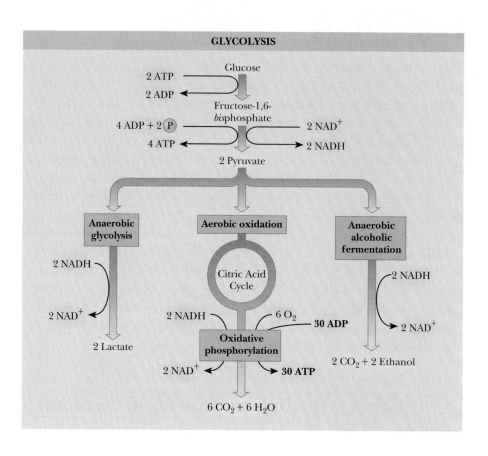

GLYCOLYSIS

▶ **FIGURE 17.1** One molecule of glucose is converted to two molecules of pyruvate. Under aerobic conditions, pyruvate is oxidized to CO_2 and H_2O by the citric acid cycle (Chapter 19) and oxidative phosphorylation (Chapter 20). Under anaerobic conditions, lactate is produced, especially in muscle. Alcoholic fermentation occurs in yeast. The NADH produced in the conversion of glucose to pyruvate is reoxidized to NAD^+ in the subsequent reactions of pyruvate.

Essential Information

In glycolysis, glucose is converted to pyruvate in a multistep pathway. When pyruvate is formed, it can be converted to carbon dioxide and water in aerobic reactions. It can also be converted to lactate under anaerobic conditions or, in some organisms, to ethyl alcohol.

In all these reactions, the conversion of glucose to product is an oxidation reaction, requiring an accompanying reduction reaction in which NAD^+ is converted to NADH, a point to which we shall return when we discuss the pathway in detail. The breakdown of glucose to pyruvate can be summarized as follows:

$$\begin{array}{l} \text{Glucose (\textbf{Six carbon atoms})} \rightarrow \text{2 Pyruvate (\textbf{Three carbon atoms})} \\ \underline{\text{2 ATP} + \text{4 ADP} + \text{2 P}_i \rightarrow \text{2 ADP} + \text{4 ATP (\textbf{Phosphorylation})}} \\ \text{Glucose} + \text{2 ADP} + \text{2 P}_i \rightarrow \text{2 Pyruvate} + \text{2 ATP (\textbf{Net reaction})} \end{array}$$

Figure 17.2 shows the reaction sequence with the names of the compounds. All sugars in the pathway have the D configuration; we shall assume this point throughout this chapter.

A Summary of the Reactions of Glycolysis

Step 1. *Phosphorylation* of glucose to give glucose-6-phosphate (ATP is the source of the phosphate group). (See Equation 17.1, page 467.)

$$\text{Glucose} + \text{ATP} \rightarrow \text{Glucose-6-phosphate} + \text{ADP}$$

Step 2. *Isomerization* of glucose-6-phosphate to give fructose-6-phosphate. (See Equation 17.2, page 470.)

$$\text{Glucose-6-phosphate} \rightarrow \text{Fructose-6-phosphate}$$

Step 3. *Phosphorylation* of fructose-6-phosphate to give fructose-1,6-*bis*phosphate (ATP is the source of the phosphate group). (See Equation 17.3, page 470.)

$$\text{Fructose-6-phosphate} + \text{ATP} \rightarrow \text{Fructose-1,6-\textit{bis}phosphate} + \text{ADP}$$

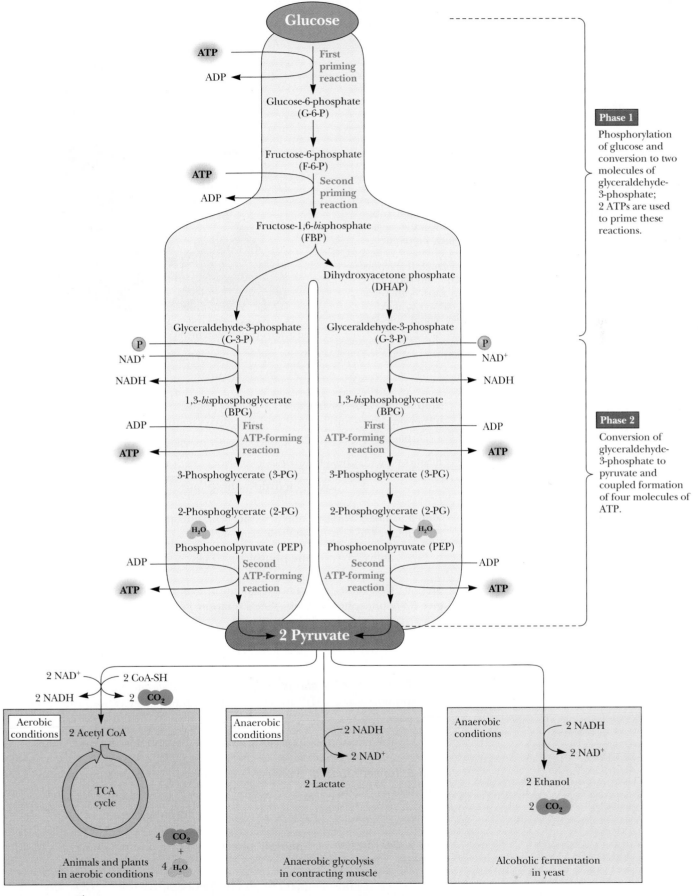

Biochemistry Now™ ▲ **ACTIVE FIGURE 17.2** The glycolytic pathway. **Watch this Active Figure at http://now.brookscole.com/campbell5**

Biochemical Connections

Louis Pasteur

Louis Pasteur, the French biologist, is famous for many studies. Perhaps the most notable was overthrowing the vital-force theory by showing that food would not decay in a sterile environment, only in one in which microorganisms were present. These studies led to the process of pasteurization, which is used for milk and for many other foods.

Among biochemists, Pasteur is well known for his studies of anaerobic and aerobic metabolism. During Pasteur's time, there was no clear concept of enzymes, so he worked only with whole organisms, namely yeast cells. He was able to show that aerobic metabolism was a much more efficient way to degrade sugars than anaerobic metabolism; that is, one gets much more energy from glucose if oxygen is available.

It is an interesting point in the history of science that, long before the existence of a biotechnology industry and most corporate research, Pasteur was supported in his work with yeast supplied by the French wine industry, whose goal was to make a better and more consistent wine, which they could sell for higher profits.

▲ Louis Pasteur (1822–1895). His research on fermentation led to important discoveries in microbiology and chemistry.

Step 4. *Cleavage* of fructose-1,6-*bis*phosphate to give two 3-carbon fragments, glyceraldehyde-3-phosphate and dihydroxyacetone phosphate. (See Equation 17.4, page 471.)

Fructose 1,6-*bis*phosphate →

Glyceraldehyde-3-phosphate + Dihydroxyacetone phosphate

Step 5. *Isomerization* of dihydroxyacetone phosphate to give glyceraldehyde-3-phosphate. (See Equation 17.5, page 471.)

Dihydroxyacetone phosphate → Glyceraldehyde-3-phosphate

Step 6. *Oxidation* (and phosphorylation) of glyceraldehyde-3-phosphate to give 1,3-*bis*phosphoglycerate. (See Equation 17.6, page 472.)

Glyceraldehyde-3-phosphate + NAD$^+$ + P$_i$ →

NADH + 1,3-*bis*phosphoglycerate + H$^+$

Step 7. *Transfer of a phosphate group* from 1,3-*bis*phosphoglycerate to ADP (phosphorylation of ADP to ATP) to give 3-phosphoglycerate. (See Equation 17.7, page 476.)

1,3-*bis*phosphoglycerate + ADP → 3-Phosphoglycerate + ATP

Step 8. *Isomerization* of 3-phosphoglycerate to give 2-phosphoglycerate. (See Equation 17.8, page 477.)

3-Phosphoglycerate → 2-Phosphoglycerate

Step 9. *Dehydration* of 2-phosphoglycerate to give phosphoenolpyruvate. (See Equation 17.9, page 477.)

2-Phosphoglycerate → Phosphoenolpyruvate + H$_2$O

Step 10. *Transfer of a phosphate group* from phosphoenolpyruvate to ADP (phosphorylation of ADP to ATP) to give pyruvate. (See Equation 17.10, page 478.)

$$\text{Phosphoenolpyruvate} + \text{ADP} \rightarrow \text{Pyruvate} + \text{ATP}$$

Note that only one of the ten steps in this pathway involves an electron-transfer reaction. We shall now look at each of these reactions in detail.

17.2 | How Is the 6-Carbon Glucose Converted to the 3-Carbon Glyceraldehyde-3-Phosphate?

The first steps of the glycolytic pathway prepare for the electron transfer and the eventual phosphorylation of ADP; these reactions make use of the free energy of hydrolysis of ATP. Figure 17.3 summarizes this part of the pathway, which is often called the *preparation phase* of glycolysis.

Step 1. Glucose is phosphorylated to give glucose-6-phosphate. The phosphorylation of glucose is an endergonic reaction.

$$\text{Glucose} + \text{P}_i \rightarrow \text{Glucose-6-phosphate} + \text{H}_2\text{O}$$

$$\Delta G^{\circ\prime} = 13.8 \text{ kJ mol}^{-1} = 3.3 \text{ kcal mol}^{-1}$$

The hydrolysis of ATP is exergonic.

$$\text{ATP} + \text{H}_2\text{O} \rightarrow \text{ADP} + \text{P}_i$$

$$\Delta G^{\circ\prime} = -30.5 \text{ kJ mol}^{-1} = -7.3 \text{ kcal mol}^{-1}$$

These two reactions are coupled, so the overall reaction is the sum of the two and is exergonic.

$$\text{Glucose} + \text{ATP} \rightarrow \text{Glucose-6-phosphate} + \text{ADP}$$

$$\Delta G^{\circ\prime} = (13.8 + -30.5) \text{ kJ mol}^{-1} = -16.7 \text{ kJ mol}^{-1} = -4.0 \text{ kcal mol}^{-1}$$

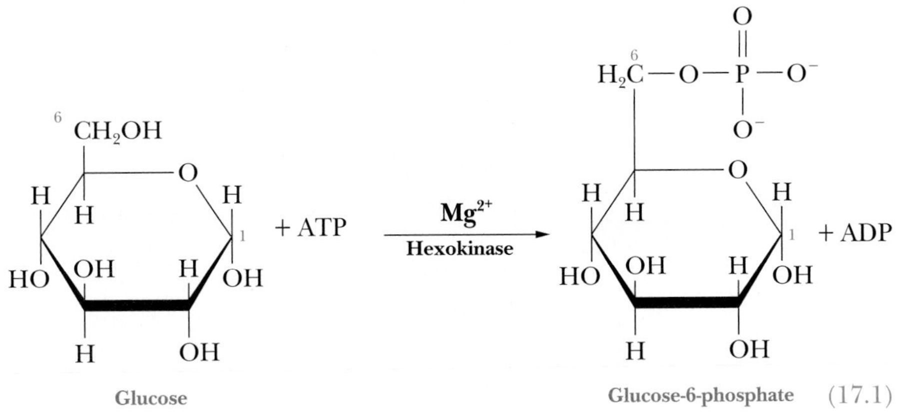

Glucose   Glucose-6-phosphate   (17.1)

Recall that the $\Delta G^{\circ\prime}$ is calculated under standard states with the concentration of all reactants and products at 1 M except hydrogen ion. If we look at the actual ΔG in the cell, the number will vary depending on cell type and metabolic state, but a typical value for this reaction is -33.9 kJ mol^{-1} or -8.12 kcal mol^{-1}. Thus the reaction is typically even more favorable under cellular conditions. Table 17.1 gives the $\Delta G^{\circ\prime}$ and the ΔG values for all the reactions of anaerobic glycolysis in erythrocytes.

This reaction illustrates the use of chemical energy originally produced by the oxidation of nutrients and ultimately trapped by phosphorylation of ADP

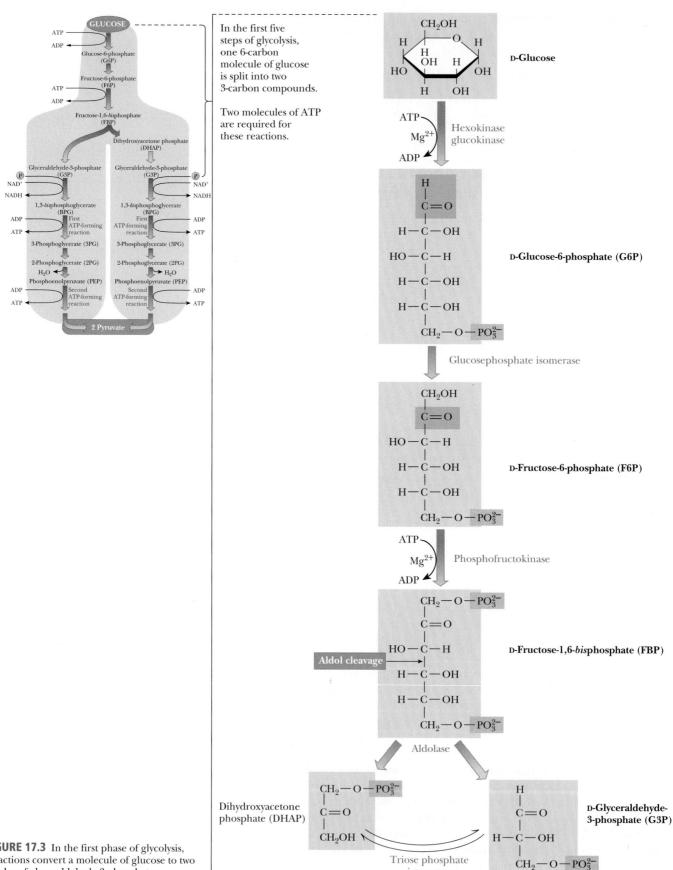

▲ **FIGURE 17.3** In the first phase of glycolysis, five reactions convert a molecule of glucose to two molecules of glyceraldehyde-3-phosphate.

In the first five steps of glycolysis, one 6-carbon molecule of glucose is split into two 3-carbon compounds.

Two molecules of ATP are required for these reactions.

Table 17.1

The Reactions of Glycolysis and Their Standard Free-Energy Changes

Step	Reaction	Enzyme	$\Delta G^{\circ\prime}$*		ΔG**
			kJ mol^{-1}	*kcal mol^{-1}*	*kJ/mol*
1	Glucose + ATP → Glucose-6-phosphate + ADP	Hexokinase/Glucokinase	−16.7	−4.0	−33.9
2	Glucose-6-phosphate → Fructose-6-phosphate	Glucose phosphate isomerase	+1.67	+0.4	−2.92
3	Fructose-6-phosphate + ATP → Fructose-1,6-*bis*phosphate + ADP	Phosphofructokinase	−14.2	−3.4	−18.8
4	Fructose-1,6-*bis*phosphate → Dihydroxyacetone phosphate + Glyceraldehyde-3-phosphate	Aldolase	+23.9	+5.7	−0.23
5	Dihydroxyacetone phosphate → Glyceraldehyde-3-phosphate	Triose phosphate isomerase	+7.56	+1.8	+2.41
6	2(Glyceraldehyde-3-phosphate + NAD$^+$ + P$_i$ → 1,3-*bis*phosphoglycerate + NADH + H$^+$)	Glyceraldehyde-3-P dehydrogenase	2(+6.20)	2(+1.5)	2(−1.29)
7	2(1,3-*bis*phosphoglycerate + ADP → 3-Phosphoglycerate + ATP)	Phosphoglycerate kinase	2(−18.8)	2(−4.5)	2(+0.1)
8	2(3-Phosphoglycerate → 2-Phosphoglycerate)	Phosphoglyceromutase	2(+4.4)	2(+1.1)	2(+0.83)
9	2(2-Phosphoglycerate → Phosphoenolpyruvate + H$_2$O)	Enolase	2(+1.8)	2(+0.4)	2(+1.1)
10	2(Phosphoenolpyruvate + ADP → Pyruvate + ATP)	Pyruvate kinase	2(−31.4)	2(−7.5)	2(−23.0)
Overall	Glucose + 2 ADP + 2 P$_i$ + NAD$^+$ → 2 Pyruvate → 2 ATP + NADH + H$^+$	Lactate dehydrogenase	−73.3	−17.5	−98.0
	2 Pyruvate + NADH + H$^+$ → Lactate + NAD$^+$		2(−25.1)	2(−6.0)	2(−14.8)
	Glucose + 2 ADP + 2 P$_i$ → 2 Lactate + 2 ATP		−123.5	−29.5	−127.6

*$\Delta G^{\circ\prime}$ values are assumed to be the same at 25°C and 37°C and are calculated for standard-state conditions (1 M concentration of reactants and products, pH 7.0).

**ΔG values are calculated at 310 K (37°C) using steady-state concentrations of these metabolites found in erythrocytes.

to ATP. Recall from Section 15.10 that ATP does not represent stored energy, just as an electric current does not represent stored energy. The chemical energy of nutrients is released by oxidation and is made available for immediate use on demand by being trapped as ATP.

The enzyme that catalyzes this reaction is **hexokinase.** The term "kinase" is applied to the class of ATP-dependent enzymes that transfer a phosphate group from ATP to a substrate. The substrate of hexokinase is not necessarily glucose; rather, it can be any one of a number of hexoses, such as glucose, fructose, and mannose. Glucose-6-phosphate inhibits the activity of hexokinase; this is a control point in the pathway. In some organisms or tissues, there are multiple isozymes of hexokinase. One isoform of hexokinase found in the human liver, called glucokinase, lowers blood glucose levels after one has eaten a meal. Liver glucokinase requires a much higher substrate level to achieve saturation than hexokinase does. Because of this, when glucose levels are high, the liver will be able to metabolize glucose via glycolysis preferentially over the other tissues. When glucose levels are low, hexokinase will still be active in all tissues (see the Biochemical Connections box on page 147).

A large conformational change takes place in hexokinase when substrate is bound. It has been shown by X-ray crystallography that, in the absence of substrate, two lobes of the enzyme that surround the binding site are quite far apart. When glucose is bound, the two lobes move closer together, and the glucose becomes almost completely surrounded by protein (Figure 17.4). This type of behavior is consistent with the induced-fit theory of enzyme

Cleft for binding
of glucose and ATP

Cleft divides
molecule into
two lobes

Glucose

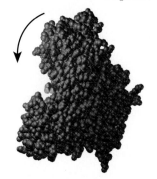

Free hexokinase

Glucose has bound in cleft,
and upper lobe has moved
with respect to lower lobe
to lie over glucose

Hexokinase–glucose complex

▲ **FIGURE 17.4** A comparison of the conformations of hexokinase and the hexokinase–glucose complex.

action (Section 6.4). In all kinases for which the structure is known, a cleft closes when substrate is bound.

Step 2. Glucose-6-phosphate isomerizes to give fructose-6-phosphate. **Glucose-phosphate isomerase** is the enzyme that catalyzes this reaction. The C-1 aldehyde group of glucose-6-phosphate is reduced to a hydroxyl group, and the C-2 hydroxyl group is oxidized to give the ketone group of fructose-6-phosphate, with no net oxidation or reduction. (Recall from Section 16.1 that glucose is an aldose, a sugar whose open-chain, noncyclic structure contains an aldehyde group, while fructose is a ketose, a sugar whose corresponding structure contains a ketone group.) The phosphorylated forms, glucose-6-phosphate and fructose-6-phosphate, are an aldose and a ketose, respectively.

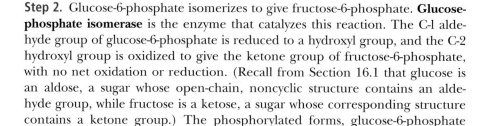

Glucose-6-phosphate ⇌ (Glucosephosphate isomerase) ⇌ Fructose-6-phosphate

(17.2)

Step 3. Fructose-6-phosphate is further phosphorylated, producing fructose-1,6-*bis*phosphate.

As in the reaction in Step 1, the endergonic reaction of phosphorylation of fructose-6-phosphate is coupled to the exergonic reaction of hydrolysis of ATP, and the overall reaction is exergonic. See Table 17.1.

Fructose-6-phosphate + ATP →(Mg^{2+}, Phosphofructokinase)→ Fructose-1,6-*bis*phosphate + ADP

(17.3)

The reaction in which fructose-6-phosphate is phosphorylated to give fructose-1,6-*bis*phosphate is the one in which the sugar is committed to glycolysis. Glucose-6-phosphate and fructose-6-phosphate can play roles in other pathways, but fructose-1,6-*bis*phosphate does not. After fructose-1,6-*bis*phosphate is formed from the original sugar, no other pathways are available, and the molecule must undergo the rest of the reactions of glycolysis. The phosphorylation of fructose-6-phosphate is highly exergonic and irreversible, and **phosphofructokinase,** the enzyme that catalyzes it, is the key regulatory enzyme in glycolysis.

Phosphofructokinase is a tetramer that is subject to allosteric feedback regulation of the type we discussed in Chapter 7. There are two types of subunits,

Biochemistry❂Now™
Go to BiochemistryNow and click on Biochemistry Interactive to learn more about the regulation of phosphofructokinase.

designated M and L, that can combine into tetramers to give different permutations (M_4, M_3L, M_2L_2, ML_3, and L_4). These combinations of subunits are referred to as **isozymes,** and they have subtle physical and kinetic differences (Figure 17.5). The subunits differ slightly in amino acid composition, so the two isozymes can be separated from each other by electrophoresis (Chapter 5). The tetrameric form that occurs in muscle is designated M_4, while that in liver is designated L_4. In red blood cells, several of the combinations can be found. Individuals who lack the gene that directs the synthesis of the M form of the enzyme can carry on glycolysis in their livers but suffer muscle weakness because they lack the enzyme in muscle.

When the rate of the phosphofructokinase reaction is observed at varying concentrations of substrate (fructose-6-phosphate), the sigmoidal curve typical of allosteric enzymes is obtained. ATP is an allosteric effector in the reaction. High levels of ATP depress the rate of the reaction, and low levels of ATP stimulate the reaction (Figure 17.6). When there is a high level of ATP in the cell, a good deal of chemical energy is immediately available from hydrolysis of ATP. The cell does not need to metabolize glucose for energy, so the presence of ATP inhibits the glycolytic pathway at this point. There is also another, more potent, allosteric effector of phosphofructokinase. This effector is fructose-2,6-*bis*phosphate; we shall discuss its mode of action in Section 18.3 when we consider general control mechanisms in carbohydrate metabolism.

Step 4. Fructose-1,6-*bis*phosphate is split into two 3-carbon fragments. The cleavage reaction here is the reverse of an aldol condensation; the enzyme that catalyzes it is called **aldolase.** In the enzyme isolated from most animal sources (the one from muscle is the most extensively studied), the basic side chain of an essential lysine residue plays the key role in catalyzing this reaction. The thiol group of a cysteine also acts as a base here.

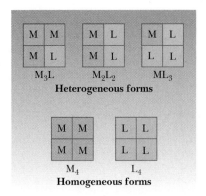

▲ **FIGURE 17.5** The possible isozymes of phosphofructokinase. The symbol M refers to the monomeric form that predominates in muscle, while the symbol L refers to the form that predominates in liver.

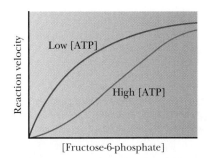

[Fructose-6-phosphate]

▲ **FIGURE 17.6** At high [ATP], phosphofructokinase behaves cooperatively, and the plot of enzyme activity versus [fructose-6-phosphate] is sigmoidal. High [ATP] thus inhibits PFK, decreasing the enzyme's affinity for fructose-6-phosphate.

Fructose-1,6-*bis*phosphate → Dihydroxyacetone phosphate + D-Glyceraldehyde-3-phosphate (17.4)

Aldolase

Step 5. The dihydroxyacetone phosphate is converted to glyceraldehyde-3-phosphate.

Dihydroxyacetone phosphate → D-Glyceraldehyde-3-phosphate (17.5)

Triosephosphate isomerase

The enzyme that catalyzes this reaction is **triosephosphate isomerase.** (Both dihydroxyacetone and glyceraldehyde are trioses.)

One molecule of glyceraldehyde-3-phosphate has already been produced by the aldolase reaction; we now have a second molecule of glyceraldehyde-3-phosphate, produced by the triosephosphate isomerase reaction. The original molecule of glucose, which contains six carbon atoms, has now been converted to two molecules of glyceraldehyde-3-phosphate, each of which contains three carbon atoms.

The ΔG value for this reaction under physiological conditions is slightly positive ($+2.41$ kJ mol^{-1} or $+0.58$ kcal mol^{-1}). It might be tempting to think that the reaction would not occur and that glycolysis would be halted at this step. We must remember that, just as coupled reactions involving ATP hydrolysis add their ΔG values together for the overall reaction, glycolysis is composed of many reactions that have very negative ΔG values that can drive the reaction to completion. A few reactions in glycolysis have small, positive ΔG values (see Table 17.1), but four reactions have very large, negative values, so that the ΔG for the whole process is negative.

17.3 | How Is Glyceraldehyde-3-Phosphate Converted to Pyruvate?

Essential Information

In the first stages of glycolysis, glucose is converted to two molecules of glyceraldehyde-3-phosphate. The key intermediate in this series of reactions is fructose-1,6-*bis*phosphate. The reaction that produces this intermediate is a key control point of the pathway, and the enzyme that catalyzes it, phosphofructokinase, is subject to allosteric regulation.

At this point, a molecule of glucose (a six-carbon compound) that enters the pathway has been converted to two molecules of glyceraldehyde-3-phosphate. We have not seen any oxidation reactions yet, but now we shall encounter them. Keep in mind that in the rest of the pathway two molecules of each of the three-carbon compounds take part in every reaction for each original glucose molecule. Figure 17.7 summarizes the second part of the pathway, which is often referred to as the *payoff phase* of glycolysis, since it is in this phase that ATP is produced instead of used.

Step 6. The oxidation of glyceraldehyde-3-phosphate to 1,3-*bis*phosphoglycerate.

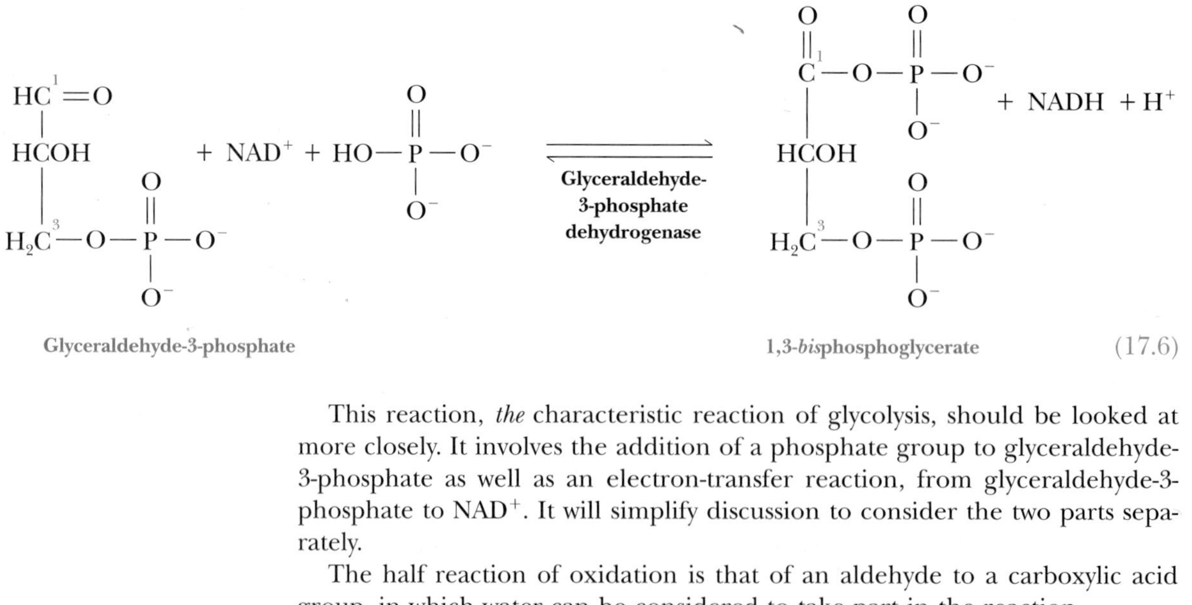

(17.6)

This reaction, *the* characteristic reaction of glycolysis, should be looked at more closely. It involves the addition of a phosphate group to glyceraldehyde-3-phosphate as well as an electron-transfer reaction, from glyceraldehyde-3-phosphate to NAD$^+$. It will simplify discussion to consider the two parts separately.

The half reaction of oxidation is that of an aldehyde to a carboxylic acid group, in which water can be considered to take part in the reaction.

$$RCHO + H_2O \rightarrow RCOOH + 2\,H^+ + 2\,e^-$$

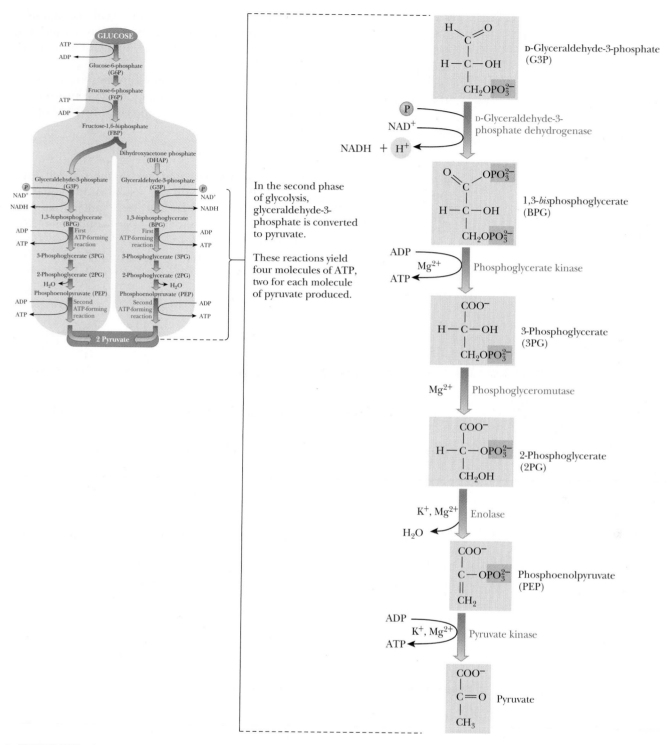

In the second phase of glycolysis, glyceraldehyde-3-phosphate is converted to pyruvate.

These reactions yield four molecules of ATP, two for each molecule of pyruvate produced.

▲ **FIGURE 17.7** The second phase of glycolysis.

The half reaction of reduction is that of NAD$^+$ to NADH (Section 15.9).

$$NAD^+ + 2 H^+ + 2 e^- \rightarrow NADH + H^+$$

The overall redox reaction is thus

$$RCHO + H_2O + NAD^+ \rightarrow RCOOH + H^+ + NADH$$

Glyceraldehyde-3-phosphate   **3-Phosphoglycerate**

in which R indicates the portions of the molecule other than the aldehyde and carboxylic acid groups, respectively. The oxidation reaction is exergonic under standard conditions ($\Delta G^{\circ\prime} = -43.1$ kJ mol^{-1} = -10.3 kcal mol^{-1}), but oxidation is only part of the overall reaction.

The phosphate group that is linked to the carboxyl group does not form an ester, since an ester linkage requires an alcohol and an acid. Instead, the carboxylic acid group and phosphoric acid form a mixed anhydride of two acids by loss of water (Section 2.2),

$$\text{3-Phosphoglycerate} + P_i \rightarrow \text{1,3-}bis\text{phosphoglycerate} + H_2O$$

in which the substances involved in the reaction are in the ionized form appropriate at pH 7. Note that ATP and ADP do not appear in the equation. The source of the phosphate group is phosphate ion itself, rather than ATP. The phosphorylation reaction is endergonic under standard conditions ($\Delta G^{\circ\prime}$ = 49.3 kJ mol^{-1} = 11.8 kcal mol^{-1}).

The overall reaction, including electron transfer and phosphorylation, is

$$RCHO + HOPO_3^{2-} + NAD^+ \rightleftharpoons RC\overset{\displaystyle O}{\overset{\displaystyle \|}{-}}OPO_3^{2-} + NADH + H^+$$

or

$$\text{Glyceraldehyde-3-phosphate} + P_i + NAD^+ \xrightarrow{\text{Glyceraldehyde-3-phosphate dehydrogenase}} \text{1,3-}bis\text{phosphoglycerate} + NADH + H^+$$

Let's show the two reactions that make up this reaction.

1. **Oxidation of glyceraldehyde-3-phosphate** ($\Delta G^{\circ\prime} = -43.1$ kJ mol^{-1} = -10.3 kcal mol^{-1})

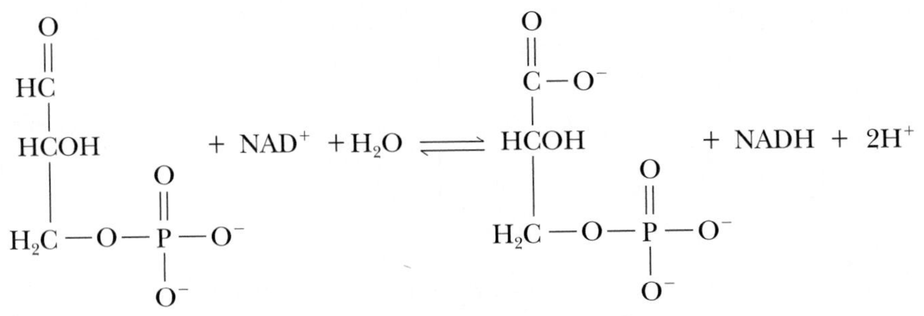

Glyceraldehyde-3-phosphate   3-Phosphoglycerate

2. **Phosphorylation of 3-phosphoglycerate** ($\Delta G^{\circ\prime}$ = 49.3 kJ mol^{-1} = 11.8 kcal mol^{-1})

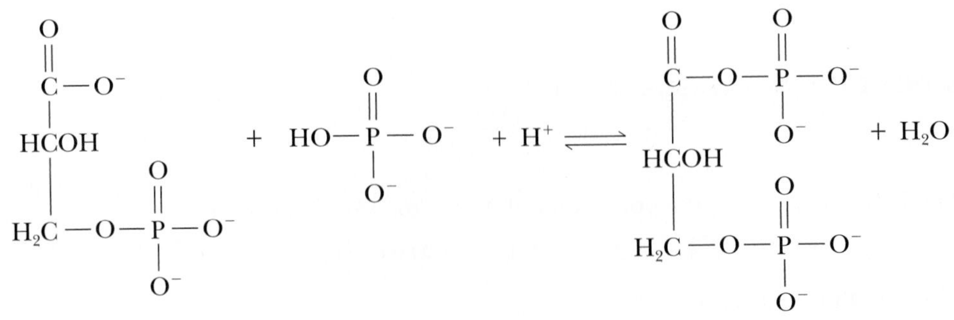

3-Phosphoglycerate   1,3-*bis*phosphoglycerate

The standard free-energy change for the overall reaction is the sum of the values for the oxidation and phosphorylation reactions. The overall reaction is not far from equilibrium, being only slightly endergonic.

$$\Delta G^{\circ\prime} \text{ overall} = \Delta G^{\circ\prime} \text{ oxidation} + \Delta G^{\circ\prime} \text{ phosphorylation}$$

$$= (-43.1 \text{ kJ mol}^{-1}) + (49.3 \text{ kJ mol}^{-1})$$

$$= 6.2 \text{ kJ mol}^{-1} = 1.5 \text{ kcal mol}^{-1}$$

This value of the standard free-energy change is for the reaction of one mole of glyceraldehyde-3-phosphate; the value must be multiplied by 2 to get the value for each mole of glucose ($\Delta G^{\circ\prime} = 12.4 \text{ kJ mol}^{-1} = 3.0 \text{ kcal mol}^{-1}$). The ΔG under cellular conditions is slightly negative ($-1.29 \text{ kJ mol}^{-1}$ or $-0.31 \text{ kcal mol}^{-1}$) (Table 17.1). The enzyme that catalyzes the conversion of glyceraldehyde-3-phosphate to 1,3-*bis*phosphoglycerate is **glyceraldehyde-3-phosphate dehydrogenase.** This enzyme is one of a class of similar enzymes, the NADH-linked dehydrogenases. The structures of a number of dehydrogenases of this type have been studied via X-ray crystallography. The overall structures are not strikingly similar, but the structure of the binding site for NADH is quite similar in all these enzymes (Figure 17.8). (The oxidizing agent is NAD$^+$; both oxidized and reduced forms of the coenzyme bind to the enzyme.) One portion of the binding site is specific for the nicotinamide ring, and one portion is specific for the adenine ring.

The molecule of glyceraldehyde-3-phosphate dehydrogenase is a tetramer, consisting of four identical subunits. Each subunit binds one molecule of NAD$^+$, and each subunit contains an essential cysteine residue. A thioester involving the cysteine residue is the key intermediate in this reaction. In the phosphorylation step, the thioester acts as a high-energy intermediate (see Chapter 15 for a discussion of thioesters). Phosphate ion attacks the thioester, forming a mixed anhydride of the carboxylic and phosphoric acids, which is also a high-energy compound (Figure 17.9). This compound is 1,3-*bis*phosphoglycerate, the product of the reaction. Production of ATP requires a high-energy compound as starting material. The 1,3-*bis*phosphoglycerate fulfills this requirement and transfers a phosphate group to ADP in a highly exergonic reaction (i.e., it has a high phosphate-group transfer potential).

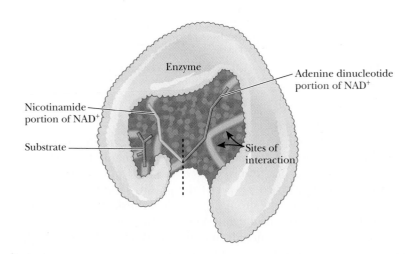

▲ **FIGURE 17.8** Schematic view of the binding site of an NADH-linked dehydrogenase. There are specific binding sites for the adenine nucleotide portion of the coenzyme (shown in red to the right of the dashed line) and for the nicotinamide portion of the coenzyme (shown in yellow to the left of the dashed line), in addition to the binding site for the substrate. Specific interactions with the enzyme hold the substrate and coenzyme in the proper positions. Sites of interaction are shown as a series of pale green lines.

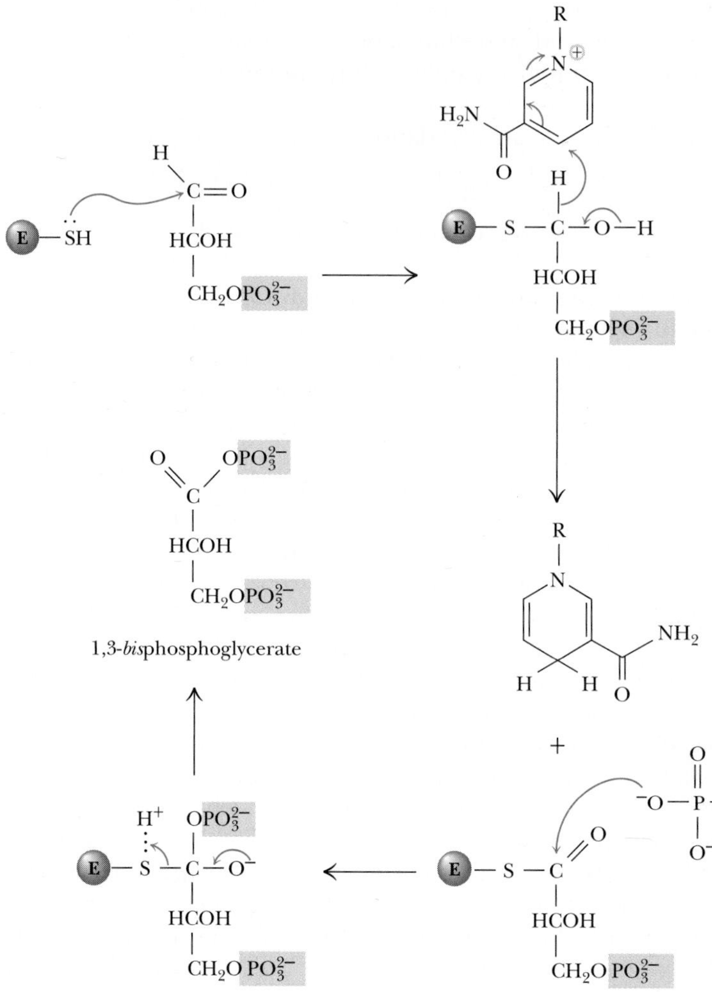

Step 7. The next step is one of the two reactions in which ATP is produced by phosphorylation of ADP.

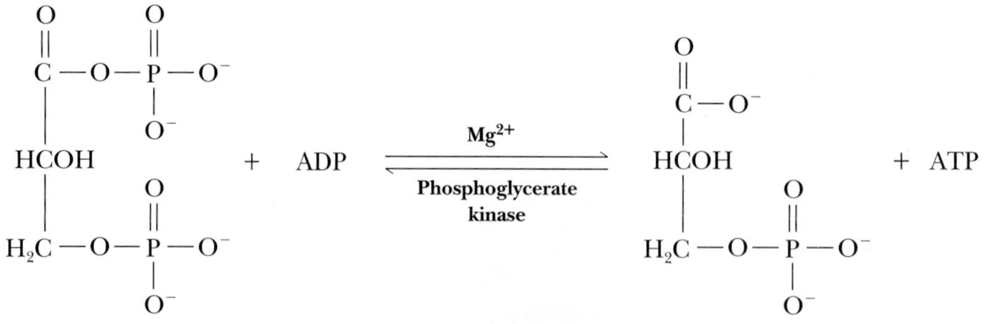

1,3-*bis*phosphoglycerate   3-Phosphoglycerate   (17.7)

The enzyme that catalyzes this reaction is **phosphoglycerate kinase.** By now the term "kinase" should be familiar as the generic name for a class of ATP-dependent phosphate-group transfer enzymes. The most striking feature of the reaction has to do with energetics of the phosphate-group transfer. In this step in glycolysis, a phosphate group is transferred from 1,3-*bis*phosphoglycer-ate to a molecule of ADP, producing ATP, the first of two such reactions in the glycolytic pathway. We already mentioned that 1,3-*bis*phosphoglycerate can easily transfer a phosphate group to other substances. Note that a substrate, namely 1,3-*bis*phosphoglycerate, has transferred a phosphate group to ADP. This transfer is typical of **substrate-level phosphorylation.** It is to be distinguished from oxidative phosphorylation (Sections 20.1 through 20.5), in which transfer of phosphate groups is linked to electron-transfer reactions in which oxygen is the ultimate electron acceptor. The only requirement for substrate-level phosphorylation is that the standard free energy of the hydrolysis reaction is more negative than that for hydrolysis of the new phosphate compound being formed. Recall that the standard free energy of hydrolysis of 1,3-*bis*phosphoglycerate is -49.3 kJ mol^{-1}. We have already seen that the standard free energy of hydrolysis of ATP is -30.5 kJ mol^{-1}, and we must change the sign of the free-energy change when the reverse reaction occurs:

$$ADP + P_i + H^+ \rightarrow ATP + H_2O$$

$$\Delta G^{\circ\prime} = 30.5 \text{ kJ mol}^{-1} = 7.3 \text{ kcal mol}^{-1}$$

The net reaction is

$$1,3\text{-}bis\text{phosphoglycerate} + ADP \rightarrow 3\text{-Phosphoglycerate} + ATP$$

$$\Delta G^{\circ\prime} = -49.3 \text{ kJ mol}^{-1} + 30.5 \text{ kJ mol}^{-1} = -18.8 \text{ kJ mol}^{-1} = -4.5 \text{ kcal mol}^{-1}$$

Two molecules of ATP are produced by this reaction for each molecule of glucose that enters the glycolytic pathway. In the earlier stages of the pathway, two molecules of ATP were invested to produce fructose-1,6-*bis*phosphate, and now they have been recovered. At this point, the balance of ATP use and production is exactly even. The next few reactions will bring about the production of two more molecules of ATP for each original molecule of glucose, leading to the net gain of two ATP molecules in glycolysis.

Step 8. The phosphate group is transferred from carbon 3 to carbon 2 of the glyceric acid backbone, setting the stage for the reaction that follows.

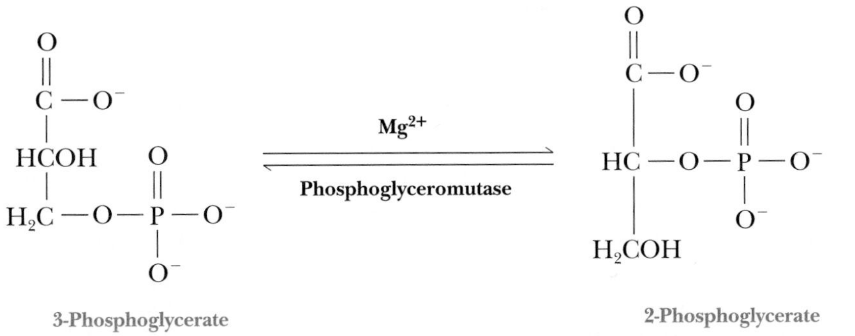

3-Phosphoglycerate   2-Phosphoglycerate   (17.8)

The enzyme that catalyzes this reaction is **phosphoglyceromutase.**

Step 9. The 2-phosphoglycerate molecule loses one molecule of water, producing phosphoenolpyruvate. This reaction does not involve electron transfer; it is a dehydration reaction. **Enolase,** the enzyme that catalyzes this reaction, requires Mg^{2+} as a cofactor. The water molecule that is eliminated binds to Mg^{2+} in the course of the reaction.

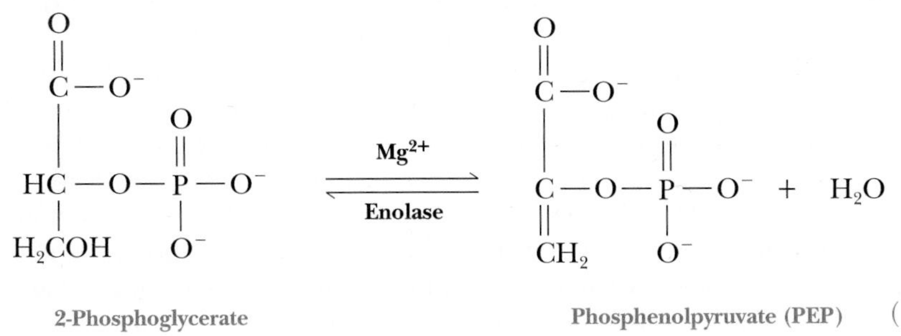

2-Phosphoglycerate   Phosphenolpyruvate (PEP)   (17.9)

Step 10. Phosphoenolpyruvate transfers its phosphate group to ADP, producing ATP and pyruvate.

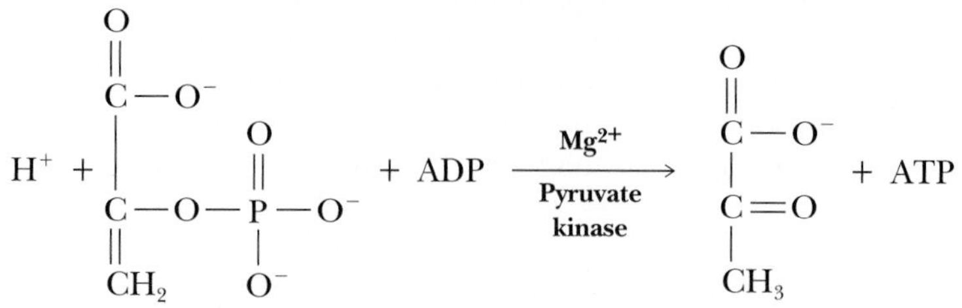

$$\text{Phosphoenolpyruvate} \qquad\qquad\qquad\qquad \text{Pyruvate} \qquad (17.10)$$

The double bond shifts to the oxygen on carbon 2 and a hydrogen shifts to carbon 3. Phosphoenolpyruvate is a high-energy compound with a high phosphate-group transfer potential. The free energy of hydrolysis of this compound is more negative than that of ATP (-61.9 kJ mol^{-1} versus -30.5 kJ mol^{-1}, or -14.8 kcal mol^{-1} versus -7.3 kcal mol^{-1}). The reaction that occurs in this step can be considered to be the sum of the hydrolysis of phosphoenolpyruvate and the phosphorylation of ADP. This reaction is another example of substrate-level phosphorylation.

$$\text{Phosphoenolpyruvate} \rightarrow \text{Pyruvate} + P_i$$

$$\Delta G^{\circ\prime} = -61.9 \text{ kJ mol}^{-1} = -14.8 \text{ kcal mol}^{-1}$$

$$\text{ADP} + P_i \rightarrow \text{ATP}$$

$$\Delta G^{\circ\prime} = 30.5 \text{ kJ mol}^{-1} = 7.3 \text{ kcal mol}^{-1}$$

The net reaction is

$$\text{Phosphoenolpyruvate} + \text{ADP} \rightarrow \text{Pyruvate} + \text{ATP}$$

$$\Delta G^{\circ\prime} = -31.4 \text{ kJ mol}^{-1} = -7.5 \text{ kcal mol}^{-1}$$

Since two moles of pyruvate are produced for each mole of glucose, twice as much energy is released for each mole of starting material.

Pyruvate kinase is the enzyme that catalyzes this reaction. Like phosphofructokinase, it is an allosteric enzyme consisting of four subunits of two different types (M and L), as we saw with phosphofructokinase. Pyruvate kinase is inhibited by ATP. The conversion of phosphoenolpyruvate to pyruvate slows down when the cell has a high concentration of ATP—that is to say, when the cell does not have a great need for energy in the form of ATP. Due to the different isozymes of pyruvate kinase found in liver versus muscle, the control of glycolysis is handled differently in these two tissues, which we will look at in detail in Chapter 18.

Control Points in the Glycolytic Pathway

One of the most important questions that we can ask about any metabolic pathway is, At which points is control exercised? Pathways can be "shut down" if an organism has no immediate need for their products, which saves energy for the organism. In glycolysis, three reactions are control points. The first is the reaction of glucose to glucose-6-phosphate, catalyzed by hexokinase; the second, which is the production of fructose-1,6-*bis*phosphate, is catalyzed by phosphofructokinase; and the last is the reaction of PEP to pyruvate, catalyzed by pyruvate kinase (Figure 17.10). It is frequently observed that control is exercised near the start and end of a pathway, as well as at points involving key intermediates such as fructose-1,6-*bis*phosphate. When we have learned more about the metabolism of carbohydrates, we can return to the role of

Essential Information

In the final stages of glycolysis, two molecules of pyruvate are produced for each molecule of glucose that entered the pathway. These reactions involve electron transfer (oxidation–reduction) and the net production of two ATP for each glucose.

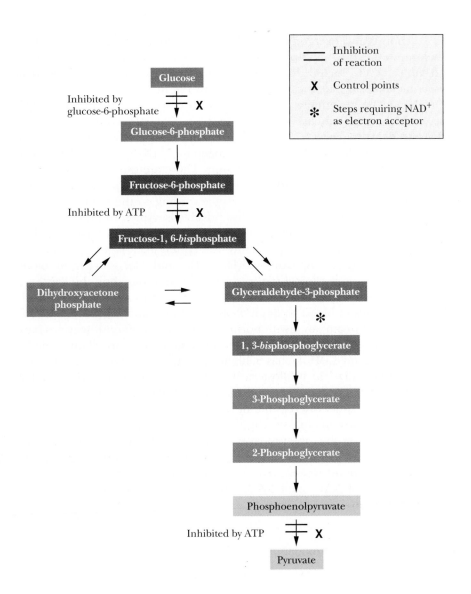

◀ **FIGURE 17.10** Control points in glycolysis.

phosphofructokinase and fructose-1,6-*bis*phosphate in the regulation of several pathways of carbohydrate metabolism (Section 18.3).

17.4 | How Is Pyruvate Metabolized Anaerobically?

The Conversion of Pyruvate to Lactate in Muscle

The final reaction of anaerobic glycolysis is the reduction of pyruvate to lactate.

$$(17.11)$$

This reaction is also exergonic ($\Delta G^{\circ\prime} = -25.1$ kJ mol^{-1} = -6.0 kcal mol^{-1}); as before, we need to multiply this value by two to find the energy yield for each molecule of glucose that enters the pathway. Lactate is a dead end in muscle metabolism, but it can be recycled in the liver to form pyruvate and even glucose by a pathway called gluconeogenesis ("new synthesis of glucose"), which we will discuss in Section 18.2.

Lactate dehydrogenase (LDH) is the enzyme that catalyzes this reaction. Like glyceraldehyde-3-phosphate dehydrogenase, LDH is an NADH-linked dehydrogenase and consists of four subunits. There are two kinds of subunits, designated M and H, which vary slightly in amino acid composition. The quaternary structure of the tetramer can vary according to the relative amounts of the two kinds of subunits, yielding five possible isozymes. In human skeletal muscle, the homogeneous tetramer of the M_4 type predominates, and in heart the other homogeneous possibility, the H_4 tetramer, is the predominant form. The heterogeneous forms—M_3H, M_2H_2, and MH_3—occur in blood serum. A very sensitive clinical test for heart disease is based on the existence of the various isozymic forms of this enzyme. The relative amounts of the H_4 and MH_3 isozymes in blood serum increase drastically after myocardial infarction (heart attack) compared with normal serum. The different isozymes have slightly different kinetic properties due to their subunit compositions. The H_4 isozyme (also called LDH 1) has a higher affinity for lactate as a substrate. The M_4 isozyme (LDH 5) is allosterically inhibited by pyruvate. These differences reflect the isozymes' general roles in metabolism. The muscle is a highly anaerobic tissue, whereas the heart is not.

At this point, one might ask why the reduction of pyruvate to lactate (a waste product in aerobic organisms) is the last step in anaerobic glycolysis, a pathway that provides energy for the organism by oxidation of nutrients. There is another point to consider about the reaction, one that involves the relative amounts of NAD$^+$ and NADH in a cell. The half reaction of reduction can be written

$$\text{Pyruvate} + 2\,H^+ + 2\,e^- \rightarrow \text{Lactate}$$

and the half reaction of oxidation is

$$\text{NADH} + H^+ \rightarrow \text{NAD}^+ + 2\,e^- + 2\,H^+$$

The overall reaction is, as we saw earlier,

$$\text{Pyruvate} + \text{NADH} + H^+ \rightarrow \text{Lactate} + \text{NAD}^+$$

The NADH produced from NAD$^+$ by the earlier oxidation of glyceraldehyde-3-phosphate is used up with no net change in the relative amounts of NADH and NAD$^+$ in the cell (Figure 17.11). This regeneration is needed under anaerobic conditions in the cell so that NAD$^+$ will be present for further glycolysis to take place. Without this regeneration, the oxidation reactions in anaerobic organisms would soon come to a halt because of the lack of NAD$^+$ to serve as an oxidizing agent in fermentative processes. The production of lactate buys time for the organism experiencing anaerobic metabolism and shifts some of the load away from the muscles and onto the liver, in which gluconeogenesis can reconvert lactate to pyruvate and glucose (Chapter 18). The same considerations apply in alcoholic fermentation (which will be discussed in the following three paragraphs). On the other hand, NADH is a frequently encountered reducing agent in many reactions, and it is lost to the organism in lactate production. Aerobic metabolism makes more efficient use of reducing agents ("reducing power") such as NADH because the conversion of pyruvate to lactate does not occur in aerobic metabolism. The NADH produced in the stages of glycolysis leading to the production of pyruvate is available for use in other reactions in which a reducing agent is needed.

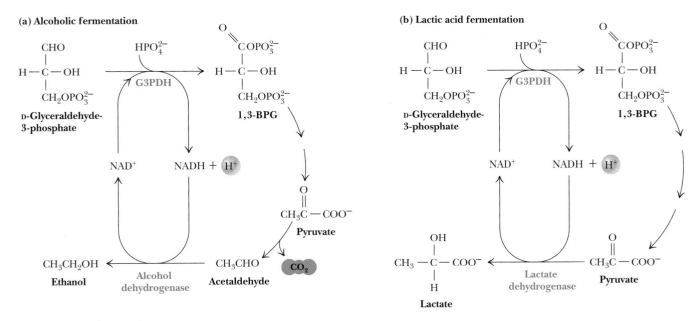

(a) Alcoholic fermentation

(b) Lactic acid fermentation

Biochemistry (Now™   ▲ **ACTIVE FIGURE 17.11** The recycling of NAD^+ and NADH in anaerobic glycolysis. **Watch this Active Figure at http://now.brookscole.com/campbell5**

Alcoholic Fermentation

Two other reactions related to the glycolytic pathway lead to the production of ethanol by *alcoholic fermentation*. This process is one of the alternative fates of pyruvate (Section 17.1). In the first of the two reactions that lead to the production of ethanol, pyruvate is decarboxylated (loses carbon dioxide) to produce acetaldehyde. The enzyme that catalyzes this reaction is *pyruvate decarboxylase.*

This enzyme requires Mg^{2+} and a cofactor we have not seen before, **thiamine pyrophosphate** (TPP). (Thiamine itself is vitamin B_1.) In TPP the carbon atom between the nitrogen and the sulfur in the thiazole ring (Figure 17.12) is highly reactive. It forms a carbanion (an ion with a negative charge on a carbon atom) quite easily, and the carbanion, in turn, attacks the carbonyl group of pyruvate to form an adduct. Carbon dioxide splits off, leaving

Biochemical Connections

Anaerobic Metabolism and Tooth Decay

Dental caries, or tooth decay, is one of the most prevalent diseases in the United States and possibly in the world, although modern treatments such as fluoride and flossing have greatly reduced its incidence in young people. Contributing factors in tooth decay are a combination of a diet high in refined sugars, the development of dental plaque, and anaerobic metabolism. The high-sugar diet allows for rapid growth of bacteria in the mouth, and sucrose is perhaps the most efficiently used sugar because the bacteria can make their polysaccharide "glue" more efficiently from this nonreducing sugar. The bacteria grow in expanding sticky colonies, forming plaque on the tooth surface. The bacteria growing under the surface of the plaque must uti-

lize anaerobic metabolism because oxygen does not diffuse readily through the waxy surface of dental plaque. The two predominant by-products, lactate and pyruvate, are relatively strong organic acids, and these acid products actually cause destruction of the enamel surface. The bacteria, of course, grow rapidly in the pock holes. If the enamel is eaten all the way through, the bacteria grow even more readily in the softer dentin layer beneath the enamel.

Fluoridation results in a much harder enamel surface, and the fluoride may actually inhibit the metabolism of the bacteria. Daily flossing disrupts the plaque, and the anaerobic conditions never get started.

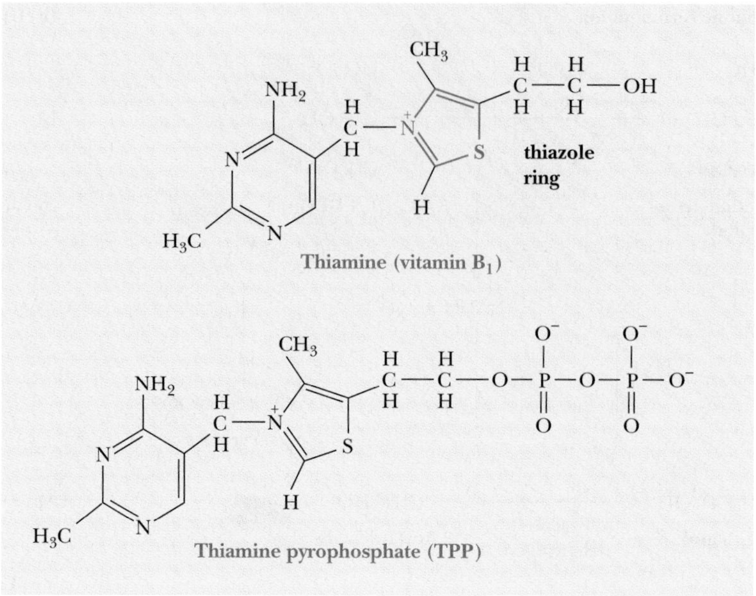

► **FIGURE 17.12** The structures of thiamine (vitamin B_1) and thiamine pyrophosphate (TPP), the active form of the coenzyme.

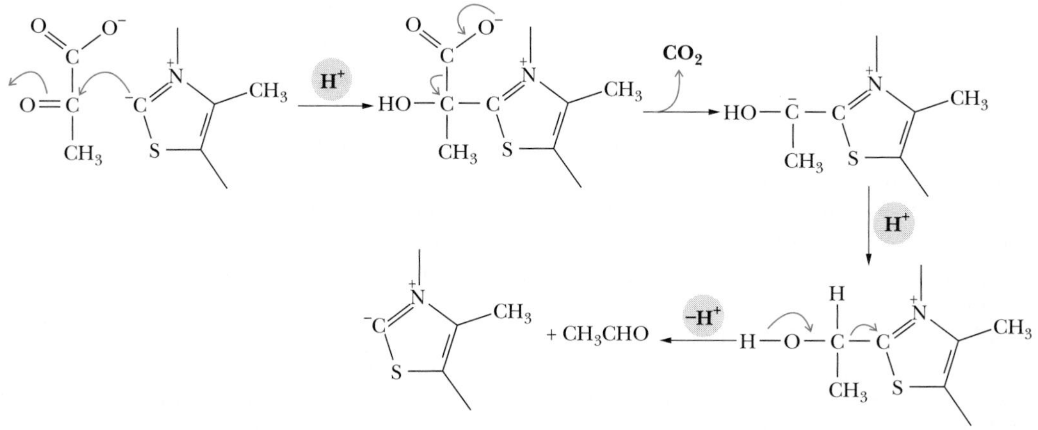

Biochemistry ⑤ Now™ ▲ **ANIMATED FIGURE 17.13** The mechanism of the pyruvate decarboxylase reaction. The carbanion form of the thiazole ring of TPP is strongly nucleophilic. The carbanion attacks the carbonyl carbon of pyruvate to form an adduct. Carbon dioxide splits out, leaving a two-carbon fragment (activated acetaldehyde) covalently bonded to the coenzyme. A shift of electrons releases acetaldehyde, regenerating the carbanion. **See this figure animated at http://now.brookscole.com/campbell**

a two-carbon fragment covalently bonded to TPP. There is a shift of electrons, and the two-carbon fragment splits off, producing acetaldehyde (Figure 17.13). The two-carbon fragment bonded to TPP is sometimes called activated acetaldehyde, and TPP can be found in several reactions that are decarboxylations.

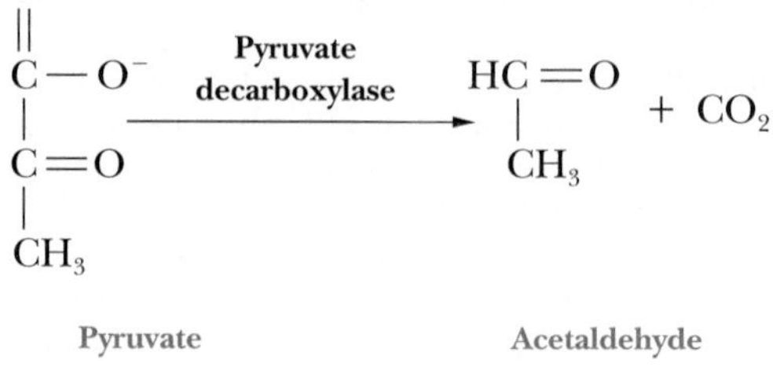

Pyruvate   Acetaldehyde   (17.12)

Biochemical Connections

Fetal Alcohol Syndrome

The variety of injuries to a fetus caused by maternal consumption of ethanol is called fetal alcohol syndrome. In catabolism of ethanol by the body, the first step is conversion to acetaldehyde—the reverse of the last reaction of alcoholic fermentation. The level of acetaldehyde in the blood of a pregnant woman is the key to detecting fetal alcohol syndrome. It has been shown that the acetaldehyde is transferred across the placenta and that it accumulates in the liver of the fetus. Acetaldehyde is toxic, and this is one of the most important factors in fetal alcohol syndrome.

In addition to the toxic effects of acetaldehyde, consumption of ethanol during pregnancy harms the fetus in other ways. It depresses transfer of nutrients to the fetus, resulting in lower levels of sugars (hypoglycemia), vitamins, and essential amino acids. Lower levels of oxygen (hypoxia) also occur. This last effect is more drastic when the mother smokes during pregnancy, as well as consuming alcohol.

The labels of alcoholic beverages now include a warning against consumption during pregnancy. The American Medical Association has issued the unequivocal warning that "there is no known safe level of alcohol during pregnancy."

The carbon dioxide produced is responsible for the bubbles in beer and in sparkling wines. Acetaldehyde is then reduced to produce ethanol, and, at the same time, one molecule of NADH is oxidized to NAD^+ for each molecule of ethanol produced.

$$\text{Acetaldehyde} + \text{NADH} \rightarrow \text{Ethanol} + NAD^+$$

The reduction reaction of alcoholic fermentation is similar to the reduction of pyruvate to lactate, in the sense that it provides for recycling of NAD^+ and thus allows further anaerobic oxidation (fermentation) reactions. The net reaction for alcoholic fermentation is

$$\text{Glucose} + 2\,\text{ADP} + 2\,P_i + 2\,H^+ \rightarrow 2\,\text{Ethanol} + 2\,\text{ATP} + 2\,CO_2 + 2\,H_2O$$

NAD^+ and NADH do not appear explicitly in the net equation. It is essential that the recycling of NADH to NAD^+ takes place here, just as it does when lactate is produced, so that there can be further anaerobic oxidation. **Alcohol dehydrogenase,** the enzyme that catalyzes the conversion of acetaldehyde to ethanol, is similar to lactate dehydrogenase in many ways. The most striking similarity is that both are NADH-linked dehydrogenases, and both are tetramers.

17.5 | How Much Energy Can Be Produced by Glycolysis?

Now that we have seen the reactions of the glycolytic pathway, we can do some bookkeeping and determine the standard free-energy change for the entire pathway by using the data from Table 17.1.

The overall process of glycolysis is exergonic. We can calculate the $\Delta G^{\circ\prime}$ for the entire reaction by adding up the $\Delta G^{\circ\prime}$ values from each of the steps. Remember that all of the reactions from triose phosphate isomerase to pyruvate kinase are doubled. This gives a final figure from glucose to two pyruvates of -74.0 kJ mol^{-1} or -17.5 kcal mol^{-1}. The energy released in the exergonic phases of the process drives the endergonic reactions. The net reaction of glycolysis explicitly includes an important endergonic process, that of phosphorylation of two molecules of ADP.

$$2\,\text{ADP} + 2\,P_i \rightarrow 2\,\text{ATP}$$

$\Delta G^{\circ\prime}$ reaction $= 61.0$ kJ mol$^{-1} = 14.6$ kcal mol^{-1} glucose consumed

Without the production of ATP, the reaction of one molecule of glucose to produce two molecules of pyruvate would be even more exergonic. Thus, *subtracting* out the synthesis of ATP:

$$
\begin{array}{ll}
\text{Glucose} + 2\,\text{ADP} + 2\,\text{P}_i \rightarrow 2\,\text{Pyruvate} + 2\,\text{ATP} & \Delta G^{\circ\prime} = -73.4 \text{ kJ mol}^{-1} -17.5 \text{ kcal mol}^{-1} \\
-(2\,\text{ATP} + 2\,\text{P}_i \rightarrow 2\,\text{ATP}) & \Delta G^{\circ\prime} = -61.0 \text{ kJ} -14.6 \text{ kcal mol}^{-1} \\
\hline
\text{Glucose} \rightarrow 2\,\text{Pyruvate} & \Delta G^{\circ\prime} = -134.4 \text{ kJ mol}^{-1} \\
& = -32.1 \text{ kcal mol}^{-1} \text{ glucose consumed}
\end{array}
$$

(The corresponding figure for the conversion of one mole of glucose to two moles of lactate is -184.6 kJ mol^{-1} = -44.1 kcal mol^{-1}.) Without production of ATP, the energy released by the conversion of glucose to pyruvate would be lost to the organism and dissipated as heat. The energy required to produce the two molecules of ATP for each molecule of glucose can be recovered by the organism when the ATP is hydrolyzed in some metabolic process. We discussed this point briefly in Chapter 15, when we compared the thermodynamic efficiency of anaerobic and aerobic metabolism. The percentage of the energy released by the breakdown of glucose to lactate that is "captured" by the organism when ADP is phosphorylated to ATP is the efficiency of energy use in glycolysis; it is $(61.0/184.6) \times 100$, or about 33%. Recall this percentage from Section 15.10. It comes from calculating the energy used to phosphorylate two moles of ATP as a percentage of the energy released by the conversion of one mole of glucose to two moles of lactate. The net release of energy in glycolysis, 123.6 kJ (29.5 kcal) for each mole of glucose converted to lactate, is dissipated as heat by the organism. Without the production of ATP to serve as a source of energy for other metabolic processes, the energy released by glycolysis would serve no purpose for the organism, except to help maintain body temperature in warm-blooded animals. A soft drink with ice can help keep you warm even on the coldest day of winter (if it is not a diet drink) because of its high sugar content.

The free-energy changes we have listed in this section are the standard values, assuming the standard conditions, such as 1 M concentrations of all solutes except hydrogen ion. Concentrations under physiological conditions can differ markedly from standard values. Fortunately, there are well-known methods (Section 15.3) for calculating the difference in the free-energy change. Also, large changes in concentrations frequently lead to relatively small differences in the free-energy change, about a few kilojoules per mole. Some of the free-energy changes may be different under physiological conditions from the values listed here for standard conditions, but the underlying principles and the conclusions drawn from them remain the same.

Summary

17.1 What Is the Overall Pathway in Glycolysis?
In glycolysis, one molecule of glucose gives rise, after a long series of reactions, to two molecules of pyruvate. Along the way, two net molecules of ATP and NADH are produced.

17.2 How Is the 6-Carbon Glucose Converted to the 3-Carbon Glyceraldehyde-3-Phosphate?
In the first half of glycolysis, glucose is phosphorylated to glucose-6-phosphate, using an ATP in the process. Glucose-6-phosphate is isomerized to fructose-6-phosphate, which is then phosphorylated again to fructose-1,6-*bis*phosphate, utilizing another ATP. Fructose-1,6-*bis*phosphate is a key intermediate, and the enzyme that catalyzes its formation, phosphofructokinase, is an important controlling factor in the pathway. Fructose 1,6-*bis*phosphate is then split into two 3-carbon compounds, glyceraldehyde-3-phosphate and dihydroxyacetone phosphate, the latter of which is then also converted to glyceraldehyde-3-phosphate. The overall reaction in the first half of the pathway is the conversion of one molecule of glucose into two molecules of glyceraldehyde-3-phosphate at the expense of two molecules of ATP.

17.3 How Is Glyceraldehyde-3-Phosphate Converted to Pyruvate?
Glyceraldehyde-3-phosphate is oxidized to 1,3-*bis*phosphoglycerate and NAD$^+$ is reduced to NADH. The 1,3-*bis*phosphoglycerate is then converted to 3-phosphoglycerate and ATP is produced. 3-Phosphoglycerate is converted in two steps to phosphoenol pyruvate, an important high-energy compound. Phosphoenol pyruvate is then converted to pyruvate and ATP is produced. The overall reaction of the second half of the pathway is that two molecules of glyceraldehyde-3-phosphate are converted to two molecules of pyruvate and four molecules of ATP are produced.

17.4 How Is Pyruvate Metabolized Anaerobically?

Several metabolic fates are possible for pyruvate. In aerobic metabolism, pyruvate loses carbon dioxide; the remaining two carbon atoms become linked to coenzyme A as an acetyl group to form acetyl-CoA, which then enters the citric acid cycle. There are two fates for pyruvate in anaerobic metabolism. In organisms capable of alcoholic fermentation, pyruvate loses carbon dioxide to produce acetaldehyde, which, in turn, is reduced to produce ethanol. In other organisms, the common fate of pyruvate in anaerobic metabolism is reduction to lactate.

17.5 How Much Energy Can Be Produced by Glycolysis?

In each of two reactions in the pathway, one molecule of ATP is hydrolyzed for each molecule of glucose metabolized. In each of two other reactions, two molecules of ATP are produced by phosphorylation of ADP for each molecule of glucose, giving a total of four ATP molecules produced. There is a net gain of two ATP molecules for each molecule of glucose processed in glycolysis. The anaerobic breakdown of glucose to lactate can be summarized as follows:

$$\text{Glucose} + 2\,\text{ADP} + 2\,\text{P}_i \rightarrow 2\,\text{Lactate} + 2\,\text{ATP}$$

The overall process of glycolysis is exergonic.

Reaction	$\Delta G^{\circ\prime}$	
	kJ mol^{-1}	kcal mol^{-1}
Glucose + 2 ADP + 2 P$_i$ → 2 Pyruvate + 2 ATP	−73.3	−17.5
2 (Pyruvate + NADH + H$^+$ → Lactate + NAD$^+$)	−50.2	−12.0
Glucose + 2 ADP + 2 P$_i$ → 2 Lactate + 2 ATP	−123.5	−29.5

Without production of ATP, glycolysis would be still more exergonic, but the energy released would be lost to the organism and dissipated as heat.

Critical Questions to Review

17.1 What Is the Overall Pathway in Glycolysis?

1. Fact Check Which reaction or reactions that we have met in this chapter require ATP? Which reaction or reactions produce ATP? List the enzymes that catalyze the reactions that require and that produce ATP.

2. Fact Check Which reaction or reactions that we have met in this chapter require NADH? Which reaction or reactions require NAD$^+$? List the enzymes that catalyze the reactions that require NADH and that require NAD$^+$.

3. Fact Check What are the possible metabolic fates of pyruvate?

17.2 How Is the 6-Carbon Glucose Converted to the 3-Carbon Glyceraldehyde-3-Phosphate?

4. Fact Check Explain the origin of the name of the enzyme aldolase.

5. Fact Check Define isozymes and give an example from the material discussed in this chapter.

6. Fact Check Why would enzymes be found as isozymes?

7. Fact Check Why is the formation of fructose-1,6-*bis*phosphate the committed step in glycolysis?

8. Thought Question Show that the reaction Glucose → 2 Glyceraldehyde-3-phosphate is slightly endergonic ($\Delta G^{\circ\prime} = 2.2$ kJ mol^{-1} = 0.53 kcal mol^{-1}); that is, it is not too far from equilibrium. Use the data in Table 17.1.

9. Thought Question What is the metabolic advantage of having both hexokinase and glucokinase to phosphorylate glucose?

10. Thought Question What are the metabolic effects of not being able to produce the M subunit of phosphofructokinase?

11. Thought Question In what way is the observed mode of action of hexokinase consistent with the induced-fit theory of enzyme action?

12. Thought Question How does ATP act as an allosteric effector in the mode of action of phosphofructokinase?

17.3 How is Glyceraldehyde-3-Phosphate Converted to Pyruvate?

13. Fact Check At what point in glycolysis are all the reactions considered doubled?

14. Fact Check Which of the enzymes discussed in this chapter are NADH-linked dehydrogenases?

15. Fact Check Define substrate-level phosphorylation and give an example from the reactions discussed in this chapter.

16. Fact Check Which reactions are the control points in glycolysis?

17. Fact Check Which molecules act as inhibitors of glycolysis? Which molecules act as activators?

18. Fact Check There are many NADH-linked dehydrogenases that have similar active sites. Which part of glyceraldehyde-3-phosphate dehydrogenase would be the most conserved between other enzymes?

19. Fact Check Several of the enzymes of glycolysis fall into classes that we will see often in metabolism. What are the reaction types catalyzed by each of the following:
(a) Kinases
(b) Isomerases
(c) Aldolases
(d) Dehydrogenases

20. Fact Check What is the difference between an isomerase and a mutase?

21. Thought Question Is the reaction of 2-phosphoglycerate to phosphoenolpyruvate a redox reaction? Give the reason for your answer.

22. Thought Question Show the carbon atom that changes oxidation state during the reaction catalyzed by glyceraldehyde-3-phosphate dehydrogenase. What is the functional group that changes during the reaction?

23. Thought Question Discuss the logic of the nature of the allosteric inhibitors and activators of glycolysis. Why would these molecules be used?

24. Thought Question Many species have a third type of LDH subunit that is found predominantly in the testes. If this subunit, called C, were expressed in other tissues and could combine with the M and H subunits, how many LDH isozymes would be possible? What would their compositions be?

25. Thought Question The M and H subunits of lactate dehydrogenase have very similar sizes and shapes but differ in amino acid composition. If the only difference between the two were that the H subunit had a glutamic acid in a position where the M subunit had a serine, how would the five isozymes of LDH separate on electrophoresis using a gel at pH 8.6? (See Chapter 5 for details on electrophoresis.)

26. Thought Question Why is the formation of fructose-1,6-*bis*phosphate a step in which control is likely to be exercised in the glycolytic pathway?

27. Thought Question High levels of glucose-6-phosphate inhibit glycolysis. If the concentration of glucose-6-phosphate decreases, activity is restored. Why?

28. Thought Question Most metabolic pathways are relatively long and appear to be very complex. For example, there are ten individual chemical reactions in glycolysis, converting glucose to pyruvate. Suggest a reason for the complexity.

29. Thought Question The mechanism involved in the reaction catalyzed by phosphoglyceromutase is known to involve a phosphorylated enzyme intermediate. If 3-phosphoglycerate is radioactively labeled with ^{32}P, the product of the reaction, 2-phosphoglycerate, does not have any radioactive label. Design a mechanism to explain these facts.

17.4 How Is Pyruvate Metabolized Anaerobically?

30. Fact Check What does the material of this chapter have to do with beer? What does it have to do with tired and aching muscles?

31. Fact Check If lactic acid is the buildup product of strenuous muscle activity, why is sodium lactate often given to hospital patients intravenously?

32. Fact Check What is the metabolic purpose of lactic acid production?

33. Thought Question Using the Lewis electron-dot notation, show explicitly the transfer of electrons in the following redox reactions.

(a) Pyruvate + NADH + H^+ → Lactate + NAD^+

(b) Acetaldehyde + NADH + H^+ → Ethanol + NAD^+

(c) Glyceraldehyde-3-phosphate + NAD^+ → 3-Phosphoglycerate + NADH + H^+ (redox reaction only)

34. Thought Question Briefly discuss the role of thiamine pyrophosphate in enzymatic reactions, using material from this chapter to illustrate your points.

35. Thought Question What is unique about TPP that makes it useful in decarboxylation reactions?

36. Biochemical Connections Beriberi is a disease caused by a deficiency of vitamin B_1 (thiamine) in the diet. Thiamine is the precursor of thiamine pyrophosphate. In view of what you have learned in this chapter, why is it not surprising that alcoholics tend to develop this disease?

37. Thought Question It is well known among hunters that meat from animals that have been run to death tastes sour. Suggest a reason for this observation.

38. Thought Question What is the metabolic advantage in the conversion of glucose to lactate, in which there is no *net* oxidation or reduction?

39. Biochemical Connections Cancer cells grow so rapidly that they have a higher rate of anaerobic metabolism than most body tissues, especially at the center of a tumor. Can you use drugs that poison the enzymes of anaerobic metabolism in the treatment of cancer? Why, or why not?

17.5 How Much Energy Can Be Produced by Glycolysis?

40. Thought Question Show how the estimate of 33% efficiency of energy use in anaerobic glycolysis is derived.

41. Fact Check What is the net gain of ATP molecules derived from the reactions of glycolysis?

42. Fact Check How does the result in Question 41 differ from the gross yield of ATP?

43. Fact Check Which of the reactions in glycolysis are coupled reactions?

44. Fact Check Which steps in glycolysis are physiologically irreversible?

45. Thought Question Show, by a series of equations, the energetics of phosphorylation of ADP by phosphoenolpyruvate.

46. Thought Question What should be the net ATP yield for glycolysis when fructose, mannose, and galactose are used as the starting compounds? Justify your answer.

47. Thought Question In the muscles, glycogen is broken down via the following reaction:

$$(\text{Glucose})_n + P_i \rightarrow \text{Glucose-1-phosphate} + (\text{Glucose})_{n-1}$$

What would be the ATP yield per molecule of glucose in the muscle if glycogen were the source of the glucose?

48. Thought Question Using Table 17.1, predict whether the following reaction is thermodynamically possible:

$$\text{Phosphoenolpyruvate} + P_i + 2\,\text{ADP} \rightarrow \text{Pyruvate} + 2\,\text{ATP}$$

49. Thought Question Does the reaction shown in Question 8 occur in nature? If not, why not?

50. Thought Question According to Table 17.1, several reactions have very positive $\Delta G°'$ values. How can this be explained, given that these reactions do occur in the cell?

51. Thought Question According to Table 17.1, four reactions have positive ΔG values. How can this be explained?

Biochemistry ⒺNow™
Assess your understanding of this chapter's topics with additional quizzing and tutorials at **http://now.brookscole.com/campbell5**

Annotated Bibliography

Bodner, G. M. Metabolism: Part I, Glycolysis, or the Embden–Meyerhoff Pathway. *J. Chem. Ed.* **63**, 566–570 (1986). [A clear, concise summary of the pathway. Part of a series on metabolism of carbohydrates and lipids.]

Boyer, P. D., ed. *The Enzymes*, Vols. 5–9. New York: Academic Press, 1972. [A standard reference with review articles on the glycolytic enzymes; lactate dehydrogenase and alcohol dehydrogenase appear in Volume 10.]

Florkin, M., and E. H. Stotz, eds. *Comprehensive Biochemistry*. New York: Elsevier, 1967. [Another standard reference. Volume 17, *Carbohydrate Metabolism*, deals with glycolysis.]

Karl, P. I., B. H. J. Gordon, C. S. Lieber, and S. E. Fisher. Acetaldehyde Production and Transfer by the Perfused Human Placental Cotyle-don. *Science* **242**, 273–275 (1988). [A report describing some of the processes involved in fetal alcohol syndrome.]

Light, W. J. *Alcoholism and Women, Genetics, and Fetal Development*. Springfield, Ill.: Charles C. Thomas Publishers, 1988. [A book that devotes a large amount of space to fetal alcohol syndrome.]

Lipmann, F. A Long Life in Times of Great Upheaval. *Ann. Rev. Biochem.* **53**, 1–33 (1984). [The reminiscences of a Nobel laureate whose research contributed greatly to the understanding of carbohydrate metabolism. Very interesting reading from the standpoints of autobiography and the author's contributions to biochemistry.]

Storage Mechanisms and Control in Carbohydrate Metabolism

We usually think of carbohydrates as "quick energy." It is certainly true that the body can mobilize carbohydrates more easily than fat, even though fats contain more energy on a gram-for-gram basis. Free glucose in the blood-stream is quickly depleted, but we have stored glucose in readily accessible form as the polymer glycogen. In sustained, high-intensity activity, glycogen in muscle cells is broken down to glucose in a quick response to the need for energy. Before long, the supply of glycogen is depleted in turn. In such cases, pyruvate, the end product of glycolysis, can be converted to lactate and exported from muscle to the liver. There, with the help of its energy reserves of ATP, the liver can convert lactate to "new glucose" and return the glucose to muscle for another round of glycolysis. This process, called gluconeogenesis, is essentially (with some notable exceptions) glycolysis in reverse. New glucose can also be made from pyruvate formed by molecules from the citric acid cycle (Chapter 19). This "homemade" glucose produced by gluconeogenesis is available to supply glucose to the brain, which has little reserve of its own. The brain requires a steady stream of energy in the form of glucose (usually its only fuel), no matter what the level of mental activity is, whether the person is awake or asleep.

Control of carbohydrate metabolism is important in physical activity of all sorts.

Critical Questions

18.1 How Is Glycogen Produced and Degraded?

18.2 How Does Gluconeogenesis Produce Glucose from Pyruvate?

18.3 How Is Carbohydrate Metabolism Controlled?

18.4 Why Is Glucose Sometimes Diverted through the Pentose Phosphate Pathway?

18.1 | How Is Glycogen Produced and Degraded?

When we digest a meal high in carbohydrates, we have a supply of glucose that exceeds our immediate needs. We store glucose as a polymer, glycogen (Section 16.4), that is similar to the starches found in plants; glycogen differs from starch only in the degree of chain branching. In fact, glycogen is sometimes called "animal starch" because of this similarity. A look at the metabolism of glycogen will give us some insights into how glucose can be stored in this form and made available on demand. In the degradation of glycogen, several glucose residues can be released simultaneously, one from each end of a branch, rather than one at a time as would be the case in a linear polymer. This feature is useful to an organism in meeting short-term demands for energy by increasing the glucose supply as quickly as possible (Figure 18.1). It has been shown by mathematical modeling that the structure of glycogen is *optimized* for its ability to store and deliver energy quickly and for the longest amount of time possible. The key to this optimization is the average chain length of the branches (13 residues). If the average chain length were much greater or much shorter, glycogen would not be as efficient a vehicle for energy storage and release on demand. Experimental results support the conclusions reached from the mathematical modeling.

Breakdown of Glycogen

Glycogen is found primarily in liver and muscle. The release of glycogen stored in the liver is triggered by low levels of glucose in blood. Liver glycogen breaks down to glucose-6-phosphate, which is hydrolyzed to give glucose. The release of glucose from the liver by this breakdown of glycogen replenishes the supply of glucose in the blood. In muscle, glucose-6-phosphate obtained from glycogen breakdown enters the glycolytic pathway directly rather than being hydrolyzed to glucose and then exported to the bloodstream.

Three reactions play roles in the conversion of glycogen to glucose-6-phosphate. In the first reaction, each glucose residue cleaved from glycogen reacts with phosphate to give glucose-1-phosphate. Note particularly that this cleavage reaction is one of **phosphorolysis** rather than hydrolysis.

$$(Glucose)_n \ + \ HO-P-O^- \ \rightleftharpoons \ (Glucose)_{n-1} \ + \ \text{Glucose-1-phosphate}$$

Glycogen   Phosphate ion   Remainder of glycogen   Glucose-1-phosphate

In a second reaction, glucose-1-phosphate isomerizes to give glucose-6-phosphate.

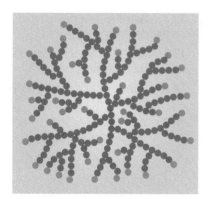

Glycogen

▲ **FIGURE 18.1** The highly branched structure of glycogen makes it possible for several glucose residues to be released at once to meet energy needs. This would not be possible with a linear polymer. The red dots indicate the terminal glucose residues that are released from glycogen. The more branch points there are, the more of these terminal residues that are available at one time.

Glucose-1-phosphate   ⇌   Glucose-6-phosphate

Complete breakdown of glycogen also requires a debranching reaction to hydrolyze the glycosidic bonds of the glucose residues at branch points in the glycogen structure. The enzyme that catalyzes the first of these reactions is **glycogen phosphorylase;** the second reaction is catalyzed by **phosphoglucomutase.**

$$\text{Glycogen} + P_i \xrightarrow{\text{Glycogen phosphorylase}} \text{Glucose-1-phosphate} + \text{Remainder of glycogen}$$

$$\text{Glucose-1-phosphate} \xrightarrow{\text{Phosphoglucomutase}} \text{Glucose-6-phosphate}$$

Glycogen phosphorylase cleaves the $\alpha(1\rightarrow4)$ linkages in glycogen. Complete breakdown requires **debranching enzymes** that degrade the $\alpha(1\rightarrow6)$ linkages. Note that no ATP is hydrolyzed in the first reaction. In the glycolytic pathway, we saw another example of phosphorylation of a substrate directly by phosphate without involvement of ATP: the phosphorylation of glyceraldehyde-3-phosphate to 1,3-*bis*phosphoglycerate. This is an alternative mode of entry to the glycolytic pathway that "saves" one molecule of ATP for each molecule of

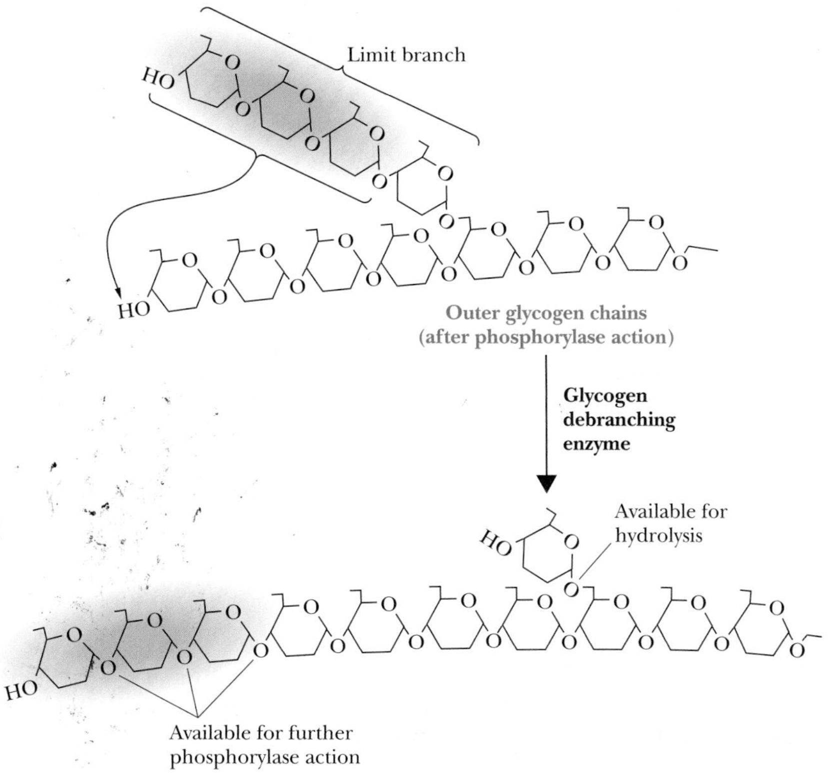

◀ **FIGURE 18.2** The mode of action of the debranching enzyme in glycogen breakdown. The enzyme transfers three $\alpha(1\rightarrow4)$-linked glucose residues from a limit branch to the end of another branch. The same enzyme also catalyzes the hydrolysis of the $\alpha(1\rightarrow6)$-linked residue at the branch point.

glucose because it bypasses the first step in glycolysis. When glycogen rather than glucose is the starting material for glycolysis, there is a net gain of three ATP molecules for each glucose monomer, rather than two ATP molecules, as when glucose itself is the starting point. Thus, glycogen is a more effective energy source than glucose. Of course, there is no "free lunch" in biochemistry and, as we shall see, it takes energy to put the glucoses together into glycogen.

The debranching of glycogen involves the transfer of a "limit branch" of three glucose residues to the end of another branch, where they are subsequently removed by glycogen phosphorylase. The same glycogen debranching enzyme then hydrolyzes the $\alpha(1\rightarrow6)$ glycosidic bond of the last glucose residue remaining at the branch point (Figure 18.2).

When an organism needs energy quickly, glycogen breakdown is important. Muscle tissue can mobilize glycogen more easily than fat and can do so anaerobically. With low-intensity exercise, such as jogging or long-distance running, fat is the preferred fuel, but, as the intensity increases, muscle and liver glycogen becomes more important. Some athletes, particularly middle-distance runners and cyclists, try to build up their glycogen reserves before a race by eating large amounts of carbohydrates. The Biochemical Connections box on page 493 goes into more detail on this subject.

Formation of Glycogen from Glucose

The formation of glycogen from glucose is not the exact reversal of the breakdown of glycogen to glucose. The synthesis of glycogen requires energy, which is provided by the hydrolysis of a nucleoside triphosphate, UTP.

In the first stage of glycogen synthesis, glucose-1-phosphate (obtained from glucose-6-phosphate by an isomerization reaction) reacts with UTP to produce uridine diphosphate glucose (also called UDP-glucose or UDPG) and pyrophosphate (PP_i).

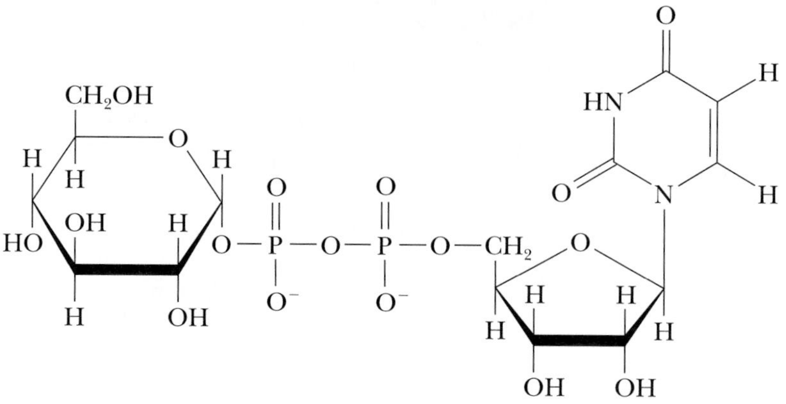

Uridine disphosphate glucose
(UDPG)

The enzyme that catalyzes this reaction is *UDP-glucose pyrophosphorylase*. The exchange of one phosphoric anhydride bond for another has a free-energy change close to zero. The release of energy comes about when the enzyme inorganic pyrophosphatase catalyzes the hydrolysis of pyrophosphate to two phosphates, a strongly exergonic reaction.

It is common in biochemistry to see the energy released by the hydrolysis of pyrophosphate combined with the free energy of hydrolysis of a nucleoside triphosphate. The coupling of these two exergonic reactions to a reaction that is not energetically favorable allows an otherwise endergonic reaction to take place. The supply of UTP is replenished by an exchange reaction with ATP, which is catalyzed by nucleoside phosphate kinase:

$$UDP + ATP \rightleftharpoons UTP + ADP$$

This exchange reaction makes the hydrolysis of any nucleoside triphosphate energetically equivalent to the hydrolysis of ATP.

The addition of UDPG to a growing chain of glycogen is the next step in glycogen synthesis. Each step involves formation of a new $\alpha(1\rightarrow4)$ glycosidic bond in a reaction catalyzed by the enzyme **glycogen synthase** (Figure 18.3). This enzyme cannot simply form a bond between two isolated glucose molecules; it must add to an existing chain with $\alpha(1\rightarrow4)$ glycosidic linkages. The initiation of glycogen synthesis requires a primer for this reason. The hydroxyl group of a specific tyrosine of the protein *glycogenin* (37,300 Da) serves this purpose. In the first stage of glycogen synthesis, a glucose residue is linked to this tyrosine hydroxyl, and glucose residues are successively added to this first one. The glycogenin molecule itself acts as the catalyst for addition of glucoses until there are about eight of them linked together. At that point, glycogen synthase takes over.

	$\Delta G^{\circ\prime}$	
	kJ mol^{-1}	kcal mol^{-1}
Glucose-1-phosphate + UTP $\rightleftharpoons$ UDPG + PP_i	$\sim$0	$\sim$0
H_2O + PP_i $\rightarrow$ 2 P_i	-30.5	-7.3
Overall:　Glucose-1-phosphate + UTP $\rightarrow$ UDPG + 2 P_i	-30.5	-7.3

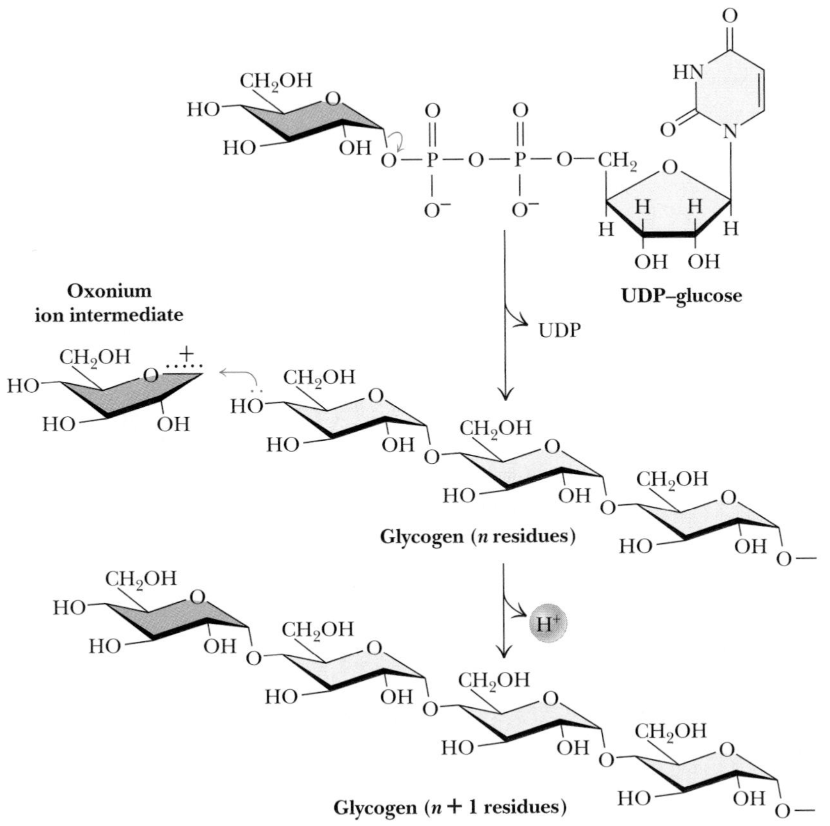

Biochemistry ⑤ Now™ **ANIMATED FIGURE 18.3**
The reaction catalyzed by glycogen synthase.
A glucose residue is transferred from UDPG to
the growing end of a glycogen chain in an α(1→4)
linkage. **See this figure animated at http://now
.brookscole.com/campbell5**

Synthesis of glycogen requires the formation of α(1→6) as well as α(1→4) glycosidic linkages. A **branching enzyme** accomplishes this task. It does so by transferring a segment about seven residues in length from the end of a growing chain to a branch point where it catalyzes the formation of the required α(1→6) glycosidic linkage (Figure 18.4). Note that this enzyme has already catalyzed the breaking of an α(1→4) glycosidic linkage in the process of transferring the oligosaccharide segment. Each transferred segment must come from a chain at least 11 residues long; each new branch point must be at least 4 residues away from the nearest existing branch point.

> **Essential Information**
>
> Glycogen is the storage form of glucose in animals, including humans. Glucose polymerizes to form glycogen when the organism has no immediate need for the energy derived from glucose breakdown. Glycogen releases glucose when energy demands are high. The whole process is subject to several different control mechanisms.

Control of Glycogen Metabolism: A Case Study in Control Mechanisms

How does an organism ensure that glycogen synthesis and glycogen breakdown do not operate simultaneously? If this were to occur, the main result would be the hydrolysis of UTP, which would waste chemical energy stored in the phosphoric anhydride bonds. A major controlling factor lies in the behavior of glycogen phosphorylase. This enzyme is subject not only to allosteric control but also to another control feature: covalent modification. We saw an earlier example of this kind of control in the sodium–potassium pump in Section 8.6. In that example, phosphorylation and dephosphorylation of an enzyme determined whether it was active, and a similar effect takes place here.

Figure 18.5 summarizes some of the salient control features that affect glycogen phosphorylase activity. The enzyme is a dimer that exists in two forms, the inactive T (taut) form and the active R (relaxed) form. In the T form (and *only* in the T form), it can be modified by phosphorylation of a specific serine residue on each of the two subunits. The esterification of the serines to phosphoric acid is catalyzed by the enzyme *phosphorylase kinase;* the

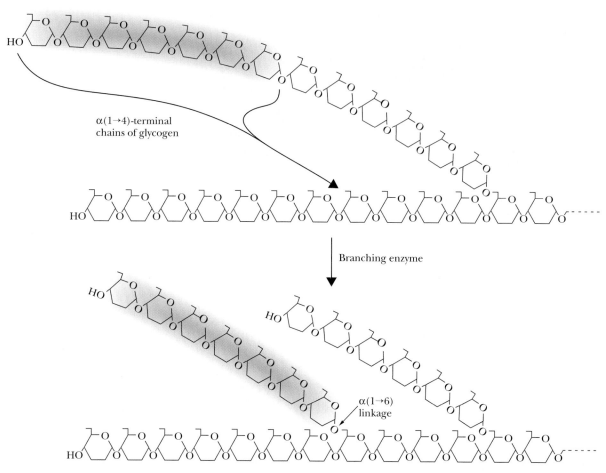

α(1→4)-terminal
chains of glycogen

Branching enzyme

α(1→6)
linkage

▲ **FIGURE 18.4** The mode of action of the branching enzyme in glycogen synthesis. A segment seven residues long is transferred from a growing branch to a new branch point, where an α(1→6) linkage is formed.

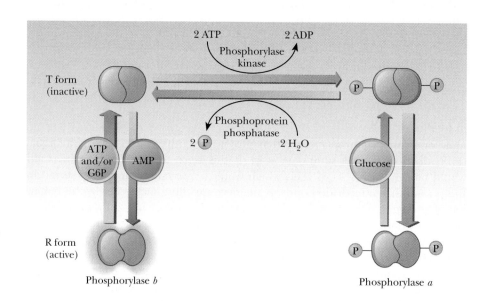

Biochemistry ⓔ**Now**™ **ACTIVE FIGURE 18.5**
Glycogen phosphorylase activity is subject to allosteric control and covalent modification. Phosphorylation of the *a* form of the enzyme converts it to the *b* form. Only the T form is subject to modification and demodification. The *a* and *b* forms respond to different allosteric effectors (see text). **Watch this Active Figure at http://now.brookscole.com/ campbell5**

dephosphorylation is catalyzed by *phosphoprotein phosphatase*. The phosphorylated form of glycogen phosphorylase is called **phosphorylase** *a*, and the dephosphorylated form is called **phosphorylase** *b*. The switch from phosphorylase *b* to phosphorylase *a* is the major form of control over the activity of

Biochemical Connections

Glycogen Loading

Glycogen is the primary energy source for a muscle that was at rest and then starts working vigorously. The energy of ATP hydrolysis derived from glycogen breakdown is initially produced *anaerobically*, with the lactic acid product being processed back to glucose in the liver. As an athlete becomes well conditioned, the muscle cells have more mitochondria, allowing for more *aerobic* metabolism of fats and carbohydrates for energy. The switch to aerobic metabolism takes a few minutes, which is why athletes must warm up before an event. In long-distance events, athletes rely more on fat metabolism than they do in short-distance events, but, in any race, there will be a final surge at the end in which the level of muscle glycogen may well determine the winner.

The idea behind glycogen loading is that, if there is more available glycogen, then a person can carry out anaerobic metabolism for a longer period of time, either at the end of a distance event or for the entire event, if the effort level is high enough. This is probably true, but several questions come to mind: How long does the glycogen last? What is the best way to "load" glycogen? Is it safe? Theoretical calculations estimate that it takes 8 to 12 minutes to use all the glycogen in the skeletal muscle, although this range varies greatly depending on the intensity level. Allowing for loading of extra glycogen, it might last half an hour. There is evidence that glycogen may be used more slowly in well-conditioned athletes because they exhibit is higher fat utilization.

Early loading methods involved glycogen depletion for three days via a high-protein diet and extreme exercise, followed by loading from a high-carbohydrate diet and resting. This method yields a marked increase of glycogen, but some of it then is stored in the heart (which usually has little or no glycogen). The practice actually stresses the heart muscle. There is clearly some danger here. There are also dangers associated with the high-protein diet because too much protein often leads to a mineral imbalance, which also stresses the heart and the kidneys. Again, there is some danger. In addition, the training was often nonoptimal during the week because the athlete had trouble performing while on the low-carbohydrate diet and didn't train much at all during the loading phase. Simple carbohydrate loading without previous extreme glycogen depletion does increase glycogen, but not as much; however, this increase does not risk potential stress to the heart.

Simple loading merely involves eating diets rich in pasta, starch, and complex carbohydrate for a few days before the strenuous exertion. It is not clear whether simple loading works. It is certainly possible to increase the amount of glycogen in muscle, but a question remains about how long it will last during vigorous exercise. Ultimately, all diet considerations for athletes are very individual, and what works for one may not work for another.

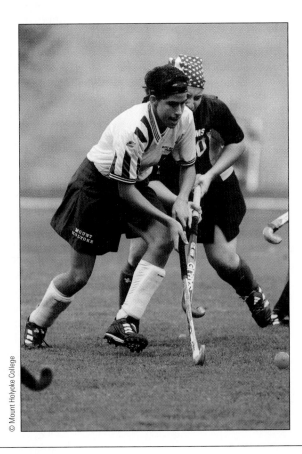

© Mount Holyoke College

phosphorylase. The response time of the changes is on the order of seconds to minutes. Phosphorylase is also controlled more quickly in times of urgency by allosteric effectors, with a response time of milliseconds.

In liver, glucose is an allosteric inhibitor of phosphorylase *a*. It binds to the substrate site and favors the transition to the T state. It also exposes the phosphorylated serines so that the phosphatase can hydrolyze them. This shifts the equilibrium to phosphorylase *b*. In muscle, the primary allosteric effectors are ATP, AMP, and glucose-6-phosphate (G6P). When the muscles use ATP to contract, AMP levels rise. This increase in AMP stimulates formation of the R state of phosphorylase *b*, which is active. When ATP is plentiful or glucose-6-phosphate builds up, these molecules act as allosteric inhibitors shifting the equilibrium back to the T form. These differences ensure that glycogen will be degraded when there is a need for energy, as is the case with high [AMP], low [G6P], and low [ATP]. When the reverse is true (low [AMP], high [G6P], and high [ATP]), the need for energy, and consequently for glycogen breakdown, is less. "Shutting down" glycogen phosphorylase activity is the appropriate response. The combination of covalent modification and allosteric control of the process allows for a degree of fine-tuning that would not be possible with either mechanism alone. Hormonal control also enters into the picture. When epinephrine is released from the adrenal gland in response to stress, it triggers a series of events, discussed more fully in Section 24.4, that suppress the activity of glycogen synthase and stimulate that of glycogen phosphorylase.

The activity of glycogen synthase is subject to the same type of covalent modification as glycogen phosphorylase. The difference is that the response is opposite. The inactive form of glycogen synthase is the phosphorylated form. The active form is unphosphorylated. The hormonal signals (glucagon or epinephrine) stimulate the phosphorylation of glycogen synthase via an enzyme called cAMP-dependent protein kinase (Chapter 24). After the glycogen synthase is phosphorylated, it becomes inactive at the same time the hormonal signal is activating phosphorylase. Glycogen synthase can also be phosphorylated by several other enzymes, including phosphorylase kinase and several enzymes called glycogen synthase kinases. Glycogen synthase is dephosphorylated by the same phosphoprotein phosphatase that removes the phosphate from phosphorylase. The phosphorylation of glycogen synthase is also more complicated in that there are multiple phosphorylation sites. As many as nine different amino acid residues have been found to be phosphorylated. As the progressive level of phosphorylation increases, the activity of the enzyme decreases.

Glycogen synthase is also under allosteric control. It is inhibited by ATP. This inhibition can be overcome by glucose-6-hosphate, which is an activator. However, the two forms of glycogen synthase respond very differently to glucose-6-phosphate. The phosphorylated (inactive) form is called **glycogen synthase D** (for glucose-6-phosphate dependent) because it is only active under very high concentrations of glucose-6-phosphate. In fact, the level necessary to give significant activity would be beyond the physiological range. The nonphosphorylated form is called **glycogen synthase I** (for glucose-6-phosphate independent) because it is active even with low concentrations of glucose-6-phosphate. Thus, even though purified enzymes can be shown to respond to allosteric effectors, the true control over the activity of glycogen synthase is by its phosphorylation state, which, in turn, is controlled by hormonal states. (See the article by Shulman in the Annotated Bibliography at the end of the chapter for a detailed description of glycogen synthase.)

The fact that two target enzymes, glycogen phosphorylase and glycogen synthase, are modified in the same way by the same enzymes links the opposing processes of synthesis and breakdown of glycogen even more intimately.

Finally, the modifying enzymes are themselves subject to covalent modification and allosteric control. This feature complicates the process considerably but adds the possibility of an amplified response to small changes in conditions. A small change in the concentration of an allosteric effector of a modifying enzyme can cause a large change in the concentration of an active, modified target enzyme; this amplification response is due to the fact that the substrate for the modifying enzyme is itself an enzyme. At this point, the situation has become very complex indeed, but it is a good example of how opposing processes of breakdown and synthesis can be controlled to the advantage of an organism. When we see in the next section how glucose is synthesized from lactate, we shall have another example, one that we can contrast with glycolysis to explore in more detail how carbohydrate metabolism is controlled.

18.2 | How Does Gluconeogenesis Produce Glucose from Pyruvate?

The conversion of pyruvate to glucose occurs by a process called **gluconeogenesis.** Gluconeogenesis is not the exact reversal of glycolysis. We first met pyruvate as a product of glycolysis, but it can arise from other sources to be the starting point of the anabolism of glucose. Some of the reactions of glycolysis are essentially irreversible; these reactions are bypassed in gluconeogenesis. An analogy is a hiker who goes directly down a steep slope but who climbs back up the hill by an alternative, easier route. We shall see that the biosynthesis and the degradation of many important biomolecules follow different pathways.

There are three irreversible steps in glycolysis, and the differences between glycolysis and gluconeogenesis are found in these three reactions. The first of the glycolytic reactions is the production of pyruvate (and ATP) from phosphoenolpyruvate. The second is the production of fructose-1,6-*bis*phosphate from fructose-6-phosphate, and the third is the production of glucose-6-phosphate from glucose. Because the first of these reactions is exergonic, the reverse reaction is endergonic. Reversing the second and third reactions would require the production of ATP from ADP, which is also an endergonic reaction. The net result of gluconeogenesis includes the reversal of these three glycolytic reactions, but the pathway is different, with different reactions and different enzymes (Figure 18.6).

Oxaloacetate Is an Intermediate in the Production of Phosphoenolpyruvate in Gluconeogenesis

The conversion of pyruvate to phosphoenolpyruvate in gluconeogenesis takes place in two steps. The first step is the reaction of pyruvate and carbon dioxide to give oxaloacetate. This step requires energy, which is available from the hydrolysis of ATP.

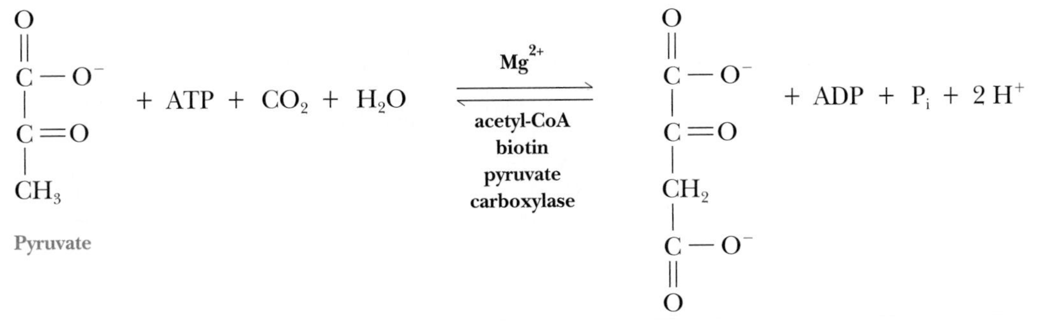

$$
\begin{array}{c}
\underset{\text{Pyruvate}}{
\begin{array}{l}
\text{O} \\
\parallel \\
\text{C}-\text{O}^- \\
\mid \\
\text{C}=\text{O} \\
\mid \\
\text{CH}_3
\end{array}}
\;+\; \text{ATP} \;+\; \text{CO}_2 \;+\; \text{H}_2\text{O}
\;\;\underset{\substack{\text{acetyl-CoA}\\\text{biotin}\\\text{pyruvate}\\\text{carboxylase}}}{\overset{\text{Mg}^{2+}}{\rightleftharpoons}}\;\;
\underset{\text{Oxaloacetate}}{
\begin{array}{l}
\text{O} \\
\parallel \\
\text{C}-\text{O}^- \\
\mid \\
\text{C}=\text{O} \\
\mid \\
\text{CH}_2 \\
\mid \\
\text{C}-\text{O}^- \\
\parallel \\
\text{O}
\end{array}}
\;+\; \text{ADP} \;+\; \text{P}_i \;+\; 2\,\text{H}^+
\end{array}
$$

The enzyme that catalyzes this reaction is *pyruvate carboxylase,* an allosteric enzyme found in the mitochondria. Acetyl-CoA is an allosteric effector that activates pyruvate carboxylase. If high levels of acetyl-CoA are present (in other words, if there is more acetyl-CoA than is needed to supply the citric acid cycle), pyruvate (a precursor of acetyl-CoA) can be diverted to gluconeogenesis. (Oxaloacetate from the citric acid cycle can frequently be a starting point for gluconeogenesis as well.) Magnesium ion (Mg^{2+}) and biotin are also

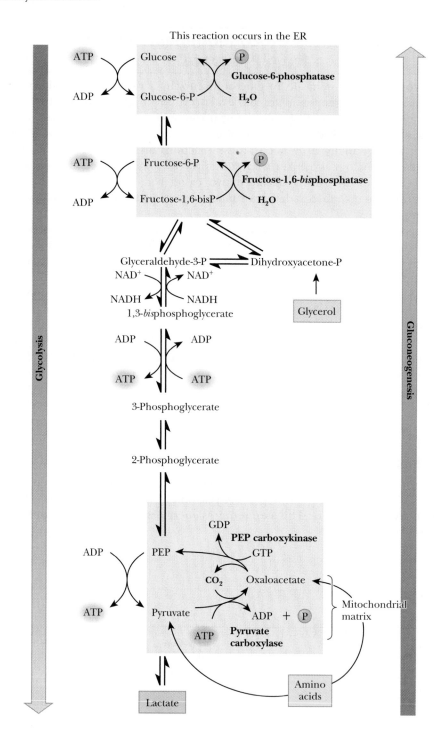

▶ **FIGURE 18.6** The pathways of gluconeogenesis and glycolysis. Species in blue, green, and pink shaded boxes indicate other entry points for gluconeogenesis (in addition to pyruvate).

required for effective catalysis. We have seen Mg^{2+} as a cofactor before, but we have not seen biotin, which requires some discussion.

Biotin is a carrier of carbon dioxide; it has a specific site for covalent attachment of CO_2 (Figure 18.7). The carboxyl group of the biotin forms an amide bond with the ε-amino group of a specific lysine side chain of pyruvate carboxylase. The CO_2 is attached to the biotin, which, in turn, is covalently bonded to the enzyme, and then the CO_2 is shifted to pyruvate to form oxaloacetate (Figure 18.8). Note that ATP is required for this reaction.

The conversion of oxaloacetate to phosphoenolpyruvate is catalyzed by the enzyme *phosphoenolpyruvate carboxykinase (PEPCK)*, which is found in the mitochondria and the cytosol. This reaction also involves hydrolysis of a nucleoside triphosphate—GTP, in this case, rather than ATP.

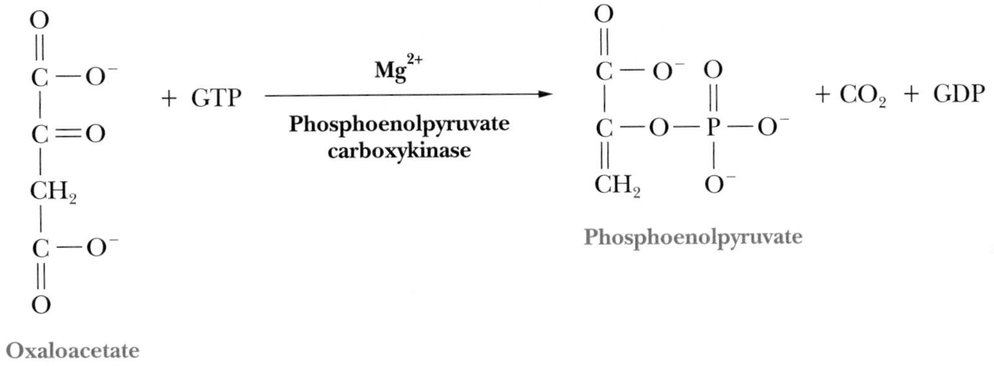

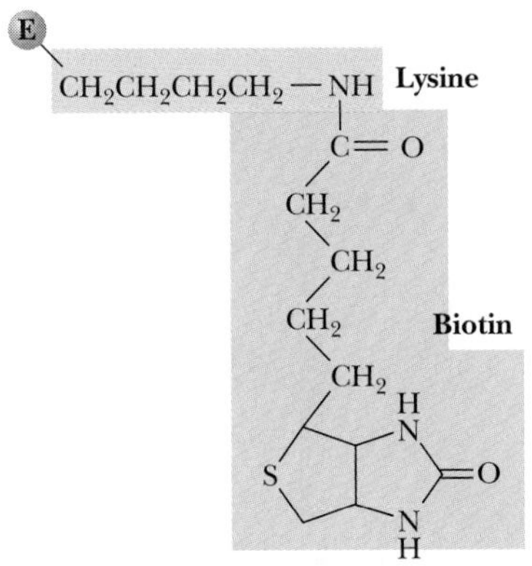

Oxaloacetate

Phosphoenolpyruvate

The successive carboxylation and decarboxylation reactions are both close to equilibrium (they have low values of their standard free energies); as a result, the conversion of pyruvate to phosphoenolpyruvate is also close to equilibrium ($\Delta G^{\circ\prime} = 2.1$ kJ mol^{-1} = 0.5 kcal mol^{-1}). A small increase in the level of oxaloacetate can drive the equilibrium to the right, and a small increase in the level of phosphoenolpyruvate can drive it to the left. A concept well known in general chemistry, the **law of mass action,** relates the concentrations of reactants and products in a system at equilibrium. Changing the concentration of reactants or products will cause a shift to reestablish equilibrium. A reaction

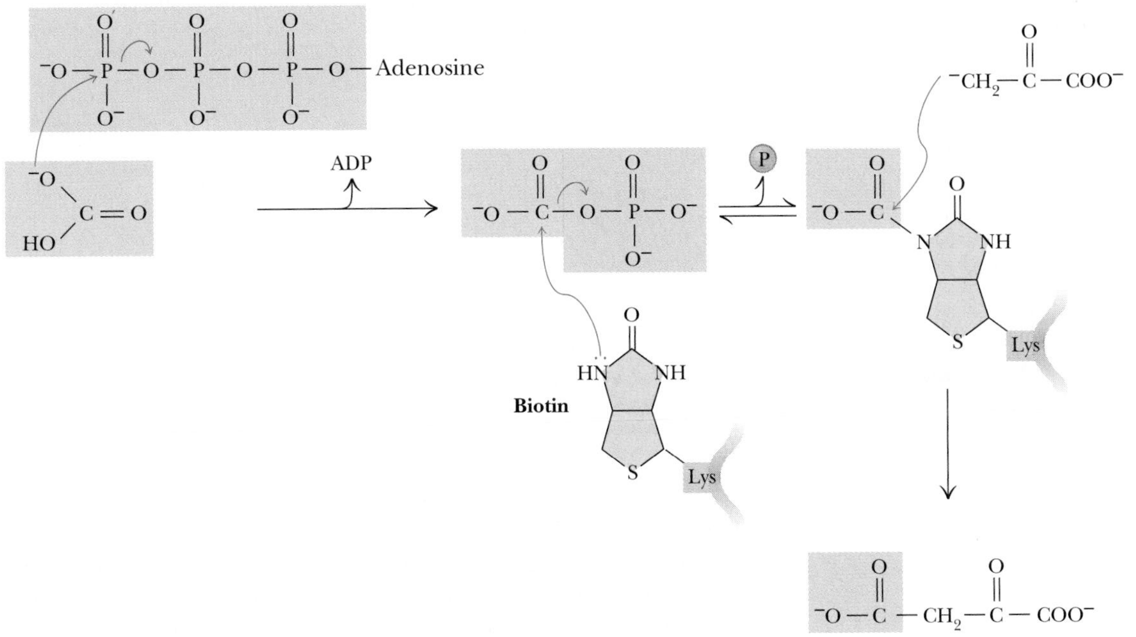

▲ **FIGURE 18.7** The structure of biotin and its mode of attachment to pyruvate carboxylase.

▲ **FIGURE 18.8** The two stages of the pyruvate carboxylase reaction. CO_2 is attached to the biotinylated enzyme. CO_2 is transferred from the biotinylated enzyme to pyruvate, forming oxaloacetate. ATP is required in the first part of the reaction.

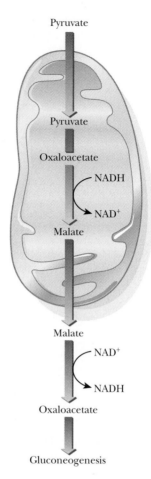

▲ **FIGURE 18.9** Pyruvate carboxylase catalyzes a compartmentalized reaction. Pyruvate is converted to oxaloacetate in the mitochondria. Because oxaloacetate cannot be transported across the mitochondrial membrane, it must be reduced to malate, transported to the cytosol, and then oxidized back to oxaloacetate before gluconeogenesis can continue.

will proceed to the right on addition of reactants and to the left on addition of products.

$$\text{Pyruvate} + \text{ATP} + \text{GTP} \rightarrow \text{Phosphoenolpyruvate} + \text{ADP} + \text{GDP} + P_i$$

The oxaloacetate formed in the mitochondria can have two fates with respect to gluconeogenesis. It can continue to form PEP, which can then leave the mitochondria via a specific transporter to continue gluconeogenesis in the cytosol. The other possibility is that the oxaloacetate can be turned into malate via mitochondrial malate dehydrogenase, a reaction that uses NADH, as shown in Figure 18.9. Malate can then leave the mitochondria, and have the reaction reversed by cytosolic malate dehydrogenase. The reason for this two-step process is that oxaloacetate cannot leave the mitochondria, but malate can. (The pathway involving malate is the one that takes place in the liver, where gluconeogenesis largely takes place.) You might wonder why these two paths exist to get PEP into the cytosol to continue gluconeogenesis. The answer brings us back to a familiar enzyme we saw in glycolysis, glyceraldehyde-3-phosphate dehydrogenase. Remember, from Chapter 17, that the purpose of lactate dehydrogenase is to reduce pyruvate to lactate so that NADH could be oxidized to form NAD^+, which is needed to continue glycolysis. This reaction must be reversed in gluconeogenesis, and the cytosol has a low ratio of NADH to NAD^+. The purpose of the roundabout way of getting oxaloacetate out of the mitochondria via malate dehydrogenase is to produce NADH in the cytosol so that gluconeogenesis can continue.

The Role of Sugar Phosphates in Gluconeogenesis

The other two reactions in which gluconeogenesis differs from glycolysis are ones in which a phosphate-ester bond to a sugar-hydroxyl group is hydrolyzed. Both reactions are catalyzed by phosphatases, and both reactions are exergonic. The first reaction is the hydrolysis of fructose-1,6-*bis*phosphate to produce fructose-6-phosphate and phosphate ion ($\Delta G^{\circ\prime} = -16.7$ kJ mol^{-1} $= -4.0$ kcal mol^{-1}).

Fructose-1,6-*bis*phosphate

Fructose-6-phosphate

This reaction is catalyzed by the enzyme *fructose-1,6-bisphosphatase,* an allosteric enzyme strongly inhibited by adenosine monophosphate (AMP) but stimulated by ATP. Because of allosteric regulation, this reaction is also a control point in the pathway. When the cell has an ample supply of ATP, the formation rather than the breakdown of glucose is favored. This enzyme is inhibited by fructose-2,6-*bis*phosphate, a compound we met in Section 17.2 as an extremely potent activator of phosphofructokinase. We shall return to this point in the next section.

The second reaction is the hydrolysis of glucose-6-phosphate to glucose and phosphate ion ($\Delta G^{\circ\prime} = -13.8$ kJ mol^{-1} = -3.3 kcal mol^{-1}). The enzyme that catalyzes this reaction is *glucose-6-phosphatase.*

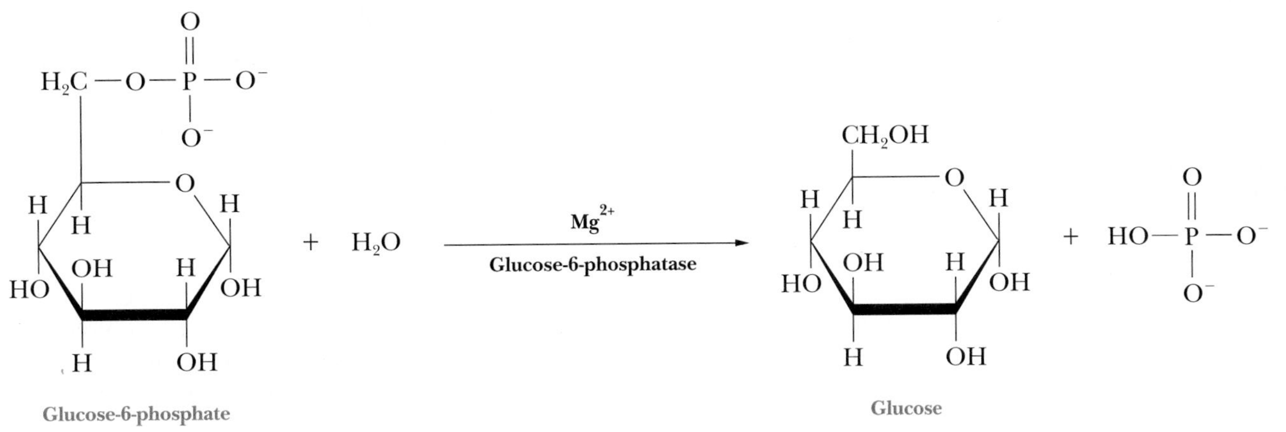

Glucose-6-phosphate   Glucose

When we discussed glycolysis, we saw that both of the phosphorylation reactions, which are the reverse of these two phosphatase-catalyzed reactions, are endergonic. In glycolysis, the phosphorylation reactions must be coupled to the hydrolysis of ATP to make them exergonic and thus energetically allowed. In gluconeogenesis, the organism can make direct use of the fact that the hydrolysis reactions of the sugar phosphates are exergonic. The corresponding reactions are not the reverse of each other in the two pathways. They differ from each other in whether they require ATP and in the enzymes involved. Hydrolysis of glucose-6-phosphate to glucose occurs in the endoplasmic reticulum. This is an example of an interesting pathway that requires three cellular locations (mitochondria, cytosol, endoplasmic reticulum).

> ### Essential Information
>
> Glucose is formed from pyruvate, which, in turn, can be obtained from lactate that accumulates in muscle during exercise. This process, called gluconeogenesis, takes place in the liver after lactate is transported there by the blood. The newly formed glucose is transported back to the muscles by the blood.

18.3 How Is Carbohydrate Metabolism Controlled?

We have now seen several aspects of carbohydrate metabolism: glycolysis, gluconeogenesis, and the reciprocal breakdown and synthesis of glycogen. Glucose has a central role in all these processes. It is the starting point for glycolysis, in which it is broken down to pyruvate, and for the synthesis of glycogen, in which many glucose residues combine to give the glycogen polymer. Glucose is also the product of gluconeogenesis, which has the net effect of reversing glycolysis; glucose is also obtained from the breakdown of glycogen. Each of the opposing pathways, glycolysis and gluconeogenesis, on the one hand, and the breakdown and synthesis of glycogen, on the other hand, is not the exact reversal of the other, even though the net results are. In other words, a different path is used to arrive at the same place. It is time to see how all these related pathways are controlled.

Control of Phosphofructokinase and Fructose-1,6-*bis*phosphatase

An important element in the control process involves fructose-2,6-*bis*phosphate (F2,6P). We mentioned, in Section 17.2, that this compound is an important allosteric activator of phosphofructokinase (PFK), the key enzyme of glycolysis; it is also an inhibitor of fructose *bis*phosphate phosphatase (FBPase), which plays a role in gluconeogenesis. A high concentration of F2,6P stimulates glycolysis, whereas a low concentration stimulates gluconeogenesis. The concentration of F2,6P in a cell depends on the balance between its synthesis,

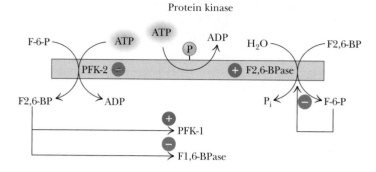

The formation and breakdown of fructose-2,6-*bis*-phosphate (F2,6P) are catalyzed by two enzyme activities on the same protein. These two enzyme activities are controlled by a phosphorylation/dephosphorylation mechanism. Phosphorylation activates the enzyme that degrades F2,6P whereas dephosphorylation activates the enzyme that produces it. **Watch this Active Figure at http://now .brookscole.com/campbell5**

catalyzed by *phosphofructokinase-2* (PFK-2), and its breakdown, catalyzed by *fructose*-bis*phosphatase-2* (FBPase-2). The enzymes that control the formation and breakdown of F2,6P are themselves controlled by a phosphorylation/dephosphorylation mechanism similar to what we have already seen in the case of glycogen phosphorylase and glycogen synthase (Figure 18.10). Both enzyme activities are located on the same protein (a dimer of about 100 kDa molecular mass). Phosphorylation of the dimeric protein leads to an increase in activity of FBPase-2 and a decrease in the concentration of F2,6P, ultimately stimulating gluconeogenesis. Dephosphorylation of the dimeric protein leads to an increase in PFK-2 activity and an increase in the concentration of F2,6P, ultimately stimulating glycolysis. The net result is similar to the control of glycogen synthesis and breakdown that we saw in Section 18.1.

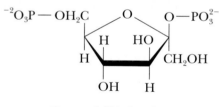

Fructose-2,6-*bis*phosphate
(F2,6P)

Figure 18.11 shows the effect of fructose-2,6-*bis*phosphate on the activity of FBPase. The inhibitor works by itself, but its effect is greatly increased by the presence of the allosteric inhibitor AMP.

Table 18.1 summarizes important mechanisms of metabolic control. Even though we discuss them here in the context of carbohydrate metabolism, they apply to all aspects of metabolism. Of the four kinds of control mechanisms listed in Table 18.1—allosteric control, covalent modification, substrate cycling, and genetic control—we have seen examples of allosteric control and covalent modification and, in Chapter 11, discussed genetic control using the *lac* operon as an example. Substrate cycling is a mechanism that we can profitably discuss here.

The term **substrate cycling** refers to the fact that opposing reactions can be catalyzed by different enzymes. Consequently, the opposing reactions can be independently regulated and have different rates. It would not be possible to have different rates with the same enzyme because a catalyst speeds up a reaction and the reverse of the reaction to the same extent (Section 6.2). We shall use the conversion of fructose-6-phosphate to fructose-1,6-*bis*phosphate and then back to fructose-6-phosphate as an example of a substrate cycle. In gly-

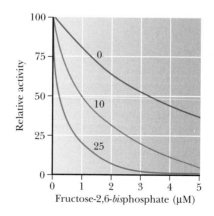

▲ **FIGURE 18.11** The effect of AMP (0, 10, and 25 μM [micromolar]) on the inhibition of fructose-1,6-*bis*phosphatase by fructose-2,6-*bis*phosphate. Activity was measured in the presence of 10 μM fructose-1,6-*bis*phosphate. (*Adapted from Van Schaftingen, E., and H. G. Hers, 1981. Inhibition of fructose-1,6-bisphosphatase by fructose-2,6-bisphosphate. Proc. Nat'l Acad. of Sci., U.S.A. 78:2861–2863.*)

Table 18.1		
Mechanisms of Metabolic Control		
Type of Control	**Mode of Operation**	**Examples**
Allosteric	Effectors (substrates, products, or coenzymes) of a pathway inhibit or activate an enzyme. (Responds rapidly to to external stimuli.)	ATCase (Section 7.2); phosphofructokinase (Section 17.2)
Covalent modification	Inhibition or activation of enzyme depends on formation or breaking of a bond, frequently by phosphorylation or dephosphorylation. (Responds rapidly to external stimuli.)	Sodium–potassium pump (Section 8.6); glycogen phosphorylase, glycogen synthase (Section 18.1)
Substrate cycles	Two opposing reactions, such as formation and breakdown of a given substance, are catalyzed by different enzymes, which can be activated or inhibited separately. (Responds rapidly to external stimuli.)	Glycolysis (Chapter 17) and gluconeogenesis (Section 18.2)
Genetic control	The amount of enzyme present is increased by protein synthesis. (Longer-term control than the other mechanisms listed here.)	Induction of β-galactosidase (Section 11.2)

colysis, the reaction catalyzed by phosphofructokinase is highly exergonic under physiological conditions ($\Delta G = -25.9$ kJ mol^{-1} = -6.2 kcal mol^{-1}).

$$\text{Fructose-6-phosphate} + \text{ATP} \rightarrow \text{Fructose-1,6-}bis\text{phosphate} + \text{ADP}$$

The opposing reaction, which is part of gluconeogenesis, is also exergonic ($\Delta G = -8.6$ kJ mol^{-1} = -2.1 kcal mol^{-1} under physiological conditions) and is catalyzed by another enzyme, namely fructose-1,6-bisphosphatase.

$$\text{Fructose-1,6-}bis\text{phosphate} + \text{H}_2\text{O} \rightarrow \text{Fructose-6-phosphate} + \text{P}_i$$

Note that the opposing reactions are not the exact reverse of one another. If we add the two opposing reactions together, we obtain the net reaction

$$\text{ATP} + \text{H}_2\text{O} \rightleftharpoons \text{ADP} + \text{P}_i$$

Hydrolysis of ATP is the energetic price that is paid for independent control of the opposing reactions.

Using combinations of these control mechanisms, an organism can set up a division of labor among tissues and organs to maintain control of glucose metabolism. A particularly clear example is found in the Cori cycle. Shown in Figure 18.12, the Cori cycle is named for Gerty and Carl Cori, who first described it. There is cycling of glucose due to glycolysis in muscle and gluconeogenesis in liver. Glycolysis in fast-twitch skeletal muscle produces lactate under conditions of oxygen debt, such as a sprint. Fast-twitch muscle has comparatively few mitochondria, so metabolism is largely anaerobic in this tissue. The buildup of lactate contributes to the muscular aches that follow strenuous exercise. Gluconeogenesis recycles the lactate that is produced (lactate is first oxidized to pyruvate). The process occurs to a great extent in the liver after the lactate is transported there by the blood. Glucose produced in the liver is transported back to skeletal muscle by the blood, where it becomes an energy store for the next burst of exercise. This is the main reason that athletes receive post-event massages and that they always warm down after the event. Warming down keeps the blood flowing through the muscles and allows the lactate and other acids to leave the cells and enter the blood. Massages increase this movement from the cells to the blood. Note that we have a division of labor between two different types of organs—muscle and liver. In the same cell (of whatever type), these two metabolic pathways—glycolysis

Essential Information

A number of control mechanisms operate in carbohydrate metabolism. In the mechanism of substrate cycling, the synthesis and the breakdown of a given compound are catalyzed by two different enzymes. Energy is required, but independent control can be exercised over the two opposing processes.

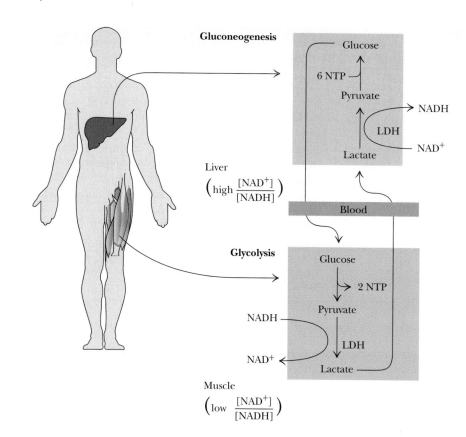

Biochemistry <u>Now</u>™ ANIMATED FIGURE 18.12
The Cori cycle. Lactate produced in muscles by glycolysis is transported by the blood to the liver. Gluconeogenesis in the liver converts the lactate back to glucose, which can be carried back to the muscles by the blood. Glucose can be stored as glycogen until it is degraded by glycogenolysis. (NTP stands for nucleoside triphosphate.) **See this figure animated at http://now.brookscole.com/campbell5**

and gluconeogenesis—are not highly active simultaneously. When the cell needs ATP, glycolysis is more active; when there is little need for ATP, gluconeogenesis is more active. Because of the hydrolysis of ATP and GTP in the reactions of gluconeogenesis that differ from those of glycolysis, the overall pathway from two molecules of pyruvate back to one molecule of glucose is exergonic ($\Delta G°' = -37.6$ kJ mol^{-1} = -9.0 kcal mol^{-1}, for one mole of glucose). The conversion of pyruvate to lactate is exergonic, which means that the reverse reaction is endergonic. The energy released by the exergonic conversion of pyruvate to glucose by gluconeogenesis facilitates the endergonic conversion of lactate to pyruvate.

Note that the Cori cycle requires the net hydrolysis of two ATP and two GTP. ATP is produced by the glycolytic part of the cycle, but the portion involving gluconeogenesis requires yet more ATP in addition to GTP.

Glycolysis:

Glucose + 2 NAD$^+$ + 2 ADP + 2 P$_i$ →

$\qquad$ 2 Pyruvate + 2 NADH + 4 H$^+$ + **2 ATP** + 2 H$_2$O

Gluconeogenesis:

2 Pyruvate + 2 NADH + 4 H$^+$ + **4 ATP + 2 GTP** + 6 H$_2$O →

$\qquad$ Glucose + 2 NAD$^+$ + 4 ADP + 2 GDP + 6 P$_i$

Overall:

$\qquad$ 2 ATP + 2 GTP + 4 H$_2$O → 2 ADP + 2 GDP + 4 P$_i$

The hydrolysis of both ATP and GTP is the price of increased simultaneous control of the two opposing pathways.

Control of Pyruvate Kinase

The final step of glycolysis is also a major control point in glucose metabolism. Pyruvate kinase (PK) is allosterically affected by several compounds. ATP and alanine both inhibit it. The ATP makes sense because there would be no reason to sacrifice glucose to make more energy if there is ample ATP. The alanine may be less intuitive. Alanine is the amino version of pyruvate. In other words, it is one reaction away from pyruvate via an enzyme called a transaminase. Therefore, a high level of alanine indicates that a high level of pyruvate is already present, so the enzyme that would make more pyruvate can be shut down. Fructose-1,6-*bis*phosphate allosterically activates PK so that the incoming products of the first reactions of glycolysis can be processed.

Pyruvate kinase is also found as isozymes with three different types of subunits, M, L, and A. The M subunit predominates in muscle; the L, in liver; and the A, in other tissues. A native pyruvate kinase molecule has four subunits, similar to lactate dehydrogenase and phosphofructokinase. In addition to the allosteric controls mentioned earlier, the liver isozymes also are subject to covalent modification, as shown in Figure 18.13. Low levels of blood sugar will trigger the release of glucagon, which will lead to the production of a protein kinase, as we saw with glycogen phosphorylase. The protein kinase will phosphorylate PK, which will render it less active. In this way, glycolysis is shut down in the liver when blood glucose is low.

▲ Gerty and Carl Cori, codiscoverers of the Cori cycle.

Control of Hexokinase

Hexokinase is inhibited by high levels of its product, glucose-6-phosphate. When glycolysis is inhibited through phosphofructokinase, glucose-6-phosphate builds up, shutting down hexokinase. This will keep glucose from being metabolized in the liver when it is needed in the blood and other tissues. However, the liver contains a second enzyme that phosphorylates glucose, glucokinase. (See the Biochemical Connections box on page 147.) Glucokinase has a higher K_M for glucose than hexokinase, so it functions only when glucose is abundant. If there is an excess of glucose in the liver, glucokinase will phosphorylate it to glucose-6-phosphate. The purpose of this phosphorylation is so that it can eventually be polymerized into glycogen.

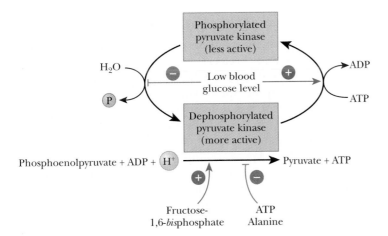

◄ **FIGURE 18.13** Control of liver pyruvate kinase by phosphorylation. When blood glucose is low, phosphorylation of pyruvate kinase is favored. The phosphorylated form is less active, thereby slowing glycolysis and allowing pyruvate to produce glucose by gluconeogenesis.

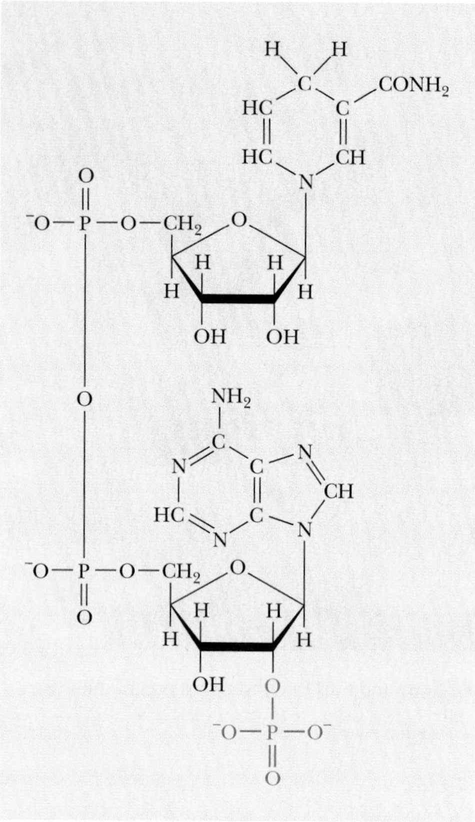

▲ **FIGURE 18.14** The structure of reduced adenine dinucleotide phosphate (NADPH).

18.4 | Why Is Glucose Sometimes Diverted through the Pentose Phosphate Pathway?

The **pentose phosphate pathway** is an alternative to glycolysis and differs from it in several important ways. In glycolysis, one of our most important concerns was the production of ATP. In the pentose phosphate pathway, the production of ATP is not the crux of the matter. As the name of the pathway indicates, five-carbon sugars, including ribose, are produced from glucose. Ribose and its derivative deoxyribose play an important role in the structure of nucleic acids. Another important facet of the pentose phosphate pathway is the production of nicotinamide adenine dinucleotide phosphate (NADPH), a compound that differs from nicotinamide adenine dinucleotide (NADH) by having one extra phosphate group esterified to the ribose ring of the adenine nucleotide portion of the molecule (Figure 18.14). A more important difference is the way these two coenzymes function. NADH is produced in the oxidative reactions that give rise to ATP. NADPH is a reducing agent in biosynthesis, which, by its very nature, is a reductive process. For example, in Chapter 21, we shall see the important role that NADPH plays in the biosynthesis of lipids.

The pentose phosphate pathway begins with a series of oxidation reactions that produce NADPH and five-carbon sugars. The remainder of the pathway involves nonoxidative reshuffling of the carbon skeletons of the sugars involved. The products of these nonoxidative reactions include substances such as fructose-6-phosphate and glyceraldehyde-3-phosphate, which play a role in glycolysis. Some of these reshuffling reactions will reappear when we look at the production of sugars in photosynthesis.

Oxidative Reactions of the Pentose Phosphate Pathway

In the first reaction of the pathway, glucose-6-phosphate is oxidized to 6-phosphogluconate (Figure 18.15, *top*). The enzyme that catalyzes this reaction is *glucose-6-phosphate dehydrogenase*. Note that NADPH is produced by the reaction.

The next reaction is an oxidative decarboxylation, and NADPH is produced once again. The 6-phosphogluconate molecule loses its carboxyl group, which is released as carbon dioxide, and the five-carbon keto-sugar (ketose) ribulose-5-phosphate is the other product. The enzyme that catalyzes this reaction is *6-phosphogluconate dehydrogenase*. In the process, the C-3 hydroxyl group of the 6-phosphogluconate is oxidized to form a β-keto acid, which is unstable and readily decarboxylates to form ribulose-5-phosphate.

Nonoxidative Reactions of the Pentose Phosphate Pathway

In the remaining steps of the pentose phosphate pathway, several reactions involve transfer of two- and three-carbon units. To keep track of the carbon backbone of the sugars and their aldehyde and ketone functional groups, we shall write the formulas in the open-chain form.

There are two different reactions in which ribulose-5-phosphate isomerizes. In one of these reactions, catalyzed by *phosphopentose-3-epimerase*, there is an inversion of configuration around carbon atom 3, producing xylulose-5-phosphate, which is also a ketose (Figure 18.15, *bottom*). The other isomerization reaction, catalyzed by *phosphopentose isomerase*, produces a sugar with an aldehyde group (an aldose) rather than a ketone. In this second reaction, ribulose-5-phosphate isomerizes to ribose-5-phosphate (Figure 18.15, *bottom*). Ribose-5-phosphate is a necessary building block for the synthesis of nucleic acids and coenzymes such as NADH.

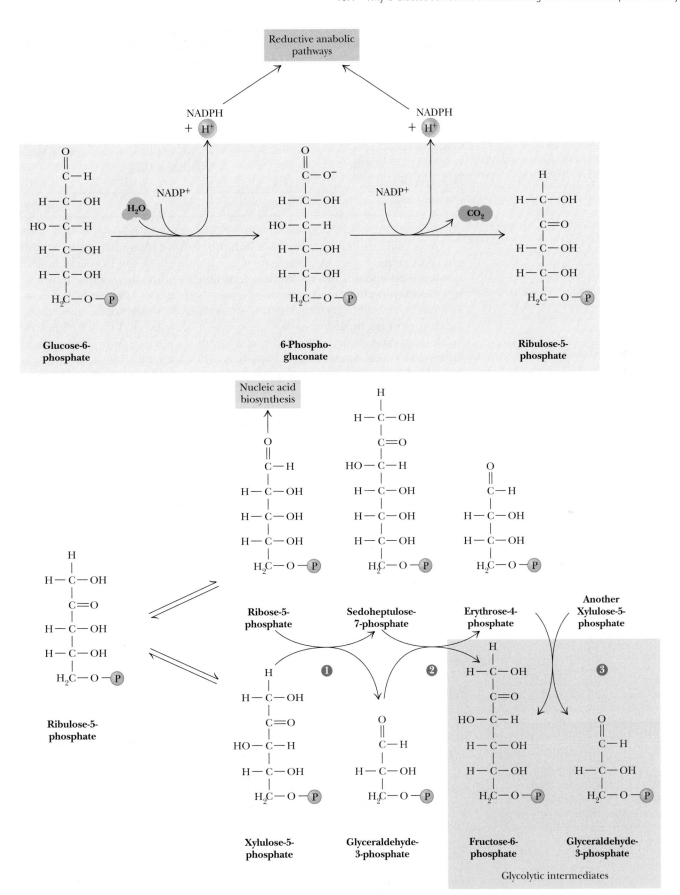

Biochemistry ⊛ Now™ ▲ **ACTIVE FIGURE 18.15** The pentose phosphate pathway. The numerals in the red circles indicate steps discussed in the text. **Watch this Active Figure at http://now.brookscole.com/campbell5**

Table 18.2

Group-Transfer Reactions in the Pentose Phosphate Pathway

	Reactant	Enzyme	Products
Two-carbon shift	$C_5 + C_5$	Transketolase $\rightleftharpoons$	$C_7 + C_3$
Three-carbon shift	$C_7 + C_3$	Transaldolase $\rightleftharpoons$	$C_6 + C_4$
Two-carbon shift	$C_5 + C_4$	Transketolase $\rightleftharpoons$	$C_6 + C_3$
Net reaction	$3\ C_5$	$\rightleftharpoons$	$2\ C_6 + C_3$

The group-transfer reactions that link the pentose phosphate pathway with glycolysis require the two 5-carbon sugars produced by the isomerization of ribulose-5-phosphate. Two molecules of xylulose-5-phosphate and one molecule of ribose-5-phosphate rearrange to give two molecules of fructose-6-phosphate and one molecule of glyceraldehyde-3-phosphate. In other words, three molecules of pentose (with five carbon atoms each) give two molecules of hexose (with six carbon atoms each) and one molecule of a triose (with three carbon atoms). The total number of carbon atoms (15) does not change, but there is considerable rearrangement as a result of group transfer.

Two enzymes, *transketolase* and *transaldolase,* are responsible for the reshuffling of the carbon atoms of sugars such as ribose-5-phosphate and xylulose-5-phosphate in the remainder of the pathway, which consists of three reactions. Transketolase transfers a two-carbon unit. Transaldolase transfers a three-carbon unit. Transketolase catalyzes the first and third reactions in the rearrangement process, and transaldolase catalyzes the second reaction. The results of these transfers can be summarized in Table 18.2. In the first of these reactions, a two-carbon unit from xylulose-5-phosphate (five carbons) is transferred to ribose-5-phosphate (five carbons) to give sedoheptulose-7-phosphate (seven carbons) and glyceraldehyde-3-phosphate (three carbons), as shown in Figure 18.15, *bottom,* red numeral 1.

In the reaction catalyzed by transaldolase, a three-carbon unit is transferred from the seven-carbon sedoheptulose-7-phosphate to the three-carbon glyceraldehyde-3-phosphate (Figure 18.15, red numeral 2). The products of the reaction are fructose-6-phosphate (six carbons) and erythrose-4-phosphate (four carbons).

In the final reaction of this type in the pathway, xylulose-5-phosphate reacts with erythrose-4-phosphate. This reaction is catalyzed by transketolase. The products of the reaction are fructose-6-phosphate and glyceraldehyde-3-phosphate (Figure 18.15, red numeral 3).

In the pentose phosphate pathway, glucose-6-phosphate can be converted to fructose-6-phosphate and glyceraldehyde-3-phosphate by a means other than the glycolytic pathway. For this reason, the pentose phosphate pathway is also called the *hexose monophosphate shunt,* and this name is used in some texts. A major feature of the pentose phosphate pathway is the production of ribose-5-phosphate and NADPH. The control mechanisms of the pentose phosphate pathway can respond to the varying needs of organisms for either or both of these compounds.

Essential Information

In the pentose phosphate pathway, two important processes take place. One is the formation of five-carbon sugars, particularly ribose, a component of RNA. The other is the formation of NADPH, a reducing agent required in many anabolic reactions.

Control of the Pentose Phosphate Pathway

As we have seen, the reactions catalyzed by transketolase and transaldolase are reversible, which allows the pentose phosphate pathway to respond to the needs of an organism. The starting material, glucose-6-phosphate, will undergo

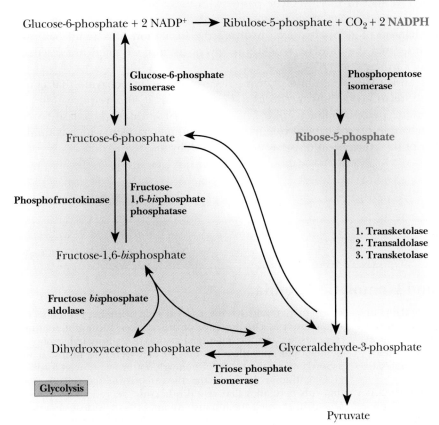

Glucose-6-phosphate + 2 NADP$^+$ $\longrightarrow$ Ribulose-5-phosphate + CO_2 + 2 **NADPH**

◀ **FIGURE 18.16** Relationships between the pentose phosphate pathway and glycolysis. If the organism needs NADPH more than ribose-5-phosphate, the entire pentose phosphate pathway is operative. If the organism needs ribose-5-phosphate more than NADPH, the nonoxidative reactions of the pentose phosphate pathway, operating in reverse, produce ribose-5-phosphate (see text).

different reactions depending on whether there is a greater need for ribose-5-phosphate or for NADPH. The operation of the oxidative portion of the pathway depends strongly on the organism's requirement for NADPH. The need for ribose-5-phosphate can be met in other ways, since ribose-5-phosphate can be obtained from glycolytic intermediates without the oxidative reactions of the pentose phosphate pathway (Figure 18.16).

If the organism needs more NADPH than ribose-5-phosphate, the reaction series goes through the complete pathway just discussed. The oxidative reactions at the beginning of the pathway are needed to produce NADPH. The net reaction for the oxidative portion of the pathway is

6 Glucose-6-phosphate + 12 NADP$^+$ + 6 H_2O $\rightarrow$

6 Ribose-5-phosphate + 6 CO_2 + 12 NADPH + 12 H$^+$

The following Biochemical Connections box discusses a clinical manifestation of an enzyme malfunction in the pentose phosphate pathway.

If the organism has a greater need for ribose-5-phosphate than for NADPH, fructose-6-phosphate and glyceraldehyde-3-phosphate can give rise to ribose-5-phosphate by the successive operation of the transketolase and transaldolase reactions, bypassing the oxidative portion of the pentose phosphate pathway (follow the red shaded path down to glyceraldehyde 3-phosphate and then go up to ribose-5-phosphate) (Figure 18.16). The reactions catalyzed by transketolase and transaldolase are reversible, and this fact plays an important role in the ability of the organism to adjust its metabolism to changes in conditions. We shall now take a look at the mode of action of these two enzymes.

Transaldolase has many features in common with the enzyme aldolase, which we met in the glycolytic pathway. Both an aldol cleavage and an aldol condensation occur at different stages of the reaction. We already saw the mechanism of aldol cleavage, involving the formation of a Schiff base, when we discussed the aldolase reaction in glycolysis, and we need not discuss this point further.

Transketolase resembles pyruvate decarboxylase, the enzyme that converts pyruvate to acetaldehyde (Section 17.4), in that it also requires Mg^{2+} and thiamine pyrophosphate (TPP). As in the pyruvate decarboxylase reaction, a carbanion plays a crucial role in the reaction mechanism, which is similar to that of the conversion of pyruvate to acetaldehyde.

Biochemical Connections

The Pentose Phosphate Pathway and Hemolytic Anemia

The pentose phosphate pathway is the only source of NADPH in red blood cells, which, as a result, are highly dependent on the proper functioning of the enzymes involved. A glucose-6-phosphate dehydrogenase deficiency leads to an NADPH deficiency, which can, in turn, lead to *hemolytic anemia* because of wholesale destruction of red blood cells.

The relationship between NADPH deficiency and anemia is an indirect one. NADPH is required to reduce the peptide glutathione from the disulfide to the free thiol form. Mammalian red blood cells lack mitochondria, which host many redox reactions. Consequently, these cells are limited in the ways in which they can deal with redox balance. A substance like glutathione, which can take part in redox reactions, assumes greater impor-

tance than would be the case in cells with large numbers of mitochondria. The presence of the reduced form of glutathione is necessary for the maintenance of the sulfhydryl groups of hemoglobin and other proteins in their reduced forms, as well as for keeping the Fe(II) of hemoglobin in its reduced form.

Glutathione also maintains the integrity of red cells by reacting with peroxides that would otherwise degrade fatty-acid side chains in the cell membrane. About 11% of African-Americans are affected by glucose-6-phosphate dehydrogenase deficiency. This condition, like the sickle-cell trait, leads to increased resistance to malaria, accounting for some of its persistence in the gene pool in spite of its otherwise deleterious consequences.

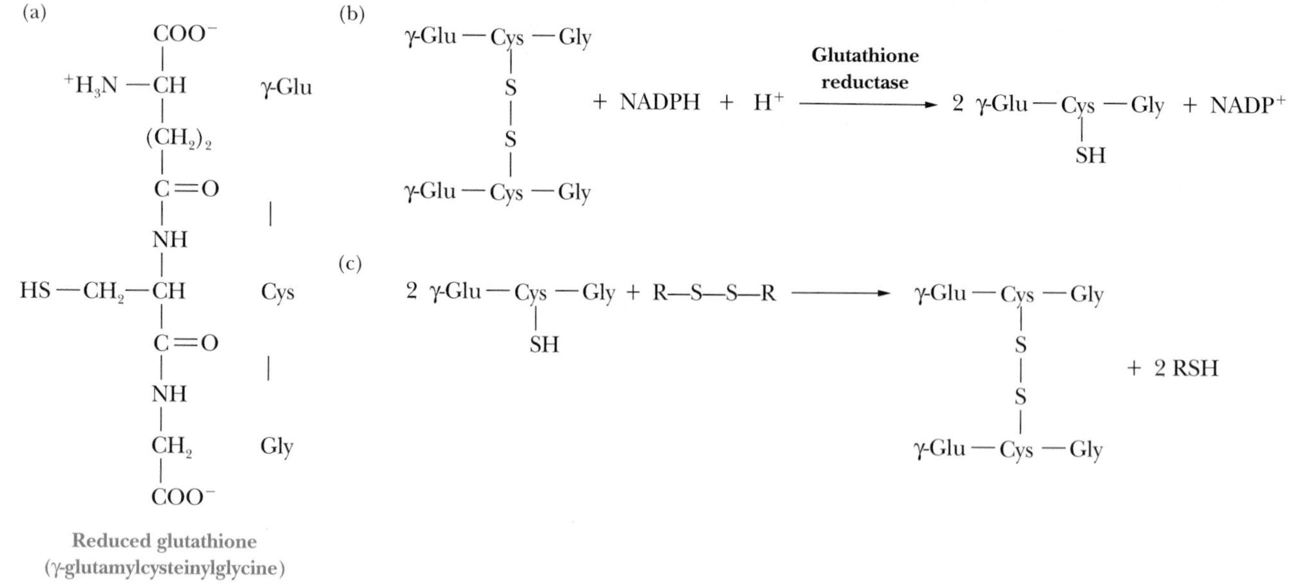

▲ Glutathione and its reactions. (a) The structure of glutathione. (b) The role of NADPH in the production of glutathione. (c) The role of glutathione in maintaining the reduced form of protein sulfhydryl groups.

Summary

18.1 How Is Glycogen Produced and Degraded?

When an organism has an available supply of extra glucose, more than is immediately needed as a source of energy extracted in glycolysis, it forms glycogen, a polymer of glucose. Glycogen synthase catalyzes the reaction between a glycogen molecule and UDP-glucose to add a glucose molecule to the glycogen via an $\alpha(1\rightarrow4)$ linkage. Branching enzyme moves sections of a chain of glucoses so that there are $\alpha(1\rightarrow6)$ branch points. Glycogen can readily be broken down to glucose in response to energy needs. Glycogen phosphorylase uses phosphate to break an $\alpha(1\rightarrow4)$ linkage, yielding glucose-1-phosphate and a glycogen molecule shorter by one glucose. Debranching enzyme aids in the degradation of the molecule around the $\alpha(1\rightarrow6)$ linkages. Control mechanisms ensure that both formation and breakdown of glycogen are not active simultaneously, a situation that would waste energy.

18.2 How Does Gluconeogenesis Produce Glucose from Pyruvate?

The conversion of pyruvate (the product of glycolysis) to glucose takes place by a process called gluconeogenesis. Gluconeogenesis is not the exact reversal of glycolysis. There are three irreversible steps in glycolysis, and it is in these three reactions that gluconeogenesis differs from glycolysis. The net result of gluconeogenesis is the reversal of these three glycolytic reactions, but the pathway is different, with different reactions and different enzymes.

18.3 How Is Carbohydrate Metabolism Controlled?

In the same cell, glycolysis and gluconeogenesis are not highly active simultaneously. When the cell needs ATP, glycolysis is more active; when there is little need for ATP, gluconeogenesis is more active. Glycolysis and gluconeogenesis play roles in the Cori cycle. The division of labor between liver and muscle allows glycolysis and gluconeogenesis to take place in different organs to serve the needs of an organism. Carbohydrate metabolism is controlled at several distinct points. Glycogen synthase and glycogen phosphorylase are reciprocally controlled by phosphorylation. Glycolysis and gluconeogenesis are controlled at several points, with phosphofructokinase and fructose *bis*phosphatase being the most important. Hexokinase and pyruvate kinase are also important control points.

18.4 Why Is Glucose Sometimes Diverted through the Pentose Phosphate Pathway?

The pentose phosphate pathway is an alternative pathway for glucose metabolism. In this pathway five-carbon sugars, including ribose, are produced from glucose. In the oxidative reactions of the pathway, NADPH is also produced. Control of the pathway allows the organism to adjust the relative levels of production of five-carbon sugars and of NADPH according to its needs.

Critical Questions to Review

18.1 How Is Glycogen Produced and Degraded?

1. **Fact Check** Why is it essential that the mechanisms that activate glycogen synthesis also deactivate glycogen phosphorylase?

2. **Fact Check** How does phosphorolysis differ from hydrolysis?

3. **Fact Check** Why is it advantageous that breakdown of glycogen gives rise to glucose-6-phosphate rather than to glucose?

4. **Fact Check** Briefly outline the role of UDPG in glycogen biosynthesis.

5. **Fact Check** Name two control mechanisms that play a role in glycogen biosynthesis. Give an example of each.

6. **Thought Question** Does the net gain of ATP in glycolysis differ when glycogen, rather than glucose, is the starting material? If so, what is the change?

7. **Thought Question** In metabolism, glucose-6-phosphate (G6P) can be used for glycogen synthesis or for glycolysis, among other fates. What does it cost, in terms of ATP equivalents, to store G6P as glycogen, rather than to use it for energy in glycolysis? *Hint:* The branched structure of glycogen leads to 90% of glucose residues being released as glucose-1-phosphate and 10% as glucose.

8. **Thought Question** How does the cost of storing glucose-6-phosphate (G6P) as glycogen differ from the answer you obtained in Question 7 if G6P were used for energy in aerobic metabolism?

9. **Biochemical Connections** You are planning to go on a strenuous hike and are advised to eat plenty of high-carbohydrate foods, such as bread and pasta, for several days beforehand. Suggest a reason for the advice.

10. **Biochemical Connections** Would eating candy bars, high in sucrose rather than complex carbohydrates, help to build up glycogen stores?

11. **Biochemical Connections** Would it be advantageous to consume a candy bar with a high refined-sugar content *immediately* before you start the strenuous hike in Question 9?

12. **Thought Question** The concentration of lactate in blood rises sharply during a sprint and declines slowly for about an hour afterward. What causes the rapid rise in lactate concentration? What causes the decline in lactate concentration after the run?

13. **Thought Question** A researcher claims to have discovered a variant form of glycogen. The variation is that it has very few branches (every 50 glucose residues or so) and that the branches are only three residues long. Is it likely that this discovery will be confirmed by later work?

14. **Thought Question** What is the source of the energy needed to incorporate glucose residues into glycogen? How is it used?

15. **Thought Question** Why is it useful to have a primer in glycogen synthesis?

16. **Thought Question** Is the glycogen synthase reaction exergonic or endergonic? What is the reason for your answer?

17. **Thought Question** What is the effect on gluconeogenesis and glycogen synthesis of (a) increasing the level of ATP, (b) decreasing the concentration of fructose-1,6-*bis*phosphate, and (c) increasing the concentration of fructose-6-phosphate?

18. **Thought Question** Briefly describe "going for the burn" in a workout in terms of the material in this chapter.

19. **Thought Question** Suggest a reason why sugar nucleotides, such as UDPG, play a role in glycogen synthesis, rather than sugar phosphates, such as glucose-6-phosphate.

18.2 How Does Gluconeogenesis Produce Glucose from Pyruvate?

20. **Fact Check** What reactions in this chapter require acetyl-CoA or biotin?

21. **Fact Check** Which steps of glycolysis are irreversible? What bearing does this observation have on the reactions in which gluconeogenesis differs from glycolysis?

22. **Fact Check** What is the role of biotin in gluconeogenesis?

23. **Fact Check** How does the role of glucose-6-phosphate in gluconeogenesis differ from that in glycolysis?

24. **Thought Question** Avidin, a protein found in egg whites, binds to biotin so strongly that it inhibits enzymes that require biotin. What is the effect of avidin on glycogen formation? On gluconeogenesis? On the pentose phosphate pathway?

25. **Thought Question** How does the hydrolysis of fructose-1,6-*bis*phosphate bring about the reversal of one of the physiologically irreversible steps of glycolysis?

18.3 How Is Carbohydrate Metabolism Controlled?

26. **Fact Check** Which reaction or reactions discussed in this chapter require ATP? Which reaction or reactions produce ATP? List the enzymes that catalyze the reactions that require and that produce ATP.

27. **Fact Check** How does fructose-2,6-*bis*phosphate play a role as an allosteric effector?

28. **Fact Check** How do glucokinase and hexokinase differ in function?

29. **Fact Check** What is the Cori cycle?

30. **Thought Question** Earlier biochemists called substrate cycles "futile cycles." Why might they have chosen such a name? Why is it something of a misnomer?

31. **Thought Question** Why is it advantageous for two control mechanisms—allosteric control and covalent modification—to be involved in the metabolism of glycogen?

32. **Thought Question** How can different time scales for response be achieved in control mechanisms?

33. **Thought Question** How do the control mechanisms in glycogen metabolism lead to amplification of response to a stimulus?

34. **Thought Question** Why would you expect to see that reactions of substrate cycles involve different enzymes for different directions?

35. **Thought Question** Suggest a reason or reasons why the Cori cycle takes place in the liver and in muscle.

36. **Thought Question** Explain how fructose-2,6-*bis*phosphate can play a role in more than one metabolic pathway.

37. **Thought Question** How can the synthesis and breakdown of fructose-2,6-*bis*phosphate be controlled independently?

38. **Thought Question** How is it advantageous for animals to convert ingested starch to glucose and then to incorporate the glucose into glycogen?

18.4 Why Is Glucose Sometimes Diverted through the Pentose Phosphate Pathway?

39. **Fact Check** List three differences in structure or function between NADH and NADPH.

40. **Fact Check** What are four possible metabolic fates of glucose-6-phosphate?

41. **Biochemical Connections** What is the connection between material in this chapter and hemolytic anemia?

42. **Fact Check** Show how the pentose phosphate pathway, which is connected to the glycolytic pathway, can do the following.

    (a) Make both NADPH and pentose phosphates, in roughly equal amounts

    (b) Make mostly or only NADPH

    (c) Make mostly or only pentose phosphates

43. **Fact Check** What is a major difference between transketolase and transaldolase?

44. **Biochemical Connections** List two ways in which glutathione functions in red blood cells.

45. **Fact Check** Does thiamine pyrophosphate play a role in the reactions of the pentose phosphate pathway? If so, what is that role?

46. **Thought Question** Using the Lewis electron-dot notation, show explicitly the transfer of electrons in the following redox reaction.

    Glucose-6-phosphate + $NADP^+ \rightarrow$

    6-Phosphoglucono-δ-lactone + NADPH + H^+

    The lactone is a cyclic ester that is an intermediate in the production of 6-phosphogluconate.

47. **Thought Question** Suggest a reason why a different reducing agent (NADPH) is used in anabolic reactions rather than NADH, which plays a role in catabolic ones.

48. **Thought Question** Explain how the pentose phosphate pathway can respond to a cell's need for ATP, NADPH, and ribose-5-phosphate.

49. **Thought Question** Why is it reasonable to expect that glucose-6-phosphate will be oxidized to a lactone (see Question 46) rather than to an open-chain compound?

50. **Thought Question** How would it affect the reactions of the pentose phosphate pathway to have an epimerase and not an isomerase to catalyze the reshuffling reactions?

Biochemistry ⓔ**Now**™

Assess your understanding of this chapter's topics with additional quizzing and tutorials at **http://now.brookscole.com/campbell5**

Annotated Bibliography

Florkin, M., and E. H. Stotz, eds. *Comprehensive Biochemistry*. New York: Elsevier, 1967. [A standard reference. Volume 17, Carbohydrate Metabolism, deals with glycolysis and related topics.]

Horecker, B. L. Transaldolase and Transketolase. *Comprehensive Biochemistry*. (1973). [Volume 15 is a review of these two enzymes and their mechanism of action.]

Lipmann, F. A Long Life in Times of Great Upheaval. *Ann. Rev. Biochem.* **53**, 1–33 (1984). [The reminiscences of a Nobel laureate whose research contributed greatly to the understanding of carbohydrate metabolism. Very interesting reading from the standpoint of autobiography and the author's contributions to biochemistry.]

Shulman, R. G., and D. L. Rothman, Enzymatic Phosphorylation of Muscle Glycogen Synthase: A Mechanism for Maintenance of Metabolic Homeostasis. *Proc. Nat. Acad. Sci.* **93**, 7491–7495 (1996). [An in-depth article about metabolic flux and covalent modification of enzymes.]

The Citric Acid Cycle

If the mitochondrion is the power plant of the cell, then the citric acid cycle operating inside the mitochondrion is its engine room. Here, metabolic fuels—glucose derived from carbohydrates, amino acids derived from proteins, and fatty acids derived from lipids—are fed into the cycle and will ultimately be oxidized to carbon dioxide and water. Their energy is transferred to electron carriers and finally to the terminal electron acceptor—oxygen. In the first stage of energy extraction from carbohydrates, glucose is catabolized to pyruvate in the ten anaerobic steps of glycolysis (see Chapter 17). In aerobic catabolism, pyruvate is converted to "high-energy" two-carbon acetyl-CoA, which then enters the citric acid cycle. As the first step of the cycle, acetyl-CoA combines with oxaloacetate to form citric acid. Each turn of the cycle releases two molecules of CO_2, regenerates the starting molecule oxaloacetate, and delivers reducing agents to the electron transport chain. In the final stage of aerobic metabolism, proton flow across the inner mitochondrial membrane is coupled to synthesis of ATP, the energy currency of the cell. The citric acid cycle, as the hub of metabolic pathways, serves to connect the breakdown and synthesis of proteins, carbohydrates, and lipids. Most of the metabolites of major nutrients can be fed into the citric acid cycle as acetyl-CoA, or as citric acid cycle intermediates, and then oxidized to produce energy. Alternatively, intermediates of the citric acid cycle can be removed and converted to other biomolecules.

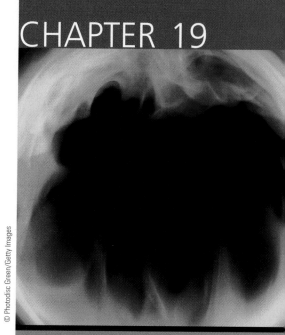

© Photodisc Green/Getty Images

The citric acid cycle is the central pathway of metabolism. It plays a pivotal role in the production of energy by a cell.

Critical Questions

19.1 What Role Does the Citric Acid Cycle Play in Metabolism?

19.2 What Is the Overall Pathway of the Citric Acid Cycle?

19.3 How Is Pyruvate Converted to Acetyl-CoA?

19.4 What Are the Individual Reactions of the Citric Acid Cycle?

19.5 What Are the Energetics of the Citric Acid Cycle, and How Is It Controlled?

19.6 What Is the Glyoxylate Cycle?

19.7 What Role Does the Citric Acid Cycle Play in Catabolism?

19.8 What Role Does the Citric Acid Cycle Play in Anabolism?

19.9 Why Isn't Oxygen Part of the Equation?

19.1	What Role Does the Citric Acid Cycle Play in Metabolism?

The evolution of aerobic metabolism, by which nutrients are oxidized to carbon dioxide and water, was an important step in the history of life on Earth. Organisms can obtain far more energy from nutrients by aerobic oxidation than by anaerobic oxidation. (Even yeast—which is usually thought of in terms of the anaerobic reactions of alcoholic fermentation and is responsible for producing bread, beer, and wine—uses the citric acid cycle and aerobically degrades glucose to carbon dioxide and water.) We saw, in Chapter 17, that glycolysis produces only two molecules of ATP for each molecule of glucose metabolized. In this chapter and the next, we shall see how 30 to 32 molecules of ATP can be produced from each molecule of glucose in complete aerobic oxidation to carbon dioxide and water. Three processes play roles in aerobic metabolism: the **citric acid cycle,** which we discuss in this chapter, and **electron transport** and **oxidative phosphorylation,** both of which we shall discuss in Chapter 20 (Figure 19.1). These three processes operate together in aerobic metabolism; separate discussion is a matter of convenience only.

Metabolism consists of catabolism, which is the oxidative breakdown of nutrients, and anabolism, which is reductive synthesis of biomolecules. The citric acid cycle is **amphibolic,** meaning that it plays a role in both catabolism and anabolism. While the citric acid cycle is a part of the pathway of aerobic oxidation of nutrients (a catabolic pathway; see Section 19.7), some of the molecules that are included in this cycle are the starting points of biosynthetic (anabolic) pathways (see Section 19.8). Metabolic pathways operate simultaneously, even though we talk about them separately. We should always keep this point in mind.

Biochemistry ⊘ Now™
Test yourself on these Critical Questions at the BiochemistryNow website at **http://now .brookscole.com/campbell5**

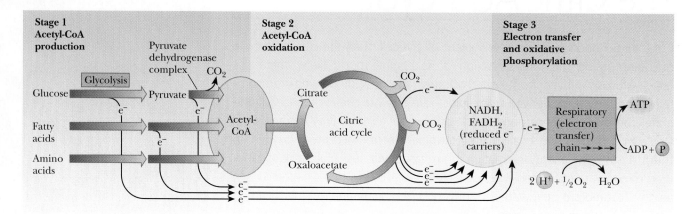

▲ **FIGURE 19.1** The central relationship of the citric acid cycle to catabolism. Amino acids, fatty acids, and glucose can all produce acetyl-CoA in stage 1 of catabolism. In stage 2, acetyl-CoA enters the citric acid cycle. Stages 1 and 2 produce reduced electron carriers (shown here as e⁻). In stage 3, the electrons enter the electron transport chain, which then produces ATP.

Essential Information

The citric acid cycle is amphibolic. It plays a role in both catabolism and anabolism. It is *the* central metabolic pathway.

There are two other common names for the citric acid cycle. One is the *Krebs cycle*, after Sir Hans Krebs, who first investigated the pathway (work for which he received a Nobel Prize in 1953). The other name is the *tricarboxylic acid cycle (or TCA cycle)*, from the fact that some of the molecules involved are acids with three carboxyl groups. We shall start our discussion with a general overview of the pathway and then go on to discuss specific reactions.

19.2 What Is the Overall Pathway of the Citric Acid Cycle?

An important difference between glycolysis and the citric acid cycle is the part of the cell in which these pathways occur. In eukaryotes, glycolysis occurs in the cytosol, while the citric acid cycle takes place in mitochondria. Most of the enzymes of the citric acid cycle are present in the mitochondrial matrix.

A quick review of some aspects of mitochondrial structure is in order here because we shall want to describe the exact location of each of the components of the citric acid cycle and the electron transport chain. Recall, from Chapter 1, that a mitochondrion has an inner and an outer membrane (Figure 19.2). The region enclosed by the inner membrane is called the **mitochondrial matrix,** and there is an **intermembrane space** between the inner and outer membranes. The inner membrane is a tight barrier between the matrix and the cytosol, and very few compounds can cross this barrier without a specific transport protein (Section 8.4). The reactions of the citric acid cycle take place in the matrix, except for the one in which the intermediate electron acceptor is FAD. The enzyme that catalyzes the FAD-linked reaction is an integral part of the inner mitochondrial membrane and is linked directly to the electron transport chain (Chapter 20).

The citric acid cycle is shown in schematic form in Figure 19.3. Under aerobic conditions, pyruvate produced by glycolysis is oxidized further, with carbon dioxide and water as the final products. First, the pyruvate is oxidized to one carbon dioxide molecule and to one acetyl group, which becomes linked to an intermediate, coenzyme A (CoA) (Section 15.7). The acetyl-CoA enters the citric acid cycle. In the citric acid cycle, two more molecules of carbon dioxide are produced for each molecule of acetyl-CoA that enters the cycle, and electrons are transferred in the process. The immediate electron acceptor in all cases but one is NAD⁺, which is reduced to NADH. In the one case in which

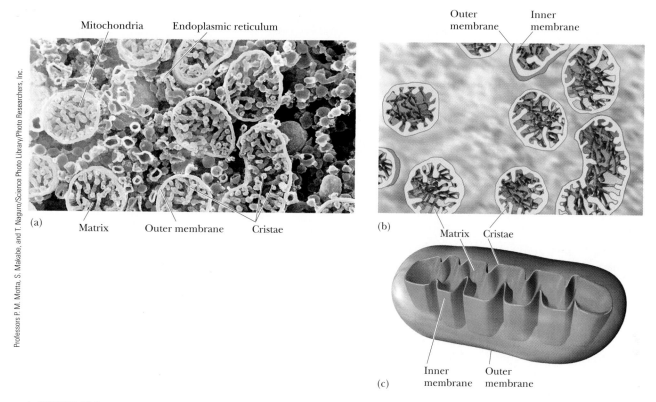

▲ FIGURE 19.2 The structure of a mitochondrion. (a) Colored scanning electron microscope image showing the internal structure of a mitochondrion (green, magnified 19,200 ×). (b) Interpretive drawing of the scanned image. (c) Perspective drawing of a mitochondrion. (For an electron micrograph of mitochondrial structure, see Figure 1.13.)

there is another intermediate electron acceptor, FAD (flavin adenine dinucleotide), which is derived from riboflavin (vitamin B_2), takes up two electrons and two hydrogen ions to produce $FADH_2$. The electrons are passed from NADH and $FADH_2$ through several stages of an electron transport chain with a different redox reaction at each step. The final electron acceptor is oxygen, with water as the product. Note that, starting from pyruvate, a three-carbon compound, three carbons are lost as CO_2 via the production of acetyl-CoA and one turn of the cycle. The cycle produces energy in the form of reduced electron equivalents (the NADH and $FADH_2$ that will enter the electron transport chain), but the carbon skeletons are effectively lost. The cycle also produces one high-energy compound directly, GTP (guanosine triphosphate).

In the first reaction of the cycle, the two-carbon acetyl group condenses with the four-carbon oxaloacetate ion to produce the six-carbon citrate ion. In the next few steps, the citrate isomerizes, and then it both loses carbon dioxide and is oxidized. This process, called **oxidative decarboxylation,** produces the five-carbon compound α-ketoglutarate, which again is oxidatively decarboxylated to produce the four-carbon compound succinate. The cycle is completed by regeneration of oxaloacetate from succinate in several steps. We shall see many of these intermediates again in other pathways, especially α-ketoglutarate, which is very important in amino acid and protein metabolism.

There are eight steps in the citric acid cycle, each catalyzed by a different enzyme. Four of the eight steps—Steps 3, 4, 6, and 8—are oxidation reactions (see Figure 19.3). The oxidizing agent is NAD^+ in all except Step 6, in which FAD plays the same role. In Step 5, a molecule of GDP (guanosine diphosphate) is phosphorylated to produce GTP. This reaction is equivalent to the production of ATP because the phosphate group is easily transferred to ADP, producing GDP and ATP.

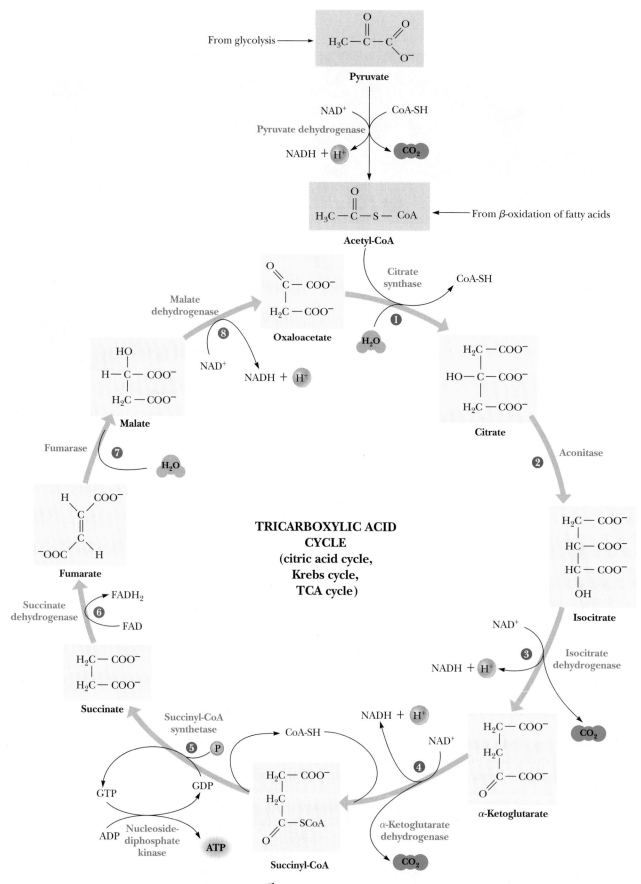

From glycolysis →

Pyruvate

NAD^+ CoA-SH

Pyruvate dehydrogenase

$NADH + H^+$ CO_2

Acetyl-CoA   ← From β-oxidation of fatty acids

Malate
dehydrogenase

Oxaloacetate

Citrate
synthase   CoA-SH

❶

H_2O

Citrate

❽

NAD^+

$NADH + H^+$

Malate

Aconitase

❷

Fumarase

❼

H_2O

Isocitrate

**TRICARBOXYLIC ACID
CYCLE**
**(citric acid cycle,
Krebs cycle,
TCA cycle)**

NAD^+

❸

Fumarate

$NADH + H^+$

Isocitrate
dehydrogenase

Succinate
dehydrogenase $FADH_2$

❻

FAD

$NADH + H^+$

NAD^+

❹

CO_2

Succinate

Succinyl-CoA
synthetase   CoA-SH

❺   P

GTP   GDP

ADP   Nucleoside-
diphosphate
kinase

ATP

Succinyl-CoA

α-Ketoglutarate
dehydrogenase

CO_2

α-Ketoglutarate

Biochemistry Now™ ▲ **ACTIVE FIGURE 19.3** An overview of the citric acid cycle. Note
the names of the enzymes. The loss of CO_2 is indicated, as is the phosphorylation of GDP to GTP.
The production of NADH and $FADH_2$ is also indicated. **Watch this Active Figure at
http://now.brookscole.com/campbell5**

19.3 | How Is Pyruvate Converted to Acetyl-CoA?

Pyruvate can come from several sources, including glycolysis, as we have seen. It moves from the cytosol into the mitochondrion via a specific transporter. There, an enzyme system called the **pyruvate dehydrogenase complex** is responsible for the conversion of pyruvate to carbon dioxide and the acetyl portion of acetyl-CoA. There is an —SH group at one end of the CoA molecule, which is the point at which the acetyl group is attached. As a result, CoA is frequently shown in equations as CoA-SH. Because CoA is a thiol (the sulfur [thio] analog of an alcohol), acetyl-CoA is a **thioester,** with a sulfur atom replacing an oxygen of the usual carboxylic ester. This difference is important, since thioesters are high-energy compounds (Chapter 15). In other words, the hydrolysis of thioesters releases enough energy to drive other reactions. An oxidation reaction precedes the transfer of the acetyl group to the CoA. The whole process involves several enzymes, all of which are part of the pyruvate dehydrogenase complex. The overall reaction

$$\text{Pyruvate} + \text{CoA-SH} + \text{NAD}^+ \rightarrow \text{Acetyl-CoA} + CO_2 + H^+ + \text{NADH}$$

is exergonic ($\Delta G^{\circ\prime} = -33.4$ kJ mol^{-1} = -8.0 kcal mol^{-1}), and NADH can then be used to generate ATP via the electron transport chain (Chapter 20).

The overall reaction of the pyruvate dehydrogenase complex

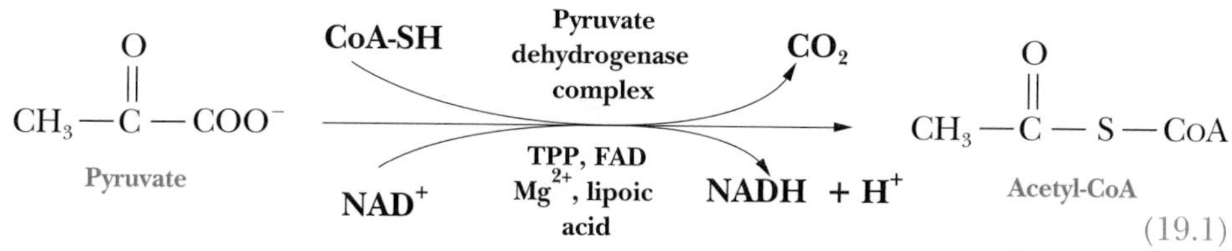

(19.1)

Five enzymes make up the pyruvate dehydrogenase complex in mammals. They are *pyruvate dehydrogenase (PDH), dihydrolipoyl transacetylase, dihydrolipoyl dehydrogenase, pyruvate dehydrogenase kinase,* and *pyruvate dehydrogenase phosphatase.* The first three are involved in the conversion of pyruvate to acetyl-CoA. The kinase and the phosphatase are enzymes used in the control of PDH (Section 19.5) and are present on a single polypeptide. The reaction takes place in five steps. Two enzymes catalyze reactions of *lipoic acid,* a compound that has a disulfide group in its oxidized form and two sulfhydryl groups in its reduced form.

Lipoic acid differs in one respect from other coenzymes. It is a vitamin, rather than a metabolite of a vitamin, as is the case with many other coenzymes (Table 7.3). (The classification of lipoic acid as a vitamin is open to question. There is no evidence of a requirement for it in the human diet, but it is required for the growth of some bacteria and protists.)

Lipoic acid can act as an oxidizing agent; the reaction involves hydrogen transfer, which frequently accompanies biological oxidation–reduction reactions (Section 15.5). Another reaction of lipoic acid is the formation of a thioester linkage with the acetyl group before it is transferred to the acetyl-CoA. Lipoic acid can act simply as an oxidizing agent, or it can simultaneously take part in two reactions—a redox reaction and the shift of an acetyl group by transesterification.

The first step in the reaction sequence that converts pyruvate to carbon dioxide and acetyl-CoA is catalyzed by pyruvate dehydrogenase, as shown in Figure 19.4. This enzyme requires thiamine pyrophosphate (TPP; a metabolite of vitamin B$_1$ or thiamine) as a coenzyme. The coenzyme is not covalently bonded to the enzyme; they are held together by noncovalent interactions.

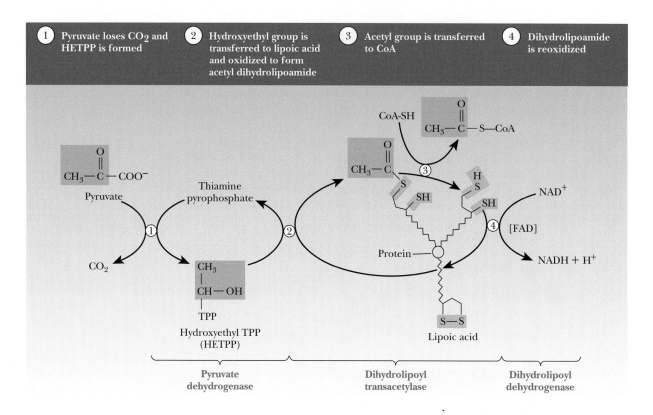

▲ **FIGURE 19.4** The mechanism of the pyruvate dehydrogenase reaction. Decarboxylation of pyruvate occurs with formation of hydroxyethyl-TPP (Step 1). Transfer of the two-carbon unit to lipoic acid in Step 2 is followed by formation of acetyl-CoA in Step 3. Lipoic acid is reoxidized in Step 4 of the reaction.

Mg^{2+} is also required. We saw the action of TPP as a coenzyme in the conversion of pyruvate to acetaldehyde, catalyzed by pyruvate decarboxylase (Section 17.4). In the pyruvate dehydrogenase reaction, an α-keto acid, pyruvate, loses carbon dioxide; the remaining two-carbon unit becomes covalently bonded to TPP.

The second step of the reaction is catalyzed by dihydrolipoyl transacetylase. This enzyme requires lipoic acid as a coenzyme. The lipoic acid is covalently bonded to the enzyme by an amide bond to the ε-amino group of a lysine side chain. The two-carbon unit that originally came from pyruvate is transferred from the thiamine pyrophosphate to the lipoic acid, and, in the process, a hydroxyl group is oxidized to produce an acetyl group. The disulfide group of the lipoic acid is the oxidizing agent, which is itself reduced, and the product of the reaction is a thioester. In other words, the acetyl group is now covalently bonded to the lipoic acid by a thioester linkage (see Figure 19.4).

The third step of the reaction is also catalyzed by dihydrolipoyl transacetylase. A molecule of CoA-SH attacks the thioester linkage, and the acetyl group is transferred to it. The acetyl group remains bound in a thioester linkage; this time it appears as acetyl-CoA rather than esterified to lipoic acid. The reduced form of lipoic acid remains covalently bound to dihydrolipoyl transacetylase (see Figure 19.4). The reaction of pyruvate and CoA-SH has now reached the stage of the products, carbon dioxide and acetyl-CoA, but the lipoic acid coenzyme is in a reduced form. The rest of the steps serve to regenerate the lipoic acid, so further reactions can be catalyzed by the transacetylase.

In the fourth step of the overall reaction, the enzyme dihydrolipoyl dehydrogenase reoxidizes the reduced lipoic acid from the sulfhydryl to the disulfide form. The lipoic acid still remains covalently bonded to the transacetylase enzyme. The dehydrogenase also has a coenzyme, FAD (Section 15.5), that is bound to the enzyme by noncovalent interactions. As a result, FAD is reduced to $FADH_2$. $FADH_2$ is reoxidized in turn. The oxidizing agent is NAD^+, and NADH is the product along with reoxidized FAD. Enzymes such as pyruvate dehydrogenase are called flavoproteins because of their attached FADs.

The reduction of NAD^+ to NADH accompanies the oxidation of pyruvate to the acetyl group, and the overall equation shows that there has been a transfer of two electrons from pyruvate to NAD^+ (Equation 19.1). The electrons gained by NAD^+ in generating NADH in this step are passed to the electron transport chain (the next step in aerobic metabolism). In Chapter 20, we shall see that the transfer of electrons from NADH ultimately to oxygen will give rise to 2.5 ATP. Two molecules of pyruvate are produced for each molecule of glucose, so that there will eventually be five ATP from each glucose from this step alone.

The reaction leading from pyruvate to acetyl-CoA is a complex one that requires three enzymes, each of which has its own coenzyme in addition to NAD^+. The spatial orientation of the individual enzyme molecules with respect to one another is itself complex. In the enzyme isolated from *E. coli,* the arrangement is quite compact, so that the various steps of the reaction can be thoroughly coordinated. There is a core of 24 dihydrolipoyl transacetylase molecules. The 24 polypeptide chains are arranged in eight trimers, with each trimer occupying the corner of a cube. There are 12 $\alpha\beta$ dimers of pyruvate dehydrogenase, and they occupy the edges of the cube. Finally, six dimers of dihydrolipoyl dehydrogenase lie on the six faces of the cube (Figure 19.5). Note that many levels of structure combine to produce a suitable environment for the conversion of pyruvate to acetyl-CoA. Each of the enzyme molecules in this array has its own tertiary structure, and the array itself has the cubical structure we have just seen.

A compact arrangement, such as the one in the pyruvate dehydrogenase multienzyme complex, has two great advantages over an arrangement in which the various components are more widely dispersed. First, the various stages of the reaction can take place more efficiently because the reactants and the enzymes are so close to each other. The role of lipoic acid is particularly important here. Recall that the lipoic acid is covalently attached to the transacetylase enzyme that occupies a central position in the complex. The lipoic acid and the lysine side chain to which it is bonded are long enough to

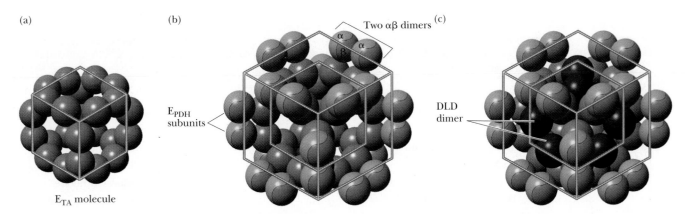

▲ FIGURE 19.5 The structure of the pyruvate dehydrogenase complex. (a) 24 dihydrolipoyl transacetylase (TA) subunits. (b) 24 $\alpha\beta$ dimers of pyruvate dehydrogenase are added to the cube (two per edge). (c) Addition of 12 dihydrolipoyl dehydrogenase subunits (two per face) completes the complex.

Essential Information

The two-carbon unit needed at the start of the citric acid cycle is obtained by converting pyruvate to acetyl-CoA. This conversion requires the three primary enzymes of the pyruvate dehydrogenase complex, as well as the cofactors TPP, FAD, NAD$^+$, and lipoic acid. The overall reaction of pyruvate dehydrogenase is the conversion of pyruvate, NAD$^+$, and CoA-SH to acetyl-CoA, NADH + H$^+$, and CO$_2$.

act as a "swinging arm," which can move to the site of each of the steps of the reaction (Figure 19.4). As a result of the swinging-arm action, the lipoic acid can move to the pyruvate dehydrogenase site to accept the two-carbon unit and then transfer it to the active site of the transacetylase. The acetyl group can then be transesterified to CoA-SH from the lipoic acid. Finally, the lipoic acid can swing to the active site of the dehydrogenase so that the sulfhydryl groups can be reoxidized to a disulfide.

A second advantage of a multienzyme complex is that regulatory controls can be applied more efficiently in such a system than in a single enzyme molecule. In the case of the pyruvate dehydrogenase complex, controlling factors are intimately associated with the multienzyme complex itself, which we shall study in Section 19.5.

19.4 | What Are the Individual Reactions of the Citric Acid Cycle?

The reactions of the citric acid cycle proper and the enzymes that catalyze them are listed in Table 19.1. We shall now discuss each of these reactions in turn.

Step 1. *Formation of Citrate*   The first step of the citric acid cycle is the reaction of acetyl-CoA and oxaloacetate to form citrate and CoA-SH. This reaction is called a condensation because a new carbon–carbon bond is formed. The condensation reaction of acetyl-CoA and oxaloacetate to form citryl-CoA takes place in the first stage of the reaction. The condensation is followed by the hydrolysis of citryl-CoA to give citrate and CoA-SH.

The condensation of acetyl-CoA and oxaloacetate to form citrate

Overall reaction

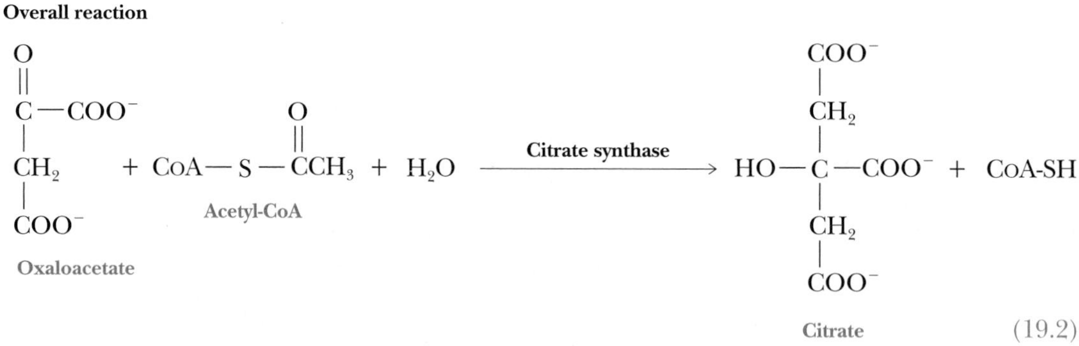

(19.2)

Table 19.1		
The Reactions of the Citric Acid Cycle		
Step	**Reaction**	**Reaction**
1	Acetyl-CoA + Oxaloacetate + H$_2$O → Citrate + CoA-SH	Citrate synthase
2	Citrate → Isocitrate	Aconitase
3	Isocitrate + NAD$^+$ → α-Ketoglutarate + NADH + CO$_2$ + H$^+$	Isocitrate dehydrogenase
4	α-Ketoglutarate + NAD$^+$ + CoA-SH → Succinyl-CoA + NADH + CO$_2$ + H$^+$	α-Ketoglutarate dehydrogenase
5	Succinyl-CoA + GDP + P$_i$ → Succinate + GTP + CoA-SH	Succinyl-CoA synthetase
6	Succinate + FAD → Fumarate + FADH$_2$	Succinate dehydrogenase
7	Fumarate + H$_2$O → L-Malate	Fumarase
8	L-Malate + NAD$^+$ → Oxaloacetate + NADH + H$^+$	Malate dehydrogenase

The reaction is catalyzed by the enzyme **citrate synthase,** originally called "condensing enzyme." A synthase is an enzyme that makes a new covalent bond during the reaction, but it does not require the direct input of ATP. It is an exergonic reaction ($\Delta G°' = -32.8 \text{ kJ mol}^{-1} = -7.8 \text{ kcal mol}^{-1}$) because the hydrolysis of a thioester releases energy. Thioesters are considered high-energy compounds.

Biochemistry Now™
Go to BiochemistryNow and click on Biochemistry Interactive to explore the citrate synthase reaction.

Step 2. ***Isomerization of Citrate to Isocitrate*** The second reaction of the citric acid cycle, the one catalyzed by aconitase, is the isomerization of citrate to isocitrate. The enzyme requires Fe^{2+}. One of the most interesting features of the reaction is that citrate, a symmetrical (achiral) compound, is converted to isocitrate, a chiral compound, a molecule that cannot be superimposed on its mirror image.

It is often possible for a chiral compound to have several different isomers. Isocitrate has four possible isomers, but only one of the four is produced by this reaction. (We shall not discuss nomenclature of the isomers of isocitrate here. See Question 28 at the end of this chapter for a question about the other isomers.) Aconitase, the enzyme that catalyzes the conversion of citrate to isocitrate, is able to select one end of the citrate molecule in preference to the other.

The formation of isocitrate (a chiral compound) from citrate (an achiral compound)

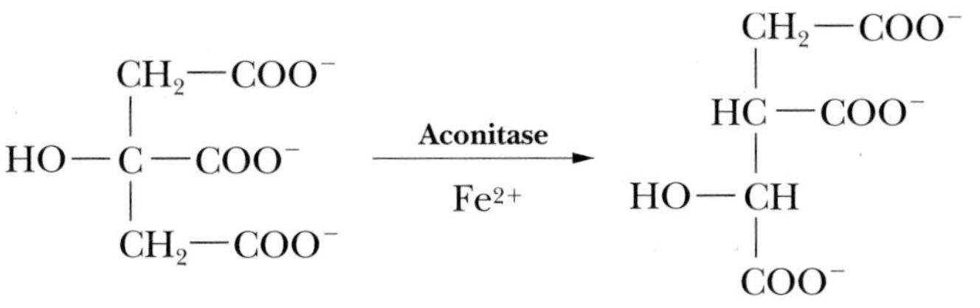

Citrate Isocitrate (19.3)

This type of behavior means that the enzyme can bind a symmetrical substrate in an unsymmetrical binding site. In Section 6.13, we mentioned that this possibility exists, and here we have an example of it. The enzyme forms an unsymmetrical three-point attachment to the citrate molecule (Figure 19.6). The reaction proceeds by removal of a water molecule from the citrate to produce *cis*-aconitate, and then water is added back to the *cis*-aconitate to give isocitrate.

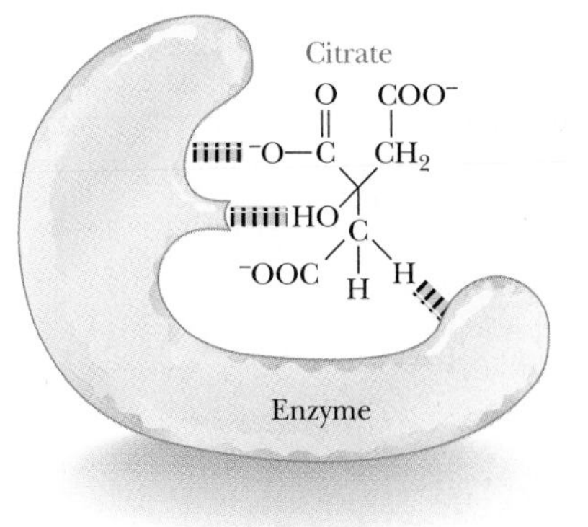

▲ **FIGURE 19.6** Three-point attachment to the enzyme aconitase makes the two —CH_2—COO^- ends of citrate stereochemically nonequivalent.

***cis*-Aconitate as an intermediate in the conversion of citrate to isocitrate**

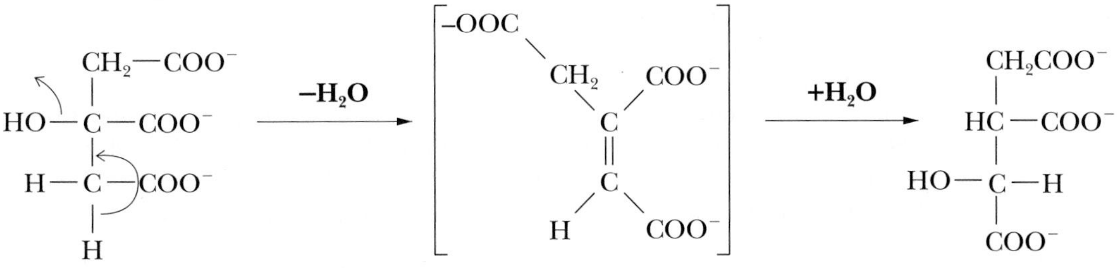

Citrate *cis*-Aconitate (enzyme-bound) Isocitrate (19.4)

Biochemical Connections

Plant Poisons and the Citric Acid Cycle

Another possible substrate for citrate synthase is *fluoroacetyl-CoA*. The source of the fluoroacetyl-CoA is fluoroacetate, which is found in the leaves of various types of poisonous plants, including locoweeds. Animals that ingest these plants form fluoroacetyl-CoA, which, in turn, is converted to fluorocitrate by their citrate synthase. Fluorocitrate, in turn, is a potent inhibitor of *aconitase,* the enzyme that catalyzes the next reaction of the citric acid cycle. These plants are poisonous because they produce a potent inhibitor of life processes.

The poison called Compound 1080 (pronounced "ten-eighty") is sodium fluoroacetate. Ranchers who want to protect their sheep from attacks by coyotes put the poison just outside the ranch fence. When the coyotes eat this poison, they die. The mechanism of poisoning by Compound 1080 is the same as that by plant poisons.

Stephen Krasemann/Photo Researchers, Inc.

The formation of fluorocitrate from fluoroacetate

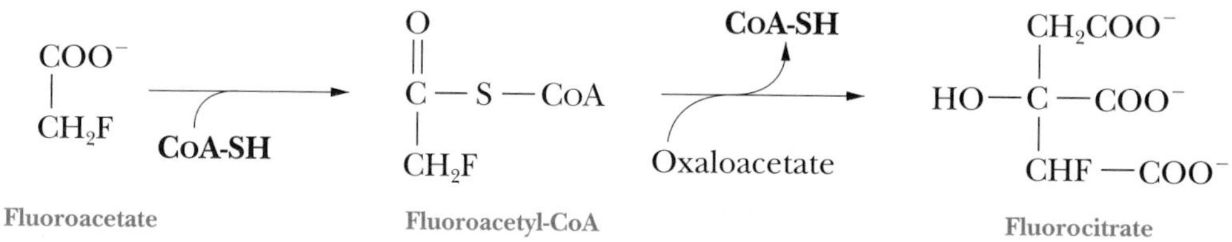

Fluoroacetate → (CoA-SH) → Fluoroacetyl-CoA → (CoA-SH, Oxaloacetate) → Fluorocitrate

The intermediate, *cis*-aconitate, remains bound to the enzyme during the course of the reaction. There is some evidence that the citrate is complexed to the Fe(II) in the active site of the enzyme in such a way that the citrate curls back on itself in a nearly circular conformation. Several authors have been unable to resist the temptation to call this situation the "ferrous wheel."

Step 3. *Formation of α-Ketoglutarate and CO$_2$—First Oxidation* The third step in the citric acid cycle is the oxidative decarboxylation of isocitrate to α-ketoglutarate and carbon dioxide. This reaction is the first of two oxidative decarboxylations of the citric acid cycle; the enzyme that catalyzes it is **isocitrate dehydrogenase.** The reaction takes place in two steps (Figure 19.7). First, isocitrate is oxidized to oxalosuccinate, which remains bound to the enzyme. Then oxalosuccinate is decarboxylated, and the carbon dioxide and α-ketoglutarate are released.

This is the first of the reactions in which NADH is produced. One molecule of NADH is produced from NAD$^+$ at this stage by the loss of two electrons in the oxidation. As we saw in our discussion of the pyruvate dehydrogenase complex, each NADH produced will lead to the production of 2.5 ATP in later stages of aerobic metabolism. Recall also that there will be two NADH, equivalent to five ATP for each original molecule of glucose.

Step 4. *Formation of Succinyl-CoA and CO$_2$—Second Oxidation* The second oxidative decarboxylation takes place in Step 4 of the citric acid cycle, in

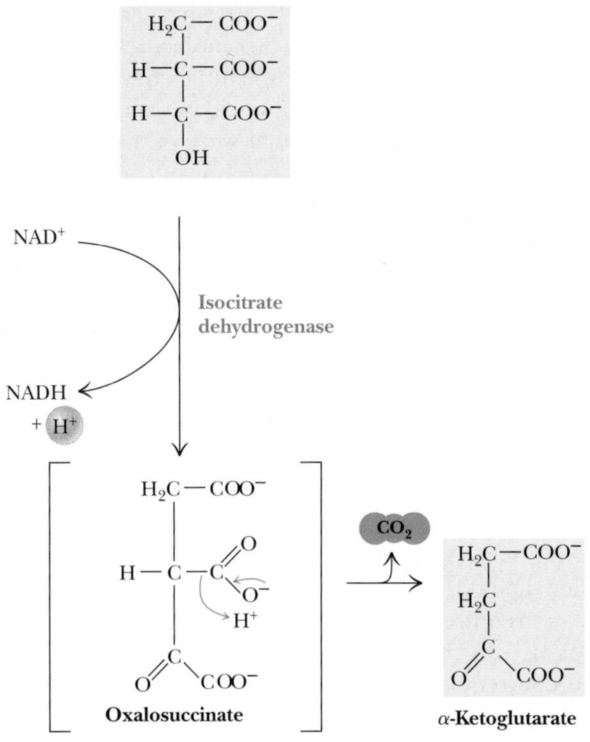

Biochemistry ⊜ Now™ **ANIMATED FIGURE 19.7** The isocitrate dehydrogenase reaction. **See this figure animated at http://now.brookscole.com/campbell5**

(19.5)

which carbon dioxide and succinyl-CoA are formed from α-ketoglutarate and CoA.

The conversion of α-ketoglutarate to succinyl-CoA

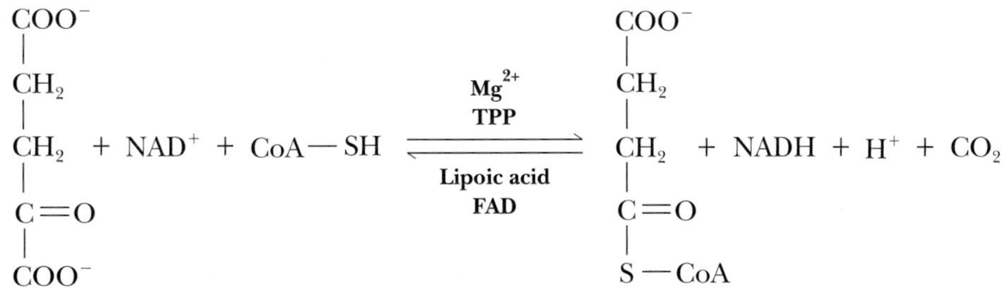

α-Ketoglutarate   Succinyl-CoA   (19.6)

This reaction is similar to the one in which acetyl-CoA is formed from pyruvate, with NADH produced from NAD^+. Once again, each NADH will eventually give rise to 2.5 ATP, with five ATP from each original molecule of glucose.

The reaction occurs in several stages and is catalyzed by an enzyme system called the **α-ketoglutarate dehydrogenase complex,** which is very similar to the pyruvate dehydrogenase complex. Each of these multienzyme systems consists of three enzymes that catalyze the overall reaction. The reaction takes place in several steps, and there is again a requirement for thiamine pyrophosphate (TPP), FAD, lipoic acid, and Mg^{2+}. This reaction is highly exergonic ($\Delta G°' = -33.4$ kJ $mol^{-1} = -8.0$ kcal mol^{-1}), as is the one catalyzed by pyruvate dehydrogenase.

At this point, two molecules of CO_2 have been produced by the oxidative decarboxylations of the citric acid cycle. Removal of the CO_2 makes the citric acid cycle irreversible in vivo, although, in vitro, each separate reaction is reversible. One might suspect that the two molecules of CO_2 arise from the two carbon atoms of acetyl-CoA. Labeling studies have shown that this is not the case, but a full discussion of this point is beyond the scope of this text. The two CO_2 arise from carbon atoms that were part of the oxaloacetate with which the acetyl group condensed. The carbons of this acetyl group are incorporated into the oxaloacetate that will be regenerated for the next round of the cycle. The release of the CO_2 molecules has a profound influence on mammalian physiology, as will be discussed later in this chapter. We should also mention that the α-ketoglutarate dehydrogenase complex reaction is the third one in which we have encountered an enzyme that requires TPP.

Step 5. *Formation of Succinate* In the next step of the cycle, the thioester bond of succinyl-CoA is hydrolyzed to produce succinate and CoA-SH; an accompanying reaction is the phosphorylation of GDP to GTP. The whole reaction is catalyzed by the enzyme **succinyl-CoA synthetase.** A synthetase is an enzyme that creates a new covalent bond and requires the direct input of energy from a high-energy phosphate. Recall that we met a synthase (citrate synthase) earlier. The difference between a synthase and a synthetase is that a synthase does not require energy from phosphate-bond hydrolysis, whereas a synthetase does. In the reaction mechanism, a phosphate group covalently bonded to the enzyme is directly transferred to the GDP. The phosphorylation of GDP to GTP is endergonic, as is the corresponding ADP-to-ATP reaction ($\Delta G°' = 30.5$ kJ mol^{-1} = 7.3 kcal mol^{-1}).

The conversion of succinyl-CoA to succinate

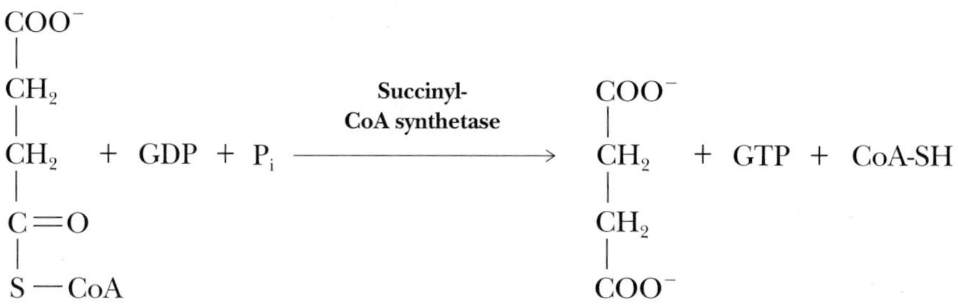

Succinyl-CoA + GDP + P$_i$ → Succinate + GTP + CoA-SH (19.7)

The energy required for the phosphorylation of GDP to GTP is provided by the hydrolysis of succinyl-CoA to produce succinate and CoA. The free energy of hydrolysis ($\Delta G°'$) of succinyl-CoA is -33.4 kJ mol^{-1} (-8.0 kcal mol^{-1}). The overall reaction is slightly exergonic ($\Delta G°' = -3.3$ kJ mol^{-1} = -0.8 kcal mol^{-1}) and, as a result, does not contribute greatly to the overall production of energy by the mitochondrion. Note that the name of the enzyme describes the reverse reaction. Succinyl-CoA synthetase would produce succinyl-CoA while spending an ATP or another high-energy phosphate. This reaction is the opposite of that.

The enzyme **nucleosidediphosphate kinase** catalyzes the transfer of a phosphate group from GTP to ADP to give GDP and ATP.

$$GTP + ADP \rightarrow GDP + ATP$$

This reaction step is called substrate-level phosphorylation to distinguish it from the type of reaction for production of ATP that is coupled to the electron transport chain. The production of ATP in this reaction is the only place in the citric acid cycle in which chemical energy in the form of ATP is made available to the cell. Except for this reaction, the generation of ATP characteristic of aerobic metabolism is associated with the electron transport chain, the subject of Chapter 20. About 30 to 32 molecules of ATP can be obtained from the oxidation of a single molecule of glucose by the combination of anaerobic and aerobic oxidation, compared with only two molecules of ATP produced by anaerobic glycolysis alone. (This variation in the stoichiometry of ATP produced is the result of differences in metabolic state and mechanisms of transport in different tissues, as will be explained in Chapter 20.) The combined reactions that occur in mitochondria are of great importance to aerobic organisms.

In the next three steps in the citric acid cycle (Steps 6 through 8), the four-carbon succinate ion is converted to oxaloacetate ion to complete the cycle.

The final stages of the citric acid cycle

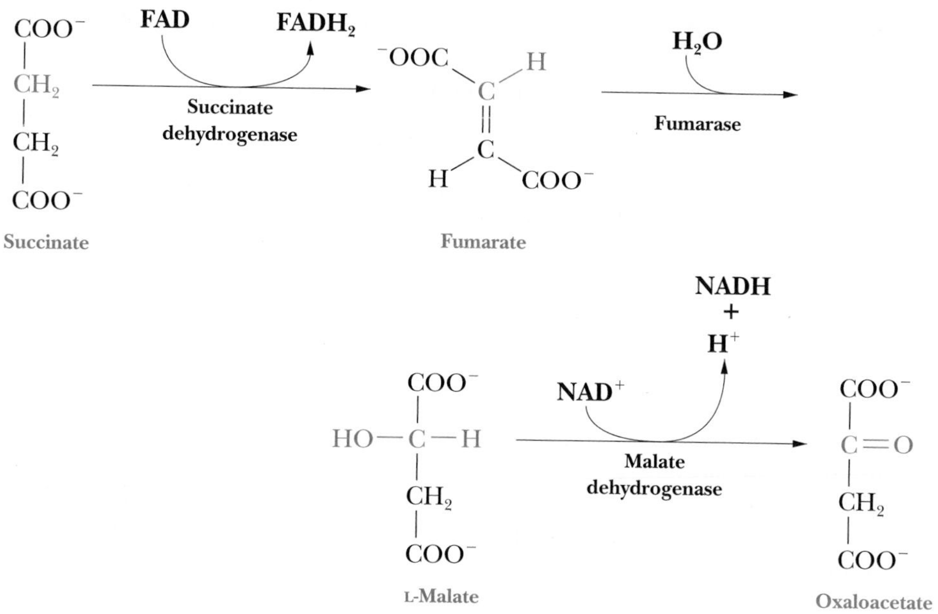

$$(19.8)$$

Step 6. *Formation of Fumarate—FAD-Linked Oxidation*

Succinate is oxidized to fumarate, a reaction that is catalyzed by the enzyme **succinate dehydrogenase.** This enzyme is an integral protein of the inner mitochondrial membrane. We shall have much more to say about the enzymes bound to the inner mitochondrial membrane in Chapter 20. The other individual enzymes of the citric acid cycle are in the mitochondrial matrix. The electron acceptor, which is FAD rather than NAD$^+$, is covalently bonded to the enzyme; succinate dehydrogenase is also called a flavoprotein because of the presence of FAD with its flavin moiety. In the succinate dehydrogenase reaction, FAD is reduced to FADH$_2$ and succinate is oxidized to fumarate.

The conversion of succinate to fumarate

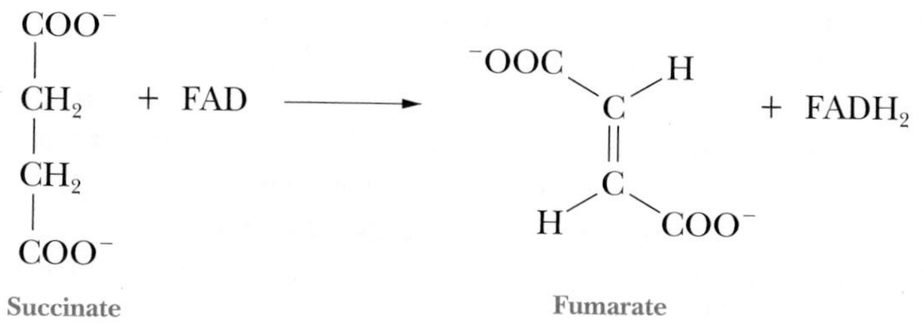

$$(19.9)$$

The overall reaction is

$$\text{Succinate} + \text{E—FAD} \rightarrow \text{Fumarate} + \text{E—FADH}_2$$

The E—FAD and E—FADH$_2$ in the equation indicate that the electron acceptor is covalently bonded to the enzyme. The FADH$_2$ group passes electrons on to the electron transport chain, and eventually to oxygen, and gives rise to 1.5 ATP, rather than 2.5, as is the case with NADH.

Succinate dehydrogenase contains iron atoms but does not contain a heme group; it is referred to as a **nonheme iron protein** or an *iron-sulfur protein*. The latter name refers to the fact that the protein contains several clusters that consist of four atoms each of iron and of sulfur.

Step 7. *Formation of L-Malate* In Step 7, which is catalyzed by the enzyme **fumarase,** water is added across the double bond of fumarate in a hydration reaction to give malate. Again, there is stereospecificity in the reaction. Malate has two enantiomers, L- and D-malate, but only L-malate is produced.

The conversion of fumarate to L-malate

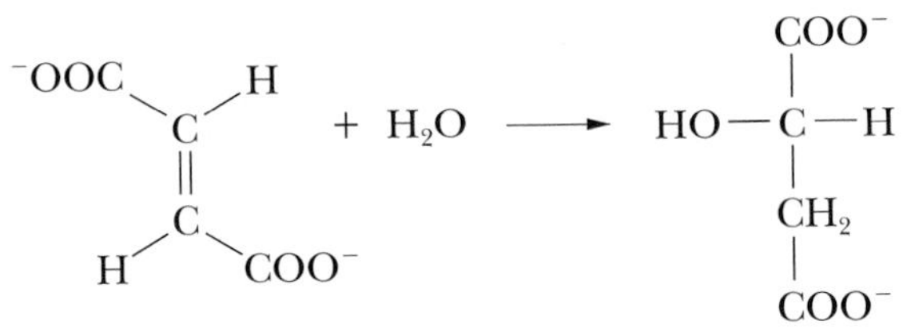

Fumarate   L-Malate   (19.10)

Step 8. *Regeneration of Oxaloacetate—Final Oxidation Step* Malate is oxidized to oxaloacetate, and another molecule of NAD$^+$ is reduced to NADH. This reaction is catalyzed by the enzyme **malate dehydrogenase.** The oxaloacetate can then react with another molecule of acetyl-CoA to start another round of the cycle.

The conversion of L-malate to oxaloacetate

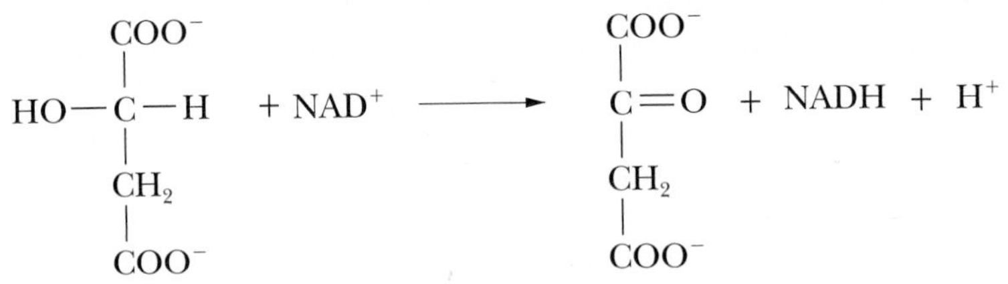

L-Malate   Oxaloacetate   (19.11)

The oxidation of pyruvate by the pyruvate dehydrogenase complex and the citric acid cycle results in the production of three molecules of CO$_2$. As a result of these oxidation reactions, one molecule of GDP is phosphorylated to GTP, one molecule of FAD is reduced to FADH$_2$, and four molecules of NAD$^+$ are reduced to NADH. Of the four molecules of NADH produced, three come from the citric acid cycle, and one comes from the reaction of the pyruvate dehydrogenase complex. The overall stoichiometry of the oxidation reactions is the sum of the pyruvate dehydrogenase reaction and the citric acid cycle. Note that only one high-energy phosphate, GTP, is produced *directly* from the citric acid cycle, but many more ATP will arise from reoxidation of NADH and FADH$_2$.

Pyruvate dehydrogenase complex:

$$\text{Pyruvate} + \text{CoA-SH} + \text{NAD}^+ \rightarrow \text{Acetyl-CoA} + \text{NADH} + \text{CO}_2 + \text{H}^+$$

Citric acid cycle:

$$\text{Acetyl-CoA} + 3\,\text{NAD}^+ + \text{FAD} + \text{GDP} + \text{P}_i + 2\,\text{H}_2\text{O} \rightarrow$$
$$2\,\text{CO}_2 + \text{CoA-SH} + 3\,\text{NADH} + 3\,\text{H}^+ + \text{FADH}_2 + \text{GTP}$$

Overall reaction:

$$\text{Pyruvate} + 4\,\text{NAD}^+ + \text{FAD} + \text{GDP} + \text{P}_i + 2\,\text{H}_2\text{O} \rightarrow$$
$$3\,\text{CO}_2 + 4\,\text{NADH} + \text{FADH}_2 + \text{GTP} + 4\,\text{H}^+$$

Eventual ATP production per pyruvate:

$$4\,\text{NADH} \rightarrow 10\,\text{ATP} \quad (2.5\,\text{ATP for each NADH})$$
$$1\,\text{FADH} \rightarrow 1.5\,\text{ATP} \quad (1.5\,\text{ATP for each FADH}_2)$$
$$1\,\text{GTP} \rightarrow 1\,\text{ATP}$$

Total 12.5 ATP per pyruvate or 25 ATP per glucose

There were also two ATP produced per glucose in glycolysis and two NADH, which will give rise to another five ATP (seven more ATP total). In the next chapter, we shall say more about the subject of ATP production from the complete oxidation of glucose.

At this point, we would do well to recapitulate what we have said about the citric acid cycle (see Figure 19.3). When studying a pathway such as this, we might learn many details but also be able to see the big picture. The entire pathway is shown with the enzyme names outside the circle. The most important reactions can be identified by those that have important cofactors ($NADH$, $FADH_2$, GTP). Also important are those steps where CO_2 is given off. These important reactions also play a large role in the cycle's contribution to our metabolism. One purpose of the cycle is to produce energy. It does that by producing GTP directly and by producing reduced electron carriers ($NADH$ and $FADH_2$). The three decarboxylations mean that, for every three carbons entering as pyruvate, three carbons are effectively lost during the cycle, a fact that has many implications to our metabolism, as we shall see later in the chapter.

> **Essential Information**
>
> In the citric acid cycle and the pyruvate dehydrogenase reaction, one molecule of pyruvate is oxidized to three molecules of carbon dioxide. In the process, four NAD^+ are reduced to NADH, and one FAD to $FADH_2$; in addition, one GDP is phosphorylated to GTP.

19.5 | What Are the Energetics of the Citric Acid Cycle, and How Is It Controlled?

The reaction of pyruvate to acetyl-CoA is exergonic, as we have seen ($\Delta G^{\circ\prime} = -33.4$ kJ mol$^{-1} = -8.0$ kcal mol^{-1}). The citric acid cycle itself is also exergonic ($\Delta G^{\circ\prime} = -44.3$ kJ mol$^{-1} = -10.6$ kcal mol^{-1}), and you will be asked in Question 38 to confirm this point. The standard free energy changes for the individual reactions are listed in Table 19.2. Of the individual reactions of the cycle, only one is strongly endergonic: the oxidation of malate to oxaloacetate ($\Delta G^{\circ\prime} = +29.2$ kJ mol$^{-1} = +7.0$ kcal mol^{-1}). This endergonic reaction is, however, coupled to one of the strongly exergonic reactions of the cycle, the condensation of acetyl-CoA and oxaloacetate to produce citrate and coenzyme A ($\Delta G^{\circ\prime} = -32.2$ kJ mol$^{-1} = -7.7$ kcal mol^{-1}). (Recall that these values for the free-energy changes refer to standard conditions. The effect of concentrations of metabolites in vivo can change matters drastically.) In addition to the energy released by the oxidation reactions, there is more release of energy to come in the electron transport chain. When the four NADH and

Table 19.2

The Energetics of Conversion of Pyruvate to CO_2

		$\Delta G^{\circ\prime}$	
Step	Reaction	kJ mol^{-1}	kcal mol^{-1}
	Pyruvate + CoA-SH + NAD$^+$ → Acetyl-CoA + NADH + CO_2	−33.4	−8.0
1	Acetyl-CoA + Oxaloacetate + H_2O → Citrate + CoA-SH + H$^+$	−32.2	−7.7
2	Citrate → Isocitrate	+6.3	+1.5
3	Isocitrate + NAD$^+$ → α-Ketoglutarate + NADH + CO_2 + H$^+$	−7.1	−1.7
4	α-Ketoglutarate + NAD$^+$ + CoA-SH → Succinyl-CoA + NADH + CO_2 + H	−33.4	−8.0
5	Succinyl-CoA + GDP + P$_i$ → Succinate + GTP + CoA-SH	−3.3	−0.8
6	Succinate + FAD → Fumarate + FADH$_2$	~0	~0
7	Fumarate + H_2O → L-Malate	−3.8	−0.9
8	L-Malate + NAD$^+$ → Oxaloacetate + NADH + H$^+$	+29.2	+7.0
Overall:	Pyruvate + 4 NAD$^+$ + FAD + GDP + P$_i$ + 2 H_2O → CO_2 + 4 NADH + FADH$_2$ + GTP + 4 H$^+$	−77.7	−18.6

single FADH$_2$ produced by the pyruvate dehydrogenase complex and citric acid cycle are reoxidized by the electron transport chain, considerable quantities of ATP are produced.

Control of the citric acid cycle is exercised at three points; that is, three enzymes within the citric acid cycle play a regulatory role (Figure 19.8). There is also control of access to the cycle via pyruvate dehydrogenase.

Control of Pyruvate Dehydrogenase

The overall reaction is part of a pathway that releases energy. It is not surprising that the enzyme that initiates it is inhibited by ATP and NADH because both compounds are abundant when a cell has a good deal of energy readily available. The end products of a series of reactions inhibit the first reaction of

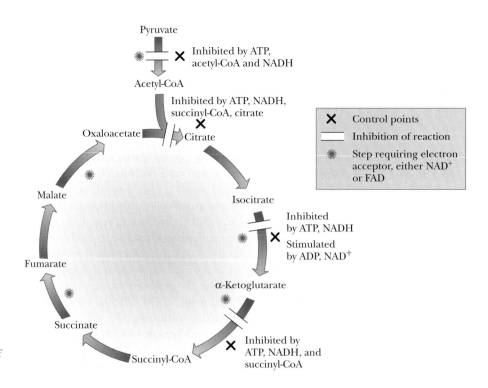

▶ **FIGURE 19.8** Control points in the conversion of pyruvate to acetyl-CoA and in the citric acid cycle.

the series, and the intermediate reactions do not take place when their products are not needed. Consistent with this picture, the pyruvate dehydrogenase (PDH) complex is activated by ADP, which is abundant when a cell needs energy. In mammals, the actual mechanism by which the inhibition takes place is the phosphorylation of pyruvate dehydrogenase. A phosphate group is covalently bound to the enzyme in a reaction catalyzed by the enzyme *pyruvate dehydrogenase kinase*. When the need arises for pyruvate dehydrogenase to be activated, the hydrolysis of the phosphate ester linkage (dephosphorylation) is catalyzed by another enzyme, *phosphoprotein phosphatase*. This latter enzyme is itself activated by Ca^{2+}. Both enzymes are associated with the mammalian pyruvate dehydrogenase complex, permitting effective control of the overall reaction from pyruvate to acetyl-CoA. The PDH kinase and PDH phosphatase are found on the same polypeptide chain. High levels of ATP activate the kinase. Pyruvate dehydrogenase is also inhibited by high levels of acetyl-CoA. This makes a great deal of metabolic sense. When fats are plentiful and are being degraded for energy, their product is acetyl-CoA (Chapter 21). Thus, if acetyl-CoA is plentiful, there is no reason to send carbohydrates to the citric acid cycle. Pyruvate dehydrogenase is inhibited, and the acetyl-CoA for the TCA cycle comes from other sources.

Control of the Citric Acid Cycle Proper

Within the citric acid cycle itself, the three control points are the reactions catalyzed by citrate synthase, isocitrate dehydrogenase, and the α-ketoglutarate dehydrogenase complex. We have already mentioned that the first reaction of the cycle is one in which regulatory control appears, as is to be expected in the first reaction of any pathway. Citrate synthase is an allosteric enzyme inhibited by ATP, NADH, succinyl-CoA, and its own product, citrate.

The second regulatory site is the isocitrate dehydrogenase reaction. In this case, ADP and NAD^+ are allosteric activators of the enzyme. We have called attention to the recurring pattern in which ATP and NADH inhibit enzymes of the pathway, and ADP and NAD^+ activate these enzymes.

The α-ketoglutarate dehydrogenase complex is the third regulatory site. As before, ATP and NADH are inhibitors. Succinyl-CoA is also an inhibitor of this reaction. This recurring theme in metabolism reflects the way in which a cell can adjust to an active state or to a resting state.

When a cell is metabolically active it uses ATP and NADH at a great rate, producing large amounts of ADP and NAD^+ (Table 19.3). In other words, when the ATP/ADP ratio is low, the cell is using energy and needs to release more energy from stored nutrients. A low $NADH/NAD^+$ ratio is also characteristic of an active metabolic state. On the other hand, a resting cell has

Table 19.3
Relationship between the Metabolic State of a Cell and the ATP/ADP and NADH/NAD$^+$ Ratios
Cells in a resting metabolic state
Need and use comparatively little energy
High ATP, low ADP levels imply high ATP/ADP ratio
High NADH, low NAD$^+$ levels imply high NADH/NAD$^+$ ratio
Cells in a highly active metabolic state
Need and use more energy than resting cells
Low ATP, high ADP levels imply low ATP/ADP ratio
Low NADH, high NAD$^+$ levels imply low NADH/NAD$^+$ ratio

fairly high levels of ATP and NADH. The ATP/ADP ratio and the NADH/NAD^+ ratio are also high in resting cells, which do not need to maintain a high level of oxidation to produce energy.

When cells have low energy requirements (that is, when they have a high "energy charge") with high ATP/ADP and $NADH/NAD^+$ ratios, the presence of so much ATP and NADH serves as a signal to "shut down" the enzymes responsible for oxidative reactions. When cells have a low energy charge, characterized by low ATP/ADP and $NADH/NAD^+$ ratios, the need to release more energy and to generate more ATP serves as a signal to "turn on" the oxidative enzymes. This relationship of energy requirements to enzyme activity is the basis for the overall regulatory mechanism exerted at a few key control points in metabolic pathways.

19.6 | What Is the Glyoxylate Cycle?

In plants and in some bacteria, but not in animals, acetyl-CoA can serve as the starting material for the biosynthesis of carbohydrates. Animals can convert carbohydrates to fats, but not fats to carbohydrates. (Acetyl-CoA is produced in the catabolism of fatty acids.) Two enzymes are responsible for the ability of plants and bacteria to produce glucose from fatty acids. **Isocitrate lyase** cleaves isocitrate, producing glyoxylate and succinate. **Malate synthase** catalyzes the reaction of glyoxylate with acetyl-CoA to produce malate.

The unique reactions of the glyoxylate cycle

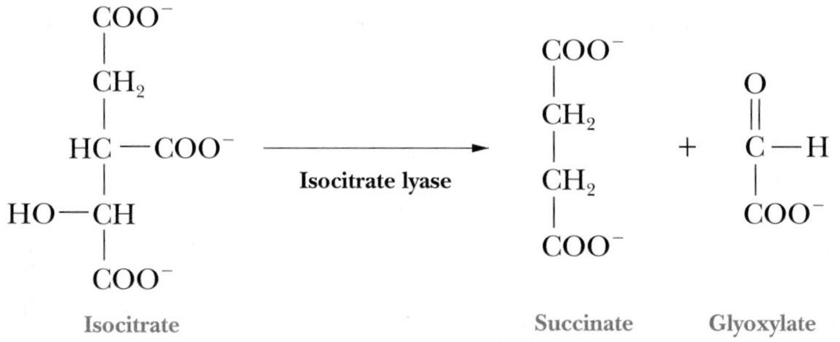

The conversion of isocitrate to glyoxylate and succinate

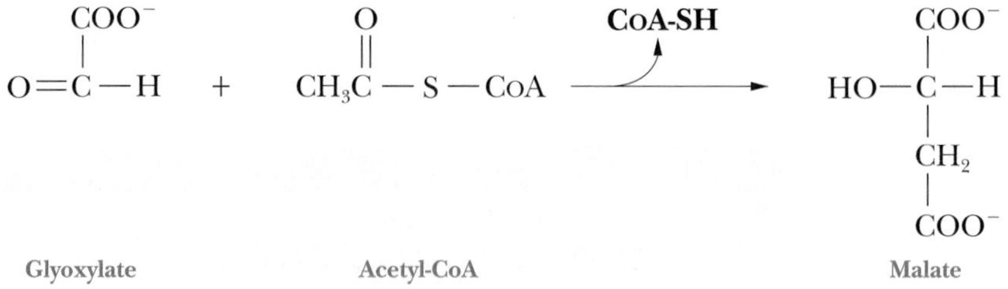

The reaction of glyoxylate with acetyl-CoA to produce malate   (19.12)

These two reactions in succession bypass the two oxidative decarboxylation steps of the citric acid cycle. The net result is an alternative pathway, the **glyoxylate cycle** (Figure 19.9). Two molecules of acetyl-CoA enter the glyoxylate cycle; they give rise to one molecule of malate and eventually to one molecule of oxaloacetate. Two 2-carbon units (the acetyl groups of acetyl-CoA) give rise to a four-carbon unit (malate), which is then converted to oxaloacetate (also a four-carbon compound). Glucose can then be produced from oxaloacetate

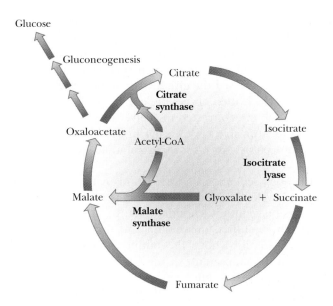

◄ **FIGURE 19.9** The glyoxylate cycle. This pathway results in the net conversion of two acetyl-CoA to oxaloacetate. All the reactions are shown in purple. The unique reactions of the glyoxylate cycle are shown with a light green highlight in the center of the circle.

by gluconeogenesis. This is a subtle, yet very important, distinction between the glyoxylate cycle and the citric acid cycle. The carbon skeletons that enter the citric acid cycle as acetyl-CoA are effectively lost by the decarboxylation steps. This means that if oxaloacetate (OAA) is drawn off to make glucose, there will be no OAA to continue the cycle. For this reason, fats cannot lead to a *net* production of glucose. With the glyoxylate cycle, the bypass reactions go around the decarboxylations, creating an *extra* four-carbon compound that can be drawn off to make glucose without depleting the citric acid cycle of its starting compound.

Specialized organelles in plants, called **glyoxysomes,** are the sites of the glyoxylate cycle. This pathway is particularly important in germinating seeds. The fatty acids stored in the seeds are broken down for energy during germination. First, the fatty acids give rise to acetyl-CoA, which can enter the citric acid cycle and go on to release energy in the ways we have already seen. The citric acid cycle and the glyoxylate cycle can operate simultaneously. Acetyl-CoA also serves as the starting point for the synthesis of glucose and any other compounds needed by the growing seedling. (Recall that carbohydrates play an important structural, as well as energy-producing, role in plants.)

The glyoxylate cycle also occurs in bacteria. This point is far from surprising because many types of bacteria can live on very limited carbon sources. They have metabolic pathways that can produce all the biomolecules they need from quite simple molecules. The glyoxylate cycle is one example of how bacteria manage this feat.

19.7 What Role Does the Citric Acid Cycle Play in Catabolism?

The nutrients taken in by an organism can include large molecules. This observation is especially true in the case of animals, which ingest polysaccharides and proteins, which are polymers, as well as lipids. Nucleic acids constitute a very small percentage of the nutrients present in foodstuffs, and we shall not consider their catabolism.

The first step in the breakdown of nutrients is the degradation of large molecules to smaller ones. Polysaccharides are hydrolyzed by specific enzymes to produce sugar monomers; an example is the breakdown of starch by amylases. Lipases hydrolyze triacylglycerols to give fatty acids and glycerol. Proteins are digested by proteases, with amino acids as the end products. Sugars, fatty acids, and amino acids then enter their specific catabolic pathways.

In Chapter 17, we discussed the glycolytic pathway by which sugars are converted to pyruvate, which then enters the citric acid cycle. In Chapter 21, we will see how fatty acids are converted to acetyl-CoA; we learned about the fate of acetyl-CoA in the citric acid cycle earlier in this chapter. Amino acids enter the cycle by various paths. We will discuss catabolic reactions of amino acids in Chapter 23.

Figure 19.10 shows schematically the various catabolic pathways that feed into the citric acid cycle. The catabolic reactions occur in the cytosol; the citric acid cycle takes place in mitochondria. Many of the end products of catabolism cross the mitochondrial membrane and then participate in the citric acid cycle. This figure also shows the outline of pathways by which amino acids are converted to components of the citric acid cycle. Be sure to notice that sugars, fatty acids, and amino acids are all included in this overall catabolic scheme. Just as "all roads lead to Rome," all pathways lead to the citric acid cycle.

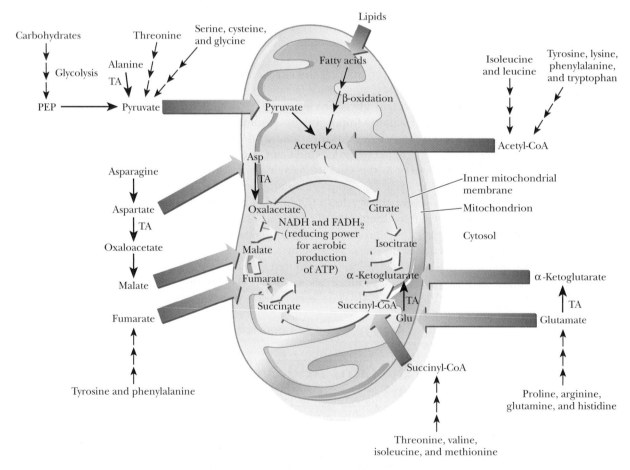

▲ **FIGURE 19.10** A summary of catabolism, showing the central role of the citric acid cycle. Note that the end products of the catabolism of carbohydrates, lipids, and amino acids all appear. (PEP is phosphoenolpyruvate; α-KG is α-ketoglutarate; TA is transamination; →→→ is a multistep pathway.)

19.8 | What Role Does the Citric Acid Cycle Play in Anabolism?

The citric acid cycle is a source of starting materials for the biosynthesis of many important biomolecules, but the supply of the starting materials that are components of the cycle must be replenished if the cycle is to continue operating. See the Biochemical Connections box on page 532. In particular, the oxaloacetate in an organism must be maintained at a level sufficient to allow acetyl-CoA to enter the cycle. A reaction that replenishes a citric acid cycle intermediate is called an **anaplerotic reaction.** In some organisms, acetyl-CoA can be converted to oxaloacetate and other citric acid cycle intermediates by the glyoxylate cycle (Section 16.6), but mammals cannot do this. In mammals, oxaloacetate is produced from pyruvate by the enzyme *pyruvate carboxylase* (Figure 19.11). We already encountered this enzyme and this reaction in the context of gluconeogenesis (see Section 18.2), and here we have another highly important role for this enzyme and the reaction it catalyzes. The supply of oxaloacetate would soon be depleted if there were no means of producing it from a readily available precursor.

This reaction, which produces oxaloacetate from pyruvate, provides a connection between the amphibolic citric acid cycle and the anabolism of sugars by gluconeogenesis. On this same topic of carbohydrate anabolism, we should note again that pyruvate cannot be produced from acetyl-CoA in mammals. Because acetyl-CoA is the end product of catabolism of fatty acids, we can see that mammals could not exist with fats or acetate as the sole carbon source. The intermediates of carbohydrate metabolism would soon be depleted. Carbohydrates are the principal energy and carbon source in animals (Figure 19.11), and glucose is especially critical in humans because it is the preferred fuel for our brain cells. Plants can carry out the conversion of acetyl-CoA to pyruvate and oxaloacetate, so they can exist without carbohydrates as a carbon

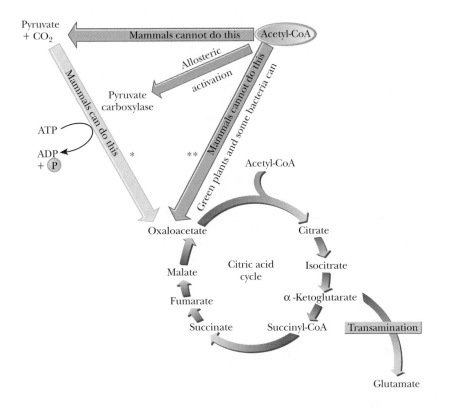

◀ **FIGURE 19.11** How mammals keep an adequate supply of metabolic intermediates. An anabolic reaction uses a citric acid cycle intermediate (α-ketoglutarate is transaminated to glutamate in our example), competing with the rest of the cycle. The concentration of acetyl-CoA rises and signals the allosteric activation of pyruvate carboxylase to produce more oxaloacetate.
* Anaplerotic reaction.
**Part of glyoxylate pathway.

Biochemical Connections

Anaplerotic Reactions

The citric acid cycle is important not only as a source of energy during aerobic metabolism but also as a key pathway in the synthesis of important metabolic intermediates. We shall see in subsequent chapters that it is a source of starting materials for the production of amino acids, carbohydrates, vitamins, nucleotides, and heme. However, if these intermediates are used for the synthesis of other molecules, then they must be replenished to maintain the catalytic nature of this cycle. The term **anaplerotic** means "filling up," and the reactions that replenish the citric acid cycle are called anaplerotic reactions. One source of needed compounds, available to all organisms, is that group of amino acids that can be converted to citric acid cycle intermedi-

ates in a single reaction. A simple reaction available to all organisms is to add carbon dioxide to the pyruvate and phosphoenolpyruvate generated from metabolism of sugars. Another source, important in bacteria and plants, is the glyoxylate cycle discussed in Section 19.6. This source is vital to the ability of plants to fix carbon dioxide to carbohydrates.

Some anaerobic organisms have developed only parts of the citric acid cycle, which they use exclusively to make the important precursors. These simple yet important reactions emphasize the truly connected nature of what we often artificially separate into "pathways." They also illustrate the convergence of evolution to a few key molecules and metabolic steps.

source. The conversion of pyruvate to acetyl-CoA does take place in both plants and animals (see Section 19.3).

The anabolic reactions of gluconeogenesis take place in the cytosol. Oxaloacetate is not transported across the mitochondrial membrane. Two mechanisms exist for the transfer of molecules needed for gluconeogenesis from mitochondria to the cytosol. One mechanism takes advantage of the fact that phosphoenolpyruvate can be formed from oxaloacetate in the mitochondrial matrix (this reaction is the next step in gluconeogenesis); phosphoenolpyruvate is then transferred to the cytosol, where the remaining reactions take place (Figure 19.12). The other mechanism relies on the fact that malate, which is another intermediate of the citric acid cycle, can be transferred to the cytosol. There is a *malate dehydrogenase* enzyme in the cytosol as well as in mitochondria, and malate can be converted to oxaloacetate in the cytosol.

$$\text{Malate} + \text{NAD}^+ \rightarrow \text{Oxaloacetate} + \text{NADH} + \text{H}^+$$

Oxaloacetate is then converted to phosphoenolpyruvate, leading to the rest of the steps of gluconeogenesis (Figure 19.12).

Gluconeogenesis has many steps in common with the production of glucose in photosynthesis, but photosynthesis also has many reactions in common with the pentose phosphate pathway. Thus, nature has evolved common strategies to deal with carbohydrate metabolism in all its aspects.

Lipid Anabolism

The starting point of lipid anabolism is acetyl-CoA. The anabolic reactions of lipid metabolism, like those of carbohydrate metabolism, take place in the cytosol; these reactions are catalyzed by soluble enzymes that are not bound to membranes. Acetyl-CoA is mainly produced in mitochondria, whether from pyruvate or from the breakdown of fatty acids. An indirect transfer mechanism exists for transfer of acetyl-CoA in which citrate is transferred to the cytosol (Figure 19.13). Citrate reacts with CoA-SH to produce citryl-CoA, which is then cleaved to yield oxaloacetate and acetyl-CoA. The enzyme that catalyzes this reaction requires ATP and is called ATP-citrate lyase. The overall reaction is

$$\text{Citrate} + \text{CoA-SH} + \text{ATP} \rightarrow \text{Acetyl-CoA} + \text{Oxaloacetate} + \text{ADP} + \text{P}_i$$

Acetyl-CoA is the starting point for lipid anabolism in both plants and animals. An important source of acetyl-CoA is the catabolism of carbohydrates. We have just seen that animals cannot convert lipids to carbohydrates, but they can convert carbohydrates to lipids. The efficiency of the conversion of

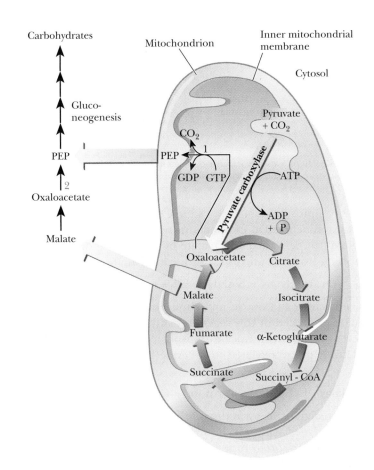

◄ **FIGURE 19.12** Transfer of the starting materials of gluconeogenesis from the mitochondrion to the cytosol. Note that phosphoenolpyruvate (PEP) can be transferred from the mitochondrion to the cytosol, as can malate. Oxaloacetate is not transported across the mitochondrial membrane. (1 is PEP carboxykinase in mitochondria; 2 is PEP carboxykinase in cytosol; other symbols are as in Figure 19.10.)

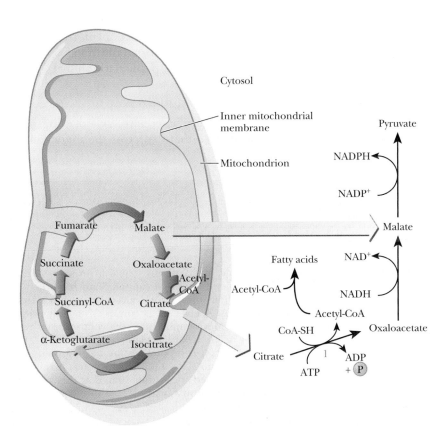

◄ **FIGURE 19.13** Transfer of the starting materials of lipid anabolism from the mitochondrion to the cytosol. (1 is ATP-citrate lyase; other symbols are as in Figure 19.10.) It is not definitely established whether acetyl-CoA is transported from the mitochondrion to the cytosol.

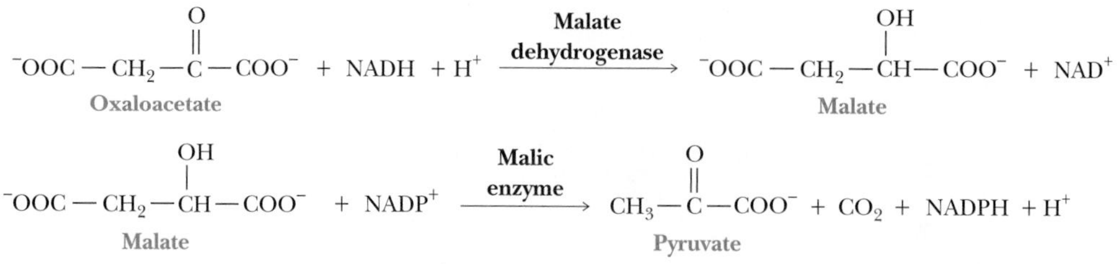

$$\underset{\text{Oxaloacetate}}{{}^-OOC-CH_2-\overset{\overset{\displaystyle O}{\parallel}}{C}-COO^-} + NADH + H^+ \xrightarrow[\text{dehydrogenase}]{\textbf{Malate}} \underset{\text{Malate}}{{}^-OOC-CH_2-\overset{\overset{\displaystyle OH}{|}}{CH}-COO^-} + NAD^+$$

$$\underset{\text{Malate}}{{}^-OOC-CH_2-\overset{\overset{\displaystyle OH}{|}}{CH}-COO^-} + NADP^+ \xrightarrow[\text{enzyme}]{\textbf{Malic}} \underset{\text{Pyruvate}}{CH_3-\overset{\overset{\displaystyle O}{\parallel}}{C}-COO^-} + CO_2 + NADPH + H^+$$

▲ **FIGURE 19.14** Reactions involving citric acid cycle intermediates that produce NADPH for fatty acid anabolism. Note that these reactions take place in the cytosol.

carbohydrates to lipids in animals is a source of considerable chagrin to many humans (see the Biochemical Connections box on page 536.)

Oxaloacetate can be reduced to malate by the reverse of a reaction we saw in the last section in the context of carbohydrate anabolism.

$$\text{Oxaloacetate} + NADH + H^+ \rightarrow \text{malate} + NAD^+$$

Malate can move into and out of mitochondria by active transport processes, and the malate produced in this reaction can be used again in the citric acid cycle. However, malate need not be transported back into mitochondria but can be oxidatively decarboxylated to pyruvate by *malic enzyme*, which requires $NADP^+$.

$$\text{Malate} + NADP^+ \rightarrow \text{Pyruvate} + CO_2 + NADPH + H^+$$

These last two reactions are a reduction reaction followed by an oxidation; there is *no net oxidation*. There is, however, a *substitution of NADPH for NADH*. This last point is an important one because many of the enzymes of fatty acid synthesis require NADPH. The pentose phosphate pathway (Section 18.4) is the principal source of NADPH in most organisms, but here we have another source (Figure 19.14).

The two ways of producing NADPH clearly indicate that all metabolic pathways are related. The exchange reactions involving malate and citryl-CoA constitute a control mechanism in lipid anabolism, while the pentose phosphate pathway is part of carbohydrate metabolism. Both carbohydrates and lipids are important energy sources in many organisms, particularly animals.

Anabolism of Amino Acids and Other Metabolites

The anabolic reactions that produce amino acids have, as a starting point, those intermediates of the citric acid cycle that can cross the mitochondrial

Biochemical Connections

Acetyl-CoA

Which molecule is arguably the most important metabolic intermediate? Acetyl-CoA is perhaps *the* central molecule of metabolism. When one plots a chart of all known metabolic pathways, acetyl-CoA ends up close to the center of that chart.

The reasons are quite simple. This important compound really links the metabolism of the three major classes of nutrients to each other. All sugars, all fatty acids, and many amino acids pass through acetyl-CoA on their way to becoming water and carbon dioxide. Equally important is the key use of this intermediate in the synthesis of the major biomolecules. Some, but not all, organisms can carry out all these conversions. Bacte-

ria provide an example of organisms that can do so, whereas humans are an example of ones that cannot. Many bacteria can live off acetic acid as their sole carbon source; however, it is first converted to acetyl-CoA. Acetyl-CoA is converted to fatty acids, terpenes, and steroids. More important is the conversion of two molecules of acetyl-CoA to malate in plants and bacteria via the glyoxylate pathway. This key compound is the starting point for the synthesis of both amino acids and carbohydrates. It is interesting to note, as mentioned in Section 19.6, that this key glyoxylate reaction is missing in mammals.

membrane into the cytosol. We have already seen that malate can cross the mitochondrial membrane and give rise to oxaloacetate in the cytosol. Oxaloacetate can undergo a transamination reaction to produce aspartate, and aspartate, in turn, can undergo further reactions to form not only amino acids but also other nitrogen-containing metabolites, such as pyrimidines. Similarly, isocitrate can cross the mitochondrial membrane and produce α-ketoglutarate in the cytosol. Glutamate arises from α-ketoglutarate as a result of another transamination reaction, and glutamate undergoes further reactions to form still more amino acids. Succinyl-CoA gives rise not to amino acids but to the porphyrin ring of the heme group. Another difference is that the first reaction of heme biosynthesis, the condensation of succinyl-CoA and glycine to form δ-aminolevulinic acid (see supplementary material on nitrogen metabolism on the website), takes place in the mitochondrial matrix, while the remainder of the pathway occurs in the cytosol.

The overall outline of anabolic reactions is shown in Figure 19.15. We used the same type of diagram in Figure 19.10 to show the overall outline of catabolism. The similarity of the two schematic diagrams points out that catabolism

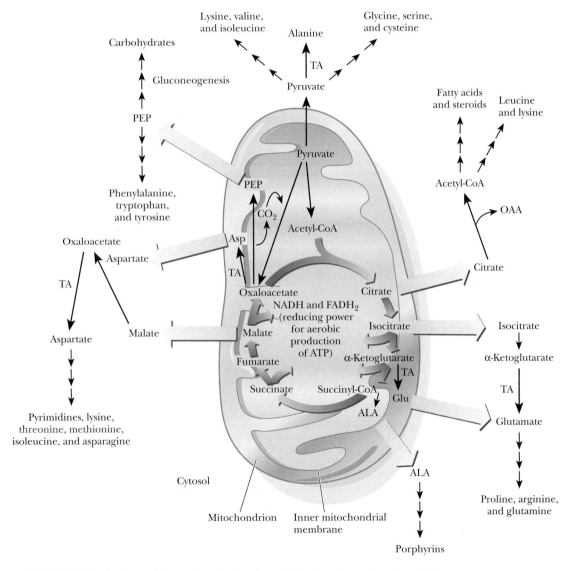

▲ **FIGURE 19.15** A summary of anabolism, showing the central role of the citric acid cycle. Note that there are pathways for the biosynthesis of carbohydrates, lipids, and amino acids. OAA is oxaloacetate, and ALA is δ-aminolevulinic acid. Symbols are as in Figure 19.10.)

and anabolism, while not exactly the same, are closely related. The operation of any metabolic pathway, anabolic or catabolic, can be "speeded up" or "slowed down" in response to the needs of an organism by control mechanisms, such as feedback control. Regulation of metabolism takes place in similar ways in many different pathways.

19.9 | Why Isn't Oxygen Part of the Equation?

The citric acid cycle is considered part of aerobic metabolism, but we have not encountered any reactions in this chapter in which oxygen takes part. The reactions of the citric acid cycle are intimately related to those of electron transport and oxidative phosphorylation, which do eventually lead to oxygen. The citric acid cycle provides a vital link between the chemical energy of nutrients and the chemical energy of ATP. Many molecules of ATP can be generated as a result of coupling to oxygen, and we shall see that the number depends on the NADH and $FADH_2$ generated in the citric acid cycle. Recall the classic equation for the aerobic oxidation of glucose:

$$\text{Glucose} + 6\,O_2 \rightarrow 6\,H_2O + 6\,CO_2$$

We have seen the metabolism of glucose through glycolysis. Now we see where the CO_2 comes from—namely, the three decarboxylation reactions associated with the citric acid cycle. In the next chapter, we will see where the water and oxygen come from.

Biochemical Connections

Why Is It So Hard to Lose Weight?

One of the great tragedies about being human is that it is far too easy to gain weight and far too difficult to lose it. If we had to analyze the specific chemical reactions that make this a reality, we would look very carefully at the citric acid cycle, especially the decarboxylation reactions.

As everybody knows, all food in excess can be stored as fat. This is true for carbohydrates, proteins, and, of course, fats. In addition, these molecules can be interconverted, with the exception that fats cannot give a net yield of carbohydrates, as we saw in Section 19.6. Why can fats not yield carbohydrates? The real answer lies in the fact that the only way a fat molecule would have to make glucose would be to enter the citric acid cycle as acetyl-CoA and then to be drawn off as oxaloacetate for gluconeogenesis. Unfortunately, the two carbons that enter are effectively lost by the decarboxylations. (We have already seen that, in one round of the citric acid cycle, it isn't really these same two carbons that are lost; nevertheless, a two-carbon loss is a two-carbon loss, regardless of which two carbons they were.) This leads to an imbalance in the catabolic pathways versus the anabolic pathways.

All roads lead to fat, but fat cannot lead back to carbohydrate. Humans are very sensitive to glucose levels in the blood because so much of our metabolism is geared toward protecting our brain cells, which prefer glucose as a fuel. If we eat more carbohydrates than we need, the excess carbohydrates will turn to fat. As we know, it is very easy to put on fat, especially as we age. What about the reverse? Why don't we just stop eating? Won't that reverse the process? The answer is yes and no. When we start eating less, fat stores will be mobilized for energy. Fat is

an excellent source of energy because it forms acetyl-CoA and gives a steady influx for the citric acid cycle. Thus, we can lose some weight by reducing our caloric intake. Unfortunately, our blood sugar will also drop as soon as our glycogen stores run out. We don't have very much stored glycogen that could maintain our blood glucose levels.

After the blood glucose drops, we become depressed, sluggish, and irritable. We start having negative thoughts like, "this dieting thing is really stupid. I should eat a pint of ice cream." If we continue the diet, and given that we cannot turn fats into carbohydrates, where will the blood glucose come from? There is only one source left, which is proteins. Proteins will be degraded to amino acids, and they will eventually be converted to pyruvate for gluconeogenesis. Thus, we will begin to lose muscle as well as fat.

There is a bright side to all of this, however. Using our knowledge of biochemistry, we can see that there is a better way to lose weight than dieting—exercise! If you exercise correctly, you can train your body to use fats to supply acetyl-CoA for the citric acid cycle. If you maintain a normal diet, you will maintain your blood glucose and not degrade proteins for that purpose; your ingested carbohydrates will be sufficient to maintain blood glucose and carbohydrate stores. With the proper balance of exercise to food intake, and the proper balance of the right types of nutrients, we can increase the breakdown of fat without sacrificing our carbohydrate stores or our proteins. In essence, it is easier and healthier to train off the weight than to diet off the weight. This has been known for a long time. Now we are in a position to see why it is so, biochemically.

Summary

19.1 What Role Does the Citric Acid Cycle Play in Metabolism?

The citric acid cycle plays a central role in metabolism. It is the first part of aerobic metabolism; it is also amphibolic (both catabolic and anabolic). Unlike glycolysis, which takes place in the cytosol, the citric acid cycle occurs in mitochondria. Most of the enzymes of the citric acid cycle are in the mitochondrial matrix. Succinate dehydrogenase, the sole exception, is localized in the inner mitochondrial membrane.

19.2 What Is the Overall Pathway of the Citric Acid Cycle?

Pyruvate produced by glycolysis is transformed by oxidative decarboxylation into acetyl-CoA in the presence of coenzyme A. Acetyl-CoA then enters the citric acid cycle by reacting with oxaloacetate to produce citrate. The reactions of the citric acid cycle include two other oxidative decarboxylations, which transform the six-carbon compound citrate into the four-carbon compound succinate. The cycle is completed by regeneration of oxaloacetate from succinate in a multistep process that includes two other oxidation reactions. The overall reaction, starting with pyruvate, is

$$Pyruvate + 4\,NAD^+ + FAD + GDP + P_i + 2\,H_2O \rightarrow$$
$$3\,CO_2 + 4\,NADH + FADH_2 + GTP + 4\,H^+$$

NAD^+ and FAD are the electron acceptors in the oxidation reactions. The cycle is strongly exergonic.

19.3 How Is Pyruvate Converted to Acetyl-CoA?

Pyruvate is produced by glycolysis in the cytosol of the cell. The citric acid cycle takes place in the matrix of the mitochondria, so the pyruvate must first pass through a transporter into this organelle. There, pyruvate will find pyruvate dehydrogenase, a large, multisubunit protein made up of three enzymes involved in the production of acetyl-CoA plus two enzyme activities involved in control of the enzymes. The reaction requires several cofactors, including FAD, lipoic acid, and TPP.

19.4 What Are the Individual Reactions of the Citric Acid Cycle?

Acetyl-CoA condenses with oxaloacetate to give citrate, a six-carbon compound. Citrate isomerizes to isocitrate, which then undergoes an oxidative decarboxylation to α-ketoglutarate, a five-carbon compound. This then undergoes another oxidative decarboxylation producing succinyl-CoA, a four-carbon compound. The two decarboxylation steps also produce NADH. Succinyl-CoA is converted to succinate with the concomitant production of GTP. Succinate is oxidized to fumarate, and $FADH_2$ is produced. Fumarate is converted to malate, which is then oxidized to oxaloacetate while another NADH is produced.

19.5 What Are the Energetics of the Citric Acid Cycle, and How Is It Controlled?

The overall pathway has a $\Delta G^{\circ\prime}$ of -77.7 kJ mol^{-1}. During the course of the cycle, starting from pyruvate, four NADH molecules and one $FADH_2$ are produced. Between the GTP formed directly and the reoxidation of the reduced electron carriers by the electron transport chain, the citric acid cycle produces 25 ATP. Control of the citric acid cycle is exercised at three points. There is also a control point outside the cycle, the reaction in

which pyruvate produces acetyl-CoA. Within the citric acid cycle, the three control points are the reactions catalyzed by citrate synthase, isocitrate dehydrogenase, and the α-ketoglutarate dehydrogenase complex. In general, ATP and NADH are inhibitors, and ADP and NAD^+ are activators of the enzymes at the control points.

19.6 What Is the Glyoxylate Cycle?

In plants and bacteria, there is a pathway related to the citric acid cycle: the glyoxylate cycle. The two oxidative decarboxylations of the citric acid cycle are bypassed. This pathway plays a role in the ability of plants to convert acetyl-CoA to carbohydrates, a process that does not occur in animals.

19.7 What Role Does the Citric Acid Cycle Play in Catabolism?

Like a giant traffic circle of life, the citric acid cycle has many routes entering it. Many members of the three basic nutrient types, proteins, fats, and carbohydrates, are metabolized to smaller molecules that can cross the mitochondrial membrane and enter the citric acid cycle as one of the intermediate molecules. In this way, the cycle allows us to get energy from the food we eat. Carbohydrates and many amino acids can enter the cycle either as pyruvate or as acetyl-CoA. Lipids enter as acetyl-CoA. Because of the transamination reaction possible with glutamate and α-ketoglutarate, almost any amino acid can be transaminated to glutamate, producing α-ketoglutarate that can enter the cycle. Several other pathways lead to amino acids entering the pathway as succinate, fumarate, or malate.

19.8 What Role Does the Citric Acid Cycle Play in Anabolism?

While the citric acid cycle takes place in mitochondria, many anabolic reactions take place in the cytosol. Oxaloacetate, the starting material for gluconeogenesis, is a component of the citric acid cycle. Malate, but not oxaloacetate, can be transported across the mitochondrial membrane. After malate from mitochondria is carried to the cytosol, it can be converted to oxaloacetate by malate dehydrogenase, an enzyme that requires NAD^+. Malate, which crosses the mitochondrial membrane, plays a role in lipid anabolism, in a reaction in which malate is oxidatively decarboxylated to pyruvate by an enzyme that requires $NADP^+$, producing NADPH. This reaction is an important source of NADPH for lipid anabolism, with the pentose phosphate pathway the only other source. In addition, most of the intermediates have anabolic pathways leading to amino acids and fatty acids, as well as some that lead to porphyrins or pyrimidines.

19.9 Why Isn't Oxygen Part of the Equation?

Glycolysis and the citric acid cycle account for some of the overall equation for the oxidation of glucose:

$$C_6H_{12}O_6 + O_2 \rightarrow 6\,CO_2 + 6\,H_2O$$

The glucose is seen in glycolysis. The decarboxylation steps of the citric acid cycle account for the CO_2. However, the oxygen in the equation does not appear until the last step of the electron transport chain. If there is insufficient oxygen available, the electron transport chain will not be able to process the reduced electron carriers from the TCA cycle, and it will have to slow down as well. Continued activity under these circumstances will cause the pyruvate produced by glycolysis to be processed anaerobically to lactate.

Critical Questions to Review

19.1 What Role Does the Citric Acid Cycle Play in Metabolism?

1. **Fact Check** Which pathways are involved in the anaerobic metabolism of glucose? Which pathways are involved in the aerobic metabolism of glucose?

2. **Fact Check** How many ATPs can be produced from one molecule of glucose anaerobically? Aerobically?

3. **Fact Check** What are the different names used to describe the pathway discussed in this chapter?

4. **Fact Check** What is meant by the statement that a pathway is amphibolic?

19.2 What Is the Overall Pathway of the Citric Acid Cycle?

5. **Fact Check** In what part of the cell does the citric acid cycle take place? Does this differ from the part of the cell where glycolysis occurs?

6. **Fact Check** How does pyruvate from glycolysis get to the pyruvate dehydrogenase complex?

7. **Fact Check** What electron acceptors play a role in the citric acid cycle?

8. **Fact Check** What three molecules produced during the citric acid cycle are an indirect or direct source of high-energy compounds?

19.3 How Is Pyruvate Converted to Acetyl-CoA?

9. **Fact Check** How many enzymes are involved in mammalian pyruvate dehydrogenase? What are their functions?

10. **Fact Check** Briefly describe the dual role of lipoic acid in the pyruvate dehydrogenase complex.

11. **Fact Check** What is the advantage to the organization of the PDH complex?

12. **Fact Check** In the PDH reaction alone, we can see cofactors that come from four different vitamins. What are they?

13. **Thought Question** Draw the structures of the activated carbon groups bound to thiamine pyrophosphate in three enzymes that contain this coenzyme. *Hint:* Keto–enol tautomerism may enter into the picture.

14. **Thought Question** Prepare a sketch showing how the individual reactions of the three enzymes of the pyruvate dehydrogenase complex give rise to the overall reaction.

19.4 What Are the Individual Reactions of the Citric Acid Cycle?

15. **Fact Check** Why is the reaction catalyzed by citrate synthase considered a condensation reaction?

16. **Fact Check** What does it mean when an enzyme has the name "synthase"?

17. **Biochemical Connections** What is fluoroacetate? Why is it used?

18. **Fact Check** With respect to stereochemistry, what is unique about the reaction catalyzed by aconitase?

19. **Fact Check** In which steps of the aerobic processing of pyruvate is CO_2 produced?

20. **Fact Check** In which steps of the aerobic processing of pyruvate are reduced electron carriers produced?

21. **Fact Check** What type of reaction is catalyzed by isocitrate dehydrogenase and α-ketoglutarate dehydrogenase?

22. **Fact Check** What are the similarities and differences between the reactions catalyzed by pyruvate dehydrogenase and α-ketoglutarate dehydrogenase?

23. **Fact Check** What does it mean when an enzyme is called a synthetase?

24. **Fact Check** Why can we say that production of a GTP is equivalent to production of an ATP?

25. **Fact Check** What are the major differences between the oxidations in the citric acid cycle that use NAD^+ as an electron acceptor and the one that uses FAD?

26. **Fact Check** ATP is a competitive inhibitor of NADH binding to malate dehydrogenase, as are ADP and AMP. Suggest a structural basis for this inhibition.

27. **Fact Check** Is the conversion of fumarate to malate a redox (electron transfer) reaction or not? Give the reason for your answer.

28. **Thought Question** We have seen one of the four possible isomers of isocitrate, the one produced in the aconitase reaction. Draw the configurations of the other three.

29. **Thought Question** Show, by Lewis electron-dot structures of the appropriate portions of the molecule, where electrons are lost in the following conversions:
    (a) Pyruvate to acetyl-CoA
    (b) Isocitrate to α-ketoglutarate
    (c) α-Ketoglutarate to succinyl-CoA
    (d) Succinate to fumarate
    (e) Malate to oxaloacetate

19.5 What Are the Energetics of the Citric Acid Cycle, and How Is It Controlled?

30. **Fact Check** Which steps of aerobic metabolism of pyruvate through the citric acid cycle are control points?

31. **Fact Check** Describe the multiple ways that PDH is controlled.

32. **Fact Check** What are the two most common inhibitors of steps of the citric acid cycle and the reaction catalyzed by pyruvate dehydrogenase?

33. **Thought Question** How does an increase in the ADP/ATP ratio affect the activity of isocitrate dehydrogenase?

34. **Thought Question** How does an increase in the $NADH/NAD^+$ ratio affect the activity of pyruvate dehydrogenase?

35. **Thought Question** Would you expect the citric acid cycle to be more or less active when a cell has a high ATP/ADP ratio and a high $NADH/NAD^+$ ratio? Give the reason for your answer.

36. **Thought Question** Would you expect the $\Delta G^{\circ\prime}$ for the hydrolysis of a thioester to be (a) large and negative, (b) large and positive, (c) small and negative, or (d) small and positive? Give the reason for your answer.

37. **Thought Question** Acetyl-CoA and succinyl-CoA are both high-energy thioesters, but their chemical energy is put to different uses. Elaborate.

38. **Thought Question** Some reactions of the citric acid cycle are endergonic. Show how the overall cycle is exergonic. (See Table 19.2.)

39. **Thought Question** How could the expression "milking it for all it's worth" relate to the citric acid cycle?

40. **Thought Question** Using the information in Chapters 17–19, calculate the amount of ATP that can be produced from one molecule of lactose metabolized aerobically through glycolysis and the citric acid cycle.

19.6 What Is the Glyoxylate Cycle?

41. **Fact Check** Which enzymes of the citric acid cycle are missing from the glyoxylate cycle?

42. **Fact Check** What are the unique reactions of the glyoxylate cycle?

43. **Biochemical Connections** Why is it possible for bacteria to survive on acetic acid as a sole carbon source, but not human beings?

19.7 What Role Does the Citric Acid Cycle Play in Catabolism?

44. **Fact Check** Describe the various purposes of the citric acid cycle.

45. **Thought Question** The intermediates of glycolysis are phosphorylated, but those of the citric acid cycle are not. Suggest a reason why.

46. **Thought Question** Discuss oxidative decarboxylation, using a reaction from this chapter to illustrate your points.

47. **Thought Question** Many soft drinks contain citric acid as a significant part of their flavor. Is this a good nutrient?

19.8 What Role Does the Citric Acid Cycle Play in Anabolism?

48. **Fact Check** NADH is an important coenzyme in catabolic processes, whereas NADPH appears in anabolic processes. Explain how an exchange of the two can be effected.

49. **Biochemical Connections** What are the anaplerotic reactions in mammals?

50. **Thought Question** Why is acetyl-CoA considered to be the central molecule of metabolism?

19.9 Why Isn't Oxygen Part of the Equation?

51. **Thought Question** Why is the citric acid cycle considered part of aerobic metabolism, even though molecular oxygen does not appear in any reaction?

Biochemistry ⊜ Now™

Assess your understanding of this chapter's topics with additional quizzing and tutorials at **http://now.brookscole.com/campbell5**

Annotated Bibliography

Bodner, C. M. The Tricarboxylic Acid (TCA), Citric Acid, or Krebs Cycle. *J. Chem. Ed.* **63,** 673–677 (1986). [A concise and well-written summary of the citric acid cycle. Part of a series on metabolism.]

Boyer, P. D., ed. *The Enzymes.* 3rd ed. New York: Academic Press, 1975. [There are reviews on aconitase in Volume 5 and on dehydrogenases in Volume 11.]

Krebs, H. A. *Reminiscences and Reflections.* New York: Oxford University Press, 1981. [A review of the citric acid cycle, along with the autobiography.]

Popjak, G. Stereospecificity of Enzyme Reactions. In Boyer, P. D., ed., *The Enzymes.* 3rd ed., vol. 2. New York: Academic Press, 1970. [A review of stereochemical aspects of the citric acid cycle.]

See also the bibliographies for Chapters 16 to 18.

CHAPTER 20

Electron Transport and Oxidative Phosphorylation

Mitochondria, shown here, are the sites of the citric acid cycle, electron transport, and oxidative phosphorylation.

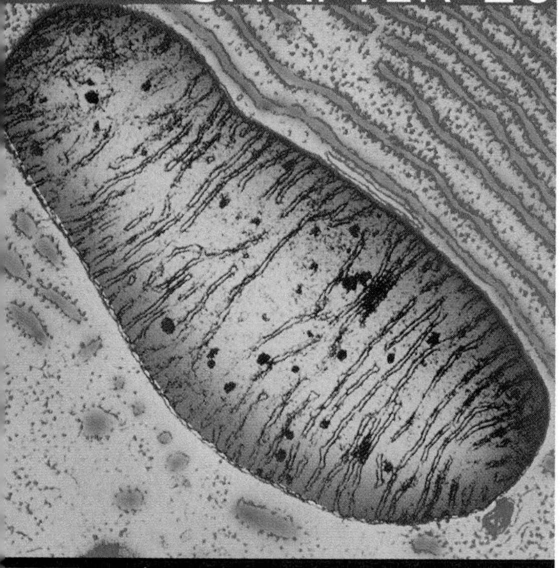

© Keith R. Porter/Photo Researchers, Inc.

Critical Questions

Biochemistry ⊜ Now™
Test yourself on these Critical Questions at the BiochemistryNow website at **http://now** **.brookscole.com/campbell5**

Energy derived from the oxidation of metabolic fuels is ultimately converted to ATP, the energy currency of the cell. In eukaryotic cells, under aerobic conditions, ATP is generated by the power of electron transport along the inner membrane of the mitochondrion coupled with proton transport across the inner membrane. The electron transport chain is actually four closely related enzyme complexes embedded in the inner mitochondrial membrane. In a series of oxidation–reduction transfers, they conduct electrons along the membrane from one complex to another until the electrons reach their final destination, where they combine with molecular oxygen to reduce O_2 to 2 H_2O. The energy of electron transport can then be used by three of these same enzyme complexes to pump protons across the inner membrane out into the intermembrane space. The reverse flow of protons back through the membrane into the inner matrix drives the production of ATP. An ATP synthase complex embedded in the inner membrane binds ADP and phosphate to catalyze the formation of ATP. The flow of protons through the ATP synthase from the intermembrane space to the inner matrix releases the new ATP that has been synthesized. This process is very similar to the production of ATP by photosynthesis (Chapter 22) in the thylakoid membrane of the chloroplast in green plants.

20.1 | What Role Does Electron Transport Play in Metabolism?

Aerobic metabolism is a highly efficient way for an organism to extract energy from nutrients. In eukaryotic cells, the aerobic processes (including conversion of pyruvate to acetyl-CoA, the citric acid cycle, and electron transport) all occur in the mitochondria, while the anaerobic process, glycolysis, takes place outside the mitochondria in the cytosol. We have not yet seen any reactions in which oxygen plays a part, but in this chapter we shall discuss the role of oxygen in metabolism as the final acceptor of electrons in the **electron transport chain.** The reactions of the electron transport chain take place in the inner mitochondrial membrane.

The energy released by the oxidation of nutrients is used by organisms in the form of the chemical energy of ATP. Production of ATP in the mitochondria is the result of **oxidative phosphorylation,** in which ADP is phosphorylated to give ATP. The production of ATP by oxidative phosphorylation (an endergonic process) is separate from electron transport to oxygen (an exergonic process), but the reactions of the electron transport chain are strongly linked to one another and are tightly coupled to the synthesis of ATP by phosphorylation of ADP. The operation of the electron transport chain leads to pumping of protons (hydrogen ions) across the inner mitochondrial membrane, creating a pH gradient (also called a **proton gradient**). This proton gradient represents stored potential energy and provides the basis of the coupling mechanism (Figure 20.1). Chemiosmotic coupling is the name given to this mechanism (Section 20.5). Oxidative phosphorylation gives rise to most of the ATP production associated with the complete oxidation of glucose.

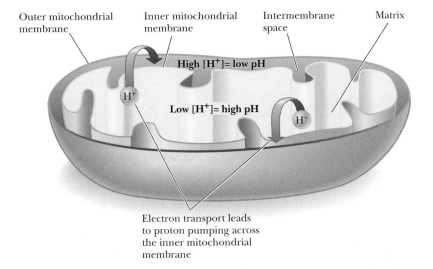

Outer mitochondrial membrane

Inner mitochondrial membrane

Intermembrane space

Matrix

High [H$^+$]= low pH

Low [H$^+$]= high pH

H$^+$

H$^+$

Electron transport leads to proton pumping across the inner mitochondrial membrane

◄ **FIGURE 20.1** A proton gradient is established across the inner mitochondrial membrane as a result of electron transport. Transfer of electrons through the electron transport chain leads to the pumping of protons from the matrix to the intermembrane space. The proton gradient (also called the pH gradient), together with the membrane potential (a voltage across the membrane), provides the basis of the coupling mechanism that drives ATP synthesis.

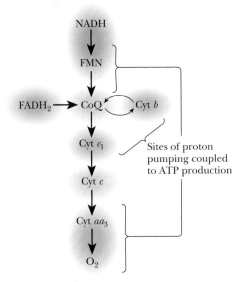

Sites of proton pumping coupled to ATP production

◄ **FIGURE 20.2** Schematic representation of the electron transport chain, showing sites of proton pumping coupled to oxidative phosphorylation. FMN is the flavin coenzyme *f*lavin *mono*nucleotide, which differs from FAD in not having an adenine nucleotide. CoQ is coenzyme Q (see Figure 20.4). Cyt b, cyt c_1, cyt c, and cyt aa_3 are the heme-containing proteins cytochrome b, cytochrome c_1, cytochrome c, and cytochrome aa_3, respectively.

The NADH and FADH$_2$ molecules generated in glycolysis and the citric acid cycle transfer electrons to oxygen in the series of reactions known collectively as the electron transport chain. The NADH and FADH$_2$ are oxidized to NAD$^+$ and FAD and can be used again in various metabolic pathways. Oxygen, the ultimate electron acceptor, is reduced to water; this completes the process by which glucose is completely oxidized to carbon dioxide and water. We have already seen how carbon dioxide is produced from pyruvate, which in turn is produced from glucose by the pyruvate dehydrogenase complex and the citric acid cycle. In this chapter, we shall see how water is produced.

The complete series of oxidation–reduction reactions of the electron transport chain is presented in schematic form in Figure 20.2. A particularly noteworthy point about electron transport is that, on average, 2.5 moles of ATP are generated for each mole of NADH that enters the electron transport chain, and, on average, 1.5 moles of ATP are produced for each mole of FADH$_2$. The general outline of the process is that NADH passes electrons to coenzyme Q, as does FADH$_2$, providing an alternative mode of entry into the electron transport chain. Electrons are then passed from coenzyme Q to a

(a) Ethanol ⟶ acetaldehyde
−0.197 V

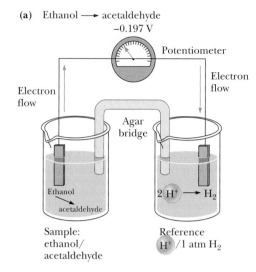

(b) Fumarate ⟶ succinate
+0.031 V

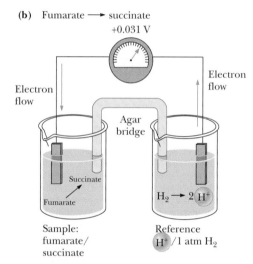

Biochemistry ⓔNow™ ACTIVE FIGURE 20.3
Experimental apparatus used to measure the standard reduction potential of the indicated redox couples: (a) the ethanol/acetaldehyde couple, (b) the fumarate/succinate couple. Figure 20.3(a) shows a sample/reference half-cell pair for measurement of the standard reduction potential of the ethanol/acetaldehyde couple. Because electrons flow toward the reference half-cell and away from the sample half-cell, the standard reduction potential is negative, specifically −0.197 V. In contrast, the fumarate/succinate couple (b) accepts electrons from the reference half-cell; that is, reduction occurs spontaneously in the system, and the reduction potential is thus positive. For each half-cell, a **half-cell reaction** describes the reaction taking place. For the fumarate/succinate half-cell coupled to a H^+/H_2 reference half-cell (b), the reaction taking place is indeed the reduction of fumarate.

Fumarate + 2 H^+ + 2 e^- ⟶ Succinate

E° = +0.031 V

However, the reaction occurring in the ethanol/acetaldehyde half-cell (a) is the oxidation of ethanol, which is the reverse of the reaction listed in Table 20.1.

Ethanol ⟶ Acetaldehyde + 2 H^+ + 2 e^-

E° = −0.197 V

Watch this Active Figure at http://now.brookscole .com/campbell5

series of proteins called cytochromes (which are designated by lowercase letters) and, eventually, to oxygen.

20.2 What Are the Reduction Potentials for the Electron Transport Chain?

Up until now, most of the energy considerations we have had concerned phosphorylation potentials. In Section 15.6, we saw how the free-energy change associated with hydrolysis of ATP could be used to drive otherwise endergonic reactions. The opposite is also true—when a reaction is highly exergonic, it can drive the formation of ATP. When we look closely at the energy changes in electron transport, a more useful approach is to consider the change in energy associated with the movement of electrons from one carrier to another. Each carrier in the electron transport chain can be isolated and studied, and each can exist in an oxidized or a reduced form (Section 15.5). If we had two potential electron carriers, such as NADH and coenzyme Q (see Section 20.3), for example, how would we know whether electrons would be more likely to be transferred from the NADH to the coenzyme Q or the other way around? This is determined by measuring a **reduction potential** for each of the carriers. A molecule with a high reduction potential will tend to be reduced if it is paired with a molecule with a lower reduction potential. This is measured by making a simple battery cell, as shown in Figure 20.3. The reference point is the half-cell on the right where hydrogen ion is in aqueous solution in equilibrium with hydrogen gas. The reduction of hydrogen ion to hydrogen gas

$$2\ H^+ + 2\ e^- \rightarrow H_2$$

is the control and is considered to have a voltage (E) of zero. The sample to be tested is in the other half-cell. The electric circuit is completed by bridge with a salt-containing agar gel.

Figure 20.3(a) shows what happens if ethanol and acetaldehyde are put into the sample half-cell. Electrons flow away from the sample cell and toward the reference cell. This means that the hydrogen ion is being reduced to hydrogen gas and the ethanol is being oxidized to acetaldehyde. Therefore, the hydrogen/H^+ pair has a higher reduction potential than the ethanol/acetaldehyde pair. If we look at Figure 20.3(b), we see the opposite. When fumarate and succinate are put into the sample half-cell, the electrons flow in the opposite direction, meaning that fumarate is being reduced to succinate while hydrogen gas is being oxidized to H^+. The direction of electron flow and the magnitude of the observed voltage allow us to make a table, as shown in Table 20.1. Because this is a table of standard reduction potentials, all the reactions are shown as reductions. The value being measured is the standard biological voltage of each half reaction $E^{\circ\prime}$. This value is calculated based on the compounds in the cells being at 1 M and the pH being 7 at the standard temperature of 25°C.

To interpret the data in this table for the purpose of electron transport, we need to look at the reduction potentials of the electron carriers involved. A reaction at the top of the table will tend to occur as written if it is paired with a reaction that is lower down on the table. For example, we have already seen that the final step of the electron transport chain is the reduction of oxygen to water. This reaction is at the top of Table 20.1 with a reduction potential of 0.816 volts, a very positive number. If this reaction were paired directly with $NAD^+/NADH$, what would happen? The standard reduction potential for

Table 20.1

Standard Reduction Potentials for Several Biological Reduction Half-Reactions

Reduction Half-Reaction	$E°'$ (V)
$\frac{1}{2} O_2 + 2 H^+ + 2 e^- \rightarrow H_2O$	0.816
$Fe^{3+} + e^- \rightarrow Fe^{2+}$	0.771
Cytochrome $a_3(Fe^{3+}) + e^- \rightarrow$ Cytochrome $a_3(Fe^{2+})$	0.350
Cytochrome $a(Fe^{3+}) + e^- \rightarrow$ Cytochrome $a(Fe^{2+})$	0.290
Cytochrome $c(Fe^{3+}) + e^- \rightarrow$ Cytochrome $c(Fe^{2+})$	0.254
Cytochrome $c_1(Fe^{3+}) + e^- \rightarrow$ Cytochrome $c_1(Fe^{2+})$	0.220
$CoQH^· + H^+ + e^- \rightarrow CoQH_2$ (coenzyme Q)	0.190
$CoQ + 2 H^+ + 2 e^- \rightarrow CoQH_2$	0.060
Cytochrome $b_H(Fe^{3+}) + e^- \rightarrow$ Cytochrome $b_H(Fe^{2+})$	0.050
Fumarate $+ 2 H^+ + 2 e^- \rightarrow$ Succinate	0.031
$CoQ + H^+ + e^- \rightarrow CoQH^·$	0.030
$[FAD] + 2 H^+ + 2 e^- \rightarrow [FADH_2]$	0.003–0.091*
Cytochrome $b_L(Fe^{3+}) + e^- \rightarrow$ Cytochrome $b_L(Fe^{2+})$	−0.100
Oxaloacetate $+ 2 H^+ + e^- \rightarrow$ Malate	−0.166
Pyruvate $+ 2 H^+ + 2 e^- \rightarrow$ Lactate	−0.185
Acetaldehyde $+ 2 H^+ + 2 e^- \rightarrow$ Ethanol	−0.197
$FMN + 2 H^+ + 2 e^- \rightarrow FMNH_2$	−0.219
$FAD + 2 H^+ + 2 e^- \rightarrow FADH_2$	−0.219
1,3-*bis*phosphoglycerate $+ 2 H^+ + 2 e^- \rightarrow$ Glyceraldehyde-3-phosphate $+ P_i$	−0.290
$NAD^+ + 2 H^+ + 2 e^- \rightarrow NADH + H^+$	−0.320
$NADP^+ + 2 H^+ + 2 e^- \rightarrow NADPH + H^+$	−0.320
α-Ketoglutarate $+ CO_2 + 2 H^+ + 2 e^- \rightarrow$ Isocitrate	−0.380
Succinate $+ CO_2 + 2 H^+ + 2 e^- \rightarrow \alpha$-Ketoglutarate $+ H_2O$	−0.670

* Typical values for reduction of bound FAD in flavoproteins such as succinate dehydrogenase.

Note that we have shown a number of components of the electron transport chain individually. We are going to see them again as parts of complexes. We have also included values for a number of reactions we saw in earlier chapters.

NAD^+ forming NADH is given near the bottom of the table. Its reduction potential is −0.320 V.

$$NAD^+ + 2 H^+ + 2 e^- \rightarrow NADH + H+ \quad E°' = -0.320 \text{ V}$$

This means that, if the two half-reactions are paired during a redox reaction, the one for the NADH will have to be reversed. NADH will give up its electrons so that oxygen can be reduced to water:

$NADH + H^+ \rightarrow NAD^+ + 2 H^+ + 2 e^-$	0.320
$\frac{1}{2} O_2 + 2 H^+ + 2 e^- \rightarrow H_2O$	0.816
Sum $NADH + \frac{1}{2} O_2 + H^+ \rightarrow NAD^+ + H_2O$	1.136

The overall voltage for this reaction will be the sum of the standard reduction potentials—in this case, 0.816 V + 0.320 V, or 1.136 V. Note that we had to change the sign on the standard reduction potential for the NADH because we had to reverse the direction of its reaction.

The $\Delta G°$ of a redox reaction is calculated using

$$\Delta G° = -nF \Delta E°'$$

where n is the number moles of electrons transferred, F is Faraday's constant (96.485 kJ V^{-1} mol^{-1}), and $\Delta E^{\circ\prime}$ is the total voltage for the two half-reactions. As we can see by this equation, the ΔG° is negative when the $\Delta E^{\circ\prime}$ is positive. Therefore, we can always calculate the direction a redox reaction will go under standard conditions by combining the two half-reactions in the way that gives the largest positive value for $\Delta E^{\circ\prime}$. For this example, the ΔG° would be calculated as follows:

$$\Delta G^{\circ} = -(2)(96.485 \text{ kJ V}^{-1} \text{ mol}^{-1})(1.136 \text{ V}) = -219 \text{ kJ mol}^{-1}$$

This would be a very large number if NADH reduced oxygen directly. As we shall see in the next section, NADH passes its electrons along a chain that eventually leads to oxygen, but it does not reduce oxygen directly.

Before moving on, it should be noted that, just as there is a difference between ΔG° and ΔG, there is a similar difference between ΔE° and ΔE. Recall from Chapter 15 that we devoted several sections to the question of standard states, including the modified standard state for biochemical reactions. The notations ΔG and ΔE refer to the free-energy change and the reduction potential under any conditions, respectively. When all components of a reaction are in their standard state (1 atm pressure, 25°C, all solutes at 1 M concentration), we write ΔG° and ΔE°, respectively, for the standard free-energy change and standard reduction potential. The modified standard state for biochemical reactions takes note of the fact that having all solutes at 1 M concentration includes the hydrogen ion concentration. That implies a pH equal to zero. Consequently, we define a modified standard state for biochemistry that differs from the usual one only in that pH = 7. Under these conditions, we write $\Delta G^{\circ\prime}$ and $\Delta E^{\circ\prime}$ for the standard free-energy change and the standard reduction potential, respectively. The true direction of electron flow in a redox reaction will also be based on the true values of the concentrations for the reactants and products, since the cellular concentrations are never 1 M.

20.3	**How Are the Electron Transport Complexes Organized?**

Intact mitochondria isolated from cells can carry out all the reactions of the electron transport chain; the electron transport apparatus can also be resolved into its component parts by a process called fractionation. Four separate **respiratory complexes** can be isolated from the inner mitochondrial membrane. These complexes are multienzyme systems. In the last chapter, we encountered other examples of such multienzyme complexes, such as the pyruvate dehydrogenase complex and the α-ketoglutarate dehydrogenase complex. Each of the respiratory complexes can carry out the reactions of a portion of the electron transport chain.

Complex I   The first complex, **NADH-CoQ oxidoreductase,** catalyzes the first steps of electron transport, namely the transfer of electrons from NADH to **coenzyme Q (CoQ).** This complex is an integral part of the inner mitochondrial membrane and includes, among other subunits, several proteins that contain an iron–sulfur cluster and the flavoprotein that oxidizes NADH. (The total number of subunits is more than 20. This complex is a subject of active research, which has proven to be a challenging task because of its complexity. It is particularly difficult to generalize about the nature of the iron–sulfur clusters because they vary from species to species.) The flavoprotein has a

flavin coenzyme, called flavin mononucleotide, or FMN, which differs from
FAD in not having an adenine nucleotide.

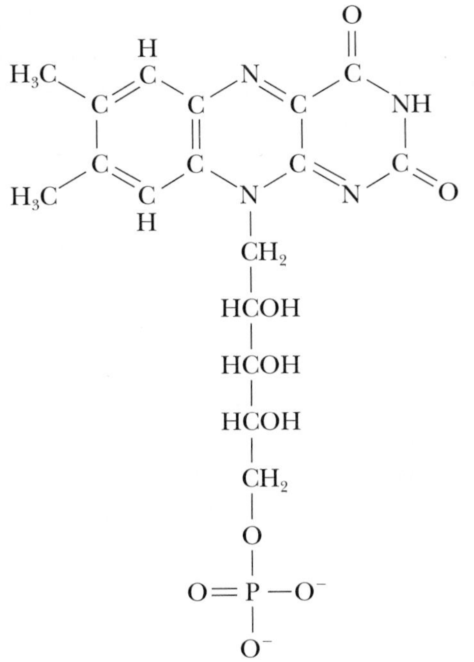

The structure of FMN
(Flavin mononucleotide)

The reaction occurs in several steps, with successive oxidation and reduction
of the flavoprotein and the iron–sulfur moiety. The first step is the transfer of
electrons from NADH to the flavin portion of the flavoprotein:

$$NADH + H^+ + E\text{—}FMN \rightarrow NAD^+ + E\text{—}FMNH_2$$

in which the notation E—FMN indicates that the flavin is covalently bonded
to the enzyme. In the second step, the reduced flavoprotein is reoxidized,
and the oxidized form of the iron–sulfur protein is reduced. The reduced
iron–sulfur protein then donates its electrons to coenzyme Q, which becomes
reduced to $CoQH_2$ (Figure 20.4). Coenzyme Q is also called ubiquinone. The
equations for the second and third steps are shown here:

$$E\text{—}FMNH_2 + 2\ Fe\text{—}S_{oxidized} \rightarrow E\text{—}FMN + 2\ Fe\text{—}S_{reduced} + 2\ H^+$$

$$2\ Fe\text{—}S_{reduced} + CoQ + 2\ H^+ \rightarrow 2\ Fe\text{—}S_{oxidized} + CoQH_2$$

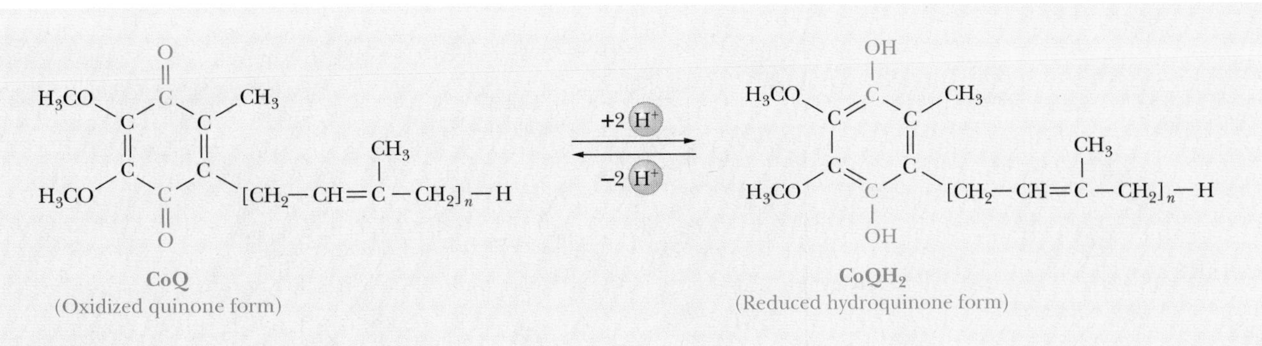

▲ FIGURE 20.4 The oxidized and reduced forms of coenzyme Q. Coenzyme Q is also called
ubiquinone.

The notation Fe—S indicates the iron–sulfur clusters. The overall equation for the reaction is

$$NADH + H^+ + CoQ \rightarrow NAD^+ + CoQH_2$$

This reaction is one of the three responsible for the proton pumping (Figure 20.5) that creates the pH (proton) gradient. The standard free-energy change ($\Delta G^{\circ\prime} = -81$ kJ mol^{-1} = -19.4 kcal mol^{-1}) indicates that the reaction is strongly exergonic, releasing enough energy to drive the phosphorylation of ADP to ATP (Figure 20.6). An important consideration about proton pumping and electron transport is the subtle differences between the electron carriers. Although they can all exist in an oxidized or reduced form, there is an order to which ones will tend to reduce the others, as we saw in Section 20.2. In other words, reduced NADH will donate its electrons to coenzyme Q, but not the other way around. Thus, there is a direction to the electron flow in the complexes we will study.

The other important subtlety is that some of the carriers, such as NADH, carry electrons and hydrogens in their reduced forms, where others, such as the iron–sulfur protein we just saw, can carry only electrons. This is the basis of the proton pumping that ultimately leads to ATP production. When a carrier

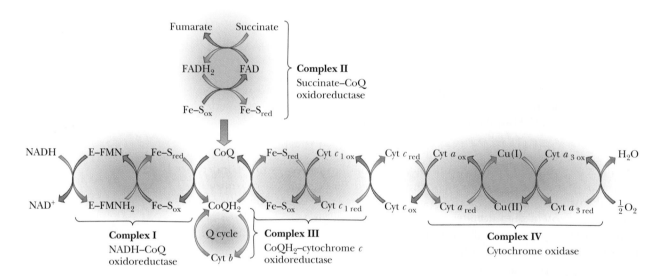

▲ **FIGURE 20.5** The electron transport chain, showing the respiratory complexes. In the reduced cytochromes, the iron is in the Fe(II) oxidation state; in the oxidized cytochromes, the oxygen is in the Fe(III) oxidation state.

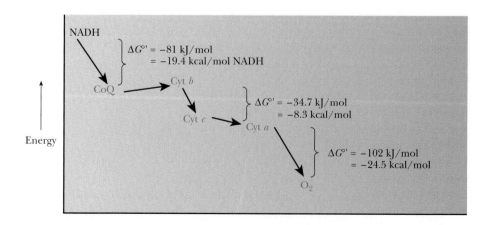

▶ **FIGURE 20.6** The energetics of electron transport.

such as NADH reduces the iron–sulfur protein, it passes along its electrons, but not its hydrogens. The architecture of the inner mitochondrial membrane and the electron carriers allows the hydrogen ions to pass out on the opposite side of the membrane. We shall look more closely at this in Section 20.5.

The final electron receptor of complex I, coenzyme Q, is mobile—that is to say, it is free to move in the membrane and to pass the electrons it has gained to the third complex for further transport to oxygen. We shall now see that the second complex also transfers electrons from an oxidizable substrate to coenzyme Q.

Complex II   The second of the four membrane-bound complexes, **succinate-CoQ oxidoreductase,** also catalyzes the transfer of electrons to coenzyme Q. However, its source of electrons (in other words, the substance being oxidized) differs from the oxidizable substrate (NADH) acted on by NADH-CoQ oxidoreductase. In this case the substrate is succinate from the citric acid cycle, which is oxidized to fumarate by a flavin enzyme (see Figure 20.5).

$$\text{Succinate} + \text{E—FAD} \rightarrow \text{Fumarate} + \text{E—FADH}_2$$

The notation E—FAD indicates that the flavin portion is covalently bonded to the enzyme. The flavin group is reoxidized in the next stage of the reaction as another iron–sulfur protein is reduced:

$$\text{E—FADH}_2 + \text{Fe—S}_{\text{oxidized}} \rightarrow \text{E—FAD} + \text{Fe—S}_{\text{reduced}}$$

This reduced iron–sulfur protein then donates its electrons to oxidized coenzyme Q, and coenzyme Q is reduced.

$$\text{Fe—S}_{\text{reduced}} + \text{CoQ} + 2\,\text{H}^+ \rightarrow \text{Fe—S}_{\text{oxidized}} + \text{CoQH}_2$$

The overall reaction is

$$\text{Succinate} + \text{CoQ} \rightarrow \text{Fumarate} + \text{CoQH}_2$$

We already saw the first step of this reaction when we discussed the oxidation of succinate to fumarate as part of the citric acid cycle. The enzyme traditionally called succinate dehydrogenase, which catalyzes the oxidation of succinate to fumarate (Section 19.3), has been shown by later work to be a part of this enzyme complex. Recall that the succinate dehydrogenase portion consists of a flavoprotein and an iron–sulfur protein. The other components of Complex II are a *b*-type cytochrome and two iron–sulfur proteins. The whole complex is an integral part of the inner mitochondrial membrane. The standard free-energy change ($\Delta G°'$) is -13.5 kJ mol^{-1} = -3.2 kcal mol^{-1}. The overall reaction is exergonic, but there is not enough energy from this reaction to drive ATP production, and no hydrogen ions are pumped out of the matrix during this step.

In further steps of the electron transport chain, electrons are passed from coenzyme Q, which is then reoxidized, to the first of a series of very similar proteins called **cytochromes.** Each of these proteins contains a heme group, and in each heme group the iron is successively reduced to Fe(II) and reoxidized to Fe(III). This situation differs from that of the iron in the heme group of hemoglobin, which remains in the reduced form as Fe(II) through the entire process of oxygen transport in the bloodstream. There are also some structural differences between the heme group in hemoglobin and the heme groups in the various types of cytochromes.

The successive oxidation–reduction reactions of the cytochromes

$$\text{Fe(III)} + e^- \rightarrow \text{Fe(II)} \quad \text{(reduction)}$$

and

$$\text{Fe(II)} \rightarrow \text{Fe(III)} + e^- \quad \text{(oxidation)}$$

differ from one another because the free energy of each reaction, $\Delta G^{\circ\prime}$, differs from the others because of the influences of the various types of hemes and protein structures. Each of the proteins is slightly different in structure, and thus each protein has slightly different properties, including the tendency to participate in oxidation–reduction reactions. The different types of cytochromes are distinguished by lowercase letters (a, b, c); further distinctions are possible with subscripts, as in c_1.

Complex III The third complex, **CoQH$_2$-cytochrome c oxidoreductase** (also called cytochrome reductase), catalyzes the oxidation of reduced coenzyme Q (CoQH$_2$). The electrons produced by this oxidation reaction are passed along to cytochrome c in a multistep process. The overall reaction is

$$CoQH_2 + 2\ Cyt\ c[Fe(III)] \rightarrow CoQ + 2\ Cyt\ c[Fe(II)] + 2\ H^+$$

Recall that the oxidation of coenzyme Q involves two electrons, whereas the reduction of Fe(III) to Fe(II) requires only one electron. Therefore, two molecules of cytochrome c are required for every molecule of coenzyme Q. The components of this complex include cytochrome b (actually two b-type cytochromes, cytochrome b_H and b_L), cytochrome c_1, and several iron–sulfur proteins (Figure 20.5). Cytochromes can carry electrons, but not hydrogens. This is another location where hydrogen ions leave the matrix. When reduced CoQH$_2$ is oxidized to CoQ, the hydrogen ions pass out on the other side of the membrane.

The third complex is an integral part of the inner mitochondrial membrane. Coenzyme Q is soluble in the lipid component of the mitochondrial membrane. It is separated from the complex in the fractionation process that resolves the electron transport apparatus into its component parts, but the coenzyme is probably close to respiratory complexes in the intact membrane (Figure 20.7). Cytochrome c itself is not part of the complex but is loosely bound to the outer surface of the inner mitochondrial membrane, facing the

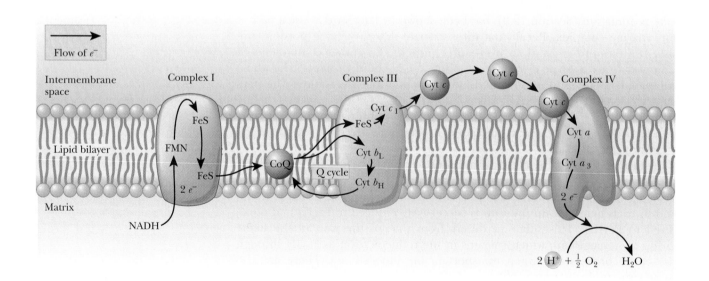

▲ **FIGURE 20.7** The compositions and locations of respiratory complexes in the inner mitochondrial membrane, showing the flow of electrons from NADH to O$_2$. Complex II is not involved and not shown. NADH has accepted electrons from substrates such as pyruvate, isocitrate, α-ketoglutarate, and malate. Note that the binding site for NADH is on the matrix side of the membrane. Coenzyme Q is soluble in the lipid bilayer. Complex III contains two b-type cytochromes, which are involved in the Q cycle. Cytochrome c is loosely bound to the membrane, facing the intermembrane space. In Complex IV, the binding site for oxygen lies on the side toward the matrix.

intermembrane space. It is noteworthy that these two important electron carriers, coenzyme Q and cytochrome c, are not part of the respiratory complexes but can move freely in the membrane. The respiratory complexes themselves move within the membrane (recall lateral motion within membranes from Section 8.3), and electron transport occurs when one complex encounters the next complex in the respiratory chain as they move.

The flow of electrons from reduced coenzyme Q to the other components of the complex does not take a simple, direct path. It is becoming clear that a cyclic flow of electrons involves coenzyme Q twice. This behavior depends on the fact that, as a quinone, coenzyme Q can exist in three forms (Figure 20.8). The semiquinone form, which is intermediate between the oxidized and reduced forms, is of crucial importance here. Because of the crucial involvement of coenzyme Q, this portion of the pathway is called the **Q cycle.**

In part of the Q cycle, *one* electron is passed from reduced coenzyme Q to the iron–sulfur clusters to cytochrome c_1, leaving coenzyme Q in the semiquinone form.

$$CoQH_2 \rightarrow Fe\text{—}S \rightarrow Cyt\ c_1$$

The notation FE—S indicates the iron–sulfur clusters. The series of reactions involving coenzyme Q and cytochrome c_1, but omitting the iron–sulfur proteins, can be written as

$$CoQH_2 + Cyt\ c_1 (oxidized) \rightarrow$$

$$Cyt\ c_1(reduced) + CoQ^- (semiquinone\ anion) + 2\ H^+$$

The semiquinone, along with the oxidized and reduced forms of coenzyme Q, participates in a cyclic process in which the two b cytochromes are reduced and oxidized in turn. A second molecule of coenzyme Q is involved, transferring a second electron to cytochrome c_1, and from there to the mobile carrier cytochrome c. We are going to omit a number of details of the process in the interest of simplicity. Each of the two molecules of coenzyme Q involved in the Q cycle loses one electron. The net result is the same as if one molecule of CoQ had lost two electrons. It is known that one molecule of $CoQH_2$ is regenerated, and one is oxidized to CoQ, which is consistent with this picture. Most importantly, the Q cycle provides a mechanism for electrons to be transferred one at a time from coenzyme Q to cytochrome c_1.

Proton pumping, to which ATP production is coupled, occurs as a result of the reactions of this complex. The Q cycle is implicated in the process, and the whole topic is under active investigation. The standard free-energy change ($\Delta G°'$) is -34.2 kJ $= -8.2$ kcal for each mole of NADH that enters the electron transport chain (see Figure 20.6). The phosphorylation of ADP requires 30.5 kJ mol^{-1} = 7.3 kcal mol^{-1}, and the reaction catalyzed by the third complex supplies enough energy to drive the production of ATP.

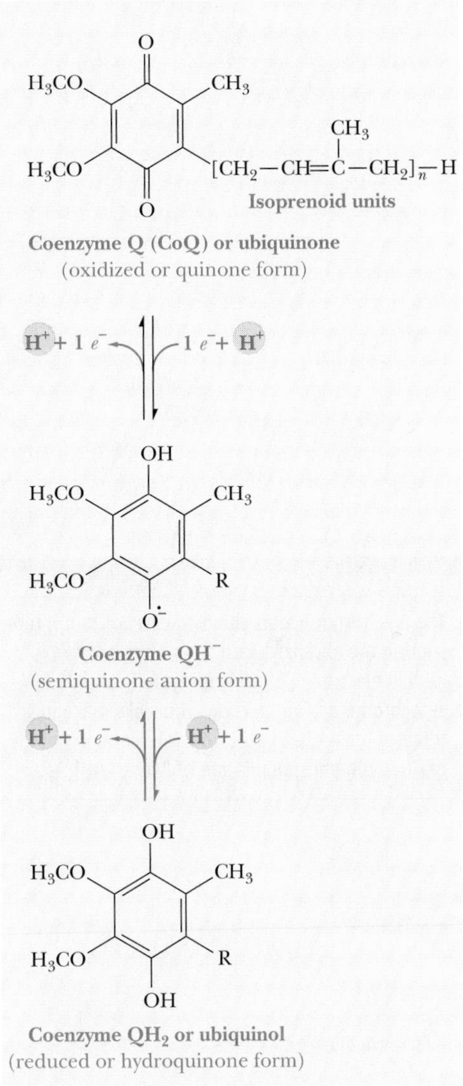

Complex IV The fourth complex, **cytochrome c oxidase,** catalyzes the final steps of electron transport, the transfer of electrons from cytochrome c to oxygen.

The overall reaction is

$$2\ Cyt\ c[Fe(II)] + 2\ H^+ + \tfrac{1}{2} O_2 \rightarrow 2\ Cyt\ c[Fe(III)] + H_2O$$

Proton pumping also takes place as a result of this reaction. Like the other respiratory complexes, cytochrome oxidase is an integral part of the inner mitochondrial membrane and contains cytochromes a and a_3, as well as two Cu^{2+} ions that are involved in the electron transport process. Taken as a whole, this complex contains about ten subunits. In the flow of electrons, the copper ions are intermediate electron acceptors that lie between the two a-type cytochromes in the sequence

$$Cyt\ c \rightarrow Cyt\ a \rightarrow Cu^{2+} \rightarrow Cyt\ a_3 \rightarrow O_2$$

Coenzyme Q (CoQ) or ubiquinone
(oxidized or quinone form)

Coenzyme QH$^-$
(semiquinone anion form)

Coenzyme QH$_2$ or ubiquinol
(reduced or hydroquinone form)

▲ **FIGURE 20.8** The oxidized and reduced forms of coenzyme Q, showing the intermediate semiquinone anion form involved in the Q cycle.

To show the reactions of the cytochromes more explicitly,

$$\text{Cyt } c \text{ [reduced, Fe(II)] + Cyt } aa_3 \text{ [oxidized, Fe(III)]} \rightarrow$$

$$\text{Cyt } aa_3 \text{ [reduced, Fe(II)] + Cyt } c \text{ [oxidized, Fe(III)]}$$

Cytochromes a and a_3 taken together form the complex known as cytochrome oxidase. The reduced cytochrome oxidase is then oxidized by oxygen, which is itself reduced to water. The half-reaction for the reduction of oxygen (oxygen acts as an oxidizing agent) is

$$\tfrac{1}{2}\,O_2 + 2\,H^+ + 2\,e^- \rightarrow H_2O$$

The overall reaction is

$$2\text{ Cyt } aa_3 \text{ [reduced, Fe(II)]} + \tfrac{1}{2}\,O_2 + 2\,H^+ \rightarrow$$

$$2\text{ Cyt } aa_3 \text{ [oxidized, Fe(III)]} + H_2O$$

Note that in this final reaction we have finally seen the link to molecular oxygen in aerobic metabolism.

The standard free-energy change ($\Delta G^{\circ\prime}$) is -110 kJ $= -26.3$ kcal for each mole of NADH that enters the electron transport chain (see Figure 20.6). We have now seen the three places in the respiratory chain where electron transport is coupled to ATP production by proton pumping. These three places are the NADH dehydrogenase reaction, the oxidation of cytochrome b, and the reaction of cytochrome oxidase with oxygen, although the mechanism for proton transfer in cytochrome oxidase remains a mystery. Table 20.2 summarizes the energetics of electron transport reactions.

Cytochromes and Other Iron-Containing Proteins of Electron Transport

In contrast to the electron carriers in the early stages of electron transport, such as NADH, FMN, and CoQ, the cytochromes are macromolecules. These proteins are found in all types of organisms and are typically located in membranes. In eukaryotes, the usual site is the inner mitochondrial membrane, but cytochromes can also occur in the endoplasmic reticulum.

All cytochromes contain the heme group, which is also a part of the structure of hemoglobin and myoglobin (Section 4.5). In the cytochromes, the iron of the heme group does not bind to oxygen; instead, the iron is involved in the series of redox reactions, which we have already seen. There are differences in the side chains of the heme group of the cytochromes involved in the various stages of electron transport (Figure 20.9). These structural differ-

Table 20.2

The Energetics of Electron Transport Reactions

Reaction	$\Delta G^{\circ\prime}$	
	kJ (mol NADH)$^{-1}$	kcal (mol NADH)$^{-1}$
$NADH + H^+ + E{-}FMN \rightarrow NAD^+ + E{-}FMNH_2$	-38.6	-9.2
$E{-}FMNH_2 + CoQ \rightarrow E{-}FMN + CoQH_2$	-42.5	-10.2
$CoQH_2 + 2\text{ Cyt } b\text{[Fe(III)]} \rightarrow CoQ + 2\,H^+ + 2\text{ Cyt } b\text{[Fe(II)]}$	$+11.6$	$+2.8$
$2\text{ Cyt } b\text{[Fe(II)]} + 2\text{ Cyt } c_1\text{[Fe(III)]} \rightarrow 2\text{ Cyt } c_1\text{[Fe(II)]} + 2\text{ Cyt } b\text{[Fe(III)]}$	-34.7	-8.3
$2\text{ Cyt } c_1\text{[Fe(II)]} + 2\text{ Cyt } c\text{[Fe(III)]} \rightarrow 2\text{ Cyt } c\text{[Fe(II)]} + 2\text{ Cyt } c_1\text{[Fe(III)]}$	-5.8	-1.4
$2\text{ Cyt } c\text{[Fe(II)]} + 2\text{ Cyt } (aa_3)\text{ [Fe(III)]} \rightarrow 2\text{ Cyt } (aa_3)\text{ [Fe(II)]} + 2\text{ Cyt } c\text{[Fe(III)]}$	-7.7	-1.8
$2\text{ Cyt } (aa_3)\text{[Fe(II)]} + \tfrac{1}{2}\,O_2 + 2\,H^+ \rightarrow 2\text{ Cyt } (aa_3)\text{[Fe(III)]} + H_2O$	-102.3	-24.5
Overall reaction: $NADH + H^+ + \tfrac{1}{2}\,O_2 \rightarrow NAD^+ + H_2O$	-220	-52.6

(a)

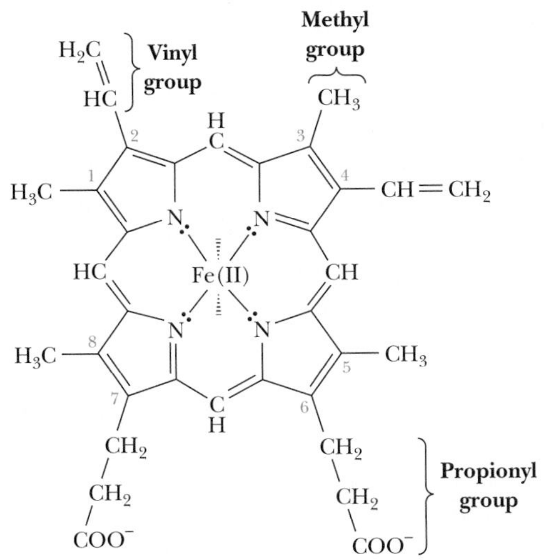

(b)

POSITION	*a* CYTOCHROMES	*c* CYTOCHROMES
1	Same	Same
2 (in *a*)	$-CH-CH_2-(CH_2-CH=C-CH_2)_3H$ with OH and CH_3 substituents	
2 (in *c*)		$-CHCH_3$ $S-$protein (Covalent attachment)
3	Same	Same
4	Same	$-CHCH_3$ $S-$protein
5	Same	Same
6	Same	Same
7	Same	Same
8	$-C=O$ (Formyl group) H	Same

▲ **FIGURE 20.9** The heme group of cytochromes. (a) Structures of the heme of all *b* cytochromes and of hemoglobin and myoglobin. The wedge bonds show the fifth and sixth coordination sites of the iron atom. (b) A comparison of the side chains of *a* and *c* cytochromes to those of *b* cytochromes.

ences, combined with the variations in the polypeptide chain and in the way the polypeptide chain is attached to the heme, account for the differences in properties among the cytochromes in the electron transport chain.

Nonheme iron proteins do not contain a heme group, as their name indicates. Many of the most important proteins in this category contain sulfur, as is the case with the iron–sulfur proteins that are components of the respiratory complexes. The iron is usually bound to cysteine or to S^{2-}. There are still many questions about the location and mode of action of iron–sulfur proteins in mitochondria.

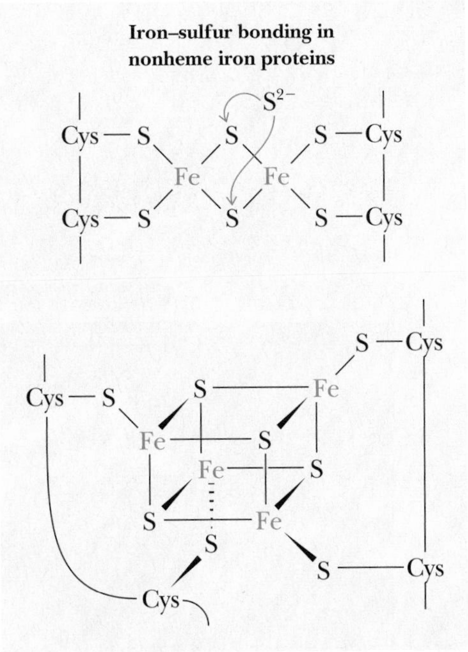

20.4 | What Is the Connection between Electron Transport and Phosphorylation?

Some of the energy released by the oxidation reactions in the electron transport chain is used to drive the phosphorylation of ADP. The phosphorylation of each mole of ADP requires 30.5 kJ = 7.3 kcal, and we have seen how each of the reactions catalyzed by three of the four respiratory complexes provides more than enough energy to drive this reaction, although it is by no means a direct usage of this energy. It is a common theme in metabolism that energy to be used by cells is converted to the chemical energy of ATP as needed. The energy-releasing oxidation reactions give rise to proton pumping and thus to the pH gradient across the inner mitochondrial membrane. In addition to the pH gradient, there is a voltage difference across the membrane generated by the concentration differences of ions inside and out. The energy of the electrochemical potential (voltage drop) across the membrane is converted to the chemical energy of ATP by the coupling process.

A coupling factor is needed to link oxidation and phosphorylation. A complex protein oligomer, separate from the electron transport complexes, serves this function; the complete protein spans the inner mitochondrial membrane and projects into the matrix as well. The portion of the protein that spans the membrane is called F_0. It consists of three different kinds of polypeptide chains (a, b, and c), and research is in progress to characterize it further. The portion that projects into the matrix is called F_1; it consists of five different kinds of polypeptide chains in the ratio $\alpha_3\beta_3\gamma\delta\varepsilon$. Electron micrographs of mitochondria show the projections into the matrix from the inner mitochondrial membrane (Figure 20.10). The schematic organization of the protein

Biochemistry *Now*™

Go to Biochemistry Now and click on Biochemistry Interactive to learn more about the complex that carries out ATP synthesis.

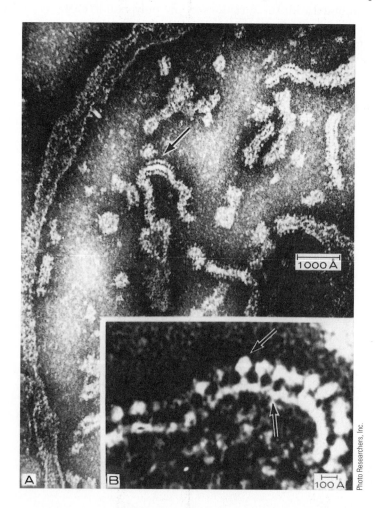

FIGURE 20.10 Electron micrograph of projections into the matrix space of a mitochondrion. Note the difference in scale between part A and part B. The top arrows indicate the matrix side and F_1 subunit. The bottom arrow in Part B indicates the intermembrane space.

1000 Å

100 Å

A   B

Photo Researchers, Inc.

(a)

(b)

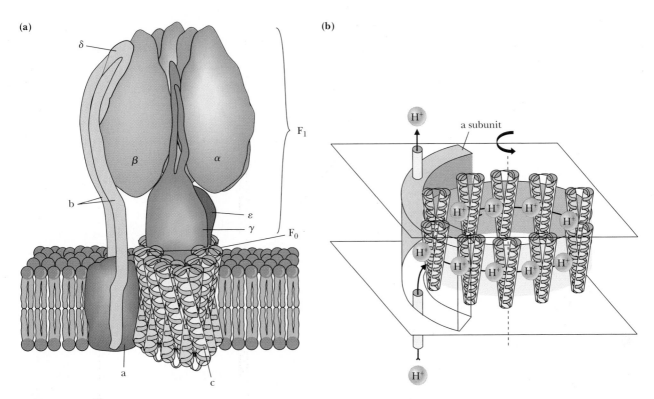

Biochemistry ⊘ Now™ ▲ **ANIMATED FIGURE 20.11** A model of the F_1 and F_0 components of the ATP synthase, a rotating molecular motor. The a, b, α, β, and δ subunits constitute the stator of the motor, and the c, γ, and ε subunits form the rotor. Flow of protons through the structure turns the rotor and drives the cycle of conformational changes in α and β that synthesize ATP. **See this figure animated at http://now.brookscole.com/campbell5**

can be seen in Figure 20.11. The F_1 sphere is the site of ATP synthesis. The whole protein complex is called **ATP synthase.** It is also known as mitochondrial ATPase because the reverse reaction of ATP hydrolysis, as well as phosphorylation, can be catalyzed by the enzyme. The hydrolytic reaction was discovered before the reaction of the synthesis of ATP, hence the name. The 1997 Nobel Prize in chemistry was shared by an American scientist, Paul Boyer of UCLA, and a British scientist, John Walker of the Medical Research Council in Cambridge, England, for their elucidation of the structure and mechanism of this enzyme. (The other half of this prize went to a Danish scientist, Jens Skou, for his work on the sodium–potassium pump [Section 8.6], which also functions as an ATPase.)

Compounds known as **uncouplers** inhibit the phosphorylation of ADP without affecting electron transport. A well-known example of an uncoupler is *2,4-dinitrophenol.* Various antibiotics, such as *valinomycin* and *gramicidin A,* are also uncouplers (Figure 20.12). When mitochondrial oxidation processes are operating normally, electron transport from NADH or $FADH_2$ to oxygen results in the production of ATP. When an uncoupler is present, oxygen is still reduced to H_2O, but ATP is not produced. If the uncoupler is removed, ATP synthesis linked to electron transport resumes.

A term called the **P/O ratio** is used to indicate the coupling of ATP production to electron transport. The P/O ratio gives the number of moles of P_i consumed in the reaction ADP + P_i → ATP for each mole of oxygen atoms consumed in the reaction $\frac{1}{2} O_2 + 2 H^+ + 2 e^- \rightarrow H_2O$. As we have already mentioned, 2.5 moles of ATP are produced when 1 mole of NADH is oxidized to NAD^+. Recall that oxygen is the ultimate acceptor of the electrons from NADH and that $\frac{1}{2}$ mole of O_2 molecules (one mole of oxygen atoms) is

Essential Information

The coupling of electron transport to oxidative phosphorylation requires a multisubunit membrane-bound enzyme, ATP synthase. This enzyme has a channel for protons to flow from the intermembrane space into the mitochondrial matrix. The proton flow is coupled to ATP production in a process that appears to involve a conformational change of the enzyme.

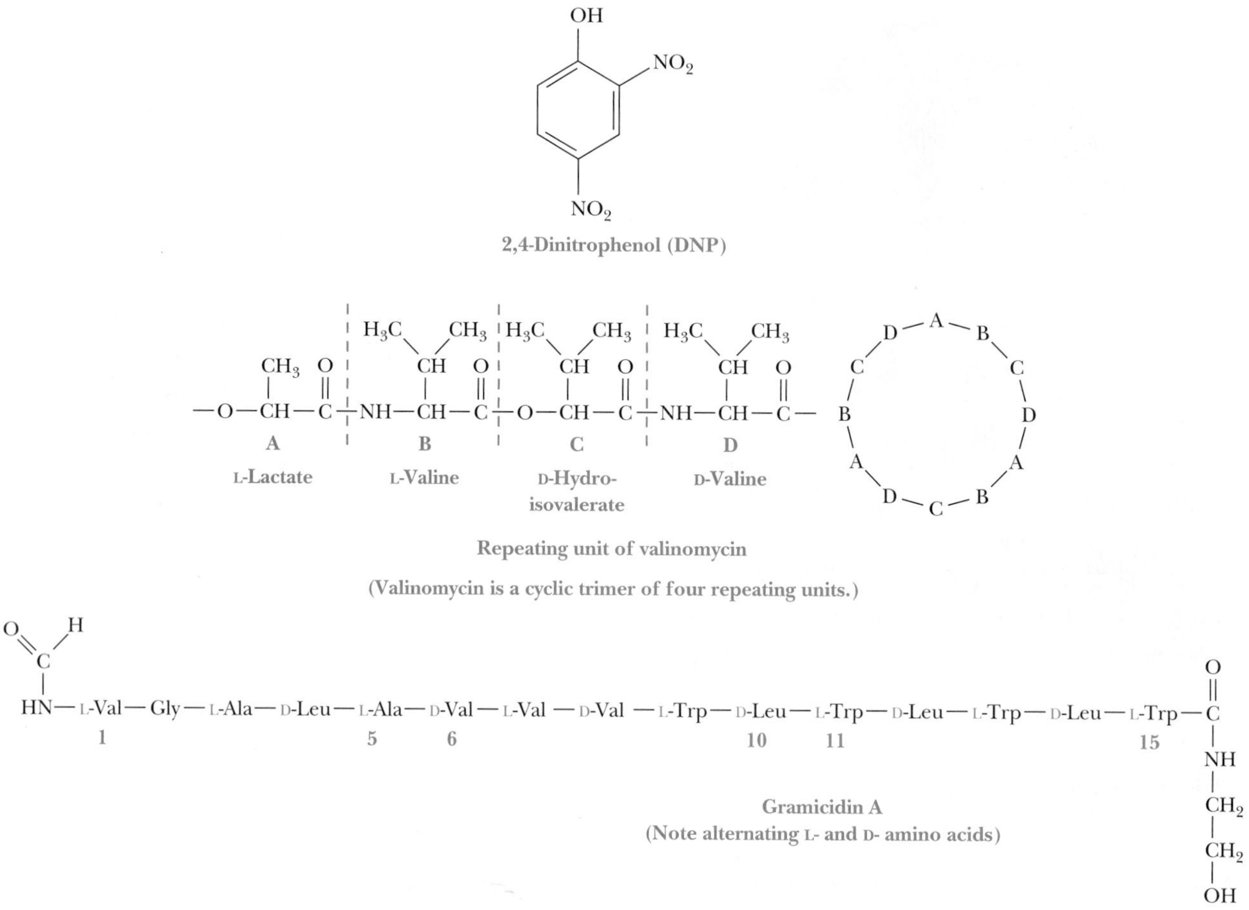
2,4-Dinitrophenol (DNP)
L-Lactate
L-Valine
D-Hydro-isovalerate
D-Valine
Repeating unit of valinomycin
(Valinomycin is a cyclic trimer of four repeating units.)
HN—L-Val—Gly—L-Ala—D-Leu—L-Ala—D-Val—L-Val—D-Val—L-Trp—D-Leu—L-Trp—D-Leu—L-Trp—D-Leu—L-Trp—C
Gramicidin A
(Note alternating L- and D- amino acids)

▲ **FIGURE 20.12** Some uncouplers of oxidative phosphorylation: 2,4-dinitrophenol, valinomycin, and gramicidin A.

reduced for each mole of NADH oxidized. The experimentally determined P/O ratio is 2.5 when NADH is the substrate oxidized. The P/O ratio is 1.5 when $FADH_2$ is the substrate oxidized (also an experimentally determined value). Until recently, biochemists had used integral values of 3 and 2 for the P/O ratios for reoxidation of NADH and $FADH_2$, respectively. The nonintegral consensus values given here clearly underscore the complexity of electron transport, oxidative phosphorylation, and the manner in which they are coupled.

20.5 What Is the Mechanism of Coupling in Oxidative Phosphorylation?

Several mechanisms have been proposed to account for the coupling of electron transport and ATP production. The mechanism that served as the point of departure in all discussions is chemiosmotic coupling, which was later modified to include a consideration of conformational coupling.

Chemiosmotic Coupling

As originally proposed, the **chemiosmotic coupling** mechanism was based entirely on the difference in proton concentration between the intermem-

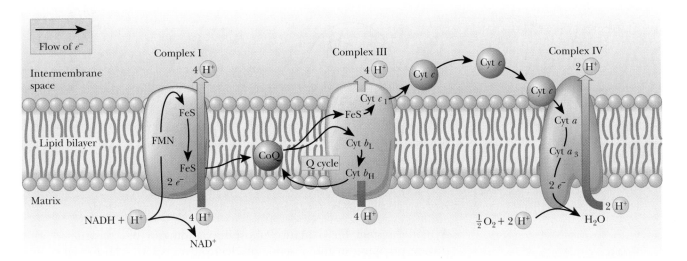

Biochemistry ⊜ **Now**™ ▲ **ACTIVE FIGURE 20.13** The creation of a proton gradient in chemiosmotic coupling. The overall effect of the electron transport reaction series is to move protons (H^+) out of the matrix into the intermembrane space, creating a difference in pH across the membrane. **Watch this Active Figure at http://now.brookscole.com/campbell5**

brane space and the matrix of an actively respiring mitochondrion. In other words, the proton (hydrogen ion, H^+) gradient across the inner mitochondrial membrane is the crux of the matter. The proton gradient exists because the various proteins that serve as electron carriers in the respiratory chain are not symmetrically oriented with respect to the two sides of the inner mitochondrial membrane, nor do they react in the same way with respect to the matrix and the intermembrane space (Figure 20.13). Note that Figure 20.13 repeats the information found in Figure 20.7, with the addition of the flow of protons. The number of protons transported by respiratory complexes is uncertain and even a matter of some controversy. Figure 20.13 shows a consensus estimate for each complex. In the process of electron transport, the proteins of the respiratory complexes take up protons from the matrix to transfer them in redox reactions; these electron carriers subsequently release protons into the intermembrane space when they are reoxidized, creating the proton gradient. As a result, there is a higher concentration of protons in the intermembrane space than in the matrix, which is precisely what we mean by a proton gradient. It is known that the intermembrane space has a lower pH than the matrix, which is another way of saying that there is a higher concentration of protons in the intermembrane space than in the matrix. The proton gradient in turn can drive the production of ATP that occurs when the protons flow back into the matrix.

Since chemiosmotic coupling was first suggested by the British scientist Peter Mitchell in 1961, a considerable body of experimental evidence has accumulated to support it.

1. A system with definite inside and outside compartments (closed vesicles) is essential for oxidative phosphorylation. The process does not occur in soluble preparations or in membrane fragments without compartmentalization.
2. Submitochondrial preparations that contain closed vesicles can be prepared; such vesicles can carry out oxidative phosphorylation, and the asymmetrical orientation of the respiratory complexes with respect to the membrane can be demonstrated (Figure 20.14).

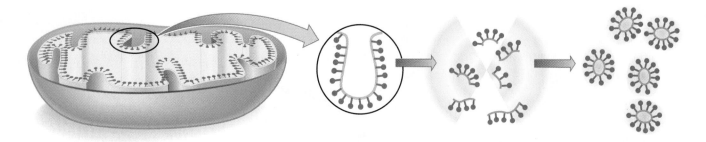

▲ **FIGURE 20.14** Closed vesicles prepared from mitochondria can pump protons and produce ATP.

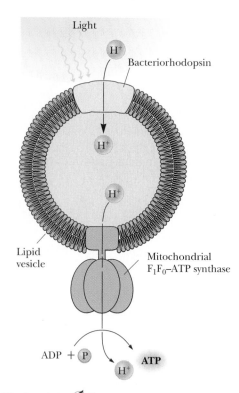

Biochemistry ⑤ Now™ ANIMATED FIGURE 20.15
ATP can be produced by closed vesicles with bacteriorhodopsin as a proton pump. **See this figure animated at http://now.brookscole.com/campbell5**

3. A model system for oxidative phosphorylation can be constructed with proton pumping in the absence of electron transport. The model system consists of reconstituted membrane vesicles, mitochondrial ATP synthase, and a proton pump. The pump is bacteriorhodopsin, a protein found in the membrane of halobacteria. The proton pumping takes place when the protein is illuminated (Figure 20.15).

4. The existence of the pH gradient has been demonstrated and confirmed experimentally.

The way in which the proton gradient leads to the production of ATP depends on ion channels through the inner mitochondrial membrane; these channels are a feature of the structure of ATP synthase. Protons flow back into the matrix through ion channels in the ATP synthase; the F_0 part of the protein is the proton channel. The flow of protons is accompanied by formation of ATP, which takes place in the F_1 unit (Figure 20.16). The unique feature of chemiosmotic coupling is the direct linkage of the proton gradient to the phosphorylation reaction. The details of the way in which phosphoryla-

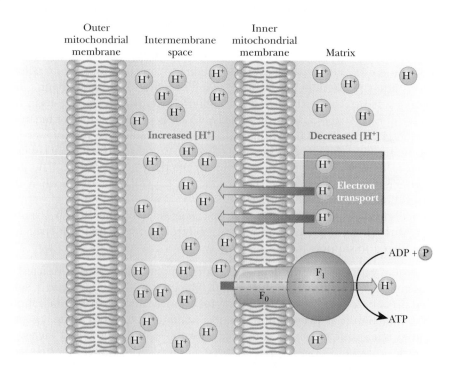

▶ **FIGURE 20.16** Formation of ATP accompanies the flow of protons back into the mitochondrial matrix.

tion takes place as a result of the linkage to the proton gradient are not explicitly specified in this mechanism.

A reasonable mode of action for uncouplers can be proposed in light of the existence of a proton gradient. Dinitrophenol is an acid; its conjugate base, dinitrophenolate anion, is the actual uncoupler because it can react with protons in the intermembrane space, reducing the difference in proton concentration between the two sides of the inner mitochondrial membrane. The antibiotic uncouplers, such as gramicidin A and valinomycin, are **ionophores,** creating a channel through which ions such as H^+, K^+, and Na^+ can pass through the membrane. The proton gradient is overcome, resulting in the uncoupling of oxidation and phosphorylation. The following Biochemical Connections box discusses a natural uncoupler.

Conformational Aspects of Coupling

In **conformational coupling,** the proton gradient is indirectly related to ATP production. The proton gradient leads to conformational changes in a number of proteins, particularly in the ATP synthase itself. It appears from recent evidence that the proton gradient is involved in the release of tightly bound ATP from the synthase as a result of the conformational change (Figure 20.17). There are three sites for substrate on the synthase and three possible conformational states: open (O), with low affinity for substrate; loose-binding (L), which is not catalytically active; and tight-binding (T), which is catalytically active. At any given time, each of the sites is in one of three different conformational states. These states interconvert as a result of the proton flux through the synthase. ATP already formed by the synthase is bound at a site in the T conformation, while ADP and P_i bind at a site in the L conformation. A proton flux converts the site in the T conformation to the O conformation, releasing the ATP. The site at which ADP and P_i are bound assumes the T conformation, which can then give rise to ATP. More recently, it has been shown that the F_1 portion of ATP synthase acts as a rotary motor. The c, γ, and ϵ subunits constitute the rotor, turning within a stationary barrel of a domain consisting of the δ subunit in association with the $\alpha_3\beta_3$ hexamer and the a and b subunits (refer to Figure 20.11 for a detailed picture of the subunits). The γ and ϵ subunits constitute the rotating "shaft" that mediates the energy exchange between the proton flow at F_0 and ATP synthesis at F_1. In essence, the chemical energy of the proton gradient is converted to mechanical energy in the form of the rotating proteins. This mechanical energy is then converted to the chemical energy stored in the high-energy phosphate bonds of ATP.

Electron micrographs have shown that the conformation of the inner mitochondrial membrane and cristae is distinctly different in the resting and active states. This evidence long supported the idea that conformational changes play a role in the coupling of oxidation and phosphorylation. See the article by Stock and Fillingame in the Annotated Bibliography of this chapter for more on the structure of ATPase.

20.6 | How Are Respiratory Inhibitors Used to Study Electron Transport?

If a pipeline is blocked, there will be a backup. Liquid will accumulate upstream of the blockage point, and there will be less liquid downstream. In electron transport, the flow of electrons is from one compound to another rather than along a pipe, but the analogy of a blocked pipeline can be useful

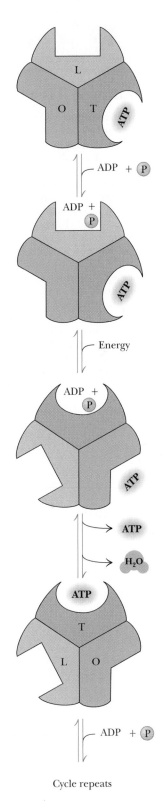

Cycle repeats

Biochemistry ☺ Now™ ANIMATED FIGURE 20.17 The role of conformational change in releasing the ATP from ATP synthase. According to the binding change mechanism, the effect of the proton flux is to cause a conformational change that leads to the release of already formed ATP from ATP synthase. **See this figure animated at http://now.brookscole .com/campbell5**

Biochemical Connections

Brown Adipose Tissue: A Case of Useful Inefficiency

When electron transport generates a proton gradient, some of the energy produced takes the form of heat. There are two situations in which dissipation of energy as heat is useful to organisms: cold-induced nonshivering thermogenesis (production of heat) and diet-induced thermogenesis. Cold-induced nonshivering thermogenesis enables animals to survive in the cold once they have become adapted to such conditions, and diet-induced thermogenesis prevents the development of obesity in spite of prolonged overeating. (Energy is dissipated as heat as food molecules are metabolized instead of being stored as fat.). These two processes may be the same biochemically. It is firmly established that they occur principally, if not exclusively, in brown adipose tissue (BAT), which is rich in mitochondria. (Brown fat takes its color from the large number of mitochondria present in it, unlike the usual white fat cells.) The key to this "inefficient" use of energy in brown adipose tissue appears to be a mitochondrial protein called thermogenin, also referred to as the "uncoupling protein." When this membrane-bound protein is activated in thermogenesis, it serves as a proton channel through the inner mitochondrial membrane. Like all other uncouplers, it "punches a hole" in the mitochondrial membrane and decreases the effect of the proton gradient. Protons flow back into the matrix through thermogenin, bypassing the ATP synthase complex.

Very little research on the biochemistry or physiology of brown adipose tissue has been done in humans. Most of the work on both obesity and adaptation to cold stress has been done on small mammals, such as rats, mice, and hamsters. What role, if any, brown fat deposits play in the development or prevention of obesity in humans is an open question for researchers. Recently, researchers have devoted much energy toward identifying the gene that encodes the uncoupling protein involved in obesity. The ultimate goal of this research is to use the protein or drugs that alter its regulation to combat obesity. Some researchers have also proposed a link between sudden infant death syndrome (SIDS), which is also known as crib death, and metabolism in brown fat tissue. They think that a lack of BAT, or a switch from BAT to normal adipose tissue too early, could lead to body-temperature cooling in a way that could affect the central nervous system.

for understanding the workings of the pathway. When a flow of electrons is blocked in a series of redox reactions, reduced compounds will accumulate before the blockage point in the pathway. Recall that reduction is a gain of electrons, and oxidation represents a loss of electrons. The compounds that come after the blockage point will be lacking electrons and will tend to be found in the oxidized form (Figure 20.18). By using **respiratory inhibitors,** we can gather additional evidence to establish the order of components in the electron transport pathway.

The use of respiratory inhibitors to determine the order of the electron transport chain depends on determining the relative amounts of oxidized and reduced forms of the various electron carriers in intact mitochondria. The logic of the experiment can be seen from the analogy of the blocked pipe. In this case, the reduced form of the carrier upstream (reduced carrier 2) will accumulate because it cannot pass electrons farther in the chain. Likewise, the oxidized form of the carrier downstream (oxidized carrier 3) will also accumulate because the supply of electrons that it could accept has been cut off (Figure 20.18). By use of careful techniques, intact mitochondria can be isolated from cells and can carry out electron transport if an oxidizable substrate is available. If electron transport in mitochondria occurs in the presence and absence of a respiratory inhibitor, there will be different relative amounts of oxidized and reduced forms of the electron carriers.

The type of experiment done to determine the relative amounts of oxidized and reduced forms of electron carriers depends on the spectroscopic properties of these substances. The oxidized and reduced forms of cytochromes can be distinguished from one another. Specialized spectroscopic techniques exist to detect the presence of electron carriers in intact

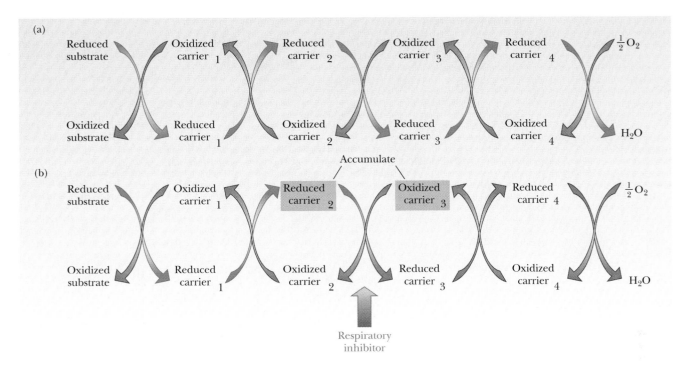

▲ **FIGURE 20.18** The effect of respiratory inhibitors. (a) No inhibitor present. Schematic view of electron transport. The red arrows indicate the flow of electrons. (b) Inhibitor present. The flow of electrons from carrier 2 to carrier 3 is blocked by the respiratory inhibitor. Reduced carrier 2 accumulates, as does oxidized carrier 3, since they cannot react with each other.

mitochondria. The individual types of cytochromes can be identified by the wavelength at which the peak appears, and the relative amounts can be determined from the intensities of the peaks.

There are three sites in the electron transport chain at which inhibitors have an effect, and we shall look at some classic examples. At the first site, barbiturates (of which amytal is an example) block the transfer of electrons from the flavoprotein NADH reductase to coenzyme Q. Rotenone is another inhibitor that is active at this site. This compound is used as an insecticide; it is highly toxic to fish, but not to humans, and is often used to kill the fish in a lake prior to introducing fish of a different species. The second site at which blockage can occur is that of electron transfer involving the b cytochromes, coenzyme Q, and cytochrome c_1. The classic inhibitor associated with this blockage is the antibiotic antimycin A (Figure 20.19). More recently developed inhibitors that are active in this part of the electron transport chain include myxothiazol and 5-n-undecyl-6-hydroxy-4,7-dioxobenzothiazol (UHDBT). These compounds played a role in establishing the existence of the Q cycle. The third site subject to blockage is the transfer of electrons from the cytochrome aa_3 complex to oxygen. Several potent inhibitors operate at this site (Figure 20.20), such as cyanide (CN^-), azide (N_3^-), and carbon monoxide (CO). Note that each of the three sites of action of respiratory inhibitors corresponds to one of the respiratory complexes. Research is continuing with some of the more recently developed inhibitors; the goal of additional work is to elucidate more of the details of the electron transport process.

Essential Information

Respiratory inhibitors block the electron transport chain at sites that correspond to each of the respiratory complexes. Experiments on these substances, many of which are highly toxic, were used to determine the path of electrons in respiration.

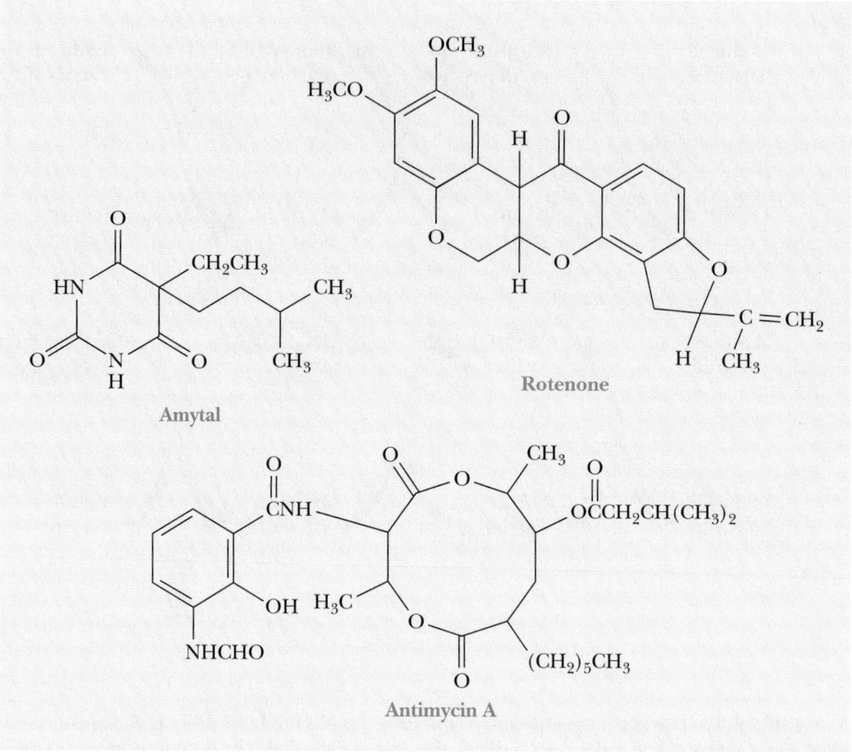

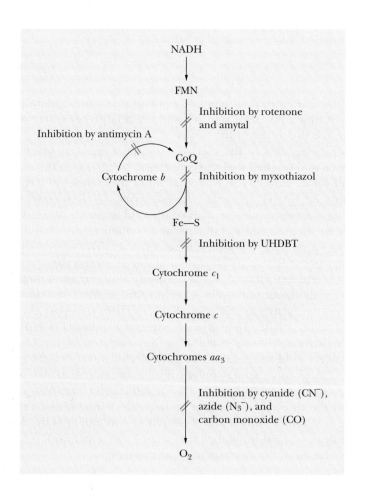

▶ **FIGURE 20.19** Structures of some respiratory inhibitors.

▶ **FIGURE 20.20** Sites of action of some respiratory inhibitors.

20.7 | **What Are Shuttle Mechanisms?**

NADH is produced by glycolysis, which occurs in the cytosol, but NADH in the cytosol cannot cross the inner mitochondrial membrane to enter the electron transport chain. However, the electrons can be transferred to a carrier that can cross the membrane. The number of ATP molecules generated depends on the nature of the carrier, which varies according to the type of cell in which it occurs.

One carrier system that has been extensively studied in insect flight muscle is the **glycerol–phosphate shuttle.** This mechanism uses the presence on the outer face of the inner mitochondrial membrane of an FAD-dependent enzyme that oxidizes glycerol phosphate. The glycerol phosphate is produced by the reduction of dihydroxyacetone phosphate; in the course of the reaction, NADH is oxidized to NAD^+. In this reaction, the oxidizing agent (which is itself reduced) is FAD, and the product is $FADH_2$ (Figure 20.21). The $FADH_2$ then passes electrons through the electron transport chain, leading to the production of 1.5 moles of ATP for each mole of cytosolic NADH. This mechanism has also been observed in mammalian muscle and brain.

A more complex and more efficient shuttle mechanism is the **malate–aspartate shuttle,** which has been found in mammalian kidney, liver, and heart. This shuttle uses the fact that malate can cross the mitochondrial membrane, while oxaloacetate cannot. The noteworthy point about this shuttle mechanism is that the transfer of electrons from NADH in the cytosol produces NADH in the mitochondrion. In the cytosol, oxaloacetate is reduced to malate by the cytosolic malate dehydrogenase, accompanied by the oxidation of cytosolic NADH to NAD^+ (Figure 20.22). The malate then crosses the

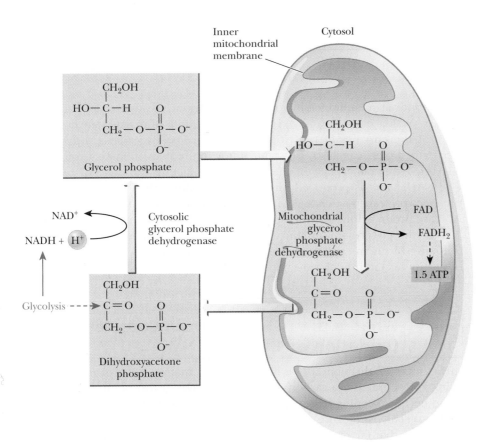

Mitochondrion

◀ **FIGURE 20.21** The glycerol–phosphate shuttle.

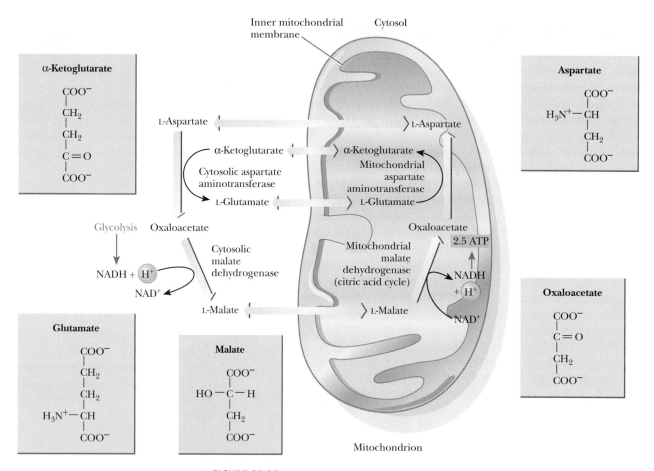

▲ **FIGURE 20.22** The malate–aspartate shuttle.

mitochondrial membrane. In the mitochondrion, the conversion of malate back to oxaloacetate is catalyzed by the mitochondrial malate dehydrogenase (one of the enzymes of the citric acid cycle). Oxaloacetate is converted to aspartate, which can also cross the mitochondrial membrane. Aspartate is converted to oxaloacetate in the cytosol, completing the cycle of reactions. The NADH that is produced in the mitochondrion thus passes electrons to the electron transport chain. With the malate–aspartate shuttle, 2.5 moles of ATP are produced for each mole of cytosolic NADH rather than 1.5 moles of ATP in the glycerol–phosphate shuttle, which uses $FADH_2$ as a carrier. The following Biochemical Connections box discusses some practical applications of our understanding of the catabolic pathways.

20.8 | What Is the ATP Yield from Complete Oxidation of Glucose?

In Chapters 17 through 20, we have discussed many aspects of the complete oxidation of glucose to carbon dioxide and water. At this point, it is useful to do some bookkeeping to see how many molecules of ATP are produced for each molecule of glucose oxidized. Recall that some ATP is produced in glycolysis, but that far more ATP is produced by aerobic metabolism. Table 20.3 summarizes ATP production and also follows the reoxidation of NADH and $FADH_2$.

Table 20.3

Yield of ATP from Glucose Oxidation

Pathway	ATP Yield per Glucose		NADH	FADH$_2$
	Glycerol–Phosphate Shuttle	Malate–Aspartate Shuttle		
Glycolysis: glucose to pyruvate (cytosol)				
Phosphorylation of glucose	−1	−1		
Phosphorylation of fructose-6-phosphate	−1	−1		
Dephosphorylation of 2 molecules of 1,3-BPG	+2	+2		
Dephosphorylation of 2 molecules of PEP	+2	+2		
Oxidation of 2 molecules of glyceraldehyde-3-phosphate yields 2 NADH			+2	
Pyruvate conversion to acetyl-CoA (mitochondria)				
2 NADH produced			+2	
Citric acid cycle (mitochondria)				
2 molecules of GTP from 2 molecules of succinyl-CoA	+2	+2		
Oxidation of 2 molecules each of isocitrate, α-ketoglutarate, and malate yields 6 NADH			+6	
Oxidation of 2 molecules of succinate yields 2 FADH$_2$				+2
Oxidative phosphorylation (mitochondria)				
2 NADH from glycolysis yield 1.5 ATP each if NADH is oxidized by glycerol–phosphate shuttle; 2.5 ATP by malate–aspartate shuttle	+3	+5	−2	
Oxidative decarboxylation of 2 pyruvate to 2 acetyl-CoA: 2 NADH produce 2.5 ATP each	+5	+5	−2	
2 FADH$_2$ from each citric acid cycle produce 1.5 ATP each	+3	+3		−2
6 NADH from citric acid cycle produce 2.5 ATP each	+15	+15	−6	
Net Yield	+30	+32	0	0

(*Note:* These P/O ratios of 2.5 and 1.5 for mitochondrial oxidation of NADH and FADH$_2$ are "consensus values." Since they may not reflect actual values, and since these ratios may change depending on metabolic conditions, these estimates of ATP yield from glucose oxidation are approximate.)

Biochemical Connections

Sports and Metabolism

Trained athletes, especially at the elite level, are more aware of the results of anaerobic and aerobic metabolism than nonathletes. Genetic endowment and training are important to the success of the athlete, but a keen understanding of physiology and metabolism is equally important. To plan nutrition for performance, a serious athlete must understand the nature of metabolism as it relates to his or her chosen sport. A working muscle has four different sources of energy available after a period of rest:

1. Creatine phosphate, which reacts directly with ADP in substrate-level phosphorylation to produce ATP;
2. Glucose from the glycogen of muscle stores, initially consumed by anaerobic metabolism;

3. Glucose from the liver, both from its glycogen stores and the gluconeogenesis from lactic acid produced in the muscle (the Cori cycle), again initially consumed by anaerobic metabolism; and
4. Aerobic metabolism in the muscle mitochondria.

Initially, all four energy sources are available to the muscle. When the creatine phosphate runs out, only the other sources are left. When muscle glycogen runs out, the anaerobic boost it provided slows down, and when the liver glycogen is gone, only aerobic metabolism to carbon dioxide and water is left.

(Continued)

Biochemical Connections (Continued)

It is difficult to make accurate calculations of how much each of these nutrients might supply to a rapidly working muscle, but it is interesting to note that simple calculations are consistent with there being less than a 1-minute supply of creatine phosphate, a figure that can be compared with the length of time for sprint events, which is typically less than a minute. It is interesting to note that creatine supplements for athletes are sold in health food stores, and results suggest that, for power lifting or short sprints, such as the 100-meter dash, this supplementation is effective. There is about 10 to 30 minutes' worth of glycogen in the muscle cells, with this figure varying dramatically based on the intensity of the exercise. Performance in running events ranging in distance from 1500 meters up to 10 kilometers can be heavily influenced by the muscle glycogen levels at the start of the event. Of course, glycogen loading (Chapter 18) could affect this last figure significantly. One reason for the difficulty in making these calculations is the uncertainty of what proportion of the liver glycogen is metabolized only to lactic acid and how much is metabolized in the liver. It is known that one rate-limiting step for aerobic metabolism is the shuttling of both NADH and pyruvate from the cytoplasm into the mitochondrion.

In this regard, it is interesting to note that well-conditioned and well-trained athletes actually have a higher number of mitochondria in their muscle cells. For long-distance events, such as the marathon or road-cycling events, aerobic metabolism certainly comes into play. "Fat burning" is the term frequently used, and it reflects metabolic fact. Fatty acids are degraded to acetyl-CoA, which then enters the citric acid cycle; marathoners and cyclists are known for their notably lean frames, with a minimal amount of body fat. It is interesting to note that running the marathon, which takes between two and three hours for very fit runners, uses more fatty acids, and is done at a lower level of oxygen uptake, than riding in a professional road-cycling event that may take up to 7 hours. Clearly, there are differences in metabolism for sports even within the category known as endurance events.

Perhaps the most studied athlete of modern times is cyclist Lance Armstrong. As a young elite rider, he was a world professional road race champion in 1993 and won a few stages of the prestigious Tour de France. He was powerfully built and excelled in time-trial events and single-day road races, but he was never considered a threat in the major stage races because he did not climb the major European mountains very well. After suffering a disappointing Olympic Games in Atlanta in 1996, he was diagnosed with testicular cancer, which had spread to several organs, including his brain. He was given little chance to live, but, after several surgeries and intense chemotherapy, he recovered and resumed his cycling career. The hospitalization and chemotherapy caused him to lose 15 to 20 pounds, and, in 1998, he became not only competitive but also a true challenger again at the World Championships and the Tour of Spain. Cycling fans were surprised at how well he climbed the Spanish Pyrenees, but few suspected that, in the next few years, he would go on to become the first cyclist to win six Tour de France titles.

Lance Armstrong was the second American to win this event. He always was an amazing aerobic machine, and, when his metab-

▲ Cancer survivor and champion cyclist Lance Armstrong on his way to a third Tour de France victory.

olism didn't have to carry as much weight up the mountains, he was able to climb with the best in the world. Of course, his true strength came from his will to win, which he credits to his ordeal with cancer.

Underscoring the importance of the electron transport chain and mitochondria to the athlete is the story of another great cycling champion, Greg LeMond, who was the first American to win the Tour de France, and he went on to win it a total of three times. Like Lance Armstrong, Greg LeMond also had a tragedy in the middle of his career, a hunting accident. He was shot in the back with buckshot shortly after his first Tour de France victory. Remarkably, he recovered and went on to win two more Tours de France. However, he never really felt well again, and he later commented that, even in his final Tour de France victory in 1990, something was definitely wrong. The next few years were disappointing for LeMond and his fans, and he never made it back into the top places of a race. He seemed to be putting on weight and didn't respond to training. Finally, in desperation, he underwent some painful muscle biopsies and discovered that he had a rare condition called mitochondrial myopathy. When he trained hard, his mitochondria began to disappear. He was essentially an aerobic athlete without the ability to process fuels aerobically. He retired from competition shortly thereafter.

Summary

20.1 What Role Does Electron Transport Play in Metabolism?
In the final stages of aerobic metabolism, electrons are transferred from NADH to oxygen (the ultimate electron acceptor) in a series of oxidation–reduction reactions known as the electron transport chain. This series of events is dependent upon the presence of oxygen in the final step. This pathway allows for the reoxidation of the reduced electron carriers produced in glycolysis, the citric acid cycle, and several other catabolic pathways, and is the true source of the ATPs produced by catabolism.

20.2 What Are the Reduction Potentials for the Electron Transport Chain?
The overall reaction of the electron transport chain shows a very large, negative $\Delta G^{\circ\prime}$ due to the large differences in reduction potentials between the reactions involving NADH and those involving oxygen. If NADH were to reduce oxygen directly, the $\Delta E^{\circ\prime}$ would be more than 1 V. In reality, there are many redox reactions in between, and the correct order of events in the electron transport chain was predicted by comparing the reduction potentials of the individual reactions long before the order was established experimentally.

20.3 How Are the Electron Transport Complexes Organized?
Four separate respiratory complexes can be isolated from the inner mitochondrial membrane. Each of the respiratory complexes can carry out the reactions of a portion of the electron transport chain. In addition to the respiratory complexes, two electron carriers, coenzyme Q and cytochrome c, are not bound to the complexes but are free to move within and along the membrane, respectively. Complex I accomplishes the reoxidation of NADH and sends electrons to Coenzyme Q. Complex II reoxidizes $FADH_2$ and also sends electrons to CoQ. Complex III involves the Q cycle and shuttles electrons to cytochrome c. Complex IV takes the electrons from cytochrome c and passes them to oxygen in the final step of electron transport.

20.4 What Is the Connection between Electron Transport and Phosphorylation?
During the process of electron transport, several reactions occur in which reduced carriers that have both electrons and protons to donate are linked to carriers that can only accept electrons. At these points, hydrogen ions are released to the other side of the inner mitochondrial membrane, causing the formation of a pH gradient. The energy inherent in the charge and chemical separation of the hydrogen ions is used to phosphorylate ADP to ATP when the hydrogen ions pass back into the mitochondria through ATP synthase.

20.5 What Is the Mechanism of Coupling in Oxidative Phosphorylation?
A complex protein oligomer is the coupling factor that links oxidation and phosphorylation. The complete protein spans the inner mitochondrial membrane and projects into the matrix as well. The portion of the protein that spans the membrane is called F_0; it consists of three different kinds of polypeptide chains (a, b, and c).

The portion that projects into the matrix is called F_1; it consists of five different kinds of polypeptide chains (α, β, γ, δ, and ϵ, in the ratio $\alpha_3\beta_3\gamma\delta\epsilon$). The F_1 sphere is the site of ATP synthesis. The whole protein complex is called ATP synthase. It is also known as mitochondrial ATPase. Two mechanisms, the chemiosmotic mechanism and the conformational coupling mechanism, have been proposed to explain the coupling.

Chemiosmotic coupling is the mechanism most widely used to explain the manner in which electron transport and oxidative phosphorylation are coupled to one another. In this mechanism, the proton gradient is directly linked to the phosphorylation process. The way in which the proton gradient leads to the production of ATP depends on ion channels through the inner mitochondrial membrane; these channels are a feature of the structure of ATP synthase. Protons flow back into the matrix through proton channels in the F_0 part of the ATP synthase. The flow of protons is accompanied by formation of ATP, which occurs in the F_1 unit. In the conformational coupling mechanism, the proton gradient is indirectly related to ATP production. It appears from recent evidence that the effect of the proton gradient is not the formation of ATP but the release of tightly bound ATP from the synthase as a result of the conformational change.

20.6 How Are Respiratory Inhibitors Used to Study Electron Transport?
Many of the workings of the electron transport chain have been elucidated by experiments using respiratory inhibitors. These inhibitors specifically block the transfer of electrons at specific points in the respiratory complexes. Examples are CO and CN^-, both of which block the final step of the electron transport chain, and rotenone, which blocks the transfer of electrons from NADH reductase to coenzyme Q. When such a blockage occurs, it causes electrons to "pile up" behind the block, giving a reduced carrier that cannot be oxidized. By noting which carriers become trapped in a reduced state and which ones are trapped in an oxidized state, the link between carriers can be established.

20.7 What Are Shuttle Mechanisms?
Two shuttle mechanisms—the glycerol–phosphate shuttle and the malate–aspartate shuttle—transfer the electrons, but not the NADH, produced in cytosolic reactions into the mitochondrion. In the first of the two shuttles, which is found in muscle and brain, the electrons are transferred to FAD; in the second, which is found in kidney, liver, and heart, the electrons are transferred to NAD^+. With the malate–aspartate shuttle, 2.5 molecules of ATP are produced for each molecule of cytosolic NADH, rather than 1.5 ATP in the glycerol–phosphate shuttle, a point that affects the overall yield of ATP in these tissues.

20.8 What Is the ATP Yield from Complete Oxidation of Glucose?
Approximately 2.5 molecules of ATP are generated for each molecule of NADH that enters the electron transport chain and approximately 1.5 molecules of ATP for each molecule of $FADH_2$. When glucose is metabolized anaerobically, the only net ATPs that are produced are those from the substrate-level phosphorylation steps. This leads to a total of only two ATPs per glucose entering glycolysis. When the pyruvate generated from glycolysis can enter the citric acid cycle, and the resulting NADH and $FADH_2$ molecules are reoxidized through the electron transport chain, a total of 30 or 32 ATPs are produced, with the difference being due to the two possible shuttles.

Critical Questions to Review

20.1 What Role Does Electron Transport Play in Metabolism?

1. **Fact Check** Briefly summarize the steps in the electron transport chain from NADH to oxygen.

2. **Fact Check** Are electron transport and oxidative phosphorylation the same process? Why or why not?

3. **Thought Question** List the reactions of electron transport from NADH to oxygen. Indicate the reactions that liberate enough energy to drive the phosphorylation of ADP.

4. **Thought Question** Show how the reactions of the electron transport chain differ from those in Question 3 when $FADH_2$ is the starting point for electron transport. Show how the reactions that

liberate enough energy to drive the phosphorylation of ADP differ from the pathway when NADH is the starting point.

5. **Thought Question** How does mitochondrial structure contribute to aerobic metabolism, particularly to the integration of the citric acid cycle and electron transport?

20.2 What Are the Reduction Potentials for the Electron Transport Chain?

6. **Fact Check** Why is it reasonable to compare the electron transport process to a battery?

7. **Fact Check** Why are all the reactions in Table 20.1 written as reduction reactions?

8. **Mathematical** Using the information in Table 20.2, calculate $\Delta G^{\circ\prime}$ for the following reaction:

$$2 \text{ Cyt } aa_3 \text{ [oxidized; Fe(III)]} + 2 \text{ Cyt } b\text{[reduced; Fe(II)]} \rightarrow$$
$$2 \text{ Cyt } aa_3 \text{ [reduced; Fe(II)]} + 2 \text{ Cyt } b \text{ [oxidized; Fe(III)]}$$

9. **Mathematical** Calculate $E^{\circ\prime}$ for the following reaction:

$$\text{NADH} + \text{H}^+ + \tfrac{1}{2} \text{O}_2 \rightarrow \text{NAD}^+ + \text{H}_2\text{O}$$

10. **Mathematical** Calculate $E^{\circ\prime}$ for the following reaction:

$$\text{NADH} + \text{H}^+ + \text{Pyruvate} \rightarrow \text{NAD}^+ + \text{Lactate}$$

11. **Mathematical** Calculate $E^{\circ\prime}$ for the following reaction:

$$\text{Succinate} + \tfrac{1}{2} \text{O}_2 \rightarrow \text{Fumarate} + \text{H}_2\text{O}$$

12. **Mathematical** For the following reaction, identify the electron donor and the electron acceptor, and calculate $E^{\circ\prime}$.

$$\text{FAD} + 2 \text{ Cyt } c \text{ (Fe}^{2+}) + 2 \text{ H}^+ \rightarrow \text{FADH}_2 + 2 \text{ Cyt } c \text{ (Fe}^{3+})$$

13. **Mathematical** Which is more favorable energetically, the oxidation of succinate to fumarate by NAD^+ or by FAD? Give the reason for your answer.

14. **Thought Question** Comment on the fact that the reduction of pyruvate to lactate, catalyzed by lactate dehydrogenase, is strongly exergonic (recall this from Chapter 15), even though the standard free-energy change for the half-reaction

$$\text{Pyruvate} + 2 \text{ H}^+ + 2 \text{ } e^- \rightarrow \text{Lactate}$$

is positive ($\Delta G^{\circ\prime} = 36.2 \text{ kJ mol}^{-1} = 8.8 \text{ kcal mol}^{-1}$), indicating an endergonic reaction.

20.3 How Are the Electron Transport Complexes Organized?

15. **Fact Check** What do cytochromes have in common with hemoglobin or myoglobin?

16. **Fact Check** How do the cytochromes differ from hemoglobin and myoglobin in terms of chemical activity?

17. **Fact Check** Which of the following does not play a role in respiratory complexes: cytochromes, flavoproteins, iron–sulfur proteins, or coenzyme Q?

18. **Fact Check** Do any of the respiratory complexes play a role in the citric acid cycle? If so, what is that role?

19. **Fact Check** Do all the respiratory complexes generate enough energy to phosphorylate ADP to ATP?

20. **Thought Question** Two biochemistry students are about to use mitochondria isolated from rat liver for an experiment on oxidative phosphorylation. The directions for the experiment specify addition of purified cytochrome c from any source to the reaction mixture. Why is the added cytochrome c needed? Why does the source not have to be the same as that of the mitochondria?

21. **Thought Question** Cytochrome oxidase and succinate-CoQ oxidoreductase are isolated from mitochondria and are incubated in the presence of oxygen, along with cytochrome c, coenzyme Q,

and succinate. What is the overall oxidation–reduction reaction that can be expected to take place?

22. **Thought Question** What are two advantages of the components of the electron transport chain being embedded in the inner mitochondrial membrane?

23. **Thought Question** Reflect on the evolutionary implications of the structural similarities and functional differences of cytochromes on the one hand and hemoglobin and myoglobin on the other.

24. **Thought Question** Experimental evidence strongly suggests that the protein portions of cytochromes have evolved more slowly (as judged by the number of changes in amino acids per million years) than the protein portions of hemoglobin and myoglobin and even more slowly than hydrolytic enzymes. Suggest a reason why.

25. **Thought Question** What is the advantage of having mobile electron carriers in addition to large membrane-bound complexes of carriers?

26. **Thought Question** What is the advantage of having a Q cycle in electron transport in spite of its complexity?

27. **Thought Question** Why do the electron-transfer reactions of the cytochromes differ in standard reduction potential, even though all the reactions involve the same oxidation–reduction reaction of iron?

28. **Thought Question** Is there a fundamental difference between the one- and two-electron reactions in the electron transport chain?

29. **Thought Question** What is the underlying reason for the differences in spectroscopic properties among the cytochromes?

30. **Thought Question** What would be some of the challenges involved in removing respiratory complexes from the inner mitochondrial membrane in order to study their properties?

20.4 What Is the Connection between Electron Transport and Phosphorylation?

31. **Fact Check** Describe the role of the F_1 portion of ATP synthase in oxidative phosphorylation.

32. **Fact Check** Is mitochondrial ATP synthase an integral membrane protein?

33. **Fact Check** Define the term P/O ratio and indicate why it is important.

34. **Fact Check** In what sense is mitochondrial ATP synthase a motor protein?

35. **Thought Question** What is the approximate P/O ratio that can be expected if intact mitochondria are incubated in the presence of oxygen, along with added succinate?

36. **Thought Question** Why is it difficult to determine an exact number for P/O ratios?

37. **Thought Question** What are some of the difficulties in determining the exact number of protons pumped across the inner mitochondrial membrane by the respiratory complexes?

20.5 What Is the Mechanism of Coupling in Oxidative Phosphorylation?

38. **Fact Check** Briefly summarize the main arguments of the chemiosmotic coupling hypothesis.

39. **Fact Check** Why does ATP production require an intact mitochondrial membrane?

40. **Biochemical Connections** Briefly describe the role of uncouplers in oxidative phosphorylation.

41. **Fact Check** What role does the proton gradient play in chemiosmotic coupling?

42. **Biochemical Connections** Why was dinitrophenol once used as a diet drug?

43. Thought Question Criticize the statement: "The role of the proton gradient in chemiosmosis is to provide the energy to phosphorylate ADP."

20.6 How Are Respiratory Inhibitors Used to Study Electron Transport?

44. Fact Check What is the effect of each of the following substances on electron transport and production of ATP? Be specific about which reaction is affected.

(a) Azide

(b) Antimycin A

(c) Amytal

(d) Rotenone

(e) Dinitrophenol

(f) Gramicidin A

(g) Carbon monoxide

45. Fact Check How can respiratory inhibitors be used to indicate the order of components in the electron transport chain?

46. Thought Question What is the fundamental difference between uncouplers and respiratory inhibitors?

20.7 What Are Shuttle Mechanisms?

47. Fact Check How does the yield of ATP from complete oxidation of one molecule of glucose in muscle and brain differ from that in liver, heart, and kidney? What is the underlying reason for this difference?

48. Thought Question The malate–aspartate shuttle yields about 2.5 moles of ATP for each mole of cytosolic NADH. Why does nature use the glycerol–phosphate shuttle, which yields only about 1.5 moles of ATP?

20.8 What Is the ATP Yield from Complete Oxidation of Glucose?

49. Mathematical What yield of ATP can be expected from complete oxidation of each of the following substrates by the reactions of glycolysis, the citric acid cycle, and oxidative phosphorylation?

(a) Fructose-1,6-*bis*phosphate

(b) Glucose

(c) Phosphoenolpyruvate

(d) Glyceraldehyde-3-phosphate

(e) NADH

(f) Pyruvate

50. Mathematical The free-energy change ($\Delta G°'$) for the oxidation of the cytochrome aa_3 complex by molecular oxygen is -102.3 kJ $= -24.5$ kcal for each mole of electron pairs transferred. What is the maximum number of moles of ATP that could be produced in the process? How many moles of ATP are actually produced? What is the efficiency of the process, expressed as a percentage?

Biochemistry ⊜ Now™

Assess your understanding of this chapter's topics with additional quizzing and tutorials at **http://now.brookscole.com/campbell5**

Annotated Bibliography

Cannon, B., and J. Nedergaard. The Biochemistry of an Inefficient Tissue: Brown Adipose Tissue. *Essays in Biochemistry* **20,** 110–164 (1985). [A review describing the usefulness to mammals of the "inefficient" production of heat in brown fat.]

Dickerson, R. E. Cytochrome *c* and the Evolution of Energy Metabolism. *Sci. Amer.* **242** (3), 136–152 (1980). [An account of the evolutionary implications of cytochrome *c* structure.]

Fillingame, R. The Proton-Translocating Pumps of Oxidative Phosphorylation. *Ann. Rev. Biochem.* **49,** 1079–1114 (1980). [A review of chemiosmotic coupling.]

Fillingame, R. H. Molecular Rotary Motors. *Science* **286,** 1687–1688 (1999). [A review of research on ATP synthase.]

Hatefi, Y. The Mitochondrial Electron Transport and Oxidative Phosphorylation System. *Ann. Rev. Biochem.* **54,** 1015–1069 (1985). [A review that emphasizes the coupling between oxidation and phosphorylation.]

Hinkle, P. C., and R. E. McCarty. How Cells Make ATP. *Sci. Amer.* **238** (3), 104–123 (1978). [Chemiosmotic coupling and the mode of action of uncouplers.]

Lane, M. D., P. L. Pedersen, and A. S. Mildvan. The Mitochondrion Updated. *Science* **234,** 526–527 (1986). [A report on an international conference on bioenergetics and energy coupling.]

Mitchell, P. Keilin's Respiratory Chain Concept and Its Chemiosmotic Consequences. *Science* **206,** 1148–1159 (1979). [A Nobel Prize lecture by the scientist who first proposed the chemiosmotic coupling hypothesis.]

Moser, C. C., et al. Nature of Biological Electron Transfer. *Nature* **355,** 796–802 (1992). [An advanced treatment of electron transfer in biological systems.]

Stock, D., A. G. W. Leslie, and J. F. Walker. Molecular Architecture of the Rotary Motor in ATP Synthase. *Science* **286,** 1700–1705 (1999). [An article about the structure and function of ATP synthase.]

Trumpower, B. The Protonmotive Q Cycle: Energy Transduction by Coupling of Proton Translocation to Electron Transfer by the Cytochrome *bc*1 Complex. *J. Biol. Chem.* **265,** 11409–11412 (1990). [An advanced article that goes into detail about the Q cycle.]

Vignais, P. V., and J. Lunardi. Chemical Probes of the Mitochondrial ATP Synthesis and Translocation. *Ann. Rev. Biochem.* **54,** 977–1014 (1985). [A review about the synthesis and use of ATP.]

CHAPTER 21

Lipid Metabolism

Lipid metabolism allows polar bears to thrive in arctic climates and to endure months of hibernation.

© Wayne R. Bilenduke/The Image Bank/Getty Images

Biochemistry ⑤ Now™

Test yourself on these Critical Questions at the BiochemistryNow website at **http://now .brookscole.com/campbell5**

In the energy economy of the cell, glucose reserves are like ready cash, whereas lipid reserves are like a fat savings account. The potential energy of lipids resides in the fatty-acid chains of triacylglycerols. When there are excess calories, fatty acids are synthesized and stored in fat cells. When energy demands are great, fatty acids are catabolized to liberate energy. The synthesis of fatty acids begins with acetyl-CoA, after which carbon atoms are added to the growing hydrocarbon chain, usually two at a time. Catabolism proceeds in the opposite direction—beginning with the carboxyl group, converting long-chain fatty acyl groups to acetyl-CoA, which is subsequently oxidized in the citric acid cycle. Acetyl-CoA is oxidized in the citric acid cycle to provide energy that is temporarily stored as ATP. Pathways of lipid synthesis and catabolism of lipids occur simultaneously, but in different parts of the cell. If these pathways were to operate at the same time in the same place, the lipids would be taken apart as soon as they were synthesized. To avoid such a futile cycle, the pathways of lipid synthesis and catabolism differ in important ways: (1) synthesis takes place in the cytosol, while catabolism takes place in the mitochondrion; (2) NADPH is the donor of high-energy electrons in lipid synthesis, whereas FAD and NAD^+ are electron acceptors in lipid catabolism; and (3) different activating ligands are used (for example, coenzyme A is used in catabolism, while the acyl carrier protein is used in anabolism).

21.1 | How Are Lipids Involved in the Generation and Storage of Energy?

In the past few chapters we have seen how energy can be released by the catabolic breakdown of carbohydrates in aerobic and anaerobic processes. In Chapter 16, we saw that there are carbohydrate polymers (such as starch in plants and glycogen in animals) that represent stored energy, in the sense that these carbohydrates can be hydrolyzed to monomers and then oxidized to provide energy in response to the needs of an organism. In this chapter, we shall see how the metabolic oxidation of lipids releases large quantities of energy through production of acetyl-CoA, NADH, and $FADH_2$ and how lipids represent an even more efficient way of storing chemical energy.

21.2 | How Are Lipids Catabolized?

The oxidation of fatty acids is the chief source of energy in the catabolism of lipids; in fact, lipids that are sterols (steroids that have a hydroxyl group as part of their structure; Section 8.2) are not catabolized as a source of energy but are excreted. Both triacylglycerols, which are the main storage form of the chemical energy of lipids, and phosphoacylglycerols, which are important components of biological membranes, have fatty acids as part of their covalently bonded structures. In both types of compounds, the bond between the fatty acid and the rest of the molecule can be hydrolyzed (Figure 21.1), with the reaction catalyzed by suitable groups of enzymes—**lipases,** in the case of triacylglycerols (Section 8.2), and **phospholipases,** in the case of phosphoacylglycerols.

O
‖
CH₂OCR

O
‖
CHOCR   → H₂O Lipases →

O
‖
CH₂OCR

Triacylglycerol

Glycerol

O
‖
RCO⁻

Free fatty acids

← Phospholipases H₂O ←

Glycerylphosphorylcholine

Reuse or oxidation

O
‖
CH₂OCR

O
‖
CHOCR

O
‖
CH₂OPOCH₂CH₂N⁺(CH₃)₃
|
O⁻

Phosphatidylcholine

▲ **FIGURE 21.1** The release of fatty acids for future use. The source of fatty acids can be a triacylglycerol (left) or a phospholipid such as phosphatidylcholine (right).

Several different phospholipases can be distinguished on the basis of the site at which they hydrolyze phospholipids (Figure 21.2). Phospholipase A₂ is widely distributed in nature; it is also being actively studied by biochemists interested in its structure and mode of action, which involves hydrolysis of phospholipids at the surface of micelles (Section 2.1). Phospholipase D occurs in spider venom and is responsible for the tissue damage that accompanies spider bites. Snake venoms also contain phospholipases; the concentration of phospholipases is particularly high in venoms with comparatively low concentrations of the toxins (usually small peptides) that are characteristic of some kinds of venom. The lipid products of hydrolysis act to lyse red blood cells, preventing clot formation. Snakebite victims bleed to death in this situation.

The release of fatty acids from triacylglycerols in adipocytes is controlled by hormones. In a scheme that will look familiar from our discussions of carbohydrate metabolism, a hormone binds to a receptor on the plasma membrane of the adipocyte (Figure 21.3). This hormone binding activates adenylate cyclase, which leads to production of active protein kinase A (cAMP-dependent protein kinase). Protein kinase phosphorylates triacylglycerol lipase, which cleaves the fatty acids from the glycerol backbone. The main hormone that has this effect is epinephrine. Caffeine also mimics epinephrine in this regard, which is one reason competitive runners often drink caffeine the morning of a race. Distance runners want to burn fat more efficiently to spare their carbohydrate stores for the later stages of the race.

Fatty-acid oxidation begins with **activation** of the molecule. In this reaction, a thioester bond is formed between the carboxyl group of the fatty acid and the thiol group of coenzyme A (CoA-SH). The activated form of the fatty acid is an acyl-CoA, the exact nature of which depends on the nature of the fatty acid itself. Keep in mind throughout this discussion that all acyl-CoA molecules are thioesters, since the fatty acid is esterified to the thiol group of CoA. The enzyme that catalyzes formation of the ester bond, an *acyl-CoA synthetase*, requires ATP for its action. In the course of the reaction, an acyl adenylate intermediate is formed. The acyl group is then transferred to CoA-SH. ATP is converted to AMP and PPᵢ, rather than to ADP and Pᵢ. The PPᵢ is hydrolyzed to 2 Pᵢ; the hydrolysis of two high-energy phosphate bonds provides energy for the activation of the fatty acid and is equivalent to the use of two ATP. The formation of the acyl-CoA is endergonic without the energy provided by the hydrolysis of the two high-energy bonds. Note also that the hydrolysis of ATP to AMP and two Pᵢ represents an increase in entropy (Figure 21.4). There are

A phosphoacylglycerol

▲ **FIGURE 21.2** Several phospolipases hydrolyze phosphoacylglycerols. They are designated A₁, A₂, C, and D. Their sites of action are shown. The site of action of phospholipase A₂ is the B site, and the name phospholipase A₂ is the result of historical accident (see text).

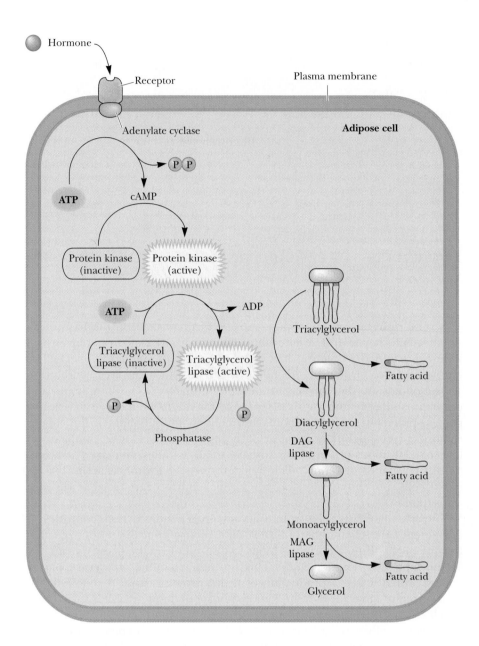

▶ **FIGURE 21.4** The formation of an acyl-CoA.

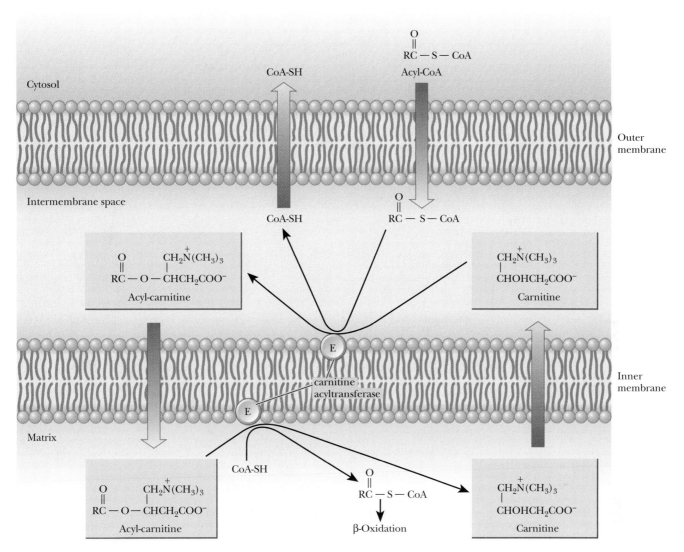

▲ **FIGURE 21.5** The role of carnitine in the transfer of acyl groups to the mitochondrial matrix.

several enzymes of this type, some specific for longer-chain fatty acids and some for shorter-chain fatty acids. Both saturated and unsaturated fatty acids can serve as substrates for these enzymes. The esterification takes place in the cytosol, but the rest of the reactions of fatty-acid oxidation occur in the mitochondrial matrix. The activated fatty acid must be transported into the mitochondrion so that the rest of the oxidation process can proceed.

The acyl-CoA can cross the outer mitochondrial membrane but not the inner membrane (Figure 21.5). In the intermembrane space, the acyl group is transferred to **carnitine** by transesterification; this reaction is catalyzed by the enzyme **carnitine acyltransferase,** which is located in the inner membrane. Acyl carnitine, a compound that can cross the inner mitochondrial membrane, is formed. This enzyme has a specificity for acyl groups between 14 and 18 carbons long and is often called **carnitine palmitoyltransferase (CPT-I)** for this reason. The acylcarnitine passes through the inner membrane via a specific carnitine/acylcarnitine transporter called **carnitine translocase.** Once in the matrix, the acyl group is transferred from carnitine to

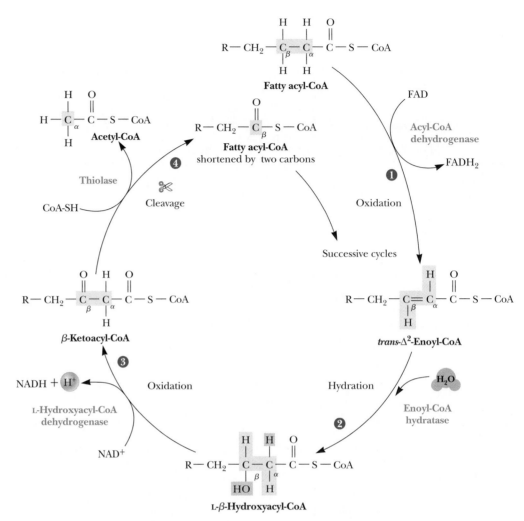

Biochemistry⊘Now™ ▲ **ACTIVE FIGURE 21.6** The β-oxidation of saturated fatty acids involves a cycle of four enzyme-catalyzed reactions. Each cycle produces one $FADH_2$ and one NADH, and it liberates acetyl-CoA, resulting in a fatty acid that is two carbons shorter. The Δ symbol represents a double bond, and the number associated with it is the location of the double bond (based on counting the carbonyl group as carbon one). **Watch this Active Figure at http://now.brookscole.com/campbell5**

mitochondrial CoA-SH by another transesterification reaction, involving a second carnitine palmitoyltransferase (**CPT-II**) located on the inner face of the membrane.

In the matrix, a repeated sequence of reactions successively cleaves two-carbon units from the fatty acid, starting from the carboxyl end. This process is called **β-oxidation,** since the oxidative cleavage takes place at the β-carbon of the acyl group esterified to CoA. The β-carbon of the original fatty acid becomes the carboxyl carbon in the next stage of degradation. The whole cycle requires four reactions (Figure 21.6).

1. The acyl-CoA is *oxidized* to an α,β unsaturated acyl-CoA (also called a β-enoyl-CoA). The product has the *trans* arrangement at the double bond. This reaction is catalyzed by an FAD-dependent acyl-CoA dehydrogenase.
2. The unsaturated acyl-CoA is *hydrated* to produce a β-hydroxyacyl-CoA. This reaction is catalyzed by the enzyme enoyl-CoA hydratase.

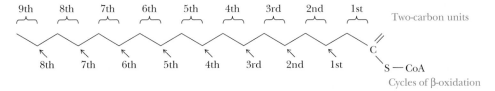

9th   8th   7th   6th   6th   5th   4th   3rd   2nd   1st   Two-carbon units

8th   7th   6th   5th   4th   3rd   2nd   1st

S — CoA

Cycles of β-oxidation

▲ **FIGURE 21.7** Stearic acid (18 carbons) gives rise to nine 2-carbon units after eight cycles of β-oxidation. The ninth 2-carbon unit remains esterified to CoA after eight cycles of β-oxidation have removed eight successive two-carbon units, starting at the carboxyl end on the right. Thus, it takes only eight rounds of β-oxidation to completely process an 18-carbon fatty acid to acetyl-CoA.

3. A second *oxidation* reaction is catalyzed by β-hydroxyacyl-CoA dehydrogenase, an NAD^+-dependent enzyme. The product is a β-ketoacyl-CoA.
4. The enzyme thiolase catalyzes the *cleavage* of the β-ketoacyl-CoA; a molecule of CoA is required for the reaction. The products are acetyl-CoA and an acyl-CoA that is two carbons shorter than the original molecule that entered the β-oxidation cycle. The CoA is needed in this reaction to form the new thioester bond in the smaller acyl-CoA molecule. This smaller molecule then undergoes another round of the β-oxidation cycle.

When a fatty acid with an even number of carbon atoms undergoes successive rounds of the β-oxidation cycle, the product is acetyl-CoA. (Fatty acids with even numbers of carbon atoms are the ones normally found in nature, so acetyl-CoA is the usual product of fatty-acid catabolism.) The number of molecules of acetyl-CoA produced is equal to half the number of carbon atoms in the original fatty acid. For example, stearic acid contains 18 carbon atoms and gives rise to nine molecules of acetyl-CoA. Note that the conversion of one 18-carbon stearic acid molecule to nine 2-carbon acetyl units requires eight, not nine, cycles of β-oxidation (Figure 21.7). The acetyl-CoA enters the citric acid cycle, with the rest of the oxidation of fatty acids to carbon dioxide and water taking place through the citric acid cycle and electron transport. Recall that most of the enzymes of the citric acid cycle are located in the mitochondrial matrix, and we have just seen that the β-oxidation cycle takes place in the matrix as well. In addition to mitochondria, other sites of β-oxidation are known. Peroxisomes and glyoxysomes (Section 1.6), organelles that carry out oxidation reactions, are also sites of β-oxidation, albeit to a far lesser extent than the mitochondria. Certain drugs, called hypolipidemic drugs, are used in an attempt to control obesity. Some of these work by stimulating β-oxidation in peroxisomes.

Biochemistry ⓔ Now™

Go to BiochemistryNow and click on Biochemistry Interactive to discover the main functions of Coenzyme A.

Essential Information

The breakdown of fatty acids takes place in the mitochondrial matrix and proceeds by successive removal of two-carbon units as acetyl-CoA. Each cleavage of a two-carbon moiety requires a four-step reaction sequence called β-oxidation. The complete oxidation of fatty acids by the citric acid cycle and the electron-transport chain releases large amounts of energy.

| 21.3 | **What Is the Energy Yield from the Oxidation of Fatty Acids?** |

In carbohydrate metabolism, the energy released by oxidation reactions is used to drive the production of ATP, with most of the ATP produced in aerobic processes. In the same aerobic processes—namely, the citric acid cycle and oxidative phosphorylation—the energy released by the oxidation of acetyl-CoA formed by β-oxidation of fatty acids can also be used to produce ATP. There are two sources of ATP to keep in mind when calculating the overall yield of ATP. The first source is the reoxidation of the NADH and $FADH_2$ produced by the β-oxidation of the fatty acid to acetyl-CoA. The second source is

ATP production from the processing of the acetyl-CoA through the citric acid cycle and oxidative phosphorylation. We shall use the oxidation of stearic acid, which contains 18 carbon atoms, as our example.

Eight cycles of β-oxidation are required to convert one mole of stearic acid to nine moles of acetyl-CoA; in the process eight moles of FAD are reduced to $FADH_2$ and eight moles of NAD^+ are reduced to NADH.

$$CH_3(CH_2)_{16}\overset{\overset{\displaystyle O}{\|}}{C} - S - CoA + 8\ FAD + 8\ NAD^+ + 8\ H_2O + 8\ CoA\text{-}SH \rightarrow$$

$$9\ CH_3 - \overset{\overset{\displaystyle O}{\|}}{C} - S - CoA + 8\ FADH_2 + 8\ NADH + 8\ H^+$$

The nine moles of acetyl-CoA produced from each mole of stearic acid enter the citric acid cycle. One mole of $FADH_2$ and three moles of NADH are produced for each mole of acetyl-CoA that enters the citric acid cycle. At the same time, one mole of GDP is phosphorylated to produce GTP for each turn of the citric acid cycle.

$$9\ CH_3\overset{\overset{\displaystyle O}{\|}}{C} - S - CoA + 9\ FAD + 27\ NAD^+ + 9\ GDP + 9\ P_i + 27\ H_2O \rightarrow$$

$$18\ CO_2 + 9\ CoA\text{-}SH + 9\ FADH_2 + 27\ NADH + 9\ GTP + 27\ H^+$$

The $FADH_2$ and NADH produced by β-oxidation and by the citric acid cycle enter the electron transport chain, and ATP is produced by oxidative phosphorylation. In our example, there are 17 moles of $FADH_2$, 8 from β-oxidation, and 9 from the citric acid cycle; there are also 35 moles of NADH, 8 from β-oxidation and 27 from the citric acid cycle. Recall that 2.5 moles of ATP are produced for each mole of NADH that enters the electron transport chain, and 1.5 moles of ATP result from each mole of $FADH_2$. Because $17 \times 1.5 = 25.5$ and $35 \times 2.5 = 87.5$, we can write the following equations:

$$17\ FADH_2 + 8.5\ O_2 + 25.5\ ADP + 25.5\ P_i \rightarrow 17\ FAD + 25.5\ ATP + 17\ H_2O$$

$$35\ NADH + 35\ H^+ + 17.5\ O_2 + 87.5\ ADP + 87.5\ P_i \rightarrow$$

$$35\ NAD^+ + 87.5\ ATP + 35\ H_2O$$

The overall yield of ATP from the oxidation of stearic acid can be obtained by adding the equations for β-oxidation, for the citric acid cycle, and for oxidative phosphorylation. In this calculation, we take GDP as equivalent to ADP and GTP as equivalent to ATP, which means that the equivalent of nine ATP must be added to those produced in the reoxidation of $FADH_2$ and NADH. There are nine ATP equivalent to the nine GTP from the citric acid cycle, 25.5 ATP from the reoxidation of $FADH_2$, and 87.5 ATP from the reoxidation of NADH, for a grand total of 122 ATP.

$$CH_3(CH_2)_{16}\overset{\overset{\displaystyle O}{\|}}{C} - S - CoA + 26\ O_2 + 122\ ADP + 122\ P_i \rightarrow$$

$$18\ CO_2 + 17\ H_2O + 122\ ATP + CoA\text{-}SH$$

Table 21.1			
The Balance Sheet for Oxidation of One Molecule of Stearic Acid			
Reaction	**NADH Molecules**	**FADH$_2$ Molecules**	**ATP Molecules**
1. Stearic acid → Stearyl-CoA (activation step)			−2
2. Stearyl-CoA → 9 acetyl-CoA (8 cycles of β-oxidation)	+8	+8	
3. 9 Acetyl-CoA → 18 CO$_2$ (citric acid cycle); GDP → GTP (9 molecules)	+27	+9	+9
4. Reoxidation of NADH from β-oxidation cycle	−8		+20
5. Reoxidation of NADH from citric acid cycle	−27		+67.5
6. Reoxidation of FADH$_2$ from β-oxidation cycle		−8	+12
7. Reoxidation of FADH$_2$ from citric acid cycle		−9	+13.5
	0	0	+120

Note that there is no net change in the number of molecules of NADH or FADH$_2$.

George Holton/Photo Researchers, Inc.

Tom McHugh/Photo Researchers, Inc.

The activation step in which stearyl-CoA was formed is not included in this calculation, and we must subtract the ATP that was required for that step. Even though only one ATP was required, two high-energy phosphate bonds are lost because of the production of AMP and PP$_i$. The pyrophosphate must be hydrolyzed to phosphate (P$_i$) before it can be recycled in metabolic intermediates. As a result, we must subtract the equivalent of two ATP for the activation step. The net yield of ATP becomes 120 moles of ATP for each mole of stearic acid that is completely oxidized. See Table 21.1 for a balance sheet. Keep in mind that these values are theoretical consensus values that not all cells attain.

As a comparison, note that 32 moles of ATP can be obtained from the complete oxidation of one mole of glucose; but glucose contains six, rather than 18, carbon atoms. Three glucose molecules contain 18 carbon atoms, and a more interesting comparison is the ATP yield from the oxidation of three glucose molecules, which is $3 \times 32 = 96$ ATP for the same number of carbon atoms. The yield of ATP from the oxidation of the lipid is still higher than that from the carbohydrate, even for the same number of carbon atoms. The reason is that a fatty acid is all hydrocarbon except for the carboxyl group; that is, it exists in a highly reduced state. A sugar is already partly oxidized because of the presence of its oxygen-containing groups. Because the oxidation of a fuel leads to the reduced electron carriers used in the electron transport chain, a more reduced fuel, such as a fatty acid, can be oxidized further than a partially oxidized fuel, such as a carbohydrate.

Another point of interest is that water is produced in the oxidation of fatty acids. We have already seen that water is also produced in the complete oxidation of carbohydrates. The production of **metabolic water** is a common feature of aerobic metabolism. This process can be a source of water for organisms that live in desert environments. Camels are a well-known example; the stored lipids in their humps are a source of both energy and water during long trips through the desert. Kangaroo rats provide an even more striking example of adaptation to an arid environment. These animals have been observed to live indefinitely without having to drink water. They live on a diet of seeds, which are rich in lipids but contain little water. The metabolic water that kangaroo rats produce is adequate for all their water needs. This metabolic response to arid conditions is usually accompanied by a reduced output of urine.

21.4 How Are Unsaturated Fatty Acids and Odd-Carbon Fatty Acids Catabolized?

Fatty acids with odd numbers of carbon atoms are not as frequently encountered in nature as are the ones with even numbers of carbon atoms. Odd-numbered fatty acids also undergo the β-oxidation process (Figure 21.8). The last cycle of β-oxidation produces one molecule of propionyl-CoA. An enzymatic pathway exists to convert propionyl-CoA to succinyl-CoA, which then enters the citric acid cycle. In this pathway, propionyl-CoA is first carboxylated to methyl malonyl-CoA in a reaction catalyzed by propionyl-CoA carboxylase, which then undergoes rearrangement to form succinyl-CoA. Because propionyl-CoA is also a product of the catabolism of several amino acids, the conversion of propionyl-CoA to succinyl-CoA also plays a role in amino acid metabolism (Section 23.4). The conversion of methyl malonyl-CoA to succinyl-CoA requires vitamin B_{12} (cyanocobalamin), which has a cobalt (III) ion in its active state.

The conversion of a monounsaturated fatty acid to acetyl-CoA requires a reaction that is not encountered in the oxidation of saturated acids, a *cis–trans* isomerization (Figure 21.9). Successive rounds of β-oxidation of oleic acid (18:1) provide an example of these reactions. The process of β-oxidation gives rise to unsaturated fatty acids in which the double bond is in the *trans* arrangement, whereas the double bonds in most naturally occurring fatty acids are in the *cis* arrangement. In the case of oleic acid, there is a *cis* double bond between carbons 9 and 10. Three rounds of β-oxidation produce a 12-carbon unsaturated fatty acid with a *cis* double bond between carbons 3 and 4. The hydratase of the β-oxidation cycle requires a *trans* double bond between carbon atoms 2 and 3 as a substrate. A ***cis–trans* isomerase** produces a *trans* double bond between carbons 2 and 3 from the *cis* double bond between carbons 3 and 4. From this point forward, the fatty acid is metabolized the same as for saturated fatty acids. When oleic acid is β-oxidized, the

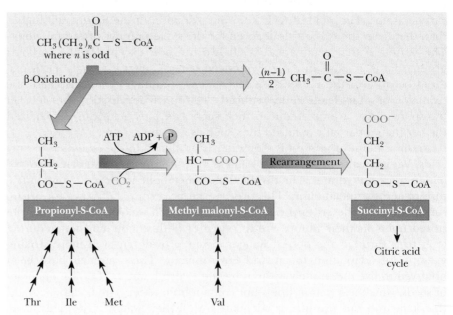

▶ **FIGURE 21.8** The oxidation of a fatty acid containing an odd number of carbon atoms.

The catabolism of some amino acids also yields propionyl-CoA and methyl malonyl-CoA

first step (fatty acyl-CoA dehydrogenase) is skipped, and the isomerase deals with the *cis* double bond, putting it into the proper position and orientation to continue the pathway.

When polyunsaturated fatty acids are β-oxidized, another enzyme is needed to handle the second double bond. Let's consider how linoleic acid (18:2) would be metabolized (Figure 21.10). This fatty acid has *cis* double bonds at positions 9 and 12 as shown in Figure 21.10, which are indicated as *cis*Δ^9 and *cis*Δ^{12}. Three normal cycles of β-oxidation occur, as in our example with oleic acid, before the isomerase must switch the position and orientation of the double bond. The cycle of β-oxidation continues until a 10-carbon fatty acyl-CoA is attained that has one *cis* double bond on its carbon 4 (*cis*Δ^4). Then the first step of β-oxidation occurs, putting in a *trans* double bond between carbons 2 and 3 (α and β). Normal β-oxidation cannot continue at this point because the fatty acid with the two double bonds so close together is a poor substrate for the hydratase. Therefore, a second new enzyme, *2,4-dienoyl-CoA reductase,* uses NADPH to reduce this intermediate. The result is a fatty acyl-CoA with a *trans* double bond between carbons 3 and 4. The isomerase then switches the *trans* double from carbon 3 to carbon 2, and β-oxidation continues.

A molecule with three double bonds, such as linolenic acid (18:3), would use the same two enzymes to handle the double bonds. The first double bond will require the isomerase. The second one will require the reductase and the isomerase, and the third will require the isomerase. For practice, you can diagram the β-oxidation of an 18-carbon molecule with *cis* double bonds at positions 9, 12, and 15 to see that this is true. Unsaturated fatty acids make up a large enough portion of the fatty acids in storage fat (40% for oleic acid alone) to make the reactions of the *cis–trans* isomerase and the epimerase of particular importance.

The oxidation of unsaturated fatty acids does not generate as many ATPs as it would for a saturated fatty acid with the same number of carbons. This is because the presence of a double bond means that the acyl-CoA dehydrogenase step will be skipped. Thus, fewer $FADH_2$ will be produced.

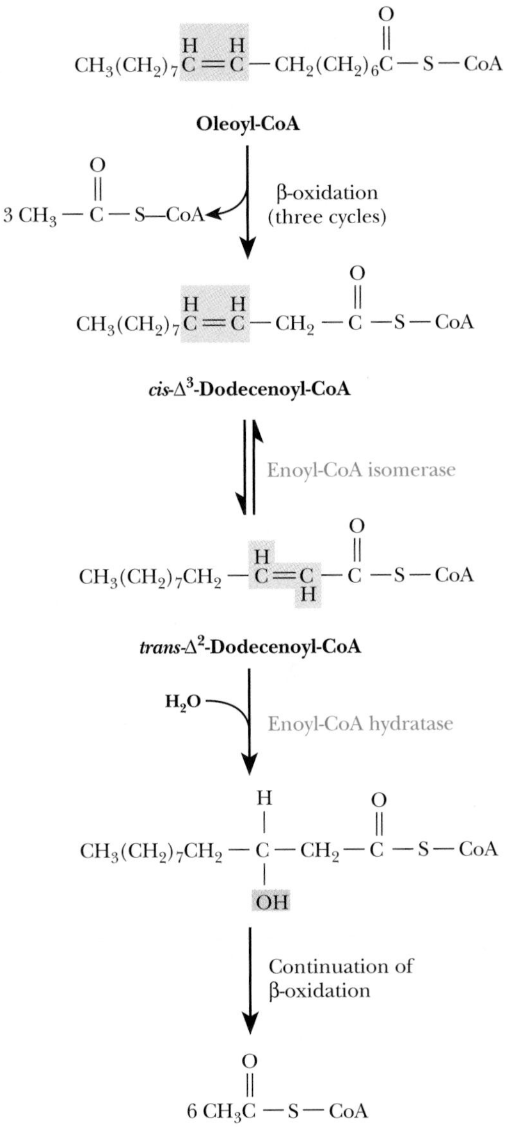

21.5 | What Are Ketone Bodies?

Substances related to acetone (**"ketone bodies"**) are produced when an excess of acetyl-CoA arises from β-oxidation. This condition occurs when not enough oxaloacetate is available to react with the large amounts of acetyl-CoA that could enter the citric acid cycle. Oxaloacetate in turn arises from glycolysis because it is formed from pyruvate in a reaction catalyzed by pyruvate carboxylase. A situation like this can come about when an organism has a high intake of lipids and a low intake of carbohydrates, but there are also other possible causes, such as starvation and diabetes. Starvation conditions cause an organism to break down fats for energy, leading to the production of large amounts of acetyl-CoA by β-oxidation. The amount of acetyl-CoA is excessive by comparison with the amount of oxaloacetate available to react with it. In the case of diabetics, the cause of the imbalance is not inadequate intake of carbohydrates but rather the inability to metabolize them.

The reactions that result in ketone bodies start with the condensation of two molecules of acetyl-CoA to produce acetoacetyl-CoA. *Acetoacetate* is produced from acetoacetyl-CoA through condensation with another acetyl-CoA to form β-hydroxy-β-methylglutaryl-CoA (HMG-CoA), a compound we will see

▲ **FIGURE 21.9** β-oxidation of unsaturated fatty acids. In the case of oleoyl-CoA, three β-oxidation cycles produce three molecules of acetyl-CoA and leave *cis*-Δ^3-dodecenoyl-CoA. Rearrangement of enoyl-CoA isomerase gives the *trans*-Δ^2 species, which then proceeds normally through the β-oxidation pathway.

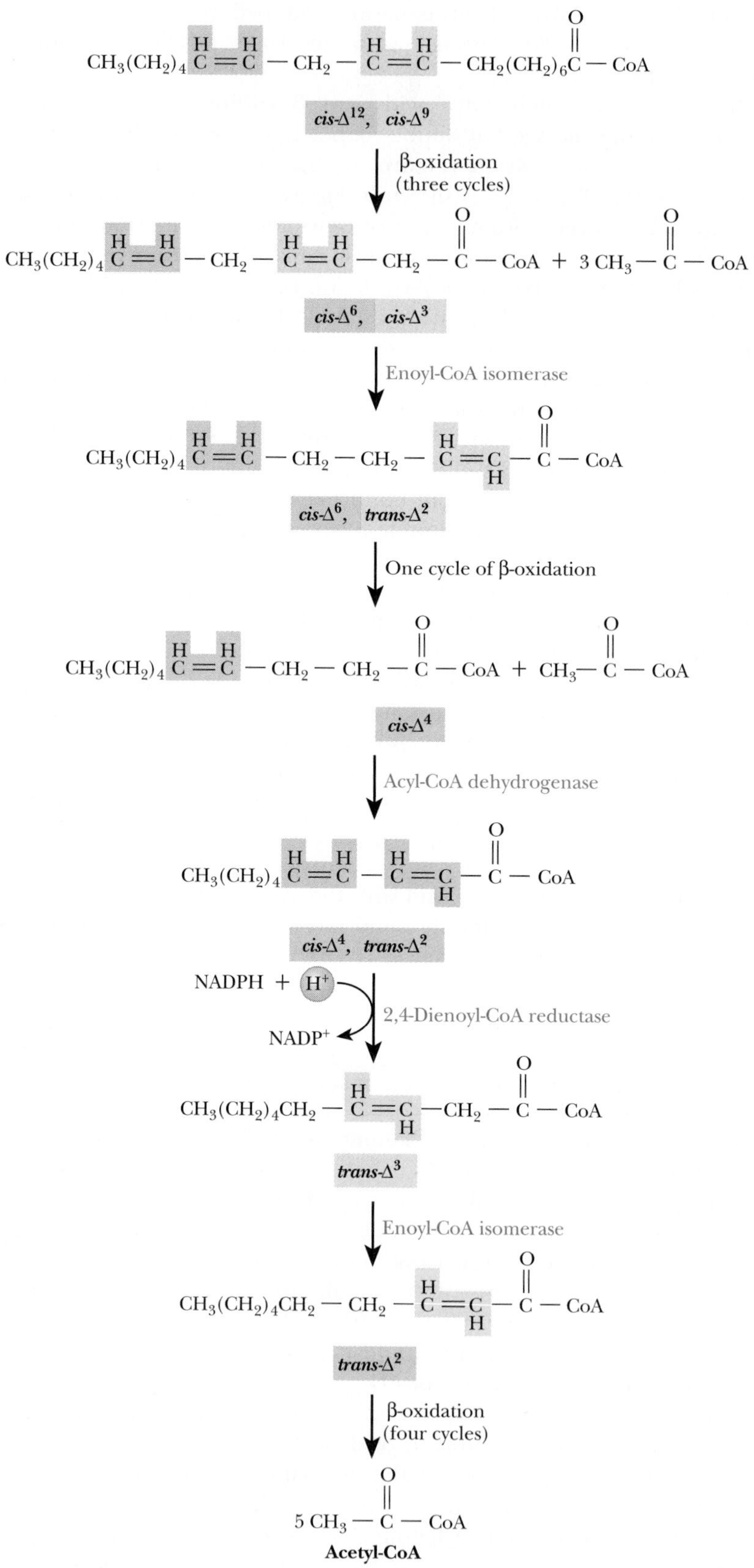

▶ **FIGURE 21.10** The oxidation pathway for polyunsaturated fatty acids, illustrated for linoleic acid. Three cycles of β-oxidation on linoleoyl-CoA yield the *cis*-Δ^3, *cis*-Δ^6 intermediate, which is converted to a *trans*-Δ^2, *cis*-Δ^6 intermediate. An additional round of β-oxidation gives *cis*-Δ^4 enoyl-CoA, which is oxidized to the *trans*-Δ^2, *cis*-Δ^4 species by acyl-CoA dehydrogenase. The subsequent action of 2,4-dienoyl-CoA reductase yields the *trans*-Δ^3 product, which is converted by enoyl-CoA isomerase to the *trans*-Δ^2 form. Normal β-oxidation then produces five molecules of acetyl-CoA.

Biochemical Connections

Ketone Bodies and Effective Weight Loss

Consistent overeating invariably results in weight gain, no matter what the source of the excess calories may be: carbohydrates, lipid, protein, or even alcohol. When you gain weight, your adipose tissue tends both to add more cells and to enlarge the ones already there. Severely obese people are unequivocally known to have added cells. Obese children, even those who are not necessarily extremely overweight, also add more adipose cells. Obese children retain those added cells for life. During weight loss, the existing cells merely shrink, so there are many little "starving" cells just waiting for a chance to grow again. The existence of these starving cells may explain why many have a tendency to regain weight so easily.

Most of the time, the body cells depend on glucose from the blood for their energy. Overnight, when carbohydrate levels in the blood decrease, blood proteins are degraded in the liver to convert them to glucose; adipose tissue is also degraded to some extent. In the course of a diet, the body begins to break down muscle mass (protein) for conversion to glucose. After a few days to a week, nitrogen depletion becomes excessive. The body turns away from sources of glucose and turns to fats in adipose tissue. These fats are metabolized to fatty acids and glycerol, with the glycerol available for conversion to glucose, taking care of specialized tissues with a specific requirement for glucose. The fatty acids are converted to short-chain oxy acids usually called ketone bodies, such as β-hydroxybutyrate and β-ketobutyrate, which are highly water soluble and are circulated easily in the blood. The ketone bodies are efficient nutrients because they enter directly into the mitochondria for aerobic metabolism. The heart uses ketone bodies all the time. Adipose tissue surrounds the heart and even permeates into it, providing a direct and efficient energy supply for this constantly working tissue. The brain will adapt to the use of ketone bodies as nutrients after about a week of fasting. This was a surprising observation because it had been believed that the brain could only use sugars. At this point, weight loss becomes very effective.

Why do you lose weight so effectively? Part of the answer depends on the fact that the ketone bodies are potentially dangerous because they are relatively strong organic acids and they lower the blood pH, possibly causing acidosis. Recall that, at low pH, hemoglobin is less able to bind oxygen. If you drink lots of fluids, your kidney will flush out the ketones in the urine to adjust body pH. You do lose weight very efficiently at this point because you are literally putting excess calories "down the drain." The water in "Dr. Stillman's Water Diet" was there for a reason.

It is important to note that prolonged ketosis/acidosis can be dangerous or even fatal. One reason is that the excess urination could produce a secondary mineral imbalance, and this is dangerous, especially if one has a weak heart. Medically supervised diets try to avoid a true ketotic state, and people on such a limited-calorie diet should be under constant physician care. You can tell you are in extreme ketosis by the metallic taste (acetone breath) in your mouth when you get up in the morning.

---

again when we look at cholesterol synthesis (Figure 21.11). HMG-CoA lyase then releases acetyl-CoA to give acetoacetate. Acetoacetate can then have two fates. A reduction reaction can produce *β-hydroxybutyrate* from acetoacetate. The other possible reaction is the spontaneous decarboxylation of acetoacetate to give *acetone*. The odor of acetone can frequently be detected on the breath of diabetics whose disease is not controlled by suitable treatment. The excess of acetoacetate, and consequently of acetone, is a pathological condition known as *ketosis*. Because acetoacetate and β-hydroxybutyrate are acidic, their presence at high concentration overwhelms the buffering capacity of the blood. The consequent lowering of blood pH (ketoacidosis) is dealt with by the body by excreting H^+ into the urine, accompanied by excretion of Na^+, K^+, and water. Severe dehydration can result (excessive thirst is a classic symptom of diabetes); diabetic coma is another possible danger.

The principal site of synthesis of ketone bodies is liver mitochondria, but they are not used there because the liver lacks the enzymes necessary to recover acetyl-CoA from ketone bodies. It is easy to transport ketone bodies in the bloodstream because, unlike fatty acids, they are water soluble and do not need to be bound to proteins, such as serum albumin. Organs other than the liver can use ketone bodies, particularly acetoacetate. Even though glucose is the usual fuel in most tissues and organs, acetoacetate can be used as a fuel. In heart muscle and the renal cortex, acetoacetate is the preferred source of energy.

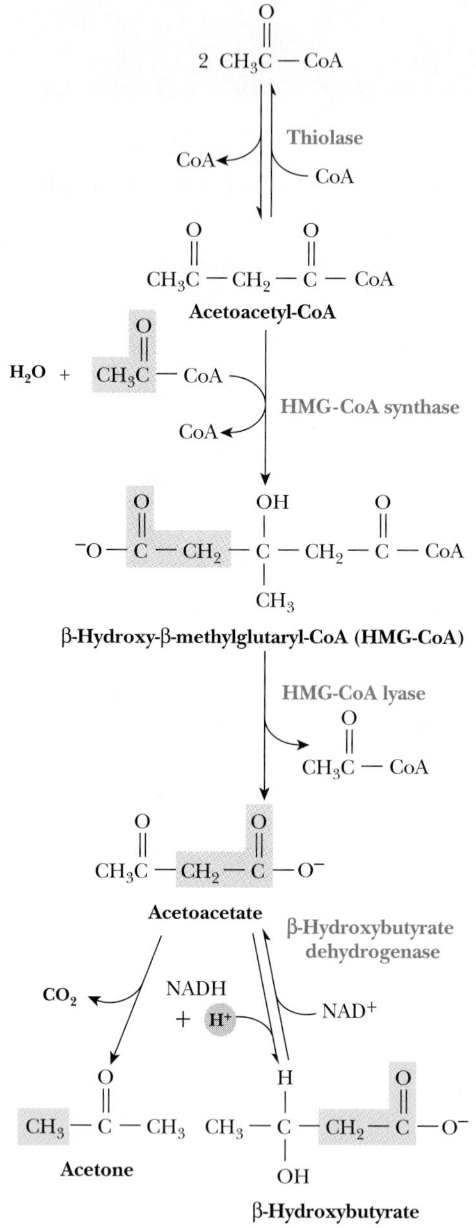

▲ **FIGURE 21.11** The formation of ketone bodies, synthesized primarily in the liver.

Even in organs such as the brain, in which glucose is the preferred fuel, starvation conditions can lead to the use of acetoacetate for energy. In this situation, acetoacetate is converted to two molecules of acetyl-CoA, which can then enter the citric acid cycle. The key point here is that starvation gives rise to long-term, rather than short-term, regulation over a period of hours to days rather than minutes. The decreased level of glucose in the blood over a period of days changes the hormone balance in the body, particularly involving insulin and glucagon (see Section 24.4). (Short-term regulation, such as allosteric interactions or covalent modification, can occur in a matter of minutes.) The rates of protein synthesis and breakdown are subject to change under these conditions. The specific enzymes involved are those involved in fatty-acid oxidation (increase in levels) and those for lipid biosynthesis (decrease in levels).

21.6 | How Are Fatty Acids Produced?

The anabolism of fatty acids is not simply a reversal of the reactions of β-oxidation. Anabolism and catabolism are not, in general, the exact reverse of each other; for instance, gluconeogenesis (Section 18.2) is not simply a reversal of the reactions of glycolysis. A first example of the differences between the degradation and the biosynthesis of fatty acids is that the anabolic reactions take place in the cytosol. We have just seen that the degradative reactions of β-oxidation take place in the mitochondrial matrix. The first step in fatty acid biosynthesis is transport of acetyl-CoA to the cytosol.

Acetyl-CoA can be formed either by β-oxidation of fatty acids or by decarboxylation of pyruvate. (Degradation of certain amino acids also produces acetyl-CoA; see Section 23.6.) Most of these reactions take place in the mitochondria, requiring a transport mechanism to export acetyl-CoA to the cytosol for fatty-acid biosynthesis. The transport mechanism is based on the fact that citrate can cross the mitochondrial membrane. Acetyl-CoA condenses with oxaloacetate, which cannot cross the mitochondrial membrane, to form citrate (recall that this is the first reaction of the citric acid cycle). The citrate that is exported to the cytosol can undergo the reverse reaction, producing oxaloacetate and acetyl-CoA (Figure 21.12). Acetyl-CoA enters the pathway for fatty-acid biosynthesis, while oxaloacetate undergoes a series of reactions in which there is a substitution of NADPH for NADH (see the discussion of lipid anabolism in Section 21.7). This substitution exercises control over the pathway because NADPH is required for fatty acid anabolism.

In the cytosol, acetyl-CoA is carboxylated, producing **malonyl-CoA,** a key intermediate in fatty-acid biosynthesis (Figure 21.13). This reaction is catalyzed by the *acetyl-CoA carboxylase* complex, which consists of three enzymes and requires Mn^{2+}, biotin, and ATP for activity. We have already seen that enzymes catalyzing reactions that take place in several steps frequently consist of several separate protein molecules, and this enzyme follows that pattern. In this case, acetyl-CoA carboxylase consists of the three proteins *biotin carboxylase,* the *biotin carrier protein,* and *carboxyl transferase.* Biotin carboxylase catalyzes the transfer of the carboxyl group to biotin. The "activated CO_2" (the carboxyl group derived from the bicarbonate ion HCO_3^-) is covalently bound to biotin. Biotin (whether carboxylated or not) is bound to the biotin carrier protein by an amide linkage to the ε-amino group of a lysine side chain. The amide linkage to the side chain that bonds biotin to the carrier protein is long enough and flexible enough to move the carboxylated biotin into position to transfer the carboxyl group to acetyl-CoA in the reaction catalyzed by carboxyl transferase, producing malonyl-CoA (Figure 21.14). In addition to its role as a starting point in fatty acid synthesis, malonyl-CoA strongly inhibits

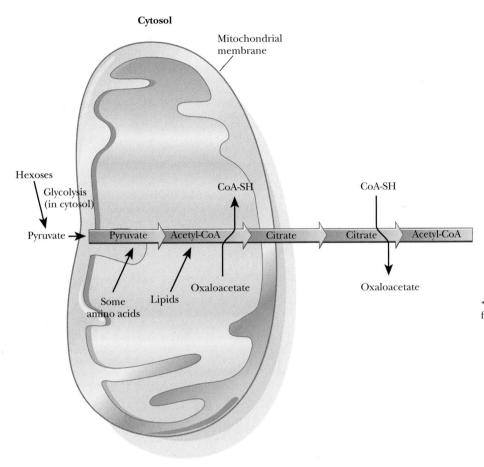

Cytosol

Mitochondrial
membrane

Hexoses

Glycolysis
(in cytosol)

Pyruvate →

Pyruvate   Acetyl-CoA   Citrate   Citrate   Acetyl-CoA

CoA-SH   CoA-SH

Oxaloacetate   Oxaloacetate

Some
amino acids   Lipids

Mitochondrion

◀ **FIGURE 21.12** The transport of acetyl groups
from the mitochondrion to the cytosol.

$$\underset{\textbf{Acetyl-CoA}}{H_3C-\overset{\overset{\textstyle O}{\|}}{C}-S-CoA}\ +\ ATP\ +\ HCO_3^-\ \xrightarrow[\textbf{Mn}^{2+}]{\textbf{Biotin}}$$

$$\underset{\textbf{Malonyl-CoA}}{^-OOC-CH_2-\overset{\overset{\textstyle O}{\|}}{C}-S-CoA}\ +\ ADP\ +\ P\ +\ H^+$$

◀ **FIGURE 21.13** The formation of malonyl-CoA,
catalyzed by acetyl-CoA carboxylase.

the carnitine acyltransferase I on the outer face of the inner mitochondrial
membrane. This avoids a futile cycle in which fatty acids are β-oxidized in the
mitochondria to make acetyl-CoA just so they can be remade into fatty acids
in the cytosol.

The biosynthesis of fatty acids involves the successive addition of two-carbon units to the growing chain. Two of the three carbon atoms of the malonyl
group of malonyl-CoA are added to the growing fatty-acid chain with each
cycle of the biosynthetic reaction. This reaction, like the formation of the
malonyl-CoA itself, requires a multienzyme complex located in the cytosol
and not attached to any membrane. The complex, made up of the individual
enzymes, is called **fatty-acid synthase.**

The usual product of fatty-acid anabolism is *palmitate,* the 16-carbon saturated fatty acid. All 16 carbons come from the acetyl group of acetyl-CoA; we
have already seen how malonyl-CoA, the immediate precursor, arises from

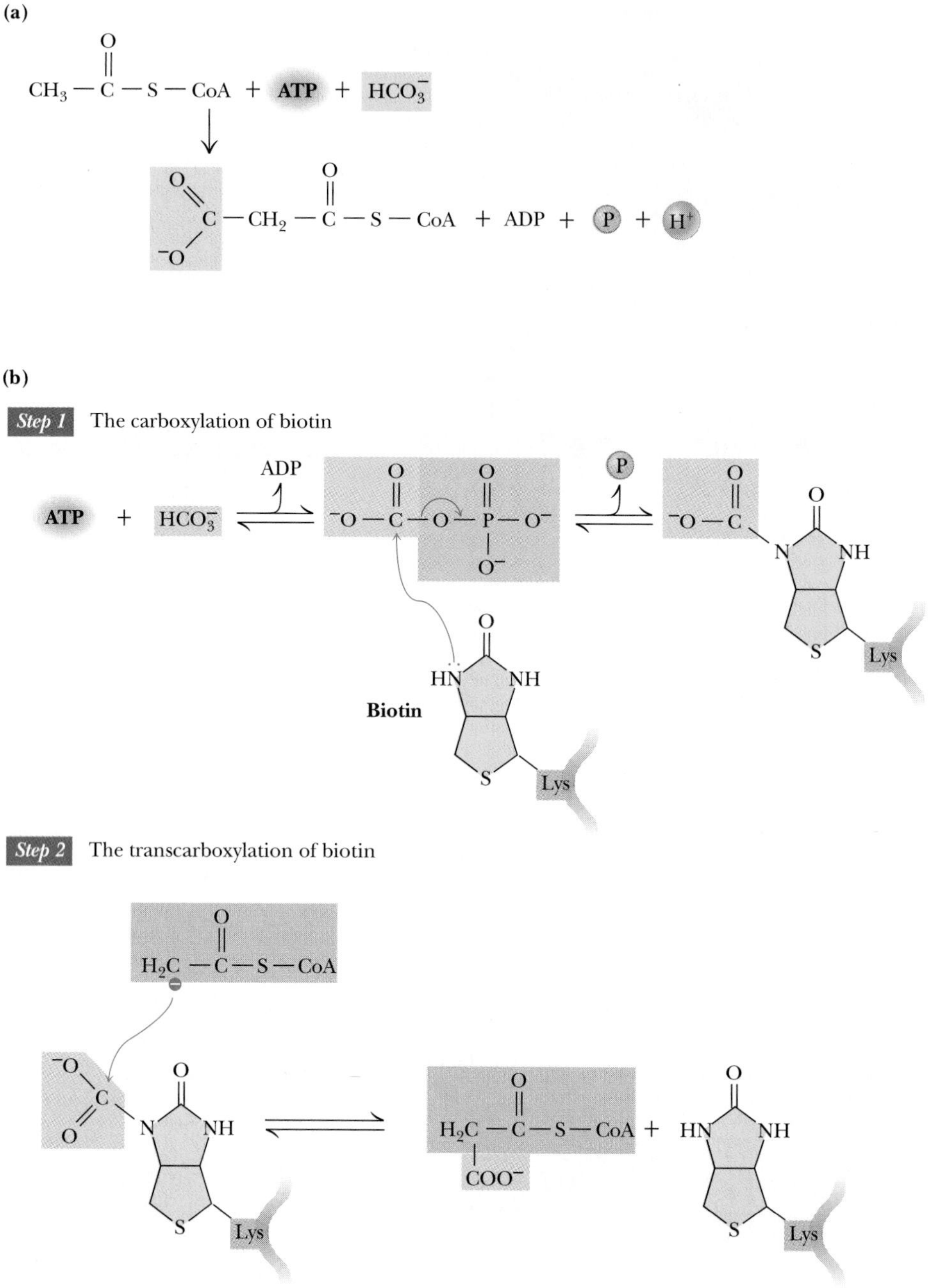

(a)

$$CH_3 - \overset{\overset{\displaystyle O}{\|}}{C} - S - CoA \ + \ \textbf{ATP} \ + \ HCO_3^-$$

$$\overset{\overset{\displaystyle O}{\|}}{\underset{\underset{\displaystyle -O}{}}{C}} - CH_2 - \overset{\overset{\displaystyle O}{\|}}{C} - S - CoA \ + \ ADP \ + \ \textcircled{P} \ + \ \textcircled{H^+}$$

(b)

<u>Step 1</u> The carboxylation of biotin

<u>Step 2</u> The transcarboxylation of biotin

Biochemistry Ⓔ**Now**™ ▲ **ACTIVE FIGURE 21.14** (a) The acetyl-CoA carboxylase reaction produces malonyl-CoA for fatty-acid synthesis. (b) A mechanism for the acetyl-CoA carboxylase reaction. Bicarbonate is activated for carboxylation reactions by formation of *N*-carboxybiotin. ATP drives the reaction forward, with transient formation of a carbonyl-phosphate intermediate (Step 1). In a typical biotin-dependent reaction, nucleophilic attack by the acetyl-CoA carbanion on the carboxyl carbon of *N*-carboxybiotin—a transcarboxylation—yields the carboxylated product (Step 2). **Watch this Active Figure at http://now.brookscole.com/campbell5**

acetyl-CoA. But first there is a priming step in which one molecule of acetyl-CoA is required for each molecule of palmitate produced. In this priming step, the acetyl group from acetyl-CoA is transferred to an **acyl carrier protein (ACP),** which is considered a part of the fatty-acid synthase complex (Figure 21.15). The acetyl group is bound to the protein as a thioester. The group on the protein to which the acetyl group is bonded is the 4-phosphopantetheine

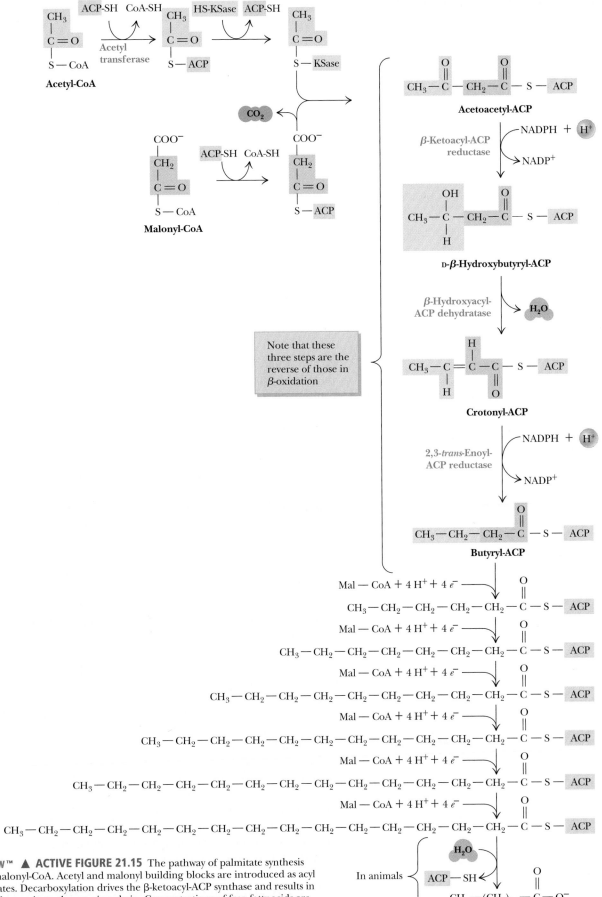

Biochemistry Now™ ▲ **ACTIVE FIGURE 21.15** The pathway of palmitate synthesis from acetyl-CoA and malonyl-CoA. Acetyl and malonyl building blocks are introduced as acyl carrier-protein conjugates. Decarboxylation drives the β-ketoacyl-ACP synthase and results in the addition of two-carbon units to the growing chain. Concentrations of free fatty acids are extremely low in most cells, and newly synthesized fatty acids exist primarily as acyl-CoA esters. **Watch this Active Figure at http://now.brookscole.com/campbell5**

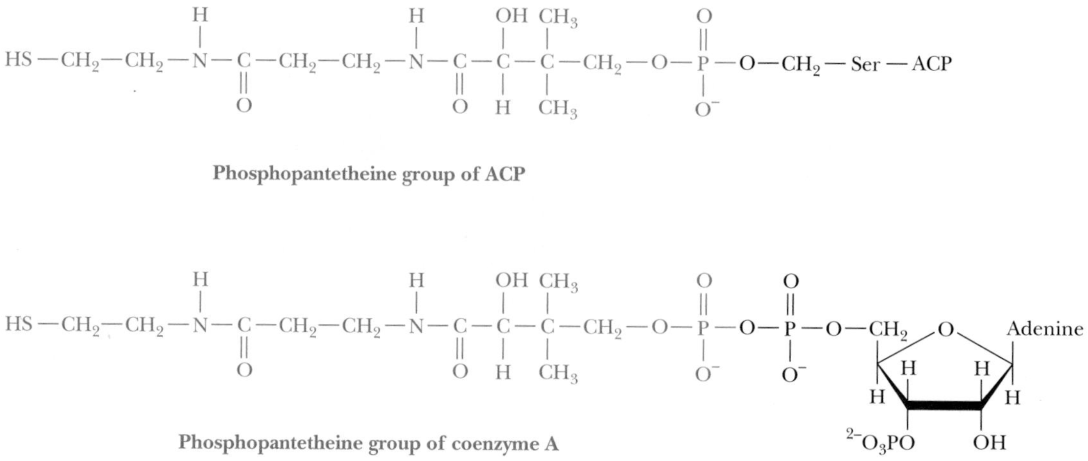

$$HS-CH_2-CH_2-\overset{\displaystyle H}{\underset{\displaystyle }{N}}-\overset{\displaystyle }{\underset{\displaystyle O}{C}}-CH_2-CH_2-\overset{\displaystyle H}{\underset{\displaystyle }{N}}-\overset{\displaystyle }{\underset{\displaystyle O}{C}}-\overset{\displaystyle OH}{\underset{\displaystyle H}{C}}-\overset{\displaystyle CH_3}{\underset{\displaystyle CH_3}{C}}-CH_2-O-\overset{\displaystyle O}{\underset{\displaystyle O^-}{P}}-O-CH_2-Ser-ACP$$

Phosphopantetheine group of ACP

$$HS-CH_2-CH_2-\overset{\displaystyle H}{\underset{\displaystyle }{N}}-\overset{\displaystyle }{\underset{\displaystyle O}{C}}-CH_2-CH_2-\overset{\displaystyle H}{\underset{\displaystyle }{N}}-\overset{\displaystyle }{\underset{\displaystyle O}{C}}-\overset{\displaystyle OH}{\underset{\displaystyle H}{C}}-\overset{\displaystyle CH_3}{\underset{\displaystyle CH_3}{C}}-CH_2-O-\overset{\displaystyle O}{\underset{\displaystyle O^-}{P}}-O-\overset{\displaystyle O}{\underset{\displaystyle O^-}{P}}-O-CH_2$$

Adenine

$^{2-}O_3PO \quad OH$

Phosphopantetheine group of coenzyme A

▲ **FIGURE 21.16** Structural similarities between coenzyme A and the phosphopantetheine group of ACP.

group, which in turn is bonded to a serine side chain; note in Figure 21.16 that this group is structurally similar to CoA-SH itself. The acetyl group is transferred from CoA-SH, to which it is bound by a thioester linkage, to the ACP; the acetyl group is bound to the ACP by a thioester linkage. Although the functional group of ACP is similar to that of CoA-SH, it is noteworthy that fatty-acid synthesis in the cytosol uses only ACP. In essence, the ACP is a label that marks acetyl groups for fatty-acid synthesis.

The acetyl group is transferred in turn from the ACP to another protein, to which it is bound by a thioester linkage to a cysteine-SH; the other protein is β-ketoacyl-S-ACP-synthase (HS-KSase) (Figure 21.15). The first of the successive additions of two of the three malonyl carbons to the fatty acid starts at this point. The malonyl group itself is transferred from a thioester linkage with CoA-SH to another thioester bond with ACP (Figure 21.15). The next step is a condensation reaction that produces acetoacetyl-ACP (Figure 21.15). In other words, the principal product of this reaction is an acetoacetyl group bound to the ACP by a thioester linkage. Two of the four carbons of acetoacetate come from the priming acetyl group, and the other two come from the malonyl group. The carbon atoms that arise from the malonyl group are the one directly bonded to the sulfur and the one in the —CH$_2$— group next to it. The CH$_3$CO— group comes from the priming acetyl group. The other carbon of the malonyl group is released as CO_2; the CO_2 that is lost is the original CO_2 that was used to carboxylate the acetyl-CoA to produce malonyl-CoA. The synthase is no longer involved in a thioester linkage. This is an example of a decarboxylation being used to drive an otherwise unfavorable condensation reaction.

Acetoacetyl-ACP is converted to butyryl-ACP by a series of reactions involving two reductions and a dehydration (Figure 21.15). In the first reduction, the β-keto group is reduced to an alcohol, giving rise to D-β-hydroxybutyryl-ACP. In the process, NADPH is oxidized to NADP$^+$; the enzyme that catalyzes this reaction is β-ketoacyl-ACP reductase (Figure 21.15). The dehydration step, catalyzed by β-hydroxyacyl-ACP dehydratase, produces crotonyl-ACP (Figure 21.15). Note that the double bond is in the *trans* configuration. A second reduction reaction, catalyzed by β-enoyl-ACP reductase, produces butyryl-ACP (Figure 21.15). In this reaction, NADPH is the coenzyme, as it was in the first reduction reaction in this series.

In the second round of fatty-acid biosynthesis, butyryl-ACP plays the same role as acetyl-ACP in the first round. The butyryl group is transferred to the synthase, and a malonyl group is transferred to the ACP. Once again there is a condensation reaction with malonyl-ACP (Figure 21.15). In this second round, the condensation produces a six-carbon β-ketoacyl-ACP. The two added carbon atoms come from the malonyl group, as they did in the first round. The reduction and dehydration reactions take place as before, giving rise to hexanoyl-ACP. The same series of reactions is repeated until palmitoyl-ACP is produced. In mammalian systems, the process stops at C_{16} because the fatty-acid synthase does not produce longer chains. Mammals produce longer-chain fatty acids by modifying the fatty acids formed by the synthase reaction.

Fatty-acid synthases from different types of organisms have markedly different characteristics. In *Escherichia coli*, the multienzyme system consists of an aggregate of separate enzymes, including a separate ACP. The ACP is of primary importance to the complex and is considered to occupy a central position in it. The phosphopantetheine group plays the role of a "swinging arm," much like that of biotin, which was discussed earlier in this chapter. This bacterial system has been extensively studied and has been considered a typical example of a fatty-acid synthase. In eukaryotes, however, fatty acid synthesis occurs on a multienzyme complex. In yeast, this complex consists of two different types of subunits, called α and β, arranged in an $\alpha_6\beta_6$ complex. In mammals, the fatty-acid synthase contains only one type of subunit, but the active enzyme is a dimer of this single subunit. Each of the subunits is a *multifunctional enzyme* that catalyzes reactions requiring several different proteins in the *E. coli* system. The growing fatty-acid chain swings back and forth between enzyme activities contained on different subunits using ACP as a "swinging arm." Like the bacterial system, the eukaryotic system keeps all the components of the reaction in proximity to one another, which shows us another example of the advantages to multienzyme complexes.

Several additional reactions are required for the elongation of fatty-acid chains and the introduction of double bonds. When mammals produce fatty acids with longer chains than that of palmitate, the reaction does not involve cytosolic fatty-acid synthase. There are two sites for the chain-lengthening reactions: the endoplasmic reticulum (ER) and the mitochondrion. In the chain-lengthening reactions in the mitochondrion, the intermediates are of the acyl-CoA type rather than the acyl-ACP type. In other words, the chain-lengthening reactions in the mitochondrion are the reverse of the catabolic reactions of fatty acids, with acetyl-CoA as the source of added carbon atoms; this is a difference between the main pathway of fatty-acid biosynthesis and these modification reactions. In the ER, the source of additional carbon atoms is malonyl-CoA. The modification reactions in the ER also differ from the biosynthesis of palmitate in that, like the mitochondrial reaction, there are no intermediates bound to ACP.

Reactions in which a double bond is introduced in fatty acids mainly take place on the ER. The insertion of the double bond is catalyzed by a mixed-function oxidase that requires molecular oxygen (O_2) and NAD(P)H. During the reaction, both NAD(P)H and the fatty acid are oxidized, while oxygen is reduced to water. Reactions linked to molecular oxygen are comparatively rare (Section 19.9). Mammals cannot introduce a double bond beyond carbon atom 9 (counting from the carboxyl end) of the fatty-acid chain. As a result, linoleate [$CH_3-(CH_2)_4-CH=CH-CH_2-CH=CH-(CH_2)_7-COO^-$], with two double bonds, and linolenate [$CH_3-(CH_2)_4-CH=CH-CH_2-CH=CH-CH_2-CH=CH-(CH_2)_4-COO^-$], with three double bonds, must be included in the diets of mammals. They are **essential fatty acids** because they are precursors of other lipids, including prostaglandins.

Essential Information

The biosynthesis of fatty acids proceeds by the addition of two-carbon units to the hydrocarbon chain. The process takes place in the cytosol and is catalyzed in many organisms by a large multienzyme complex.

Table 21.2

A Comparison of Fatty Acid Degradation and Biosynthesis

Degradation	Biosynthesis
1. Product is acetyl-CoA	Precursor is acetyl-CoA
2. Malonyl-CoA is not involved; no requirement for biotin	Malonyl-CoA is source of two-carbon units; biotin required
3. Oxidative process; requires NAD^+ and FAD and produces ATP	Reductive process; requires NADPH and ATP
4. Fatty acids form thioesters with CoA-SH	Fatty acids form thioesters with acyl carrier proteins (ACP-SH)
5. Starts at carboxyl end (CH_3CO_2—)	Starts at methyl end (CH_3CH_2—)
6. Occurs in the mitochondrial matrix, with no ordered aggregate of enzymes	Occurs in the cytosol, catalyzed by an ordered multienzyme complex
7. β-Hydroxyacyl intermediates have the L configuration	β-Hydroxyacyl intermediates have the D configuration

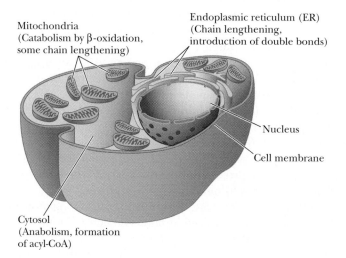

Mitochondria
(Catabolism by β-oxidation,
some chain lengthening)

Endoplasmic reticulum (ER)
(Chain lengthening,
introduction of double bonds)

Nucleus

Cell membrane

Cytosol
(Anabolism, formation
of acyl-CoA)

▶ **FIGURE 21.17** A portion of an animal cell, showing the sites of various aspects of fatty-acid metabolism. The cytosol is the site of fatty-acid anabolism. It is also the site of formation of acyl-CoA, which is transported to the mitochondrion for catabolism by the β-oxidation process. Some chain-lengthening reactions (beyond C_{16}) take place in the mitochondria. Other chain-lengthening reactions take place in the endoplasmic reticulum (ER), as do reactions that introduce double bonds.

Even though both the anabolism and the catabolism of fatty acids require successive reactions of two-carbon units, the two pathways are not the exact reversal of each other. The differences between the two pathways can be summarized in Table 21.2. The sites in the cell in which various anabolic and catabolic reactions take place are shown in Figure 21.17.

21.7	How Are Acylglycerols and Compound Lipids Produced?

Other lipids, including triacylglycerols, phosphoacylglycerols, and steroids, are derived from fatty acids and metabolites of fatty acids, such as acetoacetyl-CoA. Free fatty acids do not occur in the cell to any great extent; they are normally found incorporated in triacylglycerols and phosphoacylglycerols. The biosynthesis of these two types of compounds takes place principally on the ER of liver cells or fat cells (adipocytes).

Biochemical Connections

Acetyl-CoA Carboxylase—A New Target in the Fight against Obesity?

Malonyl-CoA has two very important functions in metabolism. First, it is the committed intermediate in fatty-acid synthesis. Second, it strongly inhibits carnitine palmitoyltransferase I and therefore fatty-acid oxidation. The level of malonyl-CoA in the cytosol can determine if the cell will be oxidizing fats or storing fats. The enzyme that produces malonyl-CoA is acetyl-CoA carboxylase, or ACC. There are two isoforms of this enzyme encoded by separate genes. ACC1 is found in the liver and adipose tissue, while ACC2 is found in cardiac and skeletal muscle. High glucose concentrations and high insulin concentrations lead to stimulation of ACC2. Exercise has the opposite effect. During exercise, an AMP-dependent protein kinase phosphorylates ACC2 and inactivates it.

Some recent studies looked at the nature of weight gain and weight loss with respect to ACC2 (see papers by Ruderman and

Flier and by Abu-Elheiga et al. in the Annotated Bibliography at the end of this chapter). The investigators created a strain of mice lacking the gene for ACC2. These mice ate more than their wild-type counterparts but had significantly lower stores of lipids (30%–40% less in skeletal muscle and 10% less in cardiac muscle). Even the adipose tissue, which still had ACC1, showed a reduction in stored triacylglycerols of up to 50%. The mice showed no other abnormalities. They grew and reproduced normally and had normal life spans. The investigators concluded that reduced pools of malonyl-CoA due to the lack of ACC2 results in increased β-oxidation via removal of the block on carnitine palmitoyltransferase I, and a decrease in fatty acid synthesis. They speculate that ACC2 would be a good target for drugs used to combat obesity.

Triacylglycerols

The glycerol portion of lipids is derived from glycerol-3-phosphate, a compound available from glycolysis. In liver and kidney, another source is glycerol released by degradation of acylglycerols. An acyl group of a fatty acid is transferred from an acyl-CoA. The products of this reaction are CoA-SH and a *lysophosphatidate* (a monoacylglycerol phosphate) (Figure 21.18). The acyl group is shown as esterified at carbon atom 2 (C-2) in this series of equations, but it is equally likely that it is esterified at C-1. A second acylation reaction takes place, catalyzed by the same enzyme, producing a *phosphatidate* (a diacylglyceryl phosphate). Phosphatidates occur in membranes and are precursors of other phospholipids. The phosphate group of the phosphatidate is removed by hydrolysis, producing a *diacylglycerol*. A third acyl group is added in a reaction in which the source of the acyl group is an acyl-CoA rather than the free fatty acid.

Phosphoacylglycerols

Phosphoacylglycerols (phosphoglycerides) are based on phosphatidates, with the phosphate group esterified to another alcohol, frequently a nitrogen-containing alcohol such as ethanolamine [see Phosphoacylglycerols (Phosphoglycerides) in Section 8.2]. The conversion of phosphatidates to other phospholipids frequently requires the presence of nucleoside triphosphates, particularly *cytidine triphosphate* (CTP). The role of CTP depends on the type of organism, because the details of the biosynthetic pathway are not the same in mammals and bacteria. We shall use a comparison of the synthesis of phosphatidylethanolamine in mammals and in bacteria (Figure 21.19) as a case study of the kinds of reactions commonly encountered in phosphoglyceride biosynthesis.

In bacteria, CTP reacts with phosphatidate to produce cytidine diphosphodiacylglycerol (a CDP diglyceride). The CDP diglyceride reacts with serine to form phosphatidylserine. Phosphatidylserine is then decarboxylated to give

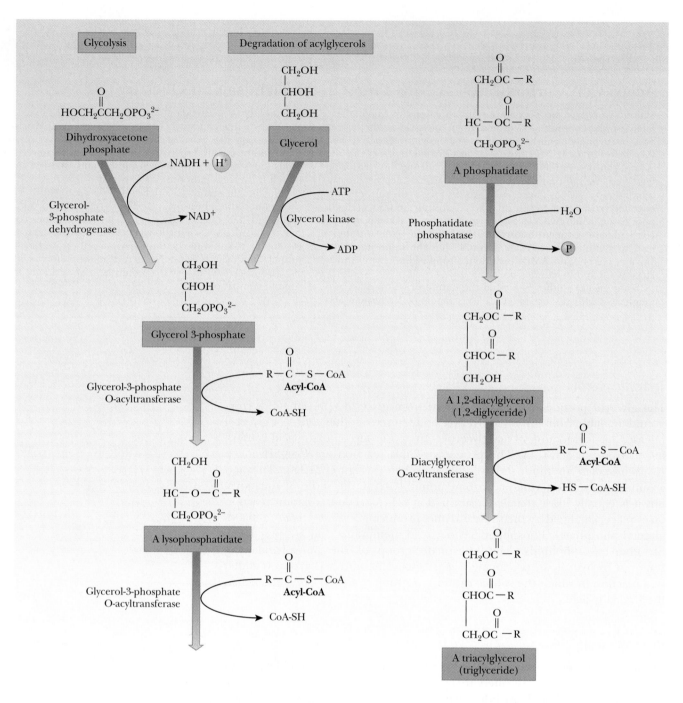

▲ **FIGURE 21.18** Pathways for the biosynthesis of triacylglycerols.

phosphatidylethanolamine. In eukaryotes, the synthesis of phosphatidylethanolamine requires two preceding steps in which the component parts are processed (Figure 21.20). The first of these two steps is the removal by hydrolysis of the phosphate group of the phosphatidate, producing a diacylglycerol; the second step is the reaction of ethanolamine phosphate with CTP to produce pyrophosphate (PP_i) and cytidine diphosphate ethanolamine (CDP-ethanolamine). The CDP-ethanolamine and diacylglycerol react to form phosphatidylethanolamine.

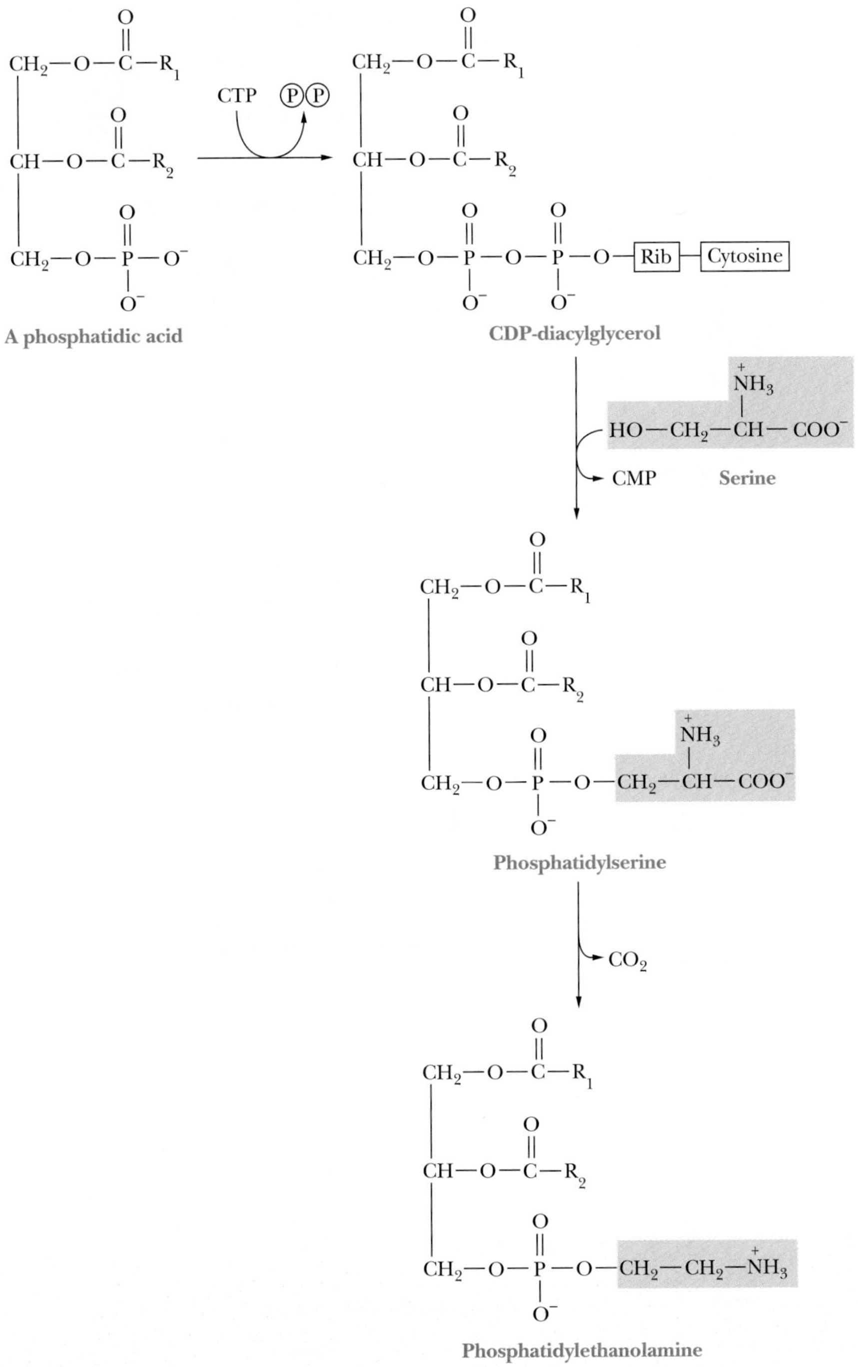

◀ **FIGURE 21.19** The biosynthesis of phosphatidylethanolamine in bacteria. See text for details about how the pathway differs in mammals.

In mammals, phosphatidylethanolamine can be produced another way. Alcohol exchange from serine to ethanolamine allows the interconversion of phosphatidylethanolamine with phosphatidylserine (Figure 21.21).

Sphingolipids

The structural basis of sphingolipids is not glycerol but *sphingosine,* a long-chain amine (see Sphingolipids in Section 8.2). The precursors of sphingosine are palmitoyl-CoA and the amino acid serine, which react to produce

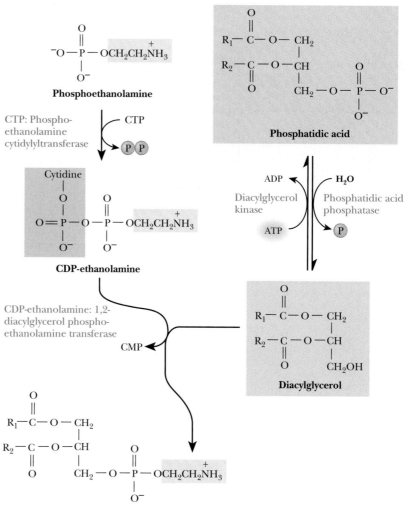

▶ **FIGURE 21.20** Production of phosphatidylethanolamine in eukaryotes.

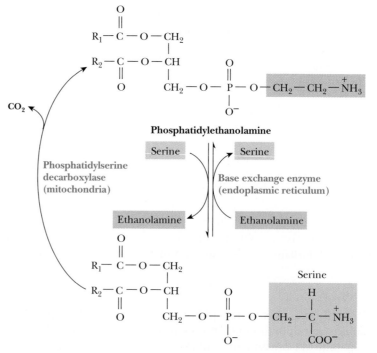

▶ **FIGURE 21.21** The interconversion of phosphatidylethanolamine and phosphatidylserine in
mammals.

Biochemical Connections

Tay–Sachs Disease

Tay–Sachs disease is an inborn error of lipid metabolism with particularly tragic consequences. In this disease, there is a blockage in the catabolism of gangliosides (see Sphingolipids in Section 8.2). The enzyme hexosaminidase A, responsible for the hydrolysis of N-acetylgalactosamine from ganglioside GM_2, is missing. (Ironically, this lipid-storage disease depends on the carbohydrate moiety of a glycolipid.)

Inclusions of accumulated GM_2 appear in the neurons of affected individuals. Those affected with Tay–Sachs disease appear normal as newborns, but, by the age of 1 year, they exhibit the characteristic symptoms of weakness, retardation, and blindness. This disease is fatal by age 3 or 4. It is possible to detect the disease during fetal development via amniocentesis, a technique based on testing of amniotic fluid or amniotic cells. An enzyme assay for the level of hexosaminidase A in the amniotic fluid depends on the use of an artificial substrate that releases a fluorescent compound when hydrolyzed. The nonappearance of fluorescence indicates the lack of hexosaminidase A that is characteristic of Tay–Sachs disease.

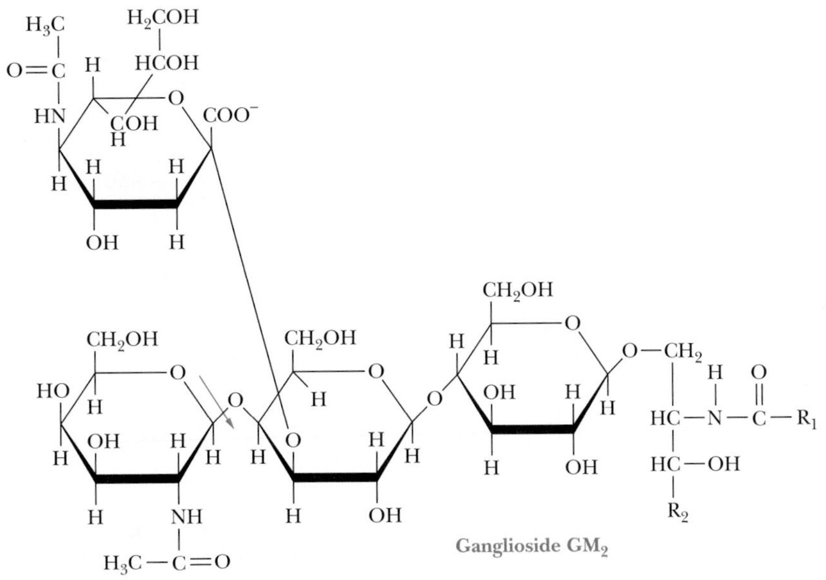

Ganglioside GM_2

▲ The enzyme that catalyzes the hydrolysis of the bond indicated by a red arrow is missing in Tay–Sachs disease.

dihydrosphingosine. The carboxyl group of the serine is lost as CO_2 in the course of this reaction (Figure 21.22). An oxidation reaction introduces a double bond, with sphingosine as the resulting compound. Reaction of the amino group of sphingosine with another acyl-CoA to form an amide bond results in an N-*acylsphingosine*, also called a **ceramide.** Ceramides in turn are the parent compounds of sphingomyelins, cerebrosides, and gangliosides. Attachment of phosphorylcholine to the primary alcohol group of a ceramide produces a *sphingomyelin*, whereas attachment of sugars such as glucose at the same site produces *cerebrosides*. *Gangliosides* are formed from ceramides by attachment of oligosaccharides that contain a sialic acid residue, also at the primary alcohol group. See Sphingolipids in Section 8.2 for the structures of these compounds.

Gangliosides play a part in *Tay–Sachs disease* (see the Biochemical Connections box). Because the enzyme hexosaminidase A is missing, the catabolism of gangliosides is blocked, creating conditions that invariably prove fatal to affected individuals.

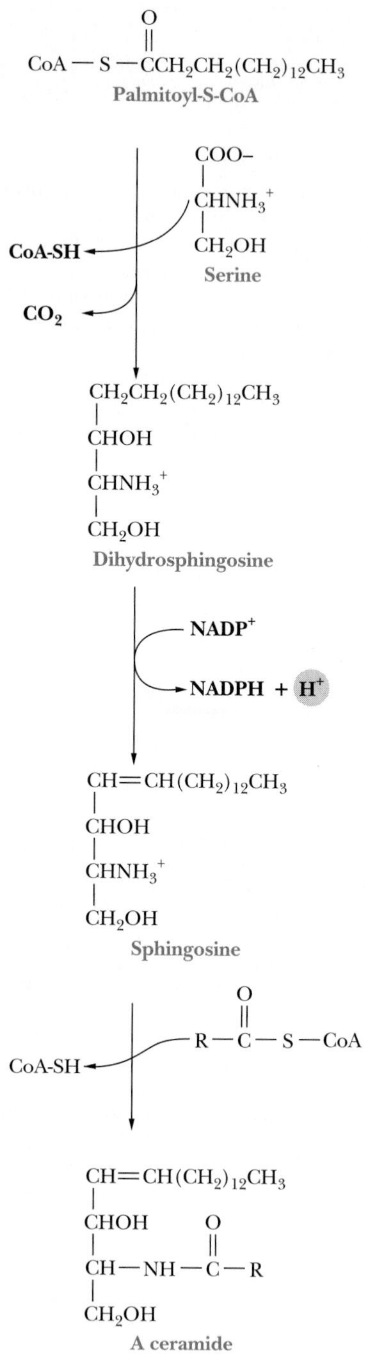

$$CoA - S - \underset{\underset{\displaystyle \text{Palmitoyl-S-CoA}}{}}{\overset{\overset{\displaystyle O}{\|}}{C}CH_2CH_2(CH_2)_{12}CH_3}$$

$$\begin{array}{c} COO^- \\ | \\ CHNH_3^+ \\ | \\ CH_2OH \\ \text{Serine} \end{array}$$

CoA-SH ←

CO₂ ←

$$\begin{array}{c} CH_2CH_2(CH_2)_{12}CH_3 \\ | \\ CHOH \\ | \\ CHNH_3^+ \\ | \\ CH_2OH \\ \textbf{Dihydrosphingosine} \end{array}$$

— **NADP⁺**

→ **NADPH + H⁺**

$$\begin{array}{c} CH{=}CH(CH_2)_{12}CH_3 \\ | \\ CHOH \\ | \\ CHNH_3^+ \\ | \\ CH_2OH \\ \textbf{Sphingosine} \end{array}$$

$$R - \overset{\overset{\displaystyle O}{\|}}{C} - S - CoA$$

CoA-SH ←

$$\begin{array}{c} CH{=}CH(CH_2)_{12}CH_3 \\ | \\ CHOH \qquad O \\ | \qquad\qquad \| \\ CH - NH - C - R \\ | \\ CH_2OH \\ \textbf{A ceramide} \end{array}$$

▶ **FIGURE 21.22** The biosynthesis of sphingolipids. When ceramides are formed, they can react (a) with choline to yield sphingomyelins, (b) with sugars to yield cerebrosides, or (c) with sugars and sialic acid to yield gangliosides.

21.8 | How Is Cholesterol Produced?

The ultimate precursor of all the carbon atoms in cholesterol and in the other steroids that are derived from cholesterol is the acetyl group of acetyl-CoA. There are many steps in the biosynthesis of steroids. The condensation of three acetyl groups produces mevalonate, which contains six carbons. Decarboxylation of mevalonate produces the five-carbon isoprene unit frequently encountered in the structure of lipids. The involvement of **isoprene units** is a key point in the biosynthesis of steroids and of many other compounds that have the generic name terpenes. Vitamins A, E, and K come from reactions involving terpenes that humans cannot carry out. That is why

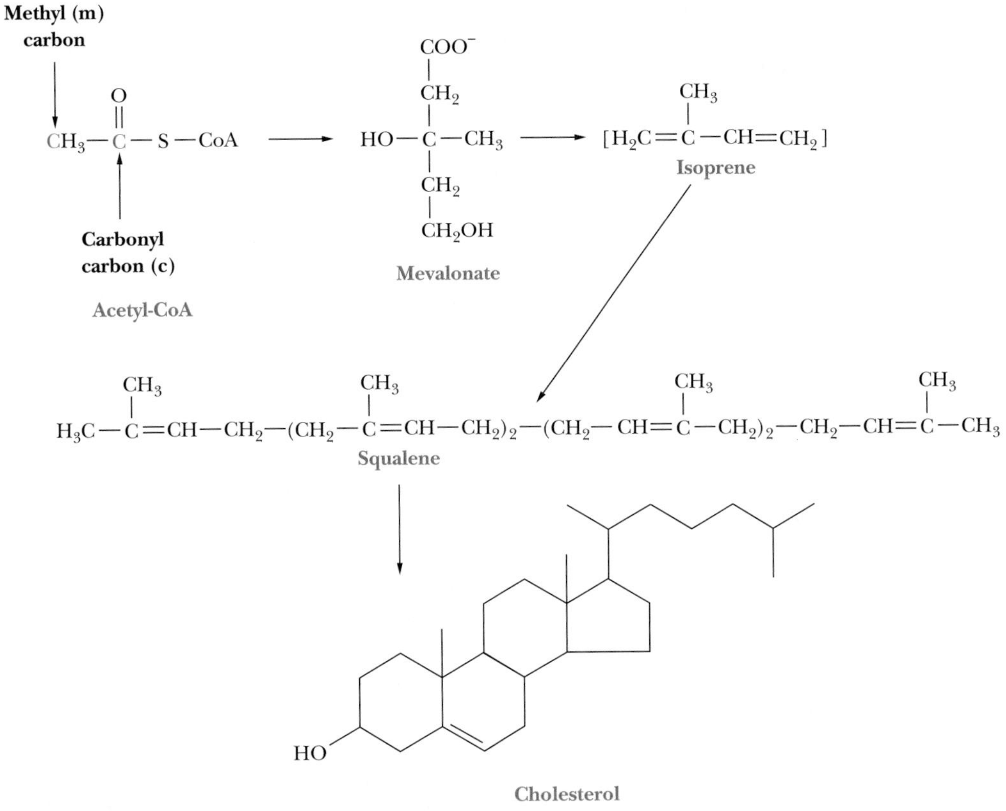

Methyl (m) carbon

$CH_3 - \overset{\overset{O}{\|}}{C} - S - CoA$ **Carbonyl carbon (c)**

Acetyl-CoA

$\longrightarrow$

$HO - \overset{\overset{COO^-}{|}\ \overset{CH_2}{|}}{\underset{\underset{CH_2OH}{|}\ \underset{CH_2}{|}}{C}} - CH_3$

Mevalonate

$\longrightarrow$

$[H_2C = \overset{\overset{CH_3}{|}}{C} - CH = CH_2]$

Isoprene

$H_3C - \overset{\overset{CH_3}{|}}{C} = CH - CH_2 - (CH_2 - \overset{\overset{CH_3}{|}}{C} = CH - CH_2)_2 - (CH_2 - CH = \overset{\overset{CH_3}{|}}{C} - CH_2)_2 - CH_2 - CH = \overset{\overset{CH_3}{|}}{C} - CH_3$

Squalene

HO

Cholesterol

▲ **FIGURE 21.23** Outline of the biosynthesis of cholesterol.

these compounds are vitamins that we must consume in our diets; vitamin D, the remaining lipid-soluble vitamin, is derived from cholesterol (Section 8.8). Isoprene units are involved in the biosynthesis of ubiquinone (coenzyme Q) and of derivatives of proteins and tRNA with specific five-carbon units attached. Isoprene units are often added to proteins to act as anchors when the protein is attached to a membrane.

Six isoprene units condense to form squalene, which contains 30 carbon atoms. Finally, squalene is converted to **cholesterol,** which contains 27 carbon atoms (Figure 21.23); squalene can also be converted to other sterols.

$$\text{Acetate} \rightarrow \text{Mevalonate} \rightarrow \text{[Isoprene]} \rightarrow \text{Squalene} \rightarrow \text{Cholesterol}$$
$$\quad\ C_2 \qquad\quad C_6 \qquad\qquad\ C_5 \qquad\qquad C_{30} \qquad\qquad C_{27}$$

It is well established that 12 of the carbon atoms of cholesterol arise from the carboxyl carbon of the acetyl group; these are the carbon atoms labeled "c" in Figure 21.24. The other 15 carbon atoms arise from the methyl carbon of the acetyl group; these are the carbon atoms labeled "m." We shall now look at the individual steps of the process in more detail.

The conversion of three acetyl groups of acetyl-CoA to *mevalonate* takes place in several steps (Figure 21.25). We already saw the first of these steps, the production of acetoacetyl-CoA from two molecules of acetyl-CoA, when we discussed the formation of ketone bodies and the anabolism of fatty acids. A third molecule of acetyl-CoA condenses with acetoacetyl-CoA to produce *β-hydroxy-β-methylglutaryl-CoA* (also called HMG-CoA and 3-hydroxy-3-methylglutaryl-CoA). This reaction is catalyzed by the enzyme hydroxymethylglutaryl-CoA synthase; one molecule of CoA-SH is released in the process. In the next reaction, the production of mevalonate from hydroxymethylglutaryl-CoA is catalyzed by the

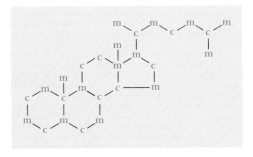

▲ **FIGURE 21.24** The labeling pattern of cholesterol. Each letter "m" indicates a methyl carbon and each letter "c" indicates a carbonyl carbon, all of which come from acetyl-CoA.

CH₃
|
C=O
|
CH₂
|
O=C—S—CoA
Acetoacetyl-CoA

$$\begin{array}{c} O \\ \parallel \\ \text{—H}_3C\text{—C—S—CoA} + H_2O \end{array}$$

Hydroxymethylglutaryl-CoA synthetase

CoA-SH + ⓗ⁺

COO⁻
|
CH₂
|
HO—C—CH₃
|
CH₂
|
O=C—S—CoA
β-Hydroxy-β-methylglutaryl-CoA

2 NADPH

Hydroxymethylglutaryl-CoA reductase

2 NADP⁺ + CoA-SH

COO⁻
|
CH₂
|
HO—C—CH₃
|
CH₂
|
CH₂OH

▲ **FIGURE 21.25** The biosynthesis of mevalonate.

enzyme hydroxymethylglutaryl-CoA reductase (HMG-CoA reductase). A carboxyl group, the one esterified to CoA-SH, is reduced to a hydroxyl group, and the CoA-SH is released. This step is inhibited by high levels of cholesterol and is the major control point of cholesterol synthesis. It is also a target for drugs to lower cholesterol levels in the body. Drugs such as *lovastatin* are inhibitors of hydroxymethyl-CoA reductase and are widely prescribed to lower blood cholesterol levels. The drug is metabolized to mevinolinic acid, which is a transition-state analogue of a tetrahedral intermediate in the reaction catalyzed by HMG-CoA reductase (Figure 21.26).

Mevalonate is then converted to an isoprenoid unit by a combination of phosphorylation, decarboxylation, and dephosphorylation reactions (Figure 21.27). Three successive reactions, each of which is catalyzed by an enzyme that requires ATP, give rise to *isopentenyl pyrophosphate,* a five-carbon isoprenoid derivative. Isopentenyl pyrophosphate and *dimethylallyl pyrophosphate,* another isoprenoid derivative, can be interconverted in a rearrangement reaction catalyzed by the enzyme isopentenyl pyrophosphate isomerase.

Condensation of isoprenoid units then leads to the production of squalene and, ultimately, cholesterol. Both of the isoprenoid derivatives we have met so far are required. Two further condensation reactions take place. As a result, *farnesyl pyrophosphate,* a 15-carbon compound, is produced. Two molecules of farnesyl pyrophosphate condense to form *squalene,* a 30-carbon compound. The reaction is catalyzed by squalene synthase, and NADPH is required for the reaction.

Figure 21.28 shows the conversion of squalene to cholesterol. The details of this conversion are far from simple. Squalene is converted to *squalene epoxide* in a reaction that requires both NADPH and molecular oxygen (O_2). This reaction is catalyzed by squalene monooxygenase. Squalene epoxide then undergoes a complex cyclization reaction to form *lanosterol.* This remarkable reaction is catalyzed by squalene epoxide cyclase. The mechanism of the reaction is a concerted reaction—that is, one in which each part is essential for any other part to take place. No portion of a concerted reaction can be left out or changed because it all takes place simultaneously rather than in a sequence of steps. The conversion of lanosterol to cholesterol is a complex process. It is known that 20 steps are required to remove three methyl groups and to move a double bond, but we shall not discuss the details of the process.

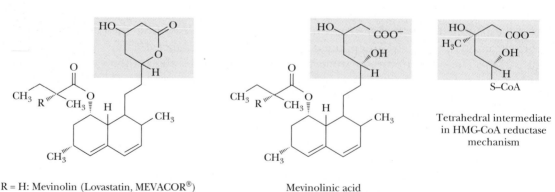

R = H: Mevinolin (Lovastatin, MEVACOR®)
R = CH₃: Synvinolin (Simnastatin, ZOCOR®)

Mevinolinic acid

Tetrahedral intermediate in HMG-CoA reductase mechanism

▲ **FIGURE 21.26** The structures of (inactive) lovastatin and synvinolin, (active) mevinolinic acid, and the tetrahedral intermediate in the HMG-CoA reductase mechanism.

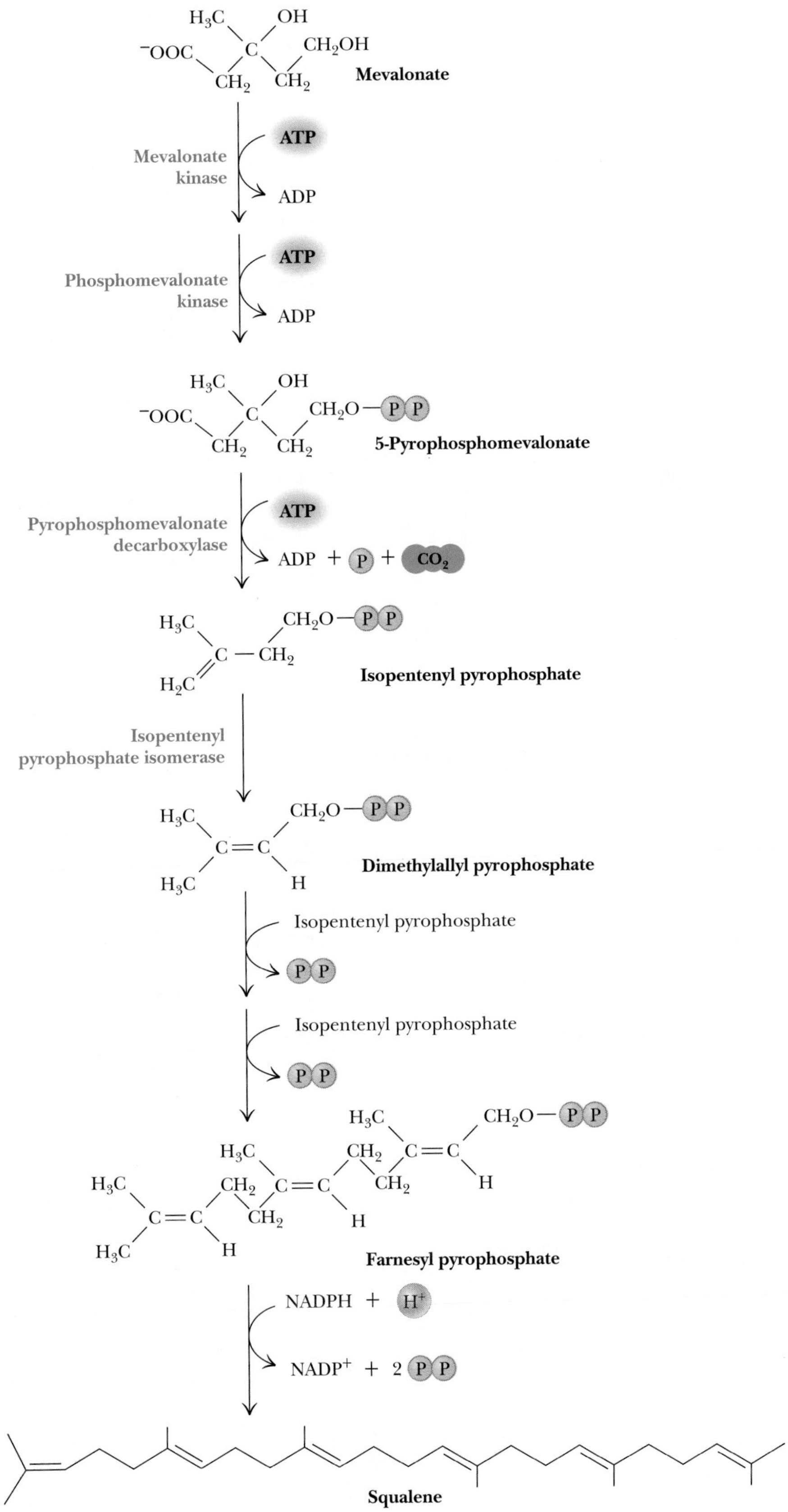

Biochemistry🔵Now™ ACTIVE FIGURE 21.27
The conversion of mevalonate to squalene. **Watch this Active Figure at http://now.brookscole.com/campbell5**

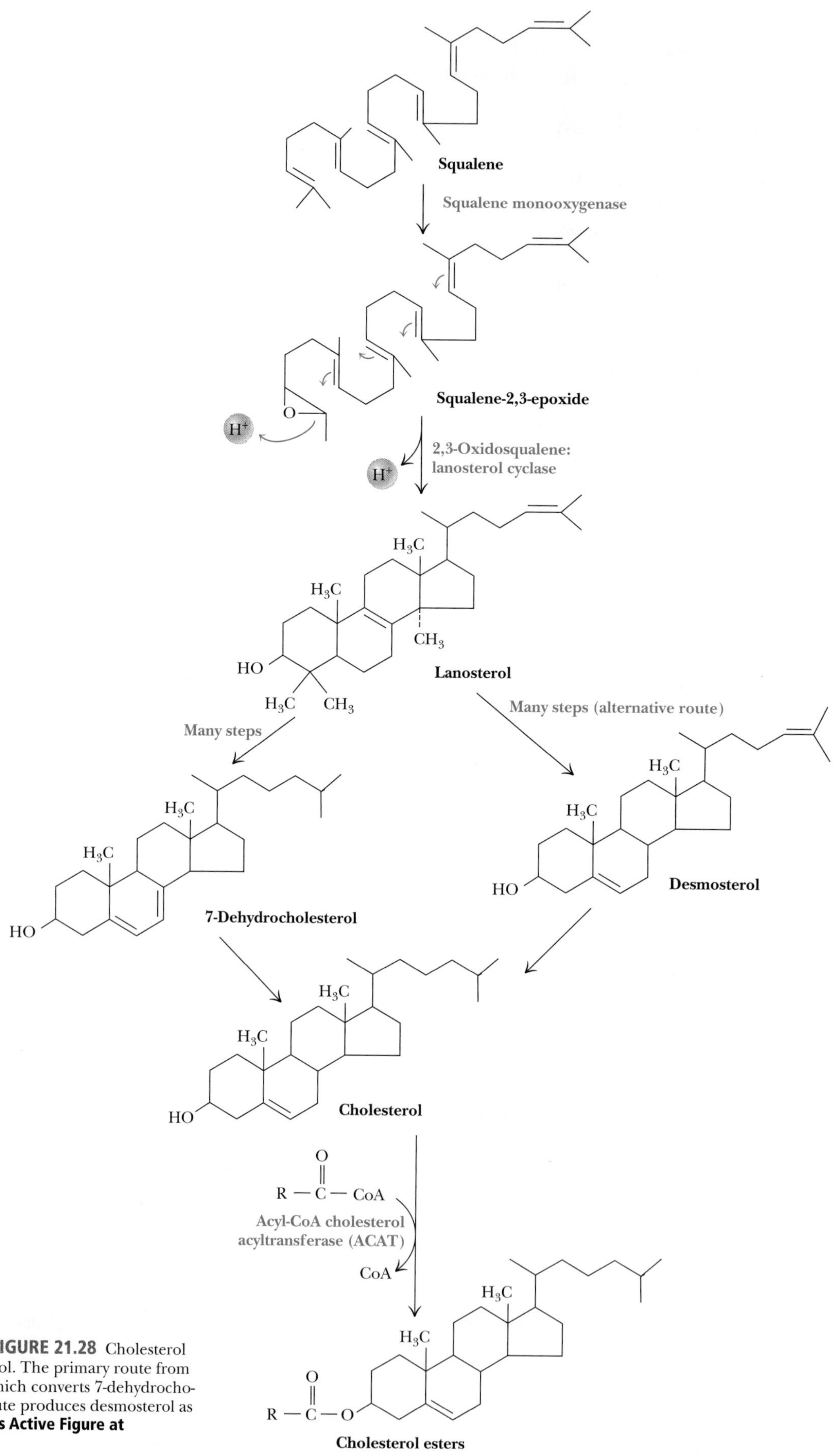

Squalene

Squalene monooxygenase

Squalene-2,3-epoxide

H⁺

2,3-Oxidosqualene:
lanosterol cyclase

H⁺

Lanosterol

Many steps

Many steps (alternative route)

7-Dehydrocholesterol

Desmosterol

Cholesterol

Acyl-CoA cholesterol
acyltransferase (ACAT)

CoA

Cholesterol esters

Biochemistry ⑤ Now™ ▲ ACTIVE FIGURE 21.28 Cholesterol
is synthesized from squalene via lanosterol. The primary route from
lanosterol involves 20 steps, the last of which converts 7-dehydrocho-
lesterol to cholesterol. An alternative route produces desmosterol as
the penultimate intermediate. **Watch this Active Figure at**
http://now.brookscole.com/campbell5

Cholesterol Is a Precursor of Other Steroids

After cholesterol is formed, it can be converted to other steroids of widely varying physiological function. The smooth ER is an important site for both the synthesis of cholesterol and its conversion to other steroids. Most of the cholesterol formed in the liver, which is the principal site of cholesterol synthesis in mammals, is converted to *bile acids,* such as cholate and glycocholate (Figure 21.29). These compounds aid in the digestion of lipid droplets by emulsifying them and rendering them more accessible to enzymatic attack.

Cholesterol is the precursor of important **steroid hormones** (Figure 21.30), in addition to the bile acids. Like all hormones, whatever their chemical nature (Section 24.3), steroid hormones serve as signals from outside a cell that regulate metabolic processes within a cell. Steroids are best known as sex hormones (they are components of birth-control pills), but they play other roles as well. *Pregnenolone* is formed from cholesterol, and *progesterone* is formed from pregnenolone. Progesterone is a sex hormone and is a precursor for other sex hormones, such as *testosterone* and *estradiol* (an estrogen). Other types of steroid hormones also arise from progesterone. The role of sex hormones in sexual maturation is discussed in Section 24.3. *Cortisone* is an

> ### Essential Information
>
> The biosynthesis of cholesterol proceeds by the condensation of five-carbon isoprenoid units to form the key intermediate squalene. Isoprenoid units in turn are derived from the reaction of three acetyl-CoA units. Once cholesterol is formed, it serves as the precursor for other steroids.

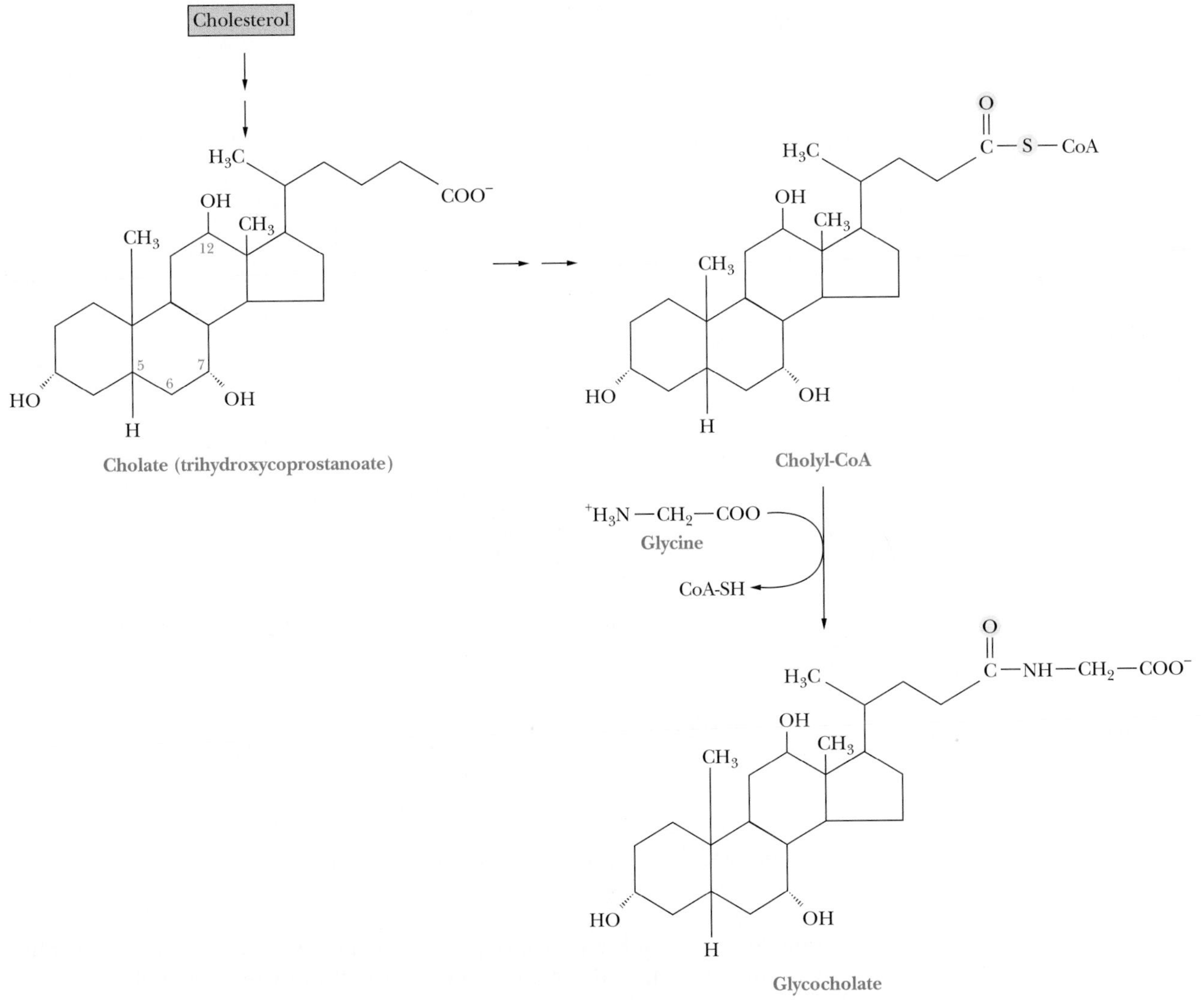

▲ **FIGURE 21.29** The synthesis of bile acids from cholesterol.

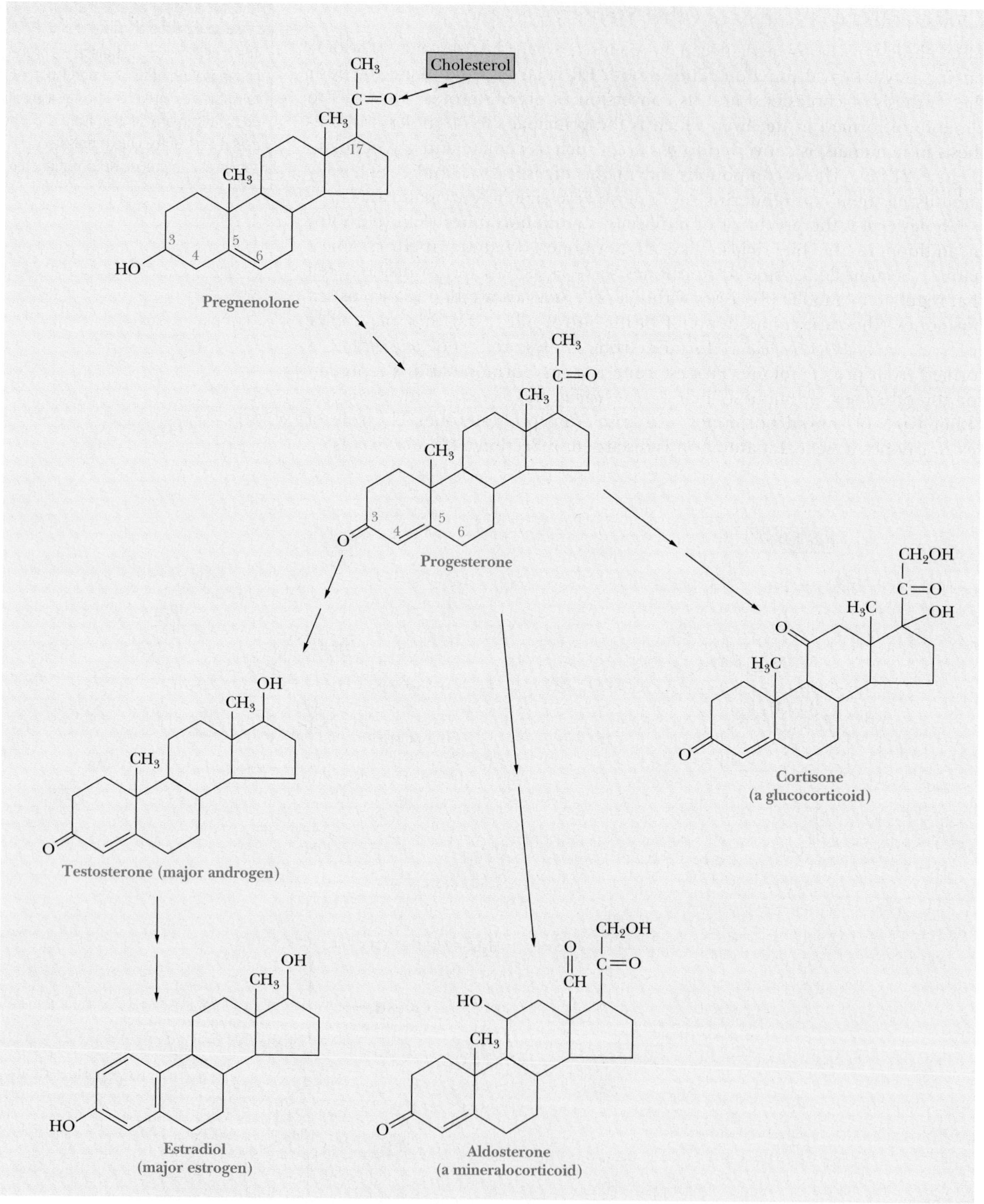

▲ **FIGURE 21.30** The synthesis of steroid hormones from cholesterol.

example of *glucocorticoids,* a group of hormones that play a role in carbohydrate metabolism, as the name implies, as well as in the metabolism of proteins and fatty acids. *Mineralocorticoids* constitute another class of hormones that are involved in the metabolism of electrolytes, including metal ions ("minerals") and water. *Aldosterone* is an example of a mineralocorticoid. In cells in which cholesterol is converted to steroid hormones, an enlarged smooth ER is frequently observed, providing a site for the process to take place.

The Role of Cholesterol in Heart Disease

Atherosclerosis is a condition in which arteries are blocked to a greater or lesser extent by the deposition of cholesterol plaques, which can lead to heart attacks. The process by which the clogging of arteries occurs is complex. Both diet and genetics are instrumental in the development of atherosclerosis. A diet high in cholesterol and fats, particularly saturated fats, will lead to a high level of cholesterol in the bloodstream. The body also makes its own cholesterol because this steroid is a necessary component of cell membranes. It is possible for more cholesterol to come from endogenous sources (synthesized within the body) than from the diet.

Cholesterol must be packaged for transport in the bloodstream; several classes of lipoproteins (summarized in Table 21.3) are involved in the transport of lipids in blood. These lipoprotein aggregates are usually classified by their densities. Besides chylomicrons, they include very-low-density lipoproteins **(VLDL),** intermediate-density lipoproteins **(IDL),** low-density lipoproteins **(LDL),** and high-density lipoproteins **(HDL).** The density increases as the protein content increases. LDL and HDL will play the major role in our discussion of heart disease. The protein portions of these aggregates can vary widely. The major lipids are generally cholesterol and its esters, in which the hydroxyl group is esterified to a fatty acid; triacylglycerols are also found in these aggregates. Chylomicrons are involved in the transport of dietary lipids, whereas the other lipoproteins primarily deal with endogenous lipids.

Figure 21.31 shows the architecture of an LDL particle. The interior consists of many molecules of cholesteryl esters (the hydroxyl group of the cholesterol is esterified to an unsaturated fatty acid, such as linoleate). On the surface, protein (apoprotein B-100), phospholipids, and unesterified cholesterol are in contact with the aqueous medium of the plasma. The protein portions of LDL particles bind to receptor sites on the surface of a typical cell. Refer to Membrane Receptors in Section 8.6 for a discussion of the process by which LDL particles are taken into the cell as one aspect of receptor action. This process is typical of the mechanism of uptake of lipids by cells, and we shall use the processing of LDL as a case study. LDL is the major player in the development of atherosclerosis.

LDL particles are degraded in the cell. LDL particles are taken into the cell by the highly regulated process of endocytosis (Section 8.6), in which a portion of the cell membrane containing the LDL particle and its receptor enters the cell. The receptor is returned to the cell surface, while the LDL particles are degraded in the lysosomes (organelles that contain degradative enzymes; see Section 1.6). The protein portion of LDL is hydrolyzed to the component amino acids, while the cholesterol esters are hydrolyzed to cholesterol and fatty acids. Free cholesterol can then be used directly as a component of membranes; the fatty acids can have any of the catabolic or anabolic fates discussed earlier in this chapter (Figure 21.32). Cholesterol not needed for membrane synthesis can be stored as oleate or palmitoleate esters in which the fatty acid is esterified to the hydroxyl group of cholesterol. The production of these esters is catalyzed by acyl-CoA:cholesterol acyltransferase (ACAT), and the presence of free cholesterol increases the enzymatic activity of ACAT. In addition, cholesterol inhibits both the synthesis and the activity of the enzyme hydroxymethylglutaryl-CoA reductase (HMG-CoA reductase). This enzyme catalyzes the production of mevalonate, the reaction that is the

Table 21.3	
Major Classes of Lipoproteins in Human Plasma	
Lipoprotein class	**Density (g mL^{-1})**
Chylomicrons	<0.95
VLDL	0.95–1.006
IDL	1.006–1.019
LDL	1.019–1.063
HDL	1.063–1.210

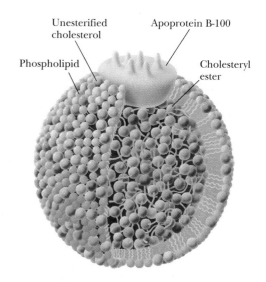

▲ **FIGURE 21.31** Schematic diagram of an LDL particle. (*From M. S. Brown and J. L. Goldstein, 1984, How LDL Receptors Influence Cholesterol and Atherosclerosis, Sci. Amer.* **251** (5), 58–66.)

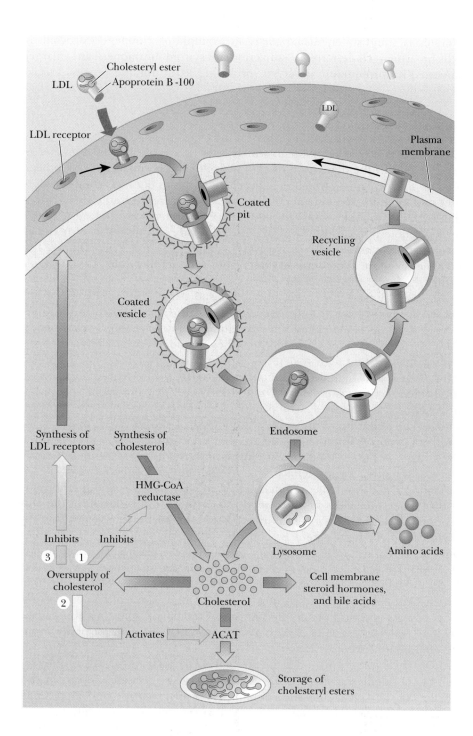

▶ **FIGURE 21.32** The fate of cholesterol in the cell (see page 599). ACAT is the enzyme that esterifies cholesterol for storage. (*From M. S. Brown and J. L. Goldstein, 1984, How LDL Receptors Influence Cholesterol and Atherosclerosis, Sci. Amer. 251 (5), 58–66.*)

committed step in cholesterol biosynthesis. This point has important implications. Dietary cholesterol suppresses the synthesis of cholesterol by the body, especially in tissues other than the liver. A third effect of the presence of free cholesterol in the cell is inhibition of synthesis of LDL receptors. As a result of reduction in the number of receptors, cellular uptake of cholesterol is inhibited, and the level of LDL in the blood increases, leading to the deposition of atherosclerotic plaques.

The crucial role of LDL receptors in maintaining the level of cholesterol in the bloodstream is especially clear in the case of *familial hypercholesterolemia,* which results from a defect in the gene that codes for the active receptors. An individual who has one gene that codes for the active receptor and one defective gene is heterozygous for this trait. Heterozygotes have blood cholesterol levels that are above average; therefore, they are at higher risk for heart disease than the general population. An individual with two defective genes, and thus with no active LDL receptor, is homozygous for the trait. Homozygotes have very high blood-cholesterol levels from birth, and there are recorded cases of heart attacks in two-year-olds with this condition. Patients who are homozygous for familial hypercholesterolemia usually die before age 20. Another genetic abnormality involved in hypercholesterolemia is the one that gives rise to a faulty apolipoprotein E, a component of IDL and VLDL, which is involved in the uptake of lipids by the cell. The unfortunate result is the same.

Before we leave this discussion, we should mention the "good" cholesterol, HDL. Unlike LDL, which transports cholesterol from the liver to the rest of the body, HDL transports it back to the liver for degradation to bile acids. It is desirable to have low levels of cholesterol and LDL in the bloodstream, but it is also desirable to have as high a proportion of total cholesterol as possible in the form of HDL. It is well known that high levels of LDL and low levels of HDL are correlated with the development of heart disease. Factors that are known to increase HDL levels, such as regular strenuous exercise, decrease the probability of heart disease. Smoking reduces the level of HDL and is highly correlated with heart disease.

Summary

21.1 How Are Lipids Involved in the Generation and Storage of Energy?
We have already seen how carbohydrates are processed catabolically and anabolically. Lipids are another class of nutrient. The catabolic oxidation of lipids releases large quantities of energy, whereas the anabolic formation of lipids represents an efficient way of storing chemical energy.

21.2 How Are Lipids Catabolized?
The oxidation of fatty acids is the chief source of energy in the catabolism of lipids. After an initial activation step in the cytosol, the breakdown of fatty acids takes place in the mitochondrial matrix by the process of β-oxidation. In this process, two-carbon units are successively removed from the carboxyl end of the fatty acid to produce acetyl-CoA, which subsequently enters the citric acid cycle. The reactions that liberate the acetyl-CoA units from a fatty acid produce NADH and $FADH_2$, which eventually produce ATP via the electron transport chain.

21.3 What Is the Energy Yield from the Oxidation of Fatty Acids?
There is a net yield of 120 ATP molecules for each molecule of stearic acid (an 18-carbon compound) that is completely oxidized to carbon dioxide and water. The source of these ATP molecules is the production of NADH and $FADH_2$ in the β-oxidation pathway, as well as the NADH, $FADH_2$, and GTP produced when the acetyl-CoA molecules are processed through the electron transport chain.

21.4 How Are Unsaturated Fatty Acids and Odd-Carbon Fatty Acids Catabolized?
The pathway of catabolism of fatty acids includes reactions in which unsaturated, as well as saturated, fatty acids can be metabolized. Odd-numbered fatty acids can also be metabolized by converting their unique breakdown product, propionyl-CoA, to succinyl-CoA, an intermediate of the citric acid cycle.

21.5 What Are Ketone Bodies?
Ketone bodies are substances related to acetone that are produced when an excess of acetyl-CoA results from β-oxidation. This situation can arise from a large intake of lipids and a low intake of carbohydrates or can occur in diabetes, in which the inability to metabolize carbohydrates causes an imbalance in the breakdown products of carbohydrates and lipids.

21.6 How Are Fatty Acids Produced?
The anabolism of fatty acids proceeds by a different pathway from β-oxidation. Some of the most important differences between the two processes are the requirement for biotin in anabolism, but not in catabolism, and the requirement for NADPH in anabolism, rather than the NAD^+ required in catabolism. Fatty-acid biosynthesis occurs in the cytosol, catalyzed by an ordered multienzyme complex; fatty-acid catabolism occurs in the mitochondrial matrix, with no ordered aggregate of enzymes.

21.7 How Are Acylglycerols and Compound Lipids Produced?
Most compound lipids, such as triacylglycerols, phosphoacylglycerols, and sphingolipids, have fatty acids as precursors. These fatty acids are linked to a backbone molecule, which is glycerol in some cases and sphingosine in others.

21.8 How Is Cholesterol Produced?
In the case of steroids, the starting material is acetyl-CoA. Isoprene units are formed from acetyl-CoA in the early stages of a lengthy process that leads ultimately to cholesterol. Cholesterol in turn is the precursor of the other steroids. Both dietary cholesterol and genetic factors influence the role of cholesterol in heart disease.

Critical Questions to Review

21.1 How Are Lipids Involved in the Generation and Storage of Energy?

1. **Thought Question** (a) The major energy storage compound of animals is fats (except in muscles). Why would this be advantageous? (b) Why don't plants use fats/oils as their *major* energy storage compound?

21.2 How Are Lipids Catabolized?

2. **Fact Check** What is the difference between phospholipase A_1 and A_2?

3. **Fact Check** How are lipases activated hormonally?

4. **Fact Check** What is the metabolic purpose of linking a fatty acid to coenzyme A?

5. **Fact Check** Outline the role of carnitine in the transport of acyl-CoA molecules into the mitochondrion. How many enzymes are involved? What are they called?

6. **Fact Check** What is the difference between the type of oxidation catalyzed by acyl-CoA dehydrogenase and that catalyzed by β-hydroxy-CoA dehydrogenase?

7. **Fact Check** Draw a six-carbon saturated fatty acid and show where the double bond is created during the first step of β-oxidation. What is the orientation of this bond?

8. **Thought Question** Why does the degradation of palmitic acid (see Question 12) to eight molecules of acetyl-CoA require seven, rather than eight, rounds of the β-oxidation process?

9. **Thought Question** Given the nature of the hormonal activation of lipases, what carbohydrate pathways would be activated or inhibited under the same conditions?

21.3 What Is the Energy Yield from the Oxidation of Fatty Acids?

10. **Fact Check** Compare the energy yields from the oxidative metabolism of glucose and of stearic acid. To be fair, calculate it on the basis of ATP equivalents per carbon and also ATP equivalents per gram.

11. **Fact Check** Which generates more ATP—the processing of the reduced electron equivalents formed during β-oxidation through the electron transport chain, or the processing of the acetyl-CoA generated from β-oxidation through the citric acid cycle and the electron transport chain?

12. **Mathematical** Calculate the ATP yield for the complete oxidation of one molecule of palmitic acid (16 carbons). How does this figure differ from that obtained for stearic acid (18 carbons)? Consider the β-oxidation steps, processing of acetyl-CoA through the citric acid cycle, and electron transport.

13. **Thought Question** It is frequently said that camels store water in their humps for long desert journeys. How would you modify this statement on the basis of information in this chapter?

21.4 How Are Unsaturated Fatty Acids and Odd-Carbon Fatty Acids Catabolized?

14. **Fact Check** Describe briefly how β-oxidation of an odd-chain fatty acid is different from that for an even-chain fatty acid.

15. **Fact Check** You hear a fellow student say that the oxidation of unsaturated fatty acids requires exactly the same group of enzymes as the oxidation of saturated fatty acids. Is the statement true or false? Why?

16. **Fact Check** What are the unique enzymes needed to β-oxidize a monounsaturated fatty acid?

17. **Fact Check** What are the unique enzymes needed to β-oxidize a polyunsaturated fatty acid?

18. **Mathematical** Calculate the net ATP yield from the complete processing of a saturated fatty acid containing 17 carbons. Consider the β-oxidation steps, processing of acetyl-CoA through the citric acid cycle, and electron transport.

19. **Mathematical** Calculate the net ATP yield from oleic acid (18:1 Δ^9). *Hint:* Remember the step that bypasses acyl-CoA dehydrogenase.

20. **Mathematical** Calculate the net ATP yield from linoleic acid (18:2 $\Delta^{9,12}$). For this calculation, assume that the loss of an NADPH is the same as the loss of an NADH.

21. **Thought Question** How many cycles of β-oxidation are required to process a fatty acid with 17 carbons?

22. **Thought Question** It has been stated many times that fatty acids cannot yield a *net* gain in carbohydrates. Why can odd-chain fatty acids be thought to break this rule to a small extent?

21.5 What Are Ketone Bodies?

23. **Fact Check** Under what conditions are ketone bodies produced?

24. **Fact Check** Briefly outline the reactions involved in ketone production.

25. **Thought Question** Why might a doctor smell the breath of a known diabetic who has just passed out?

26. **Thought Question** Why might a person who is an alcoholic have a "fatty liver"?

27. **Thought Question** A friend who is trying to lose weight complains about the odd taste in his mouth in the mornings. He says it seems like a filling has broken loose, and the metallic sensation is bothersome. What would you say?

21.6 How Are Fatty Acids Produced?

28. **Fact Check** Compare and contrast the pathways of fatty-acid breakdown and biosynthesis. What features do these two pathways have in common? How do they differ?

29. **Fact Check** Outline the steps involved in the production of malonyl-CoA from acetyl-CoA.

30. **Fact Check** What is the metabolic importance of malonyl-CoA?

31. **Fact Check** In fatty-acid degradation, we encounter coenzyme A, mitochondrial matrix, *trans* double bonds, L-alcohols, β-oxidation, NAD^+ and FAD, acetyl-CoA, and separate enzymes. What are the counterparts in fatty-acid synthesis?

32. **Fact Check** How are the two redox reactions of β-oxidation different from their counterparts in fatty-acid synthesis?

33. **Fact Check** How is ACP similar to coenzyme A? How is it different?

34. **Fact Check** What is the purpose of having ACP as a distinct activating group for fatty-acid synthesis?

35. **Fact Check** Why are linoleate and linolenate considered essential fatty acids? What step in production of polyunsaturated fatty acids are mammals unable to perform?

36. **Thought Question** Is it possible to convert fatty acids to other lipids without acyl-CoA intermediates?

37. **Thought Question** What is the role of citrate in the transport of acetyl groups from the mitochondrion to the cytosol?

38. **Thought Question** In the mitochondrion, there is a short-chain carnitine acyltransferase that can take acetyl groups from acetyl-CoA and transfer them to carnitine. How might this be related to lipid biosynthesis?

39. **Thought Question** In fatty-acid synthesis, malonyl-CoA, rather than acetyl-CoA, is used as a "condensing group." Suggest a reason for this.

40. **Thought Question** (a) Where in an earlier chapter have we encountered something comparable to the action of the acyl-carrier protein (ACP) of fatty-acid synthesis? (b) What is a critical feature of the action of the ACP?

21.7 How Are Acylglycerols and Compound Lipids Produced?

41. **Fact Check** What is the source of the glycerol in triacylglycerol synthesis?

42. **Fact Check** What is the activating group used in the formation of phosphoacylglycerols?

43. **Fact Check** What are the differences between synthesis of phosphatidylethanolamine in prokaryotes and eukaryotes?

21.8 How Is Cholesterol Produced?

44. **Fact Check** How are isoprene units important in cholesterol biosynthesis and other biochemical pathways?

45. **Fact Check** A cholesterol sample is prepared using acetyl-CoA labeled with ^{14}C at the carboxyl group as precursor. Which carbon atoms of cholesterol are labeled?

46. **Fact Check** Which molecules have cholesterol as a precursor?

47. **Thought Question** What structural feature do all steroids have in common? What are the biosynthetic implications of this common feature?

48. **Thought Question** In steroid synthesis, squalene is oxidized to squalene epoxide. This reaction is somewhat unusual, in that both a reducing agent (NADPH) and an oxidizing agent (O_2) are required. Why are both needed?

49. **Thought Question** Why must cholesterol be packaged for transport rather than occurring freely in the bloodstream?

50. **Thought Question** A drug that reduces blood cholesterol has the effect of stimulating the production of bile salts. How might this result in lower blood cholesterol? *Hint:* There are two ways.

Biochemistry ⓔ Now™

Assess your understanding of this chapter's topics with additional quizzing and tutorials at **http://now.brookscole.com/campbell5**

Annotated Bibliography

Abu-Elheiga, L., M. M. Matzuk, K. A. H. Abo-Hashema, and S. J. Wakil. Continuous Fatty Acid Oxidation and Reduced Fat Storage in Mice Lacking ACC_2. *Science* **291**, 2613–2616 (2001). [An article about metabolic effects seen in mice lacking one of the isoforms of acetyl-CoA carboxylase.]

Bodner, C. M. Lipids. *J. Chem. Ed.* **63**, 772–775 (1986). [Part of a series of concise and clearly written articles on metabolism.]

Brown, M. S., and J. L. Goldstein. How LDL Receptors Influence Cholesterol and Atherosclerosis. *Sci. Amer.* **251** (5), 58–66 (1984). [A description of the role of cholesterol in heart disease by the winners of the 1985 Nobel Prize in medicine.]

Krutch, J. W. *The Voice of the Desert*. New York: Morrow, 1975. [Chapter 7, "The Mouse That Never Drinks," is a description, primarily from a naturalist's point of view, of the kangaroo rat, but it does make the point that metabolic water is this animal's only source of water.]

Lawn, R. Lipoprotein(a) in Heart Disease. *Sci. Amer.* **266** (6), 54–60 (1992). [Relates the properties of lipids and protein structure to the blockage of arteries characteristic of heart disease.]

McCarry, J. D., and D. W. Foster. Regulation of Hepatic Fatty Acid Oxidation and Ketone Body Production. *Ann. Rev. Biochem.* **49**, 395–420 (1980). [A review.]

Ruderman, N., and J. S. Flier. Chewing the Fat—ACC and Energy Balance. *Science* **291**, 2558–2561 (2001). [A summary of information about acetyl-CoA carboxylase and lipid metabolism.]

Wakil, S. J., and E. M. Barnes. Fatty Acid Metabolism. *Compr. Biochem.* **18**, 57–104 (1971). [Extensive coverage of the topic.]

CHAPTER 22

Photosynthesis

The drama of photosynthesis, converting sunlight to energy-rich carbohydrates, is played out in the chloroplast "theater" of the green plant. In each chloroplast, there are stacks of thylakoid disks. The thylakoid membrane inside each disk is the lighted stage where the drama of Act I is performed. Here the energy of light is captured by electrons of chlorophyll molecules. The excited electrons are passed along a series of acceptors in an electron transport chain. In the process, a molecule of water is split, and oxygen is released into the atmosphere. At the same time, protons pumped out of the thylakoid membrane drive the production of ATP. Excited electrons reduce NAD^+ to NADPH, and the stored energy is used in Act II for the biosynthesis of glucose, which takes place in the dark of the stroma outside the thylakoid membrane. Carbon dioxide from the atmosphere is combined with a five-carbon sugar to produce, through an intermediate, two three-carbon sugars and, eventually, the six-carbon molecule of glucose. The energy to drive this biosynthesis comes from ATP and the reducing power of NADPH, the reduced form of nicotinamide adenine dinucleotide phosphate. Plants, at the bottom of the food chain, toil in the sun to store energy and to generate oxygen for the benefit of all animals on Earth.

© Paul Harris/Tony Stone Images/Getty

Lush rain forest vegetation. Photosynthesis linked to oxygen plays an essential role in all life, plant and animal.

Critical Questions

Biochemistry ⊘ Now™
Test yourself on these Critical Questions at the BiochemistryNow website at **http://now .brookscole.com/campbell5**

22.1	**Where Does Photosynthesis Take Place in the Cell?**

It is well known that photosynthetic organisms, such as green plants, convert carbon dioxide (CO_2) and water to carbohydrates such as glucose (written here as $C_6H_{12}O_6$) and molecular oxygen (O_2).

$$6\ CO_2 + 6\ H_2O \rightarrow C_6H_{12}O_6 + 6\ O_2$$

The equation actually represents two processes. One process, the oxidation of water to produce oxygen (the light reactions), requires light energy from the sun. The light reactions of photosynthesis in prokaryotes and eukaryotes depend on solar energy, which is absorbed by **chlorophyll** to supply the energy needed in the light reactions. The light reactions also generate NADPH, which is the reducing agent needed in the dark reactions. The other process, the fixation of CO_2 to give sugars (the dark reactions), does not use solar energy directly but rather uses it indirectly in the form of the ATP and NADPH produced in the course of the light reactions.

In prokaryotes such as cyanobacteria, photosynthesis takes place in granules bound to the plasma membrane. The site of photosynthesis in eukaryotes such as green plants and green algae is the **chloroplast** (Figure 22.1), a membrane-enclosed organelle that we discussed in Section 1.6. Like the mitochondrion, the chloroplast has inner and outer membranes and an intermembrane space. In addition, within the chloroplast are bodies called **grana,** which consist of stacks of flattened membranes called **thylakoid disks.** The grana are connected by membranes called intergranal lamellae. The thylakoid disks are formed by the folding of a third membrane within the chloroplast. The folding of the thylakoid membrane creates two spaces in the chloroplast in addition to the intermembrane space. The **stroma** lies within the inner membrane and outside the thylakoid membrane. In addition to the stroma, there is a **thylakoid space** within the thylakoid disks themselves. The

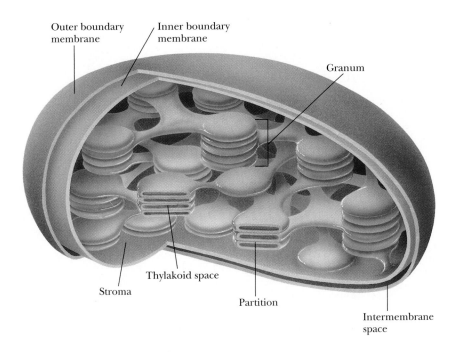

Outer boundary membrane

Inner boundary membrane

Granum

Thylakoid space

Stroma

Partition

Intermembrane space

◀ **FIGURE 22.1** Membrane structures in chloroplasts.

trapping of light and the production of oxygen take place in the thylakoid disks. The dark reactions (also called light-independent reactions), in which CO_2 is fixed to carbohydrates, take place in the stroma (Figure 22.2).

It is well established that the primary event in photosynthesis is the absorption of light by chlorophyll. The high energy states (excited states) of chlorophyll are useful in photosynthesis because the light energy can be passed along and converted to chemical energy in the light reaction. There are two principal types of chlorophyll, *chlorophyll* a and *chlorophyll* b. Eukaryotes such as green plants and green algae contain both chlorophyll *a* and chlorophyll *b*. Prokaryotes such as cyanobacteria (formerly called blue-green algae) contain only chlorophyll *a*. Photosynthetic bacteria other than cyanobacteria have

Essential Information

In eukaryotes, photosynthesis takes place in chloroplasts. The light reactions take place in the thylakoid membrane, a third membrane in chloroplasts in addition to the inner and outer membrane. The dark reactions take place in the stroma, the space between the thylakoid membrane and the inner membrane of the chloroplast.

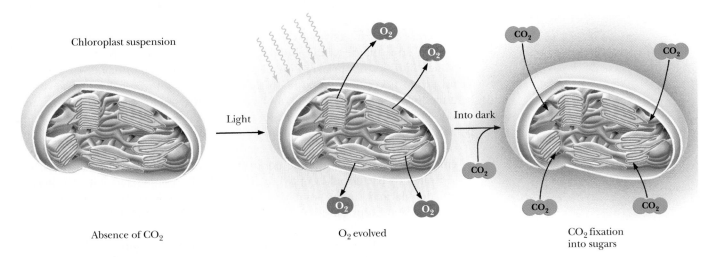

Chloroplast suspension

Light

Into dark

O_2 O_2 O_2 O_2

CO_2

CO_2 CO_2 CO_2 CO_2

Absence of CO_2

O_2 evolved

CO_2 fixation into sugars

Biochemistry☰Now™ ▲ **ANIMATED FIGURE 22.2** The light-dependent and light-independent reactions of photosynthesis. Light reactions are associated with the thylakoid membranes, and light-independent reactions are associated with the stroma. **See this figure animated at http://now.brookscole.com/campbell5**

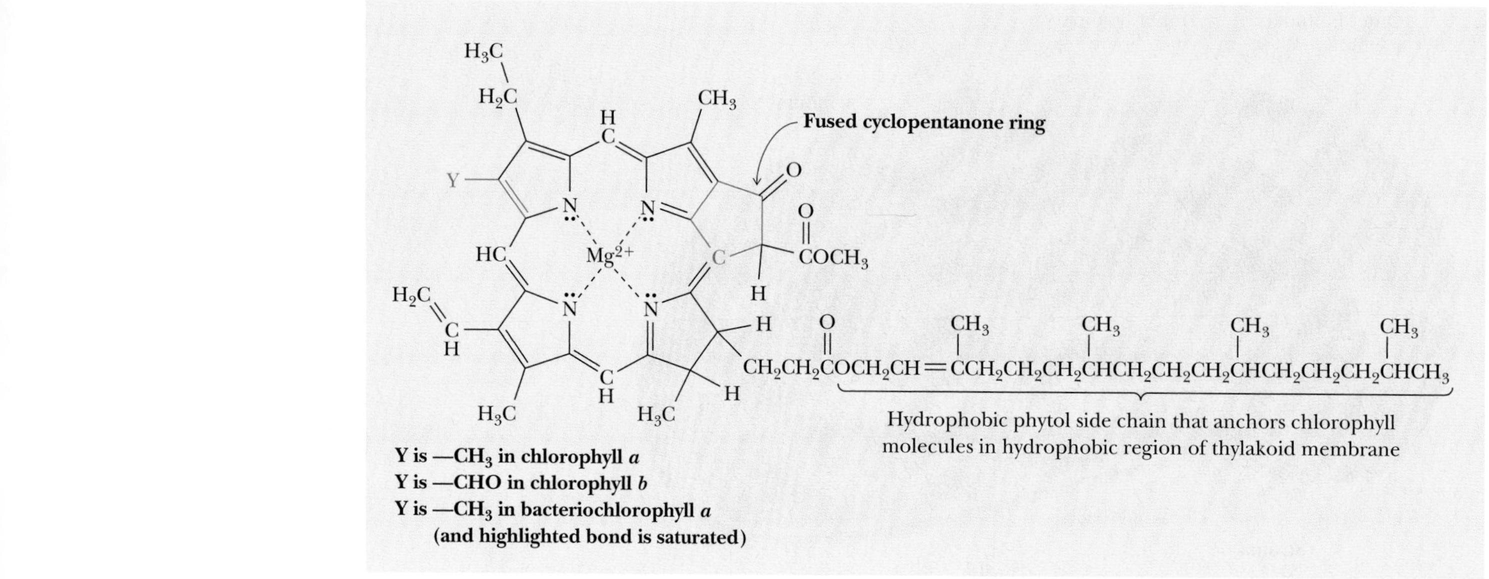

▲ **FIGURE 22.3** Molecular structures of chlorophyll *a*, chlorophyll *b*, and bacteriochlorophyll *a*.

bacteriochlorophylls, with *bacteriochlorophyll* a being the most common. Organisms such as green and purple sulfur bacteria, which contain bacteriochlorophylls, do not use water as the ultimate source of electrons for the redox reactions of photosynthesis, nor do they produce oxygen. Instead, they use other electron sources such as H_2S, which produces elemental sulfur instead of oxygen. Organisms that contain bacteriochlorophyll are anaerobic and have only one photosystem, whereas green plants have two different photosystems, as we shall see.

The structure of chlorophyll is similar to that of the heme group of myoglobin, hemoglobin, and the cytochromes in that it is based on the tetrapyrrole ring of porphyrins (Figure 22.3). (See Section 4.5.) The metal ion bound to the tetrapyrrole ring is magnesium, Mg(II), rather than the iron that occurs in heme. Another difference between chlorophyll and heme is the presence of a cyclopentanone ring fused to the tetrapyrrole ring. There is a long hydrophobic side chain, the phytol group, which contains four isoprenoid units (five-carbon units that are basic building blocks in many lipids; Section 21.8) and which binds to the thylakoid membrane by hydrophobic interactions. The phytol group is covalently bound to the rest of the chlorophyll molecule by an ester linkage between the alcohol group of the phytol and a propionic acid side chain on the porphyrin ring. The difference between chlorophyll *a* and chlorophyll *b* lies in the substitution of an aldehyde group for a methyl group on the porphyrin ring. The difference between bacteriochlorophyll *a* and chlorophyll *a* is that a double bond in the porphyrin ring of chlorophyll *a* is saturated in bacteriochlorophyll *a*. The lack of a conjugated system (alternating double and single bonds) in the porphyrin ring of bacteriochlorophylls causes a significant difference in the absorption of light by bacteriochlorophyll *a* compared with chlorophyll *a* and *b*.

The absorption spectra of chlorophyll *a* and chlorophyll *b* differ slightly (Figure 22.4). Both absorb light in the red and blue portions of the visible spectrum (600 to 700 nm and 400 to 500 nm, respectively), and the presence of both types of chlorophyll guarantees that more wavelengths of the visible spectrum are absorbed than would be the case with either one individually.

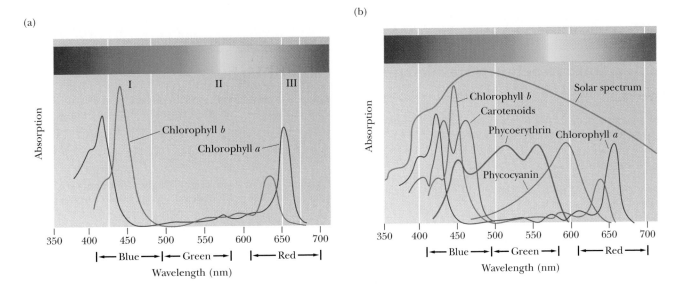

(a)

(b)

▲ **FIGURE 22.4** (a) The absorption of visible light by chlorophylls a and b. The areas marked I, II, and III are regions of the spectrum that give rise to chloroplast activity. There is greater activity in regions I and III, which are close to major absorption peaks. There are high levels of O_2 production when light from regions I and III is absorbed by chloroplasts. Lower (but measurable) activity is seen in region II, where some of the accessory pigments absorb. (b) The absorption of light by accessory pigments (superimposed on the absorption of chlorophylls a and b). The accessory pigments absorb light and transfer their energy to chlorophyll.

Recall that chlorophyll a is found in all photosynthetic organisms that produce oxygen. Chlorophyll b is found in eukaryotes such as green plants and green algae, but it occurs in smaller amounts than chlorophyll a. The presence of chlorophyll b, however, increases the portion of the visible spectrum that is absorbed and thus enhances the efficiency of photosynthesis in green plants compared with cyanobacteria. In addition to chlorophyll, various **accessory pigments** absorb light and transfer energy to the chlorophylls (Figure 22.4b). Bacteriochlorophylls, the molecular form characteristic of photosynthetic organisms that do not produce oxygen, absorb light at longer wavelengths. The wavelength of maximum absorption of bacteriochlorophyll a is 780 nm; other bacteriochlorophylls have absorption maxima at still longer wavelengths, such as 870 or 1050 nm. Light of wavelength longer than 800 nm is part of the infrared, rather than the visible, region of the spectrum. The wavelength of light absorbed plays a critical role in the light reaction of photosynthesis because the energy of light is inversely related to wavelength (see the Biochemical Connections box on page 608).

Most of the chlorophyll molecules in a chloroplast simply gather light (antennae chlorophylls). All chlorophylls are bound to proteins, either in antennae complexes or in one of two kinds of **photosystems** (membrane-bound protein complexes that carry out the light reactions). The light-harvesting molecules then pass their excitation energy along to a specialized pair of chlorophyll molecules at a **reaction center** characteristic of each photosystem (Figure 22.5). When the light energy reaches the reaction center, the chemical reactions of photosynthesis begin. The different environments of the antennae chlorophylls and the reaction-center chlorophylls give different properties to the two different kinds of molecules. In a typical chloroplast, there are several hundred light-harvesting antennae chlorophylls for each unique chlorophyll at a reaction center. The precise nature of reaction centers in both prokaryotes and eukaryotes is the subject of active research.

Essential Information

The absorption of light by chlorophyll supplies the energy required for the reactions of photosynthesis. Several different kinds of chlorophyll are known. All have a tetrapyrrole ring structure similar to that of the porphyrins of heme, but they also have differences that affect the wavelength of light they absorb. This property allows more wavelengths of sunlight to be absorbed than would be the case with a single kind of chlorophyll.

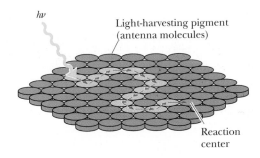

Biochemistry ⑤ Now™ ANIMATED FIGURE 22.5
Schematic diagram of a photosynthetic unit. The light-harvesting pigments, or antenna molecules (green), absorb and transfer light energy to the specialized chlorophyll dimer that constitutes the reaction center (orange). **See this figure animated at http://now.brookscole.com/campbell5**

Biochemical Connections

The Relationship between Wavelength and Energy of Light

A well-known equation relates the wavelength and energy of light, a point of crucial importance for our purposes. Max Planck established in the early 20th century that the energy of light is directly proportional to its frequency.

$$E = h\nu$$

where E is energy, h is a constant (Planck's constant), and ν is the frequency of the light. The wavelength of light is related to the frequency.

$$\nu = \frac{c}{\lambda}$$

where λ is wavelength, ν is frequency, and c is the velocity of light. We can rewrite the expression for the energy of light in terms of wavelength rather than frequency.

$$E = h\nu = \frac{hc}{\lambda}$$

Light of shorter wavelength (higher frequency) is higher in energy than light of longer wavelength (lower frequency).

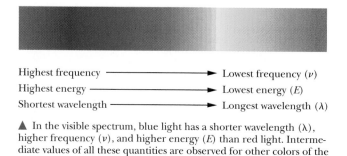

Highest frequency $\longrightarrow$ Lowest frequency (ν)

Highest energy $\longrightarrow$ Lowest energy (E)

Shortest wavelength $\longrightarrow$ Longest wavelength (λ)

▲ In the visible spectrum, blue light has a shorter wavelength (λ), higher frequency (ν), and higher energy (E) than red light. Intermediate values of all these quantities are observed for other colors of the visible spectrum.

| 22.2 | How Are Photosystems I and II Involved in the Light Reactions of Photosynthesis? |

In the light reactions of photosynthesis, water is converted to oxygen by oxidation and $NADP^+$ is reduced to NADPH. The series of redox reactions is coupled to the phosphorylation of ADP to ATP in a process called **photophosphorylation.**

$$H_2O + NADP^+ \rightarrow NADPH + H^+ + \tfrac{1}{2} O_2$$

$$ADP + P_i \rightarrow ATP$$

The light reactions consist of two parts, accomplished by two distinct but related photosystems. One part of the reaction is the reduction of $NADP^+$ to NADPH, carried out by **photosystem I (PSI).** The second part of the reaction is the oxidation of water to produce oxygen, carried out by **photosystem II (PSII).** Both photosystems carry out redox (electron transfer) reactions. The two photosystems interact with each other indirectly through an electron transport chain that links the two photosystems. The production of ATP is linked to electron transport in a process similar to that seen in the production of ATP by mitochondrial electron transport.

In the dark reactions, the ATP and NADPH produced in the light reaction provide the energy and reducing power for the fixation of CO_2. The dark reactions also constitute a redox process, since the carbon in carbohydrates is in a more reduced state than the highly oxidized carbon in CO_2. The light and dark reactions do not take place separately, but they are separated for purposes of discussion only.

The net electron transport reaction of the two photosystems taken together is, except for the substitution of NADPH for NADH, the reverse of mitochondrial electron transport. The half-reaction of reduction is that of $NADP^+$ to NADPH, whereas the half-reaction of oxidation is that of water to oxygen.

$$\begin{array}{r} NADP^+ + 2\,H^+ + 2\,e^- \rightarrow NADPH + H^+ \\ H_2O \rightarrow \frac{1}{2}\,O_2 + 2\,H^+ + 2\,e^- \\ \hline NADP^+ + H_2O \rightarrow NADPH + H^+ + \frac{1}{2}\,O_2 \end{array}$$

This is an endergonic reaction with a positive $\Delta G°' = +220 \text{ kJ mol}^{-1} = +52.6$ kcal mol¯1. The light energy absorbed by the chlorophylls in both photosystems provides the energy that allows this endergonic reaction to take place. A series of electron carriers embedded in the thylakoid membrane link these reactions. The electron carriers have an organization very similar to the carriers in the electron transport chain.

Photosystem I can be excited by light of wavelengths shorter than 700 nm, but photosystem II requires light of wavelengths shorter than 680 nm for excitation. Both photosystems must operate for the chloroplast to produce NADPH, ATP, and O_2, because the two photosystems are connected by the electron transport chain. The two systems are, however, structurally distinct in the chloroplast; photosystem I can be released preferentially from the thylakoid membrane by treatment with detergents. The reaction centers of the two photosystems provide different environments for the unique chlorophylls involved. The unique chlorophyll of photosystem I is referred to as P_{700}, where P is for pigment and the subscript 700 is for the longest wavelength of absorbed light (700 nm) that initiates the reaction. Similarly, the reaction-center chlorophyll of photosystem II is designated P_{680} because the longest wavelength of absorbed light that initiates the reaction is 680 nm. Note particularly that the path of electrons starts with the reactions in photosystem II rather than in photosystem I. The reason for the nomenclature is that photosystem I was studied extensively at an earlier date than photosystem II because it is easier to extract photosystem I from the thylakoid membrane than it is to extract photosystem II. There are two places in the reaction scheme of the two photosystems where the absorption of light supplies energy to make endergonic reactions take place (Figure 22.6).

Neither reaction-center chlorophyll is a strong enough reducing agent to pass electrons to the next substance in the reaction sequence, but the absorption of light by the chlorophylls of both photosystems provides enough energy for such reactions to take place. The absorption of light by Chl (P_{680}) allows electrons to be passed to the electron transport chain that links photosystem II and photosystem I and generates an oxidizing agent that is strong enough to split water, producing oxygen. When Chl (P_{700}) absorbs light, enough energy is provided to allow the ultimate reduction of $NADP^+$ to take place. (Note that the energy difference is shown on the vertical axis of Figure 22.6. This type of diagram is also called a Z scheme. The "Z" is rather lopsided and lies on its side, but the name is common.) In both photosystems, the result of supplying energy (light) is analogous to pumping water uphill.

Essential Information

Photosynthesis consists of two processes. The light reactions are electron transfer processes, in which water is oxidized to produce oxygen and $NADP^+$ is reduced to produce NADPH. The dark reactions are also electron transfer processes, but here carbon dioxide is reduced to carbohydrates.

Photosystem II: Water Is Split to Produce Oxygen

The oxidation of water by photosystem II to produce oxygen is the ultimate source of electrons in photosynthesis. These electrons are subsequently passed from photosystem II to photosystem I by the electron transport chain. The electrons from water are needed to "fill the hole" that is left when the

(a)

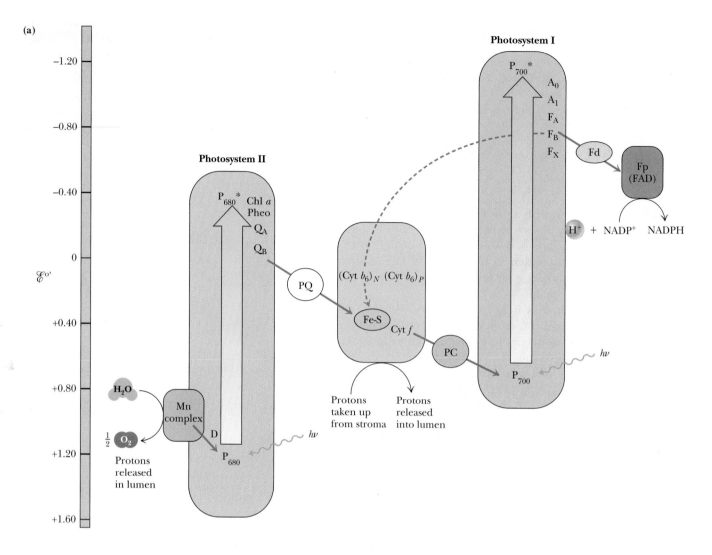

(b)

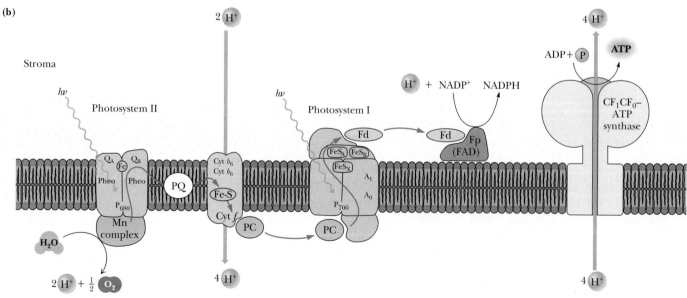

Biochemistry❂Now™ ◄ **ACTIVE FIGURE 22.6** The Z scheme of photosynthesis. (a) The Z scheme is a diagrammatic representation of photosynthetic electron flow from H_2O to $NADP^+$. The energy relationships can be derived from the $E°'$ scale beside the Z diagram, with lower standard potentials and therefore greater energy as you go from bottom to top. Energy input as light is indicated by two broad arrows, one photon appearing in P_{680} and the other in P_{700}. P_{680*} and P_{700*} represent excited states. Electron loss from P_{680*} and P_{700*} creates P_{680} and P_{700}. The representative components of the three supramolecular complexes (PSI, PSII, and the cytochrome b_6–f complex) are in shaded boxes enclosed by solid black lines. A number of components are represented by letters of the alphabet—chlorophylls and quinones by A and Q, respectively, and ferredoxins by F, and are further distinguished by subscripts. Proton translocations that establish the proton-motive force driving ATP synthesis are illustrated as well. (b) Figure showing the functional relationships among PSII, the cytochrome b_6–f complex, PSI, and the photosynthetic CF_1CF_0—ATP synthase within the thylakoid membrane. Note that e^- acceptors Q_A (for PSII) and A_1 (for PSI) are at the stromal side of the thylakoid membrane, whereas the e^- donors to P_{680} and P_{700} are situated at the lumenal side of the membrane. The consequence is charge separation (stroma, lumen) across the membrane. Also note that protons are translocated into the thylakoid lumen, giving rise to a chemiosmotic gradient that is the driving force for ATP synthesis by CF_1CF_0—ATP synthase. **Watch this Active Figure at http://now.brookscole.com/campbell5**

absorption of one photon of light leads to donation of an electron from photosystem II to the electron transport chain.

The electrons released by the oxidation of water are first passed to P_{680}, which is reduced. There are intermediate steps in this reaction because four electrons are required for the oxidation of water, and P_{680*} can accept only one electron at a time. A manganese-containing protein complex and several other protein components are required. The **oxygen-evolving complex** of photosystem II passes through a series of five oxidation states (designated as S_0 through S_4) in the transfer of four electrons in the process of evolving oxygen (Figure 22.7). One electron is passed from water to PSII for each quantum of light. In the process, the components of the reaction center go successively through oxidation states S_1 through S_4. The S_4 decays spontaneously to the S_0 state and, in the process, oxidizes two water molecules to one oxygen molecule. Note that four protons are released simultaneously. The immediate electron donor to the P_{680} chlorophyll, shown as D in Figure 22.6, is a tyrosine residue of one of the protein components that does not contain manganese. Several quinones serve as intermediate electron transfer agents to accommodate four electrons donated by one water molecule. Redox reactions of manganese also play a role here. (See the article by Govindjee and Coleman listed in the Annotated Bibliography at the end of this chapter for a discussion of the workings of this complex.)

In photosystem II, as in photosystem I, the absorption of light by chlorophyll in the reaction center produces an excited state of chlorophyll. The wavelength of light is 680 nm; the reaction-center chlorophyll of photosystem II is also referred to as P_{680}. The excited chlorophyll passes an electron to a primary acceptor. In photosystem II, the primary electron acceptor is a molecule of **pheophytin** (Pheo), one of the accessory pigments of the photosynthetic apparatus. The structure of pheophytin differs from that of chlorophyll only in the substitution of two hydrogens for the magnesium. The transfer of electrons is mediated by events that take place at the reaction center. The next electron acceptor is **plastoquinone** (PQ). The structure of plastoquinone (Figure 22.8) is similar to that of coenzyme Q (ubiquinone), a part of the respiratory electron transport chain (Section 20.2), and plastoquinone serves a very similar purpose in the transfer of electrons and hydrogen ions.

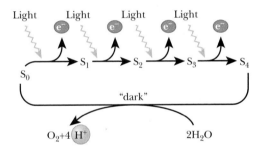

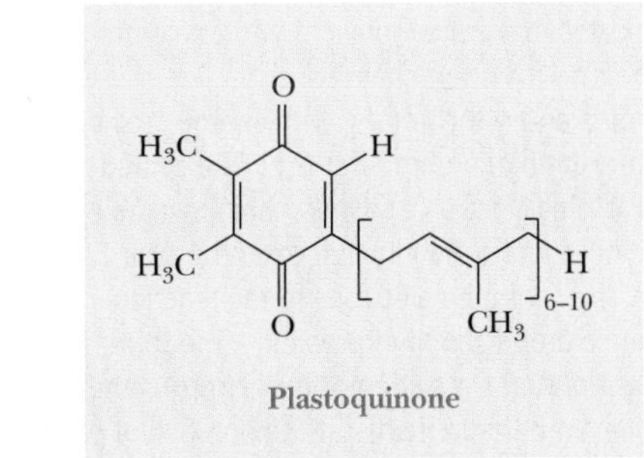

▲ **FIGURE 22.7** The PSII reaction center passes through five different oxidation states, S_0 through S_4, in the course of oxygen evolution.

▲ **FIGURE 22.8** The structure of plastoquinone. The length of the aliphatic side chain varies in different organisms.

The electron transport chain that links the two photosystems consists of pheophytin, plastoquinone, a complex of plant cytochromes (the b_6–f complex), a copper-containing protein called **plastocyanin** (PC), and the oxidized form of P_{700} (see Figure 22.6). The b_6–f complex of plant cytochromes consists of two b-type cytochromes (cytochrome b_6) and a c-type cytochrome (cytochrome f). This complex is similar in structure to the bc_1 complex in mitochondria and occupies a similar central position in an electron transport chain. This part of the photosynthetic apparatus is the subject of active research. There is a possibility that a Q cycle (recall this from Section 20.2) may operate here as well, and the object of some of this research is to establish definitely whether this is so. In plastocyanin, the copper ion is the actual electron carrier; the copper ion exists as Cu(II) and Cu(I) in the oxidized and reduced forms, respectively. This electron transport chain has another similarity to that in mitochondria, that of coupling to ATP generation.

When the oxidized chlorophyll of P_{700} accepts electrons from the electron transport chain, it is reduced and subsequently passes an electron to photosystem I, which absorbs a second photon of light. Absorption of light by photosystem II does not raise the electrons to a high enough energy level to reduce $NADP^+$; the second photon absorbed by photosystem I provides the needed energy. This difference in energy makes the "Z" of the Z scheme thoroughly lopsided, but the transfer of electrons is complete.

Photosystem I: Reduction of $NADP^+$

The absorption of light by P_{700} then leads to the series of electron transfer reactions of photosystem I. The substance to which the excited-state chlorophyll, P_{700*}, gives an electron is apparently a molecule of chlorophyll a; this transfer of electrons is mediated by processes that take place in the reaction center. The next electron acceptor in the series is bound ferredoxin, an iron–sulfur protein occurring in the membrane in photosystem I. The bound ferredoxin passes its electron to a molecule of soluble ferredoxin. Soluble ferredoxin in turn reduces an FAD-containing enzyme called ferredoxin-$NADP^+$ reductase. The FAD portion of the enzyme reduces $NADP^+$ to NADPH (Figure 22.6). We can summarize the main features of the process in two equations, in which the notation ferredoxin refers to the soluble form of the protein.

$$\text{Chl*} + \text{Ferredoxin}_{oxidized} \rightarrow \text{Chl}^+ + \text{Ferredoxin}_{reduced}$$

$$2\,\text{Ferredoxin}_{reduced} + H^+ + NADP^+ \xrightarrow{\substack{\text{Ferredoxin-NADP} \\ \text{reductase}}} 2\,\text{Ferredoxin}_{oxidized} + NADPH$$

Chl* donates one electron to ferredoxin, but the electron transfer reactions of FAD and $NADP^+$ involve two electrons. Thus, an electron from each of two ferredoxins is required for the production of NADPH.

The net reaction for the two photosystems together is the flow of electrons from H_2O to $NADP^+$ (see Figure 22.6).

$$2\,H_2O + 2\,NADP^+ \rightarrow O_2 + 2\,NADPH + 2\,H^+$$

Cyclic Electron Transport in Photosystem I

In addition to the electron transfer reactions just described, it is possible for cyclic electron transport in photosystem I to be coupled to the production of ATP (Figure 22.9). No NADPH is produced in this process. Photosystem II is not involved, and no O_2 is generated. Cyclic phosphorylation takes place when there is a high NADPH/$NADP^+$ ratio in the cell: there is not enough

Essential Information

The path of electrons in the light reactions of photosynthesis can be considered to have three parts. The first is the transfer of electrons from water to the reaction-center chlorophyll of photosystem II. The second part is the transfer of electrons from the excited-state chlorophyll of photosystem II to an electron transport chain consisting of accessory pigments and cytochromes, with energy provided by absorption of a photon of light. The components of this electron transport chain resemble those of the mitochondrial electron transport chain; they pass the electrons to the reaction-center chlorophyll of photosystem I. The third and last part of the path of the electrons is their transfer from the excited-state chlorophyll of photosystem I to the ultimate electron acceptor $NADP^+$, producing NADPH; again, energy is provided by absorption of a photon of light.

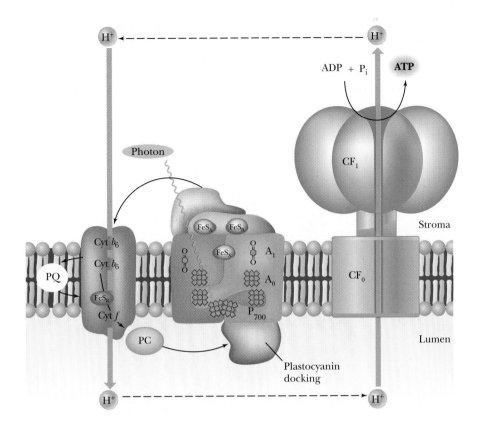

Biochemistry *Now™* **ACTIVE FIGURE 22.9**
The pathway of cyclic photophosphorylation by PSI.
Note that water is not split and that no NADPH is
produced. (*Adapted from Arnon, D. I., 1984. The discovery of
photosynthetic phosphorylation.* Trends in Biochemical Sciences
9, *258–262.*) **Watch this Active Figure at http://now
.brookscole.com/campbell5**

$NADP^+$ present in the cell to accept all the electrons generated by the excitation of P_{700}.

Structure of a Photosystem

The molecular structure of photosystems is a subject of intense interest to biochemists. The most extensively studied system is that from anaerobic phototropic bacteria of the genus *Rhodopseudomonas*. These bacteria do not produce molecular oxygen as a result of their photosynthetic activities, but enough similarities exist between the photosynthetic reactions of *Rhodopseudomonas* and photosynthesis linked to oxygen to lead scientists to draw conclusions about the nature of reaction centers in all organisms. Since the structure of this photosystem was elucidated by X-ray crystallography, the structures of PSI and PSII have also been determined and have been shown to be markedly similar. Consequently, the detailed process that goes on at the reaction center of *Rhodopseudomonas* is important enough to warrant further discussion.

It is well established that there is a pair of bacteriochlorophyll molecules (designated P_{870} from the fact that light of 870 nm is the maximum excitation wavelength) in the reaction center of *Rhodopseudomonas viridis;* the critical pair of chlorophylls is embedded in a protein complex that is in turn an integral part of the photosynthetic membrane. (We shall refer to the bacteriochlorophylls simply as chlorophylls in the interest of simplifying the discussion.) Accessory pigments, which also play a role in the light-trapping process, have specific positions close to the special pair of chlorophylls. The absorption of light by the special pair of chlorophylls raises one of their electrons to a higher energy level (Figure 22.10). This electron is passed to a series of accessory pigments. The first of these accessory pigments is pheophytin, which is structurally similar to chlorophyll, differing only in having two hydrogens in place of the magnesium. The electron is passed along to the pheophytin, raising it in turn to an excited energy level. (Note that the electron travels on only one of two possible paths, to one pheophytin but not the

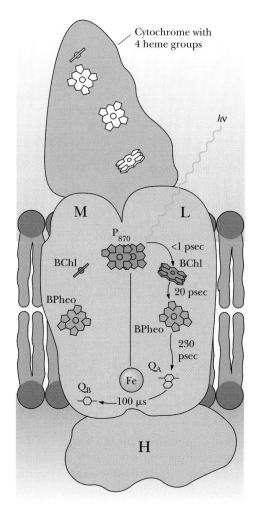

Note: The cytochrome subunit is membrane associated via a diacylglycerol moiety on its N-terminal Cys residue:

▶ **FIGURE 22.10** Model of the structure and activity of the *Rhodopseudomonas viridis* reaction center. Four polypeptides (designated cytochrome, M, L, and H) make up the reaction center, an integral membrane complex. The cytochrome maintains its association with the membrane via a diacylglyceryl group linked to its N-terminal Cys residue by a thioether bond. M and L both consist of five membrane-spanning α-helices; H has a single membrane-spanning α-helix. The prosthetic groups are spatially situated so that rapid e^- transfer from P_{870*} to Q_B is facilitated. Photoexcitation of P_{870} leads in less than 1 picosecond (psec) to reduction of the L-branch BChl only. P_{870} is re-reduced via an electron provided through the heme groups of the cytochrome.

other. Research is in progress to determine why this is so.) The next electron acceptor is menaquinone (Q_A); it is structurally similar to coenzyme Q, which plays a role in the mitochondrial electron transport chain. The final electron acceptor, which is also raised to an excited state, is coenzyme Q itself (ubiquinone, called Q_B here). The electron that had been passed to Q_B is replaced by an electron donated by a cytochrome, which acquires a positive charge in the process. The cytochrome is not bound to the membrane and diffuses away, carrying its positive charge with it. The whole process takes place in less than 10^{-3} s. The positive and negative charges have traveled in opposite directions from the chlorophyll pair and are separated from each other. This situation is similar to the proton gradient in mitochondria, where the existence of the proton gradient is ultimately responsible for oxidative phosphorylation. The separation of charge is equivalent to a battery, a form of stored energy. The reaction center has acted as a transducer, converting light energy to a form usable by the cell to carry out the energy-requiring reactions of photosynthesis. The processes that take place in *Rhodopseudomonas* serve as a model for reaction centers in photosynthesis linked to oxygen.

The structures of menaquinone and ubiquinone

22.3 | How Does Photosynthesis Produce ATP?

In Chapter 20, we saw that a proton gradient across the inner mitochondrial membrane drives the phosphorylation of ADP in respiration. The mechanism of photophosphorylation is essentially the same as that of the production of ATP in the respiratory electron transport chain. In fact, some of the strongest evidence for the chemiosmotic coupling of phosphorylation to electron transport has been obtained from experiments on chloroplasts rather than mitochondria. Chloroplasts can synthesize ATP from ADP and P_i *in the dark* if they are provided with a pH gradient.

If isolated chloroplasts are allowed to equilibrate in a pH 4 buffer for several hours, their internal pH will be equal to 4. If the pH of the buffer is raised rapidly to 8 and if ADP and P_i are added simultaneously, ATP will be produced (Figure 22.11). The production of ATP does not require the presence of light; the proton gradient produced by the pH difference supplies the driving force for phosphorylation. This experiment provides solid evidence for the chemiosmotic coupling mechanism.

Several reactions contribute to the generation of a proton gradient in chloroplasts in an actively photosynthesizing cell. The oxidation of water releases H^+ into the thylakoid space. Electron transport from photosystem II and photosystem I also helps create the proton gradient by involving plastoquinone and cytochromes in the process. Then photosystem I reduces $NADP^+$ by using H^+ in the stroma to produce NADPH. As a result, the pH of

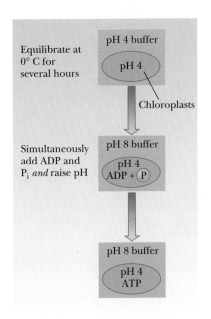

▲ **FIGURE 22.11** ATP is synthesized by chloroplasts in the dark in the presence of a proton gradient, ADP, and P_i.

Biochemical Connections

Some Herbicides Inhibit Photosynthesis

The main purpose of herbicides is to kill weeds so that they do not choke out desirable plants. One way of doing this is by selectively inhibiting photosynthesis in the weeds and not in the desired plants. A prime example is the use of 2,4-D and 2,4,5-T to kill broad-leaved weeds such as dandelions without affecting the growth of grasses. In terms of acreage, lawn grasses are the most widely grown crop in the United States.

The selectivity of herbicides is not absolute and depends on a number of factors. Transport of the herbicide to the site of action in the plant plays a role, as does absorption. One of the most important features is a higher rate of detoxification of the herbicide in the desirable plants when compared with the weeds. Some of this tolerance is genetically determined, and research is being done to enhance it by the techniques of biotechnology that we discussed in Chapter 13.

A number of other herbicides interfere with photosynthesis in specific ways. Amitrole inhibits biosynthesis of chlorophyll and carotenoids. The affected plants present a bleached appearance before they die because of the loss of their characteristic pigments. Another herbicide, atrazine, inhibits the oxidation of water to hydrogen ion and oxygen. Still other herbicides interfere with electron transfer in the two photosystems. In photosystem II, diuron inhibits electron transfer to plastoquinone, whereas bigyridylium herbicides accept electrons by competing with the electron acceptors in photosystem I. The inhibitors

David Newman/Visuals Unlimited

▲ Broad-leaved roadside weeds killed by herbicides that selectively inhibit photosynthesis. Note the unaffected corn in the background.

active in photosystem I include diquat and paraquat. The latter substance attained some notoriety when it was used to interfere with an illegal crop: it was sprayed on marijuana fields to destroy the growing plants.

the thylakoid space is lower than that of the stroma (Figure 22.12). We saw a similar situation in Chapter 20 when we discussed the pumping of protons from the mitochondrial matrix into the intermembrane space. The ATP synthase in chloroplasts is similar to the mitochondrial enzyme; in particular, it consists of two parts, CF_1 and CF_0, where the C serves to distinguish them from their mitochondrial counterparts, F_1 and F_0, respectively. Evidence exists that the components of the electron chain in chloroplasts are arranged asymmetrically in the thylakoid membrane, as is the case in mitochondria. An important consequence of this asymmetrical arrangement is the release of the ATP and NADPH produced by the light reaction into the stroma, where they provide energy and reducing power for the dark reaction of photosynthesis.

In mitochondrial electron transport, there are four respiratory complexes connected by soluble electron carriers. The electron transport apparatus of the thylakoid membrane is similar in that it consists of several large membrane-bound complexes. They are PSII (the photosystem II complex), the cytochrome b_6–f complex, and PSI (the photosystem I complex). As in mitochondrial electron transport, several soluble electron carriers form the connection between the protein complexes. In the thylakoid membrane, the soluble carriers are plastoquinone and plastocyanin, which have a role similar to that of coenzyme Q and cytochrome c in mitochondria (Figure 22.12). The proton gradient created by electron transport drives the synthesis of ATP in chloroplasts, as in mitochondria.

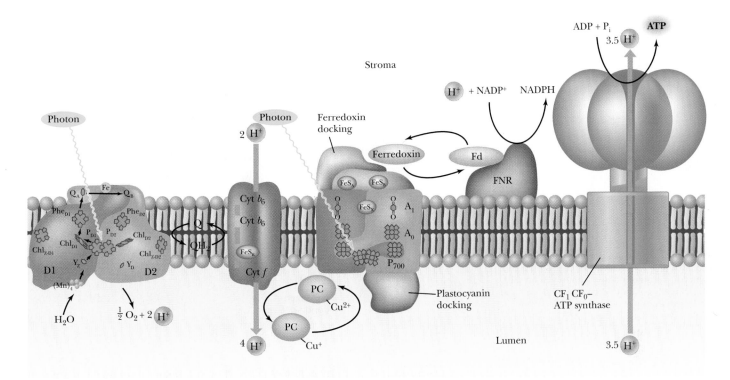

Biochemistry ⒼNow™   ▲ **ACTIVE FIGURE 22.12**   The mechanism of photophosphorylation. Photosynthetic electron transport establishes a proton gradient that is tapped by the CF_1CF_0—ATP synthase to drive ATP synthesis. Critical to this mechanism is the fact that the membrane-bound components of light-induced electron transport and ATP synthesis are asymmetric with respect to the thylakoid membrane so that directional discharge and uptake of H^+ ensue, generating the proton-motive force. **Watch this Active Figure at http://now.brookscole.com/campbell5**

22.4	What Are the Evolutionary Implications of Photosynthesis with and without Oxygen?

Photosynthetic prokaryotes other than cyanobacteria have only one photosystem and do not produce oxygen. The chlorophyll in these organisms is different from that found in photosystems linked to oxygen (Figure 22.13). Anaerobic photosynthesis is not as efficient as photosynthesis linked to oxygen, but the anaerobic version of the process appears to be an evolutionary way station. Anaerobic photosynthesis is a means for organisms to use solar energy to satisfy their needs for food and energy. Although it is efficient in the production of ATP, its efficiency is less than that of aerobic photosynthesis for carbon fixation.

A possible scenario for the development of photosynthesis starts with heterotrophic bacteria that contain some form of chlorophyll, probably bacteriochlorophyll. (*Heterotrophs* are organisms that depend on their environment for organic nutrients and for energy.) In such organisms, the light energy absorbed by chlorophyll can be trapped in the forms of ATP and NADPH. The important point about such a series of reactions is that photophosphorylation takes place, ensuring an independent supply of ATP for the organism. In addition, the supply of NADPH facilitates synthesis of biomolecules from simple sources such as CO_2. Under conditions of limited food supply, organisms that can synthesize their own nutrients have a selective advantage.

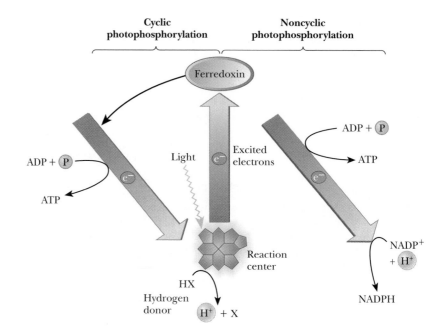

▶ **FIGURE 22.13** The two possible electron transfer pathways in a photosynthetic anaerobe. Both cyclic and noncyclic forms of photophosphorylation are shown. HX is any compound (such as H_2S) that can be a hydrogen donor. (*From L. Margulis, 1985*. Early Life, Science Books International, *Boston, p. 45.*)

Organisms of this sort are *autotrophs* (not dependent on an external source of biomolecules) but are also anaerobes. The ultimate electron source that they use is not water but some more easily oxidized substance, such as H_2S, as is the case with present-day green sulfur bacteria (and purple sulfur bacteria), or various organic compounds, as is the case with present-day purple nonsulfur bacteria. These organisms do not possess an oxidizing agent powerful enough to split water, which is a far more abundant electron source than H_2S or organic compounds. The ability to use water as an electron source confers a further evolutionary advantage.

As is frequently the case in biological oxidation–reduction reactions, hydrogens as well as electrons are transferred from a donor to an acceptor. In green plants, green algae, and cyanobacteria, the hydrogen donor and acceptor are H_2O and CO_2, respectively, with oxygen as a product. Other organisms, such as bacteria, carry out photosynthesis in which there is a hydrogen donor other than water. Some possible donors include H_2S, $H_2S_2O_3$, and succinic acid. As an example, if H_2S is the source of hydrogens and electrons, a schematic equation for photosynthesis can be written with sulfur, rather than oxygen, as a product.

$$CO_2 \;+\; 2\,H_2S \rightarrow (CH_2O) + 2\,S + H_2O$$

H-acceptor   H-donor   Carbohydrate

It is also possible for the hydrogen acceptor to be NO_2^- or NO_3^-, in which case NH_3 is a product. Photosynthesis linked to oxygen with carbon dioxide as the ultimate hydrogen acceptor is a special case of a far more general process that is widely distributed among many different organisms.

Cyanobacteria were apparently the first organisms that developed the ability to use water as the ultimate reducing agent in photosynthesis. As we have seen, this feat required the development of a second photosystem as well as a new variety of chlorophyll, chlorophyll *a* rather than bacteriochlorophyll in this case. Chlorophyll *b* had not yet appeared on the scene, since it occurs only in eukaryotes. The basic system of aerobic photosynthesis was in place with cyanobacteria. As a result of aerobic photosynthesis by cyanobacteria, the planet Earth acquired its present atmosphere with its high levels of oxygen. The existence of all other aerobic organisms depended ultimately on the activities of cyanobacteria.

22.5 | How Do the Dark Reactions of Photosynthesis Fix CO$_2$ into Glucose?

The actual storage form of the carbohydrates produced from carbon dioxide by photosynthesis is not glucose but disaccharides (e.g., sucrose in sugarcane and sugar beets) and polysaccharides (starch and cellulose). However, it is customary and convenient to write the carbohydrate product as glucose, and we shall follow this time-honored practice.

Carbon dioxide fixation takes place in the stroma. The equation for the overall reaction, like all equations for photosynthetic processes, is deceptively simple.

$$6\,CO_2 + 12\,NADPH + 18\,ATP \xrightarrow{\text{Enzymes}} C_6H_{12}O_6 + 12\,NADP^+ + 18\,ADP + 18\,P_i$$

The actual reaction pathway has some features in common with glycolysis and some in common with the pentose phosphate pathway.

The net reaction of six molecules of carbon dioxide to produce one molecule of glucose requires the carboxylation of six molecules of a five-carbon key intermediate, **ribulose-1,5-*bis*phosphate,** to form six molecules of an unstable six-carbon intermediate, which then splits to give 12 molecules of **3-phosphoglycerate.** Of these, two molecules of 3-phosphoglycerate react in turn, ultimately producing glucose. The remaining ten molecules of 3-phosphoglycerate are used to regenerate the six molecules of ribulose-1,5-*bis*phosphate. The overall reaction pathway is cyclic and is called the **Calvin cycle** (Figure 22.14) after the scientist who first investigated it, Melvin Calvin, winner of the 1961 Nobel Prize in chemistry.

The first reaction of the Calvin cycle is the condensation of ribulose-1,5-*bis*phosphate with carbon dioxide to form a six-carbon intermediate, 2-carboxy-3-ketoribitol-1,5-*bis*phosphate, which quickly hydrolyzes to give two molecules of 3-phosphoglycerate (Figure 22.15). The reaction is catalyzed by the enzyme *ribulose-1,5-bisphosphate carboxylase/oxygenase* (**rubisco**). This enzyme is located on the stromal side of the thylakoid membrane and is probably one of the most abundant proteins in nature, since it accounts for about 15% of the total protein in chloroplasts. The molecular weight of ribulose-1,5-*bis*phosphate carboxylase/oxygenase is about 560,000, and it consists of eight large subunits (molecular weight, 55,000) and eight small subunits (molecular weight, 15,000) (Figure 22.16). The sequence of the large subunit is encoded by a chloroplast gene, and that of the small subunit is encoded by a nuclear gene. The endosymbiotic theory for the development of eukaryotes (Section 1.7) is consistent with the idea of independent genetic material in organelles. The large subunit (chloroplast gene) is catalytic, whereas the small subunit (nuclear gene) plays a regulatory role, an observation that is consistent with an endosymbiotic origin for organelles such as chloroplasts.

The incorporation of CO$_2$ into 3-phosphoglycerate represents the actual fixation process; the remaining reactions are those of carbohydrates. The next two reactions lead to the reduction of 3-phosphoglycerate to form glyceraldehyde-3-phosphate. The reduction takes place in the same fashion as in gluconeogenesis, except for one unique feature (Figure 22.14): the reactions in chloroplasts require NADPH rather than NADH for the reduction of 1,3-

Biochemistry ⊜ Now™
Go to BiochemistryNow and click on Biochemistry Interactive to examine the structure of rubisco, the principal CO$_2$-fixing enzyme in nature.

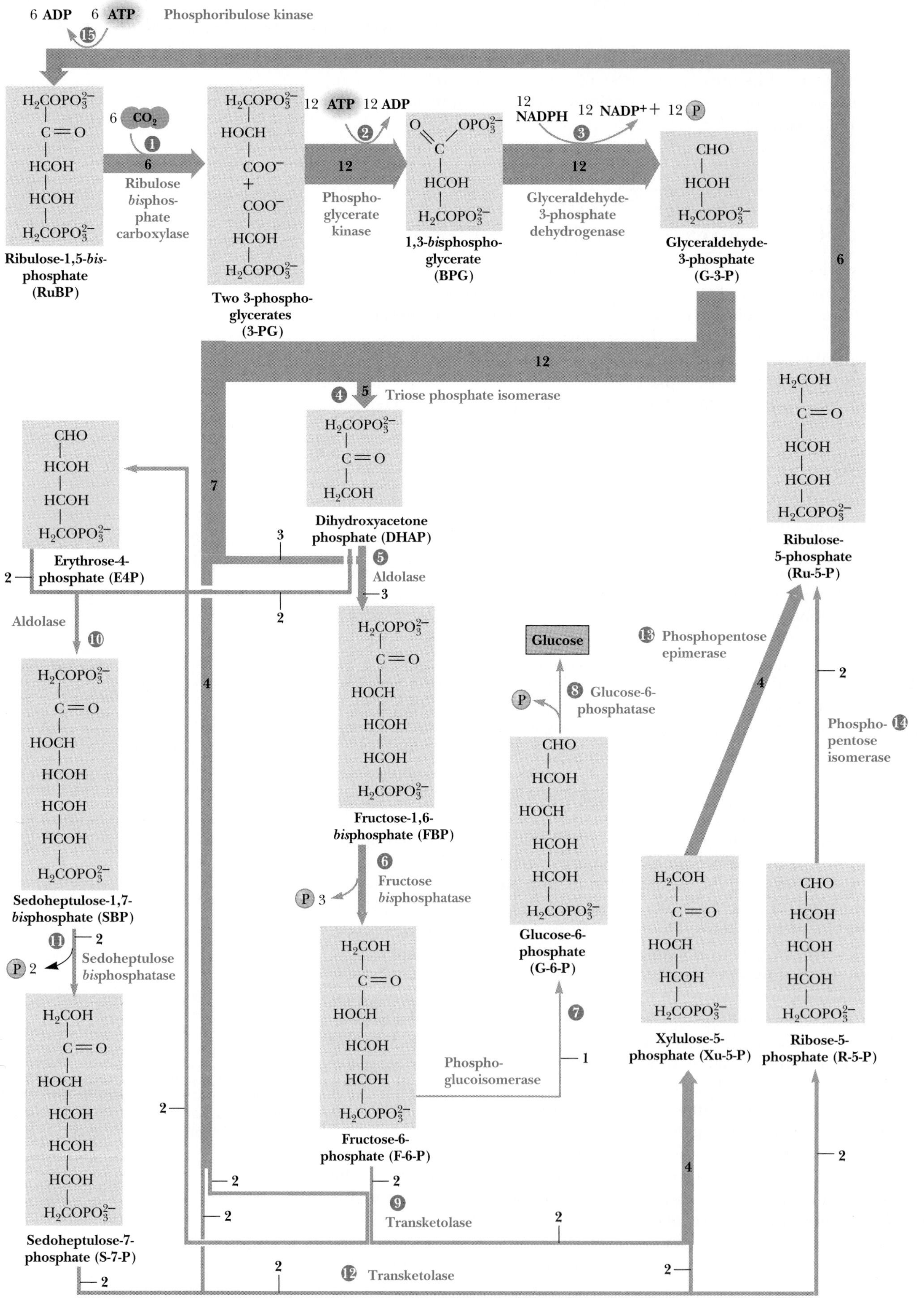

Biochemistry Now™ ▲ **ACTIVE FIGURE 22.14** The Calvin cycle of reactions. The number associated with the arrow at each step indicates the number of molecules reacting in a turn of the cycle that produces one molecule of glucose. Reactions are numbered as in Table 22.1. **Watch this Active Figure at http://now.brookscole.com/campbell5**

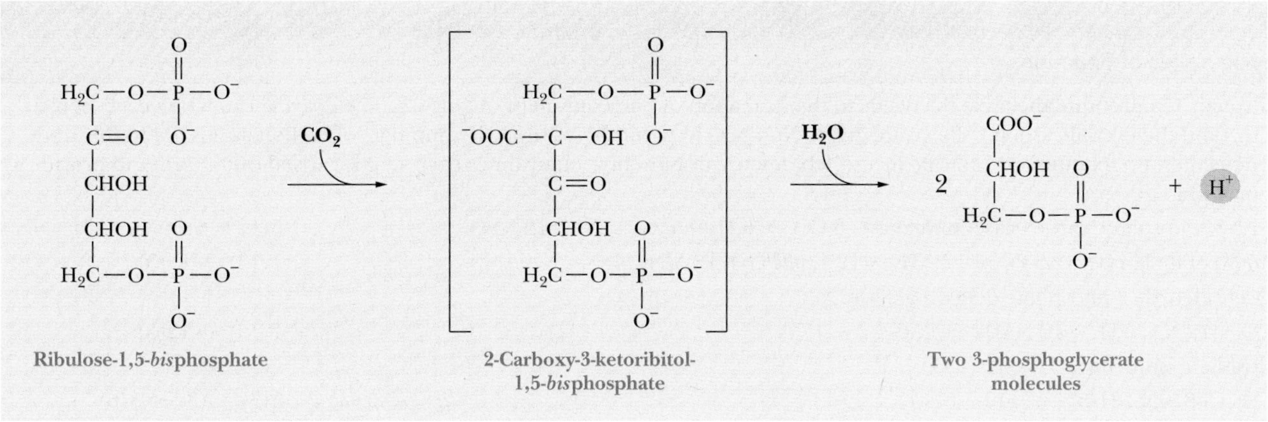

▲ **FIGURE 22.15** The reaction of ribulose-1,5-*bis*phosphate with CO$_2$ ultimately produces two molecules of 3-phosphoglycerate.

*bis*phosphoglycerate to glyceraldehyde-3-phosphate. When glyceraldehyde-3-phosphate is formed, it can have two alternative fates: one is the production of six-carbon sugars, and the other is the regeneration of ribulose-1,5-*bis*phosphate. Table 22.1 summarizes the reactions that take place and indicates their stoichiometry.

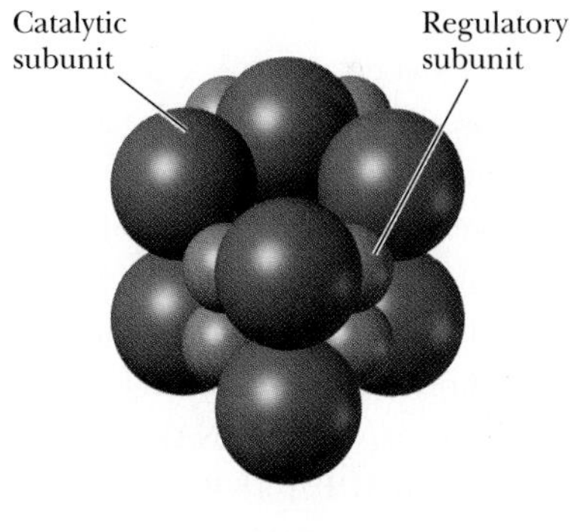

Production of Six-Carbon Sugars

The formation of glucose from glyceraldehyde-3-phosphate takes place in the same manner as in gluconeogenesis (Figure 22.14 and reactions 4 through 8 in Table 22.1). The conversion of glyceraldehyde-3-phosphate to dihydroxyacetone phosphate takes place easily (Section 17.2). Dihydroxyacetone phosphate in turn reacts with glyceraldehyde-3-phosphate, in a series of reactions we have already seen, to give rise to fructose-6-phosphate and ultimately to glucose. Because we have already seen these reactions, we shall not discuss them again.

Regeneration of Ribulose-1,5-*Bis*phosphate

This process is readily divided into four steps: *preparation, reshuffling, isomerization,* and *phosphorylation.* The preparation begins with conversion of some of the glyceraldehyde-3-phosphate to dihydroxyacetone phosphate (catalyzed by triose phosphate isomerase). This reaction also functions in the production of six-carbon sugars. Portions of both the glyceraldehyde-3-phosphate and the dihydroxyacetone phosphate are then condensed to form fructose-1,6-*bis*phosphate (catalyzed by aldolase). Fructose-1,6-*bis*phosphate is hydrolyzed to fructose-6-phosphate (catalyzed by fructose-1,6-*bis*phosphatase). (See Figure 22.14. Reactions 4 through 6 in Table 22.1 are involved here.) With a supply of glyceraldehyde-3-phosphate, dihydroxyacetone phosphate, and fructose-6-phosphate now available, the reshuffling can begin.

Most of the reactions of the reshuffling process are the same as ones we have already seen as part of the pentose phosphate pathway (Section 18.4). Consequently, we shall concentrate just on the main outline of the process later because the results are summarized in Figure 22.14 and Table 22.1. Reactions catalyzed in turn by *transketolase, aldolase,* and *sedoheptulose* bis*phosphatase* (Reactions 9 through 12 in Table 22.1) are the reactions of rearrangement of carbon skeletons in the reshuffling phase of the Calvin cycle.

▲ **FIGURE 22.16** The subunit structure of ribulose-1,5-*bis*phosphate carboxylase.

Essential Information

In the dark reactions of photosynthesis, the fixation of carbon dioxide takes place when the key intermediate ribulose-1,5-*bis*phosphate reacts with carbon dioxide to produce two molecules of 3-phosphoglycerate. This reaction is catalyzed by the enzyme ribulose-1,5-*bis*phosphate carboxylase/oxygenase (rubisco), one of the most abundant proteins in nature. The remainder of the dark reaction is the regeneration of ribulose-1,5-*bis*phosphate in the Calvin cycle.

Table 22.1

The Calvin Cycle Series of Reactions

Reactions 1 through 15 constitute the cycle that leads to the formation of one equivalent of glucose. The enzyme catalyzing each step, a concise reaction, and the overall carbon balance are given. Numbers in parentheses show the numbers of carbon atoms in the substrate and product molecules. Prefix numbers indicate in a stoichiometric fashion how many times each step is carried out in order to provide a balanced net reaction.

1. Ribulose-1,5-*bis*phosphate carboxylase/oxygenase: $6\ CO_2 + 6\ H_2O + 6\ RuBP \rightarrow 12\ 3\text{-PG}$ $6(1) + 6(5) \rightarrow 12(3)$

2. 3-Phosphoglycerate kinase: $12\ 3\text{-PG} + 12\ ATP \rightarrow 12\ 1,3\text{-BPG} + 12\ ADP$ $12(3) \rightarrow 12(3)$

3. NADP-glyceraldehyde-3-phosphate dehydrogenase:
    $12\ 1,3\text{-BPG} + 12\ NADPH \rightarrow 12\ NADP + 12\ G\text{-}3\text{-P} + 12\ P_i$ $12(3) \rightarrow 12(3)$

4. Triose phosphate isomerase: $5\ G\text{-}3\text{-P} \rightarrow 5\ DHAP$ $5(3) \rightarrow 5(3)$

5. Aldolase: $3\ G\text{-}3\text{-P} + 3\ DHAP \rightarrow 3\ FBP$ $3(3) + 3(3) \rightarrow 3(6)$

6. Fructose *bis*phosphatase: $3\ FBP + 3\ H_2O \rightarrow 3\ F\text{-}6\text{-P} + 3\ P_i$ $3(6) \rightarrow 3(6)$

7. Phosphoglucoisomerase: $1\ F\text{-}6\text{-P} \rightarrow 1\ G\text{-}6\text{-P}$ $1(6) \rightarrow 1(6)$

8. Glucose-6-phosphatase: $1\ G\text{-}6\text{-P} + 1\ H_2O \rightarrow 1\ Glucose + 1\ P_i$ $1(6) \rightarrow 1(6)$

The remainder of the pathway involves regenerating six RuBP acceptors (30 C) from the leftover two F-6-P (12 C), four G-3-P (12 C), and two DHAP (6 C).

9. Transketolase: $2\ F\text{-}6\text{-P} + 2\ G\text{-}3\text{-P} \rightarrow 2\ Xu\text{-}5\text{-P} + 2\ E4P$ $2(6) + 2(3) \rightarrow 2(5) + 2(4)$

10. Aldolase: $2\ E4P + 2\ DHAP \rightarrow 2\ SBP$ $2(4) + 2(3) \rightarrow 2(7)$

11. Sedoheptulose *bis*phosphatase: $2\ SBP + 2\ H_2O \rightarrow 2\ S\text{-}7\text{-P} + 2\ P_i$ $2(7) \rightarrow 2(7)$

12. Transketolase: $2\ S\text{-}7\text{-P} + 2\ G\text{-}3\text{-P} \rightarrow 2\ Xu\text{-}5\text{-P} + 2\ R\text{-}5\text{-P}$ $2(7) + 2(3) \rightarrow 4(5)$

13. Phosphopentose epimerase: $4\ Xu\text{-}5\text{-P} \rightarrow 4\ Ru\text{-}5\text{-P}$ $4(5) \rightarrow 4(5)$

14. Phosphopentose isomerase: $2\ R\text{-}5\text{-P} \rightarrow 2\ Ru\text{-}5\text{-P}$ $2(5) \rightarrow 2(5)$

15. Phosphoribulose kinase: $6\ Ru\text{-}5\text{-P} + 6\ ATP \rightarrow 6\ RuBP + 6\ ADP$ $6(5) \rightarrow 6(5)$

Net: $6\ CO_2 + 18\ ATP + 12\ NADPH + 12\ H^+ + 12\ H_2O \rightarrow$

   $Glucose + 18\ ADP + 18\ P_i + 12\ NADP$ $6(1) \rightarrow 1(6)$

The isomerization step (reactions 13 and 14 in Table 22.1) involves the conversion of both ribose-5-phosphate and xylulose-5-phosphate to ribulose-5-phosphate. *Ribose-5-phosphate isomerase* catalyzes the conversion of ribose-5-phosphate to ribulose-5-phosphate, and *xylulose-5-phosphate epimerase* catalyzes the conversion of xylulose-5-phosphate to ribulose-5-phosphate (Figure 22.14). The reverse of both these reactions takes place in the pentose phosphate pathway, catalyzed by the same enzymes.

In the final step (reaction 15 in Table 22.1), ribulose-1,5-*bis*phosphate is regenerated by the phosphorylation of ribulose-5-phosphate. This reaction requires ATP and is catalyzed by the enzyme *phosphoribulose kinase*. The reactions leading to the regeneration of ribulose-1,5-*bis*phosphate summarized in Table 22.1 give a net equation obtained by adding all the reactions.

Taking these points into consideration, we arrive at the *net* equation for the path of carbon in photosynthesis.

$6\ CO_2 + 18\ ATP + 12\ NADPH + 12\ H^+ + 12\ H_2O \rightarrow$

$$Glucose + 12\ NADP^+ + 18\ ADP + 18\ P_i$$

The efficiency of energy use in photosynthesis can be calculated fairly easily. The $\Delta G°'$ for the reduction of CO_2 to glucose is $+478$ kJ ($+114$ kcal) for each mole of CO_2 (see Exercise 37), and the energy of light of 600-nm wavelength is 1593 kJ mol^{-1} (381 kcal mol^{-1}). We shall not explain in detail here how this figure for the energy of the light is obtained, but it comes ultimately from

Chloroplast Genes

Chloroplasts, like mitochondria, have their own DNA. Scientists speculate that these organelles may have been independently living organisms very early in evolution. The endosymbiotic theory (Section 1.8) gives just such a scenario. Some interesting, even elaborate, interactions have evolved from the interplay of both nuclear and organelle genes. About 90 chloroplast proteins are encoded in the nucleus, and about 120 are encoded in the chloroplast. No chloroplast-coded protein is known to be used in the cytoplasm.

One of the interesting interactions involves rubisco, the principal enzyme of CO_2 fixation. The large subunit of this enzyme is coded by a chloroplast gene, and the small subunit by a nuclear gene. Mechanisms not yet understood must coordinate the two syntheses to ensure equimolar production of both subunits. The nuclear gene is translated in the cytoplasm, and the protein is then transported to the chloroplast, protected by a chaperonin, via targeting mechanisms (see Section 12.6 and the Biochemical Connections box on page 324). Special targeting sequences are used to direct various nuclear products to the appropriate chloroplast location, using reactions that require ATP hydrolysis. The chaperonin aids in formation of the final, active complex.

The chloroplast encodes its own RNA polymerase, ribosomal and transfer RNAs, and about one third of the ribosomal proteins. The DNA polymerase, aminoacyl synthetases, and the rest of the ribosomal proteins come from the nuclear genes. Different nuclear genes may be used for chloroplasts in different, specialized tissues of the plant. In different classes of plants, different genes are coded by the nucleus, even though all the same proteins constitute the chloroplast. The sequences of many of these enzymes encoded in the nucleus resemble those found in bacteria more than the sequences of proteins encoded by other nuclear genes. The chloroplast rRNAs likewise resemble bacterial rRNAs. The four subunits of the chloroplast-coded RNA polymerase are homologous to the four subunits of bacterial RNA polymerase. Furthermore, the chloroplast mRNA uses a Shine–Dalgarno sequence to bind to the ribosome, and it does not have a cap or a poly-A tail. These observations are definitely consistent with the idea that the chloroplast (and mitochondrion) originated from bacteria-like organisms, symbiotically taken into early cells, with some subsequent transfer of the organelle genes to the nucleus.

the equation $E = h\nu$. Light of wavelength 680 or 700 nm has lower energy than light at 600 nm. Thus, the efficiency of photosynthesis is at least $(478/1593) \times 100$, or 30%.

22.6 | How Is CO_2 Fixed in Tropical Plants?

In tropical plants, there is a C_4 pathway (Figure 22.17), so named because it involves four-carbon compounds. The operation of this pathway (also called the **Hatch–Slack pathway**) ultimately leads to the C_3 (based on 3-phosphoglycerate) pathway of the Calvin cycle. (There are other C_4 pathways, but this one is most widely studied. Corn [maize] is an important example of a C_4 plant, and it is certainly not confined to the tropics.)

When CO_2 enters the leaf through pores in the outer cells, it reacts first with phosphoenolpyruvate to produce oxaloacetate and P_i in the mesophyll cells of the leaf. Oxaloacetate is reduced to malate, with the concomitant oxidation of NADPH. Malate is then transported to the bundle-sheath cells (the next layer) through channels that connect the two kinds of cells.

In the bundle-sheath cells, malate is decarboxylated to give pyruvate and CO_2. In the process, $NADP^+$ is reduced to NADPH (Figure 22.18). The CO_2 reacts with ribulose-1,5-*bis*phosphate to enter the Calvin cycle. Pyruvate is transported back to the mesophyll cells, where it is phosphorylated to phosphoenolpyruvate, which can react with CO_2 to start another round of the C_4 pathway. When pyruvate is phosphorylated, ATP is hydrolyzed to AMP and PP_i. This situation represents a loss of two high-energy phosphate bonds, equivalent to the use of two ATP. Consequently, the C_4 pathway requires two

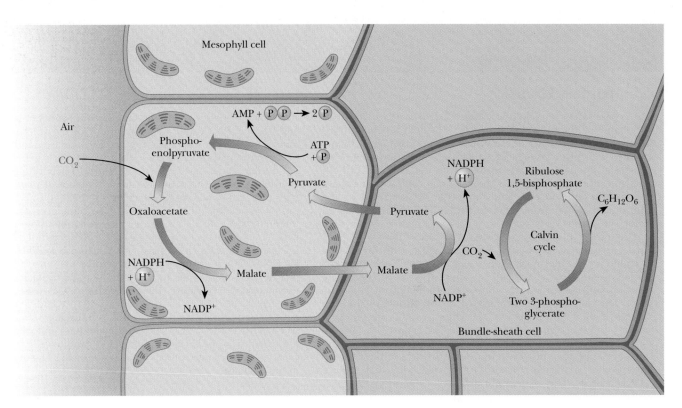

▲ **FIGURE 22.17** The C_4 pathway.

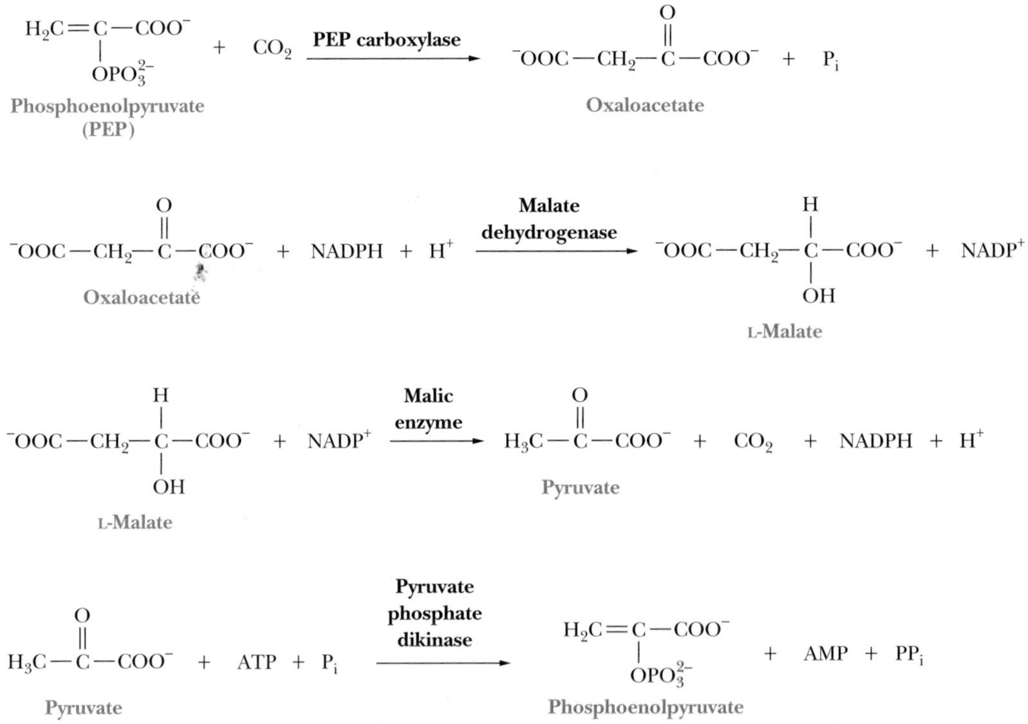

$$\underset{\substack{\text{Phosphoenolpyruvate}\\\text{(PEP)}}}{\text{H}_2\text{C}\!=\!\underset{\substack{|\\\text{OPO}_3^{2-}}}{\text{C}}\!-\!\text{COO}^-} + \text{CO}_2 \xrightarrow{\textbf{PEP carboxylase}} \underset{\text{Oxaloacetate}}{{}^-\text{OOC}\!-\!\text{CH}_2\!-\!\overset{\overset{\text{O}}{\|}}{\text{C}}\!-\!\text{COO}^-} + \text{P}_i$$

$$\underset{\text{Oxaloacetate}}{{}^-\text{OOC}\!-\!\text{CH}_2\!-\!\overset{\overset{\text{O}}{\|}}{\text{C}}\!-\!\text{COO}^-} + \text{NADPH} + \text{H}^+ \xrightarrow{\substack{\textbf{Malate}\\\textbf{dehydrogenase}}} \underset{\text{L-Malate}}{{}^-\text{OOC}\!-\!\text{CH}_2\!-\!\underset{\underset{\text{OH}}{|}}{\overset{\overset{\text{H}}{|}}{\text{C}}}\!-\!\text{COO}^-} + \text{NADP}^+$$

$$\underset{\text{L-Malate}}{{}^-\text{OOC}\!-\!\text{CH}_2\!-\!\underset{\underset{\text{OH}}{|}}{\overset{\overset{\text{H}}{|}}{\text{C}}}\!-\!\text{COO}^-} + \text{NADP}^+ \xrightarrow{\substack{\textbf{Malic}\\\textbf{enzyme}}} \underset{\text{Pyruvate}}{\text{H}_3\text{C}\!-\!\overset{\overset{\text{O}}{\|}}{\text{C}}\!-\!\text{COO}^-} + \text{CO}_2 + \text{NADPH} + \text{H}^+$$

$$\underset{\text{Pyruvate}}{\text{H}_3\text{C}\!-\!\overset{\overset{\text{O}}{\|}}{\text{C}}\!-\!\text{COO}^-} + \text{ATP} + \text{P}_i \xrightarrow{\substack{\textbf{Pyruvate}\\\textbf{phosphate}\\\textbf{dikinase}}} \underset{\text{Phosphoenolpyruvate}}{\text{H}_2\text{C}\!=\!\underset{\underset{\text{OPO}_3^{2-}}{|}}{\text{C}}\!-\!\text{COO}^-} + \text{AMP} + \text{PP}_i$$

▲ **FIGURE 22.18** The characteristic reactions of the C_4 pathway.

more ATP equivalents than the Calvin cycle alone for each CO$_2$ incorporated into glucose. Even though more ATP is required for the C$_4$ pathway than for the Calvin cycle, there is abundant light to produce the extra ATP by the light reaction of photosynthesis.

Note that the C$_4$ pathway fixes CO$_2$ in the mesophyll cells only to unfix it in the bundle-sheath cells, where CO$_2$ then enters the C$_3$ pathway. This observation raises the question of the advantage to tropical plants of using the C$_4$ pathway. The conventional wisdom on the subject focuses on the role of CO$_2$, but there is more to the situation than that. According to the conventional view, the point of the C$_4$ pathway is that it concentrates CO$_2$ and, as a result, accelerates the process of photosynthesis. Leaves of tropical plants have small pores to minimize water loss, and these small pores decrease CO$_2$ entry into the plant. Another point to consider is that the K_M for CO$_2$ of phospho-enolpyruvate carboxylase is lower than that of rubisco, allowing the outer mesophyll cells to fix CO$_2$ at a lower concentration. This also increases the concentration gradient of CO$_2$ across the leaf and facilitates the movement of CO$_2$ into the leaf through the pores. In tropical areas, where there is abundant light, the amount of CO$_2$ available to plants controls the rate of photosynthesis. The C$_4$ pathway deals with the situation, allowing tropical plants to grow more quickly and to produce more biomass per unit of leaf area than plants that use the C$_3$ pathway. A more comprehensive view of the subject includes a consideration of the role of oxygen and the process of **photorespiration**, in which oxygen is used instead of CO$_2$ during the reaction catalyzed by rubisco.

Although the actual biological role of photorespiration is not known, several points are well established. The oxygenase activity appears to be an unavoidable, wasteful activity of rubisco. Photorespiration is a salvage pathway that saves some of the carbon that would be lost due to the oxygenase activity of rubisco. In fact, the photorespiration is essential to plants even though the plant pays the price in loss of ATP and reducing power; mutations that affect this pathway can be lethal. The principal substrate oxidized in photorespiration is *glycolate* (Figure 22.19). The product of the oxidation reaction, which takes place in peroxisomes of leaf cells (Section 1.5), is *glyoxylate*. (Photorespiration is localized in peroxisomes.) Glycolate arises ultimately from the oxidative breakdown of ribulose-l,5-*bis*phosphate. The enzyme that catalyzes this reaction is ribulose-l,5-*bis*phosphate carboxylase/oxygenase, acting as an oxygenase (linked to O$_2$) rather than as the carboxylase (linked to CO$_2$) that fixes CO$_2$ into 3-phosphoglycerate.

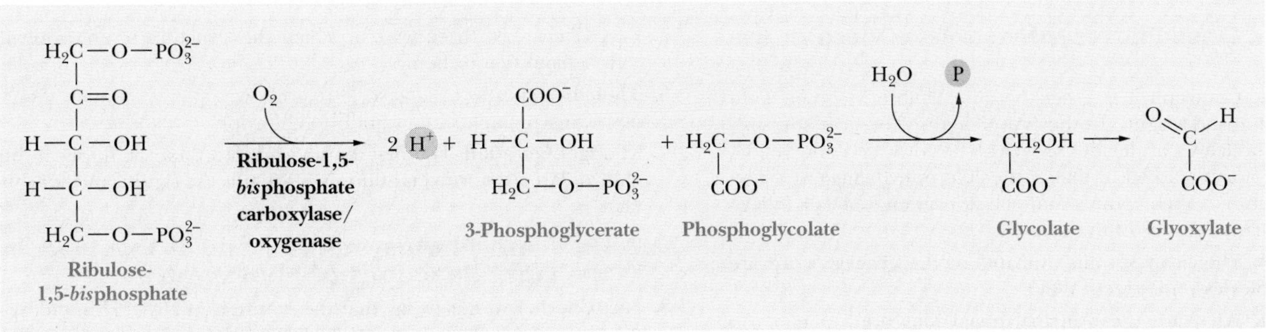

▲ **FIGURE 22.19** The characteristic reactions of photorespiration.

When levels of O_2 are high compared with those of CO_2, ribulose-1,5-*bis*phosphate is oxygenated to produce phosphoglycolate (which gives rise to glycolate) and 3-phosphoglycerate by photorespiration, rather than the two molecules of 3-phosphoglycerate that arise from the carboxylation reaction. This situation occurs in C_3 plants. In C_4 plants, the small pores decrease the entry not only of CO_2 but also of O_2 into the leaves. The ratio of CO_2 to O_2 in the bundle-sheath cells is relatively high as a result of the operation of the C_4 pathway, favoring the carboxylation reaction. C_4 plants have successfully reduced the oxygenase activity by compartmentation and thus have less need of photorespiration. This is an advantage in the hot climates in which C_4 plants are principally found.

Summary

22.1 Where Does Photosynthesis Take Place in the Cell?
In eukaryotes, the light reactions of photosynthesis take place in the thylakoid membranes of chloroplasts. A series of membrane-bound electron carriers and pigments is able to harness the light energy of the sun.

22.2 How Are Photosystems I and II Involved in the Light Reactions of Photosynthesis?
The equation for photosynthesis

$$6\,CO_2 + 6\,H_2O \rightarrow C_6H_{12}O_6 + 6\,O_2$$

actually represents two processes. One process, the oxidation of water to produce oxygen, requires light energy from the sun, and the other process, the fixation of CO_2 to give sugars, uses solar energy indirectly. In the light reactions, water is oxidized to produce oxygen, accompanied by the reduction of $NADP^+$ to NADPH. The trapping of light takes place at a reaction center within the chloroplast, and the process requires a pair of chlorophylls in a unique environment.

The light reactions consist of two parts, each carried out by a separate photosystem. The reduction of $NADP^+$ to NADPH is accomplished by photosystem I, whereas the oxidation of water to produce oxygen is done by photosystem II.

22.3 How Does Photosynthesis Produce ATP?
The two photosystems are linked by an electron transport chain coupled to the production of ATP. A proton gradient drives the production of ATP in photosynthesis, as it does in mitochondrial respiration.

22.4 What Are the Evolutionary Implications of Photosynthesis with and without Oxygen?
Some forms of bacteria also have photosynthesis, although they often have simpler systems involving only a single photosystem. Early photosynthetic bacteria probably used an electron donor other than water and did not produce oxygen. Bacteria that appeared later, as well as eukaryotes, eventually developed the dual photosystems and the ability to produce oxygen from water, which led to the Earth having an oxygen atmosphere.

22.5 How Do the Dark Reactions of Photosynthesis Fix CO_2 into Glucose?
The dark reactions of photosynthesis involve the net synthesis of one molecule of glucose from six molecules of CO_2. The net reaction of six molecules of CO_2 to produce one molecule of glucose requires the carboxylation of six molecules of a five-carbon key intermediate, ribulose-1,5-*bis*phosphate, ultimately forming 12 molecules of 3-phosphoglycerate. Of these, two molecules of 3-phosphoglycerate react to give rise to glucose. The remaining ten molecules of 3-phosphoglycerate are used to regenerate the six molecules of ribulose-1,5-*bis*phosphate. The overall reaction pathway is cyclic and is called the Calvin cycle.

22.6 How Is CO_2 Fixed in Tropical Plants?
In addition to the Calvin cycle, there is an alternative pathway for CO_2 fixation in tropical plants. It is called the C_4 pathway because it involves four carbon compounds. In this pathway, CO_2 reacts in the outer (mesophyll) cells with phosphoenolpyruvate to produce oxaloacetate and P_i. Oxaloacetate in turn is reduced to malate. Malate is transported from mesophyll cells, where it is produced, to inner (bundle-sheath) cells, where it is ultimately passed to the Calvin cycle. Plants in which the C_4 pathway operates grow more quickly and produce more biomass per unit of leaf area than C_3 plants, in which only the Calvin cycle operates.

Critical Questions to Review

22.1 Where Does Photosynthesis Take Place in the Cell?

1. **Fact Check** Chlorophyll is green because it absorbs green light less than it absorbs light of other wavelengths. The accessory pigments in the leaves of deciduous trees tend to be red and yellow, but their color is masked by that of the chlorophyll. Suggest a connection between these points and the appearance of fall foliage colors in many sections of the country.

2. **Fact Check** The bean sprouts available at the grocery store are white or colorless, not green. Why?

3. **Fact Check** What are the principal metal ions used in electron transfer in chloroplasts? Compare them to the ions found in mitochondria.

4. **Fact Check** How is the structure of chloroplasts similar to that of mitochondria? How does it differ?

5. **Fact Check** List three ways in which the structure of chlorophyll differs from that of heme.

6. **Thought Question** Suggest a reason why plants contain light-absorbing pigments in addition to chlorophylls *a* and *b*.

7. **Thought Question** The first amino acid in protein synthesis in the chloroplast is *N*-formyl methionine. What is the significance of this fact?

22.2 How Are Photosystems I and II Involved in the Light Reactions of Photosynthesis?

8. **Fact Check** Is it fair to say that the synthesis of NADPH in chloroplasts is merely the reverse of NADH oxidation in mitochondria? Explain your answer.

9. **Fact Check** Outline the events that take place at the photosynthetic reaction center in *Rhodopseudomonas*.

10. **Fact Check** What are the two places where light energy is required in the light reaction of photosynthesis? Why must energy be supplied at precisely these points?

11. **Fact Check** Do all the chlorophyll molecules in a photosynthetic reaction center play the same roles in the light reactions of photosynthesis?

12. **Fact Check** Describe some similarities between the electron transport chains in chloroplasts and in mitochondria.

13. **Thought Question** Which is likely to have evolved first, the electron transport chain in chloroplasts or in mitochondria? Explain your answer.

14. **Thought Question** Uncouplers of oxidative phosphorylation in mitochondria also uncouple photoelectron transport and ATP synthesis in chloroplasts. Give an explanation for this observation.

15. **Thought Question** A larger proton gradient is required to form a single ATP in chloroplasts than in mitochondria. Suggest a reason why. *Hint:* Ions can move across the thylakoid membrane more easily than across the inner mitochondrial membrane.

16. **Thought Question** Albert Szent-Gyorgi, a pioneer in early photosynthesis research, stated, "What drives life is a little electric current, kept up by the sunshine." What did he mean by this?

17. **Biochemical Connections** What is implied about the energy requirements of photosystems I and II by the fact that there is a difference in the minimum wavelength of light needed for them to operate (700 nm for photosystem I and 680 nm for photosystem II)?

18. **Thought Question** Is it reasonable to list standard reduction potentials (see Chapter 20) for the reactions of photosynthesis? Why or why not?

19. **Thought Question** Why is a photosynthetic reaction center comparable to a battery?

20. **Thought Question** Antimycin A is an inhibitor of photosynthesis in chloroplasts. Suggest a possible site of action, and indicate the reason for your choice.

21. **Thought Question** Would you expect H_2O or CO_2 to be the source of the oxygen produced in photosynthesis? Give the reason for your answer.

22. **Thought Question** Why do we describe the path of electrons in photosynthesis as starting in photosystem II and ending in photosystem I? In other words, why is the nomenclature "backward"?

23. **Thought Question** It has taken considerable amounts of research to establish the number of protons pumped across the mitochondrial membrane at the various stages of electron transport. Would you expect to encounter difficulties in determining the number of protons pumped in electron transport across the thylakoid membrane? Why or why not?

24. **Thought Question** The oxidation of water requires four electrons, but chlorophyll molecules can transfer only one electron at a time. Describe how these two statements can be reconciled.

25. **Thought Question** Why does a loosely bound cytochrome play a unique role in the reaction-center events in *Rhodopseudomonas*?

26. **Thought Question** What are the evolutionary implications of the similarity in structure and function of ATP synthase in chloroplasts and mitochondria?

22.3 How Does Photosynthesis Produce ATP?

27. **Fact Check** In cyclic photophosphorylation in photosystem I, ATP is produced, even though water is not split. Explain how the process takes place.

28. **Fact Check** What are the major similarities and differences between ATP synthesis in chloroplasts, as compared with mitochondria?

29. **Fact Check** How can a proton gradient be created in cyclic photophosphorylation in photosystem I?

30. **Thought Question** Can ATP production take place in chloroplasts in the absence of light? Give the reason for your answer.

31. **Thought Question** What is the advantage to plants to have the option of both cyclic and noncyclic pathways for photophosphorylation?

22.4 What Are the Evolutionary Implications of Photosynthesis with and without Oxygen?

32. **Fact Check** Is water the only possible electron donor in photosynthesis? Why or why not?

33. **Thought Question** Suppose that a prokaryotic organism that contains both chlorophyll *a* and chlorophyll *b* has been discovered. Comment on the evolutionary implications of such a discovery.

22.5 How Do the Dark Reactions of Photosynthesis Fix CO_2 into Glucose?

34. **Fact Check** Why is rubisco likely to be the most abundant protein in nature?

35. **Biochemical Connections** Is the sequence of amino acids in rubisco encoded by nuclear genes or not? Explain.

36. **Fact Check** Name some other metabolic pathways that have reactions similar to those of the dark reactions of photosynthesis.

37. **Thought Question** Using information from Section 15.3 and 15.10, show how the $\Delta G^{\circ\prime}$ of 478 kJ (114 kcal) is obtained for each mole of CO_2 fixed in photosynthesis. The reaction in question is 6 CO_2 + 6 H_2O → Glucose + 6 O_2.

38. **Thought Question** If photosynthesizing plants are grown in the presence of $^{14}CO_2$, is every carbon atom of the glucose that is produced labeled with the radioactive carbon? Why or why not?

39. **Thought Question** Rubisco has a very low turnover number, about 3 CO_2 per second. What might this low number tell us about the evolution of rubisco?

40. **Biochemical Connections** What key aspects of chloroplasts (and mitochondria) are consistent with the theory that they may have once been bacteria? List three specific features.

41. **Thought Question** Suggest a reason why the evolution of the pathway for the regeneration of ribulose-1,5-*bis*phosphate from glyceraldehyde-3-phosphate was "no big deal."

42. **Biochemical Connections** Why would nature evolve a key enzyme, rubisco, that is so sensitive to oxygen, resulting in photorespiration?

43. **Thought Question** Does the whole Calvin cycle represent carbon dioxide fixation? Why or why not?

44. **Thought Question** What is the evolutionary advantage to organisms that the Calvin cycle has a number of reactions in common with other pathways?

45. **Thought Question** Why do we refer to the conversion of six molecules of carbon dioxide (six carbon atoms) to one molecule of glucose (also six carbon atoms) as a *net* reaction?

22.6 How Is CO_2 Fixed in Tropical Plants?

46. **Fact Check** How does the production of sugars by tropical plants differ from the same reactions in the Calvin cycle?

47. **Fact Check** How does photosynthesis in C_4 plants differ from the process in C_3 plants?

48. **Fact Check** What is photorespiration?

49. **Thought Question** Why is it advantageous to the tropical plants to use the C_4 rather than the C_3 fixation pathway?

50. **Thought Question** What would be the effect on plants if photorespiration did not exist?

Biochemistry ⊜ Now™

Assess your understanding of this chapter's topics with additional quizzing and tutorials at **http://now.brookscole.com/campbell5**

Annotated Bibliography

Bering, C. L. Energy Interconversions in Photosynthesis. *J. Chem. Ed.* **62,** 659–664 (1985). [A discussion of basic concepts of photosynthesis, concentrating on the light reactions and photosystems.]

Bishop, M. B., and C. B. Bishop. Photosynthesis and Carbon Dioxide Fixation. *J. Chem. Ed.* **64,** 302–305 (1987). [Concentrates on the Calvin cycle.]

Danks, S. M., E. H. Evans, and P. A. Whittaker. *Photosynthetic Systems: Structure, Function and Assembly.* New York: Wiley, 1983. [A short book with excellent electron micrographs of chloroplasts and related structures in Chapter 1.]

Deisenhofer, J., and H. Michel. The Photosynthetic Reaction Center from the Purple Bacterium *Rhodopseudomonas viridis. Science* **245,** 1463–1473 (1989). [The authors' Nobel Prize address describing their work on the structure of the reaction center.]

Deisenhofer, J., H. Michel, and R. Huber. The Structural Basis of Photosynthetic Light Reactions in Bacteria. *Trends Biochem. Sci.* **10,** 243–248 (1985). [A discussion of the photosynthetic reaction center in bacteria.]

Dennis, D. T. *The Biochemistry of Energy Utilization in Plants.* New York: Chapman and Hall, 1987. [A short book on plant biochemistry.]

Govindjee and W. J. Coleman. How Plants Make Oxygen. *Sci. Amer.* **262** (2), 50–58 (1990). [Focuses on the water-oxidizing apparatus of photosystem II.]

Halliwell, B. *Chloroplast Metabolism: The Structure and Function of Chloroplasts in Green Leaf Cells.* New York: Oxford University Press, 1981. [A detailed description of chloroplast activity.]

Hathway, D. *Molecular Mechanisms of Herbicide Selectivity.* New York: Oxford University Press, 1989. [A short book primarily devoted to the differences in enzyme activity in weeds and desirable plants.]

Hipkins, M. F., and N. R. Baker, eds. *Photosynthesis: Energy Transduction: A Practical Approach.* Oxford, England: IRL Press, 1986. [A collection of articles about research methods used to study photosynthesis.]

Karplus, P., M. Daniels, and J. Herriott. Atomic Structure of Ferredoxin-NADP$^+$ Reductase: Prototype for a Structurally Novel Flavoenzyme Family. *Science* **251,** 60–66 (1991). [The structure of a key enzyme involved in nitrogen and sulfur metabolism, as well as in photosynthesis.]

Margulis, L. *Early Life.* Boston: Science Books International, 1982. [Chapters 2 and 3 discuss the evolutionary development of photosynthesis.]

Youvan, D. C., and B. L. Marrs. Molecular Mechanisms of Photosynthesis. *Sci. Amer.* **256** (6), 42–48 (1987). [A detailed description of a bacterial photosynthetic reaction center and the molecular events that take place there.]

Zuber, H. Structure of Light-Harvesting Antenna Complexes of Photosynthetic Bacteria, Cyanobacteria and Red Algae. *Trends Biochem. Sci.* **11,** 414–419 (1986). [Concentrates on the protein portion of the photosynthetic reaction center.]

The Metabolism of Nitrogen

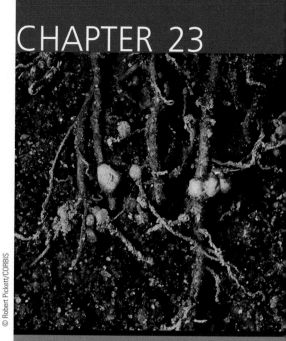

Surprisingly, nitrogen found in most living organisms does not come directly from the atmosphere, where it composes 78% of atmospheric gas. Instead, organic nitrogen enters the biological realm through bacteria living in the root nodules of leguminous plants, such as peas and beans. They convert the triple-bonded nitrogen molecule N_2 to ammonia, NH_3, which is incorporated into the amino acid glutamate, a precursor of proline and arginine. Thus, the 20 amino acids found in proteins may be formed by transforming from one amino acid into another, often by long and complex metabolic pathways. Although amino acids act principally as the building blocks of proteins, they also contribute to the synthesis of a variety of other biologically important molecules. These molecules include the nitrogenous bases of DNA and RNA, the nucleotide coenzyme electron carriers NAD^+ and $NADP^+$, and the oxygen-binding pyrrole ring of hemoglobin, as well as hormones and neurotransmitters. When amino acids are deaminated (deamination is the removal of the α-amino group, $—NH_3^+$), their carbon skeletons can be fed into the citric acid cycle to be oxidized to carbon dioxide and water. Alternatively, they may be used as precursors of other biomolecules, including glucose and fatty acids. In many ways, then, amino acids are connecting links between metabolic pathways.

Excess nitrogen in the form of ammonia is toxic. In mammals, the amino groups are concentrated in the form of glutamate, glutamine, and alanine, which are transported to the liver, where they donate their amino groups to a pathway called the urea cycle. Nitrogen is concentrated in the form of urea, which enters the bloodstream and is eventually filtered out by the kidneys. Fish excrete their excess nitrogen directly as ammonia, while birds concentrate it in the form of uric acid.

Root nodules of leguminous plants play a pivotal role in nitrogen fixation.

Critical Questions

23.1 What Processes Constitute Nitrogen Metabolism?

23.2 How Is Nitrogen Incorporated into Biologically Useful Compounds?

23.3 What Role Does Feedback Inhibition Play in Nitrogen Metabolism?

23.4 How Are Amino Acids Synthesized?

23.5 What Are the Essential Amino Acids?

23.6 How Are Amino Acids Catabolized?

23.7 How Are Purines Synthesized?

23.8 How Are Purines Catabolized?

23.9 How Are Pyrimidines Synthesized and Catabolized?

23.10 How Are Ribonucleotides Converted to Deoxyribonucleotides?

23.11 How Is dUTP Converted to dTTP?

23.1 | What Processes Constitute Nitrogen Metabolism?

We have seen the structures of many types of compounds that contain nitrogen, including amino acids, porphyrins, and nucleotides, but we have not discussed their metabolism. The metabolic pathways we have dealt with up to now have mainly involved compounds of carbon, hydrogen, and oxygen, such as sugars and fatty acids. Several important topics can be included in our discussion of the metabolism of nitrogen. The first of these is **nitrogen fixation,** the process by which inorganic molecular nitrogen from the atmosphere (N_2) is incorporated first into ammonia and then into organic compounds that are of use to organisms. Nitrate ion (NO_3^-), another kind of inorganic nitrogen, is the form in which nitrogen is found in the soil, and many fertilizers contain nitrates, frequently potassium nitrate. The process of **nitrification** (nitrate reduction to ammonia) provides another way for organisms to obtain nitrogen. Nitrate ion and nitrite ion (NO_2^-) are also involved in **denitrification** reactions, which return nitrogen to the atmosphere (Figure 23.1).

Biochemistry ⚡ Now™

Test yourself on these Critical Questions at the BiochemistryNow website at **http://now .brookscole.com/campbell5**

© Robert Pickett/CORBIS

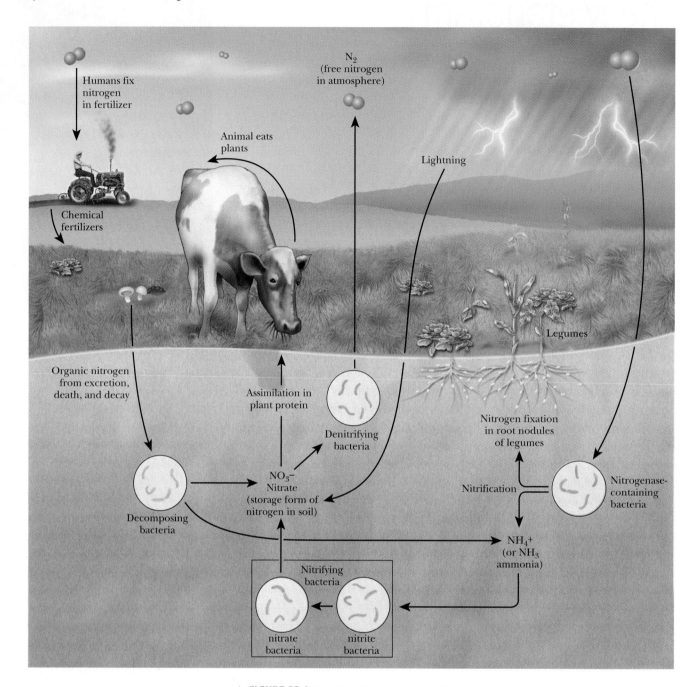

▲ **FIGURE 23.1** The flow of nitrogen in the biosphere.

Ammonia formed by either pathway, nitrogen fixation or nitrification, enters the biosphere. Ammonia is converted to organic nitrogen by plants, and organic nitrogen is passed to animals through food chains. Finally, animal waste products, such as urea, are excreted and degraded to ammonia by microorganisms. The word "ammonia" comes from sal ammoniac (ammonium chloride), which was first prepared from the dung of camels at the temple of Jupiter Ammon in North Africa. The process of death and decay releases ammonia in both plants and animals. Denitrifying bacteria reverse the conversion of ammonia to nitrate and then recycle the NO_3^- as free N_2 (Figure 23.1).

The topic of nitrogen metabolism includes the biosynthesis and breakdown of *amino acids, purines,* and *pyrimidines;* also, the metabolism of *porphyrins* is related to that of amino acids. Many of these pathways, particularly the anabolic ones, are long and complex. In discussing those pathways in which the amount of material is large and highly detailed, we shall concentrate on the most important points. Specifically, we shall concentrate on overall patterns and on interesting reactions of wide applicability. We shall also be interested in health-related aspects of this material. Other reactions will be found at the Biochemistry Interactive website for this text. It can be considered a repository of supplementary material for this chapter, and we shall refer to it a number of times.

23.2 | How Is Nitrogen Incorporated into Biologically Useful Compounds?

Bacteria are responsible for the reduction of N_2 to ammonia (NH_3). Typical nitrogen-fixing bacteria are symbiotic organisms that form nodules on the roots of leguminous plants, such as beans and alfalfa. Many free-living microbes and some cyanobacteria also fix nitrogen. Plants and animals

Biochemical Connections

Nitrogen Fertilizers

Crop production per acre in the United States is higher than in many areas of the world. In part, this is the result of extensive use of fertilizers, especially those that supply nitrogen in a form that plants can use readily. Both ammonium and nitrate ions are used; even ammonia gas can be pumped into the ground, if there is enough water available in the soil to dissolve it.

Ammonia is toxic to animals, so it often comes as a surprise that ammonia gas itself can be used for fertilization. Plants can assimilate ammonia rapidly, but they usually never get the chance to do so because the nitrifying soil bacteria, especially *Nitrosomonas* and *Nitrobacter,* rapidly convert the ammonia first to nitrite and then to nitrate. The final nitrate product is easily converted back to ammonia, but it takes energy to do it. Ammonia is especially useful as a fertilizer in the early spring and for germinating plants. In the spring, the soil is usually damp enough to dissolve the ammonia so that it can move to the plants. Because light is less available in the early spring, the young plants do not have enough energy to convert the nitrate back to ammonia until their chloroplasts develop fully. Fortunately, because of the condition of the soil, the ammonia goes directly to the plants rather than to the soil bacteria.

The genes for the enzymes for nitrogen fixation have been studied extensively. Much research is going on to determine if these genes can be incorporated into crop plants, which would reduce the amount of nitrogen fertilizer needed for maximal plant growth and crop production.

Two other sources of nitrogen fixation are often overlooked. The first is the chemical synthesis of ammonia from H_2 and N_2, called the Haber process, after its discoverer, the German chemist

Fritz Haber. This reaction is very important for the formation of chemical fertilizers, and it is responsible for a great deal of the organic nitrogen currently found in the biosphere. The second source of fixed nitrogen is that produced by lightning.

▲ Nitrogen-fixing bacteria form nodules on alfalfa roots.

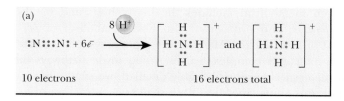

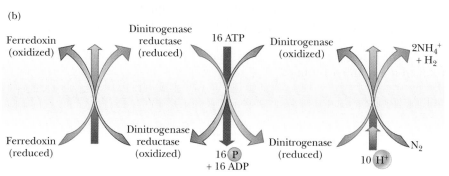

▶ **FIGURE 23.2** Some aspects of the nitrogenase reaction. (a) The reduction of N_2 to $2\,NH_4^+$. (b) The path of electrons from ferredoxin to N_2.

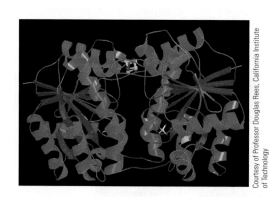

▲ **FIGURE 23.3** The X-ray structure of the *Azotobacter vinelandii* Fe–protein dimer.

Essential Information

Nitrogen enters the biosphere by the process of nitrogen fixation. Atmospheric nitrogen is converted to ammonia in its conjugate acid form, ammonium ion. A number of carbon-containing compounds form carbon–nitrogen bonds in transamination reactions.

cannot carry out nitrogen fixation. This conversion of molecular nitrogen to ammonia is the only source of nitrogen in the biosphere except for that provided by nitrates. The conjugate acid form of NH_3, ammonium ion (NH_4^+), is the form of nitrogen that is used in the first stages of the synthesis of organic compounds. Parenthetically, NH_3 obtained by chemical synthesis from nitrogen and hydrogen is the starting point for the production of many synthetic fertilizers, which frequently contain nitrates.

The **nitrogenase** enzyme complex found in nitrogen-fixing bacteria catalyzes the production of ammonia from molecular nitrogen. The half-reaction of reduction [Figure 23.2(a)] is

$$N_2 + 8\,e^- + 16\,ATP + 10\,H^+ \rightarrow 2\,NH_4^+ + 16\,ADP + 16\,P_i + H_2$$

in which six electrons are used to reduce molecular nitrogen to ammonium ion. An additional two electrons are used to reduce hydrogen ion to H_2. The total reaction catalyzed by nitrogenase is an eight-electron reduction.

The half-reaction of oxidation will vary because different organisms vary in terms of the substance oxidized to supply electrons. Several proteins are included in the nitrogenase complex. Ferredoxin is one of them (this protein also plays an important role in electron transfer in photosynthesis; Section 22.3). There are also two proteins specific to the nitrogenase reaction. One is an iron–sulfur (Fe–S) protein called *dinitrogenase reductase*. The other is an iron–molybdenum (Fe–Mo) protein, called dinitrogenase. The flow of electrons is from ferredoxin to dinitrogenase reductase to dinitrogenase to nitrogen [Figure 23.2(b)]. The nature of the nitrogenase complex is a subject of active research. Significant progress has been made in this work with the determination by X-ray crystallography of the three-dimensional structure of both the Fe protein and the Fe–Mo protein from *Azotobacter vinelandii* (Figure 23.3). The Fe protein is a dimer ("the iron butterfly"), with the iron–sulfur cluster located at the butterfly's head. The nitrogenase is even more complicated, with several types of subunits arranged into tetramers. Ferredoxin, dinitrogenase reductase, and dinitrogenase combine to perform a series of single-electron transfers, eventually transferring the eight electrons necessary to complete the reduction of N_2 to NH_4^+.

It is worth noting that the reactions of nitrogen fixation consume a great deal of energy. It is estimated that about half of the ATP produced from photosynthesis in legumes is used to fix nitrogen.

23.3 | What Role Does Feedback Inhibition Play in Nitrogen Metabolism?

The biosynthetic pathways that produce amino acids and the bases of nucleotides (purines and pyrimidines) are long and complex, requiring a large investment of energy by the organism. If there is a high level of some end product, such as an amino acid or a nucleotide, the cell saves energy by not making that compound. However, the cell needs a signal to tell it to stop producing more of that particular compound. The signal is frequently part of a **feedback inhibition** mechanism, in which the end product of a metabolic pathway inhibits an enzyme at the beginning of the pathway. We saw an example of such a control mechanism when we discussed the allosteric enzyme aspartate transcarbamoylase in Section 7.1. This enzyme catalyzes one of the early stages of pyrimidine nucleotide biosynthesis, and it is inhibited by the end product of that pathway—namely, cytidine triphosphate (CTP). Feedback inhibition is frequently encountered in the biosynthesis of amino acids and nucleotides. Another prime example of allosteric regulation by feedback inhibition is found in the activity of the enzyme glutamine synthetase, one of the key enzymes in amino acid biosynthesis (Figure 23.4). There are no fewer than

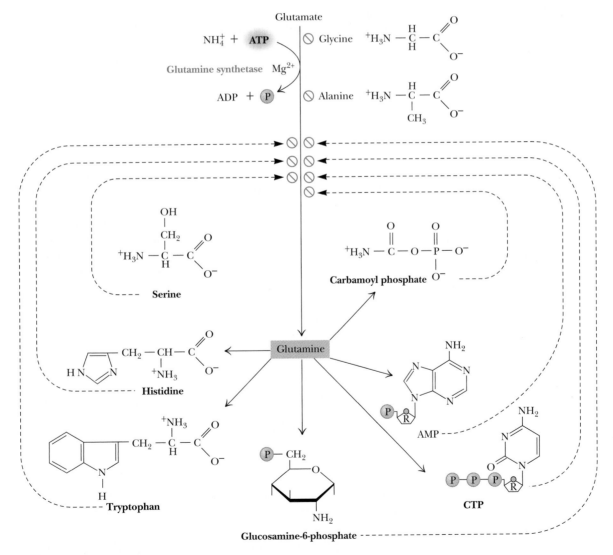

Biochemistry ⊘**Now™** ▲ **ACTIVE FIGURE 23.4** The allosteric regulation of glutamine synthetase activity by feedback inhibition. **Watch this Active Figure at http://now.brookscole.com/campbell5**

nine allosteric inhibitors involved here (glycine, alanine, serine, histidine, tryptophan, CTP, AMP, carbamoyl phosphate, and glucosamine-6-phosphate). Glycine, alanine, and serine are key indicators of amino acid metabolism in the cell. Each of the other six compounds represents an end product of a biosynthetic pathway that depends on glutamine. Feedback inhibition is very effective because a single product molecule can inhibit an enzyme capable of synthesizing many hundreds or thousands of product molecules.

23.4 | How Are Amino Acids Synthesized?

General Features

Ammonia is toxic in high concentrations, and so it must be incorporated into biologically useful compounds when it is formed by the reactions of nitrogen fixation discussed earlier in this chapter. The amino acids glutamate and glutamine are of central importance in the process. **Glutamate** arises from α-ketoglutarate, and **glutamine** is formed from glutamate (Figure 23.5). The production of glutamate is a reductive amination, and the production of glutamine is amidation. In other reactions of amino acid anabolism the α-amino group of glutamate and the side-chain amino group of glutamine are shifted to other compounds in **transamination** reactions.

The biosynthesis of amino acids involves a common set of reactions. In addition to transamination reactions, transfer of one-carbon units, such as formyl or methyl groups, occurs frequently. We are not going to discuss all the details of the reactions that give rise to amino acids. We can, however, organize this material by grouping amino acids into families based on common precursors (Figure 23.6). The reactions of some of the individual families of amino acids provide good examples of those reactions that are of general importance, such as transamination and one-carbon transfer.

We can also make some generalizations about amino acid metabolism in terms of the relationship of the carbon skeleton to the citric acid cycle and the related reactions of pyruvate and acetyl-CoA (Figure 23.7). The citric acid cycle is amphibolic; it has a part in both catabolism and anabolism. The anabolic aspect of the citric acid cycle is of interest in amino acid biosynthesis. The catabolic aspect is apparent in the breakdown of amino acids, leading to their eventual excretion, which takes place in reactions related to the citric acid cycle.

(a) $NH_4^+ + {}^-OOC-CH_2-CH_2-\overset{\overset{\displaystyle O}{\|}}{C}-COO^-$ $\xrightarrow[\text{NADP}^+]{\text{NADPH} + \text{H}^+}$ $H_2O + {}^-OOC-CH_2-CH_2-\overset{\overset{\displaystyle NH_3^+}{|}}{CH}-COO^-$

α-Ketoglutarate   Glutamate

(b) $NH_4^+ + {}^-OOC-CH_2-CH_2-\overset{\overset{\displaystyle NH_3^+}{|}}{CH}-COO^-$ $\xrightarrow{\text{ATP} \quad \text{ADP} + \text{P}}$ $H_2O + H_2N-\overset{\overset{\displaystyle O}{\|}}{C}-CH_2-CH_2-\overset{\overset{\displaystyle NH_3^+}{|}}{CH}-COO^-$

Glutamate   Glutamine

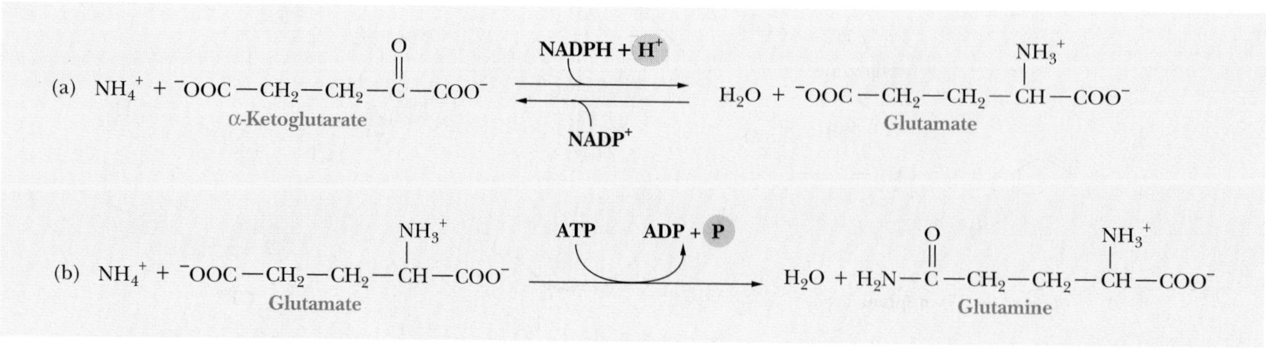

▲ **FIGURE 23.5** (a) The production of glutamate from α-ketoglutarate. (b) The production of glutamine from glutamate.

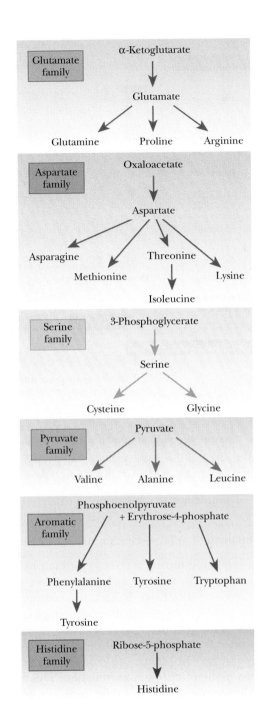

◀ **FIGURE 23.6** Families of amino acids based on biosynthetic pathways. Each family has a common precursor.

Transamination Reactions: The Role of Glutamate and Pyridoxal Phosphate

Glutamate is formed from NH_4^+ and α-ketoglutarate in a reductive amination that requires NADPH. This reaction is reversible and is catalyzed by *glutamate dehydrogenase* (GDH).

$$NH_4^+ + \text{α-ketoglutarate} + NADPH + H^+ \rightarrow \text{Glutamate} + NADP^+ + H_2O$$

Glutamate is a major donor of amino groups in reactions, and α-ketoglutarate is a major acceptor of amino groups [see Figure 23.5(a)]. Note the requirement for reducing power.

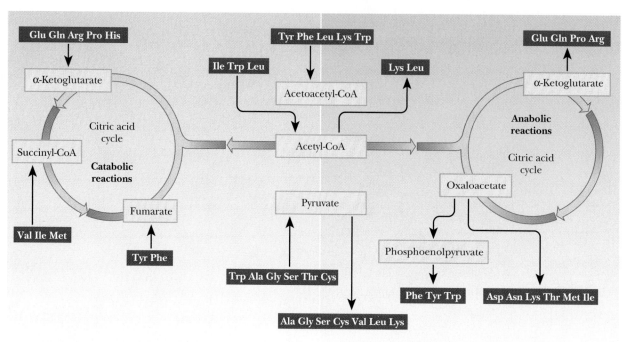

Catabolic breakdown of amino acids produces citric acid cycle intermediates

Anabolic formation of amino acids uses citric acid cycle intermediates as precursors

▲ **FIGURE 23.7** The relationship between amino acid metabolism and the citric acid cycle.

The conversion of glutamate to glutamine is catalyzed by **glutamine synthetase** (GS) in a reaction that requires ATP [see Figure 23.5(b)].

$$NH_4^+ + Glutamate + ATP \rightarrow Glutamine + ADP + P_i + H_2O$$

These reactions fix inorganic nitrogen (NH_3), forming organic (carbon-containing) nitrogen compounds, such as amino acids, but they frequently do not operate in this sequential fashion. In fact, the combination of GDH and GS is responsible for most of the assimilation of ammonia into organic compounds, especially in organisms that are rich in nitrogen sources. However, the K_M of GS is considerably lower than that of GDH. When nitrogen is limiting, as is frequently the case in plants, the conversion of glutamate to glutamine is the preferred mode of nitrogen assimilation. This means that the supply of glutamate will become depleted unless there is some way to replenish it. The reductive amination of α-ketoglutarate with the amide nitrogen of glutamine as the nitrogen source is the way this is done.

$$Reductant + \alpha\text{-Ketoglutarate} + Glutamine \rightarrow$$

$$2\ Glutamate + Oxidized\ reductant$$

The reductant can be NADH, NADPH (in yeast and bacteria), or reduced ferredoxin (in plants). The enzyme that catalyzes this reaction is glutamate synthase; it is also known as glutamate:oxoglutarate aminotransferase (GOGAT). A GS/GOGAT complex exists in plants and allows them to cope with conditions of limited nitrogen availability. Enzymes that catalyze transamination reactions require pyridoxal phosphate as a coenzyme (Figure 23.8). We discussed this compound in Section 7.8 as a typical example of a coenzyme, and here we can see its mode of action in context.

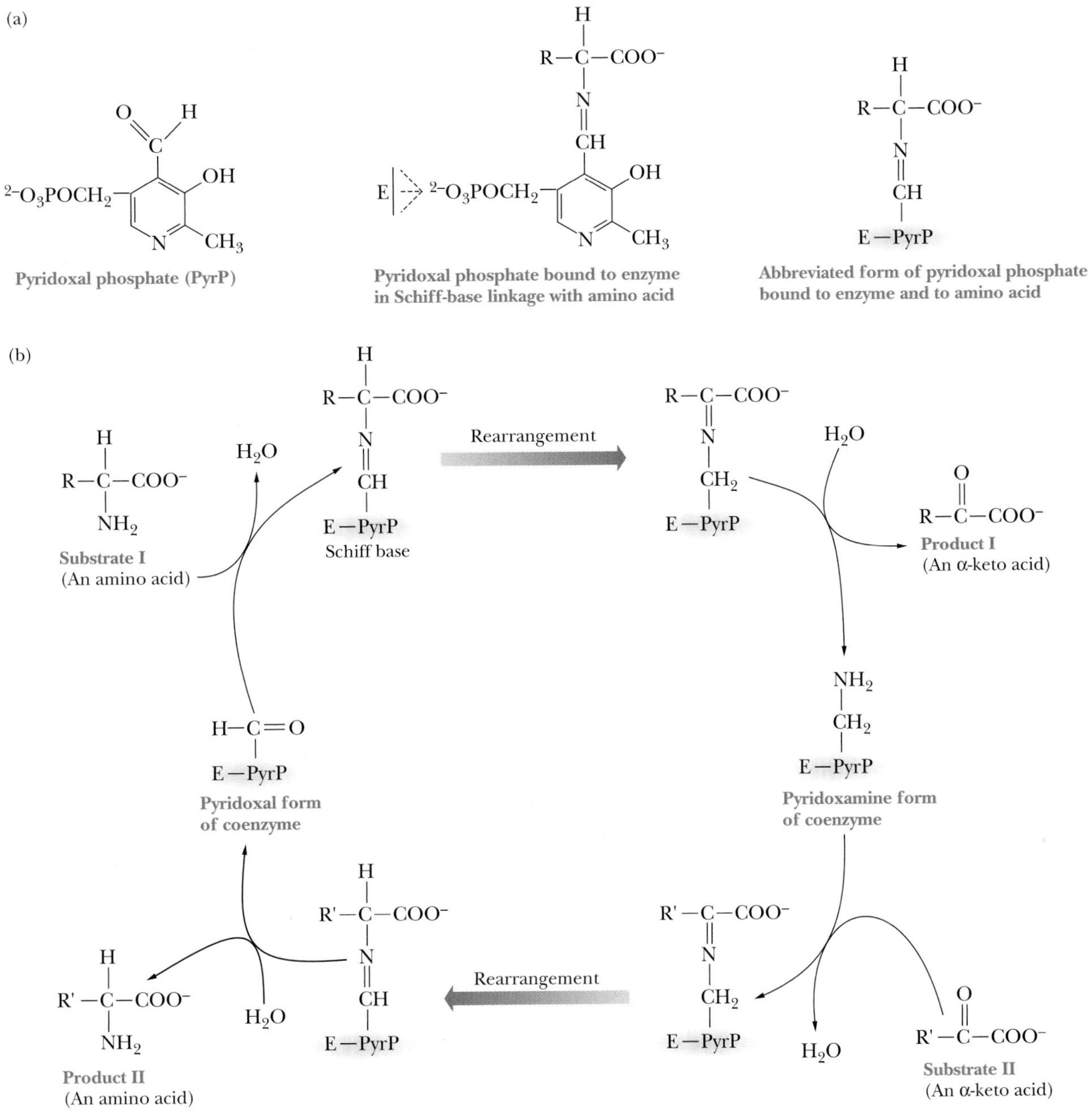

(a)

Pyridoxal phosphate (PyrP)

Pyridoxal phosphate bound to enzyme
in Schiff-base linkage with amino acid

Abbreviated form of pyridoxal phosphate
bound to enzyme and to amino acid

(b)

Biochemistry ⟨Ɛ⟩Now™ ▲ ACTIVE FIGURE 23.8 The role of pyridoxal phosphate in transamination reactions. (a) The mode of binding of pyridoxal phosphate (PyrP) to the enzyme (E) and to the substrate amino acid. (b) The reaction itself. The original substrate, an amino acid, is deaminated, while an α-keto acid is aminated to form an amino acid. The net reaction is one of transamination. Note that the coenzyme is regenerated and that the original substrate and final product are both amino acids. **Watch this Active Figure at http://now.brookscole.com/campbell5**

Pyridoxal phosphate (PyrP) forms a Schiff base with the amino group of Substrate I (the amino-group donor). The next stage is a rearrangement followed by hydrolysis, which removes Product I (the α-keto acid corresponding to Substrate I). The coenzyme now carries the amino group (pyridoxamine). Substrate II (another α-keto acid) then forms a Schiff base with pyridoxamine. Again there is a rearrangement followed by a hydrolysis, which gives rise to Product II (an amino acid) and regenerates pyridoxal phosphate. The net

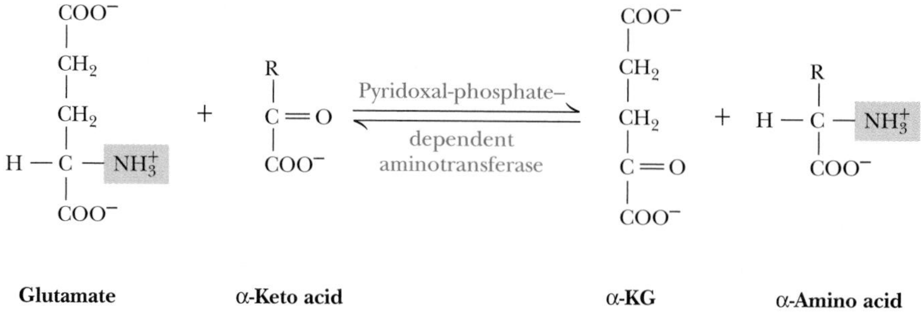

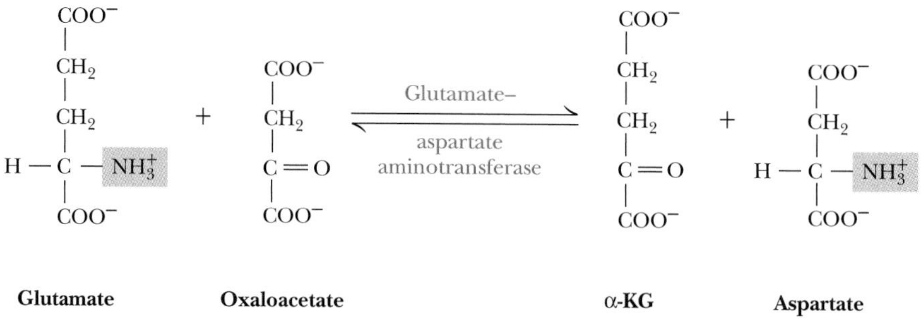

▶ **FIGURE 23.9** Transamination reactions switch an amino group from one amino acid to an α-keto acid. Glutamate and α-ketoglutarate (α-KG) are one donor/acceptor pair. *Above,* a general case. *Below,* a specific case, in which the other donor/acceptor pair is aspartate and oxaloacetate.

reaction is that an amino acid (Substrate I) reacts with an α-keto acid (Substrate II) to form an α-keto acid (Product I) and an amino acid (Product II). The amino group has been transferred from Substrate I to Substrate II, forming the amino acid, Product II. The overall reaction can be seen for a general case and for a specific case in Figure 23.9. When not involved with one of the substrates, the pyridoxal group is bound in a Schiff-base linkage to an active-site ε-NH$_2$ group of lysine. Pyridoxal phosphate is a versatile coenzyme that is also involved in other reactions, including decarboxylations, racemizations, and movement of hydroxymethyl groups, as we shall see with the conversion of serine to glycine.

One-Carbon Transfers and the Serine Family

In addition to transamination reactions, one-carbon transfer reactions occur frequently in amino acid biosynthesis. A good example of a one-carbon transfer can be found in the reactions that produce the amino acids of the serine family. This family also includes glycine and cysteine. Serine and glycine themselves are frequently precursors in other biosynthetic pathways. A discussion of the synthesis of cysteine will give us some insight into the metabolism of sulfur, as well as that of nitrogen.

The ultimate precursor of serine is 3-phosphoglycerate, which is obtainable from the glycolytic pathway. The hydroxyl group on carbon 2 is oxidized to a keto group, giving an α-keto acid. A transamination reaction in which glutamate is the nitrogen donor produces 3-phosphoserine and α-ketoglutarate. Hydrolysis of the phosphate group then gives rise to serine (Figure 23.10).

The conversion of serine to glycine involves the transfer of a one-carbon unit from serine to an acceptor. This reaction is catalyzed by *serine hydroxymethylase,* with pyridoxal phosphate as a coenzyme. The acceptor in this reaction is **tetrahydrofolate,** a derivative of folic acid and a frequently encountered carrier of one-carbon units in metabolic pathways. Its structure has three parts: a substituted pteridine ring, *p*-aminobenzoic acid, and glutamic acid (Figure 23.11).

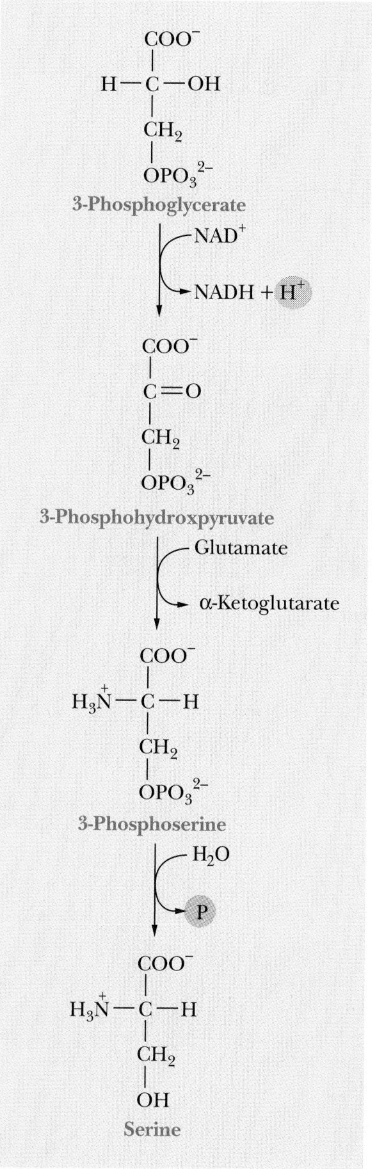

$$\text{COO}^-$$
$$|$$
$$\text{H}-\text{C}-\text{OH}$$
$$|$$
$$\text{CH}_2$$
$$|$$
$$\text{OPO}_3^{2-}$$

3-Phosphoglycerate

NAD^+

$\text{NADH} + \text{H}^+$

$$\text{COO}^-$$
$$|$$
$$\text{C}=\text{O}$$
$$|$$
$$\text{CH}_2$$
$$|$$
$$\text{OPO}_3^{2-}$$

3-Phosphohydroxpyruvate

Glutamate

α-Ketoglutarate

$$\text{COO}^-$$
$$|$$
$$\text{H}_3\overset{+}{\text{N}}-\text{C}-\text{H}$$
$$|$$
$$\text{CH}_2$$
$$|$$
$$\text{OPO}_3^{2-}$$

3-Phosphoserine

H_2O

P

$$\text{COO}^-$$
$$|$$
$$\text{H}_3\overset{+}{\text{N}}-\text{C}-\text{H}$$
$$|$$
$$\text{CH}_2$$
$$|$$
$$\text{OH}$$

Serine

◀ **FIGURE 23.10** The biosynthesis of serine.

Folic acid is a vitamin that has been identified as essential in preventing birth defects; consequently, it is now a recommended supplement for all women of child-bearing age. There is also some evidence that folic acid may prevent heart disease in both men and women over 50 years of age.

$$\text{Serine} + \text{Tetrahydrofolate} \rightarrow \text{Glycine} + \text{Methylenetetrahydrofolate} + \text{H}_2\text{O}$$

The one-carbon unit transferred in this reaction is bound to tetrahydrofolate, forming N^5,N^{10}-methylenetetrahydrofolate, in which the methylene (one-carbon) unit is bound to two of the nitrogens of the carrier (Figure 23.12). Tetrahydrofolate is not the only carrier of one-carbon units. We have already encountered biotin, a carrier of CO_2, and we have discussed the role that biotin plays in gluconeogenesis (Section 18.2) and in the anabolism of fatty acids (Section 21.6).

The conversion of serine to cysteine involves some interesting reactions. The source of the sulfur in animals differs from that in plants and bacteria. In plants and bacteria, serine is acetylated to form *O*-acetylserine. This reaction is catalyzed by *serine acyltransferase*, with acetyl-CoA as the acyl donor (Figure 23.13).

Biochemistry ❸ Now™
Go to BiochemistryNow and click on Biochemistry Interactive to see how glycine serves as a precursor in the biosynthesis of porphyrins.

(a)

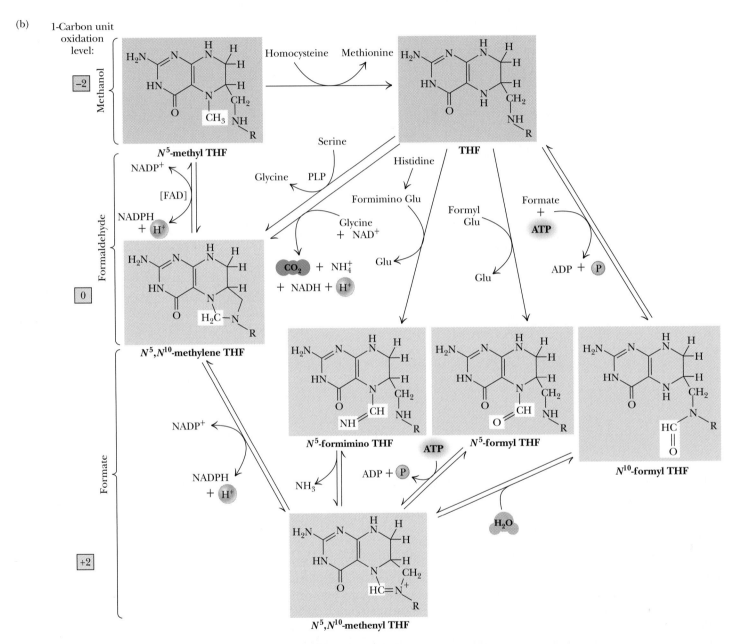

▲ **FIGURE 23.11** (a) The structure of folic acid, shown in nonionized form. (b) The reactions that introduce one-carbon units into tetrahydrofolate (THF) link seven different folate intermediates that carry one-carbon units in three different oxidation states (−2, 0, and +2). (*Adapted from T. Brody et al., in L. J. Machlin.* Handbook of Vitamins. *New York: Marcel Dekker, 1984.*)

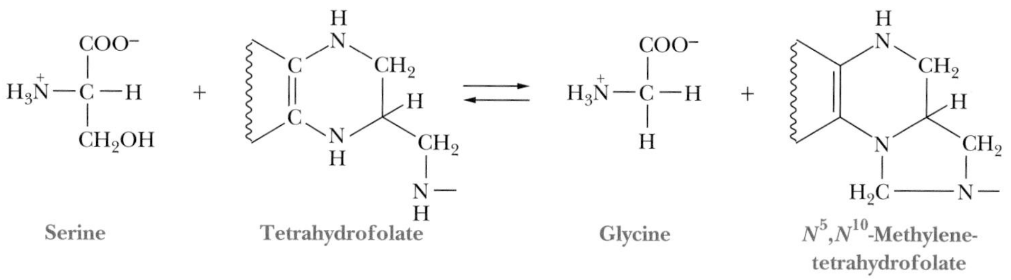

FIGURE 23.12 The conversion of serine to glycine, showing the role of tetrahydrofolate.

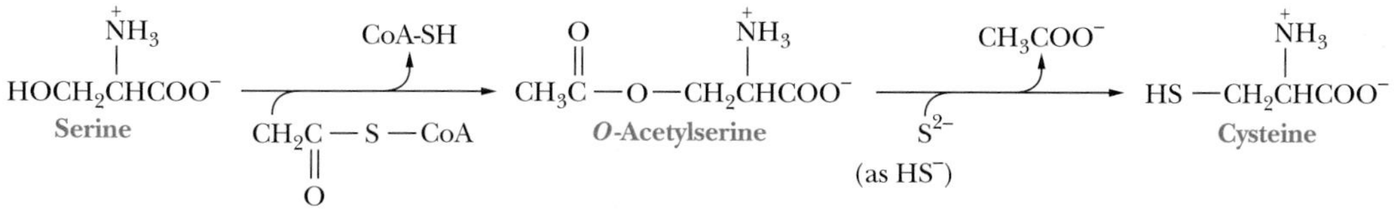

FIGURE 23.13 The biosynthesis of cysteine in plants and bacteria.

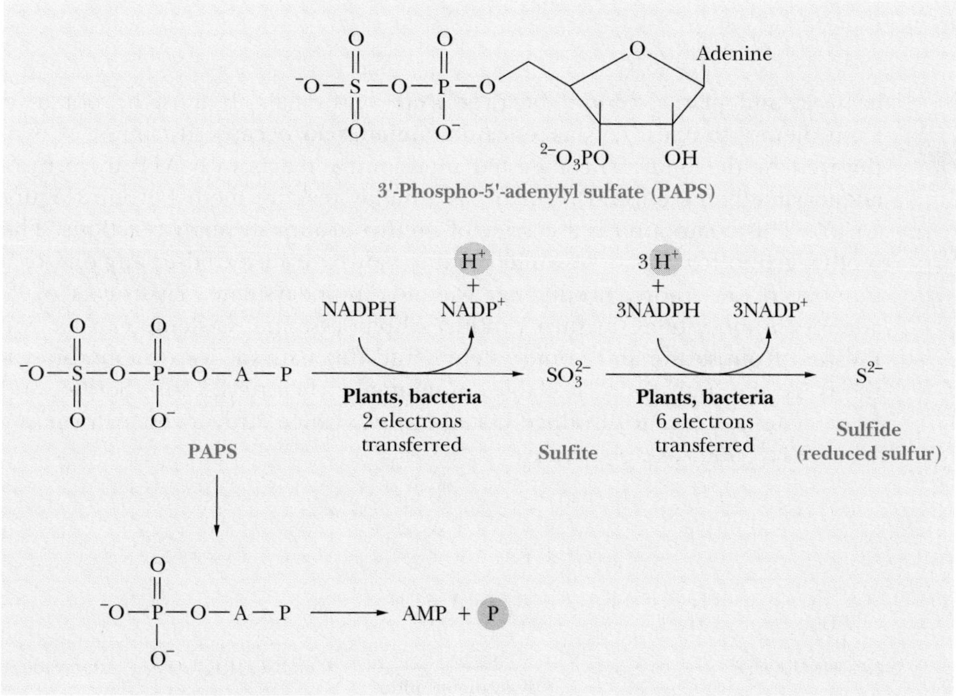

FIGURE 23.14 Electron transfer reactions of sulfur in plants and bacteria.

Conversion of O-acetylserine to cysteine requires production of sulfide by a sulfur donor. The sulfur donor for plants and bacteria is 3′-phospho-5′-adenylyl sulfate. The sulfate group is reduced first to sulfite and then to sulfide (Figure 23.14). The sulfide, in the conjugate acid form HS^-, displaces the acetyl group of the O-acetylserine to produce cysteine. Animals form cysteine from serine by a different pathway because they do not have the enzymes to carry out the sulfate-to-sulfide conversion that we have just seen. The reaction sequence in animals involves the amino acid methionine.

Methionine, which is produced by reactions of the aspartate family (see the Biochemistry Interactive website at http://now.brookscole.com/campbell5) in

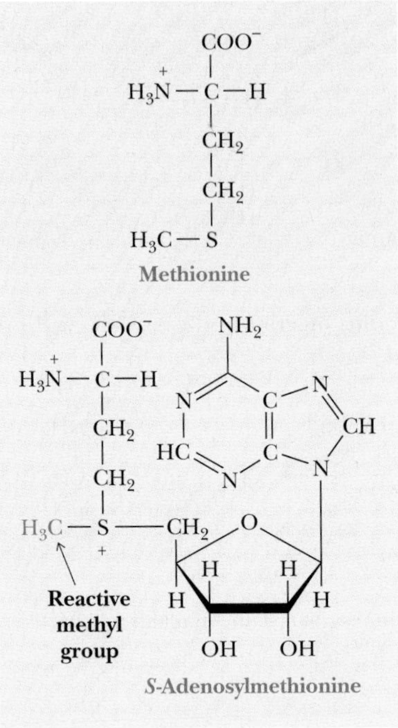

▶ **FIGURE 23.15** The structure of *S*-adenosylmethionine (SAM), with the structure of methionine shown for comparison.

Essential Information

Two of the most important classes of reactions in the biosynthesis of amino acids are transamination reactions and one-carbon transfers. The amino acids glutamate and glutamine are two of the most common donors of amino groups. Carriers of one-carbon groups include biotin, *S*-adenosylmethionine, and derivatives of folic acid.

bacteria and plants, cannot be produced by animals. It must be obtained from dietary sources. It is an **essential amino acid** because it cannot be synthesized by the body. The ingested methionine reacts with ATP to form ***S*-adenosylmethionine (SAM),** which has a highly reactive methyl group (Figure 23.15). This compound is a carrier of methyl groups in many reactions. The methyl group from *S*-adenosylmethionine can be transferred to any one of a number of acceptors, producing *S*-adenosylhomocysteine. Hydrolysis of *S*-adenosylhomocysteine in turn produces homocysteine. Cysteine can be synthesized from serine and homocysteine, and this pathway for cysteine biosynthesis is the only one available to animals (Figure 23.16). Serine and homocysteine react to produce cystathionine, which hydrolyzes to form cysteine, NH_4^+, and α-ketobutyrate.

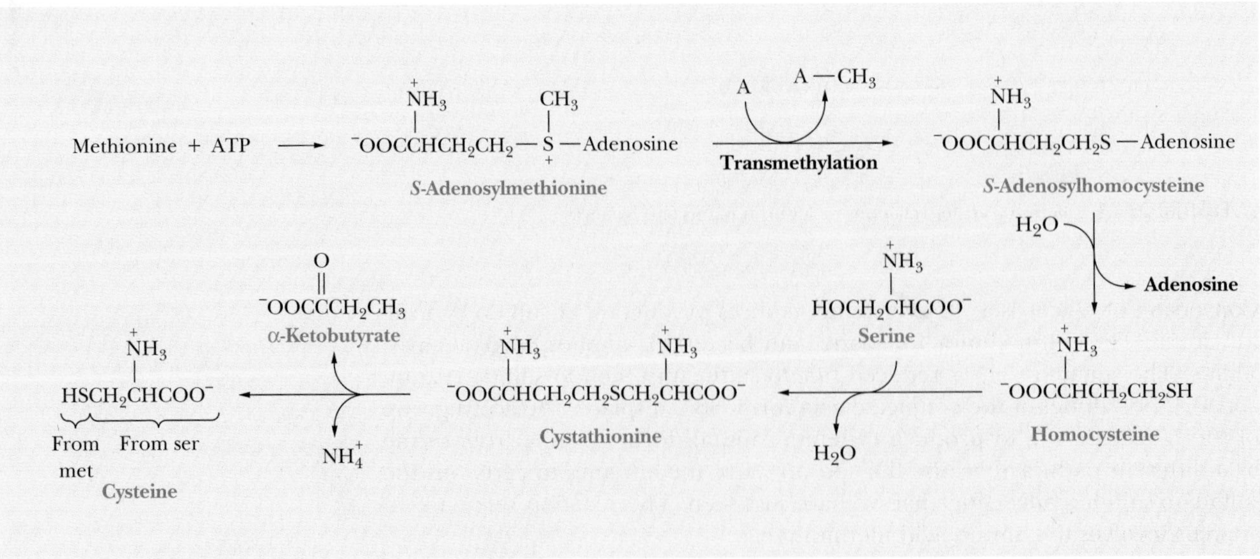

▲ **FIGURE 23.16** The biosynthesis of cysteine in animals. (A stands for acceptor.)

It is worth noting that we have now seen three important carriers of one-carbon units: biotin, a carrier of CO_2; tetrahydrofolate (FH_4), a carrier of methylene and formyl groups; and S-adenosylmethionine, a carrier of methyl groups.

23.5 What Are the Essential Amino Acids?

The biosynthesis of proteins requires the presence of all the constituent amino acids. If one of the 20 amino acids is missing or in short supply, protein biosynthesis is inhibited. Some organisms, such as *Escherichia coli*, can synthesize all the amino acids that they need. Other species, including humans, must obtain some amino acids from dietary sources. The essential amino acids in human nutrition are listed in Table 23.1. The body can synthesize some of these amino acids, but not in sufficient quantities for its needs, especially in the case of growing children. This last point applies particularly to children's requirement for arginine and histidine. Amino acids are not stored (except in proteins), and dietary sources of essential amino acids are needed at regular intervals. Protein deficiency, especially a prolonged deficiency in sources that contain essential amino acids, leads to the disease **kwashiorkor.** The problem in this disease, particularly severe in growing children, is not simply starvation but the breakdown of the body's own proteins.

23.6 How Are Amino Acids Catabolized?

When we specifically focus on the catabolism of amino acids, the first step we consider is the removal of nitrogen by transamination. Transamination reactions are also important in the anabolism of amino acids, so it is important to remind ourselves that anabolic and catabolic pathways are not the exact reverse of each other, nor do they involve exactly the same group of enzymes. In catabolism, the amino nitrogen of the original amino acid is transferred to α-ketoglutarate to produce glutamate, leaving behind the carbon skeletons. The fates of the carbon skeleton and of the nitrogen can be considered separately.

Disposition of the Carbon Skeletons

Breakdown of the carbon skeletons of amino acids follows two general pathways, the difference between the two pathways depending on the type of end product. A **glucogenic** amino acid is one that yields pyruvate or oxaloacetate on degradation. Oxaloacetate is the starting point for the production of glucose by gluconeogenesis. A **ketogenic** amino acid is one that breaks down to acetyl-CoA or acetoacetyl-CoA, leading to the formation of ketone bodies (Table 23.2; see also Section 21.5). The carbon skeletons of the amino acids give rise to metabolic intermediates such as pyruvate, acetyl-CoA, acetoacetyl-CoA, α-ketoglutarate, succinyl-CoA, fumarate, and oxaloacetate (see Figure 23.7). Oxaloacetate is a key intermediate in the breakdown of the carbon skeletons of amino acids because of its dual role in the citric acid cycle and in gluconeogenesis. The amino acids degraded to acetyl-CoA and acetoacetyl-CoA are used in the citric acid cycle, but mammals cannot synthesize glucose from acetyl-CoA. This fact is the source of the distinction between glucogenic and ketogenic amino acids. Glucogenic amino acids can be converted to glucose, with oxaloacetate as an intermediate, but ketogenic amino acids cannot be converted to glucose. Some amino acids have more than one pathway for catabolism, which explains why four of the amino acids are listed as both glucogenic and ketogenic.

Table 23.1

Amino Acid Requirements in Humans

Essential	Nonessential
Arginine*	Alanine
Histidine†	Asparagine
Isoleucine	Aspartate
Leucine	Cysteine
Lysine	Glutamate
Methionine	Glutamine
Phenylalanine	Glycine
Threonine	Proline
Tryptophan	Serine
Valine	Tyrosine

*Mammals synthesize arginine but cleave most of it to urea (Section 23.6).

†Essential for children, but not necessarily for adults.

Essential Information

Humans cannot produce some amino acids in sufficient quantities to meet their metabolic needs. Consequently, these essential amino acids must be obtained from dietary sources.

Table 23.2

Glucogenic and Ketogenic Amino Acids

Glucogenic	Ketogenic	Glucogenic and Ketogenic
Aspartate	Leucine	Isoleucine
Asparagine	Lysine	Phenylalanine
Alanine		Tryptophan
Glycine		Tyrosine
Serine		
Threonine		
Cysteine		
Glutamate		
Glutamine		
Arginine		
Proline		
Histidine		
Valine		
Methionine		

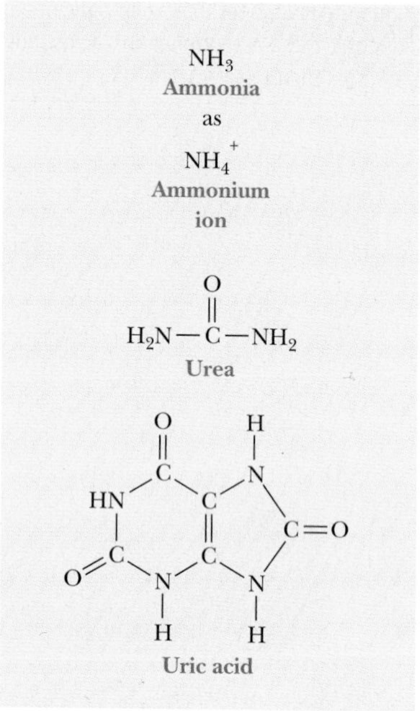

▲ **FIGURE 23.17** Nitrogen-containing products of amino acid catabolism.

Excretion of Excess Nitrogen

The nitrogen portion of amino acids is involved in transamination reactions in breakdown as well as in biosynthesis. Excess nitrogen is excreted in one of three forms: *ammonia* (as ammonium ion), *urea,* and *uric acid* (Figure 23.17). Animals, such as fish, that live in an aquatic environment excrete nitrogen as ammonia; they are protected from the toxic effects of high concentrations of ammonia not only by the removal of ammonia from their bodies but also by rapid dilution of the excreted ammonia by the water in the environment. The principal waste product of nitrogen metabolism in terrestrial animals is urea (a water-soluble compound); its reactions provide some interesting comparisons with the citric acid cycle. Birds excrete nitrogen in the form of uric acid, which is insoluble in water. They do not have to carry the excess weight of water, which could hamper flight, to rid themselves of waste products.

The Urea Cycle

A central pathway in nitrogen metabolism is the **urea cycle** (Figure 23.18). The nitrogens that enter the urea cycle come from several sources. One of the nitrogens of urea is added in the mitochondria, and its immediate precursor is glutamate, which releases ammonia via glutamate dehydrogenase. However, the ammonia nitrogens of glutamate have ultimately come from many sources as a result of transamination reactions. Mitochondrial glutaminase also provides free ammonia that can enter the cycle. A condensation reaction between

Biochemistry⌇Now™ ▶ **ACTIVE FIGURE 23.18** The urea cycle series of reactions: Transfer of the carbamoyl group of carbamoyl-P to ornithine by ornithine transcarbamoylase (OTCase, reaction 1) yields citrulline. The citrulline ureido group is then activated by reaction with ATP to give a citrullyl—AMP intermediate (reaction 2a); AMP is then displaced by aspartate, which is linked to the carbon framework of citrulline via its α-amino group (reaction 2b). The course of reaction 2 was verified using ^{18}O-labeled citrulline. The ^{18}O label (indicated by the asterisk, *) was recovered in AMP. Citrulline and AMP are joined via the ureido *O atom. The product of this reaction is argininosuccinate; the enzyme catalyzing the two steps of reaction 2 is argininosuccinate synthetase. The next step (reaction 3) is carried out by argininosuccinase, which catalyzes the nonhydrolytic removal of fumarate from argininosuccinate to give arginine. Hydrolysis of arginine by arginase (reaction 4) yields urea and ornithine, completing the urea cycle. **Watch this Active Figure at http://now.brookscole.com/campbell5**

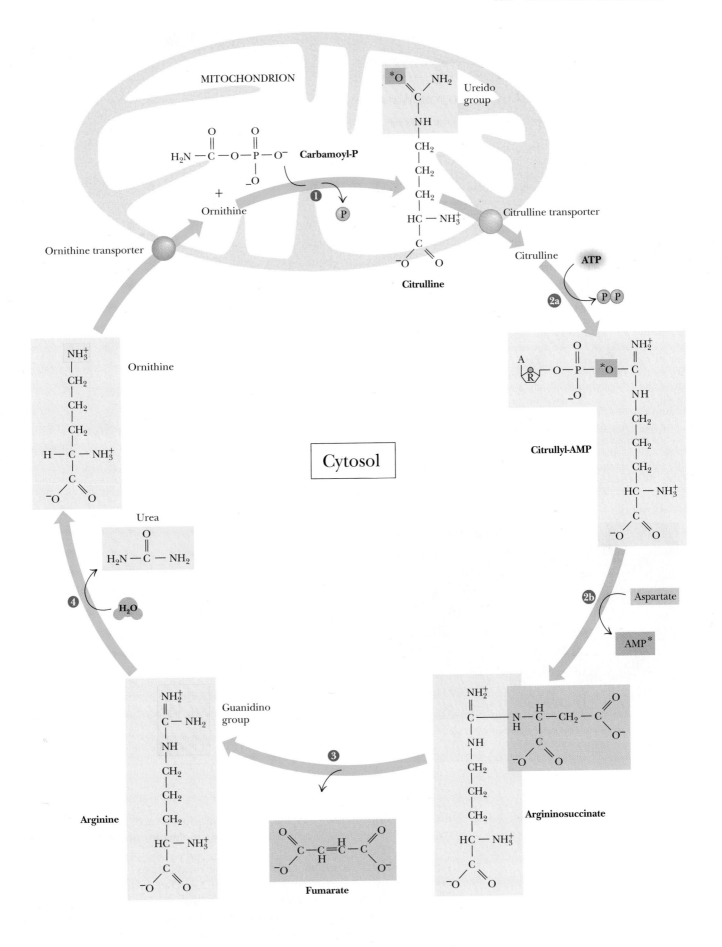

Biochemical Connections

Water and the Disposal of Nitrogen Wastes

Ammonia gas is toxic to most organisms and must usually be disposed of rapidly. In a certain sense, one can almost guess the mechanism of nitrogen-waste disposal if one knows the amount of water available to the organism in question. For example, bacteria and fish, which live in "infinite" water supplies, usually simply release ammonia into the medium, where organisms that are lower on the evolutionary scale can use it. Fish sometimes produce trimethylamine, another highly water-soluble compound, which is the characteristic "fish odor."

Most terrestrial animals do not have "infinite" water supplies, but mammals, which have bladders, usually live in conditions where adequate water is available. Their mechanism for disposal of most toxins is to prepare a water-soluble compound and then to excrete it through the urine. Thus, urea becomes a major by-product of nitrogen metabolism in mammals. Reptiles and other desert animals do not usually have much water available, and birds cannot afford to carry the weight of a fluid-filled bladder. These animals do not make urea; rather, they convert all their waste nitrogen to uric acid (Figure 23.17), the concentrated white solid so familiar in bird droppings. Some desert mammals, such as the kangaroo rat, which never drinks water but rather lives off metabolic water, also convert some of their waste nitrogen to uric acid to conserve the water used in urine.

Uric acid, the typical waste product from purines, can cause problems in primates due to its marginal water solubility. Deposits of uric acid in the joints and extremities causes gout (Section 23.8). Other mammals do not have a problem with uric acid because they convert it to allantoin, which is very water soluble.

▲ A kangaroo rat converts some of its waste nitrogen to uric acid.

▲ Catabolism of uric acid to ammonia and CO_2.

the ammonium ion and carbon dioxide produces **carbamoyl phosphate** in a reaction that requires the hydrolysis of two molecules of ATP for each molecule of carbamoyl phosphate. Carbamoyl phosphate reacts with **ornithine** (Step 1) to form **citrulline.** Citrulline is then transported to the cytosol. A second nitrogen enters the urea cycle when aspartate reacts with citrulline to form **argininosuccinate** in another reaction that requires ATP (AMP and PP_i are produced in this reaction; Step 2). The amino group of the aspartate is the source of the second nitrogen in the urea that will be formed in this series of reactions. Argininosuccinate is split to produce **arginine** and **fumarate** (Step 3). Finally, arginine is hydrolyzed to give urea and to regenerate ornithine, which is transported back to the mitochondrion (Step 4). The biosynthesis of arginine from ornithine is discussed on the Biochemistry Interactive website. Another way of looking at the urea cycle is to consider arginine as the immediate precursor of urea and to see it as producing ornithine in the process. According to this point of view, the rest of the cycle is the regeneration of arginine from ornithine.

The synthesis of fumarate is a link between the urea cycle and the citric acid cycle. Fumarate is, of course, an intermediate of the citric acid cycle, and it can be converted to oxaloacetate. A transamination reaction can convert oxaloacetate to aspartate, providing another link between the two cycles (Figure 23.19). In fact, both pathways were discovered by the same person, Hans

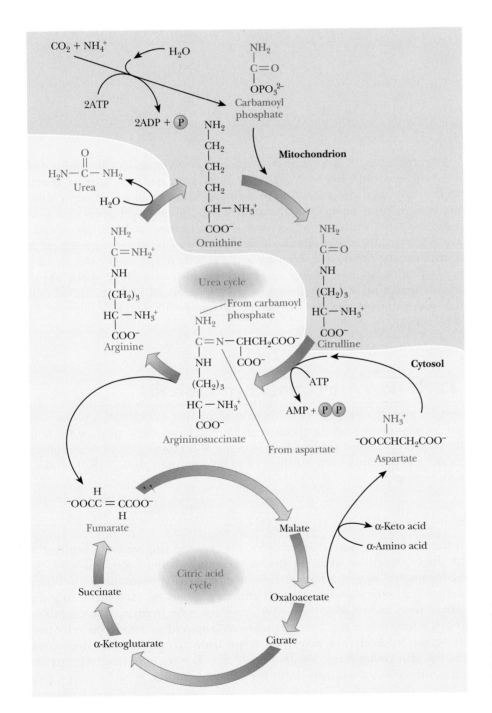

◀ **FIGURE 23.19** The urea cycle and some of its links to the citric acid cycle. Part of the cycle takes place in the mitochondrion and part in the cytosol. Fumarate and aspartate are the direct links to the citric acid cycle. Fumarate is a citric acid cycle intermediate. Aspartate comes from transamination of oxaloacetate, which is also a citric acid cycle intermediate.

Krebs. Four high-energy phosphate bonds are required because of the production of pyrophosphate in the conversion of aspartate to argininosuccinate.

In humans, urea synthesis is used to excrete excess nitrogen, such as would be found after consuming a high-protein meal. The pathway is confined to the liver. Note that arginine, the immediate precursor to urea, is the most nitrogen-rich amino acid, but the source of the nitrogen in the arginine varies. The major control point is the mitochondrial enzyme *carbamoyl-phosphate synthetase I* (CPS-I), and the formation of carbamoyl-phosphate is the committed step in the urea cycle. CPS-I is allosterically activated by *N*-acetylglutamate:

$$\underset{O^-}{\overset{O}{\|}}C - CH_2 - CH_2 - \underset{\underset{\displaystyle CH_3}{\overset{|}{\underset{|}{C=O}}}}{\overset{\overset{\displaystyle H}{|}}{\underset{|}{\overset{|}{C}}}} - \underset{O^-}{\overset{O}{\|}}C$$

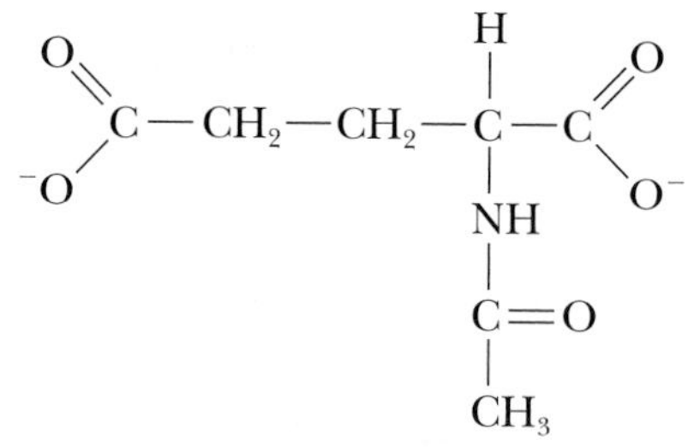

N-Acetylglutamate is formed by a reaction between glutamate and acetyl-CoA, which is catalyzed by *N*-acetylglutamate synthase. This enzyme is activated by increased concentrations of arginine. Thus, when amino acid catabolism is high, large amounts of glutamate will be present from degradation of glutamine, from synthesis via glutamate dehydrogenase, and from transamination reactions. Increased glutamate levels will lead to increased levels of *N*-acetylglutamate followed by increasing the activity of the urea cycle. In addition, any time arginine builds up, either because of protein catabolism or because ornithine is building up due to a low level of CPS-I activity, the arginine will stimulate synthesis of *N*-acetylglutamate and therefore increase the CPS-I activity.

Essential Information

The urea cycle plays a central role in nitrogen metabolism. It is involved in both the anabolism and the catabolism of amino acids and has links to the citric acid cycle.

23.7 | How Are Purines Synthesized?

We have already discussed the formation of ribose-5-phosphate as part of the pentose phosphate pathway (Section 18.4). The biosynthetic pathway for both purine and pyrimidine nucleotides makes use of preformed ribose-5-phosphate. Purines and pyrimidines are synthesized in different ways, and we shall consider them separately.

Anabolism of Inosine Monophosphate

In the synthesis of purine nucleotides, the growing ring system is bonded to the ribose phosphate while the purine skeleton is being assembled—first the five-membered ring and then the six-membered ring—eventually producing inosine-5'-monophosphate. All four nitrogen atoms of the purine ring are derived from amino acids: two from glutamine, one from aspartate, and one from glycine. Two of the five carbon atoms (adjacent to the glycine nitrogen) also come from glycine, two more come from tetrahydrofolate derivatives, and the fifth comes from CO_2 (Figure 23.20). The series of reactions producing inosine monophosphate (IMP) is long and complex.

Biochemical Connections

Chemotherapy and Antibiotics—Taking Advantage of the Need for Folic Acid

We have already seen the importance of folic acid and its derivative, tetrahydrofolate, in several reactions. This importance has been exploited in human medicine. Bacteria synthesize folic acid from *p*-aminobutyric acid (PABA). A type of antibiotic called a sulfonamide, Figure (a), works by competing with PABA in the synthesis of folic acid. Because folic acid is critical to the formation of purines, antagonists of folic acid metabolism are used to inhibit nucleic acid synthesis and cell growth. Rapidly dividing cells, such as those found in cancer and tumors, are more susceptible to these antagonists. Several related compounds, such as methotrexate, Figure (b), are used in chemotherapy to inhibit cancer cell growth.

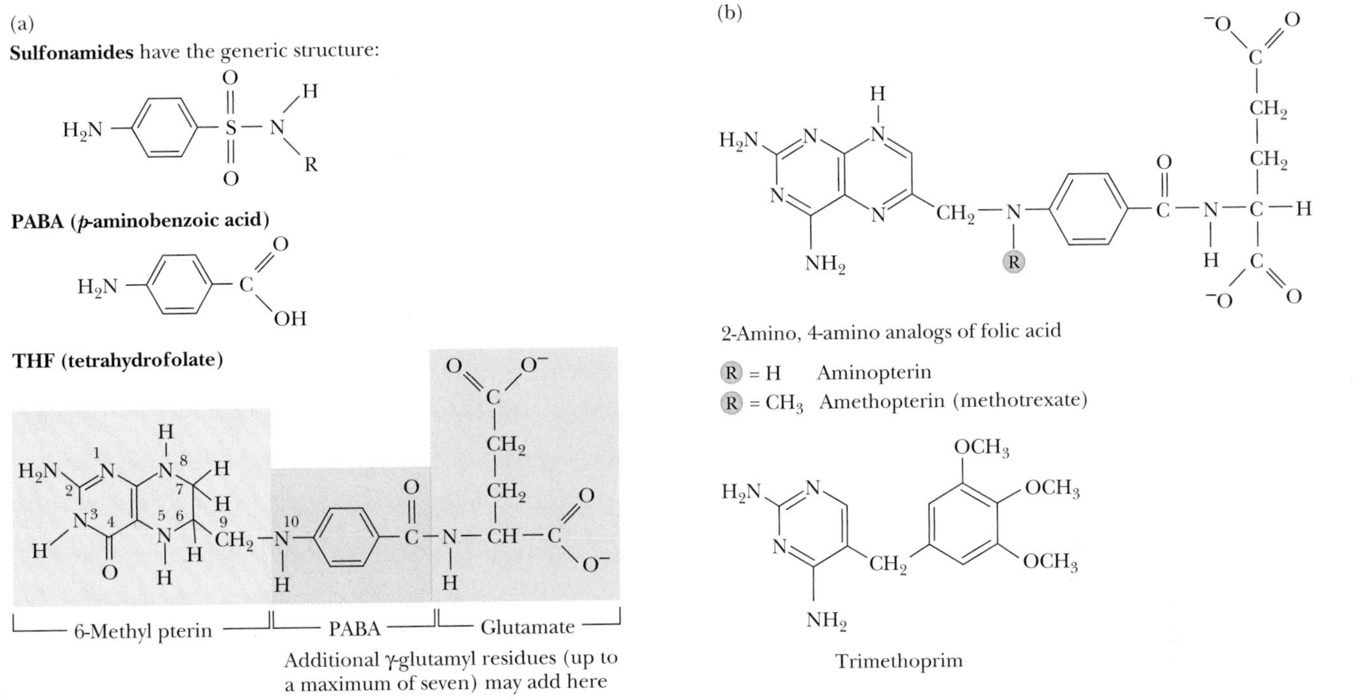

(a)
Sulfonamides have the generic structure:
PABA (p-aminobenzoic acid)
THF (tetrahydrofolate)
6-Methyl pterin
PABA
Glutamate
Additional γ-glutamyl residues (up to a maximum of seven) may add here
(b)
2-Amino, 4-amino analogs of folic acid
R = H Aminopterin
R = CH3 Amethopterin (methotrexate)
Trimethoprim

▲ (a) Sulfa drugs (sulfonamides) act as antibiotics because of their similarity to *p*-aminobenzoic acid (PABA), a precursor of folic acid synthesis. Sulfa drugs compete with PABA and stop folic acid synthesis in bacteria. (b) Three compounds that are used for chemotherapy because they interfere with folic acid metabolism. They are almost irreversible inhibitors of dihydrofolate reductase, having a 1000-fold greater affinity than dihydrofolate.

From CO2
From aspartate
From glycine
From N10-Formyl-tetrahydrofolate
From glutamine
From N10-Formyl-tetrahydrofolate

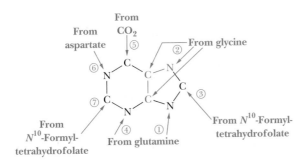

◀ **FIGURE 23.20** Sources of the atoms in the purine ring in purine nucleotide biosynthesis. The numbering system indicates the order in which each atom, or group of atoms, is added.

The Conversion of IMP to AMP and GMP

IMP is the precursor of both AMP and GMP. The conversion of IMP to AMP takes place in two stages (Figure 23.21). The first step is the reaction of

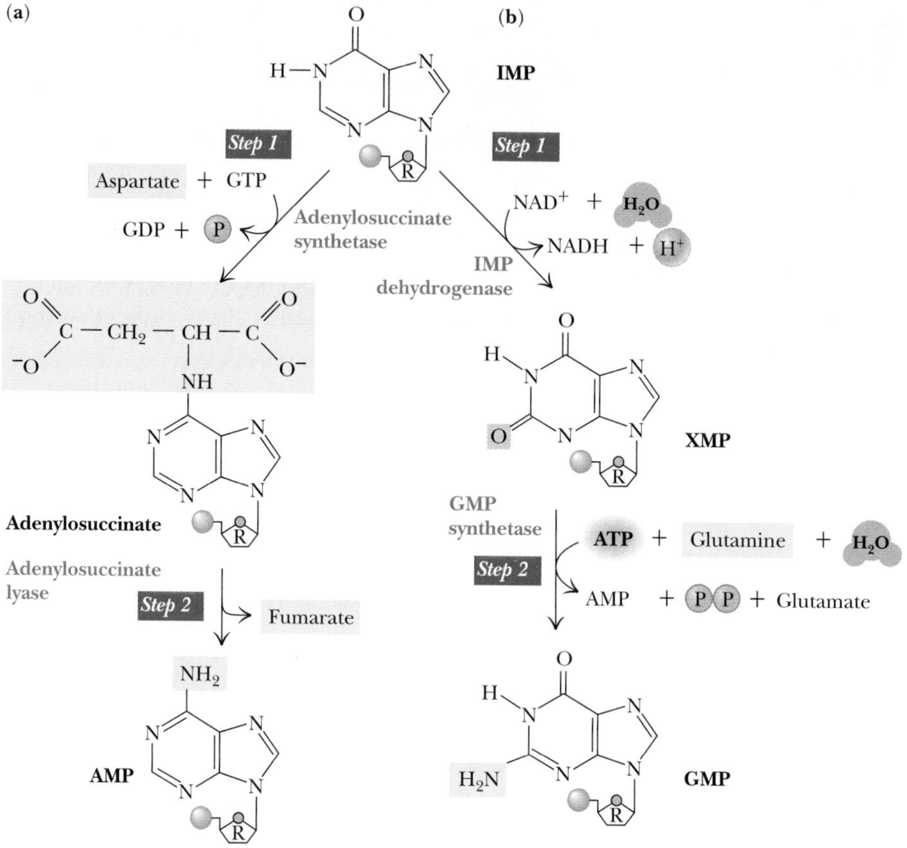

Biochemistry ⒺNow™ ANIMATED FIGURE 23.21
The synthesis of AMP and GMP from IMP. (a) AMP
synthesis: (The two reactions of AMP synthesis mimic
steps in the purine pathway leading to IMP.) In Step
1, the 6-*O* of inosine is displaced by aspartate to yield
adenylosuccinate. The energy required to drive this
reaction is derived from GTP hydrolysis. The enzyme
is adenylosuccinate synthetase. AMP is a competitive
inhibitor (with respect to the substrate IMP) of adenylosuccinate synthetase. In Step 2, adenylosuccinase
(also known as adenylosuccinate lyase, the same
enzyme that catalyzes one of the steps in the purine
pathway) carries out the nonhydrolytic removal of
fumarate from adenylosuccinate, leaving AMP.
(b) GMP synthesis: The two reactions of GMP synthesis are an NAD⁺-dependent oxidation followed by
an amidotransferase reaction. In Step 1, IMP dehydrogenase employs the substrates NAD⁺ and H₂O in
catalyzing oxidation of IMP at C-2. The products are
xanthylic acid (XMP or xanthosine monophosphate),
NADH, and H⁺. GMP is a competitive inhibitor (with
respect to IMP) of IMP dehydrogenase. In Step 2,
transfer of the amido-N of glutamine to the C-2
position of XMP yields GMP. This ATP-dependent
reaction is catalyzed by GMP synthetase. Besides GMP,
the products are glutamate, AMP, and PPᵢ. Hydrolysis
of PPᵢ to 2 Pᵢ by pyrophosphatases drives this reaction
to completion. **See this figure animated at http://
now.brookscole.com/campbell5**

aspartate with IMP to form adenylosuccinate. This reaction is catalyzed by
adenylosuccinate synthetase and requires GTP, not ATP, as an energy source
(using ATP would be counterproductive). The cleavage of fumarate from
adenylosuccinate to produce AMP is catalyzed by adenylosuccinase. This
enzyme also functions in the synthesis of the six-membered ring of IMP.

The conversion of IMP to GMP also takes place in two stages (Figure
23.21). The first of the two steps is an oxidation in which the C—H group at
the C-2 position is converted to a keto group. The oxidizing agent in the reaction is NAD⁺, and the enzyme involved is IMP dehydrogenase. The nucleotide formed by the oxidation reaction is xanthosine-5-phosphate (XMP). An
amino group from the side chain of glutamine replaces the C-2 keto group of
XMP to produce GMP. This reaction is catalyzed by GMP synthetase; ATP is
hydrolyzed to AMP and PPᵢ in the process. Note that there is some control
over the relative levels of purine nucleotides; GTP is needed for the synthesis
of adenine nucleotides, whereas ATP is required for the synthesis of guanine
nucleotides. Each of the purine nucleotides must occur at a reasonably high
level for the other to be synthesized.

Subsequent phosphorylation reactions produce purine nucleoside diphosphates (ADP and GDP) and triphosphates (ATP and GTP). The purine nucleoside monophosphates, diphosphates, and triphosphates are all feedback
inhibitors of the first stages of their own biosynthesis. Also, AMP, ADP, and
ATP inhibit the conversion of IMP to adenine nucleotides, and GMP, GDP,
and GTP inhibit the conversion of IMP to xanthylate and to guanine
nucleotides (Figure 23.22).

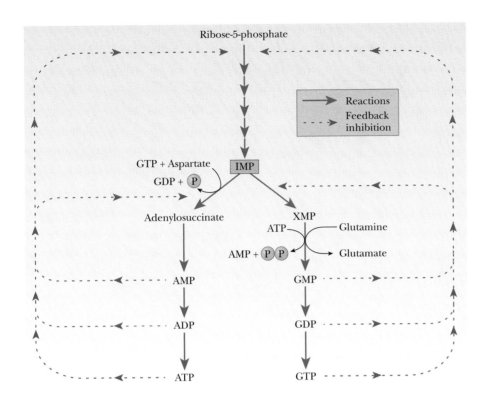

◀ **FIGURE 23.22** The role of feedback inhibition in regulation of purine nucleotide biosynthesis.

Energy Requirements for Production of AMP and GMP

The production of IMP starting with ribose-5-phosphate requires the equivalent of seven ATP (see the Biochemistry Interactive website and the article by Meyer et al. in the Annotated Bibliography at the end of this chapter). The conversion of IMP to AMP requires hydrolysis of an additional high-energy bond—in this case, that of GTP. In the formation of AMP from ribose-5-phosphate, the equivalent of eight ATP is needed. The conversion of IMP to GMP requires two high-energy bonds, given that a reaction occurs in which ATP is hydrolyzed to AMP and PP_i. For the production of GMP from ribose-5-phosphate, the equivalent of nine ATP is necessary. The anaerobic oxidation of glucose produces only two ATP for each molecule of glucose (Section 17.1). Anaerobic organisms require four molecules of glucose (which produce eight ATP) for each AMP they form, or five molecules of glucose (which produce ten ATP) for each GMP. The process is more efficient for aerobic organisms. Since 30 or 32 ATP result from each molecule of glucose, depending on the type of tissue, aerobic organisms can optimally produce four AMP (requiring 32 ATP) or three GMP (requiring 36 ATP) for each molecule of glucose oxidized. A mechanism for reuse of purines, rather than complete turnover and new synthesis, saves energy for organisms.

23.8 | How Are Purines Catabolized?

The catabolism of purine nucleotides proceeds by hydrolysis to the nucleoside and subsequently to the free base, which is further degraded. Deamination of guanine produces xanthine, and deamination of adenine produces hypoxanthine, the base corresponding to the nucleoside inosine, which is shown in Figure 23.23(a). Hypoxanthine can be oxidized to xanthine, so this

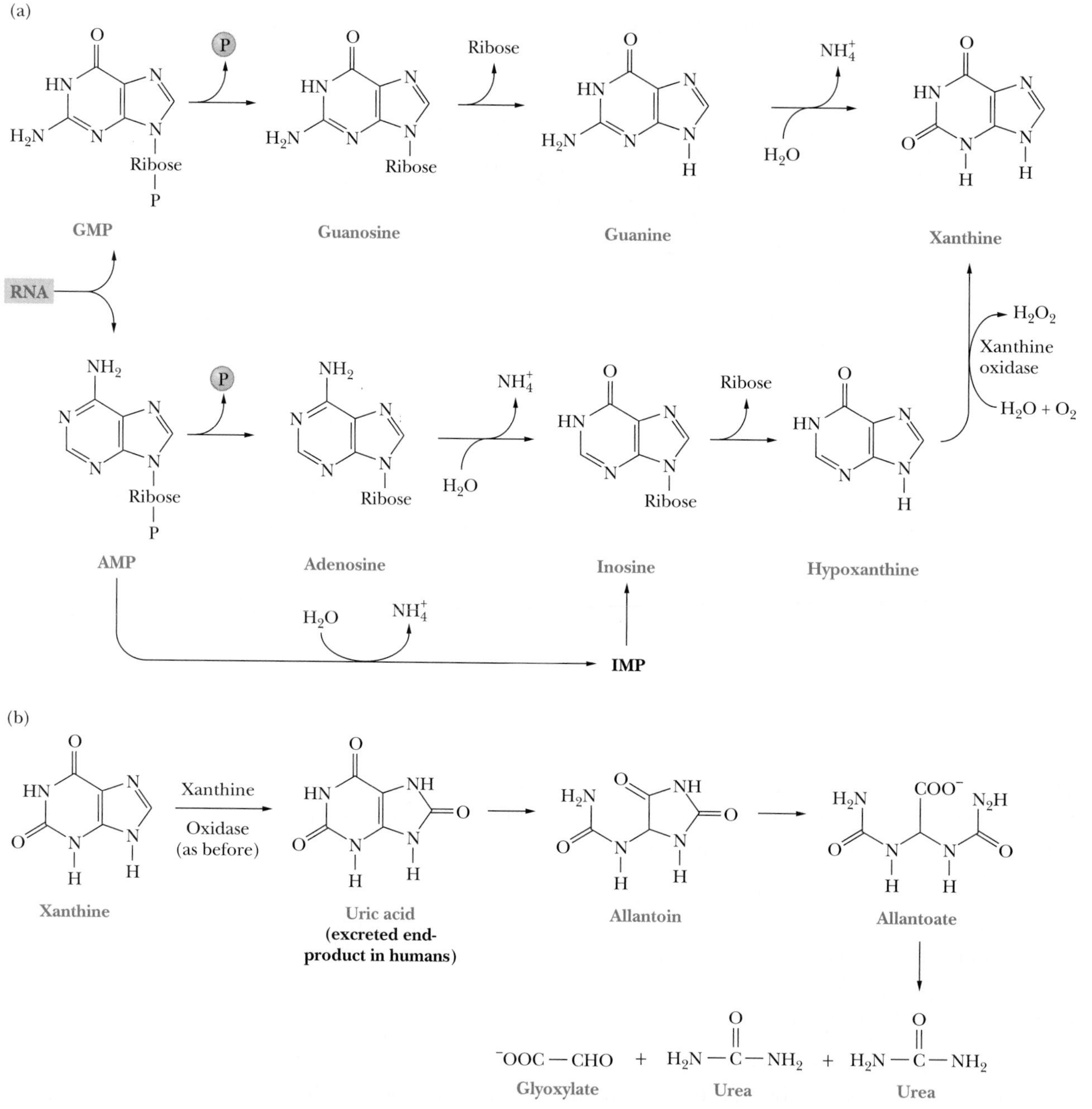

(a)

(b)

▲ **FIGURE 23.23** The reactions of purine catabolism. (a) Purine nucleotides are converted to the free base and then to xanthine. (b) Catabolic reactions of xanthine.

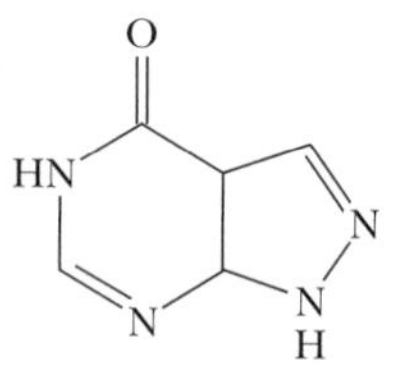

▲ Allopurinol, a substance used in the treatment of gout.

base is a common degradation product of both adenine and guanine. Xanthine is oxidized in turn to **uric acid** (Section 23.6). In birds, some reptiles, insects, Dalmatian dogs, and primates (including humans), uric acid is the end product of purine metabolism and is excreted. In all other terrestrial animals, including all other mammals, allantoin is the product excreted, whereas allantoate is the product in fish. Allantoate is further degraded to glyoxylate and urea by microorganisms and some amphibians, as shown in Figure 23.23(b). *Gout* is a disease in humans that is caused by the overproduction of uric acid. Deposits of uric acid (which is barely soluble in water) accumulate in the joints of the hands and feet. Allopurinol is a compound used to treat

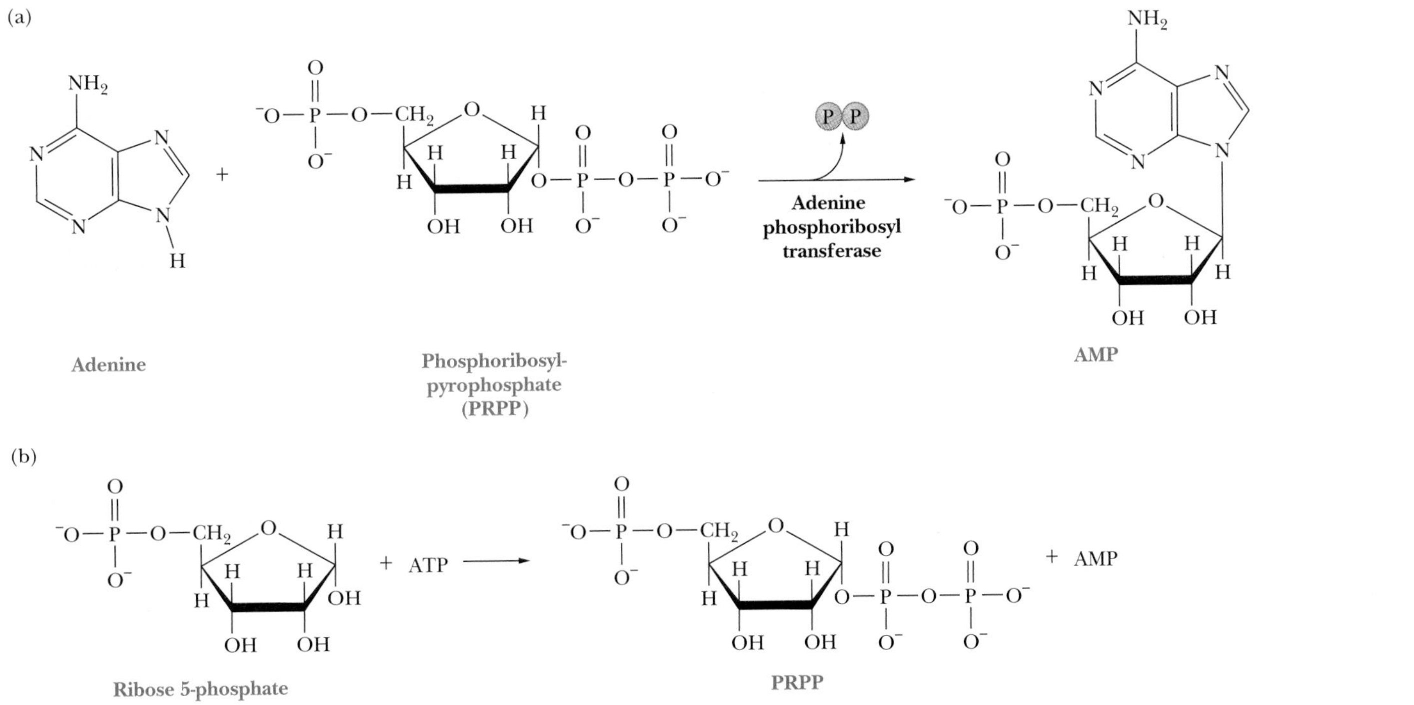

(a)

Adenine

Phosphoribosyl-
pyrophosphate
(PRPP)

Adenine
phosphoribosyl
transferase

AMP

(b)

Ribose 5-phosphate

PRPP

▲ **FIGURE 23.24** Purine salvage. (a) Adenine is the purine in this example. There are analogous reactions for salvage of guanine and hypoxanthine (see page 652). (b) The formation of phosphoribosylpyrophosphate (PRPP).

gout; it inhibits the degradation of hypoxanthine to xanthine and of xanthine to uric acid, preventing the buildup of uric acid deposits.

Salvage reactions are important in the metabolism of purine nucleotides because of the amount of energy required for the synthesis of the purine bases. A free purine base that has been cleaved from a nucleotide can produce the corresponding nucleotide by reacting with the compound phosphoribosylpyrophosphate (PRPP), formed by a transfer of a pyrophosphate group from ATP to ribose-5-phosphate (Figure 23.24).

Two different enzymes with different specificities with respect to the purine base catalyze salvage reactions. The reaction

$$\text{Adenine} + \text{PRPP} \rightarrow \text{AMP} + \text{PP}_i$$

Biochemical Connections

Lesch–Nyhan Syndrome

A deficiency of the HPRT enzyme is the cause of Lesch–Nyhan syndrome, a genetic disease. The biochemical consequences include an elevated concentration of PRPP and increased production of purines and uric acid. The accumulation of uric acid leads to kidney stones and gout, but the most striking clinical manifestations are neurological in nature. There is a compulsive tendency toward self-mutilation among patients with Lesch–Nyhan syndrome; they tend to bite off their fingertips and parts of their lips. The development of kidney stones and gouty symptoms can be prevented by administering allopurinol, but there is no real treatment for the self-destructive behavior and the mental retardation and spasticity that accompany it. The diverse manifestations of this disease show clearly that metabolism is extremely complex and that the failure of one enzyme to function can have consequences that reach far beyond the reaction that it catalyzes.

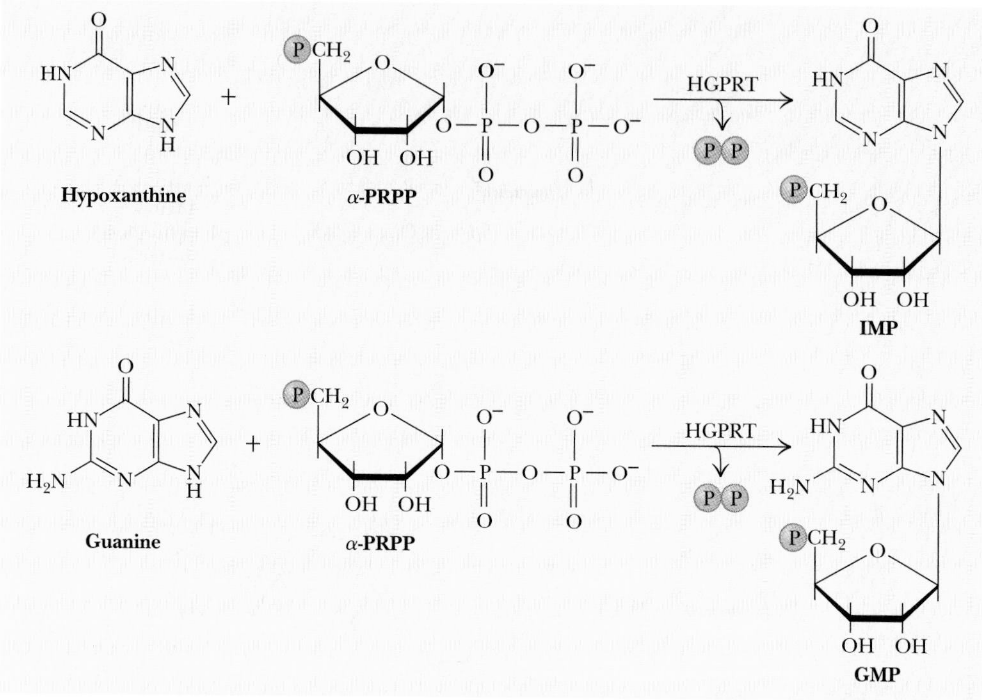

▲ **FIGURE 23.25** Purine salvage by the HGPRT reaction.

is catalyzed by adenine phosphoribosyltransferase. The corresponding reactions of guanine and hypoxanthine

$$\text{Hypoxanthine} + \text{PRPP} \xrightarrow{\text{HPRT}} \text{IMP} + \text{PP}_i$$

$$\text{Guanine} + \text{PRPP} \xrightarrow{\text{HPRT}} \text{GMP} + \text{PP}_i$$

are catalyzed by hypoxanthine-guanine phosphoribosyltransferase (HPRT) (Figure 23.25). A deficiency in HPRT can result in a serious disorder, **Lesch–Nyhan syndrome** (see the Biochemical Connections box on the previous page).

23.9 How Are Pyrimidines Synthesized and Catabolized?

The Anabolism of Pyrimidine Nucleotides

The overall scheme of pyrimidine nucleotide biosynthesis differs from that of purine nucleotides in that the pyrimidine ring is assembled before it is attached to ribose-5-phosphate. The carbon and nitrogen atoms of the pyrimidine ring come from carbamoyl phosphate and aspartate. The production of carbamoyl phosphate for pyrimidine biosynthesis takes place in the cytosol, and the nitrogen donor is glutamine. (We already saw a reaction for the production of carbamoyl phosphate when we discussed the urea cycle in Section 23.6. That reaction differs from this one because it takes place in mitochondria and the nitrogen donor is NH_4^+).

$$HCO_3^- + \text{Glutamine} + 2\,\text{ATP} + H_2O \rightarrow$$

$$\text{Carbamoyl phosphate} + \text{Glutamate} + 2\,\text{ADP} + P_i$$

The reaction of carbamoyl phosphate with aspartate to produce *N*-carbamoylaspartate is the committed step in pyrimidine biosynthesis. The compounds involved in reactions up to this point in the pathway can play other roles in metabolism; after this point, *N*-carbamoylaspartate can be used only to produce pyrimidines—thus the term "committed step."

This reaction is catalyzed by aspartate transcarbamoylase, which we discussed in detail in Chapter 7 as a prime example of an allosteric enzyme subject to feedback regulation. The next step, the conversion of *N*-carbamoylaspartate to dihydroorotate, takes place in a reaction that involves an intramolecular dehydration (loss of water) as well as cyclization. This reaction is catalyzed by dihydroorotase. Dihydroorotate is converted to orotate by dihydroorotate dehydrogenase, with the concomitant conversion of NAD$^+$ to NADH. A pyrimidine nucleotide is now formed by the reaction of orotate with PRPP to give orotidine-5′-monophosphate (OMP), which is a reaction similar to the one that takes place in purine salvage (Section 23.8). Orotate phosphoribosyltransferase catalyzes this reaction. Finally, orotidine-5′-phosphate decarboxylase catalyzes the conversion of OMP to UMP (uridine-5′-monophosphate), which is the precursor of the remaining pyrimidine nucleotides (Figure 23.26).

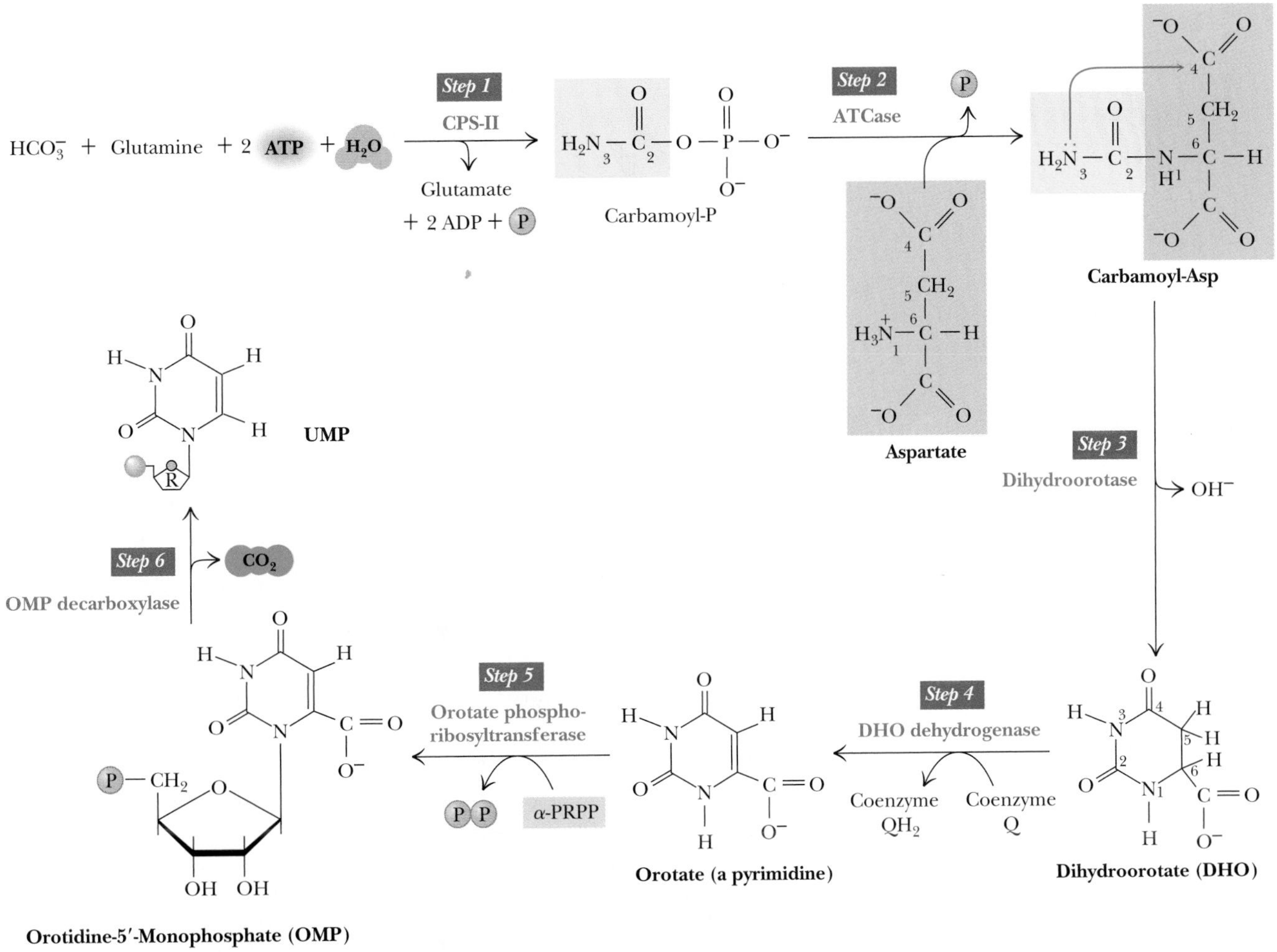

Biochemistry ⑤ Now™ ▲ ACTIVE FIGURE 23.26 The pyrimidine biosynthetic pathway. Step 1: Carbamoyl-P synthesis. Step 2: Condensation of carbamoyl phosphate and aspartate to yield carbamoyl-aspartate is catalyzed by aspartate transcarbamoylase (ATCase). Step 3: An intramolecular condensation catalyzed by dihydroorotase gives the six-membered heterocyclic ring characteristic of pyrimidines. The product is dihydroorotate (DHO). Step 4: The oxidation of DHO by dihydroorotate dehydrogenase gives orotate. (In bacteria, NAD$^+$ is the electron acceptor from DHO.) Step 5: PRPP provides the ribose-5-P moiety that transforms orotate into orotidine-5-monophosphate, a pyrimidine nucleotide. Note that orotate phosphoribosyltransferase joins N-1 of the pyrimidine to the ribosyl group in appropriate β-configuration. PP$_i$ hydrolysis renders this reaction thermodynamically favorable. Step 6: Decarboxylation of OMP by OMP decarboxylase yields UMP. **Watch this Active Figure at http://now.brookscole.com/campbell5**

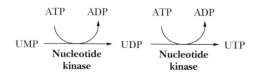

▲ FIGURE 23.27 The conversion of UMP to UTP.

Two successive phosphorylation reactions convert UMP to UTP (Figure 23.27). The conversion of uracil to cytosine takes place in the triphosphate form, catalyzed by CTP synthetase (Figure 23.28). Glutamine is the nitrogen donor, and ATP is required, as we saw earlier in similar reactions.

$$UTP + Glutamine + ATP \rightarrow CTP + Glutamate + ADP + P_i$$

Feedback inhibition in pyrimidine nucleotide biosynthesis takes place in several ways. CTP is an inhibitor of aspartate transcarbamoylase and of CTP synthetase. UMP is an inhibitor of an even earlier step, the one catalyzed by carbamoyl phosphate synthetase (Figure 23.29).

Pyrimidine Catabolism

Pyrimidine nucleotides are broken down first to the nucleoside and then to the base, as purine nucleotides are. Cytosine can be deaminated to uracil, and the double bond of the uracil ring is reduced to produce dihydrouracil. The ring opens to produce N-carbamoylpropionate, which in turn is broken down to NH_4^+, CO_2, and β-alanine (Figure 23.30).

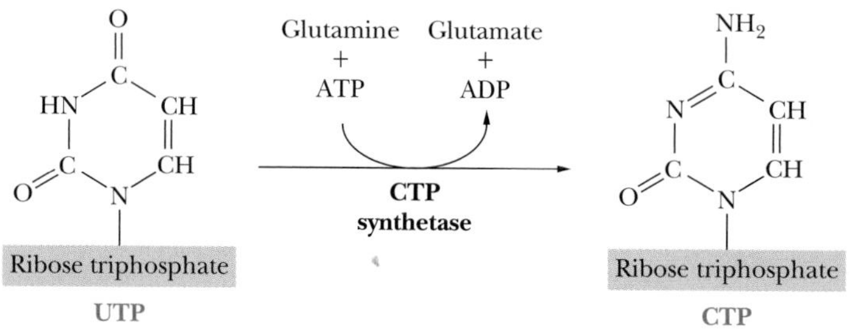

▶ FIGURE 23.28 The conversion of UTP to CTP.

▶ FIGURE 23.29 The role of feedback inhibition in the regulation of pyrimidine nucleotide biosynthesis.

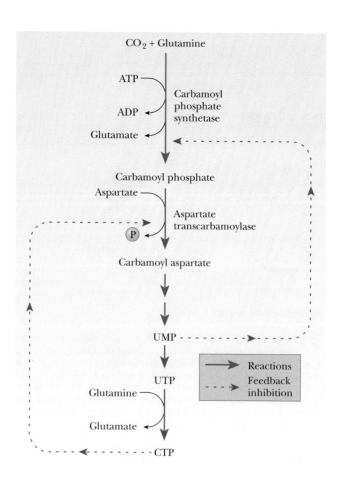

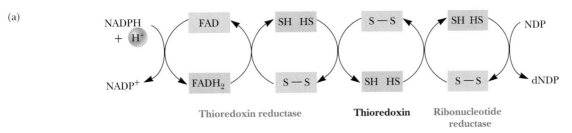

▲ FIGURE 23.30 The catabolism of pyrimidines.

23.10 | How Are Ribonucleotides Converted to Deoxyribonucleotides?

Ribonucleoside diphosphates are reduced to 2′-deoxyribonucleoside diphosphates in all organisms [Figure 23.31(a)]; NADPH is the reducing agent.

Ribonucleoside diphosphate + NADPH + H$^+$ →

Deoxyribonucleoside diphosphate + NADP$^+$ + H$_2$O

The actual process, which is catalyzed by *ribonucleotide reductase,* is more complex than the preceding equation would indicate and involves some intermediate electron carriers. The ribonucleotide reductase system from *E. coli* has been extensively studied, and its mode of action gives some clues to the nature of the process. Two other proteins are required, thioredoxin and

(a)

NADPH
+ H$^+$ FAD   SH HS   S—S   SH HS   NDP

NADP$^+$ FADH$_2$ S—S   SH HS   S—S   dNDP

Thioredoxin reductase   **Thioredoxin**   Ribonucleotide reductase

(b)

Ribonucleoside diphosphate
(NDP)

Deoxyribonucleoside diphosphate
(dNDP)

Biochemistry Now™ **▲ ANIMATED FIGURE 23.31** (a) The (—S—S—)/(—SH HS—) oxidation–reduction cycle involving ribonucleotide reductase, thioredoxin, thioredoxin reductase, and NADPH. (b) The structures of NDP and dNDP. **See this figure animated at http://now .brookscole.com/campbell5**

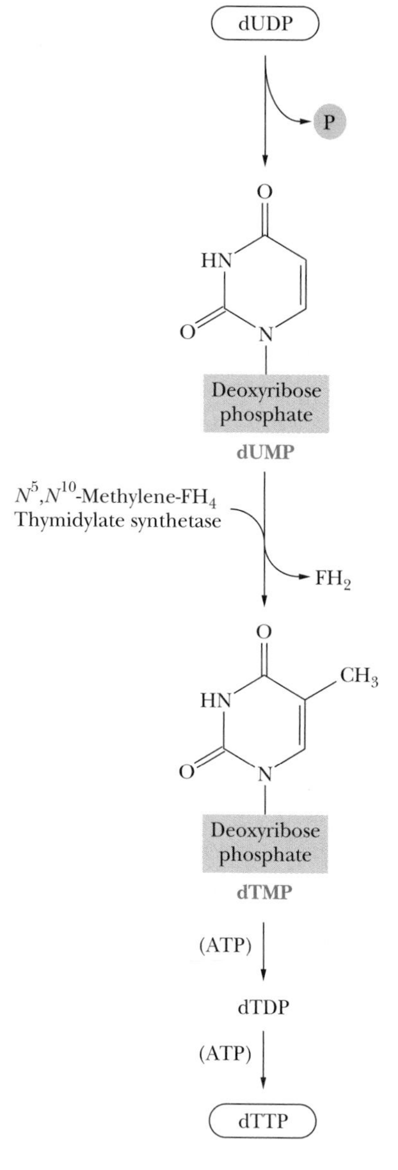

▲ **FIGURE 23.32** The conversion of dUDP to dTTP. (FH_4 is tetrahydrofolate; FH_2 is dihydrofolate.)

thioredoxin reductase. **Thioredoxin** contains a disulfide (S—S) group in its oxidized form and two sulfhydryl (—SH) groups in its reduced form. NADPH reduces thioredoxin in a reaction catalyzed by **thioredoxin reductase.** The reduced thioredoxin in turn reduces a ribonucleoside diphosphate (NDP) to a deoxyribonucleoside diphosphate (dNDP), shown in Figure 23.31(b), and this reaction is actually catalyzed by ribonucleotide reductase. Note that this reaction produces dADP, dGDP, dCDP, and dUDP. The first three are phosphorylated to give the corresponding triphosphates, which are substrates for the synthesis of DNA. Another required substrate for DNA synthesis is dTTP, and we shall now see how dTTP is produced from dUDP.

23.11 | How Is dUDP Converted to dTTP?

A one-carbon transfer is required for the conversion of uracil to thymine by attachment of the methyl group. The most important reaction in this conversion is that catalyzed by *thymidylate synthase* (Figure 23.32). The source of the one-carbon unit is N^5,N^{10}-methylenetetrahydrofolate, which is converted to dihydrofolate in the process. The metabolically active form of the one-carbon carrier is tetrahydrofolate. Dihydrofolate must be reduced to tetrahydrofolate for this series of reactions to continue, and this process requires NADPH and *dihydrofolate reductase.*

Since a supply of dTTP is necessary for DNA synthesis, inhibition of enzymes that catalyze the production of dTTP will inhibit the growth of rapidly dividing cells. Cancer cells, like all fast-growing cells, depend on continued DNA synthesis for growth. Inhibitors of thymidylate synthetase, such as fluorouracil (see Question 50), and inhibitors of dihydrofolate reductase, such as aminopterin and methotrexate (structural analogues of folate), have been used in cancer chemotherapy (Figure 23.33). The intent of such ther-

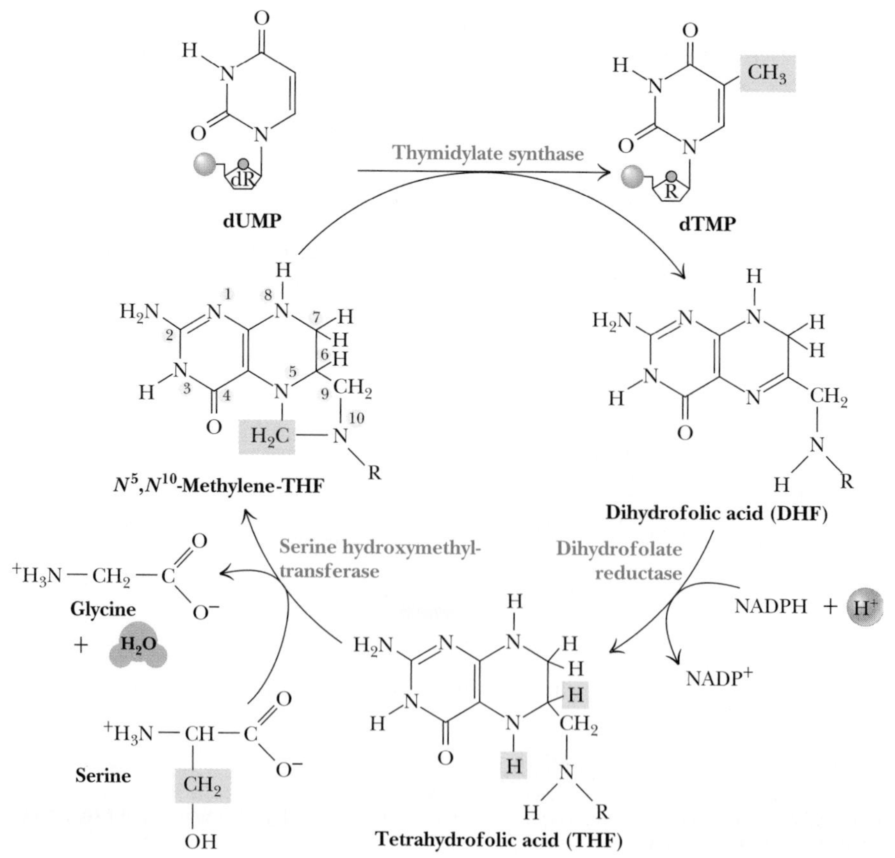

Biochemistry ⊘ Now™ **ACTIVE FIGURE 23.33**
The thymidylate synthase reaction. The 5-CH_3 group is ultimately derived from the β-carbon of serine.
Watch this Active Figure at http://now.brookscole.com/campbell5

apy is to inhibit the formation of dTTP and thus of DNA in cancer cells, causing the death of the cancer cells with minimal effect on normal cells, which grow more slowly. Chemotherapy has adverse side effects because of the highly toxic nature of most of the drugs involved; normal cells are affected to some extent, although less than the cancer cells. Enormous amounts of research are focused on finding safe and effective forms of treatment.

Summary

23.1 What Processes Constitute Nitrogen Metabolism? The metabolism of nitrogen encompasses a number of topics, including the anabolism and catabolism of amino acids, porphyrins, and nucleotides. Atmospheric nitrogen is the ultimate source of this element in biomolecules.

23.2 How Is Nitrogen Incorporated into Biologically Useful Compounds? Nitrogen fixation is the process by which molecular nitrogen from the atmosphere is made available to organisms in the form of ammonia. Nitrification reactions convert NO_3^- to NH_3 and provide another source of nitrogen.

23.3 What Role Does Feedback Inhibition Play in Nitrogen Metabolism? Feedback-inhibition control mechanisms are a unifying factor in biosynthetic pathways involving nitrogen compounds. Most of the nitrogen metabolism pathways are long and complicated and use a great deal of energy. Shutting off these processes when enough of the final product has built up is important to the energy flux of the cell.

23.4 How Are Amino Acids Synthesized? In the anabolism of amino acids, transamination reactions play an important role. Glutamate and glutamine are frequently the amino-group donors. The enzymes that catalyze transamination reactions frequently require pyridoxal phosphate as a coenzyme. One-carbon transfers also operate in the anabolism of amino acids. Carriers are required for the one-carbon groups transferred. Tetrahydrofolate is a carrier of methylene and formyl groups, and *S*-adenosylmethionine is a carrier of methyl groups.

23.5 What Are the Essential Amino Acids? Some species, including humans, cannot synthesize all the amino acids required for protein synthesis and must therefore obtain these essential amino acids from dietary sources. About half of the standard 20 amino acids are essential in humans, including arginine, histidine, isoleucine, leucine, lysine, methionine, phenylalanine, threonine, tryptophan, and valine.

23.6 How Are Amino Acids Catabolized? The catabolism of amino acids has two parts: the fate of the nitrogen and the fate of the carbon skeleton. In the urea cycle, nitrogen released by the catabolism of amino acids is converted to urea. The carbon skeleton is converted to pyruvate or oxaloacetate, in the case of glucogenic amino acids, or to acetyl-CoA or acetoacetyl-CoA, in the case of ketogenic amino acids.

23.7 How Are Purines Synthesized? The anabolic pathway of nucleotide synthesis involving purines differs from that involving pyrimidines. Both pathways use preformed ribose-5-phosphate but differ with regard to the point in the pathway at which the sugar phosphate is attached to the base. In the case of purine nucleotides, the growing base is attached to the sugar phosphate during the synthesis.

23.8 How Are Purines Catabolized? In catabolism, purine bases are frequently salvaged and reattached to sugar phosphates. Otherwise, purines are broken down to uric acid.

23.9 How Are Pyrimidines Synthesized and Catabolized? In pyrimidine biosynthesis, the base is first formed and then attached to the sugar phosphate. Pyrimidines are degraded to β-alanine.

23.10 How Are Ribonucleotides Converted to Deoxyribonucleotides? Deoxyribonucleotides for DNA synthesis are produced by the reduction of ribonucleoside diphosphates to deoxyribonucleoside diphosphates.

23.11 How Is dUTP Converted to dTTP? Another reaction specifically needed to produce substrates for DNA synthesis is the conversion of uracil to thymine. This pathway, which requires a tetrahydrofolate derivative as the carrier for one-carbon transfer, is a target for cancer chemotherapy.

Critical Questions to Review

23.1 What Processes Constitute Nitrogen Metabolism?

1. **Fact Check** What kinds of organisms can fix nitrogen? Which ones cannot?

23.2 How Is Nitrogen Incorporated into Biologically Useful Compounds?

2. **Fact Check** How is nitrogen fixed (converted from N_2 to NH_4^+)? How is it subsequently assimilated into organic compounds?

3. **Biochemical Connections** What is the Haber process?

4. **Fact Check** Write the overall reaction for the fixation of nitrogen via the nitrogenase complex.

5. **Fact Check** Describe the nitrogenase complex. How is the enzyme organized? What are its unique components?

23.3 What Role Does Feedback Inhibition Play in Nitrogen Metabolism?

6. **Fact Check** How are nitrogen-utilizing pathways controlled by feedback inhibition?

7. **Thought Question** Comment briefly on the usefulness to organisms of feedback control mechanisms in long biosynthetic pathways.

8. **Thought Question** Metabolic cycles are rather common (Calvin cycle, citric acid cycle, urea cycle). Why are cycles so useful to organisms?

23.4 How Are Amino Acids Synthesized?

9. **Fact Check** What is the relationship between α-ketoglutarate, glutamate, and glutamine in amino acid anabolism?

10. **Fact Check** Draw a transamination reaction between α-ketoglutarate and alanine.

11. **Fact Check** Diagram the reactions involving glutamate dehydrogenase and glutamine synthetase that produce glutamine from ammonia and α-ketoglutarate.

12. **Fact Check** What is the difference between glutamine synthetase and glutaminase?

13. **Fact Check** Draw the mechanism of transamination with pyridoxal phosphate.

14. **Fact Check** What cofactors are involved in one-carbon transfer reactions of amino acid anabolism?

15. **Fact Check** Sketch the structure of folic acid. Also sketch how it serves as a carrier of one-carbon groups.

16. **Fact Check** Why is there no net gain of methionine if homocysteine is converted to methionine with S-adenosylmethionine as the methyl donor?

17. **Fact Check** Show, by the equation for a typical reaction, why glutamate plays a central role in the biosynthesis of amino acids.

18. **Fact Check** By means of a structural formula, show how S-adenosylmethionine is a carrier of methyl groups.

19. **Thought Question** Sulfanilamide and related sulfa drugs were widely used to treat diseases of bacterial origin before penicillin and more advanced drugs were readily available. The inhibitory effect of sulfanilamide on bacterial growth can be reversed by p-aminobenzoate. Suggest a mode of action for sulfanilamide.

$$H_2N \text{—} \underset{\text{Sulfanilamide}}{\underset{\displaystyle\bigcirc}{}} \text{—} SO_2NH_2$$

20. **Thought Question** Proteins contain methionine but not α-amino-n-hexanoic acid. The only structural difference is the substitution of —CH_2— for —S—. Both groups are similar in size and hydrophobic character. Why is methionine more advantageous than α-amino-n-hexanoic acid?

23.5 What Are the Essential Amino Acids?

21. **Fact Check** In general, what categories of amino acids are essential in humans and which are nonessential?

22. **Fact Check** List the essential amino acids for a phenylketonuric adult and compare them with the requirements for a normal adult.

23.6 How Are Amino Acids Catabolized?

23. **Fact Check** How many α-amino acids participate directly in the urea cycle? Of these, how many can be used for protein synthesis?

24. **Fact Check** Write an equation for the net reaction of the urea cycle. Show how the urea cycle is linked to the citric acid cycle.

25. **Fact Check** Describe citrulline and ornithine based on their similarity to one of the 20 standard amino acids.

26. **Fact Check** Which amino acids in the urea cycle are the links to the citric acid cycle? Show how these links occur.

27. **Fact Check** How many ATPs are required for one round of the urea cycle? Where do these ATPs get used?

28. **Fact Check** How is carbamoyl-phosphate synthetase I (CPS-I) controlled?

29. **Fact Check** What is the logic behind high levels of arginine positively regulating N-acetylglutamate synthase?

30. **Fact Check** How does the level of glutamic acid affect the urea cycle?

31. **Fact Check** When amino acids are catabolized, what are the end products of the carbon skeletons for glucogenic amino acids? For ketogenic amino acids?

32. **Fact Check** Will an amino acid be glucogenic or ketogenic if it is catabolized to the following molecules: (a) Phosphoenolpyruvate (b) α-Ketoglutarate (c) Succinyl-CoA (d) Acetyl-CoA (e) Oxaloacetate (f) Acetoacetate

33. **Biochemical Connections** What species excrete excess nitrogen as ammonia? Which ones excrete it as uric acid?

34. **Biochemical Connections** Would you expect an ostrich to excrete excess nitrogen as uric acid, urea, or ammonia? Make an argument for your answer.

35. **Thought Question** Why is arginine an essential amino acid, when it is made in the urea cycle?

36. **Thought Question** People on high-protein diets are advised to drink lots of water. Why?

37. **Thought Question** Why is it better, when running a marathon, to drink a beverage with sugar for energy rather than one with amino acids?

38. **Thought Question** Argue logically that the urea cycle should not have evolved. Then, logically counter your argument.

23.7 How Are Purines Synthesized?

39. **Biochemical Connections** How is the importance of folic acid related to chemotherapy?

40. **Fact Check** What are the sources of the carbons and nitrogens in the purine bases?

41. **Fact Check** What is the structural difference between inosine and adenosine?

42. **Fact Check** How is tetrahydrofolate important to purine synthesis?

43. **Fact Check** Does the conversion of IMP to GMP use or produce ATP either directly or indirectly? Justify your answer.

44. **Fact Check** Discuss the role of feedback inhibition in the anabolism of purine-containing nucleotides.

23.8 How Are Purines Catabolized?

45. **Fact Check** How many high-energy phosphate bonds must be hydrolyzed in the pathway that produces GMP from guanine and PRPP by the PRPP salvage reaction, compared with the number of such bonds hydrolyzed in the pathway leading to IMP and then to GMP?

46. **Thought Question** Why do most mammals, other than primates, not suffer from gout?

23.9 How Are Pyrimidines Synthesized and Catabolized?

47. **Fact Check** What is an important difference between the biosynthesis of purine nucleotides and that of pyrimidine nucleotides?

48. **Fact Check** Compare the fates of the products of purine and pyrimidine catabolism.

23.10 How Are Ribonucleotides Converted to Deoxyribonucleotides?

49. **Fact Check** What roles do thioredoxin and thioredoxin reductase play in the metabolism of nucleotides?

23.11 How Is dUTP Converted to dTTP?

50. **Fact Check** Suggest a mode of action for fluorouracil in cancer chemotherapy.

51. **Thought Question** Chemotherapy patients receiving cytotoxic (cell-killing) agents such as FdUMP (the UMP analogue that contains fluorouracil) and methotrexate temporarily go bald. Why does this take place?

Biochemistry ⑧ Now™

Assess your understanding of this chapter's topics with additional quizzing and tutorials at **http://now.brookscole.com/campbell5**

Annotated Bibliography

Bender, D. A. *Amino Acid Metabolism.* 2nd ed. New York: John Wiley, 1985. [A general treatment of the topic, with a particularly good section on tryptophan metabolism.]

Benkovic, S. On the Mechanism of Action of Folate- and Biopterin-Requiring Enzymes. *Ann. Rev. Biochem.* **49,** 227–254 (1980). [A review of one-carbon transfers.]

Braunstein, A. E. Amino Group Transfer. In Boyer, P. D., ed. *The Enzymes.* 3rd ed., vol. 9. New York: Academic Press, 1973. [A dated, but standard, reference.]

Karplus, P., M. Daniels, and J. Herriott. Atomic Structure of Ferredoxin-NADP$^+$ Reductase: Prototype for a Structurally Novel Flavoenzyme Family. *Science* **251,** 60–66 (1991). [The structure of a key enzyme involved in nitrogen and sulfur metabolism, as well as in photosynthesis.]

Kim, J., and D. Rees. Crystallographic Structure and Functional Implications of the Nitrogenase Molybdenum-Iron Protein from *Azotobacter vinelandii. Nature* **360,** 553–560 (1992). [X-ray crystallography makes an important contribution to understanding the structure of a key protein of the nitrogen fixation process.]

Meyer, E., N. Leonard, B. Bhat, J. Stubbe, and J. Smith. Purification and Characterization of the *pur*E, *pur*K, and *pur*C Gene Products: Identification of a Previously Unrecognized Energy Requirement in the Purine Biosynthetic Pathway. *Biochemistry* **31,** 5022–5032 (1992). [The discovery of a hitherto unsuspected requirement for additional ATP in the biosynthesis of purines.]

Orme-Johnson, W. Nitrogenase Structure: Where To Now? *Science* **257,** 1639–1640 (1992). [Thoughts about nitrogen fixation based on the determination of the structure of nitrogenase by X-ray crystallography.]

Stadtman, E. R. Mechanisms of Enzyme Regulation in Metabolism. In Boyer, P. D., ed. *The Enzymes,.* 3rd ed., vol. 1. New York: Academic Press, 1970. [A review dealing with the importance of feedback control mechanisms.]

CHAPTER 24

Integration of Metabolism: Cellular Signaling

All aspects of biochemistry operate in concert, determining at the molecular level the responses that cells and whole organisms will make to the outside world. In the process of catabolism, large biopolymers are broken down to smaller molecules whereas, in anabolism, small precursors are built up into larger molecules. The citric acid cycle is amphibolic, meaning that its action is both catabolic and anabolic. To prevent overproduction of molecules in the metabolic pathways, regulatory enzymes that act by feedback inhibition control the creation of new molecules. On the other hand, hormones can send signals to speed up their production. Under stress, the hypothalamus in the brain sends hormone-releasing factors to the pituitary gland, which initiates a hormone cascade that increases the production of glucose to provide the extra energy needed to deal with a stressful situation. The body is subjected to even greater stress when infections arise from invasion by disease-causing agents. The immune system deals with such situations on both the cellular and the molecular level.

Biochemistry gives many important clues about how the body works.

© David Young-Wolff/PhotoEdit

Critical Questions

24.1 How Are the Metabolic Pathways Connected?

24.2 How Can Biochemistry Help Us Understand Nutrition?

24.3 What Are Hormones and Second Messengers?

24.4 How Are Hormones Involved in the Control of Metabolism?

24.5 What Are the Many Effects of Insulin?

Biochemistry ⊘ Now™
Test yourself on these Critical Questions at the BiochemistryNow website at **http://now.brookscole.com/campbell5**

24.1 | How Are the Metabolic Pathways Connected?

In the preceding chapters, we learned about a number of individual metabolic pathways. Some metabolites, such as pyruvate, oxaloacetate, and acetyl-CoA, appear in more than one pathway. Furthermore, reactions of metabolism can take place simultaneously, and it is important to consider control mechanisms by which some reactions and pathways are turned on and off.

All metabolism is ultimately linked to photosynthesis and the energy from the sun (Figure 24.1). The light reactions produce ATP and NADPH, which are then used to make carbohydrates in the dark reactions. These carbohydrates are the source of nutrients for other organisms. ATP and NADPH are the two consistent links between different forms of metabolism. Besides linking the light and dark reactions of photosynthesis, they are the most direct link between catabolism and anabolism (Figure 24.1). Other common molecules, such as sugars, PEP, pyruvate, and acetyl-CoA, also form a bridge between catabolic and anabolic processes. We shall now focus on some of the relationships among pathways by considering some of the physiological responses to biochemical events.

The **citric acid cycle** plays a central role in metabolism. Three main points can be considered in assigning a central role to the citric acid cycle. The first of these is its part in the catabolism of nutrients of the main types: carbohydrates, lipids, and proteins (Section 19.7). The second is the function of the citric acid cycle in the anabolism of sugars, lipids, and amino acids (Section 19.8). The third and final point is the relationship between individual metabolic pathways and the citric acid cycle. When we discuss these broader considerations, we can and should address questions that involve more than individual cells and the reactions that go on in them, such as questions of what goes on in tissues and in whole organs. In this chapter, we shall look at three such topics—nutrition, hormonal control, and the wide-reaching effects of signaling pathways. The following Biochemical Connections box describes the way in which one compound can affect an entire organism.

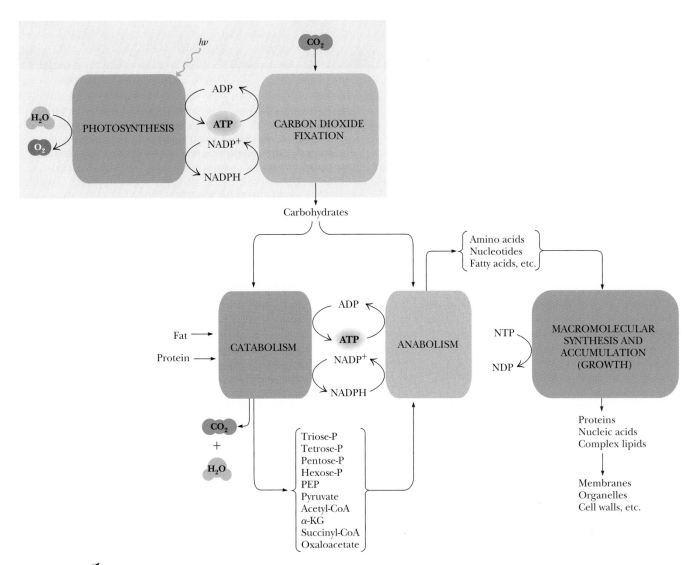

Biochemistry❋Now™ ▲ **ANIMATED FIGURE 24.1** Block diagram of intermediary metabolism showing the relationship between anabolic and catabolic processes and the common metabolites seen in many pathways. **See this figure animated at http://now.brookscole.com/campbell5**

24.2 | How Can Biochemistry Help Us Understand Nutrition?

The molecules that we process by catabolic reactions ultimately come from outside the body because we are heterotrophic organisms (dependent on external food sources). We shall devote this section to a brief look at how the foods we eat are sources of substrates for catabolic reactions. We should also bear in mind that nutrition is related to physiology as well as to biochemistry. This last point is certainly appropriate in view of the fact that many early biochemists were physiologists by training.

Required Nutrients

In humans, the catabolism of **macronutrients** (carbohydrates, fats, and proteins) to supply energy is an important aspect of nutrition. In the United States, most diets provide more than an adequate number of nutritional calories. The typical American diet is high enough in fat that essential fatty acids

Biochemical Connections

Alcohol Consumption and Addiction

Alcohol is the most abused drug in America, and alcoholism is among the most common diseases. Statistics about deaths due to drunk driving are available, but no one knows how many other accidental deaths may be indirectly caused by alcohol. Many believe that some particular biochemistry must be associated with alcoholism. There is certainly a genetic trait, shown most forcefully in a benchmark study of identical twins raised apart from each other. Attempts to find "the gene for alcoholism" have not, however, met with success. It is likely that there may be a complex genetic relationship involved.

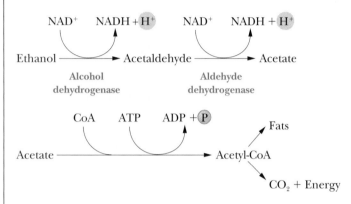

Alcohol dehydrogenase is an inducible enzyme. Its level increases in response to alcohol use. The first reaction occurs very rapidly in alcoholics, so the intoxicating effect of alcohol is actually reduced (i.e., less intoxication per ounce). Alcoholics can tolerate blood alcohol levels that would be lethal to others. For everyone, the second reaction is rate-limiting. Acetaldehyde can cause headaches, nausea, and hangovers. Malnutrition is common among alcoholics because alcohol is a source of "empty calories" without important nutrients, particularly vitamins.

Biochemical, psychological, and nutritional effects of alcohol are not the same for all people. The studies of twins indicate the possibility of a "born" alcoholic who could be totally hooked on the first drink. Fetal alcohol syndrome (see the Biochemical Connections box on page 483) is of particular concern to women. Ethanol is a teratogen; there is no "safe" level of alcohol during pregnancy. Fetal alcohol syndrome occurs in up to five of every 1000 births. Indicators include stunted growth, dysfunction of the central nervous system, and a characteristic facial shape.

Clearly, biochemistry is involved in addiction. Because many of the psychoactive drugs are structural analogs of serotonin and epinephrine (see the Biochemical Connections box on page 64), it is easy to imagine enhancement of their effects or competition with them. There is increasing interest in the effects of drugs (in general) on the production of endorphins and enkephalins (Section 3.5), the short peptides that are the brain's own opiate painkillers. In people who are not alcoholics, ethanol inhibits enkephalin synthesis because the pleasant effect of alcohol replaces the need for enkephalins. Part of the misery of a hangover is caused by the lack of enkephalins; the hangover usually lasts until the level of these compounds returns to normal.

(Section 21.6) are seldom, if ever, deficient. The only concern is that the diet contains an adequate supply of protein. If the intake of protein is sufficient, the supply of essential amino acids (Section 23.5) is normally also sufficient. Packaging on food items frequently lists the protein content in terms of both the number of grams of protein and the percentage of the daily value (DV) suggested by the Food and Nutrition Board under the auspices of the National Research Council of the National Academy of Sciences (see Table 24.1). Daily values have replaced the recommended daily allowances (RDAs) formerly seen on food packaging.

There are some key biochemical concepts to remember when analyzing a diet for protein content. First, there is no storage form for proteins. This means that proteins eaten in excess do a person no good in terms of satisfying that person's future protein requirements. All protein consumed in excess of what is needed will be turned into carbohydrate or fat, and the nitrogen from the amino group will have to be eliminated through the urea cycle (Section 23.6). Ingesting too much protein can therefore be stressful on the liver and kidneys due to the overproduction of ammonia that must be eliminated. This is the same risk faced by certain athletes who take creatine to build muscles, because creatine is a highly nitrogenated compound.

Second, the essential amino acids must be consumed daily in order for proteins to be made. It would be difficult to find a protein that did not have at least one residue of each of the common twenty amino acids. Half of these amino acids are essential, and if the diet is lacking or low in even one of these

Table 24.1

Daily Values for the Average Man and Woman, Aged 19 to 22

Nutrient	Man	Woman
Protein	56 g	44 g
Lipid-soluble vitamins		
Vitamin A	1 mg RE*	8 mg RE*
Vitamin D	7.5 μg†	7.5 μg†
Vitamin E	10 mg α-TE‡	8 mg α-TE‡
Water-soluble vitamins		
Vitamin C	60 mg	60 mg
Thiamine (vitamin B$_1$)	1.5 mg	1.1 mg
Riboflavin (vitamin B$_2$)	1.7 mg	1.3 mg
Vitamin B$_6$	3 μg	3 μg
Vitamin B$_{12}$	3 μg	3 μg
Niacin	3 μg	3 μg
Folic acid	19 mg	14 mg
Pantothenic acid (estimate)	10 mg	10 mg
Biotin (estimate)	0.3 mg	0.3 mg
Minerals		
Calcium	800 mg	800 mg
Phosphorus	800 mg	800 mg
Magnesium	350 mg	300 mg
Zinc	15 mg	15 mg
Iron	10 mg	18 mg
Copper (estimate)	3 mg	3 mg
Iodine	150 μg	150 μg

*RE = retinol equivalent, where 1 retinol equivalent = 1 μg retinol or 6 μg β-carotene. See Section 8.7.

†As cholecalciferol. See Section 8.7.

‡α-TE = α-tocopherol equivalent, where 1 α-TE = 1 mg D-α-tocopherol. See Section 8.7. Data from the Food and Nutrition Board, National Academy of Sciences–National Research Council, Washington, D.C., 1988.

essential amino acids, then protein synthesis will not be possible. Not all proteins are created equal (see the Biochemical Connections box on page 82). The protein efficiency ratio (PER) is an indication of how complete a protein is. However, mixing proteins correctly is very important, something that vegetarians know a lot about. A protein that is very low in lysine would have a low PER value. If a second protein had a low PER because it was low in tryptophan, it could be combined with the low-lysine protein to give a combination with a high PER. However, this would only work if the two were eaten together.

Third, proteins are always being degraded (Chapter 12). Because of that, even if it does not seem like a person is doing any activities that would tend to require protein replenishment, there is a constant need for quality protein in order to maintain the body's structures. Athletes are painfully aware of that fact. They must train constantly and get out of shape quickly when they stop.

Micronutrients (vitamins and minerals) are also listed on food packaging. The vitamins we require are compounds that are necessary for metabolic processes; either our bodies cannot synthesize them, or they cannot synthesize them in amounts sufficient for our needs. As a result, we must obtain vitamins from dietary sources. DVs are listed for the fat-soluble vitamins—vitamins A, D, and E (Section 8.7)—but care must be taken to avoid overdoses of these vitamins. Excesses can be toxic when large amounts of fat-soluble vitamins accumulate in adipose tissue. Excess vitamin A is especially toxic. With water-soluble vitamins, turnover is frequent enough that the danger of excess is not normally a problem.

The water-soluble vitamins with listed DVs are vitamin C, which is necessary for the prevention of scurvy (Section 4.3); and the B vitamins—niacin, pantothenic acid, vitamin B_6, riboflavin, thiamine, folic acid, biotin, and vitamin B_{12}. The B vitamins are the precursors of the metabolically important coenzymes listed in Table 7.1, where references to the reactions in which the coenzymes play a role are given. We have seen many pathways in which NADH, NADPH, FAD, TPP, biotin, pyridoxal phosphate, and coenzyme A were found, all of which came from vitamins. A summary of vitamins and their metabolic roles is given in Table 24.2. Frequently, the actual biochemical role is played by a metabolite of the vitamin rather than by the vitamin itself, but this point does not affect the dietary requirement.

Minerals, in the nutritional sense, are inorganic substances required in the ionic or free-element form for life processes. The macrominerals (those needed in the largest amounts) are sodium, potassium, chloride, magnesium, phosphorus, and calcium. The required amounts of all these minerals, except calcium, can easily be satisfied by a normal diet. Deficiencies of calcium can, and frequently do, occur. Calcium deficiencies can lead to bone fragility, with concomitant risk of fracture, which is a problem especially for elderly women. Calcium supplements are indicated in such cases. Requirements for some microminerals (trace minerals) are not always clear. It is known, for example, from biochemical evidence that chromium is necessary for glucose metabolism (a role that has recently been suggested for chromium picolinate) and that manganese is necessary for bone formation, but no deficiencies of these elements have been recorded. Requirements have been established for iron, copper, zinc, iodide, and fluoride; there are DVs for all these minerals except

Table 24.2

Vitamins: Chemical and Biochemical Facts

Vitamin	Metabolic Function	Reference
Water-Soluble		
B_1 (thiamine)	Aldehyde transfer, decarboxylation in alcoholic fermentation and citric acid cycle	Sections 17.4, 19.3
B_2 (riboflavin)	Oxidation–reduction reactions, especially in citric acid cycle and electron transport	Sections 19.3, 20.2
B_6 (pyridoxine)	Transamination reactions, especially of amino acids	Section 23.4
Niacin (nicotinic acid)	Oxidation–reduction reactions, found in many metabolic processes	Sections 17.3, 19.3, 20.2
Biotin	Carboxylation reactions in carbohydrate and lipid metabolism	Sections 18.2, 21.6
Pantothenic acid	Acyl transfer in many metabolic processes	Sections 15.7, 21.6
Folic acid	One-carbon group transfer, especially in nitrogen-containing compounds	Sections 23.4, 23.11
C (Ascorbic acid)	Hydroxylates collagen	Biochemical Connections box, p. 443.
Lipoic acid (?)	Acyl transfer, oxidation–reduction	Section 19.3
(It has been questioned whether lipoic acid is a vitamin.)		
Lipid-Soluble		
A	Isomerization mediates visual process	Section 8.7
D	Regulates calcium and phosphorus metabolism, especially in bone	Section 8.7
E	Antioxidant	Section 8.7
K	Mediates protein modification required for blood clotting	Section 8.7

fluoride. In the case of copper and zinc, needs are easily met by dietary sources, and overdoses can be toxic. A deficiency of iodide, leading to an enlarged thyroid gland (Section 24.3), has been a problem in some parts of the United States for many years. Iodized salt is used for prevention of this deficiency, and it has become unusual to find table salt without an iodine supplement. Fluoride is administered to prevent tooth decay in children and, with that end in mind, has been added to water supplies, sometimes causing considerable controversy. Iron is important because it is part of the structure of the ubiquitous heme proteins. Women of childbearing age are more susceptible to iron deficiencies than are other segments of the population, and in some cases supplements are advised. These recommended levels vary with the age of the individual and are subject to adjustment for level of activity.

The Food Pyramid

One approach to publicizing healthful food selection was the development of the Food Guide Pyramid, a graphic display that focuses on a diet sufficient in nutrients but without excesses (Figure 24.2). The goal was to use a well-chosen diet to promote good health. To avoid confusion, the development of this scheme had to take into account the fact that many people were familiar with the older recommendations about food groups. The newer recommendations pay particular attention to increasing the amount of fiber and decreasing the amount of fat in the typical diet. Variety and moderation were key concepts of the graphic presentation. From the biochemical point of view, these recommendations translate into a diet based primarily on carbohydrates, with enough protein to meet needs for essential amino acids (Section 23.5). Note that in Figure 24.2, carbohydrates are the base, with the correct amount suggested to be 6 to 11 servings of foods rich in complex carbohydrates, such as bread, cereal, rice, or pasta. Lipids should not contribute more than 30% of daily calories, but the typical American diet currently is about 45% fat. High-fat diets have been linked to heart disease and to some kinds of cancer, so the recommendation about lipid intake is of considerable importance. (See the article by Willett in the Annotated Bibliography at the end of this chapter for more on this topic.)

> ### Essential Information
>
> Human nutrition is an example of the biochemistry and physiology of heterotrophic organisms; we depend on external food sources. Not only do we require macronutrients (carbohydrates, fats, and proteins) for energy, but, in the case of essential amino acids, we also need to consume specific amino acids that our bodies cannot produce if we are to prevent protein deficiency. We also require micronutrients: vitamins and minerals. Vitamins are organic compounds for which we have no biosynthetic pathways. Vitamins function as coenzymes or precursors of coenzymes in essential reactions. The same is true of minerals, where the term refers to inorganic substances, mostly metal ions, that are required for life processes. Various metal ions function as cofactors of enzymes that catalyze essential reactions.

Biochemical Connections

Iron: An Example of a Mineral Requirement

Iron, whether in the form Fe(II) or Fe(III), is usually found in the body in association with proteins. Little or no iron can be found "free" in the blood. Because iron-containing proteins are ubiquitous, there is a dietary requirement for this mineral. Severe deficits can lead to iron-deficiency anemia.

Iron usually occurs as the Fe(III) form in food. This is also the form released from iron pots when food is cooked in them. However, iron must be in the Fe(II) state to be absorbed. Reduction from Fe(III) to Fe(II) can be accomplished by ascorbate (vitamin C) or by succinate. Factors that affect absorption include the solubility of a given compound of iron, the presence of antacids in the digestive tract, and the source of the iron. To give some examples, iron may form insoluble complexes with phosphate or oxalate, and the presence of antacids in the diges-

tive tract may decrease iron absorption. Iron from meats is more easily absorbed than iron from plant sources.

Requirements for iron vary according to age and gender. Infants and adult men need 10 mg per day; infants are born with a three- to six-month supply. Children and women (ages 16 through 50) need 15 to 18 mg per day. Women lose 20 to 23 mg of iron during each menstrual period. Pregnant and lactating women need more than 18 mg per day. After a blood loss, anyone, regardless of age or gender, needs more than these amounts. Distance runners, particularly marathoners, are also at risk of becoming anemic due to the loss of blood cells in the feet caused by the pounding of the thousands of foot falls that occur during a long run. People with iron deficiencies may experience a craving for nonfood items like clay, chalk, and ice.

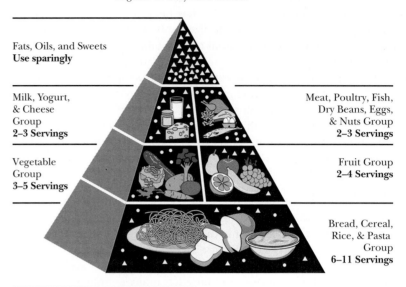

Food Guide Pyramid
A guide to daily food choices

Key

○ Fat (naturally occurring and added)
△ Sugars (added)
These symbols show fats, oils, and
added sugars in foods.

▶ **FIGURE 24.2** The Food Guide Pyramid (USDA). The recommended choices reflect a diet based primarily on carbohydrates. Smaller amounts of proteins and lipids are sufficient to meet the body's needs.

However, many scientists are now questioning some of the details of this food pyramid. It is known that certain types of fat are essential to health and actually reduce the risk of heart disease. Also, there has been little evidence to back up the claim that a high intake of carbohydrates is beneficial. Many people feel that the original food pyramid, which was published in 1992, has serious flaws. It overglorifies carbohydrates while making all fats out to be the bad guys. In addition, meat, fish, poultry, and eggs are all lumped together as if they are equivalent in terms of health. There is plenty of evidence now that does link saturated fat with high cholesterol and risk of heart disease, but monounsaturated and polyunsaturated fats have the opposite effect. While many scientists knew the distinction between the various types of fat, they felt that the average person would not understand them, and so the original pyramid was designed to send the simple message that fat was bad. The implied corollary to fat being bad was that carbohydrates were good. However, after years of study, no evidence can be shown that a diet that has 30% or fewer calories coming from fat is healthier than one with a higher level.

To further complicate matters, we have to recall the effects of the traveling forms of cholesterol—the lipoproteins. Having high levels of cholesterol traveling as high-density lipoproteins (HDL) has been correlated with a healthy heart, while having high levels of cholesterol traveling in the form of low-density lipoproteins (LDL) is related to high risk of heart disease (Chapter 21). When calories from saturated fat are replaced by carbohydrates, the levels of LDL and total cholesterol decrease, but so does the level of HDL. Since the ratio of LDL to HDL does not decrease significantly, there is little health benefit. However, the increase in carbohydrate has been shown to increase fat synthesis due to increases in insulin production. When calories from unsatu-

Multiple vitamins
FOR MOST

Alcohol in moderation
UNLESS CONTRAINDICATED

Red meat
and butter
USE SPARINGLY

White rice, white bread,
potatoes, pasta and sweets
USE SPARINGLY

Dairy or
calcium supplement
1 TO 2 SERVINGS

Fish, poultry and eggs
0 TO 2 SERVINGS

Nuts and legumes
1 TO 3 SERVINGS

Vegetables IN
ABUNDANCE

Fruit 2 to 3 SERVINGS

Whole-grain
foods AT
MOST
MEALS

Plant oils (olive, canola,
soy, corn, sunflower,
peanut and other
vegetable oils) AT
MOST MEALS

Daily exercise and weight control

NEW FOOD PYRAMID
distinguishes between healthy and unhealthy types of fat and carbohydrates. Fruits and
vegetables are still recommended, but the consumption of dairy products should be limited.

◀ **FIGURE 24.3** A new food pyramid. This version
of the recommended amounts of the different food
types reflects a distinction between unhealthy and
healthy types of carbohydrates and fats. The recommended intake of dairy products has also been
reduced compared with the original pyramid.
(*© Richard Borge. Adapted with permission.*)

rated fat are replaced with calories from carbohydrates, the results are even
worse. The LDL levels rise in comparison with the levels of HDL.

Figure 24.3 shows a more modern view of a food pyramid that takes into
account the most recent evidence and recommendations from some nutri-
tionists. Note that at the base of the pyramid is the heart and soul of good
health—exercise and weight control. There is no replacement for being active
and for restricting total calories when it comes to staying healthy. The next
level up shows that the good types of carbohydrates and the good forms of fats
occupy a prime location. Whole-grain foods are complex carbohydrates that
are digested more slowly, so they do not have the effect of raising blood glucose and causing insulin levels to rise to the extent that refined carbohydrates
like white rice and pasta do. The healthy fats come from plant oils. Vegetables
and fruits still occupy an important place in this pyramid, with nuts and
legumes just above them. Next are good sources of protein, such as fish, poultry, and eggs. Note that the recommendation says zero to two servings. This is
a change in approach, in that the type of protein is considered important and
in the fact that the guide shows that it is not necessary to eat animal protein
at all. Dairy products are found high up on the new pyramid. This is because,
despite the commercials that suggest "everybody needs milk," there are some
noted health risks in consuming dairy products. Some of the cultures that
consume large quantities of dairy products have the highest incidence of
heart disease, probably due to the high concentrations of saturated fatty acids
in milk and butter. In addition, many adults are allergic to milk proteins, and
many are unable to digest lactose. At the peak of the pyramid are the items to
be eaten only sparingly: red meat and refined carbohydrates, as well as some
natural carbohydrate sources, such as potatoes. In February 2004, the U.S.

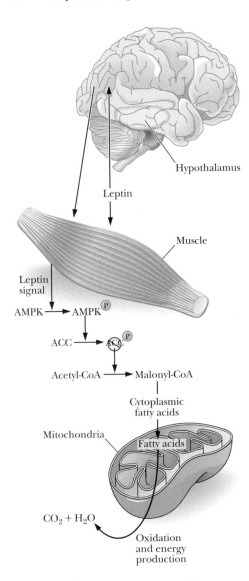

▲ **FIGURE 24.4** Leptin has multiple effects on metabolism. It affects the brain, lowering appetite. It also inactivates acetyl-CoA carboxylase (ACC). Reduced activity of ACC leads to a reduction in malonyl-CoA, which stimulates fatty-acid oxidation and reduces fatty-acid synthesis. (*From* Nature, *Vol. 415 (January 17, 2002), Fig 1, p. 268. Copyright © 2002 Nature. Reprinted with permission.*)

Department of Agriculture launched a website, the Interactive Healthy Eating Index at the URL http://209.48.219.53/. This service allows individuals to find information about foods and the role of physical activity in order to make choices for a healthful lifestyle.

Obesity

Obesity is a major public-health problem in the United States. Recent figures from the National Institutes of Health show that one-third of the population is clinically obese, defined as weighing at least 20% more than their ideal weight. Artificial sweeteners have been introduced, sometimes with great controversy (see the Biochemical Connections box on page 73), to help those who wish to control their weight. Fat substitutes have come on the market more recently, again accompanied by controversy. One thing is clear: the topic will continue to be one of great interest, with tradeoffs between palatability and health concerns providing a driving force for finding new products.

The role of the protein leptin in the control of obesity has been established in mice, and information is just appearing concerning its effect in humans. It is known that in mice leptin is a 16-kDa protein and that it is produced by the *obesity (ob)* gene. Mutations in this gene lead to a deficiency of leptin, which in turn leads to increased appetite and decreased activity, ultimately leading to weight gain. Injections of this protein into affected mice lead to decreased appetite and increased activity, with resulting weight loss. Administering leptin to leptin-deficient humans has been reported to reduce obesity; however, in clinically obese subjects, the circulating levels of leptin are often high. It is possible that some forms of obesity are caused by a lack of sensitivity to leptin rather than a lack of leptin itself.

Leptin stimulates the oxidation of fatty acids and the uptake of glucose by muscle cells. It does so by stimulating AMP-activated protein kinase, which phosphorylates an isoform of acetyl-CoA carboxylase (ACC) in muscle cells, rendering it less active (Figure 24.4). Recall from Section 21.6 that ACC plays a pivotal role in fat metabolism. When ACC activity is decreased, malonyl-CoA levels decrease and the mitochondria will be able to take up and oxidize fatty acids. Leptin also inhibits production of the mRNA for hepatic stearolyl-CoA desaturase (Section 21.6), leading to less lipid synthesis.

Leptin also works directly on the nervous system. Both leptin and insulin (Section 24.5) are long-term regulators of appetite. They circulate in the blood at concentrations roughly proportional to body-fat mass. They inhibit appetite by inhibiting specific neurons in the hypothalamus. Several laboratories have shown interest in using this information to develop treatments for human obesity. The Biochemical Connections box on page 682 gives some of the theories on diet, hormones, and obesity.

| 24.3 | **What Are Hormones and Second Messengers?** |

Hormones

The metabolic processes within a given cell are frequently regulated by signals from outside the cell. A usual means of intercellular communication takes place through the workings of the **endocrine system,** in which the ductless glands produce **hormones** as intercellular messengers. Hormones are transported from the sites of their synthesis to the sites of action by the bloodstream (Figure 24.5). In terms of their chemical structure, some typical hormones are

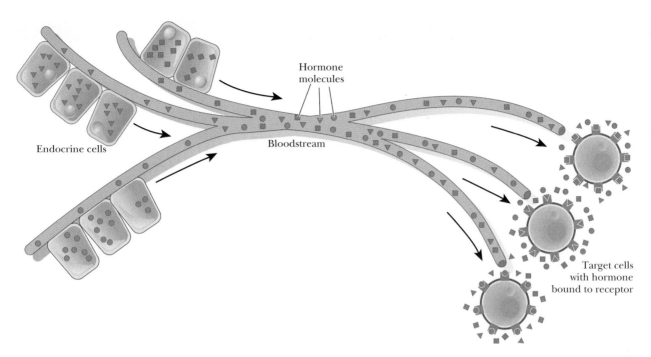

▲ **FIGURE 24.5** Endocrine cells secrete hormones into the bloodstream, which transports them to target cells.

steroids, such as estrogens, androgens, and mineralocorticoids (Section 21.8); polypeptides, such as insulin and endorphins (Section 3.5); and amino acid derivatives, such as epinephrine and norepinephrine (Table 24.3).

Hormones have several important functions in the body. They help to maintain **homeostasis,** the balance of biological activities in the body. The effect of insulin in keeping the blood glucose level within narrow limits is an example of this function. The operation of epinephrine and norepinephrine in the "fight-or-flight" response is an example of the way in which hormones mediate responses to external stimuli. Finally, hormones play roles in growth and development, as seen in the roles of growth hormone and the sex hormones. The methods and insights of biochemistry and physiology alike have helped to illuminate the workings of the endocrine system.

The release of hormones exerts control on the cells of target organs; other control mechanisms, however, determine the workings of the endocrine gland that releases the hormone in question. Simple feedback mechanisms, in which the action of the hormone leads to feedback inhibition of the release of hormone, can be postulated (Figure 24.6). The workings of the endocrine system are, in fact, much less simple, with the added complexity allowing for a greater degree of control. To illustrate with a rather restricted example, insulin is released in response to a rapid rise in the level of blood glucose. In the absence of control mechanisms, an excess of insulin can produce **hypoglycemia,** the condition of low blood glucose. In addition to negative feedback control on the release of insulin, the action of the hormone glucagon tends to increase the level of glucose in the bloodstream. The two hormones together regulate blood glucose. This example is far too restricted, as we shall see in the next section. We will look at insulin itself more closely in Section 24.5.

A more sophisticated control system involves the action of the *hypothalamus,* the *pituitary,* and specific *endocrine glands* (Figure 24.7). The central nerv-

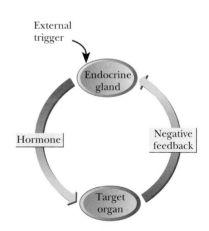

▲ **FIGURE 24.6** A simple feedback control system involving an endocrine gland and a target organ.

Table 24.3		
Selected Human Hormones		
Hormone	**Source**	**Major Effects**
Polypeptides		
Corticotropin-releasing factor (CRF)	Hypothalamus	Stimulates release of ACTH
Gonadotropin-releasing factor (GnRF)	Hypothalamus	Stimulates release of FSH and LH
Thyrotropin-releasing factor (TRF)	Hypothalamus	Stimulates release of TSH
Growth hormone–releasing factor (GRF)	Hypothalamus	Stimulates release of growth hormone
Adrenocorticotropic hormone (ACTH)	Anterior pituitary	Stimulates release of adrenocorticosteroids
Thyrotropin (TSH)	Anterior pituitary	Stimulates release of thyroxine
Follicle-stimulating hormone (FSH)	Anterior pituitary	In ovaries, stimulates ovulation and estrogen synthesis; in testes, stimulates spermatogenesis
Luteinizing hormone (LH)	Anterior pituitary	In ovaries, stimulates estrogen and progesterone synthesis; in testes, stimulates androgen synthesis
Met-enkephalin	Anterior pituitary	Has opioid effects on central nervous system
Leu-enkephalin	Anterior pituitary	Has opioid effects on central nervous system
β-Endorphin	Anterior pituitary	Has opioid effects on central nervous system
Vasopressin	Posterior pituitary	Stimulates water resorption by kidney and raises blood pressure
Oxytocin	Posterior pituitary	Stimulates uterine contractions and flow of milk
Insulin	Pancreas (β-cells of islets of Langerhans)	Stimulates uptake of glucose from bloodstream
Glucagon	Pancreas (α-cells of islets of Langerhans)	Stimulates release of glucose to bloodstream
Steroids		
Glucocorticoids	Adrenal cortex	Decrease inflammation, increase resistance to stress
Mineralocorticoids	Adrenal cortex	Maintain salt and water balance
Estrogens	Gonads and adrenal cortex	Stimulate development of secondary sex characteristics, particularly in females
Androgens	Gonads and adrenal cortex	Stimulate development of secondary sex characteristics, particularly in males
Amino acid derivatives		
Epinephrine	Adrenal medulla	Increases heart rate and blood pressure
Norepinephrine	Adrenal medulla	Decreases peripheral circulation, stimulates lipolysis in adipose tissue
Thyroxine	Thyroid	Stimulates metabolism generally

ous system sends a signal to the hypothalamus. The **hypothalamus** secretes a hormone-releasing factor, which in turn stimulates release of a trophic hormone by the anterior pituitary (Table 24.3). (The action of the hypothalamus on the posterior pituitary is mediated by nerve impulses.) **Trophic hormones** act on specific **endocrine glands,** which release the hormones to be transported to target organs. Note that feedback control is exerted at every stage of the process. Even more fine tuning is possible with zymogen activation mechanisms (Section 7.4), which exist for many well-known hormones.

The chemical natures of hormones play a predictably important role in their roles in cell signaling. Steroid hormones, for example, can enter the cell directly through the plasma membrane or can bind to plasma membrane

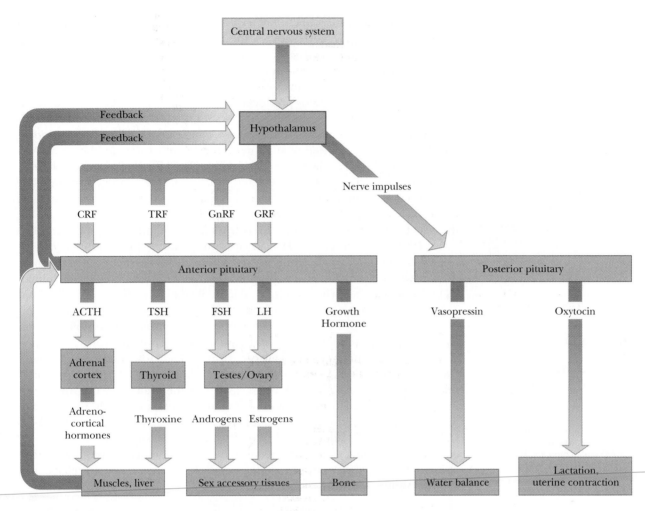

▲ **FIGURE 24.7** Hormonal control system showing the role of the hypothalamus, pituitary, and target tissues. See Table 24.3 for the names of the hormones.

receptors. Nonsteroid hormones enter the cell exclusively as a result of binding to plasma membrane receptors (Figure 24.8).

The releasing factors and trophic hormones listed in Table 24.3 tend to be polypeptides, but the chemical natures of the hormones released by specific endocrine glands show greater variation. Thyroxine, for example, produced by the thyroid, is an iodinated derivative of the amino acid tyrosine (Section 3.2). Abnormally low levels of thyroxine lead to **hypothyroidism,** characterized by lethargy and obesity, whereas increased levels produce the opposite effect *(hyperthyroidism)*. Low levels of iodine in the diet often lead to hypothyroidism and an enlarged thyroid gland *(goiter)*. This condition has largely been eliminated by the addition of sodium iodide to commercial table salt ("iodized" salt). (It is virtually impossible to find table salt that is not iodized.)

Steroid hormones (Section 21.8) are produced by the adrenal cortex and the gonads (testes in males, ovaries in females). The **adrenocortical hormones** include **glucocorticoids,** which affect carbohydrate metabolism, modulate inflammatory reactions, and are involved in reactions to stress. The **mineralo-corticoids** control the level of excretion of water and salt by the kidney. If the adrenal cortex does not function adequately, one result is *Addison's disease,* characterized by hypoglycemia, weakness, and increased susceptibility to stress. This disease is eventually fatal unless it is treated by administration of mineralocorticoids and glucocorticoids to make up for what is missing. The

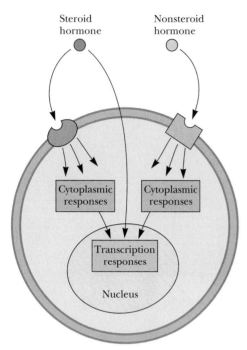

Biochemistry⊗Now™ ANIMATED FIGURE 24.8
Nonsteroid hormones bind exclusively to plasma-membrane receptors, which mediate the cellular responses to the hormone. Steroid hormones exert their effects either by binding to plasma-membrane receptors or by diffusing to the nucleus, where they modulate transcriptional events. **See this figure animated at http://now.brookscole.com/campbell5**

opposite condition, *hyperfunction of the adrenal cortex,* is frequently caused by a tumor of the adrenal cortex or of the pituitary. The characteristic clinical manifestation is *Cushing's syndrome,* marked by hyperglycemia, water retention, and the easily recognized "moon face."

The adrenal cortex produces some steroid sex hormones, the *androgens* and *estrogens,* but the main site of production is the gonads. Estrogens are required for female sexual maturation and function, but not for embryonic sexual development of female mammals. Animals that are male genetically will appear to be females if they are deprived of androgens during embryonic development. As a final example, we shall discuss growth hormone (GH), which is a polypeptide. When overproduction of GH occurs, it is usually because of a pituitary tumor. If this condition occurs while the skeleton is still growing, the result is **gigantism.** If the skeleton has stopped growing before the onset of GH overproduction, the result is **acromegaly,** characterized by enlarged hands, feet, and facial features. Underproduction of GH leads to **dwarfism,** but this condition can be treated by the injection of human GH before the skeleton reaches maturity. Animal GH is ineffective in treating dwarfism in humans. Supplies of human GH were very limited when it was possible to obtain it only from cadavers, but it can now be synthesized by recombinant DNA techniques. (Another discussion of peptide hormones can be found in the Biochemical Connections box on page 76, which treats oxytocin and vasopressin.) Human growth hormone (HGH) has recently become available to individuals who believe it will help alleviate the effects of aging. It was known that the level of HGH decreases after middle age is reached. Many have assumed that the availability of growth hormone, if one could afford it, would be a virtual fountain of youth. Even though few results are conclusive at this time, HGH is being prescribed, and the medical community has adopted rules for its use. For example, doctors will only consider prescribing it for patients over the age of 40. The same hormone is also used illegally by endurance athletes, and there is currently no reliable test to stop this illegal use.

Second Messengers

When a hormone binds to its specific receptor on a target cell, it sets off a chain of events in which the actual response within the cell is elicited. Several kinds of receptors are known. The receptors for steroid hormones tend to occur within the cell rather than as part of the membrane (steroids can pass through the plasma membrane); steroid–receptor complexes affect the transcription of specific genes. More frequently, the receptor proteins are a part of the plasma membrane. Binding of hormone to the receptor triggers a change in concentration of a second messenger. The **second messenger** brings about the changes within the cell as a result of a series of reactions.

Cyclic AMP and G Proteins

Cyclic AMP (adenosine-3′,5′-monophosphate, cAMP) is one example of a second messenger. The mode of action starts with binding of a hormone to a specific receptor called a β_1- or β_2-adrenergic receptor, which triggers the production of cAMP from ATP, catalyzed by *adenylate cyclase.* This reaction is mediated by a stimulatory G protein, a trimer consisting of three subunits—α, β, and γ. Binding of the hormone to the receptor activates the G protein; the α-subunit binds GTP while releasing GDP, giving rise to the name of the protein. The active protein has GTPase activity and slowly hydrolyzes GTP, returning the G protein to the inactive state. GDP remains bound to the α-subunit and must be exchanged for GTP when the protein is activated the next time (Figure 24.9). The G protein and adenylate cyclase are bound to the plasma membrane, while cAMP is released into the interior of the cell to act as a sec-

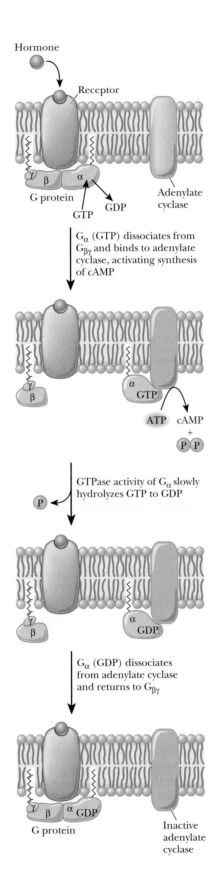

Hormone

Receptor

G protein

GTP   GDP

Adenylate cyclase

G_α (GTP) dissociates from $G_{\beta\gamma}$ and binds to adenylate cyclase, activating synthesis of cAMP

ATP   cAMP
+
P P

GTPase activity of G_α slowly hydrolyzes GTP to GDP

G_α (GDP) dissociates from adenylate cyclase and returns to $G_{\beta\gamma}$

G protein

Inactive adenylate cyclase

◀ **FIGURE 24.9** Activation of adenylate cyclase by heterotrimeric G proteins. Binding of hormone to its receptor causes a conformational change that induces the receptor to catalyze a replacement of GDP by GTP on G_α. The G_α (GTP) complex dissociates from $G_{\beta\gamma}$ and binds to adenylate cyclase, stimulating synthesis of cAMP. Bound GTP is slowly hydrolyzed to GDP by the intrinsic GTPase activity of G_α. G_α (GDP) dissociates from adenylate cyclase and reassociates with $G_{\beta\gamma}$. G_α and G_γ are lipid-anchored proteins. Adenylate cyclase is an integral membrane protein consisting of 12 transmembrane α-helical segments.

ond messenger. As we have already seen in several pathways, cAMP stimulates protein kinase A, which phosphorylates a host of enzymes and transcription factors. Some examples are known in which the binding of hormone to receptor (an α_2-receptor) inhibits rather than stimulates adenylate cyclase. A

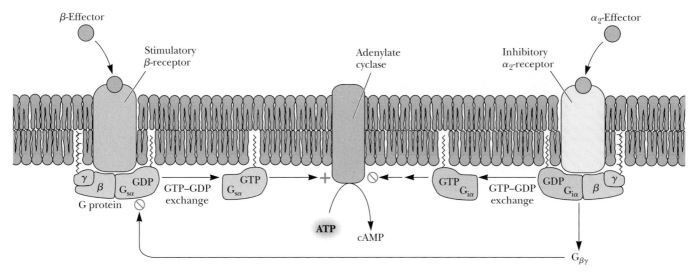

Biochemistry ⏣ Now™ ▲ **ACTIVE FIGURE 24.10** Adenylate cyclase activity is modulated by the interplay of stimulatory (G_s) and inhibitory (G_i) G proteins. Binding of hormones to β-receptors activates adenylate cyclases via G_s, whereas hormone binding to α_2 receptors leads to the inhibition of adenylate cyclase. Inhibition may occur by direct inhibition of cyclase activity by $G_{i\alpha}$ or by binding of $G_{i\beta\gamma}$ to $G_{s\alpha}$. **Watch this Active Figure at http://now.brookscole.com/campbell5**

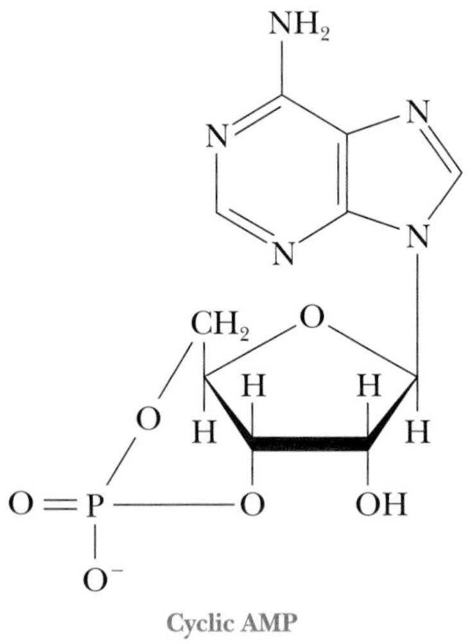

Cyclic AMP

Cyclic AMP (adenosine-3′, 5′-cyclic monophosphate, cAMP).

Biochemistry ⏣ Now™

Go to BiochemistryNow and click on Biochemistry Interactive to learn more about the heterotrimeric G protein complex.

Biochemistry ⏣ Now™

Go to BiochemistryNow and click on Biochemistry Interactive to explore the structure and function of adenylate cyclase.

G protein with a different kind of α-subunit mediates the process. The modified G protein is referred to as an inhibitory G protein to distinguish it from the kind that stimulates response to hormone binding (Figure 24.10).

In eukaryotic cells, the usual mode of action of cAMP is to stimulate a cAMP-dependent protein kinase, a tetramer consisting of two regulatory subunits and two catalytic subunits. When cAMP binds to the dimer of regulatory subunits, the two active catalytic subunits are released. The active kinase catalyzes the phosphorylation of a target enzyme or transcription factor (Figure 24.11). In the scheme shown in Figure 24.11, phosphorylation activates the enzyme. Cases are also known in which phosphorylation inactivates a target enzyme (e.g., glycogen synthase; Section 24.4). The usual site of phosphorylation is the hydroxyl group of a serine or a threonine. ATP is the source of the phosphate group that is transferred to the enzyme. The target enzyme then elicits the cellular response.

G proteins are very important signaling molecules in eukaryotes. They can be activated by combinations of hormones. For example, both epinephrine and glucagons act via a stimulatory G protein in liver cells. The effect can be cumulative so that, if both glucagons and epinephrine have been released, the cellular effect is greater. Besides the effect on cAMP, G proteins are involved in activating many other cellular processes, including stimulating phospholipase C and opening or closing membrane ion channels. They are also involved in vision and smell. There are currently more than 100 known G protein–coupled receptors and more than 20 known G proteins. See the Biochemical Connections box on page 680 for a look at a different class of G proteins.

A G protein is permanently activated by cholera toxin, leading to excessive stimulation of adenylate cyclase and chronic elevation of cAMP levels. The main danger in *cholera,* caused by the bacterium *Vibrio cholerae,* is severe dehydration as a result of diarrhea. The unregulated activity of adenylate cyclase in epithelial cells leads to the diarrhea because cAMP in epithelial cells stimulates active transport of Na^+. Excessive cAMP in the epithelial cells of the small intestine produces a large flow of Na^+ and water from the mucosal sur-

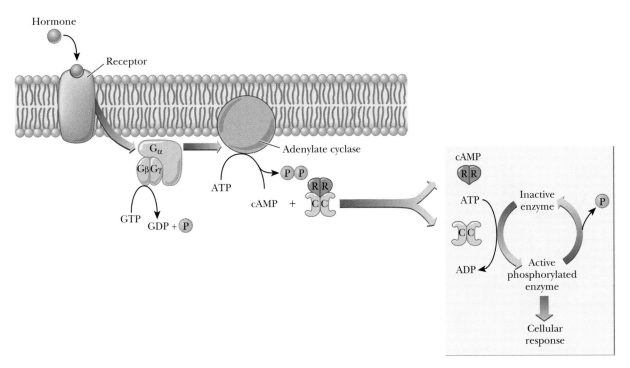

▲ **FIGURE 24.11** The activation of adenylate cyclase by the binding of hormone to the receptor and the mode of action of cAMP. The binding of hormone to the receptor leads to the production of cAMP from ATP, catalyzed by adenylate cyclase; this reaction is mediated by a G protein. Once cAMP is formed, it stimulates a protein kinase by binding to the regulatory subunits, shown as R. The active catalytic subunits, shown as C, are released and catalyze the phosphorylation of a target enzyme. The target enzyme elicits the response of the cell to the hormonal signal. This scheme applies in situations in which the phosphorylation activates the target enzyme.

face of the epithelial cells into the lumen of the intestine. If the lost fluid and salts can be replaced in cholera victims, the immune system can eliminate the actual infection within a few days.

Calcium Ion as a Second Messenger

Calcium ion (Ca^{2+}) is involved in another ubiquitous second-messenger scheme. Much of the calcium-mediated response depends on release of Ca^{2+} from intracellular reservoirs, similar to the release of Ca^{2+} from the sarcoplasmic reticulum in the action of the neuromuscular junction. A component of the inner layer of the phospholipid bilayer, *phosphatidylinositol 4,5-bisphosphate* (PIP_2), is also required in this scheme (Figure 24.12).

When the external trigger binds to its receptor on the cell membrane, it activates *phospholipase C* (Section 21.2), which hydrolyzes PIP_2 to *inositol 1,4,5-triphosphate* (IP_3) and a *diacylglycerol* (DAG), in a process mediated by a different member of the family of G proteins. The IP_3 is the actual second messenger. It diffuses through the cytosol to the endoplasmic reticulum (ER), where it stimulates the release of Ca^{2+}. A complex is formed between the calcium-binding protein calmodulin and Ca^{2+}. This calcium–calmodulin complex activates a cytosolic protein kinase, which phosphorylates target enzymes in the same fashion as in the cAMP second-messenger scheme. DAG also plays a role in this scheme; it is nonpolar and diffuses through the plasma membrane. When DAG encounters the membrane-bound protein kinase C, it too acts as a second messenger by activating this enzyme (actually a family of enzymes).

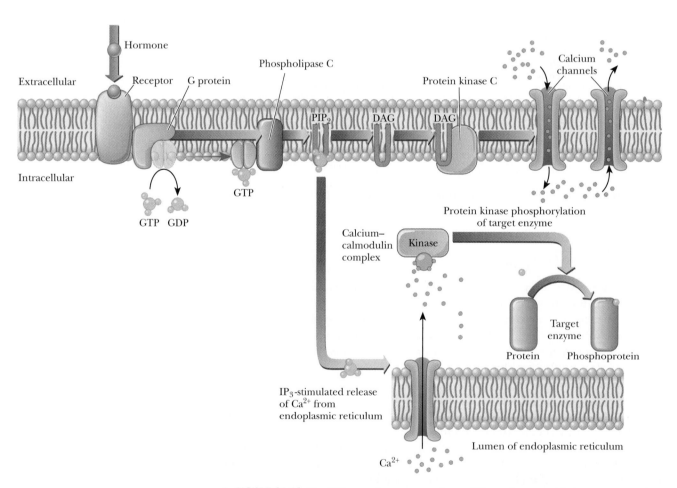

R₁ and R₂ = fatty acid residues

Ⓟ = phosphate moiety

Phosphatidylinositol 4,5-*bis*phosphate (PIP₂)

▲ **FIGURE 24.12** The PIP₂ second-messenger scheme. When a hormone binds to a receptor, it activates phospholipase C, in a process mediated by a G protein. Phospholipase C hydrolyzes PIP₂ to IP₃ and DAG. IP₃ stimulates the release of Ca^{2+} from intracellular reservoirs in the ER. A complex formed between Ca^{2+} and the calcium-binding protein calmodulin activates a cytosolic protein kinase for phosphorylation of a target enzyme. DAG remains bound to the plasma membrane, where it activates the membrane-bound protein kinase C (PKC). PKC is involved in the phosphorylation-channel proteins that control the flow of Ca^{2+} in and out of the cell. Ca^{2+} from extracellular sources can produce sustained responses even when the supply of Ca^{2+} in intracellular reservoirs is exhausted.

Protein kinase C also phosphorylates target enzymes, including channel proteins that control the flow of Ca^{2+} into and out of the cell. By controlling the flow of Ca^{2+}, this second-messenger system can produce sustained responses even when the supply of Ca^{2+} in the intracellular reservoirs becomes exhausted. (For more information on this point, see the article by Rasmussen listed in the Annotated Bibliography at the end of this chapter.)

Receptor Tyrosine Kinases

Another important type of second-messenger system involves a receptor type called a **receptor tyrosine kinase.** These receptors span the membrane of the cell and have a hormone receptor on the outside and a tyrosine kinase portion on the inside. There are several subclasses of these receptor kinases, as shown in Figure 24.13. The best known of these is class II, which includes the insulin receptor (which we will look at in more detail in Section 24.5).

These kinases are allosteric enzymes. When the hormone binds to the binding region on the outside of the cell, it induces a conformational change in the tyrosine kinase domain that activates the kinase activity. The activated tyrosine

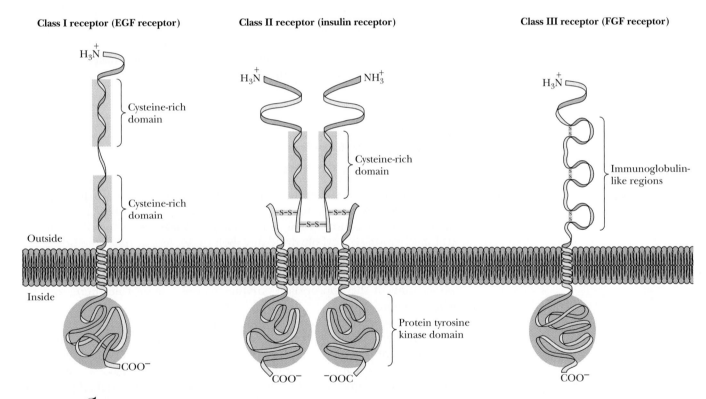

Biochemistry ⓔ Now™ ▲ **ANIMATED FIGURE 24.13** The three classes of receptor tyrosine kinases. Class I receptors are monomeric and contain a pair of Cys-rich repeat sequences. The insulin receptor, a typical class II receptor, is a glycoprotein composed of two kinds of subunits in a $\alpha_2\beta_2$ tetramer. The α- and β-subunits are synthesized as a single peptide chain, together with an N-terminal signal sequence. Subsequent proteolytic processing yields the separate α- and β-subunits. The β-subunits, of 620 residues each, are integral transmembrane proteins, with only a single transmembrane α-helix and with the amino terminus outside the cell and the carboxyl terminus inside. The α-subunits, of 735 residues each, are extracellular proteins that are linked to the β-subunits and to each other by disulfide bonds. The insulin-binding domain is located in a cysteine-rich region on the α-subunits. Class III receptors contain multiple immunoglobulin-like domains. Shown here is fibroblast growth factor (FGF) receptor, which has three immunoglobulin-like domains. (*Adapted from A. Ullrich and J. Schlessinger, 1990. Signal Transduction by Receptors with Tyrosine Kinase Activity. Cell, **61**, 203–212 (1990).*) **See this figure animated at http://now.brookscole.com/ campbell5**

Biochemical Connections

Small G Proteins and the Ras Family

A subset of the general class of G proteins is often referred to as the **small G proteins.** This is a large superfamily of proteins grouped into subfamilies called Ras (Section 14.8), Rho, Arf, Rab, and Ran. They have been found to play roles in growth and differentiation mechanisms, cell motility, and transport of material through the cell. Unlike their larger G protein brethren, the small G proteins are monomeric. However, there is strong sequence homology between the small G proteins and the G_α-subunit of the larger G proteins. A particular one that has been studied extensively is a small G protein of 21 kDa molecular mass, sometimes referred to as p21. Recall from Chapter 14 that the first gene to be found that produces this

protein was from a *rat* sarcoma virus, and was called the *ras* gene. The normal p21^{ras} protein is a GTP-binding protein with a very slow rate of GTP hydrolysis. This is natural for a protein involved in the control of slower processes, such as growth. A second protein, called **GTPase-activating protein (GAP),** increases the GTPase activity 10,000-fold.

Much research has gone into studying these small G proteins, particularly p21^{ras}. The reason is that, in about 30% of the known human tumors, there is an irregularity with this protein. The ras protein found in these tumors has an even slower hydrolysis of GTP than the normal version and is insensitive to GAPs. Thus, the process of cell division proceeds out of control.

kinases phosphorylate tyrosines on a variety of target proteins, causing alterations in membrane transport of ions and amino acids and the transcription of certain genes. Phospholipase C (seen on page 678) is one of the targets of tyrosine kinases. Another is an **insulin-sensitive protein kinase,** which phosphorylates and activates protein phosphatase 1.

24.4	How Are Hormones Involved in the Control of Metabolism?

Now that we know something about the effects of hormones in triggering responses within the cell, we can return to and expand on some earlier points about metabolic control. In Section 18.3, we discussed some points about control mechanisms in carbohydrate metabolism. We saw at that time how glycolysis and gluconeogenesis can be regulated and how glycogen synthesis and breakdown can respond to the body's needs. Phosphorylation and dephosphorylation of the appropriate enzymes played a large role there, and that whole scheme is subject to hormonal action.

Three hormones play a part in the regulation of carbohydrate metabolism: epinephrine, glucagon, and insulin. Epinephrine acts on muscle tissue to raise levels of glucose on demand, while glucagon acts on the liver, also to increase the availability of glucose. Feedback control plays a role in the process and ensures that the amount of glucose made available does not reach an excessive level (Section 24.3). The role of insulin is to trigger the feedback response that achieves this further control.

Epinephrine (also called adrenalin) is structurally related to the amino acid tyrosine. Epinephrine is released from the adrenal glands in response to stress (the fight-or-flight response). When it binds to specific receptors, it sets off a chain of events that leads to increased levels of glucose in the blood, increased glycolysis in muscle cells, and increased breakdown of fatty acids for energy. Glucagon (a peptide that contains 29 amino acid residues) is released by the α-cells of the islets of Langerhans in the pancreas, and it too binds to specific receptors to set off a chain of events to make glucose available to the organism. Each time a single hormone molecule, whether epinephrine or

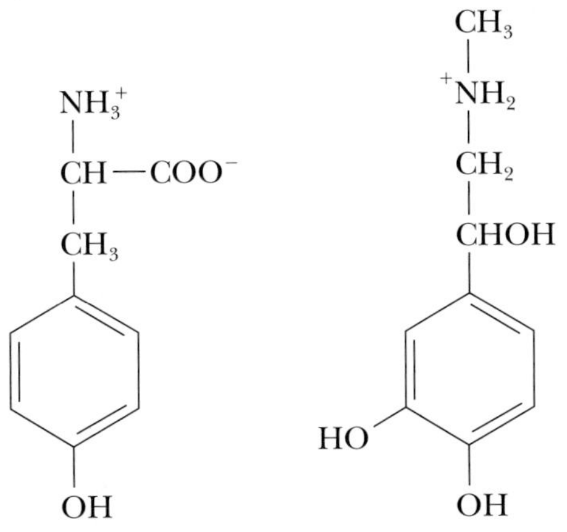

▲ Tyrosine and epinephrine. The hormone epinephrine is metabolically derived from the amino acid tyrosine.

glucagon, binds to its specific receptor, it activates a number of stimulatory G proteins. This effect starts an amplification of the hormonal signal. Each active G protein in turn stimulates adenylate cyclase several times before the G protein is inactivated by its own GTPase activity, leading to still more amplification. The cAMP produced by the increased adenylate cyclase activity allows for increased activity of the cAMP-dependent protein kinase, phosphorylating target enzymes that lead to increased glucose levels. In particular, this means an increase in the activity of the enzymes involved in gluconeogenesis and glycogen breakdown as well as a decrease in the activity of enzymes involved in glycolysis and glycogen synthesis. The series of amplifying steps is called a **cascade,** and the cumulative effect is the underlying reason why small amounts of hormones can have such marked effects.

Figure 24.14 shows how the binding of epinephrine to specific receptors leads to increased glycogen breakdown in muscle and suppression of glycogen synthesis. The hormonal stimulation leads to activation of adenylate cyclase, which in turn activates the cAMP-dependent protein kinase responsible for activating glycogen phosphorylase and inactivating glycogen synthase.

The effect of glucagon binding to receptors in stimulating gluconeogenesis in the liver and suppressing glycolysis depends on changes in the concentration of the key allosteric effector, fructose-2,6-*bis*phosphate (F2,6P). Recall from Section 18.3 that this compound is an important allosteric activator of phosphofructokinase, the key enzyme of glycolysis; it is also an inhibitor of fructose-*bis*phosphate phosphatase, which plays a role in gluconeogenesis. A high concentration of F2,6P stimulates glycolysis, whereas a low concentration stimulates gluconeogenesis. The concentration of F2,6P in a cell depends on the balance between its synthesis [catalyzed by phosphofructokinase-2 (PFK-2)] and its breakdown [catalyzed by fructose-*bis*phosphatase-2 (FBPase-2)]. The enzyme activities (on a single multifunctional protein) that control the formation and breakdown of F2,6P are themselves controlled by a phosphorylation/dephosphorylation mechanism, which in turn is subject to the same sort of hormonal control we just discussed for the enzymes of glycogen metabolism. Figure 24.15 summarizes the chain of events that leads to increased gluconeogenesis in the liver as a result of the binding of glucagon to its specific receptor. The following Biochemical Connections box discusses the role of insulin in overall metabolism and some current dieting trends.

Biochemical Connections

Insulin and Low-Carbohydrate Diets

In the 1970s, diets very high in carbohydrates became popular, both with athletes and with population at large. It was felt that a diet consisting of 60%–70% carbohydrates and 15%–20% each of fat and protein would be the healthiest (due to the high carbohydrate-to-fat ratio) and the best for athletes (due to the high levels of carbohydrates for replenishing glycogen). In the 1990s, newer diets that were based on a lower carbohydrate level became fashionable. Instead of a 70/15/15 ratio of carbohydrate/protein/fat, these diets recommended a 60/20/20 ratio or even a 50/25/25 ratio. The most notable of these diets was the one called the Zone Diet, promoted by Dr. Barry Sears. The idea behind such diets is

that a calorie is not always a calorie. In other words, the source of the calorie does matter. People had maintained high-carbohydrate diets in the belief that carbohydrates were healthier than fats, because of the myriad health problems that were attributed to too much dietary fat. Even though this seemed logical and few would argue in favor of the benefits of excessive fat, there is a possible downside to too many carbohydrates. For one thing, excess carbohydrates become fat. This may be a big consideration to nonathletes, who do not need to replenish muscle and liver glycogen as quickly and as often as an endurance athlete would. Also, a high-carbohydrate meal will stimulate the production of

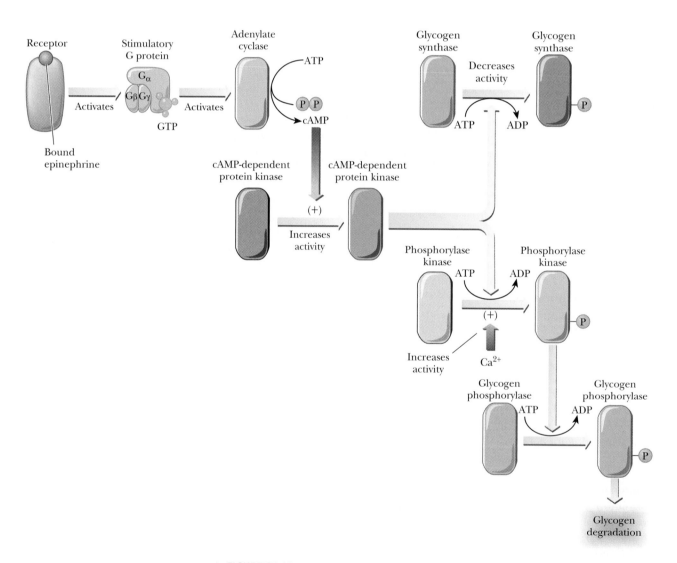

▲ **FIGURE 24.14** When epinephrine binds to its receptor, the binding activates a stimulatory G protein, which in turn activates adenylate cyclase. The cAMP thus produced activates a cAMP-dependent protein kinase. The phosphorylation reactions catalyzed by the cAMP-dependent kinase suppress the activity of glycogen synthase and enhance that of phosphorylase kinase. Glycogen phosphorylase is activated by phosphorylase kinase, leading to glycogen breakdown.

insulin. Insulin will inhibit the body's ability to use fat for energy and will stimulate the uptake of fat and its storage as triacylglycerol. A high-carbohydrate meal also has the potential of causing what is called **reactive hypoglycemia,** which occurs when high blood glucose stimulates a large insulin release, which then proceeds to clear too much glucose from the blood, causing a blood-sugar crash shortly thereafter. Many people find themselves weak, shaky, or sleepy by 10 A.M. after a high-carbohydrate breakfast. As we saw in Section 24.2, replacing fat with carbohydrates does nothing to increase the HDL/LDL ratio, either. The Zone Diet

was designed to avoid reactive hypoglycemia and the effects insulin has on fat storage. Because of the differences between fats and carbohydrates, many people may also tend to eat many more calories in the form of carbohydrates than they would in the form of fats, because carbohydrates do not give the same "filling" sensation as fats. Diet is a very personal thing. Many people have found that they feel better and actually lose weight while on a lower-carbohydrate diet. Others find just the opposite. There has been little evidence to suggest that a lower-carbohydrate diet is effective for athletes, however.

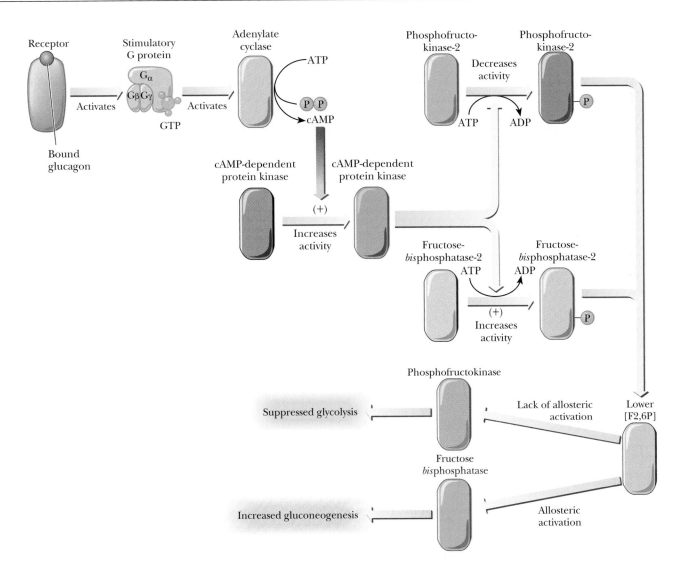

▲ **FIGURE 24.15** Binding of glucagon to its receptor sets off the chain of events that leads to the activation of a cAMP-dependent protein kinase. The enzymes phosphorylated in this case are phosphofructokinase-2, which is inactivated, and fructose-*bis*phosphatase-2, which is activated. The combined result of phosphorylating these two enzymes is to lower the concentration of fructose-2,6-*bis*phosphate (F2,6P). A lower concentration of F2,6P leads to allosteric activation of the enzyme fructose-*bis*phosphatase, thus enhancing gluconeogenesis. At the same time, the lower concentration of F2,6P implies that phosphofructokinase is lacking a potent allosteric activator, with the result that glycolysis is suppressed.

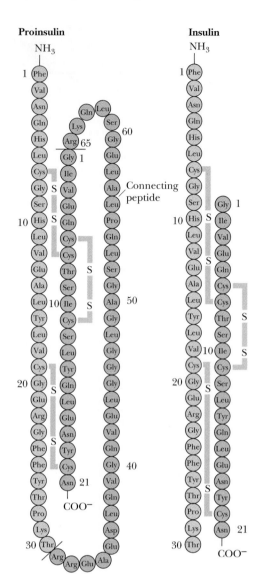

▲ FIGURE 24.16 Proinsulin is an 86-residue precursor to insulin (the sequence shown here is human proinsulin). Proteolytic removal of residues 31 through 65 yields insulin. Residues 1 through 30 (the B chain) remain linked to residues 66 through 86 by a pair of interchain disulfide bridges.

24.5 | What Are the Many Effects of Insulin?

To the average person, insulin is best known as the hormone that is deficient in people with diabetes, and it certainly was this relationship that spurred on the study of this fascinating hormone. We are just now realizing that insulin is involved in many cellular processes and in many different ways from what was previously thought.

Insulin Structure

Insulin is a peptide hormone secreted from the pancreas. In its active form, it is a 51-amino-acid peptide with two different chains, the A chain and the B chain, held together by disulfide bonds. As described in Chapter 13, insulin was one of the first proteins to be cloned and expressed for human need. Insulin is created as an 86-residue precursor, called proinsulin. Residues 31 to 65 of proinsulin are removed proteolytically to give the active form. The sequence of human insulin is shown in Figure 24.16.

Insulin Receptors

As we saw in Section 24.3, the insulin receptor is a member of the class of tyrosine receptor kinases. When insulin binds to receptor sites on the extracellular side of a cell membrane, it triggers the β-subunit to autophosphorylate a tyrosine residue on its interior portion, as shown in Figure 24.17. Once the tyrosines on the receptor are phosphorylated, the receptor then phosphorylates tyrosines on target proteins, called **insulin-receptor substrates (IRS),** which then act as the second messengers to produce a wide variety of cellular effects.

Insulin's Effect on Glucose Uptake

The body cannot tolerate great changes in the level of blood glucose. Insulin's primary job is to clear glucose out of the blood, and it does so by increasing the transport of glucose from the blood to muscle cells and adipocytes. Using mechanisms still being studied, insulin signaling leads to movement of a glucose-transporter protein called **GLUT4** from intracellular vesicles to the cell membrane. Once in the cell membrane, the GLUT4 protein allows more glucose to enter the cell, lowering the blood glucose level. It is this effect for which insulin is best known. Failure of glucose transport is the main characteristic of, and the acute risk associated with, diabetes.

Insulin Affects Many Enzymes

Insulin affects the activity of many enzymes, most of which are involved in getting rid of glucose. However, fat metabolism is also affected. Glucokinase is a liver enzyme that phosphorylates glucose to glucose-6-phosphate (see the Biochemical Connections box on page 147). It is induced by insulin, so that, when insulin is present, glucose in the liver will be sent toward catabolic pathways, such as pentose phosphate or glycolysis. Insulin also activates liver glycogen synthase and deactivates glycogen phosphorylase, causing glucose to be put into a polymeric form. In addition, insulin stimulates glycolysis through activation of phosphofructokinase and pyruvate dehydrogenase.

Insulin also has a large effect on fatty-acid metabolism. It increases fatty-acid synthesis via stimulation of acetyl-CoA carboxylase (ACC), and it increases triacylglycerol synthesis in the liver through activation of lipoprotein lipase. It also increases cholesterol synthesis via activation of hydroxymethylglutamyl-

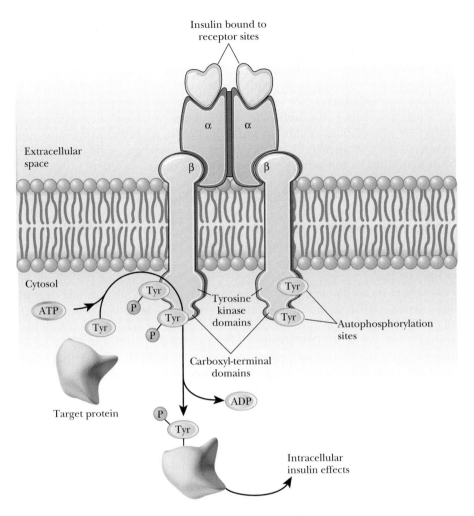

Insulin bound to
receptor sites

Extracellular
space

Cytosol

ATP

Target protein

Tyrosine
kinase
domains

Carboxyl-terminal
domains

Autophosphorylation
sites

ADP

Intracellular
insulin effects

◀ **FIGURE 24.17** The insulin receptor has two types of subunits, α and β. The α-subunit is on the extracellular side of the membrane, and it binds to insulin. The β-subunit spans the membrane. When insulin binds to the α-subunit, the β-subunits autophosphorylate on tyrosine residues. These then phosphorylate target proteins called insulin receptor substrates (IRS). These IRSs act as the second messengers in the cell. (*From Lehninger,* Principles of Biochemistry, *Second Edition, by David L. Nelson and Michael M. Cox. © 1982, 1992 by Worth Publishers. Used with permission of W. H. Freeman and Company.*)

CoA reductase (Section 21.8). Table 24.4 summarizes the effects of insulin on metabolism.

Diabetes

Much has been learned about insulin because of its relationship with diabetes. In classical, type I diabetes (or insulin-dependent diabetes), the afflicted individual does not make insulin, or at least not enough of it. This is usually caused by destruction of the beta cells of the islets of Langerhans in the pancreas from

Table 24.4			
Effect of Insulin on Metabolism			
Metabolic Process	**Location**	**Effect**	**Target**
Glucose uptake	Muscle	Increases	GLUT4 transporter
Glucose breakdown	Liver	Increases	Glucokinase
Glycolysis	Muscle and liver	Increases	PFK-1
Acetyl-CoA production	Muscle and liver	Increases	Pyruvate dehydrogenase
Glycogen synthesis	Muscle and liver	Increases	Glycogen synthase
Glycogen breakdown	Muscle and liver	Decreases	Glycogen phosphorylase
Fatty-acid synthesis	Liver and muscle	Increases	Acetyl-CoA varboxylase
Triacylglycerol dynthesis	Adipocytes	Increases	Lipoprotein lipase

a type of autoimmune disease. The only remedy for type I diabetes is regular insulin injections, and it is for this purpose that insulin is produced by recombinant DNA technology (Chapter 13).

The medical community is also concerned about the large increases in type II diabetes (or non–insulin-dependent diabetes), which is characterized by cells not responding correctly to insulin. In these cases, the person may make a normal amount of the hormone, but it does not have sufficient effect, either because it does not bind correctly to the receptor, or because the receptor does not correctly transmit the second messenger. This disease often begins later in life and is then called **adult-onset diabetes.** Where people with type I diabetes are often thin, people with type II diabetes are often obese. Evidence suggests that type II diabetes in the elderly is related to dysfunction of muscle mitochondria.

One of the most recent discoveries is that patients with type II diabetes also have an increased risk of Alzheimer's disease. In this type of diabetes, insulin is increased because it takes more insulin to accomplish the same clearance of glucose into the cells. Insulin appears to increase levels of the β-amyloid protein that forms plaques in the brain. A brain protein called **insulin degrading enzyme (IDE)** is involved in binding to and degrading insulin. This enzyme also binds to β-amyloid protein and clears it from the brain. When insulin levels are high, IDE spends more time tied up with insulin and less time clearing the β-amyloid protein. See the article by Taubes in the Annotated Bibliography for this chapter for more information about the link between insulin and Alzheimer's disease.

Insulin and Sports

Athletes must be able control their diets for maximum performance. Insulin plays a large role in the choice of a prerace breakfast for aerobic athletes. If an athlete eats a large, carbohydrate-filled breakfast in the morning, there will be a rise in blood glucose, followed by a rise in insulin. This often leads to a fall in blood glucose below the baseline level. It may take hours for the blood glucose levels to come back up. In addition, the high insulin level would cause activation of fat and glycogen synthesis and would inhibit glycogen breakdown. Thus, for a period of time after the high-carbohydrate breakfast, the athlete would be essentially running on empty. For this reason, many runners do not eat before a morning event, or, if they do, they eat little and avoid large amounts of high glycemic index foods. Athletes often drink coffee or tea in the morning. Besides the general stimulation of the central nervous system, caffeine also inhibits insulin production and stimulates fat mobilization.

Biochemical Connections

A Workout a Day Keeps Diabetes Away?

There seems to be a link between obesity and type II diabetes, although it is not clear whether diabetes leads to obesity or vice versa. The GLUT4 transporter is one of many glucose transporters, and it is the one most affected by insulin levels. It is also a protein whose levels can be affected by physical training. Studies have shown that one of the major changes associated with physical activity is an increase in the amount of GLUT4 in the muscle. In the trained state, a person will transport more glucose into the cell than when untrained. By definition, loss of function of glucose transport is type II diabetes. The training effect is such that it only takes a few days without training for the activity of GLUT4 to decrease to only half its normal level. Fortunately, the intensity of the training has less to do with the effect, at least in young to middle-aged people. With the apparent link between type II diabetes and obesity, one method of maintaining proper glucose transport appears to be staying light and fit.

Summary

24.1 How Are the Metabolic Pathways Connected?
All metabolic pathways are related, and some metabolites appear in several pathways. Furthermore, many reactions of metabolism can take place simultaneously. The citric acid cycle plays a central role in metabolism, in both catabolic and anabolic pathways. The breakdown products of sugars, fatty acids, and amino acids all enter the citric acid cycle.

24.2 How Can Biochemistry Help Us Understand Nutrition?
The sources of substrates for catabolism and for anabolism are the nutrients derived from foodstuffs. In humans, the choice of diet becomes important in the interest of obtaining enough of essential nutrients while avoiding excesses of others, such as saturated fats, where excess is known to play a role in the development of health problems. In 1992, a food guide pyramid was published to explain nutrition basics to the public. This pyramid is currently being replaced by a newer version that recognizes the differences between various types of fats and carbohydrates instead of just sending the message that all fats are bad and all carbohydrates are good.

24.3 What Are Hormones and Second Messengers?
Sophisticated fine tuning of metabolic processes in multicellular organisms is possible through the actions of hormones and second messengers. In humans, a complex hormonal system has evolved that requires releasing factors (under the control of the hypothalamus), trophic hormones (under the control of the pituitary), and specific hormones for target organs (under the control of endocrine glands). Feedback control occurs at every level of the system.

24.4 How Are Hormones Involved in the Control of Metabolism?
When a hormone binds to its receptor on the plasma membrane of a target cell, it sets off a cascade of reactions by which second messengers elicit the actual cellular response. Two of the most important second messengers, cyclic AMP (cAMP) and phosphatidylinositol-4,5-*bis*phosphate (PIP_2), activate protein kinases. Calcium ion is intimately involved in the action of PIP_2. Hormonal triggering can be added to other levels of control of metabolism, such as allosteric activation and covalent modification, to ensure an efficient response to the needs of the organism.

24.5 What Are the Many Effects of Insulin?
Insulin's primary job is to stimulate the glucose transporters in muscle—particularly the GLUT4 transporter—to take up glucose from the blood. In addition, it has a wide range of intracellular effects, such as switching off glycogen breakdown and turning on glycogen synthesis, stimulating glycolysis in the liver and muscle, turning off gluconeogenesis in the liver, and stimulating fatty-acid synthesis and storage. A recent discovery is that elevated levels of insulin in the blood may be related to Alzheimer's disease.

Critical Questions to Review

Hint: You may want to review material in Chapters 17 through 23.

24.1 How Are the Metabolic Pathways Connected?

1. **Fact Check** What are the two primary molecules that link anabolic and catabolic reactions?

2. **Fact Check** Name some of the key metabolic intermediates that are seen in more than one pathway.

3. **Fact Check** Many components of the glycolytic pathway and the citric acid cycle are direct exit or entry points to metabolic pathways of other substances. Indicate another pathway available to the following compounds.

    (a) Fructose-6-phosphate (b) Oxaloacetate (c) Glucose-6-phosphate (d) Acetyl-CoA (e) Glyceraldehyde-3-phosphate (f) α-Ketoglutarate (g) Dihydroxyacetone phosphate (h) Succinyl-CoA (i) 3-Phosphoglycerate (j) Fumarate (k) Phosphoenolpyruvate (l) Citrate (m) Pyruvate

4. **Thought Question** People who begin to lose weight often have a rapid weight loss in the first few days. Common knowledge says that this is "just" because of a loss of water from the body. Why might this be true?

5. **Thought Question** The functioning of a particular pathway often depends not only on control enzymes in that pathway but also on control enzymes of other pathways. What happens in the following pathways under the indicated conditions? Suggest what other pathway or pathways might be influenced.

    (a) High ATP or NADH concentration and the citric acid cycle. (b) High ATP concentration and glycolysis. (c) High NADPH concentration and the pentose phosphate pathway. (d) High fructose-2,6-*bis*phosphate concentration and gluconeogenesis.

6. **Thought Question** Why is it somewhat misleading to study biochemical pathways separately?

7. **Thought Question** To what extent can metabolic pathways be considered reversible? Why?

8. **Thought Question** In eukaryotic cells, metabolic pathways occur in specific locations, such as the mitochondrion or the cytosol. What sort of transport processes are required as a result?

9. **Thought Question** Why is it advantageous for a metabolic pathway to have a large number of steps?

10. **Conjectural Question** If you had your choice of doing research on any topic in this book, which would you choose? Why do you consider that topic to be interesting and important?

24.1 How Can Biochemistry Help Us Understand Nutrition?

11. **Fact Check** What is the difference between the old food pyramid and the new version?

12. **Fact Check** What do we mean when we say that there is no storage form for protein? How is this different from fats and carbohydrates?

13. **Fact Check** What is the relationship between saturated fatty acids and LDL?

14. **Fact Check** What is leptin, and how does it work?

15. **Fact Check** Many have suggested that vitamin D could be more appropriately called a hormone than a vitamin. Is this correct?

16. **Thought Question** Recent recommendations on diet suggest that the sources of calories should be distributed as follows: 50%–55% carbohydrate, 25%–30% fats, and 20% protein. Suggest some reasons for these recommendations.

17. **Biochemical Connections** People who are both alcoholic and exposed to halogen compounds often die of liver failure. Why might this be a logical ultimate result?

18. **Thought Question** It has been suggested that limits be put on the dose in vitamin A supplements sold in stores. What is a possible reason for this limitation?

19. **Fact Check** In the early 20th century, goiter was relatively common in the Midwest. Why was this so? How has it been eliminated?

20. **Thought Question** A cat named Lucullus is so spoiled that he will eat nothing but freshly opened canned tuna. Another cat, Griselda, is given only dry cat food by her far less indulgent owner. Canned tuna is essentially all protein, whereas dry cat food can be considered 70% carbohydrate and 30% protein. Assuming that these animals have no other sources of food, what can you say about the differences and similarities in their catabolic activities? (The pun is intended.)

21. **Thought Question** Immature rats are fed all the essential amino acids but one. Six hours later they are fed the missing amino acid. The rats fail to grow. Explain this observation.

22. **Thought Question** Kwashiorkor is a protein-deficiency disease that occurs most commonly in small children, who characteristically have thin arms and legs and bloated, distended abdomens due to fluid imbalance. When such children are placed on adequate diets, they tend to lose weight at first. Explain this observation.

23. **Thought Question** Why are amino acids such as arginine and histidine required in relatively large amounts by children but in smaller amounts by adults? The adult human is not able to make these amino acids.

24. **Biochemical Connections** During colonial times, iron-deficiency anemia was almost unknown in America. Why? *Hint:* The answer has nothing to do with what foods they ate.

25. **Thought Question** People on high-fiber diets often have less cancer (especially of the colon) and lower blood-cholesterol levels than people on low-fiber diets, even though fiber is not digestible. Suggest reasons for the benefits of fiber.

26. **Thought Question** Most calcium supplements have calcium carbonate as the main ingredient. Other supplements that have calcium citrate as the main ingredient are advertised as being more easily absorbed. Do you consider this a valid claim? Why?

27. **Biochemical Connections** Alcoholics tend to be malnourished, with thiamine deficiency being a particularly severe problem. Suggest a reason why this is so.

28. **Thought Question** Biologically and nutritionally important trace elements tend to be metals. What is their likely biochemical function?

29. **Thought Question** An athletic friend is preparing to run a marathon and intends to load glycogen before the race. Someone told your friend that the way to load more glycogen is by exercising excessively for two days to deplete the glycogen stores completely, and that is what your friend intends to do. What do you say about this regimen?

30. **Thought Question** Over a period of several decades, an adult human consumes tons of nutrients and more than 20,000 liters of water without significant weight gain. How is this possible? Is it an example of chemical equilibrium?

24.3 What Are Hormones and Second Messengers?

31. **Fact Check** Are all hormones closely related in their chemical structure?

32. **Fact Check** How is hormone production affected by damage to the pituitary gland? To the adrenal cortex?

33. **Fact Check** How do the actions of the hypothalamus and the pituitary gland affect the workings of endocrine glands?

34. **Fact Check** The hormone thyroxine is given as an oral dose, but insulin needs to be injected into the body. Why?

35. **Biochemical Connections** What is p21^{ras}?

36. **Fact Check** What is the difference between a G protein and a receptor tyrosine kinase? Give an example of a hormone that uses each.

37. **Fact Check** Give three examples of second messengers.

38. **Thought Question** The average male with a computer hooked up to the Internet receives thousands of spam e-mails for all kinds of things, such as Viagra, weight-loss pills, and others not fit to print. Among the spam, one can find offers for human growth hormone pills guaranteed to make one young again. Why is this unlikely?

24.4 How Are Hormones Involved in the Control of Metabolism?

39. **Fact Check** List two hormones that work through the cAMP second messenger.

40. **Fact Check** How does glucagon affect the following enzymes:
    (a) Glycogen phosphorylase (b) Glycogen synthase (c) Phosphofructokinase I

41. **Fact Check** How does epinephrine affect the enzymes listed in Question 40?

42. **Fact Check** What stops a hormone response when a G protein is involved?

43. **Thought Question** When PIP$_2$ is hydrolyzed, why does IP$_3$ diffuse into the cytosol while DAG remains in the membrane?

44. **Thought Question** Briefly describe the series of events that takes place when cAMP acts as a second messenger.

45. **Thought Question** For each of three hormones discussed in this chapter, give its source and chemical nature; also discuss the mode of action of each hormone

46. **Thought Question** Is it likely that any metabolic pathway can exist without control mechanisms?

47. **Thought Question** Cholera harms the body by its effect on a second messenger. Describe how this takes place.

48. **Thought Question** The epinephrine-mediated "amplification cascade" of Figure 24.14 has six steps, all of which are catalytic with one exception. This cascade leads to the activation of glycogen phosphorylase. This enzyme acts in turn on glycogen to yield glucose-1-phosphate (G-1-P).
    (a) Which step is not catalytic? (b) If each catalytic step had a turnover (molecules of substrate acted on per molecule of enzyme) of 10, how many molecules of G-1-P would result from one molecule of epinephrine? (c) What is the biochemical advantage of such a cascade?

49. **Thought Question** How is the amplification cascade of Question 48 reversed?

50. **Biochemical Connections** Explain what insulin and low-carbohydrate diets have to do with one another.

24.5 What Are the Many Effects of Insulin?

51. **Fact Check** What is the primary function of insulin?

52. **Fact Check** What is the second messenger for the insulin response?

53. **Fact Check** What is the link between insulin binding to the receptor and the eventual second messenger?

54. **Fact Check** What is insulin's effect on the following?
    (a) Glycogen breakdown (b) Glycogen synthesis (c) Glycolysis (d) Fatty-acid synthesis (e) Fatty-acid storage

55. **Thought Question** How is it possible that both insulin and epinephrine stimulate muscle glycolysis?

56. **Thought Question** Why would a runner who has a 5-km race to run at 9 a.m. be concerned about insulin?

57. **Biochemical Connections** Why do some people call GLUT4 the training glucose transporter?

58. **Thought Question** How are insulin, GLUT4, obesity, and type II diabetes related?

Biochemistry ⊘ Now™

Assess your understanding of this chapter's topics with additional quizzing and tutorials at **http://now.brookscole.com/campbell5**

Annotated Bibliography

Bose, A., A. Guilherme, S. I. Robida, S. M. C. Nicoloro, Q. L. Zhou, Z. Y. Jiang, D. P. Pomerleau, and M. P. Czech. Glucose Transporter Recycling in Response to Insulin Is Facilitated by Myosin Myo1c. *Nature* **420,** 821–824 (2002). [An article about the transcription factors that have been shown to affect glucose transport.]

Cohen, P., M. Miyazaki, N. D. Socci, A. Hagge-Greenberg, W. Liedtke, A. A. Soukas, R. Sharma, L. C. Hudgins, J. M. Ntambi, and J. M. Friedman. Role for Stearoyl-CoA Desaturase-1 in Leptin-Mediated Weight Loss. *Science* **297,** 240–243 (2002). [An article about one mechanism by which leptin affects weight loss.]

Cowley, M. A., J. L. Smart, M. Rubinstein, M. G. Cerdan, S. Diano, T. L. Horvath, R. D. Cone, and M. J. Low. Leptin Activates Anorexigenic POMC Neurons through a Neural Network in the Arcuate Nucleus. *Nature* **411,** 480–484 (2001). [An article about leptin and obesity.]

Friedman, J. Fat in All the Wrong Places. *Nature* **415,** 268–269 (2002). [A review article about obesity.]

Gura, T. Obesity Sheds Its Secrets. *Science* **275,** 751–753 (1997). [A Research News article on the role of the protein hormone leptin in obesity research and on the possibility of obesity therapies that might arise from this research.]

Minokoshi, Y., Y. B. Kim, D. P. Odile, L. G. D. Fryer, C. Muller, D. Carling, and B. B. Kahn. Leptin Stimulates Fatty-Acid Oxidation by Activating AMP-Activated Protein Kinase. *Nature* **415,** 339–343 (2002). [An article about another mechanism for leptin-associated weight loss.]

Rasmussen, H. The Cycling of Calcium as an Intracellular Messenger. *Sci. Amer.* **261** (4), 66–73 (1989). [An article on the role of calcium as a second messenger.]

Schwartz, M. W., and G. J. Morton. Keeping Hunger at Bay. *Nature* **418,** 595–597 (2002). [An article about hunger, hormones, and weight loss.]

Taubes, G. Insulin Insults May Spur Alzheimer's Disease. *Science* **301,** 40–41 (2003). [A recent article linking Alzheimer's disease and high insulin levels]

White, M. F. Insulin Signaling in Health and Disease. *Science* **302,** 1710–1711 (2003). [A review of insulin and its effects on health.]

Willett, W. Diet and Health: What Should We Eat? *Science* **264,** 532–537 (1994). [An excellent summary of many aspects of a complex topic.]

Willett, W. C., and M. J. Stampfer. Rebuilding the Food Pyramid. *Sci. Amer.* **288**(1), 64–71 (2003). [A thorough explanation of the changes in nutrition recommendations since the old food pyramid was published.]

See also the bibliographies for Chapters 17 through 23.

abzyme an antibody that is produced against a transition-state analog and that has catalytic activity similar to that of a naturally occurring enzyme (7.7)

accessory pigments plant pigments other than chlorophyll that play a role in photosynthesis (22.1)

acid dissociation constant a number that characterizes the strength of an acid (2.3)

acidic domain a common motif in the transcription-activation domain portion of a transcription factor (11.5)

acid strength the tendency of an acid to dissociate to a hydrogen ion and its conjugate base (2.3)

acromegaly a disease caused by an excess of growth hormone after the skeleton has stopped growing and characterized by enlarged hands, feet, and facial features (24.3)

activation energy the energy required to start a reaction (6.2)

activation step the beginning of a multistep process in which a substrate is converted to a more reactive compound (12.1)

active site the part of an enzyme to which the substrate binds and at which the reaction takes place (6.4)

active transport the energy-requiring process of moving substances into a cell against a concentration gradient (8.6)

acyl carrier protein a protein that functions in fatty-acid synthesis to carry activated carbon groups (21.6)

acyl-CoA synthetase the enzyme that catalyzes the activation step in lipid catabolism (21.6)

adenine one of the purine bases found in nucleic acids (9.2)

S-adenosylmethionine an important carrier molecule of methyl groups in amino acid metabolism (23.4)

adenylate cyclase the enzyme that catalyzes the production of cyclic AMP (24.3)

A-DNA a form of a DNA double helix characterized by having fewer residues per turn and major and minor grooves with dimensions that are more similar to each other than those of B-DNA (9.3)

ADP (adenosine diphosphate) a compound that can serve as an energy carrier when it is phosphorylated to form ATP (15.6)

adrenocortical hormones steroid hormones secreted by the adrenal cortex that have an effect on inflammation and salt and water balance (24.3)

affinity chromatography a powerful column separation procedure based on specific binding of molecules to a ligand (5.2)

agarose a complex polysaccharide used to make up resins for use in electrophoresis and in column chromatography (5.2)

alcoholic fermentation the anaerobic pathway that converts glucose to ethanol (17.1)

aldolase in glycolysis, the enzyme that catalyzes the reverse aldol condensation of fructose-1,6-*bis*phosphate (17.2)

aldose a sugar that contains an aldehyde group as part of its structure (16.1)

alleles corresponding genes on paired chromosomes (13.8)

allosteric the property of multisubunit proteins such that a conformational change in one subunit induces a change in another subunit (4.7)

allosteric effector a substance—substrate, inhibitor, or activator—that binds to an allosteric enzyme and affects its activity (7.4)

amino acid any of the fundamental building blocks of proteins; molecules that contain an amino group and a carboxyl group (3.1)

amino acid activation the formation of an ester bond between an amino acid and its specific tRNA that is catalyzed by a suitable synthetase (12.1)

amino acid analyzer an instrument that gives information on the number and kind of amino acids in a protein (5.4)

aminoacyl-tRNA synthetases enzymes that catalyze the formation of an ester linkage between an amino acid and tRNA (12.1)

amino group the NH_2 functional group (3.1)

amphibolic able to be a part both of anabolism and catabolism (19.1)

amphipathic refers to a molecule that has one end with a polar, water-soluble group and another end with a nonpolar hydrocarbon group that is insoluble in water (2.1)

amylopectin a form of starch; a branched-chain polymer of glucose (16.4)

amylose a form of starch; a linear polymer of glucose (16.4)

anabolism the synthesis of biomolecules from simpler compounds (15.3)

anaerobic glycolysis the pathway of conversion of glucose to lactate; distinguished from glycolysis, which is the conversion of glucose to pyruvate (17.1)

analytical ultracentrifugation the technique for observing the motion of particles as they sediment in a centrifuge (10.2)

anaplerotic referring to a reaction that ensures an adequate supply of an important metabolite (19.8)

anion-exchange chromatography the type of ion-exchange chromatography in which the column ligand has a net positive charge and binds to negatively charged molecules flowing through it (5.2)

anomer one of the possible stereoisomers formed when a sugar assumes the cyclic form (16.1)

anomeric carbon the chiral center created when a sugar cyclizes (16.1)

antibody a glycoprotein that binds to and immobilizes a substance that the cell recognizes as foreign (14.5)

anticodon the sequence of three bases (triplet) in tRNA that hydrogen-bonds with the mRNA triplet that specifies a given amino acid (12.2)

antigen a substance that triggers an immune response (14.5)

antigenic determinant the portion of a molecule that antibodies recognize as foreign and to which they bind (14.5)

antioxidant a strong reducing agent, which is easily oxidized and thus prevents the oxidation of other substances (8.7)

antisense strand the DNA strand that is used as a template for RNA synthesis (11.1)

2:3 antiterminator a hairpin loop that can form during transcription attenuation, allowing transcription to continue (11.2)

AP (apurinic) site a site, lacking a purine base in DNA, that is targeted by repair enzymes (10.5)

arachidonic acid a fatty acid that contains twenty carbon atoms and four double bonds; the precursor of prostaglandins and leukotrienes (8.8)

aspartate transcarbamoylase (ATCase) a classic example of an allosteric enzyme that catalyzes an early reaction in pyrimidine biosynthesis (6.5)

ATP (adenosine triphosphate) a universal energy carrier (15.6)

ATP synthase the enzyme responsible for production of ATP in mitochondria (20.4)

attenuation a type of transcription control in which the transcription is controlled after it has begun via pausing and early release of incomplete RNA sequences (11.2)

autoimmune diseases diseases in which the immune system attacks the body's own tissues (14.5)

autoradiography the technique of locating radioactively labeled substances by allowing them to expose photographic film (13.1)

bacterial plasmid a portion of circular DNA separate from the main genome of the bacterium (13.3)

bacteriophage a kind of virus that infects bacteria; bacteriophages are frequently used in molecular biology to transfer DNA between cells (13.2)

β-barrel a β-pleated sheet extensive enough to fold back on itself (4.3)

base-excision repair a process for repairing damaged DNA (10.5)

base stacking interactions between bases that are next to each other in a DNA chain (9.3)

basic-region leucine zipper a common motif found in transcription factors (11.5)

B-cell a type of white blood cell that plays an important role in the immune system and plays a role in the production of antibodies (14.5)

B-DNA the most common form of the DNA double helix (9.3)

binding assay an experimental method for selecting one molecule out of a number of possibilities by specific binding; it is used to determine the nature of many triplets of the genetic code (12.2)

bioinformatics the application of computer methods to processing large amounts of information in biochemistry (4.6)

biotin a CO_2 carrier molecule (18.2)

blotting a technique for transferring a portion of a sample to a membrane for further analysis (13.8)

blue/white screening a method for determining whether bacterial cells have incorporated a plasmid that includes a gene of interest (13.3)

Bohr effect the decrease in oxygen binding by hemoglobin caused by binding of carbon dioxide and hydrogen ion (4.7)

buffering capacity a measure of the amount of acid or base that reacts with a given buffer solution (2.6)

buffer solution a solution that resists a change in pH on addition of moderate amounts of strong acid or strong base (2.6)

Calvin cycle the pathway of carbon dioxide fixation in photosynthesis (22.5)

capsid the protein coat of a virus (14.1)

5′-cap a structure found at the 5′ end of eukaryotic mRNA (11.5)

carboxyl group the —COOH functional group that dissociates to give the carboxylate anion, —COO⁻, and a hydrogen ion (3.1)

β-carotene an unsaturated hydrocarbon; the precursor of vitamin A (8.7)

carrier protein a membrane protein to which a substance binds in passive transport into the cell (8.6)

cascade a series of steps that take place in hormonal control of metabolism, affecting a series of enzymes and amplifying the effect of a small amount of hormone (24.4)

catabolism the breakdown of nutrients to provide energy (15.3)

catabolite activator protein (CAP) a protein that can bind to a promoter when complexed with cAMP, allowing RNA polymerase to bind to its entry site on the same promoter (11.2)

catabolite repression repression of the synthesis of *lac* proteins by glucose (11.2)

catalysis the process of increasing the rate of chemical reactions (1.3)

cation-exchange chromatography the type of ion exchange chromatography in which the column resin has a net negative charge and binds to positively charged molecules flowing through the column (5.2)

cDNA complementary DNA; a form of DNA synthesized on an mRNA template, thus directly reflecting the coding sequence (13.5)

cell membrane the outer membrane of the cell that separates it from the outside world (1.5)

cellulose a polymer of glucose; an important structural material in plants (16.4)

cell wall the outer coating of bacterial and plant cells (1.5)

ceramide a lipid that contains one fatty acid linked to sphingosine by an amide bond (8.2)

cerebroside a glycolipid that contains sphingosine and a fatty acid in addition to the sugar moiety (8.2)

chaperone a protein that helps another protein fold into the correct three-dimensional structure and prevents it from associating incorrectly with other proteins (4.6)

chemiosmotic coupling the mechanism for coupling electron transport to oxidative phosphorylation; it requires a proton gradient across the inner mitochondrial membrane (20.5)

chimeric DNA DNA from more than one species covalently linked together (13.3)

chiral refers to an object that is not superimposable on its mirror image (3.1)

chlorophyll the principal photosynthetic pigment responsible for trapping light energy from the sun (22.1)

chloroplast the organelle that is the site of photosynthesis in green plants (1.6, 22.1)

cholera a disease caused by the bacterium *Vibrio cholerae* and characterized by dehydration due to excessive Na^+ transport in epithelial cells (24.3)

cholesterol a steroid that occurs in cell membranes; the precursor of other steroids (8.2, 21.8)

chromatin a complex of DNA and protein found in eukaryotic nuclei (1.6)

chromatography an experimental method for separating substances based on their molecular character (5.2)

chromosome a linear structure that contains the genetic material and associated proteins (1.6)

chymotrypsin a proteolytic enzyme that preferentially hydrolyzes amide bonds adjacent to aromatic amino acid residues (5.4)

cis–trans isomerase an enzyme that catalyzes a *cis–trans* isomerization in the catabolism of unsaturated fatty acids (21.2)

citrate synthase the enzyme that catalyzes the first step of the citric acid cycle (19.4)

citric acid cycle a central metabolic pathway; part of aerobic metabolism (19.1)

clonal selection the process by which the immune system responds selectively to antibodies actually present in an organism (14.5)

clone a genetically identical population of organisms, cells, viruses, or DNA molecules (13.3)

cloning of DNA the introduction of a section of DNA into a genome in which it can be reproduced many times (13.3)

closed complex the complex that initially forms between RNA polymerase and DNA before transcription begins (11.1)

coding strand the DNA strand that has the same sequence as the RNA that is synthesized from the template (11.1)

codon sequence of three bases on mRNA that specifies a given amino acid (12.2)

coenzyme a nonprotein substance that takes part in an enzymatic reaction and is regenerated at the end of the reaction (7.8)

coenzyme A a carrier of carboxylic acids bound to its thiol group by a thioester linkage (15.7)

coenzyme Q an oxidation–reduction coenzyme in mitochondrial electron transport (20.3)

column chromatography a form of chromatography in which the stationary phase is packed in a column (5.2)

committed step in a metabolic pathway, the formation of a substance that can play no other role in metabolism but to undergo the rest of the reactions of the pathway (17.2)

competitive inhibition a decrease in enzymatic activity caused by binding of a substrate analogue to the active site (6.7)

complementary refers to the specific hydrogen bonding of adenine with thymine (or uracil) and guanine with cytosine in nucleic acids (9.3)

complete protein a protein that contains all the essential amino acids (4.2)

concerted model a description of allosteric activity in which the conformations of all subunits change simultaneously (7.2)

configuration the three-dimensional arrangement of groups around a chiral carbon atom (16.1)

conformational coupling a mechanism for coupling electron transport to oxidative phosphorylation that depends on a conformational change in the ATP synthetase (20.5)

consensus sequences DNA sequences to which RNA polymerase binds; they are identical in many organisms (11.3)

constitutive expression the transcription and expression of genes that are not controlled by anything other than the inherent binding of the RNA polymerase to the promoter (11.2)

control sites the operator and promoter elements that modulate the production of proteins whose amino acid sequence is specified by the structural genes under their control (11.2)

cooperative binding binding to several sites such that, when the first ligand is bound, the binding of subsequent ones is easier (4.7)

cooperative transition a transition that takes place in an all-or-nothing fashion, such as the melting of a crystal (4.7)

co-repressor a substance that binds to a repressor protein making it active and able to bind to an operator gene (11.2)

core promoter in prokaryotic transcription, the portion of the DNA from the transcription start site to the −35 region (11.1)

Cori cycle a pathway in carbohydrate metabolism that links glycolysis in the liver with gluconeogenesis in the liver (18.3)

coronavirus any member of the virus family Coronaviridae, which includes the virus that causes SARS (14.2)

coupled translation in prokaryotes, the situation in which a gene is simultaneously transcribed and translated (12.4)

coupling the process by which an exergonic reaction provides energy for an endergonic one (15.6)

C-terminal the end of a protein or peptide with a carboxyl group not bonded to another amino acid (3.4)

C-terminal domain the region of a protein at the C-terminus, especially important in prokaryotic RNA polymerase B (11.3)

cut and patch a mechanism for repair of DNA by enzymatically removing incorrect nucleotides and substituting correct ones (10.5)

cyanogen bromide a reagent that cleaves proteins at internal methionine residues (5.4)

cyclic AMP a nucleotide in which the same phosphate group is esterified to the 3′ and 5′ hydroxyl groups of a single adenosine; an important second messenger (24.3)

cyclic-AMP-response element (CRE) an important eukaryotic response element that is controlled by production of cAMP in the cell (11.4)

cyclic-AMP-response-element binding protein (CREB) an important transcription factor in eukaryotes that binds to the CRE and activates transcription (11.4)

cyclins proteins that play an important role in control of the cell cycle by regulating the activity of kinases (10.6)

cytochrome any one of a group of heme-containing proteins in the electron transport chain (20.3)

cytokines soluble protein factors produced by one cell that affect another cell (14.5)

cytosine one of the pyrimidine bases found in nucleic acids (9.2)

cytoskeleton (microtrabecular lattice) a lattice of fine strands, consisting mostly of protein, that pervades the cytosol (1.6)

cytosol the portion of the cell that lies outside the nucleus and the other membrane-enclosed organelles (1.6)

debranching enzyme an enzyme that hydrolyzes the linkages in a branched-chain polymer such as amylopectin (18.1)

degenerate code a genetic code in which more than one triplet of bases can code for the same amino acid (12.2)

denaturation the unraveling of the three-dimensional structure of a macromolecule caused by the breakdown of noncovalent interactions (4.5)

dendritic cells components of the innate immune system (14.5)

denitrification the process by which nitrates and nitrites are broken down to molecular nitrogen (23.1)

density-gradient centrifugation the technique of separating substances in an ultracentrifuge by applying the sample to the top of a tube that contains a solution of varying densities (10.2)

deoxyribonucleoside a compound formed when a nucleobase and deoxyribose form a glycosidic bond (9.2)

deoxyribose a sugar that is part of the structure of DNA (9.2)

deoxy sugar a sugar in which one of the hydroxyl groups has been reduced to a hydrogen (16.2)

dextran a complex polysaccharide that is often used in column chromatography resins (5.2)

diastereomers nonsuperimposable, nonmirror-image stereoisomers (16.1)

dimer a molecule consisting of two subunits (4.7)

dipoles molecules with positive and negative ends due to an uneven distribution of electrons in bonds (2.1)

disaccharide two monosaccharides (monomeric sugars) linked by a glycosidic bond (16.3)

DNA deoxyribonucleic acid; the molecule that contains the genetic code (9.3)

DNA-binding domain the part of a transcription factor that binds to the DNA (11.5)

DNA chip a microarray of DNA samples on a single computer chip, on which many samples can be examined simultaneously (13.10)

DNA footprinting a method for studying protein–DNA interactions by comparing the digestion of DNA by nucleases in the presence and absence of the protein of interest (13.9)

DNA gyrase an enzyme that introduces supercoiling into closed circular DNA (9.4)

DNA library a collection of clones that include the total genome of an organism (13.5)

DNA ligase the enzyme that links separate stretches of DNA (10.3)

DNA polymerase the enzyme that forms DNA from deoxyribonucleotides on a DNA template (10.3)

DNase deoxyribonuclease; an enzyme that specifically hydrolyzes DNA (13.2)

domain a portion of a polypeptide chain that folds independently of other portions of the chain (4.3)

downstream in transcription, a portion of the DNA sequence that is nearer the 5′ end in the gene to be transcribed, where the DNA is read from the 3′ to the 5′ end and the RNA is formed from the 5′ to the 3′ end; in translation, a portion of the RNA sequence that is nearer the 3′ end of mRNA (13.1)

dwarfism a disease caused by a deficiency of growth hormone (24.4)

Edman degradation a method for determining the amino acid sequence of peptides and proteins (5.4)

electronegativity a measure of the tendency of an atom to attract electrons to it in a chemical bond (2.1)

electron transport to oxygen a series of oxidation–reduction reactions by which the electrons derived from oxidation of nutrients are passed to oxygen (19.1)

electrophile an electron-poor substance that tends to react with centers of negative charge or polarization (7.5)

electrophoresis a method for separating molecules on the basis of the ratio of charge to size (3.3)

electrophoretic mobility-shift assay (EMSA) a method for studying DNA–protein interactions (13.9)

−35 element (or −35 region) a portion of DNA that is 35 base pairs upstream from the start of RNA transcription that is important in control of RNA synthesis in bacteria (11.1)

elongation step in protein synthesis, the succession of reactions in which the peptide bonds are formed (12.4)

enantiomers mirror-image, nonsuperimposable stereoisomers (16.1)

endergonic energy-absorbing (1.11)

endocrine system the system of ductless glands that release hormones into the bloodstream (24.3)

endocytosis the process by which materials are brought into the cell when portions of a cell membrane are pinched off into the cell (8.6)

endonuclease an enzyme that hydrolyzes nucleic acids, attacking linkages in the middle of the polynucleotide chain (13.2)

endoplasmic reticulum (ER) a continuous single-membrane system throughout the cell (1.6)

endosymbiosis a symbiotic relationship in which a smaller organism is completely contained within a larger organism (1.8)

enhancer a DNA sequence that binds to a transcription factor and increases the rate of transcription (11.2)

enthalpy a thermodynamic quantity measured as the heat of reaction at constant pressure (1.11)

entropy a thermodynamic quantity; a measure of the energy dispersal of the universe (1.11)

enzyme a biological catalyst, usually a globular protein, with self-splicing RNA as the only exception (6.1)

epimerase an enzyme that catalyzes the inversion of configuration around a single carbon atom

epimers stereoisomers that differ only in configuration around one of several chiral carbon atoms (16.1)

epitope a binding site for an antibody on an antigen (14.5)

equilibrium the state in which a forward process and a reverse process occur at the same rate (1.11)

essential amino acids amino acids that cannot be synthesized by the body and must therefore be obtained in the diet (23.4)

essential fatty acids the polyunsaturated fatty acids (such as linoleic acid) that the body cannot synthesize; they must be obtained from dietary sources (21.6)

eukaryote an organism whose cells have a well-defined nucleus and membrane-enclosed organelles (1.4)

excision repair repair of DNA by the enzymatic removal of incorrect nucleotides and their replacement by the correct ones (10.5)

exergonic energy-releasing (1.11)

exon a DNA sequence that is expressed in the sequence of mRNA (11.6)

exonuclease an enzyme that hydrolyzes nucleic acids, starting at the end of the polynucleotide chain (13.2)

expression cassette in gene therapy, the assemblage that contains the gene being transferred (14.4)

expression vector a plasmid that has the machinery to direct the synthesis of a desired protein (13.4)

extended promoter in prokaryotic transcription, the DNA from the transcription start site to the UP element (11.1)

facilitated diffusion a process by which substances enter a cell by binding to a carrier protein; this process does not require energy (8.6)

fatty acid a compound with a carboxyl group at one end and a long, normally unbranched hydrocarbon tail at the other; the hydrocarbon tail may be saturated or unsaturated (8.1)

feedback inhibition the process by which the final product of a series of reactions inhibits the first reaction in the series (7.1, 23.3)

fibrous protein a protein whose overall shape is that of a long, narrow rod (4.3)

filter-binding assay a method used to determine the base sequence of many mRNA codons (12.2)

first-order reaction a reaction whose rate depends on the first power of the concentration of a single reactant (6.3)

Fischer projection a two-dimensional representation of the stereochemistry of three-dimensional molecules (16.1)

fis site an enhancer found in prokaryotic transcription of rRNA (11.2)

fluid-mosaic model the model for membrane structure in which proteins and a lipid bilayer exist side by side without covalent bonds between the proteins and lipids (8.5)

fluorescence a sensitive method for detection and identification of substances that absorb and reemit light (13.1)

folate reductase the enzyme that reduces dihydrofolate to tetrahydrofolate, a target for cancer chemotherapy (23.11)

fold purification a measurement of increased purity taken during a protein-purification experiment (5.1)

N-formylmethionine-tRNA^{fmet} an essential factor for the start of prokaryotic protein synthesis, the amino acid methionine formylated at the amino group and covalently bonded to its specific tRNA (12.4)

free energy a thermodynamic quantity; diagnostic for the spontaneity of a reaction at constant temperature (1.11)

functional group one of the groups of atoms that give rise to the characteristic reactions of organic compounds (1.2)

Fungi one of the five kingdoms into which living organisms are commonly classified; it includes molds and mushrooms (1.7)

furanose a cyclic sugar with a six-membered ring, named for its resemblance to the ring system in furan (16.1)

furanoside a glycoside involving a furanose (16.2)

fusion protein one that has had extra amino acids added to one of its ends (13.4)

β-galactosidase the enzyme that hydrolyzes lactose to galactose and glucose, the classic example of an inducible enzyme (11.2)

gel electrophoresis a method for separating molecules on the basis of charge-to-size ratio using a gel as a support and sieving material (5.3)

gel-filtration chromatography a type of column chromatography in which the molecules are separated according to size as they pass through the column (5.2)

gene an individual unit of inheritance (1.4)

gene therapy a method for treating a genetic disease by introducing a good copy of a defective gene (13.4)

general acid–base catalysis a form of catalysis that depends on transfer of protons (7.6)

genetic code the information for the structure and function of all living organisms (1.3)

genome the total DNA of the cell (1.4)

gigantism a disease caused by overproduction of growth hormone before the skeleton has stopped growing (24.3)

globular protein a protein whose overall shape is more or less spherical (4.3)

glucocorticoid a kind of steroid hormone involved in the metabolism of sugars (24.3)

glucogenic amino acid one that has pyruvate or oxaloacetate as a catabolic breakdown product (23.6)

gluconeogenesis the pathway of synthesis of glucose from lactate (18.2)

glucose a monosaccharide; a ubiquitous metabolite (16.1)

glyceraldehyde the simplest carbohydrate that contains a chiral carbon, the starting point of a system of describing optical isomers (16.1)

glyceraldehyde-3-phosphate a key intermediate in the reactions of sugars (17.3)

glyceraldehyde-3-phosphate dehydrogenase an important enzyme in glycolysis and gluconeogenesis (17.3)

glycerol phosphate shuttle a mechanism for transferring electrons from NADH in the cytosol to $FADH_2$ in the mitochondrion (20.7)

glycogen a polymer of glucose; an important energy storage molecule in animals (16.4)

glycolipid a lipid to which a sugar moiety is bonded (8.2)

glycolysis the anaerobic breakdown of glucose to three-carbon compounds (17.1)

glycoside a compound in which one or more sugars is bonded to another molecule (16.2)

glyoxylate cycle a pathway in plants that is an alternative to the citric acid cycle and that bypasses several citric acid cycle reactions (19.6)

glyoxysomes membrane-enclosed organelles that contain the enzymes of the glyoxylate cycle (19.6)

Golgi apparatus a cytoplasmic organelle that consists of flattened membranous sacs, usually involved in secretion of proteins (1.6)

G protein a membrane-bound protein that mediates the action of adenylate cyclase (24.3)

grana bodies within the chloroplast that contain the thylakoid disks, the site of photosynthesis (22.1)

guanine one of the purine bases found in nucleic acids (9.2)

half reaction an equation that shows either the oxidative or the reductive part of an oxidation–reduction reaction (15.4)

Haworth projection formulas a perspective representation of the cyclic forms of sugars (16.1)

helicase (rep protein) unwinds the double helix of DNA in the process of replication (10.4)

α-helix one of the most frequently encountered folding patterns in the protein backbone (4.3)

helix-turn-helix a common motif found in the DNA binding domain of transcription factors (4.3)

helper T cells components of the human immune system, the target of the AIDS virus (14.5)

heme an iron-containing cyclic compound, found in cytochromes, hemoglobin, and myoglobin (4.5)

hemiacetal a compound that is formed by reaction of an aldehyde with an alcohol and is found in the cyclic structure of sugars (16.1)

hemiketal a compound that is formed by reaction of a ketone with an alcohol and is found in the cyclic structure of sugars (16.1)

Henderson–Hasselbalch equation a mathematical relationship between the pK_a of an acid and the pH of a solution containing the acid and its conjugate base (2.4)

heteropolysaccharide polysaccharide that contains more than one kind of sugar monomer (16.4)

heterotropic effects allosteric effects that occur when different substances are bound to a protein (7.1)

heterozygous exhibiting differences in a given gene on one chromosome and the corresponding gene on the paired chromosome (13.8)

hexokinase the first enzyme of glycolysis (17.2)

hexose monophosphate shunt a synonym for the pentose phosphate pathway, in which glucose is converted to five-carbon sugars with concomitant production of NADPH (18.4)

high-performance liquid chromatography (HPLC) a sophisticated chromatography technique that gives fast and clean purifications (5.4)

histones basic proteins found complexed to eukaryotic DNA (9.3)

hnRNA heterogeneous nuclear RNA; the original form of mRNA in eukaryotes that contains intervening sequences (9.5)

holoenzyme an enzyme that has all component parts, including coenzymes and all subunits (11.1)

homeostasis the balance of biological activities in the body (24.3)

homology similarity of monomer sequences in polymers (4.6)

homopolysaccharide one that contains only one kind of sugar monomer (16.4)

homotropic effects allosteric effects that occur when several identical molecules are bound to a protein (7.1)

homozygous exhibiting no differences between a given gene on one chromosome and the corresponding gene on the paired chromosome (13.8)

hormone a substance produced by endocrine glands and delivered by the bloodstream to target cells, producing a regulatory effect (24.3)

hsp70 a protein that acts as a chaperone in *E. coli*. The letters "hsp" stand for heat shock protein, since this chaperone is produced when the bacteria are grown at elevated temperatures (12.6)

hyaluronic acid a polysaccharide found in the lubricating fluid of joints (16.4)

hydrogen bonding a noncovalent association formed between a hydrogen atom covalently bonded to one electronegative atom and a lone pair of electrons on another electronegative atom (2.2)

hydrophilic tending to dissolve in water (2.1)

hydrophobic tending not to dissolve in water (2.1)

hydrophobic bonds attractions between molecules that are nonpolar; also called hydrophobic interactions (2.1)

β-hydroxy-β-methylglutaryl-CoA an intermediate in the biosynthesis of cholesterol (21.8)

hyperbolic a characteristic of a curve on a graph such that it rises quickly and then levels off (4.7)

hyperglycemia the condition of elevated blood glucose levels (24.3)

hypoglycemia the condition of low blood glucose levels (24.3)

hypothalamus the portion of the brain that controls, among other things, many of the workings of the endocrine system (24.3)

immunoglobulins another name for antibodies, proteins that play a role in the immune system (14.5)

induced-fit model a description of substrate binding to an enzyme such that the conformation of the enzyme changes to accommodate the shape of the substrate (6.4)

inducible enzyme an enzyme whose synthesis can be triggered by the presence of some substance, which is called the inducer (11.2)

induction of enzyme synthesis the triggering of the production of an enzyme by the presence of a specific inducer (11.2)

inhibitor a substance that decreases the rate of an enzyme-catalyzed reaction (6.7)

initial rate the rate of a reaction immediately after it starts, before any significant accumulation of product (6.6)

initiation complex the aggregate of mRNA, N-formylmethione-tRNA, ribosomal subunits, and initiation factors needed at the start of protein synthesis (12.4)

initiation factor any of a large group of proteins that bind to DNA and play a role in initiation of eukaryotic RNA synthesis (12.5)

initiation step the start of protein synthesis; the formation of the initiation complex (12.4)

initiator element a loosely conserved sequence surrounding the transcription start site in eukaryotic DNA (11.3)

integral protein a protein that is embedded in a membrane (8.4)

interleukins proteins that play a role in the immune system (14.5)

intermembrane space the region between the inner and outer mitochondrial membranes (19.2)

intrinsic termination the type of transcription termination that is not dependent on the rho protein (11.1)

intron an intervening sequence in DNA that does not appear in the final sequence of mRNA (11.6)

ion-exchange chromatography a method for separating substances on the basis of charge (5.2)

ion product constant for water a measure of the tendency of water to dissociate to give hydrogen ion and hydroxide ion (2.4)

irreversible inhibition covalent binding of an inhibitor to an enzyme, causing permanent inactivation (6.7)

isoelectric focusing a method for separating substances on the basis of their isoelectric points (5.3)

isoelectric point (isoelectric pH) the pH at which a molecule has no net charge (3.3)

isoprene a five-carbon unsaturated group, which is part of the structure of many lipids (21.8)

isozymes multiple forms of an enzyme that catalyze the same overall reaction but have subtle physical and kinetic parameters (6.2)

ketogenic amino acid amino acid that has acetyl-CoA or acetoacetyl-CoA as a catabolic breakdown product (23.6)

α-ketoglutarate dehydrogenase complex one of the enzymes of the citric acid cycle; it catalyzes the conversion of α-ketoglutarate to succinyl-CoA (19.4)

ketone body one of several ketone-based molecules produced in the liver during overutilization of fatty acids when carbohydrates are limited (21.5)

ketose a sugar that contains a ketone group as part of its structure (16.1)

killer T cells components of the human immune system (14.5)

Kozak sequence the base sequence that identifies the start codon in eukaryotic protein synthesis (12.5)

Krebs cycle an alternative name for the citric acid cycle (19.1)

K system a combination of an allosteric enzyme and an inhibitor or activator, in which the presence of the inhibitor/activator changes the substrate concentration that yields one-half V_{max} (7.1)

kwashiorkor a disease caused by serious protein deficiency (23.5)

labeling covalent modification of a specific residue on an enzyme (7.5)

lac **operon** the promoter, operator, and structural genes involved in the induction of β-galactosidase and related proteins (11.2)

lactate dehydrogenase an NADH-linked dehydrogenase that catalyzes the conversion of pyruvate to lactate (17.4)

lagging strand in DNA replication, the strand that is formed in small fragments that are subsequently joined by DNA ligase (10.2)

L and D amino acids amino acids whose stereochemistry is the same as the stereochemical standards L and D glyceraldehyde, respectively (3.1)

lanosterol a precursor of cholesterol (21.8)

leading strand in DNA replication, the strand that is continuously formed in one long stretch (10.2)

Lesch–Nyhan syndrome a metabolic disease characterized by severe retardation and compulsive self-mutilation; it is caused by a deficiency of an enzyme in the purine salvage pathway (23.8)

leucine zipper (bZIP) a structural motif found in DNA-binding proteins (11.6)

leukocytes white blood cells that play an important role in the functioning of the immune system (14.5)

leukotriene a substance derived from leukocytes (white blood cells) that has three double bonds; it is of pharmaceutical importance (8.8)

lignin a polymer of coniferyl alcohol; a structural material found in woody plants (16.4)

Lineweaver–Burk double-reciprocal plot a graphical method for analyzing the kinetics of enzyme-catalyzed reactions (6.6)

lipase an enzyme that hydrolyzes lipids (8.2)

lipid a compound insoluble in water but soluble in organic solvents (8.1)

lipid bilayers aggregates of lipid molecules in which the polar head groups are in contact with water and the hydrophobic parts are not (8.3)

liposome a spherical aggregate of lipids arranged so that the polar head groups are in contact with water and the nonpolar tails are sequestered from water (8.5)

lock-and-key model a description of the binding of a substrate to an enzyme such that the active site and the substrate exactly match each other in shape (6.4)

luminescence emission of light as a result of a chemical reaction (chemiluminescence) or re-emission of absorbed light (fluorescence) (13.1)

lymphocytes a type of white blood cell; a major component of the immune system (14.5)

lymphokines soluble protein factors produced by one lymphocyte that affect another cell (14.5)

lysosomes membrane-enclosed organelles that contain hydrolytic enzymes (1.6)

macronutrients ones needed in large amounts, such as proteins, carbohydrates, or fats (24.2)

macrophages components of the innate immune system (14.5)

major histocompatability complex (MHC) protein that displays an antigen on the surface of cells of the immune system (14.5)

malate–aspartate shuttle a mechanism for transferring electrons from NADH in the cytosol to NADH in the mitochondrion (20.7)

Maloney Murine Leukemia Virus (MMLV) a vector commonly used in gene therapy (14.4)

malonyl-CoA a three-carbon intermediate important in the biosynthesis of fatty acids (21.6)

matrix (mitochondrial) the part of a mitochondrion enclosed within the inner mitochondrial membrane (1.6, 19.2)

metabolic water the water produced as a result of complete oxidation of nutrients; sometimes it is the only water source of desert-dwelling organisms (21.3)

metabolism the sum total of all biochemical reactions that take place in an organism (15.3)

metal–ion catalysis (Lewis acid–base catalysis) a form of catalysis that depends on the Lewis definition of an acid as an electron-pair acceptor and a base as an electron-pair donor (7.6)

micelle an aggregate formed by amphipathic molecules such that their polar ends are in contact with water and their nonpolar portions are on the interior (2.1)

Michaelis constant a numerical value for the strength of binding of a substrate to an enzyme; an important parameter in enzyme kinetics (6.6)

micronutrients vitamins and minerals that are needed in small amounts (24.2)

mineralocorticoid a kind of steroid hormone involved in the regulation of levels of inorganic ions "minerals" (24.3)

minerals in nutrition, inorganic substances required as the ion or free element (24.2)

miRNA (micro RNA) short stretches of RNA that affect gene expression and that play a role in growth and development (9.5)

mismatch repair a process for repairing damaged DNA (10.5)

mitochondrion an organelle that contains the apparatus responsible for aerobic oxidation of nutrients (1.6)

mitogen-activated protein kinase an enzyme that responds to cell growth and stress signals and phosphorylates key proteins that act as transcription factors (14.8)

mobile phase (eluent) in chromatography, the portion of the system in which the mixture to be separated moves (5.2)

Monera one of the five kingdoms used to classify living organisms; it includes prokaryotes (1.7)

monoclonal antibodies antibodies produced from the progeny of a single cell and specific for a single antigen (14.5)

monomer a small molecule that may bond to many others to form a polymer (1.3)

monosaccharide a compound that contains a single carbonyl group and two or more hydroxyl groups (16.1)

motif a repetitive supersecondary structure (4.3)

mRNA (messenger RNA) the kind of RNA that specifies the order of amino acids in a protein (9.5)

mucopolysaccharide a polysaccharide that has a gelatinous consistency (16.4)

multifunctional enzyme enzyme in which a single protein catalyzes several reactions (21.6)

multiple cloning site (MCS) a region of a bacterial plasmid with many restriction sites (13.3)

multiple sclerosis a disease in which the lipid sheath of nerve cells is progressively destroyed (8.2)

multipotent stem cell a stem cell that can differentiate into a number of cell types, but not all possible cell types (14.7)

mutagen an agent that brings about a mutation; such agents include radiation and chemical substances that alter DNA (10.5)

mutation a change in DNA, causing subsequent changes in the organism that can be transmitted genetically (10.5)

myelin the lipid-rich sheath of nerve cells (8.2)

native conformation a three-dimensional shape of a protein with biological activity (4.1)

natural killer (NK) cells components of the innate immune system (14.5)

negative cooperativity a cooperative effect whereby binding of the first ligand to an enzyme or protein causes the affinity for the next ligand to be lower (7.2)

nick translation a process for repairing damaged DNA (10.5)

nicotinamide adenine dinucleotide an important coenzyme in metabolism that is found in an oxidized or reduced form (7.8, 15.5)

nitrification the conversion of ammonia to nitrates (23.1)

nitrogenase the enzyme complex that catalyzes nitrogen fixation (23.2)

nitrogen fixation the conversion of molecular nitrogen to ammonia (23.1)

noncompetitive inhibition a form of enzyme inactivation in which a substance binds to a place other than the active site but distorts the active site so that the reaction is inhibited (6.7)

nonheme (iron–sulfur) protein a protein that contains iron and sulfur but no heme group (19.4)

nonoverlapping, commaless code a genetic code in which no bases are shared between the sequences of three bases (triplets) that specify an amino acid, with no intervening, noncoding bases (12.2)

nonpolar bond a bond in which two atoms share electrons evenly (2.1)

nontemplate strand the DNA strand that has the same sequence as the RNA that is synthesized from the template (11.1)

northern blotting a technique used for transferring DNA from an agarose gel after electrophoresis onto a membrane, such as nitrocellulose (13.10)

N-terminal the end of a protein or polypeptide with its amino group not linked to another amino acid by a peptide bond (3.4)

nuclear magnetic resonance (NMR) spectroscopy a method for determining the three-dimensional shape of proteins in solution (4.5)

nuclear region the portion of a prokaryotic cell that contains the DNA (1.5)

nuclease an enzyme that hydrolyzes a nucleic acid; it is specific for DNA or RNA (13.2)

nucleic acid a macromolecule formed by polymerization of nucleotides (1.3)

nucleic-acid base (nucleobase) one of the nitrogen-containing aromatic compounds that makes up the coding portion of a nucleic acid (9.2)

nucleolus a portion of the nucleus rich in RNA (1.6)

nucleophile an electron-rich substance that tends to react with sites of positive charge or polarization (7.5)

nucleophilic substitution reaction reaction in which one functional group is replaced by another as the result of nucleophilic attack (7.6)

nucleoside a purine or pyrimidine base bonded to a sugar (ribose or deoxyribose) (9.2)

nucleosome a globular structure in chromatin in which DNA is wrapped around an aggregate of histone molecules (9.3)

nucleotide a purine or pyrimidine base bonded to a sugar (ribose or deoxyribose) which in turn is bonded to a phosphate group (9.2)

nucleotide-excision repair a process for repairing damaged DNA (10.5)

nucleus the organelle that contains the main genetic apparatus in eukaryotes (1.6)

Okazaki fragments short stretches of DNA formed in the lagging strand in replication that are subsequently linked by DNA ligase (10.2)

oligomer an aggregate of several smaller units (monomers); bonding may be covalent or noncovalent (4.7)

oligosaccharide a few sugars linked by glycosidic bonds (16.1)

oncogene one that causes cancer when a triggering event takes place (14.1)

one-carbon transfers reactions in which the transfer usually involves carbon dioxide, a methyl group, or a formyl group (23.4)

open complex the form of the complex of RNA polymerase and DNA that occurs during transcription (11.1)

operator the DNA element to which a repressor of protein synthesis binds (11.2)

operon a group of operator, promoter, and structural genes (11.2)

opsin a protein in the rod and cone cells of the retina; it plays a crucial role in vision (8.7)

optical isomers (see *stereoisomers*) (16.1)

order of a reaction the experimentally determined dependence of the rate of a reaction on substrate concentrations (6.3)

organelle a membrane-enclosed portion of a cell with a specific function (1.4)

organic chemistry the study of compounds of carbon, especially of carbon and hydrogen and their derivatives (1.2)

origin of replication the point at which the DNA double helix begins to unwind at the start of replication (10.2)

origin recognition complex (ORC) a protein complex bound to DNA throughout the cell cycle that serves as an attachment site for several proteins that help control replication. (10.6)

oxidation the loss of electrons (1.9)

β-oxidation the main pathway of catabolism of fatty acids (21.2)

oxidative decarboxylation loss of carbon dioxide accompanied by oxidation (19.2)

oxidative phosphorylation a process for generating ATP; it depends on the creation of a pH gradient within the mitochondrion as a result of electron transport (19.1)

oxidizing agent a substance that accepts electrons from other substances (15.3)

oxygen-evolving complex the part of Photosystem II that splits water to produce oxygen (22.2)

palindrome a message that reads the same backward or forward (13.2)

palmitate a 16-carbon saturated fatty acid; the end product of fatty-acid biosynthesis in mammals (21.6)

passive transport the process by which a substance enters a cell without an expenditure of energy by the cell (8.6)

pause structure a hairpin loop that can form during transcription attenuation causing premature termination of transcription (11.2)

pectin a polymer of galacturonic acid; it occurs in the cell walls of plants (16.4)

pentose phosphate pathway a pathway in sugar metabolism that gives rise to five-carbon sugars and NADPH (18.4)

peptide bond an amide bond between amino acids in a protein (3.4)

peptides molecules formed by linking two to several dozen amino acids by amide bonds (3.3)

peptidoglycan a polysaccharide that contains peptide crosslinks; it is found in bacterial cell walls (16.4)

peptidyl transferase in protein synthesis, the enzyme that catalyzes formation of the peptide bond, part of the 50S ribosomal subunit (12.4)

percent recovery a measurement of the amount of an enzyme recovered at each step of a purification experiment (5.1)

peripheral proteins proteins loosely bound to the outside of a membrane (8.2)

peroxisomes membrane-bounded sacs that contain enzymes involved in the metabolism of hydrogen peroxide (H_2O_2) (1.6)

pH a measure of the acidity of a solution (2.4)

phenylketonuria a disease characterized by mental retardation in developing children; it is caused by a lack of the enzyme that converts phenylalanine to tyrosine (3.5)

pheophytin a photosynthetic pigment that differs from chlorophyll only in having two hydrogens in place of magnesium (22.2)

phosphatidic acid a compound in which two fatty acids and phosphoric acid are esterified to the three hydroxyl groups of glycerol (8.2)

phosphatidylinositol 4,5-*bis*phosphate (PIP$_2$) a membrane-bound substance that mediates the action of Ca^{2+} as a second messenger (24.3)

phosphoaclyglycerol (phosphoglyceride) a phosphatidic acid (*vide supra*) with another alcohol esterified to the phosphoric acid moiety (8.2)

3′,5′-phosphodiester bond a covalent linkage in which phosphoric acid is esterified to the 3′ hydroxyl of one nucleoside and the 5′ hydroxyl of another nucleoside; it forms the backbone of nucleic acids (9.2)

phosphofructokinase the key allosteric control enzyme in glycolysis; it catalyzes the phosphorylation of fructose-6-phosphate (17.2)

phospholipase an enzyme that hydrolyzes phospholipids (21.2)

phosphorolysis the addition of phosphoric acid across a bond, such as the glycosidic bond in glycogen, giving glucose phosphate and a glycogen remainder one residue shorter; it is analogous to hydrolysis (addition of water across a bond) (18.1)

photophosphorylation the synthesis of ATP coupled to photosynthesis (22.2)

photorespiration the process by which plants oxidize carbohydrates aerobically in the light (22.6)

photosynthesis the process of using light energy from the sun to drive the synthesis of carbohydrates (22.1)

photosynthetic unit the assemblage of chlorophylls that includes light-harvesting molecules and the special pair that actually carry out the reaction (22.2)

photosystem I the portion of the photosynthetic apparatus responsible for the production of NADPH (22.2)

photosystem II the portion of the photosynthetic apparatus responsible for the splitting of water to oxygen (22.2)

pituitary the gland that releases trophic hormones to specific endocrine glands; it is under the control of the hypothalamus (24.4)

plasma membrane another name for the cell membrane; the outer boundary of the cell (1.4)

plasmid a small, circular DNA molecule that usually contains genes for antibiotic resistance and is often used for cloning (13.2)

plastocyanin a copper-containing protein; it is part of the electron transport chain that links the two photosystems in photosynthesis (22.2)

plastoquinone a substance similar to coenzyme Q, part of the electron transport chain that links the two photosystems in photosynthesis (22.2)

β-pleated sheet one of the most important types of secondary structure, in which the protein backbone is almost fully extended with hydrogen bonding between adjacent strands (4.3)

pluripotent stem cell one that can develop into all possible cell types (14.7)

polar bond a bond in which two atoms have an unequal share in the bonding electrons (2.1)

polyacrylamide gel electrophoresis (PAGE) a form of electrophoresis in which a polyacrylamide gel serves as both a sieve and a supporting medium (5.2)

poly-A tail a long sequence of adenosine residues at the 3′ end of eukaryotic mRNA (11.6)

polylinker a region of a bacterial plasmid with many restriction sites (13.3)

polymer a macromolecule formed by the bonding together of smaller units (1.3)

polymerase chain reaction (PCR) a method for amplifying a small amount of DNA based on the reaction of isolated enzymes rather than on cloning (13.6)

polypeptide chain the backbone of a protein; it is formed by linking amino acids by peptide (amide) bonds (3.4)

polysaccharide a polymer of sugars (16.4)

polysome the assemblage of several ribosomes bound to one mRNA (12.4)

P/O ratio the ratio of ATP produced by oxidative phosphorylation to oxygen atoms consumed in electron transport (20.4)

porphyrins large-ring compounds formed by linking four pyrrole rings; they combine with iron ions to form the heme group (4.5)

positive cooperativity a cooperative effect whereby binding of the first ligand to an enzyme or protein causes the affinity for the next ligand to be higher (4.7)

pre-replication complex (pre-RC) the complex of DNA, recognition protein (ORC), activator protein (RAP), and licensing factors (RLFs) that makes DNA competent for replication in eukaryotes (10.6)

Pribnow box a DNA base sequence that is part of a prokaryotic promoter; it is located 10 bases before the transcription start site (11.1)

primary structure the order in which the amino acids in a protein are linked by peptide bonds (4.1)

primer in DNA replication, a short stretch of RNA hydrogen-bonded to the template DNA to which the growing DNA strand is bonded at the start of synthesis (10.3)

primosome the complex at the replication fork in DNA synthesis; it consists of the RNA primer, primase, and helicase (10.4)

prion a naturally occurring protein found in nervous tissue and brain that can adopt multiple forms; the abnormal form leads to prion diseases, such as mad-cow disease and human spongiform encephalopathy (Creutzfeldt-Jakob disease) (4.6)

probe a radioactively labeled strand of a nucleic acid used for selecting a complementary strand out of a mixture (13.5)

processivity the number of nucleotides incorporated in a growing DNA chain before the DNA polymerase dissociates from the template DNA (10.3)

prokaryote a microorganism that lacks a distinct nucleus and membrane-enclosed organelles (1.4)

promoter the portion of DNA to which RNA polymerase binds at the start of transcription (11.1)

proofreading the process of removing incorrect nucleotides when DNA replication is in progress (10.5)

propeller twist a twisting of the bases in a DNA double helix that allows for stronger base stacking (9.3)

prostaglandin one of a group of derivatives of arachidonic acid; it contains a five-membered ring and is of pharmaceutical importance (8.8)

prosthetic group a portion of a protein that does not consist of amino acids (4.1)

protease an enzyme that hydrolyzes proteins (7.5)

proteasome multisubunit complex of proteins that mediates degradation of other, suitably tagged proteins (12.7)

protein a macromolecule formed by polymerization of amino acids (1.3)

protein kinase a class of enzymes that modifies a protein by attaching a phosphate group to it (7.3)

proteome the total protein content of the cell (13.12)

proteomics study of interactions among all the proteins of the cell (13.12)

Protista one of the five kingdoms used to classify living organisms; it includes single-celled eukaryotes (1.7)

proton gradient the difference between the hydrogen ion concentrations in the mitochondrial matrix and that in the intermembrane space, which is the basis of coupling between oxidation and phosphorylation (20.1)

purine a nitrogen-containing aromatic compound that contains a six-membered ring fused to a five-membered ring; the parent compound of two nucleobases, adenine and guanine (9.2)

pyranose a cyclic form of a sugar containing a five-membered ring; it was named for its resemblance to pyran (16.1)

pyranoside a glycoside involving a pyranose (16.1)

pyrimidine a nitrogen-containing aromatic compound that contains a six-membered ring; the parent compound of several nucleobases (9.2)

pyrrole ring a five-membered ring that contains one nitrogen atom; part of the structure of porphyrins and heme (4.5)

pyruvate dehydrogenase complex a multienzyme complex that catalyzes the conversion of pyruvate to acetyl-CoA and carbon dioxide (19.3)

pyruvate kinase the enzyme that catalyzes the final step common to all forms of glycolysis (17.3)

Q cycle a series of reactions in the electron transport chain that provides the link between two-electron transfers and one-electron transfers (20.3)

quaternary structure the interaction of several polypeptide chains in a multisubunit protein (4.1)

rate constant a proportionality constant in the equation that describes the rate of a reaction (6.3)

rate-limiting step the slowest step in a reaction mechanism; it determines the maximum velocity of the reaction (6.3)

reaction center the site of the special pair of chlorophylls responsible for trapping light energy from the sun (22.1)

reaction order an experimentally determined number that describes the rate of a reaction in terms of the concentration of a reactant or reactants (6.3)

reading frame the starting point for reading of a genetic message (12.2)

receptor protein a protein on a cell membrane with specific binding site for extracellular substances (8.4)

recombinant DNA DNA that has been produced by linking DNA from two different sources (13.3)

reducing agent a substance that gives up electrons to other substances (15.4)

reducing sugar a sugar that has a free carbonyl group, one that can react with an oxidizing agent (16.2)

reduction the gain of electrons (1.9)

reduction potential a standard voltage that indicates the tendency of a reduction half reaction to take place (20.2)

regulatory gene gene that directs the synthesis of a repressor protein (11.2)

repair the enzymatic removal of incorrect nucleotides from DNA and their replacement by correct ones (10.5)

replication the process of duplication of DNA (10.1)

replication activator protein (RAP) the protein whose binding prepares for the start of DNA replication in eukaryotes (10.6)

replication fork in DNA replication, the point at which new DNA strands are formed (10.2)

replication licensing factors (RLFs) proteins required for DNA replication in eukaryotes (10.6)

replicator one of the multiple origins of replication in eukaryotic DNA synthesis (10.6)

replicon a part of a chromosome in which DNA synthesis is taking place (10.6)

replisome a complex of DNA polymerase, the RNA primer, primase, and helicase at the replication fork (10.4)

repressor a protein that binds to an operator gene blocking the transcription and eventual translation of structural genes under the control of that operator (11.2)

residue the portion of a monomer unit included in a polymer after splitting out of water between the linked monomers (3.4)

resonance structures structural formulas that differ from each other only in the position of electrons (3.4)

respiratory complexes the multienzyme systems in the inner mitochondrial membrane that carry out the reactions of electron transport (20.3)

response element a DNA sequence that binds to transcription factors involved in more generalized control of pathways (11.2)

restriction-fragment length polymorphism (RFLP) differences in the lengths of DNA fragments from different sources when digested with restriction enzymes; a forensic technique using DNA to identify biological samples (13.8)

restriction nuclease an enzyme that catalyzes a double-strand hydrolysis of DNA at a defined point in a specific sequence (13.2)

retinal the aldehyde form of vitamin A (8.7)

retrovirus virus in which the base sequence of RNA directs the synthesis of DNA (14.3)

reverse transcriptase the enzyme that directs the synthesis of DNA on an RNA template (14.3)

reverse turn a part of a protein where the polypeptide chain folds back on itself (4.3)

reversible inhibitor an inhibitor that is not covalently bound to an enzyme; it can be removed with restoration of activity (6.7)

R group the side chain of an amino acid that determines its identity (3.1)

rho-dependent termination the type of transcription termination that requires the rho protein (11.1)

rhodopsin a molecule crucial in vision; it is formed by the reaction of retinal and opsin (8.7)

ribonucleoside a compound formed when a nucleobase forms a glycosidic bond with ribose (9.2)

ribose a sugar that is part of the structure of RNA (9.2)

ribosome the site of protein synthesis in all organisms, consisting of RNA and protein (1.5)

ribozyme catalytic RNA (11.7)

ribulose-1,5-*bis*phosphate a key intermediate in the production of sugars in photosynthesis (22.5)

RNA ribonucleic acid (9.2)

RNA polymerase the enzyme that catalyzes the production of RNA on a DNA template (11.1)

RNA polymerase II the RNA polymerase in eukaryotes that makes mRNA; also called RNA polymerase B (11.3)

rRNA (ribosomal RNA) the kind of RNA found in ribosomes (9.5)

Rubisco (ribulose-1,5-*bis*phosphate carboxylase/oxygenase) the enzyme that catalyzes the first step in carbon dioxide fixation in photosynthesis (22.5)

salting out a purification technique for proteins based on differential solubility in salt solutions (5.1)

salvage reactions reactions that reuse compounds, such as purines, that require a large amount of energy to produce (23.8)

saponification the reaction of a triacylglycerol with base to produce glycerol and three molecules of fatty acid (8.2)

SARS (severe acute respiratory syndrome) A disease caused by a coronavirus

saturated having all carbon-carbon bonds as single bonds (8.2)

SDS polyacrylamide-gel electrophoresis (SDS–PAGE) an electrophoretic technique that separates proteins on the basis of size (5.3)

secondary structure the arrangement in space of the backbone atoms in a polypeptide chain (4.1)

second messenger a substance produced or released by a cell in response to hormone binding to a receptor on the cell surface; it elicits the actual response in the cell (24.3)

semiconservative replication the mode in which DNA reproduces itself, such that one strand comes from parent DNA and the other strand is newly formed (10.2)

sense strand the DNA strand that has the same sequence as the RNA that is synthesized from the template (11.1)

sequencer an automated instrument used in determining the amino acid sequence of a peptide or the nucleotide sequence of a nucleic acid (5.4)

sequential model a description of the action of allosteric proteins in which a conformational change in one subunit is passed along to the other subunits (7.2)

serine protease a proteolytic enzyme in which a serine hydroxyl plays an essential role in catalysis (7.5)

severe combined immune deficiency (SCID) a genetic disease that affects DNA synthesis in the cells of the immune system (14.4)

Shine–Dalgarno sequence a leader sequence in prokaryotic mRNA that precedes the start signal (12.4)

side-chain group the portion of an amino acid that determines its identity (3.1)

sigmoidal referring to an S-shaped curve on a graph, characteristic of cooperative interactions (4.7)

silencer a DNA sequence that binds to a transcription factor and reduces the level of transcription (11.2)

simple diffusion the process of passing through a pore or opening in a membrane without a requirement for a carrier or for the expenditure of energy (8.6)

single-strand binding (SSB) protein in DNA replication, a protein that protects exposed single-strand sections of DNA from nucleases (10.4)

siRNA (small interfering RNA) short stretches of RNA that control gene expression by selective suppression of genes (9.5)

site-directed mutagenesis a method for introducing specific changes in the base sequence of DNA (13.7)

SN1 unimolecular nucleophilic substitution reaction; one of the most common types of organic reactions seen in biochemistry; the rate of the reaction follows first-order kinetics (7.6)

snRNA (small nuclear RNA) an RNA type that is found in eukaryotes and is involved in splicing and some regulation of transcription (9.5)

snRNPs (small nuclear ribonucleoprotein particles) a protein-RNA complex found in the nucleus that aids in processing RNA molecules for export to the cytosol (9.5)

SN2 bimolecular nucleophilic substitution reaction; an important type of organic reaction seen in biochemistry; the rate of the reaction follows second-order kinetics (7.6)

sodium–potassium ion pump the export of sodium ion from the cell with simultaneous inflow of potassium ion, both against concentration gradients (8.6)

Southern blotting a technique used for transferring DNA from an agarose gel after electrophoresis onto a membrane, such as one made of nitrocellulose (13.8)

spacer region a region of eukaryotic DNA that is between nucleosomes (9.3)

sphingolipid a lipid whose structure is based on sphingosine (8.2)

sphingosine a long-chain amino alcohol; the basis of the structure of a number of lipids (8.2)

spliceosome a large multisubunit particle, similar in size to a ribosome, that is involved in splicing of RNA molecules (11.6)

split gene a gene that contains intervening sequences that will not be present in the mature RNA (11.6)

spontaneous in thermodynamics, characteristic of a reaction or process that will take place without outside intervention (1.10)

standard state the standard set of conditions used for comparisons of chemical reactions (15.1)

starch a polymer of glucose that plays an energy-storage role in plants (16.4)

stationary phase in chromatography, the substance that selectively retards the flow of the sample, effecting the separation (5.2)

steady state the condition in which the concentration of an enzyme–substrate complex remains constant in spite of continuous turnover (6.6)

stem cells undifferentiated cells that are precursors for all other cell types (14.7)

stereochemistry the branch of chemistry that deals with the three-dimensional shape of molecules (3.1)

stereoisomers molecules that differ from each other only in their configuration (three-dimensional shape); also called optical isomers (3.1)

stereospecific able to distinguish between stereoisomers (7.6)

steroid a lipid with a characteristic fused-ring structure (8.2)

"sticky ends" short, single-stranded stretches at the ends of double-stranded DNA; they can provide sites to which other DNA molecules with sticky ends can be linked (13.2)

(−) strand the DNA strand that is used as a template for RNA synthesis (11.1)

(+) strand the DNA strand that has the same sequence as the RNA that is synthesized from the template (11.1)

stroma in a chloroplast, the portion of the organelle that is equivalent to the mitochondrial matrix; the site of production of sugars in photosynthesis (22.1)

structural gene a gene that directs the synthesis of a protein under the control of some regulatory gene (11.2)

substrate a reactant in an enzyme-catalyzed reaction (6.4)

substrate cycling the control process in which opposing reactions are catalyzed by different enzymes (18.3)

substrate-level phosphorylation a reaction in which the source of phosphorus is inorganic phosphate ion, not ATP (17.3)

subunits the individual parts of a larger molecule (e.g., the individual polypeptide chains that make up a complete protein) (4.1)

sugar-phosphate backbone the series of ester bonds between phosphoric acid and deoxyribose (in DNA) or ribose (in RNA) (9.2)

supercoiling the presence of extra twists (over and above those of the double helix) in closed circular DNA (9.3)

supersecondary structure specific clusters of secondary structural motifs in proteins (4.3)

suppressor tRNAs tRNAes that allow translation to continue through a stop codon (12.5)

TATA box a promoter element found in eukaryotic transcription that is located 25 bases upstream of the transcription start site (11.3)

TATA-box binding protein (TBP) a part of one of the general transcription factors found in eukaryotic transcription; it binds to the TATA-box portion of the promoter (11.3)

TATA-box binding protein associated factors (TAFs) proteins that are associated with the TATA-box binding protein itself (11.3)

T cell one of two kinds of white blood cells important in the immune system—a killer T cell, which destroys infected cells, or a helper T cell, which is involved in the process of B-cell maturation (14.5)

template (antisense) strand the DNA strand that is used as a template for RNA synthesis (11.1)

termination site the area on DNA that causes termination of transcription by generating hairpin loops and a zone of weak binding between DNA and RNA (11.1)

termination step in protein synthesis, the point at which the stop signal is reached, releasing the newly formed protein from the ribosome (12.4)

tertiary structure the arrangement in space of all the atoms in a protein (4.1)

tetrahydrofolate the metabolically active form of the vitamin folic acid; a carrier of one-carbon groups (23.4)

tetramer a molecule consisting of four subunits (4.7)

thermodynamics the study of transformations and transfer of energy (1.9)

thiamine pyrophosphate a coenzyme involved in the transfer of two-carbon units (17.4)

thioester a sulfur-containing analogue of an ester (19.3)

3:4 terminator a hairpin loop that can form during transcription termination and that causes premature release of the RNA transcript (11.2)

thylakoid disks the site of the light-trapping reaction in chloroplasts (22.1)

thylakoid space the portion of the chloroplast between the thylakoid disks (22.1)

thymidylate synthetase the enzyme that catalyzes the production of thymine nucleotides needed for DNA synthesis; a target for cancer chemotherapy (23.11)

thymine one of the pyrimidine bases found in nucleic acids (9.2)

thymine dimers a defect in DNA structure caused by the action of ultraviolet light (10.5)

titration an experiment in which a measured amount of base is added to an acid (2.5)

α-tocopherol the most active form of vitamin E (8.7)

topoisomerase an enzyme that relaxes supercoiling in closed circular DNA (9.3)

torr a unit of pressure equal to that exerted by a column of mercury 1 mm high at 0°C (4.7)

transaldolase an enzyme that transfers a two-carbon unit in reactions of sugars (18.4)

transamination the transfer of amino groups from one molecule to another; an important process in the anabolism and catabolism of amino acids (23.4)

transcription the process of formation of RNA on a DNA template (9.5)

transcription-activation domain the part of a transcription factor that interacts with other proteins and complexes rather than with the DNA directly (11.4)

transcription bubble the complex of separated DNA and RNA polymerase in which transcription is actively occurring (11.1)

transcription-coupled repair a type of DNA repair that occurs during transcription (11.3)

transcription factor a protein or other complex that binds to a DNA sequence and alters the basal level of transcription (11.2)

transcription start site the location on the template DNA strand where the first ribonucleotide is used to initiate RNA synthesis (11.1)

transcriptome a group of genes that are being transcribed at a given time (13.10)

transition state the intermediate stage in a reaction in which old bonds break and new bonds are formed (6.2)

transition-state analogue a synthesized compound that mimics the form of the transition state of an enzyme reaction (7.7)

translation the process of protein synthesis in which the amino acid sequence of the protein reflects the sequence of bases in the gene that codes for that protein (9.5)

translocation in protein synthesis, the motion of the ribosome along the mRNA as the genetic message is being read (12.4)

transport protein a component of a membrane that mediates the entry of specific substances into a cell (8.4)

triacylglycerol (triglyceride) a lipid formed by esterification of three fatty acids to glycerol (8.2)

tricarboxylic acid cycle another name for the citric acid cycle (19.1)

trimer a molecule consisting of three subunits (4.7)

triosephosphate isomerase the enzyme that catalyzes the conversion of dihydroxyacetone phosphate to glyceraldehyde-3-phosphate (17.2)

triplet code a sequence of three bases (a triplet) in mRNA specifies one amino acid in a protein (12.2)

tRNA (transfer RNA) the kind of RNA to which amino acids are bonded as a preliminary step to being incorporated into a growing polypeptide chain (9.5)

trophic hormones hormones that are produced by the pituitary gland under the direction of the hypothalamus that, in turn, cause the release of specific hormones by individual endocrine glands (24.3)

trypsin a proteolytic enzyme specific for basic amino acid residues as the site of hydrolysis (5.4)

tumor suppressor a gene that encodes a protein that inhibits cell division (14.8)

turnover number the number of moles of substrate that react per second per mole of enzyme (6.6)

uncoupler a substance that overcomes the proton gradient in mitochondria, allowing electron transport to proceed in the absence of phosphorylation (20.4)

universal code the genetic code, which is the same in all organisms (12.2)

unsaturated containing some carbon–carbon double or triple bonds (8.2)

UP element a prokaryotic promoter element that is 40 to 60 bases upstream of the transcription start site (13.2)

upstream in transcription, a portion of the sequence nearer the 3′ end than the gene to be transcribed, where the DNA is read from the 3′ to the 5′ end and the RNA is formed from the 5′ to the 3′ end; in translation, nearer the 5′ end of the mRNA (13.1)

uracil one of the pyrimidine bases found in nucleic acids (9.2)

urea cycle a pathway that leads to excretion of waste products of nitrogen metabolism, especially those of amino acids (23.6)

uric acid a product of catabolism of nitrogen-containing compounds, especially purines; accumulation of uric acid in joints causes gout in humans (23.8)

vacuole a cavity within the cytoplasm of a cell, typically enclosed by a single membrane, that may serve secretory, excretory, or storage functions (1.6)

van der Waals bond a noncovalent association based on the weak attraction of transient dipoles for one another; also called a van der Waals interaction (2.1)

vector a carrier molecule for transfer of genes in DNA recombination (13.3)

virion a complete virus particle consisting of nucleic acid and coat protein (14.1)

V system a combination of an allosteric enzyme and an inhibitor or activator in which the presence of the inhibitor/activator changes the maximal velocity of the enzyme but not the substrate level that yields one-half V_{max} (7.1)

"wobble" the possible variation in the third base of a codon allowed by several acceptable forms of base pairing between mRNA and tRNA (12.2)

X-ray crystallography an experimental method for determining the three-dimensional structure of molecules, such as the tertiary or quaternary structure of proteins (4.5)

Z-DNA a form of DNA that is a left-handed helix, which has been seen to occur naturally under certain circumstances (9.3)

zero order refers to a reaction that proceeds at a constant rate, independent of the concentration of reactant (6.3)

zinc-finger motif a common motif found in the DNA binding region of transcription factors (11.5)

zwitterion a molecule that has both a positive and a negative charge (2.6)

zymogen an inactive protein that can be activated by specific hydrolysis of peptide bonds (7.4)

Chapter 1

1.1 What Are the Basic Themes for This Text?

1. A polymer is a very large molecule formed by linking smaller units (monomers) together. A protein is a polymer formed by linking amino acids together. A nucleic acid is a polymer formed by linking nucleotides together. Catalysis is the process that increases the rate of chemical reactions compared with the rate of the uncatalyzed reaction. Biological catalysts are proteins in almost all cases; the only exceptions are a few types of RNA, which can catalyze some of the reactions of their own metabolism. The genetic code is the means by which the information for the structure and function of all living things is passed from one generation to the next. The sequence of purines and pyrimidines in DNA carries the genetic code. (RNA is the coding material in some viruses).

1.2 What Is the Chemical Nature of Important Biomolecules?

2. The correct match of functional groups and the compounds containing those functional groups is given in the following list.

Amino group	$CH_3CH_2NH_2$
Carbonyl group (ketone)	CH_3COCH_3
Hydroxyl group	CH_3OH
Carboxyl group	CH_3COOH
Carbonyl group (aldehyde)	CH_3CH_2CHO
Thiol group	CH_3SH
Ester linkage	$CH_3COOCH_2CH_3$
Double bond	$CH_3CH=CHCH_3$
Amide linkage	$CH_3CON(CH_3)_2$
Ether	$CH_3CH_2OCH_2CH_3$

3. The functional groups in the compounds follow:

Glucose

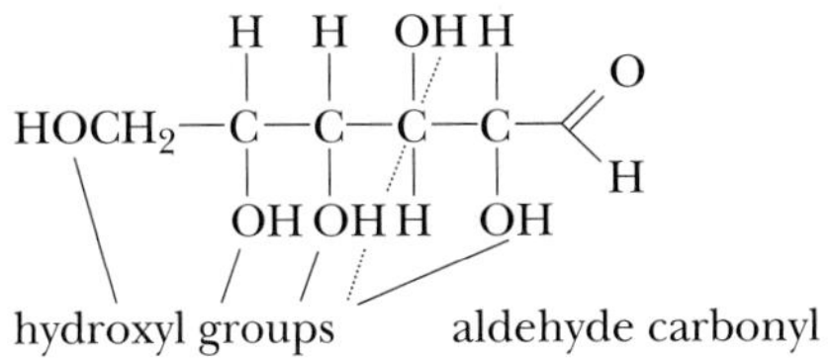

hydroxyl groups   aldehyde carbonyl

A triglyceride

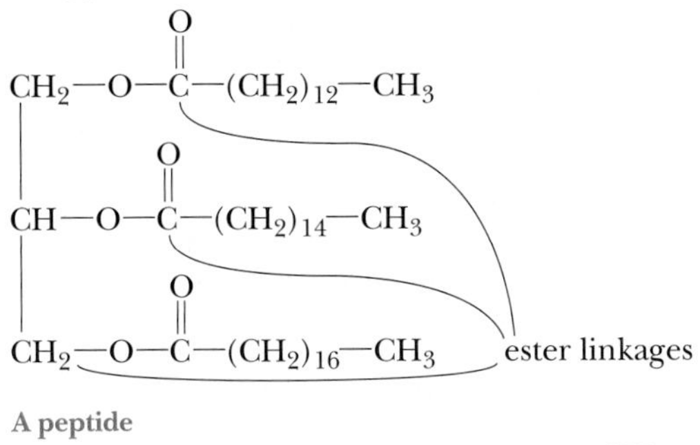

ester linkages

A peptide

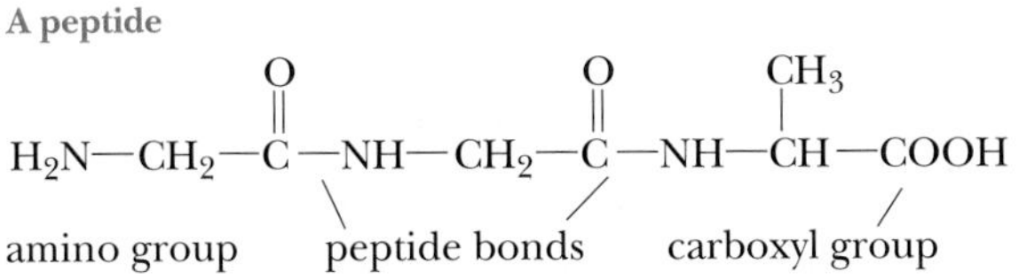

amino group   peptide bonds   carboxyl group

Vitamin A

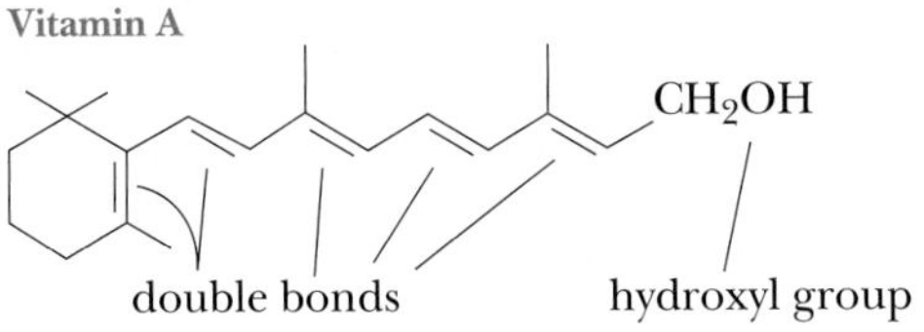

double bonds   hydroxyl group

4. Before 1828, the concept of vitalism held that organic compounds could be made only by living systems and were beyond the realm of laboratory investigations. Wöhler's synthesis showed that organic compounds, like inorganic ones, did not require a vitalistic explanation, but that, rather, they obeyed the laws of chemistry and physics and thus were subject to laboratory investigation. Subsequently, the concept was extended to the much more complex, but still testable, discipline of biochemistry.

5. Urea, like all organic compounds, has the same molecular structure, whether it is produced by a living organism or not.

6.

Item	Organic	Biochemical
Solvent	Varies (smelly)	Water (usually)
Concentrations	High	Low (mM, μM, nM)
Use catalyst?	Usually not	Almost always (enzymes)
Speed	Min, hr, day	μsec, nsec
Temp	Varies (high)	Isothermal, ambient temp
Yield	Poor–good (90%)	High (can be 100%)
Side reactions	Often*	No
Internal control	Little	Very high**—choices
Polymers (product)	Usually not	Commonly (proteins, nucleic acids, saccharides)
Bond strength	High (covalent)	High, weak (in polymers)
Bond distances	Not critical	Critical (close fit)
Compartmented	No	Yes (esp. eukaryotes)
Emphasis	One reaction	Pathways, interconnected (control** choices)†
System	Closed or open	Open (overcome $+\Delta G$)

* Example of side reactions: Glucose → G6P *or* G1P *or* G2P.
** Control levels: enzyme, hormone, gene.
† Example of choices:

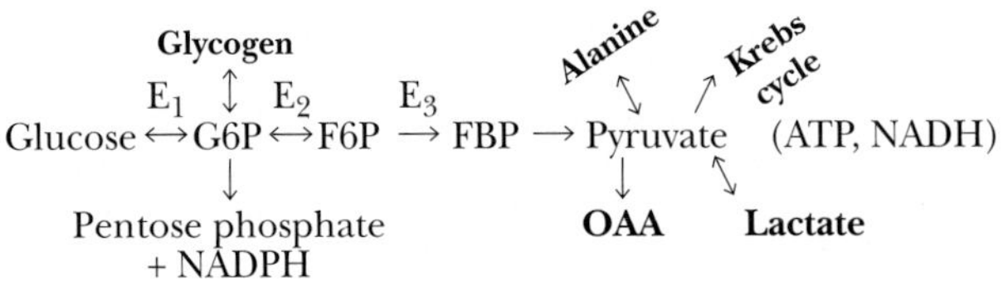

Glycogen

$E_1 \updownarrow E_2 \qquad E_3$
Glucose $\leftrightarrow$ G6P $\leftrightarrow$ F6P $\rightarrow$ FBP $\rightarrow$ Pyruvate   (ATP, NADH)

Pentose phosphate + NADPH

Alanine   Krebs cycle

OAA   Lactate

7. Five; seven if the two cyclopropane derivatives are allowed.
8. Thirteen different alcohols, 11 aldehydes/ketones, and 10 each epoxides and ethers.

1.3 What Can Biochemistry Say about Possible Origins of Life?

9. It is generally believed that carbon is the likely basis for all life forms, terrestrial or extraterrestrial.
10. Eighteen residues would give 20^{18}, or 2.6×10^{23} possibilities. Thus, 19 residues would be necessary to have at least Avogadro's number (6.022×10^{23}) of possibilities.
11. The number is 4^{40}, or 1.2×1024, which is twice Avogadro's number.
12. RNA is capable of both coding and catalysis.
13. Catalysis allows living organisms to carry out chemical reactions much more efficiently than without catalysts.
14. Two of the most obvious advantages are speed and specificity; they also work at constant temperature or produce little heat.
15. Coding allows for reproduction of cells.

16. With respect to coding, RNA-type polynucleotides have been produced from monomers in the absence of either a preexisting RNA to be copied or an enzyme to catalyze the process. The observation that some existing RNA molecules can catalyze their own processing suggests a role for RNA in catalysis. With this dual role, RNA may have been the original informational macromolecule in the origin of life.

17. It is unlikely that cells could have arisen as bare cytoplasm without a plasma membrane. The presence of the membrane protects cellular components from the environment and prevents them from diffusing away from each other. The molecules within a cell can react more easily if they are closer to each other.

1.4 How Do Prokaryotes and Eukaryotes Differ in Levels of Organization?

18. Five differences between prokaryotes and eukaryotes are: (1) Prokaryotes do not have a well-defined nucleus, but eukaryotes have a nucleus marked off from the rest of the cell by a double membrane. (2) Prokaryotes have only a plasma (cell) membrane; eukaryotes have an extensive internal membrane system. (3) Eukaryotic cells contain membrane-bounded organelles, while prokaryotic cells do not. (4) Eukaryotic cells are normally larger than those of prokaryotes. (5) Prokaryotes are single-celled organisms, while eukaryotes can be either single-celled or multicellular.

19. Protein synthesis takes place on ribosomes both in prokaryotes and in eukaryotes. In eukaryotes, ribosomes may be bound to the endoplasmic reticulum or found free in the cytoplasm; in prokaryotes, ribosomes are only found free in the cytoplasm.

1.5 What Are the Main Structural Features of Prokaryotic Cells?

20. It is unlikely that mitochondria would be found in bacteria. These eukaryotic organelles are enclosed by a double membrane, and bacteria do not have an internal membrane system. The mitochondria found in eukaryotic cells are about the same size as most bacteria.

1.6 What Are the Main Structural Features of Eukaryotic Cells?

21. See Section 1.6 for the functions of the parts of an animal cell, which are shown in Figure 1.10(a).

22. See Section 1.6 for the functions of the parts of a plant cell, which are shown in Figure 1.10(b).

23. In green plants photosynthesis takes place in the membrane system of chloroplasts, which are large membrane-enclosed organelles. In photosynthetic bacteria, there are extensions of the plasma membrane into the interior of the cell called chromatophores, which are the sites of photosynthesis.

24. Nuclei, mitochondria, and chloroplasts are all enclosed by double membranes.

25. Nuclei, mitochondria, and chloroplasts all contain DNA. The DNA found in mitochondria and in chloroplasts differs from that found in the nucleus.

26. Mitochondria carry out a high percentage of the oxidation (energy-releasing) reactions of the cell. They are the primary sites of ATP synthesis.

27. The Golgi apparatus is involved in binding carbohydrates to proteins and in exporting substances from the cell. Lysosomes contain hydrolytic enzymes, peroxisomes contain catalase (needed for the metabolism of peroxides), and glyoxysomes contain enzymes needed by plants for the glyoxylate cycle. All these organelles have the appearance of flattened sacs, and each is enclosed by a single membrane.

1.7 How Do We Classify Organisms: Five Kingdoms or Three Domains?

28. Monera includes bacteria (e.g., *E. coli*) and cyanobacteria. Protista includes such organisms as *Euglena, Volvox, Amoeba,* and *Paramecium.* Fungi includes molds and mushrooms. Plantae includes clubmosses and oak trees. Animalia includes spiders, earthworms, salmon, rattlesnakes, robins, and dogs.

29. The kingdom Monera consists of prokaryotes. Each of the other four kingdoms consists of eukaryotes.

30. The kingdom Monera is divided into the domains Eubacteria and Archaea on the basis of biochemical differences. The domain Eukarya consists of the four kingdoms of eukaryotic organisms.

1.8 Is There Common Ground for All Cells?

31. The major advantage is that of having compartments (organelles) with specialized functions (and thus division of labor). Another advantage is that cells can be much larger without surface-area-to-volume considerations being critical because of compartmentalization.

32. See the discussion of the endosymbiotic theory in Section 1.8.

33. See Question 32. The division of labor in cells gives rise to greater efficiency and a larger number of individuals. This in turn allows more opportunity for evolution and speciation.

1.9 How Do Cells Use Energy?

34. Processes that release energy are favored.

1.10 What Is the Connection between Energy and Change?

35. The term "spontaneous" means energetically favored. It does not necessarily mean fast.

1.11 What Is the Criterion for Spontaneity in Biochemical Reactions?

36. The system consists of the nonpolar solute and water, which become more disordered when a solution is formed; ΔS_{sys} is positive but comparatively small. The ΔS_{surr} is negative and comparatively large because it is a reflection of the unfavorable enthalpy change for forming the solution (ΔH_{sys}).

37. Processes (a) and (b) are spontaneous, whereas processes (c) and (d) are not. The spontaneous processes represent an increase in disorder (increase in the entropy of the universe) and have a negative $\Delta G°$ at constant temperature and pressure, while the opposite is true of the nonspontaneous processes.

38. In all cases, there is an increase in entropy, and the final state has more possible random arrangements than the initial state.

39. Since the equation involves multiplication of ΔS by T, the value of ΔG is temperature-dependent.

40. If one considers entropy to be a measure of dispersion of energy, then at higher temperatures, it is logical that molecules would have more possible arrangements due to increased molecular motion.

41. Assuming the value of ΔS is positive, an increase in temperature will increase the $-\Delta G$ contribution of the entropy component to the overall energy change.

42. The heat exchange, getting colder, reflects only the enthalpy or ΔH component of the total energy change. The entropy change must be high enough to offset the enthalpy component and to add up to an overall $-\Delta G$.

43. Entropy would increase. There are more ways that two molecules, ADP and P_i, can be randomized than a single molecule, ATP.

1.12 What Is the Connection between Thermodynamics and Life?

44. The lowering of entropy needed to give rise to organelles leads to higher entropy in the surroundings, thus increasing the entropy of the universe as a whole.

45. Compartmentalization in organelles brings components of reactions into proximity with one another. The energy change of the reaction is not affected, but the availability of components allows it to proceed more readily.

46. DNA would have higher entropy with the strands separated. There are two single strands instead of one double strand, and the single strands have more conformational mobility.

47. See the answer to Question 43. It is still unlikely that cells could have arisen as bare cytoplasm, but the question of proximity of reactants is more to the point here than the energy change of a given reaction.

48. It would be unlikely that cells of the kind we know would have evolved on a gas giant. The lack of solids and liquids on which aggregates could form would make a large difference.

49. The available materials differ from those that would have been found on Earth, and conditions of temperature and pressure are very different.

50. Mars, because of conditions more like those on Earth.

51. A number of energetically favorable interactions drive the process of protein folding, ultimately increasing the entropy of the universe.

52. Photosynthesis is endergonic, requiring light energy from the Sun. The complete aerobic oxidation of glucose is exergonic and is a source of energy for many organisms, including humans. It would be reasonable to expect the two processes to take place differently in order to provide energy for the endergonic one.

Chapter 2

2.1 What Makes Water a Polar Molecule?

1. The unique fitness of water for forming hydrogen bonds determines the properties of many important biomolecules. Water can also act as an acid and as a base, giving it great versatility in biochemical reactions.

2. If atoms did not differ in electronegativity, there would be no polar bonds. This would drastically affect all reactions that involve functional groups containing oxygen or nitrogen—that is, most biochemical reactions.

2.2 What Is a Hydrogen Bond?

3. Proteins and nucleic acids have hydrogen bonds as an important part of their structures.

4. Replication of DNA and its transcription to RNA requires hydrogen bonding of complementary bases to the DNA template strand.

5. The C—H bond is not sufficiently polar for greatly unequal distribution of electrons at its two ends. Also, there are no unshared pairs of electrons to serve as hydrogen bond acceptors.

6. Many molecules can form hydrogen bonds. Examples might be H_2O, CH_3OH, or NH_3.

7. For a bond to be called a hydrogen bond, it must have a hydrogen covalently bonded to O, N, or F. This hydrogen then forms a hydrogen bond with another O, N, or F.

8. In a hydrogen-bonded dimer of acetic acid the —OH portion of the carboxyl group on molecule 1 is hydrogen-bonded to the —C=O portion of the carboxyl group on molecule 2, and vice versa.

9. Glucose = 17 and sorbitol = 18, ribitol = 15; each alcohol group can bond to three water molecules and the ring oxygen binds to two. The sugar alcohols bind more than the corresponding sugars.

10. Positively charged ions will bind to nucleic acids as a result of electrostatic attraction to the negatively charged phosphate groups.

2.3 What Are Acids and Bases?

11. $(CH_3)_3NH^+$ (conjugate acid)
$(CH_3)_3N$ (conjugate base)
^+H_3N—CH_2—$COOH$ (conjugate acid)
^+H_3N—CH_2—COO^- (conjugate base)
^+H_3N—CH_2—COO^- (conjugate acid)
H_2N—CH_2—COO^- (conjugate base)
^-OOC—CH_2—$COOH$ (conjugate acid)
^-OOC—CH_2—COO^- (conjugate base)
^-OOC—CH_2—$COOH$ (conjugate base)
$HOOC$—CH_2—$COOH$ (conjugate acid)

12.

13. Aspirin is electrically neutral at the pH of the stomach and compares the membrane more easily than in the small intestine.

2.4 What Is pH, and What Does It Have to Do with the Properties of Water?

14. The definition of pH is $-\log[H^+]$. Due to the log function, a change in concentration of 10 will lead to a change in pH of 1. The log of 10 is 1, the log of 100 is 2, etc.

15. Blood plasma, pH 7.4 $[H^+] = 4.0 \times 10^{-8}$ M
Orange juice, pH 3.5 $[H^+] = 3.2 \times 10^{-4}$ M
Human urine, pH 6.2 $[H^+] = 6.3 \times 10^{-7}$ M
Household ammonia, pH 11.5 $[H^+] = 3.2 \times 10^{-12}$ M
Gastric juice, pH 1.8 $[H^+] = 1.6 \times 10^{-2}$ M

16. Saliva, pH 6.5 $[H^+] = 3.2 \times 10^{-7}$ M
Intracellular fluid (liver), pH 6.9 $[H^+] = 1.6 \times 10^{-7}$ M
Tomato juice, pH 4.3 $[H^+] = 5.0 \times 10^{-5}$ M
Grapefruit juice, pH 3.2 $[H^+] = 6.3 \times 10^{-4}$ M

17. Saliva, pH 6.5 $[OH^-] = 3.2 \times 10^{-8}$ M
Intracellular fluid (liver), pH 6.9 $[OH^-] = 7.9 \times 10^{-8}$ M
Tomato juice, pH 4.3 $[OH^-] = 2.0 \times 10^{-10}$ M
Grapefruit juice, pH 3.2 $[OH^-] = 1.6 \times 10^{-11}$ M

2.5 What Are Titration Curves?

18. (a) The numerical constant equal to the concentration of the products of the dissociation divided by the concentration of the undissociated acid form: $([H^+][A^-])/[HA]$.

(b) The qualitative or quantitative description of how much acid (HA) dissociates to hydrogen ion.

(c) The property of a molecule that has both a polar region and a nonpolar region.

(d) The amount of acid or base that can be added to a buffer before experiencing a sharp pH change.

(e) The point in a titration curve at which the added acid or base equals the amount of buffer originally present.

(f) The property of a molecule that is readily soluble in water (i.e., water loving).

(g) The property of a molecule that is insoluble in water (i.e., water hating).

(h) The property of a molecule that is not soluble in water. The property of a covalent bond in which there is even sharing of electrons and no dipole moments (partial charges).

(i) The property of a molecule that is soluble in water. The property of a covalent bond in which the electrons are not shared evenly and dipole moments (partial charges) exist.

(j) An experiment in which acid or base is added stepwise to a solution of a compound and the pH is measured as a function of the added substance.

2.6 What Are Buffers, and Why Are They Important?

19. The pK of the buffer should be close to the desired buffer pH, and the substance chosen should not interfere with the reaction being studied.

20. The useful pH range of a buffer is one pH unit above and below its pK_a.

21. Use the Henderson–Hasselbalch equation:

$$pH = pK_a + \log\left(\frac{[CH_3COO^-]}{[CH_3COOH]}\right)$$

$$5.00 = 4.76 + \log\left(\frac{[CH_3COO^-]}{[CH_3COOH]}\right)$$

$$0.24 = \log\left(\frac{[CH_3COO^-]}{[CH_3COOH]}\right)$$

$$\frac{[CH_3COO^-]}{[CH_3COOH]} = \text{inverse log of } 0.24 = \frac{1.7}{1}$$

22. Use the Henderson–Hasselbalch equation:

$$pH = pK_a + \log\left(\frac{[CH_3COO^-]}{[CH_3COOH]}\right)$$

$$4.00 = 4.76 + \log\left(\frac{[CH_3COO^-]}{[CH_3COOH]}\right)$$

$$-0.76 = \log\left(\frac{[CH_3COO^-]}{[CH_3COOH]}\right)$$

$$\frac{[CH_3COO^-]}{[CH_3COOH]} = \text{inverse log of } -0.76 = \frac{0.17}{1}$$

23. Use the Henderson–Hasselbalch equation:

$$pH = pK_a + \log\left(\frac{[TRIS]}{[TRIS\text{-}H^+]}\right)$$

$$8.7 = 8.3 + \log\left(\frac{[TRIS]}{[TRIS\text{-}H^+]}\right)$$

$$0.4 = \log\left(\frac{[TRIS]}{[TRISH\text{-}H^+]}\right)$$

$$\frac{[TRIS]}{[TRIS\text{-}H^+]} = \text{inverse log of } 0.4 = \frac{2.5}{1}$$

24. Use the Henderson–Hasselbalch equation:

$$pH = pK_a + \log\left(\frac{[HEPES]}{[HEPES\text{-}H^+]}\right)$$

$$7.9 = 7.55 + \log\left(\frac{[HEPES]}{[HEPES\text{-}H^+]}\right)$$

$$0.35 = \log\left(\frac{[HEPES]}{[HEPES\text{-}H^+]}\right)$$

$$\frac{[HEPES]}{[HEPES\text{-}H^+]} = \text{inverse log of } 0.35 = \frac{2.2}{1}$$

25. At pH 7.5, the ratio of $[HPO_4^{2-}]/[H_2PO_4^-]$ is 2/1 (pK_a of $H_2PO_4^-$ = 7.2), as calculated using the Henderson–Hasselbalch equation. K_2HPO_4 is a source of the base form, and HCl must be added to convert one-third of it to the acid form, according to the 2/1 base/acid ratio. Weigh out 8.7 grams of K_2HPO_4 (0.05 mol, based on a formula weight of 174 g/mol), dissolve in a small quality of distilled water, add 16.7 mL of 1 M HCl (gives 1/3 of 0.05 mol of hydrogen ion, which converts 1/3 of the 0.05 mol of HPO_4^{2-} to $H_2PO_4^-$) and dilute the resulting mixture to 1 L.

26. A 2/1 ratio of the base form to acid form is still needed, because the pH of the buffer is the same in both problems. NaH_2PO_4 is a source of the acid form, and NaOH must be added to convert two-thirds of it to the base form. Weigh out 6.0 g of NaH_2PO_4 (0.05 mol, based on a formula weight of 120 g/mol), dissolve in a small quantity of distilled water, add 33.3 mL of 1 M NaOH (gives 2/3 of 0.05 mol of hydroxide ion, which converts 2/3 of the 0.05 mol of $H_2PO_4^-$ to HPO_4^{2-}) and dilute the resulting mixture to 1 L.

27. After mixing, the buffer solution (100 mL) contains 0.75 M lactic acid and 0.25 M sodium lactate. The pK_a of lactic acid is 3.86. Use the Henderson–Hasselbalch equation

$$pH = pK_a + \log\left(\frac{[CH_3CHOHCOO^-]}{[CH_3CHOHCOOH]}\right)$$

$$pH = 3.86 + \log\left(\frac{[CH_3CHOHCOO^-]}{[CH_3CHOHCOOH]}\right)$$

$$pH = 3.86 + \log (0.25\ M/0.75\ M)$$

$$pH = 3.86 + (-0.48)$$

$$pH = 3.38$$

28. After mixing, the buffer solution (100 mL) contains 0.25 M lactic acid and 0.75 M sodium lactate. The pK_a of lactic acid is 3.86. Use the Henderson–Hasselbalch equation:

$$pH = pK_a + \log\left(\frac{[CH_3CHOHCOO^-]}{[CH_3CHOHCOOH]}\right)$$

$$pH = 3.86 + \log\left(\frac{[CH_3CHOHCOO^-]}{[CH_3CHOHCOOH]}\right)$$

$$pH = 3.86 + \log (0.75\ M/0.25\ M)$$

$$pH = 3.86 + (0.48)$$

$$pH = 4.34$$

29. Use the Henderson–Hasselbalch equation:

$$pH = pK_a + \log\left(\frac{[CH_3COO^-]}{[CH_3COOH]}\right)$$

$$pH = 4.76 + \log\left(\frac{[CH_3COO^-]}{[CH_3COOH]}\right)$$

$$pH = 4.76 + \log (0.25\ M/0.10\ M)$$

$$pH = 4.76 + 0.40$$

$$pH = 5.16$$

30. Yes, it is correct, calculate the molar amounts of the two forms and insert into the Henderson–Hasselbalch equation. (2.02 g = 0.0167 mol and 5.25 g = 0.0333 mol.)

31. The solution is a buffer because it contains equal concentrations of TRIS in the acid and free amine forms. When the two solutions are mixed, the concentrations of the resulting solution (in the absence of reaction) are 0.05 M HCl and 0.1 M TRIS because of dilution. The HCl reacts with half the TRIS present, giving 0.05 M TRIS (protonated form) and 0.05 M TRIS (free amine form).

32. Any buffer that has equal concentrations of the acid and basic forms will have a pH equal to its pK_a. Therefore, the buffer from Question 31 will have a pH of 8.3.

33. First calculate the moles of buffer that you have: 100 mL = 0.1 L, and 0.1 L of 0.1 M TRIS buffer is 0.01 mol. Since the buffer is at its pK_a, there are equal concentrations of the acid and basic form, so the amount of TRIS is 0.005 mol, and the amount of TRIS-H$^+$ is 0.005 mol. If you then add 3 mL of 1 M HCl, you will be adding 0.003 mol of H$^+$. This will react as shown:

$$TRIS + H^+ \rightarrow TRIS\text{-}H^+$$

until you run out of something, which will be the H$^+$, since it is the limiting reagent. The new amounts can be calculated as shown below:

$$\text{TRIS-H}^+ = 0.005 \text{ mol} + 0.003 \text{ mol} = 0.008 \text{ mol}$$

$$\text{TRIS} = 0.005 \text{ mol} - 0.003 \text{ mol} = 0.002 \text{ mol}$$

Now plug these values into the Henderson–Hasselbalch equation:

pH = 8.3 + log ([TRIS]/[TRIS-H$^+$]) = 8.3 + log(0.002/0.008)

pH = 7.70

34. First calculate the mol of buffer that you have (we are going to do some rounding off): 100 mL = 0.1 L, and 0.1 L of 0.1 M TRIS buffer is 0.01 mol. Since the buffer is at pH 7.70, we saw in Question 25 that the amount of TRIS is 0.002 mol, and the amount of TRIS-H$^+$ is 0.008 mol. If you then add 3 mL of 1 M HCl, you will be adding 0.003 mol of H$^+$. This will react as shown:

$$\text{TRIS} + \text{H}^+ \rightarrow \text{TRIS-H}^+$$

until you run out of something, which will be the TRIS, since it is the limiting reagent. All the TRIS is converted to TRIS-H$^+$:

$$\text{TRIS-H}^+ = 0.01 \text{ mol}$$

$$\text{TRIS} = {\sim}0 \text{ mol}$$

We have used up the buffer capacity of the TRIS. We now have 0.003 mol of H$^+$ in approximately 0.1 L of solution. This is approximately 0.3 M H$^+$.

$$\text{pH} = -\log 0.3$$

$$\text{pH} = 0.52$$

35. [H$^+$] = [A$^-$] for pure acid, thus $K_a = [\text{H}^+]^2/[\text{HA}]$

$$[\text{H}^+]^2 = K_a \times [\text{HA}] \qquad -2 \log [\text{H}^+] = \text{p}K_a - \log [\text{HA}]$$

$$\text{pH} = 1/2 \ \{\text{p}K_a - \log [\text{HA}]\}$$

36. Use the Henderson–Hasselbalch equation. [Acetate ion]/[acetic acid] = 2.3/1

37. A substance with a pK_a of 3.9 has a buffer range of 2.9 to 4.9. It will not buffer effectively at pH 7.5.

38. Use the Henderson–Hasselbalch equation. The ratio of A$^-$/HA would be 3981 to 1.

39. In all cases, the suitable buffer range covers a pH range of pK_a +/−1 pH units.

Lactic acid (pK_a = 3.86) and its sodium salt. pH 2.86–4.86
Acetic acid (pK_a = 4.76) and its sodium salt, pH 3.76–5.76
TRIS (see Table 3.4, pK_a = 8.3) in its protonated form and its free amine form, pH 7.3-9.3
HEPES (see Table 3.4, pK_a, = 7.55) in its zwitterionic and its anionic form, pH 6.55–8.55

40. Several of the buffers would be suitable, namely TES, HEPES, MOPS, and PIPES; but the best buffer would be MOPS, because its pK_a of 7.2 is closest to the desired pH of 7.3.

41. The solution is called 0.0500 M, even though the concentration of neither the free base nor the conjugate acid is 0.0500 M. Why is 0.0500 M the correct concentration to report? (Buffer concentrations are typically reported to be the sum of the two ionic forms.)

42. At the equivalence point of the titration, a small amount of acetic acid remains because of the equilibrium CH$_3$COOH $\rightleftharpoons$ H$^+$ + CH$_3$COO$^-$. There is a small, but nonzero, amount of acetic acid left.

43. Buffering capacity is based upon the amounts of the acid and base forms present in the buffer solution. A solution with a high buffering capacity can react with a large amount of added acid or base without drastic changes in pH. A solution with a low buffering capacity can react with only comparatively small amounts of acid or base before showing changes in pH. The more concentrated the buffer, the higher is its buffering capacity. The first buffer listed here has one-tenth the buffering capacity of the second, which in turn has one-tenth the buffering capacity of the third. All three buffers have the same pH, because they all have the same relative amounts of the acid and base form.

44. It would be more effective to start with the HEPES base. You want a buffer at a pH above the pK_a, which means that the base form will predominate when you have finished preparing it. It is easier to convert some of the base form to the acid form than most of the acid form to the base form.

45. In a buffer with the pH above the pK_a, the base form predominates. This would be useful as a buffer for a reaction that produces H$^+$ because there will be plenty of the base form to react with the hydrogen ion produced.

46. Zwitterions tend not to interfere with biochemical reactions.

47. It is useful to have a buffer that will maintain a stable pH even if assay conditions change. Dilution is one such possible change.

48. It is useful to have a buffer that will maintain a stable pH even if assay conditions change. Temperature variation is one such possible change.

49. The only zwitterion is $^+\text{H}_3\text{N}-\text{CH}_2-\text{COO}^-$.

50. Hypoventilation decreases the pH of blood.

Chapter 3

3.1 What Are Amino Acids, and What Is Their Three-Dimensional Structure?

1. D- and L-amino acids have different stereochemistry around the α-carbon. Peptides that contain D-amino acids are found in bacterial cell walls and in some antibiotics.

3.2 What Are the Structures and Properties of the Individual Amino Acids?

2. Proline is technically not an amino acid. Glycine contains no chiral carbon atoms.

3. An amino acid in which the R group contains the following: a hydroxyl group (serine, threonine, or tyrosine), a sulfur atom (cysteine or methionine), a second chiral carbon atom (isoleucine or threonine), an amino group (lysine), an amide group (asparagine or glutamine), an acid group (aspartate or glutamate), an aromatic ring (phenylalanine, tyrosine, or tryptophan), a branched side chain (leucine or valine).

4. In the peptide Val—Met—Ser—Ile—Phe—Arg—Cys—Tyr—Leu, the polar amino acids are Ser, Arg, Cys, and Tyr; the aromatic amino acids are Phe and Tyr; and the sulfur-containing amino acids are Met and Cys.

5. In the peptide Glu—Thr—Val—Asp—Ile—Ser—Ala, the nonpolar amino acids are Val, Ile, and Ala; the acidic amino acids are Glu and Asp.

6. Amino acids other than the usual 20 are produced by modification of one of the common amino acids. See Figure 3.5 for the structures of some modified amino acids. Hydroxyproline and hydroxylysine are found in collagen; thyroxine is found in thyroglobulin.

3.3 Do Amino Acids Have Specific Acid-Base Properties?

7. The ionized form of each of the following amino acids at pH 7—glutamic acid, leucine, threonine, histidine, and arginine:

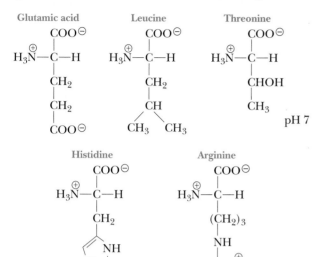

8.

9. Histidine: imidazole is deprotonated, α-amino group is predominantly deprotonated. Asparagine: α-amino group is deprotonated. Tryptophan: α-amino group is predominantly deprotonated. Proline: α-amino group is partially deprotonated. Tyrosine: α-amino group is predominantly deprotonated, phenolic hydroxyl is approximately a 50–50 mixture of protonated and deprotonated forms.

10. The isoelectric point for glutamic acid, 3.25; serine, 5.7; histidine, 7.58; lysine, 9.75; tyrosine, 5.65; arginine, 10.75.

11. Cysteine will have no net charge at pH $5.02 = (1.71 + 8.33)/2$ (see titration curve below).

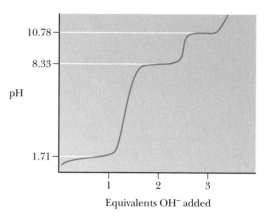

12.

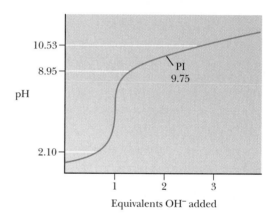

13. In all cases, the yield is 0.95^n. For 10 residues, that means 60% yield; for 50 residues, 8%; and for 100 residues, 0.6%. These are not satisfactory yields. Enzyme specificity gets around the problem.

14. The conjugate acid–base pair will act as a buffer in the pH range 1.09–3.09.

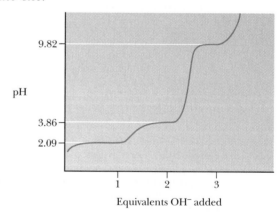

15. They have a net charge at pH extremes, and the molecules tend to repel each other. When the molecular charge is zero, the amino acids can aggregate more easily.

16. The ionic dissociation reactions of the following amino acids: aspartic acid, valine, histidine, serine, and lysine:

Aspartic acid

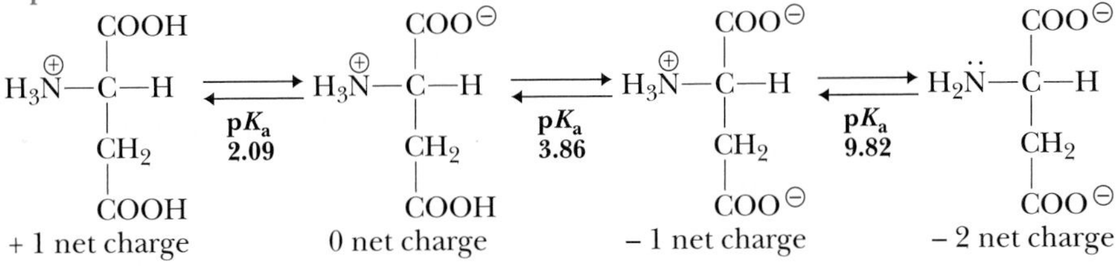

COOH
|
$H_3\overset{+}{N}$—C—H
|
CH_2
|
COOH
+ 1 net charge

$\xrightarrow{pK_a\ 2.09}$

$COO^{\ominus}$
|
$H_3\overset{+}{N}$—C—H
|
CH_2
|
COOH
0 net charge

$\xrightarrow{pK_a\ 3.86}$

$COO^{\ominus}$
|
$H_3\overset{+}{N}$—C—H
|
CH_2
|
$COO^{\ominus}$
− 1 net charge

$\xrightarrow{pK_a\ 9.82}$

$COO^{\ominus}$
|
$H_2\ddot{N}$—C—H
|
CH_2
|
$COO^{\ominus}$
− 2 net charge

Valine

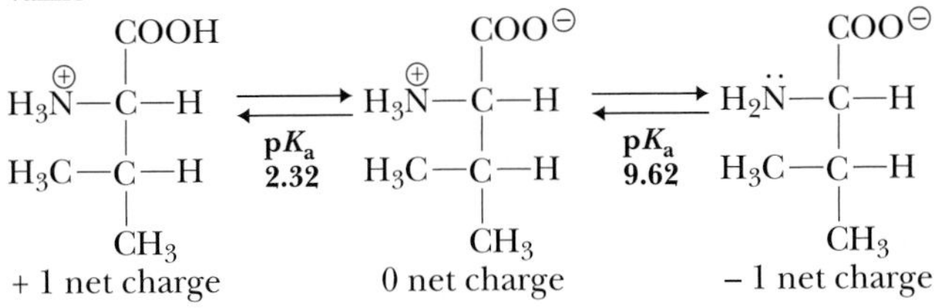

COOH
|
$H_3\overset{+}{N}$—C—H
|
H_3C—C—H
|
CH_3
+ 1 net charge

$\xrightarrow{pK_a\ 2.32}$

$COO^{\ominus}$
|
$H_3\overset{+}{N}$—C—H
|
H_3C—C—H
|
CH_3
0 net charge

$\xrightarrow{pK_a\ 9.62}$

$COO^{\ominus}$
|
$H_2\ddot{N}$—C—H
|
H_3C—C—H
|
CH_3
− 1 net charge

Histidine

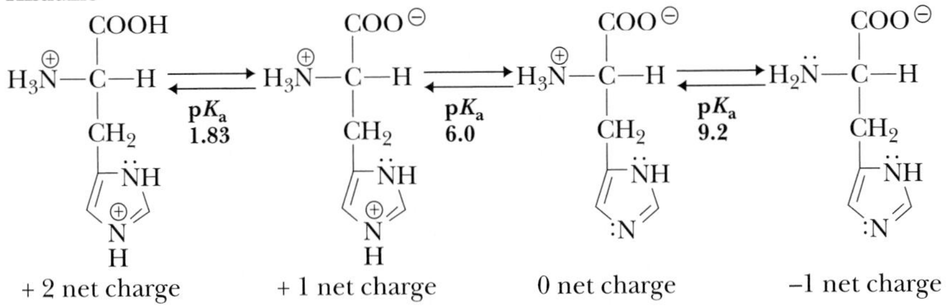

COOH
|
$H_3\overset{+}{N}$—C—H
|
CH_2
|
NH
(imidazole ring, protonated)
+ 2 net charge

$\xrightarrow{pK_a\ 1.83}$

$COO^{\ominus}$
|
$H_3\overset{+}{N}$—C—H
|
CH_2
|
NH
(imidazole ring, protonated)
+ 1 net charge

$\xrightarrow{pK_a\ 6.0}$

$COO^{\ominus}$
|
$H_3\overset{+}{N}$—C—H
|
CH_2
|
NH
(imidazole ring)
0 net charge

$\xrightarrow{pK_a\ 9.2}$

$COO^{\ominus}$
|
$H_2\ddot{N}$—C—H
|
CH_2
|
NH
(imidazole ring)
−1 net charge

Serine

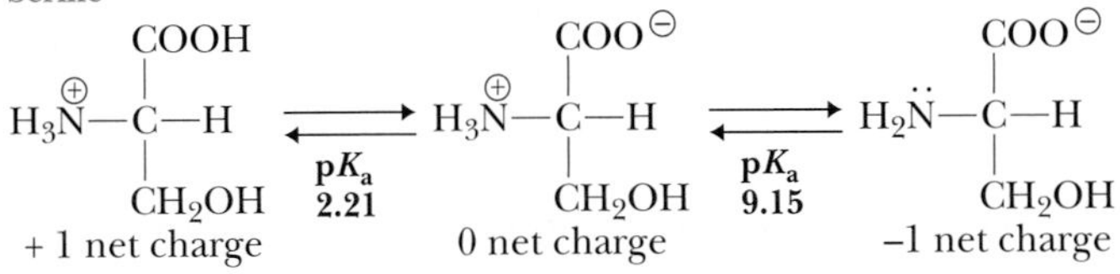

COOH
|
$H_3\overset{+}{N}$—C—H
|
CH_2OH
+ 1 net charge

$\xrightarrow{pK_a\ 2.21}$

$COO^{\ominus}$
|
$H_3\overset{+}{N}$—C—H
|
CH_2OH
0 net charge

$\xrightarrow{pK_a\ 9.15}$

$COO^{\ominus}$
|
$H_2\ddot{N}$—C—H
|
CH_2OH
−1 net charge

Lysine

COOH
|
$H_3\overset{+}{N}$—C—H
|
$(CH_2)_4$
|
$\overset{+}{N}H_3$
+ 2 net charge

$\xrightarrow{pK_a\ 2.18}$

$COO^{\ominus}$
|
$H_3\overset{+}{N}$—C—H
|
$(CH_2)_4$
|
$\overset{+}{N}H_3$
+ 1 net charge

$\xrightarrow{pK_a\ 8.95}$

$COO^{\ominus}$
|
$H_2\ddot{N}$—C—H
|
$(CH_2)_4$
|
$\overset{+}{N}H_3$
0 net charge

$\xrightarrow{pK_a\ 10.53}$

$COO^{\ominus}$
|
$H_2\ddot{N}$—C—H
|
$(CH_2)_4$
|
$\ddot{N}H_2$
−1 net charge

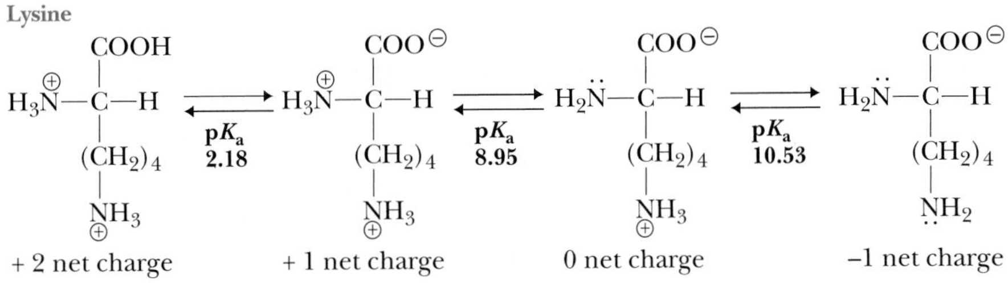

17. The pK_a for the ionization of the thiol group of cysteine is 8.33, so this amino acid could serve as a buffer in the —SH and —S— forms over the pH range 7.33–9.33. The α-amino groups of asparagine and lysine have pK'_a values of 8.80 and 8.95, respectively; these are also possible buffers, but they are both near the end of their buffer ranges.

18. At pH 4, the α-carboxyl group is deprotonated to a carboxylate, the side-chain carboxyl is still more than 50% protonated, and both amino groups are protonated. At pH 7, both the α-carboxyl group and the side-chain carboxyl group are deprotonated to a carboxylate, and both amino groups are protonated. At pH 10, both the α-carboxyl group and the side-chain carboxyl group are deprotonated to a carboxylate, one of the amino groups is primarily deprotonated, and the other amino group is a mixture of the protonated and deprotonated forms.

19. The pI refers to the form in which both carboxyl groups are deprotonated and both amino groups protonated at pH 6.96.

20. At pH 1, the charged groups are the N-terminal NH_3^+ on valine and the protonated guanidino group on arginine, giving zero net charge. The charged groups at pH 7 are the same as those at pH 1, with the addition of the carboxylate group on the C-terminal leucine, giving a net charge of -1.

21. Both peptides, Phe—Glu—Ser—Met and Val—Trp—Cys—Leu, have a charge of -1 at pH 1 because of the protonated N-terminal amino group. At pH 7, the peptide on the right has no net charge because of the protonated N-terminal amino group and the ionized C-terminal carboxylate negative charge. The peptide on the left has a net charge of -1 at pH 7 because of the side-chain carboxylate group on the glutamate in addition to the charges on the N-terminal and C-terminal groups.

22. (a) Lysine, because of the side-chain amino group; (b) Serine, because of the lack of a side-chain carboxyl; (c) Histidine, because of the presence of a titratable group in the side chain; (d) Aspartate, because of the carboxyl group in the side chain.

23. Glycine is frequently used as the basis of a buffer in the acid range near the pK of its carboxyl group. The useful buffer range is from pH 1.3 to 3.3.

3.4 What Is the Peptide Bond?

24. See Figure 3.10.

25. The resonance structures contribute to the planar arrangement by giving the C—N bond partial double-bond character.

26. Tyrosine, tryptophan, and their derivatives.

27. A monoamine oxidase is an enzyme that degrades compounds with an amino group, including neurotransmitters; consequently, they can control a person's mental state.

28. The two peptides differ in amino acid sequence but not in composition.

29. The titration curves of the two peptides will have the same general shape. The pK_a values of the α-amino and α-carboxyl groups will differ. Very careful work will show slight differences in side-chain pK_a values because of the different distances to the charged groups at the ends of the peptide. Such changes are particularly marked in proteins.

30. Asp—Leu—Phe   Leu—Asp—Phe   Phe—Asp—Leu
    Asp—Phe—Leu   Leu—Phe—Asp   Phe—Leu—Asp

31. DLF   LDF   FDL   DFL   LFD   FLD

32. You would get $20^{100} \approx 1.27 \times 10^{130}$ molecules, which is about 10^{84} Earth volumes. The same calculation for a pentapeptide gives more comprehensible results.

33. The different stereochemistry of the two peptides leads to different binding with taste receptors and therefore to the sweet taste for one and to the bitter taste for the other.

34. The high concentration of tryptophan in milk protein may mildly elevate the levels of serotonin, which relaxes the brain.

35. The tryptophan in milk might make you sleepy, whereas the tyramine in cheese should pep you up.

36. They are relatively stable because they are zwitterions. They typically have high melting points.

37. With very little doubt, no. Compare predicting the properties of water from those of hydrogen and oxygen, in either atomic or molecular form. If you knew the properties of the protein, you might be able to do the reverse to some extent.

38. The amino acids thyroxine and hydroxyproline occur in very few proteins. The genetic code does not include them, so they are produced by modification of tyrosine and proline, respectively.

39. These two peptides differ chemically. The open chain has a free C-terminal and N-terminal, but the cyclic peptide has only peptide bonds.

40. Both the C-terminal and the N-terminal of the open-chain peptide can be charged at appropriate pH values, which is not the case with the cyclic peptide. This can provide a basis for separation by electrophoresis.

41. Carbohydrates are not a source of the nitrogen needed for biosynthesis of amino acids.

42. Suggest that your friend show the carboxyl group as a charged carboxylate (—COO⁻) and the amino group in its charged form (—NH_3^+).

43. There are very few side chains that have functional groups to form crosslinks.

44. There would be many more possible conformations because of free rotation around the peptide bond.

45. There would be no possibility of disulfide crosslinks within or between peptide chains, giving more possible conformations. There would not be the possibility of oxidation–reduction reactions involving sulfhydryl and disulfide groups.

46. The big difference would be the loss of stereospecificity in the conformation of any peptide or protein. This would have drastic consequences for the kinds of reactions of the protein.

3.5 Are Small Peptides Physiologically Active?

47. Oxytocin has an isoleucine at position 3 and a leucine at position 8; it stimulates smooth muscle contraction in the uterus during labor and in the mammary glands during lactation. Vasopressin has a phenylalanine at position 3 and an arginine at position 8; it stimulates resorption of water by the kidneys, thus raising blood pressure.

48. The reduced form of glutathione consists of three amino acids with a sulfhydryl group; the oxidized form consists of six amino acids and can be considered to be the result of linking two molecules of reduced glutathione by a disulfide bridge.

49. Enkephalins are pentapeptides (Y—G—G—F—L, leucine enkephalin, and Y—G—G—F—M, methionine enkephalin), which are naturally occurring analgesics.

50. In most cases, D-amino acids are toxic. They occur in nature in antibiotics and bacterial cell walls.

Chapter 4

4.1 How Does the Structure of Proteins Determine Their Function?

1. (a) (2); (b) (4); (c) (l); (d) (3).

2. When a protein is denatured, the interactions that determine secondary, tertiary, and any quaternary structures are overcome by the presence of the denaturing agent. Only the primary structure remains intact.

3. The random portions of a protein do not contain structural motifs that are repeated within the protein, such as α-helix or β-pleated sheet, but three-dimensional features of these parts of the protein are repeated from one molecule to another. Thus, the term "random" is somewhat of a misnomer.

4.2 What Is the Primary Structure of Proteins?

4. When a protein is covalently modified, its primary structure is changed. The primary structure determines the final three-dimensional structure of the protein. The modification disrupts the folding process.

5. (a) Serine has a small side chain that can fit in any relatively polar environment. (b) Tryptophan has the largest side chain of any of the common amino acids, and it tends to require a nonpolar environment. (c) Lysine and arginine are both basic amino acids; exchanging one for the other would not affect the side-chain pK_a in a significant way. Similar reasoning applies to the substitution of a nonpolar isoleucine for a nonpolar leucine.

6. Glycine is frequently a conserved residue because its side chain is so small that it can fit into spaces that will not accommodate larger ones.

7. When alanine is replaced by isoleucine, there is not enough room in the native conformation for the larger side chain of the isoleucine. Consequently, there is a great enough change in the conformation of the protein that it loses activity. When glycine is substituted in turn for isoleucine, the presence of the smaller side chain leads to a restoration of the active conformation.

8. Meat consists largely of animal proteins and fat. The temperatures involved in cooking meat are usually more than enough to denature the protein portion of the meat.

9. Prion diseases have been linked to the immune system. It is believed that the prion proteins travel in the lymph system bound to lymphocytes and eventually arrive at the nervous tissue, where they begin to transform a normal cellular protein into an abnormal one (a prion).

10. Although there may be a strong genetic predisposition to acquire scrapie, that alone will not cause the disease. The disease must be started by ingesting a prion that already has the altered conformation, PrPsc.

4.3 What Is the Secondary Structure of Proteins?

11. Shape, solubility, and type of biological function (static, structural versus dynamic, catalytic).

12. The protein efficiency ratio is an arbitrary measurement of the essential amino acid content of a given type of protein.

13. Eggs have the highest PER.

14. The amino acids that must be consumed in the diet because the body cannot synthesize them in sufficient quantities.

15. Reasons for creating genetically modified foods include increasing their protein content, increasing their shelf life, increasing their resistance to insects or other pests, and decreasing the need for using pesticides in order to grow them economically.

16. The angles of the amide planes as they rotate about the α-carbon. The angles are both defined as zero when the two planes would be overlapping such that the carbonyl group of one contacts the N—H of the other.

17. A β-bulge is a common nonrepetitive irregularity found in antiparallel β-sheets. A misalignment occurs between strands of the β-sheet, causing one side to bow outward.

18. A reverse turn is a region of a polypeptide where the direction changes by about 180°. There are two kinds—those that contain proline and those that do not. See Figure 4.6 for examples.

19. The α-helix is not fully extended, and its hydrogen bonds are parallel to the protein fiber. The β-pleated sheet structure is almost fully extended, and its hydrogen bonds are perpendicular to the protein fiber.

20. The αα unit, the βαβ unit, the β-meander, the Greek key, and the β-barrel.

21. The geometry of the proline residue is such that it does not fit into the α-helix, but it does fit exactly for a reverse turn. See Figure 4.10(c).

22. Glycine is the only residue small enough to fit at crucial points in the collagen triple helix.

23. The principal component of wool is the protein keratin, which is a classic example of α-helical structure. The principal component of silk is the protein fibroin, which is a classic example of β-pleated sheet structure. The statement is somewhat of an oversimplification, but it is fundamentally valid.

24. Wool, which consists largely of the protein keratin, shrinks because of its α-helical conformation. It can stretch and then shrink. Silk consists largely of the protein fibroin, which has the fully extended β-sheet conformation, with far less tendency to stretch or shrink.

4.4 What Can We Say about the Thermodynamics of Protein Folding?

25. (1) Backbone H-bonds, involving the CO and NH groups of the peptide chain; (2) side-chain H-bonds, involving any possible hydrogen-bond donors or acceptors on the side chains; (3) hydrophobic interactions, involving the nonpolar groups on the protein; (4) electrostatic interactions, involving any charged groups on the protein; (5) metal ligation, involving coordination bonds between side chains and a metal ion.

26. Stabilization of any conformation depends on lowering of energy, regardless of the kind of interaction involved. With small molecule interactions, the main component of energy changes is enthalpic, with relatively small entropy changes. In the case of molecules as large as proteins, the number of possible conformations makes the entropic part of energy changes much more important. This is particularly true in the case of hydrophobic interactions.

4.5 What Is the Tertiary Structure of Proteins?

27. See Figure 4.2 for a hydrogen bond that is part of the α-helix (secondary structure). See Figure 4.13 for a hydrogen bond that is part of tertiary structure (side-chain hydrogen bonding).

28. See Figure 4.13 for electrostatic interactions, such as might be seen between the side chains of lysine and aspartate.

29. See Figure 4.13 for an example of a disulfide bond.

30. See Figure 4.13 for an example of hydrophobic bonds.

31. A chaperone is a protein that aids another protein in folding correctly and keeps it from associating with other proteins before it has reached its final, mature form.

32. *Configuration* refers to the position of groups due to covalent bonding. Examples include *cis* and *trans* isomers and optical isomers. *Conformation* refers to the positioning of groups in space due to rotation around single bonds. An example is the difference between the eclipsed and staggered conformations of ethane.

33. Five possible features limit possible protein configurations and conformations. (1) Although any one of 20 amino acids is possible at each position, only one is used, as dictated by the gene that codes for that protein. (2) Either a D- or an L-amino acid could be used at each position (except for glycine), but only L-amino acids are used. (3) The peptide group is planar, so that only *cis* and *trans* arrangements are observed. The *trans* form is more stable and is the one usually found in proteins. (4) The angles φ and ψ can theoretically take on any value from 0° to 360°, but some angles are not possible because of steric hindrance; angles that are sterically allowed may not have stabilizing interactions, such as those in the α-helix. (5) The primary structure determines an optimum tertiary structure, according to the "second half of the genetic code."

34. Technically, collagen has quaternary structure because it has multiple polypeptide chains. However, most discussions of quaternary structure involve subunits of globular proteins, not fibrous ones like collagen. Many scientists consider the collagen triple helix to be an example of a secondary structure.

4.6 Can We Predict Protein Folding from Sequence?

35. This level of sequence homology is marginal for use of comparative modeling. It is best to try that method, but then to compare the results with those obtained from the fold-recognition approach.

36. See the Protein Data Bank.

4.7 What Is the Quaternary Structure of Proteins?

37. A prion is a potentially infectious protein found in multiple forms in mammals, often concentrated in nervous tissue. It is an abnormal form of a normal cellular protein. It tends to form plaques that destroy the nervous tissue. Prions have been found to be transmissible across species.

38. A series of encephalopathies have been found to be caused by prions. In cows, the disease caused by prions is called bovine spongiform encephalopathy, or more commonly mad-cow disease. In sheep, the disease is called scrapie. In humans, it is called Creutzfeld–Jakob disease.

39. The normal form of the prion protein has a higher α-helix content compared to β-sheet. The abnormal one has an increased β-sheet content.

40. *Similarities:* both contain heme group; both are oxygen binding; secondary structure is primarily α-helix. *Differences:* hemoglobin is a tetramer, while myoglobin is a monomer; oxygen binding to hemoglobin is cooperative, but noncooperative to myoglobin.

41. The crucial residues are histidines in both proteins.

42. Myoglobin's highest level of organization is tertiary. Hemoglobin's is quaternary.

43. The function of hemoglobin is oxygen transport; its sigmoidal binding curve reflects the fact that it can bind easily to oxygen at comparatively high pressures and release oxygen at lower pressures. The function of myoglobin is oxygen storage; as a result, it is easily saturated with oxygen at low pressures, as shown by its hyperbolic binding curve.

44. In the presence of H^+ and CO_2, both of which bind to hemoglobin, the oxygen-binding capacity of hemoglobin decreases.

45. In the absence of 2,3-*bis*phosphoglycerate, the binding of oxygen by hemoglobin resembles that of myoglobin, characterized by lack of cooperativity. 2,3-*bis*phosphoglycerate binds at the center of the hemoglobin molecule, increases cooperativity, stabilizes the deoxy conformation of hemoglobin, and modulates the binding of oxygen so that it can easily be released in the capillaries.

46. Fetal hemoglobin binds oxygen more strongly than adult hemoglobin. See Figure 4.25.

47. Histidine 143 in a β-chain is replaced by a serine in a γ-chain.

48. Deoxygenated hemoglobin is a weaker acid (has a higher pK_a) than oxygenated hemoglobin. In other words, deoxygenated hemoglobin binds more strongly to H^+ than does oxygenated hemoglobin. The binding of H^+ (and of CO_2) to hemoglobin favors the change in quaternary structure to the deoxygenated form of hemoglobin.

49. The primary flaw in your friend's reasoning is a reversal of the definition of pH, which is $pH = -\log [H^+]$. If the release or binding of hydrogen ion by hemoglobin were the primary factor in the Bohr effect, the pH changes would be the opposite of those actually observed. The response of hemoglobin to changes in pH is the central point. When the pH increases, the hydrogen ion concentration decreases, and vice versa.

50. The change of a histidine to a serine in the γ-chain removes a positively charged amino acid that could have interacted with BPG. Thus there are fewer salt bridges to break, so binding is easier than it is in a β-chain.

51. Persons with sickle-cell trait have some abnormal hemoglobin. At high altitudes, there is less oxygen, and the concentration of the deoxy form of the abnormal hemoglobin increases. Less oxygen can be bound, causing the observed breathing difficulties.

52. In fetal hemoglobin, the subunit composition is $\alpha_2\gamma_2$ with replacement of the β-chains by the γ-chains. The sickle-cell mutation affects the β-chain, so the fetus homozygous for Hb S has normal fetal hemoglobin.

53. The relative oxygen affinities allow oxygen to be taken by the fetal cells from the maternal Hb.

54. Because people with sickle-cell disease are chronically anemic, some cells with fetal Hb are produced to help overcome the impaired oxygen delivery system.

55. The crystalline form changed because oxygen entered under the cover slip, transforming deoxyhemoglobin to oxyhemoglobin.

Chapter 5

5.1 How Do We Extract Pure Proteins from Cells?

1. Using a blender, a Potter-Elvejhem homogenizer, or a sonicator.

2. If you needed to maintain the structural integrity of the subcellular organelles, a Potter-Elvejhem homogenizer would be better because it is more gentle. The tissue, such as liver, must be soft enough to use with this device.

3. Salting out is a process whereby a highly ionic salt is used to reduce the solubility of a protein until it comes out of solution and can be centrifuged. The salt forms ion—dipole bonds with the water in the solution, which leaves less water available to hydrate the protein. Nonpolar side chains begin to interact between protein molecules, and they become insoluble.

4. Their amino acid content and arrangements make some proteins more soluble than others. A protein with more highly polar amino acids on the surface will be more soluble than one with more hydrophobic ones on the surface.

5. First homogenize the liver cells using a Potter-Elvejhem homogenizer. Then spin the homogenate at $500 \times g$ to sediment the unbroken cells and nuclei. Centrifuge the supernatant at 15,000 $\times g$, and collect the pellet, which contains the mitochondria.

6. No, peroxisomes and mitochondria have overlapping sedimentation characteristics. Other techniques, such as sucrose-gradient centrifugation, would have to be used to separate the two organelles.

7. If the protein were cytosolic, once the cells were broken open, you could centrifuge at $100,000 \times g$, and all the organelles would be in the pellet. Your enzyme would be in the supernatant, along with all the other cytosolic ones.

8. Isolate the mitochondria via differential or sucrose-gradient centrifugation. Use another homogenization technique, combined with a strong detergent, to release the enzyme from the membrane.

9. Tables exist to tell you how many grams of ammonium sulfate $[(NA_4)_2 SO_4]$ to add to get a certain percent saturation. A good plan would be to take the homogenate and add enough ammonium sulfate to yield a 20% saturated solution. Let the sample sit for 15 minutes on ice and then centrifuge. Separate the supernatant from the precipitate. Assay both for the protein you are working with. Add more ammonium sulfate to the supernatant to arrive at a 40% saturated solution and repeat the process. In this way, you will find out what the percent saturation in ammonium sulfate needs to be to precipitate the protein.

10. Reasonably harsh homogenization would be able to liberate the soluble protein X from the peroxisomes, which are fragile. Centrifugation at $15,000 \times g$ would sediment the mitochondria (broken or intact). The supernatant would then have protein X but no protein Y. Freeze/thaw techniques and sonication would accomplish the same thing, or the mitochondria and the peroxisomes could be separated initially by sucrose-gradient centrifugation.

5.2 What Is Column Chromatography?

11. (a) Size, (b) specific ligand-binding ability, (c) net charge

12. The largest proteins elute first; the smallest elute last. Larger proteins are excluded from the interior of the gel bead so they have less available column space to travel. Essentially, they travel a shorter distance and elute first.

13. A compound can be eluted by raising the salt concentration or by adding a mobile ligand that has a higher affinity for the bound protein than the stationary resin ligand does. Salt is cheaper but less specific. A specific ligand may be more specific, but it is likely to be expensive.

14. A compound can be eluted by raising the salt concentration or by changing the pH. Salt is cheap, but it might not be as specific for

a particular protein. Changing the pH may be more specific for a tight pI range, but extremes of pH may also denature the protein.

15. Raising the salt concentration is relatively safe. Most proteins will elute this way, and, if the protein is an enzyme, it will still be active. If necessary, the salt can be removed later via dialysis. Changing the pH enough to remove the charge can cause the proteins to become denatured. Many proteins are not soluble at the isoelectric points.

16. The basis of most resins is agarose, cellulose, dextran, or polyacrylamide.

17. See Figure 5.7.

18. Within the fractionation range of a gel-filtration column, molecules will elute with a linear relationship of log MW versus their elution volumes. A series of standards can be run to standardize the column, and then an unknown can be determined by measuring its elution volume and comparing it to a standard curve.

19. Both proteins would elute in the void volume together and would not be separated.

20. Yes, the β-amylase would come out in the void volume, but the bovine serum albumin would be included in the column bead and would elute more slowly.

21. Set up an anion-exchange column, such as Q-Sepharose (quaternary amine). Run the column at pH 8.5, a pH at which the protein X has a net negative charge. Put a homogenate containing protein X on the column and wash with the starting buffer. Protein X will bind to the column. Then elute by running a salt gradient.

22. Use a cation-exchange column, like CM-Sepharose, and run it at pH 6. Protein X will have a positive charge and will stick to the column.

23. With a quaternary amine, the column resin always has a net positive charge, and you don't have to worry about the pH of your buffer altering the form of the column. With a tertiary amine, there is a dissociable hydrogen, and the resin may be positive or neutrally charged, depending on the buffer pH.

24. The easiest way would be to use a sucrose gradient to separate the mitochondria from the peroxisomes first. Then break open the mitochondria via harsh homogenization or sonication, and then centrifuge the mitochondria. The pellet would contain protein B, while the supernatant would contain protein A. Contaminants could still exist, but they could be cleaned away by running gel filtration, on Sephadex G-75 (which would separate enzyme C from enzymes A and B), and then by running ion-exchange chromatography on Q-Sepharose at pH 7.5. Emzyme B would be neutral and would elute, while enzyme A would stick to the column.

25. Glutamic acid will be eluted first because the column pH is close to its pI. Leucine and lysine will be positively charged and will stick to the column. To elute leucine, raise the pH to around 6. To elute lysine, raise the pH to around 11.

26. A nonpolar mobile solvent will move the nonpolar amino acids fastest, so phenylalanine will be the first to elute, followed by glycine and then glutamic acid.

27. The nonpolar amino acids will stick the most to the stationary phase, so glutamic acid will move the fastest, followed by glycine and then phenylalanine.

28. A protein solution from an ammonium sulfate preparation is passed over a gel-filtration column where the proteins of interest will elute in the void volume. The salt, being very small, will move through the column slowly. In this way, the proteins will leave the salt behind and exit the column without it.

5.3 What Is Electrophoresis?

29. Size, shape, and charge.
30. Agarose and polyacrylamide.
31. Polyacrylamide.

32. DNA is the molecule most often separated on agarose electrophoresis, although proteins can also be separated.

33. Those with the highest charge/mass ratio would move the fastest. There are three variables to consider, and most electrophoreses are done in a way to eliminate two of the variables so that the separation is by size or by charge, but not by both.

34. Sodium dodecyl—sulfate polyacrylamide gel-electrophoresis. With SDS–PAGE, the charge and shape differences of proteins are eliminated so that the only parameter determining the migration is the size of the protein.

35. SDS binds to the protein in a constant ratio of 1.4 g SDS per gram of protein. It coats the protein with negative charges and puts it into a random coil shape. Thus, charge and shape are eliminated.

36. In a polyacrylamide gel used for gel-filtration chromatography, the larger proteins can travel around the beads, thereby having a shorter path to travel and therefore eluting faster. With electrophoresis, the proteins are forced to go through the matrix, so the larger ones travel more slowly because there is more friction.

37. The MW is 37,000 daltons.

5.4 How Do We Determine the Primary Structure of a Protein?

38. The Edman degradation will give the identity of the N-terminal amino acid in its first cycle, so doing a separate experiment is not necessary.

39. It might tell you if the protein were pure or if there were subunits.

40.

41. The amount of Edman reagent must exactly match the amount of N-termini in the first reaction. If there is too little Edman reagent, some of the N-termini will not react. If there is too much, some of the second amino acid will react. In either case, there will be a small amount of contaminating phenylthiohydantoin (PTH) derivatives. This error grows with the number of cycles run until the point that two amino acids are released in equal amounts, and you cannot tell which one was supposed to be the correct one.

42. In the first cycle, the first and second amino acids from the N-terminal end would be reacted and released as PTH derivatives. You would get a double signal and not know which one was the true N-terminus.

43. Val—Leu—Gly—Met—Ser—Arg—Asn—Thr—Trp—Met—Ile—Lys—Gly—Tyr—Met—Gln—Phe

44. Met—Val—Ser—Thr—Lys—Leu—Phe—Asn—Glu—Ser—Arg—Val—Ile—Trp—Thr—Leu—Met—Ile

45. It is possible that your protein is not pure and needs additional purification steps to arrive at a single polypeptide. It is also possible that the protein has subunits, so multiple polypeptide chains could be yielding the contradictory results.

46. There are two fragments that have C-termini that are not lysine or arginine, which is what trypsin is specific for. Normally there would be only one fragment ending with an amino acid that was not Arg or Lys, and we would immediately know that it was the C-terminus. Histidine is a basic amino acid, although it is usually neutral and therefore does not react with trypsin. It is possible that, in the pH environment of the reaction, the histidine was positively charged and was recognized by trypsin.

47. It would tell you a relative concentration of the various amino acids. This is important because it would help you plan your sequencing experiment better. For example, if you had a protein whose composition showed no aromatic amino acids, it would be a waste of time to use a chymotrypsin digestion.

48. Cyanogen bromide would be useless, because there is no methionine. Trypsin would be little better, because the protein is 35% basic residues. Trypsin would shred the protein into more than 30 pieces, which would be very hard to analyze.

49. Chymotrypsin would be a good choice. There are more than four residues of aromatic amino acids. The protein, containing 100 amino acids, would be cut four times, possibly yielding nice fragments roughly 20–30 amino acids long, which can be sequenced effectively by the Edman degradation.

50. It would work best if the basic residues were spread out in the protein. In that way, fragments in the proper size range would be generated. If all four of the basic residues were in the first ten amino acids, there would be one long fragment that could not be sequenced.

Chapter 6

6.1 What Makes Enzymes Such Effective Biological Catalysts?

1. Enzymes are many orders of magnitude more effective as catalysts than are nonenzymatic catalysts.

2. The majority of enzymes are proteins, but some catalytic RNAs (ribozymes) are known.

3. About 3 seconds (1 year × 1 event/10^7 events × 365 days/year × 24 hours/day × 3600 seconds/hour = 3.15 seconds).

4. Enzymes hold the substrates in favorable spatial positions, and they bind effectively to the transition state to stabilize it. Note that *all* catalysts lower the activation energy, so this is not a particular enzyme function.

6.2 What Is the Difference between the Kinetic and the Thermodynamic Aspects of Reactions?

5. The reaction of glucose with oxygen is thermodynamically favored, as shown by the negative free-energy change. The fact that glucose can be maintained in an oxygen atmosphere is a reflection of the kinetic aspects of the reaction, requiring overcoming an activation-energy barrier.

6. To the first question, most probably: local concentrations (mass-action concepts) could easily dictate the direction. To the second question, probably not: local concentrations would seldom be sufficient to overcome a relatively large $\Delta G°$ of –5.3 kcal in the reverse reaction. (See, however, the aldolase reaction in glycolysis.)

7. Heating a protein denatures it. Enzymatic activity depends on the correct three-dimensional structure of the protein. The presence of bound substrate can make the protein harder to denature.

8. The results do not prove that the mechanism is correct because results from different experiments could contradict the proposed mechanism. In that case, the mechanism would have to be modified to accommodate the new experimental results.

9. The presence of a catalyst affects the rate of a reaction. The standard free-energy change is a thermodynamic property that does not depend on the reaction rate. Consequently, the presence of the catalyst has no effect.

10. The presence of a catalyst lowers the activation energy of a reaction.

11. Enzymes, like all catalysts, increase the rate of the forward and reverse reaction to the same extent.

12. The amount of product obtained in a reaction depends on the equilibrium constant. A catalyst does not affect that.

6.3 How Can We Describe Enzyme Kinetics in Mathematical Terms?

13. The reaction is first order with respect to A, first order with respect to B, and second order overall. The detailed mechanism of the reaction is likely to involve one molecule each of A and B.

14. The easiest way to follow the rate of this reaction is to monitor the decrease in absorbance at 340 nm, reflecting the disappearance of NADH.

15. The use of a pH meter would not be a good way to monitor the rate of the reaction. You are probably running this reaction in a buffer solution to keep the pH relatively constant. If you are not running the reaction in a buffer solution, you run the risk of acid denaturation of the enzyme.

16. Enzymes tend to have fairly sharp pH optimum values. It is necessary to ensure that the pH of the reaction mixture stays at the optimum value. This is especially true for reactions that require or produce hydrogen ions.

6.4 How Do Substrates Bind to Enzymes?

17. In the lock-and-key model, the substrate fits into a comparatively rigid protein that has an active site with a well-defined shape. In the induced-fit model, the enzyme undergoes a conformational change on binding to the substrate. The active site takes shape around the substrate.

18.

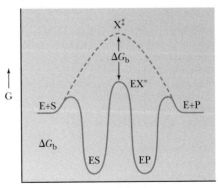

No destabilization,
thus no catalysis

19. The ES complex would be in an "energy trough," with a consequentially large activation energy to the transition state.
20. Amino acids that are far apart in the amino acid sequence can be close to each other in three dimensions because of protein folding. The critical amino acids are in the active site.
21. The overall protein structure is needed to ensure the correct arrangement of amino acids in the active site.
22. The strong inhibition indicates tight binding to the active site. Thus, the compound is very likely to be a transition-state analogue.

6.5 What Are Some Examples of Enzyme-Catalyzed Reactions?

23. See Figures 6.6 and 6.7.
24. Not all enzymes follow Michaelis–Menten kinetics. The kinetic behavior of allosteric enzymes does not obey the Michaelis–Menten equation.
25. The graph of rate against substrate concentration is sigmoidal for an allosteric enzyme but hyperbolic for an enzyme that obeys the Michaelis–Menten equation.

6.6 What Is the Michaelis–Menten Approach to Enzyme Kinetics?

26. The reaction velocity remains the same with increasing enzyme concentration. It is theoretically possible, but highly unlikely, for a reaction to be saturated with enzyme.
27. The steady-state assumption is that the concentration of the enzyme—substrate complex does not change appreciably over the time in which the experiment takes place. The rate of appearance of the complex is set equal to its rate of disappearance, simplifying the equations for enzyme kinetics.
28. Turnover number = $V_{max}/[E_T]$.
29. Use Equation 6.16. (a) $V = 0.5\ V_{max}$; (b) $V = 0.33\ V_{max}$; (c) $V = 0.09\ V_{max}$; (d) $V = 0.67\ V_{max}$; (e) $V = 0.91\ V_{max}$.
30. See graph: $V_{max} = 0.681\ \text{m}M\ \text{min}^{-1}$, $K_M = 0.421$ M.

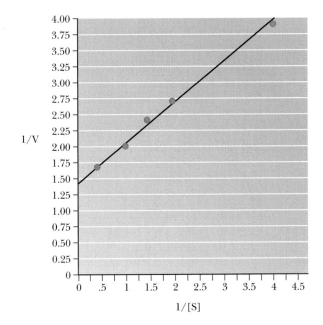

31. See graph: $V_{max} = 2.5 \times 10^{-4}\ M\ \text{sec}^{-1}$, $K_M = 1.6 \times 10^8$ M.

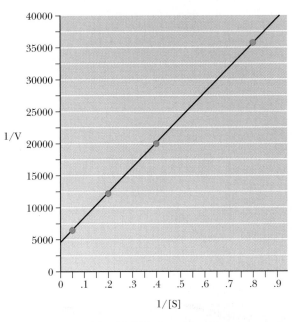

32. See graph: $K_M = 2.86 \times 10^{-2}$ M. Concentrations were not determined directly. Absorbance values were used instead as a matter of convenience.

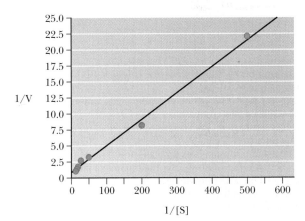

33. See graph: $V_{max} = 1.32 \times 10^{-3}\ M\ min^{-1}$, $K_M = 1.23 \times 10^{-3}$ M.

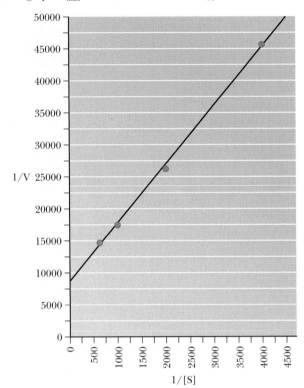

34. The turnover number is 20.43 per minute.
35. The number of moles of enzyme is 1.56×10^{-10}. Turnover number = 10,700 sec^{-1}.
36. The low K_M for the aromatic amino acids indicates that they will be oxidized preferentially.
37. It is easier to detect deviations of individual points from a straight line than from a curve.
38. The assumption that the K_M is an indication of the binding affinity between the substrate and the enzyme is valid when the rate of dissociation of the enzyme–substrate complex to product and enzyme is much smaller than the rate of dissociation of the complex to enzyme and substrate.

6.7 How Do Enzymatic Reactions Respond to Inhibitors?

39. In the case of competitive inhibition, the value of K_M increases, while the value of K_M remains unchanged in noncompetitive inhibition.
40. A competitive inhibitor blocks binding, not catalysis.
41. A noncompetitive inhibitor does not change the affinity of the enzyme for its substrate.
42. A competitive inhibitor binds to the active site of an enzyme, preventing binding of the substrate. A noncompetitive inhibitor binds at a site different from the active site, causing a conformational change, which renders the active site less able to bind substrate and convert it to product.
43. Competitive inhibition can be overcome by adding enough substrate, but this is not true for all forms of enzyme inhibition.
44. A Lineweaver–Burk plot is useful because it gives a straight line. It is easier to determine how well points fit to a straight line than to a curve.
45. In a Lineweaver–Burk plot for competitive inhibition, the lines intersect at the y-axis intercept, which is equal to $1/V_{max}$. In a Lineweaver–Burk plot for noncompetitive inhibition, the lines intersect at the x-axis intercept, which is equal to $-1/K_M$.

46. $K_M = 7.42$ mM; $V_{max} = 15.9$ mmol min^{-1}; noncompetitive inhibition.

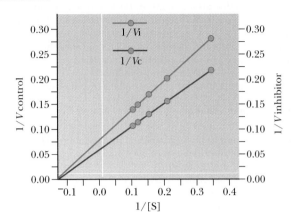

47. Competitive inhibition, $K_M = 6.5 \times 10^{-4}$. The key point here is that the V_{max} is the same within the limits of error. Some of the concentrations are given to one significant figure.

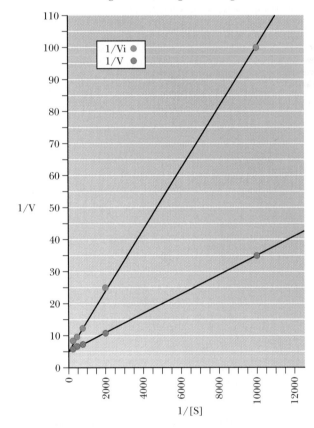

48. It is *very* good, in the case of noncompetitive inhibitors; much of metabolic control depends on feedback inhibition by downstream noncompetitive inhibitors. The question is perhaps moot in the case of competitive inhibitors, which are much less commonly encountered in vivo. Some antibiotics, however, are competitive inhibitors (good for the sick person, bad for the bacteria).
49. Both the slope and the intercepts will change. The lines will intersect above the x axis at negative values of $1/[S]$.
50. Not all AIDS drugs are enzyme inhibitors, but an important class of such drugs inhibits the HIV protease. You would need to understand the concepts of substrate binding, inhibition, and inhibitor binding.

51. An irreversible inhibitor is bound by covalent bonds. Noncovalent interactions are relatively weak and easily broken.

52. A noncompetitive inhibitor does not bind to the active site of an enzyme. Its structure need bear no relation to that of the substrate.

Chapter 7

7.1 Does the Michaelis–Menten Model Describe the Behavior of Allosteric Enzymes?

1. Allosteric enzymes display sigmoidal kinetics when rates are plotted versus substrate concentration. Michaelis–Menten enzymes exhibit hyperbolic kinetics. Allosteric enzymes usually have multiple subunits, and the binding of substrates or effector molecules to one subunit changes the binding behavior of the other subunits.

2. It is an enzyme used in the early stages of cytidine nucleotide synthesis.

3. ATP acts as a positive effector of ATCase, and CTP acts as an inhibitor.

4. The term K_M should be used for enzymes that display Michaelis–Menten kinetics. Thus, it is not used with allosteric enzymes. Technically, competitive and noncompetitive inhibition are also terms that are restricted to Michaelis–Menten enzymes, although the concepts are applicable to any enzyme. An inhibitor that binds to an allosteric enzyme at the same site as the substrate is similar to a classical competitive inhibitor. One that binds at a different site is similar to a noncompetitive inhibitor, but the equations and the graphs characteristic of competitive and noncompetitive inhibition don't work the same way with an allosteric enzyme.

5. A K system is an allosteric enzyme in which the binding of inhibitor alters the apparent substrate concentration needed to reach one-half V_{max}, $S_{0.5}$.

6. A V system is an allosteric enzyme in which the binding of inhibitor changes the V_{max} of the enzyme but not the $S_{0.5}$.

7. Homotropic effects are allosteric interactions that occur when several identical molecules are bound to a protein. The binding of substrate molecules to different sites on an enzyme, such as the binding of aspartate to ATCase, is an example of a homotropic effect. Heterotropic effects are allosteric interactions that occur when different substances (such as inhibitor and substrate) are bound to the protein. In the ATCase reaction, inhibition by CTP and activation by ATP are both heterotropic effects.

8. ATCase is made up of two different types of subunits. One of them is the catalytic subunit, and there are six of them organized into two trimers. The other is the regulatory subunit, which consists of six protein subunits organized into three dimers.

9. Enzymes that exhibit cooperativity do not show hyperbolic curves of rate versus substrate concentration. Their curves are sigmoidal. The level of cooperativity can be seen by the shape of the sigmoidal curve.

10. Inhibitors make the shape of the curve more sigmoidal.

11. Activators make the shape of the curve less sigmoidal.

12. $K_{0.5}$ is the substrate concentration that leads to half of the maximal velocity. This term is used with allosteric enzymes where the term K_M is not appropriate.

13. A mercury compound was used to separate the subunits of ATCase. When the subunits were separated, one type of subunit retained catalytic activity but was no longer allosteric and was not inhibited by CTP. The other subunit type had no ATCase activity, but it did bind to CTP and ATP.

7.2 What Are the Models for the Behavior of Allosteric Enzymes?

14. In the concerted model, all the subunits in an allosteric enzyme are found in the same form, either the T form or the R form. They are in equilibrium, with each enzyme having a characteristic ratio of the T/R. In the sequential model, the subunits change individually from T to R.

15. The sequential model can explain negative cooperativity, because a substrate binding to the T form could induce other subunits to switch to the T form, thereby reducing binding affinity.

16. Greater cooperativity is favored by having a higher ratio of the T/R form. It is also favored by having a higher dissociation constant for the substrate binding to the T form.

17. The L value is the equilibrium ratio of the T/R form. The c value is the ratio of the dissociation constants for substrate and the two forms of enzyme, such that $c = K_R/K_T$.

18. Many models are possible. We never really know for sure how the enzyme works, rather, we create a model that explains the observed behavior. It is very possible that another model would do so as well.

7.3 How Does Phosphorylation of Specific Residues Regulate Enzyme Activity?

19. A kinase is an enzyme that phosphorylates a protein using a high-energy phosphate, such as ATP, as the phosphate donor.

20. Serine, threonine, and tyrosine are the three most often phosphorylated amino acids in proteins that are acted upon by kinases. Aspartate is another one often phosphorylated.

21. The allosteric effect can be faster because it is based on simple binding equilibrium. For example, if AMP is an allosteric activator of glycogen phosphorylase, the immediate increase in AMP when muscles contract can cause muscle phosphorylase to become more active and to provide energy for the contracting muscles. The phosphorylation effect requires the hormone cascade beginning with glucagon or epinephrine. There are many steps before the glycogen phosphorylase is phosphorylated, so the response time is slower. However, the cascade effect produces many more activated phosphorylase molecules, so the effects are longer and stronger.

22. As part of the mechanism, the sodium—potassium ATPase has an aspartate residue that becomes phosphorylated. This phosphorylation alters the conformation of the enzyme and causes it to close on one side of the membrane and open on the other, moving ions in the process.

23. Glycogen phosphorylase is controlled allosterically by several molecules. In the muscle, AMP is an allosteric activator. In the liver, glucose is an allosteric inhibitor. Glycogen phosphorylase also exists in a phosphorylated form and an unphosphorylated form, with the phosphorylated form being more active.

7.4 What Are Zymogens, and How Do They Control Enzyme Activity?

24. The digestive enzymes trypsin and chymotrypsin are classic examples of regulation by zymogens. The blood-clotting protein thrombin is another.

25. Trypsin, chymotrypsin, and thrombin are all proteases. Trypsin cleaves peptide bonds where there are amino acids with positively charged side chains (Lys and Arg). Chymotrypsin cleaves peptides at amino acids with aromatic side chains. Thrombin cleaves the protein fibrinogen into fibrin.

26. The zymogen prothrombin is cleaved to give the active enzyme thrombin. The thrombin then cleaves a soluble molecule, fibrinogen, into an insoluble molecule, fibrin. Fibrin is a protein that forms part of the blood clot.

27. Chymotrypsinogen is an inactive zymogen. It is acted upon by trypsin, which cleaves peptides at basic residues, like arginine. When trypsin cleaves between the arginine and the isoleucine, chymotrypsinogen becomes semiactive, forming π-chymotrypsin. This molecule digests itself further, forming the active α-chymotrypsin. As it turns out, the α-amino group of the isoleucine produced by the first cleavage is near the active site of α-chymotrypsin and necessary for its activity.

28. Zymogens are often seen with digestive enzymes that are produced in one tissue and used in another. If the enzyme were active immediately upon production, it would digest other cell proteins, where it would cause great damage. By having it produced as a zymogen, it can be safely made and then transported to the digestive tissue, such as the stomach or small intestine, where it can then be activated.

29. This allows for a more rapid response when the hormone is needed. The hormone is already synthesized and usually just requires breaking one or two bonds to make it active. The hormone can be poised and ready to go on demand.

7.5 How Do Active-Site Events of an Enzyme Affect the Reaction Mechanism?

30. Serine and histidine are the two most critical amino acids in the active site of chymotrypsin.

31. The initial phase releases the first product and involves an acyl-enzyme intermediate. This step is faster than the second part, in which water comes into the active site and breaks the enzyme—acyl bond.

32. In the first step of the reaction, the serine hydroxyl is the nucleophile that attacks the substrate peptide bond. In the second step, water is the nucleophile that attacks the acyl-enzyme intermediate.

33. Histidine 57 performs a series of steps involving general base catalysis followed by general acid catalysis. In the first phase, it takes a hydrogen from serine 195, acting as a general base. This is followed immediately by an acid catalysis step, in which it gives the hydrogen to the amide group of the peptide bond that is breaking. A similar scheme takes place in the second phase of the reaction.

34. The first phase is faster for several reasons. The serine at position 195 is a strong nucleophile for the initial nucleophilic attack. It then forms an acyl-enzyme intermediate. In the second phase, water is the nucleophile, and it takes time for water to diffuse to the right spot to perform its nucleophilic attack. It is also not as strong a nucleophile as the serine. Therefore, it takes longer for water to perform its nucleophilic attack and break the acyl-enzyme intermediate than it takes for serine to create it.

35. Histidine 57 exists in both the protonated and unprotonated form during the chymotrypsin reaction. Its pK_a of 6.0 makes this possible in the physiological pH range.

36. Instead of a phenylalanine moiety (similar to the usual substrates of chymotrypsin), use a nitrogen-containing basic group similar to the usual substrates of trypsin.

7.6 What Types of Chemical Reactions Are Involved in Enzyme Mechanisms?

37. They act as Lewis acids (electron-pair acceptors) and can take part in enzyme catalysis mechanisms of enzymes.

38. False. The mechanisms of enzymatic catalysis are the same as those encountered in organic chemistry, operating in a complex environment.

39. General acid catalysis is the part of an enzyme mechanism in which an amino acid or other molecule donates a hydrogen ion to another molecule.

40. S_N1 stands for unimolecular nucleophilic substitution. The unimolecular part means that it obeys first-order kinetics. If the reaction is R:X + Z:→ R:Z + X:, with an S_N1 reaction, the rate is dependent on the speed with which the X breaks away from the R. The Z group comes in later and quickly, compared with the breakdown of R:X. S_N2 stands for bimolecular nucleophilic substitution. This happens with the same reaction scheme if the Z attacks the R:X molecule before it breaks down. Thus, the concentration of both R:X and Z: are important, and the rate displays second-order kinetics.

41. The S_N1 reaction leads to loss of stereospecificity as the X group leaves before the entering nucleophile. This means that the nucleophile can enter from different angles, leading to different isomers.

42. The results do not prove that the mechanism is correct, because results from different experiments could contradict the proposed mechanism. In that case, the mechanism would have to be modified to accommodate the new experimental results.

7.7 What Is the Connection between the Active Site and Transition States?

43. A good transition-state analog would have to have a tetrahedral carbon atom where the amide carbonyl group was originally found, since the transition state involves a momentary tetrahedral form. It would also have to have oxygens on the same carbon, so that there would be sufficient specificity for the active site.

44. The induced-fit model assumes that the enzyme and substrate must both move and change to conform to each other perfectly. Thus, the true fit is not between the enzyme and substrate but between the enzyme and the transition state of the substrate on its way to product. A transition-state analog will fit the enzyme nicely in this model.

45. An abzyme is created by injecting a host animal with a transition-state analog of a reaction of interest. The host animal will make antibodies to the foreign molecule, and these antibodies will have specific binding points that mimic an enzyme surrounding a transition state. The purpose is to create an antibody with catalytic activity.

46. Cocaine blocks the reuptake of the neurotransmitter dopamine at synapses. Thus, dopamine stays in the system longer, overstimulating the neuron and leading to the reward signals in the brain that lead to addiction. Using a drug to block a receptor would be of no use with cocaine addiction and would probably just make removal of dopamine even more unlikely.

47. Cocaine can be degraded by a specific enzyme that hydrolyzes an ester bond that is part of cocaine's structure. In the process of this hydrolysis, the cocaine must pass through a transition state that changes its shape. Catalytic antibodies to the transition state of the hydrolysis of cocaine hydrolyze cocaine to two harmless degradation products—benzoic acid and ecgonine methyl ester. When degraded, the cocaine cannot block dopamine reuptake. No prolongation of the neuronal stimulus occurs, and the addictive effects of the drug vanish over time.

7.8 What Are Coenzymes?

48. Nicotinamide adenine dinucleotide, oxidation—reduction; flavin adenine dinucleotide; oxidation—reduction; coenzyme A, acyl transfer; pyridoxal phosphate, transamination; biotin, carboxylation; lipoic acid, acyl transfer.

49. Most coenzymes are derivatives of compounds we call vitamins. For example, nicotinamide adenine dinucleotide is produced from the B vitamin niacin. Flavin adenine dinucleotide comes from riboflavin.

50. Vitamin B_6 is the source of pyridoxal phosphate, which is used in transamination reactions.

51. Coenzymes can accomplish the same mechanisms that the amino acids do in a reaction. For example, a metal ion may act as a general acid or base. Parts of a coenzyme, such as the reactive carbanion of thiamine pyrophosphate, may act as a nucleophile to catalyze the reaction.

52. Yes, there would be a preference. Because the coenzyme and the other substrate will be locked into the enzyme, the hydride ion would come from some functional group that had a fixed position. Therefore, the hydride would come from one side.

Chapter 8

8.1 What Is the Definition of a Lipid?

1. Solubility properties (insoluble in aqueous or polar solvents, soluble in nonpolar solvents). Some lipids are not at all structurally related.

8.2 What Are the Chemical Natures of the Lipid Types?

2. In both types of lipids, glycerol is esterified to carboxylic acids, with three such ester linkages formed in triacylglycerols and two in phosphatidyl ethanolamines. The structural difference comes in the nature of the third ester linkage to glycerol. In phosphatidyl ethanolamines, the third hydroxyl group of glycerol is esterified not to a carboxylic acid but to phosphoric acid. The phosphoric acid moiety is esterified in turn to ethanolamine. (See Figures 8.2 and 8.5.)

3.

$$
\begin{array}{l}
\text{Glycerol moiety} \\
\text{CH}_2-\text{O}-\overset{\overset{\text{O}}{\|}}{\text{C}}-(\text{CH}_2)_7\text{CH}=\text{CH}-(\text{CH}_2)_7\text{CH}_3 \quad \text{Oleic acid moiety} \\
\text{CH}-\text{O}-\overset{\overset{\text{O}}{\|}}{\text{C}}-(\text{CH}_2)_{16}\text{CH}_3 \quad \text{Stearic acid moiety} \\
\text{CH}_2-\text{O}-\overset{\overset{\text{O}}{\|}}{\underset{\underset{\text{O}^{\ominus}}{|}}{\text{P}}}-\text{O}-(\text{CH}_2)_2-\overset{\oplus}{\text{N}}(\text{CH}_3)_3 \quad \text{Choline moiety}
\end{array}
$$

4. Both sphingomyelins and phosphatidylcholines contain phosphoric acid esterified to an amino alcohol, which must be choline in the case of a phosphatidylcholine and may be choline in the case of a sphingomyelin. They differ in the second alcohol to which phosphoric acid is esterified. In phosphatidylcholines, the second alcohol is glycerol, which has also formed ester bonds to two carboxylic acids. In sphingomyelins, the second alcohol is another amino alcohol, sphingosine, which has formed an amide bond to a fatty acid. (See Figure 8.6.)

5. This lipid is a ceramide, which is one kind of sphingolipid.

6. Sphingolipids contain amide bonds, as do proteins. Both can have hydrophobic and hydrophilic parts, and both can occur in cell membranes, but their functions are different.

7. Any combination of fatty acids is possible.

$$
\begin{array}{l}
\text{Glycerol moiety} \\
\text{CH}_2-\text{O}-\overset{\overset{\text{O}}{\|}}{\text{C}}-(\text{CH}_2)_{14}\text{CH}_3 \quad \text{Palmitic acid moiety} \\
\text{CH}-\text{O}-\overset{\overset{\text{O}}{\|}}{\text{C}}-(\text{CH}_2)_7\text{CH}=\text{CH}-\text{CH}_2-\text{CH}=\text{CH}(\text{CH}_2)_4\text{CH}_3 \quad \text{Linoleic acid moiety} \\
\text{CH}_2-\text{O}-\overset{\overset{\text{O}}{\|}}{\text{C}}-(\text{CH}_2)_7(\text{CH}=\text{CHCH}_2)_3\text{CH}_3 \quad \text{Linolenic acid moiety}
\end{array}
$$

8. Steroids contain a characteristic fused-ring structure, which other lipids do not.

9. Waxes are esters of long-chain carboxylic acids and long-chain alcohols. They tend to be found as protective coatings.

10. Phospholipids are more hydrophilic than cholesterol. The phosphate group is charged, and the attached alcohol is charged or polar. These groups interact readily with water. Cholesterol has only a single polar group, an —OH.

11.

$$CH_2-O-\overset{\overset{\displaystyle O}{\|}}{C}-(CH_2)_{14}CH_3$$

$$CH-O-\overset{\overset{\displaystyle O}{\|}}{C}-(CH_2)_7CH=CH-CH_2-CH=CH-(CH_2)_4CH_3$$

$$CH_2-O-\overset{\overset{\displaystyle O}{\|}}{C}-(CH_2)_7-(CH=CH-CH_2)_3CH_3$$

$$\Big\downarrow \quad \textbf{Aqueous NaOH}$$

$$CH_2OH \qquad CH_3-(CH_2)_{14}-\overset{\overset{\displaystyle O}{\|}}{C}-O^{\ominus}Na^{\oplus}$$

$$CHOH \quad + \quad CH_3-(CH_2)_4-CH=CH-CH_2-CH=CH-(CH_2)_7-\overset{\overset{\displaystyle O}{\|}}{C}-O^{\ominus}Na^{\oplus}$$

$$CH_2OH \qquad CH_3(CH_2-CH=CH)_3-(CH_2)_7-\overset{\overset{\displaystyle O}{\|}}{C}-O^{\ominus}Na^{\oplus}$$

12. The waxy surface coating is a barrier that prevents loss of water.
13. The surface wax keeps produce fresh by preventing loss of water.
14. Cholesterol is not water-soluble, but lecithin is a good natural detergent, which is actually part of lipoproteins that transport the less soluble fats through the blood.
15. The lecithin in the egg yolks serves as an emulsifying agent by forming closed vesicles. The lipids in the butter (frequently triacylglycerols) are retained in the vesicles and do not form a separate phase.
16. The removal of the oil also removes the natural oils and waxes on the feathers. These oils and waxes must regenerate before the birds can be released.

8.3 What Is the Nature of Biological Membranes?

17. Triacylglycerols are not found in animal membranes.
18. Statements (c) and (d) are consistent with what is known about membranes. Covalent bonding between lipids and proteins [statement (e)] occurs in some anchoring motifs, but is not widespread otherwise. Proteins "float" in the lipid bilayers rather than being sandwiched between them [statement (a)]. Bulkier molecules tend to be found in the outer lipid layer [statement (b)].
19. The public is attuned to the idea of polyunsaturated fats as healthful. The *trans-* configuration gives a more palatable consistency. Recently, however, concerns have arisen about the extent to which such products mimic saturated fats.
20. Partially hydrogenated vegetable oils have the desired consistency for many foods, such as oleomargarine and components of TV dinners.
21. Many of the double bonds have been saturated. Crisco contains "partially hydrogenated vegetable oils."
22. There is less heart disease associated with diets low in saturated fatty acids.
23. The transition temperature is lower in a lipid bilayer with mostly unsaturated fatty acids compared with one with a high percentage of saturated fatty acids. The bilayer with the unsaturated fatty acids is already more disordered than the one with a high percentage of saturated fatty acids.
24. Myelin is a multilayer sheath consisting mainly of lipids (with some proteins) that insulates the axons of nerve cells, facilitating transmission of nerve impulses.

25. At the lower temperature, the membrane would tend to be less fluid. The presence of more unsaturated fatty acids would tend to compensate by increasing the fluidity of the membrane compared to one at the same temperature with a higher proportion of saturated fatty acids.
26. The higher percentage of unsaturated fatty acids in membranes in cold climates is an aid to membrane fluidity.
27. Hydrophobic interactions among the hydrocarbon tails are the main energetic driving force in the formation of lipid bilayers.

8.4 What Are Some Common Types of Membrane Proteins?

28. A glycoprotein is formed by covalent bonding between a carbohydrate and a protein, whereas a glycolipid is formed by covalent bonding between a carbohydrate and a lipid.
29. Proteins that are associated with membranes do not have to span the membrane. Some can be partially embedded in it, and some associate with the membrane by noncovalent interactions with its exterior.
30. In a 100-g sample of membrane, there are 50 g of protein and 50 g of phosphoglycerides.

$$50 \text{ g lipid} \times \frac{1 \text{ mol lipid}}{800 \text{ g lipid}} = 0.0625 \text{ mol lipid}$$

$$50 \text{ g protein} \times \frac{1 \text{ mol protein}}{50{,}000 \text{ g protein}} = 0.001 \text{ mol protein}$$

The molar ratio of lipid to protein is 0.0625/0.001 or 62.5/1.
31. Nature chooses what works. This is an efficient use of a large protein and of the energy of ATP.
32. In a protein that spans a membrane, the nonpolar residues are the exterior ones; they interact with the lipids of the cell membrane. The polar residues are in the interior, lining the channel through which the ions enter and leave the cell.

8.5 What Is the Fluid-Mosaic Model of Membrane Structure?

33. Statements (c) and (d) are correct. Transverse diffusion is only rarely observed [statement (b)], and the term "mosaic" refers to the pattern of distribution of proteins in the lipid bilayer [statement (e)]. Peripheral proteins are also considered part of the membrane [statement (a)].

8.6 What Are Some of the Functions of Membranes?

34. Biological membranes are highly nonpolar environments. Charged ions tend to be excluded from such environments rather than dissolving in them, as they would have to do to pass through the membrane by simple diffusion.

35. Statements (a) and (c) are correct; statement (b) is not correct because ions and larger molecules, especially polar ones, require channel proteins.

8.7 Which Are the Lipid-Soluble Vitamins, and What Is Their Function?

36. Cholesterol is a precursor of vitamin D_3; the conversion reaction involves ring opening.

37. Vitamin E is an antioxidant.

38. Isoprene units are five-carbon moieties that play a role in the structure of a number of natural products, including fat-soluble vitamins.

39. See Table 8.3.

40. The cis–trans isomerization of retinal in rhodopsin triggers the transmission of an impulse to the optic nerve and is the primary photochemical event in vision.

41. Vitamin D can be made in the body.

42. Lipid-soluble vitamins accumulate in fatty tissue, leading to toxic effects. Water-soluble vitamins are excreted, drastically reducing the chances of an overdose.

43. Vitamin K plays a role in the blood-clotting process. Blocking its mode of action can have an anticoagulant effect.

44. Vitamins A and E are known to scavenge free radicals, which can do oxidative damage to cells.

45. Eating carrots is good for both. Vitamin A, which is abundant in carrots, plays a role in vision. Diets that include generous amounts of vegetables are associated with a lower incidence of cancer.

8.8 What Are Prostaglandins and Leukotrienes, and What Do They Have to Do with Lipids?

46. An omega-3 fatty acid has a double bond at the third carbon from the methyl end.

47. Leukotrienes are carboxylic acids with three conjugated double bonds.

48. Prostaglandins are carboxylic acids that include a five-membered ring in their structure.

49. Prostaglandins and leukotrienes are derived from arachidonic acid. They play a role in inflammation and in allergy and asthma attacks.

50. Prostaglandins in blood platelets can inhibit their aggregation. This is one of the important physiological effects of prostaglandins.

Chapter 9

9.1 What Are the Levels of Structure in Nucleic Acids?

1. (a) Double-stranded DNA is usually thought of as having secondary structure, unless we consider its supercoiling (tertiary) or association with proteins (quaternary).
   (b) tRNA is a tertiary structure with many folds and twists in three dimensions.
   (c) mRNA is usually considered to be a primary structure, as it has little other structure.

9.2 What Is the Covalent Structure of Polynucleotides?

2. Thymine has a methyl group attached to carbon 5, while uracil does not.

3. In adenine, carbon 6 has an amino group attached; in hypoxanthine, carbon 6 is a carbonyl group.

4.

A	Adenine	Adenosine or deoxyadenosine	Adenosine-5′-triphosphate or deoxyadenosine-5′-triphosphate
G	Guanine	Guanosine or deoxyguanosine	Guanosine-5′-triphosphate or deoxyguanosine-5′-triphosphate
C	Cytosine	Cytidine or deoxycytidine	Cytidine-5′-triphosphate or deoxycytidine-5′-triphosphate
T	Thymine	Deoxythymidine	Deoxythymidine-5′-triphosphate
U	Uracil	Uridine	Uridine-5′-triphosphate

5. ATP is made from adenine, ribose, and three phosphates linked to the 5′-hydroxyl of the ribose. dATP is the same, except that the sugar is deoxyribose.

6. The sequence on the opposite strand for each of the following (all read $5' \rightarrow 3'$) is

   ACGTAT   TGCATA

   AGATCT   TCTAGA

   ATGGTA   TACCAT

7. They are DNA sequences because of the presence of thymine rather than uracil.

8. (a) Definitely yes! If there is anything that you don't want falling apart, it's your storehouse of genetic instructions. (Compare the effectiveness of a computer if all the *.exe files were deleted.) (b) In the case of messenger RNA, yes. The mRNA is the transmitter of information for protein synthesis, but it is needed only as long as a particular protein is needed. If it were long-lived, the protein would continue to be synthesized even when not needed; this would waste energy and could cause more direct detrimental effects. Thus, most mRNAs are short-lived (minutes); if more protein is needed, more mRNA is made.

9. Four different kinds of bases—adenine, cytosine, guanine, and uracil—make up the preponderant majority of the bases found in RNA, but they are not the only ones. Modified bases occur to some extent, principally in tRNA.

10. This speculation arose from the fact that ribose has three hydroxyl groups that can be esterified to phosphoric acid (at the 2′, 3′, and 5′ positions), whereas deoxyribose has free hydroxyls at the 3′ and 5′ positions alone.

11. The hydrolysis of RNA is greatly enhanced by the formation of a cyclic 2′-3′ phosphodiester intermediate. DNA, lacking the 2′-hydroxyl group, cannot form the intermediate and thus is relatively resistant to hydrolysis.

9.3 What Is the Structure of DNA?

12.

Structure	Kind of Nucleic Acid
A-form helices	Double-stranded RNA
B-form helices	DNA
Z-form helices	DNA with repeating CGCGCG sequences
Nucleosomes	Eukaryotic chromosomes
Circular DNA	Bacterial, mitochondrial, plasmid DNA

13. See Figure 9.8.

14. Statements (c) and (d) are true; statements (a) and (b) are not.

15. True. There is room for binding and access to the base pairs in both the major and minor grooves of DNA.

16. The major groove and minor groove in B-DNA have very different dimensions (width), while those in A-DNA are much closer in width.

17. Statement (c) is true. Statements (a) and (b) are false. Statement (d) is true for the B form of DNA but not for the A and Z forms.

18. (a) Supercoiling refers to twists in DNA over and above those of the double helix. (b) Positive supercoiling refers to an extra twist in DNA caused by overwinding of the helix before sealing the ends to produce circular DNA. (c) A topoisomerase is an enzyme that induces a single-strand break in supercoiled DNA, relaxes the supercoiling, and reseals the break. (d) Negative supercoiling refers to unwinding of the double helix before sealing the ends to produce circular DNA.

19. Propeller-twist is a movement of the two bases in a base pair away from being in the same plane.

20. An AG/CT step is a small section of double-stranded DNA where one strand is 5'-AG-3', and the other is 5'-CT-3'. The exact nature of such steps greatly influences the overall shape of a double helix.

21. Propeller-twist reduces the strength of the hydrogen bond but moves the hydrophobic region of the base out of the aqueous environment, thus being more entropically favorable.

22. B-DNA is a right-handed helix with specified dimensions (10 base pairs per turn, significant differences between major and minor groove, etc.). Z-DNA is a left-handed double helix with different dimensions (12 base pairs per turn, similar major and minor grooves, etc.).

23. Positive supercoils in circular DNA will be left-handed.

24. Chromatin is the complex consisting of DNA and basic proteins found in eukaryotic nuclei (see Figure 9.16).

25. See the figure in the Biochemical Connections Box on Triple Helical DNA (page 226).

26. Negative supercoiling, nucleosome winding, Z-form DNA.

27. It binds to the DNA, forming loops around itself. It then cuts both strands of DNA on one part of the loop, passes the ends across another loop, and reseals.

28. Histones are very basic proteins with many arginine and lysine residues. These residues have positively charged side chains under physiological pH. This is a source of attraction between the DNA and histones because the DNA has negatively charged phosphates: Histone—NH_3^+ attracts ^-O—P—O—DNA chain.

  When the histones become acetylated, they lose their positive charge: Histone—NH—$COCH_3$. They therefore have no attraction to the phosphates on the DNA. The situation is even less favorable if they are phosphorylated because now both the histone and the DNA carry negative charges.

29. Adenine–guanine base pairs occupy more space than is available in the interior of the double helix, whereas cytosine–thymine base pairs are too small to span the distance between the sites to which complementary bases are bonded. One would not normally expect to find such base pairs in DNA.

30. The phosphate groups in DNA are negatively charged at physiological pH. If they were grouped together closely, as in the center of a long fiber, the result would be considerable electrostatic repulsion. Such a structure would be unstable.

31. The percentage of cytosine equals that of guanine, 22%. This DNA thus has a 44% G-C content, implying a 56% A-T content. The percentage of adenine equals that of thymine, so adenine and thymine are 28% each.

32. If the DNA were not double stranded, the requirement G≡C and A≡T would no longer exist.

33. The base distribution would not have A≡T and G≡C; and total purine would not be equal to total pyrimidine.

34. The purpose of the Human Genome Project was the complete sequencing of the human genome. There are many reasons for doing this. Some are tied to basic research (i.e., the desire to know all that is knowable, especially about our own species). Some are medical in nature (i.e., a better understanding of genetic diseases and how growth and development are controlled). Some are comparative in nature, looking at the similarities and differences between genomes of other species. Our DNA is at least 95% the same as that of a chimpanzee, yet we are clearly different. An understanding of our genome will help us understand what separates mankind from other primates and nonprimates.

35. There are many legal and ethical considerations concerning human gene therapy. Some are moral and philosophical: Do we have the right to manipulate human DNA? Are we playing God? Should "tailor-made" humans be allowed? Some are more scientific: Do we have the knowledge to do it right? What happens if we make a mistake? Will a patient die that would not have died with other treatments?

36. Advantages would be that people could make informed lifestyle choices. A person with a genotype known to lead to atherosclerosis could change his or her diet and exercise habits from an early age to help fight this potential problem and could also seek preventative drug therapies. Disadvantages might involve legal issues over the right to know such information. Employers could discriminate against prospective employees based on a genotype marker that might indicate a susceptibility to drug abuse, alcoholism, or disease. A caste system based on genetics could arise.

37. Because any system involving replication of DNA by DNA polymerases must have a primer to start the reaction, the primer can be RNA or DNA, but it must bind to the template strand being read. Thus, enough of the sequence must be known to create the correct primer.

9.4 How Does the Denaturation of DNA Take Place?

38. A–T base pairs have two hydrogen bonds, whereas G–C base pairs have three. It takes more energy and higher temperature to disrupt the structure of DNA rich in G–C base pairs.

9.5 What Are the Principal Kinds of RNA, and What Are Their Structures?

39. See Figures 9.20 and 9.25.

40. Small nuclear RNA (snRNA) is found in the eukaryotic nucleus and is involved in splicing reactions of other RNA types. An snRNP is a small nuclear ribonucleoprotein particle. A complex of small nuclear RNA and protein catalyzes splicing of RNA.

41. Ribosomal RNA (rRNA) is the largest. Transfer RNA (tRNA) is the smallest.

42. Messenger RNA (mRNA) has the least amount of secondary structure (hydrogen bonding).

43. The bases in a double-stranded chain are partially hidden from the light beam of a spectrophotometer by the other bases in close proximity, as though they were in the shadow of the other bases. When the strands unwind, these bases become exposed to the light and absorb it; therefore, the absorbance increases.

44. RNA interference is the process by which small RNAs prevent the expression of genes.

45. More extensive hydrogen bonding occurs in tRNA than in mRNA. The folded structure of tRNA, which determines its binding to ribosomes in the course of protein synthesis, depends on its hydrogen-bonded arrangement of atoms. The coding sequences of mRNA must be accessible to direct the order of amino acids in proteins and should not be rendered inaccessible by hydrogen bonding.

46. They prevent intramolecular hydrogen bonding (which occurs in tRNA via the usual A–U and C–G associations), thus permitting loops that are critical for function, the most important being the anticodon loop.

47. Turnover of mRNA should be rapid to ensure that the cell can respond quickly when specific proteins are needed. Ribosomal subunits, including their rRNA component, can be recycled for many rounds of protein synthesis. As a result, mRNA is degraded more rapidly than rRNA.

48. The mistake in the DNA would be more harmful because every cell division would propagate the mistake. A mistake in transcrip-

tion would lead to one wrong RNA molecule that can be replaced with a correct version with the next transcription.

49. Eukaryotic mRNA is initially formed in the nucleus by transcription of DNA. The mRNA transcript is then spliced to remove introns, a poly-A tail is added at the $3'$ end, and a $5'$-cap is put on. This is the final mRNA, which is then transported, in most cases, out of the nucleus for translation by the ribosomes.

50. The numbers 50S, 30S, etc., refer to a relative rate of sedimentation in an ultracentrifuge and cannot be added directly. Many things besides molecular weight influence the sedimentation characteristics, such as shape and density.

Chapter 10

10.1 What Is the Flow of Genetic Information in the Cell?

1. Replication is the production of new DNA from a DNA template. Transcription is the production of RNA from a DNA template. Translation is the synthesis of proteins directed by mRNA, which reflects the base sequence of DNA.

2. False. In retroviruses, the flow of information is RNA → DNA.

3. DNA represents the permanent copy of genetic information, whereas RNA is transient. The cell could survive production of some mutant proteins, but not DNA mutation.

10.2 What Are the General Considerations in the Replication of DNA?

4. The semiconservative replication of DNA means that a newly formed DNA molecule has one new strand and one strand from the original DNA. The experimental evidence for semiconservative replication comes from density-gradient centrifugation (Figure 10.3). If replication were a conservative process, the original DNA would have two heavy strands and all newly formed DNA would have light strands.

5. A replication fork is the site of formation of new DNA. The two strands of the original DNA separate, and a new strand is formed on each original strand.

6. An origin of replication consists of a bubble in the DNA. There are two places at opposite ends where new polynucleotide chains are formed (Figure 10.4).

7. Separating the two strands of DNA requires unwinding the helix.

8. If the original Meselson–Stahl experiment had used longer pieces of DNA, the results would not have been as clear-cut. Unless the bacteria were synchronized as to their stage of development, the DNA could have represented several generations at once.

9. Replication requires separating the strands of DNA. This cannot happen unless the DNA is unwound.

10.3 How Does the DNA Polymerase Reaction Take Place?

10. The majority of DNA polymerase enzymes also have exonuclease activity.

11. DNA polymerase I is primarily a repair enzyme. DNA polymerase III is mainly responsible for the synthesis of new DNA. See Table 10.1.

12. The processivity of a DNA polymerase is the number of nucleotides incorporated before the enzyme dissociates from the template. The higher this number is, the more efficient the replication process is.

13. The reactants are deoxyribonucleotide triphosphates. They provide not only the moiety to be inserted (the deoxyribonucleotide) but also the energy to drive the reaction (dNTP → inserted NMP + PP$_i$, PP$_i$ → 2 P$_i$).

14. Hydrolysis of the pyrophosphate product prevents the reversal of the reaction by removing a product.

15. One strand of newly formed DNA uses the $3'$-to-$5'$ strand as a template. The problem arises with the $5'$-to-$3'$ strand. Nature deals with this issue by using short stretches of this strand for a number of chunks of newly formed DNA. They are then linked by DNA ligase (Figure 10.5).

16. The free $3'$ end is needed as the site to which added nucleotides will bond. A number of antiviral drugs remove the $3'$ end in some way.

17. The large negative $\Delta G°$ ensures that the back reaction of depolymerization does not occur. Energy overkill is a common strategy when it is critically important that the process does not go in the reverse direction.

18. Nucleophilic substitution is a common reaction mechanism, and the hydroxyl group at the $3'$ end of the growing DNA strand is an example of a frequently encountered nucleophile.

19. In some enzymes, there is a recognition site that is not the same as the active site. In the specific case of DNA polymerase III, the sliding clamp tethers the rest of the enzyme to the template. This ensures a high degree of processivity.

10.4 Which Proteins Are Required for DNA Replication?

20. All four deoxyribonucleoside triphosphates, template DNA, DNA polymerase, all four ribonucleoside triphosphates, primase, helicase, single-strand binding protein, DNA gyrase, DNA ligase.

21. DNA is synthesized from the $5'$ end to the $3'$ end, and the new strand is antiparallel to the template strand. One of the strands is exposed from the $5'$ end to the $3'$ end as a result of unwinding. Small stretches of new DNA are synthesized, still in an antiparallel direction from the $5'$ end to the $3'$ end and are linked by DNA ligase. See Figure 10.5.

22. DNA gyrase introduces a swivel point in advance of the replication fork. Primase synthesizes the RNA primer. DNA ligase links small, newly formed strands to produce longer ones.

23. In the replication process, the single-stranded portions of DNA are complexed to specific proteins.

24. DNA ligase seals the nicks in newly formed DNA.

25. The primer in DNA replication is a short sequence of RNA to which the growing DNA chain is bonded.

26. Specific enzymes exist to cut the DNA and give a supercoiled configuration at the replication fork that allows replication to proceed.

27. Polymerase III will not insert a deoxyribonucleotide without checking to see that the previous base is correct. It needs a previous base to check even if that base is part of a ribonucleotide.

10.5 How Do Proofreading and Repair Take Place?

28. When an incorrect nucleotide is introduced into a growing DNA chain as a result of mismatched base pairing, DNA polymerase acts as a $3'$-exonuclease, removing the incorrect nucleotide. The same enzyme then incorporates the correct nucleotide.

29. In *E. coli*, two different kinds of exonuclease activity are possible for DNA polymerase I, which functions as a repair enzyme.

30. An exonuclease nicks the DNA near the site of the thymine dimers. Polymerase I acts as a nuclease and excises the incorrect nucleotides, then acts as a polymerase to incorporate the correct ones. DNA ligase seals the nick.

31. In DNA, cytosine spontaneously deaminates to uracil. The presence of the extra methyl group is a clear indication that a thymine really belongs in that position, not a cytosine that has been deaminated.

32. About 5000 books: 10^{10} characters/error $\times$ 1 book/(2×10^6 characters) = 5×10^3 books/error.

33. 1000/characters/second $\times$ 1 word/5 characters $\times$ 60 seconds/minute = $12,000$ words/minute.

34. 1 second/1000 characters $\times$ 10^{10} characters/error = 10^7 seconds/error = 16.5 weeks/error nonstop.

35. Prokaryotes methylate their DNA soon after replication. This aids the process of mismatch repair. The enzymes that carry out the process can recognize the correct strand by its methyl groups.

The newly formed strand, which contains the incorrect base, does not have methyl groups.

36. DNA is constantly being damaged by environmental factors and by spontaneous mutations. If these mistakes accumulate, deleterious amino acid changes or deletions can arise. As a result, essential proteins, including those that control cell division and programmed cell death, will be inactive or overactive, eventually leading to cancer.

37. Prokaryotes have a last-resort mechanism for dealing with drastic DNA damage. This mechanism, called the SOS response, includes the crossing over of DNA. Replication becomes highly error prone, but it serves the need of the cell to survive.

10.6 How Is DNA Replicated in Eukaryotes?

38. Eukaryotes usually have several origins of replication, whereas prokaryotes have only one.

39. The general features of DNA replication are similar in prokaryotes and eukaryotes. The main differences are that eukaryotic DNA polymerases do not have exonuclease activity. After synthesis, eukaryotic DNA is complexed with proteins, while prokaryotic DNA is not.

40. Histones are proteins complexed to eukaryotic DNA. Their synthesis must take place at the same rate as DNA synthesis. The proteins and DNA must then assemble in proper fashion.

41. (a) Eukaryotic DNA replication must deal with histones; the linear DNA molecule in eukaryotes is a much larger molecule and requires special treatment at ends.
    (b) Special polymerases are used in the organelles.

42. Eukaryotes have more DNA polymerases, which tend to be larger molecules. Eukaryotic DNA polymerases tend not to have exonuclease activity. There are more origins of replication in eukaryotes and shorter Okazaki fragments. See Table 10.5.

43. Mechanisms exist to ensure that DNA synthesis takes place only once in the eukaryotic cell cycle, during the S phase. Preparation for DNA synthesis can and does take place in the G_1 phase, but the timing of actual synthesis is strictly controlled.

44. If the telomerase enzyme were inactivated, DNA synthesis would eventually stop. This enzyme maintains the 3′ template end strand so that it does not undergo degradation with each round of DNA synthesis. The degradation in turn arises from the removal of the RNA primer with each round of DNA synthesis.

45. If histone synthesis took place faster than DNA synthesis, it would be highly disadvantageous to invest the energy required for protein synthesis. The histones would have no DNA with which to bind.

46. Replication licensing factors (RLFs) are proteins that bind to eukaryotic DNA. They get their name from the fact that replication cannot proceed until they are bound. Some of the RLF proteins have been found to be cytosolic. They only have access to the chromosome when the nuclear membrane dissolves during mitosis. Until they are bound, replication cannot occur. This property links eukaryotic DNA replication and the cell cycle. Once RLFs have bound, the DNA is then competent for replication.

47. It will be faster in prokaryotes. The DNA is smaller, and the lack of compartmentalization within the cell facilitates the process. DNA replication in eukaryotes is linked to the cell cycle, and prokaryotic cells proliferate more quickly than those of eukaryotes.

48. In reverse transcriptase action, the single RNA strand serves as a template for the synthesis of a single DNA strand. The DNA strand, in turn, serves as the template for synthesis of the second strand of DNA.

49. Circular DNA does not have ends. This removes the necessity for maintaining the 3′ template end on removal of the RNA primer. Telomeres and telomerase are not needed with circular DNA.

50. The presence of a DNA polymerase that operates only in mitochondria is consistent with the view that these organelles are derived from bacteria incorporated by endosymbiosis. The bacteria were originally free-living organisms earlier in evolutionary history.

Chapter 11

11.1 How Does Transcription Take Place in Prokaryotes?

1. There is no primer required for transcription of DNA into RNA.

2. RNA polymerase from *E. coli* has a molecular weight of about 500,000 and four different kinds of subunits. It uses one strand of the DNA template to direct RNA synthesis. It catalyzes polymerization from the 5′ end to the 3′ end.

3. The subunit composition for the holoenzyme is $\alpha_2\beta\beta'\sigma$.

4. The core enzyme lacks the σ subunit, while the holoenzyme has it.

5. The strand that the RNA polymerase uses as a template for its RNA is called the template strand, the noncoding strand, the antisense strand, and the (−) strand. The other strand, whose sequence matches the RNA produced except for the T–U change, is called the nontemplate strand, the coding strand, the sense strand, and the (+) strand.

6. The promoter region is the portion of DNA to which RNA polymerase binds at the start of transcription. This region lies upstream (nearer the 3′ end of the template DNA) of the actual gene for the RNA. The promoter regions of DNA from many organisms have sequences in common (consensus sequences). The consensus sequences frequently lie 10 base pairs and 35 base pairs upstream of the start of transcription.

7. Moving from 5′ to 3′ on the coding strand, the order is the following: Fis site, UP element, −35 element, Pribnow box, and TSS.

8. Intrinsic termination of transcription involves the formation of a hairpin loop in the RNA being formed, which stalls the RNA polymerase over a region rich in A–U base pairs. This causes termination of transcription and release of the transcript. Rho-dependent termination often involves a similar hairpin loop, but, in addition, a Rho protein binds to the RNA and moves along it toward the transcription bubble. When the Rho protein reaches the transcription bubble, it causes termination.

9. See Figure 11.1. The top DNA strand is the nontemplate strand because it is not used to create the RNA. It is called the coding strand because it has the same sequence as the RNA produced, except for the change of T for U. It is called the sense strand because its sequence would give the correct amino acid sequence of the protein product. It is called the (+) strand again because it has the correct sequence. The bottom strand is called the template strand because it is the one used to make the RNA. It is also the noncoding strand because its sequence does not match the RNA produced. It is the antisense and the (−) strand for the same reason.

11.2 How Is Transcription Regulated in Prokaryotes?

10. An inducer is a substance that leads to transcription of the structural genes in an operon. A repressor is a substance that prevents transcription of the structural genes in an operon.

11. The σ factor is a subunit of prokaryotic RNA polymerase. It directs the polymerase to specific promoters and is one of the ways the gene expression is controlled in prokaryotes.

12. $σ^{70}$ is the normal σ subunit for RNA polymerase in *E. coli*. It directs RNA polymerase to most of the genes that are transcribed under normal circumstances. $σ^{32}$ is an alternate subunit that is produced when the cells are grown at higher temperatures. It directs the RNA polymerase to other genes that need to be expressed during heat shock conditions.

13. The catabolite activator protein is a transcription factor in *E. coli* that stimulates transcription of the *lac* operon structural genes. It responds to cAMP levels such that the *lac* operon is transcribed only when the cells must use lactose as a fuel source.

14. Transcription attenuation is the process found in prokaryotes in which transcription can continue or be prematurely aborted based on the concurrent translation of the mRNA produced. This is often seen in genes whose protein products lead to amino acid synthesis.

15. An operon consists of an operator gene, a promoter gene, and structural genes. When a repressor is bound to the operator, RNA polymerase cannot bind to the promoter to start transcription of the structural genes. When an inducer is present, it binds to the repressor, rendering it inactive. The inactive repressor can no longer bind to the operator. As a result, RNA polymerase can bind to the promoter, leading to the eventual transcription of the structural genes.

16. See Figure 11.5.

17. With phage SPO1, which infects the bacteria *B. subtilis*, the virus has a set of genes called the early genes that are transcribed by the host's RNA polymerase, using its regular σ subunit. One of the viral early genes codes for a protein called gp28. This protein is another σ subunit, which directs the RNA polymerase to preferentially transcribe more of the viral genes during the middle phase. Products of the middle phase transcription are gp33 and gp34, which together make up another σ factor that directs the transcription of the late genes.

18. See Figure 11.14. When the level of tryptophan is low, the trp-$tRNA^{trp}$ becomes limiting. This stalls the ribosome over the tryptophan codons on the mRNA. By stalling the ribosome there, the antitermination loop can form, transcription is not aborted, and the full mRNA is produced. If the ribosome does not stall there, the termination loop forms, and the leader mRNA dissociates.

11.3 How Does Transcription Take Place in Eukaryotes?

19. Exons are the portions of DNA that are expressed, which means that they are reflected in the base sequence of the final mRNA product. Introns are the intervening sequences that do not appear in the final product, but are removed during the splicing of mRNA.

20. There are three RNA polymerases in eukaryotes, compared with one in prokaryotes. There are many more transcription factors in eukaryotes, including complexes of them necessary for polymerase recruitment. RNA is extensively processed after transcription in eukaryotes, and, in most cases, the mRNA must leave the nucleus to be translated, whereas translation and transcription can occur at the same time in prokaryotes.

21. RNA polymerase I produces most of the rRNA. RNA polymerase II produces mRNA, and RNA polymerase III produces tRNA, the 5S ribosomal subunit, and snRNA.

22. The first component includes a variety of upstream elements, which act as enhancers and silencers. Two common ones are close to the core promoter and are the GC box (−40), which has a consensus sequence of GGGCGG, and the CAAT box (extending to −110), which has a consensus sequence of GGCCAATCT. The second component, found at position −25, is the TATA box, which has a consensus sequence of TATAA(T/A). The third component includes the transcription start site at position +1 and is surrounded by a sequence called the initiator element (*Inr*). The final component is a possible downstream regulator.

23. TFIIA, TFIIB, TFIID, TFIIE, TFIIF, and TFIIH are the general transcription factors. TFIID is also the TATA-box binding protein and is associated with TAFs (TBP associated factors).

24. Its primary function is as a general transcription factor involved in the formation of the open complex for transcription initiation. It binds to the basal unit and is involved in DNA melting through a helicase activity as well as promoter clearance via phosphorylation of the CTD of RNA polymerase. In addition, it also has a cyclin-dependent kinase activity. Thus, TFIIH is involved in tying transcription and cell division together. It is also involved in DNA repair mechanisms.

11.4 How Is Transcription Regulated in Eukaryotes?

25. The heat-shock element responds to increased temperature. The metal-response element responds to the presence of heavy metals, such as cadmium, and the cyclic-AMP-response element controls a wide variety of genes based on cAMP levels in the cell.

26. CREB is a transcription factor that binds to the cAMP-response element. It is involved with the transcription of hundreds of genes based on the cAMP levels of the cell. When there is cAMP, CREB is phosphorylated, which allows it to bind the CREB binding protein, which connects the CRE to the basal transcription machinery, stimulating transcription.

27. Regulation in eukaryotes is much more complicated. Prokaryotic regulation is controlled by the choice of σ subunit, the nature of the promoters, and the use of repressors/inducers. In eukaryotes, there are many more promoter elements, transcription factors, and coactivators. In addition, the DNA must be released from histone proteins, so transcription of DNA is linked to histone modifications.

28. As the mRNA is being produced, ribosomes are bound and begin to translate. A leader sequence on the mRNA leads to a leader peptide. Loops can form in the mRNA in different ways. Some loop combinations lead to transcription termination. The speed with which the ribosome is able to move on the mRNA controls which loop combinations form, and this speed is usually governed by the level of a specific tRNA that is available for the translation.

29. Assuming that there is a basal transcription rate for a particular gene, an enhancer would bind to a transcription factor and lead to a greater level of transcription, while a silencer would bind to a transcription factor and reduce the level of transcription below the basal rate.

30. A response element is an enhancer element that will bind to a specific transcription factor and increase the level of transcription of target genes. In the case of response elements, however, this is in response to a more general cell signal, such as the presence of cAMP, glucocorticoids, or heavy metals. Response elements may control a large set of genes, and a given gene may be under the control of more than one response element.

31. As seen here, CREB binds to the CRE. When phosphorylated, it also binds to CBP and bridges to the basal transcription complex.

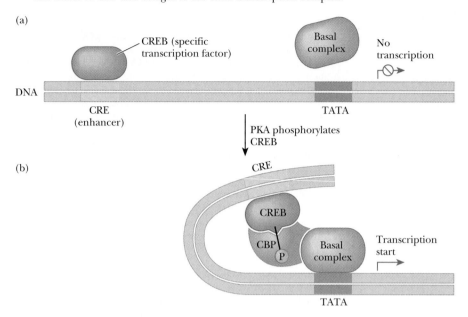

(a)

(b)

32. TFIID is one of the general transcription factors for RNA polymerase II. Part of it is a protein that binds to the TATA box in eukaryotic promoters. Associated in complex with the TATA box and the TBP are many proteins called TAFs, for TBP associated factors.

33. The statement is untrue. Many eukaryotic promoters do have TATA boxes, but there are also genes that lack one.

34. Transcription elongation in eukaryotes is controlled in several ways. There are pause sites at which RNA polymerase tends to hesitate. There is also antitermination at which RNA polymerase can transcribe past a normal termination point. The general transcription factor TFIIF stimulates elongation as well as initiation by helping RNA polymerase II read through pause sites. A separate elongation factor, TFIIS, is called an arrest-release factor because it stimulates RNA polymerase to resume transcription once it has hesitated at a pause site. Separate proteins also exist, called P-TEF and N-TEF, that act to positively or negatively affect elongation.

35. CREB is a ubiquitous transcription factor that has been found involved in many genes. It is phosphorylated when cAMP levels are high, which triggers the activation of the genes. CREB-mediated transcription has been implicated in cell proliferation, cell differentiation, spermatogenesis, release of somatostatin, development of mature T-cells, protection of nerve cells under hypoxic conditions, circadian rhythms, adaptation to exercise, regulation of gluconeogenesis, transcription regulation of phosphoenolpyruvate carboxykinase and lactate dehydrogenase, and learning and storage in long-term memory.

36. Acidic domains, glutamine-rich domains, and proline-rich domains.

11.5 What Are Some Structural Motifs in DNA-Binding Proteins?

37. Helix–turn–helix motifs, zinc fingers, and basic-region leucine zippers.

38. The major DNA binding protein motifs are helix–turn–helix, zinc fingers, and basic region leucine zippers. The helix–turn–helix motifs are organized so that the two helices of the protein fit into the major groove of the DNA. Zinc fingers are formed by combinations of cysteine and/or histidine complexed with zinc ions. A loop of protein forms around this complex, and these loops fit into the major groove of the DNA. Several such loops can be found spiraling around the DNA with the major groove. The basic-region leucine zipper has two domains. One is an area of leucines spaced out every seven amino acids. This puts them on

the same side of an α-helix, which allows them to dimerize with another such protein. The basic region is high in lysine and arginine, which bind tightly to the DNA backbone via electrostatic attraction.

11.6 How Is RNA Modified after Transcription?

39. Introns are spliced out. Bases are modified. A poly-A tail is put on the 3′ end of mRNA. A 5′-cap is put on mRNA.

40. One of the snURPs, U-1, is the target for destruction by the body's own immune system.

41. They both have multiple isoforms created by differential splicing of mRNA.

42. Trimming is necessary to obtain RNA transcripts of the proper size. Frequently, several tRNAs are transcribed in one long RNA molecule and must be trimmed to obtain active tRNAs.

43. Capping, polyadenylation, and splicing of coding sequences take place in the processing of eukaryotic mRNA.

44. The snRNPs are small nuclear ribonucleoprotein particles. They are the site of mRNA splicing.

45. Besides its traditional role in mRNA, tRNA, and rRNA, RNA serves other functions, such as splicing reactions, trimming reactions, and the peptide synthesis reaction of peptidyl transferase. It also has been shown that some small RNAs are produced; they act as gene silencers by binding to specific DNA sequences and blocking their transcription.

46. See Figure 11.34.

47. The Human Genome Project concluded that humans had far fewer genes than previously thought, yet we seem to be more biologically and biochemically complex. One possibility suggested to explain how so few genes could lead to so many proteins is that more proteins may be produced via differential splicing of mRNA. Thus, the same amount of DNA could lead to more gene products.

11.7 How Does RNA Act as an Enzyme?

48. A ribozyme is RNA that has catalytic activity without the intervention of protein at the active site. The catalytic portion of RNase P is a ribozyme. The self-splicing rRNA of *Tetrahymena* is the classic example, and it has recently been shown that the peptidyl transferase activity of the ribosome is actually a ribozyme.

49. Two mechanisms for RNA self-splicing are known. In Group I ribozymes, an external guanosine is covalently bonded at the splice site, releasing one end of the intron. The free end of the

exon thus produced attacks the end of the other exon to splice the two. The intron cyclizes in the process. (See Figure 11.34.) Group II ribozymes display a lariat mechanism. The 2'-OH of an internal adenosine attacks the splice site. (See Figure 11.36.)

50. Proteins are more efficient catalysts than RNA because they have wider variations in structure and thus can tailor the active site for maximum efficiency for a given reaction.

Chapter 12

12.1 What Is the Overall Process of Translating the Genetic Message?

1. See Figure 12.1.

12.2 What Is the Genetic Code?

2. A code in which two bases code for a single amino acid allows for only 16 (4×4) possible codons, which is not adequate to code for 20 amino acids.

3. A degenerate code is one in which more than one triplet can specify a given amino acid.

4. In the binding assay technique, various tRNA molecules, one of which is radioactively labeled with ^{14}C, are mixed with ribosomes and synthetic trinucleotides bound to a filter. If the radioactive label is detected on the filter, then it is known that the particular tRNA bound to that triplet. The binding experiments can be repeated until all the triplets are assigned.

5. The wobble base can be uracil, guanine, or hypoxanthine.

6. The codons UAA, UAG, and UGA are the stop signals. These codons are not recognized by any tRNAs, but they are recognized by proteins called release factors. A release factor not only blocks the binding of a new aminoacyl-tRNA but also affects the activity of the peptidyl transferase, so that the bond between the carboxyl end of the peptide and the tRNA is hydrolyzed.

7. Note that the sequence in the codon of mRNA is reversed because mRNA synthesis is antiparallel.

   (a) Position 1 has an intermediate effect. For purine changes, a different amino acid results in all cases. The changes tend to be conservative, with only four of the 16 possible changes leading to hydrophobic—hydrophilic differences. For our purposes, glycine is considered neither hydrophobic nor hydrophilic. The resulting protein would have a better chance of functioning than in a second-base change, but a lesser probability than in a third-base change.

   (b) Position 2 is the most informational: a different amino acid results from any change. In this case, however, the chances are high (75%) that the mutation would be a conservative one, with one hydrophobic amino acid replacing another one, so the protein would still have a good chance of functioning. A change involving serine or threonine (25% chance) would alter the polarity but would not introduce a charge on the side chain; the protein might still function.

   (c) There is a high probability of a change in the type of amino acid, including differences in charge; the probability of the resulting protein having proper function is considerably lower.

   (d) Position 3 is the least informational. There is a high probability of getting the same amino acid. The protein thus has a very good chance of functioning.

8. The concept of "wobble" specifies that the first two bases of a codon remain the same, while there is room for variation in the third base. This is precisely what is observed experimentally.

9. Hypoxanthine is the most versatile of the "wobble" bases; it can base pair with adenine, cytosine, or uracil.

10. It is quite reasonable. When codons for a given amino acid have one or two nucleotides in common, a mutation is less likely to give rise to a nonfunctional protein. The survival value of such a feature guarantees its selection in evolution.

11. An ambiguous code would allow for variation in the amino acid sequence of proteins. Consequently, there would be variation in function, including a number of nonfunctional proteins.

12. Variations in the genetic code in mitochondria support the idea of their existence as free-living bacteria early in evolutionary history.

12.3 What Is the Role of Aminoacyl-tRNA Synthetases in Amino Acid Activation?

13. The hydrolysis of ATP to AMP and PP_i provides the energy to drive the activation step.

14. Proofreading in amino acid activation takes place in two stages. The first requires a hydrolytic site on the aminoacyl-tRNA synthetase; incorrect amino acids that have become esterified to the tRNA are removed here. The second stage of proofreading requires the recognition site on the aminoacyl-tRNA synthetase for the tRNA itself. The incorrect tRNA does not bind tightly to the enzyme.

15. The following factors ensure fidelity in protein synthesis. Aminoacyl-tRNA formation includes a high degree of enzyme specificity to connect the right amino acid to the right tRNA, proofreading in the formation of some aminoacyl-adenylates, and energy "overkill." Other factors include proper hydrogen bonding of mRNA to the ribosome and between codon and anti-codon. (The latter is a relatively slow association, allowing time for mismatches to dissociate before the peptide bond is formed.) The fidelity of protein synthesis is low compared with DNA synthesis, which has proofreading procedures in addition to energy overkill and proper base pairing. The fidelity of protein synthesis is relatively high compared with RNA synthesis, which has only energy overkill and proper base pairing.

16. A separate synthetase exists for each amino acid, and this synthetase functions for all of the different tRNA molecules for that amino acid.

17. The linkage of amino acids to tRNA is as an aminoacyl ester.

18. Proofreading at the activation step allows for selection of both the amino acid and the tRNA. If proofreading took place at the level of codon–anticodon recognition, there would not be a mechanism to ensure that the correct amino acid has been esterified to the tRNA.

19. The overall process of amino acid activation is energetically favored because of the energy contributed by the hydrolysis of two phosphate bonds. Without that input of energy, it would not be favorable.

12.4 How Does Translation Take Place in Prokaryotes?

20. Peptidyl transferase catalyzes the formation of a new peptide bond in protein synthesis. The elongation factors, EF-Tu and EF-Ts, are required for binding of aminoacyl tRNA to the A site. The third elongation factor, EF-G, is needed for the translocation step in which the mRNA moves with respect to the ribosome, exposing the codon for the next amino acid.

21. The initiation complex in *E. coli* requires mRNA, the 30S ribosomal subunit, fmet-tRNA^{fmet}, GTP, and three protein-initiation factors, called IF-1, IF-2, and IF-3. The IF-3 protein is needed for the binding of mRNA to the ribosomal subunit. The other two protein factors are required for the binding of fmet-tRNA^{fmet} to the mRNA—30S complex.

22. Attachment of the 50S ribosomal subunit to the 30S subunit in the initiation complex is needed for protein synthesis to proceed to the elongation phase.

23. The A site and the P site on the ribosome are both binding sites for charged tRNAs taking part in protein synthesis. The P (peptidyl) site binds a tRNA to which the growing polypeptide chain is bonded. The A (aminoacyl) site binds to an aminoacyl tRNA. The amino acid moiety will be the next added to the nascent protein. The E (exit) site binds the uncharged tRNA until it is released from the ribosome.

24. Puromycin terminates the growing polypeptide chain by forming a peptide bond with its C-terminus, which prevents the formation of new peptide bonds (see Figure 12.14).

25. The stop codons bind to release factors, proteins that block binding of aminoacyl tRNAs to the ribosome, and to release the newly formed protein.

26. In the course of protein synthesis, mRNA binds to the smaller ribosomal subunit.

27. The Shine–Dalgarno sequence is a purine-rich leader segment of prokaryotic mRNA. It binds to a pyrimidine-rich sequence on the 16S rRNA part of the 30S ribosomal subunit and aligns it for proper translation, beginning with the AUG start codon.

28. Your friend is mistaken. The hydrogen-bonded regions contribute to the overall shape of the tRNA. Hydrogen-bonded regions are also important in the recognition of tRNAs by aminoacyl-tRNA synthetases.

29. Methionine bound to tRNAf^{met} can be formylated, but methionine bonded to tRNA^{met} cannot be.

30. Different tRNAs and different factors are involved. Initiation requires IF-2, which recognizes fmet-tRNA^{fmet} but not met-tRNA^{fmet}. Conversely, in elongation, EF-Tu recognizes metRNA^{met} but not fmet-tRNA^{fmet}.

31. The methionine anticodon (UAC) on the tRNA base pairs with the methionine codon AUG in the mRNA sequence that signals the start of protein synthesis.

32. The fidelity of protein synthesis is assured twice during protein synthesis—the first time during activation of the amino acids and the second time during the matching of the codon to the anticodon on the mRNA.

33. (a) Activation cycles needed for a protein with 150 AA: 150.
    (b) Initiation cycles needed for a protein with 150 AA: 1.
    (c) Elongation cycles needed for a protein with 150 AA: 149.
    (d) Termination cycles needed for a protein with 150 AA: 1.

34. Four high-energy phosphate bonds per amino acid: two in aminoacyl-tRNA formation, one in elongation with EF-Tu, and one in translocation from the A to the P site, involving EF-G. Forming a peptide bond requires about 5 kcal/mol. This is an expenditure of about 30 kcal/mol peptide bonds. This is the price of low entropy and high fidelity.

35. Not very precisely. Ignoring any editing or proofreading costs, a maximum value can be calculated in terms of high-energy phosphate bonds. We will designate each phosphate bond as ~P. Four are needed per amino acid, and two are needed per ribonucleotide or deoxyribonucleotide. Therefore, four ~P per amino acid × six ~P per codon × six ~P per DNA triplet = 144 ~P per amino acid (approximately 1050 kcal per mole of amino acid. However, the actual value would be much less because of several factors. A single mRNA molecule can be involved in the synthesis of several to many protein molecules before it is degraded. One gene can be involved in the synthesis of many mRNA molecules, with replication taking place only once per cell generation. In addition, rRNA and tRNA are relatively long-lived and available for repeated protein syntheses.

36. The fact that peptidyl transferase is one of the most conserved sequences in all of biology may indicate that it evolved very early in evolution and that it is so critical for all living organisms that it cannot be modified.

37. The less highly purified ribosome preparations contained polysomes, which are more active in protein synthesis than single ribosomes.

38. At first, peptide-bond formation was catalyzed by RNA. In time, as proteins catalysts developed and became more efficient, proteins became an integral part of the ribosome.

39. Electron microscopy can give information about ribosomal structure and function, but X-ray crystallography has given far more detailed information.

40. Because the tRNAs are bound in proximity to each other on the ribosome, the growing polypeptide chain and the amino acid to be added are also close to each other. This will facilitate formation of the next peptide bond.

41. A virus takes over the protein-synthesizing machinery of the cell. It uses its own nucleic acids and the cell's ribosomes.

12.5 How Does Translation Take Place in Eukaryotes?

42. Similarities between protein synthesis in bacteria and protein synthesis in eukaryotes: same start and stop codons; same genetic code; same chemical mechanisms of synthesis; interchangeable tRNAs. Major differences: In prokaryotes, the Shine–Dalgarno sequence and no introns; in eukaryotes, the 5′-cap and 3′-tail on mRNA and introns have been spliced out.

43. The original N-terminal methionine can be removed by post-translational modification.

44. Puromycin would be useful for treatment of a viral infection, but chloramphenicol would not. Viral mRNAs are translated by eukaryotic translation systems, so one must use an antibiotic active on eukaryotic systems.

45. Protein synthesis in prokaryotes takes place as a coupled process with simultaneous transcription of mRNA and translation of the message in protein synthesis. This is possible because of the lack of compartmentalization in prokaryotic cells. In eukaryotes, mRNA is transcribed and processed in the nucleus and only then exported to the cytoplasm to direct protein synthesis.

46. Some mutations can introduce stop codons. It is useful to a cell to have some mechanism to suppress the formation of incomplete proteins.

12.6 How Does Posttranslational Modification of Proteins Take Place?

47. Hydroxyproline is formed from proline, an amino acid for which there are four codons, by posttranslational modification of the collagen precursor.

12.7 How Are Proteins Degraded?

48. Ubiquitin is a small polypeptide (76 amino acids) that is highly conserved in eukaryotes. When ubiquitin is linked to a protein, it marks that protein for degradation in a proteasome.

49. If proteins to be degraded did not have some signal marking them, the process would take place more randomly and thus be less efficient.

50. If protein degradation took place at any location in a cell, indiscriminate breakdown of functional proteins could take place, so this is an unlikely occurrence. It is much more useful to the cell to have a mechanism for tagging proteins to be degraded and to do so at a specific location in the cell.

Chapter 13

13.1 How Do We Purify and Detect Nucleic Acids?

1. Safety, no need for special licensing, and convenience of disposal.

2. DNA is labeled with ^{32}P and run on a gel. The gel is placed next to X-ray paper, which is then developed. The radioactivity shows up as black bands on the X-ray paper. This is called an autoradiogram.

3. The DNA run on electrophoresis gels is usually cleaved with restriction enzymes to give linear pieces, thus the shape is uniform for DNA. The charge is a constant for DNA, in that every nucleotide has the same charge due to the phosphate groups; thus, DNA has a uniform shape and a uniform charge-to-mass ratio, so it separates solely on size, with the shorter fragments traveling fastest through the gel.

13.2 What Makes Restriction Endonucleases an Important Tool for DNA Research?

4. The use of restriction endonucleases with different specificities gives overlapping sequences that can be combined to give an overall sequence.

5. Restriction endonucleases do not hydrolyze a methylated restriction site.

6. The restriction site of the DNA of the organism that produces a restriction endonuclease is modified, usually by methylation.

7. The restriction fragments of different sizes (restriction-fragment length polymorphisms, or RFLPs) that come about as a result of different base sequences on paired chromosomes were used as genetic markers to determine the exact position of the cystic fibrosis gene on chromosome 7.

8. An endonuclease is an enzyme that cuts nucleic acid chains in the middle, as opposed to cleaving from the ends inward. The term *restriction* came from the restricted growth seen in host cells that are infected by bacteriophages when the bacteria have restriction enzymes that can cleave the viral DNA.

9. They are all palindromes (ignoring punctuation and spacing in the latter two cases), analogous to palindromic sequences of bases in DNA. Just as the five examples are distinguished by being pronounced differently, different palindromes in DNA are distinguished and acted on by different, very specific restriction endonucleases.

10. GGATCC, GAATTC, AAGCTT (remember that these are listed 5′ to 3′, so you must read the complementary strand 5′ to 3′ to see that the sequence is the same).

11. *Hae*III cuts at a sequence of four bases, cuts in the middle of the sequence, and leaves blunt ends. *Bam*HI cuts at a sequence of six bases, cuts on the second base from the 5′ end, and leaves sticky ends.

12. Sticky ends are short regions of single-stranded DNA extending from the ends of double-stranded DNA molecules. These are produced by some restriction enzymes or can be added chemically to blunt-ended double-stranded DNA. They are important because they provide a means for DNA from different sources (e.g., "foreign" gene and plasmid, both containing sticky ends) to find each other by hydrogen bonding between complementary bases. A ligase is then used to covalently link the two molecules.

13. An advantage of using *Hae*III is that it yields blunt ends. Thus, one could combine DNA cut with this enzyme with any other DNA that also had blunt ends. Enzymes exist that quickly remove the sticky overhangs from other restriction enzymes. The disadvantage is that *Hae*III is specific for a four-base sequence that is likely to occur many times in a genome, so the target DNA may also be cleaved somewhere in the middle. Also, the blunt ends make it more difficult to get specific ligation of two DNA types.

13.3 What Is Cloning?

14. A portion of exogenous DNA is introduced into a suitable vector, frequently a bacterial plasmid, and many copies of the DNA are produced when the bacteria grow. Viruses are also commonly used as vectors.

15. The most common vectors are bacterial plasmids. Viruses and cosmids can also be used, depending on the size of the foreign DNA that must be inserted.

16. The plasmid to be used as a vector needs markers both for uptake of the target DNA sequence into the plasmid and for insertion of the plasmid into host cells. Typically, a plasmid has a gene for ampicillin resistance. Only cells that have taken up a plasmid can grow on ampicillin plates. The foreign DNA is usually inserted into a second marker to select for those plasmids that took up the target DNA. This second marker may be another antibiotic-resistant gene or some other gene, such as the β-galactosidase gene.

17. The key feature of a plasmid capable of blue/white screening is the gene for the α-subunit of the enzyme β-galactosidase. These plasmids are used with a strain of *E. coli* that are deficient in the α-subunit of this enzyme. β-galactosidase can convert a colorless sugar-derivative, called X-gal, to a blue one. The site for cleavage of the plasmid by a restriction endonuclease lies within the β-galactosidase gene. Cells that have acquired a plasmid will be able to grow on ampicillin. If the plasmid reclosed on itself without the target DNA, the colonies that took up that plasmid will grow blue. Cells that have acquired the DNA insert will not be able to produce a blue color.

18. Restriction enzymes to cut DNA, DNA ligase to rejoin DNA, a suitable vector to carry the foreign DNA, a cell line to accept the vector, and a way of selecting for the correct transformants.

19. Since most recombinant DNA occurs with bacterial and viral vectors, a big concern is that a mutated virus or bacteria will be released that can infect other species and that may be resistant to drugs, thereby creating a new, potentially lethal disease. Precautions include frequent sterilization of cultures to make sure that they are all dead before disposal, working in laminar hoods that isolate the recombinant DNA from the outside, and care in the choice of vectors. Some vectors that are replication-deficient outside of certain cell types are used so that they cannot replicate outside of the lab environment.

13.4 What Is Genetic Engineering, and Why Do We Do It?

20. To increase disease resistance, resistance to pests, shelf life, level of nitrogen fixation (protein content), and resistance to temperature extremes.

21. Insulin, human growth hormone, tissue plasminogen activator, enterokinase, erythropoietin, and interferon.

22. The corn being grown in the field has been genetically engineered. The gene that was introduced came from the bacterium *Bacillus thuringensis.*

23. LDH 3 has the subunit composition H_2M_2. Each of the subunits is coded for by a separate gene, so, in order to clone LDH 3, one would have to clone the gene for the M subunit and the gene for the H subunit. These would be separate cloning experiments. Each gene would be cloned into an expression cell line, and the proteins would be expressed. The individual subunits could then be combined, and they would form tetramers, some of which would be LDH 3. This could be verified by native gel electrophoresis.

24. An expression vector, such as pET 5 plasmid, has the components of any normal cloning vector (e.g., origin of replication, selectable marker, multiple cloning site), but it also has the ability to have the inserted DNA be transcribed. It has a promoter for RNA polymerase, such as T7 polymerase, and a termination sequence. These border the multiple cloning site. These vectors are used with a cell line that will make T7 RNA polymerase when induced.

25. A fusion protein is a combination of a protein coded for by an expression vector and the target gene. A common one is a histidine tag and enterokinase, which will be linked to the target protein when transcribed and translated. They are used to help with the eventual purification of the target protein. The overexpressed target protein can be quickly separated from the rest of the host's proteins by purifying the fusion protein, which will have characteristics that make it easy to purify.

26. The bovine growth hormone is a protein that will be denatured and digested in the intestinal tract. Also, all cow's milk contains some of the hormone.

27. The DNA sequence to be inserted in the bacterial plasmid to direct the production of α-globin should be cDNA, which is a sequence complementary to the mRNA for α-globin. The cDNA can be produced on the mRNA template in a reaction catalyzed by reverse transcriptase.

28. Isolate the DNA that codes for the growth factor by means of suitable probes. Introduce the DNA into a bacterial genome. Allow the bacteria to grow and to produce human growth hormone.

29. The public is concerned about contamination with prions, which come from mammalian sources. If a mammalian protein can be expressed in large quantities in bacteria, there will be no risk of prion contamination.

13.5 What Are DNA Libraries?

30. A DNA library is a collection of cells that carry cloned pieces of the entire DNA genome of an organism. A cDNA library is made by taking the mRNA from an organism, converting it to cDNA,

and cloning that for the library. In this way, the active DNA sequence is stored.

31. If a DNA library is to represent the total genome of an organism, it must contain at least one clone for each DNA sequence. This requires several hundred thousand separate clones to ensure that every sequence is represented.

32. The amount of work involved in constructing a DNA library makes it desirable to have such libraries available to the entire scientific community, thus avoiding duplication of effort.

13.6 What Is the Polymerase Chain Reaction?

33. The polymerase chain reaction depends on repeated cycles of separation of DNA strands followed by annealing of primers. The first step requires a significantly higher temperature than the second, giving rise to the requirement for strict temperature control.

34. Part of the procedure of the polymerase chain reaction requires the use of high temperatures. When a temperature-stable RNA polymerase is used, there is no need to add fresh batches of enzyme for each round of amplification. This would need to be the case, however, if the RNA polymerase could not withstand the high temperatures.

35. Good primers have similar G–C contents for the forward and reverse reactions, have minimal secondary structure possibilities with each other or with themselves, and are long enough to give sufficient specificity for the gene to be duplicated without costing too much.

36. The contaminating DNA as well as the desired DNA is amplified at each stage of the polymerase chain reaction, giving rise to an impure product.

37. (a) Very different G–C contents; (b) forward primer will have significant secondary structure with itself (hairpin loop due to inverted Gs and Cs on end); (c) forward and reverse primers will bind to each other.

13.7 What Is Site-Directed Mutagenesis?

38. PCR would be used to create DNA with a different sequence. A primer would be designed to match the region of the DNA where the *Bam*HI restriction site was (GGATCC). The sequence of DNA on both sides would be determined and a suitable primer that would bind the restriction site to seven or eight bases above and below it would be used. This primer would have the sequence GAATTC instead of GGATCC. The sequences match at four of the six positions, so the primer would still bind. After a few cycles, the product sequence will have been changed to the *Eco*RI.

39. Because it is suspected that a critical serine and histidine are involved in the mechanism, PCR would be used to make the DNA for chymotrypsin with altered bases. These bases would be chosen so that the serine 195 and histidine 57 were changed to different amino acids. The loss of activity would support the proposed mechanism.

13.8 What Is DNA Fingerprinting?

40. The polymerase chain reaction can increase the amount of a desired DNA sample by a considerable factor, making possible definite identification of DNA samples that were too small to be characterized by other means. It can be used on hair and blood samples found at the scene of a crime to establish the presence of a suspect. This method can also be used to identify remains of possible murder victims.

41. It is easier to show that two DNA samples do not match than to prove that they are identical.

13.9 How Can We Study DNA–Protein Interactions?

42. A DNA footprint is an experiment where DNA is digested with DNase I to yield a random set of fragment sizes. A potential DNA-binding protein is then added to the DNA and the digestion repeated. If the protein binds to the DNA, it protects the DNA

from the DNase I and a different banding pattern is seen. An electrophoretic mobility-shift assay simply compares the migration of DNA on a gel with and without a potential binding protein. If the DNA binds to the protein, it moves slower in the gel, "shifting" the band to a higher position.

43. Recall that CREB is a transcription factor that binds to DNA. It links to the basal transcription unit via a mediator protein called CBP. If you ran DNA alone, and compared it to DNA plus CBP, there would be no shift. However, if you ran DNA and CREB, there would be a shift. DNA plus CREB plus CBP would show a supershift.

13.10 What Are Some Methods for Studying Transcription?

44. A DNA microarray chip can have oligonucleotides representing the known genes in a species. Cells or the whole organism could be subjected to two different metabolic conditions, and the mRNA could be collected. The mRNA can be converted to cDNA with fluorescent labels. This labeled cDNA is then reacted with the oligonucleotides on the chip, and the data are analyzed to see which genes are expressed differently.

13.11 How Do We Determine the Base Sequences of Nucleic Acids?

45. 5′GATGCCTACG3′

46. Two factors are involved here. First, large polymers must be cleaved into smaller, manageable fragments for sequencing. Enzymes (endoproteases) that cleave proteins, while showing some specificity, are far from absolutely specific, and messy mixtures result. On the other hand, restriction endonucleases are absolutely specific for palindromic base sequences in DNA, and "clean" cuts result, allowing easier purification. (Note that, if the gene for a protein isn't available but the mRNA is, the resulting cDNA can be made using reverse transcriptase.) A second factor is that only relatively short fragments of protein can be sequenced without additional internal cleavage. For example, the Edman degradation is limited to peptides of about 50 amino acids or less. With DNA, the dideoxy method coupled with poly-acrylamide-gel separation can handle DNA fragments 10 to 20 times longer.

47. DNA often has introns in the gene, so knowing the DNA sequence may give the wrong answer for the final protein sequence. Also, proteins are modified posttranslationally, so there may be modifications to the protein sequence not reflected in the DNA.

48. Open-ended answer.

49. *Benefits:* A person at risk for future heart disease could be more careful with diet and exercise. Such a person might also take a drug beforehand that would help prevent the condition from developing. Doctors with access to such information would be able to make better diagnoses and to suggest quicker treatments. *Detriments:* Employment could be based on a preconceived idea of what a good genotype is. Health and life insurance could be denied to people considered to have a risky genotype. A new type of prejudice against the "genotypically challenged" could arise.

13.12 How Can We Use Bioinformatics to Study Genomics and Proteomics?

50. The genome is the total DNA of a cell, containing all the genes of that organism. The proteome is the total complement of proteins.

51. A proteomic analysis has been done on the fruit fly *Drosophila melanogaster*.

52. The two-hybrid analysis depends on using a gene for which the transcriptional activator has two domains, one for DNA binding and one for transcription activation. DNA for each of the domains is introduced into a plasmid along with the gene for a protein to be tested. Each plasmid produces a fusion protein. If the two proteins being tested interact with each other, both the DNA-binding domain and the transcription-interaction domain can act in concert to allow for expression of a reporter gene.

Chapter 14

14.1 What Are Viruses?

1. Some viruses have DNA and some have RNA. In some cases, a viral genome is single-stranded and in others it is double-stranded.

2. (a) The virion is the entire virus particle.
   (b) The capsid is the protein coat that surrounds the viral nucleic acid.
   (c) The nucleocapsid is the combination of the nucleic acid and the capsid.
   (d) A protein spike is a membrane-bound protein that is used to help the virus attach to its host.

3. The main factors determining the family of a virus is whether its genome is DNA or RNA and whether it has a membrane envelope or not. Whether the nucleic acid is single- or double-stranded and the method of incorporation of the virus are also considered.

4. The virus attaches to a specific protein on the host cell's membrane and injects its nucleic acid inside the cell.

5. In the lytic pathway, the viral nucleic acid is replicated in the host cell and packaged into new virus particles that lyse the host cell. In the lysogenic pathway, the viral DNA is incorporated into the host DNA.

6. The influenza virus has been with us for at least 2400 years, as the first report was from 400 BC.

7. The influenza virus is an enveloped virus with a single-stranded RNA template strand genome. The most prominent features of the virus envelope are the two spike proteins, hemagglutinin and neuraminidase.

8. There is no correlation. Some viruses, such as ebola virus, are fast acting and very lethal, while others, such as HIV, are slow and just as lethal. The influenza virus is fast acting, but it is rarely lethal these days.

9. One good choice would be a drug that attacks one of the specific protein spikes on the virus. This may be an antibody that attacks it, or a drug that blocks its ability to attach to the host cell. Another choice would be a drug that inhibits a key viral enzyme, such as the reverse transcriptase of a retrovirus, or the enzymes involved in repackaging the viruses.

10. Viruses can often switch from one pathway to another, based on the condition of the host cells. If the host is healthy, then there will be sufficient material to allow the virus to replicate and to produce new virions. If the host cell is starved or unhealthy, there may be insufficient energy and material to do so. In this case, lysogeny will allow the DNA to incorporate in the host cell, where it can wait until the cell's health improves.

11. One example would be someone who had helper T cells lacking a CD4 receptor. The HIV virus must bind to the CD4 receptor as part of its attachment process.

14.2 What Virus Causes Severe Acute Respiratory Syndrome (SARS)?

12. The SARS virus is a corona virus. As such, it has a unique set of protein spikes on its membrane envelope that make it resemble a crown.

13. The first reported cases came from the province of Guangdong in southeast China.

14. Being coding-strand RNA, the RNA is ready to be translated once in a host cell. Since it has the 5′-cap and a poly-A tail, it resembles mammalian mRNA and is therefore perfect for infecting mammalian cells.

15. The modes of infection are completely different, although both can be deadly. SARS is not a retrovirus, so it has no stage where its RNA is turned into DNA that is then incorporated into the host cell's genome. This means that SARS cannot lie dormant for years before becoming active, as HIV can. Also, SARS is faster act-ing, which makes it easier to find it quickly in the population. Since infected people get sick quickly, they are diagnosed before they have the opportunity to spread it to as many people.

14.3 What Is Unique About Retroviruses?

16. A retrovirus has an RNA genome that must pass through a stage in which it is reverse transcribed to DNA, and this DNA must recombine with the host's DNA.

17. Reverse transcriptase.

18. The first is that retroviruses have been linked to cancer. The second is that human immunodeficiency virus (HIV) is a retrovirus. The third is that retroviruses can be used in gene therapy.

14.4 How Are Viruses Used in Gene Therapy?

19. Gene therapy is the process of introducing a gene into the cells of an organism that was missing functional copies of the gene.

20. Ex vivo gene therapy, in which the cells are removed from the patient before being infected with the virus carrying the therapeutic gene, and in vivo gene therapy, in which the patient is directly infected with the virus carrying the gene.

21. The two most common are the Maloney murine leukemia virus (MMLV) and adenovirus. Both must be manipulated so that the critical genes for replication are removed and replaced with an expression cassette containing the therapeutic gene.

22. When retroviruses, such as MMLV, are used, there is the danger that the therapeutic gene will incorporate in a place that will disrupt another gene. In more cases than would be predicted by random chance, this seems to occur in a place that disrupts a tumor-suppressor gene, causing cancer. There is also the danger that the patient will have a strong reaction to the virus used to introduce the therapeutic gene. In at least one case, this has had fatal consequences.

23. The biggest consideration is where the therapeutic gene has to go. Some viruses are very specific to their target cells, so if the problem is in the lungs, then a virus that is good at attacking lung cells, such as adenovirus, is a good choice. In this case, in vivo delivery would be superior, because lung cells cannot be removed from the body and then replaced. However, if the problem is in an immune cell, then bone-marrow cells can be removed and transformed and later given back to the patient, making ex vivo delivery an option.

24. There are dangers inherent to all forms of gene therapy. People who have SCID have such compromised immune systems that they cannot lead normal lives, and there are few other remedies that allow them to lead normal lives. That made SCID a prime candidate for experimental techniques. Diabetes can be controlled effectively by other techniques that are well established and not as risky.

14.5 How Does the Immune System Defend the Body?

25. AIDS is the most well-known problem of a malfunctioning immune system. SCID is also high on the list. All allergies are immune-system problems, as are autoimmune diseases. Many forms of diabetes are caused by an autoimmune disease in which a person's pancreatic cells are attacked by the immune system.

26. Innate immunity refers to a variety of protective processes, including skin, mucous, and tears as a first line of defense, and dendritic cells, phagocytes, macrophages, and natural killer cells as a second line of defense. These are always present, and the innate-immunity cells are always circulating in the body. Acquired immunity refers to the processes involving B cells and T cells, in which specific sets of them are activated in response to an antigen challenge, and these subsets then multiply.

27. One part includes physical barriers, such as skin, mucous, and tears. The cells of the innate immune system are dendritic cells, macrophages, and natural killer (NK) cells.

28. B cells, which make antibodies, killer T cells, which attack infected cells, and helper T cells, which help to activate B cells.

29. MHC's are receptors on antigen-presenting cells. They bind to fragments of antigens that have been degraded by the infected cell and display it on their surface. T cells then bind to the infected cells.

30. Clonal selection refers to the process in which a particular T cell or B cell is stimulated to divide. Only the one bearing the correct receptor for the antigens being presented is selected.

31. The cells of the innate system initially attack a pathogen, such as a virus, bacteria, or even a cancerous cell. They then present antigens from the pathogen on their surfaces via their MHC proteins. The acquired immunity cells then recognize the MHC/antigen complex, bind to it, and begin the involvement of the acquired immunity system.

32. Interferon is a cytokine produced in very small quantities that stimulates natural killer cells, which attack cancerous cells. One of the first treatments for cancer was to give the patient interferon to stimulate NK cells. Having a large supply of cloned interferon is helpful, therefore, in fighting cancer.

33. When T cells and B cells are developing, they are, in a sense, "trained." If they contain receptors that recognize self-antigens, they are eliminated when they are still young. If they don't ever see any antigens they recognize, then they die by neglect. This leaves a set of precursors to T cells and B cells that have receptors that will recognize foreign antigens but not self-antigens.

14.6 How Does Human Immunodeficiency Virus Cause AIDS?

34. The HIV virus is the causative agent of AIDS. It attacks helper T cells because it binds to the CD4 receptor. It eventually causes the death of some of the T cells, which weakens the immune system.

35. One reason is that the reverse transcriptase is inaccurate, so it makes many mutated versions of the virus. For this reason, it is difficult to make an effective vaccine. Another is a conformational change of the gp120 protein when it binds to the CD4 receptor on the T cell. The gp120 protein forms a complex with gp41 and changes shape when it binds to CD4. It also binds to a secondary site on the T cell that normally binds to a cytokine. This change exposes a part of the gp120 that was previously hidden and therefore unable to elicit antibodies. HIV is also good at evading the innate-immunity system. Natural killer cells attempt to attack the virus, but HIV binds a particular cell protein, called cyclophilin, to its capsid, which blocks the antiviral agent restriction factor 1. HIV hides from the immune system by cloaking its outer membrane in sugars that are very close to the natural sugars found on most of its host's cells.

36. Researchers have tried many vaccines to the outer coat proteins or to specific enzymes, such as the viral protease. Another is antiviral therapy designed to inhibit the activity of reverse transcriptase or protease. A combination of antiviral therapy and vaccines has been shown to be promising.

37. A vaccine acts by stimulating the immune system to produce B cells and T cells directed against particular antigens. One reason this has been largely ineffective is that the virus mutates so quickly that the antigens are never the same for very long. Even within a given individual, there may be many versions of the virus. Some vaccines have been made against the gp120 protein that is involved in docking the virus to the T cell. However, there is a conformation shift during docking that hides portions of the protein from antibodies, so that docking is still successful.

38. Neutralizing antibody b12 has a different shape from a normal immunoglobulin. It has sections of long tendrils that fit into a fold in gp120. This makes it more effective than most other antibodies. Neutralizing antibody 2G12 recognizes some of the sugars on the HIV outer membrane that are unique to HIV. Normal vaccines produce many useless antibodies, but the two neutralizing antibodies that were discovered are able to completely wipe out their targets.

14.7 Why Are Stem Cells Special?

39. Stem cells can be maintained in an undifferentiated state in culture for long time periods, essentially being immortal. They also will adapt to the tissue they are put into and will then differentiate into that type of tissue.

40. Stem cells can be found in adult tissue, but they are found in higher concentrations in fetuses, umbilical cords, or in therapeutically cloned embryos.

41. Embryos have the highest concentration of usable stem cells, making them attractive choices for this purpose. However, the fact that the stem cells come from an embryo, whether it was produced naturally or by therapeutic cloning, is controversial, owing to the questions regarding the legal status of an embryo. Many people who oppose abortion are against the use of embryos for this purpose.

42. The epigenetic state is important because the DNA in a cell usually remembers where it came from. When sperm and oocytes unite, it takes DNA from two different parents to be able to divide into daughter cells and eventually to make an organism. In whole-animal cloning, DNA from only one adult is used, and the epigenetic states must be reprogrammed; otherwise, the cloned embryo will never grow.

14.8 What Is the Biochemistry of Cancer?

43. Cancer cells will continue to grow and divide in situations in which normal cells would not, such as when they are not receiving growth signals from surrounding cells. They will also continue to grow even if surrounding tissues are sending out "stop growth" signals. Cancer cells will continue to grow and divide in situations in which normal cells would not. Cancer cells are able to co-opt the body's vascular system, causing the growth of new blood vessels to supply the cancerous cells with nutrients. Cancer cells are essentially immortal. They can continue to grow and to divide indefinitely. Cancer cells have the ability to break loose, to travel to other parts of the body, and to create new cancer areas, a process known as metastasis.

44. A tumor suppressor is a molecule that restricts the ability of a cell to grow and to divide. An oncogene is a gene whose product stimulates a cell to grow and to divide.

45. The protein called p53 is a tumor suppressor. Mutations of p53 have been found in more than half of all human cancers. Ras is involved in cell division, and mutations in this protein are involved in 30% of human tumors.

46. Viruses have been implicated in many cancers. Retroviruses are particularly dangerous because they insert their DNA into the host's DNA. When this happens in a tumor-suppressor gene, the tumor suppressor is inactivated, causing cancer. Also, the homology between proto-oncogenes and oncogenes makes it likely that the infection cycle of viruses may be responsible for some proto-oncogenes becoming oncogenic.

47. Virotherapy is the process of using a virus to attempt to treat cancer. There are two strategies for virotherapy. One is to use the virus to attack and kill cancer cells directly. In this case, the virus has a protein on its surface that is specific for a cancer cell. Once inside, it kills the cancer cell. The second is to have the virus ferry a gene into the cancer cell that will make the cell more susceptible to a chemotherapeutic agent.

48. If smoking causes cancer, then everyone who smokes would have cancer, but this is not true. Smoking has been linked to cancer, and it is a strong predictor of future cancer; but cancer is the result of many things going wrong in a cell, and there is no single, definitive cause.

49. A tumor suppressor is a protein that helps control cell growth and division. It is like the brakes on a car, trying to slow down a process. Many cancers are related to mutation of tumor suppressors. An oncogene produces something that stimulates growth and division. This is like the accelerator of the car. Many other cancers are caused ultimately by overactivation of an oncogene.

50. Ras, Jun, and Fos are all considered to be oncogenes. In the process of cell division, Ras is a necessary component, but it is usually active only when the cell should be dividing. Oncogenic forms of Ras are overactive and lead to too much cell division. Ras is an early step in the process. Jun and Fos are transcription factors that together make up AP-1, which is involved in the transcription activation pathway involving CBP.

Chapter 15

15.1 What Are Standard States for Free-Energy Changes?

1. There is a connection, and it is one of the most important points in this chapter. It can be expressed in the equation $\Delta G^{\circ\prime} = -RT \ln K_{eq}$.

2. Statements (a) and (c) indicate that the reaction will not proceed as written in the absence of coupling to an exergonic reaction. Statement (b) is ambiguous. It indicates that the reaction will not proceed as written under those conditions, but it could proceed as written at other concentrations or at another temperature.

3. The information given here deals with the thermodynamics of the reaction, not the kinetics. It is not possible to predict the rate of the reaction.

15.2 What Is a Modified Standard State for Biochemical Applications?

4. The usual thermodynamic standard state refers to pH = 0. This is not very useful in biochemistry.

5. The first statement is true, but the second is not. The standard state of solutes is normally defined as unit activity (1 M for all but the most careful work). In biological systems, the pH is frequently in the neutral range (i.e., H^+ is close to 10^{-7} M); the modification is a matter of convenience. Water is the solvent, not a solute, and its standard state is the pure liquid.

6. The designation $\Delta G^{\circ\prime}$ indicates a biological standard state. If the prime is omitted, then it is for chemical standard states.

7. No, there is no relationship between the thermodynamic quantity, ΔG° and the speed. The ΔG° reflects the thermodynamic possibility under standard states. Speed is a kinetic quantity that is based on the ability of an enzyme to catalyze the reaction and the real substrate concentrations in the cell.

8. Assuming one significant figure, 20 kJ mol^{-1}, 0 kJ mol^{-1}, +30 kJ mol^{-1}.

9. $\Delta G^{\circ\prime} = \Delta H^{\circ\prime} - T\Delta S^{\circ\prime}$ and $\Delta S^{\circ\prime} = 34.9$ J mol^{-1} $K^{-1} = 8.39$ cal mol^{-1} K^{-1}. There are two particles on the reactant side of the equation and three on the product side, representing an increase in disorder.

10. Assuming 298 K and one significant figure, (a) −50 kJ, (b) −20 kJ, and (c) −20 kJ.

11. The levels of substrates and products can affect the true ΔG of a reaction, changing it from zero to a high number as in part (a). The ΔG is negative when there is a larger amount of substrate than product.

12. The overall $\Delta G^{\circ\prime} = -260.4$ kJ mol^{-1} or −62.3 kcal mol^{-1}. The reaction is exergonic, because it has a large, negative $\Delta G^{\circ\prime}$.

13. Greater than 3333 to 1.

14. Reaction (a) will not proceed as written; $\Delta G^{\circ\prime} = +12.6$ kJ. Reaction (b) will proceed as written; $\Delta G^{\circ\prime} = -20.8$ kJ. Reaction (c) will not proceed as written; $\Delta G^{\circ\prime} = +31.4$ kJ. Reaction (d) will proceed as written; $\Delta G^{\circ\prime} = -18.0$ kJ.

15. Yes, *if* you correct for the difference in temperature and concentrations from the standard values.

16. Two aspects are involved here. (a) Very rarely, if ever, are in vivo concentrations standard concentrations; actual ΔG (not ΔG°) values are very dependent on local concentrations, especially if the number of reactant molecules and product molecules is not the same. (b) Values of ΔG° rigorously apply only to *closed* systems that can reach equilibrium. Biochemical systems, however, are *open* systems and do not reach equilibrium. If you were at equilib-

rium, you would be dead. Metabolic pathways involve series of reactions, and the metabolic pathways themselves are interconnected, including processes that take in materials from the surroundings and release waste products to the surroundings.

15.3 What Is Metabolism?

17. Group 1: Catabolism, oxidative, energy yielding. Group 2: Anabolism, reductive, energy requiring.

18. The local decrease in entropy associated with living organisms is balanced by the increase in the entropy of the surroundings caused by their presence. Coupling of reactions leads to overall dispersal of energy in the universe.

19. The synthesis of sugars by plants in photosynthesis is endergonic and requires light energy from the Sun.

20. The biosynthesis of proteins is endergonic and is accompanied by a large decrease in entropy.

21. The ATP constantly generated by living organisms is used as a source of chemical energy for endergonic processes. There is a good deal of turnover of molecules, but no net change.

15.4 How Are Oxidation and Reduction Involved in Metabolism?

22. (a) NADH is oxidized, $H^+ + NADH \rightarrow NAD^+ + 2e^- + 2H^+$. Acetaldehyde is reduced, $CH_3CH_2CHO + 2e^- + 2H^+ \rightarrow CH_3CH_2CH_2OH$
    (b) Fe^{2+} is oxidized, $Fe^{2+} \rightarrow Fe^{3+} + e^-$. Cu^{2+} is reduced, $Cu^{2+} + e^- \rightarrow Cu^+$

23. (a) Acetaldehyde is the oxidizing agent; NADH is the reducing agent.
    (b) Cu^{2+} is the oxidizing agent; Fe^{2+} is the reducing agent.

15.5 How Are Coenzymes Used in Biologically Important Oxidation–Reduction Reactions?

24. NAD^+, $NADP^+$, and FAD all contain an ADP moiety.

25. In NADPH, the 2′ hydroxyl of the ribose attached to the adenine has a phosphate attached.

26. There is little effect in the reactions. Both are coenzymes involved in oxidation–reduction reactions. The presence of the phosphate serves to distinguish two separate pools of coenzymes so that different ratios of $NADPH/NADP^+$ versus $NADH/NAD^+$ can be maintained.

27. None of these statements is true. Some coenzymes are involved in group-transfer reactions (recall this from Chapter 7). Many coenzymes contain phosphate groups, and CoA contains sulfur. ATP does not represent stored energy, but is generated on demand.

28. Redox reactions. NAD^+, or NADPH in an anabolic process, would likely be used. FAD probably would not be used because its free-energy change is too low.

29. The second half reaction (the one involving NADH) is that of oxidation, while the first half reaction (the one involving O_2) is that of reduction. The overall reaction is $\frac{1}{2} O_2 + NADH + H^+ \rightarrow H_2O + NAD^+$. O_2 is the oxidizing agent and NADH is the reducing reagent.

30. See Figures 15.3 and 15.4.

31. Glucose-6-phosphate is oxidized, and $NADP^+$ is reduced. $NADP^+$ is the oxidizing agent, and glucose-6-phosphate is the reducing agent.

32. FAD is reduced, and succinate is oxidized. FAD is the oxidizing agent, and succinate is the reducing agent.

33. It is important to have two different pools of redox coenzymes. In the cytosol, the $NAD^+/NADH$ ratio is high, but the $NADPH/NADP^+$ ratio is also high. This means that anabolic reactions can take place in the cytosol, while catabolic reactions, such as glycolysis, can also take place. If there were not two different pools of these coenzymes, no single cell location could have both catabolism and anabolism. Having two different, but structurally related, reducing agents helps to keep anabolic and catabolic reactions distinct from each other.

15.6 How Are the Production and Use of Energy Coupled?

34. The ratio of substrates to products would have to be 321, 258 to 1.

35. Creatine phosphate + ADP → Creatine + ATP;

$$\Delta G^{\circ\prime} = -12.6 \text{ kJ}$$

ATP + Glycerol → ADP + Glycerol-3-phosphate;

$$\Delta G^{\circ\prime} = -20.8 \text{ kJ}$$

Creatine phosphate + Glycerol →
Creatine + Glycerol-3-phosphate;

$$\Delta G^{\circ\prime} \text{ overall} = -33.4 \text{ kJ}$$

36. Glucose-1-phosphate → Glucose + P_i;

$$\Delta G^{\circ\prime} = -20.9 \text{ kJ mol}^{-1}$$

Glucose + P_i → Glucose-6-phosphate;

$$\Delta G^{\circ\prime} = +12.5 \text{ kJ mol}^{-1}$$

Glucose-1-phosphate → Glucose 6-phosphate;

$$\Delta G^{\circ\prime} = -8.4 \text{ kJ mol}^{-1}$$

37. In both pathways, the overall reaction is ATP + 2 H_2O → AMP + 2 P_i. Thermodynamic parameters, such as energy, are additive. The overall energy is the same because the overall pathway is the same.

38. Phosphoarginine + ADP → Arginine + ATP;

$$\Delta G^{\circ\prime} = -1.7 \text{ kJ}$$

ATP + H_2O → ADP + P_i;

$$\Delta G^{\circ\prime} = -30.5 \text{ kJ}$$

Phosphoarginine + H_2O → Arginine + P_i;

$$\Delta G^{\circ\prime} = -32.2 \text{ kJ}$$

39. ATP is less stable than ADP and P_i because of the charge distribution and loss of the resonance stabilization in the phosphate ion. There is stabilization (dispersal of energy) when ATP is hydrolyzed, leading to a negative free-energy change.

40. It is intermediate; thus, ATP is ideally positioned to serve as a phosphate donor or (as ADP) a phosphate acceptor, depending on local concentrations.

41. Creatine phosphate can phosphorylate ADP to ATP. There is a biochemical "germ of truth" here, but the effectiveness of such a supplement is another matter.

42. There is a large increase in entropy accompanying the hydrolysis of one molecule to five separate molecules.

43. PEP is a high-energy compound because energy is released upon its hydrolysis, owing to the resonance stabilization of the inorganic phosphate released and the possible keto–enol tautomerization of its product, pyruvate. See Figure 15.8.

44. The fact that a reaction is thermodynamically favorable does not mean that it will occur biologically. Even though there appears to be ample energy to catalyze the production of 2 ATPs from PEP, there is no enzyme that catalyzes this reaction.

45. Sprints and similar short periods of exercise rely on anaerobic metabolism as a source of energy, producing lactic acid. Longer periods of exercise also draw on aerobic metabolism.

15.7 How Is Coenzyme A Involved in Activation of Metabolic Pathways?

46. An activation step leads to an exergonic next step in a pathway. It is similar to the way in which organic chemists want to attach a good leaving group for the next step in a series of reactions.

47. Small energy changes will generally involve mild conditions. Also, such reactions will be more sensitive to relatively small changes in concentration and thus be easier to control.

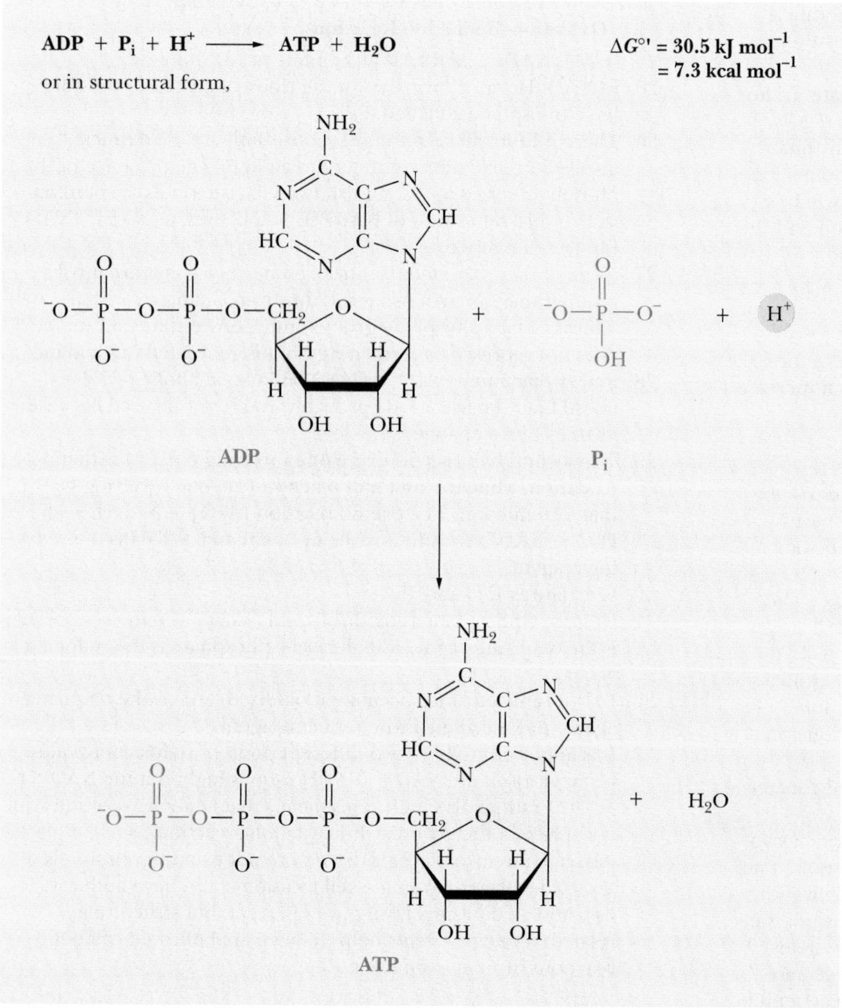

$$ADP + P_i + H^+ \longrightarrow ATP + H_2O \qquad \Delta G^{\circ\prime} = 30.5 \text{ kJ mol}^{-1} = 7.3 \text{ kcal mol}^{-1}$$

or in structural form,

48. Thioesters are high-energy compounds. The possible dissociation of the products after hydrolysis and resonance structures of the products facilitate reaction.

49. Coenzyme A serves several purposes. It is a high-energy compound, activating the initial steps of the metabolic pathway. It is used as a tag to "earmark" a molecule for a particular pathway. It is large and cannot cross membranes, so compartmentalization of pathways can be affected by binding metabolites to coenzyme A.

50. The size and complexity of the molecule make it more specific for particular enzyme-catalyzed reactions. In addition, it cannot cross membranes, so acyl-CoA molecules and other CoA derivatives can be segregated.

Chapter 16

16.1 What Are the Structures and the Stereochemistry of Monosaccharides?

1. A polysaccharide is a polymer of simple sugars, which are compounds that contain a single carbonyl group and several hydroxyl groups. A furanose is a cyclic sugar that contains a five-membered ring similar to that in furan. A pyranose is a cyclic sugar that contains a six-membered ring similar to that in pyran. An aldose is a sugar that contains an aldehyde group; a ketose is a sugar that contains a ketone group. A glycosidic bond is the acetal linkage that joins two sugars. An oligosaccharide is a compound formed by the linking of several simple sugars (monosaccharides) by glycosidic bonds. A glycoprotein is formed by the covalent bonding of sugars to a protein.

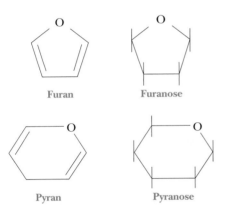

Furan   Furanose

Pyran   Pyranose

2. D-Mannose and D-galactose are both epimers of D-glucose, with inversion of configuration around carbon atoms 2 and 4, respectively; D-ribose has only five carbons, but the rest of the sugars named in this question have six.

3. All groups are aldose–ketose pairs.

4. Enantiomers are nonsuperimposable, mirror-image stereoisomers. Diastereomers are nonsuperimposable, nonmirror-image stereoisomers.

5. Four epimers of D-glucose exist, with inversion of configuration at a single carbon. The possible carbons at which this is possible are those numbered two through five.

6. Furanoses and pyranoses have five-membered and six-membered rings, respectively. It is well known from organic chemistry that rings of this size are the most stable and the most readily formed.

7. There are four chiral centers in the open-chain form of glucose (carbons two through five). Cyclization introduces another chiral center at the carbon involved in hemiacetal formation, giving a total of five chiral centers in the cyclic form.

8. Enantiomers: (a) and (f), (b) and (d). Epimers: (a) and (c), (a) and (d), (a) and (e), (b) and (f).

(a)
CHO
H—C—OH
H—C—OH
H—C—OH
CH₂OH

(b)
CHO
H—C—OH
HO—C—H
HO—C—H
CH₂OH

(c)
CHO
H—C—OH
H—C—OH
HO—C—H
CH₂OH

(d)
CHO
HO—C—H
H—C—OH
H—C—OH
CH₂OH

(e)
CHO
H—C—OH
HO—C—H
H—C—OH
CH₂OH

(f)
CHO
HO—C—H
HO—C—H
HO—C—H
CH₂OH

9. L-Sorbitol was named early in biochemical history as a derivative of L-sorbose. Reduction of D-glucose gives a hydroxy sugar that could easily be named D-glucitol, but it was originally named L-sorbitol and the name stuck.

10. Arabinose is an epimer of ribose. Nucleosides in which arabinose is substituted for ribose act as inhibitors in reactions of ribonucleosides.

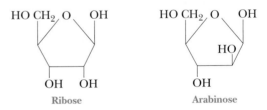

Ribose   Arabinose

11. Converting a sugar to an epimer requires inversion of configuration at a chiral center. This can only be done by breaking and re-forming covalent bonds.

12. Two different orientations with respect to the sugar ring are possible for the hydroxyl group at the anomeric carbon. The two possibilities give rise to the new chiral center.

16.2 How Do Monosaccharides React?

13. This compound contains a lactic acid side chain.

14. In a sugar phosphate, an ester bond is formed between one of the sugar hydroxyls and phosphoric acid. A glycosidic bond is an acetal, which can be hydrolyzed to regenerate the two original sugar hydroxyls.

15. A reducing sugar is one that has a free aldehyde group. The aldehyde is easily oxidized, thus reducing the oxidizing agent.

16. Vitamin C is a lactone (a cyclic ester) with a double bond between two of the ring carbons. The presence of the double bond makes it susceptible to air oxidation.

16.3 What Are Some Important Oligosaccharides?

17. Similarities: Sucrose and lactose are both disaccharides, and both contain glucose. Differences: Sucrose contains fructose, whereas lactose contains galactose. Sucrose has an α,β (1→2) glycosidic linkage, whereas lactose has a β (1→4) glycosidic linkage.

18.

Structure of gentibiose

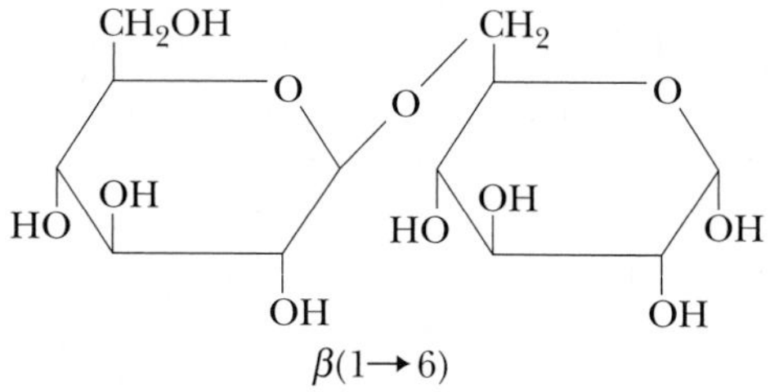

$\beta(1 \rightarrow 6)$

19. In some cases, the enzyme that degrades lactose (milk sugar) to its components—glucose and galactose—is missing. In other cases, the enzyme isomerizes galactose to glucose for further metabolic breakdown.

20.

(a)

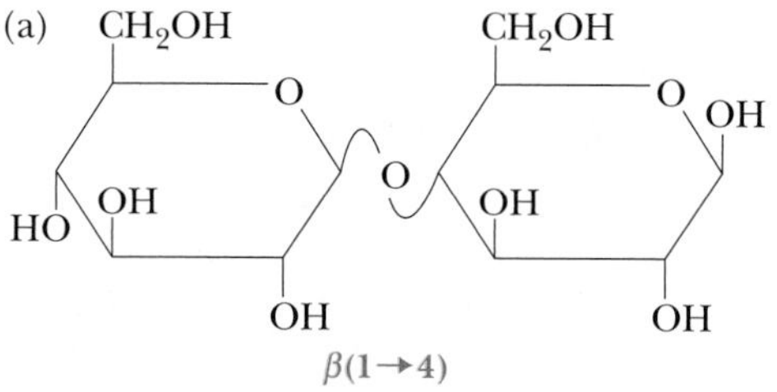

$\beta(1 \rightarrow 4)$

(b)

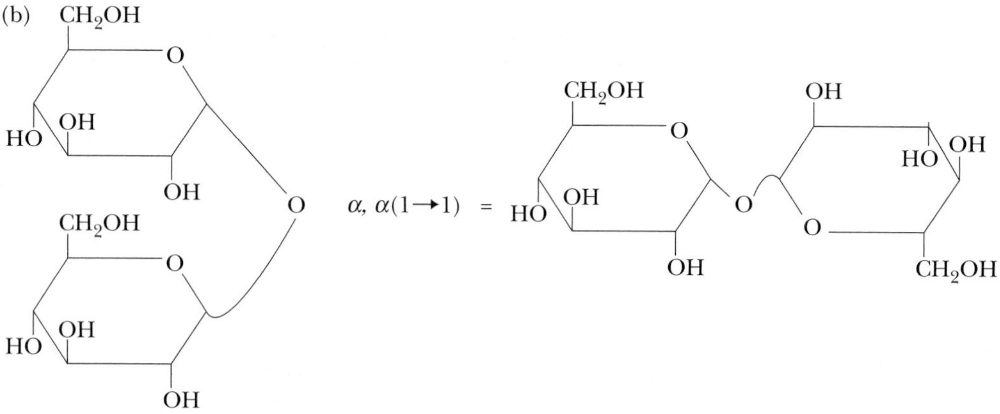

$\alpha, \alpha(1 \rightarrow 1)$ =

(c)

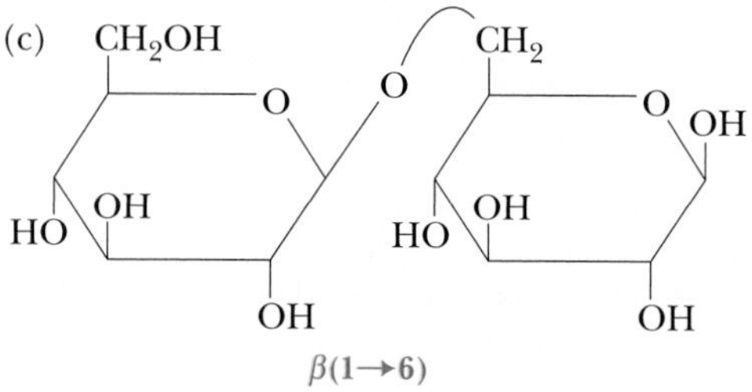

$\beta(1 \rightarrow 6)$

21. Milk contains lactose. Many people are sensitive to lactose and require an alternative beverage.

16.4 What Are the Structures and Functions of Polysaccharides?

22. The cell walls of plants consist mainly of cellulose, whereas those of bacteria consist mainly of polysaccharides with peptide crosslinks.

23. Chitin is a polymer of N-acetyl-β-D-glucosamine, whereas cellulose is a polymer of D-glucose. Both polymers play a structural role, but chitin occurs in the exoskeletons of invertebrates and cellulose primarily in plants.

24. Glycogen and starch differ mainly in the degree of chain branching. Both polymers serve as vehicles for energy storage, glycogen in animals and starch in plants.

25. Both cellulose and starch are polymers of glucose. In cellulose, the monomers are joined by a β-glycosidic linkage, whereas in starch they are joined by an α-glycosidic linkage.

26. Glycogen exists as a highly branched polymer. Starch can have both a linear and a branched form, which is not as highly branched as that of glycogen.

27. Plant cell walls consist almost exclusively of carbohydrates, whereas bacterial cell walls contain peptides.

28. Repeating disaccharide of pectin:

Galacturonic acid (α form)

unmethylated   methylated

Repeating disaccharide

$\alpha(1 \rightarrow 4)$ α-anomeric end

29. Glucose and fructose.

30. Differences in structure: Cellulose consists of linear fibers, but starch has a coil form. Differences in function: Cellulose has a structural role, but starch is used for energy storage.

31. The concentration of reducing groups is too small to detect.

32. To 2500, one place (0.02%). To 1000, four places (0.08%). To 200, 24 places (0.48%).

33. This polymer would be expected to have a structural role. The presence of the β-glycosidic linkage makes it useful as food only to termites or to ruminants, such as cows and horses; these animals harbor bacteria capable of attacking the β-linkage in their digestive tracts.

34. Because of the branching, the glycogen molecule gives rise to a number of available glucose molecules at a time when it is being hydrolyzed to provide energy. A linear molecule could produce only one available glucose at a time.

35. The digestive tract of these animals contains bacteria that have the enzyme to hydrolyze cellulose.

36. Humans lack the enzyme to hydrolyze cellulose. In addition, the fibrous structure of cellulose makes it too insoluble to digest, even if humans had the necessary enzyme.

37. The enzyme β-amylase is an exoglycosidase, degrading polysaccharides from the ends. The enzyme α-amylase is an endoglycosidase, cleaving internal glycosidic bonds.

38. Fiber binds many toxic substances in the gut and decreases the transit time of ingested food in the digestive tract, so that harmful compounds, such as carcinogens, are removed from the body more quickly than would be the case with a low-fiber diet.

39. A cellulase (an enzyme that degrades cellulose) needs an active site that can recognize glucose residues joined in a β-glycosidic linkage and hydrolyze that linkage. An enzyme that degrades starch has the same requirements with regard to glucose residues joined in an α-glycosidic linkage.

40. Cross-linking can be expected to play a role in the structures of polysaccharides where mechanical strength is an issue. Examples include cellulose and chitin. These crosslinks can be readily formed by extensive hydrogen bonding. (See Figure 16.19.)

41. The sequence of monomers in a polysaccharide is not genetically coded, and, in this sense, it does not contain information.

42. It can be useful for polysaccharides to have a number of ends, characteristic of a branched polymer, rather than the two ends of a linear polymer. This would be the case when it is necessary to release residues from the ends as quickly as possible. Polysaccharides achieve this by having 1→4 and 1→6 glycosidic linkages to a residue at a branch point.

43. Chitin is a suitable material for the exoskeleton of invertebrates because of its mechanical strength. Individual polymer strands are cross-linked by hydrogen bonding, accounting for the strength. Cellulose is another polysaccharide cross-linked in the same way, and it can play a similar role.

44. Bacterial cell walls are not likely to consist largely of protein. Polysaccharides are easily formed and confer considerable mechanical strength. They are likely to play a large role.

45. Athletes try to increase their stores of glycogen before an event. The most direct way to increase the amount of this polymer of glucose is to eat carbohydrates.

46. Iodine is the reagent that will be added to the reaction mixture in the titration. When the end point is reached, the next drop of iodine will produce a characteristic blue color in the presence of the indicator.

47. Heparin is an anticoagulant. Its presence prevents blood clotting.

48. Glycosidic bonds can be formed between the side-chain hydroxyls of serine or threonine residues and the sugar hydroxyls. In addition, there is the possibility of ester bonds forming between the side-chain carboxyl groups of aspartate or glutamate and the sugar hydroxyls.

$\alpha(1 \rightarrow 4)$ $\beta(1 \rightarrow 4)$ $\alpha(1 \rightarrow 4)$ $\beta(1 \rightarrow 4)$ $\alpha(1 \rightarrow 4)$

16.5 What Are Glycoproteins?

49. Glycoproteins are ones in which carbohydrates are covalently bonded to proteins. They play a role in eukaryotic cell membranes, frequently as recognition sites for external molecules. Antibodies (immunoglobulins) are glycoproteins.

50. The sugar portions of the blood-group glycoproteins are the source of the antigenic difference.

Chapter 17

17.1 What Is the Overall Pathway in Glycolysis?

1. Reactions that require ATP: phosphorylation of glucose to give glucose-6-phosphate and phosphorylation of fructose-6-phosphate to give fructose-1,6-*bis*phosphate. Reactions that produce ATP: transfer of phosphate group from 1,3-*bis*phosphoglycerate to ADP and transfer of phosphate group from phosphoenolpyruvate to ADP. Enzymes that catalyze reactions requiring ATP: hexokinase, glucokinase, and phosphofructokinase. Enzymes that catalyze reactions producing ATP: phosphoglycerate kinase and pyruvate kinase.

2. Reactions that require NADH: reduction of pyruvate to lactate and reduction of acetaldehyde to ethanol. Reactions that require NAD$^+$: oxidation of glyceraldehyde-3-phosphate to give 1,3-diphosphoglycerate. Enzymes that catalyze reactions requiring NADH: lactate dehydrogenase and alcohol dehydrogenase. Enzymes that catalyze reactions requiring NAD$^+$: glyceraldehyde-3-phosphate dehydrogenase.

3. Pyruvate can be converted to lactate, ethanol, or acetyl- CoA.

17.2 How Is the 6-Carbon Glucose Converted to the 3-Carbon Glyceraldehyde-3-Phosphate?

4. Aldolase catalyzes the reverse aldol condensation of fructose-1,6-*bis*phosphate to glyceraldehyde-3-phosphate and dihydroxyacetone phosphate.

5. Isozymes are oligomeric enzymes that have slightly different amino acid compositions in different organs. Lactate dehydrogenase is an example, as is phosphofructokinase.

6. Isozymes allow for subtle control of the enzyme to respond to different cellular needs. For example, in the liver, lactate dehydrogenase is most often used to convert lactate to pyruvate, but the reaction is often reversed in the muscle. Having a different isozyme in the muscle and liver allows for those reactions to be optimized.

7. Fructose-1,6-*bis*phosphate can only undergo the reactions of glycolysis. The components of the pathway up to this point can have other metabolic fates.

8. Add the $\Delta G^{\circ'}$ mol^{-1} values for the reactions from glucose to glyceraldehyde-3-phosphate. The result is 2.5 kJ mol^{-1} = 0.6 kcal mol^{-1}.

9. The two enzymes can have different tissue locations and kinetic parameters. The glucokinase has a higher K_M for glucose than hexokinase. Thus, under conditions of low glucose, the liver will not convert glucose to glucose-6-phosphate, using the substrate that is needed elsewhere. When the glucose concentration is much higher, however, glucokinase will function to help phosphorylate glucose so that it can be stored as glycogen.

10. Individuals who lack the gene that directs the synthesis of the M form of the enzyme can carry on glycolysis in their livers but suffer muscle weakness because they lack the enzyme in muscle.

11. The hexokinase molecule changes shape drastically on binding to substrate, consistent with the induced-fit theory of an enzyme adapting itself to its substrate.

12. ATP inhibits phosphofructokinase, consistent with the fact that ATP is produced by later reactions of glycolysis.

17.3 How Is Glyceraldehyde-3-Phosphate Converted to Pyruvate?

13. From the point at which aldolase splits fructose-1,6-*bis*phosphate into dihydroxyacetone phosphate and glyceraldehyde-3-phosphate; all reactions of the pathway are doubled (only the path from one glyceraldehyde-3-phosphate is usually shown).

14. NADH-linked dehydrogenases: Glyceraldehyde-3-phosphate dehydrogenase, lactate dehydrogenase, and alcohol dehydrogenase.

15. The free energy of hydrolysis of a substrate is the energetic driving force in substrate-level phosphorylation. An example is the conversion of glyceraldehyde-3-phosphate to 1,3-*bis*phosphoglycerate.

16. The control points in glycolysis are the reactions catalyzed by hexokinase, phosphofructokinase, and pyruvate kinase.

17. Hexokinase is inhibited by glucose-6-phosphate. Phosphofructokinase is inhibited by ATP and citrate. Pyruvate kinase is inhibited by ATP, acetyl-CoA, and alanine. Phosphofructokinase is stimulated by AMP and fructose-2,6-*bis*phosphate. Pyruvate kinase is stimulated by AMP and fructose-1,6-*bis*phosphate.

18. The part of the active site that binds to NADH would be the part that is most conserved, since many dehydrogenases use that coenzyme.

19. (a) Using a high-energy phosphate to phosphorylate a substrate. (b) Changing the form of a molecule without changing its empirical formula (i.e., replacing one isomer with another). (c) Performing an aldol cleavage of a sugar to yield two smaller sugars or sugar derivatives. (d) Changing the oxidation state of a substrate by removing hydrogens while simultaneously changing the oxidation state of a coenzyme (NADH, FADH$_2$, etc.).

20. An isomerase is a general term for an enzyme that changes the form of a substrate without changing its empirical formula. A mutase is an enzyme that moves a functional group, such as a phosphate, to a new location in a substrate molecule.

21. The reaction of 2-phosphoglycerate to phosphoenolpyruvate is a dehydration (loss of water) rather than a redox reaction.

22. Carbon-1 of glyceraldehyde is the aldehyde group. It changes oxidation state to a carboxylic acid, which is phosphorylated simultaneously.

23. ATP is an inhibitor of several steps of glycolysis as well as other catabolic pathways. The purpose of catabolic pathways is to produce energy, and high levels of ATP mean the cell already has sufficient energy. Glucose-6-phosphate inhibits hexokinase and is an example of product inhibition. If the glucose-6-phosphate level is high, it may indicate that sufficient glucose is available from glycogen breakdown or that the subsequent enzymatic steps of glycolysis are going slowly. Either way, there is no reason to produce more glucose-6-phosphate. Phosphofructokinase is inhibited by a special effector molecule, fructose-2,6-*bis*phosphate, whose levels are controlled by hormones. It is also inhibited by citrate, which indicates that there is sufficient energy from the citric acid cycle, probably from fat and amino acid degradation. Pyruvate kinase is also inhibited by acetyl-CoA, the presence of which indicates that fatty acids are being used to generate energy for the citric acid cycle. The main function of glycolysis is to feed carbon units to the citric acid cycle. When these carbon skeletons can come from other sources, glycolysis is inhibited to spare glucose for other purposes.

24. There would be 15 possible isozymes of LDH, combining three different subunits into combinations of four. Besides the five isozymes containing only M and H, there would also be C$_4$, CH$_3$, C$_2$H$_2$, C$_3$H, CH$_2$M, C$_2$HM, C$_3$M, CHM$_2$, C$_2$M$_2$, and CM$_3$.

25. Glutamic acid has an acidic side chain with a pK_a of 4.25. Therefore, it would be negatively charged at pH 8.6, and the H subunit would move more toward the anode (+) than the M subunit. Thus, LDH 1, which is H$_4$, would move the farthest. LDH 5, which is M$_4$, would move the least, with the other isozymes

migrating between those two extremes proportional to their H content.

26. The formation of fructose-1,6-*bis*phosphate is the committed step in the glycolytic pathway. It is also one of the energy-requiring steps of the pathway.

27. Glucose-6-phosphate inhibits hexokinase, the enzyme responsible for its own formation. Because G-6-P is used up by additional reactions of glycolysis, the inhibition is relieved.

28. With few exceptions, a biochemical reaction typically results in only one chemical modification of the substrate. Accordingly, several to many steps are needed to reach the ultimate goal.

29. The enzyme contains a phosphate group on a suitable amino acid, such as serine, threonine, and histidine. The substrate donates its phosphate group from the C-3 position to another amino acid on the enzyme, subsequently receiving the one that started out on the enzyme. Thus, the ^{32}P that was on the substrate is transferred to the enzyme, while an unlabeled phosphorus is put on the C-2 position.

17.4 How Is Pyruvate Metabolized Anaerobically?

30. The bubbles in beer are CO_2, produced by alcoholic fermentation. Tired and aching muscles are caused in part by a buildup of lactic acid, a product of anaerobic glycolysis.

31. The problem with lactic acid is that it is an acid. The H^+ produced from lactic acid formation causes the burning muscle sensation. Sodium lactate is the conjugate weak base of lactic acid. It is reconverted to glucose by gluconeogenesis in the liver. Giving sodium lactate intravenously is a good way to supply an indirect source of blood glucose.

32. The purpose of the step that produces lactic acid is to reduce pyruvate so that NADH can be oxidized to NAD^+, which is needed for the step catalyzed by glyceraldehyde-3-phosphate dehydrogenase.

33.

34. Thiamine pyrophosphate is a coenzyme in the transfer of two-carbon units. It is required for catalysis by pyruvate decarboxylase in alcoholic fermentation.

35. The important part of TPP is the five-membered ring, in which a carbon is found between a nitrogen and a sulfur. This carbon forms a carbanion and is extremely reactive, making it able to perform a nucleophilic attack on carbonyl groups, leading to decarboxylation of several compounds in different pathways.

36. Thiamine pyrophosphate is a coenzyme required in the reaction catalyzed by pyruvate carboxylase. Because this reaction is a part of the metabolism of ethanol, less will be available to serve as a coenzyme in the reactions of other enzymes that require it.

37. Animals that have been run to death have accumulated large amounts of lactic acid in their muscle tissue, accounting for the sour taste of the meat.

38. Conversion of glucose to lactate rather than pyruvate recycles NADH.

39. This is possible, and it is done. These poisons also affect other tissues, including skin, hair, cells of the intestinal lining, and especially the immune system and red blood cells. People on chemotherapy are usually more susceptible to infectious diseases than healthy people and are often somewhat anemic.

17.5 How Much Energy Can Be Produced by Glycolysis?

40. The energy released by all the reactions of glycolysis is 184.5 kJ mol glucose^{-1}. The energy released by glycolysis drives the phosphorylation of two ADP to ATP for each molecule of glucose, trapping 61.0 kJ mol glucose^{-1}. The estimate of 33% efficiency comes from the calculation $(61.0/184.5) \times 100 = 33\%$.

41. There is a net gain of two ATP molecules per glucose molecule consumed in glycolysis.

42. The gross yield is four ATP molecules per glucose molecule, but the reactions of glycolysis require two ATP per glucose.

43. The reactions catalyzed by hexokinase, phosphofructokinase, glyceraldehyde-3-phosphate dehydrogenase, phosphoglycerokinase, and pyruvate kinase.

44. The steps catalyzed by hexokinase, phosphofructokinase, and pyruvate kinase.

45. Phosphoenolpyruvate $\rightarrow$ pyruvate + P_i

$\Delta G^{\circ\prime} = -61.9$ kJ mol^{-1} = -14.8 kcal mol^{-1}

ADP + $P_i \rightarrow$ ATP

$\Delta G^{\circ\prime} = 30.5$ kJ mol^{-1} = 7.3 kcal mol^{-1}

Phosphoenolpyruvate + ADP $\rightarrow$ Pyruvate + ATP

$\Delta G^{\circ\prime} = -31.4$ kJ mol^{-1} = -7.5 kcal mol^{-1}

46. The net yield of ATP from glycolysis is the same, two ATP, when either of the three substrates is used. The energetics of the conversion of hexoses to pyruvate are the same, regardless of hexose type.

47. Starting with glucose-1-phosphate, the net yield is three ATP, because one of the priming reactions is no longer used. Thus, glycogen is a more efficient fuel for glycolysis than free glucose.

48. Phosphoenolpyruvate + ADP $\rightarrow$ Pyruvate + ATP

	$\Delta G^{\circ\prime} = -31.4$ kJ/mol
ADP + $P_i \rightarrow$ ATP	$\Delta G^{\circ\prime} = 30.5$ kJ/mol
Sum	$\Delta G^{\circ\prime} = -0.9$ kJ/mol

Thus, the reaction is thermodynamically possible under standard conditions.

49. No, the reaction shown in Question 48 does not occur in nature. We can assume that no enzyme evolved that could catalyze it. Nature is not 100% efficient.

50. A positive $\Delta G^{\circ\prime}$ does not necessarily mean that the reaction has a positive ΔG. Substrate concentrations can make a negative ΔG out of a positive $\Delta G^{\circ\prime}$.

51. The entire pathway can be looked at as a large coupled reaction. Thus, if the overall pathway has a negative ΔG, an individual step may be able to have a positive ΔG, and the pathway can still continue.

Chapter 18

18.1 How Is Glycogen Produced and Degraded?

1. These two pathways occur in the same cellular compartment, and, if both are on at the same time, a futile ATP hydrolysis cycle results. Using the same mechanism to turn them on/off or off/on is highly efficient.

2. In phosphorolysis, a bond is cleaved by adding the elements of phosphoric acid across that bond, whereas, in hydrolysis, the cleavage takes place by adding the elements of water across the bond.

3. Glucose-6-phosphate is already phosphorylated. This saves one ATP equivalent in the early stages of glycolysis.

4. Each glucose residue is added to the growing glycogen molecule by transfer from UDPG.

5. Glycogen synthase is subject to covalent modification and to allosteric control. The enzyme is active in its phosphorylated form and inactive when dephosphorylated. AMP is an allosteric inhibitor of glycogen synthase, whereas ATP and glucose-6-phosphate are allosteric activators.

6. There is a net gain of three, rather than two, ATP when glycogen, not glucose, is the starting material of glycolysis.

7. It "costs" one ATP equivalent (UTP to UDP) to add a glucose residue to glycogen. In degradation, about 90% of the glucose residues do not require ATP to produce glucose-1-phosphate. The other 10% require ATP to phosphorylate glucose. On average, this is another 0.1 ATP. Thus, the overall "cost" is 1.1 ATP, compared with the three ATP that can be derived from glucose-6-phosphate by glycolysis.

8. The ATP cost is the same, but more than 30 ATP can be derived from aerobic metabolism.

9. Eating high-carbohydrate foods for several days prior to strenuous activity is intended to build up glycogen stores in the body. Glycogen will be available to supply required energy.

10. The disaccharide sucrose can be hydrolyzed to glucose and fructose, which can both be readily converted to glucose-1-phosphate, the immediate precursor of glycogen. This is not the usual form of "glycogen loading."

11. Probably not, because the sugar spike initially will result in a rapid increase in insulin levels, which results in lowering blood glucose levels and increased glycogen storage in the liver.

12. The sprint is essentially anaerobic and produces lactate from glucose by glycolysis. Lactate is then recycled to glucose by gluconeogenesis.

13. It is unlikely that this finding will be confirmed by other researchers. The highly branched structure of glycogen is optimized for release of glucose on demand.

14. Each glucose residue added to a growing phosphate chain comes from uridine diphosphate glucose. The cleavage of the phosphate ester bond to the nucleoside diphosphate moiety supplies the needed energy.

15. The enzyme that catalyzes addition of glucose residues to a growing glycogen chain cannot form a bond between isolated glucose residues, thus the need for a primer.

16. The glycogen synthase reaction is exergonic overall because it is coupled to phosphate ester hydrolysis.

17. (a) Increasing the level of ATP favors both gluconeogenesis and glycogen synthesis. (b) Decreasing the level of fructose-1,6-*bis*phosphate would tend to stimulate glycolysis, rather than gluconeogenesis or glycogen synthesis. (c) Levels of fructose-6-phosphate do not have a marked regulatory effect on these pathways of carbohydrate metabolism.

18. "Going for the burn" in a workout refers to the sensation that accompanies lactic acid buildup. This in turn arises from anaerobic metabolism of glucose in muscle.

19. Sugar nucleotides are diphosphates. The net result is hydrolysis to two phosphate ions, releasing more energy and driving the addition of glucose residues to glycogen in the direction of polymerization.

18.2 How Does Gluconeogenesis Produce Glucose from Pyruvate?

20. Reactions that require acetyl-CoA: none. Reactions that require biotin: carboxylation of pyruvate to oxaloacetate.

21. Three reactions of glycolysis are irreversible under physiological conditions. They are the production of pyruvate and ATP from phosphoenolpyruvate, the production of fructose-1,6-*bis*phosphate from fructose-6-phosphate, and the production of glucose-6-phosphate from glucose. These reactions are bypassed in gluconeogenesis; the reactions of gluconeogenesis differ from those of glycolysis at these points and are catalyzed by different enzymes.

22. Biotin is the molecule to which carbon dioxide is attached to the process of being transferred to pyruvate. The reaction produces

oxaloacetate, which then undergoes further reactions of gluconeogenesis.

23. In gluconeogenesis, glucose-6-phosphate is dephosphorylated to glucose (the last step of the pathway); in glycolysis, it isomerizes to fructose-6-phosphate (an early step in the pathway).

24. Of the three processes—glycogen formation, gluconeogenesis, and the pentose phosphate pathway—only one, gluconeogenesis, involves an enzyme that requires biotin. The enzyme in question is pyruvate carboxylase, which catalyzes the conversion of pyruvate to oxaloacetate, an early step in gluconeogenesis.

25. The hydrolysis of fructose-1,6-*bis*phosphate is a strongly exergonic reaction. The reverse reaction in glycolysis, phosphorylation of fructose-6-phosphate, is irreversible because of the energy supplied by ATP hydrolysis.

18.3 How Is Carbohydrate Metabolism Controlled?

26. Reactions that require ATP: formation of UDP-glucose from glucose-1-phosphate and UTP (indirect requirement, because ATP is needed to regenerate UTP), regeneration of UTP, and carboxylation of pyruvate to oxaloacetate. Reactions that produce ATP: none. Enzymes that catalyze ATP-requiring reactions: UDP-glucose phosphorylase (indirect requirement), nucleoside phosphate kinase, and pyruvate carboxylase. Enzymes that catalyze ATP-producing reactions: none.

27. Fructose-2,6-*bis*phosphate is an allosteric activator of phosphofructokinase (a glycolytic enzyme) and an allosteric inhibitor of fructose *bis*phosphate phosphatase (an enzyme in the pathway of gluconeogenesis).

28. Hexokinase can add a phosphate group to any of several six-carbon sugars, whereas glucokinase is specific for glucose. Glucokinase has a lower affinity for glucose than does hexokinase. Consequently, glucokinase tends to deal with an excess of glucose, particularly in the liver. Hexokinase is the usual enzyme for phosphorylating six-carbon sugars.

29. The Cori cycle is a pathway in which there is cycling of glucose due to glycolysis in muscle and gluconeogenesis in liver. The blood transports lactate from muscle to liver and glucose from liver to muscle.

30. Substrate cycles are futile in the sense that there is no net change except for the hydrolysis of ATP. However, substrate cycles allow for increased control over opposing reactions when they are catalyzed by different enzymes.

31. Having two control mechanisms allows for fine tuning of control and for the possibility of amplification. Both mechanisms are capable of rapid response to conditions, milliseconds in the case of allosteric control and seconds to minutes in the case of covalent modification.

32. Different control mechanisms have inherently different time scales. Allosteric control can take place in milliseconds, whereas covalent control takes seconds to minutes. Genetic control has a longer time scale than either.

33. The most important aspect of the amplification scheme is that the control mechanisms affect agents that are catalysts themselves. An enhancement by several powers of ten is itself increased by several powers of ten.

34. Enzymes, like all catalysts, speed up the forward and reverse reaction to the same extent. Having different catalysts is the only way to ensure independent control over the rates of the forward and reverse process.

35. Muscle tissue uses large quantities of glucose, producing lactate in the process. The liver is an important site of gluconeogenesis to recycle the lactate to glucose.

36. Fructose-2,6-*bis*phosphate is an allosteric activator of phosphofructokinase (a glycolytic enzyme) and an allosteric inhibitor of fructose *bis*phosphate phosphatase (an enzyme in the pathway of gluconeogenesis). It thus plays a role in two pathways that are not exactly the reverse of each other.

37. The concentration of fructose-2,6-*bis*phosphate in a cell depends on the balance between its synthesis (catalyzed by phosphofructokinase-2) and its breakdown (catalyzed by fructose *bis*phosphatase-2). The separate enzymes that control the formation and breakdown of fructose-2,6-*bis*phosphate are themselves controlled by a phosphorylation/dephosphorylation mechanism.

38. Glycogen is more extensively branched than starch. It is a more useful storage form of glucose for animals because the glucose can be mobilized more easily when there is a need for energy.

18.4 Why Is Glucose Sometimes Diverted through the Pentose Phosphate Pathway?

39. NADPH has one more phosphate group than NADH (at the 2′ position of the ribose ring of the adenine nucleotide portion of the molecule). NADH is produced in oxidative reactions that give rise to ATP. NADPH is a reducing agent in biosynthesis. The enzymes that use NADH as a coenzyme are different from those that require NADPH.

40. Glucose-6-phosphate can be converted to glucose (gluconeogenesis), glycogen, pentose phosphates (pentose phosphate pathway), or pyruvate (glycolysis).

41. Hemolytic anemia is caused by defective working of the pentose phosphate pathway. There is a deficiency of NADPH, which indirectly contributes to the integrity of the red blood cells. The pentose phosphate pathway is the only source of NADPH in red blood cells.

42. (a) By using only the oxidative reactions. (b) By using the oxidative reactions, the transaldolase and transketolase reactions, and gluconeogenesis. (c) By using glycolytic reactions and the transaldolase and transketolase reactions in reverse.

43. Transketolase catalyzes the transfer of a two-carbon unit, whereas transaldolase catalyzes the transfer of a three-carbon unit.

44. In red blood cells, the presence of the reduced form of glutathione is necessary for the maintenance of the sulfhydryl groups of hemoglobin and other proteins in their reduced forms, as well as for keeping the Fe(II) of hemoglobin in its reduced form. Glutathione also maintains the integrity of red cells by reacting with peroxides that would otherwise degrade fatty-acid side chains in the cell membrane.

45. Thiamine pyrophosphate is a cofactor necessary for the function of transketolase, an enzyme that catalyzes one of the reactions in the nonoxidative part of the pentose phosphate pathway.

46.

Glucose-6-phosphate   6-Phosphoglucono-δ-lactone

47. Having different reducing agents for anabolic and catabolic pathways serves to keep the pathways separate metabolically. Thus, they are subject to independent control and do not waste energy.

48. If a cell needs NADPH, all the reactions of the pentose phosphate pathway take place. If a cell needs ribose-5-phosphate, the oxidative portion of the pathway can be bypassed; only the nonoxidative reshuffling reactions take place. The pentose phosphate pathway does not have a significant effect on the cell's supply of ATP.

49. The ester bond is more easily broken than any of the other bonds that form the sugar ring. Hydrolysis of that bond is the next step in the pathway.

50. The reshuffling reactions of the pentose phosphate pathway have both an epimerase and an isomerase. Without an isomerase, all the sugars involved are keto sugars, which are not substrates for transaldolase, one of the key enzymes in the reshuffling process.

Chapter 19

19.1 What Role Does the Citric Acid Cycle Play in Metabolism?

1. Anaerobic glycolysis is the principal pathway for the anaerobic metabolism of glucose. The pentose phosphate pathway can also be considered. Aerobic glycolysis and the citric acid cycle are responsible for the aerobic metabolism of glucose.

2. Anaerobically, two ATPs can be produced from one glucose molecule. Aerobically, this figure is 30 to 32, depending on in which tissue it is occurring.

3. The citric acid cycle is also called the Krebs cycle, the tricarboxylic acid cycle, and the TCA cycle.

4. Amphibolic means that the pathway is involved in both catabolism and anabolism.

19.2 What Is the Overall Pathway of the Citric Acid Cycle?

5. The citric acid cycle takes place in the mitochondrial matrix. Glycolysis takes place in the cytosol.

6. There is a transporter on the inner mitochondrial matrix that allows pyruvate from the cytosol to pass into the mitochondria.

7. NAD^+ and FAD are the primary electron acceptors of the citric acid cycle.

8. NADH and $FADH_2$ are indirect sources of energy produced in the TCA cycle. GTP is a direct source of energy.

19.3 How Is Pyruvate Converted to Acetyl-CoA?

9. There are five enzymes involved in the pyruvate dehydrogenase complex of mammals. Pyruvate dehydrogenase transfers a two-carbon unit to TPP and releases CO_2. Dihydrolipoyl transacetylase transfers the two-carbon acetyl unit to lipoic acid and then to coenzyme A. Dihydrolipoyl dehydrogenase reoxidizes lipoic acid and reduces NAD^+ to NADH. Pyruvate dehydrogenase kinase phosphorylates PDH. PDH phosphatase removes the phosphate.

10. Lipoic acid plays a role both in redox and in acetyl-transfer reactions.

11. There are five enzymes all in close proximity for efficient shuttling of the acetyl unit between molecules and efficient control of the complex by phosphorylation.

12. Thiamine pyrophosphate comes from the B vitamin thiamine. Lipoic acid is a vitamin. NAD^+ comes from the B vitamin niacin. FAD comes from the B vitamin riboflavin.

13.

14. See Figure 19.4.

19.4 What Are the Individual Reactions of the Citric Acid Cycle?

15. A condensation reaction is one in which a new carbon–carbon bond is formed. The reaction of acetyl-CoA and oxaloacetate to produce citrate involves formation of such a carbon–carbon bond.

16. It means that the reaction catalyzed by the enzyme produces the product that is part of the name and does not require a direct input of energy from a high-energy phosphate. Thus, citrate synthase catalyzes the synthesis of citrate without using ATP to do it.

17. Fluoroacetate is a poison that is produced naturally in some plants and is also used as a poison against undesirable pests. It is poisonous because it is used by citrate synthase to make fluorocitrate, which is an inhibitor of the citric acid cycle.

18. The reaction involves an achiral molecule (citrate) being converted to a chiral one (isocitrate).

19. Conversion of pyruvate to acetyl-CoA, conversion of isocitrate to α-ketoglutarate, and conversion of α-ketoglutarate to succinyl-CoA.

20. Conversion of pyruvate to acetyl-CoA, conversion of isocitrate to α-ketoglutarate, conversion of α-ketoglutarate to succinyl-CoA, conversion of succinate to fumarate, and conversion of malate to oxaloacetate.

21. These enzymes catalyze oxidative decarboxylations.

22. The reactions proceed by the same mechanism and use the same cofactors. The difference is the initial substrate, which is pyruvate or α-ketoglutarate. During the course of the reaction, pyruvate dehydrogenase shuttles an acetyl unit through the reaction while α-ketoglutarate dehydrogenase shuttles a succinyl unit.

23. A synthetase is an enzyme that synthesizes a molecule and uses a high-energy phosphate in the process.

24. GTP is equivalent to ATP because an enzyme, nucleoside diphosphate kinase, is able to interconvert GTP and ATP.

25. The enzymes that reduce NAD^+ are all soluble, matrix enzymes, while succinate dehydrogenase is membrane bound. The NAD^+-linked dehydrogenases all catalyze oxidations that involve carbons

and oxygens, such as an alcohol group being oxidized to an alde-hyde or aldehyde to carboxylic acid. The FAD-linked dehydrogenase oxidizes a carbon–carbon single bond to a double bond.

26. There is an adenine nucleotide portion in the structure of NADH, with a specific binding site on NADH-linked dehydrogenases for this portion of NADH.

27. The conversion of fumarate to malate is a hydration reaction, not a redox reaction.

28.

$$CH_2-COO^-$$
$$^-OOC-C-H$$
$$H-C-OH$$
$$COO^-$$

$$CH_2-COO^-$$
$$H-C-COO^-$$
$$H-C-OH$$
$$COO^-$$

$$CH_2-COO^-$$
$$^-OOC-C-H$$
$$HO-C-H$$
$$COO^-$$

19.5 What Are the Energetics of the Citric Acid Cycle, and How Is It Controlled?

30. The reactions are catalyzed by pyruvate dehydrogenase, citrate synthase, isocitrate dehydrogenase, and α-ketoglutarate dehydrogenase.

31. PDH is controlled allosterically. It is inhibited by ATP, acetyl-CoA, and NADH. In addition, it is subject to control by phosphorylation. When PDH kinase phosphorylates PDH, it becomes inactive. Removing the phosphate with the PDH phosphatase reactivates it.

32. ATP and NADH are the two most common inhibitors.

33. If the amount of ADP in a cell increases relative to the amount of ATP, the cell needs energy (ATP). This situation not only favors the reactions of the citric acid cycle, which release energy, activating isocitrate dehydrogenase, but also stimulates the formation of NADH and $FADH_2$ for ATP production by electron transport and oxidative phosphorylation.

34. If the amount of NADH in a cell increases relative to the amount of NAD^+, the cell has completed a number of energy-releasing reactions. There is less need for the citric acid cycle to be active; as a result, the activity of pyruvate dehydrogenase is decreased.

29. (a)

$$CH_3-\overset{O}{\overset{\|}{C}}\overset{\ddot{O}}{\overset{\ddots}{:}}\overset{\ddot{O}}{\overset{\ddots}{:}}^- + \text{CoA-SH} \longrightarrow CH_3-\overset{O}{\overset{\|}{C}}-S-\text{CoA} + H^+ + :\ddot{O}::C::\ddot{O}: + 2e^-$$

Pyruvate   **Acetyl-CoA**   **Carbon dioxide**

(b)

$$\begin{array}{c} COO^- \\ | \\ CH_2 \\ | \\ H-C-COO^- \\ | \\ :\ddot{O}:C:H \\ | \\ H \;\; COO^- \end{array} \longrightarrow \begin{array}{c} COO^- \\ | \\ CH_2 \\ | \\ H-C-H \\ | \\ C::\ddot{O} \\ | \\ COO^- \end{array} + CO_2 + 2e^- + H^+$$

Isocitrate   **α-Ketoglutarate**

(c)

$$\begin{array}{c} \ddot{O}: \\ :: \\ C:\ddot{O}:^- \\ | \\ C=O \\ | \\ CH_2 \\ | \\ CH_2 \\ | \\ COO^- \end{array} + \text{CoA-SH} \longrightarrow \begin{array}{c} S-\text{CoA} \\ | \\ C=O \\ | \\ CH_2 \\ | \\ CH_2 \\ | \\ COO^- \end{array} + :\ddot{O}::C::\ddot{O}: + 2e^- + H^+$$

α-Ketoglutarate   **Succinyl-CoA**

(d)

$$\begin{array}{c} COO^- \\ | \\ H:C-H \\ | \\ H-C:H \\ | \\ COO^- \end{array} \longrightarrow \begin{array}{c} COO^- \;\; H \\ \diagdown \; C \; \diagup \\ :: \\ C \\ \diagup \;\; \diagdown \\ H \qquad COO^- \end{array} + 2e^- + 2H^+$$

Succinate   **Fumarate**

(e)

$$\begin{array}{c} COO^- \\ | \\ H:\ddot{O}:C:H \\ | \\ CH_2 \\ | \\ COO^- \end{array} \longrightarrow \begin{array}{c} COO^- \\ | \\ C::\ddot{O} \\ | \\ CH_2 \\ | \\ COO^- \end{array} + 2e^- + 2H^+$$

Malate   **Oxaloacetate**

35. The citric acid cycle is less active when a cell has a high ATP/ADP ratio and a high $NADH/NAD^+$ ratio. Both ratios indicate a high "energy charge" in the cell, indicating less of a need for the energy-releasing reactions of the citric acid cycle.

36. Thioesters are "high-energy" compounds that play a role in group-transfer reactions; consequently, their $\Delta G^{\circ\prime}$ of hydrolysis is large and negative to provide energy for the reaction.

37. The energy released by hydrolysis of acetyl-CoA is needed for the condensation reaction that links the acetyl moiety to oxaloacetate, yielding citrate. The energy released by hydrolysis of succinyl-CoA drives the phosphorylation of GDP, yielding GTP.

38. Table 19.2 shows that the sum of the energies of the individual reactions is -44.3 kJ (-10.6 kcal) for each mole of acetyl-CoA that enters the cycle.

39. The expression would relate to the intensive extraction of energy from intermediate compounds by redox reactions. Including the pyruvate dehydrogenase reaction, five of nine reactions are redox reactions (in contrast with only one of ten in glycolysis). Accordingly, energy is rapidly extracted from carbon compounds (yielding the energy-less CO_2) and is transferred to NAD^+ and FAD for subsequent utilization.

40. Lactose is a disaccharide of glucose and galactose. There is no energy cost in the hydrolysis of the bond between the two monosaccharides, so essentially there are two hexoses to consider. Because the processing of any of the hexoses yields the same amount of energy, the aerobic processing of lactose would lead to 60 to 64 ATPs, depending on the tissue and on the shuttle system used.

19.6 What Is the Glyoxylate Cycle?

41. Isocitrate dehydrogenase, α-ketoglutarate dehydrogenase, and succinyl-CoA synthetase.

42. The conversion of isocitrate to succinate and glyoxylate catalyzed by isocitrate lyase and the conversion of glyoxylate and acetyl-CoA to malate catalyzed by malate synthase.

43. Bacteria that have a glyoxylate cycle can convert the acetic acid to amino acids, carbohydrates, and lipids, but humans can only use the acetic acid as an energy source or to make lipids.

19.7 What Role Does the Citric Acid Cycle Play in Catabolism?

44. The citric acid cycle is the central metabolic pathway and indirect producer of energy. It receives fuels from the other pathways at many points and generates reduced electron carriers that go into the electron transport chain. It is also involved in anabolism, as many of its intermediates can be drawn off to synthesize other compounds.

45. The citric acid cycle occurs in the mitochondrial matrix, which is more selective in its permeability than the plasma membrane.

46. In oxidative decarboxylation, the molecule that is oxidized loses a carboxyl group as carbon dioxide. Examples of oxidative decarboxylation include the conversion of pyruvate to acetyl-CoA, isocitrate to α-ketoglutarate, and α-ketoglutarate to succinyl-CoA.

47. Yes, not only is citric acid completely degraded to carbon dioxide and water, but it is also readily absorbed into the mitochondrion.

19.8 What Role Does the Citric Acid Cycle Play in Anabolism?

48. The following series of reactions exchanges NADH for NADPH.

Oxaloacetate + NADH + H^+ → Malate + NAD^+

Malate + $NADP^+$ → Pyruvate + CO_2 + NADPH + H^+

49. A variety of reactions in which amino acids are converted to citric acid cycle intermediates are considered to be anaplerotic. In addition, pyruvate + CO_2 can form oxaloacetate via pyruvate carboxylase.

50. Many compounds can form acetyl-CoA, such as fats, carbohydrates, and many amino acids. Acetyl-CoA can also form fats and ketone bodies, as well as feed directly into the citric acid cycle.

19.9 Why Isn't Oxygen Part of the Equation?

51. The NADH and $FADH_2$ produced by the citric acid cycle are the electron donors in the electron transport chain linked to oxygen. Because of this connection, the citric acid cycle is considered part of aerobic metabolism.

Chapter 20

20.1 What Role Does Electron Transport Play in Metabolism?

1. Electrons are passed from NADH to a flavin-containing protein to coenzyme Q. From coenzyme Q, the electrons pass to cytochrome b, then to cytochrome c, via the Q cycle, followed by cytochromes a and a_3. From the cytochrome aa_3 complex, the electrons are finally passed to oxygen.

2. Electron transport and oxidative phosphorylation are different processes. Electron transport requires the respiratory complexes of the inner mitochondrial membrane, whereas oxidative phosphorylation requires ATP synthase, also located on the inner mitochondrial membrane. Electron transport can take place in the absence of oxidative phosphorylation.

3. In all reactions, electrons are passed from the reduced form of one reactant to the oxidized form of the next reactant in the chain. The notation [Fe—S] refers to any one of a number of iron–sulfur proteins.

Reactions of Complex I

$NADH + E\text{-}FMN \rightarrow NAD^+ + E\text{-}FMNH_2$

$E\text{-}FMNH_2 + 2[Fe\text{—}S]_{ox} \rightarrow E\text{-}FMN + 2[Fe\text{—}S]_{red}$ **Liberation of enough energy to produce ATP**

Transfer to Coenzyme Q

$2[Fe\text{—}S]_{red} + CoQ \rightarrow 2[Fe\text{—}S]_{ox} + CoQH_2$

Reactions of Complex III
Q cycle reactions

$[Fe\text{—}S]_{red} + cyt\ c_{1ox} \rightarrow [Fe\text{—}S]_{ox} + cyt\ c_{1red}$ **Liberation of enough energy to produce ATP**

Transfer to cytochrome c

$cyt\ c_{1red} + cyt\ c_{ox} \rightarrow cyt\ c_{1ox} + cyt\ c_{red}$

Reactions of Complex IV

$cyt\ c_{red} + cyt\ a\text{-}a_{3ox} \rightarrow cyt\ c_{ox} + cyt\ a\text{-}a_{3red}$ **Liberation of enough energy to produce ATP**

$cyt\ a\text{-}a_{3red} + \frac{1}{2} O_2 \rightarrow cyt\ a\text{-}a_{3ox} + H_2O$

4. When $FADH_2$ is the starting point for electron transport, electrons are passed from $FADH_2$ to coenzyme Q in a reaction carried out by Complex II that bypasses Complex I.

$$FADH_2 + 2[Fe\text{—}S]_{ox} \rightarrow FAD + 2[Fe\text{—}S]_{red}$$

$$2[Fe\text{—}S]_{red} + CoQ \rightarrow 2[Fe\text{—}S]_{ox} + CoQH_2$$

5. Mitochondrial structure confines the reduced electron carriers produced by the citric acid cycle to the matrix. There they are close to the respiratory complexes of the electron transport chain that will pass the electrons from the carriers produced by the citric acid cycle to oxygen, the ultimate recipient of electrons and hydrogens.

20.2 What Are the Reduction Potentials for the Electron Transport Chain?

6. The electron transport chain translocates charged particles by chemical means. Interconversion of chemical and electrical energy is exactly what a battery does.

7. The reactions are all written in the same direction for purposes of comparison. By convention, they are written as reduction, rather than oxidation, reactions.

8. $\Delta G^{\circ\prime} = -60$ kJ/mol

9. We fundamentally add the half reactions in Table 20.1.

$$E^{\circ\prime} \ (V)$$

$$NAD^+ + 2\,H^+ + 2e^- \rightarrow NADH + H^+ \qquad -0.320$$

This is the wrong direction, so we reverse the equation and the sign of the potential difference.

	$E^{\circ\prime} \ (V)$
$NADH + H^+ \rightarrow NAD^+ + 2\,H^+ + 2e^-$	0.320
$\frac{1}{2}O_2 + 2\,H^+ + 2e^- \rightarrow H_2O$	0.816
$NADH + H^+ + \frac{1}{2}O_2 \rightarrow NAD^+ + H_2O$	1.136

10. We fundamentally add the half reactions in Table 20.1.

$$E^{\circ\prime} \ (V)$$

$$NAD^+ + 2\,H^+ + 2e^- \rightarrow NADH + H^+ \qquad -0.320$$

This is the wrong direction, so we reverse the equation and the sign of the potential difference.

	$E^{\circ\prime} \ (V)$
$NADH + H^+ \rightarrow NAD^+ + 2\,H^+ + 2e^-$	0.320
$Pyruvate + 2\,H^+ + 2e^- \rightarrow Lactate$	−0.185
$NADH + H^+ + Pyruvate \rightarrow NAD^+ + Lactate$	0.135

11. We fundamentally add the half reactions in Table 20.1.

$$E^{\circ\prime} \ (V)$$

$$Fumarate + 2\,H^+ + 2e^- \rightarrow Succinate \qquad 0.031$$

This is the wrong direction, so we reverse the equation and the sign of the potential difference.

	$E^{\circ\prime} \ (V)$
$Succinate \rightarrow Fumarate + 2\,H^+ + 2e^-$	−0.031
$\frac{1}{2}O_2 + 2\,H^+ + 2e^- \rightarrow H_2O$	0.816
$Succinate + \frac{1}{2}O_2 \rightarrow Fumarate + H_2O$	0.785

12. The cytochrome is the electron donor, and the flavin moiety is the electron acceptor. Once again, we add the half reactions in Table 20.1.

$$E^{\circ\prime} \ (V)$$

$$Cytochrome \ c(Fe^{3+}) + e^- \rightarrow Cytochrome \ c(Fe^{2+}) \qquad 0.254$$

This is the wrong direction, so we reverse the equation and the sign of the potential difference.

	$E^{\circ\prime} \ (V)$
$2\,Cytochrome \ c(Fe^{2+}) + 2e^- \rightarrow 2\,Cytochrome \ c(Fe^{3+})$	−0.254
$[FAD] + 2\,H^+ + 2\,e^- \rightarrow [FADH_2]$	0.091
$[FAD] + 2\,Cyt \ c(Fe^{2+}) + 2\,H^+ \rightarrow [FADH_2] + 2\,Cyt \ c(Fe^{3+})$	−0.063

This was the maximum value for a bound flavin. The negative sign indicates that this reaction will not take place as written because it is not energetically favorable.

13. Here is an illustration based on standard reduction potentials.

$$E^{\circ\prime} \ (V)$$

$$Fumarate + 2\,H^+ + 2e^- \rightarrow Succinate \qquad 0.031$$

This is the wrong direction, so we reverse the equation and the sign of the potential difference.

	$E^{\circ\prime} \ (V)$
$Succinate \rightarrow Fumarate + 2\,H^+ + 2e^-$	−0.031
$FAD + 2\,H^+ + 2e^- \rightarrow FADH_2$	−0.219
$Succinate + FAD \rightarrow Fumarate + FADH_2$	−0.250

The other possibility can be calculated in the same way.

	$E^{\circ\prime} \ (V)$
$Succinate \rightarrow Fumarate + 2\,H^+ + 2e^-$	−0.031
$NADH + H^+ \rightarrow NAD^+ + 2\,H^+ + 2e^-$	0.320
$Succinate + NAD^+ \rightarrow Fumarate + NADH$	−0.351

Both reduction potentials indicate a reaction that is not energetically favorable, less so with FAD than with NAD^+. Other factors

enter into consideration, however, in a living cell. The first is that the reactions do not take place under standard conditions, altering the values of reduction potentials. The second is that the reduced electron carriers (NADH and $FADH_2$) are reoxidized. Coupling the reactions we have looked at here to others also makes them less unfavorable.

14. The half reaction of oxidation $NADH + H^+ \rightarrow NAD^+ + 2\,H^+ + 2e^-$ is strongly exergonic ($\Delta G^{\circ\prime} = -61.3 \ kJ \ mol^{-1} = -14.8 \ kcal \ mol^{-1}$), as is the overall reaction Pyruvate $+ NADH + H^+ \rightarrow$ Lactate $+ NAD^+$ ($\Delta G^{\circ\prime} = -25.1 \ kJ \ mol^{-1} = -6.0 \ kcal \ mol^{-1}$)

20.3 How Are the Electron Transport Complexes Organized?

15. They all contain the heme group, with minor differences in the heme side chains in most cytochromes.

16. Cytochromes are proteins of electron transport; the heme ion alternates between the Fe(II) and Fe(III) states. The function of hemoglobin and myoglobin is oxygen transport and storage, respectively. The iron remains in the Fe(II) state.

17. Coenzyme Q is not bound to any of the respiratory complexes. It moves freely in the inner mitochondrial membrane.

18. A part of Complex II catalyzes the conversion of succinate to fumarate in the citric acid cycle.

19. Three of the four respiratory complexes generate enough energy to phosphorylate ADP to ATP. Complex II is the sole exception.

20. Cytochrome c is not tightly bound to the mitochondrial membrane and can easily be lost in the course of cell fractionation. This protein is so similar in most aerobic organisms that cytochrome c from one source can easily be substituted for that from another source.

21. Succinate $+ \frac{1}{2}O_2 \rightarrow$ Fumarate $+ H_2O$.

22. The components are in the proper orientation for the electrons to be transferred rapidly from one component to the next; if the components were in solution, speed would be limited to the rate of diffusion. A second advantage, which is actually a necessity, is that the components are properly positioned to facilitate the transport of protons from the matrix to the intermembrane space.

23. From an evolutionary standpoint, two different functions can be performed by identical structures or by structures that are close to identical, with only minor differences in the protein moieties. The organism saves a considerable amount of energy by not having to evolve—and to operate—two pathways.

24. The key point here is not the active site, which has a low tolerance for mutations, but the molecules with which the proteins in question are associated. Cytochromes are membrane bound and must associate with other members of the electron transport chain; most mutations are apt to interfere with the close fit, and thus they are not preserved (because they are lethal). Globins, while soluble, still form some associations, so more mutations can be tolerated, with some limits. Hydrolytic enzymes are soluble and not likely to associate with other polypeptides except substrates. They can tolerate a higher proportion of mutations.

25. Having mobile electron carriers in addition to membrane-bound respiratory complexes allows electron transport to use the most readily available complex rather than to use the same one all the time.

26. The Q cycle allows for a smooth transition from two-electron carriers (NADH and $FADH_2$) to one-electron carriers (cytochromes).

27. The protein environment of the iron differs in each of the cytochromes, causing differences in the reduction potential.

28. All the reactions in the electron transport chain are electron transfer reactions, but some of the reactants and products inherently transfer either one or two electrons, as the case may be.

29. The heme groups differ slightly in the various kinds of cytochromes. This is the main difference, with some modification due to the different protein environments.

30. Respiratory complexes contain a number of proteins, some of them quite large. This is the first difficulty. Like most proteins bound to membranes, the components of respiratory complexes are easily denatured on removal from their environment.

20.4 What Is the Connection between Electron Transport and Phosphorylation?

31. The F_1 portion of the mitochondrial ATP synthase, which projects into the matrix, is the site of ATP synthesis.

32. The F_0 portion of mitochondrial ATP synthase lies within the inner mitochondrial membrane, but the F_1 portion projects into the matrix.

33. The P/O ratio gives the number of moles of P_i consumed in the reaction $ADP + P_i \rightarrow ATP$ for each mole of oxygen atoms consumed in the reaction $\frac{1}{2}O_2 + 2H^+ + 2e^- \rightarrow H_2O$. It is a measure of the coupling of ATP production to electron transport.

34. The F_1 part of mitochondrial ATP synthase has a stationary domain (the $\alpha_3\beta_3\delta$ domain) and a domain that rotates (the $\gamma\varepsilon$ domain). This is exactly the arrangement needed for a motor.

35. A P/O ratio of 1.5 can be expected because oxidation of succinate passes electrons to coenzyme Q via a flavoprotein intermediate, bypassing the first respiratory complex.

36. Exact values for P/O ratios are difficult to determine because of the complexity of the systems that pump protons and phosphorylate ADP. The number of ADP molecules phosphorylated is directly related to the number of protons pumped across the membrane. This figure has been a matter of some controversy. It has been difficult for chemists and biochemists to accept uncertain stoichiometry.

37. The difficulties in determining the number of protons pumped across the inner mitochondrial membrane by respiratory complexes are those inherent in working with large assemblies of proteins that must be bound in a membrane environment to be active. As experimental methods improve, the task becomes less difficult.

20.5 What Is the Mechanism of Coupling in Oxidative Phosphorylation?

38. The chemiosmotic coupling mechanism is based on the difference in hydrogen ion concentration between the intermembrane space and the matrix of actively respiring mitochondria. The hydrogen ion gradient is created by the proton pumping that accompanies the transfer of electrons. The flow of hydrogen ions back into the matrix through a channel in the ATP synthase is directly coupled to the phosphorylation of ADP.

39. An intact mitochondrial membrane is necessary for compartmentalization, which in turn is necessary for proton pumping.

40. Uncouplers overcome the proton gradient on which oxidative phosphorylation depends.

41. In chemiosmotic coupling, the proton gradient is related to ATP production. The proton gradient leads to conformational changes in a number of proteins, releasing tightly bound ATP from the synthase as a result of the conformational change.

42. Dinitrophenol is an uncoupler of oxidative phosphorylation. The rationale was to dissipate energy as heat.

43. The energy released as protons pass through the F particles is actually used to cause conformational changes in the F_1 proteins, thereby releasing ATP. The "tight" conformation (one of three) provides a hydrophobic environment in which ADP is phosphorylated by adding P_i without requiring *immediate* energy.

20.6 How Are Respiratory Inhibitors Used to Study Electron Transport?

44. (a) Azide inhibits the transfer of electrons from cytochrome aa_3 to oxygen. (b) Antimycin A inhibits the transfer of electrons from cytochrome b to coenzyme Q in the Q cycle. (c) Amytal inhibits the transfer of electrons from NADH reductase to coenzyme Q.

(d) Rotenone inhibits the transfer of electrons from NADH reductase to coenzyme Q. (e) Dinitrophenol is an uncoupler of oxidative phosphorylation. (f) Gramicidin A is an uncoupler of oxidative phosphorylation. (g) Carbon monoxide inhibits the transfer of electrons from cytochrome aa_3 to oxygen.

45. Methods exist to determine the amounts of the oxidized and reduced components of the electron transport chain present in a sample. If a respiratory inhibitor is added, the reduced form of the component before the blockage point in the chain will accumulate, as will the oxidized form of the component immediately after the blockage point.

46. Uncouplers overcome the proton gradient created by electron transport, whereas respiratory inhibitors block the flow of electrons.

20.7 What Are Shuttle Mechanisms?

47. The complete oxidation of glucose produces 30 molecules of ATP in muscle and brain and 32 ATP in liver, heart, and kidney. The underlying reason is the different shuttle mechanisms for transfer to mitochondria of electrons from the NADH produced in the cytosol by glycolysis.

48. The transport "product" (in the matrix) of the malate–aspartate shuttle is NADH, whereas that of the glycerol-phosphate shuttle is $FADH_2$. The latter shuttle can thus go *against* a transmembrane NADH concentration gradient, whereas the former cannot.

20.8 What Is the ATP Yield from Complete Oxidation of Glucose?

49. (a) 34; (b) 32; (c) 13.5; (d) 17; (e) 2.5; (f) 12.5.

50. The maximum yield of ATP, to the nearest whole number, is 3.

$$102.3 \text{ kJ released} \times \frac{1 \text{ ATP}}{30.5 \text{ kJ}} = 3.35 \text{ ATP}$$

One ATP is actually produced, so the efficiency of the process is

$$\frac{1 \text{ ATP}}{3 \text{ ATP}} \times 100 = 33.3\%$$

Chapter 21

21.1 How Are Lipids Involved in the Generation and Storage of Energy?

1. (a) For mobile organisms—e.g., a migrating hummingbird—weight can be a critical factor, and packing the most energy into the least weight is decidedly advantageous. A 2.5-g hummingbird needs to add about 2 g of fat for migration energy, which would increase body weight by 80%. The equivalent amount of energy stored as glycogen would be about 5 g, which would increase its body weight by 200%; the bird would never get off the ground!

(b) For immobile plants, weight is not a critical factor, and it takes more energy to make fat or oil than it does to make starch. (The second law of thermodynamics would dictate that the energy obtained from oil would be less than that expended making oil. You can verify this numerically if you wish.) In the case of plant *seeds*, "compact" energy is beneficial, because the seed must be self-sufficient until enough growth has occurred to permit photosynthesis.

21.2 How Are Lipids Catabolized?

2. Phospholipase A_1 hydrolyzes the ester bond to carbon-1 of the glycerol backbone, while phospholipase A_2 hydrolyzes the ester bond to carbon-2 of the backbone.

3. A hormone signal activates adenylate cyclase, which makes cAMP. This activates protein kinases, which phosphorylate the lipases, thereby activating them.

4. Acyl-CoAs are high-energy compounds. An acyl-CoA has sufficient energy to initiate the β-oxidation process. The CoA is also a tag indicating that the molecule is destined for oxidation.

5. Acyl groups are esterified to carnitine to cross the inner mitochondrial membrane. There are transesterification reactions from the acyl-CoA to carnitine and from acylcarnitine to CoA (see Figure 21.5).

6. Acyl-CoA dehydrogenase removes hydrogens from adjacent carbons creating a double bond and using FAD as coenzyme. β-Hydroxy-CoA dehydrogenase oxidizes an alcohol group to a ketone group and uses NAD^+ as a coenzyme.

7.

$$CH_3CH_2\textbf{CH}_2\textbf{CH}_2CH_2—\overset{\displaystyle O}{\underset{\displaystyle \|}{C}}—S—CoA$$

The two carbons shown in boldface type are the ones that will have the double bond between them. The orientation will be *trans*.

8. Seven carbon–carbon bonds are broken in the course of β-oxidation (see Figure 21.6).

9. In the liver, glycogen breakdown and gluconeogenesis would occur. In the muscle, glycogen breakdown and glycolysis would occur.

21.3 What Is the Energy Yield from the Oxidation of Fatty Acids?

10. One obtains 6.7 ATP per carbon and 0.42 ATP per gram for stearic acid versus 5 ATP per carbon and 0.17 ATP per gram for glucose. There is more energy available from stearic acid than from glucose.

11. The processing of the acetyl-CoA through the citric acid cycle and the electron transport chain produces more energy than the processing of the NADH and $FADH_2$ produced during β-oxidation.

12. From seven cycles of β-oxidation: 8 acetyl-CoA, 7 $FADH_2$, and 7 NADH. From the processing of 8 acetyl-CoA in the citric acid cycle: 8 $FADH_2$, 24 NADH, and 8 GTP. From reoxidation of all $FADH_2$ and NADH: 22.5 ATP from 15 $FADH_2$, 77.5 ATP from 31 NADH. From 8 GTP: 8 ATP. Subtotal: 108 ATP. A 2-ATP equivalent was used in the activation step. Grand total: 106 ATP. The grand total for stearic acid was 120 ATP.

13. The humps of camels contain lipids that can be degraded as a source of metabolic water, rather than water as such.

21.4 How Are Unsaturated Fatty Acids and Odd-Carbon Fatty Acids Catabolized?

14. For an odd-chain fatty acid, β-oxidation proceeds normally until the last round. When there are five carbons left, that round of β-oxidation releases one acetyl-CoA and one propionyl-CoA. Propionyl-CoA cannot be further metabolized by β-oxidation; however, a separate set of enzymes converts propionyl-CoA into succinyl-CoA, which can then enter the citric acid cycle.

15. False. The oxidation of unsaturated fatty acids to acetyl-CoA requires a cis–trans isomerization and an epimerization, reactions that are not found in the oxidation of saturated fatty acids.

16. For a monounsaturated fatty acid, an additional enzyme is needed, the enoyl-CoA isomerase.

17. For a polyunsaturated fatty acid, two additional enzymes are needed, the enoyl-CoA isomerase and 2,4-dienoyl-CoA reductase.

18. From seven cycles of β-oxidation: 7 acetyl-CoA, 1 propionyl-CoA, 7 $FADH_2$, 7 NADH. From the processing of 7 acetyl-CoA in the citric acid cycle: 7 $FADH_2$, 21 NADH, and 7 GTP. From the processing of the propionyl-CoA: −1 ATP for conversion to succinyl-CoA, + 1 GTP from the citric acid cycle, and 1 NADH and 1 $FADH_2$ from the citric acid cycle. From reoxidation of all $FADH_2$ and NADH: 22.5 ATP from 15 $FADH_2$, and 72.5 ATP from 29 NADH. From 8 GTP: 8 ATP. Subtotal: 103 ATP. Subtract a 2-ATP

equivalent used in activation step and a 1-ATP equivalent used in the conversion to succinyl-CoA for a grand total of 100 ATP.

19. An 18-carbon saturated fatty acid yields 120 ATP. For a monounsaturated fatty acid, the double bond eliminates the step that produces $FADH_2$, so there would be 1.5 ATP less for oleic acid, or 118.5 ATP total.

20. An 18-carbon saturated fatty acid yields 120 ATP. For a diunsaturated fatty acid with the bonds in the Δ^9 and Δ^{12} positions, the first double bond eliminates an $FADH_2$. The second double bond uses an NADPH, which we are guessing is the same cost as using an NADH. Thus a total of 4 ATP are lost, compared with a saturated fatty acid, so the total is 116 ATP.

21. It would take seven cycles of β-oxidation to release 14 carbons as acetyl-CoA, with the last three being released as propionyl-CoA.

22. Fats cannot produce a net yield of glucose because they must enter the citric acid cycle as the two-carbon unit acetyl-CoA. In the first few steps, two carbons are released as CO_2. However, an odd-chain fatty acid can be considered to be partially glucogenic because the final three carbons become succinyl-CoA and enter the citric acid cycle after the decarboxylation steps. Thus, if an extra succinyl-CoA is added, it can then be drawn off later as malate and used for gluconeogenesis without removing the steady-state level of citric acid cycle intermediates.

21.5 What Are Ketone Bodies?

23. Ketones are produced when there is an imbalance in lipid catabolism, compared with carbohydrate catabolism. If fatty acids are being β-oxidized to produce acetyl-CoA, but there is insufficient oxaloacetate because it is being drawn off for gluconeogenesis, the acetyl-CoA molecules will combine to form ketone bodies.

24. Two acetyl-CoA molecules combine to form acetoacetyl-CoA. This can then release coenzyme A to yield acetoacetate, which can be converted either to β-hydroxybutyrate or to acetone.

25. If the reason for passing out is uncontrolled diabetes, the doctor expects to smell acetone on the breath, since the otherwise unused sugars are being converted to fats and ketone bodies.

26. Ethanol is converted to acetaldehyde and then to acetic acid. Humans can use that acetic acid only for energy, or they can convert it to fatty acids and other lipids.

27. The metallic taste may be due to acetone, which means that your friend may have a mild state of ketosis. Ask if your friend has consulted a doctor about the diet regimen, and perhaps recommend either backing off from such a low-calorie diet or drinking more water to flush the system more thoroughly.

21.6 How Are Fatty Acids Produced?

28. The two pathways have in common the involvement of acetyl-CoA and thioesters, and each round of breakdown or synthesis involves two-carbon units. The differences are many: malonyl-CoA is involved in biosynthesis, not in breakdown; thioesters involve CoA in breakdown, acyl carrier proteins in biosynthesis; biosynthesis occurs in the cytosol, breakdown occurs in the mitochondrial matrix; breakdown is an oxidative process that requires NAD^+ and FAD and produces ATP by electron transport and oxidative phosphorylation, whereas biosynthesis is a reductive process that requires NADPH and ATP.

29. Step 1: Biotin is carboxylated using bicarbonate ion (HCO_3^-) as the source of the carboxyl group. Step 2: The carboxylated biotin is brought into proximity with enzyme-bound acetyl-CoA by a biotin carrier protein. Step 3: The carboxyl group is transferred to acetyl-CoA, forming malonyl-CoA.

30. It is a molecule that commits itself to fatty-acid synthesis. It is also a potent inhibitor of carnitine acyltransferase I, thereby shutting down β-oxidation.

31. ACP, citrate, cytosol, *trans* double bonds, D-alcohols, β-reduction, NADPH, malonyl-CoA (except for one acetyl-CoA), and a multifunctional enzyme complex.

32. In β-oxidation, FAD is the coenzyme for the first oxidation reaction, while NAD⁺ is the coenzyme for the second. In fatty-acid synthesis, NADPH is the coenzyme for both. The β-hydroxy-acyl group in β-oxidation has the L-configuration, while it has the D-configuration in fatty acid synthesis.

33. Both have a phosphopantetheine group at the active end. In coenzyme A, this group is attached to 2′-phospho-AMP, while in ACP it is attached to a serine residue of a protein.

34. ACP is a molecule that earmarks acyl groups for fatty-acid synthesis. It can be managed separately from acyl-CoA groups. Also, the ACP attaches to the acyl groups like a "swinging arm" that tethers it to the fatty-acid synthase complex.

35. Linoleate and linolenate cannot be synthesized by the body and must therefore be obtained from dietary sources. Mammals cannot produce a double bond beyond carbon atom 9 of fatty acids.

36. Acyl-CoA intermediates are essential in the conversion of fatty acids to other lipids.

37. Acetyl groups condense with oxaloacetate to form citrate, which can cross the mitochondrial membrane. Acetyl groups are regenerated in the cytosol by the reverse reaction.

38. If acetyl-carnitine forms in the matrix of the mitochondrion, it can be translocated to the cytosol via the carnitine translocase. Thus, this could represent another way of shuttling acetyl units out of the mitochondria for synthesis.

39. Energy is needed to condense an acetyl group to the growing fatty acid. In theory, such could be done with acetyl-CoA, using ATP. In practice, the ATP is used to convert acetyl-CoA to malonyl-CoA; the condensation of the acetyl moiety of malonyl-CoA is driven in part by the accompanying decarboxylation and requires no additional energy. A possible reason for this is to avoid a metabolic confusion of pathways, perhaps particularly important in (uncompartmented) prokaryotes; one could envision an acetyl-CoA from degradation being used immediately for synthesis. Malonyl-CoA says "synthesis"; acetyl-CoA says "degradation."

40. (a) The lipoate "swinging arm" of the pyruvate dehydrogenase complex. (b) The "arm" or ACP carries the group to be acted on from one enzyme to another (avoiding a diffusion-limited process and also positioning key groups correctly). In the case of the ACP, the group to be acted on (β-carbon) is always the same distance from the ACP, regardless of the length of the growing fatty acid, and thus the critical group is always in proximity to the active sites of the several pertinent enzymes.

21.7 How Are Acylglycerols and Compound Lipids Produced?

41. The glycerol comes from degradation of other acylglycerols or from glycerol-3-phosphate derived from glycolysis.

42. The activating group found on the acylglycerol is cytidine diphosphate.

43. In prokaryotes, CTP reacts with phosphatidic acid to give a CDP-diacylglycerol. This reacts with serine to give phosphatidylserine, which decarboxylates to phosphatidylethanolamine. In eukaryotes, CDP-ethanolamine reacts with a diacylglycerol to give phosphatidylethanolamine.

21.8 How Is Cholesterol Produced?

44. In steroid biosynthesis, three acetyl-CoA molecules condense to form the six-carbon mevalonate, which then gives rise to a five-carbon isoprenoid unit. A second and then a third isoprenoid unit condense, giving rise to a 10-carbon and then a 15-carbon unit. Two of the 15-carbon units condense, forming a 30-carbon precursor of cholesterol.

45. See Figure 21.24.

46. Bile acids and steroid hormones.

47. All steroids have a characteristic fused-ring structure, implying a common biosynthetic origin.

48. One oxygen atom from O_2 is needed to form the epoxide. The NADPH is needed to reduce the other oxygen atom to water.

49. Cholesterol is nonpolar and cannot dissolve in blood, which is an aqueous medium.

50. Bile salts are made from cholesterol, and cholesterol is taken from the body into the intestine in the bile fluid.

Chapter 22

22.1 Where Does Photosynthesis Take Place in the Cell?

1. In the fall, the chlorophyll in leaves is lost, and the red and yellow colors of the accessory pigments become visible, accounting for fall foliage colors.

2. The bean sprouts are grown in the dark to prevent them from turning green; most customers will not purchase green sprouts.

3. Iron and manganese in chloroplasts; iron and copper in mitochondria. Note that all these are transition metals, which can easily undergo redox reactions.

4. Both chloroplasts and mitochondria have an inner and outer membrane. Both have their own DNA and ribosomes. Chloroplasts, however, have a third membrane, the thylakoid membrane.

5. Chlorophyll has a cyclopentanone ring fused to the tetrapyrrole ring, a feature that does not exist in heme. Chlorophyll contains magnesium, whereas heme contains iron. Chlorophyll has a long side chain based on isoprenoid units, which is not found in heme.

6. Only a relatively small portion of the visible spectrum is absorbed by chlorophylls. The accessory pigments absorb light at additional wavelengths. As a result, most of the visible spectrum can be harnessed in light-dependent reactions.

7. It is one more piece of evidence that is consistent with the evolution of chloroplasts from independent bacterial organisms.

22.2 How Are Photosystems I and II Involved in the Light Reactions of Photosynthesis?

8. By and large, the synthesis of NADPH in chloroplasts is the reverse of NADH oxidation in mitochondria. The net electron flow in chloroplasts is the reverse of that in the mitochondria, although different carriers are involved.

9. When light impinges on the reaction center of *Rhodopseudomonas*, the special pair of chlorophylls there is raised to an excited energy level. An electron is passed from the special pair to accessory pigments, first pheophytin, then menaquinone, and finally to ubiquinone. The electron lost by the special pair of chlorophylls is replaced by a soluble cytochrome, which diffuses away. The separation of charge represents stored energy (see Figure 21.9).

10. In photosystem I and in photosystem II, light energy is needed to raise the reaction-center chlorophylls to a higher energy level. Energy is needed to generate strong enough reducing agents to pass electrons to the next of the series of components in the pathway.

11. No. Most chlorophylls are light-harvesting molecules that transfer energy to the special pair that takes part in the light reactions.

12. The electron transport chain in chloroplasts, like that in mitochondria, consists of proteins, such as plastocyanin, and protein complexes, such as the cytochrome b_6-f complex. It also contains mobile electron carriers, such as pheophytin and plastoquinone (equivalent to coenzyme Q), which is also true of the mitochondrial electron transport chain.

13. Probably the electron transport chain in chloroplasts. Chloroplasts generate molecular oxygen; mitochondria use it. The early atmosphere almost certainly lacked molecular oxygen. Only

when photosynthesis introduced oxygen into the atmosphere would oxygen be needed.

14. Electron transport and ATP production are coupled to each other by the same mechanism in mitochondria and chloroplasts. In both cases, the coupling depends on the generation of a proton gradient across the inner mitochondrial membrane or across the thylakoid membrane, as the case may be.

15. In mitochondria, both a proton gradient (chemical) and an electrochemical gradient (based on charge) are formed, both contributing to the total potential energy. In chloroplasts, only a proton gradient is formed, because ions move across the thylakoid membrane and neutralize charge. The proton gradient alone is considerably less efficient.

16. With very few exceptions, life directly or indirectly depends on photosynthesis. The electric current is the flow of electrons from water to $NADP^+$, a light-requiring process. The "current" continues in the light-independent reactions, with electrons flowing from NADPH to *bis*phosphoglycerate, which ultimately yields glucose.

17. Photosystem II requires more energy than photosystem I. The shorter wavelength of light means a higher frequency. Frequency, in turn, is directly proportional to energy.

18. It is quite reasonable to list reduction potentials for the electron-transfer reactions of photosynthesis. They are entirely analogous to the electron-transfer reactions in mitochondria, for which we listed standard reduction potentials in Chapter 20.

19. A photosynthetic reaction center is analogous to a battery because its reactions produce a charge separation. The charge separation is comparable to the stored energy of the battery.

20. The electron transport chains of mitochondria and chloroplasts are similar. In mitochondria, antimycin A inhibits electron transfer from cytochrome *b* to coenzyme Q in the Q cycle. By analogy, it can be argued that antimycin A inhibits electron flow from plastoquinone to cytochrome b_6-*f*. A Q cycle may also operate in chloroplasts.

21. Oxygen produced in photosynthesis comes from water. The oxygen-evolving complex is part of the series of electron-transfer reactions from water to NADPH. Carbon dioxide is involved in the dark reactions, which are different reactions that take place in another part of the chloroplast.

22. It is well established that the path of electrons in photosynthesis goes from photosystem II to photosystem I. The reason for the nomenclature is that photosystem I is easier to isolate than photosystem II and was studied more extensively at an earlier date.

23. It would take much work to establish the number of protons pumped across the thylakoid membrane. This is partly the result of experience with mitochondria and partly a prediction based on the greater complexity of structure in the chloroplast.

24. The oxygen-evolving complex of photosystem II passes through a series of five oxidation states (designated as S_0 through S_4) in the transfer of four electrons in the process of evolving oxygen (Figure 22.6). One electron is passed from water to photosystem II for each quantum of light. In the process, the components of the reaction center go successively through oxidation states S_1 through S_4. The S_4 decays spontaneously to the S_0 state and, in the process, oxidizes two water molecules to one oxygen molecule. Four protons are released simultaneously.

25. When the loosely bound cytochrome diffuses away, a charge separation is induced. This separation of charge represents stored energy.

26. The similarity of ATP synthase in chloroplasts and mitochondria supports the idea that both may have arisen from free-living bacteria.

22.3 How Does Photosynthesis Produce ATP?

27. In cyclic photophosphorylation, the excited chlorophyll of photosystem I passes electrons directly to the electron transport chain that normally links photosystem II to photosystem I. This electron transport chain is coupled to ATP production (see Figure 22.8).

28. Both depend on a proton gradient, resulting from the flow of electrons. In chloroplasts, protons come from the splitting of water to produce oxygen. In mitochondria, protons come from the oxidation of NADH and ultimately consume oxygen and produce water.

29. The proton gradient is created by the operation of the electron transport chain that links the two photosystems in noncyclic photophosphorylation.

30. ATP can be produced by chloroplasts in the absence of light if some way exists to form a proton gradient.

31. Cyclic photophosphorylation can take place when the plant needs ATP but does not have a great need for NADPH. Noncyclic photophosphorylation can take place when the plant needs both.

22.4 What Are the Evolutionary Implications of Photosynthesis with and without Oxygen?

32. Many electron donors other than water are possible in photosynthesis. This is especially the case in bacteria, whose photosystems do not have strong enough oxidizing agents to oxidize water. Some of the alternative electron donors are H_2S and organic compounds.

33. A prokaryotic organism that contains both chlorophyll *a* and chlorophyll *b* could be a relic of an evolutionary way station in the development of chloroplasts.

22.5 How Do the Dark Reactions of Photosynthesis Fix CO_2 into Glucose?

34. Rubisco is the principal protein in chloroplasts in all green plants. This wide distribution makes it likely to be the most abundant protein in nature.

35. The amino acid sequence of the catalytic subunits of Rubisco is encoded by chloroplast genes, whereas that of the regulatory subunits is encoded by nuclear genes.

36. Gluconeogenesis and the pentose phosphate pathway have a number of reactions similar to those of the dark reactions of photosynthesis.

37. From the standpoint of thermodynamics, the production of sugars in photosynthesis is the reverse of the complete oxidation of a sugar such as glucose to CO_2 and water. The complete oxidation reaction produces six moles of CO_2 for each mole of glucose oxidized. To get the energy change for the fixation of one mole of CO_2, change the sign of the energy for the complete oxidation of glucose and divide by 6.

38. Glucose synthesized by photosynthesis is not uniformly labeled because only one molecule of CO_2 is incorporated into each molecule of ribulose-1,5-*bis*phosphate, which then goes on to give rise to sugars.

39. If Rubisco was one of the first protein enzymes to arise early in the evolution of life, it may not have the efficiency of protein enzymes that evolved later, when evolution was more dependent on modifying and adapting existing proteins.

40. Their DNA is circular. Their ribosomes are more like those of bacteria than those of eukaryotes. Their aminoacyl-tRNA synthetases use bacterial tRNAs but not eukaryotic tRNAs. In general, they do not have introns in their genomes. Their mRNA uses a Shine—Dalgarno sequence.

41. The pathway borrows heavily from the nonoxidative branch of the pentose phosphate pathway and from gluconeogenesis. Without doubt, the pathways yield sugars as well as NADPH for reductive biosynthesis. Thus, only a few new enzymes would have to evolve through mutations to enable the complete Calvin cycle to function.

42. Atmospheric oxygen is a consequence of photosynthesis. Rubisco evolved before there was a significant amount of oxygen in the atmosphere.

43. The condensation of ribulose-1,5-*bis*phosphate with carbon dioxide to form two molecules of 3-phosphoglycerate is the actual carbon dioxide fixation. The rest of the Calvin cycle regenerates ribulose-1,5-*bis*phosphate.

44. Organisms would need only a few mutations giving rise to the enzymes unique to the Calvin cycle. The rest of the pathway is already in place.

45. Six molecules of carbon dioxide fixed in the Calvin cycle do not end up in the same glucose molecule. However, labeling experiments show that six carbon atoms are incorporated into sugars for every six carbon dioxide molecules that enter the Calvin cycle.

22.6 How Is CO_2 Fixed in Tropical Plants?

46. In tropical plants, the C_4 pathway is operative in addition to the Calvin cycle.

47. In C_4 plants, when CO_2 enters the leaf through pores in the outer cells, it reacts first with phosphoenolpyruvate to produce oxaloacetate and P_i in the mesophyll cells of the leaf. Oxaloacetate is reduced to malate, with the concomitant oxidation of NADPH. Malate is then transported to the bundle-sheath cells (the next layer) through channels that connect two kinds of cells. These reactions do not take place in C_3 plants.

48. Photorespiration is a pathway in which glycolate is a substrate oxidized by Rubisco acting as an oxygenase, rather than as a carboxylase. Photorespiration is not completely understood.

49. Three reasons come to mind. (1) Light energy is usually not limiting. (2) The plants have small pores to prevent water loss, but this also limits CO_2 uptake. (3) The C_4 pathway allows for increasing the CO_2 concentration in the inner chloroplast, which would not be otherwise possible with the small pores.

50. Most plants would be more productive in the absence of photorespiration. There is another side to this picture, however. The oxygenase activity appears to be an unavoidable, wasteful activity of Rubisco. Photorespiration is a salvage pathway that saves some of the carbon that would be lost due to the oxygenase activity of Rubisco. Photorespiration is essential to plants even though the plant pays the price in loss of ATP and reducing power; mutations that affect this pathway can be lethal.

Chapter 23

23.1 What Processes Constitute Nitrogen Metabolism?

1. Nitrogen-fixing bacteria (symbiotic organisms that form nodules on the roots of leguminous plants, such as beans and alfalfa) and some free-living microbes and cyanobacteria can fix nitrogen. Plants and animals cannot.

23.2 How Is Nitrogen Incorporated into Biologically Useful Compounds?

2. Nitrogen is fixed by the nitrogenase reaction, in which N_2 is converted to NH_4^+. Very few organisms have this enzyme, which can catalyze the breaking of the triple bond in molecular nitrogen. The glutamate dehydrogenase reaction and the glutamine synthase reactions assimilate nitrogen:

$$NH_4^+ + \alpha\text{-Ketoglutarate} \rightleftharpoons$$
$$\text{Glutamate} + \text{Water (NADPH required)}$$

$$NH_4^+ + \text{Glutamate} \rightleftharpoons \text{Glutamine (ATP required)}$$

3. The chemical synthesis of ammonia from H_2 and N_2.

4. $N_2 + 8e^- + 16ATP + 10H^+ \rightarrow 2NH_4^+ + 16ADP + 16P_i + H_2$ is the half reaction for reduction via nitrogenase. The oxidation reaction varies with species.

5. The nitrogenase complex is made up of ferredoxin, dinitrogenase reductase, and nitrogenase. Dinitrogenase reductase is an iron–sulfur protein, whereas nitrogenase is an iron–molybdenum protein. The Fe—S protein is a dimer ("the iron butterfly"), with the iron–sulfur cluster located at the butterfly's head. The nitrogenase is even more complicated, with several types of subunits arranged into tetramers.

23.3 What Role Does Feedback Inhibition Play in Nitrogen Metabolism?

6. Pathways that use nitrogen to make amino acids, purines, and pyrimidines are controlled by feedback inhibition. The final product, such as CTP, inhibits the first or an early step in its synthesis.

7. Feedback control mechanisms slow down long biosynthetic pathways at or near their beginnings, saving energy for the organism.

8. Because all the components of a cycle are regenerated, only small amounts ("catalytic quantities") are needed. This is important from an energy standpoint and, perhaps with some compounds, because of insolubility problems.

23.4 How Are Amino Acids Synthesized?

9. They are all interrelated. α-Ketoglutarate can be changed to glutamate via transamination or glutamate dehydrogenase. Glutamine synthetase makes glutamine out of glutamate.

10.

α-Ketoglutarate $\qquad$ L-Alanine $\qquad$ L-Glutamate $\qquad$ Pyruvate

11.

$$NH_4^+ + {}^-OOC-CH_2-CH_2-\overset{O}{\overset{\|}{C}}-COO^- \underset{NADP^+}{\overset{NADPH + H^+}{\rightleftharpoons}} H_2O + {}^-OOC-CH_2-CH_2-\overset{NH_3^+}{\underset{}{CH}}-COO^-$$

α-Ketoglutarate $\qquad\qquad$ Glutamate

$$NH_4^+ + {}^-OOC-CH_2-CH_2-\overset{NH_3^+}{\underset{}{CH}}-COO^- \xrightarrow{\overset{ATP \quad ADP + P_i}{\qquad}} H_2O + H_2N-\overset{O}{\overset{\|}{C}}-CH_2-CH_2-\overset{NH_3^+}{\underset{}{CH}}-COO^-$$

Glutamate $\qquad\qquad$ Glutamine

12. Glutamine synthetase catalyzes the following reaction and uses energy: NH_4^+ + Glutamine + ATP → Glutamine + ADP + P_i + H_2O. Glutaminase catalyzes the following reaction and does not use energy directly: Glutamine + H_2O → Glutamate + NH_4^+.

13. See Figure 23.8.

14. The principal ones are tetrahydrofolate and *S*-adenosylmethionine.

15. See Figure 23.11.

16. Conversion of homocysteine to methionine using *S*-adenosylmethionine as the methyl donor gives no net gain; one methionine is needed to produce another methionine.

17. Glutamate + α-Keto acid → α-Ketoglutarate + Amino acid

18. See the *S*-adenosylmethionine structure in Figure 23.15. The reactive methyl group is indicated.

19. Sulfanilamide inhibits folic acid biosynthesis.

20. Methionine can play a dual role. In addition to providing a hydrophobic group, methionine (in the form of *S*-adenosylmethionine) can act as a methyl group donor.

23.5 What Are the Essential Amino Acids?

21. The essential amino acids are those with branched chains, aromatic rings, or basic side chains.

22. In both cases, the requirements are those given in Table 23.1.

23.6 How Are Amino Acids Catabolized?

23. Five α-amino acids are involved directly in the urea cycle (ornithine, citrulline, aspartate, arginosuccinate, and arginine). Of those, only aspartate and arginine are also found in proteins.

24. $H^+ + HCO_3^- + 2NH_3 + 3ATP → NH_2CONH_2 + 2ADP + 2P_i + AMP + PP_i + 2H_2O$
    The urea cycle is linked to the citric acid cycle by fumarate and by aspartate, which can be converted to malate by transamination (see Figure 23.19).

25. Ornithine is similar to lysine, but it has one fewer methylene group in the side chain. Citrulline is a keto version of arginine with a side chain C=NH_2^+ replaced by C=O.

26. Aspartate and arginosuccinate are the amino acids that link the two pathways. Aspartate is made by transamination of OAA. The aspartate then combines with citrulline to form arginosuccinate, which then releases a fumarate to go back to the TCA cycle.

27. Each round of the urea cycle costs 4 ATP, two to make carbamoyl-phosphate and effectively two (ATP → AMP) to make arginosuccinate.

28. It is controlled by a special effector molecule, *N*-acetylglutamate, which is itself controlled by levels of arginine.

29. When arginine levels build up, it means that the urea cycle is going too slow and not enough carbamoyl-phosphate is available to react with ornithine.

30. Glutamate brings ammonia groups to the matrix of the mitochondria for the urea cycle. High levels of glutamate stimulate the urea cycle.

31. Glucogenic amino acids are degraded to pyruvate or one of the citric acid cycle intermediates found after the decarboxylation steps, such as succinate or malate. Ketogenic amino acids are degraded to acetyl-CoA or acetoacetyl-CoA.

32. (a) glucogenic; (b) glucogenic; (c) glucogenic; (d) ketogenic; (e) glucogenic; (f) ketogenic.

33. Fish excrete excess nitrogen as ammonia, and birds excrete it as uric acid. Mammals excrete it as urea.

34. Because ostriches don't fly, one could argue that they would excrete their excess nitrogen as urea. On the other hand, they are birds, and as such probably have the same metabolism of their lighter counterparts, and might likely excrete it as uric acid.

35. The amounts of arginine necessary in the urea cycle are only catalytic. If arginine from the cycle is used for protein synthesis, the cycle will become depleted.

36. A high-protein diet leads to increased production of urea. Drinking more water increases the volume of urine, ensuring elimination of the urea from the body with less strain on the kidneys than if urea were at a higher concentration.

37. The metabolism of amino acids will encourage urine formation and actually a greater thirst and need for water.

38. Several enzymes, resulting from mutations, are needed for the urea cycle. Most mutations tend to be lost unless they provide some survival value. It seems improbable that all the mutations needed for all the enzymes of the cycle would arise nearly simultaneously. However, the origin of the cycle can be rather easily explained on the premise that only one new enzyme (arginase) was needed. The other enzymes of the cycle are needed for the biosynthesis of arginine. As a component of proteins, arginine was presumably needed before there was a need for a urea cycle. This is an example of nature using features already available to bring about a new function.

23.7 How Are Purines Synthesized?

39. Since folic acid is critical to the formation of purines, antagonists of folic acid metabolism are used as chemotherapy drugs to inhibit nucleic acid synthesis and cell growth. Rapidly dividing cells, such as those found in cancer and tumors, are more susceptible to these antagonists.

40. All four nitrogen atoms of the purine ring are derived from amino acids: two from glutamine, one from aspartate, and one from glycine. Two of the five carbon atoms (adjacent to the glycine nitrogen) also come from glycine, two more come from tetrahydrofolate derivatives, and the fifth comes from CO_2.

41. In inosine, carbon-6 of the ring is a ketone group, while in adenosine carbon-6 is bound to an amino group.

42. Tetrahydrofolate is a carrier of carbon groups. Two of the carbons in the purine ring are donated by tetrahydrofolate.

43. The conversion of IMP to GMP produces one NADH and uses the equivalent of 2 ATP because an ATP is converted to AMP. Because NADH gives rise to 2.5 ATP if it goes into the electron transport chain, we can say that the conversion results in a net production of ATP.

44. There is a complicated system of feedback inhibition for the production of purine-containing nucleotides. The final products, ATP and GTP, feed back to inhibit the first steps starting from ribose-5-phosphate. In addition, each intermediate, such as AMP or ADP, can also inhibit the first step. Also, each of the three forms for each nucleotide inhibit the committed reaction from IMP that eventually decides which purine nucleotide will be made.

23.8 How Are Purines Catabolized?

45. The purine salvage reaction that produces GMP requires the equivalent of 2 ATP. The pathway to IMP and then to GMP requires the equivalent of 8 ATP.

46. In most mammals, uric acid is converted to allantoic acid, which is much more water soluble than uric acid.

23.9 How Are Pyrimidines Synthesized and Catabolized?

47. In purine nucleotide biosynthesis, the growing purine ring is covalently bonded to ribose, while the ribose is added after the ring is synthesized in pyrimidine nucleotide biosynthesis.

48. Purines break down to various products, depending on the species. These products are then excreted, representing a major means of nitrogen excretion for many organisms. Pyrimidine catabolism yields, in addition to NH_4^+ and CO_2, the salvageable product β-alanine, which is a breakdown product of both cytosine and uracil.

23.10 How Are Ribonucleotides Converted to Deoxyribonucleotides?

49. Both thioredoxin and thioredoxin reductase are proteins involved in the conversion of ribonucleotides to deoxyribonucleotides. Thioredoxin is an intermediate carrier of electrons and hydrogens, and thioredoxin reductase is the enzyme that catalyzes the process.

23.11 How Is dUTP Converted to dTTP?

50. Fluorouracil substitutes for thymine in DNA synthesis. In rapidly dividing cells, such as cancer cells, the result is the production of defective DNA, causing cell death.

51. The DNA of fast-growing cells, such as those of the hair follicles, is damaged by chemotherapeutic agents.

Chapter 24

24.1 How Are the Metabolic Pathways Connected?

1. ATP and NADPH are the two molecules that link the most pathways.

2. Acetyl-CoA, pyruvate, PEP, α-ketoglutarate, succinyl-CoA, oxaloacetate, and several sugar phosphates, such as glucose-6-phosphate and fructose-6-phosphate.

3. (a) Fructose-6-phosphate—from pentose phosphate pathway (PPP)
   (b) Oxaloacetate—to phosphoenolpyruvate in gluconeogenesis, to and from aspartate, to glyoxylate cycle via citrate.
   (c) Glucose-6-phosphate—to PPP, to and from glycogen in animals, to starch in plants.
   (d) Acetyl-CoA—to and from fatty acids, to steroids (and isoprenoids), some amino acid degradations, to glyoxylate cycle via citrate.
   (e) Glyceraldehyde-3-phosphate—to reverse PPP.
   (f) α-Ketoglutarate—to and from glutamate.
   (g) Dihydroxyacetone phosphate—to and from glycerol moiety of triacylglycerols and phosphoacylglycerols.
   (h) Succinyl-CoA—degradation of fatty acids with odd numbers of carbon atoms, some amino acid degradation.
   (i) 3-Phosphoglycerate—appears in Calvin cycle.
   (j) Fumarate—some amino acid degradations.
   (k) Phosphoenolpyruvate—from oxaloacetate in gluconeogenesis.
   (l) Citrate—to glyoxylate cycle, transport across mitochondrial membrane for fatty acid and steroid synthesis.
   (m) Pyruvate—fermentation, to gluconeogenesis, also to and from alanine.

4. When the body breaks down proteins to supply material for gluconeogenesis, the increased urea output results in greater urine production, which uses water stored in the body. Fat metabolism also produces much metabolic water.

5. (a) High ATP or NADH concentration and the citric acid cycle: Isocitrate dehydrogenase (and the citric acid cycle) would be inhibited. The resulting pileup of acetyl-CoA (or citrate) would stimulate fatty acid and steroid synthesis, gluconeogenesis, and (in plants and some microorganisms) the glyoxylate cycle.
(b) High ATP concentration and glycolysis: Phosphofructokinase-1 (and glycolysis) would be inhibited. Glucose-6-phosphate would pile up, stimulating glycogen (or starch) synthesis, the oxidative pentose phosphate pathway, or glucose formation. (c) High NADPH concentration and the pentose phosphate pathway: the oxidative branch of the pentose phosphate pathway would be inhibited, thus making glucose-6-phosphate available for other purposes. These include glycolysis, glycogen synthesis, glucose synthesis, and the "reverse" pentose phosphate pathway (yielding only pentose phosphate). (d) High fructose-2,6-*bis*phosphate concentration and gluconeogenesis: fructose-2,6-*bis*phosphate inhibits fructose-1,6-*bis*phosphatase and activates phosphofructokinase-1. Gluconeogenesis would thus be inhibited and glycolysis would be stimulated, as would the reverse pentose phosphate pathway and the production of glycerol phosphate for lipids.

6. Many compounds, such as oxaloacetate, pyruvate, and acetyl-CoA, play a role in a number of reactions. More to the point, the end products of some pathways are the starting points of others. Each pathway is one aspect of an overall metabolic scheme.

7. The *effect* of biochemical pathways can be reversed. Examples include glycolysis and gluconeogenesis, glycogen formation and synthesis, and the pentose phosphate pathway. The details are not completely reversible. An irreversible step in one pathway will tend to be replaced with another reaction, catalyzed by another enzyme.

8. Transport processes are especially important for substances, such as oxaloacetate, that cannot cross the mitochondrial membrane. The same is true for electrons. Shuttle mechanisms must exist to transport electrons as the reduced form of important compounds. Compounds that cannot cross the membrane must be converted to ones that can, and then must be converted back to their original form on the other side of the membrane.

9. When a pathway has a number of steps, it is possible for energy changes to take place in steps of manageable size. It also allows for control of a pathway to be exercised at more points than would be the case if there were only a few steps.

10. The possibilities are limitless. Even more to the point, some discovery that no one expects can open even more possibilities.

24.2 How Can Biochemistry Help Us Understand Nutrition?

11. The old pyramid assumed that all carbohydrates and fats were the same and that carbohydrates were good and all fats were bad. The new pyramid recognizes that not all carbohydrates are good and not all fats are bad. Complex carbohydrates are placed lower down on the new pyramid, whereas processed ones are placed higher. Essential fats and oils are included as necessary food types. Also, dairy consumption recommendations have been reduced.

12. Fats and carbohydrates can be stored when they are consumed in excess. Fats are stored as triacylglycerols and carbohydrates are stored as glycogen. However, proteins consumed in excess are not stored. The extra protein will be broken down. The amino groups will be released as urea and the carbon skeletons will be stored as carbohydrate or fat.

13. Saturated fatty acids have been correlated with increased levels of LDL, which have been shown to be an indicator of high risk for heart disease.

14. Leptin is a hormone that affects metabolism. It affects the brain to suppress appetite and it affects metabolism directly by stimulating fatty-acid oxidation and inhibiting fatty-acid synthesis.

15. Yes, cholecalciferol is made in the body, and many of its functions are hormonelike in nature.

16. Carbohydrates are the main energy source. Excess fat consumption can lead to the formation of "ketone bodies" and to atherosclerosis. Diets extremely high in protein can put a strain on the kidneys.

17. The liver is the primary organ for alcohol metabolism and for disposing of drugs (legal, illegal, and accidental) and halocarbon compounds. When the liver spends its time dealing with these other tasks, it may not be able to carry out its other normal functions; in essence, prolonged exposure to any such "toxin" overworks the liver.

18. Vitamin A is a lipid-soluble vitamin, which can accumulate in the body. Overdoses of this vitamin can be toxic.

19. Low levels of iodine in the diet often lead to hypothyroidism and an enlarged thyroid gland (goiter). This condition has largely been eliminated by the addition of sodium iodide to commercial table salt.

20. Lucullus breaks down the protein in the tuna to amino acids, which in turn undergo the urea cycle and the breakdown of the carbon skeleton described in Chapter 23, eventually leading to the citric acid cycle and electron transport. In addition to protein catabolism, Griselda breaks down the carbohydrates to sugars, which then undergo glycolysis and enter the citric acid cycle. (Gratuitous information: Lucullus was a notorious Roman gourmand. In medieval literature, Griselda was the name usually given to a forbearing, long-suffering woman.)

21. All amino acids must be present at the same time for protein synthesis to occur. Newly synthesized proteins are necessary for growth in the immature rats.

22. The weight loss is due to correction of the bloating caused by retention of liquids.

23. After a person is fully grown, many amino acids are scavenged and recycled by the body. Because all proteins contain at least some of these two amino acids, there is enough to maintain the body. It should be noted that both again become essential if there is disease or tissue damage and that arginine is required for sperm production in males.

24. The early colonists always cooked in iron pots; enough iron is leached out to supply required amounts, as long as the body is able to absorb it. (Glass cookware did not become available until after World War I, and aluminum cookware, not until after World War II.)

25. Diets high in fiber are usually lower in fats, especially saturated fats; fiber adsorbs many potentially toxic substances, such as cholesterol and halocarbons, preventing their absorption into the body; fiber decreases transit time through the intestine, so any toxic materials in food remain in the body for less time and have a smaller chance of being absorbed or otherwise causing problems.

26. This claim has a chemical basis. Calcium carbonate will dissolve in stomach acid, releasing calcium ion in its usual hydrated form. Calcium citrate is likely to have the calcium ion bound to the citrate in a manner similar to iron in heme. Consequently, the charge of the calcium ion is effectively decreased. Calcium bound to citrate can pass a cell membrane more easily than a hydrated calcium ion.

27. Alcohol provides calories but does not provide vitamins. This is one of the leading causes of malnutrition. Metabolizing alcohol involves an enzyme (alcohol dehydrogenase) with thiamine pyrophosphate (TPP) as a cofactor. The cofactor, in turn, is a metabolite of vitamin B_1, leading to severe deficiencies.

28. Metal ions play a role in the structure and function of proteins and some coenzymes. They tend to do so because they operate as Lewis acids.

29. Severe depletion of glycogen often results in a rebound effect, in which so much is made that some is stored in inappropriate tissues, including the heart, and mineral imbalances often occur. It is best to exercise moderately before the glycogen loading because then the glycogen is stored more effectively and safely in the liver and muscle tissue where it is most needed.

30. Nutrients and water turn over in the body, sometimes very frequently. This implies that an organism is an open system. Equilibrium requires a closed system. Consequently, an organism can reach a steady state, but never equilibrium.

24.3 What Are Hormones and Second Messengers?

31. Hormones can have several different kinds of chemical structures, including steroids, polypeptides, and amino acid derivatives.

32. The anterior pituitary stimulates release of trophic hormones, which in turn stimulate specific endocrine glands; the workings of the adrenal cortex, the thyroid, and the gonads can all be affected as a result. The adrenal cortex produces adrenocortical hormones, including glucocorticoids (involved in carbohydrate metabolism, inflammatory reactions, and reaction to stress) and mineralocorticoids, which control the level of excretion of water and salt by the kidney. If the adrenal cortex does not function adequately, one result is Addison's disease, characterized by hypoglycemia, weakness, and increased susceptibility to stress. The opposite condition, hyperadrenocorticism, is Cushing's syndrome.

33. The hypothalamus secretes hormone-releasing factors. Under the influence of these factors, the pituitary secretes trophic hormones, which act on specific endocrine glands. Individual hormones are then released by the specific endocrine glands.

34. Thyroxine is an amino acid derivative and is absorbed directly from the gut into the bloodstream. If insulin were taken orally, it would be hydrolyzed to amino acids in the stomach and intestine.

35. p21^{ras} is a 21-kDa protein of the small G protein family. It is involved in cell division, and mutations in the gene for this protein are associated with many human cancers.

36. G proteins get their name because they bind GTP as part of their effect. An example is the G protein that is linked to the epinephrine receptor and leads to the production of cAMP as a second messenger. Receptor tyrosine kinases have a different mode of action. When they bind their hormone, they phosphorylate tyrosine residues on themselves and other target proteins, which then act as a second messenger. Insulin is an example of a hormone that binds to a receptor tyrosine kinase.

37. cAMP, Ca^{2+}, insulin receptor substrate.

38. Human growth hormone is a peptide hormone. If it were taken orally, the peptide would be degraded to its component amino acids in the small intestine and would be rendered useless.

24.4　How Are Hormones Involved in the Control of Metabolism?

39. Epinephrine and glucagon are the two that were discussed the most in this book.

40. Glucagon causes the activation of glycogen phosphorylase, inhibition of glycogen synthase, and inhibition of phosphofructokinase-1.

41. Epinephrine has the same affect on glycogen phosphorylase and glycogen synthase, but it has the opposite effect on phosphofructokinase-1.

42. The G protein is bound to GTP. Eventually, the GTP is hydrolyzed to GDP, which causes it to dissociate from adenylate cyclase. This stops the hormone response until the hormone dissociates from the receptor, the G protein trimers are rejoined, and the process starts over again.

43. IP_3 is a polar compound and can dissolve in the aqueous environment of the cytosol; DAG is nonpolar and interacts with the side chains of the membrane phospholipids.

44. When a stimulatory hormone binds to its receptor on a cell surface, it stimulates the action of adenylate cyclase, mediated by the G protein. The cAMP that is produced elicits the desired effect on the cell by stimulating a kinase that phosphorylates a target enzyme.

45. See Table 24.2.

46. It is most unlikely that a metabolic pathway could exist without control mechanisms. Many pathways require energy, so it is advantageous for an organism to shut down a pathway when its products are not needed. Even if a pathway does not require large amounts of energy, the many connections among pathways make it likely that control is established over the levels of important metabolites.

47. In cholera, adenylate cyclase is permanently "turned on." This in turn stimulates active transport of Na^+ and water from epithelial cells, leading to diarrhea.

48. (a) Stoichiometric amounts of cAMP are required to activate cAMP-dependent protein kinase. (b) Six catalytic steps, including the reaction catalyzed by glycogen phosphorylase, with ten molecules acted on in each step, would result in 10^6 (one million) G-1-P molecules for each epinephrine. (c) A major factor is speed. It is important to be able to use stored energy rapidly in "fight or flight" situations. A second factor is control. Note that glycogen phosphorylase is activated by kinases. The competing process of glycogen storage, catalyzed by glycogen synthetase, is inactivated by kinases. A third factor is economy. A single molecule of epinephrine activates many molecules of glycogen phosphorylase and yet more molecules of G-1-P.

49. A phosphatase dephosphorylates glycogen phosphorylase and glycogen synthetase, inactivating and activating them, respec-

tively. The phosphatase becomes active in response to high concentrations of glucose.

50. Low-carbohydrate diets are designed to prevent the high blood-sugar levels that arise when large quantities of carbohydrates are consumed. High blood sugar will lead to a rapid rise in insulin. Insulin is known to stimulate fat synthesis and to inhibit fatty-acid oxidation. Thus, low carbohydrate diets are thought to help fight weight gain.

24.5　What Are the Many Effects of Insulin?

51. Insulin's primary function is to stimulate the transport of glucose out of the blood and into the cell.

52. The second messenger is a protein called the insulin receptor substrate, which is phosphorylated on a tyrosine by the insulin receptor kinase.

53. When insulin binds to its receptor, the β-subunit of the receptor kinase autophosphorylates. When this happens, the receptor kinase is able to phosphorylate tyrosines on the insulin receptor kinase.

54. Insulin causes the following effects: (a) Glycogen breakdown is decreased. (b) Glycogen synthesis is increased. (c) Glycolysis is increased. (d) Fatty-acid synthesis is increased. (e) Fatty-acid storage is increased.

55. Insulin and epinephrine normally have opposite effects, but they both stimulate muscle glycolysis. Epinephrine is the hormone that signals the need for quick energy, which means the muscle cells must be able to use glucose via glycolysis. Insulin stimulates pathways that use up glucose so that the blood glucose will lower, so it makes sense for it to stimulate glycolysis as well. Epinephrine stimulates muscle glycolysis by activating adenylate cyclase, which makes cAMP; cAMP then activates protein kinase A, which phosphorylates phosphofructokinase-2 and fructose-*bis*phosphatase-2. In the muscle, phosphorylation of phosphofructokinase-2 activates it, producing more fructose-2,6-*bis*phosphate, which activates phosphofructokinase-1 and glycolysis. In muscle, insulin stimulates glycolysis by activating phosphofructokinase and pyruvate dehydrogenase.

56. Prerace diet can be critical to a runner. If the race is at 9 A.M., and the runner gets up at 7 A.M. and then eats a typical American breakfast of cereal, toast, or pancakes, she will have a high blood-sugar level within half an hour, which will lead to a high insulin level shortly thereafter. In that scenario, by the time the runner gets to the starting line, she will have a metabolism dedicated to fat and glycogen synthesis and will not be burning fat or carbohydrates. The runner will be like a car with a full tank of gas and a clogged fuel line.

57. It has been shown that the GLUT4 transporter responds to physical activity. When a person is active, the transporter is active and responds well to insulin. After a few days of detraining, this transporter shows only half of the activity it did before.

58. GLUT4 is one of the glucose transporters on muscle cells. It responds to insulin by moving glucose out of the blood and into the cell. In type II diabetes, insulin is present, but it does not have the same effect. It takes more insulin to accomplish the same movement out of the blood and into the cell. People with type II diabetes often show classical signs of obesity, and there is a correlation between diminishing GLUT4 activity, obesity, and diabetes.

. γ

. for a

The Standard Genetic Code					
First Position (5′ End)	Second Position				Third Position (3′ End)
	U	C	A	G	
U	UUU Phe	UCU Ser	UAU Tyr	UGU Cys	U
	UUC Phe	UCC Ser	UAC Tyr	UGC Cys	C
	UUA Leu	UCA Ser	UAA Stop	UGA Stop	A
	UUG Leu	UCG Ser	UAG Stop	UGG Trp	G
C	CUU Leu	CCU Pro	CAU His	CGU Arg	U
	CUC Leu	CCC Pro	CAC His	CGC Arg	C
	CUA Leu	CCA Pro	CAA Gln	CGA Arg	A
	CUG Leu	CCG Pro	CAG Gln	CGG Arg	G
A	AUU Ile	ACU Thr	AAU Asn	AGU Ser	U
	AUC Ile	ACC Thr	AAC Asn	AGC Ser	C
	AUA Ile	ACA Thr	AAA Lys	AGA Arg	A
	AUG Met*	ACG Thr	AAG Lys	AGG Arg	G
G	GUU Val	GCU Ala	GAU Asp	GGU Gly	U
	GUC Val	GCC Ala	GAC Asp	GGC Gly	C
	GUA Val	GCA Ala	GAA Glu	GGA Gly	A
	GUG Val	GCG Ala	GAG Glu	GGG Gly	G

*AUG forms part of the initiation signal as well as coding for internal methionine residues.